To my wife, Betsy and my mother, Vicky

Mircea Grigoriu

Stochastic Calculus

Applications in Science and Engineering

Springer Science+Business Media, LLC

Mircea Grigoriu
Cornell University
School of Civil
and Environmental Engineering
Ithaca, NY 14853
U.S.A.

Library of Congress Cataloging-in-Publication Data

Grigoriu, Mircea.
Stochastic calculus : applications in science and engineering / Mircea Grigoriu.
p. cm.
Includes bibliographical references and index.
ISBN 978-1-4612-6501-6
1. Stochastic analysis. I. Title.

QA274.2.G75 2002
519.2–dc21 2002074386
CIP

ISBN 978-1-4612-6501-6 ISBN 978-0-8176-8228-6 (eBook)
DOI 10.1007/978-0-8176-8228-6

AMS Subject Classifications: 37A50, 60GXX, 35SXX, 35QXX

Printed on acid-free paper.

Originally published by Birkhäuser Boston in 2002
Softcover reprint of the hardcover 1st edition 2002

Birkhäuser

SPIN 10835902

Typeset by the author.

9 8 7 6 5 4 3 2 1

Contents

Stochastic Calculus

Applications in Science and Engineering

Chapter 1

Introduction

Algebraic, differential, and integral equations are used in the applied sciences, engineering, economics, and the social sciences to characterize the current state of a physical, economic, or social system and forecast its evolution in time. Generally, the coefficients of and/or the input to these equations are not precisely known because of insufficient information, limited understanding of some underlying phenomena, and inherent randomness. For example, the orientation of the atomic lattice in the grains of a polycrystal varies randomly from grain to grain, the spatial distribution of a phase of a composite material is not known precisely for a particular specimen, bone properties needed to develop reliable artificial joints vary significantly with individual and age, forces acting on a plane from takeoff to landing depend in a complex manner on the environmental conditions and flight pattern, and stock prices and their evolution in time depend on a large number of factors that cannot be described by deterministic models. Problems that can be defined by algebraic, differential, and integral equations with random coefficients and/or input are referred to as **stochastic problems**.

The main objective of this book is the solution of stochastic problems, that is, the determination of the probability law, moments, and/or other probabilistic properties of the state of a physical, economic, or social system. It is assumed that the operators and inputs defining a stochastic problem are specified. We do not discuss the mathematical formulation of a stochastic problem, that is, the selection of the functional form of the equation and of the probability laws of its random coefficients and input for a particular stochastic problem.

The book is addressed to researchers and graduate students. It is intended to serve as a bridge between heuristic arguments used at times in the applied sciences and the very rich mathematical literature that is largely inaccessible to many applied scientists. Mathematicians will find interesting unresolved technical problems currently solved by heuristic assumptions.

Organization

The book is largely self-contained and has two parts. The first part includes Chapters 2-5, and develops the probabilistic tools needed for the analysis of the stochastic problems considered in the book. Essentials of probability theory are reviewed in Chapter 2. An extensive review of stochastic processes, elements of stochastic integrals, Itô's formula, and a primer on stochastic differential equations and their numerical solution is presented in Chapters 3 and 4. Numerous references are provided that contain details and material not included in Chapters 2, 3, and 4. Methods of Monte Carlo simulation for random variables, stochastic processes, and random fields are discussed in Chapter 5. These methods provide useful illustrations for some of the theoretical concepts in Chapters 2-4 and are essential for the solution of many of the stochastic problems in this book.

The second part of the book, Chapters 6-9, develops methods for solving stochastic problems. In this book stochastic problems are characterized by type

Table 1.1: Stochastic problems

INPUT	SYSTEM	
	Deterministic	*Stochastic*
Deterministic	**Chapter 6**	**Chapter 8**
Stochastic	**Chapter 7**	**Chapter 9**

rather than by the particular field of application. Table 1 shows the four types of problems considered in the book. Chapter 6 is concerned with deterministic problems, that is, problems in which both the governing differential equations are deterministic, but whose solution utilizes concepts of the theory of probability and stochastic processes. Such problems arise in physics, mechanics, material science, heat conduction, and many other fields. For example, consider a Laplace equation $\sum_{i=1}^{d} \partial^2 u(\boldsymbol{x})/\partial x_i^2 = 0$ with Dirichlet boundary conditions defined on an open bounded subset D of $\mathbb{R}^d$. The solution of this equation at an arbitrary point $\boldsymbol{x} \in D$ is equal to the expectation of $u(\boldsymbol{Y})$, where $\boldsymbol{Y}$ denotes the exit point from D of an $\mathbb{R}^d$-valued Brownian motion starting at $\boldsymbol{x}$. Numerical algorithms based on this approach are simpler and more efficient, though less general, than current finite element and finite difference based algorithms. Chapter 7 discusses problems in which the equations describing a system have deterministic coefficients and the input to these equations is stochastic. Such problems arise in the description of various types of physical, economic, and ecological systems. Chapter 8 covers problems defined by equations with random coefficients and deterministic input. Examples include the derivation of macroscopic properties of a material from the attributes of its constituents and phenomena, such as localization and pattern formation, relevant in physics and mechanics. Chapter 9 is concerned with problems characterized by equations with random coefficients and inputs. An example is the system describing the propagation of pollutants through a soil deposit with uncertain properties for which no realistic deterministic description is possible.

We now briefly describe the presentation of the material. Essential facts are in boxes throughout the book. The book contains numerous examples that have two purposes: to state consequences of the essential facts and to illustrate the use of the stated facts in the solution of stochastic problems. The statements of the facts and examples are followed by proofs or notes printed in smaller characters. Complete proofs are given if they are not very technical. The notes include the idea and/or the essential steps of technical proofs and references where complete proofs can be found.

The advanced topics on probability theory and stochastic processes in the first part of the book are essential for Chapters 6 and 7. Many of the developments in Chapters 8 and 9 are largely based on the second moment calculus. However, extensions of the methods in Chapters 6 and 7 to the problems considered in Chapters 8 and 9 require most of the advanced topics in Chapters 2-4.

Classroom use

The book can be used as a text for four different one-semester courses or one two-semester course. The four one-semester courses can differ depending on the applications they emphasize. The first, second, third, and fourth courses emphasize the applications in Chapter 6, 7, 8, and 9, respectively, and require the following background.

• **The first course:** properties of the conditional expectation (Chapter 2), stopping times and martingales (Chapter 3), the Itô calculus and diffusion processes (Chapter 4), and Monte Carlo techniques (Chapter 5).

• **The second course:** facts on stochastic processes (Chapter 3), the stochastic integral, the Itô calculus, and the theory of stochastic differential equations (Chapter 4), and Monte Carlo techniques (Chapter 5).

• **The third course:** a review of probabilistic concepts (Chapter 2), second moment calculus for stochastic processes (Chapter 3), and Monte Carlo techniques (Chapter 5).

• **The fourth course:** essential probability concepts (Chapter 2), properties of Brownian motion and second moment calculus for stochastic processes (Chapter 3), use of Itô 's formula (Chapter 4), and Monte Carlo simulation techniques (Chapter 5).

Owing to time limitations, the one-semester courses need to be focused on applications. The presentation of the book, summarizing facts on probability theory and stochastic processes that are essential for calculations and illustrating these facts by numerous examples, facilitates the course development.

A **two-semester course** can include most of the topics discussed in this book. A good part of the first semester needs to be spent on a review of the topics in Chapters 2-4. Depending on students' interest and background, the lectures can focus on understanding and use of the essential facts given in these chapters or they may incorporate proofs of some of these facts based on the material and references in the book. The Monte Carlo simulation techniques in Chapter 5 can be used during lectures on Chapters 2-4 to illustrate various probabilistic concepts. Most students in my classes on stochastic problems are in favor of this approach, which facilitates the understanding of some technical concepts of probability theory. This approach also contributes to the development of computational skills needed for solving realistic stochastic problems that, generally, do not have analytical solutions. The remainder of the first semester can be used to discuss the solution of problems defined by deterministic equations and inputs (Chapter 6). The second semester of the course can include analysis of the stochastic problems discussed in Chapters 7-9 and/or other problems that are relevant to the instructor's field. The addition of field specific applications is strongly recommended and is consistent with the goal of this book: to use a unified framework for the solution of stochastic problems arising in applications.

Acknowledgements

This book could not have been completed without the contributions of many individuals. In particular, I wish to express my deepest appreciation to Professors S. I. Resnick and G. Samorodnitsky of Cornell University for their technical advice and comments on various topics in the book, Dr. E. Simiu of National Institute of Standards and Technology for reviewing the entire manuscript, Professor S. T. Ariaratnam of the University of Waterloo, Canada, for numerous stimulating discussions, as well as Professors P. R. Dawson, J. Jenkins, S. Mukherjee, and Dr. C. Myers of Cornell University and Professor O. Ditlevsen of Denmark Technical University, Lyngby, for their useful comments. I also wish to thank my doctoral students S. Arwade, E. Mostafa, and C. Roth for their many contributions, and Mr. C. Willkens for his enthusiastic and professional support of the computer hardware and software used in this project. Finally, I am grateful to my wife Betsy for understanding, encouragement, and support. During this project, she became an accomplished sculptor and pilot.

My research on stochastic problems, some of which is incorporated in this book, has been supported by the National Science Foundation, National Institute of Standards and Technology, Electric Power Research Institute, Jet Propulsion Laboratory, Air Force Office of Scientific Research, Federal Aviation Administration, AON Financial Products, and other institutions. I am indebted to these organizations and their continued support.

Chapter 2

Probability Theory

2.1 Introduction

Essential concepts of probability theory needed in this text are reviewed and are illustrated by examples. The review includes the concepts of events, sample space, σ-field, measure, probability measure, probability space, conditional probability, independence, random variable and vector, integral of random variables, expectation, distribution, density, and characteristic functions, second moment properties, convergence of sequences of random variables, conditional expectation, and martingales. The readers familiar with these concepts can skip this chapter entirely. However, some of those readers may benefit from using this chapter as a summary of facts and examples needed in the rest of the book.

2.2 Probability space

The three components of probability space, the sample space Ω, the σ-field $\mathcal{F}$, and the probability measure P, are defined and illustrated.

2.2.1 Sample space

Consider an experiment and let Ω be the set of outcomes of this experiment, called the **sample space**. For example, $\Omega = \{1, 2, 3, 4, 5, 6\}$ for the experiment of rolling a die and $\Omega = [a, b] \subset [0, \infty)$, $0 < a < b < \infty$, for the experiment consisting of strength tests of nominally identical steel specimens. The first and second sample spaces have a finite and an uncountable number of elements, respectively. If Ω has a finite, countable, or uncountable number of elements, it is referred to as a **finite**, **countable**, or **uncountable** sample space, respectively. The elements of Ω are denoted by ω.

Example 2.1: Consider the experiment of rolling a die and two games associated with this experiment. In the first game, one loses \$10 if $\omega \le 3$ and wins \$10 if $\omega > 3$. In the second game, one loses \$10 if $\omega \le 3$ and wins \$5, \$10, or \$15 if $\omega = 4$, 5, or 6. The relevant information for the first and second games is given by the collections of subsets $\mathcal{A}_1 = \{\{1,2,3\},\{4,5,6\}\}$ and $\mathcal{A}_2 = \{\{1,2,3\},\{4\},\{5\},\{6\}\}$ of Ω, respectively. Playing these games does not require knowing in the finest detail the outcome of each roll of the die. Coarser information suffices. To play the first game we need to know only if $\omega \le 3$ or $\omega > 3$. The second game requires a more refined description of the outcomes of the experiment because it has more options of interest. ◇

Note: Similar considerations apply to the experiment of testing steel specimens for strength. Suppose that each steel specimen is subjected to the same force of magnitude $x > 0$ and the objective is to design a safe steel structure. There are two relevant outcomes for this "game", survival $A = \{\omega \ge x\}$ and failure $B = A^c$. The precise value of the strength ω of a particular specimen is not essential. To assess the likelihood of survival of steel specimens subjected to action x it is sufficient to know whether $\omega < x$ or $\omega \ge x$. ▲

2.2.2 σ-field

Let $\mathcal{F}$ be a collection of subsets of Ω relevant to a particular experiment. It seems natural to require that $\mathcal{F}$ has at least two properties: (1) If $A \in \mathcal{F}$, then the outcomes corresponding to the non-occurrence of A should also be in $\mathcal{F}$, that is, $A \in \mathcal{F}$ implies $A^c \in \mathcal{F}$, so that $\mathcal{F}$ is closed to complements, and (2) If $A_i \in \mathcal{F}$, $i \in I$, that is, if the subsets A_i occur individually, then $\cup_{i \in J} A_i$ should also be in $\mathcal{F}$ for any subset J of the index set I. These heuristic considerations are consistent with the following properties defining $\mathcal{F}$.

A non-empty collection of subsets $\mathcal{F}$ of Ω is called a σ-**field** on Ω if

1. $\emptyset \in \mathcal{F}$.
2. $A \in \mathcal{F} \Longrightarrow A^c \in \mathcal{F}$.
3. $A_i \in \mathcal{F}, \quad i \in I, I =$ a countable set $\Longrightarrow \cup_{i \in I} A_i \in \mathcal{F}$. (2.1)

Note: The first condition in the definition of the σ-field $\mathcal{F}$ can be replaced with $\Omega \in \mathcal{F}$ because $\mathcal{F}$ is closed to complements. The last conditions imply that countable intersections of events are events. We have $\cup_{i \in I} A_i \in \mathcal{F}$ for $A_i \in \mathcal{F}$ (condition **3**) so that $(\cup_{i \in I} A_i)^c \in \mathcal{F}$ (condition **2**) and $(\cup_{i \in I} A_i)^c = \cap_{i \in I} A_i^c \in \mathcal{F}$ by De Morgan's formulas.

If the last condition in the definition of a σ-field $\mathcal{F}$ is replaced with the requirement that $\mathcal{F}$ is closed under finite union, $\mathcal{F}$ is said to be a **field**. Hence, a σ-field is a field. However, a field may not be a σ-field. There is no difference between a field and a σ-field for finite sample spaces. ▲

The members of $\mathcal{F}$ are called **events**, or $\mathcal{F}$-**measurable** subsets of Ω, or just **measurable** subsets of Ω if there is no confusion regarding the reference σ-field. The pair $(\Omega, \mathcal{F})$ is said to be a **measurable space**.

Example 2.2: The collections $\mathcal{A}_i$, $i = 1, 2$, defined in Example 2.1 are not σ-fields. However,

$$\mathcal{F}_1 = \{\text{members of } \mathcal{A}_1, \emptyset, \Omega\} \quad \text{and}$$
$$\mathcal{F}_2 = \{\text{members of } \mathcal{A}_2, \{4,5,6\}, \{1,2,3,5,6\}, \{1,2,3,4,6\}, \{1,2,3,4,5\}, \emptyset, \Omega\}$$

are σ-fields. They are the smallest σ-fields including $\mathcal{A}_i$. ◇

Note: $\mathcal{A}_1$ is not closed to union while $\mathcal{A}_2$ is not closed to both union and complements so that neither is a σ-field. ▲

Let $\mathcal{A}$ be a collection of subsets of Ω and define

$$\sigma(\mathcal{A}) = \bigcap_{\mathcal{G} \supseteq \mathcal{A}} \mathcal{G}, \tag{2.2}$$

where $\mathcal{G}$ are σ-fields on Ω. Then $\sigma(\mathcal{A})$ is a unique σ-field, called the **σ-field generated by** $\mathcal{A}$. There is no σ-field smaller than $\sigma(\mathcal{A})$ that includes $\mathcal{A}$.

• The **Borel σ-field** is generated by the collection of open sets of a topological space. The members of this σ-field are called **Borel sets**.

• The Borel σ-fields on $\mathbb{R}^d$, $d > 1$, and $\mathbb{R}$ are generated by the intervals in these spaces and are denoted by $\mathcal{B}(\mathbb{R}^d) = \mathcal{B}^d$ and $\mathcal{B} = \mathcal{B}^1 = \mathcal{B}(\mathbb{R})$, respectively. The Borel σ-fields on the intervals $[a, b]$, $[a, b)$, $(a, b]$, and (a, b) of the real line are denoted by $\mathcal{B}([a, b])$, $\mathcal{B}([a, b))$, $\mathcal{B}((a, b])$, and $\mathcal{B}((a, b))$, respectively.

Note: The Borel σ-field constitutes a special case of $\sigma(\mathcal{A})$ in Eq. 2.2 that is generated by the collection of open sets $\mathcal{D}$ of a space $\mathcal{X}$. This collection is a **topology** on $\mathcal{X}$ if it (1) contains the empty set and the entire space, (2) is closed to finite intersections, and (3) is closed to uncountable unions. There are notable similarities between a topology $\mathcal{D}$ and a σ-field $\mathcal{F}$ defined on a space $\mathcal{X}$. Both $\mathcal{D}$ and $\mathcal{F}$ include the entire space and the empty set and are closed to countable unions and finite intersections. However, $\mathcal{D}$ is closed to uncountable unions while $\mathcal{F}$ is closed under complements and countable intersections.

If $\mathcal{X} = \mathbb{R}$, $\mathcal{B}$ can be generated by any of the intervals (a, b), $(a, b]$, $[a, b)$, or $[a, b]$ ([151], Section 1.7, p. 17), that is,

$$\begin{aligned}\mathcal{B} &= \sigma((a, b), -\infty \le a \le b \le +\infty) = \sigma([a, b), -\infty < a \le b \le +\infty) \\ &= \sigma([a, b], -\infty < a \le b < +\infty) = \sigma((-\infty, x], x \in \mathbb{R}) = \sigma(\text{open subsets of } \mathbb{R}).\end{aligned}$$

The Borel σ-field $\mathcal{B}^d$, $d > 1$, can be generated, for example, by the open intervals $\times_{i=1}^d (a_i, b_i)$ of $\mathbb{R}^d$. ▲

Example 2.3: Let A be a subset of Ω. The σ-field generated by A is $\mathcal{F} = \{\emptyset, A, A^c, \Omega\}$. The σ-fields $\mathcal{F}_i$ in Example 2.2 coincide with the σ-fields generated by the collections of events $\mathcal{A}_i$. ◇

Example 2.4: Let $\Omega = \mathbb{R}$ and $\mathcal{A}$ be the collection of finite unions of intervals of the form $(-\infty, a]$, $(b, c]$, and (d, ∞). This collection is a field but is not a σ-field on the sample space Ω. $\diamond$

Proof: The empty set is a member of $\mathcal{A}$ because the interval $(b, c]$ is in $\mathcal{A}$ and $(b, b] = \emptyset$. The complements, (a, ∞), $(-\infty, b] \cup (c, \infty)$, and $(-\infty, d]$, of $(-\infty, a]$, $(b, c]$, and (d, ∞), respectively, are in $\mathcal{A}$. The collection $\mathcal{A}$ is also closed under finite unions by definition. Hence, $\mathcal{A}$ is a field. However, $\mathcal{A}$ is not closed under countable intersections so that it is not a σ-field. For example, the intersection $\cap_n (b - 1/n, c] = [b, c]$ is not in $\mathcal{A}$ although the intervals $(b - 1/n, c]$ are in $\mathcal{A}$ for each n. ■

Example 2.5: The intervals $(a, b]$, $[a, b)$, and $[a, b]$, $a < b$, are Borel sets although they are not open intervals, that is, they are not members of the usual topology $\mathcal{D}$ on the real line. Singletons, that is, isolated points of the real line, are also Borel sets. $\diamond$

Proof: The intervals $(a, b + 1/n)$, $(a - 1/n, b)$, and $(a - 1/n, b + 1/m)$ are open intervals that are in $\mathcal{B}$ for each $n = 1, 2, \ldots$, so that $\cap_{n,m \geq 1} (a - 1/n, b + 1/m) = [a, b]$ is in $\mathcal{B}$. Similar calculations show that $[a, b)$, and $(a, b]$ are Borel sets. That singletons are Borel sets follows from the equality $\{a\} = \cap_{n,m \geq 1} (a - 1/n, a + 1/m)$ since $(a - 1/n, a + 1/m) \in \mathcal{B}$ holds for each n, m, and $a \in \mathbb{R}$. ■

Example 2.6: Let $\mathcal{F}$ be a σ-field on a sample space Ω. A member A of this field is called an **atom** if $A \neq \emptyset$ and if $B \subseteq A$ implies that either $B = \emptyset$ or $B = A$. Hence, atoms are the finest members of a field. For example, the atoms of the σ-fields associated with the games of Example 2.1 are $\{1, 2, 3\}, \{4, 5, 6\} \in \mathcal{F}_1$ and $\{1, 2, 3\}, \{4\}, \{5\}, \{6\} \in \mathcal{F}_2$. $\diamond$

2.2.3 Probability measure

Consider a measurable space $(\Omega, \mathcal{F})$. We define two real-valued functions on $\mathcal{F}$, a measure μ and a probability measure P, and give some of their properties.

A set function $\mu : \mathcal{F} \to [0, \infty]$ is said to be a **measure** if it is countably additive, that is,

$$\mu(\cup_{i=1}^{\infty} A_i) = \sum_{i=1}^{\infty} \mu(A_i) \quad \text{for } A_i \in \mathcal{F}, \quad A_i \cap A_j = \emptyset, \ i \neq j. \tag{2.3}$$

The triple $(\Omega, \mathcal{F}, \mu)$ is called a **measure space**.

We now give four examples of measures.

1. The **Lebesgue measure**, denoted by λ^d and defined on $(\mathbb{R}^d, \mathcal{B}^d)$, is

$$\lambda^d([a_1, b_1] \times \cdots \times [a_d, b_d]) = \prod_{i=1}^{d} (b_i - a_i),$$

that is, the volume of the interval $[a_1, b_1] \times \ldots \times [a_d, b_d]$.

2. The **counting measure** is used later in the book in conjunction with the Poisson process. Suppose that $\mathcal{F}$ is the **power set** of the sample space Ω, that is, the collection of all subsets of Ω. The counting measure of $A \in \mathcal{F}$ is the cardinal of A if A is finite and infinity otherwise.

3. The σ-**finite measure** has the property that for every $A \in \mathcal{F}$ there exists a sequence of disjoint events A_i, $i = 1, 2, \ldots$, such that $\mu(A_i) < \infty$ for every i, $\Omega = \cup_i A_i$, and $\mu(A) = \sum_{i=1}^{\infty} \mu(A \cap A_i)$. For example, the Lebesgue measure λ on $\mathbb{R}$ is σ-finite because $\lambda(A) = \sum_{i \in \mathbb{Z}} \lambda(A \cap [i, i+1))$, where $A \in \mathcal{B}$, $\mathbb{Z}$ denotes the set of integers, and $\lambda = \lambda^1$ is the Lebesgue measure on the real line.

4. The **finite measure** is a measure with the property $\mu(\Omega) < \infty$. The scaled version of this measure, $\mu(A)/\mu(\Omega)$, takes values in [0, 1].

A set function $P : \mathcal{F} \to [0, 1]$ with the properties

1. $P(\Omega) = 1$,

2. $$P(\cup_{i=1}^{\infty} A_i) = \sum_{i=1}^{\infty} P(A_i), \quad A_i \in \mathcal{F}, \quad A_i \cap A_j = \emptyset, \ i \neq j, \tag{2.4}$$

is said to be a **probability measure** or a **probability**. The triple $(\Omega, \mathcal{F}, P)$ is called a **probability space**.

Note: A probability space can be obtained from a measure space $(\Omega, \mathcal{F}, \mu)$ with a finite measure μ by setting $P(\cdot) = \mu(\cdot)/\mu(\Omega)$. A set N in $\mathcal{F}$ with probability zero is called a **null set**. A property that is true on $\Omega \setminus N$ is said to hold **almost everywhere** (a.e.), **almost surely** (a.s.), for **almost every** ω, or **with probability one** (w.p.1). Because of the second condition in Eq. 2.4, we say that the probability measure is **countably additive**.

The definition of P is consistent with our intuition. Let A and B be two events associated with an experiment defining Ω and $\mathcal{F}$. Suppose that the experiment is performed n times and denote by n_A and n_B the number of outcomes in which A and B are observed, respectively. For example, n_A and n_B may denote the number of outcomes $\{\omega \leq 3\}$ and $\{\omega > 3\}$ when a die is rolled n times. The probabilities of A and B are given by the limits as $n \to \infty$ of the relative frequencies of occurrence, n_A/n and n_B/n, of the events A and B. The relative frequencies have the properties (1) $P(\Omega) = \lim_{n\to\infty}(n_\Omega/n) = 1$ since $n_\Omega = n$ for each n and (2) if A and B are disjoint events, $P(A \cup B) = \lim_{n\to\infty}((n_A + n_B)/n) = \lim_{n\to\infty}(n_A/n) + \lim_{n\to\infty}(n_B/n) = P(A) + P(B)$, consistent with Eq. 2.4. We also note that Eqs. 2.3 and 2.4 are meaningful because $\mathcal{F}$ is closed to countable unions. Also, $A \in \mathcal{F}$ implies that A^c is in the domain of P since $\mathcal{F}$ is a σ-field. ▲

We assume throughout the book that all probability spaces are complete. A probability space $(\Omega, \mathcal{F}, P)$ is **complete** if for every $A \subset B$ such that $B \in \mathcal{F}$ and $P(B) = 0$ we have $A \in \mathcal{F}$ so that $P(A) = 0$ (Eq. 2.5). The following result shows that this assumption is not restrictive.

For any probability space $(\Omega, \mathcal{F}, P)$ there exists a complete space $(\Omega, \bar{\mathcal{F}}, \bar{P})$ such that $\mathcal{F} \subseteq \bar{\mathcal{F}}$ and $P = \bar{P}$ on $\mathcal{F}$ ([40], Theorem 2.2.5, p. 29).

The formulas in Eqs. 2.5-2.8 give properties of the probability measure P that are useful for calculations.

$$\begin{aligned} &P(A) \leq P(B), \quad A \subseteq B, \quad A, B \in \mathcal{F}. \\ &P(A) = 1 - P(A^c), \quad A \in \mathcal{F}. \\ &P(A \cup B) = P(A) + P(B) - P(A \cap B), \quad A, B \in \mathcal{F}. \end{aligned} \tag{2.5}$$

Proof: The first relationship follows since $B = A \cup (B \setminus A)$, A and $B \setminus A$ are disjoint events, $B \setminus A = B \cap A^c$, and the probability measure is a countably additive positive set function.

We have $1 = P(\Omega) = P(A \cup A^c) = P(A) + P(A^c)$ by Eq. 2.4. This also shows that the probability of the impossible event is zero because $P(\emptyset) = 1 - P(\Omega) = 0$.

The last relationship results from Eq. 2.4 and the observations that $A \cup B$ is the union of the disjoint sets $\{A \cap B^c, A \cap B, A^c \cap B\}$, A is the union of the disjoint sets $\{A \cap B^c, A \cap B\}$, and B is the union of the disjoint sets $\{A^c \cap B, A \cap B\}$. ■

If $A_i, B \in \mathcal{F}$ and A_i, $i = 1, \ldots, n$, partition Ω, we have

$$P(B) = \sum_{i=1}^{n} P(B \cap A_i). \tag{2.6}$$

Proof: This equality follows from Eq. 2.4 since $\Omega = \cup_{i=1}^{n} A_i$, $A_i \in \mathcal{F}$, $A_i \cap A_j = \emptyset$, $i \neq j$, and $B = \cup_i (B \cap A_i)$ is a union of disjoint events. ■

If $A_i \in \mathcal{F}$, $i = 1, 2, \ldots$, then

$$P(\cup_{i=1}^{\infty} A_i) \leq \sum_{i=1}^{\infty} P(A_i). \tag{2.7}$$

Proof: Since

$$\cup_{i=1}^{\infty} A_i = A_1 \cup \left(A_1^c \cap A_2\right) \cup \left(A_1^c \cap A_2^c \cap A_3\right) \cup \cdots$$

and the events $A_1, A_1^c \cap A_2, A_1^c \cap A_2^c \cap A_3, \ldots$ are disjoints, we have

$$P\left(\cup_{i=1}^{\infty} A_i\right) = P\left(A_1\right) + P\left(A_1^c \cap A_2\right) + P\left(A_1^c \cap A_2^c \cap A_3\right) + \cdots,$$

which gives Eq. 2.7 by using the first relation in Eq. 2.5. For example, $P(A_1^c \cap A_2) \leq P(A_2)$, $P(A_1^c \cap A_2^c \cap A_3) \leq P(A_3)$, and so on. A set function P satisfying Eq. 2.7 is said to be **subadditive.** ■

The probability measure satisfies the **inclusion-exclusion** formula

$$P\left(\cup_{i=1}^n A_i\right) = \sum_{i=1}^n P(A_i) - \sum_{i=2}^n \sum_{j=1}^{i-1} P(A_i \cap A_j)$$
$$+ \sum_{i=3}^n \sum_{j=2}^{i-1} \sum_{k=1}^{j-1} P(A_i \cap A_j \cap A_k) - \cdots + (-1)^{n+1} P\left(\cap_{q=1}^n A_q\right), \tag{2.8}$$

where $A_i \in \mathcal{F}$, $i = 1, \ldots, n$, are arbitrary.

Proof: This formula extends the last equation in Eq. 2.5 and results from a repeated application of this equation. For example, the probability of $P(A_1 \cup A_2 \cup A_3)$ is equal to the probability of the union of $A_1 \cup A_2$ and A_3 so that

$$\begin{aligned} P((A_1 \cup A_2) \cup A_3) &= P(A_1 \cup A_2) + P(A_3) - P((A_1 \cup A_2) \cap A_3) \\ &= P(A_1 \cup A_2) + P(A_3) - P((A_1 \cap A_3) \cup (A_2 \cap A_3)) \\ &= P(A_1) + P(A_2) - P(A_1 \cap A_2) + P(A_3) \\ &- P(A_1 \cap A_3) - P(A_2 \cap A_3)) + P(A_1 \cap A_2 \cap A_3), \end{aligned}$$

which is the inclusion-exclusion formula for $n = 3$. ■

Example 2.7: Let F_i be the failure event of component i of a series system with n components, that is, a system that fails if at least one of its components fails. The system probability of failure is $P_f = P(\cup_{i=1}^n F_i)$. If the events F_i are disjoint, the failure probability is $P_f = \sum_{i=1}^n P(F_i)$. Otherwise, $\sum_{i=1}^n P(F_i)$ gives an upper bound on P_f. ◇

Example 2.8: The probability of failure $P_f = P(\cup_{i=1}^n F_i)$ of the series system in the previous example can be calculated exactly by the inclusion-exclusion formula. However, the use of this formula is prohibitive for large values of n or even impossible if the probability of the events $F_{i_1} \cap \cdots \cap F_{i_m}$ is not known for $m \geq 2$. The calculation of the bounds

$$P_f \leq P_{f,u} = \sum_{i=1}^n P(F_i) - \sum_{i=2}^n \max_{j=1,\ldots,i-1} P(F_j \cap F_i) \quad \text{and}$$
$$P_f \geq P_{f,l} = P(F_1) + \sum_{i=2}^n \max\left(0, P(F_i) - \sum_{j=1}^{i-1} P(F_j \cap F_i)\right) \tag{2.9}$$

on P_f is relatively simple since it involves only the probability of the events F_i and $F_i \cap F_j$, $i \neq j$.

If $P(F_i) = p$ and $P(F_i \cap F_j) = p^2, i \neq j$, the upper and lower bounds on the probability of failure P_f of this system are

$$P_{f,u} = n\,p - (n-1)\,p^2 \quad \text{and} \quad P_{f,l} = p + \sum_{i=2}^{n} \max\left(0, p - (i-1)\,p^2\right).$$

These bounds are shown in Fig. 2.1 for $p = 0.1$ as a function of the system size n. The bounds are relatively wide and deteriorate as n increases. ◇

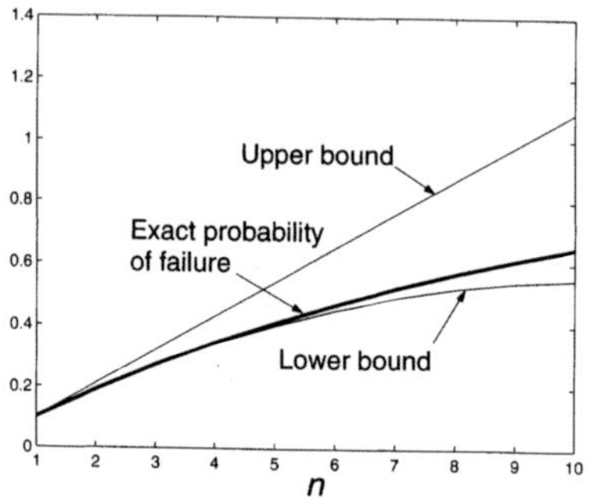

Figure 2.1. Bounds on the probability of failure for a series system

Proof: The equality $P(F_i \cap F_j) = p^2, i \neq j$, is valid if F_i and F_j are independent events, as we will see later in this section (Eq. 2.54). We have (Eq. 2.7)

$$P_f = P(\cup_{i=1}^{n} F_i) = P(F_1) + P(F_1^c \cap F_2) + P\left((F_1 \cup F_2)^c \cap F_3\right) + \cdots$$
$$+ P\left((F_1 \cup \cdots \cup F_{n-1})^c \cap F_n\right)$$

so that

$$P_f = P(F_1) + \sum_{i=2}^{n} P(S_1 \cap \cdots \cap S_{i-1} \mid F_i)\, P(F_i), \quad n \geq 2,$$

where $S_i = F_i^c$. The conditional probabilities $P(S_1 \cap \cdots \cap S_{i-1} \mid F_i)$ are smaller and larger than $P(S_j \mid F_i)$, $j = 1, \ldots, i-1$, and $1 - \sum_{j=1}^{i-1} P(F_j \mid F_i)$, respectively ([53], pp. 236-237). These inequalities give the bounds of Eq. 2.9. ■

2.3 Construction of probability spaces

The starting point in many applications involving probabilistic models is an experiment rather than a probability space. We need to construct a probability space based on the available information. This section illustrates three constructions of a probability space.

2.3.1 Countable sample space

Let $\Omega = \{\omega_1, \omega_2, \ldots\}$ be a sample space and consider a collection of numbers $p_i \geq 0$, $i = 1, 2, \ldots$, such that $\sum_{i=1}^{\infty} p_i = 1$. Take $\mathcal{F}$ to be the power set of Ω, that is, the collection of all subsets of Ω.

The function $P : \mathcal{F} \to [0, 1]$ defined by

$$P(A) = \sum_{\omega_i \in A} p_i, \quad A \in \mathcal{F} \tag{2.10}$$

is a probability measure on $(\Omega, \mathcal{F})$.

Proof: The set function P is positive for any $A \in \mathcal{F}$, $P(\Omega) = \sum_{i=1}^{\infty} p_i = 1$ by definition, and is countably additive since

$$P\left(\cup_{j\geq 1} A_j\right) = \sum_{\omega_i \in \cup_{j\geq 1} A_j} p_i = \sum_{j\geq 1} \sum_{\omega_i \in A_j} p_i = \sum_{j\geq 1} P(A_j)$$

provided that A_j, $j = 1, 2, \ldots$, are mutually disjoint events ([151], p. 41). ■

2.3.2 Product probability space

Let $p > 0$ be the probability that maximum yearly wind speed $V \geq 0$ at a site exceeds a value v_{cr} deemed critical for a facility, that is, the probability of the event $A = \{V > v_{\text{cr}}\}$. The associated sample space, σ-field, and probability measure are $\Omega = [0, \infty)$, $\mathcal{F} = \{\emptyset, A, A^c, \Omega\}$, and $P(A) = p$, respectively. To evaluate the probability that this facility performs satisfactorily during its design life of n years we need to construct a new probability space for an "experiment" consisting of n repetitions of the maximum yearly wind speed. This engineering application resembles the experiment of tossing a loaded coin n times ([151], p. 41).

Consider two probability spaces $(\Omega_k, \mathcal{F}_k, P_k)$, $k = 1, 2$, describing two experiments. These two experiments can be characterized jointly by the **product probability space** $(\Omega, \mathcal{F}, P)$ with the following components.

Product sample space:

$$\Omega = \Omega_1 \times \Omega_2 = \{(\omega_1, \omega_2) : \omega_k \in \Omega_k, k = 1, 2\}. \tag{2.11}$$

Product σ-field:

$$\mathcal{F} = \mathcal{F}_1 \times \mathcal{F}_2 = \sigma(\mathcal{R}), \quad \text{where} \tag{2.12}$$

$$\mathcal{R} = \{A_1 \times A_2 : A_1 \in \mathcal{F}_1, A_2 \in \mathcal{F}_2\} = \text{ measurable rectangles.} \tag{2.13}$$

Product probability measure, $P = P_1 \times P_2$, on the measurable space $(\Omega, \mathcal{F})$: The probability P is unique and has the property

$$P(A_1 \times A_2) = P_1(A_1)\, P(A_2), \quad A_1 \in \mathcal{F}_1, A_2 \in \mathcal{F}_2. \tag{2.14}$$

Note: The **product sample space** Ω contains the outcomes of both experiments generating the sample spaces Ω_1 and Ω_2. Let

$$\mathcal{G}_1 = \{A_1 \times \Omega_2 : A_1 \in \mathcal{F}_1\} \quad \text{and} \quad \mathcal{G}_2 = \{\Omega_1 \times A_2 : A_2 \in \mathcal{F}_2\}$$

be two collections of subsets of Ω. These collections are σ-fields on Ω, are included in $\mathcal{F}$, and $\mathcal{F} = \sigma(\mathcal{R}) = \sigma(\mathcal{G}_1, \mathcal{G}_2)$ since every member of $\mathcal{R}$ is the intersection of sets from $\mathcal{G}_1$ and $\mathcal{G}_2$.

The construction of the **product probability measure** is less simple. It can be shown that there exists a unique probability P on $(\Omega, \mathcal{F})$ such that it satisfies Eq. 2.14 ([40], Theorem 3.3.5, p. 59). ▲

Example 2.9: Let $\Omega_k = \{1, 2, 3, 4, 5, 6\}$, $k = 1, 2$, be sample spaces associated with the experiment of rolling two dice. The σ-fields $\mathcal{F}_k$ on these spaces consist of all subsets of Ω_k. The probability measures on Ω_k are $P_k(\{i\}) = 1/6$, $i = 1, \ldots, 6$.

The product sample space Ω, the product σ-field $\mathcal{F}$, and the product probability measure P are

$$\Omega = \Omega_1 \times \Omega_2 = \{\omega = (i, j)\} = \begin{bmatrix} (1,1) & (1,2) & \cdots & (1,6) \\ (2,1) & (2,2) & \cdots & (2,6) \\ (3,1) & (3,2) & \cdots & (3,6) \\ (4,1) & (4,2) & \cdots & (4,6) \\ (5,1) & (5,2) & \cdots & (5,6) \\ (6,1) & (6,2) & \cdots & (6,6) \end{bmatrix},$$

all subsets of Ω, and $P(\{\omega\}) = 1/36$, respectively. ◇

Note: The sample space Ω corresponds to the experiment of rolling two dice. The product σ-field consists of all subsets of Ω since the members of $\mathcal{R}$ are (i, j), $\cup_{i \in I_1}(i, j)$, $\cup_{j \in I_2}(i, j)$, $\cup_{i \in I_1, j \in I_2}(i, j)$, where $I_1, I_2 \subseteq \{1, 2, 3, 4, 5, 6\}$. The product probability measure can be defined for each outcome $\omega = (i, j)$ and is $P(\{\omega\}) = 1/36$ because the members of Ω are equally likely. Also, $P(\{\omega\} = (i, j))$ is equal to $P_1(\{i\})\, P_2(\{j\}) = (1/6)\,(1/6)$. ▲

Example 2.10: Consider the same experiment as in the previous example but assume that the σ-fields on $\Omega_1 = \Omega_2 = \{1, 2, 3, 4, 5, 6\}$ are

$$\mathcal{F}_1 = \{A_1 = \{1, 2\}, A_1^c, \emptyset, \Omega_1\} \quad \text{and} \quad \mathcal{F}_2 = \{A_2 = \{1, 2, 3\}, A_2^c, \emptyset, \Omega_2\}.$$

The probabilities of A_1 and A_2 are $P_1(A_1) = 2/6$ and $P_2(A_2) = 3/6$.

The product probability space is $\Omega = \{\omega = (i, j), i, j = 1, \ldots, 6\}$. The product σ-field coincides with the collection of measurable sets

$$\{A_1 \times A_2, A_1 \times A_2^c, A_1^c \times A_2, A_1^c \times A_2^c\},$$

unions of the members of this collection, and sets in which at least one component is the empty set, for example, the set $\emptyset \times A_2$. The product probability measure is given by Eq. 2.14. ◇

Note: The product sample space Ω is as in Example 2.9 but the product σ-field is much coarser than in this example because fewer events are relevant. The definition of the product probability measure is consistent with our intuition. For example, $P(A_1 \times A_2) = 6/36$ and coincides with the product of $P_1(A_1)$ and $P_2(A_2)$. ▲

The formulas of Eqs. 2.11-2.14 can be generalized to define a product probability space for three or more probability spaces. Consider the probability spaces $(\Omega_k, \mathcal{F}_k, P_k)$, $k = 1, 2, \ldots, n$.

- If n is finite, the product probability space $(\Omega, \mathcal{F}, P)$ is defined by

$$\begin{aligned} \Omega &= \Omega_1 \times \cdots \times \Omega_n, \\ \mathcal{F} &= \mathcal{F}_1 \times \cdots \times \mathcal{F}_n, \\ P &= P_1 \times \cdots \times P_n. \end{aligned} \tag{2.15}$$

- If n is infinity, the definition of Eq. 2.15 applies with n replaced by ∞.

Note: The last equalities in Eq. 2.15 are commonly used notations. If the probability spaces $(\Omega_k, \mathcal{F}_k, P_k)$ are identical, the product sample space, σ-field, and probability are denoted by Ω_1^n, $\mathcal{F}_1^n$, and P_1^n for $n < \infty$ and Ω_1^∞, $\mathcal{F}_1^\infty$, and P_1^∞ for $n = \infty$. ▲

Example 2.11: A loaded coin with sides $\{1\}$ and $\{0\}$ and probabilities $p \in (0, 1)$ and $q = 1 - p$, respectively, is tossed n times. The probability space for a single toss is $\Omega = \{0, 1\}$, $\mathcal{F} = \{\emptyset, \Omega, \{0\}, \{1\}\}$, $P(\{1\}) = p$, and $P(\{0\}) = q$. The corresponding elements of the product probability space for n tosses are

$$\Omega^n = \{\omega = (\omega_1, \ldots, \omega_n) : \omega_i = 0 \text{ or } 1\},$$

$\mathcal{F}^n$ = all subsets of Ω^n, and

$$P^n(A) = \sum_{\omega \in A} P(\{\omega\}) = \sum_{\omega \in A} p^{n_\omega} q^{n - n_\omega}, \quad A \in \mathcal{F}^n,$$

where $n_\omega = \sum_{i=1}^n \omega_i$ gives the number of 1's in $\omega = (\omega_1, \ldots, \omega_n)$. ◇

Note: We have

$$P^n(\Omega^n) = \sum_{\omega_1, \ldots, \omega_n} p^{n_\omega} q^{n - n_\omega} = \sum_{\omega_1, \ldots, \omega_{n'}} p^{n'_\omega} q^{n' - n'_\omega} (p^1 q^0 + p^0 q^1) = \cdots = 1$$

since $p^1 q^0 + p^0 q^1 = 1$, where $n' = n - 1$ and $n'_\omega = \sum_{i=1}^{n'} \omega_i$ so that $P^n(\Omega^n) = 1$. The set function P^n is also positive and countably additive ([151], Section 2.3). ▲

2.3.3 Extension of probability measure

Consider an experiment with a sample space Ω. Assume that a set function R defined on a collection $\mathcal{C}$ of subsets of Ω is available and that this function is real-valued, positive, countably additive, and $R(\Omega) = 1$. For calculation purposes we need to extend R to $\mathcal{F} = \sigma(\mathcal{C})$ such that its extension is a probability measure on the measurable space $(\Omega, \mathcal{F})$. The following theorem states conditions under which the extension of R to $\mathcal{F}$ is a probability measure. For proof and additional information on this topic, see [151] (Section 2.4).

If (1) $\mathcal{C}$ is a field on Ω and (2) R is a real-valued, positive, and countably additive function defined on $\mathcal{C}$ such that $R(\Omega) = 1$, then there exists a unique probability P on $\mathcal{F} = \sigma(\mathcal{C})$ such that $P(A) = R(A)$ for each $A \in \mathcal{C}$, that is, the restriction of P to $\mathcal{C}$ is equal to R ([66], Theorem 14, p. 94).

Example 2.12: Let $\Omega = \mathbb{R}$ and let $\mathcal{C}$ be the collection of all finite unions of intervals of the type $(a, b]$ for $a < b$, $(-\infty, a]$, (a, ∞), and $(-\infty, \infty)$ to which we add the empty set. Let $F : \mathbb{R} \to [0, 1]$ be a continuous increasing function such that $\lim_{x\to-\infty} F(x) = 0$ and $\lim_{x\to\infty} F(x) = 1$. Define $R : \mathcal{C} \to [0, 1]$ by $R((a, b]) = F(b) - F(a)$, $R((-\infty, a]) = F(a)$, $R((a, \infty)) = 1 - F(a)$, and $R((\infty, \infty)) = 1$. This set function can be extended uniquely to a probability measure on $(\mathbb{R}, \mathcal{B})$. ◇

Note: The collection of subsets $\mathcal{C}$ is a field and $\sigma(\mathcal{C}) = \mathcal{B}$. The set function R is well defined because, for example, $R((a, b] \cup (b, c]) = R((a, c])$ and $R(\mathbb{R}) = 1$, and it is finitely additive. Moreover, R is countably additive ([66], Proposition 9, p. 90). The above theorem implies the stated result. ▲

2.3.4 Conditional probability

Let $(\Omega, \mathcal{F}, P)$ be a probability space and $B \in \mathcal{F}$ an event with $P(B) > 0$. We define a new probability measure on $(\Omega, \mathcal{F})$ under the assumption that B has occurred.

The **probability of** A **conditional on** B is

$$P(A \mid B) = \frac{P(A \cap B)}{P(B)} \quad \text{for } A, B \in \mathcal{F} \text{ and } P(B) > 0. \tag{2.16}$$

Note: The set function $P(\cdot \mid B)$ is a probability because it is defined on $\mathcal{F}$, takes positive values, $P(\Omega \mid B) = P(B)/P(B) = 1$, and is countably additive, that is,

$$\begin{aligned} P(\cup_{i\in I} A_i \mid B) &= P((\cup_{i\in I} A_i) \cap B)/P(B) = P(\cup_{i\in I} (A_i \cap B))/P(B) \\ &= \sum_{i\in I} P(A_i \cap B)/P(B) = \sum_{i\in I} P(A_i \mid B) \end{aligned}$$

for any countable set I and disjoint events A_i, $i \in I$. ▲

If B has not occurred in an experiment, then B^c has occurred. The probability of A given that B^c has been observed is given by Eq. 2.16 with B^c in place of B provided that $P(B^c) > 0$. Hence, the conditional probability of A can take the values $P(A \cap B)/P(B)$ or $P(A \cap B^c)/P(B^c)$ depending on whether B or B^c has been observed. In Example 2.86 we extend the definition of Eq. 2.16 to give the probability of A given the information content of a sub-σ-field $\mathcal{G}$ of $\mathcal{F}$.

Let $A_i \in \mathcal{F}$, $i = 1, \ldots, n$, be a partition of Ω such that $P(A_i) > 0$, and let $B \in \mathcal{F}$ be an arbitrary event. The following two properties of the conditional probability are very useful for calculations.

The **law of total probability**:

$$P(B) = \sum_{i=1}^{n} P(B \cap A_i) = \sum_{i=1}^{n} P(B \mid A_i)\, P(A_i), \tag{2.17}$$

and

The **Bayes formula**:

$$P(A_j \mid B) = \frac{P(A_j)\, P(B \mid A_j)}{P(B)} = \frac{P(A_j)\, P(B \mid A_j)}{\sum_{i=1}^{n} P(A_i)\, P(B \mid A_i)}, \quad P(B) > 0. \tag{2.18}$$

Proof: The events $A_i \cap B$, $i = 1, \ldots, n$, are disjoint so that Eqs. 2.17 and 2.18 result from Eq. 2.6 and the definition of the conditional probability (Eq. 2.16). To apply the Bayes formula, the probability $P(B)$ must be strictly positive. The probabilities $P(A_i)$ and $P(A_j \mid B)$ in Eq. 2.18 can be interpreted as the **prior probability** of A_i and the **posterior probability** of A_j given the occurrence of B, respectively. $P(B \mid A_i)$ is the probability of B conditional on A_i. ■

Example 2.13: Consider the experiment of rolling two dice. The sample space, the σ-field $\mathcal{F}$, and the probability measure are $\Omega = \{\omega = (i, j) : i, j = 1, \ldots, 6\}$, the collection of all parts of Ω, and $P(\omega) = 1/36$, respectively. Consider the events $A_1 = (6, 2)$ and $A_2 = \{(6, 2), (4, 4), (1, 6)\}$. The probabilities of these events given that $B = \{\omega \in \Omega : i + j = 8\}$ has occurred are $P(A_1 \mid B) = 1/5$ and $P(A_2 \mid B) = 2/5$. ◇

Proof: The event B is $\{(6, 2), (5, 3), (4, 4), (3, 5), (2, 6)\}$. The probability of A_1 conditional on B is $P(A_1 \mid B) = 1/5$ since A_1 is one of the five equally likely members of B. The same result can be obtained from Eq. 2.16 since $P(A_1 \cap B) = P(A_1) = 1/36$ and $P(B) = 5/36$. The conditional probability of A_1 given B^c is zero. Hence, the probability of occurrence of A_1 is 1/5 and zero if B and B^c has occurred, respectively. The conditional probabilities of A_2 under B and B^c are $P(A_2 \mid B) = (2/36)/(5/36) = 2/5$ and $P(A_2 \mid B^c) = (1/36)/(31/36) = 1/31$ since $P(B) = 5/36$ and $P(B^c) = 1 - P(B) = 31/36$. We will see in Example 2.86 that the probabilities $P(A_2 \mid B)$, $P(A_2 \mid B^c)$, $P(B)$, and $P(B^c)$ are sufficient to specify the conditional probability $P(A_2 \mid \mathcal{G})$, where $\mathcal{G} = \{\emptyset, \Omega, B, B^c\}$ is the σ-field generated by B. ■

Example 2.14: Let X be the unknown strength of a physical system, that is, the maximum load that the system can sustain. Suppose that the system survives without any damage if subjected to a proof load test of intensity x_{pr}. Consider the events $A = \{X > x\}$ and $B = \{X > x_{\text{pr}}\}$ that the system strength exceeds x and x_{pr}, respectively. The probability that the system can sustain a load x given that B has occurred,

$$P_s(x) = P(X > \max\{x_{\text{pr}}, x\})/P(X > x_{\text{pr}}),$$

is larger than the system reliability $P(X > x)$ calculated when the information provided by the proof load test is not available or is ignored. ◇

Example 2.15: A system is subjected to two actions described by the events B_1 and B_2, where $P(B_1) = 0.8$ and $P(B_2) = 0.2$. Let $P(A \mid B_1) = 0.9$ and $P(A \mid B_2) = 0.7$ be the system reliability under action B_1 and B_2, respectively, where A is the event that the system survives. By the law of total probability (Eq. 2.17), the system reliability is $P_s = P(A) = P(A \mid B_1)P(B_1) + P(A \mid B_2)P(B_2)$ so that $P_s = (0.9)(0.8) + (0.7)(0.2) = 0.86$. ◇

2.3.5 Sequence of sets

Let A_i, $i = 1, 2, \ldots$, be subsets of a set Ω. The **limit supremum** and the **limit infimum** of this sequence are

$$\limsup_{i\to\infty} A_i = \cap_{j=1}^{\infty} \cup_{i=j}^{\infty} A_i = \cap_{j=1}^{\infty} B_j, \quad B_j = \cup_{i=j}^{\infty} A_i \downarrow, \tag{2.19}$$

$$\liminf_{i\to\infty} A_i = \cup_{j=1}^{\infty} \cap_{i=j}^{\infty} A_i = \cup_{j=1}^{\infty} C_j, \quad C_j = \cap_{i=j}^{\infty} A_i \uparrow. \tag{2.20}$$

Note: The limits in Eqs. 2.19-2.20 are subsets of Ω. The sequence of events $B_j = \cup_{i=j}^{\infty} A_i$ and $C_j = \cap_{i=j}^{\infty} A_i$ are monotone, B_j is a decreasing sequence, that is, $B_j \supseteq B_{j+1}$, while C_j is an increasing sequence, that is, $C_j \subseteq C_{j+1}$, for all j. These properties are denoted by $B_j \downarrow$ and $C_j \uparrow$, respectively. ▲

The limits in Eqs. 2.19 and 2.20 have some notable properties.

$$\limsup_{i\to\infty} A_i = \{A_i \text{ i.o.}\} = \left\{\omega : \sum_{i=1}^{\infty} 1_{A_i}(\omega) = \infty\right\}$$
$$= \{\omega : \omega \in A_{i_j}, j = 1, 2, \ldots\} \tag{2.21}$$

for some subsequence i_j depending on ω, where i.o. means **infinitely often**.

Proof: Recall that the function $1_A : \Omega \to \{0, 1\}$, called the **indicator function**, is defined by $1_A(\omega) = 1$ for $\omega \in A$ and $1_A(\omega) = 0$ for $\omega \notin A$.

If $\omega \in \limsup_i A_i$, then $\omega \in B_j = \cup_{i=j}^{\infty} A_i$, $\forall j$, so that there exists $i_j \geq j$ such that $\omega \in A_{i_j}$, $j = 1, 2, \ldots$, and $\sum_{i=1}^{\infty} 1_{A_i}(\omega) \geq \sum_j 1_{A_{i_j}}(\omega) = \infty$ implying that $\omega \in \{\omega : \sum_{i=1}^{\infty} 1_{A_i}(\omega) = \infty\}$. Therefore, we have $\limsup_i A_i \subset \{\omega : \sum_{i=1}^{\infty} 1_{A_i}(\omega) = \infty\}$.

If $\omega \in \{\omega : \sum_{i=1}^{\infty} 1_{A_i}(\omega) = \infty\}$, there exists $i_j \to \infty$, $j = 1, 2, \ldots$, such that $\omega \in A_{i_j}$. Hence, $\omega \in \cup_{k \geq j} A_k$ for all j's so that $\omega \in \limsup_i$ or $\{\omega : \sum_{i=1}^{\infty} 1_{A_i(\omega)} = \infty\} \subset \limsup_i A_i$ ([151], Lemma 1.3.1, p. 6). ■

$$\liminf_{i \to \infty} A_i = \{\omega : \omega \in A_n \text{ for all } n \text{ except for a finite number}\}$$
$$= \left\{\omega : \sum_{i=1}^{\infty} 1_{A_i^c}(\omega) < \infty\right\} = \{\omega : \omega \in A_i, \forall i \geq j_0(\omega)\}. \quad (2.22)$$

Proof: If $\omega \in \liminf_i A_i$, there exists $j_0(\omega)$ such that $\omega \in C_{j_0} = \cap_{i=j_0}^{\infty} A_i$, that is, ω belongs to all sets A_i for $i \geq j_0(\omega)$, or ω does not belong to the sets $A_1, \ldots, A_{j_0(\omega)}$ so that $\sum_{i=1}^{\infty} 1_{A_i^c}(\omega)$ is finite.

If ω is such that $\sum_{i=1}^{\infty} 1_{A_i^c}(\omega)$ is finite, then ω does not belong to a finite number of subsets A_i. Hence, there exists $j_0(\omega)$ such that $\omega \in A_i$ for $i \geq j_0$ or $\omega \in \cap_{i=j_0}^{\infty} A_i$. ■

$$\liminf_i A_i = (\limsup_i A_i^c)^c \subseteq \limsup_i A_i. \quad (2.23)$$

Proof: If $\omega \in \liminf_i A_i$, there exists j_0 such that $\omega \in C_{j_0}$, that is, $\omega \in A_i$ for $i \geq j_0$. Hence, $\omega \in A_i$ infinitely often, that is, $\omega \in \limsup_i A_i$. De Morgan's law $\left(\cup_{j=1}^{\infty} \cap_{i=j}^{\infty} A_i\right)^c = \cap_{j=1}^{\infty} \cup_{i=j}^{\infty} A_i^c$ yields Eq. 2.23. ■

If the limits in Eqs. 2.19 and 2.20 coincide, the sequence A_i is said to be a **convergent sequence** with the **limit**

$$\lim_{i \to \infty} A_i = \liminf_{i \to \infty} A_i = \limsup_{i \to \infty} A_i. \quad (2.24)$$

Example 2.16: Let $A_i = \{(x_1, x_2) \in \mathbb{R}^2 : x_1 \in [0, i), x_2 \in [0, 1/i)\}$, $i = 1, 2, \ldots$, be a sequence of sets in $\mathbb{R}^2$. The sequences of events $B_j = \cup_{i \geq j} A_i$ and $C_j = \cap_{i \geq j} A_i$ have the same limit $[0, \infty) \times \{0\}$ as $j \to \infty$. Hence, $\{A_i\}$ is a convergent sequence and $\lim_{i \to \infty} A_i = [0, \infty) \times \{0\}$. ◇

Example 2.17: Consider a monotone sequence of subsets A_i. This sequence is convergent with $\lim_{i \to \infty} A_i = \cup_{i=1}^{\infty} A_i$ and $\lim_{i \to \infty} A_i = \cap_{i=1}^{\infty} A_i$ if $A_i \uparrow$ and $A_i \downarrow$, respectively. ◇

Proof: We need to show that $\limsup A_i$ and $\liminf A_i$ are equal. If $A_i \uparrow$, then $B_j = \cup_{i \geq j} A_i = \cup_{i \geq 1} A_i$ so that $\limsup A_i = \cup_{i \geq 1} A_i$. The sequence $C_j = \cap_{i \geq j} A_i$ in Eq. 2.20 is A_j so that $\liminf A_i = \cup_{i \geq 1} A_i = \lim A_i$. Similar considerations give the limit of a decreasing sequence of subsets. ■

2.3.6 Sequence of events

Let $(\Omega, \mathcal{F}, P)$ be a probability space and consider a sequence of events $A_i \in \mathcal{F}$, $i = 1, 2, \ldots$, on this space. The following three statements relate to the continuity of the probability measure. The last result in this section is the Borel-Cantelli lemma, a useful tool for proving almost sure convergence.

Continuity of probability measure. If $A_i \uparrow A$ or $A_i \downarrow A$, then $P(A_i) \uparrow P(A)$ or $P(A_i) \downarrow P(A)$, respectively, where $A = \lim_{i \to \infty} A_i$.

Proof: If A_i is an increasing sequence, then $D_1 = A_1$ and $D_i = A_i \setminus A_{i-1}, i = 1, 2, \ldots$, provide a partition of $\cup_{i=1}^{\infty} A_i$ so that

$$
\begin{aligned}
P(\lim_{i \to \infty} A_i) &= P(A) = P\left(\cup_{i=1}^{\infty} D_i\right) = \sum_{i=1}^{\infty} P(D_i) = \lim_{n \to \infty} \sum_{i=1}^{n} P(D_i) \\
&= \lim_{n \to \infty} P\left(\cup_{i=1}^{n} D_i\right) = \lim_{n \to \infty} P(A_n)
\end{aligned}
$$

since P is a countably additive function. By the first property in Eq. 2.5, the numerical sequence $P(A_n)$ is increasing. Similar consideration can be applied for the case $A_i \downarrow$. If A_i is a decreasing sequence such that $\cap_{i=1}^{\infty} A_i = \emptyset$, we have $\lim_{i \to \infty} P(A_i) = P(\cap_{i=1}^{\infty} A_i) = P(\emptyset) = 0$. ■

If a sequence of events A_i is convergent with limit $A = \lim_i A_i$, then

$$
\lim_{i \to \infty} P(A_i) = P(\lim_{i \to \infty} A_i) = P(A), \tag{2.25}
$$

that is, for a convergent sequence of events, **probability and limit operations can be interchanged.**

Proof: A direct consequence of Eq. 2.25 is that for any sequence of events A_i the equalities

$$
\begin{aligned}
P(\limsup_{i \to \infty} A_i) &= \lim_{j \to \infty} P(B_j) = \lim_{j \to \infty} P(\cup_{i=j}^{\infty} A_i) \quad \text{and} \\
P(\liminf_{i \to \infty} A_i) &= \lim_{j \to \infty} P(C_j) = \lim_{j \to \infty} P(\cap_{i=j}^{\infty} A_i)
\end{aligned}
$$

hold because $B_j = \cup_{i=j}^{\infty} A_i$ and $C_j = \cap_{i=j}^{\infty} A_i$ are decreasing and increasing sequences, respectively, $\limsup A_i = \lim B_j$, and $\liminf A_i = \lim C_j$.

Since A_i is a convergent sequence of events, the decreasing and increasing sequences $D_j = \sup_{i \geq j} A_i$ and $E_j = \inf_{i \geq j} A_i$ respectively, converge to A. The inequalities $P(E_j) \leq P(A_j) \leq P(D_j)$ hold for each j because $E_j \subseteq A_j \subseteq D_j$. By the continuity of the probability measure, $P(E_j)$ and $P(D_j)$ converge to $P(A)$ as $j \to \infty$ so that the limit of $P(A_j)$ is $P(A)$. ■

Fatou's lemma. The probability measure satisfies the inequalities

$$P(\liminf_i A_i) \le \liminf_i P(A_i) \le \limsup_i P(A_i) \le P(\limsup_i A_i) \qquad (2.26)$$

for any sequence of events A_i.

Proof: Because $\limsup_i A_i$ and $\liminf_i A_i$ are events as countable unions and intersections of events (Eqs. 2.19 and 2.20), we can calculate $P(\limsup_i A_i)$ and $P(\liminf_i A_i)$. The inequality $P(\liminf_i A_i) \le P(\limsup_i A_i)$ follows from Eqs. 2.5 and 2.23. Because $P(A_i)$ is a numerical series, it satisfies the inequality $\liminf_i P(A_i) \le \limsup_i P(A_i)$. The proof of the remaining inequalities is left to the reader.

If the sequence A_i is convergent, $\limsup_i A_i = \liminf_i A_i = \lim_i A_i$ so that the left and right terms of Eq. 2.26 coincide. This observation provides an alternative proof for Eq. 2.25. ■

Borel-Cantelli Lemma. If $\sum_i P(A_i) < \infty$, then

$$P(\limsup_i A_i) = P(A_i \text{ i.o.}) = 0. \qquad (2.27)$$

Proof: Because $\limsup_i A_i \subseteq B_j$ and $B_j = \cup_{i \ge j} A_i$ (Eq. 2.19), we have

$$P(\limsup_i A_i) \le P\left(\cup_{i=j}^{\infty} A_i\right) \le \sum_{i=j}^{\infty} P(A_i), \quad j = 1, 2, \ldots,$$

by the monotonicity and subadditivity of the probability measure. The extreme right term in these inequalities converges to zero as $j \to \infty$ because it is assumed that $\sum_i P(A_i) < \infty$. Hence, $P(\limsup_i A_i) = 0$. ■

2.4 Measurable functions

Let $(\Omega, \mathcal{F})$ and $(\Psi, \mathcal{G})$ be two measurable spaces and consider a function $h : \Omega \to \Psi$ with domain Ω and range Ψ.

The function h is said to be **measurable from** $(\Omega, \mathcal{F})$ **to** $(\Psi, \mathcal{G})$ if

$$h^{-1}(B) = \{\omega : h(\omega) \in B\} \in \mathcal{F}, \quad \forall B \in \mathcal{G}. \qquad (2.28)$$

Note: It is common to indicate this property of h by the **notation** $h : (\Omega, \mathcal{F}) \to (\Psi, \mathcal{G})$, $h \in \mathcal{F}/\mathcal{G}$, or even $h \in \mathcal{F}$ provided that there can be no confusion about the σ-field $\mathcal{G}$. If the σ-fields $\mathcal{F}$ and $\mathcal{G}$ are not in doubt, it is sufficient to say that h is measurable. ▲

2.4.1 Properties

If $h : (\Omega, \mathcal{F}) \to (\Psi, \mathcal{G})$ is measurable and P is a probability measure on $(\Omega, \mathcal{F})$, then $Q : \mathcal{G} \to [0, 1]$, defined by

$$Q(B) = P(h^{-1}(B)), \quad B \in \mathcal{G}, \tag{2.29}$$

is a probability measure on $(\Psi, \mathcal{G})$, called the **probability induced by** h or the **distribution** of h.

Proof: The function Q has the properties $Q(\Psi) = 1$ by definition and $Q(\cup_{i \in I} B_i) = \sum_{i \in I} Q(B_i)$ for any countable set I and disjoint events $B_i \in \mathcal{G}$, since the sets $h^{-1}(B_i)$ are disjoint events in $\mathcal{F}$ and P is a probability. Hence, the set function Q is a probability measure on $(\Psi, \mathcal{G})$. ■

If $\mathcal{C}$ is a collection of subsets of Ψ, then $h^{-1}(\sigma(\mathcal{C})) = \sigma(h^{-1}(\mathcal{C}))$, where $h^{-1}(\sigma(\mathcal{C}))$ and $\sigma(h^{-1}(\mathcal{C}))$ are the inverse image of $\sigma(\mathcal{C})$ and the σ-field generated by $h^{-1}(\mathcal{C})$ in Ω, respectively ([151], Proposition 3.1.2, p. 73).

Let $\mathcal{C}$ be a collection of subsets of Ψ such that $\sigma(\mathcal{C}) = \mathcal{G}$. The function h is measurable if and only if $h^{-1}(\mathcal{C}) \subseteq \mathcal{F}$ ([151], Proposition 3.2.1, p. 76).

If $g : \mathbb{R}^d \to \mathbb{R}^q$ is a continuous function, then g is $\mathcal{B}^d/\mathcal{B}^q$-measurable.

Proof: Let $\mathcal{D}^r$ denote the topology on $\mathbb{R}^r$ generated by the open intervals of $\mathbb{R}^r$. Because g is a continuous function, we have $g^{-1}(D) \in \mathcal{D}^d$ for all $D \in \mathcal{D}^q$. This function is measurable since the Borel fields $\mathcal{B}^d$ and $\mathcal{B}^q$ are generated by the open sets in $\mathbb{R}^d$ and $\mathbb{R}^q$, respectively. ■

2.4.2 Definition of random variable

If a probability measure P is attached to the measurable space $(\Omega, \mathcal{F})$, the function h in Eq. 2.28 is called a **random variable**. We consider primarily random variables corresponding to $(\Psi, \mathcal{G}) = (\mathbb{R}^d, \mathcal{B}^d)$ and denote them by **upper case** letters, for example, $\boldsymbol{X} : \Omega \to \mathbb{R}^d$ for $d > 1$ and $X : \Omega \to \mathbb{R}$ for $d = 1$. Random variables with values in $\mathbb{R}^d$, $d > 1$, are also called **random vectors**.

The σ-**field generated by** a random variable $\boldsymbol{X}$ is

$$\sigma(\boldsymbol{X}) = \mathcal{F}^{\boldsymbol{X}} = \boldsymbol{X}^{-1}(\mathcal{B}^d) = \{\boldsymbol{X}^{-1}(B) : \forall B \in \mathcal{B}^d\} \tag{2.30}$$

and represents the smallest σ-field with respect to which $\boldsymbol{X}$ is measurable.

Proof: The collection $\sigma(\boldsymbol{X})$ of subsets of Ω is a σ-field because (1) $\mathbb{R}^d \in \mathcal{B}^d$ so that $\Omega = \boldsymbol{X}^{-1}(\mathbb{R}^d) \in \sigma(\boldsymbol{X})$, (2) if $B \in \mathcal{B}^d$, then $\boldsymbol{X}^{-1}(B) \in \sigma(\boldsymbol{X})$ and $(\boldsymbol{X}^{-1}(B))^c = \boldsymbol{X}^{-1}(B^c) \in$

$\sigma(X)$, and (3) if $B_i \in \mathcal{B}^d$, $i \in I$, we have $X^{-1}(B_i) \in \sigma(X)$ and $\cup_{i \in I} X^{-1}(B_i) = X^{-1}(\cup_{i \in I} B_i) \in \sigma(X)$, where I is a countable index set.

If $\mathcal{H}$ is another σ-field relative to which X is measurable, this field must include the subsets $X^{-1}(\mathcal{B}^d)$ of Ω so that $\mathcal{H}$ includes $\sigma(X)$. ■

The **distribution** of X is the probability induced by the mapping $X : \Omega \to \mathbb{R}^d$ on the measurable space $(\mathbb{R}^d, \mathcal{B}^d)$, that is, (Eq. 2.29)

$$Q(B) = P(X^{-1}(B)), \quad B \in \mathcal{B}^d. \tag{2.31}$$

The mapping $X : (\Omega, \mathcal{F}) \to (\mathbb{R}, \mathcal{B})$ is a random variable if and only if $X^{-1}((-\infty, x]) \in \mathcal{F}$, $x \in \mathbb{R}$ ([151], Corollary 3.2.1, p. 77).

Note: This result is used extensively in applications to determine whether a function is measurable and to find properties of random variables. ▲

The $\mathbb{R}^d$-valued function $X : (\Omega, \mathcal{F}) \to (\mathbb{R}^d, \mathcal{B}^d)$ is a random vector if and only if its coordinates are random variables.

Note: If X is a random vector, its coordinates $X_i = \pi_i \circ X$, $i = 1, \ldots, d$, are random variables because the projection map $\pi_i(x) = x_i$ is continuous.

Suppose now that X_i are random variables. Let $\mathcal{R}$ be the collection of open rectangles in $\mathbb{R}^d$ with members $R = I_1 \times \cdots \times I_d$, where I_i are open intervals in $\mathbb{R}$. The σ-field $\mathcal{B}^d$ is generated by these rectangles, that is, $\mathcal{B}^d = \sigma(\mathcal{R})$ ([151], Proposition 3.2.4, p. 81), so that it is sufficient to show that $X^{-1}(\mathcal{R})$ is in $\mathcal{F}$. We have $X^{-1}(R) = \cap_{i=1}^d X_i^{-1}(I_i)$ so that $X^{-1}(R) \in \mathcal{F}$ since X_i are random variables. This property shows that X is a random vector if and only if $\{\omega \in \Omega : X_i(\omega) \leq x_i\} \in \mathcal{F}$ for all $x_i \in \mathbb{R}$, $i = 1, \ldots, d$.

These considerations can be extended to a countable collection $(X_1, X_2, \ldots)$ of random variables, referred to as a **time series** or a **discrete time stochastic process**. A time series $(X_1, X_2, \ldots)$ takes values in $\mathbb{R}^\infty$. The measurable spaces $(\Omega, \mathcal{F})$ and $(\mathbb{R}^\infty, \mathcal{B}^\infty)$ need to be considered to establish whether $(X_1, X_2, \ldots)$ is a measurable function. ▲

Example 2.18: Consider the experiment of rolling a die. Let $X, Y : \Omega \to \mathbb{R}$ be defined by $X(\omega) = 1$ and -1 for $\omega \leq 3$ and $\omega > 3$, respectively, and $Y(\omega) = \omega$, where $\Omega = \{1, 2, 3, 4, 5, 6\}$ and $P(\{\omega\}) = 1/6$. The σ-fields generated by X and Y are $\mathcal{F}^X = \{\emptyset, \{1, 2, 3\}, \{4, 5, 6\}, \Omega\}$ and $\mathcal{F}^Y = \mathcal{P}(\Omega)$, where $\mathcal{P}(\Omega)$ is the power set of Ω, that is, the collection of all subsets of the sample space. We have $X \in \mathcal{F}^X$, $Y \in \mathcal{F}^Y$, and $X \in \mathcal{F}^Y$ since $\mathcal{F}^X \subset \mathcal{F}^Y$. However, $Y \notin \mathcal{F}^X$ since the information content of $\mathcal{F}^X$ is too coarse to describe Y. ◇

Example 2.19: Let $X : (\mathbb{R}, \mathcal{B}) \to (\mathbb{R}, \mathcal{B})$ be an increasing function. Then X is measurable because $\mathcal{B}$ can be generated by the collection of intervals $(a, b]$, $a \leq b$, and $X^{-1}((a, b]) = (X^{-1}(a), X^{-1}(b)]$ is in $\mathcal{B}$. ◇

Example 2.20: Let $\Omega = [0, 1]$, $\mathcal{F} = \mathcal{B}([0, 1])$, and $P(d\omega) = d\omega$. If α and $\beta > \alpha$ are constants, the function $\omega \mapsto X(\omega) = \alpha + (\beta - \alpha)\,\omega$ is a random variable on $(\Omega, \mathcal{F}, P)$ since it is continuous. The distribution of X is $Q((x_1, x_2]) = (x_2 - x_1)/(\beta - \alpha)$. ◇

Proof: Take $B = (x_1, x_2]$ in the range of X. The distribution $Q((x_1, x_2])$ is equal to the probability $P((\omega_1, \omega_2])$ of the interval $X^{-1}((x_1, x_2])$, where $\omega_i = (x_i - \alpha)/(\beta - \alpha)$. ■

2.4.3 Measurable transformations

The determination of the output of a physical system subjected to a random input involves transformations of random variables. For example, suppose that the input to a system is modeled by an $\mathbb{R}^d$-valued random variable $\boldsymbol{X}$ and the objective is to characterize the system output $Y = g(\boldsymbol{X})$, where $g : \mathbb{R}^d \to \mathbb{R}$ depends on the system properties. The mapping

$$(\Omega, \mathcal{F}) \xrightarrow{\boldsymbol{X}} (\mathbb{R}^d, \mathcal{B}^d) \xrightarrow{g} (\mathbb{R}, \mathcal{B})$$

from $(\Omega, \mathcal{F})$ to $(\mathbb{R}, \mathcal{B})$ defines the output. A first question we need to pose is whether Y is a random variable.

If $(\Omega, \mathcal{F})$, $(\Psi, \mathcal{G})$, and $(\Phi, \mathcal{H})$ are measurable spaces and

$$h : (\Omega, \mathcal{F}) \to (\Psi, \mathcal{G}), \quad g : (\Psi, \mathcal{G}) \to (\Phi, \mathcal{H})$$

are measurable functions, then

$$g \circ h : (\Omega, \mathcal{F}) \to (\Phi, \mathcal{H}) \quad \text{is measurable.}$$

Proof: For $B \in \mathcal{H}$, $g^{-1}(B)$ is in $\mathcal{G}$ since g is measurable. We have $h^{-1}(A) \in \mathcal{F}$ for any $A \in \mathcal{G}$ since h is measurable, so that $h^{-1}(g^{-1}(B)) \in \mathcal{F}$. Hence, $g \circ h$ is measurable from $(\Omega, \mathcal{F})$ to $(\Phi, \mathcal{H})$. ■

Example 2.21: If $X : (\Omega, \mathcal{F}) \to (\mathbb{R}, \mathcal{B})$ is a real-valued measurable function and $g : (\mathbb{R}, \mathcal{B}) \to (\mathbb{R}, \mathcal{B})$ is a **Borel measurable** function, that is $g^{-1}(\mathcal{B}) \subset \mathcal{B}$, then $Y = g(X) = g \circ X : (\Omega, \mathcal{F}) \to (\mathbb{R}, \mathcal{B})$ is a real-valued measurable function. ◇

Proof: We need to show that $Y^{-1}(B) = X^{-1}(g^{-1}(B)) \in \mathcal{F}$ for any $B \in \mathcal{B}$. This condition is satisfied because $B' = g^{-1}(B) \in \mathcal{B}, \forall B \in \mathcal{B}$ and $X^{-1}(B') \in \mathcal{F}, \forall B' \in \mathcal{B}$, by the properties of g and X (Fig. 2.2). ■

Example 2.22: If X is a random variable on a probability space $(\Omega, \mathcal{F}, P)$, $r > 0$ is an integer, and λ, u are some real constants, then

$$X^r, \quad |X|^r, \quad e^{-\lambda X}, \quad \text{and} \quad e^{\sqrt{-1}\,u\,X}$$

are random variables on the same probability space. ◇

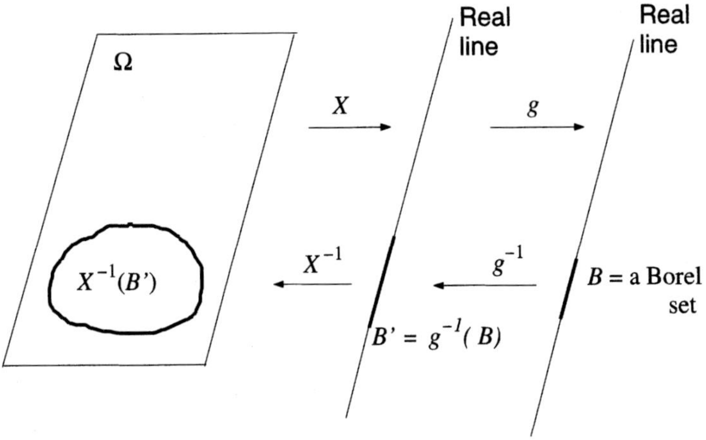

Figure 2.2. Mappings X and g

Note: A continuous function $g : \mathbb{R} \to \mathbb{R}$ is Borel measurable because the open intervals of $\mathbb{R}$ generate the σ-field $\mathcal{B}$ and g is continuous if $g^{-1}(I) \in \mathcal{D}$ for any $I \in \mathcal{D}$, where $\mathcal{D}$ denotes the topology generated by the open intervals of $\mathbb{R}$. Hence, $g \circ X$ is a random variable if g is a continuous function and X is a random variable.

The function $\omega \to e^{\sqrt{-1}\, u\, X(\omega)} \in \mathbb{C}$ defined by the last transformation is complex-valued. A complex-valued function defined on Ω is measurable if its real and imaginary parts are measurable functions from $(\Omega, \mathcal{F})$ to $(\mathbb{R}, \mathcal{B})$. Hence, $e^{\sqrt{-1}\, u\, X}$ is a measurable function because its real and imaginary parts, $\cos(u\, X)$ and $\sin(u\, X)$, are $\mathcal{F}$-measurable as continuous transformations of X. ▲

Example 2.23: If $(X, Y) : (\Omega, \mathcal{F}) \to (\mathbb{R}^2, \mathcal{B}^2)$ and $g : (\mathbb{R}^2, \mathcal{B}^2) \to (\mathbb{R}, \mathcal{B})$ are measurable functions, then $g(X, Y) \in \mathcal{F}/\mathcal{B}$. For example, the functions $X \vee Y = \max(X, Y)$, $X \wedge Y = \min(X, Y)$, $X + Y$, $X - Y$, $X\, Y$, and X/Y are measurable from $(\Omega, \mathcal{F})$ to $(\mathbb{R}, \mathcal{B})$. The transformation X/Y is defined for $Y \neq 0$ ([40], Theorem 3.1.5, p. 36)). ◇

Example 2.24: If X_i, $i = 1, 2, \ldots$, is a sequence of random variables defined on a measurable space $(\Omega, \mathcal{F})$, then $\inf_i X_i$, $\sup_i X_i$, $\liminf_i X_i$, and $\limsup_i X_i$ are measurable. ◇

Proof: The event $\{\inf_i X_i > x\} = \cap_i \{X_i > x\}$, $x \in \mathbb{R}$, is in $\mathcal{F}$ since X_i are random variables. Hence, $\inf_i X_i$ is a measurable function. Similarly, $\sup_i X_i$ is a random variable since $\{\sup_i X_i \leq x\} = \cap_i \{X_i \leq x\}$. Because $\liminf_i X_i = \sup_{i \geq 1} \inf_{j \geq i} X_j$ and $\inf_{j \geq i} X_j$ are random variables, $\liminf_i X_i$ is measurable. A similar argument shows that $\limsup_i X_i$ is a random variable. ■

Example 2.25: Consider a series $X = (X_1, X_2, \ldots)$, where X_i are measurable functions from $(\Omega, \mathcal{F})$ to $(\Psi, \mathcal{G})$. Let $\mathcal{K}$ be the collection of all subsets of $\mathbb{Z}^+ =$

$\{1, 2, \ldots\}$. The function $(m, \omega) \mapsto X_m(\omega)$ depending on the arguments m and ω is measurable from $(\mathbb{Z}^+ \times \Omega, \mathcal{K} \times \mathcal{F})$ to $(\Psi, \mathcal{G})$. Generally, this property does not hold if the discrete index m in this example is allowed to take values in an uncountable set, as we will see in the next chapter. $\diamond$

Proof: Let $A = \{(m, \omega) : X_m(\omega) \in B\}$ be the inverse image of the function $(m, \omega) \mapsto X_m(\omega)$ in $\mathbb{Z}^+ \times \Omega$ corresponding to an arbitrary member B of $\mathcal{G}$. Because X_m is measurable, this set can be expressed as the countable union $\cup_m\{\omega : X_m(\omega) \in B\}$ of sets $\{\omega : X_m(\omega) \in B\}$ that are in $\mathcal{F}$ for each $m \geq 1$. Hence, A is in $\mathcal{K} \times \mathcal{F}$. We also note that the function $m \mapsto X_m(\omega)$ is $\mathcal{K}$-measurable for a fixed $\omega \in \Omega$ since $\{m : X_m(\omega) \in B\}$ is a subset of $\mathbb{Z}^+$ so that it is in $\mathcal{K}$. ■

Example 2.26: Let $M : (\Omega, \mathcal{F}) \to (\mathbb{Z}^+, \mathcal{K})$ and $X_m : (\Omega, \mathcal{F}) \to (\Psi, \mathcal{G})$ for $m \in \mathbb{Z}^+$ be measurable functions, where $\mathcal{K}$ consists of all parts of $\mathbb{Z}^+$. The function $\omega \mapsto X_{M(\omega)}(\omega)$ is measurable from $(\Omega, \mathcal{F})$ to $(\Psi, \mathcal{G})$. $\diamond$

Proof: We need to show that $A = \{\omega : X_{M(\omega)}(\omega) \in B\}$ is in $\mathcal{F}$ for $B \in \mathcal{G}$. The sets $A_m = \{\omega : M(\omega) = m\}$, $m = 1, 2, \ldots$, provide a countable partition of Ω and $A_m \in \mathcal{F}$ for each m because M is $\mathcal{F}/\mathcal{K}$-measurable. Hence, A is equal to a countable union of disjoint parts, that is,

$$A = \cup_{m=1}^{\infty}\{A_m \cap A\} = \cup_{m=1}^{\infty}\{\omega : X_m(\omega) \in B\} \cap A_m.$$

Since X_m is an $\mathcal{F}/\mathcal{G}$-measurable function for each m, A is in $\mathcal{F}$. ■

2.5 Integration and expectation

Let $(\Omega, \mathcal{F}, P)$ be a probability space. We define the expectation for a special class of random variables taking a finite number of values and then extend this definition to arbitrary random variables.

If I is a finite index set and $A_i \in \mathcal{F}$, $i \in I$, is a partition of Ω, then

$$X = \sum_{i \in I} x_i \, 1_{A_i}, \quad x_i \in \mathbb{R}, \tag{2.32}$$

defines a **simple random variable**.

Note: Because X takes a constant value x_i in A_i and the subsets A_i partition Ω, we have $P(A_i) = P(X = x_i)$, where $\{X_i = x_i\}$ is an abbreviated notation for the event $\{\omega \in \Omega : X(\omega) = x_i\}$.

The collection of simple random variables is a vector space. Moreover, if X and Y are two simple random variables defined on a probability space $(\Omega, \mathcal{F}, P)$, then $X\,Y$, $X \wedge Y = \min(X, Y)$ and $X \vee Y = \max(X, Y)$ are simple random variables on the same space ([151], Section 5.1). ▲

If $A \in \mathcal{F}$, the **indicator function**

$$1_A(\omega) = \begin{cases} 1 & \text{if } \omega \in A \\ 0 & \text{if } \omega \notin A \end{cases} \quad \text{is a simple random variable.} \tag{2.33}$$

Proof: Because $1_A^{-1}(B) = \{\omega : 1_A(\omega) \in B\}$ is $\emptyset$, A, A^c, and Ω if $0, 1 \notin B$, $1 \in B$ but $0 \notin B$, $0 \in B$ but $1 \notin B$, and $0, 1 \in B$, respectively, we have $1_A^{-1}(B) \in \mathcal{F}$ for any $B \in \mathcal{B}$ so that 1_A is $\mathcal{F}/\mathcal{B}$-measurable. ■

The following **measurability theorem** suggests that we may be able to extend properties established for simple random variables to arbitrary random variables.

Let $X : \Omega \to \mathbb{R}$ be such that $X(\omega) \geq 0$ for all $\omega \in \Omega$. Then $X \in \mathcal{F}/\mathcal{B}$ if and only if there exists an increasing sequence of simple random variables $X_n \geq 0$, $n = 1, 2, \ldots$, converging to X, referred to as an **approximating sequence** for X ([151], Theorem 5.1.1, p. 118).

Proof: If there is an approximating sequence of simple random variables X_n and $X(\omega) = \lim_{n\to\infty} X_n(\omega)$, then X is a random variable since limits preserve measurability (Example 2.24).

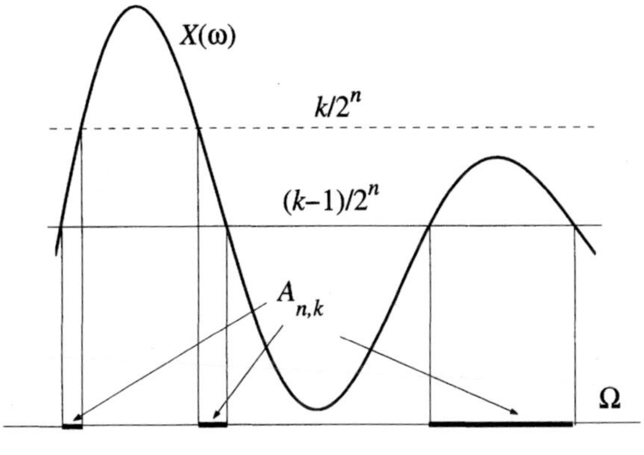

Figure 2.3. Construction of the approximating sequence X_n

If $X \in \mathcal{F}/\mathcal{B}$, construct the sequence

$$X_n = \sum_{k=1}^{n\,2^n} X\left(\frac{k-1}{2^n}\right) 1_{A_{n,k}} + n\,1_{B_n} \geq 0$$

of simple random variables, where $A_{n,k} = \{\omega : (k-1)/2^n \leq X(\omega) < k/2^n\}$ and $B_n = \{\omega : X(\omega) \geq n\}$ (Fig. 2.3). The sequence X_n, $n = 1, 2, \ldots$, is increasing, that is, $X_n \leq X_{n+1}$, and its members are measurable functions that are smaller than X. If $X(\omega) < \infty$, then $|X_n(\omega) - X(\omega)| \leq 2^{-n} \to 0$ as $n \to \infty$. If $X(\omega) = +\infty$, then $X_n(\omega) = n$ so that it approaches infinity as n increases. ■

2.5.1 Expectation operator

Consider a probability space $(\Omega, \mathcal{F}, P)$. We define the expectation for finite-valued simple random variables and then extend this definition to arbitrary real-valued random variables.

2.5.1.1 Finite-valued simple random variables

If X is a simple random variable (Eq. 2.32) such that $|x_i| < \infty$, $i \in I$, its **expectation** is

$$E[X] = \sum_{i \in I} x_i \, P(A_i). \tag{2.34}$$

Note: The expectation of the indicator function in Eq. 2.33 is $E[1_A] = P(A)$. ▲

Example 2.27: Consider a loaded die with $m < \infty$ sides showing the numbers x_i with probabilities p_i, $i = 1, \ldots, m$. Suppose we roll the die n times and $n_i \geq 0$ denotes the number of times we see side i. The arithmetic average, $\bar{x} = \sum_{i \in I} x_i \, (n_i/n)$, of the observed values approximates $E[X]$ in Eq. 2.34 for large values of n because n_i/n converges to p_i as $n \to \infty$. ◇

The following properties are direct consequences of the definition in Eq. 2.34.

If X and Y are finite-valued simple random variables, then

- $X \geq 0 \Longrightarrow E[X] \geq 0$.
- $E[\alpha X + \beta Y] = \alpha E[X] + \beta E[Y]$, $\alpha, \beta =$ constants $\Longrightarrow E[\cdot]$ is **linear**.
- $X \leq Y$ implies $E[X] \leq E[Y] \Longrightarrow E[\cdot]$ is **monotone**. (2.35)

Proof: Let $X = \sum_{i \in I} x_i \, 1_{A_i}$ and $Y = \sum_{j \in J} y_j \, 1_{B_j}$, where I and J are finite sets. $X \geq 0$ implies $x_i \geq 0$ so that $E[X]$ is positive (Eq. 2.34), that is, the first property. To prove the second property, note that $\alpha X + \beta Y$ is a finite-valued simple random variable corresponding to the partition $A_i \cap B_j$ of Ω ([151], pp. 120-121). If $X \leq Y$, the first property and the fact that $Y - X \geq 0$ is a simple random variable imply $E[Y - X] \geq 0$ or $E[Y] \geq E[X]$ (Eq. 2.34). ■

If X and X_n are finite-valued simple random variables and $X_n \uparrow X$ or $X_n \downarrow X$, then ([151], Property 5, p. 121)

$$E[X_n] \uparrow E[X] \quad \text{or} \quad E[X_n] \downarrow E[X]. \tag{2.36}$$

2.5.1.2 Positive random variables

If $P(X = \infty) > 0$, set $E[X] = \infty$. Otherwise,

$$E[X] = \lim_{n\to\infty} E[X_n], \tag{2.37}$$

where X_n is an approximating sequence of finite-valued simple random variables such that $X_n \uparrow X$ and $E[X_n]$ is finite for each n.

Note: The measurability theorem guarantees the existence of an approximating sequence. The expectation in Eq. 2.37 is well defined since the value of $E[X]$ does not change if another approximating sequence $Y_m \uparrow X$ is used in this equation in place of X_n ([151], Proposition 5.2.1, p. 122). ▲

The properties of the expectation for finite-valued simple random variables in Eqs. 2.35 and 2.36 can be extended to positive random variables with the range $[0, \infty]$.

If X, Y, and X_n, $n = 1, 2, \ldots$, are positive random variables with range $[0, \infty]$, then ([151], Section 5.2.3)

- $E[X] \in [0, \infty]$.
- $E[\alpha\, X + \beta\, Y] = \alpha\, E[X] + \beta\, E[Y], \; \alpha, \beta > 0$.
- $X_n \uparrow X \implies E[X_n] \uparrow E[X]$. (2.38)

Note: The last property, called the **monotone convergence theorem**, gives conditions under which limits and expectations can be interchanged ([151], Section 5.2.3). A similar result is given in Eq. 2.36. ▲

2.5.1.3 Arbitrary random variables

An arbitrary random variable X has the representation $X = X^+ - X^-$, where

$$X^+ = X \vee 0 = \max(X, 0) \quad \text{and} \quad X^- = (-X) \vee 0 = \max(-X, 0) \tag{2.39}$$

are positive random variables.

- If X is an arbitrary random variable such that at least one of $E[X^+]$ and $E[X^-]$ is finite, we define

$$E[X] = E[X^+] - E[X^-]. \tag{2.40}$$

- If $E[X^+]$ and $E[X^-]$ are both finite, then $E[X]$ in Eq. 2.40 exists and is finite. We say that X has a **finite expectation** or is P-**integrable**.

- If both $E[X^+]$ and $E[X^-]$ are unbounded, $E[X]$ does not exist.

Note: If the expectations $E[X^+]$ and $E[X^-]$ are unbounded, $E[X]$ given by Eq. 2.40 does not exist. If $E[X^+]$ and $E[X^-]$ are finite, then both $E[X] < \infty$ and $E[|X|] < \infty$ since $X = X^+ - X^-$ and $|X| = X^+ + X^-$. If one of the expectations $E[X^+]$ and $E[X^-]$ is finite and the other infinite, $E[X]$ is defined but is unbounded. For example, $E[X]$ exists and $E[X] = +\infty$ if $E[X^+] = +\infty$ and $E[X^-]$ is bounded. ▲

If the expectation of a random variable X exists, it is also denoted by

$$E[X] = \int_\Omega X(\omega)\, P(d\omega) \quad \text{or } E[X] = \int X\, dP. \tag{2.41}$$

- The **integral of X with respect to P over set $A \in \mathcal{F}$** is

$$E[X\, 1_A] = \int_\Omega 1_A(\omega)\, X(\omega)\, P(d\omega) = \int_A X(\omega)\, P(d\omega). \tag{2.42}$$

- If $E[X\, 1_A]$ exists and is finite, X is said to be **integrable with respect to P over A** or **P-integrable over A**.

Note: The integrals in Eqs. 2.41 and 2.42 are special cases of the **Lebesgue-Stieltjes integral** $\int h\, d\mu$, where h is a measurable function on a measure space $(\Omega, \mathcal{F}, \mu)$. The random variable X and the probability measure P in Eqs. 2.41 and 2.42 correspond to the measurable function h and the measure μ, respectively. Hence, $\int X\, dP$ has properties similar to $\int h\, d\mu$. ▲

Example 2.28: Let $\Omega = \{\omega = (i, j) : i, j = 1, 2, \ldots, 6\}$ be the sample space associated with the experiment of rolling two dice and $\mathcal{F}$ be the collection of all subsets of Ω. Define a random variable $X(\omega) = i^2 + j^2$. This function is positive and takes on a finite number of values. The integral of X with respect to P over A is (Eq. 2.42)

$$E[X\, 1_A] = \sum_{\omega \in A} (i^2 + j^2)\, \frac{1}{36} = [(2)(40) + (2)(34) + 32]\, \frac{1}{36} = 5,$$

where $A = \{\omega = (i, j) : i + j = 8\} = \{(2, 6), (3, 5), (4, 4), (5, 3), (6, 2)\}$. ◇

2.5.2 Properties of integrals of random variables

We review essential properties of integrable random variables. These properties are divided into three groups depending on the number of random variables in the integrand and on the domain of integration.

2.5.2.1 Finite number of random variables

Let X and Y be random variables defined on a probability space $(\Omega, \mathcal{F}, P)$ and let A be an event on this space, that is, $A \in \mathcal{F}$.

$\int_A X\,dP$ is finite if and only if $\int_A |X|\,dP < \infty$ ([40], Section 3.2).

Note: We have $|E[X]| < \infty$ if and only if $E[|X|] < \infty$ by setting $A = \Omega$. This result is consistent with an earlier comment (Eq. 2.40). ▲

If X and Y are P-integrable over A, $a\,X + b\,Y$ is P-integrable over A and

$$\int_A (a\,X + b\,Y)\,dP = a \int_A X\,dP + b \int_A Y\,dP, \tag{2.43}$$

where a, b are constants ([40], Section 3.2).

Note: If $A = \Omega$, Eq. 2.43 yields $E[a\,X + b\,Y] = a\,E[X] + b\,E[Y]$ showing that E is a **linear operator** in agreement with Eq. 2.35 and 2.38. This property implies that $\int_A \sum_{i=1}^n a_i\,X_i\,dP = \sum_{i=1}^n a_i \int_A X_i\,dP$ if the random variables X_i are P-integrable over A and a_i denote constants. ▲

If X is P-integrable and $A_i \in \mathcal{F}$ is a partition of Ω, then ([40], Section 3.2)

$$\int_{\cup_i A_i} X\,dP = \sum_i \int_{A_i} X\,dP. \tag{2.44}$$

If $X \geq 0$ a.s., then $\int_A X\,dP \geq 0$ ([40], Section 3.2).

If X is P-integrable, then X is finite a.s., that is, $\{\omega : X(\omega) = \pm\infty\}$ is in $\mathcal{F}$ and has probability zero ([40], Section 3.2).

If $Y \leq X$ a.s. and the integrals $\int X\,dP$ and $\int Y\,dP$ exist ([40], Section 3.2),

$$\int_A Y\,dP \leq \int_A X\,dP. \tag{2.45}$$

Note: Because $X - Y \geq 0$, we have $\int_A (X - Y)\,dP \geq 0$ by one of the above properties.

The **mean value** theorem, $a\,P(A) \leq \int_A X\,dP \leq b\,P(A)$ if $a \leq X \leq b$ a.s., and the **modulus inequality**,

$$\left|\int_A X\,dP\right| \leq \int_A |X|\,dP, \tag{2.46}$$

are direct consequences of Eq. 2.45. For example, $a\,P(A) \leq \int_A X\,dP$ follows from Eq. 2.45 with $Y = a$ and $X \geq a$ a.s. The modulus inequality can be obtained from Eq. 2.45 with $(-|X|, X)$ and $(X, |X|)$ in place of (Y, X). ▲

2.5.2.2 Sequence of random variables

Let X_n, $n = 1, 2, \ldots$, and Y be random variables defined on a probability space $(\Omega, \mathcal{F}, P)$. Let A be an event on this space, that is, $A \in \mathcal{F}$.

If $\lim_{n\to\infty} X_n = X$ a.s. and if, for each n,

- $|X_n| \le Y$ a.s., $Y \ge 0$, and $\int_A Y\,dP < \infty$, or
- $|X_n| \le c$ a.s. for a positive constant c, or
- $X_n \ge 0$ a.s. is an increasing sequence that can take on the value $+\infty$,

$$\text{then} \quad \lim_{n\to\infty} \int_A X_n\,dP = \int_A (\lim_{n\to\infty} X_n)\,dP = \int_A X\,dP. \tag{2.47}$$

Note: Under the conditions of this theorem we can interchange the limit and the integration operations. The a.s. convergence of X_n to X means that the numerical series $X_n(\omega)$ converges to $X(\omega)$ as $n \to \infty$ for each $\omega \in \Omega \setminus N$, where $N \in \mathcal{F}$ and $P(N) = 0$ (Section 2.13). An event N with this property is called a **null set**.

The statements in Eq. 2.47 corresponding to the conditions (1) $|X_n| \le Y$ a.s and $\int_A Y\,dP < \infty$, (2) $|X_n| \le c$ a.s., and (3) $X_n \ge 0$ a.s. is an increasing sequence with $+\infty$ being an allowed value are referred to as the **dominated convergence**, **bounded convergence**, and **monotone convergence** theorems, respectively. Slightly weaker versions of the dominated, bounded, and monotone convergence conditions under which Eq. 2.47 holds can be found in [40] (Section 3.2). ▲

If $\sum_n \int_A |X_n|\,dP < \infty$, then integration can be performed term by term ([40], Section 3.2)

$$\int_A \sum_n X_n\,dP = \sum_n \int_A X_n\,dP. \tag{2.48}$$

If $X_n \ge 0$ a.s. on A for each n, then ([40], Section 3.2)

$$\int_A (\liminf_n X_n)\,dP \le \liminf_n \int_A X_n\,dP \quad \textbf{(Fatou's lemma)}. \tag{2.49}$$

Note: Recall that

$$\liminf_{n\to\infty} X_n = \sup_{n\ge 1} \inf_{k\ge n} X_k = \lim_{n\to\infty} \wedge_{k\ge n} X_k \quad \text{and}$$

$$\limsup_{n\to\infty} X_n = \inf_{n\ge 1} \sup_{k\ge n} X_k = \lim_{n\to\infty} \vee_{k\ge n} X_k,$$

where $\wedge_{k\ge n} X_k$ and $\vee_{k\ge n} X_k$ are increasing and decreasing sequences. Because $\vee_{k\ge n} X_k = -\wedge_{k\ge n}(-X_k)$, we have $\limsup_{n\to\infty} X_n = -\liminf_{n\to\infty}(-X_n)$. The above limits of X_n resembles the limits for sequences of subsets in Ω (Eqs. 2.19 and 2.20). Also note that

$$\{\vee_n X_n \le x\} = \cap_n \{X_n \le x\} \in \mathcal{F} \quad \text{and} \quad \{\wedge_n X_n > x\} = \cap_n \{X_n > x\} \in \mathcal{F}$$

if X_n are random variables on $(\Omega, \mathcal{F}, P)$. ▲

If the sequence of random variables X_n is such that $|X_n| \leq Z$, where $Z \geq 0$ a.s. and P-integrable over A, then

$$\int_A (\liminf_n X_n)\, dP \leq \liminf_n \int_A X_n\, dP$$
$$\leq \limsup_n \int_A X_n\, dP \leq \int_A (\limsup_n X_n)\, dP \quad \textbf{(Lebesgue theorem)}. \tag{2.50}$$

Proof: Since Z is P-integrable, Z is finite a.s. so that $P(\{\omega \in A : Z(\omega) = +\infty\}) = 0$. Note that $X_n + Z \geq 0$ a.s. so that the Fatou lemma yields

$$\int (\liminf X_n)\, dP + \int Z\, dP \leq \liminf \int X_n\, dP + \int Z\, dP$$

or $\int (\liminf X_n)\, dP \leq \liminf \int X_n\, dP$ since $\liminf(X_n + Z) = \liminf X_n + Z$. The last inequality in Eq. 2.50 results from $\sup\{X_n, X_{n+1}, \ldots\} = -\inf\{-X_n, -X_{n+1}, \ldots\}$. The middle inequality is valid for any numerical sequence, in particular for the sequence $\int_A X_n\, dP$. ■

Example 2.29: The inequalities in Eq. 2.26 can be obtained from the Lebesgue theorem in Eq. 2.50 with $A = \Omega$, $X_n = 1_{A_n}$, and $A_n \in \mathcal{F}$. ◇

Proof: Take $A = \Omega$ and $X_n = 1_{A_n}$, $A_n \in \mathcal{F}$, in Eq. 2.50. Then

$$\int (\liminf_{n\to\infty} X_n)\, dP = \int (\sup_{n\geq 1} \inf_{k\geq n} 1_{A_k})\, dP = P\left(\sup_{n\geq 1} 1_{\cap_{k\geq n} A_k}\right) = P(\liminf_{n\to\infty} A_n)$$

because $\inf_{k\geq n} 1_{A_k} = 1_{\cap_{k\geq n} A_k}$ is an increasing sequence so that $\sup_{n\geq 1} 1_{\cap_{k\geq n} A_k}$ is equal to $\cup_{n\geq 1} 1_{\cap_{k\geq n} A_k}$ or $1_{\cup_{n\geq 1} \cap_{k\geq n} A_k}$. Similar considerations give the last inequality in Eq. 2.26. The integral $\int_A X_n\, dP$ is $\int_A 1_{A_n}\, dP = P(A_n)$ so that we have

$$\liminf_n P(A_n) \leq \limsup_n P(A_n)$$

from the middle inequality in Eq. 2.50. ■

2.5.2.3 Expectation

The properties in Sections 2.5.2.1 and 2.5.2.2 hold for any $A \in \mathcal{F}$. If $A = \Omega$, the above integrals become expectations. We give here a list of properties of the expectation operator that are useful for calculations. It is assumed that all expectations exist and are finite.

- The expectation is a **linear operator**.
- If $X \geq 0$ a.s., then $E[X] \geq 0$.
- If $Y \leq X$ a.s., then $E[Y] \leq E[X]$.
- $|E[X]| \leq E[|X|]$.
- If $\lim_{n\to\infty} X_n = X$ a.s., and there exists $Z \geq 0$ a.s. such that $E[Z] < \infty$ and $|X_n| \leq Z$, then $\lim_{n\to\infty} E[X_n] = E[\lim_{n\to\infty} X_n] = E[X]$.
- If $\sum_n E[|X_n|] < \infty$, then $E\left[\sum_n X_n\right] = \sum_n E[X_n]$.

Proof: The first four properties follow from Eq. 2.43, the fact that $X \geq 0$ a.s. implies $\int_A X\,dP \geq 0$ for $A \in \mathcal{F}$, Eq. 2.45, and Eq. 2.46, respectively. The Lebesgue theorem gives $\int_A (\lim X_n)\,dP = \lim \int_A X_n\,dP$ if $\lim X_n = X$. The last property follows from Eq. 2.48. ■

2.6 The $L_q(\Omega, \mathcal{F}, P)$ space

Let $(\Omega, \mathcal{F}, P)$ be a probability space. Denote by $L_q(\Omega, \mathcal{F}, P)$ the collection of real-valued random variables X defined on $(\Omega, \mathcal{F}, P)$ such that $E[|X|^q] < \infty$ for $q \geq 1$. If X is in $L_q(\Omega, \mathcal{F}, P)$ we write $X \sim L_q(\Omega, \mathcal{F}, P)$ or $X \sim L_q$ provided that there is no confusion about the probability space. The case $q = 2$ is most relevant for applications because it relates to second moment calculus and estimation theory discussed later in this and subsequent chapters. The L_2 space has some notable properties.

- L_2 is a vector space.
- L_2 is a **Hilbert space** with the **inner product** $\langle X, Y\rangle = E[X\,Y]$ for $X, Y \in L_2$ and the norm $\| X \|_{L_2} = \left(E[X^2]\right)^{1/2}$, referred to as the L_2-**norm**.

Proof: We need to show that (1) $\langle X, Y\rangle = E[X\,Y]$ is an inner product on L_2, (2) L_2 is a vector space, and (3) L_2 with the metric $d(X, Y) = \langle X - Y, X - Y\rangle^{1/2} = \| X - Y \|_{L_2}$ is complete.

The expectation $E[X\,Y]$ defines an **inner product** on L_2 because

$$\begin{aligned}
&\langle 0, X\rangle = 0, \quad X \in L_2,\\
&\langle X, X\rangle > 0, \quad X \in L_2, X \neq 0,\\
&\langle X, Y\rangle = \langle Y, X\rangle, \quad X, Y \in L_2,\\
&\langle X + Y, Z\rangle = \langle X, Y\rangle + \langle Y, Z\rangle, \quad X, Y, Z \in L_2,\\
&\lambda\,\langle X, Y\rangle = \langle \lambda\,X, Y\rangle, \quad X, Y \in L_2, \lambda \in \mathbb{R}.
\end{aligned}$$

The function $d : L_2 \times L_2 \to [0, \infty)$ defined by $d(X, Y) = \| X - Y \|_{L_2}$ is a **metric on** L_2 because

$$\begin{aligned}
&d(X, Y) = 0 \quad \text{if and only if } X = Y,\\
&d(X, Y) = d(Y, X) \quad \text{for each } X, Y \in L_2,\\
&d(X, Y) \leq d(X, Z) + d(Z, Y) \quad \text{for each } X, Y, Z \in L_2.
\end{aligned}$$

The first condition in the above equation is not satisfied in a strict sense. The condition holds if we do not distinguish between random variables that differ on a set of probability zero.

That L_2 is a vector space follows from the properties (a) $X \sim L_2$ and $\lambda \in \mathbb{R}$ imply $\lambda X \sim L_2$ since $E[(\lambda X)^2] = \lambda^2 E[X^2] < \infty$ and (b) $X, Y \sim L_2$ implies $X + Y \sim L_2$ since $E[(X+Y)^2] = E[X^2] + E[Y^2] + 2\,E[X\,Y]$, $E[X^2]$ and $E[Y^2]$ are finite, and $|E[X\,Y]| < \infty$ by the Cauchy-Schwarz inequality discussed later in this chapter (Eq. 2.112).

It remains to show that L_2 endowed with the metric $(X, Y) \mapsto d(X, Y) = \| X - Y \|_{L_2}$ is complete, that is, that any sequence $X_n \in L_2$, $n = 1, 2, \ldots$, with $d(X_n, X_m) \to 0$ as $n, m \to \infty$ is convergent and its limit is in L_2. The proof that L_2 is a Hilbert space can be found in [66] (Proposition 4, p. 399). We only show here that, if $X, X_n \in L_2$ and $\| X_n - X \|_{L_2} \to 0$ as $n \to \infty$, that is, X_n converges in **mean square (m.s.)** to X, then $\| X_n - X_m \|_{L_2} \to 0$ as $n, m \to \infty$, that is, X_n is a **Cauchy sequence in the mean square sense** (Section 2.13). Take $\varepsilon > 0$ and an index $\bar{n}$ such that $\| X_n - X \|_{L_2} < \varepsilon/2$ and $\| X_m - X \|_{L_2} < \varepsilon/2$ for $n, m \geq \bar{n}$. This is possible since $\| X_n - X \|_{L_2}$ converges to zero as $n \to \infty$. We have $\| X_n - X_m \|_{L_2} \leq \| X_n - X \|_{L_2} + \| X_m - X \|_{L_2} < \varepsilon$ for $n, m \geq \bar{n}$, that is, $\| X_n - X_m \|_{L_2}$ converges to zero as $n, m \to \infty$. ■

Example 2.30: The sequence of spaces $L_q(\Omega, \mathcal{F}, P)$ is decreasing, that is, $L_{q'} \subseteq L_q$ for $q \leq q'$. However, the spaces $L_q(\Omega, \mathcal{F}, \mu)$ corresponding to a measure space $(\Omega, \mathcal{F}, \mu)$ may not decrease with q. ◇

Proof: The first part follows from $(E[|X|])^p \leq E[|X|^p]$, $1 < p < \infty$, $X \in L_p$, derived from Hölder's inequality with $Y = 1$. This inequality is proved later in this chapter (Eq. 2.113). Hölder's inequality with $|X|^q$ in place of X gives $\left(E[|X|^q]\right)^p \leq E[|X|^{p\,q}]$ or $\left(E[|X|^q]\right)^{q'/q} \leq E[|X|^{q'}]$ for $p = q'/q \geq 1$. Hence, $E[|X|^{q'}] < \infty$ implies $E[|X|^q] < \infty$, that is, $L_{q'} \subseteq L_q$ for $q' \geq q$.

Let $h : (\mathbb{R}, \mathcal{B}, \lambda) \to (\mathbb{R}, \mathcal{B}, \lambda)$ be a measurable function defined by $h(\omega) = 0$ for $\omega < 1$ and $h(\omega) = 1/\omega$ for $\omega \geq 1$, where $\lambda(d\omega) = d\omega$ is the Lebesgue measure. The integrals $\int_{\mathbb{R}} h\, d\lambda$ and $\int_{\mathbb{R}} h^2\, d\lambda$ are equal to $+\infty$ and one, respectively. Hence, h is an element of $L_2(\mathbb{R}, \mathcal{B}, \lambda)$ but is not in $L_1(\mathbb{R}, \mathcal{B}, \lambda)$, so that $L_2 \not\subset L_1$. ■

Let $(\Omega, \mathcal{F}, P)$ be a probability space and $\mathcal{G}$ a sub-σ-field of $\mathcal{F}$. Consider an m.s. convergent sequence X_n in $L_2(\Omega, \mathcal{G}, P)$. If the m.s. limit X of X_n is in $L_2(\Omega, \mathcal{G}, P)$, we say that $L_2(\Omega, \mathcal{G}, P)$ is a **closed subspace** of $L_2(\Omega, \mathcal{F}, P)$. The following theorem, referred to as the **orthogonal projection theorem**, gives an essential property of the L_2 spaces.

If $\mathcal{G}$ is a sub-σ-field of $\mathcal{F}$, then for any $X \in L_2(\Omega, \mathcal{F}, P)$ there is a unique $\hat{X} \in L_2(\Omega, \mathcal{G}, P)$ such that ([156], Theorem 4.11, p. 84)

$$\| X - \hat{X} \|_{L_2} = \min\{\| X - Z \|_{L_2} : Z \in L_2(\Omega, \mathcal{G}, P)\} \quad \text{and}$$
$$\langle X - \hat{X}, Z \rangle = 0, \quad \forall Z \in L_2(\Omega, \mathcal{G}, P). \tag{2.51}$$

Note: The random variable $\hat{X} \in L_2(\Omega, \mathcal{G}, P)$ has the smallest mean square error of all members of $L_2(\Omega, \mathcal{G}, P)$, and is called the **best m.s. estimator** of X. The second equality in the above equation can be used to calculate $\hat{X}$ and shows that the error $X - \hat{X}$ is orthogonal to $L_2(\Omega, \mathcal{G}, P)$. ▲

2.7 Independence

We define independent σ-fields and apply this definition to events and random variables. We also discuss the Borel Zero-One Law giving the probability of lim sup of independent events.

2.7.1 Independence of σ-fields

Consider a probability space $(\Omega, \mathcal{F}, P)$ and a collection of sub-σ-fields, $\mathcal{F}_i$, $i \in I$, of $\mathcal{F}$.

- If I is finite and

$$P\left(\cap_{i \in I} A_i\right) = \prod_{i \in I} P(A_i), \quad \forall A_i \in \mathcal{F}_i, \tag{2.52}$$

then the σ-fields $\mathcal{F}_i$, $i \in I$, are independent.

- If I is infinite and Eq. 2.52 holds for all finite subsets of I, then the σ-fields $\mathcal{F}_i$ are independent.

Note: The above condition implies that any sub-collection of events $A_i \in \mathcal{F}_i$, $i \in I$, must satisfy Eq. 2.52 since some of the events A_i may coincide with the sample space Ω. This requirement is consistent with Eq. 2.54. If the σ-fields $\mathcal{F}_i$, $i \in I$, are on different sample spaces, the above independence condition has to be applied on the corresponding product measure space. ▲

Example 2.31: The sub-σ-fields $\mathcal{F}_1$ and $\mathcal{F}_2$ generated by the collections of events $\mathcal{A}_1 = \{\{1, 2, 3\}, \{4, 5, 6\}\}$ and $\mathcal{A}_2 = \{\{1, 2, 3\}, \{4\}, \{5\}, \{6\}\}$ in Example 2.1 are not independent. For example, the subset $\{1, 2, 3\}$ belongs to both $\mathcal{F}_1$ and $\mathcal{F}_2$, but the probability $P(\{1, 2, 3\} \cap \{1, 2, 3\}) = P(\{1, 2, 3\}) = 1/2$ differs from $P(\{1, 2, 3\})\, P(\{1, 2, 3\}) = 1/4$. ◇

2.7.2 Independence of events

The definition in Eq. 2.52 applied to events is consistent with the classical definitions of independence between events.

The events $A_i \in \mathcal{F}$, $i \in I$, of a probability space $(\Omega, \mathcal{F}, P)$ are independent if the σ-fields $\sigma(A_i)$ generated by these events are independent, where the index set I may be finite or infinite.

Example 2.32: Let A and B be two events in $\mathcal{F}$. These events are independent if the σ-fields $\sigma(A) = \{\emptyset, A, A^c, \Omega\}$ and $\sigma(B) = \{\emptyset, B, B^c, \Omega\}$ are independent or, equivalently, if $P(A \cap B) = P(A)\,P(B)$. $\diamond$

Proof: The independence of the σ-fields $\sigma(A)$ and $\sigma(B)$ requires that $P(A_i \cap B_j) = P(A_i)\,P(B_j)$ for all $A_i \in \sigma(A)$ and $B_j \in \sigma(B)$ (Eq. 2.52). The resulting non-trivial conditions of independence are $P(A \cap B) = P(A)\,P(B)$, $P(A^c \cap B) = P(A^c)\,P(B)$, $P(A \cap B^c) = P(A)\,P(B^c)$, and $P(A^c \cap B^c) = P(A^c)\,P(B^c)$. These conditions are equivalent with the classical requirement $P(A \cap B) = P(A)\,P(B)$ for the independence of two events (Eq. 2.53). For example, $P(A^c \cap B) = P(B) - P(A \cap B)$ since $A^c \cap B = B \setminus (A \cap B)$ so that $P(A^c \cap B) = P(A^c)\,P(B)$ if $P(A \cap B) = P(A)\,P(B)$. Similar considerations apply to show that $P(A \cap B^c) = P(A)\,P(B^c)$ follows from $P(A \cap B) = P(A)\,P(B)$. ■

We give now classical definitions for the independence of two or more events and for families of events.

Two events, A and B, are said to be independent if

$$P(A \cap B) = P(A)\,P(B). \tag{2.53}$$

Note: Let A and B be two events such that $P(A) > 0$ and $P(B) > 0$. If A and B are independent, the occurrence of B does not affect the probability of A so that $P(A \mid B) = P(A)$ implying $P(A \cap B) = P(A)\,P(B)$. An equivalent condition of independence for the events A and B is $P(B \mid A) = P(B)$. Note that the events A and $B = A^c$ are not independent because A cannot occur if B is observed. ▲

A finite collection of events, A_i, $i = 1, \ldots, n$, is independent if

$$P(A_{i_1} \cap A_{i_2} \cap \cdots \cap A_{i_m}) = \prod_{k=1}^{m} P(A_{i_k}) \tag{2.54}$$

holds for any subset $\{i_1, \ldots, i_m\} \subset \{1, \ldots, n\}$.

Example 2.33: The conditions of Eq. 2.54 are essential to assure the independence of three or more events. It is not sufficient to require that Eq. 2.54 be satisfied for the entire collection of events. For example, consider the sample space $\Omega = \{1, 2, 3, 4\}$ with $\mathcal{F}$ given by all parts of Ω and a probability measure P defined by $P(\{1\}) = \sqrt{2}/2 - 1/4$, $P(\{2\}) = 1/4$, $P(\{3\}) = 3/4 - \sqrt{2}/2$, and $P(\{4\}) = 1/4$. Let $A_1 = \{1, 3\}$, $A_2 = \{2, 3\}$, and $A_3 = \{3, 4\}$ be some events on $(\Omega, \mathcal{F})$. The probability of $A_1 \cap A_2 \cap A_3 = \{3\}$ is $P(\{3\}) = 3/4 - \sqrt{2}/2$ and is equal to $P(A_1)P(A_2)P(A_3)$. However, $P(A_1 \cap A_2) \neq P(A_1)P(A_2)$ ([106], p. 2). $\diamond$

Example 2.34: Let S_i, $i = 1, 2, \ldots, n$, be the event that the maximum flow in a river during year i does not exceed the height of a flood protection system. The probability that there will be no flood in n years is $P_s(n) = P(\cap_{i=1}^n S_i)$. If the events S_i are independent, the reliability of the flood protection system in n years is $P_s(n) = \prod_{i=1}^n P(S_i)$ so that $P_s(n) = p^n$ for the special case $P(S_i) = p$. $\diamond$

The families $\mathcal{C}_i \subset \mathcal{F}$, $i = 1, \ldots, n$, are independent, if for any $A_1 \in \mathcal{C}_1, \ldots, A_n \in \mathcal{C}_n$, the events $A_1, \ldots, A_n$ are independent.

The families $\mathcal{C}_t$, $t \in T$, where T is an arbitrary index set, are independent if $\mathcal{C}_t$, $t \in I$, are independent families for each finite subset I of T.

The above definitions and the following criterion can be used to prove independence of σ-fields. The criterion uses classes of events forming a π**-system**. A collection $\mathcal{C}$ of subsets of Ω is said to be a π-system if it is closed to finite intersection, that is, $A, B \in \mathcal{C}$ implies $A \cap B \in \mathcal{C}$.

If $\mathcal{C}_i$ is a non-empty class of events in $\mathcal{F}$ for each $i = 1, \ldots, n$, such that (1) $\mathcal{C}_i$ is a π-system and (2) $\mathcal{C}_i$, $i = 1, \ldots, n$, are independent, then the σ-fields $\sigma(\mathcal{C}_i)$, $i = 1, \ldots, n$, are independent ([151], Theorem 4.1.1, p. 92).

Borel Zero-One Law. If A_i is a sequence of independent events, then

$$P(\limsup_i A_i) = P(A_i \text{ i.o.}) = \begin{cases} 0 & \text{if and only if } \sum_i P(A_i) < \infty, \\ 1 & \text{if and only if } \sum_i P(A_i) = \infty. \end{cases} \quad (2.55)$$

Proof: For proof see [151] (Proposition 4.5.2, p. 103). We only show that $\sum_i P(A_i) = \infty$ implies $P(A_i \text{ i.o.}) = 1$. Note that, by the definition of the event $\{A_i \text{ i.o.}\}$ and the independence of A_i,

$$\begin{aligned} 1 - P(A_i \text{ i.o.}) &= P(\cup_{j=1}^\infty \cap_{i=j}^\infty A_i^c) = P(\lim_{j\to\infty} \cap_{i=j}^\infty A_i^c) \\ &= \lim_{j\to\infty} P(\cap_{i=j}^\infty A_i^c) = \lim_{j\to\infty} \prod_{i=j}^\infty [1 - P(A_i)]. \end{aligned}$$

The inequality $1 - x \le e^{-x}$, $0 < x < 1$, applied for $x = P(A_i)$ gives $1 - P(A_i) \le \exp[-P(A_i)]$ so that $1 - P(A_i \text{ i.o.}) \le \lim_{j\to\infty} e^{-\sum_{i=j}^\infty P(A_i)}$. Because $\sum_{i=j}^\infty P(A_i) = \infty$ for all j, $\exp[-\sum_{i=j}^\infty P(A_i)]$ is zero so that $1 - P(A_i \text{ i.o.}) = 0$. ■

2.7.3 Independence of random variables

The random variables X_i, $i \in I$, defined on a probability space $(\Omega, \mathcal{F}, P)$ are independent if the σ-fields $\sigma(X_i)$ generated by these random variables are independent, where the index set I is finite or not.

Example 2.35: Let X and Y be two real-valued random variables defined on $(\Omega, \mathcal{F}, P)$. If the σ-fields $\sigma(X)$ and $\sigma(Y)$ are independent, the random variables X and Y are said to be independent. This definition is equivalent to a classical definition for the independence for random variables given in Section 2.11.2. ◇

Note: The condition of independence of the events A and B in Example 2.32 is equivalent to the requirement that the random variables 1_A and 1_B be independent. ▲

Example 2.36: Let X_k, $k = 1, 2, \ldots$, be a sequence of independent random variables and φ_k be real-valued measurable functions defined on the real line. The random variables $\varphi_k \circ X_k$ are independent. For example, the collection of functions $\sin(X_k)$ are independent since the "sin" function is continuous and, hence, measurable. ◇

Proof: We have $\varphi_k^{-1}(\mathcal{B}) \subseteq \mathcal{B}$ and $X_k^{-1}(\varphi_k^{-1}(\mathcal{B})) \subseteq \mathcal{F}$ because φ_k is a Borel measurable function and X_k is a random variable. Because $X_k^{-1}(\varphi_k^{-1}(\mathcal{B})) \subseteq X_k^{-1}(\mathcal{B})$ and the σ-fields $X_k^{-1}(\mathcal{B})$, $k = 1, 2, \ldots$, are independent by hypothesis, the random variables $\varphi_k \circ X_k$ are independent. ■

2.8 The Fubini theorem

The definition of the integrals in Eq. 2.42 is also valid for random variables defined on a product probability space. The Fubini theorem gives conditions under which the integration on the product space can be performed sequentially.

Let $(\Omega_k, \mathcal{F}_k, P_k)$, $k = 1, 2$, be two probability spaces and denote their completions by $(\Omega_k, \bar{\mathcal{F}}_k, \bar{P}_k)$. Denote by $\Omega = \Omega_1 \times \Omega_2$, $\mathcal{F} = \mathcal{F}_1 \times \mathcal{F}_2$, and $P = P_1 \times P_2$ the product sample space, σ-field, and probability measure. Let $(\bar{\mathcal{F}}_1, \bar{P}_1)$, $(\bar{\mathcal{F}}_2, \bar{P}_2)$, and $(\bar{\mathcal{F}}, \bar{P})$ be the completions of $(\mathcal{F}_1, P_1)$, $(\mathcal{F}_2, P_2)$, and $(\mathcal{F}, P)$, respectively. The following statement is the **Fubini theorem**.

If $(\omega_1, \omega_2) \mapsto X(\omega_1, \omega_2)$ is $\bar{\mathcal{F}}$-measurable and $\bar{P}$-integrable, then

1. $X(\omega_1, \cdot)$ is $\bar{\mathcal{F}}_2$-measurable and $\bar{P}_2$-integrable for each $\omega_1 \in \Omega_1 \setminus N_1$, $N_1 \in \mathcal{F}_1$, and $P_1(N_1) = 0$.

2. $\int_{\Omega_2} X(\cdot, \omega_2)\, \bar{P}_2(d\omega_2)$ is $\bar{\mathcal{F}}_1$-measurable and $\bar{P}_1$-integrable.

3. The equality

$$\int_{\Omega} X(\omega)\, \bar{P}(d\omega) = \int_{\Omega_1} \left[\int_{\Omega_2} X(\omega_1, \omega_2)\, \bar{P}_2(d\omega_1) \right] \bar{P}_2(d\omega_1) \qquad (2.56)$$

holds. If in addition X is positive and either side of Eq. 2.56 exists and is finite or infinite, so does the other side and Eq. 2.56 is valid ([40], p. 59).

Note: We have seen in Section 2.2.3 that for any probability space $(\Psi, \mathcal{G}, Q)$ there exists a complete probability space $(\Psi, \bar{\mathcal{G}}, \bar{Q})$ such that $\mathcal{G} \subset \bar{\mathcal{G}}$ and $Q = \bar{Q}$ on $\mathcal{G}$. Hence, the assumption that X is defined on a complete probability space is not restrictive. The last statement of the theorem considers the case in which X is positive but may not be $\bar{P}$-integrable, that is, the integral $\int_\Omega X\, d\bar{P}$ may not be finite. ▲

Example 2.37: Let $X : ([0, 1] \times \Omega, \mathcal{B}([0, 1]) \times \mathcal{F}) \longrightarrow (\mathbb{R}, \mathcal{B})$ be a measurable function. The product measure space on which X is defined is endowed with the product measure $\lambda \times P$, where λ denotes the Lebesgue measure and P is a probability measure on $(\Omega, \mathcal{F})$. It is common to interpret the first argument of X as time. The integral $\chi(A, \omega) = \int_A 1_B(X(s, \omega))\, ds$, $A \in \mathcal{B}([0, 1])$, $B \in \mathcal{B}$, represents the time $X(\cdot, \omega)$, $\omega \in \Omega$, spends in B during a time interval A. The expectation of this occupation time is $E[\chi(A, \omega)] = \int_A P(X(s) \in B)\, ds$. ◇

Proof: The measurable mapping $(s, \omega) \mapsto X(s, \omega)$ generalizes the time series considered in Example 2.25 because the index s takes values in $[0, 1]$ rather than in a countable set. This mapping is a **stochastic process** (Chapter 3). The function $X(\cdot, \omega)$, called the **sample path** ω of X, is defined on $[0, 1]$ for each ω.

The indicator function, $1_B : (\mathbb{R}, \mathcal{B}) \longrightarrow (\{0, 1\}, \mathcal{K})$, is measurable because $B \in \mathcal{B}$ and $\mathcal{K} = \{\emptyset, \{0, 1\}, \{0\}, \{1\}\}$ so that

$$1_B \circ X : ([0, 1] \times \Omega, \mathcal{B}([0, 1]) \times \mathcal{F}) \longrightarrow (\{0, 1\}, \mathcal{K})$$

is measurable. The expectation of the occupation time is

$$\begin{aligned} E[\chi(A, \omega)] &= \int_\Omega \chi(A, \omega) P(d\omega) = \int_\Omega \left[\int_A 1_B(X(s, \omega))\, ds\right] P(d\omega) \\ &= \int_A \left[\int_\Omega 1_B(X(s, \omega))\, P(\omega)\right] ds = \int_A P(X(s) \in B)\, ds, \end{aligned}$$

where we have used Fubini's theorem ([151], Example 5.9.1, p. 153). ■

Example 2.38: If X is a positive random variable, its expectation can be calculated from $E[X] = \int_{[0,\infty)} P(X > x)\, dx$. ◇

Proof: The mapping $(x, \omega) \mapsto 1_{\{X(\omega)>x\}}$ is measurable from $([0, \infty) \times \Omega, \mathcal{B}([0, \infty)) \times \mathcal{F})$ to $(\{0, 1\}, \mathcal{K})$, where $\mathcal{K} = \{\emptyset, \{0\}, \{1\}, \{0, 1\}\}$. We have

$$\begin{aligned} &\int_{[0,\infty)} P(X > x)\, dx = \int_{[0,\infty)} \left[\int_\Omega 1_{\{X(\omega)>x\}}\, P(d\omega)\right] dx \\ &= \int_\Omega \left[\int_{[0,\infty)} 1_{\{X(\omega)>x\}}\, dx\right] P(d\omega) = \int_\Omega X(\omega)\, P(d\omega) = E[X] \end{aligned}$$

by Fubini's theorem and the equality $\int_{[0,\infty)} 1_{\{X(\omega)>x\}}\, dx = \int_{[0,X(\omega))} dx = X(\omega)$. ■

2.9 Radon-Nikodym derivative

We will primarily use the Radon-Nikodym derivative to improve the efficiency of the Monte Carlo simulation (Section 5.4.2).

Let $(\Omega, \mathcal{F})$ be a measurable space and $\mu, \nu : \Omega \to [0, \infty]$ be two measures on this space. If $\mu(A) = 0$, $A \in \mathcal{F}$, implies $\nu(A) = 0$, we say that ν is **absolutely continuous with respect to** μ and indicate this property by the notation $\nu \ll \mu$. If $\nu \ll \mu$ and $\mu \ll \nu$, ν and μ are said to be **equivalent** measures.

Example 2.39: Consider a measure space $(\Omega, \mathcal{F}, \mu)$ and a measurable function $h : (\Omega, \mathcal{F}) \longrightarrow ([0, \infty), \mathcal{B}([0, \infty)))$. The set function

$$\nu(A) = \int_A h \, d\mu, \quad A \in \mathcal{F}, \tag{2.57}$$

is a measure that is absolutely continuous with respect to μ. ◇

Proof: The set function ν is positive by definition. This function is countably additive since, for any disjoint sets $A_n \in \mathcal{F}$, $n = 1, 2, \ldots$, we have

$$\begin{aligned} \nu(\cup_{n=1}^\infty A_n) &= \int_{\cup_{n=1}^\infty A_n} h \, d\mu = \int \sum_{n=1}^\infty h \, 1_{A_n} \, d\mu \\ &= \sum_{n=1}^\infty \int h \, 1_{A_n} \, d\mu = \sum_{n=1}^\infty \int_{A_n} h \, d\mu = \sum_{n=1}^\infty \nu(A_n). \end{aligned}$$

The above term by term integration is valid whether $\sum_{n=1}^\infty \int h \, 1_{A_n} \, d\mu$ is or is not finite since h is positive. The measure ν is absolutely continuous with respect to μ because

$$\nu(A) = \int_A h \, d\mu = \int_\Omega h \, 1_A \, d\mu \le \sup_{\omega \in A} (h(\omega)) \, \mu(A)$$

so that $\mu(A) = 0$ implies $\nu(A) = 0$ with the usual convention $0 \cdot \infty = 0$

We also note that if μ is σ-finite so is ν, that is, there exists a sequence of disjoint events $B_j \in \mathcal{F}$ such that $\nu(B_j) < \infty$ for every j and $\nu(B) = \sum_{j=1}^\infty \nu(B \cap B_j)$, $B \in \mathcal{F}$. Take $B_j = \{\omega \in \Omega : j - 1 \le h(\omega) < j\} \in \mathcal{F}$ and note that $\nu_j(A) = \int_A (h \, 1_{B_j}) \, d\mu$ is finite for each j and $A \in \mathcal{F}$ such that $\mu(A) < \infty$. Also, $\nu_j(A)$ is bounded by $j \, \mu(A)$. The monotone convergence theorem for the sequence of measurable, increasing, and positive functions $X_n = \sum_{j=1}^n h \, 1_{B_j}$ gives (Eq. 2.47)

$$\lim_{n\to\infty} \int_A X_n \, d\mu = \int_A \left(\lim_{n\to\infty} X_n \right) d\mu.$$

The left and the right sides of this equation are

$$\lim_{n\to\infty} \sum_{j=1}^n \int_A h \, 1_{B_j} \, d\mu = \sum_{j=1}^\infty \int_A h \, 1_{B_j} \, d\mu = \sum_{j=1}^\infty \nu_j(A) \quad \text{and}$$

$$\int_A \left(\sum_{j=1}^\infty h \, 1_{B_j} \right) d\mu = \nu(A),$$

respectively, so that ν is σ-finite. ■

The **Radon-Nikodym theorem** can be interpreted as the converse of Example 2.39. The statement of this theorem is given in Eq. 2.58 without proof.

If μ and ν are σ-finite measures on a measurable space $(\Omega, \mathcal{F})$ such that $\nu \ll \mu$, then there exists a measurable function

$$h = \frac{d\nu}{d\mu} : (\Omega, \mathcal{F}) \to ([0, \infty), \mathcal{B}([0, \infty))), \tag{2.58}$$

called the **Radon-Nikodym derivative** of ν with respect to μ, such that Eq. 2.57 holds ([66], Theorem 18, p. 116).

Example 2.40: Consider a probability space $(\Omega, \mathcal{F}, P)$, a partition $A_i \in \mathcal{F}$, $i = 1, \ldots, n$, of Ω such that $P(A_i) > 0$, and a random variable $X = \sum_{i=1}^n x_i \, 1_{A_i}$ defined on this space, where $x_i \in \mathbb{R}$. Let Q be a probability measure on $(\Omega, \mathcal{F})$ such that $Q(A_i) > 0$ and $h_i = P(A_i)/Q(A_i)$, $i = 1, \ldots, n$. Denote expectations under the probability measures P and Q by E_P and E_Q, respectively. The expectations $E_P[X]$ and $E_Q[X\,h]$ of the random variables X and $X\,h$ with respect to the measures P and Q coincide. ◇

Proof: The expectation of $X\,h$ under Q is

$$E_Q[X\,h] = \sum_{i=1}^{n} (x_i\, h_i)\, Q(A_i) = \sum_{i=1}^{n} \left(x_i \, \frac{P(A_i)}{Q(A_i)} \right) Q(A_i) = E_P[X].$$

Since both P and Q are zero only on the impossible event $\emptyset$, the function h is measurable and Q is absolutely continuous with respect to P on the σ-field $\sigma(A_i, i = 1, \ldots, n)$. The function h is the Radon-Nikodym derivative of P with respect to Q. ▲

Example 2.41: Let $X \sim L_1(\Omega, \mathcal{F}, P)$ be a random variable. If Q is a probability on a measurable space $(\Omega, \mathcal{F})$ such that $P \ll Q$, we have

$$E_P[X] = \int X\, dP = \int X \frac{dP}{dQ}\, dQ = \int X\, h\, dQ = E_Q[X\, h], \tag{2.59}$$

where $h = dP/dQ$ is the Radon-Nikodym derivative. This result extends our discussion in Example 2.40 to an arbitrary random variable. The importance sampling technique uses Eq. 2.59 to improve the efficiency of the Monte Carlo solution (Section 5.4.2). ◇

2.10 Random variables

Consider a probability space $(\Omega, \mathcal{F}, P)$ and a real-valued random variable X defined on this space (Section 2.4). This section defines the distribution, density, and characteristic functions of X, and illustrates the use of these functions in calculations.

2.10.1 Distribution function

The cumulative distribution function or the distribution function of a random variable X is defined by the mapping $X : \Omega \to \mathbb{R}$ and the probability measure P.

The **distribution function** of X is

$$F(x) = P(X^{-1}((-\infty, x])) = P(\{\omega : X(\omega) \leq x\}) = P(X \leq x). \quad (2.60)$$

Note: F is the probability induced by X on $(\mathbb{R}, \mathcal{B})$ and constitutes a special case of the distribution in Eq. 2.29 for $B = (-\infty, x]$ and $h = X$. The definition is meaningful because X is $\mathcal{F}$-measurable so that $\{\omega : X(\omega) \leq x\} \in \mathcal{F}$. The notation $P(X \leq x)$ used in Eq. 2.60 is an abbreviation for $P(\{\omega : X(\omega) \leq x\})$. ▲

The distribution function F has the following properties.

- F is a right continuous, increasing function with range [0, 1].
- F can have only jump discontinuities and the set of these jumps is countable. The number of jumps of F exceeding $\varepsilon \in (0, 1)$ is smaller than $1/\varepsilon$.
- $\lim_{x\to\infty} F(x) = 1$ and $\lim_{x\to-\infty} F(x) = 0$.
- $P(a < X \leq b) = F(b) - F(a) \geq 0, \quad a \leq b$.
- $P(a \leq X < b) = F(b) - F(a) + P(X = a) - P(X = b), \quad a \leq b$.

Proof: Since F is a probability measure, its range is [0, 1]. Also, F is an increasing function because $\{X \leq x_1\} \subseteq \{X \leq x_2\}$ for $x_1 \leq x_2$ (Eq. 2.5). That F is right continuous follows from the continuity of the probability measure and the definition of F. Let $\{x_n\}$ be a decreasing numerical series converging to x, $B_n = \{\omega : X(\omega) \leq x_n\}$, and $B = \{\omega : X(\omega) \leq x\}$. The sequence of events B_n is decreasing so that $\lim_{n\to\infty} B_n = \cap_{n=1}^{\infty} B_n = B$ and (Eq. 2.25)

$$\lim_{n\to\infty} F(x_n) = \lim_{n\to\infty} P(B_n) = P(\lim_{n\to\infty} B_n) = P(B) = F(x).$$

A function $g : I \to \mathbb{R}$, $I \subset \mathbb{R}$, is discontinuous at $c \in I$ if it is not continuous at this point. The discontinuity of g at c can be of four types:
(a) Infinite discontinuity if the left limit $\lim_{x\uparrow c} g(x)$ and/or the right limit $\lim_{x\downarrow c} g(x)$ of g at c is infinity.
(b) Jump discontinuity if the left and right limits of g at c are finite but not equal.
(c) Removable discontinuity if $\lim_{x\to c} g(x)$ exists and is finite but differs from the value of g at c.
(d) Oscillating discontinuity if g is bounded and the left and right limits of g at c do not exist.

F can have only jump discontinuities because it is a bounded, increasing, and right continuous function. We need to show that F has at most a countable number of jump discontinuities. Let $\xi < \xi'$ be two distinct jump points of F and consider the open intervals $I_\xi = (F(\xi-), F(\xi))$ and $I_{\xi'} = (F(\xi'-), F(\xi'))$ associated with these jumps. Because the jump points ξ and ξ' are distinct, there exists $\tilde{\xi} \in (\xi, \xi')$ such that $F(\xi) \leq F(\tilde{\xi}) \leq F(\xi'-)$, showing that I_ξ and $I_{\xi'}$ are disjoint intervals. The collection of the intervals I_ξ is countable since each I_ξ contains a rational number and the set of rational number is countable.

The sum of all jumps of F is $\sum_{\xi \in J}[F(\xi+) - F(\xi-)] = \sum_{\xi \in J}[F(\xi) - F(\xi-)] \leq 1$, where J denotes the collection of jump points of F. Hence, $\varepsilon\, n_\varepsilon \leq 1$ so that $n_\varepsilon \leq 1/\varepsilon$, where n_ε denotes the number of jumps of F larger than ε.

The third property results from the equalities

$$\lim_{x \to +\infty} P((-\infty, x]) = P(\Omega) = 1 \quad \text{and} \quad \lim_{x \to -\infty} P((-\infty, x]) = P(\emptyset) = 0.$$

The fourth property follows from the equality $P(B \setminus A) = P(B) - P(A)$, which holds since the event $A = \{\omega : X(\omega) \leq a\}$ is included in $B = \{\omega : X(\omega) \leq b\}$ for $a \leq b$.

The last property holds because the event $\{a \leq X < b\}$ is the union of the disjoint events $\{X = a\}$ and $\{a < X \leq b\} \setminus \{X = b\}$ so that the probability of $\{a \leq X < b\}$ is $P(X = a) + P(\{a < X \leq b\} \setminus \{X = b\})$, and $P(\{a < X \leq b\} \setminus \{X = b\})$ is equal to $P(a < X \leq b) - P(X = b)$. ■

Example 2.42: Consider the system in Example 2.14 and let F denote the distribution of the system strength X. The system reliability without and with a proof test at x_{pr} is $P_s(x) = 1 - F(x)$ and $P_s(x) = (1 - F(x \vee x_{\text{pr}}))/(1 - F(x_{\text{pr}}))$, respectively.

Figure 2.4 shows the system reliability disregarding and accounting for the

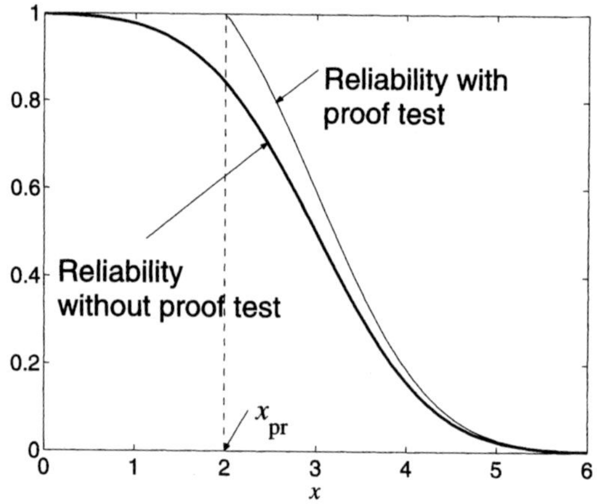

Figure 2.4. Reliability with and without proof test information

information provided by a proof test at $x_{\text{pr}} = 2$ for a lognormal random variable $X = \exp(Y)$, where Y is a Gaussian variable with mean $\mu = 1$ and variance $\sigma^2 = 9$ (Eq. 2.68). The figure suggests that the additional information provided by the proof test can be beneficial in practical applications. ◇

Example 2.43: Let F be the distribution of a random variable X defined on a probability space $(\Omega, \mathcal{F}, P)$. The distribution F is continuous at $x \in \mathbb{R}$ if and only if $P(\{X = x\}) = 0$. ◇

Proof: Let $\{x_n\}$ be a positive sequence such that $x_n \downarrow 0$ as $n \to \infty$. Define the sequence of intervals $A_n = \{\omega \in \Omega : X(\omega) \in (x - x_n, x]\}$ for $x \in \mathbb{R}$. Since this sequence is decreasing, we have $P(\{X = x\}) = P\left(\cap_{n=1}^{\infty} A_n\right) = \lim_{n\to\infty} P(A_n)$ and $P(A_n) = F(x) - F(x_n)$. If F is continuous at x, then $F(x) - F(x_n) \to 0$ as $n \to \infty$ so that $P(\{x\}) = 0$. Conversely, if $P(\{X = x\}) = 0$, we have $\lim_{n\to\infty}[F(x) - F(x_n)] = 0$ so that F is continuous at x since it is right continuous. ■

2.10.2 Density function

Consider a distribution function F that is **absolutely continuous** in $\mathbb{R}$, that is, there exists an integrable function f, called a probability density function or a density function, such that Eq. 2.61 is satisfied.

The **density function** f is an integrable function such that

$$F(b) - F(a) = \int_a^b f(x)\,dx, \quad a \le b. \tag{2.61}$$

The density function has the following properties.

- $f(x) = F'(x)$ so that $\int_{-\infty}^{x} f(\xi)\,d\xi = F(x)$.
- $f \ge 0$ because F is an increasing function.
- f is not a probability measure.
- The area under f is $\int_{-\infty}^{\infty} f(x)\,dx = 1$.

Consider a random variable X defined on a probability space $(\Omega, \mathcal{F}, P)$ and a measurable function $g : (\mathbb{R}, \mathcal{B}) \to (\mathbb{R}, \mathcal{B})$. Then $Y = g \circ X$ is a random variable defined on $(\Omega, \mathcal{F}, P)$ (Section 2.4.3).

If X and $Y = g \circ X$ are integrable and $Q(B) = P(X^{-1}(B))$, $B \in \mathcal{B}$, then

$$E[Y] = \int_\Omega Y(\omega)\,P(d\omega) = \int_\Omega g(X(\omega))\,P(d\omega) \quad \text{and} \tag{2.62}$$

$$E[Y] = \int_\mathbb{R} g(x)\,Q(dx) = \int_\mathbb{R} g(x)\,dF(x) = \int_\mathbb{R} g(x)\,f(x)\,dx. \tag{2.63}$$

Proof: If $g = 1_B$, $B \in \mathcal{B}$, then Eqs. 2.62 and 2.63 give $E[Y] = P(X \in B) = Q(B)$. If I is a finite index set and $g = \sum_{i\in I} b_i\,1_{B_i}$, the subsets $B_i \in \mathcal{F}$ partition $\mathbb{R}$, and b_i are real constants, then Eqs. 2.62 and 2.63 give $\sum_{i\in I} b_i\,P(B_i)$ because the integration is a linear operator.

If g is an arbitrary positive Borel function, there exists a sequence of simple, increasing, and measurable functions g_n, $n = 1, 2, \ldots$, converging to g as $n \to \infty$. We have seen that the expectations of $g_n(X)$ in Eqs. 2.62 and 2.63 coincide. The monotone convergence theorem shows that the expectations of $g(X)$ in Eqs. 2.62 and 2.63 coincide. If g is an arbitrary Borel function, it can be represented by $g = g^+ - g^-$, where g^+ and g^- are positive Borel functions. Because $g(X)$ is integrable and the expectation is a linear operator, the formulas in Eqs. 2.62 and 2.63 give the same result.

That Eqs. 2.62 and 2.63 give the same result can be indicated by $E_P[g \circ X] = E_Q[g]$, where E_P and E_Q denote expectations under the probabilities P and Q, respectively. We note that the probabilities P and Q live on different measurable spaces, P is defined on $(\Omega, \mathcal{F})$ while Q is the probability measure induced by X on $(\mathbb{R}, \mathcal{B})$ (Eq. 2.29). Generally, the last two formulas in Eq. 2.63 involving Riemann-Stieltjes and Riemann integrals are used to calculate $E[Y]$. ■

Example 2.44: The expectation of the **Cauchy** random variable X with density $f(x) = a/[\pi\,(a^2 + x^2)]$, $a > 0$, $x \in \mathbb{R}$, does not exist. ◇

Proof: The expectation of X^+ is (Eq. 2.63)

$$E[X^+] = E\left[1_{X \geq 0}\, X\right] = \int_0^\infty x\, \frac{a}{\pi\,(a^2 + x^2)}\, dx = \frac{a}{2\pi} \log(a^2 + x^2) \,|_0^\infty = +\infty.$$

Similarly, $E[X^-]$ is infinity. Hence, $E[X]$ is not defined. ■

Example 2.45: Let $g : (\mathbb{R}, \mathcal{B}) \to (\mathbb{R}, \mathcal{B})$ be a measurable function and X be a random variable defined on a probability space $(\Omega, \mathcal{F}, P)$. The expectation of the random variable $g(X)$ can be calculated by the last integral in Eq. 2.63 or $E_f[g(X)] = E_q\,[g(X)\, f(X)/q(X)]$, where q is a probability density function such that $I_f = \{x \in \mathbb{R} : f(x) > 0\} \subseteq I_q = \{x \in \mathbb{R} : q(x) > 0\}$. The expectations E_f and E_q correspond to f and q, respectively. The requirement on the subsets I_q and I_f of $\mathbb{R}$ is consistent with the absolute continuity condition for the existence of the Radon-Nikodym derivative (Eq. 2.58). ◇

Proof: The expectation of $g(X)$ is

$$E_f[g(X)] = \int_{I_f} g(x)\, f(x)\, dx = \int_{I_q} \left[g(x)\, \frac{f(x)}{q(x)}\right] q(x)\, dx,$$

where the last equality holds since f is zero in $I_q \setminus I_f$. The integral on I_q is well defined because q is strictly positive on I_q, and gives the expectation of the random variable $g(X)\, f(X)/q(X)$ with respect to the density function q. ■

Example 2.46: Let X be a random variable with density f. It is assumed that the expectation of X and of all functions of X considered in this example exist and are finite. Let X_i, $i = 1, \ldots, n$, be independent copies of X. The classical estimator $\bar{X} = (1/n) \sum_{i=1}^n X_i$ of $E[X]$ has mean $E[X]$ and variance $\mathrm{Var}[X]/n$. The importance sampling estimator of $E[X]$ is $\bar{X}_{\mathrm{is}} = (1/n) \sum_{i=1}^n X_i\, f(X_i)/q(X_i)$, where q is a density such that $q(x) = 0$ implies $f(x) = 0$. The mean and variance of $\bar{X}_{\mathrm{is}}$ are $E[X]$ and $\mathrm{Var}[X\, f(X)/q(X)]/n$, respectively.

The success of the importance sampling technique depends on the functional form of q. For example, $E[\bar{X}] = 1/2$ and $\mathrm{Var}[\bar{X}] = 0.05$ for a random variable X with density $f(x) = 6x(1-x)\,1_{[0,1]}(x)$. The variance of $\bar{X}_{\text{is}}$ is 0.0929 and 0.0346 for $q(x) = 1_{[0,1]}(x)$ and $q(x) = (20/7)\,x\,1_{[0,0.7]} + [2 - (20/3)(x - 0.7)]\,1_{(0.7,1]}$, respectively. This shows that $\bar{X}_{\text{is}}$ may or may not be superior to $\bar{X}$ depending on q. ◇

Note: The importance sampling technique in this illustration is a special case of Example 2.45 for $g(x) = x$. If we set $q(x) = x\,f(x)/E[X]$, the variance of $\bar{X}_{\text{is}}$ is zero. However, this estimator cannot be used because it depends on $E[X]$, that is, on the parameter to be estimated. ▲

2.10.3 Characteristic function

Let X be a real-valued random variable with distribution F and density f. Define a complex-valued function $g(x,u) = \exp\left(\sqrt{-1}\,u\,x\right)$, $u, x \in \mathbb{R}$.

The **characteristic function** of X is

$$\varphi(u) = E\left[e^{\sqrt{-1}\,u\,X}\right] = \int_{\mathbb{R}} e^{\sqrt{-1}\,u\,x}\,dF(x) = \int_{\mathbb{R}} e^{\sqrt{-1}\,u\,x}\,f(x)\,dx. \quad (2.64)$$

Note: We define the expectation for real-valued random variables and $e^{\sqrt{-1}\,u\,X}$ is complex-valued. The expectation of $e^{\sqrt{-1}\,u\,X}$ is complex-valued with real and imaginary parts $E[\cos(u\,X)]$ and $E[\sin(u\,X)]$, respectively. ▲

Example 2.47: The characteristic function exists for any random variable. For example, the characteristic function of the Cauchy random variable in Example 2.44 is $\varphi(u) = \exp(-a\,|u|)$, $u \in \mathbb{R}$. We have seen that the expectation of this random variable does not exist.

The **moment generating function,** $m(u) = E[\exp(u\,X)]$, $u \in \mathbb{R}$, is also used in calculations. However, $m(\cdot)$ may not be bounded, for example, the moment generating function of a Cauchy random variable. ◇

2.10.3.1 Properties

- $|\varphi(u)| \le \varphi(0) = 1$.
- $\varphi(-u) = \varphi(u)^*$, where z^* denotes the complex conjugate of $z \in \mathbb{C}$.
- The characteristic function is **positive definite**.
- The characteristic and the density functions are **Fourier pairs**.
- φ is **uniformly continuous** in $\mathbb{R}$.
- If $X \in L_q$, then $\varphi \in C^q(\mathbb{R})$ and $\varphi^{(k)}(0) = (\sqrt{-1})^k\,E[X^k]$, $k = 1, \ldots, q$.

Proof: Eq. 2.64 gives $\varphi(0) = 1$, $|\varphi(u)| = \left|\int e^{\sqrt{-1}\,u\,x}\,f(x)\,dx\right| \le \int |e^{\sqrt{-1}\,u\,x}|\,f(x)\,dx = 1$, and $\varphi(u) = \varphi(-u)^*$.

Let $Z = \sum_{k=1}^{n} z_k \exp(\sqrt{-1}\,u_k\,X)$ be a random variable. Because $Z\,Z^* = |Z|^2$ is a positive random variable for any $z_k \in \mathbb{C}$ and $u_k \in \mathbb{R}$,

$$0 \le E[Z\,Z^*] = \sum_{k,l=1}^{n} z_k\,z_l^*\,E\left[e^{\sqrt{-1}\,(u_k-u_l)\,X}\right] = \sum_{k,l=1}^{n} z_k\,z_l^*\,\varphi(u_k - u_l),$$

so that φ is positive definite.

For the fourth property we need to show that

$$\varphi(u) = \int_{\mathbb{R}} e^{\sqrt{-1}\,u\,x}\,f(x)\,dx \quad \text{and} \quad f(x) = \frac{1}{2\pi}\int_{\mathbb{R}} e^{-\sqrt{-1}\,u\,x}\,\varphi(u)\,du. \tag{2.65}$$

The first equation is the definition of the characteristic function. The second equation follows from the fact that, if $\beta : \mathbb{R} \to \mathbb{C}$ is a continuous, positive definite function such that $\beta(0) = 1$ and $\int |\beta(u)|\,du < \infty$, then β is a characteristic function of a distribution F on $\mathbb{R}$ that is absolutely continuous with respect to the Lebesgue measure and has a continuous density f given by ([66], Theorem 14, p. 231)

$$f(x) = \frac{1}{2\pi}\int_{\mathbb{R}} e^{-\sqrt{-1}\,u\,x}\,\beta(u)\,du.$$

It can also be shown that the characteristic function defines uniquely the density and the distribution of X ([66], Theorem 3, p. 211) and that

$$F(x_2) - F(x_1) = \frac{1}{2\pi}\lim_{\tau\to\infty}\int_{-\tau}^{\tau} \frac{e^{-\sqrt{-1}\,u\,x_2} - e^{-\sqrt{-1}\,u\,x_1}}{-\sqrt{-1}\,u}\,\varphi(u)\,du, \tag{2.66}$$

where and x_1 and $x_2 > x_1$ are points of continuity of F ([142], Theorem 3, p. 12).

To prove that φ is uniformly continuous in $\mathbb{R}$, we have to show that for an arbitrary $\varepsilon > 0$ there exists $\delta > 0$ such that $|\varphi(u+h) - \varphi(u)| < \varepsilon$ if $|h| < \delta$ for all $u \in \mathbb{R}$. The increment of the characteristic function in $(u, u+h)$ satisfies the inequality

$$\begin{aligned}
|\varphi(u+h) - \varphi(u)| &= \left|E[e^{\sqrt{-1}\,(u+h)\,X}] - E[e^{\sqrt{-1}\,u\,X}]\right| \\
&\le E\left[\left|e^{\sqrt{-1}\,u\,X}\left(e^{\sqrt{-1}\,h\,X} - 1\right)\right|\right] \le E\left[\left|e^{\sqrt{-1}\,h\,X} - 1\right|\right] \\
&= \int_{|x|>a}\left|e^{\sqrt{-1}\,h\,x} - 1\right| dF(x) + \int_{|x|\le a}\left|e^{\sqrt{-1}\,h\,x} - 1\right| dF(x)
\end{aligned}$$

for any $a > 0$. The first integral is smaller than $2\,P(|X| > a)$. We can take a sufficiently large such that $P(|X| > a) < \varepsilon/4$. The second integral can be made smaller than $\varepsilon/2$ for the selected value of a by taking $|h|$ smaller than a real $\delta > 0$ that is independent of u. Hence, the increment $|\varphi(u+h) - \varphi(u)|$ of the characteristic function does not exceed an arbitrary $\varepsilon > 0$ provided that $|h| < \delta$ irrespective of the value of u.

The proof of the relationship between $\varphi^{(k)}(u) = d^k\varphi(u)/du^k$ and moments of X can be based on the inequalities ([151], p. 298)

$$\left|e^{\sqrt{-1}\,x} - \sum_{k=0}^{n}\frac{(\sqrt{-1}\,x)^k}{k!}\right| \le \frac{|x|^{n+1}}{(n+1)!} \quad \text{and} \quad \left|e^{\sqrt{-1}\,x} - \sum_{k=0}^{n}\frac{(\sqrt{-1}\,x)^k}{k!}\right| \le \frac{2\,|x|^n}{n!}.$$

For example, suppose that $X \sim L_1$ and consider the difference

$$\frac{\varphi(u+h)-\varphi(u)}{h} - E\left[\sqrt{-1}\,X\,e^{\sqrt{-1}\,u\,X}\right] = E\left[e^{\sqrt{-1}\,u\,X}\,\frac{e^{\sqrt{-1}\,h\,X}-1-\sqrt{-1}\,h\,X}{h}\right].$$

The latter expectation is finite because

$$\left|e^{\sqrt{-1}\,u\,X}\right| \left|\frac{e^{\sqrt{-1}\,h\,X}-1-\sqrt{-1}\,h\,X}{h}\right| \le 2\,|X| \in L_1$$

by the second inequality with $n = 1$ and x replaced by $h\,X$. The first inequality gives

$$\left|\frac{e^{\sqrt{-1}\,h\,X}-1-\sqrt{-1}\,h\,X}{h}\right| \le \frac{h^2\,X^2}{2\,h} = h\,\frac{X^2}{2} \to 0, \quad \text{as } h \downarrow 0,$$

so that

$$\lim_{h\downarrow 0}\left|\frac{\varphi(u+h)-\varphi(u)}{h} - E\left[\sqrt{-1}\,X\,e^{\sqrt{-1}\,u\,X}\right]\right| = 0$$

by dominated convergence. Hence, we have $d\varphi(u)/du = E[\sqrt{-1}\,X\,e^{\sqrt{-1}\,u\,X}]$ so that $\varphi'(0) = \sqrt{-1}\,E[X]$. Higher order moments can be found in the same way. ■

Example 2.48: Let X be a random variable with characteristic function φ. The characteristic function of $a\,X + b$ is $\varphi_{a\,X+b}(u) = \varphi(a\,u)\,e^{\sqrt{-1}\,u\,b}$, where a and b are constants. ◇

Proof: The characteristic function of $a\,X + b$ is $E[\exp(\sqrt{-1}\,u(a\,X + b))]$ by Eq. 2.64 or $E[\exp(\sqrt{-1}\,u\,a\,X)]\,\exp(\sqrt{-1}\,u\,b)$. ■

Example 2.49: Let X be a random variable with the distribution function $F(x) = \sum_{i\in I} p_i\,1_{[x_i,\infty)}(x)$, where I is a finite index set, $p_i \ge 0$ such that $\sum_{i\in I} p_i = 1$, and $\{x_i\}$ is an increasing sequence of real numbers. The density and the characteristic functions, $f(x) = \sum_{i\in I} p_i\,\delta(x - x_i)$ and $\varphi(u) = \sum_{i\in I} p_i\,[\cos(u\,x_i) + \sqrt{-1}\,\sin(u\,x_i)]$, of X are Fourier pairs. The characteristic function consists of a superposition of harmonics with amplitudes p_i and frequencies coinciding with the locations x_i of the jumps of F. ◇

Example 2.50: Let X be a real-valued random variable on a probability space $(\Omega, \mathcal{F}, P)$ and $Y = g(X) = X^r$, where $r \ge 1$ is an integer. Because $g(x) = x^r$ is a continuous function, it is Borel measurable and $Y = X^r$ is a random variable. If $X \sim L_r(\Omega, \mathcal{F}, P)$, the expectation of Y exists and is finite. The **moment of order** r of X is

$$\mu(r) = E[X^r] = \int_{-\infty}^{\infty} x^r\,dF(x) = \int_{-\infty}^{\infty} x^r\,f(x)\,dx. \tag{2.67}$$

If the functions $g(x) = (x - \mu(1))^r$ and $g(x) = |x - \mu(1)|^r$ are considered, the resulting expectations define the **central moments of order** r and **absolute**

central moments of order r of X. The moments $\mu = \mu(1)$, $\sigma^2 = E(X-\mu)^2$, $v = \sigma/\mu$, $\gamma_3 = E[(X-\mu)^3]/\sigma^3$, and $\gamma_4 = E[(X-\mu)^4]/\sigma^4$ are called **mean, variance, coefficient of variation, coefficient of skewness,** and **coefficient of kurtosis**, respectively. These moments can be calculated by direct integration (Eq. 2.67) or from $\varphi^{(r)}(0) = (\sqrt{-1})^r E[X^r]$. The positive square root of the variance is called **standard deviation** and is denoted by σ. ◇

Example 2.51: A random variable X is said to be **Gaussian** with mean μ and variance σ^2, in short $X \sim N(\mu, \sigma^2)$, if it has the density

$$f(x) = \frac{1}{\sqrt{2\pi}\,\sigma} \exp\left[-\frac{1}{2}\left(\frac{x-\mu}{\sigma}\right)^2\right], \quad x \in \mathbb{R}. \tag{2.68}$$

The skewness and kurtosis coefficients of X are $\gamma_3 = 0$ and $\gamma_4 = 3$. The skewness coefficient is zero because f is symmetric about $x = \mu$.

The random variable X follows a **gamma** distribution with parameters $k > 0$ and $\lambda > 0$ if it has the density

$$f(x) = \frac{x^{k-1}\,\lambda^k\,e^{-\lambda x}}{\Gamma(k)}, \quad x \geq 0. \tag{2.69}$$

The first four moments of this random variable, denoted by $X \sim \Gamma(k,\lambda)$, are $\mu = k/\lambda$, $\sigma^2 = k/\lambda^2$, $\gamma_3 = 2/\sqrt{k}$, and $\gamma_4 = 3(1+2/k)$.

The characteristic functions of the random variables with densities given by Eqs. 2.68 and 2.69 are

$$\varphi(u) = \exp\left(\sqrt{-1}\,\mu\,u - \frac{1}{2}\sigma^2\,u^2\right) \quad \text{for} \quad X \sim N(\mu,\sigma^2) \quad \text{and} \tag{2.70}$$

$$\varphi(u) = \frac{1}{(1-\sqrt{-1}\,u/\lambda)^k} \quad \text{for} \quad X \sim \Gamma(k,\lambda). \tag{2.71}$$

Figure 2.5 shows the density and the characteristic functions of a gamma and a Gaussian random variable with mean $\mu = 1.5$ and variance $\sigma^2 = 1$. Because these densities are not even functions, their characteristic functions are complex-valued. ◇

Example 2.52: Let N be a discrete random variable taking values in $\{0, 1, 2, \ldots\}$ with the probability

$$P(N = n) = \frac{\lambda^n}{n!}\,e^{-\lambda}, \quad n = 0, 1, \ldots, \tag{2.72}$$

where $\lambda > 0$ is an intensity parameter. The probability measure in Eq. 2.72 defines a **Poisson** random variable. The characteristic function of $X = a\,N + b$ is

$$\varphi(u) = \exp\left(\lambda\,(e^{\sqrt{-1}\,a\,u} - 1) + \sqrt{-1}\,b\,u\right), \tag{2.73}$$

where a and b are constants. The random variable X is referred to as Poisson with parameters (a, b, λ). ◇

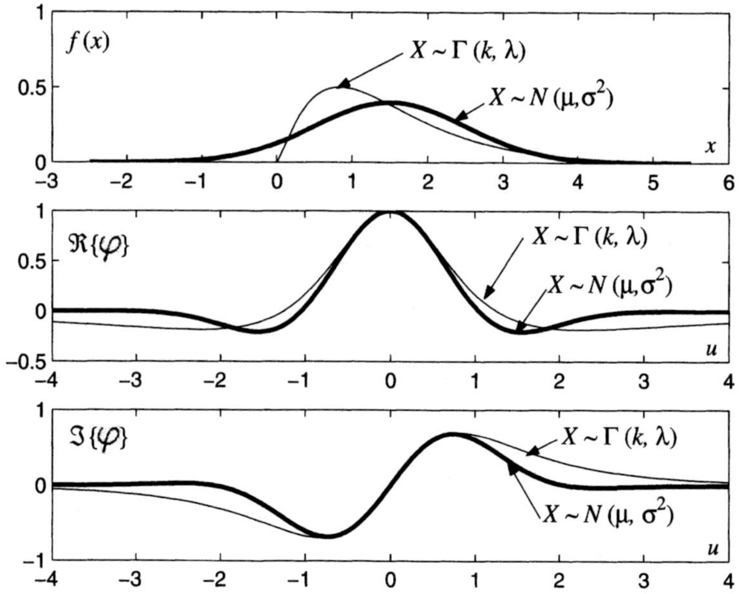

Figure 2.5. The density and characteristic functions of a gamma and a Gaussian variable with the same mean and variance

Note: The characteristic function of N is (Eq. 2.64)

$$\varphi_n(u) = E\left[e^{\sqrt{-1}\,u\,N}\right] = \sum_{k=0}^{\infty} e^{\sqrt{-1}\,u\,k}\,\frac{\lambda^k}{k!}\,e^{-\lambda} = e^{-\lambda}\sum_{k=0}^{\infty}\frac{\left(\lambda\,e^{\sqrt{-1}\,u}\right)^k}{k!}$$
$$= \exp\left[\lambda\,(e^{\sqrt{-1}\,u} - 1)\right].$$

This function and $\varphi_x(u) = \exp(\sqrt{-1}\,u\,b)\,\varphi_n(u\,a)$ give Eq. 2.73, where φ_x denotes the characteristic functions of X. ■

Example 2.53: Let N_j, $j = 1, \ldots, m$, be independent Poisson variables with intensity parameters $\lambda_j > 0$. The characteristic function of the random variable $X = \sum_{j=1}^{m} x_j\,N_j$ is

$$\varphi(u) = \exp\left[\sum_{j=1}^{m}\lambda_j\,\left(e^{\sqrt{-1}\,u\,x_j} - 1\right)\right], \tag{2.74}$$

where x_j are real numbers. ◇

Proof: We have $\varphi(u) = E\left[e^{\sqrt{-1}\,u\,X}\right] = \prod_{j=1}^{m} E\left[e^{\sqrt{-1}\,u\,x_j\,N_j}\right]$ by the independence of the random variables N_j. The result in Eq. 2.74 follows from this expression and Example 2.52. ■

Example 2.54: Let X be a **compound Poisson** random variable defined by

$$X = \sum_{k=1}^{N} Y_k \quad \text{for } N > 0 \quad \text{and } X = 0 \text{ for } N = 0, \tag{2.75}$$

where N is a Poisson variable with intensity $\lambda > 0$ (Eq. 2.72) and Y_k, $k = 1, 2, \ldots$, are independent copies of Y_1. The random variables Y_k are independent of N. The characteristic function of X is

$$\varphi(u) = \exp\left(-\lambda \int_{\mathbb{R}} (1 - e^{\sqrt{-1}\, u\, y})\, dF_y(y)\right), \tag{2.76}$$

where F_y denotes the distribution function of Y_1. ◇

Proof: We have

$$\begin{aligned}
\varphi(u) &= E\left[e^{\sqrt{-1}\, u\, X}\, 1_{\{N \ge 1\}} + e^{\sqrt{-1}\, u\, X}\, 1_{\{N=0\}}\right] = E\left[e^{\sqrt{-1}\, u \sum_{k=1}^{N} Y_k}\right] + P(N = 0) \\
&= E\left\{E\left[\prod_{k=1}^{N} e^{\sqrt{-1}\, u\, Y_k} \mid N\right]\right\} + P(N = 0) \\
&= \sum_{k=1}^{\infty} (\varphi_y(u))^k\; P(N = k) + P(N = 0) \\
&= \sum_{k=0}^{\infty} \frac{(\lambda\, \varphi_y(u))^k}{k!}\, e^{-\lambda} = \exp[-\lambda\, (1 - \varphi_y(u))],
\end{aligned}$$

where φ_y denotes the characteristic function of Y_1. The expression of φ in Eq. 2.76 is an alternative form of the above result. ■

2.10.3.2 Infinitely divisible characteristic function

Let X be a random variable with characteristic and distribution functions φ and F, respectively.

> If there is a characteristic function φ_n for every integer $n \ge 1$ such that
>
> $$\varphi(u) = (\varphi_n(u))^n, \quad u \in \mathbb{R}, \tag{2.77}$$
>
> then φ is said to be **infinitely divisible (i.d.)**.

Note: This definition shows that a random variable X with an infinitely divisible characteristic function has the representation $X = \sum_{i=1}^{n} X_i^{(n)}$ for each $n \ge 1$, where $X_i^{(n)}$, $i = 1, \ldots, n$, are independent identically distributed (iid) random variables with the characteristic function φ_n. ▲

Example 2.55: The characteristic function of $X \sim N(\mu, \sigma^2)$ is infinitely divisible. ◇

Proof: The characteristic functions $\varphi_n(u) = \exp\left(\sqrt{-1}\,\mu\,u/n - \sigma^2\,u^2/(2\,n)\right)$ and φ in Eq. 2.70 satisfy Eq. 2.77 for each n. Hence, X can be represented by a sum of n independent Gaussian variables with mean μ/n and variance σ^2/n. ■

Example 2.56: Let N denote a Poisson random variable with intensity $\lambda > 0$ and let a, b be some constants (Example 2.52). The characteristic function of $a\,N + b$ is infinitely divisible. ◇

Proof: $\varphi_n(u) = \exp\left((\lambda/n)\,(e^{\sqrt{-1}\,a\,u} - 1) + \sqrt{-1}\,(b/n)\,u\right)$ and φ in Eq. 2.73 satisfy the condition $\varphi = (\varphi_n)^n$ for any $n \geq 1$ integer. Hence, a Poisson variable X with parameters (a, b, λ) can be represented by a sum of n independent Poisson variables with parameters $(a, b/n, \lambda/n)$ for each n (Example 2.52). ■

Infinitely divisible characteristic functions are encountered in the analysis of the Brownian motion and other processes with stationary independent increments introduced in the next chapter and used extensively in this book. We review here essential properties, methods of construction, and canonical representations of infinitely divisible characteristic functions.

Properties of i.d. characteristic functions.

If φ is an i.d. characteristic function, then φ has no real zeros, that is, $\varphi(u) \neq 0$ for every $u \in \mathbb{R}$ ([124], Theorem 5.3.1, p. 80).

Proof: This property can be used as a criterion for finding whether a particular characteristic function is not infinitely divisible. For example, the characteristic function in Example 2.49 is not infinitely divisible.

If φ is infinitely divisible, there exist characteristic functions $\varphi_n = \varphi^{1/n}$ for each integer $n > 0$ so that $g(u) = \lim_{n\to\infty} \varphi_n(u) = \lim_{n\to\infty}(\varphi(u))^{1/n}$ takes only two values, zero for $\varphi(u) = 0$ and one for $\varphi(u) \neq 0$. Because φ and φ_n are characteristic functions, there is an interval $I \subset \mathbb{R}$ containing zero in which both φ and φ_n are not zero so that $\log(\varphi_n(u)) = (1/n)\,\log(\varphi(u))$, $u \in I$. The right side of this equation approaches zero as $n \to \infty$ so that $\varphi_n(u) \to 1$, $u \in I$, as n increases indefinitely.

The function g is a characteristic function as a limit of characteristic functions ([124], Chapter 3), can be either zero or one, and $g(u) = 1$ for $u \in I$. Hence, $g(u) = 1$ everywhere by the continuity of the characteristic function so that $\varphi(u) \neq 0$, $\forall u \in \mathbb{R}$. ■

Finite products of i.d. characteristic functions are i.d. characteristic functions ([124], Theorem 5.3.2, p. 81).

Proof: Let φ_k, $k = 1, \ldots, q$, be i.d. characteristic functions so that $\varphi_k = (\varphi_{k,n})^n$, $n \geq 1$, where $\varphi_{k,n}$ are characteristic functions. The function $\varphi = \prod_{k=1}^q \varphi_k$ is a characteristic function as a product of characteristic functions and

$$\varphi(u) = \prod_{k=1}^{q} \varphi_k(u) = \prod_{k=1}^{q} [\varphi_{k,n}(u)]^n = \left[\prod_{k=1}^{q} \varphi_{k,n}(u)\right]^n = \varphi_n(u)^n$$

for any $n \geq 1$, where $\varphi_n = \prod_{k=1}^q \varphi_{k,n}$. Hence, φ is infinitely divisible. ■

If φ is an i.d. characteristic function so is $|\varphi|$ ([124], Corollary to Theorem 5.3.2, p. 81).

Proof: The function $|\varphi(u)|^2 = \varphi(u)\,(\varphi(u))^* = \varphi(u)\,\varphi(-u)$ is an i.d. characteristic function by the previous property so that $(|\varphi(u)|^2)^{1/(2n)} = |\varphi(u)|^{1/n}$ is a characteristic function for each integer $n > 0$. ■

A characteristic function that is the limit of a sequence of i.d. characteristic functions is infinitely divisible ([124], Theorem 5.3.3, p. 82).

If φ is an i.d. characteristic function, φ^α is a characteristic function for each real number $\alpha > 0$. The converse is also true ([124], Corollary to Theorem 5.3.3, p. 82).

Example 2.57: The characteristic function of $X \sim N(0, 1)$ is $\varphi(u) = e^{-u^2/2}$. We have seen in Example 2.55 that φ is infinitely divisible. This function has no real zeros consistent with a property of i.d. characteristic functions. The function $\varphi(u)^\alpha = e^{-u^2\alpha/2}$, $\alpha > 0$, is also an i.d. characteristic function because it corresponds to the variable $N(0, \alpha)$, a result consistent with the last property. ◇

Construction of an i.d. characteristic function.

If $g : \mathbb{R} \to \mathbb{C}$ is a characteristic function and $p > 0$ denotes a real number,

$$\varphi(u) = \exp\{p\,[g(u) - 1]\}, \quad u \in \mathbb{R}, \tag{2.78}$$

is an i.d. characteristic function ([124], Lemma 5.4.1, p 83).

Note: This fact can be used to construct infinitely divisible characteristic functions. For example, the characteristic function in Eq. 2.73 with $b = 0$ can be obtained from Eq. 2.78 with $p = \lambda$ and $g(u) = \exp(\sqrt{-1}\,a\,u)$, $a \in \mathbb{R}$. ▲

A characteristic function φ is infinitely divisible if and only if

$$\varphi(u) = \lim_{n\to\infty} \exp[p_n\,(g_n(u) - 1)], \tag{2.79}$$

where $p_n > 0$ are real numbers and g_n denote characteristic functions ([124], Theorem 5.4.1, p. 83).

The limit of a sequence of finite products of Poisson type characteristic functions is an i.d. characteristic function. Moreover, every i.d. characteristic function can be written as the limit of a sequence of finite products of Poisson type characteristic functions ([124], Theorem 5.4.2, p. 83).

Proof: We only prove the stated representation of i.d. characteristic functions.

Suppose that φ is an i.d. characteristic function. According to the previous theorem, we have $\varphi(u) = \lim_{n\to\infty} \exp[p_n\,(g_n(u) - 1)]$, where $p_n > 0$ and g_n are characteristic functions. Take $g_n(u) = \int_{\mathbb{R}} e^{\sqrt{-1}\,u\,x}\, dG_n(x)$ in Eq. 2.79, where G_n, $n = 1, 2, \ldots$, is a collection of distributions. For $a > 0$ consider a partition $-a = a_0 \le a_1 \le \cdots \le a_m = a$ of $(-a, a)$ such that $\max_i (a_i - a_{i-1}) \to 0$ as $m \to \infty$. The function $p_n\,(g_n(u) - 1)$ can be approximated by

$$p_n \int_{-a}^{a} \left(e^{\sqrt{-1}\,u\,x} - 1\right) dG_n(x) = \lim_{m\to\infty} \sum_{k=1}^{m} b_k \left(e^{\sqrt{-1}\,u\,a_k} - 1\right) \quad \text{so that}$$

$$\exp\left[p_n \int_{-a}^{a} \left(e^{\sqrt{-1}\,u\,x} - 1\right) dG_n(x)\right] = \lim_{m\to\infty} \prod_{k=1}^{m} \exp\left[b_k \left(e^{\sqrt{-1}\,u\,a_k} - 1\right)\right],$$

where $b_k = p_n\,[G_n(a_k) - G_n(a_{k-1})]$. For any $a > 0$ the last expression is the limit of a finite product of characteristic functions corresponding to Poisson random variables (Eq. 2.73). The stated result follows by taking the limit as $a \to \infty$. ■

Canonical representations of i.d. characteristic functions.

Let $a \in \mathbb{R}$ be a constant, θ be a real-valued, bounded, and increasing function defined on the real line such that $\theta(-\infty) = 0$. If a function φ admits the representation

$$\log(\varphi(u)) = \sqrt{-1}\,a\,u + \int_{\mathbb{R}} \left(e^{\sqrt{-1}\,u\,x} - 1 - \frac{\sqrt{-1}\,u\,x}{1+x^2}\right) \frac{1+x^2}{x^2}\, d\theta(x) \quad (2.80)$$

for all $u \in \mathbb{R}$, then φ is an i.d. characteristic function. The constant a and the function θ are uniquely determined by φ ([124], Lemma 5.5.1, p. 85).

Note: The integrand of the integral in Eq. 2.80 is defined by continuity at $x = 0$ so that it is equal to $-u^2/2$ at this point. ▲

If φ is an i.d. characteristic function, there exists a sequence of functions $\xi_n(u)$ such that $\lim_{n\to\infty} \xi_n(u) = \log(\varphi(u))$, $u \in \mathbb{R}$, where

$$\xi_n(u) = \sqrt{-1}\,a_n\,u + \int_{\mathbb{R}} \left(e^{\sqrt{-1}\,u\,x} - 1 - \frac{\sqrt{-1}\,u\,x}{1+x^2}\right) \frac{1+x^2}{x^2}\, d\theta_n(x),$$

$$a_n = n \int_{\mathbb{R}} \frac{x}{1+x^2}\, dF_n(x), \quad \theta_n(x) = n \int_{-\infty}^{x} \frac{y}{1+y^2}\, dF_n(y), \quad (2.81)$$

and F_n denotes the distribution function of $\varphi^{1/n}$ ([124], Lemma 5.5.3, p. 88).

Lévy-Khinchine representation. A function φ is an i.d. characteristic function if and only if $\log(\varphi(u))$ is given by Eq. 2.80. The representation is unique and the integrand defined at $x = 0$ by continuity is $-u^2/2$ ([124], Theorem 5.5.1, p. 89).

Note: If the function θ has a jump $\sigma^2 = \theta(0+) - \theta(0-)$ at $x = 0$, the Lévy-Khinchine representation can be given in the form

$$\log(\varphi(u)) = \sqrt{-1}\, a\, u - \frac{\sigma^2 u^2}{2} + \int_{\mathbb{R}\setminus\{0\}} \left(e^{\sqrt{-1}\, u\, x} - 1 - \frac{\sqrt{-1}\, u\, x}{1+x^2} \right) \frac{1+x^2}{x^2}\, d\theta(x) \tag{2.82}$$

for each $u \in \mathbb{R}$ because the integrand is $-u^2/2$ at $x = 0$ (Eq. 2.80). An alternative form of this equation is

$$\log(\varphi(u)) = \sqrt{-1}\, a\, u - \frac{\sigma^2 u^2}{2} + \int_{\mathbb{R}\setminus\{0\}} \left(e^{\sqrt{-1}\, u\, x} - 1 - \frac{\sqrt{-1}\, u\, x}{1+x^2} \right) d\zeta(x), \tag{2.83}$$

where

$$\zeta(x) = \begin{cases} \int_{-\infty}^{x} \frac{1+y^2}{y^2}\, d\theta(y) & \text{for } x < 0, \\ -\int_{x}^{\infty} \frac{1+y^2}{y^2}\, d\theta(y) & \text{for } x > 0. \end{cases} \tag{2.84}$$

The function ζ is defined on $\mathbb{R} \setminus \{0\}$, is increasing in $(-\infty, 0)$ and $(0, \infty)$, satisfies the conditions $\zeta(-\infty) = 0$, $\zeta(\infty) = 0$, and the integral $\int_{(-\varepsilon,\varepsilon)\setminus\{0\}} x^2\, d\zeta(x) < \infty$ for any $\varepsilon > 0$. The representation in Eq. 2.83 is unique and is referred to as the **Lévy representation** for the i.d. characteristic function φ. ▲

An alternative version of the **Lévy-Khinchine representation** is

$$\log(\varphi(u)) = \sqrt{-1}\, a\, u - \frac{\sigma^2 u^2}{2} + \int_{\mathbb{R}\setminus\{0\}} \left(e^{\sqrt{-1}\, u\, x} - 1 - \sqrt{-1}\, u\, \chi(x) \right) \lambda_L(dx), \quad u \in \mathbb{R}, \tag{2.85}$$

where $\chi(y) = -1_{(-\infty,1]}(y) + y\, 1_{(-1,1)}(y) + 1_{[1,\infty)}(y)$ and λ_L is a **Lévy measure**, that is, a measure defined on $\mathbb{R} \setminus \{0\}$ such that $\int_{\mathbb{R}\setminus\{0\}} (y^2 \wedge 1)\, \lambda_L(dy) < \infty$ ([66], Theorem 13, p. 299).

Example 2.58: The characteristic functions of the Gaussian, Poisson, and compound Poisson random variables are special cases of the Lévy-Khinchine representation given by Eq. 2.85. ◇

Proof: The Lévy-Khinchine representation in Eq. 2.85 with $\lambda_L = 0$ gives the characteristic function of a Gaussian variable (Eq. 2.70). The characteristic function in Eq. 2.85 with $\sigma^2 = 0$ and $\lambda_L(B) = \lambda\, 1_{\{\xi\}}(B)$ for $\xi \in \mathbb{R} \setminus \{0\}$, $\lambda > 0$, and $B \in \mathcal{B}$, is

$$\log(\varphi(u)) = \sqrt{-1}\, u\, (a - \lambda\, \chi(\xi)) + \lambda\, (e^{\sqrt{-1}\, u\, \xi} - 1)$$

and has the form of the characteristic function for a Poisson variable (Eq. 2.73). If $\sigma^2 = 0$ and $\lambda_L(dx) = \lambda\, dF(x)$, $\lambda > 0$, Eq. 2.85 becomes

$$\begin{aligned}\log(\varphi(u)) &= \sqrt{-1}\, a\, u + \lambda \int_{\mathbb{R}\setminus\{0\}} \left(e^{\sqrt{-1}\, u\, x} - 1 - \sqrt{-1}\, u\, \chi(x)\right) dF(x) \\ &= \sqrt{-1}\, a\, u - \sqrt{-1}\, \lambda\, u \int_{\mathbb{R}\setminus\{0\}} \chi(x)\, dF(x) + \lambda \int_{\mathbb{R}\setminus\{0\}} \left(e^{\sqrt{-1}\, u\, x} - 1\right) dF(x)\end{aligned}$$

and resembles the characteristic function of a compound Poisson variable (Eq. 2.76). ■

2.10.3.3 α-Stable random variable

A random variable X is α-**stable** if, for any $a_i > 0$ and independent copies X_i, $i = 1, 2$, of X, there exist $c > 0$ and b such that

$$c\,(a_1\, X_1 + a_2\, X_2) + b \stackrel{d}{=} X, \tag{2.86}$$

that is, $c\,(a_1\, X_1 + a_2\, X_2) + b$ and X have the same distribution.

Note: Because $a_1\, X_1 + a_2\, X_2$ and $(X - b)/c$ have the same distribution, X_k, $k = 1, 2$, are independent copies of X, and $\varphi(u) = E[\exp(\sqrt{-1}\, u\, X)]$ is the characteristic function of X, Eq. 2.86 gives $\varphi(a_1\, u)\, \varphi(a_2\, u) = \varphi(u/c) \exp(-\sqrt{-1}\, u\, b/c)$.

There are alternatives to the definition in Eq. 2.86. For example, X has an α-stable distribution if for each $n \geq 2$ there are two real numbers, $c_n > 0$ and d_n, such that $\sum_{i=1}^n X_i$ and $c_n\, X + d_n$ have the same distribution, where X_i are independent copies of X ([162], Definition 1.1.4, p. 3). ▲

We give here two properties of α-stable random variables that are particularly useful for calculations. An extensive discussion on α-stable random variables is in [161, 162].

The characteristic function of an α-stable variable is infinitely divisible.

Proof: Let φ denote the characteristic function of an α-stable variable X and let $X_1, \ldots, X_n$ denote independent copies of this random variable. Because X is an α-stable variable, we have $\sum_{i=1}^n X_i \stackrel{d}{=} c_n\, X + d_n$, where $c_n > 0$ and d_n are real numbers. This equality gives $[\varphi(u)]^n = \varphi(c_n\, u)\, e^{\sqrt{-1}\, u\, d_n}$ or

$$\varphi(u) = [\varphi(u/c_n)]^n\, e^{-\sqrt{-1}\, u\, d_n/c_n}$$

if we replace u with u/c_n. We have $\varphi_n(u) = \varphi(u/c_n)\, e^{\sqrt{-1}\, u\, d_n/(n\, c_n)}$ and $\varphi(u) = [\varphi_n(u)]^n$. Hence, φ is an i.d. characteristic function. ■

This property shows that the class of α-stable variables is a subset of the class of random variables with i.d. characteristic functions. Hence, α-stable variables have all the properties of random variables with i.d. characteristic functions. For example, the characteristic functions of α-stable variables have no real zeros.

The characteristic function of any α-stable variable has the form

$$\varphi(u) = \exp\{\sqrt{-1}\,\mu\, u - \sigma^\alpha\, |u|^\alpha\, [1 + \sqrt{-1}\,\beta\ \mathrm{sign}(u)\,\omega(|u|, \alpha)]\}, \quad (2.87)$$

$$\text{where } \omega(|u|, \alpha) = \begin{cases} -\tan(\pi\alpha/2), & \alpha \neq 1, \\ (2/\pi)\log|u|, & \alpha = 1, \end{cases} \quad (2.88)$$

$\mu, \sigma \geq 0$, $|\beta| \leq 1$, $0 < \alpha \leq 2$ are real constants ([124], Theorem 5.7.3, p. 102).

Note: The characteristic function of an α-stable random variable can be obtained from Eqs. 2.83 and 2.84 with

$$\zeta(u) = \begin{cases} c_1\, |x|^{-\alpha} & \text{for } x < 0, \\ -c_2\, |x|^{-\alpha} & \text{for } x < 0, \end{cases}$$

where $c_1, c_2 \geq 0$ are real numbers such that $c_1 + c_2 > 0$ ([124], Theorem 5.7.2, p. 101). The integrals

$$\int_{-\infty}^{0} \left(e^{\sqrt{-1}\,u\,x} - 1 - \frac{\sqrt{-1}\,u\,x}{1+x^2} \right) \frac{dx}{|x|^{\alpha+1}} \quad \text{and}$$

$$\int_{0}^{\infty} \left(e^{\sqrt{-1}\,u\,x} - 1 - \frac{\sqrt{-1}\,u\,x}{1+x^2} \right) \frac{dx}{|x|^{\alpha+1}}$$

can be calculated for $\alpha < 1$, $\alpha = 1$, and $\alpha > 1$ and give Eqs. 2.87 and 2.88.

The parameters α, β, σ, and μ, referred to as **stability index**, **skewness**, **scale**, and **shift** or **location**, respectively, control the distribution type, the departure from a symmetric distribution about μ, the range of likely values, and the shift from zero. We use the notation $X \sim S_\alpha(\sigma, \beta, \mu)$ to indicate that X is a real-valued α-stable random variable with parameters $(\alpha, \sigma, \beta, \mu)$. The density of an α-stable variable with $\beta = 0$ is symmetric about its location parameter μ. ▲

Example 2.59: The characteristic functions of Gaussian and Cauchy variables with parameters $S_2(1, 0, 0)$ and $S_1(1, 0, 0)$ are shown in the left graph of Fig. 2.6. The imaginary parts of these characteristic functions are zero. The right graph in the figure shows the real and imaginary parts of the characteristic function,

$$\varphi(u) = \exp\left[-|u|\left(1 - 2\sqrt{-1}\ \mathrm{sign}(u)\ \log|u|/\pi\right)\right],$$

of the Cauchy random variable $S_1(1, -1, 0)$. This characteristic function is complex-valued since $\beta \neq 0$. ◇

2.11 Random vectors

Let $(\Omega, \mathcal{F}, P)$ be a probability space and

$$X : (\Omega, \mathcal{F}) \to (\mathbb{R}^d, \mathcal{B}^d), \quad (2.89)$$

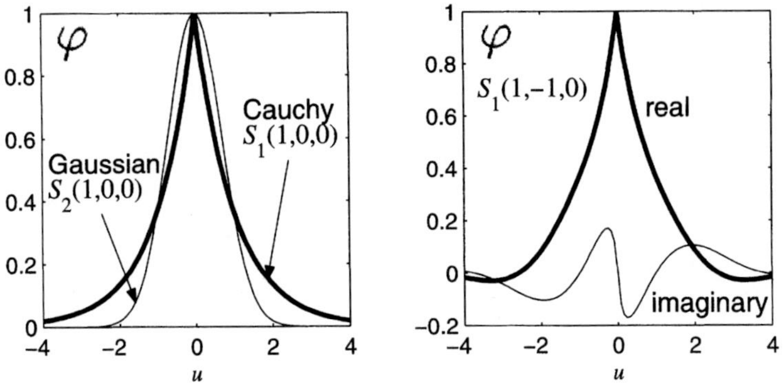

Figure 2.6. Characteristic function of some α-stable random variables

be a measurable function, that is, $X^{-1}(B) \in \mathcal{F}$ for every Borel set $B \in \mathcal{B}^d$. This function, called a **random vector** in $\mathbb{R}^d$ or $\mathbb{R}^d$**-valued random variable**, induces the probability measure $Q(B) = P(X \in B) = P(X^{-1}(B))$, $B \in \mathcal{B}^d$, on the measurable space $(\mathbb{R}^d, \mathcal{B}^d)$.

2.11.1 Joint distribution and density functions

Let $B = \times_{i+1}^d (-\infty, x_i]$ be a rectangle in $\mathbb{R}^d$ for $x_i \in \mathbb{R}$, $i = 1, \ldots, d$. This rectangle is in $\mathcal{B}^d$ so that $X^{-1}(B) \in \mathcal{F}$ because X is measurable.

The **joint distribution function** of X is

$$F(x) = P(\cap_{i=1}^d \{X_i \le x_i\}), \quad x = (x_1, \ldots, x_d) \in \mathbb{R}^d. \tag{2.90}$$

The domain and the range of this function are $\mathbb{R}^d$ and [0, 1], respectively.

The following properties of F result directly from its definition in Eq. 2.90.

- $\lim_{x_k \to \infty} F(x)$, $1 \le k \le d$, is the joint density of the $\mathbb{R}^{d-1}$-valued random variable $(X_1, \ldots, X_{k-1}, X_{k+1}, \ldots, X_d)$.
- $\lim_{x_k \to -\infty} F(x) = 0$ for $k \in \{1, \ldots, d\}$.
- The function $x_k \mapsto F(x)$ is increasing for each $k \in \{1, \ldots, d\}$.
- The function $x_k \mapsto F(x)$ is right continuous for each $k \in \{1, \ldots, d\}$.

If F is such that

$$f(x) = \frac{\partial^d F(x)}{\partial x_1 \cdots \partial x_d} \tag{2.91}$$

exists, then f is called the **joint density** function of X.

Note: X takes values in the infinitesimal rectangle $\times_{i=1}^{d}(x_i, x_i + dx_i]$ with probability $P\left(\cap_{i=1}^{d}\{X_i \in (x_i, x_i + dx_i]\}\right) \simeq f(\boldsymbol{x})\,d\boldsymbol{x}$. The distribution of one or more coordinates of $\boldsymbol{X}$ can be obtained from the joint distribution or the joint density of $\boldsymbol{X}$. For example, the **marginal distribution** and **marginal density** of X_1 are $F_1(x_1) = F(x_1, \infty, \ldots, \infty)$ and $f_1(x_1) = \int_{\mathbb{R}^{d-1}} f(\boldsymbol{x})\,dx_2 \cdots dx_d$ or $f_1(x_1) = dF_1(x_1)/dx_1$, respectively. ▲

Suppose that the last $d_2 < d$ coordinates of a random vector $\boldsymbol{X}$ have been measured and are equal to $\boldsymbol{z} = (z_1, \ldots, z_{d_2})$. Let $\boldsymbol{X}^{(1)}$ and $\boldsymbol{X}^{(2)}$ be vectors consisting of the first $d_1 = d - d_2$ and the last d_2 coordinates of $\boldsymbol{X}$.

The **conditional density** $f^{(1|2)}$ of $\boldsymbol{X}^{(1)}$ given $\boldsymbol{X}^{(2)} = \boldsymbol{z}$ is

$$f^{(1|2)}(\boldsymbol{x}^{(1)} \mid \boldsymbol{z}) = \frac{f(\boldsymbol{x}^{(1)}, \boldsymbol{z})}{f^{(2)}(\boldsymbol{z})}, \tag{2.92}$$

where $f^{(i)}$ denotes the density of $\boldsymbol{X}^{(i)}$, $i = 1, 2$, and $\boldsymbol{x}^{(1)} = (x_1, \ldots, x_{d_1})$.

Note: The definition of the conditional probability $P(A \mid B)$ in Eq. 2.16 with $A = \{X_1 \in (x_1, x_1 + dx_1], \ldots, X_{d_1} \in (x_{d_1}, x_{d_1} + dx_{d_1}]\}$ and $B = \{X_{d_1+1} \in (z_1, z_1 + dz_1], \ldots, X_d \in (z_{d_2}, z_{d_2} + dz_{d_2}]\}$ provides a heuristic justification for Eq. 2.92. We can view $P(A \mid B)$ as $[(f(\boldsymbol{x}^{(1)}, \boldsymbol{z})/f^{(2)}(\boldsymbol{z}))]\,d\boldsymbol{x}^{(1)}$ and gives the probability content of the infinitesimal rectangle $(x_1, x_1 + dx_1] \times \cdots \times (x_{d_1}, x_{d_1} + dx_{d_1}]$ under the condition $\boldsymbol{X}^{(2)} = \boldsymbol{z}$. For a rigorous discussion, see [66] (Section 21.3, pp. 416-417). ▲

If $\boldsymbol{X}$ and $\boldsymbol{Y}$ are $\mathbb{R}^d$-valued random variables and there is a one-to-one correspondence between these variables given by $\boldsymbol{y} = \boldsymbol{g}(\boldsymbol{x})$ and $\boldsymbol{x} = \boldsymbol{h}(\boldsymbol{y})$, then the densities f_x and f_y of $\boldsymbol{X}$ and $\boldsymbol{Y}$ are related by

$$f_y(\boldsymbol{y}) = f_x(\boldsymbol{h}(\boldsymbol{y})) \left| \frac{\partial x_i}{\partial y_j} \right|, \quad \boldsymbol{x}, \boldsymbol{y} \in \mathbb{R}^d. \tag{2.93}$$

Proof: The postulated correspondence implies that the Jacobian

$$\frac{\partial(x_1, \ldots, x_d)}{\partial(y_1, \ldots, y_d)} = \begin{vmatrix} \frac{\partial x_1}{\partial y_1} & \cdots & \frac{\partial x_1}{\partial y_d} \\ \vdots & \ddots & \vdots \\ \frac{\partial x_d}{\partial y_1} & \cdots & \frac{\partial x_d}{\partial y_d} \end{vmatrix}$$

is non-zero and finite everywhere. The Jacobian of the inverse transformation has the same properties since $\frac{\partial(x_1,\ldots,x_d)}{\partial(y_1,\ldots,y_d)} \frac{\partial(y_1,\ldots,y_d)}{\partial(x_1,\ldots,x_d)} = 1$. Let D_x be a neighborhood of $\boldsymbol{x} \in \mathbb{R}^d$ and $D_y = \{\boldsymbol{\eta} \in \mathbb{R}^d : \boldsymbol{\eta} = \boldsymbol{g}(\boldsymbol{\xi}), \boldsymbol{\xi} \in D_x\}$ denote the image of D_x by the transformation $\boldsymbol{x} \mapsto \boldsymbol{y} = \boldsymbol{g}(\boldsymbol{x})$. The equality $P(\boldsymbol{X} \in D_x) = P(\boldsymbol{Y} \in D_y)$ can be written as $\int_{D_x} f_x(\boldsymbol{\xi})\,d\boldsymbol{\xi} = \int_{D_y} f_y(\boldsymbol{\eta})\,d\boldsymbol{\eta}$, and implies

$$\int_{D_y} f_x(\boldsymbol{h}(\boldsymbol{\eta})) \left| \frac{\partial(\xi_1, \ldots, \xi_d)}{\partial(\eta_1, \ldots, \eta_d)} \right| d\boldsymbol{\eta} = \int_{D_y} f_y(\boldsymbol{\eta})\,d\boldsymbol{\eta}$$

by change of variables. The last equality gives Eq. 2.93.

Suppose now that the mapping $\boldsymbol{y} \mapsto \boldsymbol{x} = \boldsymbol{h}(\boldsymbol{y})$ does not have a unique solution. Let $\{A_\nu\}$ be a partition of the $\boldsymbol{x}$-space so that the mapping $\boldsymbol{y} \mapsto \boldsymbol{x}$ is unique in each A_ν. The probability mass of D_y is equal to the sum $\sum_\nu f_\nu \, |J_\nu|$ of the corresponding contributions in the subsets A_ν of the $\boldsymbol{x}$-space, where each term $f_\nu \, |J_\nu|$ is equal to the right side of Eq. 2.93 for the restriction of the mapping $\boldsymbol{y} \mapsto \boldsymbol{x} = \boldsymbol{h}(\boldsymbol{y})$ to A_ν. ■

Example 2.60: Let X be a real-valued random variable with a continuous density f_x and define $Y = X^2$. The density of Y is

$$f_y(y) = \left(f_x(y^{1/2}) + f_x(-y^{1/2}) \right) / (2\, y^{1/2})$$

for $y > 0$. ◇

Proof: The mapping from y to x is unique on $A_1 = (-\infty, 0)$ and $A_2 = (0, \infty)$ and is given by $y \mapsto x = h(y) = -y^{1/2}$ and $y \mapsto x = h(y) = y^{1/2}$, respectively. The contribution of $y \mapsto x = -y^{1/2}$ to f_y is $f_x(-y^{1/2}) \, |d(-y^{1/2})/dy|$ (Eq. 2.93).

This result can also be obtained by direct calculations from the relationship $P(Y \leq y) = P(-y^{1/2} < X \leq y^{1/2})$. ■

Example 2.61: Let X_1 and X_2 be real-valued random variables and define $Y = X_1 + X_2$. The density of Y is $f_y(y) = \int_{\mathbb{R}} f_x(\eta, y - \eta) \, d\eta$. ◇

Proof: It is convenient to augment Y to a vector $\tilde{\boldsymbol{Y}} \in \mathbb{R}^d$, which is related to $\boldsymbol{X}$ by a one-to-one mapping. We take $\tilde{Y}_1 = X_1$ and $\tilde{Y}_2 = Y = X_1 + X_2$ so that $|\partial x_i / \partial y_j| = 1$ and $f_{\tilde{y}}(\tilde{y}_1, \tilde{y}_2) = f_x(\tilde{y}_1, \tilde{y}_2 - \tilde{y}_1)$. The result follows by integration.

Direct calculations can also be used for solution. From

$$P(Y \leq y) = P(X_1 + X_2 \leq y) = \int_{-\infty}^{\infty} d\xi_1 \int_{-\infty}^{y-\xi_1} d\xi_2 \, f_x(\xi_1, y - \boldsymbol{x}_1)$$

we can calculate f_y by differentiation. ■

2.11.2 Independence

Consider a family of random variables X_i, $i \in I$, defined on a probability space $(\Omega, \mathcal{F}, P)$, where the index set I is finite or infinite. We say that X_i are independent random variables if the σ-fields $\sigma(X_i)$ generated by these random variables are independent (Section 2.7.3).

Example 2.62: Let X_1 and X_2 be random variables defined on a probability space $(\Omega, \mathcal{F}, P)$. Denote by F and F_i the joint distribution of (X_1, X_2) and the distribution of X_i, respectively. If X_1 and X_2 are independent, then $F(x_1, x_2) = F_1(x_1) \, F_2(x_2)$ for all $x_1, x_2 \in \mathbb{R}$. ◇

Note: The independence of X_1 and X_2 implies that the σ-fields $\sigma(X_1)$ and $\sigma(X_2)$ generated by X_1 and X_2 are independent. Hence, $P(A_1 \cap A_2) = P(A_1) \, P(A_2)$ for any $A_i \in \sigma(X_i)$, $i = 1, 2$. This property implies $F(x_1, x_2) = F_1(x_1) \, F_2(x_2)$ if we take $A_i = X_1^{-1}((-\infty, x_i]) \in \mathcal{F}$, $i = 1, 2$. ▲

The following equation gives an alternative definition of independence for random variables. This new definition is equivalent with the above definition of independence based on the σ-fields generated by the coordinates of $\boldsymbol{X}$ ([151], Corollary 4.2.2, p. 94).

A family of random variables X_i, $i \in I$, is independent if and only if for all finite $J \subset I$ ([151], Theorem 4.2.1, p. 94)

$$P(X_i \le x_i, i \in J) = \prod_{i \in J} P(X_i \le x_i), \quad x_i \in \mathbb{R}. \tag{2.94}$$

Note: If $I = \{1, \ldots, d\}$ is finite, the condition in Eq. 2.94 becomes $P(\cap_{i=1}^d \{X_i \le x_i\}) = \prod_{i=1}^d P(X_i \le x_i)$, $x_i \in \mathbb{R}$, or $F(\boldsymbol{x}) = \prod_{i=1}^d F_i(x_i)$, where $\boldsymbol{x} = (x_1, \ldots, x_d)$, F denotes the joint distribution of $\boldsymbol{X} = (X_1, \ldots, X_d)$, and F_i is the distribution of X_i. The independence condition $F(\boldsymbol{x}) = \prod_{i=1}^d F_i(x_i)$ is also satisfied by all subsets of $I = \{1, \ldots, d\}$. For example, if we set $x_d = \infty$, this condition applies to the first $d-1$ coordinates of $\boldsymbol{X}$. If the distributions F and F_i have densities f and f_i, Eq. 2.94 implies $f(\boldsymbol{x}) = \prod_{i \in J}^d f_i(x_i)$. ▲

2.11.3 Characteristic function

Consider a Borel measurable function $\boldsymbol{g} : (\mathbb{R}^d, \mathcal{B}^d) \to (\mathbb{R}^q, \mathcal{B}^q)$ and an $\mathbb{R}^d$-valued random variable $\boldsymbol{X} = (X_1, \ldots, X_d)$ defined on a probability space $(\Omega, \mathcal{F}, P)$. The function $\boldsymbol{g}(\boldsymbol{X})$ is measurable from $(\Omega, \mathcal{F})$ to $(\mathbb{R}^q, \mathcal{B}^q)$ and its expectation can be calculated from the following formulas.

$$E[\boldsymbol{g}(\boldsymbol{X})] = \int_\Omega \boldsymbol{g}(\boldsymbol{X}(\omega))\, P(d\omega) \quad \text{and} \tag{2.95}$$

$$E[\boldsymbol{g}(\boldsymbol{X})] = \int_{\mathbb{R}^d} \boldsymbol{g}(\boldsymbol{x})\, Q(d\boldsymbol{x}) = \int_{\mathbb{R}^d} \boldsymbol{g}(\boldsymbol{x})\, dF(\boldsymbol{x}) = \int_{\mathbb{R}^d} \boldsymbol{g}(\boldsymbol{x})\, f(\boldsymbol{x})\, d\boldsymbol{x}. \tag{2.96}$$

Note: That Eqs. 2.95 and 2.96 give the same result can be shown by extending the arguments used to prove the equivalence of Eqs. 2.62 and 2.63. The chain of equalities in Eq. 2.96 holds because $Q(d\boldsymbol{x}) = dF(\boldsymbol{x}) = f(\boldsymbol{x})\, d\boldsymbol{x}$. If $q = 1$, a_i denote some constants, and $g(\boldsymbol{x}) = \sum_{i=1}^d a_i x_i$, then the expectation of $g(\boldsymbol{X})$ is $\sum_{i=1}^d a_i E[X_i]$ showing that expectation is a linear operator. ▲

The formulas in Eqs. 2.95 and 2.96 can be extended to complex-valued continuous functions of $\boldsymbol{X}$, for example, we can take $g(\boldsymbol{X}) = \exp(\sqrt{-1}\, \boldsymbol{u}^T \boldsymbol{X})$, where $\boldsymbol{u} = (u_1, \ldots, u_d) \in \mathbb{R}^d$ and $\boldsymbol{u}^T$ denotes the transpose of $\boldsymbol{u}$.

The **joint characteristic function** of $\boldsymbol{X}$ is

$$\varphi(\boldsymbol{u}) = E\left[e^{\sqrt{-1}\, \boldsymbol{u}^T \boldsymbol{X}}\right] = \int_{\mathbb{R}^d} e^{\sqrt{-1}\, \boldsymbol{u}^T \boldsymbol{x}}\, dF(\boldsymbol{x}) = \int_{\mathbb{R}^d} e^{\sqrt{-1}\, \boldsymbol{u}^T \boldsymbol{x}}\, f(\boldsymbol{x})\, d\boldsymbol{x}. \tag{2.97}$$

The following properties of φ are useful for calculations.

- $|\varphi(\boldsymbol{u})| \leq 1$ for all $\boldsymbol{u} \in \mathbb{R}^d$.
- The joint characteristic and the joint density functions are Fourier pairs.
- If $\boldsymbol{X}$ has independent coordinates, $\varphi(\boldsymbol{u}) = \prod_{k=1}^{d} \varphi_k(u_k)$, where φ_k is the characteristic function of X_k.
- φ is uniformly continuous.

Proof: The first property results from the definition of the characteristic function since

$$|\varphi(\boldsymbol{u})| \leq \int_{\mathbb{R}^d} \left| e^{\sqrt{-1}\,\boldsymbol{u}^T \boldsymbol{x}} \right| dF(\boldsymbol{x}) = 1.$$

The characteristic and the density functions of $\boldsymbol{X}$ are Fourier pairs related by Eq. 2.97 and ([62], p. 524)

$$f(\boldsymbol{x}) = \frac{1}{(2\,\pi)^d} \int_{\mathbb{R}^d} e^{-\sqrt{-1}\,\boldsymbol{u}^T \boldsymbol{x}}\, \varphi(\boldsymbol{u})\, d\boldsymbol{u}. \tag{2.98}$$

If the coordinates of the random vector $\boldsymbol{X}$ are independent, the characteristic function becomes (Eq. 2.97)

$$\begin{aligned}
\varphi(\boldsymbol{u}) &= E\left[e^{\sqrt{-1}\,\boldsymbol{u}^T \boldsymbol{X}}\right] = \int_{\mathbb{R}^d} \prod_{k=1}^{d} \left(e^{\sqrt{-1}\,u_k X_k}\, f_k(x_k)\, dx_k \right) \\
&= \prod_{k=1}^{d} E\left[e^{\sqrt{-1}\,u_k X_k}\right] = \prod_{k=1}^{d} \varphi_k(u_k),
\end{aligned}$$

where $\varphi_k(u_k) = E[e^{\sqrt{-1}\,u_k X_k}]$ is the characteristic function of the coordinate k of $\boldsymbol{X}$. The Fourier transform of $\varphi(\boldsymbol{u})$ shows that the density of $\boldsymbol{X}$ is equal to the product of the densities of its coordinates. Hence, the above equality provides an alternative way of checking whether a random vector has independent coordinates.

For the last property we need to show that for any $\varepsilon > 0$ there is $\delta > 0$ such that $\| \boldsymbol{h} \| < \delta$ implies $|\varphi(\boldsymbol{u} + \boldsymbol{h}) - \varphi(\boldsymbol{u})| < \varepsilon$, where $\| \cdot \|$ denotes the usual norm in $\mathbb{R}^d$. The increment of the characteristic function from $\boldsymbol{u}$ to $\boldsymbol{u} + \boldsymbol{h}$ is

$$\begin{aligned}
|\varphi(\boldsymbol{u} + \boldsymbol{h}) - \varphi(\boldsymbol{u})| &= \left| E\left[e^{\sqrt{-1}\,(\boldsymbol{u}+\boldsymbol{h})^T \boldsymbol{X}} - e^{\sqrt{-1}\,\boldsymbol{u}^T \boldsymbol{X}} \right] \right| \\
&\leq E\left[\left| e^{\sqrt{-1}\,\boldsymbol{u}^T \boldsymbol{X}} \left(e^{\sqrt{-1}\,\boldsymbol{h}^T \boldsymbol{X}} - 1 \right) \right| \right] \leq E\left[\left| e^{\sqrt{-1}\,\boldsymbol{h}^T \boldsymbol{X}} - 1 \right| \right] \\
&= \int_{D^c} \left| e^{\sqrt{-1}\,\boldsymbol{h}^T \boldsymbol{x}} - 1 \right| dF(\boldsymbol{x}) + \int_{D} \left| e^{\sqrt{-1}\,\boldsymbol{h}^T \boldsymbol{x}} - 1 \right| dF(\boldsymbol{x}),
\end{aligned}$$

where $D = \times_{i=1}^{d} [-a, a]$ and $0 < a < \infty$. The integral on D^c is smaller than $2\,P(\boldsymbol{X} \in D^c)$ so that it can be made smaller than $\varepsilon/2$ for a sufficiently large a. For this value of a, the integral on D can also be made smaller than $\varepsilon/2$ by choosing an adequate value of δ since $\max_{\boldsymbol{x} \in D} |\exp(\sqrt{-1}\,\boldsymbol{h}^T \boldsymbol{x}) - 1|$ converges to zero as $\| \boldsymbol{h} \| \to 0$. ■

2.11.4 Moments

Let $\boldsymbol{X}$ be an $\mathbb{R}^d$-valued random variable on a probability space $(\Omega, \mathcal{F}, P)$ and consider the function $g(\boldsymbol{x}) = \prod_{i=1}^{d} x_i^{s_i}$, where $s_i \geq 0$ are integers. Because g is continuous, $g(\boldsymbol{X})$ is a real-valued random variable.

If $\boldsymbol{X} \sim L_s$, that is, $X_i \sim L_s$ for each $i = 1, \ldots, d$, the **moments of order** $s = \sum_{i=1}^{d} s_i$ of $\boldsymbol{X}$ exist, are finite, and are given by

$$\mu(s_1, \ldots, s_d) = E[g(\boldsymbol{X})] = E\left[\prod_{i=1}^{d} X_i^{s_i}\right]. \tag{2.99}$$

Note: The characteristic function can be used to calculate moments of any order of $\boldsymbol{X}$ provided that they exist. For example,

$$\mu(s_1, \ldots, s_d) = (\sqrt{-1})^s \frac{\partial^s \varphi(\boldsymbol{u})}{\partial^{s_1} u_1 \ldots \partial^{s_d} u_d} |_{\boldsymbol{u}=\boldsymbol{0}} . \tag{2.100}$$

If the coordinates of $\boldsymbol{X}$ are independent random variables, the moments of $\boldsymbol{X}$ can be calculated from $\mu(s_1, \ldots, s_d) = \prod_{i=1}^{d} E[X_i^{s_i}]$. ▲

Moments of any order of a random vector $\boldsymbol{X}$ can be obtained from Eq. 2.99 by selecting adequate values for the exponents s_i. For example,

- **Mean** of $X_i : \mu_i = E[X_i] = \mu(s_1, \ldots, s_d)$, for $s_i = 1, s_j = 0, j \neq i$.
- **Correlation** of $(X_i, X_j) : r_{i,j} = E[X_i X_j] = \mu(s_1, \ldots, s_d)$, for $s_i = s_j = 1, s_k = 0, k \neq i, j$.
- **Covariance** of $(X_i, X_j) : c_{i,j} = E[(X_i - \mu_i)(X_j - \mu_j)] = r_{i,j} - \mu_i \mu_j$.
- **Variance** of $X_i : \sigma_i^2 = c_{i,i} = E[(X_i - \mu_i)^2] = r_{i,i} - \mu_i^2$. (2.101)

Note: The relationship between $c_{i,j}$ and $r_{i,j}$ holds by the linearity of the expectation operator. If $c_{i,j} = 0$ for $i \neq j$, then X_i and X_j are said to be **uncorrelated**. If $r_{i,j} = 0$ for $i \neq j$, then X_i and X_j are said to be **orthogonal**. If $\mu_i = \mu_j = 0$, the coordinates X_i and X_j of $\boldsymbol{X}$ are uncorrelated if and only if they are orthogonal. ▲

It is common to list the means, correlations, and covariances of the coordinates of $\boldsymbol{X}$ in vectors and matrices. We view all vectors as column vectors and denote the transpose of a vector $\boldsymbol{a}$ by $\boldsymbol{a}^T$. The **mean vector**, the **correlation matrix**, and the **covariance matrix** of $\boldsymbol{X}$ are, respectively,

$$\begin{aligned} &\boldsymbol{\mu} = \{\mu_i\} = E[\boldsymbol{X}], \quad \boldsymbol{r} = \{r_{i,j}\} = E[\boldsymbol{X}\,\boldsymbol{X}^T], \quad \text{and} \\ &\boldsymbol{c} = \{c_{i,j}\} = E[(\boldsymbol{X} - \boldsymbol{\mu})(\boldsymbol{X} - \boldsymbol{\mu})^T]. \end{aligned} \tag{2.102}$$

The pair $(\boldsymbol{\mu}, \boldsymbol{r})$ or $(\boldsymbol{\mu}, \boldsymbol{c})$ gives the **second moment properties** of $\boldsymbol{X}$. A short hand notation for these properties is $\boldsymbol{X} \sim (\boldsymbol{\mu}, \boldsymbol{r})$ or $\boldsymbol{X} \sim (\boldsymbol{\mu}, \boldsymbol{c})$. The information content of $(\boldsymbol{\mu}, \boldsymbol{r})$ and $(\boldsymbol{\mu}, \boldsymbol{c})$ is equivalent (Eq. 2.101).

The **correlation coefficient** of X_i and X_j is

$$\rho_{i,j} = \frac{c_{i,j}}{\sigma_i \, \sigma_j} = \frac{E[(X_i - \mu_i)\,(X_j - \mu_j)]}{\sqrt{c_{i,i}\, c_{j,j}}}. \tag{2.103}$$

The correlation coefficient $\rho_{i,j}$ has the following remarkable properties.

- $|\rho_{i,j}| \leq 1$.
- $\rho_{i,j} = \pm 1$ if and only if X_i and X_j are linearly related.
- If X_i and X_j are independent, $\rho_{i,j} = 0$. The converse is not generally true.

Proof: These properties can be obtained from classical inequalities discussed later in this chapter or by direct calculations as shown here. Let $\tilde{X}_i = (X_i - \mu_i)/\sigma_i$ denote a scaled version of X_i with mean zero and variance one. Because $E[(\tilde{X}_i \pm \tilde{X}_j)^2] = 2\,(1 \pm \rho_{i,j}) \geq 0$, we have $|\rho_{i,j}| \leq 1$.

If a, b are constants and $X_i = a\,X_j + b$, then $\rho_{i,j} = \pm 1$ results by direct calculations. If $\rho_{i,j} = \pm 1$, we have $E[(\tilde{X}_i \mp \tilde{X}_j)^2] = 0$. One of the inequalities in the next section shows that the last equality implies that $\tilde{X}_i - \tilde{X}_j = 0$ a.s. so that X_i and X_j are linearly related a.s. (Example 2.65).

If X_i and X_j are independent, we have

$$\begin{aligned} E[(X_i - \mu_i)\,(X_j - \mu_j)] &= \int_{\mathbb{R}^2} (\xi - \mu_i)\,(\eta - \mu_j)\, f(\xi, \eta)\, d\xi\, d\eta \\ &= E[X_i - \mu_i]\, E[X_j - \mu_j] = 0, \end{aligned}$$

because the joint density f of (X_i, X_j) is equal to the product of the densities of X_i and X_j. Generally, the converse of this property is not true. For example, let X_i be a Gaussian random variable with mean zero and variance one and let $X_j = X_i^2$. These variables are uncorrelated because $E[X_i\,(X_j - 1)] = E[X_i^3 - X_i] = 0$ and X_i has zero odd moments. However, X_i and X_j are not independent since X_j is perfectly known if X_i is specified. ■

2.11.5 Gaussian vector

Let $\boldsymbol{X}$ be an $\mathbb{R}^d$-valued random variable with mean $\boldsymbol{\mu}$ and covariance matrix $\boldsymbol{c}$. This vector is **Gaussian** with mean $\boldsymbol{\mu}$ and covariance matrix $\boldsymbol{c}$ if its density and characteristic functions are given by Eqs. 2.104 and 2.105, respectively.

$$f(\boldsymbol{x}) = [(2\,\pi)^d \det(\boldsymbol{c})]^{-1/2} \exp\left[-\frac{1}{2}(\boldsymbol{x}-\boldsymbol{\mu})^T\, \boldsymbol{c}^{-1}\,(\boldsymbol{x}-\boldsymbol{\mu})\right] \quad \text{and} \tag{2.104}$$

$$\varphi(\boldsymbol{u}) = \exp\left(\sqrt{-1}\,\boldsymbol{u}^T\,\boldsymbol{\mu} - \frac{1}{2}\boldsymbol{u}^T\,\boldsymbol{c}\,\boldsymbol{u}\right). \tag{2.105}$$

Note: The functions f and φ are defined on $\mathbb{R}^d$ and are real and complex-valued, respectively. If the covariance matrix $\boldsymbol{c}$ is diagonal, the coordinates of $\boldsymbol{X}$ are not only uncorrelated but also independent because the joint density of $\boldsymbol{X}$ is equal to the product of the marginal densities of this vector. Hence, the independence and the lack of correlation are equivalent concepts for Gaussian vectors.

To indicate that $\boldsymbol{X}$ is an $\mathbb{R}^d$-valued Gaussian variable with second moment properties $(\boldsymbol{\mu}, \boldsymbol{c})$, we write $\boldsymbol{X} \sim N_d(\boldsymbol{\mu}, \boldsymbol{c})$ or just $\boldsymbol{X} \sim N(\boldsymbol{\mu}, \boldsymbol{c})$ if there can be no confusion about the dimension of $\boldsymbol{X}$. ▲

Example 2.63: If $d = 1$, $\mu = 0$, and $c = 1$, $\boldsymbol{X} = X$ is called the **standard Gaussian variable**. The density and the distribution of this variable are

$$\phi(x) = \frac{1}{\sqrt{2\,\pi}} \exp\left(-\frac{x^2}{2}\right) \quad \text{and} \quad \Phi(x) = \int_{-\infty}^{x} \phi(y)\,dy. \tag{2.106}$$

The **standard bivariate Gaussian vector** corresponding to $d = 2$, $\boldsymbol{\mu} = \boldsymbol{0}$, $c_{1,1} = c_{2,2} = 1$, and $c_{1,2} = c_{2,1} = \rho$ has the density

$$\phi(x_1, x_2; \rho) = \frac{1}{2\,\pi\,\sqrt{1-\rho^2}} \exp\left[-\frac{x_1^2 - 2\,\rho\,x_1\,x_2 + x_2^2}{2(1-\rho^2)}\right] \tag{2.107}$$

and the characteristic function

$$\varphi(u_1, u_2; \rho) = \exp\left(-\frac{u_1^2 + 2\,\rho\,u_1\,u_2 + u_2^2}{2}\right). \tag{2.108}$$

Figure 2.7 shows the density and the characteristic functions in Eqs. 2.107 and 2.108 for $\rho = \pm 0.8$. ◇

We give two essential properties satisfied by any Gaussian vector $\boldsymbol{X} \sim N(\boldsymbol{\mu}, \boldsymbol{c})$. One of these properties uses the notations $\boldsymbol{X}^{(1)}$ and $\boldsymbol{X}^{(2)}$ for the first $d_1 < d$ and the last $d_2 = d - d_1$ coordinates of $\boldsymbol{X}$, $\boldsymbol{\mu}^{(p)} = E[\boldsymbol{X}^{(p)}]$, and $\boldsymbol{c}^{(p,q)} = E[(\boldsymbol{X}^{(p)} - \boldsymbol{\mu}^{(p)})(\boldsymbol{X}^{(p)} - \boldsymbol{\mu}^{(p)})^T]$, $p, q = 1, 2$.

- Linear transformations of a Gaussian vector are Gaussian vectors.
- The conditional vector $\hat{\boldsymbol{X}} = \boldsymbol{X}^{(1)} \mid (\boldsymbol{X}^{(2)} = \boldsymbol{z})$ is an $\mathbb{R}^{d'}$-valued Gaussian variable with mean and covariance matrices

$$\hat{\boldsymbol{\mu}} = \boldsymbol{\mu}^{(1)} + \boldsymbol{c}^{(1,2)}\,(\boldsymbol{c}^{(2,2)})^{-1}(\boldsymbol{z} - \boldsymbol{\mu}^{(2)}) \quad \text{and}$$
$$\hat{\boldsymbol{c}} = \boldsymbol{c}^{(1,1)} - \boldsymbol{c}^{(1,2)}\,(\boldsymbol{c}^{(2,2)})^{-1}\,\boldsymbol{c}^{(2,1)}. \tag{2.109}$$

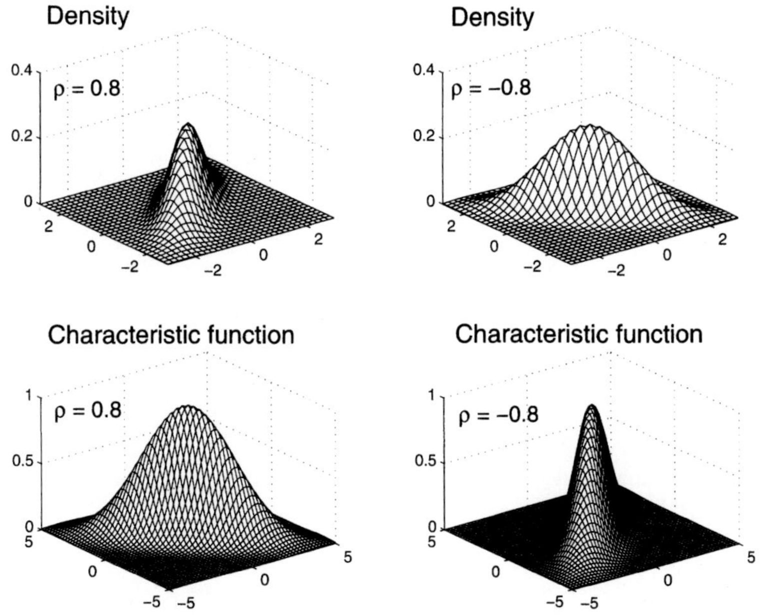

Figure 2.7. The density and the characteristic function of the standard bivariate Gaussian vector with $\rho = 0.8$ and $\rho = -0.8$

Proof: Let $\boldsymbol{Y} = \boldsymbol{a}\,\boldsymbol{X} + \boldsymbol{b}$, where $\boldsymbol{a}$ and $\boldsymbol{b}$ are real-valued (q, d) and $(q, 1)$ constant matrices. The characteristic function of $\boldsymbol{Y}$ at $\boldsymbol{v} \in \mathbb{R}^q$ is

$$E\left[e^{\sqrt{-1}\,\boldsymbol{v}^T\,\boldsymbol{Y}}\right] = E[\exp(\sqrt{-1}\,\boldsymbol{v}^T\,(\boldsymbol{a}\,\boldsymbol{X} + \boldsymbol{b})] = E[\exp(\sqrt{-1}\,\boldsymbol{v}^T\,\boldsymbol{a}\,\boldsymbol{X})]\,\exp(\sqrt{-1}\,\boldsymbol{v}^T\,\boldsymbol{b}).$$

The first term is the characteristic function of $\boldsymbol{X}$ for $\boldsymbol{u} = \boldsymbol{a}^T\,\boldsymbol{v}$ so that (Eq. 2.105)

$$\begin{aligned} E\left[e^{\sqrt{-1}\,\boldsymbol{v}^T\,\boldsymbol{Y}}\right] &= \exp\left(\sqrt{-1}(\boldsymbol{a}^T\,\boldsymbol{v})^T\,\boldsymbol{\mu} - \frac{1}{2}(\boldsymbol{a}^T\,\boldsymbol{v})^T\,\boldsymbol{c}\,(\boldsymbol{a}^T\,\boldsymbol{v})\right)\exp(\sqrt{-1}\,\boldsymbol{v}^T\,\boldsymbol{b}) \\ &= \exp\left(\sqrt{-1}\,\boldsymbol{v}^T(\boldsymbol{a}\,\boldsymbol{\mu} + \boldsymbol{b}) - \frac{1}{2}\boldsymbol{v}^T\,(\boldsymbol{a}^T\,\boldsymbol{c}\,\boldsymbol{a})\boldsymbol{v}\right). \end{aligned}$$

Hence, $\boldsymbol{Y}$ is a Gaussian vector with mean $\boldsymbol{\mu}_y = \boldsymbol{a}\,\boldsymbol{\mu} + \boldsymbol{b}$ and covariance $\boldsymbol{c}_y = \boldsymbol{a}^T\,\boldsymbol{c}\,\boldsymbol{a}$.

The properties of $\hat{\boldsymbol{X}}$ can be obtained by direct calculations based on the density of $\boldsymbol{X}$ in Eq. 2.104 and the definition of the conditional density (Eq. 2.92). If $d_1 = d_2 = 1$, $\boldsymbol{\mu} = \boldsymbol{0}$, $c_{1,1} = c_{2,2} = 1$, and $c_{1,2} = c_{2,1} = \rho$, then $\hat{X} = X_1 \mid (X_2 = z)$ is a Gaussian variable with mean $\hat{\mu} = \rho\,z$ and variance $\hat{c} = 1 - \rho^2$. If $\rho = 0$, that is, X_1 and X_2 are independent, $\hat{X}$ and X_1 have the same distribution. Otherwise, $\hat{X}$ has a smaller variance than X_1. The probabilistic characteristics of the conditional vector $\hat{X}$ have useful applications in reliability studies. For example, suppose that X_1 and X_2 control the performance of a system and we can measure only X_2. If the correlation coefficient ρ is not zero, the measured value of X_2 can be used to reduce the uncertainty in X_1. ■

Example 2.64: Let X be an $\mathbb{R}^2$-valued random variable with mean vector $\boldsymbol{\mu} = (\mu_1, \mu_2)$ and covariance matrix $\boldsymbol{c}$ with entries $c_{1,1} = \sigma_1^2$, $c_{1,2} = c_{2,1} = \rho\,\sigma_1\,\sigma_2$, and $c_{2,2} = \sigma_2^2$. The random variable,

$$\hat{X}_1 = \frac{\rho\,\sigma_1}{\sigma_2}\,(X_2 - \mu_2) + \mu_1,$$

is the optimal, mean square, linear estimator of X_1 corresponding to an observation of X_2. ◇

Proof: Let $Z = a\,X_2 + b$, $a, b \in \mathbb{R}$, be a linear estimator of X_1. We require that the estimator is unbiased and minimizes the mean square error $E[(Z - X_1)^2]$. The first condition implies $E[Z] = \mu_1$ so that $\mu_1 = a\,\mu_2 + b$ and $Z = a\,(X_2 - \mu_2) + \mu_1$. The mean square error of the estimator Z of X_1 is

$$E[(Z - X_1)^2] = E[(a\,(X_2 - \mu_2) - (X_1 - \mu_1))^2] = a^2\,\sigma_2^2 + \sigma_1^2 - 2\,a\,\rho\,\sigma_1\,\sigma_2,$$

and takes its minimum value at $a = \rho\,\sigma_1/\sigma_2$. This optimal value of a is the solution of $\partial E[(Z - X_1)^2]/\partial a = 0$. ■

2.12 Useful inequalities

The Chebyshev, Cauchy-Schwarz, Hölder, and Jenssen inequalities are useful for both applications and theoretical considerations. These inequalities are stated and proved. Let X and Y be random variables defined on a probability space $(\Omega, \mathcal{F}, P)$.

Chebyshev's inequality. If $r : \mathbb{R} \to (0, \infty)$ is a strictly positive, even function that increases in $(0, \infty)$ and $E[r(X)] < \infty$, then

$$P(|X| > a) \le \frac{E[r(X)]}{r(a)}, \quad a > 0. \tag{2.110}$$

Proof: The function $r(X)$ is a random variable because $\{\omega : r(X(\omega)) \le y\}$, $y \ge 0$, is equal to $\{\omega : X(\omega) \in [-r^{-1}(y), r^{-1}(y)]\}$ and X is a random variable. We have

$$E[r(X)] = \int_\Omega r(X)\,dP \ge \int_{|X| \ge a} r(X)\,dP \ge r(a)\,P(|X| > a)$$

because $r > 0$ and $r(x) \ge r(a)$ for $|x| \ge a$. The common form of the Chebyshev inequality corresponds to $r(x) = |x|^p$, where $p \ge 1$ is an integer. ■

Jensen's inequality. If $g : \mathbb{R} \to \mathbb{R}$ is a convex function and X and $g(X)$ are integrable random variables, then

$$g(E[X]) \le E[g(X)]. \tag{2.111}$$

Proof: We use two facts to prove Eq. 2.111. First, convex functions are continuous so that $g(X)$ is a random variable. Second, a convex function g has the property

$$g(x) = \sup\{l(x) : l(u) \leq g(u), \forall u \in \mathbb{R}\},$$

where l denotes a linear function. We have

$$E[g(X)] = E[\sup\{l(X)\}] \geq \sup\{E[l(X)]\} = \sup\{l(E[X])\} = g(E[X]),$$

where the above inequality and the above last equality hold by Eq. 2.50 and the linearity of the expectation operator, respectively.

The inequalities,

$$|E[X]| \leq E[|X|] \quad \text{and} \quad (E[X])^{2q} \leq E[X^{2q}] \quad \text{for an integer } q > 0,$$

follow from Eq. 2.111 with $g(x) = |x|$ and $g(x) = x^{2q}$, respectively. ■

Cauchy-Schwarz's inequality . If $E[X^2]$ and $E[Y^2]$ are finite, then $E[X\,Y]$ exists and

$$|E[X\,Y]| \leq (E[X^2])^{1/2}\,(E[Y^2])^{1/2}. \tag{2.112}$$

Proof: The expectation $E[(X + \lambda\,Y)^2] = E[Y^2]\,\lambda^2 + 2\,E[X\,Y]\,\lambda + E[X^2]$ is positive for any $\lambda \in \mathbb{R}$. Hence, the polynomial of λ has no real roots so that we must have $(E[X\,Y])^2 - E[X^2]\,E[Y^2] \leq 0$. ■

Hölder's inequality. If $1 < p < \infty$ and q is given by $1/p + 1/q = 1$, $E[|X|^p] < \infty$, and $E[|Y|^q] < \infty$, then

$$|E[X\,Y]| \leq E[|X\,Y|] \leq E[|X|^p]^{1/p}\,E[|Y|^q]^{1/q}. \tag{2.113}$$

Proof: The inequality $|a\,b| \leq |a|^p/p + |b|^q/q$ holds for $a, b \in \mathbb{R}$ and p, q as in Eq. 2.113. The expectation of this inequality with $a = X/(E[|X|^p])^{1/p}$ and $b = Y/(E[|Y|^q])^{1/q}$ gives Eq. 2.113. The Hölder inequality gives $E[|X|] \leq E[|X|^p]^{1/p}$ for $Y = 1$ and the Cauchy-Schwarz inequality for $p = q = 2$.

The inequality $E[|X|] \leq E[|X|^p]^{1/p}$ shows that $X \sim L_p$ implies $X \sim L_1$, where $p > 1$ is an integer. ■

Minkowski's inequality. If $1 \leq p < \infty$ and $X, Y \in L_p$, then $X + Y \in L_p$ and

$$E[|X + Y|^p]^{1/p} \leq E[|X|^p]^{1/p} + E[|Y|^p]^{1/p}. \tag{2.114}$$

Proof: That $X, Y \in L_p$ implies $X + Y \in L_p$ follows from the inequality

$$|X + y|^p \leq 2\left(|X|^p \vee |Y|^p\right) \leq 2\left(|X|^p + |Y|^p\right).$$

If $p = 1$, the inequality in Eq. 2.114 is trivial. If $p > 1$, the expectation of

$$|X+Y|^p = |X+Y|\,|X+Y|^{p-1} \le (|X|+|Y|)\,|X+Y|^{p-1} \quad \text{is}$$

$$E[|X+Y|^p] \le E[|X|\,|X+Y|^{p-1}] + E[|Y|\,|X+Y|^{p-1}].$$

The Hölder inequality gives the upper bounds

$$E[|X|^p]^{1/p}\, E[|X+Y|^{(p-1)q}]^{1/q} \quad \text{and} \quad E[|Y|^p]^{1/p}\, E[|X+Y|^{(p-1)q}]^{1/q}$$

on the right side terms of the above inequality, where $1/p + 1/q = 1$, so that we have

$$E[|X+Y|^p] \le (E[|X|^p]^{1/p} + E[|Y|^p]^{1/p})\, E[|X+Y|^{(p-1)q}]^{1/q}.$$

Eq. 2.114 results by division with $E[|X+Y|^{(p-1)\,q}]^{1/q}$ since $(p-1)\,q = p$. ■

Example 2.65: Consider an $\mathbb{R}^2$-valued random variable $\boldsymbol{X} = (X_1, X_2)$ with the second moment properties

$$\boldsymbol{\mu} = \begin{bmatrix} \mu_1 \\ \mu_2 \end{bmatrix} \quad \text{and} \quad \boldsymbol{c} = \begin{bmatrix} \sigma_1^2 & \rho\,\sigma_1\,\sigma_2 \\ \rho\,\sigma_1\,\sigma_2 & \sigma_2^2 \end{bmatrix}.$$

The above inequalities can be used to prove that the correlation coefficient ρ takes values in $[-1, 1]$ and $\rho = \pm 1$ if and only if X_1 and X_2 are linearly related. ◇

Proof: Set $\tilde{X}_i = (X_i - \mu_i)/\sigma_i \sim (0, 1)$. The Cauchy-Schwarz inequality applied to the random variables $\tilde{X}_1$ and $\tilde{X}_2$ gives $|\rho|^2 \le 1$. If $X_1 = a\,X_2 + b$, then $\rho = a/\sqrt{a^2}$ so that $\rho = \pm 1$. If $\rho = 1$, the expectation of $[\tilde{X}_1 - \tilde{X}_2]^2$ is zero so that $P(|\tilde{X}_1 - \tilde{X}_2| > \varepsilon) = 0$, $\forall \varepsilon > 0$, by the Chebyshev inequality. Hence, $\tilde{X}_1 = \tilde{X}_2$ a.s. or $X_1 = a\,X_2 + b$ a.s. ■

2.13 Convergence of random variables

Let X and X_n, $n = 1, 2, \ldots$, be real-valued random variables defined on a probability space $(\Omega, \mathcal{F}, P)$. The distribution functions of X and X_n are F and F_n, respectively. The convergence of the sequence X_n to X has various definitions depending on the way in which the difference between X_n and X is measured. We now define the **almost sure** convergence ($X_n \xrightarrow{\text{a.s.}} X$), the convergence in **probability** ($X_n \xrightarrow{\text{pr}} X$), the convergence in **distribution** ($X_n \xrightarrow{\text{d}} X$), and the **convergence in** L_p or the L_p **convergence** ($X_n \xrightarrow{\text{m.p.}} X$), where $p \ge 1$ is an integer. The L_p convergence for $p = 2$, called convergence in **mean square** , is denoted by $X_n \xrightarrow{\text{m.s.}} X$ or $\text{l.i.m.}_{\,n\to\infty} X_n = X$.

- $X_n \xrightarrow{\text{a.s.}} X$ if $\lim_{n\to\infty} X_n(\omega) = X(\omega)$, $\forall \omega \in \Omega \setminus N$, $P(N) = 0$.
- $X_n \xrightarrow{\text{pr}} X$ if $\lim_{n\to\infty} P(|X_n - X| > \varepsilon) = 0$, $\forall \varepsilon > 0$.
- $X_n \xrightarrow{\text{d}} X$ if $\lim_{n\to\infty} F_n(x) = F(x)$, $\forall x \in \mathbb{R}$.
- $X_n \xrightarrow{\text{m.p.}} X$ if $\lim_{n\to\infty} E[|X_n - X|^p] = 0$.

Note: The a.s. convergence has some equivalent definitions. For example, it can be shown that $X_n \xrightarrow{\text{a.s.}} X$ if and only if for any $\varepsilon > 0$ we have

$$\lim_{m\to\infty} P(|X_n - X| \le \varepsilon \text{ for all } n \ge m) = 1 \quad \text{or}$$
$$\lim_{m\to\infty} P(|X_n - X| > \varepsilon \text{ for some } n \ge m) = 0.$$

An alternative form of the latter condition is $P(|X_n - X| > \varepsilon \text{ i.o.}) = 0$ since $A_m(\varepsilon) = \cup_{n=m}^{\infty}\{|X_n - X| > \varepsilon\}$, $\forall \varepsilon > 0$, is a decreasing sequence of events, so that

$$\begin{aligned}\lim_{m\to\infty} P(A_m(\varepsilon)) &= P(\lim_{m\to\infty} A_m(\varepsilon)) = P(\limsup_{n\to\infty}\{|X_n - X| > \varepsilon\}) \\ &= P(\{|X_n - X| > \varepsilon\} \text{ i.o.}),\end{aligned}$$

where the last equality holds by Eq. 2.21. ▲

We summarize some useful **properties of convergent sequences** of random variables and the relationships between different types of convergence ([151], Section 6.3). These relationships are illustrated in Fig. 2.8.

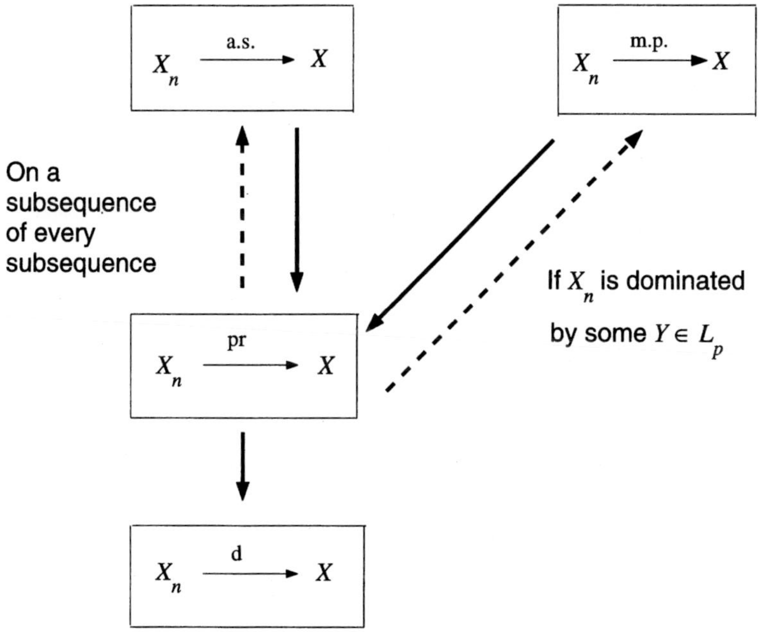

Figure 2.8. Types of convergence for random sequences and their relationship

The convergence $X_n \xrightarrow{\text{pr}} X$ implies:

- X is unique with probability one.
- X_n is Cauchy in probability. The converse is also true.
- $X_n \xrightarrow{\text{d}} X$.
- $X_n \xrightarrow{\text{m.p.}} X$ if X_n is dominated by a random variable in L_p, that is, there exists $Y \in L_p$ and $|X_n| \leq Y$ a.s. for $n \geq 1$.

Note: A sequence of random variables X_n is **Cauchy in probability** if for arbitrary $\varepsilon > 0$ and $\eta > 0$ there exists $n(\varepsilon, \eta)$ such that $P(|X_m - X_n| > \varepsilon) < \eta$ for $m, n \geq n(\varepsilon, \eta)$.

The limits of sequences of random variables have similar properties as the limits of numerical series. For example, $X_n \xrightarrow{\text{pr}} X$ and $Y_n \xrightarrow{\text{pr}} Y$ imply $\alpha X_n + \beta Y_n \xrightarrow{\text{pr}} \alpha X + \beta Y$, where $\alpha, \beta \in \mathbb{R}$, and $X_n Y_n \xrightarrow{\text{pr}} X Y$.

If $X_n \xrightarrow{\text{pr}} X$ and $Y \geq 0$ is a random variable such that $|X_n| < Y$ a.s. for each n, then $|X| \leq Y$ a.s.

$X_n \xrightarrow{\text{pr}} X$ if and only if each subsequence $\{X_{n_k}\}$ of $\{X_n\}$ contains a subsequence $\{X_{n_{k_i}}\}$ such that $X_{n_{k_i}} \xrightarrow{\text{a.s}} X$.

$X_n \xrightarrow{\text{a.s}} X$ implies $X_n \xrightarrow{\text{pr}} X$.

$X_n \xrightarrow{\text{m.p.}} X$ implies $X_n \xrightarrow{\text{pr}} X$ and $X_n \xrightarrow{\text{m.r.}} X$ if $r < p$.

Note: Proofs of the above statements can be found, for example, in [40] (Chapter 4) and [150] (Section 6.3). ▲

Example 2.66: Let $([0,1], \mathcal{B}([0,1]), P)$ be a probability space with $P(d\omega) = d\omega$ and let $X_n = 2^n \, 1_{(0,1/n)}$ be a sequence of random variables defined on this space. The sequence X_n converges to zero in probability as $n \to \infty$ but does not converge in L_p, $p \geq 1$. ◇

Proof: For any $\varepsilon > 0$, we have $P(|X_n| > \varepsilon) = P((0, 1/n)) = 1/n \to 0$ as $n \to \infty$. Hence, X_n converges in probability to zero. On the other hand,

$$E[|X_n|^p] = (2^n) \, P((0, 1/n)) = 2^n/n \to \infty \quad \text{as } n \to \infty.$$

Hence, convergence in probability does not imply L_p convergence ([151], p. 182). ■

Example 2.67: Consider a sequence of random variables X_n defined on the probability space $([0,1], \mathcal{B}([0,1]), \lambda)$ by $X_1 = 1_{[0,1/2]}$, $X_2 = 1_{[1/2,1]}$, $X_3 = 1_{[1,1/3]}$, $X_4 = 1_{[1/3,2/3]}$, $X_5 = 1_{[2/3,1]}$, and so on, where λ is the Lebesgue measure. This sequence converges in L_p to zero but does not converge a.s. to zero. ◇

Proof: The sequence converges to zero in L_p, $p > 0$, since

$$E[|X_1|^p] = E[|X_2|^p] = 1/2, \; E[|X_3|^p] = E[|X_4|^p] = E[|X_5|^p] = 1/3, \ldots,$$

so that $E[|X_n|^p] \to 0$ as $n \to \infty$. However, the numerical sequence $X_n(\omega)$ does not converge to zero ([151], p. 182). ▲

Example 2.68: If X_n is a sequence of uncorrelated random variables with finite mean μ and variance σ^2, then

$$S_n = \sum_{k=1}^{n} \frac{X_n - \mu}{n} \xrightarrow{\text{pr}} 0, \quad n \to \infty. \tag{2.115}$$

This result is referred to as the **weak law of large numbers** ([106], p. 36). ◇

Proof: The mean and variance of S_n are $\mu_n = 0$ and $\sigma_n^2 = \sigma^2/n$. The Chebyshev inequality (Eq. 2.110) gives $P(|S_n| > \varepsilon) \leq \sigma_n^2/\varepsilon^2 = \sigma^2/(n\,\varepsilon^2)$ for arbitrary $\varepsilon > 0$ and each $n \geq 1$. Hence, $P(|S_n| > \varepsilon) \to 0$ as $n \to \infty$ so that $S_n \xrightarrow{\text{pr}} 0$. This convergence indicates that most of the probability mass of S_n is concentrated in a small vicinity of zero for a sufficiently large n. The convergence $S_n \xrightarrow{\text{pr}} 0$ does not provide any information on the behavior of the numerical sequence $S_n(\omega)$ for an arbitrary but fixed $\omega \in \Omega$. Other versions of the weak law of large numbers can be found in [151] (Section 7.2). ■

Example 2.69: If X_n is a sequence of independent identically distributed random variables such that $E[|X_1|]$ exists and is finite, then

$$\frac{1}{n}\sum_{k=1}^{n} X_k \xrightarrow{\text{a.s.}} E[X_1], \quad n \to \infty. \tag{2.116}$$

This result is known as the **strong law of large numbers** [151] (Sections 7.4 and 7.5). Figure 2.9 shows five samples of $(1/n)\sum_{k=1}^{n} X_k$, where X_k are independent Gaussian variables with mean zero and variance one. All samples approach the mean of X_1 as n increases in agreement with Eq. 2.116. ◇

Note: The strong law of large numbers characterizes the behavior of the numerical sequence $(1/n)\sum_{k=1}^{n} X_k(\omega)$. The a.s. convergence of $(1/n)\sum_{k=1}^{n} X_k$ to $\mu = E[X_1]$ means that the numerical sequence $(1/n)\sum_{k=1}^{n} X_k(\omega)$ converges to μ for each $\omega \in \Omega \setminus N$, where $N \in \mathcal{F}$ and $P(N) = 0$. ▲

Example 2.70: If X_k, $k = 1, 2, \ldots$, is a sequence of iid random variables with finite mean $\mu = E[X_1]$ and variance σ^2, then

$$S_n^* = \frac{1}{\sqrt{n}} \sum_{k=1}^{n} \frac{X_k - \mu}{\sigma} \xrightarrow{\text{d}} N(0, 1), \quad n \to \infty. \tag{2.117}$$

This result is known as the **central limit theorem** ([151], Section 8.2). ◇

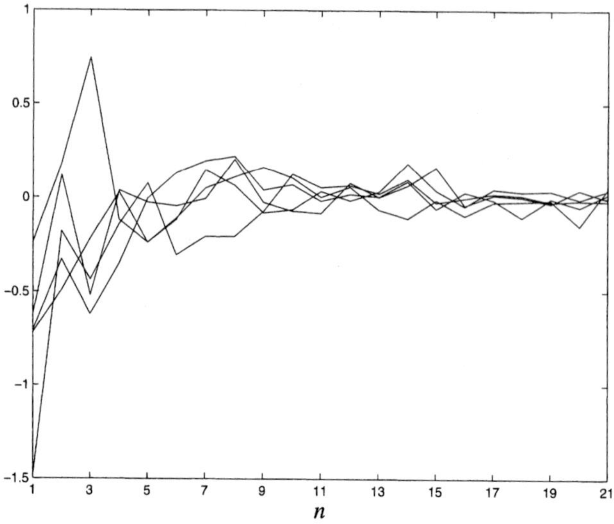

Figure 2.9. Five sample paths of $(1/n)\sum_{k=1}^{n} X_k$ for $X_k \sim N(0,1)$

Proof: The mean and variance of S_n^* are zero and one for each n so that this sequence has the required second moment properties. We show that the characteristic function of S_n^* approaches the corresponding function of the standard Gaussian variable as $n \to \infty$. The characteristic function of S_n^* is

$$\varphi_n(u) = \exp\left[-\sqrt{-1}\, u\, \mu\, \sqrt{n}/\sigma\right] \varphi(u/(\sqrt{n}\,\sigma))^n,$$

where φ denotes the characteristic function of X_1. An alternative form of this characteristic function is

$$\varphi_n(u) = \exp\left\{-\sqrt{-1}\, u\, \mu\, \frac{\sqrt{n}}{\sigma} + n\left[\sqrt{-1}\, u\, \frac{\mu}{\sqrt{n}\,\sigma} + \frac{(\sqrt{-1}\, u)^2}{2\, n\, \sigma^2} + O(n^{-3/2})\right]\right\}$$

and results from $\varphi(u/(\sqrt{n}\,\sigma))^n = \exp\left[\ln\left(\varphi(u/(\sqrt{n}\,\sigma))^n\right)\right]$ by expanding the function $\ln\left(\varphi(u/(\sqrt{n}\,\sigma))^n\right)$ in a Taylor series ([79], p. 376). The limit of the characteristic function of S_n^* as $n \to \infty$ is equal to $\exp(-u^2/2)$ which is the characteristic function of the standard Gaussian variable. ■

Example 2.71: Consider a sequence of Bernoulli variables X_n taking the values one and zero with probabilities $P(X_n = 1) = p_n$ and $P(X_n = 0) = 1 - p_n$, $p_n \in (0, 1)$. If $\sum_{n=1}^{\infty} p_n < \infty$, then $P(\lim_{n\to\infty} X_n = 0) = 1$. ◇

Proof: The assumption $\sum_{n=1}^{\infty} p_n = \sum_{n=1}^{\infty} P(X_n = 1) < \infty$ and the Borel-Cantelli lemma yield $P((X_n = 1) \text{ i.o.}) = 0$ or

$$1 = P(\{(X_n = 1) \text{ i.o.}\}^c) = P(\{\limsup_{n\to\infty}(X_n = 1)\}^c) = P(\liminf_{n\to\infty}(X_n = 0)),$$

as stated ([151], Example 4.5.1, p. 103). ■

2.14 Random walk

Let $X = (X_1, X_2, \ldots)$ be a sequence of iid random variables defined on a probability space $(\Omega, \mathcal{F}, P)$ with values in a measurable space $(\Psi, \mathcal{G})$. The sequence $R = (R_0, R_1, \ldots)$ defined by

$$R_n = \sum_{i=1}^{n} X_i, \quad n = 1, 2, \ldots, \quad \text{and} \quad R_0 = 0 \tag{2.118}$$

is called a **random walk**.

The random variables $(R_1, \ldots, R_n)$ and $(X_1, \ldots, X_n)$ are related by a measurable mapping for each $n \geq 1$ so that $(R_1, \ldots, R_n)$ is $\sigma(X_1, \ldots, X_n)$-measurable. The random variables $(R_0, R_1, \ldots, R_n)$ are also $\sigma(X_1, \ldots, X_n)$-measurable because the σ-field $\{\emptyset, \Omega\}$ generated by the random variable $R_0 : \Omega \to \mathbb{R}$ is included in $\sigma(X_1, \ldots, X_n)$ for each $n \geq 1$. We will think of the index n of the random walk as time.

Example 2.72: Suppose that the random variables X_i in Eq. 2.118 are real-valued with finite mean μ and variance σ^2 and that the measurable space $(\Psi, \mathcal{G})$ is $(\mathbb{R}, \mathcal{B})$. The mean, variance, and coefficient of variation of the random walk R_n corresponding to these random variables are

$$E[R_n] = n\,\mu, \quad \text{Var}[R_n] = n\,\sigma^2, \quad \text{and c.o.v.}[R_n] = \frac{\sqrt{\text{Var}[R_n]}}{E[R_n]} = \frac{v}{\sqrt{n}},$$

where $v = \sigma/\mu$ and c.o.v.$[R_n]$ are defined if $\mu \neq 0$.

The variance of the random walk increases linearly with the index n, a property shared by the Brownian motion process (Section 3.13). If $\mu = 0$, the random walk has mean zero and oscillates about its mean value with an amplitude that increases with n. If $\mu \neq 0$ and $n \to \infty$, the average and the coefficient of variation of R_n approach $\pm\infty$ depending on the sign of μ and zero, respectively. This observation suggests that R_n approaches $+\infty$ or $-\infty$ as time n increases. ◇

The heuristic considerations in Example 2.72 are correct in the sense of the following two results, which are stated without proof.

If $P(X_1 = 0) < 1$, the asymptotic behavior of the random walk as $n \to \infty$ is one and only one of the following ([150], Proposition 7.2.2, p. 562):

- $R_n \to +\infty$ a.s.,
- $R_n \to -\infty$ a.s., or
- $-\infty = \liminf R_n < \limsup R_n = +\infty$ a.s. (2.119)

If $P(X_1 = 0) < 1$ and $E[|X_1|] < \infty$, the asymptotic behavior of the random walk R_n as $n \to \infty$ is ([150], Proposition 7.2.3, p. 563):

- If $E[X_1] > 0$, $\quad R_n \to +\infty \quad$ a.s.
- If $E[X_1] < 0$, $\quad R_n \to -\infty \quad$ a.s.
- If $E[X_1] = 0$, $\quad -\infty = \liminf R_n < \limsup R_n = +\infty$. (2.120)

Note: Eq. 2.119 shows that R_n converges to an infinity or oscillates between $-\infty$ and $+\infty$ as $n \to \infty$. The additional information on the mean of X_1 in Eq. 2.120 allows us to predict the behavior of the random walk more precisely than in Eq. 2.119. For example, if the random variables X_k take the values one and zero with probabilities p, $0 < p < 1$, and $1 - p$, the corresponding random walk R_n converges to $+\infty$ a.s. since $E[X_1] = p$. ▲

Example 2.73: Let $X_1, X_2, \ldots$ be a sequence of iid Gaussian random variables with mean μ and standard deviation σ. Figure 2.10 shows samples of the random walk R_n in Eq. 2.118 for $X_1 \sim N(\mu, \sigma^2)$, where $\mu = 1$, $\mu = 0$, and $\mu = -1$ and

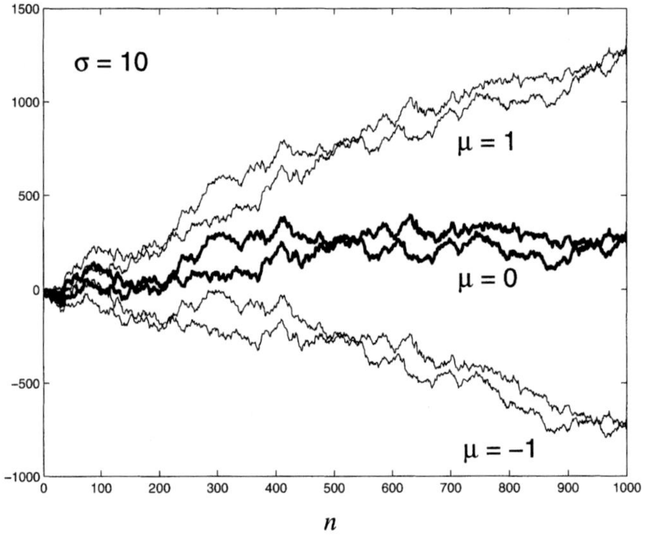

Figure 2.10. Sample paths of the random walk for $X_k \sim N(\mu, 10^2)$ with $\mu = 1$, $\mu = 0$, and $\mu = -1$

$\sigma = 10$. The samples corresponding to $\mu = 1$ and $\mu = -1$ exhibit an increasing and decreasing trend, respectively, while the samples for $\mu = 0$ oscillate about zero in agreement with Eq. 2.120. ◇

Example 2.74: Consider a sequence of iid random variables A_n, $n = 1, 2, \ldots$, defined on a probability space $(\Omega, \mathcal{F}, P)$. Let X_n be another sequence of random variables defined on the same probability space generated by the recurrence formula $X_n = A_n X_{n-1}$, $n = 1, 2, \ldots$, with $X_0 = x \in \mathbb{R} \setminus \{0\}$. If $\ln|A_1| \sim L_1(\Omega, \mathcal{F}, P)$ and $E[\ln|A_1|] < 0$, then X_n converges to zero a.s. as $n \to \infty$. The limit

$$\lambda_{\mathrm{LE}} = \lim_{n\to\infty} \frac{1}{n} \sum_{k=1}^{n} \ln|A_k| = E[\ln|A_1|]$$

is called the **Lyapunov exponent** in the analysis of dynamic systems (Section 8.7). The solution X_n is stable a.s., that is, it converges a.s. to zero, if $\lambda_{\mathrm{LE}} < 0$ and diverges if $\lambda_{\mathrm{LE}} > 0$. ◇

Proof: The recurrence formula for X_n yields $X_n/x = \prod_{k=1}^{n} A_k$ for $X_0 = x$ so that we have $\ln|X_n/x| = \sum_{k=1}^{n} \ln|A_k|$. It follows that $\ln|X_n/x|$ is a random walk since $\ln|A_k|$ are iid random variables (Eq. 2.118). If $E[\ln|A_1|] < 0$, the random walk $\ln|X_n/x|$ converges to $-\infty$ a.s. as $n \to \infty$ (Eq. 2.120) so that the solution $|X_n/x| = \exp\left(\sum_{k=1}^{n} \ln|A_k|\right)$ converges to zero a.s. If $E[\ln|A_1|] > 0$, $\ln|X_n/x|$ tends to $+\infty$ a.s. as $n \to \infty$ so that $|X_n/x| = \exp\left(\sum_{k=1}^{n} \ln|A_k|\right)$ approaches $+\infty$ a.s. ■

Example 2.75: Suppose that the random variables A_n in Example 2.74 are exponential with distribution $F(x) = 1 - \exp(-\rho\, x)$, $\rho > 0$, $x \geq 0$. The solution X_n is stable a.s. for $\rho > \rho^* \simeq 0.56$, that is, $X_n \xrightarrow{\text{a.s.}} 0$ as $n \to \infty$. Figure 2.11 shows the dependence of the Lyapunov exponent λ_{LE} on ρ in the range [0.4, 0.7]. ◇

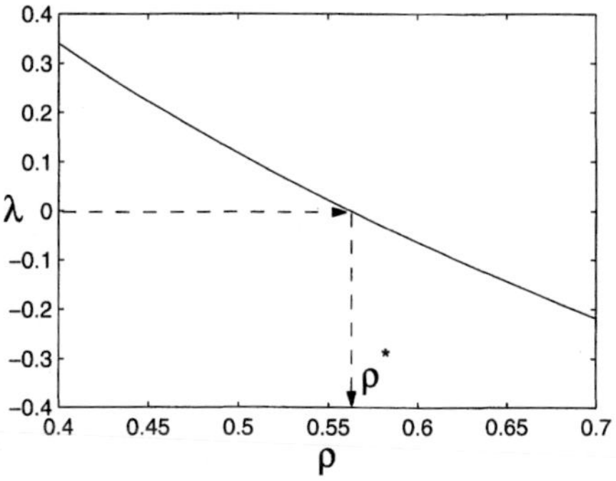

Figure 2.11. Stability condition for $X_n = A_n X_{n-1}$ with A_n iid exponential variables

Proof: The mean of $\ln|A_k| = \ln(A_k)$ is finite and can be calculated from

$$\lambda_{\mathrm{LE}} = \int_0^\infty \ln(u)\, \rho \exp(-\rho\, u)\, du = -\rho \int_{-\infty}^{\infty} \xi \exp[-\xi - \rho \exp(-\xi)]\, d\xi$$

by numerical integration. The Lyapunov exponent $\lambda_{\mathrm{LE}} = E[\ln|A_1|]$ decreases with ρ and is negative for $\rho > \rho^* \simeq 0.56$. ■

2.15 Filtration

Let $(\Omega, \mathcal{F})$ be a measurable space. An increasing collection $\mathcal{F}_0 \subseteq \mathcal{F}_1 \subseteq \cdots \mathcal{F}_n \subseteq \cdots \subseteq \mathcal{F}$ of sub-σ-fields of $\mathcal{F}$ is said to be a **filtration** in $(\Omega, \mathcal{F})$.

A probability space $(\Omega, \mathcal{F}, P)$ endowed with a filtration $(\mathcal{F}_n)_{n\geq 0}$ is called a **filtered probability space** and is denoted by $(\Omega, \mathcal{F}, (\mathcal{F}_n)_{n\geq 0}, P)$. We assume that $\mathcal{F}_0$ contains all the P-null sets of $\mathcal{F}$.

Let $(\Omega, \mathcal{F})$ and $(\Psi, \mathcal{G})$ be measurable spaces, $(\mathcal{F}_n)_{n\geq 0}$ denote a filtration on $(\Omega, \mathcal{F})$, and $X = (X_0, X_1, \ldots)$ be a sequence of measurable functions from $(\Omega, \mathcal{F})$ to $(\Psi, \mathcal{G})$.

The sequence X is said to be **adapted to the filtration** $(\mathcal{F}_n)_{n\geq 0}$ or $\mathcal{F}_n$**-adapted** if X_n is $\mathcal{F}_n$-measurable for each $n \geq 0$. The **minimal** or **natural filtration** of $X = (X_0, X_1, \ldots)$, that is, the smallest σ-field with respect to which X is adapted, is $\sigma(X_0, X_1, \ldots, X_n)$.

Example 2.76: Let $X = (X_1, X_2, \ldots)$ be outcomes of a sequence of coin tosses. The σ-field $\mathcal{F}_n = \sigma(X_1, \ldots, X_n)$ represents the knowledge at time $n \geq 1$. The information content of $\mathcal{F}_n$ is sufficient to decide whether an event related to the first n tosses has or has not occurred. For example, the event

$$A = \{\text{at least 2 heads in the first five tosses}\}$$

is $\mathcal{F}_5$-measurable because we can decide after five tosses whether A has or has not occurred. If {tail, tail, head, tail} is a sample $(X_1(\omega), X_2(\omega), X_3(\omega), X_4(\omega))$ of the first four tosses, the event A remains undecided so that $A \notin \mathcal{F}_4$. If $(X_1(\omega), X_2(\omega), X_3(\omega), X_4(\omega))$ is {tail, head, head, tail}, A has been observed in the first four tosses for this sample irrespective of the fifth outcome. However, we cannot conclude that $A \in \mathcal{F}_4$. This conclusion would be correct if we could tell whether or not A has occurred by watching only the first four outcomes irrespective of the particular sample of (X_1, X_2, X_3, X_4). ◇

Example 2.77: Let $\mathcal{F}_n = \sigma(R_1, \ldots, R_n)$, $n \geq 1$, and $\mathcal{F}_0 = \{\emptyset, \Omega\}$ be the natural filtration of the random walk sequence $R = (R_0, R_1, \ldots)$ in Eq. 2.118. The sequence R is $\mathcal{F}_n$-adapted. ◇

Note: The sequence of σ-fields $\mathcal{F}_n$ contains the information accumulated by observing the random walk R_n up to and including time $n \geq 0$. The σ-field $\mathcal{F}_{n+1}$ includes $\mathcal{F}_n$ and other sets such that $R_{n+1} = R_n + X_{n+1}$ is measurable. Hence, the sequence of sub-σ-fields $\mathcal{F}_n = \sigma(R_0, R_1, \ldots, R_n)$ is a filtration in $(\Omega, \mathcal{F})$ and R is $\mathcal{F}_n$-adapted. ▲

2.16 Stopping time

In a game of chance we may decide to quit or continue to play after each round depending on what happened up to that point and on the available resources, for example, we may be out of money. Hence, the time T at which we stop playing the game is a random variable that takes values in $\{1, 2, \ldots\}$ and depends on the entire game history. Let $\mathcal{F}_n$ represent the knowledge accumulated at time $n \geq 1$. The event $\{T = n\}$ to quit the game after n rounds should be in $\mathcal{F}_n$. A similar "game" can be envisioned between a physical or biological system and the environment. The system quits the game when its damage state reaches a critical level, which is the analogue to running out of cash. Depending on the situation, the time T at which such a system quits the game may be called failure time or death time.

We define in this section stopping times and prove some of their properties. Additional information on stopping times can be found in [59] (Chapter 2) and [151] (Section 10.7).

Let $(\Omega, \mathcal{F})$ be a measurable space and $(\mathcal{F}_n)_{n \geq 0}$ be a filtration of $\mathcal{F}$. A $\{0, 1, 2, \ldots, \infty\}$-valued random variable T defined on $(\Omega, \mathcal{F})$ is a **stopping time** with respect to $(\mathcal{F}_n)_{n \geq 0}$ or an $\mathcal{F}_n$**-stopping time** if $\{\omega : T(\omega) \leq n\} \in \mathcal{F}_n$ for each $n \geq 0$.

Note: This definition states that T is an $\mathcal{F}_n$-stopping time if $\mathcal{F}_n$ contains sufficient information to determine whether T is smaller or larger than n for each time $n \geq 0$. For example, $T = \inf\{n \geq 0 : |R_n| \geq a\}$, $a > 0$, gives the first time when the random walk R in Eq. 2.118 exits $(-a, a)$. We can determine whether $T \leq n$ or $T > n$ by observing the random walk up to time n.

We also note that a constant random variable $t \in \{0, 1, \ldots\}$ is a stopping time since $\{\omega : t \leq n\}$ is either $\emptyset$ or Ω so that the event $\{t \leq n\}$ is in $\mathcal{F}_n$ for each $n \geq 0$. If T is an $\mathcal{F}_n$-stopping time so is $T + t$ since $\{\omega : T(\omega) + t \leq n\} = \{\omega : T(\omega) \leq n - t\} \in \mathcal{F}_{(n-t)\vee 0}$ and $\mathcal{F}_{(n-t)\vee 0} \subset \mathcal{F}_n$ for $t \geq 0$. ▲

T is a stopping time if and only if $\{T = n\} \in \mathcal{F}_n$ for each $n \geq 0$.

Proof: If T is a stopping time, we have $\{T \leq n\} \in \mathcal{F}_n$ and $\{T \leq n - 1\}^c \in \mathcal{F}_{n-1} \subseteq \mathcal{F}_n$. Hence, $\{T = n\} = \{T \leq n\} \cap \{T \leq n - 1\}^c \in \mathcal{F}_n$ since $\mathcal{F}_n$ is a σ-field.

Suppose now that $\{T = n\} \in \mathcal{F}_n$ for each $n \geq 0$. Because $\{T = m\} \in \mathcal{F}_m \subseteq \mathcal{F}_n$ for $m \leq n$ and $\{T \leq n\} = \cup_{m=0}^{n} \{T = m\}$, we have $\{T \leq n\} \in \mathcal{F}_n$. ■

If T is an $\mathcal{F}_n$-stopping time,

$$\begin{aligned} \mathcal{F}_T &= \{A \in \mathcal{F} : A \cap \{T \leq n\} \in \mathcal{F}_n \text{ for all } n \geq 0\} \\ &= \{A \in \mathcal{F} : A \cap \{T = n\} \in \mathcal{F}_n \text{ for all } n \geq 0\}, \end{aligned} \tag{2.121}$$

is a sub-σ-field of $\mathcal{F}$ including events in $\mathcal{F}$ that "occur up to time T".

Proof: We show first that the collection of sets in the second definition of $\mathcal{F}_T$ is a σ-field, and then that the above two definitions of $\mathcal{F}_T$ are equivalent.

We have (1) $\Omega \cap \{T = n\} = \{T = n\} \in \mathcal{F}_n$ since T is a stopping time so that $\Omega \in \mathcal{F}_T$, (2) $A \in \mathcal{F}_T$ implies $A^c \in \mathcal{F}_T$ since $A^c \cap \{T = n\} = \{T = n\} \cap (A \cap \{T = n\})^c \in \mathcal{F}_n$, and (3) $A_i \in \mathcal{F}_T, i = 1, 2, \ldots$, implies $\cup_{i=1}^{\infty} A_i \in \mathcal{F}_T$ since

$$\left(\cup_{i=1}^{\infty} A_i\right) \cap \{T = n\} = \cup_{i=1}^{\infty} (A_i \cap \{T = n\}) \in \mathcal{F}_n.$$

Hence, the collection of sets in the second definition of $\mathcal{F}_T$ is a σ-field.

We show now that the above two definitions of $\mathcal{F}_T$ coincide. Take $A \in \mathcal{F}$ and assume that $A \cap \{T = n\} \in \mathcal{F}_n$ for all $n \geq 0$. Hence, $A \cap \{T \leq n\} = \cup_{k=1}^{n} (A \cap \{T = k\})$ is in $\mathcal{F}_n$. It remains to show that $A \cap \{T \leq n\} \in \mathcal{F}_n$ for all $n \geq 0$ implies $A \cap \{T = n\} \in \mathcal{F}_n$ for all $n \geq 0$. This follows from the equality $A \cap \{T = n\} = (A \cap \{T \leq n\}) \cap (A \cap \{T \leq n-1\})^c$ since $A \cap \{T \leq n\} \in \mathcal{F}_n$ and $A \cap \{T \leq n-1\} \in \mathcal{F}_{n-1} \subseteq \mathcal{F}_n$. ■

If $X = (X_0, X_1, \ldots)$ is a sequence of functions measurable from $(\Omega, \mathcal{F})$ to $(\Psi, \mathcal{G})$ and $\mathcal{F}_n = \sigma(X_0, X_1, \ldots, X_n)$, then the **hitting time**

$$T(\omega) = \inf\{n \geq 0 : X_n(\omega) \in B\}, \quad B \in \mathcal{G}, \tag{2.122}$$

is an $\mathcal{F}_n$-stopping time.

Proof: Take $B \in \mathcal{G}$. The random variable $T(\omega) = \inf\{n \geq 0 : X_n(\omega) \in B\}$ satisfies the condition $\{T \leq n\}$ if at least one of the events $\{X_0 \in B\}$, $\{X_0 \notin B, \ldots, X_{k-1} \notin B, X_k \in B\}$, $k = 1, \ldots, n$, occurs, that is,

$$\{T \leq n\} = \{X_0 \in B\} \cup \left(\cup_{k=1}^{n} \{X_0 \notin B, \ldots, X_{k-1} \notin B, X_k \in B\}\right)$$

so that $\{T \leq n\} \in \mathcal{F}_n$ because it consists of finite intersections and unions of events in $\mathcal{F}_k$ and $\mathcal{F}_k \subseteq \mathcal{F}_n$ for $k \leq n$. ■

A stopping time T is $\mathcal{F}_T$-measurable.

Proof: We need to show that $\{T \leq m\}$ is in $\mathcal{F}_T$ for each $m \geq 0$, that is, $\{T \leq m\} \cap \{T \leq n\} \in \mathcal{F}_n$ for all $n \geq 0$. This condition is satisfied because $\{T \leq m\} \cap \{T \leq n\} = \{T \leq m \wedge n\}$ and $\{T \leq m \wedge n\} \in \mathcal{F}_{m \wedge n} \subseteq \mathcal{F}_n$. ■

If S and T are $\mathcal{F}_n$-stopping times such that $S \leq T$, then $\mathcal{F}_S \subseteq \mathcal{F}_T$.

Proof: If $A \in \mathcal{F}_S$, then $A \cap \{S \leq n\} \in \mathcal{F}_n$ for all $n \geq 0$. We need to show that $A \cap \{S \leq n\} \in \mathcal{F}_n$ implies $A \cap \{T \leq n\} \in \mathcal{F}_n$ for all $n \geq 0$. This implication is true since the sets $A \cap \{S \leq n\}$ and $\{T \leq n\}$ are in $\mathcal{F}_n$, and $A \cap \{T \leq n\} = (A \cap \{S \leq n\}) \cap \{T \leq n\}$ holds if $S \leq T$. ■

If S and T are $\mathcal{F}_n$-stopping times, then $S \wedge T$ and $S \vee T$ are stopping times.

Proof: We have

$$\{S \wedge T \leq n\} = \{S \leq n\} \cup \{T \leq n\} \quad \text{and} \quad \{S \vee T \leq n\} = \{S \leq n\} \cap \{T \leq n\}$$

so that the events $\{S \wedge T \leq n\}$ and $\{S \vee T \leq n\}$ are in $\mathcal{F}_n$ for each $n \geq 0$ since S and T are stopping times. ■

If S and T are $\mathcal{F}_n$-stopping times, then $\mathcal{F}_{S \wedge T} = \mathcal{F}_S \cap \mathcal{F}_T$.

Proof: Because $S \wedge T \leq S, T$, we have $\mathcal{F}_{S \wedge T} \subset \mathcal{F}_S, \mathcal{F}_T$ by one of the above properties so that $\mathcal{F}_{S \wedge T} \subset \mathcal{F}_S \cap \mathcal{F}_T$. To prove the opposite inclusion, take $A \in \mathcal{F}_S \cap \mathcal{F}_T$ and note that, if $A \in \mathcal{F}_S \cap \mathcal{F}_T$, then

$$A \cap \{S \wedge T \leq n\} = A \cap (\{S \leq n\} \cup \{T \leq n\}) = (A \cap \{S \leq n\}) \cup (A \cap \{T \leq n\})$$

belongs to $\mathcal{F}_n$ so that $A \in \mathcal{F}_{S \wedge T}$. ■

If $X = (X_0, X_1, \ldots)$ is a sequence of functions measurable from $(\Omega, \mathcal{F})$ to $(\Psi, \mathcal{G})$, $\mathcal{F}_n = \sigma(X_0, X_1, \ldots, X_n)$, and T denotes an $\mathcal{F}_n$-stopping time, then X_T is $\mathcal{F}_T$-measurable.

Proof: Take $B \in \mathcal{G}$. Then

$$\{X_T \in B\} = \cup_{k=0}^{\infty} \{T = k\} \cap \{X_T \in B\} = \cup_{k=0}^{\infty} \{T = k\} \cap \{X_k \in B\} \in \mathcal{F}$$

since $\{T = k\} \in \mathcal{F}_k \subset \mathcal{F}$ and $\{X_k \in B\} \in \mathcal{F}$. It remains to show that X_T is $\mathcal{F}_T$-measurable, that is, that $A = \{X_T \in B\} \in \mathcal{F}_T$ for each $B \in \mathcal{G}$ or $A \cap \{T \leq n\}$ is in $\mathcal{F}_n$ for all $n \geq 0$. We have

$$\begin{aligned} A \cap \{T \leq n\} &= \{X_T \in B\} \cap \{T \leq n\} = \cup_{k=0}^{n} \{T = k\} \cap \{X_T \in B\} \\ &= \cup_{k=0}^{n} \{T = k\} \cap \{X_k \in B\} \in \mathcal{F}_n \end{aligned}$$

since $\mathcal{F}_n$ is a filtration on $\mathcal{F}$, X is $\mathcal{F}_n$-adapted, and $\mathcal{F}_k \subseteq \mathcal{F}_n$ for $k \leq n$. ■

If (1) $X = (X_1, X_2, \ldots)$ is a sequence of functions measurable from $(\Omega, \mathcal{F})$ to $(\Psi, \mathcal{G})$ that are iid, (2) P is the probability measure on $(\Omega, \mathcal{F})$, and (3) T is an a.s. finite $\mathcal{F}_n = \sigma(X_1, \ldots, X_n)$-stopping time, $n \geq 1$, then the sequence $Y = (Y_1, Y_2, \ldots)$, where

$$Y_n(\omega) = X_{T(\omega)+n}(\omega), \quad n \geq 1, \tag{2.123}$$

is independent of $\mathcal{F}_T$ and has the same distribution as X.

Proof: Because X_k are independent random variables, $(X_{n+1}, X_{n+2}, \ldots)$, $n \geq 1$, obtained from X by translating the time with n units, is independent of the past history, that is, of the σ-fields $\mathcal{F}_m$, $m \leq n$. We show that this property is preserved even if time is translated by a random amount T, provided that T is a finite stopping time, that is, that the sequence $Y = (X_{T+1}, X_{T+2}, \ldots)$ is independent of the σ-field $\mathcal{F}_T$. The Brownian motion and other processes considered in this book have this property, called the **strong Markov property**.

We show first that the measurable functions $X, Y : (\Omega, \mathcal{F}) \to (\Psi^\infty, \mathcal{G}^\infty)$ have the same distribution. For $B \in \mathcal{G}^\infty$, we have

$$\begin{aligned} P(Y \in B) &= P((X_{T+1}, X_{T+2}, \ldots) \in B) \\ &= \sum_{k=1}^{\infty} P((X_{k+1}, X_{k+2}, \ldots) \in B)\, P(T = k) \\ &= \sum_{k=1}^{\infty} P((X_1, X_2, \ldots) \in B)\, P(T = k) = P(X \in B). \end{aligned}$$

To show that Y is independent of $\mathcal{F}_T$, we need to prove that $P(A \cap \{Y \in B\}) = P(A)\,P(Y \in B)$, where $A \in \mathcal{F}_T$ and $B \in \mathcal{G}^\infty$. We have

$$\begin{aligned} P(A \cap \{Y \in B\}) &= \sum_{k=1}^{\infty} P(A \cap \{Y \in B\} \mid T = k)\, P(T = k) \\ &= \sum_{k=1}^{\infty} P(\tilde{A}_k \cap \{(X_{k+1}, X_{k+2}, \dots) \in B\}) \\ &= \sum_{k=1}^{\infty} P(\tilde{A}_k)\, P((X_{k+1}, X_{k+2}, \dots) \in B) = P(A)\, P(Y \in B) \end{aligned}$$

because the events $\tilde{A}_k = A \cap \{T = k\}$ and $(X_{k+1}, X_{k+2}, \dots) \in B$ are independent for each k and the sequences X and Y have the same distribution. Since A is an arbitrary member of $\mathcal{F}_T$, we conclude that Y is independent of $\mathcal{F}_T$. ■

Example 2.78: Let R be the random walk in Eq. 2.118, where X_i are iid Gaussian random variables with mean μ and standard deviation $\sigma > 0$. Let $T = \inf\{n \geq 0 : R_n \notin (-a, a)\}$ be the first time R leaves $(-a, a)$, $a > 0$. Figure 2.12 shows 1000 samples of the stopping time T and the corresponding histogram of these samples for $\mu = 0.2$, $\sigma = 1$, and $a = 5$. ◇

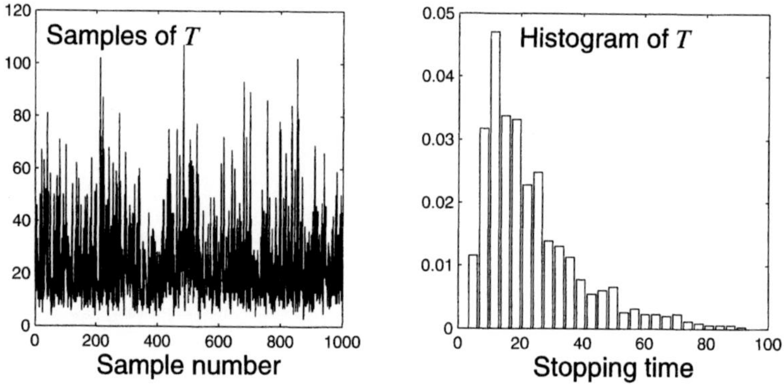

Figure 2.12. Samples and a histogram of T for $R_n = \sum_{i=1}^{n} X_i$, $X_1 \sim N(0.2, 1)$, and $a = 5$

2.17 Conditional expectation

We have defined the conditional density $f^{(1|2)}$ of a random vector $\boldsymbol{X}^{(1)}$ given that another vector $\boldsymbol{X}^{(2)}$ has a known value $\boldsymbol{z}$ (Eq. 2.92). This density can be used to calculate the expectation

$$E[\boldsymbol{X}^{(1)} \mid \boldsymbol{X}^{(2)} = \boldsymbol{z}] = \int_{\mathbb{R}^{d_1}} \boldsymbol{y}\, f^{(1|2)}(\boldsymbol{y} \mid \boldsymbol{z})\, d\boldsymbol{y} \qquad (2.124)$$

of the conditional vector $\boldsymbol{X}^{(1)} \mid (\boldsymbol{X}^{(2)} = z)$, that is, the expectation of $\boldsymbol{X}^{(1)}$ given the information $\boldsymbol{X}^{(2)} = z$.

We extend the definition of the conditional expectation $E[\boldsymbol{X}^{(1)} \mid \boldsymbol{X}^{(2)}]$ in Eq. 2.124 by considering information more general than $\boldsymbol{X}^{(2)} = z$. This section defines the conditional expectation $E[\cdot \mid \mathcal{G}]$, where $(\Omega, \mathcal{F}, P)$ is a probability space and $\mathcal{G}$ is a sub-σ-field of $\mathcal{F}$. The conditional expectation $E[\cdot \mid \mathcal{G}]$ is needed in many applications involving stochastic processes, as we will see in this and the subsequent chapters.

Example 2.79: Consider the experiment of rolling two dice. The sample space, σ-field, and probability measure for this experiment are $\Omega = \{\omega = (i, j) : i, j = 1, \ldots, 6\}$, all subsets of Ω, and $P(\{\omega\}) = 1/36$ for $\omega \in \Omega$, respectively. Define the random variables $X(\omega) = i + j$ and $Z(\omega) = i \wedge j$. The expected value of X is $E[X] = (1/36)[(1+1) + (1+2) + \cdots + (6+6)] = 7$.

Suppose we are told the value of Z and asked to determine the expectation of X given Z, that is, the expectation of the conditional variable $X \mid Z$ denoted by $E[X \mid Z]$. The conditional expectation $E[X \mid Z]$ is a simple random variable with density in Fig. 2.13 so that

$$E\{E[X \mid Z]\} = \sum_z E[X \mid Z = z] \, P(Z = z) = (12) P(Z = 6) \\ + (32/3) P(Z = 5) + \cdots + (52/11) P(Z = 1) = 7 = E[X].$$

is the expected value of $E[X \mid Z]$. ◇

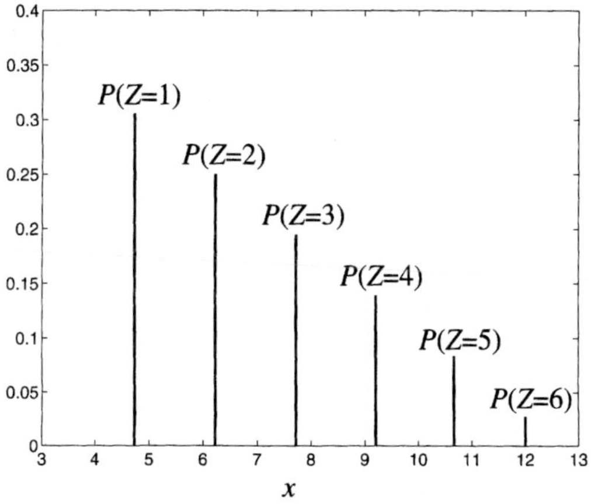

Figure 2.13. The density of $E[X \mid Z]$

Proof: If $Z = i \wedge j = \min(i, j) = 6$, there is a single outcome $(6, 6)$ so that $X \mid (Z = 6) = 6 + 6 = 12$. The density and the expectation of the conditional random variable

$X \mid (Z = 6)$ are $f^{(1|2)}(x) = f_{X|Z}(x \mid 6) = \delta(x - 12)$ and $E[X \mid Z = 6] = 12$, where δ denotes the Dirac delta function. The probability of the event $\{Z = 6\}$ is 1/36. If $Z = 5$, there are three equally likely outcomes, (6, 5), (5, 5), and (5, 6), so that $f_{X|Z}(x \mid 5) = (1/3)\delta(x - 10) + (2/3)\delta(x - 11)$ and $E[X \mid Z = 5] = (10)(1/3) + (11)(2/3) = 32/3$. The probability of the event $\{Z = 5\}$ is 3/36.

Note that the random variable $E[X \mid Z]$ does not have a density. The above expressions of $f^{(1|2)}$ are formal. They are used here and in some of the following examples for simplicity. ■

2.17.1 σ-field generated by a countable partition of Ω

Let $(\Omega, \mathcal{F}, P)$ be a probability space and let Λ_n be a countable collection of measurable sets that partition Ω, that is, $\cup_n \Lambda_n = \Omega$, $\Lambda_n \cap \Lambda_{n'} = \emptyset$ for $n \neq n'$, and $\Lambda_n \in \mathcal{F}$. Denote by $\mathcal{G}$ the σ-field generated by $\{\Lambda_n\}$. The members of $\mathcal{G}$ are unions of sub-collections of $\{\Lambda_n\}$. Let X be an integrable random variable on $(\Omega, \mathcal{F}, P)$.

If $P(\Lambda_n) > 0$, define the function $E[X \mid \mathcal{G}] : \Omega \to \mathbb{R}$ by

$$E[X \mid \mathcal{G}](\omega) = \sum_n E[X \mid \Lambda_n]\, 1_{\Lambda_n}(\omega), \quad \text{where} \tag{2.125}$$

$$E[X \mid \Lambda_n] = \int_\Omega X(\omega)\, P_{\Lambda_n}(d\omega) = \frac{1}{P(\Lambda_n)} \int_{\Lambda_n} X(\omega)\, P(d\omega). \tag{2.126}$$

Note: $E[X \mid \Lambda_n]$ is the conditional expectation of X with respect to Λ_n and $P_{\Lambda_n}(A) = P(A \mid \Lambda_n) = P(A \cap \Lambda_n)/P(\Lambda_n)$. ▲

The function $E[X \mid \mathcal{G}] : \Omega \to \mathbb{R}$ has the following properties.

- $E[X \mid \mathcal{G}]$ is $\mathcal{G}$-**measurable**.
- $E[X \mid \mathcal{G}]$ can be viewed as an **approximation** of X.
- $E[X \mid \mathcal{G}]$ satisfies the **defining relation**

$$\int_\Lambda X\, dP = \int_\Lambda E[X \mid \mathcal{G}]\, dP, \quad \forall \Lambda \in \mathcal{G}. \tag{2.127}$$

Proof: Since $1_{\Lambda_n} \in \mathcal{G}$ and $E[X \mid \Lambda_n]$ are some constants, the conditional expectation in Eqs. 2.125 and 2.126 is a $\mathcal{G}$-measurable function. We also note that $E[X \mid \mathcal{G}]$ is a discrete random variable on $(\Omega, \mathcal{G})$ whose probability density function is $f(x) = \sum_n P(\Lambda_n)\, \delta(x - E[X \mid \Lambda_n])$.

Since $E[X \mid \Lambda_n]$ can be viewed as a local average of X over the subset Λ_n, $E[X \mid \mathcal{G}]$ provides an approximation for the random variable X. The accuracy of this approximation depends on the refinement of the partition Λ_n. If the partition has a single element that necessarily coincides with the sample space, $E[X \mid \mathcal{G}]$ is equal to $E[X]$. If the sample space is countable and the partition is given by the elements $\{\omega\}$ of the sample space, $E[X \mid \mathcal{G}]$ is equal to X.

It remains to prove Eq. 2.127. If Λ coincides with the sample space Ω, then

$$\int_\Omega E[X \mid \mathcal{G}]\, dP = E\{E[X \mid \mathcal{G}]\} = \sum_n E[X \mid \Lambda_n]\, P(\Lambda_n) = \int_\Omega X\, dP = E[X]$$

because $E[X \mid \mathcal{G}]$ is equal to $E[X \mid \Lambda_n]$ on the atoms Λ_n of $\mathcal{G}$ with probabilities $P(\Lambda_n)$. If Λ is an arbitrary element of $\mathcal{G}$, that is, $\Lambda = \cup_{k\in J} \Lambda_k$ is a union of some of the sets partitioning Ω, we have (Eqs. 2.44 and 2.126)

$$\int_\Lambda X\, dP = \sum_{k\in J} \int_{\Lambda_k} X\, dP = \sum_{k\in J} E[X \mid \Lambda_k]\, P(\Lambda_k) \quad \text{and}$$

$$\begin{aligned}\int_\Lambda E[X \mid \mathcal{G}]\, dP &= \sum_{k\in J} \int_{\Lambda_k} E[X \mid \mathcal{G}]\, dP = \sum_{k\in J} \int_{\Lambda_k} \sum_n E[X \mid \Lambda_n]\, 1_{\Lambda_n}(\omega)\, P(d\omega) \\ &= \sum_{k\in J} \sum_n \int_{\Lambda_k \cap \Lambda_n} E[X \mid \Lambda_n]\, P(d\omega) = \sum_{k\in J} \int_{\Lambda_k} E[X \mid \Lambda_k]\, P(d\omega) \\ &= \sum_{k\in J} E[X \mid \Lambda_k]\, P(\Lambda_k).\end{aligned}$$

The above chain of equalities is based on properties of integrals of random variables and Eqs. 2.125 and 2.126. ■

Example 2.80: Let $(\Omega, \mathcal{F}, P)$ be a probability space, where $\Omega = [0, 1]$, $\mathcal{F} = \mathcal{B}([0, 1])$, and $P(d\omega) = d\omega$. Consider a random variable $X(\omega) = 2 + \sin(2\pi\,\omega)$, $\omega \in \Omega$, defined on this space. Let $\Lambda_1 = [0, 1/4)$, $\Lambda_2 = [1/4, 3/4)$, $\Lambda_3 = [3/4, 1)$, and $\Lambda_4 = \{1\}$ be a partition of Ω and denote by $\mathcal{G}$ the σ-field generated by this partition. We have

$$E[X \mid \mathcal{G}] = 2\,(1 + 1/\pi)\, 1_{\Lambda_1} + 2\, 1_{\Lambda_2} + 2\,(1 - 1/\pi)\, 1_{\Lambda_3} + x\, 1_{\Lambda_4},$$

where x can be any real number. The lack of uniqueness is of no concern since $P(\Lambda_4) = 0$. The conditional expectation will be defined as a class of random variables that are equal a.s. (Eq. 2.128). Figure 2.14 shows the measurable functions X and $E[X \mid \mathcal{G}]$ for $x = 3$. The conditional expectation $E[X \mid \mathcal{G}]$ represents an approximation of X whose accuracy depends on the partition $\{\Lambda_n\}$ of Ω. ◇

Proof: The conditional expectation $E[X \mid \mathcal{G}]$ is given by Eq. 2.125 with $E[X \mid \Lambda_n]$ in Eq. 2.126. For example, we have $E[X \mid \Lambda_1] = \frac{1}{1/4} \int_0^{1/4} [2 + \sin(2\pi\,\omega)]\, d\omega = 2(1 + 1/\pi)$ for Λ_1. ■

The **defining relation** (Eq. 2.127) is essential for calculations and reveals some useful properties of the conditional expectation.

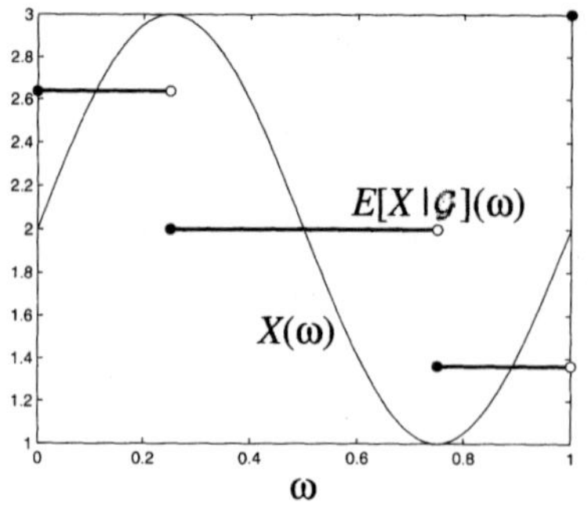

Figure 2.14. The conditional expectation $E[X \mid \mathcal{G}]$ of $X(\omega) = 2+\sin(2\pi\,\omega)$ for $\Lambda_1 = [0, 1/4)$, $\Lambda_2 = [1/4, 3/4)$, $\Lambda_3 = [3/4, 1)$, and $\Lambda_4 = \{1\}$

The integrals $\int_\Lambda X\,dP$ and $\int_\Lambda E[X \mid \mathcal{G}]\,dP$ are equal for every $\Lambda \in \mathcal{G}$ although their integrands, X and $E[X \mid \mathcal{G}]$, are $\mathcal{F}$-measurable and $\mathcal{G}$-measurable, respectively. These integrals represent local averages of the random variables X and $E[X \mid \mathcal{G}]$ over Λ.

$E[X \mid \mathcal{G}]$ is integrable because X is an integrable random variable.

$E[X \mid \mathcal{G}]$ is unique up to an equivalence.

Proof: Let h_i, $i = 1, 2$, be two $\mathcal{G}$-measurable functions satisfying the defining relation $\int_\Lambda X\,dP = \int_\Lambda h_i\,dP$, $\forall\Lambda \in \mathcal{G}$. Hence, h_1 and h_2 are two versions of the conditional expectation. The subset $\Lambda' = \{\omega : h_1(\omega) > h_2(\omega)\}$ is in $\mathcal{G}$ since the functions h_i are $\mathcal{G}$-measurable. The integral $\int_{\Lambda'}(h_1 - h_2)\,dP$ is zero by hypothesis so that $P(\Lambda') = 0$ since $h_1 - h_2 > 0$ on Λ'. Similar arguments show that the subset $\Lambda'' = \{\omega : h_1(\omega) < h_2(\omega)\}$ has probability zero so that $P(\{\omega : h_1(\omega) \neq h_2(\omega)\}) = 0$ or $h_1 = h_2$ a.s., that is, h_1 and h_2 can differ only on a set of probability zero ([40], Section 9.1). The conditional expectation $E[X \mid \mathcal{G}]$ is represented by all $\mathcal{G}$-measurable random variables satisfying the defining relation in Eq. 2.127. The random variables in this class are equal a.s. and represent versions of the conditional expectation. ■

Example 2.81: Let $\Lambda_n = \{Z = n\}$, $n = 1, \ldots, 6$, be subsets of the sample space Ω of the experiment in Example 2.79. The σ-field $\mathcal{G} = \sigma(Z)$ consists of unions of Λ_n's. The conditional expectation, $E[X \mid Z]$, calculated previously coincides with $E[X \mid \mathcal{G}]$ in Eq. 2.125, where $X(\omega) = i + j$. ◇

Proof: The conditional expectation $E[X \mid \mathcal{G}]$ is constant on Λ_n and is equal to $E[X \mid \Lambda_n]$. For example, $E[X \mid \Lambda_6]$ is equal to (1/(1/36))(12)(1/36)=12 and $E[X \mid \Lambda_5]$ is (1/(3/36))(11+10+11)(1/36)=32/3. Hence, the conditional expectations $E[X \mid Z]$ and $E[X \mid \sigma(Z)]$ coincide. ■

Example 2.82: Let X be a random variable with distribution function F and finite mean defined on a probability space $(\Omega, \mathcal{F}, P)$. If $a \in \mathbb{R}$ is such that $F(a) \in (0, 1)$, then

$$E[X \mid \mathcal{G}] = \frac{\int_{-\infty}^{a} x\, dF(x)}{F(a)} 1_{(-\infty,a]} + \frac{\int_{a}^{\infty} x\, dF(x)}{1 - F(a)} 1_{(-\infty,a]^c}$$

for $\mathcal{G} = \{\emptyset, \Omega, A, A^c\}$, where $A = X^{-1}((-\infty, a]) \in \mathcal{F}$. ◇

Proof: The events A and A^c partition Ω so that $E[X \mid \mathcal{G}]$ is given by Eq. 2.126 with

$$E[X \mid A] = \frac{1}{P(A)} \int_A X\, dP = \frac{1}{F(a)} \int_{-\infty}^{a} x\, dF(x) \quad \text{and}$$
$$E[X \mid A^c] = \frac{1}{P(A^c)} \int_{A^c} X\, dP = \frac{1}{1 - F(a)} \int_{a}^{\infty} x\, dF(x).$$

The expectation of $E[X \mid \mathcal{G}]$ is

$$E\{E[X \mid \mathcal{G}]\} = E[X \mid A]\, P(A) + E[X \mid A^c]\, P(A^c) = E[X]$$

so that the random variables $E[X \mid \mathcal{G}]$ and X have the same expectation. ■

2.17.2 General σ-field

Let X be an integrable random variable defined on a probability space $(\Omega, \mathcal{F}, P)$ and $\mathcal{G}$ denote a sub-σ-field of $\mathcal{F}$.

The **conditional expectation** $E[X \mid \mathcal{G}]$ **of** X **with respect to** $\mathcal{G}$ is defined to be the class of $\mathcal{G}$-**measurable** functions satisfying the **defining relation** (Eq. 2.127)

$$\int_\Lambda X\, dP = \int_\Lambda E[X \mid \mathcal{G}]\, dP, \quad \forall \Lambda \in \mathcal{G}. \qquad (2.128)$$

Note: The definition is meaningful because, if $E[|X|] < \infty$ and $\mathcal{G}$ is a sub-σ-field of $\mathcal{F}$, there exists a unique equivalence class of integrable random variables, denoted by $E[X \mid \mathcal{G}]$, that is $\mathcal{G}$-measurable and satisfies the defining condition of Eq. 2.128 for all $\Lambda \in \mathcal{G}$ ([40], Theorem 9.1.1, p. 297). ▲

The following properties result directly from Eq. 2.128.

- If $\mathcal{G} = \{\emptyset, \Omega\}$, then $E[X \mid \mathcal{G}] = E[X]$.
- If $\mathcal{G} = \mathcal{F}$, then $E[X \mid \mathcal{F}] = X$ a.s.
- $E\{E[X \mid \mathcal{G}]\} = E[X]$.

Proof: If $\mathcal{G} = \{\emptyset, \Omega\}$, then $E[X \mid \mathcal{G}]$ is constant on Ω so that Eq. 2.128 gives $E[X] = E[X \mid \mathcal{G}]$ for $\Lambda = \Omega$.

If $\mathcal{G} = \mathcal{F}$, then $E[X \mid \mathcal{G}] = E[X \mid \mathcal{F}]$ is $\mathcal{F}$-measurable and the defining relation yields $\int_\Lambda (X - E[X \mid \mathcal{F}])\, dP = 0$ for all $\Lambda \in \mathcal{F}$ so that $X = E[X \mid \mathcal{F}]$ a.s.

The defining relation gives $E\{E[X \mid \mathcal{G}]\} = E[X]$ for $\Lambda = \Omega$. ■

We list here some properties of the conditional expectation $E[X \mid \mathcal{G}]$ that are used frequently in calculations.

$$E\left[(X - E[X \mid \mathcal{G}])\, Z\right] = 0, \quad \forall Z \in \mathcal{G}. \tag{2.129}$$

Proof: If $Z = 1_\Lambda$, $\Lambda \in \mathcal{G}$, then Eq. 2.129 holds because $E\{(X - E[X \mid \mathcal{G}])\, 1_\Lambda\} = \int_\Lambda (X - E[X \mid \mathcal{G}])\, dP$ and this integral is zero by the defining relation. Eq. 2.129 is also valid for a simple random variable $Z = \sum_n b_n\, 1_{\Lambda_n}$, $\Lambda_n \in \mathcal{G}$, by the linearity of expectation. The extension to an arbitrary random variable Z results from the representation of Z by a difference of two positive random variables, which can be defined as limits of simple random variables ([40], Section 9.1). ■

The conditional expectation $E[X \mid \mathcal{G}]$ is **the projection of** X **on** $\mathcal{G}$ and $X - E[X \mid \mathcal{G}]$ **is orthogonal to** $\mathcal{G}$ (Eq. 2.129, Fig. 2.15). Moreover, $E[X \mid \mathcal{G}]$ represents the **best mean square (m.s.) estimator** of X with respect to the information content of $\mathcal{G}$ ([98], Sections 4.3 and 4.4).

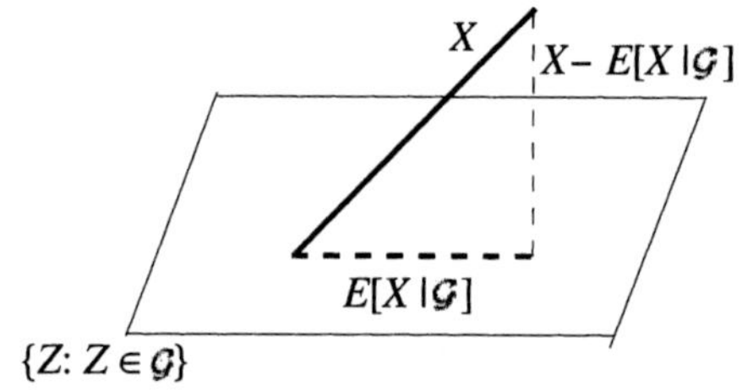

Figure 2.15. A geometrical interpretation of conditional expectation

Note: The function $(X_1, X_2) \mapsto \langle X_1, X_2 \rangle = E[X_1\, X_2]$ is the inner product defined on L_2 (Section 2.6). If $\langle X_1, X_2 \rangle = 0$, X_1 and X_2 are said to be **orthogonal**. A random variable $Y \in L_2(\Omega, \mathcal{F}, P)$ is orthogonal to $L_2(\Omega, \mathcal{G}, P)$ if $\langle Y, Z \rangle = 0$ for all $Z \in L_2(\Omega, \mathcal{G}, P)$. Eq. 2.129 states that $X - E[X \mid \mathcal{G}]$ is orthogonal to $L_2(\Omega, \mathcal{G}, P)$. According to the orthogonal projection theorem in Eq. 2.51, $E[X \mid \mathcal{G}]$ is the best m.s. estimator of X given the information content of $\mathcal{G}$. ▲

$$E[X\, Z \mid \mathcal{G}] = Z\, E[X \mid \mathcal{G}] \quad \text{a.s.}, \quad \forall Z \in \mathcal{G}. \tag{2.130}$$

Proof: The random variables $E[X\, Z \mid \mathcal{G}]$ and $Z\, E[X \mid \mathcal{G}]$ are $\mathcal{G}$-measurable by the definition of the conditional expectation and the fact that the product of two $\mathcal{G}$-measurable functions, the functions Z and $E[X \mid \mathcal{G}]$, is $\mathcal{G}$-measurable. If $Z = 1_\Delta$, $\Delta \in \mathcal{G}$, Eq. 2.130 holds a.s. because for $\Lambda \in \mathcal{G}$ the left and the right sides of this equation are

$$\int_\Lambda E[X\, 1_\Delta \mid \mathcal{G}]\, dP = \int_\Lambda X\, 1_\Delta\, dP = \int_{\Lambda \cap \Delta} X\, dP \quad \text{and}$$
$$\int_\Lambda 1_\Delta\, E[X \mid \mathcal{G}]\, dP = \int_{\Lambda \cap \Delta} E[X \mid \mathcal{G}]\, dP = \int_{\Lambda \cap \Delta} X\, dP,$$

respectively, by the defining relation. Hence, Eq. 2.130 holds for a simple random variable Z because the conditional expectation is a linear function (Eq. 2.131). The extension to general random variables results from the representation of an arbitrary random variable by the difference of two positive random variables, which can be defined as limits of simple random variables ([40], Theorem 9.1.3, p. 300). ■

We now list essential properties of the conditional expectation . These properties hold a.s., and are similar to the properties of the expectation operator.

If X and X_n are integrable random variables, then ([151], Section 10.3):

- $X \in \mathcal{G}$ implies $E[X \mid \mathcal{G}] = X$.
- $E[a\,X_1 + b\,X_2 \mid \mathcal{G}] = a\,E[X_1 \mid \mathcal{G}] + b\,E[X_2 \mid \mathcal{G}]$ (**linearity**).
- $X_1 \leq X_2$ implies $E[X_1 \mid \mathcal{G}] \leq E[X_2 \mid \mathcal{G}]$ (**monotonicity**).
- $|E[X \mid \mathcal{G}]| \leq E[|X| \mid \mathcal{G}]$ (**modulus inequality**).
- $X_n \uparrow (\downarrow) X$ implies $E[X_n \mid \mathcal{G}] \uparrow (\downarrow) E[X \mid \mathcal{G}]$ (**monotone convergence**).
- $|X_n| \leq Y$, $E[Y] < \infty$, and $X_n \to X$ imply $E[X_n \mid \mathcal{G}] \to E[X \mid \mathcal{G}]$ (**dominated convergence**). (2.131)

Cauchy-Schwarz and **Jensen** inequalities are, respectively,

$$(E[X\,Y \mid \mathcal{G}])^2 \leq E[X^2 \mid \mathcal{G}]\,E[Y^2 \mid \mathcal{G}] \quad \text{and} \tag{2.132}$$

$$g(E[X \mid \mathcal{G}]) \leq E[g(X) \mid \mathcal{G}] \quad \text{for a convex function } g. \tag{2.133}$$

If $\mathcal{G}_1$ and $\mathcal{G}_2$ are sub-σ-fields of $\mathcal{F}$ such that $\mathcal{G}_1 \subset \mathcal{G}_2$, then we can perform the following **change of fields in the conditional expectation**

$$E[X \mid \mathcal{G}_1] = E[X \mid \mathcal{G}_2] \Longleftrightarrow E[X \mid \mathcal{G}_2] \in \mathcal{G}_1 \quad \text{and} \tag{2.134}$$

$$E\{E[X \mid \mathcal{G}_2] \mid \mathcal{G}_1\} = E[X \mid \mathcal{G}_1] = E\{E[X \mid \mathcal{G}_1] \mid \mathcal{G}_2\}. \tag{2.135}$$

Proof: If $E[X \mid \mathcal{G}_1] = E[X \mid \mathcal{G}_2]$ holds (Eq. 2.134), the defining relation gives $\int_\Lambda X\,dP = \int_\Lambda E[X \mid \mathcal{G}_1]\,dP$ for any $\Lambda \in \mathcal{G}_1$ and the latter integral is equal to $\int_\Lambda E[X \mid \mathcal{G}_2]\,dP$ by hypothesis. Hence, we have $\int_\Lambda X\,dP = \int_\Lambda E[X \mid \mathcal{G}_2]\,dP$ for all $\Lambda \in \mathcal{G}_1$, so that $E[X \mid \mathcal{G}_2]$ is $\mathcal{G}_1$-measurable. Conversely, if $E[X \mid \mathcal{G}_2]$ is $\mathcal{G}_1$-measurable, then $E\{E[X \mid \mathcal{G}_2] \mid \mathcal{G}_1\} = E[X \mid \mathcal{G}_2]$ by the first statement in Eq. 2.131. We also have $E[X \mid \mathcal{G}_1] = E\{E[X \mid \mathcal{G}_2] \mid \mathcal{G}_1\}$ by Eq. 2.135.

For Eq. 2.135 take $\Lambda \in \mathcal{G}_1 \subset \mathcal{G}_2$ and apply the defining relation in Eq. 2.128 twice, first for the random variable $E[X \mid \mathcal{G}_2]$ and second for the random variable X, that is,

$$\int_\Lambda E\{E[X \mid \mathcal{G}_2] \mid \mathcal{G}_1\}\,dP = \int_\Lambda E[X \mid \mathcal{G}_2]\,dP = \int_\Lambda X\,dP = \int_\Lambda E[X \mid \mathcal{G}_1]\,dP.$$

This gives the first equality of Eq. 2.135. The second equality of this equation results since $E[X \mid \mathcal{G}_1]$ is $\mathcal{G}_1$-measurable and, therefore, $\mathcal{G}_2$-measurable. Hence, $E\{E[X \mid \mathcal{G}_1] \mid \mathcal{G}_2\} = E[X \mid \mathcal{G}_1]$ by the first property of Eq. 2.131. ■

Example 2.83: Let $R = (R_0, R_1, \ldots)$ be the random walk in Eq. 2.118 and $\mathcal{F}_m$ be the σ-field generated by $(R_0, R_1, \ldots, R_m)$, $m \geq 0$. The average of the random walk at a time $n > m$ given the past history $(R_0, R_1, \ldots, R_m)$, that is, the σ-field $\mathcal{F}_m$, is $E[R_n \mid \mathcal{F}_m] = R_m + (n-m)\, E[X_1]$, where X_k are independent identically distributed random variables in L_1. ◇

Proof: We have

$$E[R_n \mid \mathcal{F}_m] = E\left[R_m + \sum_{k=m+1}^{n} X_k \mid \mathcal{F}_m\right] = E[R_m \mid \mathcal{F}_m] + E\left[\sum_{k=m+1}^{n} X_k \mid \mathcal{F}_m\right] = R_m + \sum_{k=m+1}^{n} E[X_k]$$

because the conditional expectation is a linear operator, R_m is $\mathcal{F}_m$-measurable, and X_k is independent of $\mathcal{F}_m$ for $k > m$. ■

Example 2.84: Let $(\Omega, \mathcal{F}, P)$ be a probability space, $\{\Lambda_n\}$ denote a countable partition of Ω such that $\Lambda_n \in \mathcal{F}$ for each n, and $\mathcal{G}$ be the σ-field generated by $\{\Lambda_n\}$. Then the random variable $X - E[X \mid \mathcal{G}]$ is orthogonal to $\mathcal{G}$, where the conditional expectation $E[X \mid \mathcal{G}]$ is equal to $\sum_n E[X \mid \Lambda_n]\, 1_{\Lambda_n}$. ◇

Proof: The conditional expectation $E[X \mid \mathcal{G}]$ is given by Eqs. 2.125 and 2.126, where the members of $\mathcal{G}$ are $\cup_{n \in J} \Lambda_n\}$ with $J \subseteq \{1, 2, \ldots\}$. It remains to show that $X - E[X \mid \mathcal{G}]$ is orthogonal to $\mathcal{G}$, that is, that $E[(X - E[X \mid \mathcal{G}])\, Z] = 0$ for all $Z \in \mathcal{G}$. Because $Z \in \mathcal{G}$, we have $Z = \sum_n \alpha_n\, 1_{\Lambda_n}$ so that

$$E[X\,Z] - E\{E[X \mid \mathcal{G}]\, Z\} = \sum_n \alpha_n\, E[X\, 1_{\Lambda_n}] - \sum_{m,n} E[X \mid \Lambda_n]\, \alpha_m\, E[1_{\Lambda_m}\, 1_{\Lambda_n}] = 0.$$

The above expression is zero since $E[X\, 1_{\Lambda_n}] = E[X \mid \Lambda_n]\, P(\Lambda_n)$ (Eq. 2.126) and $E[1_{\Lambda_m}\, 1_{\Lambda_n}] = P(\Lambda_m \cap \Lambda_n) = \delta_{mn}\, P(\Lambda_n)$ so that the second summation becomes $\sum_n E[X \mid \Lambda_n]\, \alpha_n\, P(\Lambda_n)$. ■

Example 2.85: Let X and Z be random variables in $L_2(\Omega, \mathcal{F}, P)$. We have seen that the conditional expectation $E[X \mid \sigma(Z)] = E[X \mid Z]$ is the best m.s. estimator of X with respect to the information content of $\sigma(Z)$. The best m.s. linear estimator of X is

$$\hat{X} = E[X] + \frac{E[X\,Z] - E[X]\,E[Z]}{E[Z^2] - E[Z]^2}\,(Z - E[Z])\,.$$

This estimator has the functional form of the conditional Gaussian variable in Example 2.64, represents the linear regression of X with respect to Z, and becomes $\hat{X} = E[X]$ if X and Z are not correlated. ◇

Proof: The general form of a linear estimator of X given Z is $\hat{X} = a\,Z + b$, where a, b are some constants. Hence, $\hat{X}$ is $\sigma(Z)$-measurable. We require that X satisfies the defining and the orthogonality conditions. These conditions give $E[X] = E[\hat{X}]$ for $\Lambda = \Omega$ (Eq. 2.128) and $\langle X - \hat{X}, Z\rangle = 0$ (Eq. 2.129) or $E[X] = a\,E[Z] + b$ and $E[X\,Z] = a\,E[Z^2] + b\,E[Z]$. The solution a, b of these equations introduce in $a\,Z + b$ gives $\hat{X}$. The resulting estimator has the property that its m.s. error $E[(\hat{X} - X)^2]$ is smaller than the error $E[(a\,Z + b - X)^2]$ of any other linear estimator. ■

The conditional expectation can be used to generalize the definition of conditional probability in Eq. 2.16. Let $(\Omega, \mathcal{F}, P)$ be a probability space and consider a sub-σ-field $\mathcal{G}$ of $\mathcal{F}$.

The **conditional probability with respect to** $\mathcal{G}$ is

$$P(A \mid \mathcal{G}) = E[1_A \mid \mathcal{G}], \quad A \in \mathcal{F}. \tag{2.136}$$

Note: The definition is meaningful since 1_A is an $\mathcal{F}$-measurable function. Because conditional probability is a conditional expectation, $P(A \mid \mathcal{G})$ is $\mathcal{G}$-measurable and

$$\int_\Lambda P(A \mid \mathcal{G})\,dP = \int_\Lambda E[1_A \mid \mathcal{G}]\,dP = \int_\Lambda 1_A\,dP = P(A \cap \Lambda), \quad \forall \Lambda \in \mathcal{G},$$

where $\int_\Lambda E[1_A \mid \mathcal{G}]\,dP = \int_\Lambda 1_A\,dP$ holds by the defining relation of conditional expectation. ▲

Example 2.86: Let $(\Omega, \mathcal{F}, P)$ be a probability space and $A, B \in \mathcal{F}$ such that $P(B) > 0$ and $P(B^c) > 0$. The conditional probability in Eq. 2.136 is

$$P(A \mid \mathcal{G}) = \frac{P(A \cap B)}{P(B)}\,1_B + \frac{P(A \cap B^c)}{P(B^c)}\,1_{B^c} \tag{2.137}$$

for the σ-field $\mathcal{G} = \{\emptyset, \Omega, B, B^c\}$ generated by B. Hence, $P(A \mid \mathcal{G})$ is a random variable taking the values $P(A \cap B)/P(B)$ and $P(A \cap B^c)/P(B^c)$ with probabilities $P(B)$ and $P(B^c)$, respectively. This observation is consistent with the definition of conditional probability in Eq. 2.16. ◇

Proof: We have $P(A \mid \mathcal{G}) = E[1_A \mid \mathcal{G}]$ by Eq. 2.136 and (Eqs. 2.125 and 2.126)

$$\begin{aligned} E[1_A \mid \mathcal{G}] &= E[1_A \mid B]\,1_B + E[1_A \mid B^c]\,1_{B^c} \\ &= \left(\frac{1}{P(B)}\int_B 1_A\,dP\right)1_B + \left(\frac{1}{P(B^c)}\int_{B^c} 1_A\,dP\right)1_{B^c}. \end{aligned}$$

If $\mathcal{G} = \mathcal{F}$, the corresponding conditional probability is $P(A \mid \mathcal{F}) = E[1_A \mid \mathcal{F}] = 1_A$, $A \in \mathcal{F}$, because 1_A is $\mathcal{F}$-measurable. Hence, $P(A \mid \mathcal{F})$ is equal to one and zero for $\omega \in A$ and $\omega \in A^c$, respectively. ■

2.18 Martingales

Let $(\Omega, \mathcal{F}, P)$ be a probability space endowed with a filtration $(\mathcal{F}_n)_{n\geq 0}$ such that $\mathcal{F}_\infty = \cup_{n=1}^{\infty} \mathcal{F}_n \subseteq \mathcal{F}$ and let X_n, $n = 0, 1, \ldots$, be random variables defined on $(\Omega, \mathcal{F}, P)$. The sequence $X = (X_0, X_1, X_2, \ldots)$ is referred to as a **discrete time stochastic process** or just a **stochastic process**. The numerical sequence $(X_0(\omega), X_1(\omega), \ldots)$, $\omega \in \Omega$, is called a **sample** or **sample path** of X. Continuous time stochastic processes are discussed in Chapter 3.

The sequence X is an $\mathcal{F}_n$-**martingale** if

1. $E[|X_n|] < \infty, \quad n = 0, 1, \ldots,$
2. X is **adapted** to $\mathcal{F}_n$, that is, $X_n \in \mathcal{F}_n$ for each $n \geq 0$, and
3. $E[X_n \mid \mathcal{F}_m] = X_m, \quad 0 \leq m \leq n.$ (2.138)

Note: If the equality in Eq. 2.138 is replaced by $\geq$ and $\leq$, then X is said to be an $\mathcal{F}_n$-**submartingale** and $\mathcal{F}_n$-**supermartingale**, respectively. The filtration $\mathcal{F}_n$ need not be mentioned if there is no confusion about it.

If the random variables X_n are in $L_p(\Omega, \mathcal{F}, P)$, X is called a p-**integrable** martingale, submartingale, or supermartingale depending on the last condition of Eq. 2.138. If $p = 2$, then X is said to be a **square integrable** martingale, submartingale, or supermartingale. ▲

Suppose $m \geq 1$ rounds have been completed in a game with unit stake. We can think of X_n as our total winnings (or losses) after n rounds of a game so that $X_n - X_m$, $n > m$, gives our net total winnings in the future rounds $m + 1, \ldots, n$. The best m.s. estimator of $X_n - X_m$, $n > m$, given our knowledge $\mathcal{F}_m$ after m rounds is $E[X_n - X_m \mid \mathcal{F}_m]$, where $\mathcal{F}_m = \sigma(X_1, \ldots, X_m)$, $m \geq 1$, and $\mathcal{F}_0 = \{\emptyset, \Omega\}$. If X is a martingale, then $E[X_n - X_m \mid \mathcal{F}_m] = 0$, that is, our average fortune $E[X_n \mid \mathcal{F}_m]$ at a future time n is equal to our current fortune X_m. This game can be generalized by allowing stakes other then one. Let A_i be our stake at game $i = 0, 1, \ldots$, where $A_0 = 0$. Because we decide our stake for round $m + 1$ on our knowledge $\mathcal{F}_m$ after m games, A_{m+1} is $\mathcal{F}_m$-measurable. Processes with this property are said to be **predictable** processes. The sequence $A_1, A_2, \ldots$ is called gambling strategy. The summation,

$$M_n = \sum_{i=1}^{n} A_i \,(X_i - X_{i-1}), \tag{2.139}$$

gives our total winnings after $n \geq 1$ games, where $X_0 = 0$. The sequence $M = (M_0, M_1, \ldots)$ defined by Eq. 2.139 for $n \geq 1$ and $M_0 = 0$ is a discrete version of the stochastic integral considered in Chapter 4. The integrand A_i is a predictable process and the integrator X_i is a martingale.

Example 2.87: Consider the random walk in Eq. 2.118. We have shown that

$$(R_0, R_1, \ldots, R_n) \quad \text{and} \quad (R_1, \ldots, R_n)$$

are $\sigma(X_1, \ldots, X_n)$-measurable for $n \geq 1$ so that $R = (R_0, R_1, R_2, \ldots)$ is adapted to the filtration $(\mathcal{F}_n)_{n\geq 0}$ given by $\mathcal{F}_n = \sigma(X_1, \ldots, X_n)$, $n = 1, 2, \ldots$, and $\mathcal{F}_0 = \{\emptyset, \Omega\}$. If the random variables X_i have finite mean, R is an $\mathcal{F}_n$-supermartingale, martingale, or submartingale depending on the value of $E[X_1]$. ◇

Proof: It remains to prove the last condition in Eq. 2.138. For $n \geq m \geq 0$, we have

$$E[R_n \mid \mathcal{F}_m] = R_m + \sum_{i=m+1}^{n} E[X_i] = R_m + (n-m)\, E[X_1]$$

by Example 2.83. If $E[X_1]$ is negative, zero, or positive, then R is a supermartingale, martingale, or submartingale, respectively. ■

Example 2.88: Let $R = (R_0, R_1, \ldots)$ be the random walk in Eq. 2.118. If the random variables X_i are in $L_2(\Omega, \mathcal{F}, P)$ and have mean zero, the sequence $S_n = R_n^2 = \sum_{i,j=1}^{n} X_i\, X_j$, $n \geq 1$, with $S_0 = 0$ is an $\mathcal{F}_n$-submartingale and $S_n - n\, E[X_1^2]$ is an $\mathcal{F}_n$-martingale. ◇

Proof: The sequences S_n and $S_n - n\, E[X_1^2]$ have the first two properties in Eq. 2.138. We also have

$$\begin{aligned} E[S_{n+1} \mid \mathcal{F}_n] &= E[(R_n + X_{n+1})^2 \mid \mathcal{F}_n] = E[R_n^2 + 2\, R_n\, X_{n+1} + X_{n+1}^2 \mid \mathcal{F}_n] \\ &= R_n^2 + 2\, R_n\, E[X_{n+1}] + E[X_{n+1}^2] = S_n + E[X_1^2] \geq S_n \end{aligned}$$

since R_n is $\mathcal{F}_n$-measurable, X_{n+1} is independent of $\mathcal{F}_n$, and $E[X_{n+1}] = 0$. Hence, $S = (S_0, S_1, \ldots)$ is a submartingale. We also have

$$E[S_{n+1} - (n+1)\, E[X_1^2] \mid \mathcal{F}_n] = (S_n + E[X_1^2]) - (n+1)\, E[X_1^2] = S_n - n\, E[X_1^2]$$

so that $S_n - n\, E[X_1^2]$ is a martingale. ■

Example 2.89: Let $Y_n = \exp(\sqrt{-1}\, u\, R_n)/\varphi_n(u)$, $u \in \mathbb{R}$, where $R = (R_0, R_1, \ldots)$ is the random walk in Eq. 2.118 and $\varphi_n(u) = E[\exp(\sqrt{-1}\, u\, R_n)]$, $u \in \mathbb{R}$, denotes the characteristic function of R_n. Then Y_n is a martingale. ◇

Proof: The characteristic function of R_n is $\varphi_n(u) = \varphi(u)^n$, where φ denotes the characteristic function of X_1. The random variables Y_n are $\mathcal{F}_n = \sigma(R_1, \ldots, R_n)$-measurable for each $n \geq 0$ and have unit mean. For $n > m$, we have

$$\begin{aligned} E[Y_n \mid \mathcal{F}_m] &= \frac{1}{\varphi_n(u)}\, E\left[e^{\sqrt{-1}\, u\, (R_m + \sum_{k=m+1}^{n} X_k} \mid \mathcal{F}_m\right] \\ &= \frac{e^{\sqrt{-1}\, u\, R_m}}{\varphi_n(u)}\, E\left[e^{\sqrt{-1}\, u\, \sum_{k=m+1}^{n} X_k}\right] = \frac{e^{\sqrt{-1}\, u\, R_m}}{\varphi_n(u)}\, \varphi(u)^{n-m} = Y_m \end{aligned}$$

since $R_m \in \mathcal{F}_m$ and the random variables X_k, $k \geq m+1$, are independent of $\mathcal{F}_m$. ■

2.18.1 Properties

Most of the properties given in this section follow directly from the definition of submartingale, martingale, and supermartingale processes.

- The expectation of a submartingale, martingale, and supermartingale $X = (X_0, X_1, \ldots)$ is an increasing, constant, and decreasing function of time.
- X is a martingale if it is both a submartingale and a supermartingale.
- If X is a submartingale, then $-X$ is a supermartingale.

Proof: The expectation of Eq. 2.138 is $E\{E[X_n \mid \mathcal{F}_m]\} = E[X_m]$, $n \geq m \geq 0$. We also have $E\{E[X_n \mid \mathcal{F}_m]\} = E[X_n]$ (Eq. 2.128) so that $E[X_n]$ is constant. If X is a submartingale, then $E[X_n \mid \mathcal{F}_m] \geq X_m$ (Eq. 2.138) and the expectation of this inequality is $E[X_n] \geq E[X_m]$. The opposite inequality results for a supermartingale. Martingales model fair games because the average fortune $E[X_n]$ of a player does not change in time. Submartingales and supermartingales correspond to super-fair and unfair games, respectively.

If X is both a submartingale and a supermartingale, it must satisfy simultaneously the conditions $E[X_n \mid \mathcal{F}_m] \geq X_m$ and $E[X_n \mid \mathcal{F}_m] \leq X_m$ for $n \geq m$, which imply $E[X_n \mid \mathcal{F}_m] = X_m$.

The condition $E[X_n \mid \mathcal{F}_m] \geq X_m$, $n \geq m$, implies $E[(-X_n) \mid \mathcal{F}_m] \leq (-X_m)$ which gives the last property. ■

The third condition in Eq. 2.138 can be replaced by

$$E[X_{n+1} \mid \mathcal{F}_n] = X_n, \quad n \geq 0. \tag{2.140}$$

Proof: We have $E[X_{n+2} \mid \mathcal{F}_n] = E\{E[X_{n+2} \mid \mathcal{F}_{n+1}] \mid \mathcal{F}_n\} = E[X_{n+1} \mid \mathcal{F}_n] = X_n$ by Eq. 2.135 and the fact that X is an $\mathcal{F}_n$-martingale. ■

If $X = (X_0, X_1, \ldots)$ is an $\mathcal{F}_n$-square integrable martingale, the sequence $Y = (Y_0, Y_1, \ldots)$ defined by $Y_0 = X_0 - E[X_0]$ and $Y_n = X_n - X_{n-1}$, $n \geq 1$, has the orthogonality property

$$E[Y_i \, Y_j] = 0 \quad \text{for } i \neq j. \tag{2.141}$$

Proof: We have $Y_n \in L_1$, $Y_n \in \mathcal{F}_n$, and

$$E[Y_{n+1} \mid \mathcal{F}_n] = E[X_{n+1} - X_n \mid \mathcal{F}_n] = X_n - X_n = 0$$

since $E[X_{n+1} \mid \mathcal{F}_n] = X_n$ and $E[X_n \mid \mathcal{F}_n] = X_n$. For $n > m$,

$$E[Y_n \, Y_m] = E\{E[Y_n \, Y_m \mid \mathcal{F}_m]\} = E\{Y_m \, E[Y_n \mid \mathcal{F}_m]\} = 0$$

since $Y_m \in \mathcal{F}_m$ and $E[Y_n \mid \mathcal{F}_m] = 0$. ■

Doob decomposition. If $X = (X_0, X_1, \ldots)$ is an $\mathcal{F}_n$-submartingale, then there is an $\mathcal{F}_n$-martingale M_n and a process A_n such that $A_0 = 0$, $A_n \uparrow$, $A_n \in \mathcal{F}_{n-1}$, $n \geq 1$, and the representation

$$X_n = A_n + M_n, \quad n \geq 0, \quad \text{is valid and unique.} \tag{2.142}$$

Proof: The notation $A_n \uparrow$ means $A_n \geq A_{n-1}$. We say that A is **predictable** because A_n is $\mathcal{F}_{n-1}$-measurable. The representation in Eq. 2.142 shows that submartingales have a predictable part A that can be told ahead of time and an unpredictable part M.

We first show that if the representation in Eq. 2.142 exists, it is unique. We have $E[X_n \mid \mathcal{F}_{n-1}] = E[A_n \mid \mathcal{F}_{n-1}] + E[M_n \mid \mathcal{F}_{n-1}] = A_n + M_{n-1}$ for $n \geq 1$. Substituting M_{n-1} in this equation with its expression from Eq. 2.142, we obtain the recurrence formula $A_n = A_{n-1} + E[X_n \mid \mathcal{F}_{n-1}] - X_{n-1}$, which defines A uniquely with $A_0 = 0$.

We now show that the decomposition in Eq. 2.142 exists, that is, that there are processes A and M with the claimed properties. Let A_n, $n = 0, 1, \ldots$, be defined by the above recurrence formula with $A_0 = 0$. Note that $A_n \in \mathcal{F}_{n-1}$ and $A_n \geq A_{n-1}$ since X_n is a submartingale. Hence, A_n, $n = 0, 1, \ldots$, has the stated properties. We also have

$$\begin{aligned} &E[M_n \mid \mathcal{F}_{n-1}] = E[X_n - A_n \mid \mathcal{F}_{n-1}] = E[X_n \mid \mathcal{F}_{n-1}] \\ &- E\left[A_{n-1} + E[X_n \mid \mathcal{F}_{n-1}] - X_{n-1} \mid \mathcal{F}_{n-1}\right] = -A_{n-1} + X_{n-1} = M_{n-1} \end{aligned}$$

so that M is an $\mathcal{F}_n$-martingale. ■

Example 2.90: Let X_n, $n \geq 0$, be a square integrable martingale. Then X_n^2 is a submartingale with Doob decomposition in Eq. 2.142, where $M_n = X_n^2 - A_n$ and $A_n = \sum_{i=1}^n E\left[X_i^2 - X_{i-1}^2 \mid \mathcal{F}_{i-1}\right]$.

Proof: The process X_n^2 satisfies the first two conditions in Eq. 2.138. Also,

$$\begin{aligned} E\left[X_n^2 \mid \mathcal{F}_{n-1}\right] &= E\left[(X_n - X_{n-1})^2 + 2\, X_n\, X_{n-1} - X_{n-1}^2 \mid \mathcal{F}_{n-1}\right] \\ &= E\left[(X_n - X_{n-1})^2 \mid \mathcal{F}_{n-1}\right] + X_{n-1}^2 \geq X_{n-1}^2 \end{aligned}$$

since $E\left[(X_n - X_{n-1})^2 \mid \mathcal{F}_{n-1}\right] \geq 0$, X_{n-1} is $\mathcal{F}_{n-1}$-measurable, and X_n is a martingale. Hence, X_n^2 is a submartingale so that it admits the decomposition in Eq. 2.142. The recurrence formula in the Doob decomposition defining A becomes $A_n = A_{n-1} + E\left[X_n^2 \mid \mathcal{F}_{n-1}\right] - X_{n-1}^2$. ■

Let $X = (X_0, X_1, \ldots)$ be a square integrable $\mathcal{F}_n$-martingale and $A = (A_0, A_1, \ldots)$ be an $\mathcal{F}_n$-predictable process such that $A_0 = 0$ and $E[A_n^2] < \infty$. Then M_n in Eq. 2.139 is an $\mathcal{F}_n$-martingale.

Proof: The expectation of $|M_n|$ satisfies the condition

$$E[|M_n|] \leq \sum_{i=1}^{n} E\left[|A_i|\,|X_i - X_{i-1}|\right] \leq \sum_{i-1}^{n} \left(E[A_i^2]\, E[(X_i - X_{i-1})^2]\right)^{1/2}$$

so that it is finite since A_i and X_i have finite second moments. That $M_n \in \mathcal{F}_n$ follows from its definition and the properties of A_n and X_n. Recall that $A_n \in \mathcal{F}_{n-1}, n \geq 1$, since A is $\mathcal{F}_n$-predictable. For $n > m$, we have

$$
\begin{aligned}
E[M_n \mid \mathcal{F}_m] &= E\left[M_m + \sum_{i=m+1}^{n} A_i\,(X_i - X_{i-1}) \mid \mathcal{F}_m\right] \\
&= M_m + \sum_{i=m+1}^{n} E[A_i\,(X_i - X_{i-1}) \mid \mathcal{F}_m] \\
&= M_m + \sum_{i=m+1}^{n} E\left\{E[A_i\,(X_i - X_{i-1}) \mid \mathcal{F}_{i-1}] \mid \mathcal{F}_m\right\} = M_m
\end{aligned}
$$

since $A_i \in \mathcal{F}_{i-1}$ and X_i is a martingale so that $E[A_i\,(X_i - X_{i-1}) \mid \mathcal{F}_{i-1}] = A_i\, E[X_i - X_{i-1} \mid \mathcal{F}_{i-1}] = 0$.

If the integrator X of the discrete stochastic integral in Eq. 2.139 is not a martingale, the integral $M = (M_0, M_1, \ldots)$ may not be a martingale either. For example, we have $E[M_n \mid \mathcal{F}_{n-1}] = M_{n-1} + E[A_n\,(X_n - X_{n-1}) \mid \mathcal{F}_{n-1}]$ for $m = n-1$ or $M_{n-1} + Z_n\, A_n$, where $Z_n = E[X_n - X_{n-1} \mid \mathcal{F}_{n-1}]$. ■

2.18.2 Stopped martingales

Let $X = (X_0, X_1, \ldots)$ be an $\mathcal{F}_n$-submartingale, martingale, or supermartingale (Eq. 2.138) and let T denote an $\mathcal{F}_n$-stopping time. We call

$$X_n^T(\omega) = X_{n \wedge T(\omega)}(\omega) \quad \text{for } n = 0, 1, \ldots \text{ and } \omega \in \Omega, \tag{2.143}$$

the sequence X **stopped** at T. Figure 2.16 shows five samples of a random walk $R_n = \sum_{i=1}^{n} X_i$ for $n \geq 1$ and $R_0 = 0$ (Example 2.73, Fig. 2.10) stopped at the time T when it leaves the interval $(-a, a)$, $a = 100$, for the first time, where X_i are independent Gaussian variables with mean zero and variance 100.

> If $X_n, n \geq 0$, is adapted to a filtration $\mathcal{F}_n$, so is X_n^T in Eq. 2.143.

Proof: Note that

$$X_{n \wedge T} = \sum_{k=0}^{n-1} X_k\, 1_{\{T=k\}} + X_n\, 1_{\{T \geq n\}},$$

$X_k \in \mathcal{F}_k \subseteq \mathcal{F}_n$ for $k \leq n$ since X is adapted, and $1_{\{T=k\}} \in \mathcal{F}_k \subseteq \mathcal{F}_n$ and $1_{\{T \geq n\}} \in \mathcal{F}_n$ since T is a stopping time. Hence, $X_{n \wedge T}$ is $\mathcal{F}_n$-measurable for all $n \geq 0$. ■

> If $X_n, n \geq 0$, is an $\mathcal{F}_n$-submartingale, martingale, or supermartingale, so is X_n^T in Eq. 2.143.

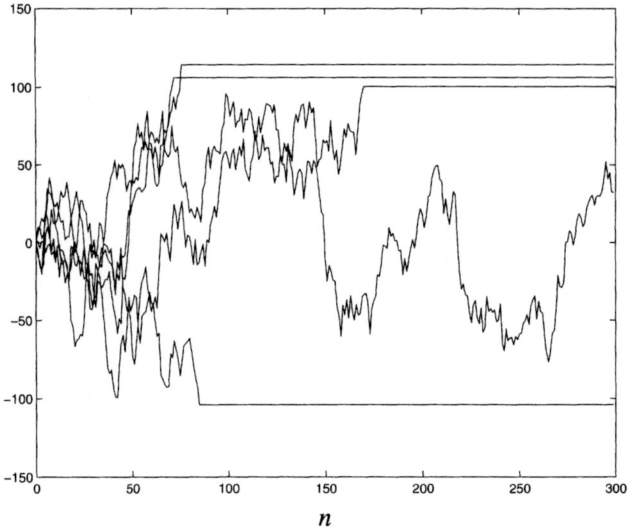

Figure 2.16. Samples of a stopped random walk

Proof: We have shown that X_n^T is $\mathcal{F}_n$-adapted. It remains to prove that X_n^T has also the first and third properties in Eq. 2.138. That X_n^T is integrable follows from

$$E[|X_{n\wedge T}|] = \int_{\{T\leq n\}} |X_{n\wedge T}|\, dP + \int_{\{T>n\}} |X_{n\wedge T}|\, dP$$
$$= \sum_{k=0}^{n} \int_{\{T=k\}} |X_k|\, dP + \int_{\{T>n\}} |X_n|\, dP,$$

$E[|X_n|] < \infty$, and the inequalities $\int_{\{T=k\}} |X_k|\, dP \leq E[|X_k|]$ and $\int_{\{T>n\}} |X_n|\, dP \leq E[|X_n|]$. For the last property in Eq. 2.138 we use $X_{n\wedge T} = \sum_{k=0}^{n-1} X_k\, 1_{\{T=k\}} + X_n\, 1_{\{T\geq n\}}$ so that $X_{(n+1)\wedge T} - X_{n\wedge T} = (X_{n+1} - X_n)\, 1_{\{T>n\}}$. The expectation $E[X_{(n+1)\wedge T} - X_{n\wedge T} \mid \mathcal{F}_n]$ is equal to $1_{\{T>n\}}\, E[X_{n+1} - X_n \mid \mathcal{F}_n]$ since $1_{\{T>n\}}$ is $\mathcal{F}_n$-measurable. If X is a submartingale, martingale, or supermartingale, $E[X_{n+1} - X_n \mid \mathcal{F}_n]$ is positive, zero, or negative so that X_n^T is a submartingale, martingale, or supermartingale, respectively. ■

> If (1) X is an $\mathcal{F}_n$-martingale, (2) T is a stopping time with respect to $\mathcal{F}_n$ such that $T < \infty$ a.s., (3) X_T is integrable, and (4) $E[X_n\, 1_{\{T>n\}}] \to 0$ as $n \to \infty$, then $E[X_T] = \mu$, where $\mu = E[X_n]$. This statement is referred to as the **optional stopping theorem.**

Proof: The value of the martingale X at T can be written as $X_T = X_{n\wedge T} + (X_T - X_n)\, 1_{\{T>n\}}$ so that $E[X_T] = \mu + E[X_T\, 1_{\{T>n\}}] - E[X_n\, 1_{\{T>n\}}]$ since X_n^T is a martingale. The expectation $E[X_n\, 1_{\{T>n\}}]$ converges to zero as $n \to \infty$ by hypothesis. The expectation $E[X_T\, 1_{\{T>n\}}] = \sum_{k=n+1}^{\infty} E[X_k\, 1_{\{T=k\}}]$ also converges to zero as $n \to \infty$

because X_T is integrable so that the series $\sum_{k=0}^{\infty} E[X_k \, 1_{\{T=k\}}]$ is convergent. Hence, the expectation of X_T is μ. ■

2.18.3 Inequalities

We present two inequalities that are most useful for applications. Additional inequalities and convergence theorems can be found in, for example, [66] (Chapter 24) and [151] (Chapter 10).

Doob maximal inequality. If $X = (X_0, X_1, \ldots)$ is a positive $\mathcal{F}_n$-submartingale and $\lambda > 0$ is an arbitrary constant, then

$$P\left(\max_{0 \le k \le n} X_k \ge \lambda\right) \le \frac{1}{\lambda} \, E\left[X_n \, 1_{\{\max_{0\le k\le n} X_k \ge \lambda\}}\right]. \tag{2.144}$$

Proof: Set $X_n^* = \max_{0 \le k \le n} X_k$. For $\lambda > 0$ let $T = \min\{k \le n : X_k \ge \lambda\}$ if there exists $k \le n$ such that $X_k \ge \lambda$ and let $T = n$ otherwise.

Because X is a positive submartingale and T is a stopping time, we have $E[X_n] \ge E[X_T]$ since $T \le n$. Also,

$$E[X_T] = E[X_T \, 1_{\{X_n^* \ge \lambda\}}] + E[X_T \, 1_{\{X_n^* < \lambda\}}] \ge \lambda \, P(X_n^* \ge \lambda) + E[X_n \, 1_{\{X_n^* < \lambda\}}]$$

since $X_n^* \ge \lambda$ implies $X_T \ge \lambda$ and $X_n^* < \lambda$ implies $X_T = X_n$. The above results give

$$E[X_n] \ge E[X_T] \ge \lambda \, P(X_n^* \ge \lambda) + E[X_n \, 1_{\{X_n^* < \lambda\}}]$$

or $\lambda \, P(X_n^* \ge \lambda) \le E[X_n] - E[X_n \, 1_{\{X_n^* < \lambda\}}] = E[X_n \, 1_{\{X_n^* \ge \lambda\}}]$. ■

Doob maximal L_2 inequality. If $X = (X_0, X_1, \ldots)$ is a square integrable positive $\mathcal{F}_n$-submartingale, then

$$E\left[\left(\max_{0 \le k \le n} X_k\right)^2\right] \le 4 \, E[X_n^2]. \tag{2.145}$$

Proof: The equality $E[Y^2] = 2 \int_0^{\infty} y \, P(Y > y) \, dy$ holding for any positive square integrable random variable Y applied to $X_n^* = \max_{0 \le k \le n} X_k$ gives

$$E[(X_n^*)^2] = 2 \int_0^{\infty} x \, P(X_n^* > x) \, dx \le 2 \int_0^{\infty} E[X_n \, 1_{\{X_n^* \ge x\}}] \, dx,$$

where the above inequality follows from Eq. 2.144. We also have

$$\begin{aligned}\int_0^{\infty} E[X_n \, 1_{\{X_n^* \ge x\}}] \, dx &= \int_0^{\infty} \left(\int_{\Omega} X_n \, 1_{\{X_n^* \ge x\}} \, dP\right) dx = \int_{\Omega} \left(\int_0^{X_n^*} X_n \, dx\right) dP \\ &= \int_{\Omega} X_n^* \, X_n \, dP = E[X_n^* \, X_n] \le \left(E[(X_n^*)^2] \, E[X_n^2]\right)^{1/2},\end{aligned}$$

where the Fubini theorem and the Cauchy-Schwarz inequality were used. We have found $E[(X_n^*)^2] \le 2\left(E[(X_n^*)^2]\,E[X_n^2]\right)^{1/2}$, which gives the Doob inequality in Eq. 2.145 following division by $\left(E[(X_n^*)^2]\right)^{1/2}$. ■

Example 2.91: The sequence $X_n = \sum_{k=0}^{n} Z_k$, $n = 0, 1, \ldots$, with $Z_0 = 0$ and $Z_k \ge 0$ independent random variables is a positive $\mathcal{F}_n$-submartingale, where $\mathcal{F}_n = \sigma(Z_1, \ldots, Z_n)$ for $n \ge 1$ and $\mathcal{F}_0 = \{\emptyset, \Omega\}$. Figure 2.17 shows ten samples

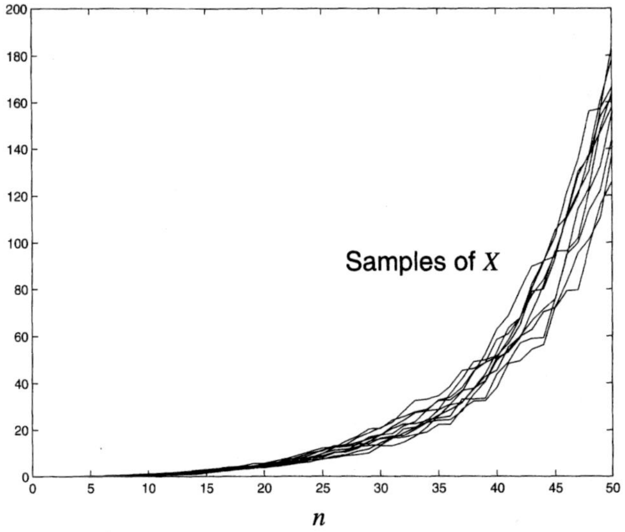

Figure 2.17. Ten samples of a positive submartingale X

of X for random variables Z_k uniformly distributed in the range $(0, 2\,\mu_k)$, where $\mu_k = \rho\,(e^{\rho\,k} - 1)$ and $\rho = 0.1$. The samples resemble the growth of a crack in a the plate with the number $n \ge 0$ of stress cycles ([79], p. 17). ◇

Note: That X is a submartingale follows from $E[X_n \mid \mathcal{F}_m] = E[X_m + \sum_{k=m+1}^{n} Z_k \mid \mathcal{F}_m] = X_m + \sum_{k=m+1}^{n} E[Z_k] \ge X_m$. The expectation $E[X_n] = \sum_{k=0}^{n} E[Z_k]$ of X_n is an increasing function of time.

The submartingale X in this example is square integrable so that the inequality in Eq. 2.145 applies but gives a trivial result since X has increasing samples so that $X_n^* = \max_{0 \le k \le n} X_k = X_n$. Estimates of $P(X_n^* \ge \lambda)$ and $E[X_n\, 1_{X_n^* \ge \lambda}]/\lambda$ in Eq. 2.144 based on 1,000 samples of X are 0.2910 and 0.3136, respectively for $\lambda = 160$. ▲

2.19 Problems

2.1: Let Ω be a sample space and let $\mathcal{A}$ be a collection of subsets of Ω. Show that $\sigma(\mathcal{A}) = \cap_{\mathcal{G} \supseteq \mathcal{A}} \mathcal{G}$ is a σ-field, where $\mathcal{G}$ are σ-fields.

2.2: Let $\Omega = \{1, 2, 3, 4, 5, 6\}$ be a sample space and consider the collection

$$\mathcal{F} = \{\emptyset, \Omega, \{1, 2, 3\}, \{2, 4, 6\}, \{2, 4\}, \{1, 3, 5, 6\}, \{6\}, \{1, 2, 3, 4, 5\}\}$$

of subsets of Ω. Is $\mathcal{F}$ a σ-field?

2.3: Let $\mathcal{F}$ be a σ-field on a sample space Ω. Show that $\mathcal{F}$ is closed to countable intersections.

2.4: Complete the proof of Eq. 2.8

2.5: Suppose that A and B are independent events. Show that A^c and B are independent events.

2.6: Prove that any monotone function is measurable.

2.7: Suppose that $h : (\mathbb{R}, \mathcal{B}) \to (\mathbb{R}, \mathcal{B})$ is measurable and let a be a constant. Show that $h^a(x) = h(x)$ if $h(x) \leq a$ and $h^a(x) = a$ if $h(x) > a$ is measurable.

2.8: Let X be a random variable defined on a probability space $(\Omega, \mathcal{F}, P)$. Show that the σ-field generated by X is the smallest field with respect to which this random variable is measurable.

2.9: Complete the details of the proof of Eq. 2.26.

2.10: Show that $\inf_i X_i$, $\sup_i X_i$, $\liminf_i X_i$, and $\limsup_i X_i$ are random variables on $(\Omega, \mathcal{F}, P)$, where $X_i, i = 1, 2, \ldots$, are random variables on this space.

2.11: Prove the statements in Section 2.5.2.1.

2.12: Prove the statements in Section 2.5.2.2.

2.13: Let $h_n = 1_{[n,n+1]}$, $n = 1, 2, \ldots$, and let λ denote the Lebesgue measure. We have $\int_{\mathbb{R}} h_n \, d\lambda = 1$ for all n. Show that $\int (\lim h_n) \, d\lambda \neq \lim \int h_n \, d\lambda$.

2.14: Use the dominated convergence theorem to find the limit as $n \to \infty$ of the integral $\int_1^\infty \sqrt{x}/(1 + n\, x^3) \, dx$.

2.15: Find the characteristic function of a random variable X that is uniformly distributed in a bounded interval (a, b).

2.16: Let $X_i \sim S_\alpha(\sigma_i, \beta_i, \mu_i)$, $i = 1, \ldots, n$, be independent α-stable random variables. Find the characteristic function of $\sum_{i=1}^n X_i$. Comment on your result.

2.17: Let (Y_1, Y_2) be an $\mathbb{R}^2$-valued Gaussian variable with the density in Eq. 2.107. Find the joint density of $(X_1 = g_1(Y_1), X_2 = g_2(Y_2))$, where g_i, $i = 1, 2$, are increasing functions. Specialize your results for $g_i(y) = \exp(y)$.

2.18: Calculate the mean and variance of the estimator $\hat{X}$ in Example 2.64.

2.19: Show that an $\mathbb{R}^d$-valued random variable $X \in L_2$ is Gaussian if and only if $\sum_{i=1}^{d} \alpha_i X_i$ is a real-valued Gaussian variable for every collection of constants $\alpha_1, \ldots, \alpha_d$.

2.20: Show that $X_n \xrightarrow{\text{m.s.}} X$ and $Y_n \xrightarrow{\text{m.s.}} Y$ imply $X_n + Y_n \xrightarrow{\text{m.s.}} X + Y$.

2.21: Let $g : \mathbb{R} \to \mathbb{R}$ be a continuous function. Let X and $X_1, X_2, \ldots$ be random variables on a probability space $(\Omega, \mathcal{F}, P)$. Show that $X_n \xrightarrow{\text{a.s.}} X$ and $X_n \xrightarrow{\text{pr}} X$ imply $g(X_n) \xrightarrow{\text{a.s.}} g(X)$ and $g(X_n) \xrightarrow{\text{pr}} g(X)$, respectively.

2.22: Take a collection of random variables X_i, $i \in I$, with the same mean and variance but different distributions. Plot the probabilities $P(|X_i| > x)$ as a function of $x > 0$ and a Chebyshev bound. Can the bound be used to approximate $P(|X_i| > x)$?

2.23: Let $R_n = \sum_{i=1}^{n} X_i$ for $n \geq 1$ and $R_0 = 0$, where $X_i > 0$ are iid random variables. Find the probability law of the stopping time $T = \inf\{n \geq 0 : R_n > a\}$, where $a > 0$ is a constant.

2.24: Let X and X_i, $i = 1, \ldots, n$, be random variables on a probability space $(\Omega, \mathcal{F}, P)$. Show that the best mean square estimator of X given $(X_1, \ldots, X_n)$ is $\hat{X} = E[X \mid \sigma(X_1, \ldots, X_n)]$.

2.25: Let X_i, $i = 0, 1, \ldots$, be a sequence of random variables generated by $X_{i+1} = \rho X_i + W_i$, where ρ is a constant and W_i are iid random variables, and X_0 is independent of W_i. Is X_i a martingale, submartingale, or supermartingale?

2.26: Let X be an integrable random variable defined on a probability space $(\Omega, \mathcal{F}, P)$ and let $\mathcal{F}_n$, $n \geq 0$, be a filtration on this space. Show that $X_n = E[X \mid \mathcal{F}_n]$ is an $\mathcal{F}_n$-martingale.

2.27: Let X_i, $i \geq 1$, be iid random variables defined on a probability space $(\Omega, \mathcal{F}, P)$ such that $P(X_i = 1) = P(X_i = -1) = 1/2$. Consider the filtration $\mathcal{F}_n = \sigma(X_1, \ldots, X_n)$ on this space and the random walk $R_n = \sum_{i=1}^{n} X_i$. Show that $R_n^2 - n$ and $(-1)^n \cos(\pi R_n)$ are $\mathcal{F}_n$-martingales.

2.28: Let X_n be an $\mathcal{F}_n$-martingale and let $B \in \mathcal{B}$. Show that $T = \min\{n : X_n \in B\}$ is an $\mathcal{F}_n$-stopping time.

Chapter 3

Stochastic Processes

3.1 Introduction

In the previous chapter we defined a time series or a discrete time stochastic process as a countable family of random variables $X = (X_1, X_2, \ldots)$. Time series provide adequate models in many applications. For example, X_n may denote the damage of a physical system after n loading cycles or the value of a stock at the end of day n. However, there are situations in which discrete time models are too coarse. For example, consider the design of an engineering system subjected to wind, wave, and other random forces over a time interval I. To calculate the system dynamic response, we need to know these forces at each time $t \in I$. The required collection of force values is an uncountable set of random variables indexed by $t \in I$, referred to as a **continuous time stochastic process** or just a **stochastic process**. We use upper case letters for all random quantities. A real-valued stochastic process is denoted by $\{X(t), t \in I\}$ or X. If the process takes on values in $\mathbb{R}^d$, $d > 1$, we use the notation $\{\boldsymbol{X}(t), t \in I\}$ or $\boldsymbol{X}$.

If X is indexed by a space coordinate $\boldsymbol{\xi} \in D \subset \mathbb{R}^q$ rather than time $t \in I$, then $\{\boldsymbol{X}(\boldsymbol{\xi}), \boldsymbol{\xi} \in D\}$ is called a **random field**. There are notable differences between random processes and fields. For example, the concept of past and future has a clear meaning for stochastic processes but not for random fields. However, stochastic processes and random fields share many properties. There are also situations in which a random function depends on both space and time parameters. The evolution in time of (1) the wave height in the North Sea, (2) the temperature everywhere in North America, (3) the ground acceleration in California during an earthquake, and (4) the Euler angles and the dislocation density in a material subjected to plastic deformation are some examples. These functions are random fields at each fixed time $t \in I$ and stochastic processes at each site $\boldsymbol{\xi} \in D$. They are referred to as **space-time stochastic processes** and are denoted by $\{\boldsymbol{X}(t, \boldsymbol{\xi}), t \geq 0, \boldsymbol{\xi} \in D\}$

This chapter reviews essential concepts on stochastic processes and illustrates them by numerous examples. The review includes properties of stochastic processes, filtration, stopping time, and finite dimensional distribution. Second moment calculus is considered in detail because of its extensive use in applications. The concluding sections of the chapter deal with martingales, Poisson, Brownian motion, and Lévy processes. These processes are needed subsequently to develop the Itô formula and apply it to the solution of a wide class of stochastic problems. Properties of the random fields used in the book are also reviewed.

3.2 Definitions

Let $X : I \times \Omega \mapsto \mathbb{R}^d$ be a function of two arguments, $t \in I$ and $\omega \in \Omega$, where I is a subset of $\mathbb{R}$ or $[0, \infty)$ and $(\Omega, \mathcal{F}, P)$ denotes a probability space.

If $X(t)$ is an $\mathbb{R}^d$-valued random variable on the probability space $(\Omega, \mathcal{F}, P)$ for each $t \in I$, that is, $X(t) \in \mathcal{F}$ for each $t \in I$, then X is said to be an $\mathbb{R}^d$-valued **stochastic process**.

Note: We will also refer to an $\mathbb{R}^d$-valued stochastic process with $d > 1$ as a **vector stochastic process**. The function $X(\cdot, \omega)$ for a fixed $\omega \in \Omega$ is called a **sample path, path, sample**, or **realization** of X. The function $X(t, \cdot)$ for a fixed t is by definition an $\mathbb{R}^d$-valued random variable. The measurable space $(\mathbb{R}^d, \mathcal{B}^d)$ used in the definition of X is sufficiently general for our discussion. The definition of a stochastic process X does not require that the functions $X(\cdot, \omega) : [0, \infty) \mapsto \mathbb{R}^d$ and $X(\cdot, \cdot) : [0, \infty) \times \Omega \mapsto \mathbb{R}^d$ be measurable. ▲

Random fields are defined in the same way. Let $(\Omega, \mathcal{F}, P)$ denote a probability space and let $X : D \times \Omega \mapsto \mathbb{R}^d$ be a function of two arguments, $\xi \in D$ and $\omega \in \Omega$, where $D \subset \mathbb{R}^q$ and $q \geq 1$ is an integer.

If $X(\xi)$ is an $\mathbb{R}^d$-valued random variable on the probability space $(\Omega, \mathcal{F}, P)$ for each $\xi \in D$, that is, $X(\xi) \in \mathcal{F}$ for each $\xi \in D$, then X is said to be an $\mathbb{R}^d$-valued **random field** in D.

Example 3.1: Figure 3.1 shows daily average temperatures measured in Central Park, New York City, during two different years starting on the first of January and ending on the last day of December. These records can be interpreted as two samples of a stochastic process $\{X(t), t = 1, \ldots, 365\}$ giving daily average temperatures in Central Park. A collection of samples of X of the type shown in this figure can be used to estimate the probability law of X. ◇

Example 3.2: Let Φ denote one of the three Euler angles of the atomic lattice orientation in aluminum AL 7075. Figure 3.2 shows 14,000 measurements of Φ performed on an aluminum AL 7075 plate D with dimensions 540×540 μm. The left plot shows the spatial variation of Φ over the plate. The right plot presents contour lines for Φ. These lines partition D in subsets characterized by nearly constant values of Φ, called grains or crystals. The plots in this figure can be

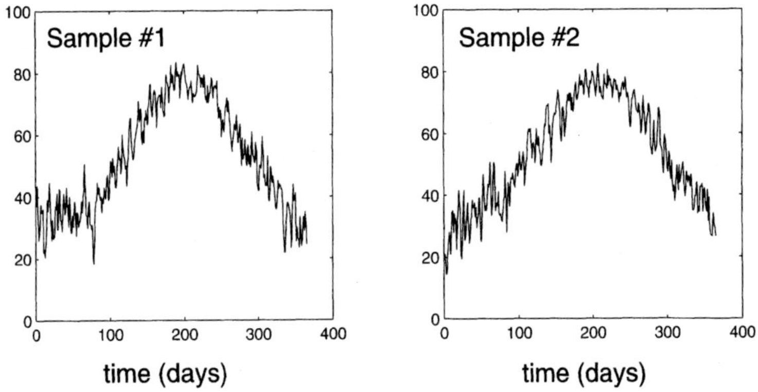

Figure 3.1. Two records of daily temperatures in Central Park, New York City in two consecutive years

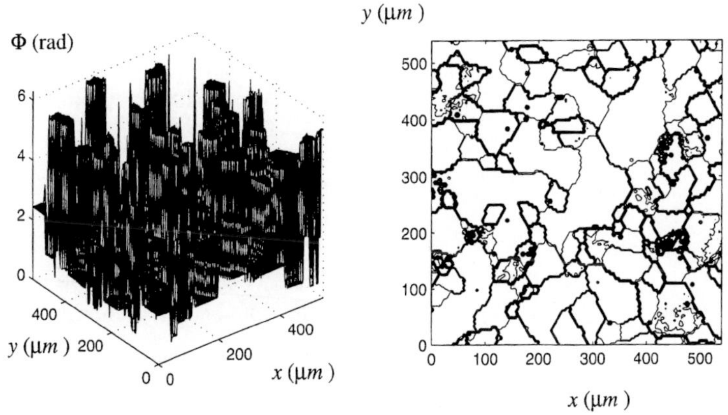

Figure 3.2. Three dimensional and contour lines for an Euler angle of atomic lattice orientation in aluminum AL 7075

interpreted as a sample of the random field $\Phi : D \times \Omega \to \mathbb{R}$, $D \subset \mathbb{R}^2$. Properties of this field can be inferred from measurements on a collection of aluminum specimens. $\diamond$

Example 3.3: Let $X : I \times \Omega \to \mathbb{R}$ be defined by $X(t, \omega) = Y(\omega) \cos(t)$, where $I = [0, 10]$ and Y is a random variable defined by $Y(\omega) = \omega$ on the probability space $([0, 1], \mathcal{B}([0, 1]), P)$, where $P(d\omega) = d\omega$. Figure 3.3 shows the mapping $(t, \omega) \mapsto X(t, \omega)$ and 10 samples of X. $\diamond$

Note: $X(t, \cdot)$ is $\mathcal{B}([0, 1])$-measurable for a fixed t, since $\cos(t)$ is a constant and $Y(\omega)$ is a continuous function of ω so that it is $\mathcal{B}([0, 1])$-measurable. Hence, X is a stochastic process. $X(\cdot, \omega)$ is $\mathcal{B}([0, 10])$-measurable for a fixed ω because $Y(\omega)$ is a constant and $\cos(t)$

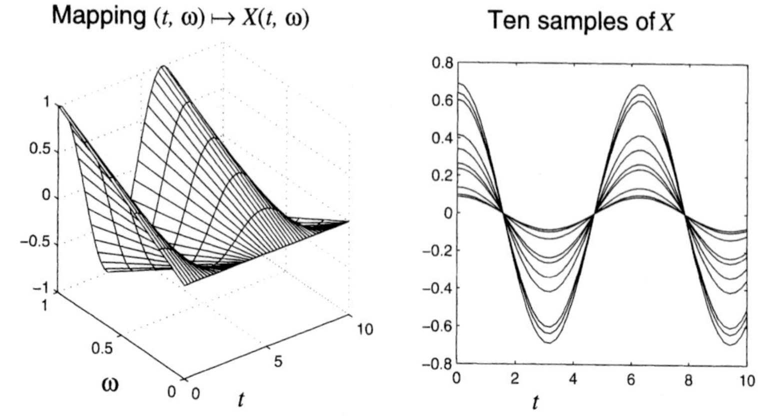

Figure 3.3. The mapping $(t, \omega) \mapsto X(t, \omega) = Y(\omega)\cos(t)$ and ten samples of X

is continuous. It is natural to ask whether the function $(t, \omega) \mapsto X(t, \omega)$ is measurable from $([0, 10] \times [0, 1], \mathcal{B}([0, 10]) \times \mathcal{B}([0, 1]))$ to $([-1, 1], \mathcal{B}([-1, 1]))$, where $[-1, 1]$ is the range of X. The answer to this question is in the affirmative (Example 3.4). ▲

We have seen that a time series $X = (X_1, X_2, \ldots)$ viewed as a function $(m, \omega) \mapsto X_m(\omega)$ is measurable from $(\mathbb{Z}^+ \times \Omega, \mathcal{K} \times \mathcal{F})$ to $(\mathbb{R}, \mathcal{B})$, where $\mathbb{Z}^+ = \{1, 2, \ldots\}$, $\mathcal{K}$ consists of all subsets of $\mathbb{Z}^+$, and $(\Omega, \mathcal{F})$ denotes the measurable space on which the random variables X_n, $n = 1, 2, \ldots$, are defined (Section 2.4.3). Generally, this property is not preserved when the time becomes continuous, that is, for stochastic processes and random fields.

A stochastic process $\boldsymbol{X}$ is **measurable** if the function $\boldsymbol{X} : I \times \Omega \mapsto \mathbb{R}^d$ is measurable from $(I \times \Omega, \mathcal{B}(I) \times \mathcal{F})$ to $(\mathbb{R}^d, \mathcal{B}^d)$.

Note: If $\boldsymbol{X}$ is a measurable stochastic process, then $\boldsymbol{X}(t, \cdot)$ is $\mathcal{F}$-measurable and $\boldsymbol{X}(\cdot, \omega)$ is $\mathcal{B}(I)$-measurable by Fubini's theorem (Section 2.8). The first statement is valid even if X is not measurable because X is a stochastic process.

The assumption that a process is measurable may not be very restrictive since many of the stochastic processes used in applications are measurable (Example 3.4). ▲

The requirements in the definitions of stochastic processes and random fields are quite weak so that the samples of these random functions can be quite erratic. We can impose additional conditions to control the behavior of these functions, for example, we may require that a stochastic process satisfy some measurability conditions and/or its samples are continuous. A stochastic process $X(t)$, $t \geq 0$, is said to be **right/left continuous** or **continuous** if the function $X(t, \omega)$, $t \geq 0$, is right/left continuous or continuous, respectively, for every $\omega \in \Omega$. If the above properties hold for every $\omega \in \Omega \setminus N$ such that $N \in \mathcal{F}$ and $P(N) = 0$, we say that X is **right/left continuous** a.s. or **continuous** a.s. Requirements on the sample properties may imply some measurability properties, as illustrated by the following examples.

Example 3.4: Let $\{X(t), t \in [0, 1]\}$ be a real-valued stochastic process. If X (1) starts at zero, that is, $X(0) = 0$, (2) has a.s. continuous samples, (3) has stationary independent Gaussian increments, that is, the random variable $X(t) - X(s)$, $t \geq s$, is Gaussian with mean zero and variance $t - s$, and (4) the increments of X are independent of the past, that is, $X(t) - X(s)$, $t > s$, is independent of $\mathcal{F}_s = \sigma(B(u), 0 \leq u \leq s)$, then X is called a **Brownian motion** or a **Wiener** process and is denoted by B. The Brownian motion process is measurable. ◇

Proof: Let $B^{(n)}(t, \omega) = B(k\,2^{-n}, \omega)$, $t \in ((k-1)\,2^{-n}, k\,2^{-n}]$, $k = 1, \ldots, 2^n$, and $B^{(n)}(0, \omega) = 0$ be an approximating sequence for B in the time interval $[0, 1]$. The process $B^{(n)}$ has piece-wise constant sample paths, depends on a countable number of values of B, and can be viewed as a random walk with time step $\Delta t = 2^{-n}$ defined by a sum of independent Gaussian variables with mean zero and variance 2^{-n} for each n (Section 2.14). Hence, the function $(t, \omega) \mapsto B^{(n)}(t, \omega)$ is measurable from $\mathcal{B}([0, 1]) \times \mathcal{F}$ to $(\mathbb{R}, \mathcal{B})$. The measurable mapping $(t, \omega) \mapsto B^{(n)}(t, \omega)$ converges to $(t, \omega) \mapsto B(t, \omega)$ point-wise by the continuity of the sample paths of B so that the function $B(t, \omega) = \lim_{n \to \infty} B^{(n)}(t, \omega)$ is measurable, since limits of measurable functions are measurable. We will see in Section 3.13 that B has continuous samples.

We also note that the random variables $X(t) - X(s)$ and $X(v) - X(u)$, $u < v \leq s < t$, are independent since $X(t) - X(s)$ is independent of $\mathcal{F}_s$ by property (4). Hence, the Brownian motion process has independent increments. ■

In Section 2.15 we have defined filtration and adapted time series. These concepts extend directly to the case in which the time index is continuous.

A collection of sub-σ-fields $(\mathcal{F}_t)_{t \geq 0}$ of $\mathcal{F}$ such that $\mathcal{F}_s \subseteq \mathcal{F}_t \subset \mathcal{F}$, $0 \leq s \leq t$, is called a **filtration** in $(\Omega, \mathcal{F})$. A probability space $(\Omega, \mathcal{F}, P)$ endowed with a filtration $(\mathcal{F}_t)_{t \geq 0}$ is said to be a **filtered probability space** and is denoted by $(\Omega, \mathcal{F}, (\mathcal{F}_t)_{t \geq 0}, P)$. A stochastic process $\boldsymbol{X}$ defined on $(\Omega, \mathcal{F}, P)$ is **adapted to the filtration** $(\mathcal{F}_t)_{t \geq 0}$, or $\mathcal{F}_t$**-adapted**, or just **adapted** if $\boldsymbol{X}(t) \in \mathcal{F}_t$ for each $t \geq 0$.

Note: The definition is given for the most common case considered in our discussions when t takes values in $I = [0, \infty)$. A similar definition holds for an arbitrary index set I. If $\boldsymbol{X}$ is $\mathcal{F}_t$-adapted, then $\boldsymbol{X}$ is a stochastic process since $\boldsymbol{X}(t) \in \mathcal{F}_t$ implies $\boldsymbol{X}(t) \in \mathcal{F}$ for each $t \geq 0$. Generally, the converse is not true. ▲

Let $\{\boldsymbol{X}(t), t \geq 0\}$ be a stochastic process defined on a probability space $(\Omega, \mathcal{F}, P)$. The **natural filtration**

$$\mathcal{F}_t^{\boldsymbol{X}} = \sigma\left(\boldsymbol{X}(s), 0 \leq s \leq t\right) = \sigma\left(\cup_{0 \leq s \leq t} \sigma(\boldsymbol{X}(s))\right) \tag{3.1}$$

of $\boldsymbol{X}$ is the smallest filtration with respect to which $\boldsymbol{X}$ is adapted.

Note: In Section 2.15 we have defined the natural filtration $\mathcal{F}_n^X = \sigma(X_1, \ldots, X_n), n \geq 1$, for a time series $X = (X_1, X_2, \ldots)$. The definition of $\mathcal{F}_t^X$ in Eq. 3.1 is a direct extension of the definition of $\mathcal{F}_n^X$ to a continuous time index. If $(\mathcal{F}_t)_{t\geq 0}$ is a filtration on $(\Omega, \mathcal{F}, P)$ such that $\mathcal{F}_t^X \subset \mathcal{F}_t$ for each $t \geq 0$, then X is $\mathcal{F}_t$-adapted. ▲

Example 3.5: Let $([0, 1], \mathcal{B}([0, 1]), P)$ be a probability space and a random variable $Y(\omega) = \omega$ defined on this space, where $P(d\omega) = d\omega$. Consider also a stochastic process $X : [0, 1] \times \Omega \to \mathbb{R}$ defined by $X(t, \omega) = Y(\omega)$. The natural filtration of X is $\mathcal{F}_t^X = \mathcal{B}([0, 1]), t \geq 0$. ◇

Note: We have $\sigma(X(t)) = \sigma(Y)$ for all $t \in [0, 1]$ so that $\mathcal{F}_t^X = \sigma(\cup_{0\leq s\leq t}\sigma(X(s))) = \sigma(Y)$. Because $\omega \mapsto Y(\omega)$ is the identity function, the σ-field generated by the random variable Y is $\mathcal{B}([0, 1])$. ▲

Example 3.6: Let $\{B(t), t \geq 0\}$ be a Brownian motion, $\mathcal{F}_t = \sigma(B(s), 0 \leq s \leq t)$ denote the natural filtration of B, and let $g : [0, \infty) \times \mathbb{R} \to \mathbb{R}$ be a continuous function. The stochastic process $Y(t) = g(t, B(t))$ is $\mathcal{F}_t$-adapted. For example, $B(t)^2$, $B(t)^2 - t$, $B(t)^q$ for an arbitrary integer $q \geq 1$, and $\max_{0\leq s\leq t} B(s)^q$ are $\mathcal{F}_t$-adapted. We say that these processes are **adapted to the Brownian motion**. On the other hand, the process $Z(t) = B(t + \tau)$, $\tau > 0$, is not adapted to the Brownian motion. ◇

Note: The natural filtration of some of the processes derived from B may be smaller than $\mathcal{F}_t$. For example, the natural filtration $\mathcal{F}_t' = \sigma(B(s)^2, 0 \leq s \leq t)$ is smaller than $\mathcal{F}_t$, $t \geq 0$, because $\mathcal{F}_t'$ contains the entire information on B^2 and $|B|$ but not on B. Hence, the stochastic process B^2 is both $\mathcal{F}_t'$ and $\mathcal{F}_t$-adapted ([131], Example 1.5.2, p. 78).

The random variable $B(t + \tau)$ is $\mathcal{F}_{t+\tau}$-measurable but it is not $\mathcal{F}_t$-measurable for $\tau > 0$. Hence, Z is not $\mathcal{F}_t$-adapted. ▲

A stochastic process X defined on a filtered probability space $(\Omega, \mathcal{F}, (\mathcal{F}_t)_{t\geq 0}, P)$ is **progressively measurable** or **progressive** with respect to $(\mathcal{F}_t)_{t\geq 0}$ if the function $X(s, \omega) : [0, t] \times \Omega \to \mathbb{R}$ is $\mathcal{B}([0, t)) \times \mathcal{F}_t$-measurable for every $t \geq 0$.

Note: A progressively measurable process X is measurable since $\mathcal{B}([0, t)) \times \mathcal{F}_t$ is included in $\mathcal{B}([0, \infty)) \times \mathcal{F}$. The process X is also $\mathcal{F}_t$-adapted because $X(t, \cdot)$ is $\mathcal{F}_t$-measurable for each $t \geq 0$ (Fubini's theorem). ▲

Example 3.7: If X is an $\mathcal{F}_t$-adapted stochastic processes with continuous samples, then X is progressively measurable. Hence, the Brownian motion is progressively measurable with respect to its natural filtration. ◇

Proof: Take a time t in $[0, 1]$. Let $X^{(n)}(t, \omega) = X(k\,2^{-n}, \omega)$ if $t \in ((k-1)\,2^{-n}, k\,2^{-n}]$, $k = 1, \ldots, 2^n$, and $X^{(n)}(t, \omega) = 0$ if $t = 0$, $n = 1, 2, \ldots$, define a sequence approximating $\{X(t), t \in [0, 1]\}$. The mapping $(t, \omega) \mapsto X^{(n)}(t, \omega)$ is $\mathcal{B}([0, k\,2^{-n}]) \times \mathcal{F}_{k\,2^{-n}}$-measurable if k and n are such that $t \in ((k-1)\,2^{-n}, k\,2^{-n}]$ so that it is also $\mathcal{B}([0, s]) \times \mathcal{F}_s$-measurable for $s \geq k\,2^{-n}$. That the process X is progressively measurable follows from the continuity of the sample paths of X and the convergence of the mapping $(t, \omega) \mapsto X^{(n)}(t, \omega)$ to $(t, \omega) \mapsto X(t, \omega)$. ■

Figure 3.4 illustrates the relationship between adapted, measurable, and progressive processes. We have seen that a progressive process is measurable

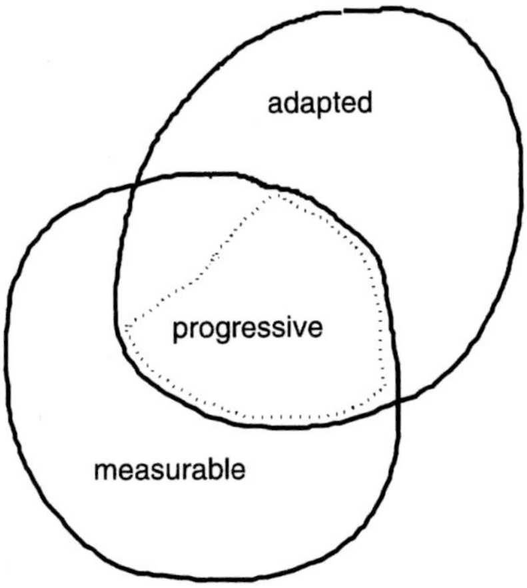

Figure 3.4. Relationship between adapted, measurable, and progressive processes

and adapted. An adapted process may or may not be measurable. An adapted and measurable process may not be progressive. The following statement generalizes Example 3.7.

Any right or left continuous adapted process $X(t)$, $t \geq 0$, is progressive.

Proof: Suppose that X is right continuous and define the sequence of processes

$$X_{n,t}(s, \omega) = \begin{cases} X(i\,t/n, \omega), & \text{if } (i-1)\,t/n < s \leq i\,t/n,\ i = 1, \ldots, n, \\ X(s, \omega), & \text{if } s = 0 \end{cases}$$

for $n = 1, 2, \ldots$ and $t > 0$. For any $B \in \mathcal{B}$, the set

$$\begin{aligned} \{(s, \omega) : 0 \leq s \leq t, \omega \in \Omega, X_{n,t}(s, \omega) \in B\} &= \{0\} \times \{\omega : X(0, \omega) \in B\} \\ &\cup \left(\cup_{i=1}^{n} ((i-1)\,t/n, i\,t/n] \times \{\omega : X(i\,t/n, \omega) \in B\}\right) \end{aligned}$$

is in $\mathcal{B}([0, t]) \times \mathcal{F}_t$ since X is adapted. Hence, $X_{n,t}(s, \omega)$ with $0 \leq s \leq t$ and $\omega \in \Omega$ is $\mathcal{B}([0, t]) \times \mathcal{F}_t$-measurable for each $n \geq 1$ so that $X_{n,t}$ is progressive. By right continuity we have $X_{n,t}(s, \omega) \to X(s, \omega)$ as $n \to \infty$ for all $0 \leq s \leq t$ and $\omega \in \Omega$. We conclude that X is progressive since limits of measurable functions are measurable. ■

We assume throughout the book that the filtration $\mathcal{F}_t$, $t \geq 0$, on $(\Omega, \mathcal{F}, P)$ has two properties, referred to as the **usual hypotheses**:

1. $\mathcal{F}_t$ is **complete** meaning that (a) $\mathcal{F}$ is complete, that is, for every $A \subset B$ such that $B \in \mathcal{F}$ and $P(B) = 0$ we have $A \in \mathcal{F}$ so that $P(A) = 0$ (Section 2.2.3), and (b) $\mathcal{F}_0$ contains all null subsets of $\mathcal{F}$, that is, $A \in \mathcal{F}$ such that $P(A) = 0$ implies $A \in \mathcal{F}_0$.

2. $\mathcal{F}_t$ is **right continuous**, that is, the σ-field $\mathcal{F}_{t+} = \cap_{s>t}\mathcal{F}_s$ is equal to $\mathcal{F}_t$ for all $t \geq 0$. The filtration $\mathcal{F}_{t+}$ provides an "infinitesimal peek into the future" in the sense that $A \in \mathcal{F}_{t+}$ implies $A \in \mathcal{F}_{t+\varepsilon}$ for all $\varepsilon > 0$. Note that a right continuous filtration $\mathcal{F}_n$ indexed by $n = 0, 1, \ldots$, is constant since $\mathcal{F}_{n+} = \mathcal{F}_{n+1}$.

Example 3.8: Let $(\Omega, \mathcal{F}, (\mathcal{F}_t)_{t\geq 0}, P)$ be a filtered probability space and X be an $\mathcal{F}_t$-adapted stochastic process. If $\mathcal{F}_t$ is right continuous, then the random variable $\dot{X}(t) = \limsup_{u\downarrow t}(X(u) - X(t))/(u - t)$ is $\mathcal{F}_t$-measurable provided it exists. ◇

Proof: Define the random variables $Y_n = n\,(X(t + 1/n) - X(t))$ and $Z_n = \sup_{m\geq n} Y_m$, take $p > n$. Note that $Z_p \in \mathcal{F}_{t+1/p} \subset \mathcal{F}_{t+1/n}$ so that $\inf_p Z_p \in \mathcal{F}_{t+1/n}$ for any n and $\inf_p Z_p \in \cap_{n=1}^{\infty}\mathcal{F}_{t+1/n} = \mathcal{F}_{t+}$. If the filtration $\mathcal{F}_t$ is right continuous, then $\dot{X}(t)$ is $\mathcal{F}_t$-measurable for each $t \geq 0$. ■

3.3 Continuity

We have already defined right/left continuity, continuity, a.s. right/left continuity, and a.s. continuity. Generally, it is difficult to establish whether a process is sample continuous from these definitions. The following **Kolmogorov criterion** provides a simple alternative. If $\{X(t), t \in I\}$ is a real-valued stochastic process and $I \subset [0, \infty)$ is a closed interval, and there exist three constants $\alpha, \beta, \zeta > 0$ such that

$$E[|X(t+h) - X(t)|^{\alpha}] \leq \zeta\, h^{1+\beta}, \tag{3.2}$$

then $\lim_{h\to 0}\sup_{s,t\in I, |s-t|<h}|X(s) - X(t)| = 0$ a.s., that is, almost every sample path of X is uniformly continuous in I. An $\mathbb{R}^d$-valued process, $d > 1$, is sample continuous if each of its coordinates satisfies the condition in Eq. 3.2 ([197], Proposition 4.2, p. 57).

We give in this section some of the most common definitions of continuity for stochastic processes. Both continuity at a point and in an interval are discussed. Let $\{\boldsymbol{X}(t), t \geq 0\}$ be an $\mathbb{R}^d$-valued stochastic process defined on a probability space $(\Omega, \mathcal{F}, P)$.

X is **continuous in probability at** $t \geq 0$ if

$$\lim_{s\to t} P(\| \boldsymbol{X}(s) - \boldsymbol{X}(t) \| \geq \varepsilon) = 0, \quad \forall \varepsilon > 0. \tag{3.3}$$

$\boldsymbol{X}$ is **continuous in the p'th mean at** t if

$$\lim_{s \to t} E[\| \boldsymbol{X}(s) - \boldsymbol{X}(t) \|^p] = 0. \tag{3.4}$$

$\boldsymbol{X}$ is **almost surely (a.s.) continuous at** t if

$$P\left(\{\omega : \lim_{s \to t} \| \boldsymbol{X}(s, \omega) - \boldsymbol{X}(t, \omega) \| = 0\}\right) = 1 \quad \text{or}$$
$$P(\Omega_t) = P(\{\omega : \lim_{s \to t} \boldsymbol{X}(s, \omega) \neq \boldsymbol{X}(t, \omega)\}) = 0. \tag{3.5}$$

Note: The Euclidean norm $\| \boldsymbol{x} \| = \left(\sum_{i=1}^{d} x_i^2\right)^{1/2}$ is used in the above definition to assess the magnitude of the difference $\boldsymbol{X}(t) - \boldsymbol{X}(s)$.

If $p = 2$ in Eq. 3.4, $\boldsymbol{X}$ is said to be **continuous in the mean square sense** or **m.s. continuous** at t. The mean square continuity is used extensively in the second moment calculus considered later in this and the following chapters.

The subset Ω_t of Ω in Eq. 3.5, consisting of the sample paths of $\boldsymbol{X}$ that are not continuous at time t, is in $\mathcal{F}$ because it is defined by the condition $\| \boldsymbol{X}(t+, \omega) - \boldsymbol{X}(t-, \omega) \| > 0$ and $\| \boldsymbol{X}(t+, \omega) - \boldsymbol{X}(t-, \omega) \|$ is a measurable function of the random variables $\boldsymbol{X}(t+)$ and $\boldsymbol{X}(t-)$.

The process $\boldsymbol{X}$ is continuous in probability, continuous in the p'th mean, and almost surely **continuous in an interval** $I \subset [0, \infty)$ if it is continuous in probability, continuous in the p'th mean, and almost surely continuous at each $t \in I$, respectively. ▲

Example 3.9: A process $\{C(t) \in \mathbb{R}, t \geq 0\}$, with $C(0) = 0$ and

$$C(t) = \sum_{k=1}^{N(t)} Y_k, \quad t > 0, \tag{3.6}$$

is said to be a **compound Poisson** process, where Y_k, $k = 1, 2, \ldots$, are independent identically distributed (iid) real-valued random variables and N is a **Poisson counting** process with intensity parameter $\lambda > 0$. The compound Poisson process in Eq. 3.6 and the Brownian motion process in Example 3.4 are continuous in probability at each time $t \geq 0$. These processes are also continuous in probability in $[0, \infty)$. Figure 3.5 shows a sample of a Brownian motion and a sample of a compound Poisson process with jumps $Y_1 \sim N(0, \sigma^2)$ and intensity λ such that $\lambda\, E[Y_1^2] = \lambda\, \sigma^2 = 1$. The two processes have identical first two moments but their samples differ significantly. Processes with the same first two moments are said to be equal in the second moment sense (Section 3.8). ◇

Proof: The process N has **stationary independent increments**, that is, (1) the distribution of the random variable $N(t) - N(s)$, $s < t$, depends only on the time lag $t - s$ rather than the times t and s and (2) the increments of N over non-overlapping time intervals are

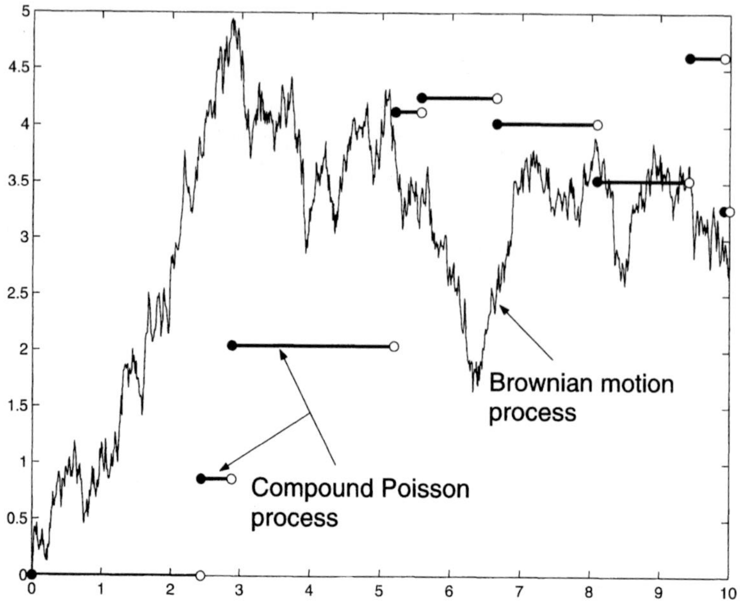

Figure 3.5. Samples of a Brownian motion and a compound Poisson process with $Y_1 \sim N(0, \sigma^2)$ and $\lambda \sigma^2 = 1$

independent random variables, for example, $N(t)-N(u)$ and $N(v)-N(s)$ are independent for $s < v \leq u < t$ (Sections 3.6.4 and 3.12). The probability,

$$P(N(t) - N(s) = n) = \frac{(\lambda (t-s))^n}{n!} e^{-\lambda (t-s)}, \quad n = 0, 1, \ldots, \tag{3.7}$$

of $N(t)-N(s)$, $s < t$, depends only on the time lag $t-s$ and an intensity parameter $\lambda > 0$. The compound Poisson process C in Eq. 3.6 has similar properties. If the random variables Y_k are equal to unity, then C is a Poisson process.

B is continuous in probability at time t because $P(|B(t) - B(s)| > \varepsilon)$ is equal to $2\,\Phi(-\varepsilon/\sqrt{|t-s|})$ so that it converges to zero as $s \to t$ for any $\varepsilon > 0$. The process C is continuous in probability at $t \geq 0$ because the event $|C(t) - C(s)| > \varepsilon$ can occur if there is at least one jump in the time interval $(s, t]$, and the probability $P(N(t-s) > 0) = 1 - \exp[-\lambda (t-s)]$ of this event approaches zero as $s \to t$. Because these processes are continuous in probability at each time $t \geq 0$, they are also continuous in probability in $[0, \infty)$. ■

Example 3.10: The Brownian motion B is m.s. continuous at each $t \geq 0$. If $Y_1 \in L_2$, then the compound Poisson process C in Example 3.9 is m.s. continuous at each $t \geq 0$. ◇

Proof: The second moment of the increments of B and C are $E[(B(t) - B(s))^2] = t - s$ and $\lambda (t-s)\, E[Y_1^2] + (\lambda (t-s)\, E[Y_1])^2$, respectively. The second moment of $B(t) - B(s)$

follows from the definition of B. The expression of the expectation $E[(C(t) - C(s))^2]$ can be found in [79] (Section 3.3). The limits of the second moments of the increments of B and C are zero as $s \to t$. ■

Example 3.11: The Brownian motion process (Example 3.4) and the compound Poisson process (Example 3.9) are a.s. continuous at each $t \geq 0$.

Proof: The random variable $B(t) - B(t - 1/n)$ is Gaussian with mean zero and variance $1/n$ for any integer $n \geq 1$. For an $\varepsilon > 0$ and $n = 1, 2, \ldots$ define the sequence of events $A_n(\varepsilon) = \{|B(t) - B(t - 1/n)| > \varepsilon\}$. By Fatou's lemma (Section 2.3.6) we have

$$P(\liminf_{n\to\infty} A_n(\varepsilon)) \leq \liminf_{n\to\infty} P(A_n(\varepsilon)) = \liminf_{n\to\infty} 2\,(1 - \Phi(\sqrt{n}\,\varepsilon)) = 0.$$

Consider now the compound Poisson process C and define the sequence of events $A_n = \{|C(t) - C(t-1/n)| > 0\}, n \geq 1$. We have $P(\liminf_n A_n) \leq \liminf_n P(A_n)$, where $P(A_n)$ is the probability $1 - e^{-\lambda/n}$ that C has at least a jump in the interval $(t - 1/n, t]$. Because $P(A_n)$ converges to zero as $n \to \infty$, we have $P(\Omega_t) = 0$ (Eq. 3.5). ■

Example 3.12: Almost sure continuity at each time t does not imply sample continuity. ◇

Proof: If X is a.s. continuous at each $t \geq 0$, the probability of the set Ω_t in Eq. 3.5 is zero at each time. However, the probability of $\cup_{t\geq 0}\Omega_t$ may not be zero or may not even be defined since this set is an uncountable union of measurable sets, which is not necessarily measurable.

For example, let $(\Omega, \mathcal{F}, P)$ be a probability space with $\Omega = [0, 1]$, $\mathcal{F} = \mathcal{B}([0, 1])$, and $P(d\omega) = d\omega$ and let X be a real-valued stochastic process defined by $X(t, \omega) = 0$ and $X(t, \omega) = 1$ for $t < \omega$ and $t \geq \omega$, respectively. Because Ω_t in Eq. 3.5 is $\{t\}$, we have $P(\Omega_t) = 0$ so that X is continuous a.s. at each $t \in [0, 1]$. However, the probability of $\cup_{t\in[0,1]}\Omega_t = [0, 1] = \Omega$ is equal to 1 so that the process is not sample continuous. Another example is the compound Poisson process that has a.s. continuous samples at each time t but its samples are piece-wise constant functions. The Poisson process has the same property. ■

Example 3.13: The Brownian motion B is sample continuous while the compound Poisson process C does not have continuous samples (Fig. 3.5). ◇

Proof: Because $B(t+h) - B(t) \sim N(0, h), h > 0$, for any $t \geq 0$, we have $E[(B(t+h) - B(t))^4] = 3\,h^2$ so that the Kolmogorov condition in Eq. 3.2 is satisfied for $\alpha = 4$, $\beta = 1$, and $\zeta \geq 3$. That C is not sample continuous can be seen from Fig. 3.5. ■

A process X is called **càdlàg** if its sample paths are a.s. right continuous with left limits. X is called **càglàd** if its sample paths are a.s. left continuous with right limits. The names are acronyms for the French "continu à droite, limites à gauche" and "continu à gauche, limites à droite", respectively.

Note: The Brownian motion is both càdlàg and càglàd because it has continuous samples almost surely. The samples of the compound Poisson process C are right continuous with left limit so that C is càdlàg. The difference $C(t+) - C(t-) = C(t) - C(t-)$ at time t is one of the random variables Y_k if C has a jump at t and zero otherwise. ▲

3.4 Stopping times

In Section 2.16 we have defined and illustrated stopping times associated with time series. We consider now stopping times related to stochastic processes. For example, let $X(t)$, $t \geq 0$, be the state of a system and assume that X has continuous samples. Denote by T the first time when X exceeds a critical value causing system failure and by $\tilde{T}$ the time when X reaches its largest value during a time interval $[0, \tau]$, $\tau > 0$. To determine whether T is smaller or larger than a time $t \geq 0$, it is sufficient to monitor X up to this time. Hence, the event $\{T \leq t\}$ is in $\mathcal{F}_t = \sigma(X(s), 0 \leq s \leq t)$. On the other hand, we need to have complete information on X in $[0, \tau]$ to tell whether $\tilde{T}$ is smaller than $t < \tau$. The information content of $\mathcal{F}_t$ is insufficient to make this assertion. Hence, $\{\tilde{T} \leq t\}$ is not in $\mathcal{F}_t$.

Let $(\Omega, \mathcal{F}, (\mathcal{F}_t)_{t \geq 0}, P)$ be a filtered probability space and $T : \Omega \to [0, \infty]$ be a random variable defined on this space. If $\{\omega : T(\omega) \leq t\} \in \mathcal{F}_t$ for all $t \geq 0$, then T is said to be an $\mathcal{F}_t$-**stopping time** or a **stopping time**.

Note: A stopping time T is said to be **finite** a.s. if $P(T < \infty) = 1$ or $P(T = \infty) = 0$. ▲

Example 3.14: Any positive constant is a stopping time on an arbitrary filtered probability space $(\Omega, \mathcal{F}, (\mathcal{F}_t)_{t \geq 0}, P)$. ◇

Proof: Let $T = a$, where $a \geq 0$ is a real number. Then $\{\omega : T(\omega) \leq t\}, t \geq 0$, is $\emptyset$ if $a > t$ and the sample space Ω if $a \leq t$ so that it is in $\mathcal{F}_t, t \geq 0$. ■

- If T is an $\mathcal{F}_t$-stopping time, then $\{T < t\} \in \mathcal{F}_t$ and $\{T = t\} \in \mathcal{F}_t, t \geq 0$.
- If $\{T < t\} \in \mathcal{F}_t$ for all $t \geq 0$, then T is an $\mathcal{F}_t$-stopping time.

Proof: We have $\{T < t\} = \cup_{n=1}^{\infty} \{T \leq t - 1/n\}$ and $\{T \leq t - 1/n\} \in \mathcal{F}_{t-1/n} \subset \mathcal{F}_t$ so that $\{T < t\}$ is in $\mathcal{F}_t$. The event $\{T = t\}$ is equal to $\{T \leq t\} \cap \{T < t\}^c$ so that it is in $\mathcal{F}_t$.

Suppose now that $\{T < t\} \in \mathcal{F}_t$ for all $t \geq 0$. For an arbitrary $t \geq 0$ and integer $m \geq 1$ we have $\{T \leq t\} = \cap_{n=m}^{\infty} \{T < t + 1/n\}$ so that $\{T \leq t\} \in \mathcal{F}_{t+1/m}$ since $\{T < t + 1/n\} \in \mathcal{F}_{t+1/n} \subset \mathcal{F}_{t+1/m}$. Because m is arbitrary we have $\{T \leq t\} \in \cap_{m=1}^{\infty} \mathcal{F}_{t+1/m} = \mathcal{F}_{t+} = \mathcal{F}_t$, where the latter equality holds since $\mathcal{F}_t$ is assumed to be a right continuous filtration. ■

Let $(\Omega, \mathcal{F}, (\mathcal{F}_t)_{t \geq 0}, P)$ be a filtered probability space, $T_1, T_2, \ldots$ be stopping times on this space, and $c \in [0, \infty]$ a constant. Then

- $c + T_1$ and $c \wedge T_1$ are $\mathcal{F}_t$-stopping times.
- $\sup_{n \geq 1} T_n$ and $\inf_{n \geq 1} T_n$ are $\mathcal{F}_t$-stopping times.
- $\liminf_{n \to \infty} T_n$, $\limsup_{n \to \infty} T_n$ are $\mathcal{F}_t$-stopping times.

Proof: If $c = \infty$, then $c + T_1 = \infty$ is a constant, which is a stopping time. Also, the random variable $c \wedge T_1 = T_1$ is a stopping time. If $c < \infty$, then $\{c + T_1 \leq t\} = \{T_1 \leq t - c\}$ is $\emptyset$

for $t < c$ and $\{T_1 \leq t - c\} \in \mathcal{F}_{t-c} \subset \mathcal{F}_t$ for $t \geq c$. The event $\{c \wedge T_1 \leq t\}$ is Ω for $t \geq c$ and $\{T_1 \leq t\}$ for $t < c$. Hence, $c + T_1$ and $c \wedge T_1$ are stopping times.

The statements in the second bullet results from $\{\sup_{n\geq 1} T_n \leq t\} = \cap_{n\geq 1}\{T_n \leq t\}$, $\{\inf_{n\geq 1} T_n < t\} = \cup_{n=1}^{\infty}\{T_n < t\}$, and $\{T_n < t\} \in \mathcal{F}_t$ by the right continuity of the filtration $\mathcal{F}_t$, $t \geq 0$.

The statements in the last bullet follow from the definitions,

$$\liminf_{n\to\infty} T_n = \sup_{m\geq 1} \inf_{n\geq m} T_n \quad \text{and} \quad \limsup_{n\to\infty} T_n = \inf_{m\geq 1} \sup_{n\geq m} T_n,$$

of " lim inf " and " lim sup ". ■

Let $X(t)$, $t \geq 0$, be a right continuous adapted process and B denote an open set. Then the **first entrance time** $T = \inf\{t \geq 0 : X(t) \in B\}$ is an $\mathcal{F}_t$-stopping time.

Proof: Since X is a right continuous process and B is an open set, we have $\{T < t\} = \cup_{0\leq s<t,\ s=\text{rational}}\{X(s) \in B\}$ and $\{X(s) \in B\} \in \mathcal{F}_s \subset \mathcal{F}_t$. Hence, T is an $\mathcal{F}_t$-stopping time because $\mathcal{F}_t = \mathcal{F}_{t+}$ by assumption. ■

Example 3.15: Let $T = \inf\{t \geq 0 : B(t) > a\}$ be the first time when a Brownian motion B exceeds a level $0 < a < \infty$, that is, the first entrance time of B in (a, ∞), and let $\mathcal{F}_t$, $t \geq 0$, denote a right continuous filtration such that $\mathcal{F}_t \supset \sigma(B(s), 0 \leq s \leq t)$. Then T is an $\mathcal{F}_t$-stopping time. Figure 3.6 shows the density of T for $a = 1, 2, 3$ calculated from $P(T \leq t) = 2\,[1 - \Phi(a/\sqrt{t})]$ by differentiation. ◇

Proof: T is an $\mathcal{F}_t$-stopping time by the previous property. We have $P(B(t) > a) = 1 - \Phi(a/\sqrt{t})$ and

$$\begin{aligned} P(B(t) > a) &= P(B(t) > a \mid T > t)\, P(T > t) + P(B(t) > a \mid T \leq t)\, P(T \leq t) \\ &= (1/2)\, P(T \leq t), \end{aligned}$$

where Φ denotes the distribution of the standard Gaussian variable. The last equality in the above formula holds because $P(B(t) > a \mid T > t) = 0$ by the definition of T, and $P(B(t) > a \mid T \leq t) = 1/2$ because $B(T) = a$ prior to t for $T \leq t$ and the Brownian motion is symmetric with respect to its starting point. ■

Example 3.16: Let T be an $\mathcal{F}_t$-stopping time and $0 = t_0 < t_1 < \cdots < t_k < \cdots$ a sequence of numbers such that $t_k \to \infty$ as $k \to \infty$. Define an approximation $\hat{T}$ of T by $\hat{T} = t_{k+1}$ if $T \in [t_k, t_{k+1})$ and $\hat{T} = \infty$ if $T = \infty$. Then $\hat{T}$ is an $\mathcal{F}_t$-stopping time. ◇

Proof: If $t \in [t_k, t_{k+1})$, we have $\{\hat{T} \leq t\} = \{T < t_k\}$ and the last event is in $\mathcal{F}_{t_k} \subset \mathcal{F}_t$. Note that the random variable $\tilde{T} = t_k$ if $T \in [t_k, t_{k+1})$ and $\tilde{T} = \infty$ if $T = \infty$ is not an $\mathcal{F}_t$-stopping time. ■

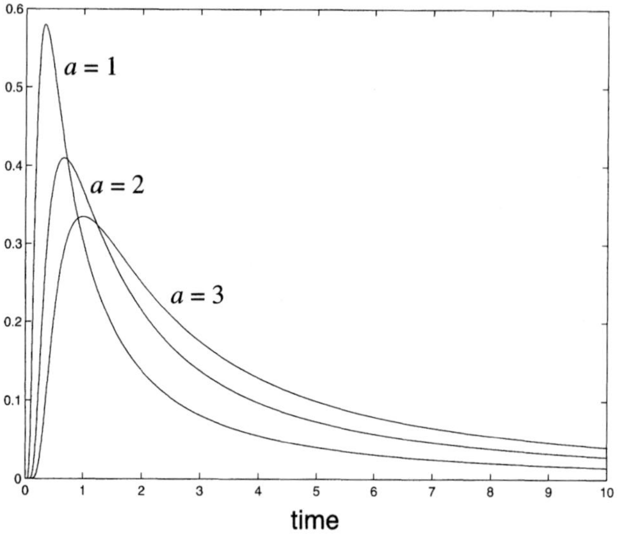

Figure 3.6. Density of the stopping time $T = \inf\{t \geq 0 : B(t) \geq a\}$ for $a = 1, 2, 3$

We define now two σ-fields including events that occur up to a stopping time T using similar considerations as in Section 2.16.

Let T be an $\mathcal{F}_t$-stopping time. Then

$$\begin{aligned} \mathcal{F}_T &= \{A \in \mathcal{F} : A \cap \{T \leq t\} \in \mathcal{F}_t, \text{ for all } t \geq 0\}, \\ \mathcal{F}_{T+} &= \{A \in \mathcal{F} : A \cap \{T \leq t\} \in \mathcal{F}_{t+}, \text{ for all } t \geq 0\} \end{aligned} \tag{3.8}$$

are sub-σ-fields of $\mathcal{F}$ including events in $\mathcal{F}$ "occurring up to time T."

Note: $\mathcal{F}_T$ and $\mathcal{F}_{T+}$ include all events in $\mathcal{F}$ occurring prior to T. If the filtration $(\mathcal{F}_t)_{t\geq 0}$ is right continuous as considered in our discussion, then $\mathcal{F}_T$ and $\mathcal{F}_{T+}$ coincide. The proof that $\mathcal{F}_T$ and $\mathcal{F}_{T+}$ are sub-σ-fields of $\mathcal{F}$ can be based on arguments as in Section 2.16 ([61], Proposition 1.4, p. 52). ▲

In Section 2.18.2 we have defined stopped martingales in the discrete time setting. This definition can be extended to stochastic processes. Let $\boldsymbol{X}$ be a stochastic process defined on a probability space $(\Omega, \mathcal{F}, P)$ endowed with a filtration $(\mathcal{F}_t)_{t\geq 0}$ and let T, T' be $\mathcal{F}_t$-stopping times. Then

$$\boldsymbol{X}^T(t) = \boldsymbol{X}(t \wedge T) \tag{3.9}$$

is said to be the process $\boldsymbol{X}$ **stopped** at T. The following properties are valid.

- T and $T \wedge T'$ are $\mathcal{F}_T$-measurable.
- If $T \leq T'$, then $\mathcal{F}_T \subseteq \mathcal{F}_{T'}$.
- If $\boldsymbol{X}$ is $\mathcal{F}_t$-progressive, then $\boldsymbol{X}(T)$ is $\mathcal{F}_T$-measurable on $\{T < \infty\}$.
- Let $\mathcal{G}_t$ denote $\mathcal{F}_{T \wedge t}$ for each $t \geq 0$. Then $(\mathcal{G}_t)_{t \geq 0}$ is a filtration and the stopped process $\boldsymbol{X}^T$ is both $\mathcal{G}_t$ and $\mathcal{F}_t$-progressively measurable.

Note: Some of these properties will be used, for example, in Chapters 6 and 7 to solve deterministic partial differential equations and assess the evolution of mechanical, ecological, or other systems. The proof of these and related properties can be found in [61] (Proposition 1.4, p. 52). ▲

3.5 Finite dimensional distributions and densities

Let $\{\boldsymbol{X}(t), t \geq 0\}$ be an $\mathbb{R}^d$-valued stochastic process defined on a probability space $(\Omega, \mathcal{F}, P)$. Let $n \geq 1$ be an integer, $t_i \geq 0$, $i = 1, \ldots, n$, be arbitrary distinct times, and set $\mathcal{X}_n = (\boldsymbol{X}(t_1), \ldots, \boldsymbol{X}(t_n))$.

The **finite dimensional distributions (f.d.d.'s) of order** n of $\boldsymbol{X}$ are the distributions of the random vectors $\mathcal{X}_n$ (Section 2.11.1), that is,

$$F_n(\boldsymbol{x}^{(1)}, \ldots, \boldsymbol{x}^{(n)}; t_1, \ldots, t_n) = P(\cap_{i=1}^n \{\boldsymbol{X}(t_i) \in \times_{k=1}^d (-\infty, x_{i,k}]\}), \quad (3.10)$$

where $\boldsymbol{x}^{(i)} = (x_{i,1}, \ldots, x_{i,d}) \in \mathbb{R}^d$ and $n \geq 1$ is an integer.

Note: This definition is meaningful since $\times_{k=1}^d (-\infty, x_{i,k}]$ are Borel sets so that the sets $\{\boldsymbol{X}(t_i) \in \times_{k=1}^d (-\infty, x_{i,k}]\}$ are in $\mathcal{F}$. The finite dimensional distributions F_n in Eq. 3.10 are probability measures induced by $\boldsymbol{X}$ on the measurable space $(\mathbb{R}^{n\,d}, \mathcal{B}^{n\,d})$ (Sections 2.4.1 and 2.10.1). The **marginal distribution** $F_1(\cdot; t)$ of $\boldsymbol{X}$ at time t is also denoted by $F(\cdot; t)$. ▲

If X is a real-valued stochastic process ($d = 1$), its finite dimensional distributions are

$$F_n(x_1, \ldots, x_n; t_1, \ldots, t_n) = P(X(t_1) \leq x_1, \ldots, X(t_n) \leq x_n), \quad (3.11)$$

where $x_i \in \mathbb{R}$, $i = 1, \ldots, n$. The corresponding **finite dimensional densities** of X can be obtained from

$$f_n(x_1, \ldots, x_n; t_1, \ldots, t_n) = \frac{\partial^n}{\partial x_1 \cdots \partial x_n} F(x_1, \ldots, x_n; t_1, \ldots, t_n) \quad (3.12)$$

provided that these derivatives exist. The **marginal density** of X is the density of the random variable $X(t)$ and is denoted by $f_1(\cdot; t)$ or $f(\cdot; t)$.

Generally, the information available on a stochastic process $\boldsymbol{X}$ in applications is sufficient to estimate at most its first and second order finite dimensional distributions. Hence, assumptions need to be made to specify all finite dimensional distributions of X. Two difficulties may be encountered. First, the postulated finite dimensional distributions can be inconsistent with the physics of the phenomenon modeled by X. For example, the Gaussian distribution is an inadequate model for the strength of a physical system because strength is positive and bounded. Second, an arbitrary collection of distributions F_n, $n = 1, 2, \ldots$, may not define a stochastic process.

If a collection of finite dimensional distributions F_n, $n = 1, 2, \ldots$, satisfies the **consistency** and the **symmetry** conditions, then it can be used to define a stochastic process. A similar statement holds for random fields ([1], Section 1.5).

Note: A collection of distributions F_n, $n = 1, 2, \ldots$, of the type in Eq. 3.10 satisfies the **consistency** condition if any F_m of order $m < n$ can be derived from F_n. For example,

$$F_m(x_1, \ldots, x_m; t_1, \ldots, t_m) = F_n(x_1, \ldots, x_m, \infty, \ldots, \infty; t_1, \ldots, t_m, \ldots, t_n). \quad (3.13)$$

The family of distributions F_n, $n = 1, 2, \ldots$, satisfies the **symmetry** condition if the functions F_n are invariant when their arguments x_i and t_i are subjected to the same permutation. ▲

Example 3.17: Denote by $\Omega = C([0, 1])$ the collection of real-valued continuous functions defined on $[0, 1]$ and let $X : [0, 1] \times C([0, 1]) \to \mathbb{R}$ be given by $X(t, \omega) = \omega(t)$, where ω is a member of $C([0, 1])$. The family of finite dimensional distributions

$$F_n(x_1, \ldots, x_n; t_1, \ldots, t_n) = \prod_{k=1}^{n} \Phi(x_k), \qquad n = 1, 2, \ldots,$$

satisfies the conditions of consistency and symmetry. Hence, this family can be used to define a stochastic process but not the process X in this example. ◇

Proof: That the family of distributions F_n satisfies the conditions of consistency and symmetry follows from its definition.

However, this family cannot be used to define the process X since (1) the sequence of events $A_n = \{\omega : X(t, \omega) > \varepsilon, X(t + 1/n, \omega) < -\varepsilon\}$ converges to the empty set as $n \to \infty$ by the continuity of the samples of X so that $P(A_n) \to P(\emptyset) = 0$ as $n \to \infty$ for every $\varepsilon > 0$ and (2) $P(X(t) > \varepsilon, X(s) < -\varepsilon) = (\Phi(-\varepsilon))^2$ for any $t \neq s$ and $\varepsilon > 0$ by the definition of F_n so that

$$P(A_n) = P(X(t) > \varepsilon, X(t + 1/n) < -\varepsilon) = (\Phi(-\varepsilon))^2, \quad n = 1, 2, \ldots,$$

does not approach zero as $n \to \infty$, a contradiction. The conclusion is somewhat obvious in this example since the assumptions that $X(t)$ is independent of $X(s)$ for $t \neq s$ and that X has continuous samples are at variance. ■

The finite dimensional distributions of a process X do not provide information on all of its statistics. For example, the probability of the set $\{|X(t)| < a, 0 \le t \le 1\} = \cap_{0\le t\le 1}\{|X(t)| < a\}$, $a > 0$, depends on an uncountable number of values of X so that it cannot be calculated from its f.d.d.'s. Moreover, this set may not even be in $\mathcal{F}$ since it is the intersection of an uncountable number of events. Additional conditions are needed in the definition of X to assure that $\{|X(t)| < a, 0 \le t \le 1\}$ and other similar sets are in $\mathcal{F}$.

3.6 Classes of stochastic processes

We define stationary, ergodic, Markov, independent increment, Gaussian, translation, and mixture of translation processes, and illustrate them by examples. These processes are used frequently in applications [79, 175].

3.6.1 Stationary processes

An $\mathbb{R}^d$-valued stochastic process $\boldsymbol{X}$ is said to be **stationary in the strict sense** or **stationary** if

$$(\boldsymbol{X}(t_1), \ldots, \boldsymbol{X}(t_n)) \stackrel{d}{=} (\boldsymbol{X}(t_1+\tau), \ldots, \boldsymbol{X}(t_n+\tau)) \tag{3.14}$$

for any $n \ge 1$, distinct times t_i, $i = 1, \ldots, n$, and time shift τ.

Note: The condition in Eq. 3.14 is equivalent to the requirement that all finite dimensional distributions of $\boldsymbol{X}$ are invariant under an arbitrary time shift. If $\boldsymbol{X}$ is stationary, then the finite dimensional distributions of $\boldsymbol{X}$ depend on only the lag between the times $\{t_1, \ldots, t_n\}$ rather than their values. Hence, the marginal distribution of $\boldsymbol{X}$ is time invariant. ▲

Example 3.18: The Brownian motion B is not a stationary process since $B(t) \sim N(0, t)$ so that its marginal distribution, $F(x; t) = P(B(t) \le x) = \Phi(x/\sqrt{t})$, changes in time. ◇

Example 3.19: Let Y_k be independent identically distributed random variables that are independent of a random variable Y. The time series $X_k = Y + Y_k$, $k = 1, 2, \ldots$, is stationary. ◇

Proof: The finite dimensional distributions of the process $X = (X_1, X_2, \ldots)$ are

$$F_n(x_1, \ldots, x_n; k_1, \ldots, k_n) = P(\cap_{i=1}^n \{Y + Y_{k_i} \le x_i\})$$
$$= \int_{\mathbb{R}} \left[\prod_{i=1}^n P(Y_{k_i} \le x_i - y)\right] dF_Y(y) = \int_{\mathbb{R}} \left[\prod_{i=1}^n F_1(x_i - y)\right] dF_Y(y),$$

where F_1 and F_y denote the distributions of Y_1 and Y, respectively. The expression of F_n shows that X is a stationary time series.

Note that the time series $Z_k = kY + Y_k$ is not stationary because its f.d.d.'s $F_n(x_1, \ldots, x_n; k_1, \ldots, k_n)$ depend on the times k_i. ■

3.6.2 Ergodic processes

An $\mathbb{R}^d$-valued stochastic process $\boldsymbol{X}$ is said to be **ergodic** if ensemble averages equal time averages, that is, if we have

$$E[g(\boldsymbol{X}(t))] = \lim_{\tau \to \infty} \frac{1}{\tau} \int_{-\tau/2}^{\tau/2} g(\boldsymbol{X}(s))\, ds \quad \text{a.s.} \tag{3.15}$$

for any real-valued measurable function g such that $E[|g(\boldsymbol{X}(t))|] < \infty$.

Note: The condition in Eq. 3.15 states that the ensemble average $E[g(\boldsymbol{X}(t))]$ can be calculated from the temporal average $(1/\tau)\int_{-\tau/2}^{\tau/2} g(\boldsymbol{X}(s))\, ds$ on a sample path of $\boldsymbol{X}$ of length $\tau \to \infty$ ([176], Section 3.2.3, [197], Section 2.3). ▲

Example 3.20: The time series $X = (X_1, X_2, \ldots)$ in Example 3.19 is ergodic if $Y = c$ is a constant. Otherwise, X is not an ergodic time series. ◇

Proof: If $Y = c$, the random variables $g(c + Y_k)$ are independent so that the sum

$$(1/n) \sum_{k=1}^{n} g(c + Y_k)$$

converges a.s. to $E[g(c+Y_1)]$ as $n \to \infty$ by the strong law of large numbers (Section 2.13), provided that $E[|g(c + Y_1)|]$ is finite.

If Y is not a degenerated random variable, X is not ergodic because Eq. 3.15 does not hold for all measurable functions g. For example, if $g(x) = x$ the left and right sides of Eq. 3.15 are $E[Y] + E[Y_1]$ and $Y(\omega) + E[Y_1]$, respectively, where $Y(\omega)$ is a sample of the random variable Y. ■

Weaker ergodicity conditions can be defined by restricting the form of the function g in Eq. 3.15. For example, this equation gives the conditions for **ergodicity of the mean** and **ergodicity of the marginal distribution** for $g(\boldsymbol{x}) = \boldsymbol{x}$ and $g(\boldsymbol{x}) = 1_{\times_{i=1}^d (-\infty, \xi_i]}(\boldsymbol{x})$ respectively, where $\boldsymbol{\xi} = (\xi_1, \ldots, \xi_d) \in \mathbb{R}^d$ ([141], Section 9.8).

3.6.3 Markov processes

We limit our discussion to real-valued processes for simplicity. The extension to $\mathbb{R}^d$-valued processes is straightforward.

A real-valued stochastic process X is **Markov** if for every integer $n \geq 1$ and times $t_1 < \cdots < t_n$ the conditional random variables $X(t_n) \mid X(t_{n-1})$ and $(X(t_{n-2}), \ldots, X(t_1)) \mid X(t_{n-1})$ are independent, that is, the past of X is independent of its future conditional on the present.

Note: This definition implies

$$\begin{aligned} & f_{X(t_n),X(t_{n-2}),\ldots,X(t_1)|X(t_{n-1})}(x_n, x_{n-2}, \ldots, x_1 \mid x_{n-1}) \\ & \quad = f_{X(t_{n-2}),\ldots,X(t_1)|X(t_{n-1})}(x_{n-2}, \ldots, x_1 \mid x_{n-1})\, f_{X(t_n)|X(t_{n-1})}(x_n \mid x_{n-1}), \end{aligned} \tag{3.16}$$

where the subscripts of f indicate the reference random variables. For example, the function $f_{X(t_n)|X(t_{n-1})}$ is the density of $X(t_n) \mid X(t_{n-1})$. The alternative forms of this definition,

$$\begin{aligned} & f_n(x_1, \ldots, x_n; t_1, \ldots, t_n) \\ & \quad = f_{n-1}(x_1, \ldots, x_{n-1}; t_1, \ldots, t_{n-1})\, f_{X(t_n)|X(t_{n-1})}(x_n \mid x_{n-1}) \end{aligned} \tag{3.17}$$

and

$$f_{X(t_n)|\{X(t_{n-1}),\ldots,X(t_1)\}}(x_n \mid x_{n-1}, \ldots, x_1) = f_{X(t_n)|X(t_{n-1})}(x_n \mid x_{n-1}), \tag{3.18}$$

can be obtained from Eq. 3.16 by multiplying with the density of $X(t_{n-1})$ and from Eq. 3.17 by dividing with the density of $(X(t_{n-1}), \ldots, X(t_1))$, respectively. A Markov process can also be defined by the condition $E[g(X(t+s)) \mid \mathcal{F}_t^X] = E[g(X(t+s)) \mid X(t)]$, where $t, s \geq 0$ are arbitrary times and $g : \mathbb{R} \to \mathbb{R}$ is a Borel function ([61], p. 156). ▲

The density of the conditional random variable $X(t) \mid X(s)$, $t > s$, called the **transition density** of X, has the following remarkable properties.

If X is a Markov process, then all its finite dimensional densities are specified completely by its marginal and transition densities.

Proof: Repeated application of Eq. 3.17 gives

$$f_n(x_1, \ldots, x_n; t_1, \ldots, t_n) = f(x_1; t_1) \prod_{i=2}^{n} f_{X(t_i)|X(t_{i-1})}(x_i \mid x_{i-1}),$$

where $f(x_1; t_1) = f_1(x_1; t_1)$ is the density of $X(t_1)$. ■

Chapman-Kolmogorov equation. The conditional densities $f_{X(t)|X(t_0)}$, $f_{X(t)|X(s)}$, and $f_{X(s)|X(t_0)}$, $t_0 < s < t$, are related by

$$f_{X(t)|X(t_0)}(x \mid x_0) = \int_{\mathbb{R}} f_{X(t)|X(s)}(x \mid y)\, f_{X(s)|X(t_0)}(y \mid x_0)\, dy. \tag{3.19}$$

Proof: The density of the conditional random variable $X(t) \mid X(t_0)$, $t_0 < s < t$, is

$$\begin{aligned} f_{X(t)|X(t_0)}(x \mid x_0) &= \frac{f_2(x_0, x; t_0, t)}{f_1(x_0; t_0)} = \frac{1}{f_1(x_0; t_0)} \int_{\mathbb{R}} f_3(x_0, y, x; t_0, s, t)\, dy \\ &= \int_{\mathbb{R}} \frac{f_3(x_0, y, x; t_0, s, t)}{f_2(x_0, y; t_0, s)} \frac{f_2(x_0, y; t_0, s)}{f_1(x_0; t_0)}\, dy, \end{aligned}$$

which gives Eq. 3.19. ■

Example 3.21: Let $X(t) = \sum_{k=1}^{t} Y_k$ be a discrete time process, where $t \geq 1$ is an integer and Y_k, $k = 1, 2, \ldots$, are iid random variables. The process X is Markov and $\varphi_{X(t)|(X(s)=y)}(u) = e^{\sqrt{-1}\,u\,y} \prod_{k=s+1}^{t} \varphi_{Y_1}(u)$ is the characteristic function of $X(t) \mid X(s)$, $t > s$. ◇

Proof: For $t > s$ we have $X(t) \mid (X(s) = y) = y + \sum_{k=s+1}^{t} Y_k$ so that future values $X(t)$ of X are independent of its past values conditional on $X(s) = y$. The characteristic function results from the expression of $X(t) \mid (X(s) = y)$, where φ_{Y_1} denotes the characteristic function of Y_1. ■

3.6.4 Independent increment processes

• An $\mathbb{R}^d$-valued stochastic process $\boldsymbol{X}$ is said to have **independent increments** if the random variables $\boldsymbol{X}(t) - \boldsymbol{X}(v)$ and $\boldsymbol{X}(u) - \boldsymbol{X}(s)$, $s < u \leq v < t$, are independent.

• If $\boldsymbol{X}$ has independent increments and the distribution of $\boldsymbol{X}(t) - \boldsymbol{X}(s)$, $s \leq t$, depends on only the time lag $t - s$ rather than the values of the times s and t, then $\boldsymbol{X}$ is said to have **stationary independent increments**.

Note: The Brownian motion and the compound Poisson processes in Examples 3.4 and 3.9 have stationary independent increments. For example, by the definition of the Brownian motion process B, the increments of B over non-overlapping intervals are independent and $B(t) - B(s) \sim N(0, t - s)$, $t \geq s$. ▲

The finite dimensional densities of a real-valued process X with $X(0) = 0$ and independent increments are

$$f_n(x_1, \ldots, x_n; t_1, \ldots, t_n) = f_{Y_1}(x_1)\, f_{Y_2}(x_2 - x_1) \ldots f_{Y_n}(x_n - x_{n-1}), \quad (3.20)$$

where f_{Y_i} denotes the density of $Y_i = X(t_i) - X(t_{i-1})$, $i = 1, \ldots, n$.

Proof: Let $0 = t_0 < t_1 < t_2 < \cdots < t_n$ be n arbitrary times. Then

$$P(X(t_1) \leq x_1, \ldots, X(t_n) \leq x_n) = P\left(Y_1 \leq x_1, Y_1 + Y_2 \leq x_2, \ldots, \sum_{k=1}^{n} Y_k \leq x_n\right)$$

so that the f.d.d. of order n of X is

$$P(X(t_1) \leq x_1, \ldots, X(t_n) \leq x_n) = \int_{-\infty}^{x_1} dy_1\, f_{Y_1}(y_1) \int_{-\infty}^{x_2 - y_1} dy_2\, f_{Y_2}(y_2) \cdots$$

$$\cdots \int_{-\infty}^{x_{n-1} - \sum_{k=1}^{n-2} y_k} dy_{n-1}\, f_{Y_{n-1}}(y_{n-1}) \int_{-\infty}^{x_n - \sum_{k=1}^{n-1} y_k} dy_n\, f_{Y_n}(y_n)$$

where f_{Y_k} denotes the density of Y_k. The corresponding finite dimensional density of order n of X can be obtained from the above equation by differentiation or from the density of the random vector $(Y_1, \ldots, Y_n)$ and its relation to $(X(t_1), \ldots, X(t_n))$. ■

Let X be a real-valued process with stationary independent increments. The following three properties of X are useful in applications.

If X is a real-valued process with stationary independent increments, then X is a Markov process.

Proof: The densities of the conditional variables $X(t_n) \mid \{X(t_{n-1}), \ldots, X(t_1)\}$ and $X(t_n) \mid X(t_{n-1})$ coincide (Eq. 3.20). This property shows that the Brownian motion and the compound Poisson processes are Markov. ■

If X is a real-valued process with stationary independent increments, then

$$P(X(t+\tau_1+\tau_2) - X(t) \le x) = \int_{\mathbb{R}} F_{\tau_2}(x-y)\, f_{\tau_1}(y)\, dy,$$

$$f_{\tau_1+\tau_2}(x) = \int_{\mathbb{R}} f_{\tau_2}(x-y)\, f_{\tau_1}(y)\, dy, \quad \text{and}$$

$$\varphi_{\tau_1+\tau_2}(u) = \varphi_{\tau_1}(u)\, \varphi_{\tau_2}(u) \tag{3.21}$$

for $\tau_1, \tau_2 > 0$, where F_τ, f_τ, φ_τ denote the distribution, the density, and the characteristic function of $X(t+\tau) - X(t)$, $\tau > 0$.

Proof: Set $\Delta_1 = X(t+\tau_1) - X(t)$ and $\Delta_2 = X(t+\tau_1+\tau_2) - X(t+\tau_1)$ for $\tau_1, \tau_2 > 0$. The distribution of $X(t+\tau_1+\tau_2) - X(t) = \Delta_1 + \Delta_2$ is

$$P(X(t+\tau_1+\tau_2) - X(t) \le x) = P(\Delta_1 + \Delta_2 \le x) = \int_{\mathbb{R}} F_{\tau_2}(x-y)\, f_{\tau_1}(y)\, dy$$

since the random variables Δ_1 and Δ_2 are independent. The density of $X(t+\tau_1+\tau_2) - X(t)$ results from the differentiation of the above equation. The characteristic function of $X(t+\tau_1+\tau_2) - X(t)$ is the product of the characteristic functions of Δ_1 and Δ_2, that is, $\varphi_{\tau_1+\tau_2}(u) = \varphi_{\tau_1}(u)\, \varphi_{\tau_2}(u)$. This relationship becomes $\varphi_{\sum_{k=1}^m \tau_k}(u) = \prod_{k=1}^m \varphi_{\tau_k}(u)$ for m non-overlapping time intervals of length τ_k, $k = 1, \ldots, m$. ■

The mean $\mu(t) = E[X(t)]$ and variance $\sigma(t)^2 = E[(X(t)-\mu(t))^2]$ of a process $X(t)$, $t \ge 0$, with $X(0) = 0$ and stationary independent increments are linear functions of time.

Proof: Properties of the expectation give

$$\begin{aligned}
\mu(t+\tau) &= E[(X(t+\tau) - X(t)) + X(t)] = \mu(\tau) + \mu(t),\\
\sigma(t+\tau)^2 &= E\{[(X(t+\tau) - X(t)) - (\mu(t+\tau) - \mu(t)) + (X(t) - \mu(t))]^2\}\\
&= E\{[(X(t+\tau) - X(t)) - (\mu(t+\tau) - \mu(t))]^2\} + E\{[X(t) - \mu(t)]^2\}\\
&= \sigma(\tau)^2 + \sigma(t)^2
\end{aligned}$$

for $t, \tau \ge 0$ so that $\mu(t) = t\,\mu(1)$ and $\sigma(t)^2 = t\,\sigma(1)^2$. This result is consistent with the evolution of mean and variance of the Brownian motion B and the compound Poisson process C. These moments are $E[B(t)] = 0$, $\mathrm{Var}[B(t)] = t$, $E[C(t)] = \lambda t\, E[Y_1]$, and $\mathrm{Var}[C(t)] = \lambda t\, E[Y_1^2]$ provided that $Y_1 \in L_2$, where $\lambda > 0$ is the intensity of the Poisson process N in the definition of C (Example 3.9). ■

Example 3.22: The finite dimensional density of order n of a Brownian motion B is

$$f_n(x_1, \ldots, x_n; t_1, \ldots, t_n) = \phi\left(\frac{x_1}{\sqrt{t_1}}\right) \phi\left(\frac{x_2 - x_1}{\sqrt{t_2 - t_1}}\right) \ldots \phi\left(\frac{x_n - x_{n-1}}{\sqrt{t_n - t_{n-1}}}\right), \tag{3.22}$$

where ϕ denotes the density of the standard Gaussian variable. The characteristic function of the increment $C(t) - C(s)$, $t \geq s$, of a compound Poisson process $C(t) = \sum_{k=1}^{N(t)} Y_k$ is

$$E\left[e^{\sqrt{-1}\, u\, (C(t) - C(s))}\right] = e^{\lambda\,(t-s)\,[1 - \varphi_y(u)]}, \tag{3.23}$$

where $\lambda > 0$ is the intensity of the underlying Poisson process N and φ_y is the characteristic function of Y_1. ◇

Proof: Properties of Brownian motion and Eq. 3.20 give Eq. 3.22. The characteristic function of $C(t) - C(s) \stackrel{d}{=} C(t-s)$, $t \geq s$, is

$$E\left[e^{\sqrt{-1}\, u\, C(t-s)}\right] = \sum_{n=0}^{\infty} E\left[e^{\sqrt{-1}\, u\, C(t-s)} \mid N(t-s) = n\right] \frac{[\lambda\,(t-s)]^n}{n!} e^{-\lambda\,(t-s)}$$

$$= e^{-\lambda\,(t-s)} \left(1 + \sum_{n=1}^{\infty} \left(\varphi_y(u)\right)^n \frac{[\lambda\,(t-s)]^n}{n!}\right) = e^{-\lambda\,(t-s)\,[1 - \varphi_y(u)]},$$

where $\varphi_y(u) = E[e^{\sqrt{-1}\, u\, Y_1}]$. ■

3.6.5 Gaussian processes

• A stochastic process $\boldsymbol{X}$ is said to be **Gaussian** if all its finite dimensional distributions are Gaussian.

• A Gaussian process $\boldsymbol{X}$ satisfying the condition in Eq. 3.14 is called a **stationary Gaussian** process.

Note: To prove that a stochastic process $\boldsymbol{X}$ is Gaussian, we need to show that $\mathcal{X}_n = (\boldsymbol{X}(t_1), \ldots, \boldsymbol{X}(t_n)) \in \mathbb{R}^{n\,d}$ is a Gaussian vector for any integer $n \geq 1$ and times t_i, $i = 1, \ldots, n$, (Section 2.11.5). ▲

Example 3.23: Let $\boldsymbol{Z} = (Z_1, \ldots, Z_m) \in \mathbb{R}^m$ be a Gaussian vector and

$$X(t) = \sum_{k=1}^{m} Z_k\, w_k(t) = \boldsymbol{w}(t)\, \boldsymbol{Z}, \quad t \geq 0, \tag{3.24}$$

where $w_k(t)$, $k = 1, \ldots, m$, are real-valued, deterministic, and continuous functions of time and $\boldsymbol{w}(t)$ denotes a row vector with coordinates $w_k(t)$. Then X is a Gaussian process. ◇

Proof: Let $\mathcal{X}_n = (X(t_1), \ldots, X(t_n))$, where $n \geq 1$ is an integer and $(t_1, \ldots, t_n)$ denote arbitrary times. The vector $\mathcal{X}_n$ can be expressed as a linear transformation of the Gaussian vector $\boldsymbol{Z}$ so that it is Gaussian (Section 2.11.5). ■

Let $\boldsymbol{g} : [0, \infty) \times \mathbb{R}^m \to \mathbb{R}^d$ be a measurable function and $\boldsymbol{Z}$ be an $\mathbb{R}^m$-valued random variable, where $d, m \geq 1$ are integers. The function $\boldsymbol{X}(t) = \boldsymbol{g}(t, \boldsymbol{Z})$ depends on a finite number of random variables, the variables $Z_1, \ldots, Z_m$, and is called a **parametric stochastic process**. Parametric processes are used, for example, to approximate general stochastic processes for Monte Carlo simulation (Chapter 5). The process in Eq. 3.24 is a parametric Gaussian process.

3.6.6 Translation processes

An $\mathbb{R}^d$-valued stochastic process $\boldsymbol{X}$ is said to be a **translation** process if its coordinates are memoryless, measurable, nonlinear functions of an $\mathbb{R}^d$-valued Gaussian process $\boldsymbol{G}$, that is, $X_i(t) = g_i(\boldsymbol{G}(t))$ and $g_i : \mathbb{R}^d \to \mathbb{R}$, $i = 1, \ldots, d$, are measurable functions.

Note: Generally, we define the coordinates of $\boldsymbol{X}$ by $X_i(t) = g_i(G_i(t))$, $i = 1, \ldots, d$, so that $g_i : \mathbb{R} \to \mathbb{R}$ are real-valued functions defined on $\mathbb{R}$. If the processes G_i are independent and $X_i(t) = g_i(G_i(t)), i = 1, \ldots, d$, then the non-Gaussian processes X_i are independent of each other. Otherwise, the coordinates of $\boldsymbol{X}$ are dependent non-Gaussian processes. ▲

Example 3.24: Let

$$X(t) = g(G(t)) = F^{-1} \circ \Phi(G(t)), \quad t \geq 0, \tag{3.25}$$

be a real-valued translation process, where F is an arbitrary distribution with density f, Φ is the distribution of $N(0, 1)$, and G is a Gaussian process such that $E[G(t)] = 0$ and $E[G(t)^2] = 1$. The marginal distribution of X is F. The finite dimensional density of order n of X is

$$\begin{aligned} & f_n(x_1, \ldots, x_n; t_1, \ldots, t_n) \\ & = [(2\pi)^n \det(\boldsymbol{\rho})]^{-1/2} \prod_{p=1}^{n} \frac{f(x_p)}{\phi(y_p)} \exp\left(-\frac{1}{2}\, \boldsymbol{y}^T \boldsymbol{\rho}^{-1}\, \boldsymbol{y}\right), \end{aligned} \tag{3.26}$$

where $\boldsymbol{\rho} = \{E[G(t_p)\, G(t_q)]\}$, $y_p = \Phi^{-1} \circ F(x_p)$, $p, q = 1, \ldots, n$, ϕ is the density of $N(0, 1)$, and $\boldsymbol{y} = (y_1, \ldots, y_n)$. ◇

Note: The marginal distribution of X is $P(X(t) \leq x) = P(F^{-1} \circ \Phi(G(t))) = P(G(t) \leq \Phi^{-1}(F(x))) = F(x)$. The finite dimensional distribution F_n of order n of X is the probability of the event $\{X(t_1) \leq x_1, \ldots, X(t_n) \leq x_n\}$ or $\{G(t_1) \leq y_1, \ldots, G(t_n) \leq y_n\}$. The density f_n in Eq. 3.26 can be obtained by differentiating F_n. Alternatively, f_n can be calculated from the density of the Gaussian vector $(G(t_1), \ldots, G(t_n))$ in Section 2.11.5 and the change of variable in Eq. 3.25. ▲

3.6.7 Mixture of translation processes

Let $X_k(t) = g_k(G_k(t)) = \left(F^{(k)}\right)^{-1} \circ \Phi(G_k(t))$, $t \geq 0$, $k = 1, \ldots, m$, be real-valued translation processes, where $F^{(k)}$ are absolutely continuous distributions and G_k denote stationary Gaussian processes with mean zero, variance one, correlation function $\rho_k(\tau) = E[G_k(t)\, G_k(t+\tau)]$, and spectral density s_k. The processes G_k are independent of each other. The finite dimensional densities of the translation processes X_k are (Eq. 3.26)

$$f_n^{(k)}(x_1, \ldots, x_n; t_1, \ldots, t_n)$$
$$= [(2\pi)^n \det(\boldsymbol{\rho}_k)]^{-1/2} \prod_{p=1}^{n} \frac{f^{(k)}(x_p)}{\phi(y_{k,p})} \exp\left(-\frac{1}{2}\, \boldsymbol{y}_k^T\, \boldsymbol{\rho}_k^{-1}\, \boldsymbol{y}_k\right), \tag{3.27}$$

where $\boldsymbol{\rho}_k = \{E[G_k(t_p)\, G_k(t_q)]\}$, $y_{k,p} = \Phi^{-1} \circ F^{(k)}(x_p)$, $p, q = 1, \ldots, n$, ϕ is the density of $N(0,1)$, $f_1^{(k)}(x) = f^{(k)}(x) = dF^{(k)}(x)/dx$, and $\boldsymbol{y}_k = (y_{k,1}, \ldots, y_{k,n})$.

Let $\mathcal{X}$ be the collection of real-valued stochastic processes with finite dimensional densities

$$f_n(x_1, \ldots, x_n; t_1, \ldots, t_n)$$
$$= \sum_{k=1}^{m} p_k\, [(2\pi)^n \det(\boldsymbol{\rho}_k)]^{-1/2} \prod_{i=1}^{n} \frac{f^{(k)}(x_p)}{\phi(y_{k,p})} \exp\left(-\frac{1}{2}\, \boldsymbol{y}_k^T\, \boldsymbol{\rho}_k^{-1}\, \boldsymbol{y}_k\right), \tag{3.28}$$

where $p_k \geq 0$, $k = 1, \ldots, m$, are such that $\sum_{k=1}^{m} p_k = 1$. The processes in $\mathcal{X}$ are referred to as **mixtures of translation processes**. A member of $\mathcal{X}$ is said to be non-degenerate if $m > 1$ and $p_k \in (0,1)$ for all $k = 1, \ldots, m$. We now give some of the properties of these processes.

- Translation processes are included in $\mathcal{X}$. The non-degenerate members of $\mathcal{X}$ are not translation processes.
- The members of $\mathcal{X}$ are stationary non-Gaussian processes.
- The moments $\mu_k(q_1, \ldots, q_n; t_1, \ldots, t_n) = E\left[\prod_{i=1}^{n} X_k(t_i)^{q_i}\right]$ and $\mu(q_1, \ldots, q_n; t_1, \ldots, t_n) = E\left[\prod_{i=1}^{n} X(t_i)^{q_i}\right]$ of order $q = \sum_{i=1}^{n} q_i$ of X_k and X with finite dimensional densities in Eqs. 3.27 and 3.28, respectively, are related by

$$\mu(q_1, \ldots, q_n; t_1, \ldots, t_n) = \sum_{k=1}^{m} p_k\, \mu_k(q_1, \ldots, q_n; t_1, \ldots, t_n). \tag{3.29}$$

Proof: Take $m = 2$ and $n = 2$ in Eq. 3.28. Suppose that f_2 in this equation defines a non-degenerate process in $\mathcal{X}$, the densities $f_2^{(k)}$ satisfy the condition $f_2^{(k)}(x_1, x_2) =$

$f^{(k)}(x_1)\, f^{(k)}(x_2)$, $k = 1, 2$, and $\mu_k(1; t) = 0$, $k = 1, 2$. Hence, the $\mathbb{R}^2$-valued random variable $\boldsymbol{X}$ with density $f_2 = p\, f_2^{(1)} + (1-p)\, f_2^{(2)}$, $p \in (0, 1)$, has uncorrelated coordinates (Eq. 3.29). If $\boldsymbol{X}$ were a translation vector, then we would have

$$p\, f_2^{(1)}(x_1, x_2) + (1-p)\, f_2^{(2)}(x_1, x_2) = p\, f^{(1)}(x_1)\, f^{(1)}(x_2) + (1-p)\, f^{(2)}(x_1)\, f^{(2)}(x_2)$$
$$= \left(p\, f^{(1)}(x_1) + (1-p)\, f^{(2)}(x_1)\right)\left(p\, f^{(1)}(x_2) + (1-p)\, f^{(2)}(x_2)\right)$$

since uncorrelated translation variables are independent. The above equality gives

$$\left(f^{(1)}(x_1) - f^{(2)}(x_1)\right)\left(f^{(1)}(x_2) - f^{(2)}(x_2)\right) = 0$$

for all $(x_1, x_2) \in \mathbb{R}^2$. This implies that $f^{(1)}$ coincides with $f^{(2)}$ so that the mixture is degenerate in contradiction with the initial assumption.

The last two properties follow from the properties of the Gaussian processes G_k and the definitions of the processes X_k and X by elementary calculations. ■

Example 3.25: Let $\zeta(\tau, \sigma) = E[X(t)\, X(t+\tau)\, X(t+\sigma)]$ be the second order correlation function of $X \in \mathcal{X}$ with finite dimensional density in Eq. 3.28. Then $\zeta(\tau, \sigma) = \sum_{k=1}^{m} p_k\, \zeta_k(\tau, \sigma)$, where $\zeta_k(\tau, \sigma) = E[X_k(t)\, X_k(t+\tau)\, X_k(t+\sigma)]$. ◇

Note: This is a special case of Eq. 3.29. The second order correlation functions ζ and ζ_k depend on only two arguments because the processes X and X_k are stationary. ▲

3.7 Second moment properties

Many of the definitions and results in this section relate to the second moment properties of random vectors (Section 2.11.4). Let $\boldsymbol{X}$ be an $\mathbb{R}^d$-valued process in $L_2(\Omega, \mathcal{F}, P)$, that is, the coordinates of $\boldsymbol{X}$ have finite second moments at all times.

- Mean function : $\boldsymbol{\mu}(t) = E[\boldsymbol{X}(t)]$
- Correlation function : $\boldsymbol{r}(t, s) = E[\boldsymbol{X}(t)\, \boldsymbol{X}(s)^T]$
- Covariance function : $\boldsymbol{c}(t, s) = E[(\boldsymbol{X}(t) - \boldsymbol{\mu}(t))\, (\boldsymbol{X}(s) - \boldsymbol{\mu}(s))^T]$

(3.30)

Note: The moments in Eq. 3.30 exist and are finite because $\boldsymbol{X}(t) \in L_2$. For example, the entry (i, j) of $\boldsymbol{r}(t, s)$ giving the correlation function $r_{i,j}(t, s) = E[X_i(t)\, X_j(s)]$ is finite since $(E[X_i(t) X_j(s)])^2 \le E[X_i(t)^2]\, E[X_j(s)^2]$ by the Cauchy-Schwarz inequality and the expectations $E[X_i(t)^2]$ exist and are finite by hypothesis. Similar considerations show that the mean and covariance functions of $\boldsymbol{X}$ exist and are finite.

Generally, the condition $\boldsymbol{X}(t) \in L_2$, $t \ge 0$, is insufficient for the existence of moments of $\boldsymbol{X}$ of order higher than two. The class of Gaussian processes is an exception because their moments of order three and higher are related to the first two moments by algebraic equations ([79], Appendix B). ▲

The **second moment properties** of X are given by the pair of functions $(\boldsymbol{\mu}, \boldsymbol{r})$ or $(\boldsymbol{\mu}, \boldsymbol{c})$ in Eq. 3.30.

Note: The functions $(\boldsymbol{\mu}, \boldsymbol{r})$ and $(\boldsymbol{\mu}, \boldsymbol{c})$ contain the same information since the correlation and covariance functions are related by $\boldsymbol{c}(t,s) = \boldsymbol{r}(t,s) - \boldsymbol{\mu}(t)\,\boldsymbol{\mu}(t)^T$.

If $r_{i,j}(t,s) = 0$ at all times, the processes X_i and X_j are said to be **orthogonal.** If $c_{i,j}(t,s) = 0$ at all times, the processes X_i and X_j are said to be **uncorrelated**. If the processes X_i and X_j have mean zero, that is, $\mu_i(t) = 0$ and $\mu_j(t) = 0$ at all times, they are orthogonal if and only if they are uncorrelated. ▲

Example 3.26: Generally, the second moment properties of a process X are insufficient to define its finite dimensional distributions. The Gaussian process is a notable exception. ◇

Proof: Let $n \geq 1$ be an integer and t_i, $i = 1, \ldots, n$, denote arbitrary times. The finite dimensional distribution of order n of X is the distribution of the vector $\mathcal{X}_n = (X(t_1), \ldots, X(t_n))$. If X is a Gaussian process, $\mathcal{X}_n$ is a Gaussian vector whose second moment properties can be obtained from the mean and correlation functions of X. The distribution of $\mathcal{X}_n$ is completely defined by its first two moments (Section 2.11.5). ■

If a process X is stationary, then its finite dimensional distributions are invariant to a time shift so that (1) the mean function $\boldsymbol{\mu}$ is a constant and (2) the correlation and covariance functions $\boldsymbol{r}$ and $\boldsymbol{c}$ depend only on the time lag. These observations suggest the following definition.

A stochastic process X is said to be **weakly stationary** or **stationary in the weak sense** if

1. The mean function $\boldsymbol{\mu}(t) = \boldsymbol{\mu}$ is time invariant and

2. The correlation and the covariance functions $\boldsymbol{r}(t,s)$ and $\boldsymbol{c}(t,s)$ depend on only the time lag $\tau = t - s$, that is, $\boldsymbol{r}(t,s) = \boldsymbol{r}(\tau)$ and $\boldsymbol{c}(t,s) = \boldsymbol{c}(\tau)$.

Note: A stationary process in L_2 is weakly stationary but the converse is not generally true. A notable exception is the Gaussian process that is stationary if and only if it is weakly stationary. For example, let X be a real-valued weakly stationary Gaussian process with mean μ and covariance function $c(t,s) = E[(X(t) - \mu)(X(s) - \mu)]$. The vector $\mathcal{X}_n = (X(t_1), \ldots, X(t_n))$ is Gaussian with mean $\mu\,\mathbf{1}_n$ and covariance $\boldsymbol{c} = \{c(t_i - t_j)\}$, $i, j = 1, \ldots, n$, for any integer $n \geq 1$ and times $0 \leq t_1 < \cdots < t_n$, where $\mathbf{1}_n$ is an n-dimensional vector with unit entries. The finite dimensional density of order n of X at $\boldsymbol{x} = (x_1, \ldots, x_n)$, that is, the density

$$f(\boldsymbol{x}; t_1, \ldots, t_n) = [(2\pi)^n \det(\boldsymbol{c})]^{-1/2} \exp\left[-\frac{1}{2}(\boldsymbol{x} - \mu\,\mathbf{1}_n)^T\, \boldsymbol{c}^{-1}\,(\boldsymbol{x} - \mu\,\mathbf{1}_n)\right]$$

of $\mathcal{X}_n$ is invariant to a time shift.

The above notations $\boldsymbol{r}(t,s) = \boldsymbol{r}(\tau)$ and $\boldsymbol{c}(t,s) = \boldsymbol{c}(\tau)$, $\tau = t - s$, are used for convenience when dealing with weakly stationary stochastic processes. They may cause some confusion since the functions $\boldsymbol{r}(t,s)$, $\boldsymbol{c}(t,s)$ and $\boldsymbol{r}(\tau)$, $\boldsymbol{c}(\tau)$ are defined on $\mathbb{R}^2$ and $\mathbb{R}$, respectively. ▲

Example 3.27: Let X be a real-valued discrete time stochastic process defined by

$$X_{k+1} = \rho\, X_k + W_k, \quad k = 0, 1, \ldots,$$

where $\rho \in (0, 1)$, W_k are uncorrelated random variables with mean $E[W_k] = 0$ and variance $\mathrm{Var}[W_k] = 1$ that are uncorrelated to the initial value X_0 of X. The process X is weakly stationary as $k \to \infty$ but may or may not be stationary depending on the higher order properties of W_k. ◇

Proof: The recurrence formulas $\mu_{k+1} = \rho\, \mu_k$ and $\sigma_{k+1}^2 = \rho^2\, \sigma_k^2 + 1$ for the mean $\mu_k = E[X_k]$ and the variance $\sigma_k^2 = E[(X_k - \mu_k)^2]$ of X_k result by averaging the defining equation for X and its square, respectively. The covariance function of X satisfies the equation $c(k+p, k) = E[\tilde{X}_{k+p}\, \tilde{X}_k] = \rho^p\, \sigma_k^2$, where $\tilde{X}_k = X_k - \mu_k$. This equation results by taking the expectation of the product of $\tilde{X}_{k+p} = \rho^p\, \tilde{X}_k + \sum_{s=1}^{p} \rho^{s-1}\, W_{k+p-s}$ with $\tilde{X}_k$ because X_u and W_v, $v \geq u$, are not correlated. The asymptotic values of the mean, variance, and covariance functions of X as $k \to \infty$ are zero, $1/(1-\rho^2)$, and $\rho^p/(1-\rho^2)$, respectively, so that X becomes a weakly stationary process for large values of k.

However, X may not be stationary. For example, $\varphi_k(u) = E[\exp(\sqrt{-1}\, u\, X_k)]$ satisfies the recurrence formula $\varphi_{k+1}(u) = \varphi_k(\rho\, u)\, \varphi_{W_k}(u)$ if the random variables W_k and X_k are independent for all k's, where $\varphi_{W_k}(u) = E[\exp(\sqrt{-1}\, u\, W_k)]$. Suppose that X becomes stationary for large values of k and denote the stationary characteristic function of X by φ. For these values of k the recurrence formula becomes $\varphi(u) \simeq \varphi(\rho\, u)\, \varphi_{W_k}(u)$ or $\varphi(u)/\varphi(\rho\, u) \simeq \varphi_{W_k}(u)$. This equality is not possible because its left side is time invariant while its right side depends on time if, for example, W_k has the above second moment properties but different characteristic functions at each time k. ■

Example 3.28: Let $X(t) = g(G(t)) = F^{-1} \circ \Phi(G(t))$ be the translation process in Eq. 3.25. Suppose that F is such that $\int x\, dF(x) = 0$ and $\int x^2\, dF(x) = 1$ and that G is a stationary Gaussian process with mean zero, variance one, and covariance function $\rho(\tau) = E[G(t)\, G(t+\tau)]$. The covariance function of X is

$$\xi(\tau) = E[g(G(t))\, g(G(t+\tau))] = \int_{\mathbb{R}^2} g(y_1)\, g(y_2)\, \phi(y_1, y_2; \rho(\tau))\, dy_1\, dy_2, \tag{3.31}$$

where ϕ denotes the joint density of $(G(t), G(t+\tau))$ (Section 2.11.5). The covariance function ξ of X takes on values in the range $[\xi^*, 1]$, where $\xi^* = E[g(Z)\, g(-Z)]$ and Z is a standard Gaussian variable. The translation process X is weakly stationary and stationary. ◇

Proof: The formula in Eq. 3.31 results from the definition of the translation process X or a theorem by Price ([79], Section 3.1.1). This theorem also shows that $\xi(\tau)$ increases with $\rho(\tau)$. The covariance function of X is $\xi(\tau) = 0$ and 1 for $\rho(\tau) = 0$ and 1, respectively. However, $\rho(\tau) = -1$ does not imply $\xi(\tau) = -1$. If we have $\rho(\tau) = -1$, that is, $G(t) = Z$ and $G(t+\tau) = -Z$, the value of the covariance $\xi^* = E[g(Z)\, g(-Z)]$ of $(X(t), X(t+\tau))$ depends on g. For example, $\xi^* = -1$ if $g(y) = y^3$ but $\xi^* > -1$ if $g(y) = \alpha + \exp(\beta + \gamma\, G(t))$, where α, β, γ are some constants ([79], Section 3.1.1). This property of the covariance function of X is particularly relevant in applications when we attempt to fit a non-Gaussian translation model to a marginal distribution $\hat{F}$ and covariance function $\hat{\xi}$ estimated from data. If $\hat{\xi}$ takes on values outside the range $[\hat{\xi}^*, 1]$, it is not possible to fit a translation process to $\hat{F}$ and $\hat{\xi}$. Additional conditions for the existence of a translation process with specified $\hat{F}$ and $\hat{\xi}$ are given in [79] (Section 3.1.1).

That X is weakly stationary follows from Eq. 3.31. Also, the finite dimensional density f_n of translation processes in Eq. 3.26 depends on time through the covariance matrix $\boldsymbol{\rho} = \{E[G(t_p)\, G(t_q)] = \rho(t_p - t_q)\}$ so that it is invariant to a time shift. ■

3.7.1 Properties of the correlation function

Let $\boldsymbol{X}$ be an $\mathbb{R}^d$-valued stochastic process in $L_2(\Omega, \mathcal{F}, P)$ with correlation functions $r_{i,j}(t,s) = E[X_i(t)\, X_j(s)]$. We divide the properties of these functions in two groups corresponding to $r_{i,i}$ and $r_{i,j}$, $i \neq j$.

- $r_{i,i}$ is symmetric, that is, $r_{i,i}(t,s) = r_{i,i}(s,t)$. If $\boldsymbol{X}$ is weakly stationary, then $r_{i,i}$ is an even function, that is, $r_{i,i}(\tau) = r_{i,i}(-\tau)$, where $\tau = t - s$.
- The inequality $|r_{i,i}(t,s)|^2 \le r_{i,i}(t,t)\, r_{i,i}(s,s)$ holds. If $\boldsymbol{X}$ is weakly stationary, then $|r_{i,i}(\tau)| \le r_{i,i}(0)$.
- $r_{i,i}$ is positive definite.

Proof: The first property follows from the definition of the correlation function and the fact that this function depends only on the time lag for weakly stationary processes. The Cauchy-Schwarz inequality gives the second property. That $r_{i,i}$ is positive definite, that is, $\sum_{k,l}^n a_k\, a_l\, r_{i,i}(t_k, t_l) \ge 0$ for any integer $n \ge 1$, constants $(a_1, \ldots, a_n)$, and times $(t_1, \ldots, t_n)$, results from the fact that $\left(\sum_{k=1}^n a_k\, X_i(t_k)\right)^2$ is a positive random variable so that its expectation $\sum_{k,l}^n a_k\, a_l\, r_{i,i}(t_k, t_l)$ must be positive.

The notations $r_{i,i}(t,s)$ and $r_{i,i}(\tau)$, $\tau = t - s$, are somewhat inconsistent but are common in the second moment analysis of stochastic processes, and we will use them. The same comment applies to the correlation function $r_{i,j}$ of distinct coordinates of $\boldsymbol{X}$. ■

- $r_{i,j}$, $i \neq j$, satisfies the condition $r_{i,j}(t,s) = r_{j,i}(s,t)$. If $\boldsymbol{X}$ is weakly stationary, then $r_{i,j}(\tau) = r_{j,i}(-\tau)$, where $\tau = t - s$.
- The inequality $|r_{i,j}(t,s)| \le \sqrt{r_{i,i}(t,t)\, r_{j,j}(s,s)}$ holds. If $\boldsymbol{X}$ is weakly stationary, then $|r_{i,j}(\tau)| \le \sqrt{r_{i,i}(0)\, r_{j,j}(0)}$.

Proof: The first property follows from the definition of the correlation function because $r_{i,j}(t,s) = E[X_i(t)\, X_j(s)] = E[X_j(s)\, X_i(t)] = r_{j,i}(s,t)$. If $\boldsymbol{X}$ is weakly stationary, we have $r_{i,j}(t+\tau, t) = r_{j,i}(t, t+\tau)$ so that $r_{i,j}(\tau) = r_{j,i}(-\tau)$. The Cauchy-Schwarz inequality gives the second property. We also have $r_{i,i}(t,t)\, r_{j,j}(s,s) \le [r_{i,i}(t,t) + r_{j,j}(s,s)]^2/4$ since $4\,a\,b \le (a+b)^2$ for any $a, b \in \mathbb{R}$. ■

Example 3.29: Let $\boldsymbol{X}$ be a $\mathbb{C}^d$-valued process in $L_2(\Omega, \mathcal{F}, P)$, that is, its coordinates are complex-valued processes such that $E[X_i(t)\, X_i(t)^*] < \infty$, $i = 1, \ldots, d$, at all times. The correlation functions of $\boldsymbol{X}$ are

$$r_{i,j}(t,s) = E[X_i(t)\, X_j(s)^*], \quad i,j = 1, \ldots, d, \tag{3.32}$$

and satisfy the condition $r_{i,j}(t,s) = r_{j,i}(s,t)^*$, where z^* denotes the complex conjugate of $z \in \mathbb{C}$. ◇

Proof: Let U_i and V_i denote the real and imaginary parts of X_i. Then

$$\begin{aligned} r_{i,j}(t,s) &= E\left[\left(U_i(t) + \sqrt{-1}\, V_i(t)\right)\left(U_j(s) - \sqrt{-1}\, V_j(s)\right)\right] \\ &= E\left[U_i(t)\, U_j(s) + V_i(t)\, V_j(s)\right] + \sqrt{-1}\, E\left[V_i(t)\, U_j(s) - U_i(t)\, V_j(s)\right] \end{aligned}$$

and is equal to the complex conjugate of $r_{j,i}(s,t)$ that can be obtained by similar calculations. Alternatively, note that $r_{j,i}(s,t)^* = \left(E[X_j(s)\, X_i(t)^*]\right)^* = E[X_j(s)^*\, X_i(t)] = r_{i,j}(t,s)$. Other properties of the correlation functions for complex-valued processes can be derived in the same manner. ■

Example 3.30: Let A_k, B_k be uncorrelated random variables with mean zero and unit variance. The function

$$X(t) = \sum_{k=1}^{n} \sigma_k\, [A_k \cos(\nu_k\, t) + B_k \sin(\nu_k\, t)], \quad t \geq 0, \tag{3.33}$$

is a real-valued parametric stochastic process with mean zero and covariance function

$$c(t,s) = \sum_{k=1}^{n} \sigma_k^2 \cos(\nu_k\, (t-s)), \tag{3.34}$$

where $\sigma_k, \nu_k > 0$, $k = 1, 2, \ldots, n$, are some constants. The process X is weakly stationary. If in addition A_k, B_k are Gaussian variables, then X is a stationary Gaussian process. ◇

Proof: The mean is zero because X is a linear function of A_k and B_k, and these random variables have mean zero. The covariance function of X is

$$\begin{aligned} c(t,s) &= \sum_{k,l=1}^{n} \sigma_k\, \sigma_l\, E[(A_k \cos(\nu_k\, t) + B_k \sin(\nu_k\, t))(A_l \cos(\nu_l\, s) + B_l \sin(\nu_l\, s))] \\ &= \sum_{k=1}^{n} \sigma_k^2\, (\cos(\nu_k\, t) \cos(\nu_k\, s) + \sin(\nu_k\, t) \sin(\nu_k\, s)) \\ &= \sum_{k=1}^{n} \sigma_k^2 \cos(\nu_k\, (t-s)) = c(t-s) \end{aligned}$$

by the linearity of the expectation operator and the properties of the random variables A_k, B_k. Because $c(t,s)$ depends on the time lag $t - s$ rather than the times t and s, X is a weakly stationary process.

If in addition the random variables A_k, B_k are Gaussian, then X is stationary since it is a weakly stationary Gaussian process. ■

3.7.2 Power spectral density

Let X be a weakly stationary stochastic process in $L_2(\Omega, \mathcal{F}, P)$. We have seen that the mean and the correlation functions or the mean and covariance functions provide equivalent second moment properties for weakly stationary processes. The spectral density of a process X provides an alternative way for specifying the second moment properties of this process.

3.7.2.1 Bochner's theorem

A continuous function $r : \mathbb{R} \mapsto \mathbb{C}$ is positive definite if and only if it has the representation

$$r(\tau) = \int_{-\infty}^{\infty} e^{\sqrt{-1}\,\nu\,\tau}\, d\mathcal{S}(\nu), \tag{3.35}$$

where $\mathcal{S}$ is a real-valued, increasing, and bounded function.

Note: The function r is said to be positive definite if $\sum_{k,l=1}^{n} \zeta_k\, \zeta_l^*\, r(t_k - t_l) \geq 0$ for any integer $n \geq 1$, arguments $t_k \in \mathbb{R}$, and complex constants ζ_k, $k = 1, \ldots, n$, where z^* denotes the complex conjugate of $z \in \mathbb{C}$.

The proof of Bochner's theorem can be found in [45] (Section 7.4). We only show that $r(\cdot)$ given by Eq. 3.35 is positive definite. Note that

$$\sum_{k,l=1}^{n} \zeta_k\, \zeta_l^*\, r(t_k - t_l) = \sum_{k,l=1}^{n} \zeta_k\, \zeta_l^* \int_{-\infty}^{\infty} e^{\sqrt{-1}\,\nu\,(t_k - t_l)}\, d\mathcal{S}(\nu)$$

$$= \int_{-\infty}^{\infty} \left(\sum_{k=1}^{n} \zeta_k\, e^{\sqrt{-1}\,\nu\,t_k} \right) \left(\sum_{l=1}^{n} \zeta_l\, e^{\sqrt{-1}\,\nu\,t_l} \right)^* d\mathcal{S}(\nu)$$

$$= \int_{-\infty}^{\infty} \left| \sum_{k=1}^{n} \zeta_k\, e^{\sqrt{-1}\,\nu\,t_k} \right|^2 d\mathcal{S}(\nu)$$

is positive for any integer $n \geq 1$, arguments t_k, and complex constants ζ_k. ▲

3.7.2.2 $\mathbb{R}$-valued stochastic processes

If X is a real-valued weakly stationary process, its correlation function $r(\tau) = E[X(t+\tau)\, X(t)]$ is a positive definite function. The Bochner theorem states that, if the correlation function $r : \mathbb{R} \to \mathbb{R}$ is continuous, then there exists a real-valued, increasing, and bounded function $\mathcal{S}$, called the **spectral distribution** function of X, such that r has the representation in Eq. 3.35. The spectral distribution function is defined up to a constant that can be eliminated by setting $\mathcal{S}(-\infty) = 0$. If $\mathcal{S}$ is **absolutely continuous**, there exists a function $s(\nu) = d\mathcal{S}(\nu)/d\nu$, $\nu \in \mathbb{R}$, called the **spectral density** or the **mean power spectral density** function of X.

The correlation and spectral density functions are Fourier pairs, that is,

$$r(\tau) = \int_{-\infty}^{\infty} e^{\sqrt{-1}\,\nu\,\tau}\, s(\nu)\, d\nu \quad \text{and} \quad s(\nu) = \frac{1}{2\pi}\int_{-\infty}^{\infty} e^{-\sqrt{-1}\,\nu\,\tau}\, r(\tau)\, d\tau. \tag{3.36}$$

Note: If $\mu = E[X(t)]$ is not zero, the spectral density function of X has an atom at $\nu = 0$ equal to μ^2 since $r(\tau) = \mu^2 + c(\tau)$ (Eq. 3.36). Formal calculations give

$$s(\nu) = \mu^2\,\delta(\nu) + \tilde{s}(\nu), \quad \text{where } \tilde{s}(\nu) = \frac{1}{2\pi}\int_{-\infty}^{\infty} e^{-\sqrt{-1}\,\nu\,\tau}\, c(\tau)\, d\tau, \tag{3.37}$$

where $\delta(\cdot)$ denotes the Dirac delta function. In engineering applications the **one-sided spectral density** function

$$g(\nu) = \mu^2\,\delta(\nu) + 2\,\tilde{s}(\nu), \quad \nu \geq 0, \tag{3.38}$$

is preferred since $\nu \geq 0$ in Eq. 3.38 can be interpreted as frequency. By analogy with the spectral distribution $\mathcal{S}$ of X, we can define the **one-sided spectral distribution** function of this process by

$$\mathcal{G}(\nu) = \int_0^{\nu} g(\xi)\, d\xi = \mu^2 + 2\int_{0+}^{\nu} s(\xi)\, d\xi = \mu^2 + 2\int_0^{\nu} \tilde{s}(\xi)\, d\xi, \quad \nu \geq 0,$$

where the notation $\int_{0+}^{\nu}$ indicates that the integral does not include the value of the integrand at zero. ▲

We summarize now some properties of the spectral and the one-sided spectral density functions.

- $s(\cdot)$ or $g(\cdot)$ provide the second moment properties of X.
- $s(\cdot)$ and $g(\cdot)$ are positive functions and $s(\nu) = s(-\nu)$ for all $\nu \in \mathbb{R}$.
- An alternative to Eq. 3.36 involving only real-valued functions is

$$r(\tau) = \int_0^{\infty} g(\nu)\,\cos(\nu\,\tau)\, d\nu \quad \text{and} \quad g(\nu) = \frac{1}{\pi}\int_0^{\infty} r(\tau)\,\cos(\nu\,\tau)\, d\tau. \tag{3.39}$$

- The area under the spectral densities $s(\cdot)$ and $g(\cdot)$ is $r(0)$. If $E[X(t)] = 0$, this area is the variance of X.

Proof: The Fourier transform of the spectral density is the correlation function of X and the mass of the atom of its spectral density at $\nu = 0$ is μ^2. Hence, the functions $s(\cdot)$ and $g(\cdot)$ deliver the second moment properties of X. Because $\mathcal{S}$ is an increasing function, the spectral densities $s(\cdot)$ and $g(\cdot)$ are positive. The spectral density $s(\cdot)$ is an even function since the correlation function of X is even (Eq. 3.36). The last two properties result from Eq. 3.36, Eq. 3.38, and the fact that the spectral density $s(\cdot)$ is an even function. ■

Table 3.1. Examples of correlation and one-sided spectral density functions for weakly stationary processes with mean zero (adapted from [175])

Process X	$r(\tau)$	$g(\nu)$
WN	$\pi\, g_0\, \delta(\tau)$	g_0
BLWN	$g_0 \sin(\nu_c \tau)/\tau$	$\begin{cases} g_0, & 0 < \nu \le \nu_c \\ 0, & \nu > \nu_c \end{cases}$
RP	$g_0(\nu_b - \nu_a) \sin(\rho) \cos(\gamma)/\rho,$	$\begin{cases} g_0, & \nu_a < \nu \le \nu_b \\ 0, & \nu \notin (\nu_a, \nu_b] \end{cases}$
BN	$\begin{cases} \sigma^2 (1 - \|\tau\|/\Delta), & \|\tau\| \le \Delta \\ 0, & \|\tau\| > \Delta \end{cases}$	$\sigma^2 \frac{\Delta}{\pi} \frac{\sin^2(\nu\,\Delta/2)}{(\nu\,\Delta/2)^2}$
1M	$\sigma^2 \exp(-\lambda\|\tau\|)$	$\frac{2\sigma^2\lambda}{\pi\,(\nu^2+\lambda^2)}$
2M	$\sigma^2 \exp(-\lambda\,\|\tau\|)\,(1 + \lambda\,\|\tau\|)$	$\frac{4\sigma^2\lambda^3}{\pi\,(\nu^2+\lambda^2)^2}$

Example 3.31: The spectral density and the one-sided spectral density functions of the stochastic process in Eq. 3.33 are

$$s(\nu) = \frac{1}{2} \sum_{k=1}^{n} \sigma_k^2 \left[\delta(\nu - \nu_k) + \delta(\nu + \nu_k)\right] \quad \text{and} \quad g(\nu) = \sum_{k=1}^{n} \sigma_k^2\, \delta(\nu - \nu_k), \tag{3.40}$$

respectively, where $\delta(\cdot)$ denotes the Dirac delta function. ◇

Note: The spectral distribution function S is a piece-wise constant with jumps of magnitude $\sigma_k^2/2$ at $-\nu_k$ and ν_k, $k = 1, \ldots, n$. The definition of the spectral density is based on formal operations since $dS(\nu)/d\nu$ does not exist in $\mathbb{R}$. This operation can be made rigorous by considering a sequence of processes whose spectral densities are defined and converge to the spectral density of X. ▲

Table 3.1 gives examples of covariance and spectral density functions for real-valued weakly stationary processes, where the WN, BLWN, RP, BN, 1M, and 2M denote white noise, band-limited white noise, rectangular pulse, binary noise, first order Markov, and second order Markov process, respectively, and g_0, ν_c, ν_a, ν_b, Δ, λ, and σ^2 are positive constants. The notations $\rho = (\nu_b - \nu_a)\,\tau/2$ and $\gamma = (\nu_b + \nu_a)\,\tau/2$ are also used in the above table.

3.7.2.3 $\mathbb{C}$-valued stochastic processes

We have defined complex-valued processes in Example 3.29. If X is a weakly stationary complex-valued process, its correlation function $r(t - s) = E[X(t)\, X(s)^*]$ is complex-valued (Eq. 3.32). This correlation function can be obtained from its real and imaginary parts (Example 3.29) and has the following properties.

- The correlation function of X has the property $r(t-s) = r(s-t)^*$.
- The correlation function of X is positive definite.

Proof: The equality $r(t,s) = r(s,t)^*$ (Example 3.29) implies the first property. The second property results from the observation that the random variable

$$\left(\sum_{k=1}^{n} \zeta_k \, X(t_k)\right) \left(\sum_{l=1}^{n} \zeta_l \, X(t_l)\right)^* = \left|\sum_{k=1}^{n} \zeta_k \, X(t_k)\right|^2$$

is positive for any integer $n \geq 1$, times t_k, and complex-valued constants ζ_k so that its expectation $\sum_{k,l=1}^{n} \zeta_k \, \zeta_l^* \, r(t_k - t_l)$ is positive. The Bochner theorem (Eq. 3.35) states that X has a spectral distribution function. ■

Example 3.32: The correlation and spectral density of complex-valued processes can be real-valued. For example, let X be a stationary Gaussian process with mean zero and covariance function $c(\tau) = E[X(t+\tau)\,X(t)] = \exp(-\lambda\,|\tau|)$, $\lambda > 0$. The correlation function and the spectral density of the complex-valued process $Y(t) = \exp(\sqrt{-1}\,X(t))$ are real-valued functions. ◇

Proof: We have $r_Y(\tau) = E[Y(t+\tau)\,Y(t)^*] = E[\exp(\sqrt{-1}\,(X(t+\tau) - X(t)))]$ so that $r_Y(\tau) = \exp\left(-\frac{1}{2}\,e^{-\lambda\,|\tau|}\right)$ because $X(t+\tau) - X(t)$ is a Gaussian variable with mean zero and variance $e^{-\lambda\,|\tau|}$. Hence, $r_Y(\tau)$ is an even real-valued function so that the spectral density of Y has the same properties. ■

3.7.2.4 $\mathbb{R}^d$-valued stochastic processes

If $\boldsymbol{X}$ is an $\mathbb{R}^d$-valued weakly stationary stochastic process with mean zero and continuous correlation function $\boldsymbol{r}(\tau) = E[\boldsymbol{X}(t+\tau)\,\boldsymbol{X}(t)^T]$, then

$$r_{i,j}(\tau) = \int_{\mathbb{R}} e^{\sqrt{-1}\,\nu\,\tau}\, d\mathcal{S}_{i,j}(\nu) = \int_{\mathbb{R}} e^{\sqrt{-1}\,\nu\,\tau}\, s_{i,j}(\nu)\,d\nu, \quad \text{where}$$

$$\mathcal{S}_{i,i} = \text{spectral distribution function of } X_i,$$

$$\mathcal{S}_{i,j} = \mathcal{S}_{j,i}^* = \frac{1}{2}\left[\mathcal{S}_1 - \sqrt{-1}\,\mathcal{S}_2 - (1-\sqrt{-1})\,(\mathcal{S}_{i,i} + \mathcal{S}_{j,j})\right], \quad i \neq j, \tag{3.41}$$

$\mathcal{S}_p$, $p = 1, 2$, is defined by Eq. 3.42 below, and $s_{i,j}(\nu) = d\mathcal{S}_{i,j}(\nu)/d\nu$.

Proof: The representation of the entries $r_{i,i}$ of the correlation function follows from results in Section 3.7.2.2. We show now that the correlation functions $r_{i,j}$, $i \neq j$, have a similar representation. The integral representations of $r_{i,j}$ involving the spectral densities $s_{i,j}$ are valid if the functions $s_{i,j}$ exist.

The complex-valued process $Y(t) = \sum_{i=1}^{d} \zeta_i X_i(t)$, $\zeta_i \in \mathbb{C}$, is weakly stationary because it has mean zero and correlation function $r_y(\tau) = E[Y(t+\tau) Y(t)^*] = \sum_{k,l=1}^{d} \zeta_k \zeta_l^* r_{k,l}(\tau)$. If r_y is continuous, then r_y has the representation in Eq. 3.36 since it is positive definite. Let $i, j \in \{1, \ldots, d\}$ be some fixed, distinct indices and let Y_1 and Y_2 be special cases of Y corresponding to (1) $\zeta_i = \zeta_j = 1$ and $\zeta_k = 0$ for $k \neq i, j$ and (2) $\zeta_i = \sqrt{-1}$, $\zeta_j = 1$, and $\zeta_k = 0$ for $k \neq i, j$, respectively. The correlation functions of these processes,

$$r_{y_1}(\tau) = r_{i,i}(\tau) + r_{j,j}(\tau) + r_{i,j}(\tau) + r_{j,i}(\tau) \quad \text{and}$$
$$r_{y_2}(\tau) = r_{i,i}(\tau) + r_{j,j}(\tau) + \sqrt{-1}\, r_{i,j}(\tau) - \sqrt{-1}\, r_{j,i}(\tau),$$

have the representations

$$r_{y_p}(\tau) = \int_{-\infty}^{\infty} e^{\sqrt{-1}\,\nu\,\tau}\, d\mathcal{S}_p(\nu), \quad p = 1, 2, \tag{3.42}$$

where $\mathcal{S}_p$ denotes the spectral distribution of Y_p, $p = 1, 2$ (Eq. 3.36). The last two equations and the representation of the correlation functions $r_{i,i}$ in Bochner's theorem give

$$r_{i,j}(\tau) = \int_{\mathbb{R}} e^{\sqrt{-1}\,\nu\,\tau}\, d\mathcal{S}_{i,j}(\nu), \quad \text{where}$$
$$\mathcal{S}_{i,j} = \mathcal{S}_{j,i}^* = \frac{1}{2}\left[\mathcal{S}_1 - \sqrt{-1}\,\mathcal{S}_2 - (1 - \sqrt{-1})\,(\mathcal{S}_{i,i} + \mathcal{S}_{j,j})\right]$$

for the correlation functions $r_{i,j}$, $i \neq j$. The spectral distributions $\mathcal{S}_{i,j}$, $i \neq j$, are bounded complex-valued functions and the matrix $\{\mathcal{S}_{i,j}\}$ is a Hermitian since $\mathcal{S}_{i,j} = \mathcal{S}_{j,i}^*$. The spectral distributions $\mathcal{S}_{i,j}$, $i \neq j$, are not monotone functions.

It remains to show that it is possible to define a spectral density for distinct pairs of coordinates of $\boldsymbol{X}$. Consider a bounded interval $[\nu, \nu + \Delta\nu)$ and set $\Delta\mathcal{S}_{i,j}(\nu) = \mathcal{S}_{i,j}(\nu + \Delta\nu) - \mathcal{S}_{i,j}(\nu)$. The Hermitian matrix $\{\Delta\mathcal{S}_{i,j}(\nu)\}$, $i, j = 1, \ldots, d$, has the property $\sum_{i,j=1}^{d} \zeta_i\, \zeta_j^*\, \Delta\mathcal{S}_{i,j}(\nu) \geq 0$, $\zeta_i \in \mathbb{C}$, since

$$r_y(\tau) = \int_{-\infty}^{\infty} e^{\sqrt{-1}\,\nu\,\tau}\, d\mathcal{S}_y(\nu) = \sum_{i,j=1}^{n} \zeta_i\, \zeta_j^* \int_{-\infty}^{\infty} e^{\sqrt{-1}\,\nu\,\tau}\, d\mathcal{S}_{i,j}(\nu),$$

and the spectral distribution $\mathcal{S}_y$ of $Y(t) = \sum_{k=1}^{d} \zeta_k\, X(t_k)$ is an increasing function. For a fixed pair of distinct indices $i, j \in \{1, \ldots, d\}$ and $\zeta_k = 0$ for $k \neq i, j$ the sum

$$\sum_{i,j=1}^{d} \zeta_i\, \zeta_j^*\, \Delta\mathcal{S}_{i,j}(\nu)$$

becomes

$$\Delta\mathcal{S}_{i,i}(\nu)\,|\zeta_i|^2 + \Delta\mathcal{S}_{j,j}(\nu)\,|\zeta_j|^2 + 2\,\Re(\Delta\mathcal{S}_{i,j}(\nu)\,\zeta_i\,\zeta_j^*) \geq 0, \quad \text{or}$$
$$\Delta\mathcal{S}_{i,i}(\nu)\,|\zeta_i|^2 + \Delta\mathcal{S}_{j,j}(\nu)\,|\zeta_j|^2 + 2\,|\Delta\mathcal{S}_{i,j}(\nu)|\,|\zeta_i|\,|\zeta_j^*| \geq 0,$$

because the real part of $\Delta\mathcal{S}_{i,j}(\nu)\,\zeta_i\,\zeta_j^*$ is smaller than its absolute value. The latter expression divided by $|\zeta_j|^2$ is a polynomial of $\eta = |\zeta_i|/|\zeta_j|$ that has no real roots so that

$|\Delta\mathcal{S}_{i,j}(\nu)|^2 \leq |\Delta\mathcal{S}_{i,i}(\nu)|\,|\Delta\mathcal{S}_{j,j}(\nu)|$. Hence, $s_{i,j}(\nu)$ exists and satisfies the condition $|s_{i,j}(\nu)|^2 \leq s_{i,i}(\nu)\, s_{j,j}(\nu)$ at any $\nu \in \mathbb{R}$, where $s_{i,i}(\nu) = d\mathcal{S}_{i,i}/d\nu$ and $s_{j,j}(\nu) = d\mathcal{S}_{j,j}/d\nu$ ([45], Section 8.1). ■

Example 3.33: Let $X_i(t) = \sqrt{1-\rho}\, W_i(t) + \sqrt{\rho}\, W(t)$, $i = 1, 2$, be the coordinates of an $\mathbb{R}^2$-valued process $\boldsymbol{X}$, where $\rho \in (0, 1)$ and (W_1, W_2, W) are zero-mean, weakly stationary processes that are uncorrelated with each other. Then $\boldsymbol{X}$ is a weakly stationary process with spectral densities

$$s_{i,i}(\nu) = (1-\rho)\, s_{W_i}(\nu) + \rho\, s_W(\nu), \qquad i = 1, 2,$$
$$s_{1,2}(\nu) = s_{2,1}(\nu) = \rho\, s_W(\nu)$$

where s_{W_i} and s_W denote the spectral densities of W_i and W, respectively.

Proof: The covariance functions of $\boldsymbol{X}$ are $c_{i,j}(\tau) = (1-\rho)\,\delta_{ij}\, c_{W_i}(\tau) + \rho\, c_W(\tau)$, where c_{W_i} and c_W denote the covariance functions of W_i and W. The spectral densities $s_{i,i}$ are the Fourier transform of the covariance functions $c_{i,i}$ by Bochner's theorem (Eq. 3.35). The correlation functions in Eq. 3.42 are

$$r_{y_1}(\tau) = (1-\rho)\,\left(c_{W_1}(\tau) + c_{W_2}(\tau)\right) + 4\,\rho\, c_W(\tau),$$
$$r_{y_2}(\tau) = (1-\rho)\,\left(c_{W_1}(\tau) + c_{W_2}(\tau)\right) + 2\,\rho\, c_W(\tau),$$

so that

$$s_1(\nu) = \frac{1}{2\pi}\int_{\mathbb{R}} e^{-\sqrt{-1}\,\nu\,\tau}\, r_{y_1}(\tau)\, d\tau = (1-\rho)\,\left(s_{W_1}(\nu) + s_{W_2}(\nu)\right) + 4\,\rho\, s_W(\nu),$$
$$s_2(\nu) = \frac{1}{2\pi}\int_{\mathbb{R}} e^{-\sqrt{-1}\,\nu\,\tau}\, r_{y_1}(\tau)\, d\tau = (1-\rho)\,\left(s_{W_1}(\nu) + s_{W_2}(\nu)\right) + 2\,\rho\, s_W(\nu).$$

The spectral densities $s_{1,2}$ and $s_{2,1}$ result from Eq. 3.41. ■

3.8 Equivalent stochastic processes

In both applications and theoretical studies we need to assess differences and similarities between two stochastic processes $\boldsymbol{X}$ and $\boldsymbol{Y}$ that may or may not be defined on the same probability space. If $\boldsymbol{X}$ and $\boldsymbol{Y}$ have some common properties, we say that they are equivalent in a sense defined by the shared properties.

> The stochastic processes $\boldsymbol{X}$ and $\boldsymbol{Y}$ defined on the same probability space $(\Omega, \mathcal{F}, P)$ are said to be **indistinguishable** if their samples coincide a.s.

Note: Let Ω_0 be the subset of Ω collecting the samples $\boldsymbol{X}(\cdot, \omega)$ and $\boldsymbol{Y}(\cdot, \omega)$ of $\boldsymbol{X}$ and $\boldsymbol{Y}$ that do not coincide. The processes $\boldsymbol{X}$ and $\boldsymbol{Y}$ are indistinguishable if $P(\Omega_0) = 0$. ▲

> The stochastic processes $\boldsymbol{X}$ and $\boldsymbol{Y}$ defined on the same probability space $(\Omega, \mathcal{F}, P)$ are said to be **modifications** if $P(\{\omega : \boldsymbol{X}(t, \omega) = \boldsymbol{Y}(t, \omega)\}) = 1$ at each value of t.

Note: Indistinguishable processes are modifications but modifications may not be indistinguishable. Let $\Omega_t = \{\omega : \boldsymbol{X}(t,\omega) \neq \boldsymbol{Y}(t,\omega)\}$ be the subset of Ω on which the modifications $\boldsymbol{X}$ and $\boldsymbol{Y}$ differ at t. The set Ω_t is an event and $P(\Omega_t) = 0$. If the probability of $\cup_t \Omega_t$ is zero, then $\boldsymbol{X}$ and $\boldsymbol{Y}$ will be indistinguishable. Because $\cup_t \Omega_t$ is an uncountable union of members of $\mathcal{F}$, it may not be in $\mathcal{F}$, in which case $P(\cup_t \Omega_t)$ is not even defined. ▲

The following conditions of equivalence for $\boldsymbol{X}$ and $\boldsymbol{Y}$ are weaker and do not require that these processes be defined on the same probability space.

> The stochastic processes $\boldsymbol{X}$ and $\boldsymbol{Y}$ are said to be **versions** if they have the same finite dimensional distributions.

> The stochastic processes $\boldsymbol{X}$ and $\boldsymbol{Y}$ are said to be equal in the **second moment sense** if they have the same second moment properties.

Note: Generally, processes with the same second moment properties have different finite dimensional distributions. An exception is the class of Gaussian processes. Recall that two Gaussian processes with the same second moment properties are versions.

Other types of equivalence between stochastic processes can be defined. For example, processes with the same marginal distribution and second moment properties or processes with the same finite dimensional distributions of order two can be viewed as equivalent. ▲

Example 3.34: Suppose that the stochastic processes X and Y are modifications with right continuous sample. Then X and Y are indistinguishable. ◇

Proof: The event $\{X(r) \neq Y(r)\}$ has zero measure for each $r \in \mathbb{Q}$ so that the event $A = \cup_{r \in \mathbb{Q}} \{X(r) \neq Y(r)\}$ has zero probability since the set of rational numbers $\mathbb{Q}$ is countable. If $r_n \downarrow t$, $r_n \in \mathbb{Q}$, and $t \in \mathbb{R}$, then $\{X(r_n) \neq Y(r_n)\} \subset A$ for each n so that we also have $\{X(t) \neq Y(t)\} \subset A$ at any time $t \in \mathbb{R}$ by right continuity. Hence, X and Y are indistinguishable. ■

Example 3.35: Let $X(t)$ and $Y(t)$, $t \in [0,1]$, be real-valued processes defined on the probability space $([0,1], \mathcal{B}([0,1]), P)$, where $P(d\omega) = d\omega$ such that $X(t,\omega) = 0$ and $Y(t,\omega) = 0$ or 1 for $t \neq \omega$ or $t = \omega$, respectively. The processes X and Y are not indistinguishable but are versions and modifications. ◇

Proof: The samples of X and Y differ for all $\omega \in \Omega$, for example, $\sup_{t\in[0,1]} X(t,\omega) = 0$ while $\sup_{t\in[0,1]} Y(t,\omega) = 1$ for each $\omega \in [0,1]$.

The processes X and Y are modifications because $\Omega_t = \{\omega : X(t,\omega) \neq Y(t,\omega)\} = \{t\}$ and $P(\{t\}) = 0$ for any $t \in [0,1]$. These processes are also versions since the vectors $(X(t_1), \ldots, X(t_n))$ and $(Y(t_1), \ldots, Y(t_n))$ differ on a subset of Ω with measure zero for any times $t_1, \ldots, t_n$ in $[0,1]$. ■

Example 3.36: The mean and covariance functions of the Brownian motion and the compound Poisson processes B and C are

$$\mu_B(t) = 0, \quad c_B(t,s) = \min(t,s) \tag{3.43}$$

and

$$\mu_C(t) = \lambda\, t\, E[Y_1], \quad c_C(t, s) = \lambda \min(t, s)\, E[Y_1^2], \tag{3.44}$$

with the notations in Eq. 3.6 and the assumption $E[Y_1^2] < \infty$. If $E Y_1 = 0$ and $\lambda\, E[Y_1^2] = 1$, the Brownian motion and the compound Poisson process are equal in the second moment sense. However, the samples of these processes differ significantly (Fig 3.5) showing that the second moment equivalence provides little information on the sample properties. ◇

Proof: The Brownian motion has mean zero by definition. The expectation $E[B(t)\, B(s)]$, $t > s$, is

$$E[(B(t) - B(s) + B(s))\, B(s)] = E[(B(t) - B(s))\, B(s)] + E[B(s)^2] = s$$

since $E[(B(t) - B(s))\, B(s)] = E[B(t) - B(s)]\, E[B(s)] = 0$ by the independence of the increments of B and $E[B(s)^2] = s$.

We have $E[C(t)] = \sum_{n=0}^{\infty} E[C(t) \mid N(t) = n]\, P(N(t) = n)$, $E[C(t) \mid N(t) = n] = n\, E[Y_1]$, and $P(N(t) = n) = e^{-\lambda t}\, (\lambda\, t)^n / n!$ so that

$$E[C(t)] = \sum_{n=0}^{\infty} n\, E[Y_1]\, e^{-\lambda t}\, (\lambda\, t)^n / n! = E[Y_1]\, e^{-\lambda t} \sum_{n=0}^{\infty} n\, (\lambda\, t)^n / n!$$

$$= E[Y_1]\, e^{-\lambda t}\, (\lambda\, t) \sum_{n=1}^{\infty} (\lambda\, t)^{n-1} / (n-1)! = \lambda\, t\, E[Y_1].$$

Similar calculations can be performed to find the mean and the covariance functions of C in Eq. 3.44. ■

Many physical phenomena are modeled in applications by stochastic processes $\boldsymbol{X}$ whose properties have to be inferred from the available records of these phenomena and/or other type of information, for example, restrictions imposed by the physics of the modeled phenomena. Generally, the available information is insufficient to find uniquely the probability law of $\boldsymbol{X}$. There may be many stochastic processes that are consistent with the available information. This lack of uniqueness in modeling can have significant practical implications. For example, if the available information on the input to a linear dynamic system is limited to the mean and covariance functions, the second moment properties of the state of this system are defined uniquely. However, the extreme, range, and other sample properties of the output can differ significantly depending on the particular model used for $\boldsymbol{X}$ ([79], Chapter 5).

3.9 Second moment calculus

The Chebyshev inequality, the Cauchy-Schwarz inequality, the mean square (m.s.) convergence of sequences of random variables (Sections 2.12 and 2.13) and the following facts are essential tools for second moment calculations.

• If $X_n, X \in L_2$ and $X_n \xrightarrow{\text{m.s.}} X$, then

$$\lim_{n\to\infty} E[X_n] = E[\text{l.i.m.}_{\,n\to\infty} X_n] = E[X]. \tag{3.45}$$

• If $X_n, X, Y_n, Y \in L_2$ and $X_n \xrightarrow{\text{m.s.}} X$, $Y_n \xrightarrow{\text{m.s.}} Y$, then

$$\lim_{m,n\to\infty} E[X_m Y_n] = E[(\text{l.i.m.}_{\,m\to\infty} X_m)\,(\text{l.i.m.}_{\,n\to\infty} Y_n)] = E[X\,Y]. \tag{3.46}$$

• If a sequence X_n has two m.s. limits X and Y, then $X = Y$ a.s., that is,

$$X_n \xrightarrow{\text{m.s.}} X \text{ and } X_n \xrightarrow{\text{m.s.}} Y \quad \text{implies } P(X \neq Y) = 0. \tag{3.47}$$

Proof: The above statements show that the expectations and the m.s. limits can be interchanged under some conditions (Eqs. 3.45-3.46) and that the m.s. limit is unique with probability one, that is, the subset of Ω in which the m.s. limits of a sequence X_n may differ has zero measure (Eq. 3.47).

We have

$$0 \leq |E[X] - E[X_n]| = |E[X - X_n]| \leq \left(E[(X - X_n)^2]\right)^{1/2}$$

for each $n \geq 1$ by the Cauchy-Schwarz inequality so that $\lim_{n\to\infty} E[X_n] = E[X]$ since $X_n \xrightarrow{\text{m.s.}} X$ by assumption.

Note that

$$\begin{aligned}
|E[X_m Y_n - X\,Y]| &= |E[X_m Y_n - X_m Y + X_m Y - X\,Y]| \\
&\leq |E[X_m (Y_n - Y)]| + |E[(X_m - X)Y]| \\
&\leq \left(E[X_m^2]\,E[(Y_n - Y)^2]\right)^{1/2} + \left(E[(X_m - X)^2]\,E[Y^2]\right)^{1/2},
\end{aligned}$$

where the Cauchy-Schwarz inequality has been used to obtain the final upper bound on $|E[X_m Y_n - X\,Y]|$. The property in Eq. 3.46 follows from the above inequality since $X_m \xrightarrow{\text{m.s.}} X$ and $Y_n \xrightarrow{\text{m.s.}} Y$ by hypothesis.

The uniqueness with probability 1 of the m.s. limit in Eq. 3.47 follows from the Chebyshev inequality $P(|X - Y| > \varepsilon) \leq E[(X - Y)^2]/\varepsilon^2$, $\varepsilon > 0$, and the inequalities

$$\begin{aligned}
0 &\leq E[(X - Y)^2] \leq E[((X - X_n) + (X_n - Y))^2] \\
&\leq 2\,E[(X - X_n)^2] + 2\,E[(X_n - Y)^2],
\end{aligned}$$

where the last inequality holds since $(a + b)^2 \leq 2\,a^2 + 2\,b^2$. Because $X_n \xrightarrow{\text{m.s.}} X, Y$, we have $P(|X - Y| > \varepsilon) = 0$, $\forall \varepsilon > 0$. ■

The remainder of this section gives definitions and basic properties of m.s. continuity, differentiation, and integration for real-valued processes in $L_2(\Omega, \mathcal{F}, P)$. We also extend these results to $\mathbb{R}^d$-valued stochastic processes.

3.9.1 Continuity

We have defined in Section 3.3 several types of continuity for stochastic processes. Generally, it is difficult to show that a process X is sample continuous, that is, that almost all samples of X are continuous functions of time. Weaker definitions of continuity may be adequate in many applications. We define again m.s. continuity and give simple criteria for determining whether a process is m.s. continuous at a time t and in a time interval.

A real-valued stochastic process X in L_2 is **m.s. continuous** or **continuous in the mean square sense** at time t if $\text{l.i.m.}_{s \to t} X(s) = X(t)$, that is,

$$\lim_{s \to t} E[(X(s) - X(t))^2] = \lim_{s \to t} \| X(t) - X(s) \|_2^2 = 0. \tag{3.48}$$

Note: This definition states that X is m.s. continuous at t if the distance between $X(t)$ and $X(s)$ defined by the second moment of the difference $X(t) - X(s)$ approaches zero as $s \to t$. The process X is said to be **m.s. continuous in an interval** I if it is m.s. continuous at each $t \in I$. The norm in Eq. 3.48 is the norm in L_2.

An $\mathbb{R}^d$-valued stochastic process X is m.s. continuous at a time t if and only if its coordinates X_i, $i = 1, \ldots, d$ are m.s. continuous at t. ▲

The definition in Eq. 3.48 is not very useful for checking m.s. continuity. We give a simple criterion for assessing whether a process is m.s. continuous.

- A real-valued process X is **m.s. continuous at a time** t if and only if its correlation function $r(u, v) = E[X(u)\, X(v)]$ is continuous at $u = v = t$, that is, $\lim_{u,v \to t} r(u, v) = r(t, t)$.
- A real-valued weakly stationary process X is **m.s. continuous at a time** t if and only if $\lim_{\tau \to 0} r(\tau) = r(0)$.

Proof: If X is m.s. continuous at t, the following chain of equalities and inequalities

$$\begin{aligned}
0 &\leq |r(u, v) - r(t, t)| \\
&= |E[X(u)\, X(v)] - E[X(t)\, X(u)] + E[X(t)\, X(u)] - E[X(t)\, X(t)]| \\
&\leq |E[X(u)\, (X(v) - X(t))]| + |E[X(t)\, (X(u) - X(t))]| \\
&\leq \left(E[X(u)^2]\, E[(X(v) - X(t))^2]\right)^{1/2} + \left(E[X(t)^2]\, E[(X(u) - X(t))^2]\right)^{1/2}
\end{aligned}$$

shows that $r(u, v) \to r(t, t)$ as $u, v \to t$. The last relationship follows by the Cauchy-Schwarz inequality.

Conversely, the relationship $E[(X(s) - X(t))^2] = r(s, s) + r(t, t) - 2\, r(s, t)$ and the convergence $r(u, v) \to r(t, t)$, $u, v \to t$, imply the m.s. continuity of X at time t. ■

A real-valued process X is m.s. continuous in an interval I if and only if its correlation function $r(u, v) = E[X(u)\, X(v)]$ is continuous on $I \times I$.

Proof: We have $E[(X(t) - X(s))^2] = r(t,t) + r(s,s) - 2\,r(t,s)$ for $t, s \in I$ arbitrary so that X is m.s. continuous in I if $r(t,t)+r(s,s)-2\,r(t,s)$ converges to zero as $|t-s| \to 0$. Hence, the process X is m.s. continuous in I if r is continuous in $I \times I$.

A process X is **uniformly m.s. continuous in** I if for each $\varepsilon > 0$ there exists $\delta > 0$ such that $\| X(u) - X(v) \|_2 < \varepsilon$ for $|u - v| < \delta$, $u, v \in I$, where δ depends on ε but not on the arguments of X. Hence, the distance between $X(u)$ and $X(v)$ can be bounded uniformly by ε if $|u - v|$ is smaller than a fixed value δ. It can also be shown that if X is m.s. continuous in a closed and bounded interval I, then X is uniformly m.s. continuous in I ([157], Theorem 2.5, p. 34). A similar result holds for deterministic functions. ■

Example 3.37: The Brownian motion B, the compound Poisson process C, and the process X in Eq. 3.33 are m.s. continuous. Also, the processes B and X are sample continuous while C is not. ◇

Proof: The correlation functions of B and C at two times t and s are proportional with $t \wedge s$. Because $t \wedge s$ is a continuous function, B and C are m.s. continuous. The process X is weakly stationary so that it is sufficient to observe that its correlation function is continuous at the origin (Eq. 3.34).

The differences between the sample properties of B, C, and X show that m.s. continuity is a relatively weak requirement that does not provide much information on the sample continuity of a process. ■

3.9.2 Differentiation

The derivative $\dot{X}(t) = dX(t)/dt$ of the process X in Eq. 3.33 is

$$\dot{X}(t,\omega) = \sum_{k=1}^{n} \sigma_k \, \nu_k \, [-A_k(\omega) \, \sin(\nu_k \, t) + B_k(\omega) \, \cos(\nu_k \, t)]$$

and can be calculated sample by sample. Generally, such calculations are not possible since analytical expressions are rarely available for the samples of arbitrary stochastic processes.

A real-valued process X in L_2 is **mean square differentiable** at t if

$$\text{l.i.m.}_{h \to 0} \frac{X(t+h) - X(t)}{h} \tag{3.49}$$

exists. This limit, denoted by $\dot{X}(t)$ or $dX(t)/dt$, is called the **m.s. derivative** of X at t.

Note: Because the limit in Eq. 3.49 is not known, it is not possible to calculate the second moment of the difference $\dot{X}(t) - (X(t+h) - X(t))/h$. To prove the existence of $\dot{X}(t)$, we need to show that $\{(X(t+h_n) - X(t))/h_n\}$ is a Cauchy sequence in L_2 if $h_n \to 0$ as $n \to \infty$. Because L_2 is complete, $\{(X(t+h_n) - X(t))/h_n\}$ has a limit in L_2 denoted by $\dot{X}(t)$. Recall that a sequence $\{Y_n\}$ in L_2 is Cauchy if $\| Y_n - Y_m \|_2 \to 0$ as $m, n \to \infty$.

An $\mathbb{R}^d$-valued process $\boldsymbol{X}$ is m.s. differentiable at t if and only if its coordinates, that is, the real-valued stochastic processes X_i, $i = 1, 2, \ldots$, are m.s. differentiable at t. The process $\boldsymbol{X}$ is said to be **m.s. differentiable** in an interval I if it has an m.s. derivative at each time $t \in I$. ▲

As for m.s. continuity, we give a criterion for assessing whether a real-valued process X is m.s. differentiable. We also give some useful formulas involving the m.s. differentiation and the expectation operators.

• $\dot{X}(t)$ exists in the mean square sense and $E[\dot{X}(t)^2] < \infty$ if and only if $\partial^2 r(u, v)/(\partial u\, \partial v)$ exists and is finite at $u = v = t$.

• If X is a weakly stationary process, then $\dot{X}(t)$ exists in the mean square sense and $E[\dot{X}(t)^2] < \infty$ if and only if $r''(\tau) = d^2 r(\tau)/d\tau^2$ exists and is finite at $\tau = 0$.

Note: The proof of these statements considers a sequence $Y_n = [X(t + h_n) - X(t)]/h_n$, where $h_n \downarrow 0$ as $n \to \infty$. The existence of $\dot{X}$ in the m.s. sense implies that Y_n is an m.s. Cauchy sequence, that is $\| Y_n - Y_m \|_2 \to 0$ as $m, n \to \infty$. This property can be used to show that $\partial^2 r(u, v)/(\partial u\, \partial v)$ exists and is finite at $u = v = t$. The converse results by considering a finite difference approximation of $\partial^2 r(u, v)/(\partial u\, \partial v)$ and showing that the existence of this partial derivative implies $E[\dot{X}(t)^2] < \infty$ ([175], Section 4.4). ▲

If X is a real-valued m.s. differentiable stochastic process, then

$$\frac{d}{dt} E[X(t)] = E[\dot{X}(t)], \quad \frac{\partial}{\partial t} E[X(t)\, X(s)] = E[\dot{X}(t)\, X(s)], \quad \text{and}$$

$$\frac{\partial^2}{\partial t\, \partial s} E[X(t)\, X(s)] = E[\dot{X}(t)\, \dot{X}(s)] \quad \text{or } r_{\dot{X},\dot{X}}(t, s) = \frac{\partial^2 r(t, s)}{\partial t\, \partial s}. \tag{3.50}$$

Additional properties are in Table 3.2.

Proof: The formulas in Eq. 3.50 show that differentiation and expectation can be interchanged. The proof of these formulas is based on Eq. 3.45. For example, the derivative of $E[X(t)]$ is given by the limit of $E[(X(t + h) - X(t))/h]$ as $h \to 0$. Because $\dot{X}$ exists, the expectation and the limit operations can be interchanged (Eq. 3.45) so that $\lim_{h\to 0} E[(X(t+h) - X(t))/h]$ is equal to $E[\text{l.i.m.}_{h\to 0}(X(t+h) - X(t))/h] = E[\dot{X}(t)]$.

The existence of derivatives of X of order two and higher require additional conditions. For example, $\ddot{X}(t)$ exists and $E[\ddot{X}(t)^2] < \infty$ if and only if $\partial r_{\dot{X},\dot{X}}(u, v)/\partial u\, \partial v$ exists and is finite at $u = v = t$. ■

If X is a real-valued, weakly stationary, and m.s. differentiable process, then

$$r_{\dot{X},\dot{X}}(\tau) = -\frac{d^2 r(\tau)}{d\tau^2} = -r''(\tau) \quad \text{and } r_{\dot{X},\dot{X}}(\tau) = -\int_{-\infty}^{\infty} \nu^2\, e^{\sqrt{-1}\,\nu\,\tau} dS(\nu). \tag{3.51}$$

Proof: The first formula in Eq. 3.51 follows from the last formula in Eq. 3.50. The second formula in Eq. 3.51 can be obtained by differentiating Eq. 3.35. These results

Table 3.2. Properties of mean square differentiation (adapted from [175]).

Non-stationary Processes	Weakly Stationary Processes
$\frac{d}{dt}[a\,X(t)+b\,Y(t)] = a\,\frac{dX(t)}{dt}+b\,\frac{dX(t)}{dt}$, $a, b =$ constants	
$\frac{d}{dt}[g(t)\,X(t)] = \frac{dg(t)}{dt}\,X(t)+g(t)\,\frac{dX(t)}{dt}$, $g =$ a differentiable function	
$E\left[X^{(n)}(t)\right] = \frac{d^n}{dt^n}\,E\,[X(t)]$	$E\left[X^{(n)}(t)\right] = 0, \quad n \geq 1$
$E\left[X^{(n)}(t)\,X^{(m)}(s)\right] = \frac{\partial^{n+m} r(t,s)}{\partial t^n\,\partial s^m}$	$E\left[X^{(n)}(t)\,X^{(m)}(s)\right] = (-1)^n\,\frac{d^{n+m} r(\tau)}{d\tau^{n+m}}$
Note: $X^{(n)}(t) = \frac{d^n}{dt^n} X(t)$ and $\tau = s - t$.	

show that $r''(\tau) = -\int_{-\infty}^{\infty} \nu^2\, e^{\sqrt{-1}\,\nu\,\tau}\, dS(\nu)$ so that the variance of $\dot{X}$ is $E[\dot{X}(t)^2] = \int_{-\infty}^{\infty} \nu^2\, dS(\nu)$. The latter result can also be obtained by direct calculations. We have

$$\begin{aligned} r''(\tau) &= \frac{d}{d\tau} E[\dot{X}(t+\tau)\,X(t)] = \frac{d}{d\tau} E[\dot{X}(t)\,X(t-\tau)] \\ &= -E[\dot{X}(t)\,\dot{X}(t-\tau)] = -E[\dot{X}(t+\tau)\,\dot{X}(t)] \end{aligned}$$

since X is a weakly stationary process. ■

Example 3.38: Let B denote a Brownian motion and C be a compound Poisson process with $E[Y_1] = 0$ and $\lambda\, E[Y_1^2] = 1$ (Eq. 3.6). The correlation function of these two processes is $r(u, v) = u \wedge v$ so that they are not m.s. differentiable since $\partial^2 r(u, v)/(\partial u\, \partial v)$ does not exist.

Let us attempt to calculate the mixed derivative of $u \wedge v$ anyway. Formal calculations give $\partial^2 r(u, v)/(\partial u\, \partial v) = \delta(u - v)$, where $\delta(\cdot)$ denotes the Dirac delta function. Also, the one-sided spectral density of $\dot{B}$ and $\dot{C}$ is $g(\nu) = 1/\pi$, $\nu \geq 0$. Processes with constant spectral density and delta correlation functions are called **white noise** processes in the engineering and physics literature. These calculations defining white noise processes are meaningless because $\dot{B}$ and $\dot{C}$ do not exist in the m.s. sense. We will consider alternative ways of defining white noise processes based on properties of the increments of the Brownian, compound Poisson, and other processes. ◇

Note: The differentiation of $u \wedge v$ relative to u gives a step function that is 1 for $v \geq u$ and zero otherwise. The differentiation of this step function with respect to v gives the Dirac delta function that is zero for $u \neq v$ and has unit mass at $u = v$. Continuing with formal calculations we find that the one-sided spectral density function of B is constant and equal to $1/\pi$ (Eq. 3.39 and 3.38). ▲

Example 3.39: If $X \in L_2$ is m.s. differentiable at a time t, then X is m.s. continuous at this time. ◇

Proof: We need to show that $\| X(t+h) - X(t) \|_2$ converges to zero as $h \to 0$. Because X is m.s. differentiable, the right side of the equation

$$\| X(t+h) - X(t) \|_2 = |h| \; \| [X(t+h) - X(t)]/h \|_2$$

converges to (0) $\| \dot{X}(t) \|_2$ as $h \to 0$ so that, X is m.s. continuous at t. ■

3.9.3 Integration

Let $\{X(t), t \in [a,b]\}$ be a real-valued stochastic process and let $h : [a,b] \to \mathbb{R}$ be a real-valued function, where $[a,b] \subset \mathbb{R}$. Our objective is to define the integrals

$$\int_a^b h(t)\, dX(t) \quad \text{and} \quad \int_a^b X(t)\, dh(t) \tag{3.52}$$

in the mean square sense. These integrals are random variables because their values depend on the particular sample of X used for calculations. If X is a parametric process, the integrals in Eq. 3.52 can be written explicitly. For example, if X is in Eq. 3.33 and h is an integrable function in $[0,t]$, we have

$$Y(t) = \int_0^t h(s)\, dX(s) = \sum_{k=1}^n \sigma_k \, \nu_k \left[A_k \, h_{s,k}(t) + B_k \, h_{c,k}(t) \right],$$

where $h_{c,k}(t) = \int_0^t h(s) \cos(\nu_k s)\, ds$ and $h_{s,k}(t) = -\int_0^t h(s) \sin(\nu_k s)\, ds$ are Riemann integrals. The process Y has mean zero, covariance function

$$E[Y(t)\, Y(s)] = \sum_{k=1}^n \sigma_k^2 \, \nu_k^2 \left[h_{c,k}(t)\, h_{c,k}(s) + h_{s,k}(t)\, h_{s,k}(s) \right]$$

and can be interpreted as the output of a linear filter with transfer function h to a random input X.

Generally, it is not possible to obtain analytical solutions for the integrals in Eq. 3.52. We define these integrals in the m.s. sense. The definition of these integrals requires some notation, that we introduce now. A finite set of points $p = (a = t_0 < t_1 < \cdots < t_m = b)$ is called a **partition** of $[a,b]$ and $t'_k \in [t_{k-1}, t_k]$ is an **intermediate point** of p. The mesh of p is $\Delta(p) = \max_{1 \le k \le m}(t_k - t_{k-1})$. A partition p' is a **refinement** of p if $p' \supseteq p$. Refining partitions are not necessary for the definition of the m.s. integrals in this section. We will use refining partitions to define some of the stochastic integrals in Chapter 4.

Let $p_n = (a = t_0^{(n)} < t_1^{(n)} < \cdots < t_{m_n}^{(n)} = b)$ be a sequence of partitions of $[a,b]$ with intermediate points $t_k'^{(n)} \in [t_{k-1}^{(n)}, t_k^{(n)}]$ such that $\Delta(p_n) \to 0$ as

$n \to \infty$. Define the sums

$$S_{h,X}(p_n) = \sum_{k=1}^{m_n} h(t_k'^{(n)}) \left(X(t_k^{(n)}) - X(t_{k-1}^{(n)}) \right) \quad \text{and}$$
$$S_{X,h}(p_n) = \sum_{k=1}^{m_n} X(t_k'^{(n)}) \left(h(t_k^{(n)}) - h(t_{k-1}^{(n)}) \right). \tag{3.53}$$

If $S_{h,X}(p_n)$ and $S_{X,h}(p_n)$ are Cauchy sequences in L_2, we say that h is **m.s. Riemann-Stieltjes integrable on** $[a, b]$ **with respect to** X and that X is **m.s. Riemann-Stieltjes integrable on** $[a, b]$ **with respect to** h, respectively.

Note: The m.s. limits of the Cauchy sequences $S_{h,X}$ and $S_{X,h}$ define the integrals in Eq. 3.52. These definitions are admissible because the limits of the Cauchy sequences in Eq. 3.53 are in L_2 and are independent of the intermediate points $t_k'^{(n)}$ ([157], Theorem 2.16, p. 41). The sequences $S_{h,X}$ and $S_{X,h}$ are similar to the Riemann-Stieltjes sums used to define classical integrals.

The independence of the limits of the Cauchy sequences in Eq. 3.53 on the intermediate points $t_k'^{(n)}$ of the partitions of $[a, b]$ is essential for the definition of the m.s. integrals in Eq. 3.52. If the limits were dependent on $t_k'^{(n)}$, the above definitions would not be admissible. This condition is not always satisfied (Section 4.3).

The definition of the m.s. integrals in Eq. 3.52 can be extended to the integrals $\int_a^b h(t)\, d\boldsymbol{X}(t)$ and $\int_a^b \boldsymbol{X}(t)\, dh(t)$ depending on an $\mathbb{R}^d$-valued process $\boldsymbol{X}$. The coordinates $\int_a^b h(t)\, dX_i(t)$ and $\int_a^b X_i(t)\, dh(t)$, $i = 1, \ldots, d$, of these integrals are given by limits of Cauchy sequences as in Eq. 3.53. ▲

3.9.3.1 Variation functions

Consider a real-valued process X in L_2, an interval $[a, b]$ of the real line, and a function $h : [a, b] \to \mathbb{R}$.

- $v_h(p) = \sum_{k=1}^{m} |h(t_k) - h(t_{k-1})|$ = the **variation of** h and
- $v_X(p) = \sum_{k=1}^{m} \| X(t_k) - X(t_{k-1}) \|_2$ = the **variation of** X (3.54)

on $[a, b]$ **relative to the partition** $p = (a = t_0 < t_1 < \cdots < t_m = b)$.

$$\tilde{v}_r(p,q) = \sum_{k=1}^{m}\sum_{l=1}^{n} |r(t_k, s_l) - r(t_k, s_{l-1}) - r(t_{k-1}, s_l) + r(t_{k-1}, s_{l-1})|$$
$$= \text{the } \textbf{variation of } r \qquad (3.55)$$

on $[a, b] \times [a, b]$ **relative to the partitions** $p = (a = t_0 < t_1 < \cdots < t_m = b)$ **and** $q = (a = s_0 < s_1 < \cdots < s_m = b)$ of $[a, b]$.

Note: The variation of r on $[a, b] \times [a, b]$ is defined by

$$\tilde{v}_r(p,q) = \sum_{k=1}^{m}\sum_{l=1}^{n} \left|E[(X(t_k) - X(t_{k-1}))(X(s_l) - X(s_{l-1}))]\right|$$

so that

$$\tilde{v}_r(p,q) \le \sum_{k=1}^{m}\sum_{l=1}^{n} \left(E[(X(t_k) - X(t_{k-1}))^2]\, E[(X(s_l) - X(s_{l-1}))^2]\right)^{1/2} = v_X(p)\, v_X(q)$$

for any partitions p and q of $[a, b]$. ▲

- $v_h = \sup_p\{v_h(p)\}$ = the **total variation of** h,
- $v_X = \sup_p\{v_X(p)\}$ = the **total variation of** X (3.56)

on $[a, b]$, where the supremum is taken over the set of all partitions p of $[a, b]$.

$$\tilde{v}_r = \sup_{p,q} \tilde{v}_r(p,q) = \text{the } \textbf{total variation} \text{ of } r \textbf{ on } [a,b] \times [a,b], \qquad (3.57)$$

where the supremum is calculated over all partitions (p, q) of $[a, b] \times [a, b]$.

- If $v_h < \infty$, we say that h is of **bounded variation on** $[a, b]$.
- If $v_X < \infty$ ($\tilde{v}_r < \infty$), we say that X is of **bounded variation in the strong (weak) sense on** $[a, b]$.

Note: It can be shown that a monotone or a differentiable function h is of bounded variation. The total variation of h on an interval $[a, c]$, $a < b < c$, is the sum of the total variations of this function on $[a, b]$ and $[b, c]$. Moreover, $v_h(p)$ increases as the partition p is refined ([157], Section 2.4.2).

It can be shown that (1) it is sufficient to consider identical partitions for the two arguments of r, that is, we can take $p = q$ in Eq. 3.57, (2) the variation $\tilde{v}_r(p, q)$ increases as the partition of $[a, b] \times [a, b]$ is refined, (3) the total variation of r over finite unions of disjoint intervals is the sum of the total variations of r on these intervals, and (4) if X is of bounded variation in the strong sense, then X is of bounded variation in the weak sense (see note following Eq. 3.55 and [157], Section 2.4.4). ▲

Example 3.40: If X is an m.s. differentiable stochastic process defined on $[a, b]$, then X is of bounded variation in the strong sense on this interval. ◇

Proof: Let $p = (a = t_0 < t_1 < \cdots < t_m = b)$ be a partition of $[a, b]$. The variation of X on $[a, b]$ relative to p is (Eq. 3.54)

$$v_X(p) = \sum_{k=1}^{m} \| \int_{t_{k-1}}^{t_k} \dot{X}(t)\, dt \|_2 \leq \sum_{k=1}^{m} \int_{t_{k-1}}^{t_k} \| \dot{X}(t) \|_2 \, dt \leq \alpha\, (b-a),$$

where $\alpha = \max_{t\in[a,b]} \| \dot{X}(t) \|_2$. The equality and inequality in this equation follow from Eq. 3.58 and 3.61 in a subsequent section. ■

Example 3.41: The Brownian motion B is of bounded variation in the weak sense on any bounded interval $[0, \tau]$, $0 < \tau < \infty$, but is not of bounded variation in the strong sense on $[0, \tau]$. ◇

Proof: Let $p = (0 = t_0 < t_1 < \cdots < t_m = \tau)$ be a partition of $[0, \tau]$. The variation of the correlation function $r(t, s) = E[B(s)\, B(t)] = s \wedge t$ on $[0, \tau] \times [0, \tau]$ with respect to p is

$$\begin{aligned}\tilde{v}_r(p, p) &= \sum_{k,l=1}^{m} \left| E\left[(B(t_k) - B(t_{k-1}))\, (B(t_l) - B(t_{l-1})) \right] \right| \\ &= \sum_{k=1}^{m} \left| E\left[(B(t_k) - B(t_{k-1}))^2 \right] \right| = \sum_{k=1}^{m} \left(t_k - t_{k-1} \right) = \tau < \infty.\end{aligned}$$

The variation of B relative to a partition p_n of $[0, \tau]$ with points $t_k^{(n)} = k\,\tau/n$, $k = 0, 1, \ldots, n$, is

$$v_B(p_n) = \sum_{k=1}^{n} \| B(t_k^{(n)}) - B(t_{k-1}^{(n)}) \|_2 = \sum_{k=1}^{n} \sqrt{\tau/n} = \sqrt{n\,\tau}$$

so that $v_B(p_n) \to \infty$ as $n \to \infty$. ■

Example 3.42: The compound Poisson process C in Eq. 3.6 with jumps $Y_k \sim L_2$ such that $E[Y_k] = 0$ is of bounded variation in the weak sense on $[0, \tau]$ but is not of bounded variation in the strong sense. ◇

Proof: Let $p = (0 = t_0 < t_1 < \cdots < t_m = \tau)$ be a partition of $[0, \tau]$. The variation of the correlation function of C on $[0, \tau] \times [0, \tau]$ relative to p is

$$\tilde{v}_r(p, p) = \sum_{k=1}^{m} E[(C(t_k) - C(t_{k-1}))^2] = \lambda\, \tau\, E[Y_1^2] < \infty.$$

The variation of C on $[0, \tau]$ relative to a partition p_n defined by the points $t_k^{(n)} = k\,\tau/n$, $k = 0, 1, \ldots, n$, is

$$v_C(p_n) = \sum_{k=1}^{n} \| C(t_k^{(n)}) - C(t_{k-1}^{(n)}) \|_2 = \sum_{k=1}^{n} \left(\lambda\, (\tau/n)\, E[Y_1^2] \right)^{1/2} = \left(\lambda\, \tau\, n\, E[Y_1^2] \right)^{1/2}$$

so that $v_C(p_n) \to \infty$ as $n \to \infty$. The properties $E[C(t)] = \lambda\, t\, E[Y_1]$, $\mathrm{Var}[C(t)] = \lambda\, t\, E[Y_1^2]$, and $E[C(t)^2] = (\lambda\, t\, E[Y_1])^2 + \lambda\, t\, E[Y_1^2]$ of C have been used in the above equations. ■

3.9.3.2 Conditions of existence

We summarize conditions for the existence of the integrals $\int_a^b X(t)\, dh(t)$ and $\int_a^b h(t)\, dX(t)$ in Eq. 3.52 and some relationships between these integrals.

If X is m.s. continuous and h is of bounded variation on $[a, b]$, then

- $\int_a^b X(t)\, dh(t)$ exists and $\| \int_a^b X(t)\, dh(t) \|_2 \leq \alpha\, v_h$,
- $\| \int_a^b X(t)\, dt \|_2 \leq \int_a^b \| X(t) \|_2\, dt \leq \alpha(b-a)$, and
- $$E\left[\int_a^b X(t)\, dt \int_a^b Z(t)\, dt\right] = \int_{[a,b]^2} E[X(u)\, Z(v)]\, du\, dv, \tag{3.58}$$

where $\alpha = \max_{t\in[a,b]} \| X(t) \|_2$ and Z is an m.s. continuous stochastic process.

If X is m.s. continuous and h is continuously differentiable on $[a, b]$, then

- $\int_a^b X(t)\, dh(t)$, $\int_a^b X(t)\, \dot{h}(t)\, dt$ exist and
- $$\int_a^b X(t)\, dh(t) = \int_a^b X(t)\, \dot{h}(t)\, dt. \tag{3.59}$$

If X is m.s. continuous on $[a, b]$, then the integral $\int_a^t X(s)\, ds$ exists, is m.s. differentiable on $[a, b]$, and

$$\frac{d}{dt} \int_a^t X(s)\, ds = X(t), \quad t \in [a, b]. \tag{3.60}$$

If $\dot{X}$ is m.s. continuous on $[a, b]$, then the following integral exists and

$$\int_a^b \dot{X}(t)\, dt = X(b) - X(a). \tag{3.61}$$

Note: The existence of the integral $\int_a^b X(t)\, dt$ in the last two properties in Eq. 3.58 follows from the first property in this equation applied for the special case $h(t) = t$. The continuity of the function $t \mapsto \| X(t) \|_2$ on $[a, b]$ implies the existence of $\int_a^b \| X(t) \|_2\, dt$.

The proof of the statements in Eqs. 3.58, 3.59, 3.60, and 3.61 can be found in [157] (Theorem 2.22, p. 46, Theorem 2.23, p. 50, Theorem 2.24, p. 51, and Theorem 2.25, p. 51, respectively). ▲

If h_i are continuous and $X_i \in L_2$ are processes of bounded variation in the weak sense on $[a_i, b_i]$, $i = 1, 2$, then

- $\int_{a_1}^{b_1} \int_{a_2}^{b_2} h_1(s)\, h_2(t) \frac{\partial^2 E[X_1(s)\, X_2(t)]}{\partial s\, \partial t}\, ds\, dt, \quad \int_{a_i}^{b_i} h_i(t)\, dX_i(t)$ exist and
- $E\left[\int_{a_1}^{b_1} h_1(s)\, dX_1(s) \int_{a_2}^{b_2} h_2(t)\, dX_2(t)\right]$

$$= \int_{a_1}^{b_1} \int_{a_2}^{b_2} h_1(s)\, h_2(t) \frac{\partial^2 E[X_1(s)\, X_2(t)]}{\partial s\, \partial t}\, ds\, dt. \tag{3.62}$$

If h is continuous and X is m.s. differentiable with m.s. continuous derivative $\dot{X}$ on $[a, b]$, then

- $\int_a^b h(t)\, dX(t), \quad \int_a^b h(t)\, \dot{X}\, dt$ exist and
- $\int_a^b h(t)\, dX(t) = \int_a^b h(t)\, \dot{X}(t)\, dt. \quad (3.63)$

If $h, k : [a, b] \to \mathbb{R}$ are continuous and $X \in L_2$ is a process of bounded variation in the weak sense on $[a, b]$, then

- $Y(t) = \int_a^t h(s)\, dX(s)$ exists, is of bounded variation in the weak sense on $[a, b]$ for $t \in [a, b]$, and
- $\int_a^b h(t)\, dY(t)$ exists, and $\int_a^b h(t)\, dY(t) = \int_a^b h(t)\, k(t)\, dX(t). \quad (3.64)$

Note: If we take $a_i = a$, $b_i = b$, $h_i = h$, and $X_i = X$, $i = 1, 2$, in Eq. 3.62, and assume as in this equation that h is continuous and $X \in L_2$ is of bounded variation in the weak sense on $[a, b]$, then

- $\int_a^b \int_a^b h(s)\, h(t) \frac{\partial^2 r(s,t)}{\partial s\, \partial t}\, ds\, dt, \quad \int_a^b h(t)\, dX(t)$ exist and
- $E\left[\left(\int_a^b h(t)\, dX(t)\right)^2\right] = \int_a^b \int_a^b h(s)\, h(t) \frac{\partial^2 r(s,t)}{\partial s\, \partial t}\, ds\, dt \leq \beta^2\, \tilde{v}_r, \quad (3.65)$

where $\beta = \max_{t\in[a,b]} |h(t)|$ ([157], Theorem 2.29, p. 59). This special case also shows that the integral $\int_a^b h(t)\,dX(t)$ exists if h is continuous and X is of bounded variation in the strong sense on $[a, b]$ ([157], Theorem 2.26, p. 52). The proof of the statements in Eqs. 3.62, 3.63, and 3.64 can be found in [157] (Theorem 2.30, p. 60, Theorem 2.27, p. 53, and Theorem 2.31, p. 61, respectively).

We also note that the m.s. integrals in Eqs. 3.58, 3.59, 3.60, 3.62, and 3.64 with X and Z being equal to a Brownian motion process B are defined in the m.s. sense. For example, we have $\| \int_a^b B(t)\,dt \|_2 \le \int_a^b \| B(t) \|_2\, dt \le b^{1/2}\,(b-a)$ and $E\left[\left(\int_a^b B(t)\,dt\right)^2\right] = \int_{[a,b]^2} (u \wedge v)\,du\,dv$. ▲

3.9.3.3 Properties for calculations

We summarize some properties of the m.s. integrals in Eq. 3.52 that are relevant for calculations. These properties resemble features of the Riemann-Stieltjes integral.

Integration by parts. The m.s. integral $\int_a^b h(t)\,dX(t)$ exists if and only if the m.s. integral $\int_a^b X(t)\,dh(t)$ exists and

$$\int_a^b h(t)\,dX(t) = h(t)\,X(t)\,\Big|_a^b - \int_a^b X(t)\,dh(t). \tag{3.66}$$

If $c \in [a, b]$ and the following m.s. integrals exist, then

- $$\int_a^b h(t)\,dX(t) = \int_a^c h(t)\,dX(t) + \int_c^b h(t)\,dX(t),$$
- $$\int_a^b X(t)\,dh(t) = \int_a^c X(t)\,dh(t) + \int_c^b X(t)\,dh(t). \tag{3.67}$$

The m.s. integral is linear. If h, k and $X, Y \in L_2$ are such that $\int_a^b h(t)\,dX(t)$, $\int_a^b k(t)\,dX(t)$, and $\int_a^b h(t)\,dY(t)$ exist in m.s., then

- $$\int_a^b [\alpha\, h(t) + \beta\, k(t)]\,dX(t) = \alpha \int_a^b h(t)\,dX(t) + \beta \int_a^b k(t)\,dX(t),$$
- $$\int_a^b h(t)\,d[\alpha\, X(t) + \beta\, Y(t)] = \alpha \int_a^b h(t)\,dX(t) + \beta \int_a^b h(t)\,dY(t), \tag{3.68}$$

where $\alpha, \beta \in \mathbb{R}$. Similar properties hold for the m.s. integral $\int_a^b X(t)\,dh(t)$.

Expectations and m.s. integrals interchange. If the m.s. integral $\int_a^b h(t)\, dX(t)$ or $\int_a^b X(t)\, dh(t)$ exists, then the integrals $\int_a^b h(t)\, dE[X(t)]$ and $\int_a^b E[X(t)]\, dh(t)$ exist, respectively, and

- $E\left[\int_a^b h(t)\, dX(t)\right] = \int_a^b h(t)\, dE[X(t)]$,
- $E\left[\int_a^b X(t)\, dh(t)\right] = \int_a^b E[X(t)]\, dh(t)$. (3.69)

Note: The proof of the statements in Eqs. 3.66, 3.67, 3.68, and 3.69 can be found in [157] (Theorem 2.17, p. 42, Theorem 2.18, p. 43, Theorem 2.19, p. 44, and Theorem 2.20, p. 44, respectively) and are based on the definition of the m.s. integrals in Eq. 3.52. For example, the expectation of $S_{h,X}(p_n)$ is

$$E[S_{h,X}(p_n)] = \sum_{k=1}^{m_n} h(t_k'^{(n)}) \left(E[X(t_k^{(n)})] - E[X(t_{k-1}^{(n)})] \right).$$

The left side of this equation converges to $E\left[\int_a^b h(t)\, dX(t)\right]$ because $S_{h,X}(p_n)$ is a Cauchy sequence in L_2 approximating $\int_a^b h(t)\, dX(t)$ as $\Delta(p_n) \to 0$ (Eq. 3.45). The right side of the above equation converges as $\Delta(p_n) \to 0$ to the classical Riemann-Stieltjes integral $\int_a^b h(t)\, dE[X(t)]$. ▲

Example 3.43: Let $h : [a, b] \to \mathbb{R}$ be a differentiable function on $[a, b]$ and let X be an m.s. continuous process. The m.s. integral $Y = \int_a^b X(t)\, dh(t)$ is well defined and $E[Y^2] = \int_a^b \int_a^b r(t, s)\, g(t)\, g(s)\, dt\, ds$ exists and is bounded, where r denotes the correlation function of X and $g(t) = dh(t)/dt$. ◇

Proof: The integral $Y = \int_a^b X(t)\, dh(t)$ exists in the mean square sense by one of the properties in Eq. 3.58 since h being differentiable in $[a, b]$ is also of bounded variation in this interval. The expectation $E[Y^2]$ results from the last property in Eq. 3.58 since the process $g\, X$ is m.s. continuous. ■

Example 3.44: Let h and X be as in Example 3.43. The mean and variance of the m.s. integral $Y = \int_a^b X(t)\, dh(t)$ are

$$\mu_y = \int_a^b g(t)\, \mu(t)\, dt \quad \text{and} \quad \sigma_y^2 = \int_a^b \int_a^b g(t)\, g(s)\, c(t, s)\, dt\, ds$$

where μ and c are the mean and covariance functions of X. If X is a Gaussian process, Y is a Gaussian variable with the characteristic function $\varphi_y(u) = \exp(\sqrt{-1}\, u\, \mu_y - u^2\, \sigma_y^2/2)$. If X is a Brownian motion process, then $\mu_y = 0$ and $\sigma_y^2 = \int_a^b \int_a^b (t \wedge s)\, g(t)\, g(s)\, dt\, ds$. ◇

Proof: The properties in Eq. 3.58 can be used to confirm the above statements. We present here a direct proof. The sum $S_{g\,X,1}(p_n) = \sum_{k=1}^{m_n} g(t_k'^{(n)})\, X(t_k'^{(n)})\,(t_k^{(n)} - t_{k-1}^{(n)})$ is a Cauchy sequence in L_2 with expectation $\mu_{y,n} = \sum_{k=1}^{n} g(t_k'^{(n)})\, \mu(t_k'^{(n)})\,(t_k^{(n)} - t_{k-1}^{(n)})$ that converges to μ_y as $\Delta(p_n) \to 0$. Similar considerations show that the expectation of the square of the centered sums $S_{g\,X,1}(p_n) - \mu_{y,n}$ converges to σ_y^2 as $\Delta(p_n) \to 0$. If X is Gaussian, then Y is a Gaussian variable with the specified characteristic function. ■

Example 3.45: Let $h : [0, \infty) \to \mathbb{R}$ be a differentiable function and let X be a m.s. continuous process. Then $Y(t) = \int_0^t g(s)\, X(s)\, ds$, $t \geq 0$, is a m.s. differentiable process with mean and correlation functions

$$E[Y(t)] = \int_0^t g(s)\, \mu(s)\, ds \quad \text{and}$$
$$E[Y(t)\, Y(s)] = \int_0^t du\, g(u) \int_0^s dv\, g(v)\, r(t, s),$$

where $g(t) = dh(t)/dt$. The process $Y(t)$, $t \geq 0$, can represent the output of a linear dynamic system with transfer function g subjected to an input X (Section 7.2). If X is Gaussian, Y is a Gaussian process so that the above moments define all finite dimensional distributions of Y. ◇

Proof: The existence of the process Y and the expressions of its first two moments result from Eq. 3.58 since $Y(t) = \int_0^t X(s)\, dh(s)$ and h is differentiable so that it is of bounded variation on compact sets.

The process Y is m.s. differentiable since $g\, X$ is m.s. continuous (Eq. 3.60) so that Y is also m.s. continuous. Note that Y satisfies the differential equation $\dot{Y}(t) = g(t)\, X(t)$, $t \geq 0$, with the initial condition $Y(0) = 0$. The right side of this equation is m.s. continuous so that its left side must also be m.s. continuous. Hence, $\int_0^t \dot{Y}(s)\, ds = Y(t) - Y(0)$ by Eq. 3.61, which gives $Y(t) = \int_0^t g(s)\, X(s)\, ds$, that is, the definition of Y.

We give the definition of a compact set here for convenience. Consider a topological space (Section 2.2.2) and a subset K of this space. The set K is **compact** if every open cover of K has a finite subcover, that is, if any collection D_α, $\alpha \in I$, of open sets such that $K \subset \bigcup_{\alpha \in I} D_\alpha$ has a finite subcollection $D_{\alpha_1}, \ldots, D_{\alpha_n}$ with the property $K \subset \bigcup_{k=1}^{n} D_{\alpha_k}$. It can be shown that a set $K \subset \mathbb{R}$ is compact if and only if K is closed and bounded ([26], Theorem 13.6, p. 95). Hence, the intervals $[a, \infty)$ and $[a, b)$ of the real line are not compact sets, but $[a, b]$ is a compact set. The same conclusions follows from the definition. For example, $[a, \infty)$ is not a compact set since it is not possible to extract from the open cover $(a - 1/n, n)$, $n = 1, 2, \ldots$, of this interval a finite cover. ■

3.9.4 Spectral representation

We show that any weakly stationary stochastic process can be represented by a superposition of harmonics with random amplitude and phase. The representation is very useful in many applications.

Example 3.46: Let X be the real-valued stochastic process in Eq. 3.33 with covariance and spectral density functions given by Eqs. 3.34 and 3.40, respectively. An alternative representation of X and its covariance and spectral density functions is

$$X(t) = \sum_{p=-n}^{n} C_p \, e^{\sqrt{-1}\, \nu_p \, t}, \quad c(\tau) = \sum_{p=-n}^{n} \frac{\sigma_p^2}{2} \, e^{\sqrt{-1}\, \nu_p \, \tau}, \quad \text{and}$$

$$s(\nu) = \sum_{p=-n}^{n} \frac{\sigma_p^2}{2} \, [\delta(\nu - \nu_p) + \delta(\nu + \nu_p)], \tag{3.70}$$

where $\nu_{-k} = -\nu_k$, $\sigma_{-k} = \sigma_k$, $k = 1, \ldots, n$, the random coefficients C_p are complex-valued with means $E[C_p] = 0$ and correlations $E[C_p \, C_q^*] = (\sigma_p^2/2)\, \delta_{pq}$, $C_0 = 0$, and $\sigma_0 = 0$. ◇

Note: The representation in Eq. 3.70 results from Eqs. 3.33, 3.34, and 3.40 and the relationships $\cos(u) = [e^{\sqrt{-1}\, u} + e^{-\sqrt{-1}\, u}]/2$, $\sin(u) = [e^{\sqrt{-1}\, u} - e^{-\sqrt{-1}\, u}]/(2\sqrt{-1})$. The coefficients of the harmonics of X are $C_p = \sigma_p \, [A_p - \sqrt{-1}\, B_p]/2$, where $A_{-p} = A_p$ and $B_{-p} = -B_p$. The second moment properties of C_p are $E[C_p] = 0$ and $E[C_p \, C_q^*] = (\sigma_p \, \sigma_q/4)\, E[(A_p - \sqrt{-1}\, B_p)\,(A_q + \sqrt{-1}\, B_q)] = (\sigma_p^2/2)\, \delta_{pq}$, where $A_{-k} = A_k$ and $B_{-k} = -B_k$ for $k = 1, \ldots, n$. ▲

The representation in Eq. 3.70 shows that X consists of a superposition of harmonics $C_p \, e^{\sqrt{-1}\, \nu_p \, t}$ with frequencies ν_p. The coefficients C_p of these harmonics are orthogonal, have mean zero, and variance $s(\nu_p) = \sigma_p^2/2$. This observation suggests the representation $X(t) = \int_{-\infty}^{\infty} e^{\sqrt{-1}\, \nu \, t} \, dZ(\nu)$ for an arbitrary weakly stationary process X, where the random complex-valued coefficients $dZ(\nu)$ associated with the harmonics of frequencies $\nu \in \mathbb{R}$ are such that $E[dZ(\nu)] = 0$ and $E[dZ(\nu)\, dZ(\nu')^*] = \delta(\nu - \nu')\, s(\nu)\, d\nu$. It turns that our intuition is correct.

3.9.4.1 $\mathbb{C}$- and $\mathbb{R}$-valued stochastic processes

Spectral representation theorem. If X is a complex-valued, weakly stationary, and m.s. continuous process with spectral distribution $\mathcal{S}$ and spectral density s, then there exists a complex-valued process Z with orthogonal increments defined up to an additive constant such that the m.s. integral

$$X(t) = \int_{-\infty}^{\infty} e^{\sqrt{-1}\, \nu \, t} \, dZ(\nu) \tag{3.71}$$

exists at any time t and, for $Z(-\infty) = 0$, we have

$$E[Z(\nu)] = 0, \quad E[|Z(\nu)|^2] = \mathcal{S}(\nu), \quad \text{and} \quad E[|dZ(\nu)|^2] = d\mathcal{S}(\nu) = s(\nu)\, d\nu. \tag{3.72}$$

Note: This theorem shows that a complex-valued weakly stationary stochastic process can be constructed as a superposition of orthogonal harmonics $e^{\sqrt{-1}\,\nu t}\,dZ(\nu)$ with frequency ν and random amplitude and phase defined by the increments $dZ(\nu)$ of a process Z, called the **spectral process** associated with X ([45], Section 7.5). The representation in Eq. 3.71 is the continuous version of Eq. 3.70.

Since X is m.s. continuous, its correlation function is continuous so that the Bochner theorem in Eq. 3.35 can be applied and gives $E[X(t+\tau)\,X(t)] = \int_{-\infty}^{\infty} e^{\sqrt{-1}\,\nu\tau}\,d\mathcal{S}(\nu)$.

The m.s. integral in Eq. 3.71 generalizes the m.s. integral defined by Eq. 3.52 because both its integrand and integrator are complex-valued. The definition, properties, and calculation formulas established for the m.s. integral in Eq. 3.52 can be extended to this type of integral by applying results of the previous section to the real and imaginary parts of the integral in Eq. 3.71. These considerations show, for example, that the integral in Eq. 3.71 exists in the mean square sense because its integrand $e^{\sqrt{-1}\,\nu t}$ is a continuous function and the integrator Z is of bounded variation in the weak sense. The variation of the correlation function of Z in a bounded interval $[-a,a]\times[-a,a]$, $a>0$, for a partition $p = (-a = \nu_0 < \nu_1 < \cdots < \nu_m = a)$ of this interval is (Eq. 3.55)

$$\begin{aligned}\tilde{v}_r(p,p) &= \sum_{k,l=1}^{m} \left|E[(Z(\nu_k)-Z(\nu_{k-1}))(Z(\nu_l)-Z(\nu_{l-1}))^*]\right| \\ &= \sum_{k=1}^{m} E[|Z(\nu_k)-Z(\nu_{k-1})|^2] = \sum_{k=1}^{m}(\mathcal{S}(\nu_k)-\mathcal{S}(\nu_{k-1}))) = \mathcal{S}(a)-\mathcal{S}(-a) \qquad (3.73)\end{aligned}$$

so that Z is of bounded variation in the weak sense on $[-a,a]$ for any $a>0$. The process Z is also of bounded variation in the weak sense on $\mathbb{R}$ because $\lim_{a\to\infty}[\mathcal{S}(a)-\mathcal{S}(-a)] = \mathcal{S}(\infty) < \infty$. ▲

Spectral representation theorem. If X is a real-valued, weakly stationary, and m.s. continuous process with one-sided spectral density g, then there exist two real-valued processes U, V with orthogonal increments such that the m.s. integral

$$X(t) = \int_0^{\infty} [\cos(\nu t)\,dU(\nu) + \sin(\nu t)\,dV(\nu)] \qquad (3.74)$$

exists at any time t and we have

$$\begin{aligned}&E[U(\nu)] = E[V(\nu)] = 0, \quad E[dU(\nu)\,dV(\nu')] = 0, \quad \text{and} \\ &E[dU(\nu)^2] = E[dV(\nu)^2] = g(\nu)\,d\nu \quad \text{for } \nu,\nu' \ge 0. \qquad (3.75)\end{aligned}$$

Note: The properties of the processes U, V are similar to the properties of the coefficients A_k, B_k in Eq. 3.33. The moments of U and V in Eq. 3.75 can be obtained from Eqs. 3.71 and 3.72 under the assumption that X is real-valued ([45], Section 7.6). ▲

Example 3.47: The correlation functions of the processes in Eqs. 3.71 and 3.74

are

$$E[X(t)\,X(s)^*] = \int_{-\infty}^{\infty} e^{\sqrt{-1}\,(t-s)\,\nu}\, s(\nu)\, d\nu \quad \text{and}$$

$$E[X(t)\,X(s)] = \int_{0}^{\infty} \cos(\nu(t-s))\, g(\nu)\, d\nu,$$

respectively, if X takes complex and real values, respectively. Both processes have mean zero. ◇

Proof: We have seen in Eq. 3.69 that expectation and integration can be interchanged under some conditions that are satisfied by the m.s. integrals in Eqs. 3.71 and 3.74 (Section 3.9.3.2 and 3.9.3.3). This result and the properties of the processes Z, U, and V give the above correlation functions. ■

The following two examples illustrate linear operations on weakly stationary processes. These operations are commonly used in applications, for example, to find properties of the solution of linear differential equations with random input (Chapter 7).

Example 3.48: Let $Y(t) = \sum_{k=1}^{n} c_k\, X(t+t_k)$, where X is a complex-valued process with spectral representation in Eq. 3.71, t_k are times, $c_k \in \mathbb{C}$ denote some constants, and $n \geq 1$ is an integer. The stochastic process Y has the spectral representation $Y(t) = \int_{-\infty}^{\infty} e^{\sqrt{-1}\,\nu\, t}\, d\tilde{Z}(\nu)$, where $d\tilde{Z}(\nu) = h(\nu)\, dZ(\nu)$ and $h(\nu) = \sum_{k=1}^{n} c_k\, e^{\sqrt{-1}\,\nu\, t_k}$. ◇

Proof: We have

$$Y(t) = \int_{-\infty}^{\infty} \left[\sum_{k=1}^{n} c_k\, e^{\sqrt{-1}\,\nu\, t_k}\right] e^{\sqrt{-1}\,\nu\, t}\, dZ(\nu) = \int_{-\infty}^{\infty} h(\nu)\, e^{\sqrt{-1}\,\nu\, t}\, dZ(\nu)$$

so that Y consists of a superposition of harmonics $e^{\sqrt{-1}\,\nu\, t}$ with amplitudes $h(\nu)\, dZ(\nu)$. If h, referred to as **gain**, is such that $\int_{-\infty}^{\infty} |h(\nu)|^2\, dS(\nu) < \infty$, then Y is in L_2. ■

Example 3.49: Let X be a real-valued weakly stationary process. If $\dot{X}$ exists in an m.s. sense, then

$$\dot{X}(t) = \int_{-\infty}^{\infty} \sqrt{-1}\,\nu\, e^{\sqrt{-1}\,\nu\, t}\, dZ(\nu) \tag{3.76}$$

with the notation in Eqs. 3.71-3.72. ◇

Proof: Take $a, \tau > 0$ and let

$$X^{(a)}(t) = \int_{-a}^{a} e^{\sqrt{-1}\,\nu\, t}\, dZ(\nu) \quad \text{and} \quad \dot{X}^{(a)}(t) = \int_{-a}^{a} \sqrt{-1}\,\nu\, e^{\sqrt{-1}\,\nu\, t}\, dZ(\nu).$$

We need to show that $E\left[\left((X^{(a)}(t+\tau) - X^{(a)}(t))/\tau - \dot{X}^{(a)}(t)\right)^2\right]$ converges to zero as $\tau \to 0$ for any $a > 0$. The stated result follows by letting a approach infinity. That the above expectation converges to zero as $\tau \to 0$ follows by straightforward calculations. ■

Example 3.50: Let X be the process in Eq. 3.71 and define

$$Y(t) = \int_{-\infty}^{\infty} e^{\sqrt{-1}\,\nu\,t}\,h(\nu)\,dZ(\nu),$$

where $h(\nu) = 0$, 1, and 2 for $\nu < -\xi$, $\nu \in [-\xi, \xi]$, and $\nu > \xi$, respectively, and $\xi > 0$. We have

$$Y(t) = \int_{-\xi}^{\xi} e^{\sqrt{-1}\,\nu\,t}\,dZ(\nu) + 2\int_{\xi}^{\infty} e^{\sqrt{-1}\,\nu\,t}\,dZ(\nu) \tag{3.77}$$

so that

$$\bar{Y}(t) = \text{l.i.m.}_{\xi \to 0} Y(t) = \Delta Z(0) + 2\int_{0+}^{\infty} e^{\sqrt{-1}\,\nu\,t}\,dZ(\nu), \tag{3.78}$$

where $\Delta Z(0) = \text{l.i.m.}_{\xi \to 0} \int_{-\xi}^{\xi} e^{\sqrt{-1}\,\nu\,t}\,dZ(\nu)$ denotes the jump of Z at $\nu = 0$. If the spectral distribution $\mathcal{S}$ is continuous at $\nu = 0$, the jump $\Delta Z(0)$ vanishes.

The **Hilbert transform** $\hat{X}$ of X is defined by $\bar{Y}(t) = X(t) + \sqrt{-1}\,\hat{X}(t)$. If X is the real-valued process in Eq. 3.74, its Hilbert transform has the representation

$$\hat{X}(t) = \int_0^{\infty} [\sin(\nu\,t)\,dU(\nu) - \cos(\nu\,t)\,dV(\nu)] \tag{3.79}$$

and is real-valued with the notation in Eqs. 3.74 and 3.75. ◇

Note: The definition of $\hat{X}$ shows that this process can be obtained from X by a linear operation with the gain $h(\nu) = \sqrt{-1}$, 0, and $-\sqrt{-1}$ for $\nu < 0$, $\nu = 0$, and $\nu > 0$, respectively, that is, $\hat{X}(t) = \int_{\mathbb{R}} h(\nu)\,e^{\sqrt{-1}\,\nu\,t}\,dZ(\nu)$ ([45], p. 142). The expression of the gain results from Eq. 3.78 and the definition of $\hat{X}$. ▲

3.9.4.2 $\mathbb{R}^d$-valued stochastic processes

Spectral representation theorem. If $\boldsymbol{X}$ is an $\mathbb{R}^d$-valued, weakly stationary, and m.s. continuous process with spectral density functions $s_{i,j}$, there exist two $\mathbb{R}^d$-valued processes $\boldsymbol{U}$, $\boldsymbol{V}$ with orthogonal increments and mean zero such that

$$\boldsymbol{X}(t) = \int_0^{\infty} [\cos(\nu t)\,d\boldsymbol{U}(\nu) + \sin(\nu t)\,d\boldsymbol{V}(\nu)] \tag{3.80}$$

exists in the m.s. sense at any time and

$$E[dU_i(\nu)\,dU_j(\nu')] = E[dV_i(\nu)\,dV_j(\nu')] = \delta(\nu - \nu')\,g_{i,j}(\nu)\,d\nu,$$
$$E[dU_i(\nu)\,dV_j(\nu')] = -E[dV_i(\nu)\,dU_j(\nu')] = \delta(\nu - \nu')\,h_{i,j}(\nu)\,d\nu, \quad \text{where}$$
$$g_{i,j}(\nu) = s_{i,j}(\nu) + s_{i,j}(-\nu) \quad \text{and} \quad h_{i,j}(\nu) = -\sqrt{-1}\,[s_{i,j}(\nu) - s_{i,j}(-\nu)]. \tag{3.81}$$

Proof: We have shown that the correlation and spectral density functions of the coordinates of X are Fourier pairs (Eq. 3.41). An alternative form of the correlation function in Eq. 3.41 is

$$r_{i,j}(\tau) = \int_{\mathbb{R}} e^{\sqrt{-1}\,\nu\,\tau}\, s_{i,j}(\nu)\,d\nu = \int_0^\infty \Big[\big(s_{i,j}(\nu) + s_{i,j}(-\nu)\big)\cos(\nu\,\tau)$$
$$+ \sqrt{-1}\,\big(s_{i,j}(\nu) - s_{i,j}(-\nu)\big)\sin(\nu\,\tau)\Big]\,d\nu.$$

The correlation function $r_{i,j}$ based on the representation of X in Eq. 3.80 is

$$r_{i,j}(\tau) = E\left[X_i(t+\tau)\,X_j(t)\right] = \int_0^\infty \left[g_{i,j}(\nu)\cos(\nu\,\tau) - h_{i,j}(\nu)\sin(\nu\,\tau)\right]d\nu.$$

The above expressions of $r_{i,j}$ yield the definitions of the second moment properties of U and V in Eq. 3.81. ■

Example 3.51: Let X be the $\mathbb{R}^2$-valued process in Example 3.33 with coordinates $X_i(t) = \sqrt{1-\rho}\,W_i(t) + \sqrt{\rho}\,W(t)$, where (W_1, W_2, W) are zero-mean, weakly stationary processes that are mutually uncorrelated and $\rho \in (0,1)$. The process $X = (X_1, X_2)$ has the spectral representation

$$X(t) = \int_0^\infty \left[\cos(\nu\,t)\,dU(\nu) + \sin(\nu\,t)\,dV(\nu)\right],$$

where U and V are uncorrelated with each other, $E[dU_i(t)] = E[dV_i(t)] = 0$, $E[dU_i(\nu)\,dU_j(\nu')] = E[dV_i(\nu)\,dV_j(\nu')] = [(1-\rho)\,g^{(i)}(\nu)\,\delta_{ij} + \rho\,g(\nu)]\,\delta(\nu - \nu')\,d\nu$ and $g^{(i)}$, g denote the one-sided spectral densities of W_i, W. The one-sided spectral densities of X are $g_{i,j}(\nu) = [(1-\rho)\,g^{(i)}(\nu)\,\delta_{ij} + \rho\,g(\nu)]$. ◇

Proof: The spectral representations,

$$W_i(t) = \int_0^\infty \left[\cos(\nu\,t)\,dU^{(i)}(\nu) + \sin(\nu\,t)\,dV^{(i)}(\nu)\right], \quad i = 1, 2,$$
$$W(t) = \int_0^\infty \left[\cos(\nu\,t)\,dU(\nu) + \sin(\nu\,t)\,dV(\nu)\right],$$

of processes W_i and W are given by Eqs. 3.74 and 3.75. These processes and the definition of X show that the coordinates of U and V in the spectral representation of X are $U_i(\nu) = \sqrt{1-\rho}\,U^{(i)}(\nu) + \sqrt{\rho}\,U(t)$ and $V_i(\nu) = \sqrt{1-\rho}\,V^{(i)}(\nu) + \sqrt{\rho}\,V(t)$. We have $h_{i,j}(\nu) = 0$ because $E[dU_i(\nu)\,dV_j(\nu')] = 0$ (Eq. 3.81) so that $s_{i,j}(\nu) = s_{i,j}(-\nu)$. The equality $g_{i,j}(\nu) = s_{i,j}(\nu) + s_{i,j}(-\nu)$ shows that the ordinates of $g_{i,j}$ are twice the ordinates of $s_{i,j}$. The same result can be obtained by taking the Fourier transform of the covariance functions $c_{i,j}(\tau) = (1-\rho)\,\delta_{ij}\,c_{W_i}(\tau) + \rho\,c_W(\tau)$ of X. ■

3.9.4.3 Random fields

Let $(\Omega, \mathcal{F}, P)$ be a probability space. A function $X : D \times \Omega \to \mathbb{R}^d$ is said to be an $\mathbb{R}^d$-valued **random field** if $X(t)$ is an $\mathbb{R}^d$-valued random variable on

$(\Omega, \mathcal{F}, P)$ for each $\boldsymbol{t} \in D$, where $D \subset \mathbb{R}^{d'}$ and $d, d' \geq 1$ are integers. We limit our discussion to real-valued random fields ($d = 1$) and denote these fields by X or $\{X(\boldsymbol{t}), \boldsymbol{t} \in D\}$.

The f.d.d. of order n of X is the probability of $\cap_{i=1}^{n} \{X(\boldsymbol{t}_i) \leq x_i\}$, where $\boldsymbol{t}_i \in D$ and $x_i \in \mathbb{R}$. The random field X is said to be **strictly stationary/homogeneous** or just **stationary/homogeneous** if its finite dimensional distributions are invariant under a space shift, that is, if the points $\boldsymbol{t}_i$, $i = 1, \ldots, n$, are mapped into $\boldsymbol{t}_i + \boldsymbol{\tau} \in D$, where $\boldsymbol{\tau} \in \mathbb{R}^{d'}$ and the integers $n \geq 1$ are arbitrary. The marginal and the finite dimensional distributions of X are space invariant and depend only on the "space lag" $(\boldsymbol{t}_1 - \boldsymbol{t}_2)$, respectively.

Suppose that $X \in L_2(\Omega, \mathcal{F}, P)$ and define by

$$\begin{aligned}
\mu(\boldsymbol{t}) &= E[X(\boldsymbol{t})], \\
r(\boldsymbol{t}, \boldsymbol{s}) &= E[X(\boldsymbol{t})\, X(\boldsymbol{s})], \quad \text{and} \\
c(\boldsymbol{t}, \boldsymbol{s}) &= E[(X(\boldsymbol{t}) - \mu(\boldsymbol{t}))\, (X(\boldsymbol{s}) - \mu(\boldsymbol{s}))]
\end{aligned} \tag{3.82}$$

the **mean**, **correlation**, and **covariance** functions, respectively. The pair (μ, r) or (μ, c) defines the **second moment properties** of X. The properties of these functions result from Section 3.7.1. If X is stationary, then $\mu(\boldsymbol{t}) = \mu$ is constant and $r(\boldsymbol{t}, \boldsymbol{s})$, $c(\boldsymbol{t}, \boldsymbol{s})$ depend only on $(\boldsymbol{t} - \boldsymbol{s})$. If X is such that

$$\begin{aligned}
&\mu(\boldsymbol{t}) = \text{constant and} \\
&r(\boldsymbol{t}, \boldsymbol{s}),\ c(\boldsymbol{t}, \boldsymbol{s}) = \text{functions of only } \boldsymbol{t} - \boldsymbol{s},
\end{aligned} \tag{3.83}$$

then X is called a **weakly stationary/homogeneous** random field. A stationary field is weakly stationary but the converse is not generally true.

If the correlation function of a weakly stationary random field X is continuous, it has the representation (Bochner's theorem in Eq. 3.35)

$$r(\boldsymbol{\tau}) = \int_{\mathbb{R}^{d'}} e^{\sqrt{-1}\, \boldsymbol{\tau} \cdot \boldsymbol{\nu}}\, d\mathcal{S}(\boldsymbol{\nu}) = \int_{\mathbb{R}^{d'}} e^{\sqrt{-1}\, \boldsymbol{\tau} \cdot \boldsymbol{\nu}}\, s(\boldsymbol{\nu})\, d\boldsymbol{\nu}, \tag{3.84}$$

where $\mathcal{S}$ is a bounded, real-valued function such that $\int_R d\mathcal{S}(\boldsymbol{\nu}) \geq 0$ for all $R \in \mathcal{B}^{d'}$, $\mathcal{S}(-\infty, \ldots, -\infty) = 0$, and $\boldsymbol{\tau} \cdot \boldsymbol{\nu} = \sum_{k=1}^{d'} \tau_i\, \nu_i$ ([1], Theorem 2.1.2, p. 25). The function $\mathcal{S}$ is called **spectral distribution**. If $\mathcal{S}$ is absolutely continuous, then $d\mathcal{S}(\boldsymbol{\nu}) = [\partial^{d'} \mathcal{S}(\boldsymbol{\nu}) / \partial \nu_1 \cdots \partial \nu_{d'}]\, d\boldsymbol{\nu} = s(\boldsymbol{\nu})\, d\boldsymbol{\nu}$, where s denotes the **spectral density** function. The last equality in Eq. 3.84 is valid if X has a spectral density function.

The second moment calculus developed in Section 3.9 can be extended directly to random fields. We can define m.s. continuity and differentiability for random fields as in Sections 3.9.1 and 3.9.2. The extension of the m.s. integration in Section 3.9.3 to random fields involves minor changes because the integration is performed over a subset of $\mathbb{R}^{d'}$ rather then an interval of the real line. A very useful discussion on the second moment calculus for random fields can be found in [1] (Chapters 1 and 2). If X is a weakly stationary m.s. continuous random field with spectral distribution $\mathcal{S}$ and spectral density s, then X has the spectral representation (Eq. 3.71)

$$X(\boldsymbol{t}) = \int_{\mathbb{R}^{d'}} e^{\sqrt{-1}\,\boldsymbol{\nu}\cdot\boldsymbol{t}}\, dZ(\boldsymbol{\nu}), \tag{3.85}$$

where the random field Z is such that $E[Z(\boldsymbol{\nu})] = 0$ and $E[dZ(\boldsymbol{\nu})\, dZ(\boldsymbol{\nu}')^*] = \delta(\boldsymbol{\nu} - \boldsymbol{\nu}')\, d\mathcal{S}(\boldsymbol{\nu}) = \delta(\boldsymbol{\nu} - \boldsymbol{\nu}')\, s(\boldsymbol{\nu})\, d\boldsymbol{\nu}$. If X is a real-valued field, then (Eq. 3.84)

$$\int_{\mathbb{R}^{d'}} \sin(\boldsymbol{\nu}\cdot\boldsymbol{t})\, s(\boldsymbol{\nu})\, d\boldsymbol{\nu} = 0 \tag{3.86}$$

because the correlation function of X must be real-valued. The condition in Eq. 3.86 is satisfied if the spectral density of X is such that $s(\boldsymbol{\nu}) = s(-\boldsymbol{\nu})$ for all $\boldsymbol{\nu} \in \mathbb{R}^{d'}$.

Example 3.52: Let X be a real-valued random field defined by

$$X(\boldsymbol{t}) = \sum_{k=1}^{n} \sigma_k \left[A_k \cos(\boldsymbol{\nu}_k\cdot\boldsymbol{t}) + B_k \sin(\boldsymbol{\nu}_k\cdot\boldsymbol{t})\right], \quad \boldsymbol{t} \in \mathbb{R}^2, \tag{3.87}$$

where $E[A_k] = E[B_k] = 0$, $E[A_k\, B_l] = 0$, $E[A_k\, A_l] = E[B_k\, B_l] = \delta_{kl}$, $\sigma_k > 0$, and $\boldsymbol{\nu}_k \in \mathbb{R}^2$ for all $k, l = 1, \ldots, n$. The field has mean zero. The correlation and spectral density functions of X are

$$r(\boldsymbol{\tau}) = E[X(\boldsymbol{t}+\boldsymbol{\tau})\, X(\boldsymbol{t})] = \sum_{k=1}^{n} \sigma_k^2 \cos(\boldsymbol{\nu}_k\cdot\boldsymbol{\tau}) \quad \text{and}$$

$$s(\boldsymbol{\nu}) = \sum_{k=1}^{n} \frac{\sigma_k^2}{2} \left[\delta(\boldsymbol{\nu} - \boldsymbol{\nu}_k) + \delta(\boldsymbol{\nu} + \boldsymbol{\nu}_k)\right] \tag{3.88}$$

so that X is a weakly stationary random field. ◇

Proof: The expectation of the defining equations of X gives $E[X(\boldsymbol{t})] = 0$. The correlation function of X is

$$\begin{aligned} E[X(\boldsymbol{t})\, X(\boldsymbol{s})] &= \sum_{k,l=1}^{n} \sigma_k\, \sigma_l\, \delta_{kl} \left[\cos(\boldsymbol{\nu}_k\cdot\boldsymbol{t})\, \cos(\boldsymbol{\nu}_k\cdot\boldsymbol{s}) + \sin(\boldsymbol{\nu}_k\cdot\boldsymbol{t})\, \sin(\boldsymbol{\nu}_k\cdot\boldsymbol{s})\right] \\ &= \sum_{k=1}^{n} \sigma_k^2 \cos\left(\boldsymbol{\nu}_k\cdot(\boldsymbol{t} - \boldsymbol{s})\right) \end{aligned}$$

by the properties of the random coefficients (A_k, B_k). The Fourier transform of this function is the spectral density of X. The random field X is closely related to the stochastic process in Example 3.30. Both random functions are constructed by superposing n harmonics with random amplitude and phase.

The waves generating the random field X in Eq. 3.87 can be visualized simply since the domain of definition of X is $\mathbb{R}^2$. The zeros of $\cos(\boldsymbol{\nu}_k\cdot\boldsymbol{t})$ are the lines $\boldsymbol{\nu}_k\cdot\boldsymbol{t} = \pi/2 + q\,\pi$,

$q = 0, 1, \ldots$, in the (t_1, t_2)-space so that the wave associated with the frequency $\boldsymbol{\nu}_k$ has length $2\pi/\sqrt{\nu_{k,1}^2 + \nu_{k,2}^2}$ and evolves in a direction perpendicular to the zero lines at an angle $\theta_k = \tan^{-1}(\nu_{k,2}/\nu_{k,1})$ relative to the coordinate t_1. ■

3.9.4.4 Karhunen-Loéve representation

We show that a complex-valued process in L_2 can be represented by a linear combination of a countable number of deterministic functions with random coefficients. This representation, referred to as the **Karhunen-Loéve representation,** generalizes the expansion in Example 3.30 and holds also for random fields ([1], Section 3.3). The Karhunen-Loéve representation can be useful in applications since it provides alternative definitions for stochastic processes and random fields as functions of a countable number of random variables. However, the Karhunen-Loéve representation is relatively difficult to obtain and captures only the second moment properties of a random function.

Let $\{X(t), t \in [a, b]\}$ be a complex-valued stochastic process with mean zero and correlation function $r(t, s) = E[X(t)\, X(s)^*]$. Recall that the correlation function of X is symmetric, that is, $r(t, s) = r(s, t)^*$ (Example 3.29) and satisfies the condition $\int_a^b \int_a^b |r(t, s)|^2\, dt\, ds < \infty$ since X is in L_2. Consider the integral equation

$$\int_a^b r(t, s)\, \phi(s)\, ds = \lambda\, \phi(t), \quad t \in [a, b]. \tag{3.89}$$

A number $\lambda \neq 0$ and a function ϕ satisfying Eq. 3.89 such that $\int_a^b |\phi(t)|^2\, dt < \infty$ is called an **eigenvalue** and an **eigenfunction** for Eq. 3.89, respectively.

It can be shown that (1) Eq. 3.89 has at least one eigenvalue and eigenfunction, (2) the collection of eigenvalues for Eq. 3.89 is at most countable and strictly positive, (3) for each eigenvalue λ_k there exists at most a finite number of linearly independent eigenfunctions, (4) the eigenfunctions ϕ_k corresponding to distinct eigenvalues are orthogonal, that is, $\int_a^b \phi_k(t)\, \phi_l(t)^*\, dt = 0$ for $k \neq l$ and $\lambda_k \neq \lambda_l$, and (5) if the correlation function r of X is square integrable and continuous in $[a, b] \times [a, b]$, the Mercer theorem holds, that is, $r(t, s) = \sum_{k=1}^{\infty} \lambda_k\, \phi_k(t)\, \phi_k(s)^*$, and this series representation converges absolutely and uniformly in $[a, b] \times [a, b]$ ([1], Section 3.3, [49], Appendix 2, [98], Section 6.2).

Karhunen-Loéve representation. Let $\{X(t), t \in [a,b]\}$ be a complex-valued stochastic process with mean zero and correlation function $r(t,s) = E[X(t)\,X(s)^*]$. If $r(\cdot,\cdot)$ is square integrable and continuous in $[a,b] \times [a,b]$, then X has the representation

$$X(t) = \underset{n\to\infty}{\text{l.i.m.}} \sum_{k=1}^{n} \sigma_k\, X_k\, \phi_k(t) \quad \text{at each } t \in [a,b], \quad \text{where}$$

$$\sigma_k = \sqrt{\lambda_k}, \quad X_k = \frac{1}{\sigma_k} \int_a^b X(t)\,\phi_k(t)^*\,dt, \tag{3.90}$$

$E[X_k] = 0$, and $E[X_k\, X_l^*] = \delta_{kl}$.

Note: We only show that X_k in Eq. 3.90 has the stated properties. The mean of X_k is zero since $E[X(t)] = 0$. The correlation of these random variables is

$$E[X_k\, X_l^*] = \frac{1}{\sigma_k\,\sigma_l} \int_a^b \int_a^b E[X(t)\,X(s)^*]\,\phi_k(t)^*\,\phi_l(s)\,dt\,ds$$
$$= \frac{1}{\sigma_k\,\sigma_l} \int_a^b \left[\int_a^b r(t,s)\,\phi_l(s)\,ds\right] \phi_k(t)^*\,dt = \frac{\sigma_l}{\sigma_k} \int_a^b \phi_l(t)\phi_k(t)^*\,dt = \delta_{kl}.$$

It can be shown that the m.s. error,

$$E\left[\left(X(t) - X^{(n)}(t)\right)\left(X(t) - X^{(n)}(t)\right)^*\right] = r(t,t) - \sum_{k=1}^{n} |\sigma_k|^2\,|\phi_k(t)|^2,$$

of $X^{(n)}(t) = \sum_{k=1}^{n} \sigma_k\, X_k\, \phi_k(t)$ converges to zero as $n \to \infty$ ([49], Section 6-4). ▲

Example 3.53: Let $\{X(t), t \in [-\xi, \xi]\}$, $\xi > 0$, be a real-valued stochastic process with mean zero and correlation function $r(t,s) = (1/4)\,e^{-2\alpha|\tau|}$, where $\tau = t - s$ denotes the time lag and $\alpha > 0$. The Karhunen-Loéve representation of X in $[-\xi, \xi]$ is

$$X(t) = \sum_{k=1}^{\infty} \left[\sqrt{\lambda_k}\, X_k\, \phi_k(t) + \sqrt{\hat{\lambda}_k}\, \hat{X}_k\, \hat{\phi}_k(t)\right],$$

where

$$\lambda_k = \frac{1}{4\alpha\,(1 + b_k^2)}, \quad \phi_k(t) = \frac{\cos(2\,\alpha\, b_k\, t)}{\sqrt{\xi + \sin(4\,\alpha\, b_k\, \xi)/(4\,\alpha\, b_k)}},$$
$$\hat{\lambda}_k = \frac{1}{4\alpha\,(1 + \hat{b}_k^2)}, \quad \hat{\phi}_k(t) = \frac{\sin(2\,\alpha\, \hat{b}_k\, t)}{\sqrt{\xi - \sin(4\,\alpha\, \hat{b}_k\, \xi)/(4\,\alpha\, \hat{b}_k)}},$$

and $b_k, \hat{b}_k$ are the solutions of $b_k\, \tan(2\,\alpha\,\xi\, b_k) = 1$, $\hat{b}_k\, \cot(2\,\alpha\,\xi\,\hat{b}_k) = 1$. ◇

Note: The eigenvalues and eigenfunction in the representation of X are the solution of the integral equation $\int_{-\xi}^{\xi} e^{-2\alpha|t-s|}\,\phi(s)\,ds = 4\,\lambda\,\phi(t)$ defined by Eq. 3.89 ([49], Example 6-4.1, p. 99). ▲

Example 3.54: Let X be a real-valued process defined in $[0, 1]$ with mean zero and correlation function $r(t, s) = t \wedge s$. The Karhunen-Loéve representation of this process is

$$X(t) = \underset{n\to\infty}{\text{l.i.m.}}\, \frac{\sqrt{2}}{\pi} \sum_{k=0}^{n} \frac{\sin(k + 1/2)\,\pi\,t}{k + 1/2}\, Z_k,$$

where $E[Z_k] = 0$, $E[Z_k^2] = 1$, and the sequence $Z_0, Z_1, \ldots$ is uncorrelated. ◇

Proof: The condition in Eq. 3.89 becomes

$$\int_0^t s\,\phi(s)\,ds + t \int_t^1 \phi(s)\,ds = \lambda\,\phi(t), \quad t \in (0, 1),$$

and gives $\int_t^1 \phi(s)\,ds = \lambda\,\phi'(t)$ by differentiation with respect to t. We can perform this operation since the left side of the above equation is differentiable so that ϕ on its right side is also differentiable. Differentiating one more time with respect to time, we have $\lambda\,\phi''(t) + \phi(t) = 0$ in $(0, 1)$ with the boundary conditions $\phi(0) = 0$ and $\phi'(1) = 0$, which result from Eq. 3.89 with $r(t, s) = t \wedge s$ and the equation $\int_t^1 \phi(s)\,ds = \lambda\,\phi'(t)$, respectively. The solution of the above differential equation gives the eigenvalues and eigenfunctions in the representation of X.

We note that the Brownian motion and the compound Poisson processes have correlation functions similar to the correlation function of the process X in this example. Hence, the above Karhunen-Loéve representation can be used to characterize these two processes. This observation also demonstrates a notable limitation of the Karhunen-Loéve representation. The representation cannot distinguish between processes that are equal in the second moment sense. ■

Example 3.55: Let $\{W(t), t \in [a, b]\}$ be a white noise defined formally as a process with mean zero and correlation function $E[W(t)\,W(s)] = \gamma\,\delta(t - s)$, where $t, s \in [a, b]$, $\gamma > 0$ is interpreted as the noise intensity, and $\delta(\cdot)$ denotes the Dirac delta function. Then $W(t) = \sqrt{\gamma}\,\sum_{k=1}^{\infty} W_k\,\phi_k(t)$ is the Karhunen-Loéve representation of W, where W_k are random variables with $E[W_k] = 0$ and $E[W_k\,W_l] = \delta_{kl}$ and $\{\phi_k\}$ is a collection of orthonormal functions spanning the class of square integrable functions defined in $[a, b]$. ◇

Proof: Eq. 3.89 gives $\gamma\,\phi(t) = \lambda\,\phi(t)$ so that all eigenvalues are equal to γ and the eigenfunctions are arbitrary provided that they are orthonormal and span the space of square integrable functions. The first two moments of the random variables $\{W_k\}$ result from the definition of X_k in Eq. 3.90. These calculations are formal since the process W does not exists in the second moment sense. However, this representation of the white noise is frequently used in engineering applications [175]. ■

Example 3.56: Suppose that a weakly stationary process X with mean zero and unknown correlation function $E[X(t)\,X(s)]$ is observed through an imperfect device. The observation equation is $Y(t) = X(t) + W(t)$, $0 \le t \le \tau$, where W is the white noise process in the previous example. The processes X and W are not correlated. It is assumed that the observed process Y has mean zero and a known correlation function $r_y(t-s) = E[Y(t)\,Y(s)]$, $t, s \in [0, \tau]$. Then the best m.s. estimator of X is

$$\hat{X}_{\text{opt}}(t) = \sum_{k=1}^{\infty} \frac{1}{1+\gamma/\lambda_k}\, Y_k\, \phi_k(t)$$

where γ/λ_k is the noise to signal ratio, (λ_k, ϕ_k) are the eigenvalues and eigenfunctions given by Eq. 3.89 with $r(t,s) = r_y(t-s)$, and the properties of the random variables Y_k result from Eq. 3.90 written for the process Y. If the noise intensity γ decreases to zero, the optimal estimator $\hat{X}_{\text{opt}}$ has the same Karhunen-Loéve representation as X. $\diamond$

Proof: Let $X(t) = \sum_{k=1}^{\infty} \sqrt{\lambda_k}\, X_k\, \phi_k(t)$ be the Karhunen-Loéve representation of X, where ϕ_k and λ_k are the solutions of Eq. 3.89 with r being the correlation function of X. The previous example has shown that W admits the expansion $W(t) = \sqrt{\gamma}\,\sum_{k=1}^{\infty} W_k\,\phi_k(t)$ so that

$$Y(t) = X(t) + W(t) = \sum_{k=1}^{\infty} (\sqrt{\lambda_k}\, X_k + \sqrt{\gamma}\, W_k)\, \phi_k(t) = \sum_{k=1}^{\infty} Y_k\, \phi_k(t),$$

where $Y_k = \sqrt{\lambda_k}\, X_k + \sqrt{\gamma}\, W_k$ (Eq. 3.90 and Example 3.55). The above equation shows that the Karhunen-Loéve representations of X and Y involve the same eigenfunctions ϕ_k and these functions can be obtained from Eq. 3.89 with $r(t,s) = r_y(t-s)$. The Karhunen-Loéve representation of Y can be calculated since this process is assumed to have known second moment properties.

Consider the estimator $\hat{X}(t) = \sum_{k=1}^{\infty} h_k\, Y_k\, \phi_k(t)$, $t \in [0, \tau]$ of X, where the coefficients h_k need to be determined. The m.s. error, $e = E\left[\int_0^\tau \left(X(t) - \hat{X}(t)\right)^2 dt\right]$, of $\hat{X}$ is equal to $\sum_{k=1}^{\infty} E\left[\left|h_k\, Y_k - \sqrt{\lambda_k}\, X_k\right|^2\right]$ so that it is minimized for $h_k = \lambda_k/(\lambda_k + \gamma)$. The expression of $\hat{X}$ with $h_k = \lambda_k/(\lambda_k+\gamma)$ is the stated optimal estimator. In some of the above operations we interchange summations with integrals and expectations. It can be shown that these operations are valid ([49], Appendix 2). ■

The results in Eqs. 3.89-3.90 can be extended directly to random fields in L_2 provided that their correlation functions satisfy similar conditions. For example, let $X(\boldsymbol{t}) \in L_2$, $\boldsymbol{t} \in R$, be a real-valued random field, where R is a bounded rectangle in $\mathbb{R}^d$. If the correlation function $r(\boldsymbol{t}, \boldsymbol{s}) = E[X(\boldsymbol{t})\,X(\boldsymbol{s})]$ of X is square integrable and continuous in $R \times R$, then X admits a Karhunen-Loéve representation similar to the one in Eq. 3.90 ([1], Section 3.3, [98], Section 6.2).

3.10 Extremes of stochastic processes

Let $\{X(t), t \geq 0\}$ be the state of a physical system and suppose that the system performance is adequate as long as X does not leave an open subset D of $\mathbb{R}^d$. The stopping time $T = \inf\{t \geq 0 : X(t) \notin D, X(0) \in D\}$ is defined to be the system failure time. The probabilities $P_s(t) = P(T > t)$ and $P_f(t) = 1 - P_s(t)$ are referred to as reliability and failure probability in $(0, t]$, respectively.

Let $N(D; t)$ or $N(D; I)$ denote the number of times X exits D during a time interval $I = [0, t]$. Then

$$P_f(t) \leq E[N(D; t)] + P(X(0) \notin D). \tag{3.91}$$

Proof: We have $P_f(t) = P(\{N(D; t) > 0\} \cup \{X(0) \notin D\})$ so that $P_f(t) \leq P(N(D; t) > 0) + P(X(0) \notin D)$. Also,

$$P(N(D; t) > 0) = \sum_{k=1}^{\infty} P(N(D; t) = k) \leq \sum_{k=1}^{\infty} k\, P(N(D; t) = k) = E[N(D; t)].$$

Numerical results show that the upper bound on $P_f(t)$ in Eq. 3.91 is relatively tight for highly reliable systems, that is, systems characterized by very small values of $P_f(t)$ and $E[N(D; t)]$ ([175], Chapter 7). If $E[N(D; t)]$ is not finite or is finite but larger than 1, the upper bound in Eq. 3.91 provides no information on $P_f(t)$. ■

The following two sections give conditions under which $E[N(D; t)]$ is finite and formulas for calculating $E[N(D; t)]$ and the density of T.

3.10.1 Mean crossing rate

Let $\{X(t), t \geq 0\}$ be a real-valued stationary stochastic process with a.s. continuous samples and continuous marginal distribution. We say that a sample $X(\cdot, \omega)$ of X upcrosses a level x at time t or has an x-**upcrossing** at t if for each $\varepsilon > 0$ there exist $t_1 \in (t - \varepsilon, t)$ and $t_2 \in (t, t + \varepsilon)$ such that $X(t_1, \omega) < x < X(t_2, \omega)$. Suppose that $D = (-\infty, x)$ and $I = (0, t]$. Since X can exit D only through its right end, $N(D; I)$ in Eq. 3.91 represents the number of x-upcrossings of X in I, and is denoted by $N(x; I)$ or $N(x; t)$.

Let X_h be a piece-wise linear approximation of X so that $X_h(j\,h, \omega) = X(j\,h, \omega)$, $j = 1, 2, \ldots$, where $h > 0$ is a small time step. The scaled probability $J_h(x) = P(X(0) < x < X(h))/h$ is equal to the average number of x-upcrossings of X_h per unit of time or the **mean x-upcrossing rate** of X_h since X_h has one x-upcrossing in h with probability $P(X(0) < x < X(h))$ and no x-upcrossing in this interval if the event $\{X(0) < x < X(h)\}$ does not occur. Let $N_h(x; I)$ denote the number of x-upcrossings of X_h in I. It can be shown that $N_h(x; I) \leq N(x; I)$ and that $N_h(x; I) \to N(x; I)$ a.s. and $E[N_h(x; I)] \to E[N(x; I)]$ as $h \to 0$ ([121], Lemma 7.2.2, p. 148).

The following result, referred to as the Rice formula, is used extensively in engineering. The proof of Rice's formula is based on the approximation X_h of X. Let $Z_h = (X(h) - X(0))/h$ be the slope of X_h in a time step and let $f_h(\cdot, \cdot)$ denote the joint density of $(X(0), Z_h)$.

Rice's formula. If (1) X is a real-valued stationary process with a.s. continuous samples and continuous marginal distribution, (2) the density $f_{X(t),\dot{X}(t)}$ of $(X(t), \dot{X}(t))$ exists, and (3) $f_h(\xi, z)$ is continuous in ξ for all z and small $h > 0$ and converges to $f_{X(t),\dot{X}(t)}(\xi, z)$ uniformly in ξ for fixed z as $h \to 0$, and (4) $\int_0^\infty z\, f_h(\xi, z)\, dz < \infty$, then the mean x-upcrossing rate of X is

$$\lambda(x)^+ = \lim_{h \to 0} J_h(x) = \int_0^\infty z\, f_{X(t),\dot{X}(t)}(x, z)\, dz$$
$$= E\left[\dot{X}(t)^+ \mid X(t) = x\right] f_{X(t)}(x). \tag{3.92}$$

Proof: We first note that the mean rate at which X crosses with negative slope a level $x \in \mathbb{R}$ at a time t, that is, the **mean x-downcrossing rate** of X at t, is

$$\lambda(x)^- = E\left[\dot{X}(t)^- \mid X(t) = x\right] f_{X(t)}(x).$$

If X is not a stationary process, its mean x-upcrossing and x-downcrossing rates will depend on time and will be denoted by $\lambda(x; t)^+$ and $\lambda(x; t)^-$, respectively.

The event $X(0) < x < X(h)$ that X_h has an x-upcrossing in h can be written as $\{X(0) < x < X(0) + h\, Z_h\} = \{X(0) < x\} \cap \{Z_h > (x - X(0))/h\}$ so that

$$J_h(x) = h^{-1} \int_0^\infty dz \int_{x - z\,h}^{x} du\, f_h(u, z) = h^{-1} \int_0^\infty dz \int_0^1 dv\, (h\, z)\, f_h(x - h\, z\, v, z)$$

where the last equality results by the change of variables $u = x - h\, z\, v$. The limit of $J_h(x)$ as $h \to 0$ gives Eq. 3.92. Details on this proof and on crossings of stochastic processes can be found in [121] (Chapter 7).

If $\boldsymbol{X}$ is an $\mathbb{R}^d$-valued stochastic process and $D \in \mathcal{B}^d$, it can be shown that under similar assumptions the mean rate at which $\boldsymbol{X}$ exits D at a time t or the **mean D-outcrossing rate** of $\boldsymbol{X}$ at t is ([175], Chapter 7)

$$\lambda(D; t) = \int_{\partial D} E[\dot{X}_{(\boldsymbol{n})}(t)+ \mid \boldsymbol{X}(t) = \boldsymbol{x}]\, f_{\boldsymbol{X}(t)}(\boldsymbol{x})\, d\sigma(\boldsymbol{x}), \tag{3.93}$$

where $\partial D = \bar{D} \setminus D$ denotes the boundary of D and $\dot{X}_{(\boldsymbol{n})}(t) = \dot{\boldsymbol{X}}(t) \cdot \boldsymbol{n}(\boldsymbol{x})$ is the projection of $\dot{\boldsymbol{X}}(t)$ on the outer normal $\boldsymbol{n}(\boldsymbol{x})$ to D at $\boldsymbol{x} \in \partial D$. ■

Example 3.57: Let X be an m.s. differentiable Gaussian process with mean and covariance functions $\mu(t)$ and $c(t, s)$, respectively. The mean x-upcrossing rate of X at time t is given by Eq. 3.92 with

$$E[\dot{X}(t)+ \mid X(t) = x] = \hat{\sigma}(t) \left[\phi\left(\frac{\hat{\mu}(t)}{\hat{\sigma}(t)}\right) + \frac{\hat{\mu}(t)}{\hat{\sigma}(t)}\, \Phi\left(\frac{\hat{\mu}(t)}{\hat{\sigma}(t)}\right)\right] \quad \text{and}$$

$$f(x; t) = \frac{1}{\breve{\sigma}(t)}\, \phi\left(\frac{x - \breve{\mu}(t)}{\breve{\sigma}(t)}\right), \quad \text{where}$$

$\hat{\mu}(t) = \dot{\mu}(t) + [c_{,t}(t,s)\,|_{t=s}\,/c(t,t)]\,(x - \mu(t))$, $c_{,t}(t,s) = \partial c(t,s)/\partial t$, $\hat{\sigma}(t)^2 = c_{,ts}(t,s)\,|_{t=s} - \left(c_{,t}(t,s)\,|_{t=s}\right)^2/c(t,t)$, and $c_{,ts}(t,s) = \partial^2 c(t,s)/(\partial t\,\partial s)$.

Figure 3.7 shows the variation in time of the mean x-upcrossing rate of

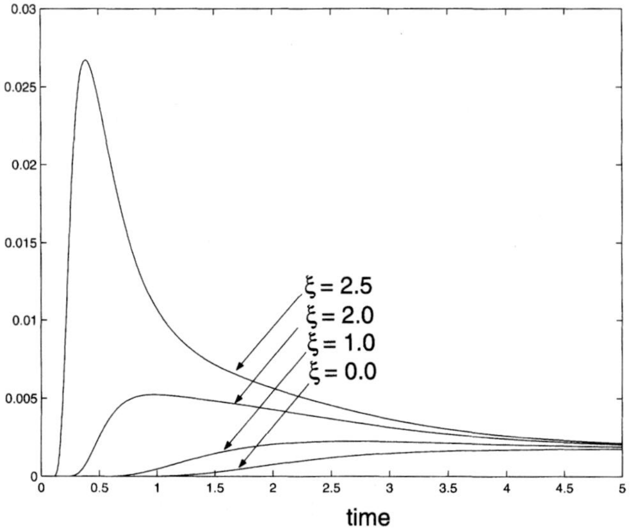

Figure 3.7. Mean x-upcrossing rate for a non-stationary Gaussian process $X(t) = Y(t) \mid (Y(0) = \xi)$, $x = 3.0$, and $t \in [0, 5]$

$X(t) = Y(t) \mid (Y(0) = \xi)$ for $x = 3$ and $t \in [0, 5]$, where Y is a stationary Gaussian process with mean zero and covariance function $\rho(\tau) = E[Y(t)\,Y(t+\tau)] = (1 + \alpha\,|\tau|)\,e^{-\alpha|\tau|}$ for $\alpha = 1$ and several values of ξ. We note that the mean x-upcrossing rate $\lambda(x;t)^+$ (1) converges to the stationary value $\lambda(x)^+ = (\alpha/(2\pi))\exp(-x^2/2)$ as $t \to \infty$ and (2) is an increasing function of ξ at each time $t > 0$. The first property is a consequence of the reduced effect of the initial value of X on its future values as time increases. The increased likelihood that X upcrosses a level x for an initial value $X(0) = \xi$ closer to this level explains the second property of $\lambda(x;t)^+$. ◇

Proof: The $\mathbb{R}^2$-valued Gaussian variable $(X(t), \dot{X}(t))$ has mean $\boldsymbol{\mu} = (\mu(t), \dot{\mu}(t))$ and covariance matrix $\boldsymbol{c}$ with entries $c_{11} = 1 - \rho(t)^2$, $c_{12} = c_{21} = \rho(t)\,\rho'(t)$, and $c_{22} = -\rho''(0) - \rho'(t)^2$. Hence, $\dot{X}(t) \mid (X(t) = x)$ is a Gaussian random variable with mean and standard deviation $\hat{\mu}(t)$ and $\hat{\sigma}(t)$ (Section 2.11.5).

The mean x-upcrossing rate of Y is $\lambda(x)^+ = (\alpha/\sqrt{2\pi})\,\phi(x)$ since $Y(t)$ and $\dot{Y}(t)$ are independent Gaussian variables so that $E[\dot{Y}(t)+ \mid Y(t)] = E[\dot{Y}(t)+]$ and $E[\dot{Y}(t)+] = \alpha/\sqrt{2\pi}$. ■

3.10.2 First passage time density

Let $\{X(t), t \geq 0\}$ be a real-valued Gaussian process with mean zero and covariance function $c(t, s) = E[X(t)\, X(s)]$. Suppose that $X(0) < x$ and denote by $T = \inf\{t > 0 : X(t) \geq x\}$ the first time X exceeds a level x.

Durbin formula. If $c(\cdot, \cdot)$ has continuous first order partial derivatives and $\lim_{s \uparrow t} \mathrm{Var}[X(t) - X(s)]/(t - s) < \infty$, then

$$f_{T,\dot{X}(T)}(t, z) = E\left[1_{\{X(s)<x,\, 0 \leq s<t\}}\, \dot{X}(t)^+ \mid X(t) = x,\, \dot{X}(t) = z\right] f_{X(t),\dot{X}(t)}(x, z),$$
$$f_T(t) = E\left[1_{\{X(s)<x,\, 0 \leq s<t\}}\, \dot{X}(t)^+ \mid X(t) = x\right] f_{X(t)}(x). \tag{3.94}$$

Note: The density f_T of T is equal to the mean x-upcrossing rate of the samples of X that have not ever left $D = (-\infty, x)$ in $[0, t]$. Because $1_{\{X(s)<x,\, 0 \leq s<t\}}$ is zero on the subset of the sample space corresponding to samples of X that upcross x at least once in $[0, t]$, we have $f_T(t) \leq \lambda(x; t)^+$. Details on Durbin's formula in Eq. 3.94, extensions of this formula, and computer codes for calculating the density of T can be found in [56, 158, 159]. ▲

Example 3.58: Let $\{X(t), t = 0, 1, \ldots\}$ be a time series with independent values following the same distribution F. Denote by $T = \inf\{t \geq 0 : X(t) > x\}$ the first time the series X exceeds a specified level x. The exact and approximate expressions of the distributions of T are

$$F_T(t) = 1 - F(t)^{t+1} \quad \text{and} \quad F_T(t) \simeq F(x)\, e^{-\lambda(x)^+ t},$$

where $\lambda(x)^+$ denotes the mean x-upcrossing rate of X.

Figure 3.8 shows the probability $P(T > t)$ for $F(x) = \Phi(x)$ and $x = 1.5$.

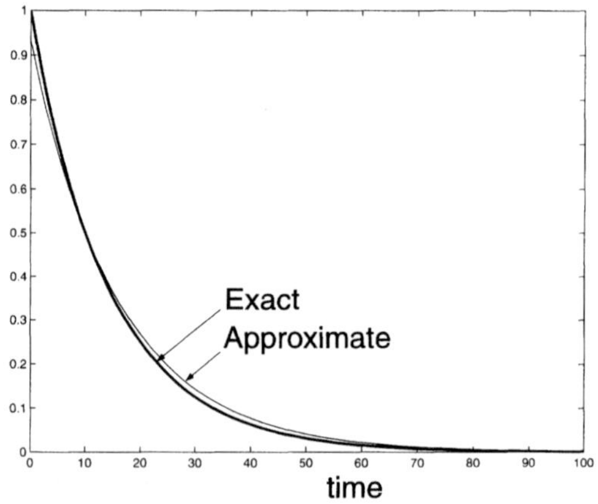

Figure 3.8. Exact and approximate probabilities $P(T > t)$ for $F(x) = \Phi(x)$ and $x = 1.5$

The approximation of $P(T > t)$ based on the mean x-upcrossing rate of X is accurate in this case. ◇

Proof: Because the events $\{T > t\}$ and $\cap_{k=0}^{t}\{X(k) \leq x\}$ have the same probability, we have $P(T > t) = (P(X_1 \leq x))^{t+1} = F(x)^{t+1}$.

Note also that the event $\{T > t\}$ occurs if $X(0) \leq x$ and X has no x-upcrossing in t (Eq. 3.91) so that

$$P(T > t) = P(\{X(0) \leq x\} \cap \{N(x; t) = 0\}) \simeq P(X(0) \leq x)\, P(N(x; t) = 0).$$

The above approximate equality is based on the assumption that $\{X(0) \leq x\}$ and $\{N(x; t) = 0\}$ are independent events. The approximation of $P(T > t)$ in Fig. 3.8 is given by the above equation in which the probability of $\{N(x; t) = 0\}$ is taken to be $\exp(-\lambda(x)^+ t)$, where $\lambda(x)^+ = F(x)\,(1 - F(x))$ is the mean x-upcrossing rate of X. This expression of $P(N(x; t) = 0)$ is based on an asymptotic result stating that the x-upcrossings of X approach a Poisson process with intensity $\lambda(x)^+$ as x increases indefinitely ([45], Section 12.2). ■

3.11 Martingales

We have examined in Section 2.18 discrete time martingales. Here we consider continuous time stochastic processes that are martingales.

A real-valued stochastic process X defined on a filtered probability space $(\Omega, \mathcal{F}, (\mathcal{F}_t)_{t\geq 0}, P)$ is an $\mathcal{F}_t$**-martingale** if

$$\begin{aligned} &1.\ E[|X(t)|] < \infty, \quad \forall t \geq 0,\\ &2.\ X \text{ is } \mathcal{F}_t - \text{adapted}, \quad \text{and}\\ &3.\ E[X(t) \mid \mathcal{F}_s] = X(s) \ \text{ a.s.}, \quad \forall s \leq t. \end{aligned} \tag{3.95}$$

Note: The last condition in Eq. 3.95 can be replaced with $E[X(t) \mid \mathcal{F}_s] = X(s \wedge t)$, $t, s \geq 0$, since $E[X(t) \mid \mathcal{F}_s] = X(t) = X(t \wedge s)$ for $t < s$ and $E[X(t) \mid \mathcal{F}_s] = X(s)$ for $t \geq s$.

If the equality $E[X(t) \mid \mathcal{F}_s] = X(s)$, $s \leq t$, is replaced by $\geq$ and $\leq$, X is said to be an $\mathcal{F}_t$**-submartingale** and $\mathcal{F}_t$**-supermartingale**, respectively.

If X is in $L_2(\Omega, \mathcal{F}, P)$ and satisfies the last two conditions in Eq. 3.95, then X is said to be an $\mathcal{F}_t$**-square integrable martingale**. Similarly, a submartingale/supermartingale $X \in L_2(\Omega, \mathcal{F}, P)$ is called a square integrable submartingale/supermartingale. If the first condition in Eq. 3.95 is replaced by $E[|X(t)|^p] < \infty$, $p > 1$, then X is said to be a p-integrable martingale. ▲

We have seen in Section 2.18 that the martingale, submartingale, and supermartingale can be viewed as models for fair, super-fair, and unfair games by interpreting $t \geq s$, s, $E[X(t) \mid \mathcal{F}_s]$, and $X(s)$ as future, present, average future fortune, and current fortune, respectively. For example, the average future fortune of a player $E[X(t) \mid \mathcal{F}_s]$ is equal to his or her current fortune $X(s)$ for martingales but it is larger/smaller then $X(s)$ for submartingales/supermartingales. The conditions that a stochastic process X must satisfy to be a martingale, submartingale, and supermartingale add structure to this process, as we will see in this section.

Example 3.59: Consider a filtered probability space $(\Omega, \mathcal{F}, (\mathcal{F}_t)_{t\geq 0}, P)$ and a random variable $Y \in \mathcal{F}$ such that $E[|Y|] < \infty$. The process $X(t) = E[Y \mid \mathcal{F}_t]$, $t \geq 0$, is an $\mathcal{F}_t$-martingale. ◇

Proof: We have $E[|X(t)|] = E[|E(Y \mid \mathcal{F}_t)|] \leq E[E[|Y| \mid \mathcal{F}_t]] = E[|Y|] < \infty$ by the Jensen inequality for conditional expectation (Section 2.17.2) and hypothesis. X is $\mathcal{F}_t$-adapted because $E[Y \mid \mathcal{F}_t]$ is $\mathcal{F}_t$-measurable for each $t \geq 0$. Properties of the conditional expectation give $E[X(t) \mid \mathcal{F}_s] = E[E(Y \mid \mathcal{F}_t) \mid \mathcal{F}_s] = E[Y \mid \mathcal{F}_s] = X(s)$ for all $s \leq t$. Hence, X is a martingale, referred to as the martingale **closed** by the random variable Y. ■

Example 3.60: The Brownian motion process B is an $\mathcal{F}_t$-martingale, where $\mathcal{F}_t = \sigma(B(s) : 0 \leq s \leq t)$ denotes the natural filtration of B. ◇

Proof: We have $E[|B(t)|] \leq \left(E[B(t)^2]\right)^{1/2} = t^{1/2}$ by the Cauchy-Schwarz inequality and properties of B. Since $\mathcal{F}_t = \sigma(B(s) : 0 \leq s \leq t)$ for all $t \geq 0$, B is $\mathcal{F}_t$-adapted. For $t \geq s$ we have

$$\begin{aligned} E[B(t) \mid \mathcal{F}_s] &= E[(B(t) - B(s)) + B(s) \mid \mathcal{F}_s] \\ &= E[B(t) - B(s) \mid \mathcal{F}_s] + E[B(s) \mid \mathcal{F}_s] = B(s) \end{aligned}$$

since the conditional expectation is a linear operator, the increment $B(t) - B(s)$ is independent of $B(u)$, $u \leq s$, so that $E[B(t) - B(s) \mid \mathcal{F}_s] = E[B(t) - B(s)]$, B has mean zero, and $B(s)$ is $\mathcal{F}_s$-measurable. Hence, B is a martingale. ■

Example 3.61: The compound Poisson process C in Eq. 3.6 can be an $\mathcal{F}_t$-martingale, submartingale, or supermartingale depending on the sign of $E[Y_1]$, where $\mathcal{F}_t$ denotes the natural filtration of C and Y_1 is assumed to be in L_1. ◇

Proof: The mean of the absolute value of $C(t)$ is smaller than $E\left[\sum_{k=1}^{N(t)} |Y_k|\right] = \lambda\, t\, E[|Y_1|]$, which is finite because $E[|Y_1|] < \infty$ by hypothesis. The process is $\mathcal{F}_t$-adapted. For $t \geq s$ we have

$$\begin{aligned} E[C(t) \mid \mathcal{F}_s] &= E[(C(t) - C(s)) + C(s) \mid \mathcal{F}_s] \\ &= E[C(t) - C(s) \mid \mathcal{F}_s] + E[C(s) \mid \mathcal{F}_s] = \lambda\,(t - s)\, E[Y_1] + C(s) \end{aligned}$$

since C has stationary independent increments, $E[C(t) - C(s)] = \lambda\,(t - s)\, E[Y_1]$, and $C(s)$ is $\mathcal{F}_s$-measurable. If $E[Y_1]$ is zero, positive, or negative, then C is a martingale, submartingale, or supermartingale, respectively. The above equalities also show that the **compensated compound Poisson** process $C(t) - \lambda\, t\, E[Y_1]$ is an $\mathcal{F}_t$-martingale. ■

Example 3.62: Let $\{B_i(\nu), 0 \leq \nu \leq \bar{\nu}\}$, $i = 1, 2$, be two independent Brownian motions, $0 < \nu_1 < \nu_2 < \cdots < \nu_n < \nu_{n+1} = \bar{\nu} < \infty$ be some frequencies, and $\sigma_k > 0$, $k = 1, \ldots, n$, denote arbitrary numbers. Then

$$X(t) = \sum_{k=1}^{n} \frac{\sigma_k}{\sqrt{\nu_k}} \left[B_1(\nu_k)\, \Delta B_1(\nu_k)\, \cos(\nu_k\, t) + B_2(\nu_k)\, \Delta B_2(\nu_k)\, \sin(\nu_k\, t)\right] \tag{3.96}$$

is a weakly stationary process with one-sided spectral density function $g(\nu) = \sum_{k=1}^{n} \sigma_k^2\, \delta(\nu - \nu_k)$, where $\Delta B_i(\nu_k) = B(\nu_{k+1}) - B(\nu_k)$, $i = 1, 2$. The definition of X constitutes a discrete version of Eq. 3.74.

Figure 3.9 shows a sample of X scaled to have unit variance for $n = 100$, $\Delta\nu = \nu_{k+1} - \nu_k = 1/10$, $\sigma_k^2 = g(\nu_k)\,\Delta\nu$, and $g(\nu) = (2/\pi)/(\nu^2 + 1)$. The estimated value of the coefficient of kurtosis of X is 2.18, indicating that X is not a Gaussian process. ◇

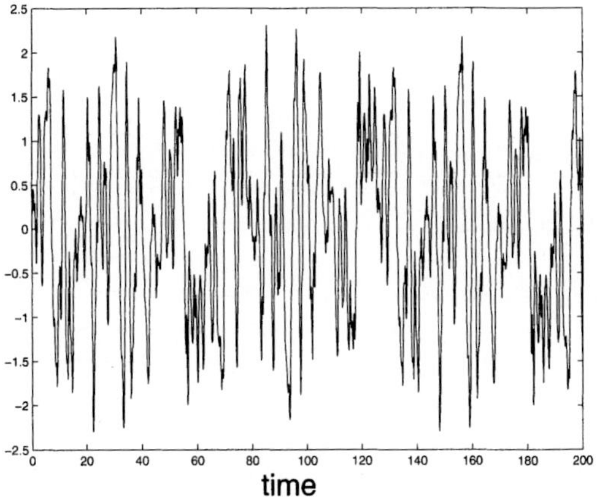

Figure 3.9. A sample of X in Eq. 3.96

Proof: The mean and variance of the increments $B_i(\nu_k)\,\Delta B_i(\nu_k)$ are zero and $\nu_k\,\Delta\nu_k$, respectively. The correlation of these increments for $k < l$ is

$$\begin{aligned} E[B_i(\nu_k)\,\Delta B_i(\nu_k)\,B_i(\nu_l)\,\Delta B_i(\nu_l)] &= E\left\{E[B_i(\nu_k)\,\Delta B_i(\nu_k)\,B_i(\nu_l)\,\Delta B_i(\nu_l) \mid \mathcal{F}_l]\right\} \\ &= E\left\{B_i(\nu_k)\,\Delta B_i(\nu_k)\,B_i(\nu_l)\,E[\Delta B_i(\nu_l) \mid \mathcal{F}_l]\right\} = 0 \end{aligned}$$

since $B_i(\nu_k)$, $\Delta B_i(\nu_k)$, and $B_i(\nu_l)$ are $\mathcal{F}_l$-measurable and $E[\Delta B_i(\nu_l) \mid \mathcal{F}_l] = 0$. Hence, the increments $B_i(\nu)\,\Delta B_i(\nu)$ have the same second moment properties as the increments of U and V in the spectral representation of a real-valued process (Eq. 3.74). ■

A real-valued process X defined on a filtered probability space $(\Omega, \mathcal{F}, (\mathcal{F}_t)_{t\geq 0}, P)$ is an $\mathcal{F}_t$-**local martingale**, **submartingale**, or **supermartingale** if there exists an increasing sequence T_n, $n = 1, 2, \ldots$, of $\mathcal{F}_t$-stopping times, that is, $T_n \leq T_{n+1}$, such that (1) $\lim_{n\to\infty} T_n = +\infty$ a.s. and (2) the stopped process $X^{T_n}(t) = X(t \wedge T_n)$ is an $\mathcal{F}_t$-martingale, submartingale, or supermartingale, respectively, for each n.

Note: The sequence T_n is said to **reduce** X, and is referred to as a **localizing sequence**. A martingale, submartingale, or supermartingale can have two or more localizing sequences. In many applications the stopping times T_n are related to sample properties of X, for example, the sequence of stopping times $T_n = \inf\{t \geq 0 : |X(t)| > n\}$ gives the first time when X leaves the interval $[-n, n]$. This sequence of stopping times is increasing but may or may not converge to infinity depending on the properties of X. If X is a Brownian motion B, then T_n is a localizing sequence for B (Example 3.64).

We also note that every martingale, submartingale, and supermartingale is a local martingale, submartingale, and supermartingale, respectively. For example, we can take $T_n = \infty$ for all $n \geq 1$. ▲

Example 3.63: Let N be a Poisson process with intensity $\lambda > 0$. The random variables $T_n = \inf\{t \geq 0 : N(t) = n\}$, $n = 1, 2, \ldots$, define an increasing sequence of stopping times such that $T_n \to \infty$ a.s. as $n \to \infty$. The processes N^{T_n}, $n = 1, 2, \ldots$, are $\mathcal{F}_t$-submartingales so that N is an $\mathcal{F}_t$-local submartingale with the localizing sequence T_n. Note also that N is an $\mathcal{F}_t$-submartingale. ◇

Proof: The random variable T_n is an $\mathcal{F}_t$-stopping time since the events $\{T_n \leq t\}$ and $\{N(t) \geq n\}$ coincide and $\{N(t) \geq n\} \in \mathcal{F}_t$. We also have $T_n = \sum_{i=1}^{n} Z_i$, where Z_i are independent exponential random variables with mean $1/\lambda$ so that $T_{n+1} = T_n + Z_{n+1} \geq T_n$ a.s. for all $n \geq 1$ and $T_n \to \infty$ a.s. as $n \to \infty$ since $E[Z_1] = 1/\lambda > 0$ (Section 2.14).

We will see in Section 3.11.2 that right continuous stopped martingales, submartingales, and supermartingales are martingales, submartingales, and supermartingales, respectively. Hence, N^{T_n} is an $\mathcal{F}_t$-submartingale for each $n \geq 1$. ■

Example 3.64: Let $\{\boldsymbol{B}(t), t \geq 0\}$ be a Brownian motion in $\mathbb{R}^d$ such that $\boldsymbol{B}(0) = \boldsymbol{x} \in \mathbb{R}^d$ and let $S(\boldsymbol{x}, n) = \{\boldsymbol{y} \in \mathbb{R}^d : \| \boldsymbol{y} - \boldsymbol{x} \| \leq n\}$, $n = 1, 2, \ldots$, be a sphere of radius n centered at $\boldsymbol{x} = (x_1, \ldots, x_d)$. The coordinates B_i of $\boldsymbol{B}$ are independent real-valued Brownian motions starting at x_i rather than zero. The random variables $T_n = \inf\{t \geq 0 : \boldsymbol{B}(t) \notin S(\boldsymbol{x}, n)\}$ define an increasing sequence of stopping times with respect to the natural filtration of $\boldsymbol{B}$ and $T_n \to \infty$ a.s. as $n \to \infty$. ◇

Proof: The random variables T_n satisfy the condition $T_n \leq T_{n+1}$ and are stopping times since we can tell whether T_n exceeds an arbitrary time $t \geq 0$ from properties of $\boldsymbol{B}$ up to this time. Since the Brownian motions B_i have continuous samples, they cannot reach infinity in a finite time so that $T_n \to \infty$ a.s. as $n \to \infty$. ■

3.11.1 Properties

Many of the properties of the martingales considered here are counterparts of the properties of discrete time martingales in Section 2.18.1.

• The expectation of a martingale, submartingale, and supermartingale X is a constant, increasing, and decreasing function, respectively.

• X is a martingale if it is both a submartingale and a supermartingale.

• If X is a submartingale, then $-X$ is a supermartingale.

Proof: These properties follow from the definition of the martingale, submartingale, and supermartingale processes. For example, the expectation of a submartingale X is an increasing function of time because the condition $X(s) \leq E[X(t) \mid \mathcal{F}_s]$, $s \leq t$, implies $E[X(s)] \leq E[E(X(t) \mid \mathcal{F}_s)] = E[X(t)]$. ■

Increments of a martingale X over non-overlapping intervals are orthogonal provided that $X(t) \in L_2$ for all $t \geq 0$.

Proof: For $u \leq v \leq s \leq t$ we have

$$\begin{aligned} E[(X(t) - X(s))\,(X(v) - X(u))] &= E\left\{E\left[(X(t) - X(s))\,(X(v) - X(u)) \mid \mathcal{F}_s\right]\right\} \\ &= E\left\{(X(v) - X(u))\,E\left[X(t) - X(s) \mid \mathcal{F}_s\right]\right\} = 0 \end{aligned}$$

since $X(v) - X(u)$ is $\mathcal{F}_v$-measurable, $\mathcal{F}_v \subset \mathcal{F}_s$, and $E[X(t) - X(s) \mid \mathcal{F}_s] = 0$ by the martingale property. The above expectations are finite since $X(t) \in L_2$, $t \geq 0$. ■

If X is an $\mathcal{F}_t$-martingale, then X is also an $\mathcal{F}_t^X$-martingale, where $\mathcal{F}_t^X = \sigma(X(s) : s \leq t)$ denotes the natural filtration of X.

Proof: We have $E[|X(t)|] < \infty$ because X is a martingale. The random variable $X(t)$ is $\mathcal{F}_t^X$-measurable by the definition of the natural filtration. For $t \geq s$ we have

$$E[X(t) \mid \mathcal{F}_s^X] = E\{E[X(t) \mid \mathcal{F}_s] \mid \mathcal{F}_s^X\} = E[X(s) \mid \mathcal{F}_s^X] = X(s)$$

since X is an $\mathcal{F}_t$-martingale and $X(s) \in \mathcal{F}_s^X$. ■

If X is an $\mathcal{F}_t$-adapted process with finite mean and has increments independent of the past, then the **compensated process** $X(t) - E[X(t)]$ is an $\mathcal{F}_t$-martingale.

Proof: The compensated process $X(t) - E[X(t)]$ satisfies the first two conditions in Eq. 3.95 by hypothesis. We also have

$$E[X(t) \mid \mathcal{F}_s] = E[(X(t) - X(s)) + X(s) \mid \mathcal{F}_s] = E[(X(t) - X(s))] + X(s)$$

for $t \geq s$ since $X(s) \in \mathcal{F}_s$ and X has increments independent of the past, that is, $X(t) - X(s)$ is independent of $\mathcal{F}_s$. The above equation gives $E[X(t) - E[X(t)] \mid \mathcal{F}_s] = X(s) - E[X(s)]$ and shows that $X(t) - E[X(t)]$ is a martingale. This result is consistent with Example 3.61 showing that a compensated compound Poisson process with jumps in L_1 is a martingale. ■

Every martingale has an optional modification ([61], Lemma 4.1, p. 72)).

Note: Two definitions are needed to clarify the meaning of the above statement. Let $\mathcal{P}$ and $\mathcal{O}$ be the smallest σ-fields on $[0, \infty) \times \Omega$ with respect to which all left and right continuous $\mathcal{F}_t$-adapted processes are measurable, respectively. A process X is called **predictable** if the mapping $(t, \omega) \mapsto X(t, \omega)$ is $\mathcal{P}$-measurable. If this mapping is $\mathcal{O}$-measurable, then X is said to be **optional**. Predictable processes allow "a peek into the future" because they are left continuous. We note two important properties:

1. Predictable and optional processes are measurable because the σ-fields $\mathcal{P}$ and $\mathcal{O}$ are included in $\mathcal{B}([0, \infty)) \times \mathcal{F}$ and

2. $\mathcal{P} \subseteq \mathcal{O}$.

We only show that $\mathcal{P}$ is included in $\mathcal{O}$. Let X be an $\mathcal{F}_t$-adapted process with left continuous paths and define $X^{(n)}(t) = X(k/2^n)$ for $t \in [k/2^n, (k+1)/2^n)$ (Fig. 3.10). The process $X^{(n)}$ is $\mathcal{F}_t$-adapted, has right continuous samples, and approaches X as $n \to \infty$.

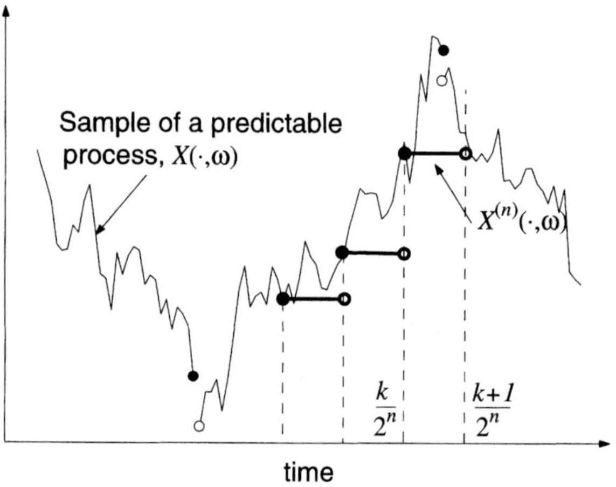

Figure 3.10. A hypothetical sample of a predictable process and the corresponding sample of an optional process $X^{(n)}$ approximating X

Hence, X is also an optional process so that $\mathcal{P} \subseteq \mathcal{O}$. The use of a similar construction showing the other inclusion fails. Let us assume that X is an $\mathcal{F}_t$-adapted process with right continuous paths. The sequence of processes $\tilde{X}^{(n)}(t) = X((k+1)/2^n)$, $t \in (k/2^n, (k+1)/2^n]$, is left continuous and approaches X as $n \to \infty$ but is not $\mathcal{F}_t$-adapted. ▲

Every martingale has a unique càdlàg modification ([147], Corollary 1, p. 8).

Note: This statement augments the previous property and shows that martingales have well-behaved paths. If a martingale is specified by a collection of finite dimensional distributions, we can assume that its sample paths are càdlàg. ▲

Doob-Meyer decomposition. If X is a right continuous $\mathcal{F}_t$-submartingale of class DL, then there exists a unique (up to indistinguishability) right continuous $\mathcal{F}_t$-adapted increasing process A with $A(0) = 0$ such that $M = X - A$ is an $\mathcal{F}_t$-martingale and

$$E\left[\int_0^{t\wedge T} Y(s-)\,dA(s)\right] = E\left[\int_0^{t\wedge T} Y(s)\,dA(s)\right] = E[Y(t\wedge T)\,A(t\wedge T)]$$

for every non-negative right continuous $\mathcal{F}_t$-martingale Y, every $t \geq 0$, and $\mathcal{F}_t$-stopping time T ([61], Theorem 5.1, p. 74).

Note: We have proved a similar decomposition for discrete time submartingales (Section 2.18.1). A right continuous $\mathcal{F}_t$-submartingale X is said to be of class DL if the family $X(t \wedge T)$ with T ranging over the collection of $\mathcal{F}_t$-stopping times is uniformly integrable for each $t \geq 0$. A family of random variables Z_α, $\alpha \in I$, is uniformly integrable if $\sup_{\alpha\in I}\int_{|Z_\alpha|\geq n}|Z_\alpha|\,dP$ converges to zero as $n \to \infty$.

We say that A is an **increasing process** if $A(t, \omega)$, $t \geq 0$, are increasing functions for all $\omega \in \Omega$. For example, the Poisson process is increasing. ■

3.11.2 Stopped martingales

If X is an $\mathcal{F}_t$-submartingale and T_1, T_2 are $\mathcal{F}_t$-stopping times assuming values in a finite set $\{t_1, t_2, \ldots, t_n\} \subseteq [0, \infty)$, then ([61], Lemma 2.2, p. 55)

$$E[X(T_2) \mid \mathcal{F}_{T_1}] \geq X(T_1 \wedge T_2). \tag{3.97}$$

Note: This inequality is similar to the last condition in Eq. 3.95 with s, t replaced by the stopping times T_1, T_2. By the defining properties of the conditional expectation, Eq. 3.97 is satisfied if and only if $\int_A X(T_2)\,dP \geq \int_A X(T_1 \wedge T_2)\,dP, \forall A \in \mathcal{F}_{T_1}$. ▲

Optional sampling theorem. If X is a right continuous $\mathcal{F}_t$-submartingale and T_1, T_2 are $\mathcal{F}_t$-stopping times, then ([61], Theorem 2.13, p. 61)

$$E[X(T_2 \wedge t) \mid \mathcal{F}_{T_1}] \geq X(T_1 \wedge T_2 \wedge t) \quad \text{for each } t > 0. \tag{3.98}$$

If, in addition, $T_2 < \infty$ a.s., $\lim_{t\to\infty} E[|X(t)|\,1_{\{T_2>t\}}] = 0$, and $E[|X(T_2)|] < \infty$, then $E[X(T_2) \mid \mathcal{F}_{T_1}] \geq X(T_1 \wedge T_2)$.

Note: The inequality in Eq. 3.98 does not follow from Eq. 3.97 since the values of T_1 and T_2 are not restricted to a finite set. This generalization can be achieved by requiring X to be right continuous. The results in Eqs. 3.97 and 3.98 hold also for martingales and supermartingales with " $\geq$ " replaced by " $=$ " and " $\leq$ ", respectively. ▲

Stopped right continuous (local) submartingales, martingales, and supermartingales are (local) submartingales, martingales, and supermartingales, respectively.

Proof: Suppose that $X(t)$, $t \geq 0$, is a right continuous submartingale and T is a stopping time. We have $E[|X(T \wedge t)|] < \infty$ since X is a submartingale. Since X is right continuous and adapted, it is progressive (Section 3.2) so that $X(T \wedge t) \in \mathcal{F}_{T \wedge t} \subset \mathcal{F}_t$. Also, Eq. 3.98 with $T_1 = s \leq t$ and $T_2 = T$ yields $E[X(T \wedge t) \mid \mathcal{F}_s] \geq X(T \wedge s)$. Hence, X^T is a submartingale.

Suppose now that $X(t)$, $t \geq 0$, is a right continuous local submartingale, T is a stopping time, and that T_n, $n = 1, 2, \ldots$, is a localizing sequence for X. The above considerations applied to the submartingale X^{T_n} show that $X^{T_n}(T \wedge t)$, $t \geq 0$, is a submartingale. Hence, $X^{T_n}(T \wedge t) = X(T_n \wedge T \wedge t) = X^T(T_n \wedge t)$, $t \geq 0$, is a submartingale for all $n \geq 1$ so that T_n is a localizing sequence for X^T.

Similar arguments can be used to show that stopped right continuous local martingales and supermartingales are local martingales and supermartingales, respectively. ■

If X is an $\mathcal{F}_t$-right continuous submartingale which is bounded from below, then X is of class DL. Hence, X admits the Doob-Meyer decomposition.

Proof: The inequality in Eq. 3.98 with $T_2 = T$, $T_1 = T \wedge t$, $t \geq 0$, gives

$$E[X(T \wedge t) \mid \mathcal{F}_{T \wedge t}] \geq X(T \wedge t).$$

The family $E[X(T \wedge t) \mid \mathcal{F}_{T \wedge t}]$ of random variables indexed by the stopping times T is uniformly integrable for each $t \geq 0$ ([40], Theorem 4.5.3, p. 96). Since $X(T \wedge t)$ is bounded from below, then $X(T \wedge t)$ indexed by T is uniformly integrable. Hence, X is of class DL. For example, the Poisson process N is right continuous and is bounded from below since $N(t) \geq 0$, $t \geq 0$. Hence, N is of class DL. ■

3.11.3 Inequalities

Jensen inequality. If X is an $\mathcal{F}_t$-martingale and $\varphi : \mathbb{R} \mapsto \mathbb{R}$ is a convex function such that $E[|\varphi(X(t))|] < \infty$, $t \geq 0$, then

$$E[\varphi(X(t)) \mid \mathcal{F}_s] \geq \varphi(X(s)), \quad s \leq t. \tag{3.99}$$

Proof: We first note that Eq. 3.99 implies that $\varphi(X)$ is an $\mathcal{F}_t$-submartingale. The Jensen inequality can be used to show, for example, that $|B|$ and B^2 are submartingales, where B is a Brownian motion. We have also established a Jensen inequality for random variables (Section 2.12).

Because φ is a convex function, we have $\varphi(x) = \sup\{l(x) : l(u) \leq \varphi(u), \forall u\}$, where $l(u) = a\,u + b$ is a linear function with a, b real constants. Also,

$$\begin{aligned} E[\varphi(X(t)) \mid \mathcal{F}_s] &= E[\sup\{l(X(t))\} \mid \mathcal{F}_s] \geq \sup\{E[l(X(t)) \mid \mathcal{F}_s]\} \\ &= \sup\{l(E[X(t) \mid \mathcal{F}_s])\} = \sup\{l(X(s))\} = \varphi(X(s)) \end{aligned}$$

since $E[\sup\{l(X(t))\} \mid \mathcal{F}_s] \geq E[l(X(t)) \mid \mathcal{F}_s]$ for any linear function l, the conditional expectation is a linear operator, and X is an $\mathcal{F}_t$-martingale. ■

If X is an $\mathcal{F}_t$-submartingale, $0 < \tau < \infty$, and $F \subset [0, \tau]$ is a finite set, then for each $x > 0$ and bounded interval (a, b) we have ([61], Lemma 2.3, p. 56 and Lemma 2.5, p. 57)

$$
\begin{aligned}
P(\max_{t \in F}\{X(t)\} \geq x) &\leq E[X(\tau)^+]/x, \\
P(\min_{t \in F}\{X(t)\} \leq -x) &\leq \left(E[X(\tau)^+] - X(0)\right)/x, \quad \text{and} \\
E[U(a, b, F)] &\leq \left(E[(X(\tau) - a)^+]\right)/(b - a).
\end{aligned}
\tag{3.100}
$$

Note: If $T_1 = \min\{t \in F : X(t) \leq a\}$ and $T_2 = \min\{t > T_1 : t \in F : X(t) \geq b\}$, we say that X has an oscillation in (T_1, T_2) larger than $b - a$. This event relates to the event of an a-downcrossing followed by a b-upcrossing defined for differentiable processes (Section 3.10.1). The number of oscillations of X in F larger than $b - a$ is denoted by $U(a, b, F)$.

The bounds in Eq. 3.100 show that the samples of X cannot be very rough because (1) $E[X(\tau)^+]/x$ and $(E[X(\tau)^+] - X(0))/x$ converge to zero as $|x| \to \infty$ so that the maximum and the minimum values of X in F are finite with probability one and (2) the average number of oscillations of X in F that are larger than $b - a$ is finite. ▲

If X is a right continuous submartingale, the first two inequalities in Eq. 3.100 can be extended to

$$
\begin{aligned}
P(\sup_{0 \leq t \leq \tau}\{X(t)\} \geq x) &\leq E[X(\tau)^+]/x, \\
P(\inf_{0 \leq t \leq \tau}\{X(t)\} \leq -x) &\leq \left(E[X(\tau)^+] - E[X(0)]\right)/x,
\end{aligned}
\tag{3.101}
$$

for each $x > 0$ and $\tau > 0$ ([61], Proposition 2.16, p. 63).

Example 3.65: A Brownian motion B is a continuous martingale with respect to its natural filtration so that the inequalities in Eq. 3.101 can be applied. The probability $P(\max_{t \in [0,\tau]}\{B(t)\} \geq x)$ of the largest value of the standard Brownian motion B in $[0, \tau]$ and its upper bound $E[B(\tau)^+]/x$ given by the first inequality in Eq. 3.101 are shown in Fig. 3.11 for $\tau = 10$ and $x \in [1, 5]$. The upper bound $E[B(\tau)^+]/x$ is relatively wide for all values of x so that it does not provide a very useful approximation for $P(\max_{t \in [0,\tau]}\{B(t)\} \geq x)$. ◇

Note: The expectation $E[B(\tau)^+]$ is $(2\pi\,\tau)^{-1/2} \int_0^\infty x\, e^{x^2/(2\tau)}\, dx = \sqrt{\tau/(2\pi)}$. Let $T = \inf\{t \geq 0 : B(t) \geq x\}$ be a stopping time with respect to the natural filtration of B. The probability that the largest value of B exceeds x in $[0, \tau]$ is equal to $P(T \leq \tau)$ and the expression of this probability is given in Example 3.15. ▲

Example 3.66: Let $\{X_n, n = 0, 1, \ldots\}$ be an $\mathcal{F}_n$-martingale and define a process $(H \cdot X)_n = \sum_{k=1}^n H_k(X_k - X_{k-1})$, where $\{H_n, n = 1, 2, \ldots\}$ is an arbitrary sequence of random variables. The process $H \cdot X$ is an $\mathcal{F}_n$-martingale if H is predictable, that is, $H_n \in \mathcal{F}_{n-1}$ (Section 2.18). ◇

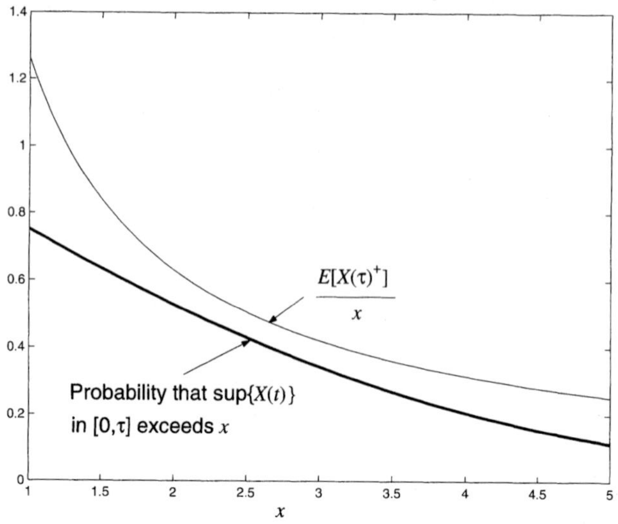

Figure 3.11. Probability of the largest value of a Brownian motion and a bound on this probability for $\tau = 10$

Proof: If $H_n \in \mathcal{F}_{n-1}$, then

$$\begin{aligned} E[(H \cdot X)_{n+1} \mid \mathcal{F}_n] &= (H \cdot X)_n + E[H_{n+1}(X_{n+1} - X_n) \mid \mathcal{F}_n] \\ &= (H \cdot X)_n + H_{n+1} E[X_{n+1} - X_n \mid \mathcal{F}_n] = (H \cdot X)_n \end{aligned}$$

so that $(H \cdot X)$ is an $\mathcal{F}_n$-martingale. If H is not predictable, the process $(H \cdot X)$ may not be a martingale since the random variables $E[H_{n+1}(X_{n+1} - X_n) \mid \mathcal{F}_n]$ and $H_{n+1} E[X_{n+1} - X_n \mid \mathcal{F}_n]$, generally, differ.

The formula of $(H \cdot X)$ resembles an integral, referred to as a stochastic integral and defined in Chapter 4. In Section 2.18 we have interpreted H_n, $H_n\, D_n = H_n(X_n - X_{n-1})$, and $(H \cdot X)_n$ as the amount of money we bet at round n, the amount of money we would win or lose for a bet H_n, and our fortune after n bets, respectively. Because our bet H_n at time n can be based on only the outcomes $\{X_1, X_2, \ldots, X_{n-1}\}$, we require that H be a predictable process, that is, $H_n \in \mathcal{F}_{n-1}$. The condition $H_n \in \mathcal{F}_{n-1}$ is consistent with the game rule that the bets H_n must be placed after completion of round $n - 1$.

If the time is continuous, the characterization of a predictable process is less obvious. However, a similar interpretation can be constructed. For example, consider the game of roulette. Let T be the random time when the ball comes to rest measured from the moment at which the wheel is spun and the ball is released. The players can bet at any time $t < T$ but not at $t \geq T$. One way of requiring that the bets be placed strictly before T is to impose the condition that the amount of money we bet at time t is a left continuous function so that it is not possible to take advantage of a jump in the process we are betting on ([58], Sections 2.1 and 2.2) ■

3.11.4 Quadratic variation and covariation processes

We define variation and covariation processes for martingales and present some useful properties. These processes are also discussed in Section 4.5 for semimartingales. The variation and covariation processes are needed to establish the Itô formula (Chapter 4) and solve various stochastic problems (Chapters 5-9).

If X is a right continuous $\mathcal{F}_t$-local martingale, then there exists a right continuous increasing process $[X]$ such that, for each $t \geq 0$ and each sequence of partitions $p_n = (0 = t_0^{(n)} \leq t_1^{(n)} \leq \cdots \leq t_{m_n}^{(n)} = t)$ of $[0, t]$ with $\Delta(p_n) \to 0$ as $n \to \infty$, we have

$$S_n = \sum_{k=1}^{m_n} \left(X(t_k^{(n)}) - X(t_{k-1}^{(n)})\right)^2 \xrightarrow{\text{pr}} [X](t), \quad \text{as } n \longrightarrow \infty. \tag{3.102}$$

If X is a right continuous square integrable martingale, then the convergence in Eq. 3.102 is in L_1 ([61], Proposition 3.4, p. 67).

Note: The process $[X]$ is called the **square bracket** or **quadratic variation** process. This process is also denoted by $[X, X]$. The latter notation is used almost exclusively in the later chapters of the book. ▲

Example 3.67: For a Brownian motion $B(t)$, $t \geq 0$, the quadratic variation process is $[B](t) = t$.

Proof: Take $t > 0$ and consider the partition $p_n = k\,t/n$, $k = 0, 1, \ldots, n$, of $[0, t]$. The sequence of sums in Eq. 3.102 is $S_n = \sum_{k=1}^{n} (B(k\,t/n) - B((k-1)\,t/n))^2$ so that $S_n \stackrel{d}{=} (t/n) \sum_{k=1}^{n} G_k^2$, where G_k, $k = 1, \ldots, n$, are independent $N(0, 1)$, since the increments $B(k\,t/n) - B((k-1)\,t/n)$ of B are independent Gaussian variables with mean zero and variance t/n. Hence, $S_n \to t\, E[G_1^2] = t$ as $n \to \infty$ by the law of large numbers. We will revisit this property of the Brownian motion later in this chapter (Eq. 3.112). ■

Example 3.68: Let N be a Poisson process with unit intensity. The quadratic variation process of $M(t) = N(t) - t$, $t \geq 0$, is $[M](t) = N(t)$.

Proof: Take $t > 0$ and consider the partition $p_n = k\,t/n$, $k = 0, 1, \ldots, n$, of $[0, t]$. For a sufficiently large n, the intervals $((k-1)\,t/n, k\,t/n]$, $k = 1, \ldots, n$, of the partition contain at most one jump of N. Let $J_n(t)$ be the collection of intervals $((k-1)\,t/n, k\,t/n]$ containing a jump of N. The sum S_n in Eq. 3.102 is

$$\begin{aligned} S_n &= \sum_{k \notin J_n(t)} (M(k\,t/n) - M((k-1)\,t/n))^2 + \sum_{k \in J_n(t)} (M(k\,t/n) - M((k-1)\,t/n))^2 \\ &= \sum_{k \notin J_n(t)} \frac{1}{n^2} + \sum_{k \in J_n(t)} \left(1 - \frac{1}{n}\right)^2 = \frac{1}{n^2}(n - N(t)) + N(t)\left(1 - \frac{1}{n}\right)^2, \end{aligned}$$

and the last expression converges to $N(t)$ as $n \to \infty$. ■

If X is a continuous $\mathcal{F}_t$-local martingale, then $[X]$ can be taken as a continuous process [61] (Proposition 3.6, p. 71).

If X is a (local) square integrable, right continuous $\mathcal{F}_t$-martingale, then

$$X^2 - [X] \quad \text{is a (local) martingale.} \tag{3.103}$$

Proof: The expectation of the absolute value of $X(t)^2 - [X](t)$ is finite for each $t \geq 0$ since X is a square integrable martingale and $[X]$ is in L_1 (Eq. 3.102). The process $X^2 - [X]$ is $\mathcal{F}_t$-adapted because $[X]$ is adapted by definition. It remains to show the last property in Eq. 3.95 holds. For $t \geq s$ we have

$$\begin{aligned} E[X(t)^2 - X(s)^2 \mid \mathcal{F}_s] &= E[(X(t) - X(s))^2 + 2\,X(t)\,X(s) - 2\,X(s)^2 \mid \mathcal{F}_s] \\ &= E[(X(t) - X(s))^2 \mid \mathcal{F}_s] + 2\,E[X(t)\,X(s) \mid \mathcal{F}_s] \\ &\quad - 2\,E[X(s)^2 \mid \mathcal{F}_s] = E[(X(t) - X(s))^2 \mid \mathcal{F}_s], \end{aligned}$$

since X is a martingale. The above equality gives

$$\begin{aligned} E\left[\sum_{k=1}^{m_n} \left(X(t_k^{(n)}) - X(t_{k-1}^{(n)})\right)^2 \mid \mathcal{F}_s\right] &= E\left[\sum_{k=1}^{m_n} \left(X(t_k^{(n)})^2 - X(t_{k-1}^{(n)})^2\right) \mid \mathcal{F}_s\right] \\ &= E[X(t)^2 - X(s)^2 \mid \mathcal{F}_s] \end{aligned}$$

for any sequence of partitions $s = t_0^{(n)} \leq t_1^{(n)} \leq \cdots \leq t_{m_n}^{(n)} = t$ of the interval $[s, t]$. The left side of the above equalities converges in L_1 to $[X](t) - [X](s)$ as $\Delta(p_n) \to 0$ (Eq. 3.102) so that

$$E[X(t)^2 - [X](t) \mid \mathcal{F}_s] = E[X(s)^2 - [X](s) \mid \mathcal{F}_s] = X(s)^2 - [X](s) \quad \text{a.s.}$$

showing that $X^2 - [X]$ is a martingale.

It can also be shown that, if X is a local square integrable martingale, $X^2 - [X]$ is a local martingale ([61], Proposition 6.1, p. 79). ■

If X and Y are right continuous, square integrable martingales, then $X\,Y - [X, Y]$ is a martingale, where

$$[X, Y] = \frac{1}{2}\left([X + Y] - [X] - [Y]\right) \tag{3.104}$$

is the **covariation** of X and Y.

Note: If X and Y are right continuous square integrable martingales so are the processes $X + Y$ and $X - Y$. Hence, the definition in Eq. 3.104 is meaningful.

The covariation process $[X, Y](t)$ can be calculated as the limit of the sums

$$\sum_k (X(t_k^{(n)}) - X(t_{k-1}^{(n)}))(Y(u_k^{(n)}) - Y(u_{k-1}^{(n)}))$$

as the meshes of the partitions $\{t_k^{(n)}\}$ and $\{u_k^{(n)}\}$ of $[0, t]$ converge to zero (Eq. 3.102). The convergence is in probability. If the martingales X and Y are square integrable, the convergence is in L_1. Similar statements hold for local martingales ([61], Proposition 6.2, p. 79).

If the covariation of the martingales X and Y is zero, that is, $[X, Y] = 0$, we say that X and Y are **orthogonal.** ▲

If X is a right continuous, square integrable martingale, then there exists a unique right continuous adapted increasing process $< X >$ satisfying the conditions for the Doob-Meyer decomposition so that

$$X^2 - < X > \quad \text{is a martingale.} \tag{3.105}$$

Note: The process X^2 is a submartingale which is bounded from below since $X^2(t) \geq 0$ for all $t \geq 0$. We have seen that such a process is of class DL. Hence, X^2 admits the Doob-Meyer decomposition, where A in the Doob-Meyer decomposition is denoted by $< X >$.

Consider the decomposition of X^2 in Eqs. 3.103 and 3.105. Since $X^2 - [X]$ and $X^2 - < X >$ are martingales, so is $(X^2 - < X >) - (X^2 - [X]) = [X] - < X >$. Generally, $[X]$ and $< X >$ differ. If X is continuous, the processes $[X]$ and $< X >$ are indistinguishable ([61], p. 79). ▲

Example 3.69: Let $M(t) = N(t) - \lambda\, t$, $t \geq 0$, be a compensated Poisson process, where N is a Poisson process with intensity $\lambda > 0$. We have $[M](t) = N(t)$ and $< M > (t) = \lambda\, t$, $t \geq 0$. ◇

Proof: We have seen in Example 3.68 that $[M](t) = N(t)$. We show here that the process A in the Doob-Meyer decomposition is $< M > (t) = \lambda\, t$, $t \geq 0$.

Since M is a square integrable martingale and $< M >$ is continuous, we only need to show that $M^2(t) - < M > (t) = M^2(t) - \lambda\, t$ is a martingale. For $s \leq t$ we have

$$\begin{aligned}
E[M^2(t) \mid \mathcal{F}_s] &= E\left[(N(t) - N(s) - \lambda\,(t-s) - (N(s) - \lambda\, s))^2 \mid \mathcal{F}_s\right] \\
&= E\left[(N(t) - N(s) - \lambda\,(t-s))^2 \mid \mathcal{F}_s\right] \\
&\quad + 2\, E\,[(N(t) - N(s) - \lambda\,(t-s))\,(N(s) - \lambda\, s) \mid \mathcal{F}_s] \\
&\quad + E\left[(N(s) - \lambda\, s)^2 \mid \mathcal{F}_s\right] = \lambda\,(t-s) + (N(s) - \lambda\, s)^2 \\
&= M(s)^2 + \lambda\,(t-s)
\end{aligned}$$

so that $M^2(t) - \lambda\, t$, $t \geq 0$, is a martingale. ■

In Section 3.9.3.1 we have defined the variation and the total variation for deterministic functions and stochastic processes (Eqs. 3.54-3.56). In the following sections we will extend the concept of variation for deterministic functions to samples $X(\cdot, \omega)$, $\omega \in \Omega$, of a stochastic process $X(t)$, $t \geq 0$. The **variation** of $X(\cdot, \omega)$ in a time interval $[a, b]$ is defined by $V_x(\omega) = \sup_{\{\text{all } p\}} V_x(\omega, p)$, where $p = (a = t_0 < t_1 < \cdots < t_m = b)$ is a partition of $[a, b]$, $V_x(\omega, p) = \sum_{k=1}^{m} |X(t_k, \omega) - X(t_{k-1}, \omega)|$, and the supremum is calculated over all partitions p of $[a, b]$. We say that X is of **finite variation** if almost all samples of X are of finite variation on each compact of $[0, \infty)$. Otherwise, X is said to be of unbounded variation.

3.12 Poisson processes

We have defined the Poisson and the compound Poisson processes in Example 3.9. In this section we consider these processes in more detail. The processes and random variables considered in this section are defined on the same filtered probability space $(\Omega, \mathcal{F}, (\mathcal{F}_t)_{t\geq 0}, P)$.

If $\{T_n, n = 0, 1, 2, \ldots, \infty\}$ with $T_0 = 0$ a.s. is a strictly increasing sequence of positive random variables, then

$$N(t) = \sum_{n\geq 1} 1_{\{t\geq T_n\}} \tag{3.106}$$

is the **counting** process associated with the sequence T_n. If $T = \sup_n\{T_n\} = \infty$ a.s., then N is said to be **without explosions**.

Note: The counting process N (1) has right continuous, piece-wise constant samples with unit jumps at the random times T_n and starts at zero with probability one if T_1 has no probability mass at zero, (2) takes on positive integer values including $\{+\infty\}$, (3) has càdlàg samples if $T = \infty$ a.s., and (4) is $\mathcal{F}_t$-adapted if and only if T_n are $\mathcal{F}_t$-stopping times. ▲

A counting process N is said to be a **Poisson** process if it (1) is $\mathcal{F}_t$-adapted, (2) is without explosions, and (3) has stationary increments that are independent of the past, that is, $N(t) - N(s)$, $t > s$, has the same distribution as $N(t - s)$ and is independent of $\mathcal{F}_s$.

Note: The requirement of increments independent of the past is stronger than the requirement of independent increments stated in Section 3.6.4. Consider a process X adapted to a filtration $\mathcal{F}_t$, $t \geq 0$, and take $0 \leq u < v \leq s < t$. Since $X(t) - X(s)$ is independent of $\mathcal{F}_s$ by hypothesis, it is also independent of $X(v) - X(u) \in \mathcal{F}_v \subseteq \mathcal{F}_s$. ▲

The probability that N has n **jumps** in $(0, t]$ is

$$P(N(t) = n) = \frac{(\lambda\, t)^n}{n!}\, e^{-\lambda\, t}, \quad n = 0, 1, 2, \ldots, \tag{3.107}$$

for some $\lambda > 0$. Hence, the random variable $N(t)$ follows a Poisson distribution with parameter $\lambda\, t$ for each $t > 0$.

Note: The probability in Eq. 3.107 results from the defining properties of the Poisson process ([147], Theorem 23, p. 14). The samples of N look like staircases with steps of unit height and random width. The Poisson process is a special case of the compound Poisson process (Fig. 3.5). We also note that N is continuous in probability, m.s., and a.s. at each time $t > 0$ but does not have continuous samples.

The probability in Eq. 3.107 shows that (1) the period between consecutive jumps of N is an exponential random variable with distribution $P(T_1 > t) = P(N(t) = 0) = e^{-\lambda t}$ and mean $1/\lambda$ and (2) N starts at zero a.s. because $P(T_1 > t) \to 1$ as $t \to 0$. ■

The mean, variance, covariance function, cumulant of order q, and characteristic function of a Poisson process N with intensity $\lambda > 0$ are

$$E[N(t)] = \lambda t, \quad \text{Var}[N(t)] = \lambda t, \quad \text{Cov}[N(s), N(t)] = \lambda \min(s, t),$$
$$\chi_q = \lambda t, \quad \text{and } \varphi(u; t) = \exp\left[-\lambda t \left(1 - e^{\sqrt{-1}\, u}\right)\right]. \tag{3.108}$$

Note: These properties can be obtained from the probability law of N in Eq. 3.107 ([79], pp. 79-83). The relationship between cumulants and moments can be found in [79] (Appendix B, p. 377), where $q \geq 1$ is an integer. Because $E[N(t)]$ gives the expected number of jumps of N in time t, $\lambda = E[N(t)]/t$ represents the average number of jumps per unit of time and is called the **intensity** or the **mean arrival rate** of N. ▲

Example 3.70: If N is a Poisson counting process with intensity λ, then N is a submartingale and $N(t) - \lambda t$, $(N(t) - \lambda t)^2 - \lambda t$ are martingales. ◇

Proof: The compensated process $N(t) - \lambda t$ is a martingale by one of the properties in Section 3.11.1.

To prove that $(N(t)-\lambda t)^2-\lambda t$ is a martingale, we note that $E[(N(t)-\lambda t)^2]-\lambda t < \infty$ and $(N(t) - \lambda t)^2 - \lambda t$ is adapted. Also, for $t \geq s$,

$$E\left[(N(t) - \lambda t)^2 \mid \mathcal{F}_s\right] = E\left[((N(t) - \lambda t) - (N(s) - \lambda s) + (N(s) - \lambda s))^2 \mid \mathcal{F}_s\right]$$
$$= E\left[(N(t) - N(s) - \lambda (t - s))^2 \mid \mathcal{F}_s\right] + E\left[(N(s) - \lambda s)^2 \mid \mathcal{F}_s\right]$$
$$+ 2\, E\left[(N(t) - N(s) - \lambda (t - s))\, (N(s) - \lambda s) \mid \mathcal{F}_s\right] = \lambda (t - s) + (N(s) - \lambda s)^2$$

since $N(t) - N(s)$ is independent of $\mathcal{F}_s$, $N(s) \in \mathcal{F}_s$, and $N(t) - N(s) - \lambda (t - s)$ is a martingale. Hence, $(N(t) - \lambda t)^2 - \lambda t$ is a martingale. We have performed similar calculations in Example 3.69. ■

Example 3.71: Let $C(t) = \sum_{n \geq 1} Y_n\, 1_{\{t \geq T_n\}} = \sum_{n=1}^{N(t)} Y_n$ be the compound Poisson process in Example 3.9 with $Y_1 \in L_q$, $q \geq 2$. The mean, variance, covariance function, cumulant of order q, and characteristic function of C are

$$E[C(t)] = \lambda t\, E[Y_1], \quad \text{Var}[C(t)] = \lambda t\, E[Y_1^2],$$
$$\text{Cov}[C(s), C(t)] = \lambda \min(s, t)\, E[Y_1^2], \quad \chi_q = \lambda t\, E[Y_1^q], \quad \text{and}$$
$$\varphi(u; t) = \exp\left[-\lambda t \left(1 - \varphi_{Y_1}(u)\right)\right], \tag{3.109}$$

where $\lambda > 0$ is the intensity of the Poisson process N and φ_{Y_1} denotes the characteristic function of Y_1. ◇

Note: These properties of C can be obtained from the definition of this process and the probability law of the Poisson process in Eq. 3.107 ([79], Section 3.3). ▲

Example 3.72: Let C be a compound Poisson process with $E[|Y_1|] < \infty$. Denote by $\mathcal{F}_t$ the natural filtration of C. Then C is an $\mathcal{F}_t$-submartingale, martingale, or supermartingale if $E[Y_1] \geq 0, = 0$, or ≤ 0, respectively. The **compensated compound Poisson** process $\tilde{C}(t) = C(t) - \lambda\, t\, E[Y_1]$ is an $\mathcal{F}_t$-martingale. If $Y_1 \geq 0$ a.s., then $A(t) = \lambda\, t\, E[Y_1]$ is the process in the Doob-Meyer decomposition of the submartingale C. ◇

Proof: The first part of the statement has been proved in Example 3.61, where it was shown that $E[C(t) \mid \mathcal{F}_s] = \lambda\,(t-s)\, E[Y_1] + C(s)$ for $s \leq t$.

If $Y_1 \geq 0$ a.s., then C is a submartingale such that $C(t) \geq 0$ for all $t \geq 0$. Hence, C is of class DL (Section 3.11.2) so that it admits the Doob-Meyer decomposition in Section 3.11.1. Since $\tilde{C}(t) = C(t) - \lambda\, t\, E[Y_1]$ is a martingale and $\lambda\, t\, E[Y_1]$ is an adapted continuous increasing process, then the process A in the Doob-Meyer decomposition is $A(t) = \lambda\, t\, E[Y_1]$. ■

Consider the compound Poisson process $C(t) = \sum_{k=1}^{N(t)} Y_k$ in Examples 3.9 and 3.71 and let T_k denote the time of jump k. Figure 3.12 shows a hypothetical

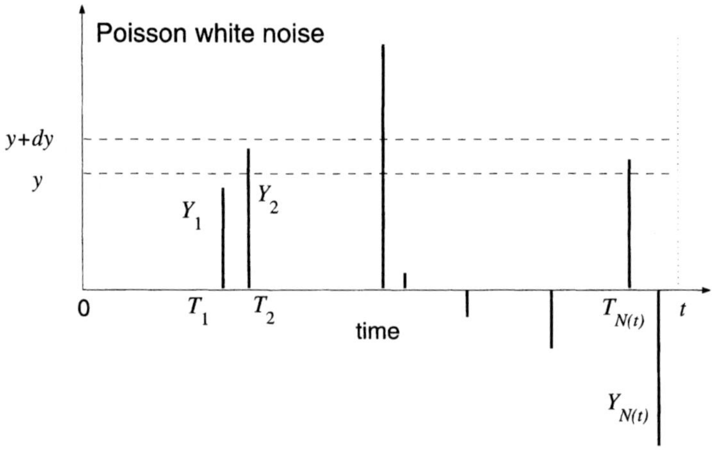

Figure 3.12. A hypothetical sample of the jumps of a compound Poisson process

sample of the sequence of jumps of C up to a time $t > 0$. This sequence, referred to as **Poisson white noise**, can be interpreted as the formal derivative of C and is $W_P(t) = \sum_{k=1}^{N(t)} Y_k\, \delta(t - T_k)$, where $\delta(\cdot)$ denotes the Dirac delta function. Let

$\mathcal{M}(t, dy)$ be a random measure giving the number of jumps of C in $(y, y + dy]$ during a time interval $(0, t]$. This measure is equal to 2 for the sample of W_P in Fig. 3.12. The measure $\mathcal{M}(t, dy)$ (1) is random because its value depends on the sample of C, (2) has the expectation $E[\mathcal{M}(t, dy)] = \lambda\, t\, dF(y) = \mu(dy)\, t$, where $\mu(dy) = \lambda\, dF(y)$ and F denotes the distribution of Y_1, and (3) provides the alternative definition,

$$C(t) = \int_{\mathbb{R}} y\, \mathcal{M}(t, dy), \tag{3.110}$$

for the compound Poisson process C ([172], Theorem 3.3.2, p. 145).

Let C be a compound Poisson process, Λ be a Borel set in $\mathbb{R}$, and $\mathcal{M}(t, dy)$ be a random measure giving the number of jumps of C in $(y, y + dy]$ during $(0, t]$. Then

$$C^{\Lambda}(t) = \sum_{k=1}^{N(t)} Y_k\, 1_{\Lambda}(Y_k) = \sum_{0<s\le t} \Delta C(s)\, 1_{\Lambda}(\Delta C(s)) = \int_{\Lambda} y\, \mathcal{M}(t, dy) \tag{3.111}$$

is a compound Poisson process, where $\Delta C(s) = C(s) - C(s-)$ and $C(s-) = \lim_{s \uparrow t} C(s)$.

Note: Because C^{Λ} retains only the jumps of C in Λ, it is called a **thinned** version of C. If $\Lambda = \mathbb{R}$, the processes C^{Λ} and C coincide. Figure 3.13 shows a hypothetical sample of a

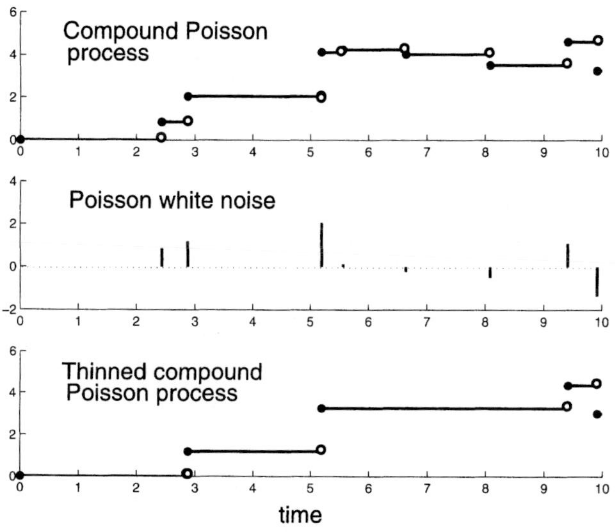

Figure 3.13. Samples of a compound Poisson process C, its formal derivative $W_P(t) = dC(t)/dt$, and a thinned version C^{Λ} of C for $\Lambda = (-a, a)^c$, $a = 1$

compound Poisson process C, the corresponding samples of the jumps of C, referred to as **Poisson white noise**, and the thinned version C^Λ for $\Lambda = (-a, a)^c$, $a = 1$.

Denote by $\tilde{Y}_k = Y_k \, 1_\Lambda(Y_k)$ the jumps of C taking values in Λ. The characteristic function of $\tilde{Y}_1$ is

$$\begin{aligned}
\varphi_{\tilde{Y}_1}(u) &= E\left[e^{\sqrt{-1}\, u\, Y_1\, 1_\Lambda(Y_1)}\right] \\
&= \int_\Lambda e^{\sqrt{-1}\, u\, y\, 1_\Lambda(y)} f_{Y_1}(y)\, dy + \int_{\mathbb{R}\setminus\Lambda} e^{\sqrt{-1}\, u\, y\, 1_\Lambda(y)} f_{Y_1}(y)\, dy \\
&= \int_\Lambda e^{\sqrt{-1}\, u\, y} f_{Y_1}(y)\, dy + \int_{\mathbb{R}\setminus\Lambda} f_{Y_1}(y)\, dy = \varphi_{Y_1}^{(\Lambda)}(u) + P(Y_1 \notin \Lambda)
\end{aligned}$$

so that $\lambda\, t \left(1 - \varphi_{\tilde{Y}_1}(u)\right) = \lambda\, t \left(P(Y_1 \in \Lambda) - \varphi_{Y_1}^{(\Lambda)}(u)\right) = \tilde{\lambda}\, t\, (1 - \tilde{\varphi}(u))$, where $\tilde{\lambda} = \lambda\, P(Y_1 \in \Lambda)$, $\varphi_{Y_1}^{(\Lambda)}(u) = \int_\Lambda e^{\sqrt{-1}\, u\, y} f_{Y_1}(y)\, dy$, and $\tilde{\varphi}(u) = \varphi_{Y_1}^{(\Lambda)}(u)/P(Y_1 \in \Lambda)$ denotes the characteristic function of $Y_1 \mid (Y_1 \in \Lambda)$. Hence, the characteristic function of C^Λ is $\varphi_{C^\Lambda}(u) = \exp\left[\tilde{\lambda}\, t\, (1 - \tilde{\varphi}(u))\right]$, C^Λ has jumps arriving at an average rate $\tilde{\lambda}$, and the characteristic function of these jumps is $\tilde{\varphi}$.

The definitions in Eq. 3.111 can be modified to consider other transformations of C. For example, the processes

$$\begin{aligned}
C_h(t) &= \sum_{k=1}^{N(t)} h(Y_k) = \int_{\mathcal{Y}} h(y)\, \mathcal{M}(t, dy) \quad \text{and} \\
C_h^\Lambda(t) &= \sum_{k=1}^{N(t)} h(Y_k)\, 1_\Lambda(Y_k) = \sum_{0<s\le t} h(\Delta C(s))\, 1_\Lambda(\Delta C(s)) = \int_\Lambda h(y)\, \mathcal{M}(t, dy)
\end{aligned}$$

are compound Poisson processes with jumps $h(Y_k)$ and $h(Y_k)\, 1_\Lambda(Y_k)$ arriving at the mean rates λ and $\tilde{\lambda}$, respectively, where $h : \mathbb{R} \to \mathbb{R}$ is a Borel measurable function. ▲

Example 3.73: Let C be a compound Poisson process with jumps satisfying the condition $E[Y_1^2] < \infty$. The quadratic variation of the compensated compound Poisson process $\tilde{C}(t) = C(t) - \lambda\, t\, E[Y_1]$ is $[\tilde{C}](t) = \sum_{k=1}^{N(t)} Y_k^2$. ◇

Proof: Since $\tilde{C}$ is a right continuous square integrable martingale, the sequence $S_n = \sum_{k=1}^{m_n} \left(\tilde{C}(t_k^{(n)}) - \tilde{C}(t_{k-1}^{(n)})\right)^2$ converges to $[\tilde{C}](t)$ in probability and L_1 as the mesh of the sequence of partitions $p_n = (0 = t_0^{(n)} \le t_1^{(n)} \le \cdots \le t_{m_n}^{(n)} = t)$ of $[0, t]$ converges to zero (Eq. 3.102).

The expression of $[\tilde{C}]$ results from the observation that the terms of the sum S_n in Eq. 3.102 are not zero only in intervals $(t_{k-1}^{(n)}, t_k^{(n)}]$ containing at least one jump of C and considerations in Example 3.68. ■

3.13 Brownian motion process

A **Brownian motion** $\boldsymbol{B}(t)$, $t \ge 0$, in $\mathbb{R}^d$ consists of d independent real-valued Brownian motions B_i, $i = 1, \ldots, d$. It is assumed unless stated otherwise that $\boldsymbol{B}$ starts at $\mathbf{0} \in \mathbb{R}^d$. If $\boldsymbol{B}(0) = \boldsymbol{x}$, we say that the Brownian motion starts at $\boldsymbol{x}$. Because the properties of $\boldsymbol{B}$ are determined by the properties of its coordinates B_i, we consider here only real-valued Brownian motions, and denote these processes by B. We have defined the Brownian motion process B in Example 3.4. This section gives essential properties of B.

There exists a modification of the Brownian motion with a.s. continuous paths ([147], Theorem 26, p. 17).

Note: This theorem is consistent with a previous result based on the Kolmogorov criterion (Eq. 3.2). We will work exclusively with continuous modifications of the Brownian motion B. Therefore, it will be assumed that B has continuous samples. ▲

If $p_n = (0 = t_0^{(n)} \le t_1^{(n)} \le \cdots \le t_{m_n}^{(n)} = t)$ is a sequence of **refining partitions** of $[0, t]$ such that $\Delta(p_n) \to 0$ as $n \to \infty$, then ([147], Theorem 28, p. 18)

$$\lim_{n\to\infty} \sum_{k=1}^{m_n} \left(B(t_k^{(n)}) - B(t_{k-1}^{(n)}) \right)^2 = t \quad \text{a.s.} \tag{3.112}$$

Note: This property justifies the notation $(dB(t))^2 = dt$ used extensively in applications. If the mesh of the sequence of partitions p_n converges to zero as $n \to \infty$ but is not refining, then the limit in Eq. 3.112 exists in m.s. (Example 3.74). We have also seen that the sequence $S_n = \sum_{k=1}^{m_n} \left(B(t_k^{(n)}) - B(t_{k-1}^{(n)}) \right)^2$ converges to t as $n \to \infty$ in probability and L_1 (Eq. 3.102).

Figure 3.14 shows three samples of $Z_{m_n}(s) = \sum_{k=1}^{m_n} \left(B(t_k^{(n)} \wedge s) - B(t_{k-1}^{(n)} \wedge s) \right)^2$

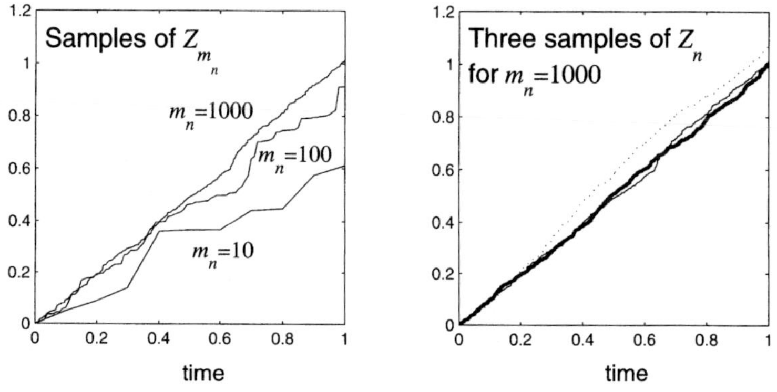

Figure 3.14. Samples of $Z_{m_n}(s) = \sum_{k=1}^{m_n} \left(B(t_k^{(n)} \wedge s) - B(t_{k-1}^{(n)} \wedge s) \right)^2$ for $s \le t$, $t = 1$, and $t_k^{(n)} = k/m_n$

for $s \le t$, $t = 1$, $t_k^{(n)} = k/m_n$, and $m_n = 10, 100$, and 1,000 corresponding to a single sample of the Brownian motion B. The figure also shows three samples of Z_{m_n} corresponding to the same partition of $[0, 1]$ ($m_n = 1,000$) but different samples of the Brownian motion. If m_n is small, the samples of Z_{m_n} can differ significantly from the identity function $t \mapsto t$ and from one another. However, these samples nearly coincide with $t \mapsto t$ as the partition is refined such that $\Delta(p_n) \to 0$ as $n \to \infty$. ▲

> The samples of a Brownian motion process B are of unbounded variation a.s. in any bounded interval.

Proof: Let $p_n = (0 = t_0^{(n)} \le t_1^{(n)} \le \cdots \le t_{m_n}^{(n)} = t)$ be a sequence of refining partitions of $[0, t]$ such that $\Delta(p_n) \to 0$ as $n \to \infty$. The left side of the inequality

$$\sum_{k=1}^{m_n} [B(t_k^{(n)}) - B(t_{k-1}^{(n)})]^2 \le \max_k |B(t_k^{(n)}) - B(t_{k-1}^{(n)})| \sum_{k=1}^{m_n} |B(t_k^{(n)}) - B(t_{k-1}^{(n)})|$$

and $\max_k |B(t_k^{(n)}) - B(t_{k-1}^{(n)})|$ converge to t and zero a.s. as $n \to \infty$ by Eq. 3.112 and the continuity of the samples of B, respectively. To satisfy the above inequality, the summation$\sum_{k=1}^{m_n} |B(t_k^{(n)}) - B(t_{k-1}^{(n)})|$ must approach infinity a.s. as $n \to \infty$ ([147], Theorem 29, p. 19).

> Brownian motion is a square integrable martingale with respect to its natural filtration.

Proof: We have shown that B is an $\mathcal{F}_t = \sigma(B(s) : 0 \le s \le t)$-martingale. Because $E[B(t)^2] = t < \infty$ for each $t > 0$, B is an $\mathcal{F}_t$-square integrable martingale. ■

Example 3.74: The equality in Eq. 3.112 holds also in the mean square sense if the mesh of the sequence of partitions p_n in this equation converges to zero as $n \to \infty$. The sequence of partitions p_n does not have to be refining. ◇

Proof: The first two moments of $Y_{m_n} = \sum_{k=1}^{m_n} \left[B(t_k^{(n)}) - B(t_{k-1}^{(n)})\right]^2 - t$ are $E[Y_{m_n}] = 0$ and $E[Y_{m_n}^2] = 2\sum_{k=1}^{m_n} (t_k^{(n)} - t_{k-1}^{(n)})^2$ since the Brownian motion has stationary independent Gaussian increments and the fourth moment of $N(0, \sigma^2)$ is $3\,\sigma^4$. Hence,

$$E[Y_{m_n}^2] \le 2 \max_k (t_k^{(n)} - t_{k-1}^{(n)}) \sum_{k=1}^{m_n} (t_k^{(n)} - t_{k-1}^{(n)}) = 2\,t\,\Delta(p_n) \to 0 \quad \text{as } n \to \infty$$

so that $Y_{m_n} \to 0$ in m.s. as $n \to \infty$. ■

Example 3.75: Let B be a Brownian motion, N denote a Poisson process with intensity $\lambda > 0$, and $\tilde{N}(t) = N(t) - \lambda\,t$. Assume that B and N are independent of each other. The square bracket or quadratic variation process of $B + \tilde{N}$ is $[B + \tilde{N}](t) = t + N(t)$. ◇

Proof: Let p_n be a sequence of partitions of a time interval $[0, t]$ and consider the sums

$$S_n = \sum_k \left[(B(t_k^{(n)}) + \tilde{N}(t_k^{(n)}) - (B(t_{k-1}^{(n)}) + N(t_{k-1}^{(n)})\right]^2$$
$$= \sum_k (\Delta B_k^{(n)})^2 + \sum_k (\Delta \tilde{N}_k^{(n)})^2 + 2 \sum_k (\Delta B_k^{(n)})(\Delta \tilde{N}_k^{(n)}),$$

where $\Delta B_k^{(n)} = B(t_k^{(n)}) - B(t_{k-1}^{(n)})$ and $\Delta \tilde{N}_k^{(n)} = \tilde{N}(t_k^{(n)}) - \tilde{N}(t_{k-1}^{(n)})$. If the sequence of partitions p_n is refining and $\Delta(p_n) \to 0$ as $n \to \infty$, then $\sum_k (\Delta B_k^{(n)})^2$ and $\sum_k (\Delta \tilde{N}_k^{(n)})^2$ converge to t and $N(t)$ a.s., while $\sum_k (\Delta B_k^{(n)})(\Delta \tilde{N}_k^{(n)})$ converges to zero since the Brownian motion has continuous samples so that

$$\left|\sum_k (\Delta B_k^{(n)})(\Delta \tilde{N}_k^{(n)})\right| \le \sum_k |\Delta B_k^{(n)}|\,|\Delta \tilde{N}_k^{(n)}| \le \max_k |\Delta B_k^{(n)}| \sum_k |\Delta \tilde{N}_k^{(n)}|$$

and $\max_k |\Delta B_k^{(n)}| \to 0$ as $n \to \infty$. Hence, S_n converges to $t + N(t)$ a.s. as $n \to \infty$. If the sequence of partitions p_n is not refining, then S_n converges to $t + N(t)$ in m.s. and probability as $n \to \infty$.

Similar arguments can be applied to show that $[B + \tilde{C}](t) = t + \sum_{k=1}^{N(t)} Y_k^2$, where B is a Brownian motion, $C(t) = \sum_{k=1}^{N(t)} Y_k$ is a compound Poisson process with jumps $Y_1, Y_2, \ldots$ in L_2, C is independent of B, and $\tilde{C}(t) = C(t) - \lambda\, t\, E[Y_1]$. ■

3.14 Lévy processes

This section defines Lévy processes, presents some of their essential properties, and gives the Lévy decomposition and the Lévy-Khinchine formulas. It also shows that Lévy processes are more complex than the superposition of a Brownian motion and compound Poisson processes.

> Let $(\Omega, \mathcal{F}, (\mathcal{F}_t)_{t \ge 0}, P)$ be a filtered probability space. A process $\{X(t), t \ge 0\}$ defined on this space is **Lévy** if it
>
> **1.** is $\mathcal{F}_t$-adapted and starts at zero,
>
> **2.** has stationary increments that are independent of the past, that is, $X(t) - X(s)$, $t > s$, has the same distribution as $X(t - s)$ and is independent of $\mathcal{F}_s$, and
>
> **3.** is continuous in probability.

Table 3.3 summarizes the **three defining properties** for the Poisson, Brownian motion, and Lévy processes. The third property differentiates these processes. Yet, the Poisson, Brownian motion, and Lévy processes are closely related. For example, the Poisson and Brownian motion processes are Lévy processes since they are continuous in probability. The compound Poisson process C

Table 3.3. Defining properties for Poisson, Brownian motion, and Lévy processes

Poisson	Brownian motion	Lévy
$\mathcal{F}_t$-adapted starting at zero		
stationary increments that are independent of the past		
counting process without explosions	Gaussian increments $B(t) - B(s) \sim N(0, t-s)$	continuous in probability

in Eq. 3.6 is also a Lévy process because it is $\mathcal{F}_t$-adapted, starts at zero, has stationary increments that are independent of the past, and $P(|C(t) - C(s)| > 0) \to 0$ as $|t - s| \to 0$.

We have shown that the Poisson, compound Poisson, and Brownian motion processes are not m.s. differentiable. It will be seen that the samples of the Lévy process are also too rough to be differentiable. Yet, it is common in the engineering literature to use the formal derivative of the compound Poisson, Brownian motion, and Lévy processes, called the **Poisson, Gaussian**, and **Lévy white noise** process, respectively, to model the input to various physical systems. Because the Poisson, Gaussian, and Lévy white noise processes do not exist, calculations involving these processes are formal so that the resulting findings may be questionable. We will see in the following chapters how white noise processes can be incorporated in the theory of stochastic differential equations in a rigorous manner.

Example 3.76: If X_1 and X_2 are two independent Lévy processes defined on the same filtered probability space $(\Omega, \mathcal{F}, (\mathcal{F}_t)_{t\geq 0}, P)$, then $X = X_1 + X_2$ is also a Lévy process. ◇

Proof: X is $\mathcal{F}_t$-adapted and has stationary independent increments, since it is the sum of two processes with these properties. It remains to show that X is continuous in probability. Set $A_i = \{|X_i(t) - X_i(s)| > \varepsilon/2\}$, $i = 1, 2$, and $A = \{|X(t) - X(s)| > \varepsilon\}$ for some $\varepsilon > 0$. Because $A^c \supseteq A_1^c \cap A_2^c$, we have $A \subseteq A_1 \cup A_2$ so that $P(A) \leq P(A_1) + P(A_2)$ implies $P(|X(t) - X(s)| > \varepsilon) \to 0$ as $|t - s| \to 0$ since the processes X_1, X_2 are continuous in probability. ■

Example 3.77: The characteristic function of a Lévy process X has the form

$$\varphi(u; t) = E[e^{\sqrt{-1}\, u\, X(t)}] = e^{-t\,\psi(u)}, \tag{3.113}$$

where ψ is a continuous function with $\psi(0) = 0$. ◇

Proof: Lévy processes have stationary independent increments, that is, the random variables $X(t+s) - X(s)$ and $X(s) - X(0)$, $s, t \geq 0$, are independent, so that

$$\varphi(u; t+s) = E\left[e^{\sqrt{-1}\,u\,[(X(t+s)-X(s))+(X(s)-X(0))]}\right] = \varphi(u; t)\,\varphi(u; s),$$

which implies $\varphi(u; t) = [\varphi(u; 1)]^t = \exp(-t\,\psi(u))$ for any integer $t \geq 1$. Hence, φ is an infinitely divisible characteristic function (Section 2.10.3.2). A similar argument shows that the above relationship can be extended to $t \geq 0$ rational. It can be shown that the mapping $t \mapsto \varphi(u; t)$ is right continuous since X has a càdlàg modification (Section 3.14.1) so that $\varphi(u; t) = \exp(-t\,\psi(u))$ holds for any time $t \geq 0$ ([147], Section 4).

The characteristic functions of the Poisson and compound Poisson processes in Eqs. 3.108 and 3.109 have the functional form in Eq. 3.113. If X is a Brownian motion B, then $\varphi(u; t) = E[e^{\sqrt{-1}\,u\,B(t)}] = \exp(-t\,u^2/2)$ since $B(t) \sim N(0, t)$ so that $\psi(u) = u^2/2$. The functional form of ψ in Eq. 3.113 for an arbitrary Lévy process is given at the end of this section (Eq. 3.122). ■

Example 3.78: Let X be an $\mathcal{F}_t$-adapted process starting at zero that has stationary increments that are independent of the past. Suppose that the characteristic function of the random variable $X(t)$ for some $t \geq 0$ is $\varphi(u; t) = e^{-t\,|u|^\alpha}$, where $u \in \mathbb{R}$ and $\alpha \in (0, 2)$. Then X is a Lévy process that may or may not be an $\mathcal{F}_t$-martingale depending on the value of α. ◇

Proof: Let f denote the density of $X(t) - X(s)$, $t > s$. For every $\varepsilon > 0$ the probability $P(|X(t) - X(s)| > \varepsilon) = \int_{(-\varepsilon,\varepsilon)^c} f(x)\,dx$ converges to zero as $(t-s) \to 0$ since $f(x) = \frac{1}{2\pi}\int_{-\infty}^{\infty} e^{\sqrt{-1}\,u\,x}\, e^{-(t-s)\,|u|^\alpha}\,du$ approaches a delta function centered at the origin as $(t-s) \to 0$ so that its integral over the interval $(-\varepsilon, \varepsilon)^c$ converges to zero.

It can be shown that $E[|X(t)|^p]$ is finite if and only if $p \in (0, \alpha)$ ([162], Proposition 1.2.16, p. 18). Hence, X is not a martingale for $\alpha \leq 1$ because the expectation $E[|X(t)|]$ is not bounded but this process is an $\mathcal{F}_t$-martingale for $\alpha > 1$. If $\alpha = 2$, the characteristic function of $X(t)$ is $\varphi(u; t) = e^{-t\,u^2}$ so that $X(t) \sim N(0, 2\,t)$ is a square integrable martingale that has the same distribution as $\sqrt{2}\,B$, where B denotes a Brownian motion process. ■

3.14.1 Properties

Each Lévy process has a unique modification which is Lévy and càdlàg ([147], Theorem 30, p. 21).

Note: Only this modification is considered in our discussion so that we assume that the Lévy process has càdlàg samples, that is, right continuous samples with left limits. The Brownian motion and the compound Poisson are examples of processes with càdlàg samples that are also Lévy processes. ▲

A Lévy process preserves its properties if it is restarted at a stopping time ([147], Theorem 32, p. 23).

Note: Let X be a Lévy process, T denote a stopping time, and $Y(t) = X(T+t) - X(T)$, $t \geq 0$, be a new process derived from X. The process Y (1) is a Lévy process adapted to $\mathcal{F}_{T+t}$, (2) is independent of $\mathcal{F}_T$, and (3) has the same distribution as X. Hence, the Brownian motion, Poisson, and compound Poisson processes have these properties because they are Lévy processes. ▲

A Lévy process X can have only jump discontinuities, where

$$\Delta X(t) = X(t) - X(t-) \text{ is the **jump** of } X \text{ at } t, \tag{3.114}$$

$X(t-) = \lim_{s \uparrow t} X(s)$, and $X(t) = X(t+) = \lim_{s \downarrow t} X(s)$.

Note: X has only jump discontinuities because it has càdlàg samples. The random variables $X(t-)$ and $X(t+) = X(t)$ denote the left and the right limits of X at time t, respectively, (Fig. 3.15). ▲

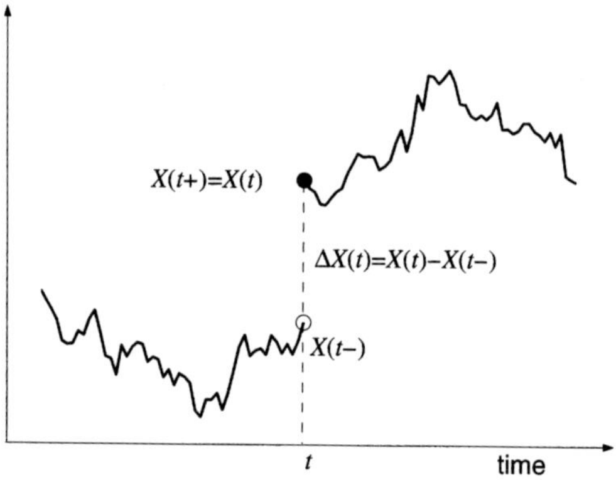

Figure 3.15. A jump discontinuity of X at time t

If a Lévy process X has bounded jumps, that is, if there exists a finite constant $c > 0$ such that $\sup_t |\Delta X(t)| \leq c < \infty$ a.s., then X has finite absolute moments of any order, that is, $E[|X(t)|^n] < \infty$ for $n = 1, 2, \ldots$ ([147], Theorem 34, p. 25).

Proof: Because Lévy processes have càdlàg samples, they can have only jump discontinuities ([147], p. 6).

Let $0 < c < \infty$ be a constant such that $\sup_t |\Delta X(t)| \leq c$ a.s. and define the stopping times

$$T_1 = \inf\{t > 0 : |X(t)| \geq c\} \quad \text{and}$$
$$T_{n+1} = \inf\{t > T_n : |X(t) - X(T_n)| \geq c\} \quad \text{for } n \geq 1.$$

The sequence $(T_1, T_2, \ldots)$ is strictly increasing since X is right continuous. Moreover, the stopped process X^{T_n} must satisfy the condition $\sup_t |X^{T_n}(t)| \le 2\,n\,c < \infty$ since $|X(t) - X(T_k)|$ cannot exceed $2\,c$ for $t \in (T_k, T_{k+1}]$ by the definition of the stopping times T_k. Because a Lévy process preserves its properties if it is restarted at a stopping time, the random variable $T_n - T_{n-1}$ is independent of $\mathcal{F}_{T_{n-1}}$ and has the same distribution as T_1. Hence,

$$E[e^{-T_n}] = \prod_{k=1}^{n} E[e^{-(T_k - T_{k-1})}] = \left(E[e^{-T_1}]\right)^n$$

since $T_n = (T_n - T_{n-1}) + (T_{n-1} - T_{n-2}) + \cdots + (T_1 - T_0)$ and $T_0 = 0$. Because $\{|X(t)| > 2\,n\,c\} \subseteq \{T_n < t\}$, we have

$$P(|X(t)| > 2\,n\,c) \le P(T_n < t) = P(e^{-T_n} > e^{-t}) \le E[e^{-T_n}]\,e^t = (E[e^{-T_1}])^n\,e^t$$

by the Chebyshev inequality. The last formula giving the tail behavior of the distribution of $|X(t)|$ shows that $X(t)$ has moments of any order since $0 \le E[e^{-T_1}] < 1$. ■

3.14.2 The Lévy decomposition

The Lévy decomposition and the Lévy-Khinchine formulas show that a Lévy process X can be represented by the sum of three processes (Fig. 3.16). We also show that X can be represented by the sum of three independent processes, a Brownian motion, a compound Poisson process, and a limit of compound Poisson processes (Example 3.84).

If X is a Lévy process and $\Lambda \in \mathcal{B}$ is a Borel set such that $0 \notin \bar{\Lambda}$, then

$$N^{\Lambda}(t) = \sum_{0<s\le t} 1_{\Lambda}(\Delta X(s)) = \sum_{n=1}^{\infty} 1_{\{T_n^{\Lambda} \le t\}} \quad \text{is a Poisson process} \qquad (3.115)$$

with intensity $\lambda_L(\Lambda) = E[N^{\Lambda}(1)]$, called the **Lévy measure**, where $T_1^{\Lambda} = \inf\{t > 0 : \Delta X(t) \in \Lambda\}$ and $T_{n+1}^{\Lambda} = \inf\{t > T_n^{\Lambda} : \Delta X(t) \in \Lambda\}$, $n \ge 1$.

Proof: The notation $\bar{\Lambda}$ is used for the closure of Λ, for example, the closure of an open interval $\Lambda = (\alpha, \beta)$ is $\bar{\Lambda} = [\alpha, \beta]$. The process N^{Λ} is defined by the jumps of X in Λ, where $T_1^{\Lambda} = \inf\{t > 0 : \Delta X(t) \in \Lambda\}$ denotes the first time X has a jump in Λ and $T_{n+1}^{\Lambda} = \inf\{t > T_n^{\Lambda} : \Delta X(t) \in \Lambda\}$, $n \ge 1$, is the time of jump $n+1$ of X in Λ. The process N^{Λ} is similar to C^{Λ} in Eq. 3.111.

The definition in Eq. 3.115 is meaningful because Lévy processes have càdlàg samples so that they can have only jump discontinuities. The process N^{Λ} (1) is a counting process giving the number of jumps of X in Λ during a time interval, (2) is adapted by definition, (3) is without explosions since X has càdlàg paths so that the random variables T_n^{Λ} are stopping times with the property $\lim_{n\to\infty} T_n^{\Lambda} = \infty$, and (4) has stationary increments that are independent of the past. The distribution of this increments depends on only the time lag $t - s$ because X has stationary increments. Hence, N^{Λ} is a Poisson counting process with intensity $\lambda_L(\Lambda) = E[N^{\Lambda}(1)] < \infty$ ([147], p. 26).

Let $\mathcal{M}(t, dy, \omega) = \#\{s \le t : \Delta X(s, \omega) \in (y, y + dy]\}$ be a random measure counting the number of jumps of X in $(y, y + dy]$ during $(0, t]$ for each fixed (t, ω). The Poisson process N^Λ can be defined by $N^\Lambda(t, \omega) = \int_\Lambda \mathcal{M}(t, dy, \omega)$ for each ω, where $\lambda_L(dy) = E[\mathcal{M}(1, dy)]$ is the Lévy measure of N^Λ for $\Lambda = (y, y + dy]$. ■

Example 3.79: If Λ is a Borel set such that $0 \notin \bar{\Lambda}$, then

$$E\left[\int_\Lambda h(y)\,\mathcal{M}(t, dy)\right] = t\int_\Lambda h(y)\,\lambda_L(dy) \quad \text{and}$$
$$E\left[\left(\int_\Lambda h(y)\,\mathcal{M}(t, dy) - t\int_\Lambda h(y)\,\lambda_L(dy)\right)^2\right] = t\int_\Lambda h(y)^2\,\lambda_L(dy), \quad (3.116)$$

where $\lambda_L(dy) = E[\mathcal{M}(1, dy)]$ denotes a Lévy measure and $h\,1_\Lambda$ is a function in L_2 relative to the Lévy measure λ_L. ◇

Note: If Λ_i are disjoint Borel sets and $h = \sum_i a_i\,1_{\Lambda_i}$, then

$$E\left[\int_\Lambda h(y)\,\mathcal{M}(t, dy)\right] = E\left[\sum_i a_i\,N^{\Lambda\cap\Lambda_i}(t)\right] = \sum_i a_i\,E\left[N^{\Lambda\cap\Lambda_i}(t)\right]$$
$$= t\sum_i a_i\,\lambda_L(\Lambda\cap\Lambda_i)$$

since $N^{\Lambda\cap\Lambda_i}$ is a Poisson process with intensity $\lambda_L(\Lambda\cap\Lambda_i)$. Similar considerations can be used to prove the second equality in Eq. 3.116. The extension to an arbitrary function $h\,1_\Lambda \in L_2$ can be found in [147] (Theorem 38, p. 28). ▲

Let X be a Lévy process and Λ denote a Borel set such that $0 \notin \bar{\Lambda}$. The **associated jump process,**

$$J^\Lambda(t) = \sum_{0<s\le t} \Delta X(s)\,1_\Lambda(\Delta X(s)) = \int_\Lambda y\,\mathcal{M}(t, dy), \quad (3.117)$$

and $X - J^\Lambda$ are Lévy processes ([147], Theorem 37, p. 27).

Proof: J^Λ has piece-wise constant càdlàg sample paths with jumps $\Delta X(T_n^\Lambda)$ arriving at the times T_n^Λ defining the Poisson process N^Λ (Eq. 3.115). J^Λ is a compound Poisson process defined by the large jumps of X, that is, the jumps of X in Λ. Therefore, almost all samples of J^Λ are of bounded variation on compacts.

Since J^Λ is a compound Poisson process, it is also Lévy. The process $X - J^\Lambda$ is Lévy as the difference of two Lévy processes. ■

If X is a Lévy process and $\Lambda = (-\infty, -a) \cup (a, \infty)$, $a > 0$, then

$$Y^a(t) = X(t) - \sum_{0<s\leq t} \Delta X(s)\, 1_{\{|\Delta X(s)|>a\}} = X(t) - \int_{|x|>a} y\, \mathcal{M}(t, dy) \tag{3.118}$$

has bounded absolute moments of any order.

Proof: The process in Eq. 3.118 can also be written as $Y^a(t) = X(t) - J^\Lambda(t)$ with the notation in Eq. 3.117. The jumps of Y^a have magnitude smaller than a so that $E[|Y^a(t)|^n] < \infty$ for any order n, by one of the properties in Section 3.14.1. It is common to define the process Y^a for $a = 1$, but any other value of $a > 0$ is acceptable. ■

The jump processes J^{Λ_i} in Eq. 3.117 corresponding to Borel sets Λ_1 and Λ_2 such that $0 \notin \bar{\Lambda}_i$, $i = 1, 2$, and $\Lambda_1 \cap \Lambda_2 = \emptyset$ are independent Lévy processes ([147], Theorem 39, p. 30).

Note: A similar property holds for processes obtained from a compound Poisson process by retaining its jumps in two disjoint sets. ▲

A Lévy process X can be represented by $X = Y + Z$, where Y and Z are independent Lévy processes, Y is a zero mean martingale with bounded jumps including all small jumps of X, and Z has paths of finite variation on compacts including all the large jumps of X ([147], Theorem 40, p. 31).

Proof: Recall that a process is said to have samples of **finite variation on compacts** if almost all its samples are of bounded variation on each compact of $[0, \infty)$.

Take $\Lambda = (-1, 1)^c$. The processes J^Λ and $X - J^\Lambda$ are Lévy (Eq. 3.117). Any sample function $s \mapsto X(s, \omega)$ has a finite number of jumps in Λ during a time interval $[0, t]$ so that J^Λ is of finite variation on compacts. The jumps of the Lévy process $X - J^\Lambda$ are smaller than 1 so that this process has bounded absolute moments of any order. Because $X - J^\Lambda$ has stationary independent increments and $X(0) - J^\Lambda(0) = 0$, we have $E[X(t) - J^\Lambda(t)] = \beta\, t$, where $\beta = E[X(1) - J^\Lambda(1)]$. Hence, $Y(t) = (X(t) - J^\Lambda(t)) - \beta\, t$ is a martingale. The claimed representation of X results by setting $Z(t) = J^\Lambda(t) + \beta\, t$.

That Y and Z are independent follows from the observations that (1) the processes J^Λ and $J^{\Lambda(\varepsilon)}$ are independent Lévy processes according to the previous property applied for $\Lambda_1 = \Lambda$ and $\Lambda_2 = \Lambda(\varepsilon) = (-1, -\varepsilon) \cup (\varepsilon, 1)$, $\varepsilon \in (0, 1)$, and (2) the process $J^{\Lambda(\varepsilon)}$ approaches $X - J^\Lambda$ as $\varepsilon \to 0$. ■

A Lévy process X with bounded jumps, that is, $\sup_t |\Delta X(t)| \le a$ a.s., $a > 0$, has the representation

$$X(t) - E[X(t)] = Z^c(t) + \int_{\{|y| \le a\}} y\, \tilde{\mathcal{M}}(t, dy) = Z^c(t) + Z^d(t), \quad (3.119)$$

where $\tilde{\mathcal{M}}(t, dy) = \mathcal{M}(t, dy) - t\,\lambda_L(dy)$, Z^d and Z^c are independent Lévy processes, Z^c is a martingale with continuous path, and Z^d is a martingale ([147], Theorem 41, p. 31).

Note: The representation in Eq. 3.119 shows that the process Y in the decomposition $X = Y + Z$ of a general Lévy process defined by the previous statement can be represented by the sum of two independent Lévy processes, the processes Z^c and Z^d. ■

The last two statements show that any Lévy process X has the representation (Fig. 3.16)

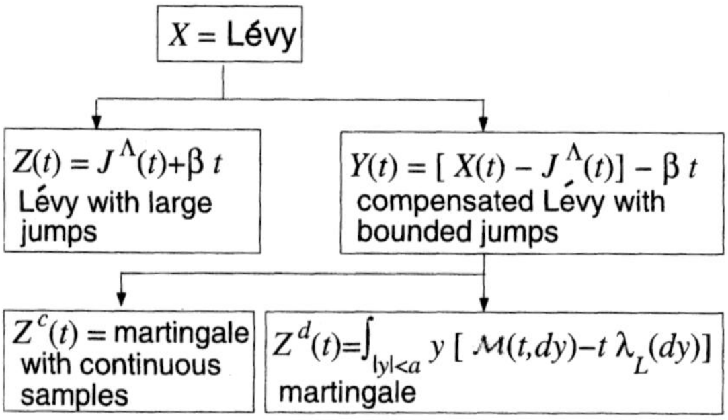

Figure 3.16. Components of a general Lévy process X

$$X(t) = [(X(t) - J^\Lambda(t)) - \beta\, t] + [J^\Lambda(t)) + \beta\, t] = Z^c(t) + Z^d(t) + Z(t), \quad (3.120)$$

where Z_c, Z^d and Z are independent Lévy processes, Z^c is a martingale with continuous samples, Z^d is a martingale, and Z has paths of finite variation on compacts (Fig. 3.16). The processes $Z^c + Z^d$ and Z include the small and large jumps of X, respectively.

The following theorem, referred to as the Lévy decomposition, provides additional information on the representation in Fig. 3.16 and Eq. 3.120 for an arbitrary Lévy process X. The representation depends on (1) a random measure $\mathcal{M}$ defining a Poisson process $N^\Lambda(t) = \int_\Lambda \mathcal{M}(t, dy)$ with intensity $\lambda_L(\Lambda)$ for any Borel set Λ, $0 \notin \bar{\Lambda}$ such that, if Γ is another Borel set with the properties $0 \notin \bar{\Gamma}$ and $\Gamma \cap \Lambda = \emptyset$, then N^Λ and N^Γ are independent, (2) $\beta = E[X(1) - J^\Lambda(1)] \in \mathbb{R}$, and (3) a Brownian motion process B that is independent of N^Λ.

Lévy decomposition. If X is a Lévy process and λ_L is a measure on $\mathbb{R} \setminus \{0\}$ such that $\int \min(1, y^2)\, \lambda_L(dy) < \infty$, then ([147], Theorem 42, p. 32)

$$X(t) = B(t) + \int_{|y|<1} y(\mathcal{M}(t, dy) - t\, \lambda_L(dy))$$
$$+ t\, E\left[X(1) - \int_{|y|\geq 1} y\, \mathcal{M}(1, dy)\right] + \int_{|y|\geq 1} y\, \mathcal{M}(t, dy)$$
$$= B(t) + \int_{|y|<1} y(\mathcal{M}(t, dy) - t\, \lambda_L(dy)) + \beta\, t + \sum_{0<s\leq t} \Delta X(s)\, 1_{\{|\Delta X(s)|\geq 1\}}. \tag{3.121}$$

Lévy-Khinchine formula. A Lévy process X is defined by its characteristic function $E[e^{\sqrt{-1}\, u\, X(t)}] = e^{-t\, \psi(u)}$, where

$$\psi(u) = \frac{\sigma^2}{2} u^2 + \int_{|y|<1} (1 - e^{\sqrt{-1}\, u\, y} + \sqrt{-1}\, u\, y)\, \lambda_L(dy)$$
$$- \sqrt{-1}\, \beta\, u + \int_{|y|\geq 1} (1 - e^{\sqrt{-1}\, u\, y})\, \lambda_L(dy), \tag{3.122}$$

and λ_L, σ^2, and β need to be specified. The parameters λ_L, σ^2, and β define a Lévy process uniquely in distribution ([147], Theorem 43, p. 32).

Example 3.80: Let B denote a Brownian motion that is independent of a compound Poisson process $C(t) = \sum_{k=1}^{N(t)} Y_k$. Then $X = B + C$ is a Lévy process. The processes B and C correspond to the components Y and Z, respectively, in the representation of a general Lévy process (Fig. 3.16). ◇

Note: B is a martingale with a.s. continuous samples and $E[|B(t)|^p] < \infty$, $p \geq 1$, and C has paths of finite variation on compacts. The variation $\sum_{k=1}^{N(t)} |Y_k|$ of C in $(0, t]$, $0 < t < \infty$, is bounded a.s. since $P(N(t) > n) \to 0$ as $n \to \infty$ for each time $t \geq 0$. ▲

Example 3.81: Let $X(t)$, $t \geq 0$, be a Lévy process with characteristic function $\varphi(u; t) = E[e^{\sqrt{-1}\, u\, L(t)}] = e^{-t\, |u|^\alpha}$, where $\alpha \in (0, 2]$. Figure 3.17 shows a sample of a Brownian motion B, its variation, and its quadratic variation, respectively. The figure also shows similar results for a Lévy process X of the type considered in Example 3.78 with $\alpha = 1.5$. The sample of X exhibits jump discontinuities whose effects are most visible in the sample of the quadratic variation process $[X]$ of X. The steady increase in the sample of $[X]$ resembling the sample of $[B]$ is interrupted by jumps characteristic of the quadratic variation of a compound Poisson process. ◇

Note: The sample of X in Fig. 3.17 has been generated by the recurrence formula $X(t + \Delta t) = X(t) + \Delta X(t)$, $t = 0, \Delta t, 2\,\Delta t, \ldots$, where $X(0) = 0$ and $\Delta X(t)$ are independent

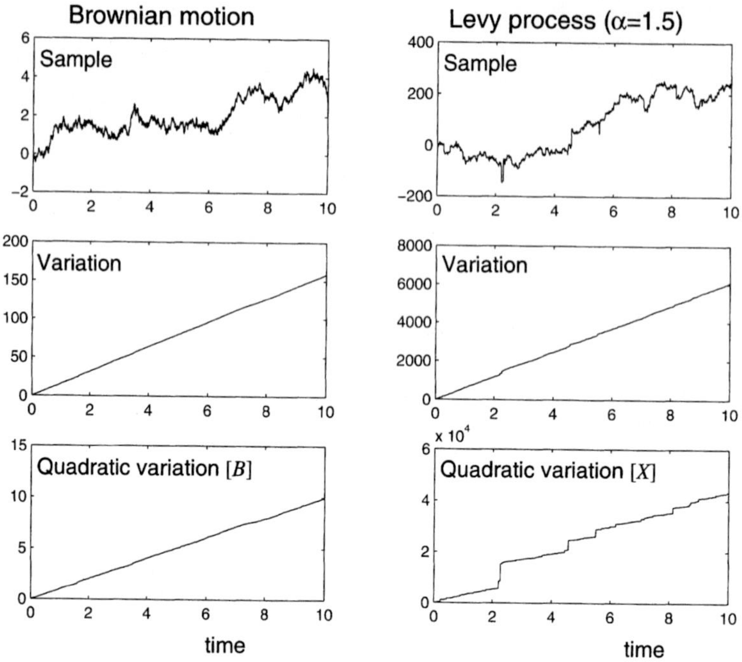

Figure 3.17. Samples of a Brownian motion and a Lévy process and their variation and quadratic variation

random variables with the characteristic function $\varphi(u) = e^{-\Delta t\,|u|^{\alpha}}$. Additional information on the generation of samples of X and other processes is given in Chapter 5. ▲

Example 3.82: Let $\lambda_L(dy) = dy/|y|^{\alpha+1}$ for $y \in \mathbb{R} \setminus \{0\}$ and $\alpha \in (0, 2)$ be a measure defining a Lévy process X. Consider also a compound Poisson process $C(t) = \sum_{k=1}^{N(t)} Y_k$, where $\lambda > 0$ denotes the intensity of the Poisson process N and Y_k are independent random variables with the distribution F defined by $\lambda_L(dy) = \lambda\, dF(y)$.

There are notable similarities and essential differences between X and C. Both processes are defined by random measures $\mathcal{M}(dt, dy)$ giving the number of jumps in $(t, t + dt] \times (y, y + dy]$. However, the average number of jumps in a Borel set Λ per unit of time is finite for C but may be infinite for X. ◇

Proof: The average number of jumps in $\Lambda \in \mathcal{B}$ per unit time is $\int_\Lambda \lambda_L(dy) = \lambda \int_\Lambda dF(y) \leq \lambda < \infty$ for C. However, the average number of jumps of X per unit of time $\int_\Lambda \lambda_L(dy)$

may or may not be finite depending on Λ. For example,

$$\int_{(-a,a)^c} \lambda_L(dy) = 2 \int_a^\infty y^{-(\alpha+1)}\, dy = \frac{1}{\alpha\, a^\alpha} < \infty \quad \text{and}$$
$$\int_{(-a,a)} \lambda_L(dy) = 2 \int_{0+}^a y^{-(\alpha+1)}\, dy = \infty$$

for $\Lambda = (-a, a)^c$ and $\Lambda = (-a, a)$, $0 < a < \infty$, respectively. The large jumps of X generate the compound Poisson process $J^{(-a,a)^c}$ defined by Eq. 3.117. This process has iid jumps $|\Delta X(t)| \geq a$ arriving in time at the mean rate $1/(\alpha\, a^\alpha)$. The small jumps of X do not define a compound Poisson process because the mean arrival rate of these jumps is not finite. Additional considerations on the jumps of X and measures needed to count the small jumps of this process can be found in [162] (Section 3.12). ■

Example 3.83: Let π be a measure defined on $\mathbb{R}^d \setminus \{\mathbf{0}\}$ satisfying the condition $\int \left(1 \wedge \| \boldsymbol{y} \|^2\right) \pi(d\boldsymbol{y}) < \infty$, and define the function

$$\begin{aligned}\psi(\boldsymbol{u}) &= \sqrt{-1}\, \boldsymbol{a}^T \boldsymbol{u} + \frac{1}{2} q(\boldsymbol{u}) \\ &+ \int_{\mathbb{R}^d} \left(1 - e^{\sqrt{-1}\, \boldsymbol{u}^T \boldsymbol{y}} + \sqrt{-1}\, \boldsymbol{u}^T \boldsymbol{y}\, 1_{\{\|\boldsymbol{y}\|<1\}}\right) \pi(d\boldsymbol{y}) \qquad (3.123)\end{aligned}$$

for every $\boldsymbol{u} \in \mathbb{R}^d$, where $\boldsymbol{a} \in \mathbb{R}^d$ and $q : \mathbb{R}^d \to \mathbb{R}$ is a quadratic form of $\boldsymbol{u}$. Then there exists a unique Lévy process $\boldsymbol{X}$ in distribution such that its characteristic function is $\varphi(\boldsymbol{u}; t) = E\left[e^{\sqrt{-1}\, \boldsymbol{u}^T \boldsymbol{X}(t)}\right] = e^{-t\, \psi(\boldsymbol{u})}$. Moreover, the jump process $\{\Delta \boldsymbol{X}(t), t \geq 0\}$ of $\boldsymbol{X}$ generates a Poisson process with measure π ([20], Theorem 1, p. 13). The measure $\pi(d\boldsymbol{y})$ gives the number of jumps of $\boldsymbol{X}$ in $(\boldsymbol{y}, \boldsymbol{y} + d\boldsymbol{y}]$ per unit of time.

Figure 3.18 shows samples of a real-valued Lévy X process with $a = 0.1$, $q(u) = \sigma^2 u^2$, $\sigma = 1$, $\varepsilon = 0.1$, $\pi(dy) = |y|^{-(\alpha+1)}\, dy$, and several values of α. The representation $X \simeq X^{(1)} + X^{(2)} + X^{(3,\varepsilon)}$ was used to generate the samples in this figure, where $X^{(1)}$, $X^{(2)}$, and $X^{(3,\varepsilon)}$ are mutually independent processes, $X^{(1)} = \sigma\, B(t) - a\, t$, $X^{(2)}$ and $X^{(3,\varepsilon)}$ are compound Poisson processes with measures $\pi^{(2)}(dy) = \pi(dy)\, 1_{\{|y|\geq 1\}}$ and $\pi^{(3,\varepsilon)}(dy) = \pi(dy)\, 1_{\{\varepsilon<|y|<1\}}$, and $\varepsilon \in (0, 1)$. ◇

Proof: We show that $\boldsymbol{X} = \boldsymbol{X}^{(1)} + \boldsymbol{X}^{(2)} + \boldsymbol{X}^{(3)}$, where $\boldsymbol{X}^{(1)}(t) = \boldsymbol{\sigma}\, \boldsymbol{B}(t) - \boldsymbol{a}\, t$, $\boldsymbol{\sigma}$ is a (d, d) matrix, the coordinates of $\boldsymbol{B} \in \mathbb{R}^d$ are independent Brownian motion, $\boldsymbol{X}^{(2)}$ is a compound Poisson process with jumps of magnitude larger than 1, and $\boldsymbol{X}^{(3)}$ is a process including all jumps of $\boldsymbol{X}$ strictly smaller than 1.

The characteristic function of $\boldsymbol{X}^{(1)}(t) \sim N(-\boldsymbol{a}\, t, \boldsymbol{\sigma}\, \boldsymbol{\sigma}^T\, t)$ is $\varphi^{(1)}(\boldsymbol{u}; t) = e^{-t\, \psi^{(1)}(\boldsymbol{u})}$, where $\psi^{(1)}(\boldsymbol{u}) = \sqrt{-1}\, \boldsymbol{u}^T \boldsymbol{a} + \frac{1}{2} \boldsymbol{u}^T \boldsymbol{\sigma}\, \boldsymbol{\sigma}^T \boldsymbol{u}$.

Consider first the large jumps of $\boldsymbol{X}$, that is, jumps with magnitude larger than 1 and denote $\pi^{(2)}(d\boldsymbol{y}) = \pi(d\boldsymbol{y})\, 1_{\{\|\boldsymbol{y}\|\geq 1\}}$. The total mass of $\pi^{(2)}$ is finite and $\boldsymbol{X}^{(2)}(t) = \sum_{s\leq t} \Delta \boldsymbol{X}^{(2)}(s)$ is a compound Poisson process with characteristic function $\varphi^{(2)}(\boldsymbol{u}; t) =$

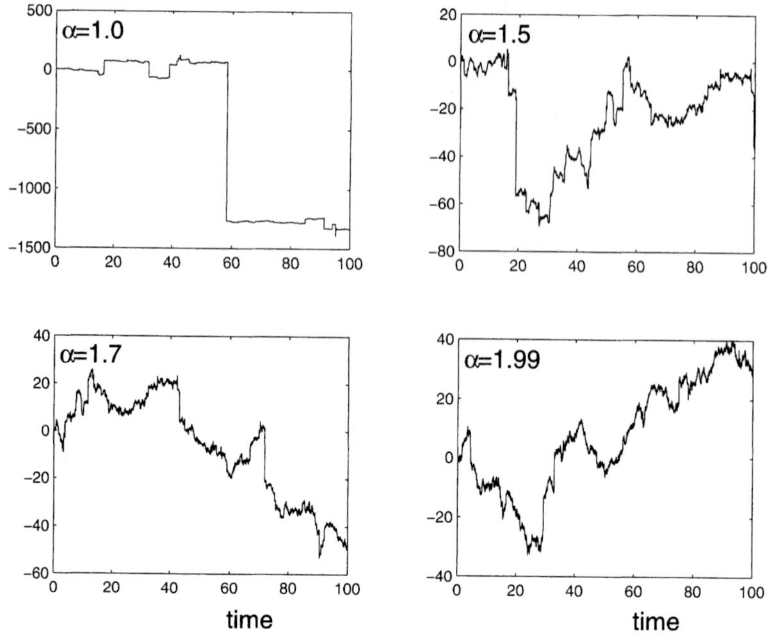

Figure 3.18. Samples of a Lévy process X with $a = 0.1$, $\sigma = 1$, $\varepsilon = 0.1$, $\pi(dy) = |y|^{-(\alpha+1)}\,dy$, and several values of α

$e^{-t\,\psi^{(2)}(\boldsymbol{u})}$, where $\psi^{(2)}(\boldsymbol{u}) = \int_{\mathbb{R}^d}\left(1 - e^{\sqrt{-1}\,\boldsymbol{u}^T\,\boldsymbol{y}}\right)\pi^{(2)}(d\boldsymbol{y})$, $\Delta\boldsymbol{X}^{(2)}(s) = \Delta\boldsymbol{X}(s)$ if $\|\,\Delta\boldsymbol{X}(s)\,\| \geq 1$, and $\Delta\boldsymbol{X}^{(2)}(s) = \boldsymbol{0}$ if $\|\,\Delta\boldsymbol{X}(s)\,\| < 1$.

Consider now the process $\boldsymbol{X}^{(3)}(t) = \sum_{s\leq t}\Delta\boldsymbol{X}^{(3)}(s)$ corresponding to the small jumps of $\boldsymbol{X}$ defined by $\Delta\boldsymbol{X}^{(3)}(s) = \Delta\boldsymbol{X}(s)$ if $\|\,\Delta\boldsymbol{X}(s)\,\| < 1$ and zero otherwise. Define the compensated compound Poisson process

$$\boldsymbol{X}^{(3,\varepsilon)}(t) = \sum_{s\leq t}\Delta\boldsymbol{X}(s)\,1_{\{\varepsilon<\|\Delta\boldsymbol{X}(s)\|<1\}} - t\int_{\mathbb{R}^d}\boldsymbol{y}\,\pi^{(3,\varepsilon)}(d\boldsymbol{y}), \quad t \geq 0,$$

for every $\varepsilon > 0$, where $\pi^{(3,\varepsilon)}(d\boldsymbol{y}) = \pi(d\boldsymbol{y})\,1_{\{\varepsilon<\|\boldsymbol{y}\|<1\}}$. This process is independent of $\boldsymbol{X}^{(2)}$ and has the characteristic function $\varphi^{(3,\varepsilon)}(\boldsymbol{u};t) = e^{-t\,\psi^{(3,\varepsilon)}(\boldsymbol{u})}$, where

$$\psi^{(3,\varepsilon)}(\boldsymbol{u}) = \int_{\mathbb{R}^d}\left(1 - e^{\sqrt{-1}\,\boldsymbol{u}^T\,\boldsymbol{y}} + \sqrt{-1}\,\boldsymbol{u}^T\,\boldsymbol{y}\right)\pi^{(3,\varepsilon)}(d\boldsymbol{y}).$$

It can be shown that $\{\boldsymbol{X}^{(3,\varepsilon)}, \varepsilon > 0\}$ is a Cauchy sequence with the norm defined by the expectation $E\left[\sup_{0\leq s\leq t}\|\,\boldsymbol{Y}(s)\,\|^2\right]$ and its limit denoted by $\boldsymbol{X}^{(3)}$ is a Lévy process with the characteristic function $\varphi^{(3)}(\boldsymbol{u};t) = e^{-t\,\psi^{(3)}(\boldsymbol{u})}$, where ([20], Theorem 1, p. 13)

$$\psi^{(3)}(\boldsymbol{u}) = \int_{\mathbb{R}^d}\left(1 - e^{\sqrt{-1}\,\boldsymbol{u}^T\,\boldsymbol{y}} + \sqrt{-1}\,\boldsymbol{u}^T\,\boldsymbol{y}\right)\pi^{(3)}(d\boldsymbol{y}).$$

Hence, $X^{(3)}$ is a limit of compound Poisson processes with bounded jumps.

The processes $X^{(1)}$, $X^{(2)}$, and $X^{(3)}$ correspond to the processes Z^c, Z, and Z^d in Eq. 3.120 and Fig. 3.16 giving the decomposition of a general Lévy process. ■

In the following chapter we will define an integral, referred to as a stochastic integral, with integrand and integrator given by a process with càglàd samples and a semimartingale, respectively. This integral is sufficiently general to establish the Itô formula and find the solution of a variety of stochastic problems. We will define semimartingales and give some of their essential properties in Section 4.4.1. For example, we will see that an adapted process X with càdlàg samples is a semimartingale X if it admits the representation $X(t) = X(0) + M(t) + A(t)$, where $M(0) = A(0) = 0$, M is a local martingale, and A is an adapted process with càdlàg samples of finite variation on compacts.

Example 3.84: Any Lévy process is a semimartingale. ◇

Proof: If X is a Lévy process, then it has the representation $X = Y + Z$, where (1) Y and Z are independent Lévy processes, (2) Y is a martingale with bounded jumps so that it is a square integrable martingale (Section 3.10.2), and (3) Z has paths of finite variation on compacts. The process X is adapted and càdlàg since it is Lévy and so are the processes Y and Z. ■

Example 3.85: The Poisson process N and the square of a Brownian motion B are semimartingales. ◇

Proof: The processes N and B^2 admit the representation $M + A$, where $(M(t) = N(t) - \lambda\, t, A(t) = \lambda\, t)$ and $(M(t) = B(t)^2 - t, A(t) = t)$, respectively. The properties of M and A show that N and B^2 are semimartingales. For example, $B(t)^2 - t$ is a martingale since $E[B(t)^2] = t < \infty$, B^2 is $\mathcal{F}_t = \sigma(B(s), 0 \leq s \leq t)$-adapted, and for $t \geq s$ we have

$$E[B(t)^2 \mid \mathcal{F}_s] = E\left[(B(t) - B(s))^2 + 2\, B(t)\, B(s) - B(s)^2 \mid \mathcal{F}_s\right] = (t - s) + B(s)^2.$$

The martingale $B(t)^2 - t$ has finite moments of any order and starts at zero. The process $A(t) = t$ is continuous, adapted, and $A(0) = 0$. We also note that the filtration $\mathcal{F}_t = \sigma(B(s), 0 \leq s \leq t)$ includes the natural filtration of B^2. ■

3.15 Problems

3.1: Let Y be a real-valued random variable on a probability space $(\Omega, \mathcal{F}, P)$. Find the natural filtration of the stochastic process $X(t) = Y\, h(t)$, $t \geq 0$, where $h : [0, \infty) \to \mathbb{R}$ is a continuous function.

3.2: Complete the proofs of the properties of stopping times in Section 3.4.

3.3: Prove the first three statements following Eq. 3.9.

3.4: Write the finite dimensional densities for an $\mathbb{R}^d$-valued Gaussian process X with specified mean and covariance functions.

3.5: Is the process in Eq. 3.33 ergodic in the first two moments, that is, is Eq. 3.15 satisfied with $g(x) = x^p$, $p = 1, 2$?

3.6: Let $X(t)$, $t \geq 0$, be a real-valued Gaussian process with mean zero and covariance function $E[X(t)\, X(s)] = t \wedge s$. Show that this process has independent increments.

3.7: Show that a compound Poisson process has the second moment properties in Eq. 3.109.

3.8: Prove the results given in Table 3.1.

3.9: Consider the process X in Eq. 3.33 with $A_k = A + U + k$ and $B_k = B + V_k$, where A, B, U_k, V_k are mutually uncorrelated random variables with zero-mean zero and unit-variance. Is X a weakly stationary process?

3.10: Complete the proof of the m.s. continuity criterion in Section 3.9.1.

3.11: Show that if X is m.s. continuous on a closed bounded interval I, then X is uniformly m.s. continuous on I.

3.12: Prove that the m.s. derivative of a process X exists and $E[\dot{X}(t)^2] < \infty$ if and only if $\partial^2 r(u, v)/(\partial u\, \partial v)$ exists and is finite at $u = v = t$.

3.13: Consider an m.s. differentiable process X defined on an interval I. Show that $\dot{X}(t) = 0$ for all $t \in I$ if and only if X is constant on I.

3.14: Prove that the m.s. derivative of a weakly stationary process X exists and $E[\dot{X}(t)^2] < \infty$ if and only if $d^2 r(\tau)/d\tau^2$ exists and is finite at $\tau = 0$.

3.15: Show that a monotone or differentiable function $h : [a, b] \to \mathbb{R}$ is of bounded variation on $[a, b]$.

3.16: Show that the variation $v_h(p)$ defined by Eq. 3.54 increases as the partition p is refined.

3.17: Prove the statements in Eqs. 3.58, 3.59, 3.60, and 3.61.

3.18: Prove Eqs. 3.62, 3.63, and 3.66.

3.19: Find the second moment properties of a process $Y(t) = \int_0^t \zeta(u)\, dX(u)$, where $X(t) = \int_{\mathbb{R}} \exp(\sqrt{-1}\, t\, \nu)\, dZ(\nu)$ (Eq. 3.71) and Z has the properties in Eq. 3.72.

3.20: Develop the Karhunen-Loéve representation for a band limited white noise process with mean zero (Table 3.1).

3.21: Find the mean crossing rate of a process $Y(t) = a(t)\,X(t)$, where X is given by Eq. 3.33 and a is a differentiable function.

3.22: Find the mean upcrossing rate of the translation process in Eq. 3.25, where G satisfies the conditions in Eq. 3.92. Specialize your results for a lognormal and exponential distribution function F.

3.23: Develop an algorithm for calculating the density f_T in Eq. 3.94 approximately.

3.24: Show that the expectation of a submartingale and supermartingale is an increasing and decreasing function of time, respectively.

3.25: Prove that a counting process N associated with the strictly increasing sequence of positive random variables $\{T_n, n = 1, 2, \ldots\}$ is adapted to the natural filtration of N if and only if $\{T_n, n = 1, 2, \ldots\}$ are stopping times.

3.26: Complete the proof of Eq. 3.116 in Example 3.79.

3.27: Show that the process J^{Λ} defined by Eq. 3.117 has paths of finite variation on compacts.

3.28: Let X be a Lévy process in L_1. Show that $X(t) - E[X(t)]$ is a martingale.

3.29: Show that the sum of two independent compound Poisson processes is a compound Poisson process.

Chapter 4

Itô's Formula and Stochastic Differential Equations

4.1 Introduction

The probabilistic concepts reviewed in the previous chapters are applied to develop one of the most useful tools for the solution of stochastic problems, the Itô calculus. Our objectives are to:

1. Define the **stochastic integral** or the **Itô integral**, that is, an integral involving càglàd (left continuous with right limits) integrands and semimartingale integrators. The Itô integral cannot be defined in the Riemann-Stieltjes sense because many semimartingales do not have sufficiently smooth samples. The stochastic integral considered in this chapter is an extension of the original Itô integral defined for Brownian motion integrators. We also define an alternative to the Itô integral, called the **Fisk-Stratonovich integral** or the **Stratonovich integral**, that has properties similar to the Riemann-Stieltjes integral. The relationship between the Itô and the Stratonovich integrals is also established.

2. Develop a change of variable formula for functions of semimartingales, called the **Itô formula**. This formula differs from the change of variable formula of classical calculus and is essential for the solution of many stochastic problems in engineering, physics, and other fields.

3. Introduce a new type of differential equation, called a **stochastic differential equation**, giving the propagation in time of the state of a physical system driven by semimartingales, and establish conditions under which the solution of a stochastic differential equation exists and is unique. The semimartingale noise includes the Gaussian, Poisson, and Lévy white noise processes and is sufficiently general for our developments. The theory of stochastic differential equations provides essential tools for the analysis of deterministic partial differential equations, random

vibration, and other stochastic problem in sciences and engineering, as we will see in the following chapters.

Examples are presented to clarify the calculation of the Itô and Fisk-Stratonovich integrals, illustrate the relationship between these integrals, and demonstrate the use of the Itô formula to find probabilistic characteristics of the solution of stochastic differential equations driven by white noise.

4.2 Riemann-Stieltjes integrals

A finite set of points $p = \{t_0, t_1, \ldots, t_m\}$ such that $0 = t_0 < \cdots < t_m = t$ is called a **partition** of $[0, t]$. A point $t_k' \in [t_{k-1}, t_k]$ is an **intermediate point** of p. The **mesh** of the partition p is $\Delta(p) = \max_{1 \le k \le m}(t_k - t_{k-1})$. A partition p' is a **refinement** of p or **refines** p if $p' \supseteq p$. A real-valued function h is of **bounded** r**-variation** on $[0, t]$ if $\sup\left\{\sum_i |h(t_i) - h(t_{i-1})|^r\right\} < \infty$, where $r > 0$ and the supremum is taken over **all partitions** of the interval $[0, t]$. If $r = 1$, h is said to be of **bounded variation** on $[0, t]$ (Section 3.9.3.1).

Consider two real-valued functions f and g defined on a bounded interval $[0, t] \subset \mathbb{R}$. Define the **Riemann-Stieltjes sum of** f **with respect to** g corresponding to a partition p of $[0, t]$ by

$$s_{f,p}(g) = \sum_{k=1}^{m} f(t_k')\,[g(t_k) - g(t_{k-1})]. \tag{4.1}$$

We say that f is **Riemann-Stieltjes integrable with respect to** g **on** $[0, t]$ if there is a number a such that, for every $\varepsilon > 0$, there is a partition p_ε of $[0, t]$ with the property $|s_{f,p}(g) - a| < \varepsilon$ for every $p \supseteq p_\varepsilon$ and every choice of intermediate points ([5], p. 141). The number a is called the **Riemann-Stieltjes integral of** f **with respect to** g **on** $[0, t]$ and is denoted by $\int_0^t f(s)\,dg(s)$ or $\int_0^t f\,dg$.

It can be shown that $\int_0^t f\,dg$ exists if ([131], p. 94)

1. f and g have no discontinuities at the same point $s \in [0, t]$ and

2. f and g are of bounded p-variation and q-variation, respectively, for some $p > 0$ and $q > 0$ such that $1/p + 1/q > 1$.

Note: The Riemann-Stieltjes integral $\int_0^t f\,dg$ can be obtained as the limit of a sequence of sums s_{f,p_n} (Eq. 4.1), where p_n is a sequence of refining partitions with arbitrary intermediate points $t_k'^{(n)} \in [t_{k-1}^{(n)}, t_k^{(n)}]$ such that $\Delta(p_n) \to 0$ as $n \to \infty$. ▲

Let f_1, f_2 be Riemann-Stieltjes integrable with respect to g on $[0, t]$, f be Riemann-Stieltjes integrable with respect to g, g_1, g_2 on $[0, t]$, c_1, c_2 some constants, and $s \in (0, t)$. The Riemann-Stieltjes integral has the following properties ([5], Theorem 7.2, p. 142, Theorem 7.3, p. 142, and Theorem 7.4, p. 143).

- $\int_0^t (c_1 f_1 + c_2 f_2)\, dg = c_1 \int_0^t f_1\, dg + c_2 \int_0^t f_2\, dg,$
- $\int_0^t f\, d(c_1 g_1 + c_2 g_2) = c_1 \int_0^t f\, dg_1 + c_2 \int_0^t f\, dg_2,$ and
- $\int_0^t f\, dg = \int_0^s f\, dg + \int_s^t f\, dg.$ (4.2)

Note: The last equality in Eq. 4.2 holds if $\int_0^t f\, dg$ exists. The integral $\int_s^t f\, dg$ in this equation is $\int_0^t 1_{[s,t]}\, f\, dg$. ▲

Let B be a Brownian motion process. The sample paths $s \mapsto B(s,\omega)$ of B are of bounded q-variation on any finite interval $[0,t]$ for $q > 2$ [182]. Hence, $\int_0^t f(s)\, dB(s,\omega)$ exists as a Riemann-Stieltjes integral for almost all sample paths of B if f is of bounded variation since $1/p + 1/q > 1$ for $p = 1$ and $q > 2$. The definition of $\int_0^t f\, dB$ as a Riemann-Stieltjes integral corresponding to the sample paths of the Brownian motion is referred to as **path by path** definition. The integrals

$$\int_0^t e^s\, dB(s,\omega), \quad \int_0^t \cos(s)\, dB(s,\omega), \quad \text{and} \quad \int_0^t s^k\, dB(s,\omega)$$

exist as Riemann-Stieltjes integrals for almost all sample paths of the Brownian motion because the functions e^s, $\cos(s)$, and s^k are of bounded variation.

These observations may suggest that the path by path definition can be extended to more general integrals, for example, $\int_0^t B\, dB$ and $\int_0^t f\, dB$, where f is an arbitrary continuous function. However, the Riemann-Stieltjes integrals $\int_0^t B(s,\omega)\, dB(s,\omega)$ and $\int_0^t f(s)\, dB(s,\omega)$ may not exist because:

1. The condition $1/p + 1/q > 1$ is not satisfied by the integrand and the integrator of the path by path integral $\int_0^t B(s,\omega)\, dB(s,\omega)$ and

2. The Riemann-Stieltjes integral $\int_0^t f(s)\, dg(s)$ does not exist for all continuous functions f on $[0,t]$ unless g is of bounded variation ([147], Theorem 52, p. 40) so that the integral $\int_0^t f(s)\, dB(s,\omega)$ cannot be defined as a path by path Riemann-Stieltjes integral for an arbitrary continuous function f.

We need an alternative definition for $\int_0^t B\, dB$, $\int_0^t f\, dB$, and other more general integrals involving random integrands and integrators. The definition considered in the following sections allows a broader class of integrands and integrators than the path by path definition but is less intuitive than the Riemann-Stieltjes integral.

4.3 Preliminaries on stochastic integrals

Consider two real-valued stochastic processes, X and Y. Our objective is to define the integral of X with respect to Y on a bounded interval I of the real line, referred to as the **stochastic integrals of X with respect to Y on I**. We will clarify in the first part of this chapter the precise meaning of the stochastic integral and establish conditions that X and Y must satisfy so that this integral exists in some sense. The definition of the stochastic integral is based on sequences of sums resembling the Riemann-Stieltjes sums of classical calculus (Eq. 4.1). However, there are two notable differences between the definition of stochastic integrals and Riemann-Stieltjes integrals.

1. The sequence of sums defining stochastic integrals is random. Convergence criteria for sequences of random variables need to be used to define the stochastic integral. One commonly used convergence criterion is mean square convergence.

2. The limit of the sequence of sums defining a stochastic integral may depend on the selection of the intermediate points $t_k'^{(n)}$ of the sequence of partitions p_n of $[0, t]$. In contrast, the Riemann-Stieltjes integral has the same value irrespective of the intermediate points used to define the sequence of sums in Eq. 4.1.

We illustrate these differences by examples calculating the stochastic integrals $\int_0^t B\, dB$ and $\int_0^t N\, dN$, where B and N denote the Brownian motion and Poisson processes, respectively. These integrals are defined as limits of random sequences of sums corresponding to partitions of the interval of integration $[0, t]$. Generally, the result of the integration depends on the selection of the intermediate points of the partitions of $[0, t]$. Two stochastic integrals are defined for particular selections of the intermediate points, the **Itô** and the **Stratonovich** integrals. We show that the Itô integral of the Brownian motion with respect to itself is a martingale while the corresponding Stratonovich integral is not a martingale. We also show that the Itô and the pathwise Riemann-Stieltjes definitions of the integral $\int_0^t N(s-)\, dN(s)$ coincide.

Example 4.1: Consider the random sequence

$$J_{B,n}(B) = \sum_{k=1}^{m_n} B(t_{k-1}^{(n)}) \left[B(t_k^{(n)}) - B(t_{k-1}^{(n)}) \right],$$

where $p_n = (0 = t_0^{(n)} < t_1^{(n)} < \cdots < t_{m_n}^{(n)} = t)$ is a sequence of partitions of $[0, t]$ with intermediate points $t_k'^{(n)} = t_{k-1}^{(n)}$ such that $\Delta(p_n) \to 0$ as $n \to \infty$. The limit of $J_{B,n}(B)$ as $n \to \infty$ exists in m.s. and in probability. This limit, denoted by $\int_0^t B(s)\, dB(s)$ or $\int_0^t B\, dB$ and called the **Itô integral** of B with respect to B on $[0, t]$, is

$$\int_0^t B(s)\, dB(s) = \frac{1}{2}(B(t)^2 - t), \quad t \geq 0. \tag{4.3}$$

If, in addition, the sequence of partitions of $[0, t]$ is refining, that is, $p_n \subset p_{n+1}$, then $J_{B,n}(B)$ converges also a.s. to the limit in Eq. 4.3. The Itô integral differs from $B(t)^2/2$, that is, the expression of the integral $\int_0^t B\,dB$ obtained by the formal use of classical calculus. $\diamond$

Proof: The notation $J_{B,n}(B)$ is used for simplicity although it is slightly inconsistent with the definition of $s_{f,p}(g)$ in Eq. 4.1. We have

$$J_{B,n}(B) = \frac{1}{2}\sum_{k=1}^{m_n}\left(B(t_k^{(n)})^2 - B(t_{k-1}^{(n)})^2\right) - \frac{1}{2}\sum_{k=1}^{m_n}\left(B(t_k^{(n)}) - B(t_{k-1}^{(n)})\right)^2$$

$$= \frac{1}{2}B(t)^2 - \frac{1}{2}\sum_{k=1}^{m_n}\left(B(t_k^{(n)}) - B(t_{k-1}^{(n)})\right)^2,$$

where the last equality holds since $\sum_{k=1}^{m_n}\left((B(t_k^{(n)})^2 - B(t_{k-1}^{(n)})^2)\right)$ is a telescopic series whose sum is $B(t)^2$. The m.s. limit of $J_{B,n}(B)$ is $(B(t)^2 - t)/2$ since

$$E\left[\left(J_{B,n}(B) - \frac{1}{2}\left(B(t)^2 - t\right)\right)^2\right] = E\left[\left(-\frac{1}{2}\sum_{k=1}^{m_n}\left(B(t_k^{(n)}) - B(t_{k-1}^{(n)})\right)^2 + \frac{t}{2}\right)^2\right]$$

$$= \frac{1}{4}\sum_{k,l} E\left[(\Delta B_k)^2\,(\Delta B_l)^2\right] - \frac{t}{2}\sum_{k} E\left[(\Delta B_k)^2\right] + \frac{t^2}{4}$$

approaches zero as $n \to \infty$, where $\Delta B_k = B(t_k^{(n)}) - B(t_{k-1}^{(n)})$. Hence, the Itô integral $\int_0^t B\,dB$ can be defined as the m.s. limit of $J_{B,n}(B)$. The integral can also be defined as the limit in probability of $J_{B,n}(B)$ because the m.s. convergence implies convergence in probability (Section 2.13).

Because $\lim_{n\to\infty}\sum_{k=1}^{m_n}\left[B(t_k^{(n)}) - B(t_{k-1}^{(n)})\right]^2 = t$ holds a.s. for a sequence of refining partitions of $[0, t]$ (Section 3.13 in this book, [147], Theorem 28, p. 18), $J_{B,n}(B)$ converges a.s. to $(B(t)^2 - t)/2$ for refining partitions so that in this setting the Itô integral can be defined as the a.s. limit of $J_{B,n}(B)$. ■

Example 4.2: Consider the random sequence

$$\tilde{J}_{B,n}(B) = \sum_{k=1}^{m_n} B(t_k^{\prime\,(n)})\left[B(t_k^{(n)}) - B(t_{k-1}^{(n)})\right]$$

that differs from $J_{B,n}(B)$ in Example 4.1 by the choice of the intermediate points, $t_k^{\prime\,(n)} = (1-\theta)t_{k-1}^{(n)} + \theta\, t_k^{(n)}$, $\theta \in [0, 1]$, instead of $t_{k-1}^{(n)}$. The limit of $\tilde{J}_{B,n}(B)$ as $n \to \infty$ exists in m.s. and in probability and is

$$\underset{n\to\infty}{\text{l.i.m.}}\ \tilde{J}_{B,n}(B) = \frac{1}{2}\,B(t)^2 + (\theta - 1/2)\,t. \tag{4.4}$$

If $\theta = 1/2$, this limit, denoted by $\int_0^t B(s) \circ dB(s)$ or $\int_0^t B \circ dB$ and called the **Stratonovich integral** of B with respect to B on $[0, t]$, is

$$\int_0^t B(s) \circ dB(s) = \int_0^t B \circ dB = \frac{1}{2} B(t)^2. \tag{4.5}$$

The Stratonovich integral coincides with the result obtained by the formal use of the classical calculus. The difference between the Itô and Stratonovich integrals is caused by the relationship between the integrands $B(t_{k-1}^{(n)})$ and $B(t_k'^{(n)})$ and the integrators $B(t_k^{(n)}) - B(t_{k-1}^{(n)})$ of these integrals. They are independent for the Itô integral but are dependent for the Stratonovich integral. ◇

Proof: Take a fixed $\theta \in [0, 1]$ and denote by

$$\begin{aligned} \Delta B_k &= B(t_k^{(n)}) - B(t_{k-1}^{(n)}) \sim N(0, \Delta t_k), \\ \Delta B_k' &= B(t_k'^{(n)}) - B(t_{k-1}^{(n)}) \sim N(0, \theta\, \Delta t_k), \quad \text{and} \\ \Delta B_k'' &= B(t_k^{(n)}) - B(t_k'^{(n)}) \sim N(0, (1-\theta)\, \Delta t_k) \end{aligned}$$

the increments of B in $(t_{k-1}^{(n)}, t_k^{(n)})$, $(t_{k-1}^{(n)}, t_k'^{(n)})$, and $(t_k'^{(n)}, t_k^{(n)})$, respectively, where $\Delta t_k = t_k^{(n)} - t_{k-1}^{(n)}$. With this notation we have

$$\tilde{J}_{B,n}(B) = \sum_{k=1}^{m_n} B(t_{k-1}^{(n)})\, \Delta B_k + \sum_{k=1}^{m_n} \Delta B_k'\, \Delta B_k$$

since $B(t_k'^{(n)}) = B(t_{k-1}^{(n)}) + \Delta B_k'$. We have shown that the first term of $\tilde{J}_{B,n}(B)$ converges in m.s. to $(B(t)^2 - t)/2$ as $n \to \infty$ and its limit defines the Itô integral (Eq. 4.3). It remains to show that the second term in $\tilde{J}_{B,n}(B)$, that is, $\sum_{k=1}^{m_n} \Delta B_k'\, \Delta B_k$, converges in m.s. to $\theta\, t$ since $X_n \xrightarrow{\text{m.s.}} X$ and $Y_n \xrightarrow{\text{m.s.}} Y$ imply $X_n + Y_n \xrightarrow{\text{m.s.}} X + Y$ (Section 2.13). The first moment of the second term of $\tilde{J}_{B,n}(B)$ is

$$E\left[\sum_{k=1}^{m_n} \Delta B_k'\, \Delta B_k\right] = \sum_k E\left[(\Delta B_k')^2\right] + \sum_k E\left[\Delta B_k'\, \Delta B_k''\right] = \sum_k \theta\, \Delta t_k = \theta\, t$$

since $\Delta B_k = \Delta B_k' + \Delta B_k''$ and $E[\Delta B_k'\, \Delta B_k''] = E[\Delta B_k']\, E[\Delta B_k''] = 0$. The second moment of this term,

$$\begin{aligned} E\left[\left(\sum_{k=1}^{m_n} \Delta B_k'\, \Delta B_k\right)^2\right] &= E\left[\left(\sum_{k=1}^{m_n} (\Delta B_k')^2 + \sum_{k=1}^{m_n} \Delta B_k'\, \Delta B_k''\right)^2\right] \\ &= \sum_{k \neq l} E[(\Delta B_k')^2\, (\Delta B_l')^2] + \sum_k E[(\Delta B_k')^4] + \sum_{k,l} E[\Delta B_k'\, \Delta B_k''\, \Delta B_l'\, \Delta B_l''] \\ &\quad + 2 \sum_{k,l} E[(\Delta B_k')^2\, \Delta B_l'\, \Delta B_l''] = \theta^2 \sum_{k \neq l} \Delta t_k\, \Delta t_l + 3\theta^2 \sum_k (\Delta t_k)^2 \\ &\quad + \theta\,(1-\theta) \sum_k (\Delta t_k)^2 = \theta^2 \sum_{k,l} \Delta t_k\, \Delta t_l + 2\theta^2 \sum_k (\Delta t_k)^2 + \theta\,(1-\theta) \sum_k (\Delta t_k)^2, \end{aligned}$$

converges to $(\theta t)^2$ as $n \to \infty$ since $\sum_{k,l} \Delta t_k \, \Delta t_l$ approaches t^2 and $\sum_k (\Delta t_k)^2 \leq \max_l \{\Delta t_l\} \sum_k \Delta t_k$ converges to zero as $\Delta(p_n) \to 0$. Hence, $\sum_{k=1}^{m_n} \Delta B'_k \, \Delta B_k$ converges in m.s. to $\theta\, t$ so that the m.s. limit of $\tilde{J}_{B,n}(B)$ exists and is

$$\underset{n\to\infty}{\text{l.i.m.}}\ \tilde{J}_{B,n}(B) = \frac{1}{2}\,(B(t)^2 - t) + \theta\, t = \frac{1}{2}\, B(t)^2 + (\theta - 1/2)\, t.$$

This limit coincides with the Itô integral for $\theta = 0$ but differs from this integral for $\theta \neq 0$. The Stratonovich integral results for $\theta = 1/2$. ■

The stochastic integrals in Examples 4.1 and 4.2 can be extended to define stochastic processes. Take a fixed $t > 0$ and set

$$M(s) = \int_0^t 1_{[0,s]}(u)\, B(u)\, dB(u) = \int_0^s B(u)\, dB(u) \quad \text{and} \tag{4.6}$$

$$\tilde{M}(s) = \int_0^t 1_{[0,s]}(u)\, B(u) \circ dB(u) = \int_0^s B(u) \circ dB(u), \tag{4.7}$$

where $s \in [0, t]$. These integrals can also be defined, for example, as the m.s. limits of the sums

$$J_{B,n}(B)(s) = \sum_{k=1}^{m_n} B(t_{k-1}^{(n)}) \left[B(t_k^{(n)} \wedge s) - B(t_{k-1}^{(n)} \wedge s) \right] \quad \text{and} \tag{4.8}$$

$$\tilde{J}_{B,n}(B)(s) = \sum_{k=1}^{m_n} B(t_k'^{(n)}) \left[B(t_k^{(n)} \wedge s) - B(t_{k-1}^{(n)} \wedge s) \right]. \tag{4.9}$$

Calculations as in Examples 4.1 and 4.2 can be used to show that $J_{B,n}(B)(s)$ and $\tilde{J}_{B,n}(B)(s)$ converge in m.s. as $n \to \infty$ to $(B(s)^2 - s)/2$ and $B(s)^2 + (\theta - 1/2)\, s$, respectively.

Example 4.3: Let $t > 0$ be a fixed time and M be the stochastic process in Eq. 4.6. Then (1) $E[M(s)] = 0$, (2) $E[M(s)^2] = s^2/2$, (3) $M(s) = M(\sigma) + \int_\sigma^s B(u)\, dB(u)$ for $\sigma \in (0, s)$, (4) M has continuous samples, and (5) M is a square integrable martingale with respect to the natural filtration $\mathcal{F}_t = \sigma(B(s), 0 \leq s \leq t)$ of the Brownian motion B. ◇

Proof: The stated properties of M follow from Eq. 4.3 properties of the Brownian motion. However, we present here a direct proof of these properties.

With the notations $B_{k-1} = B(t_{k-1}^{(n)})$, $\Delta B_k = B(t_k^{(n)} \wedge s) - B(t_{k-1}^{(n)} \wedge s)$, and $\mathcal{F}_k = \mathcal{F}_{t_k^{(n)}}$, the first moment of $J_{B,n}(B)(s)$ is

$$E\left[\sum_k B_{k-1}\, \Delta B_k \right] = \sum_k E\left\{ E[B_{k-1}\, \Delta B_k \mid \mathcal{F}_{k-1}] \right\} = E\left\{ B_{k-1}\, E[\Delta B_k] \right\} = 0$$

for each n and $s \in [0, t]$ because $\Delta B_k = 0$ for $s \le t_{k-1}^{(n)}$ and $E[\Delta B_k \mid \mathcal{F}_{k-1}] = E[\Delta B_k] = 0$ for $s > t_{k-1}^{(n)}$. The second moment of $J_{B,n}(B)(s)$ is

$$\begin{aligned}
E[(J_{B,n}(B)(s))^2] &= E\left[\sum_{k,l} B_{k-1}\, B_{l-1}\, \Delta B_k\, \Delta B_l\right] \\
&= \sum_k E[B_{k-1}^2\, (\Delta B_k)^2] + 2 \sum_{k>l} E[B_{k-1}\, B_{l-1}\, \Delta B_k\, \Delta B_l] \\
&= \sum_k E\left\{E[B_{k-1}^2\, (\Delta B_k)^2 \mid \mathcal{F}_{k-1}]\right\} + 2 \sum_{k>l} E\left\{E[B_{k-1}\, B_{l-1}\, \Delta B_k\, \Delta B_l \mid \mathcal{F}_{k-1}]\right\} \\
&= \sum_k E\left\{B_{k-1}^2\, E[(\Delta B_k)^2]\right\} + 2 \sum_{k>l} E\left\{B_{k-1}\, B_{l-1}\, \Delta B_l\, E[\Delta B_k]\right\} \\
&= \sum_k t_{k-1}^{(n)} \left(t_k^{(n)} \wedge s - t_{k-1}^{(n)} \wedge s\right)
\end{aligned}$$

since B_{k-1} and B_{l-1} are $\mathcal{F}_{k-1}$-measurable for $k > l$, ΔB_k is independent of $\mathcal{F}_{k-1}$, $E[\Delta B_k] = 0$, and $E[(\Delta B_k)^2] = t_k^{(n)} \wedge s - t_{k-1}^{(n)} \wedge s$. Hence,

$$E\left[\left(J_{B,n}(B)(s)\right)^2\right] = \sum_k t_{k-1}^{(n)} \left(t_k^{(n)} \wedge s - t_{k-1}^{(n)} \wedge s\right)$$

for each n and $s \in [0, t]$ and converges to $\int_0^s u\, du = s^2/2$ as the partition mesh approaches zero.

The third property of M follows from

$$\int_0^s B(u)\, dB(u) = \int_0^t 1_{[0,\sigma]}(u)\, B(u)\, dB(u) + \int_0^t 1_{[\sigma,s]}(u)\, B(u)\, dB(u)$$

and the linearity of $J_{B,n}(B)(\cdot)$ with respect to the integrand, where $0 \le \sigma \le s \le t$.

A proof that M is sample continuous can be found in [42] (Theorem 2.6, p. 38). The proof starts with the observation that $J_{B,n}(B)(\cdot)$ is a sample continuous process since

$$J_{B,n}(B)(s) = J_{B,n}(B)(t_{k-1}^{(n)}) + B(t_{k-1}^{(n)})\, [B(s) - B(t_{k-1}^{(n)})]$$

for $s \in (t_{k-1}^{(n)}, t_k^{(n)})$.

It is left to show that M is a square integrable martingale. This result also follows from Eq. 4.3. Here we only sketch a direct proof of this property. We show that $J_{B,n}(B)$ is a square integrable, that is, $E[(J_{B,n}(B)(s))^2] < \infty$, $s \in [0, t]$, $J_{B,n}(B)(s) \in \mathcal{F}_s$, $0 \le s \le t$, and $E[J_{B,n}(B)(s) \mid \mathcal{F}_\sigma] = J_{B,n}(B)(\sigma)$ for $s \ge \sigma$ (Section 3.11). We have already shown that $E\left[\left(J_{B,n}(B)(s)\right)^2\right]$ is finite. That $J_{B,n}(B)(s)$ is $\mathcal{F}_s$-adapted follows from the definition of $J_{B,n}(B)(s)$. The process $J_{B,n}(B)(s)$ has the martingale property $E[J_{B,n}(B)(s) \mid \mathcal{F}_\sigma] = J_{B,n}(B)(\sigma)$, $s \ge \sigma$, since

$$E[J_{B,n}(B)(s) \mid \mathcal{F}_\sigma] = E[J_{B,n}(B)(s) - J_{B,n}(B)(\sigma) \mid \mathcal{F}_\sigma] + J_{B,n}(B)(\sigma)$$

and

$$E[J_{B,n}(B)(s) - J_{B,n}(B)(\sigma) \mid \mathcal{F}_\sigma] = \sum_q E[B_{q-1}\,\Delta B_q \mid \mathcal{F}_\sigma]$$
$$= \sum_q E\left\{E[B_{q-1}\,\Delta B_q \mid \mathcal{F}_{q-1}] \mid \mathcal{F}_\sigma\right\} = \sum_q E\left\{B_{q-1}\,E[\Delta B_q] \mid \mathcal{F}_\sigma\right\} = 0,$$

where B_{q-1} and ΔB_q have the same meaning as in the previous equations but are restricted to the interval $[\sigma, s]$. The above equalities hold by properties of the Brownian motion. Because the martingale property is preserved by L_2-limits, M given by Eq. 4.6 is a square integrable martingale ([42], Proposition 1.3, p. 13). ■

Example 4.4: Consider a fixed time t and the stochastic process $\tilde{M}$ in Eq. 4.7. This process, that is, the Stratonovich integral of B with respect to B on $[0, s]$, is not a martingale with respect to the natural filtration of B. ◇

Proof: Properties of the sequence of sums defining the Stratonovich integral can be used to show that $\tilde{M}$ is not a martingale, as was done in the previous example. An alternative approach is taken here. We note that $\tilde{M}(t) = \int_0^t B \circ dB = B(t)^2/2$ (Eq. 4.5) cannot be an $\mathcal{F}_t = \sigma(B(s), 0 \le s \le t)$-martingale since its expectation $E[B(t)^2/2] = t/2$ is not constant. ■

Example 4.5: The sample paths of the Itô and Stratonovich integrals, $\int_0^s B\,dB$ and $\int_0^s B \circ dB$, differ and the difference between these integrals increases in time. Figure 4.1 shows a sample of $\int_0^s B\,dB$ and a sample of $\int_0^s B \circ dB$ corresponding to the same sample path $B(\cdot, \omega)$ of a Brownian motion B, that is, the functions $(B(t, \omega)^2 - t)/2$ and $B(t, \omega)^2/2$, respectively.

Samples of the Itô and the Stratonovich integrals corresponding to the same sample $B(\cdot, \omega)$ of a Brownian motion B are also shown in Fig. 4.2 with solid lines. The dotted lines in this figure give sample paths of $J_{B,n}(B)(\cdot)$ and $\tilde{J}_{B,n}(B)(\cdot)$ for two partitions p_n of $[0, t]$ with points $t_k^{(n)} = k\,t/m_n, k = 0, 1, \ldots, m_n, m_n = 10$, and $m_n = 100$ obtained from the same sample $B(\cdot, \omega)$ of B. These plots suggest that the sample paths of $J_{B,n}(B)(\cdot)$ and $\tilde{J}_{B,n}(B)(\cdot)$ approach the sample paths of the Itô and Stratonovich integrals, respectively, as n increases. ◇

Example 4.6: Consider the random sequence

$$J_{N,n}(N) = \sum_{k=1}^{m_n} N(t_{k-1}^{(n)})\,[N(t_k^{(n)}) - N(t_{k-1}^{(n)})],$$

where $p_n = (0 = t_0^{(n)} < t_1^{(n)} < \cdots < t_{m_n}^{(n)} = t)$ is a sequence of partitions of $[0, t]$ such that $\Delta(p_n) \to 0$ as $n \to \infty$ and N is a Poisson process with intensity $\lambda > 0$. Then $J_{N,n}(N)$ converges in m.s. and a.s. as $n \to \infty$. The limit of $J_{N,n}(N)$,

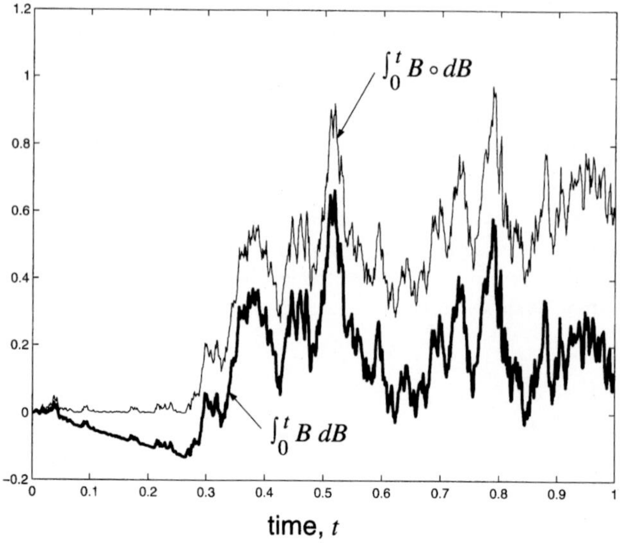

Figure 4.1. Sample paths of Itô and Stratonovich integrals $\int_0^t B\,dB$ and $\int_0^t B \circ dB$

denoted by $\int_0^t N(s-)\,dN(s)$ or $\int_0^t N_-\,dN$ and called the **Itô integral** of N_- with respect to N on $[0, t]$, is

$$\int_0^t N_-(s)\,dN(s) = \int_0^t N(s-)\,dN(s) = \frac{1}{2}\left(N(t)^2 - N(t)\right). \tag{4.10}$$

Because N has càdlàg samples, $N_-(s) = N(s-) = \lim_{u \uparrow s} N(u)$ is a càglàd process. The Itô integral $\int N_-\,dN$ coincides with the path by path Riemann-Stieltjes integral and is a martingale.

Figure 4.3 shows a sample of a Poisson process N and the corresponding sample of the stochastic integral $\int_0^t N_-\,dN$. The values of the sample of $\int N_-\,dN$ at the jump times $T_1 = 3$, $T_2 = 4$, and $T_3 = 8$ of the sample of N are $N(3-)\,\Delta N(3) = 0$, $0 + N(4-)\,\Delta N(4) = 0 + (1)(1) = 1$, $1 + N(8-)\,\Delta N(8) = 1 + (2)(1) = 3$, respectively, where $\Delta N(t) = N(t) - N(t_-)$. ◇

Proof: We first show that the Itô and the path by path Riemann-Stieltjes integrals of N_- with respect to N coincide. Consider a classical Riemann-Stieltjes integral $\int_a^b \alpha(x)\,d\beta(x)$, where β is a step function with jumps β_k at x_k, $k = 1, 2, \ldots, n$. If α and β are such that they are not discontinuous from the right or from the left at each x_k simultaneously, then ([5], Theorem 7.11, p. 148)

$$\int_a^b \alpha(x)\,d\beta(x) = \sum_{k=1}^{n} \alpha(x_k)\,\beta_k.$$

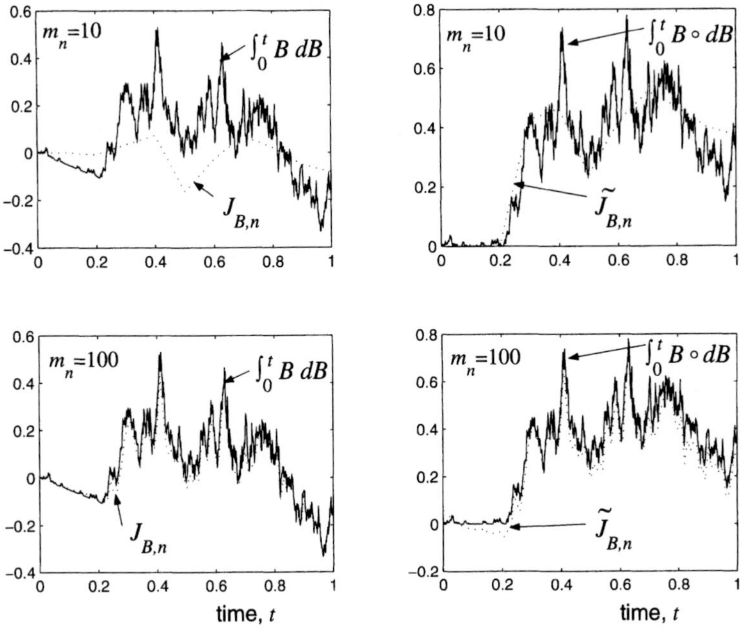

Figure 4.2. Samples of the approximations $J_{B,n}(B)$ and $\tilde{J}_{B,n}(B)$ of $\int B\, dB$ and $\int B \circ dB$ for $m_n = 10$ and $m_n = 100$

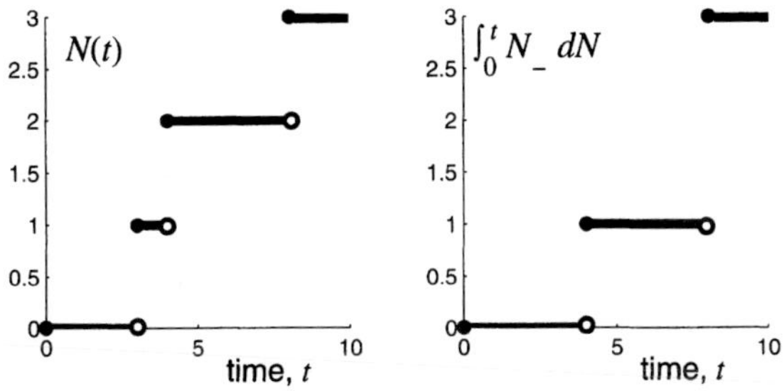

Figure 4.3. A hypothetical sample of the Poisson process N and the corresponding sample of the Itô integral $\int N_-\, dN$

This result implies

$$\int_0^t N(s-, \omega)\, dN(s, \omega) = \sum_{k=1}^{N(t,\omega)} N(T_{k-1}(\omega), \omega)\, [N(T_k(\omega), \omega) - N(T_{k-1}(\omega), \omega)]$$
$$= \sum_{k=1}^{N(t,\omega)} (k-1) = \frac{1}{2}(N(t, \omega)^2 - N(t, \omega)),$$

where T_k, $k = 1, 2, \ldots$, denote the jump times of N and $T_0 = 0$. Hence, the path by path integral $\int_0^t N(s-, \omega)\, dN(s, \omega)$ coincides with the Itô integral in agreement with a result in [147] (Theorem 17, p. 54), and $J_{N,n}(N)$ converges a.s. to $\int_0^t N_-\, dN$ as $n \to \infty$.

It remains to show that $J_{N,n}(N)$ converges in m.s. to $\int_0^t N_-\, dN$ as $n \to \infty$. Note that $J_{N,n}(N)$ can be written as

$$J_{N,n}(N) = \frac{1}{2} \sum_{k=1}^{m_n} \left(N(t_k^{(n)})^2 - N(t_{k-1}^{(n)})^2 \right) - \frac{1}{2} \sum_{k=1}^{m_n} \left(N(t_k^{(n)}) - N(t_{k-1}^{(n)}) \right)^2$$
$$= \frac{1}{2} N(t)^2 - \frac{1}{2} \sum_{k=1}^{m_n} \left(N(t_k^{(n)}) - N(t_{k-1}^{(n)}) \right)^2$$

so that $J_{N,n}(N)$ is an increasing sequence such that $0 \leq J_{N,n}(N) \leq N(t)^2/2$ a.s. for each n. Hence, we have $J_{N,n}(N) \stackrel{\text{m.s.}}{\longrightarrow} \int_0^t N_-\, dN$ as $n \to \infty$ by dominated convergence and the a.s. convergence of $J_{N,n}(N)$ to $\int_0^t N_-\, dN$ (Section 2.5.2.2)

We also present here an alternative proof for the m.s. convergence of $J_{N,n}(N)$ to $\int_0^t N_-\, dN = (N(t)^2 - N(t))/2$ based on direct calculations and elementary arguments. We need to show that the expectation of the square of

$$J_{N,n}(N) - \frac{1}{2}(N(t)^2 - N(t)) = \frac{1}{2} \sum_k \left[\Delta N_k - (\Delta N_k)^2 \right]$$

converges to zero as $n \to \infty$, where $\Delta N_k = N(t_k^{(n)}) - N(t_{k-1}^{(n)})$, that is, the sum

$$\sum_{k,l} E\left[\left(\Delta N_k - (\Delta N_k)^2 \right) \left(\Delta N_l - (\Delta N_l)^2 \right) \right]$$
$$= \sum_{k,l} E\left[\Delta N_k\, \Delta N_l - \Delta N_k\, (\Delta N_l)^2 - \Delta N_l\, (\Delta N_k)^2 + (\Delta N_k)^2\, (\Delta N_l)^2 \right]$$
$$= \lambda^4 \sum_{k,l} (\Delta t_k)^2\, (\Delta t_l)^2 - 2\,\lambda^2 \sum_k (\Delta t_k)^2 + 4\,\lambda^3 \sum_k (\Delta t_k)^3$$

converges to zero as $\Delta t_k \to 0$, where $\Delta t_k = t_k^{(n)} - t_{k-1}^{(n)}$. The first four moments of the increments of N used to establish the above expression can be obtained from the observation that the cumulants χ_r of any order r of ΔN_k are $\lambda\, \Delta t_k$ and the relationships $\mu_1 = \chi_1$, $\mu_2 = \chi_2 + \chi_1^2$, $\mu_3 = \chi_3 + 3\, \chi_1\, \chi_2 + \chi_1^3$, and $\mu_4 = \chi_4 + 3\, \chi_2^2 + 4\, \chi_1\, \chi_3 + 6\, \chi_1^2\, \chi_2 + \chi_1^4$ between the moments μ_r of order r and the cumulants of a random variable ([79], p. 83 and p. 377). The cumulant of order r of a random variable with characteristic function φ is equal to the value of $(\sqrt{-1})^{-r}\, d^r \varphi(u)/du^r$ at $u = 0$. ■

4.4 Stochastic integrals

This section (1) provides additional information on semimartingales because of their close relation to stochastic integrals, (2) defines the stochastic integral for a class of simple predictable integrands and semimartingale integrators,

(3) extends this definition to stochastic integrals with càglàd integrands and semimartingale integrators, and (4) gives properties of the stochastic integral that are useful for calculations. The presentation in this and the following two sections is based on [147].

The Itô integrals $\int_0^t B\,dB$ and $\int_0^t N_-\,dN$ in the previous section are special cases of the stochastic integral defined here. The stochastic integral differs significantly from the integrals in Section 3.9.3. The integrals in Section 3.9.3 (1) have random integrand and deterministic integrator or vice versa and (2) the random integrand/integrator of these integrals are in L_2. In contrast, both the integrand and integrator of the stochastic integral considered here are stochastic processes whose expectations may not exist.

4.4.1 Semimartingales

Let $(\Omega, \mathcal{F}, (\mathcal{F}_t)_{t\geq 0}, P)$ be a filtered probability space. We denote by $\mathcal{S}$, $\mathcal{D}$, and $\mathcal{L}$ the class of **simple predictable, $\mathcal{F}_t$-adapted with càdlàg paths**, and **$\mathcal{F}_t$-adapted with càglàd paths** processes, respectively.

Let $0 = T_0 \leq T_1 \leq \cdots \leq T_{n+1} < \infty$ be a finite sequence of $\mathcal{F}_t$-stopping times and H_i denote random variables such that $H_i \in \mathcal{F}_{T_i}$ and $|H_i| < \infty$ a.s., $i = 0, 1, \ldots, n$, that is, $|H_i(\omega)| < \infty$ for $\omega \in \Omega \setminus \Omega_0$ and $P(\Omega_0) = 0$.

A process $\{H(t), t \geq 0\}$ is said to be **simple predictable** if

$$H(t) = H_0\, 1_{\{0\}}(t) + \sum_{i=1}^{n} H_i\, 1_{(T_i, T_{i+1}]}(t). \tag{4.11}$$

Note: We have $\mathcal{S} \subset \mathcal{L}$ since the processes in $\mathcal{S}$ are $\mathcal{F}_t$-adapted with càglàd, piece-wise constant sample paths.

The classes of processes $\mathcal{S}$, $\mathcal{D}$, and $\mathcal{L}$ are vector spaces. For example, the sum of two $\mathcal{F}_t$-adapted processes with càdlàg paths is an $\mathcal{F}_t$-adapted process with càdlàg paths. Also, multiplying an $\mathcal{F}_t$-adapted process with càdlàg paths with a real constant does not change its type, so that $\mathcal{D}$ is a vector space. ▲

Consider a real-valued stochastic process X defined on a filtered probability space $(\Omega, \mathcal{F}, (\mathcal{F}_t)_{t\geq 0}, P)$ and let L_0 denote the collection of real-valued random variables defined on this space.

Let $I_X : \mathcal{S} \to L_0$ be a mapping defined by

$$I_X(H) = H_0\, X(0) + \sum_{i=1}^{n} H_i\, [X(T_{i+1}) - X(T_i)]. \tag{4.12}$$

Note: The mapping I_X is linear and its range is L_0 since it is a finite sum of random variables H_i weighted by increments of X. We also note that $I_X(H)$ resembles $J_{B,n}(B)$

and $J_{N,n}(N)$ used in Examples 4.1 and 4.6 to define the stochastic integrals $\int B\,dB$ and $\int N_-\,dN$, respectively. ▲

To check whether I_X is a continuous mapping, we need to define the meaning of convergence in $\mathcal{S}$ and L_0. A sequence H^n in $\mathcal{S}$ is said to **converge uniformly in** (t, ω) to $H \in \mathcal{S}$ if, for any $\varepsilon > 0$, there is an integer $n_\varepsilon \geq 1$ independent of (t, ω) such that $|H^n(t, \omega) - H(t, \omega)| < \varepsilon$ for $n \geq n_\varepsilon$ or, equivalently, $\sup_{\omega \in \Omega, t \geq 0} |H^n(t, \omega) - H(t, \omega)| \to 0$, $n \to \infty$. The convergence in probability is used in L_0.

- A stochastic process X is a **total semimartingale** if (1) $X \in \mathcal{D}$ and (2) I_X in Eq. 4.12 is continuous from $\mathcal{S}$ to L_0.
- A process X is a **semimartingale** if X^t is a total semimartingale for each $t \in [0, \infty)$.

Note: The continuity of I_X means that the uniform convergence $H^n \to H$ in $\mathcal{S}$ implies the convergence $I(H^n) \xrightarrow{\text{pr}} I_X(H)$ in L_0 as $n \to \infty$.

The definition of the (total) semimartingale shows that the collection of (total) semimartingales is a vector space because, if X_1 and X_2 are (total) semimartingales and c_1, c_2 are some constants, then $c_1 X_1 + c_2 X_2$ is a (total) semimartingale. ▲

Example 4.7: Let

$$C(t) = \sum_{k=1}^{N(t)} Y_k, \quad t \geq 0, \tag{4.13}$$

be a compound Poisson process, where N is a Poisson process with respect to a filtration $\mathcal{F}_t$ and the iid random variables Y_k are $\mathcal{F}_0$-measurable. Then $C_-(t) = C(t-) = \lim_{s \uparrow t} C(s)$ is a simple predictable process. ◇

Proof: Recall that a Poisson process N is an adapted counting process without explosions that has stationary independent increments (Section 3.12). We have

$$\{\omega : T_n(\omega) \leq t\} = \{\omega : N(t, \omega) \geq n\} \in \mathcal{F}_t \quad \text{for each } t,$$

so that the jump times T_i of N are $\mathcal{F}_t$-stopping times (see also [147], Theorem 22, p. 14). We have $H_i = \sum_{j=1}^{i} Y_j \in \mathcal{F}_0 \subseteq \mathcal{F}_{T_i}$, $i \geq 1$, $|H_i| < \infty$ a.s., and $T_{n+1} < \infty$ a.s. for each n. The jump times are finite because $P(T_{n+1} > t) \leq E[T_{n+1}]/t$ by Chebyshev's inequality and $E[T_{n+1}] = (n+1)/\lambda$ so that $P(T_{n+1} > t)$ approaches zero as $t \to \infty$, where λ denotes the intensity of the Poisson process. Hence, C_- is in $\mathcal{S}$. ■

Example 4.8: Set $H = C_-$ and $X = B$ in Eq. 4.12, where C_- is defined in the previous example and B is a Brownian motion assumed to be independent of C. Figure 4.4 shows a sample of C_- and B. The random variable in Eq. 4.12 is $I_X(H) = H_1\,[B(T_2) - B(T_1)] + H_2\,[B(T_3) - B(T_2)]$ for $n = 2$. If C has unit jumps, that is, C is a Poisson process, then

$$P(I_X(H) \leq \xi) = \lambda^2 \int_0^\infty \int_0^\infty \Phi\left(\frac{\xi}{\sqrt{\eta_1 + 4\,\eta_2}}\right) e^{-\lambda\,(\eta_1 + \eta_2)}\,d\eta_1\,d\eta_2,$$

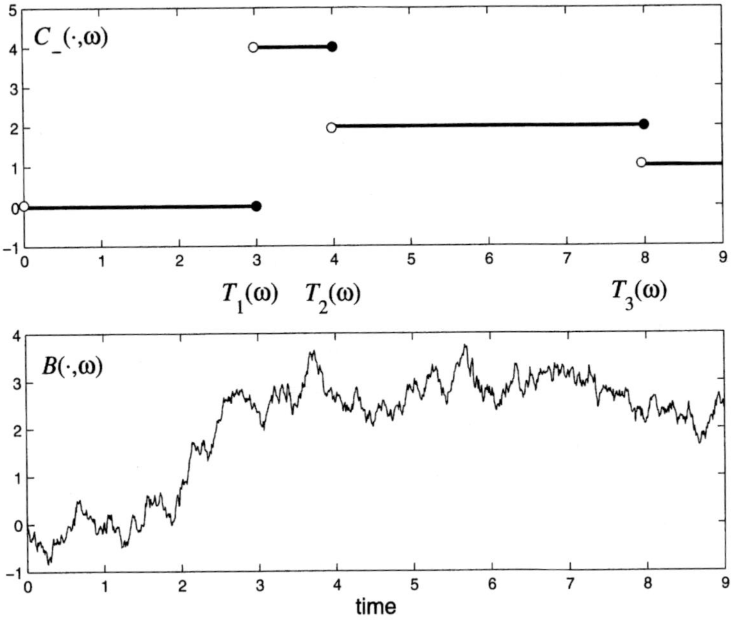

Figure 4.4. A sample of C_- and a sample of B used to calculate $I_X(H) = I_B(C_-)$

where Φ denotes the distribution of the standard Gaussian variable. Figure 4.5 shows the distribution of $I_X(H)$ for $n = 2$ and two values of the intensity parameter λ of the Poisson process. ◇

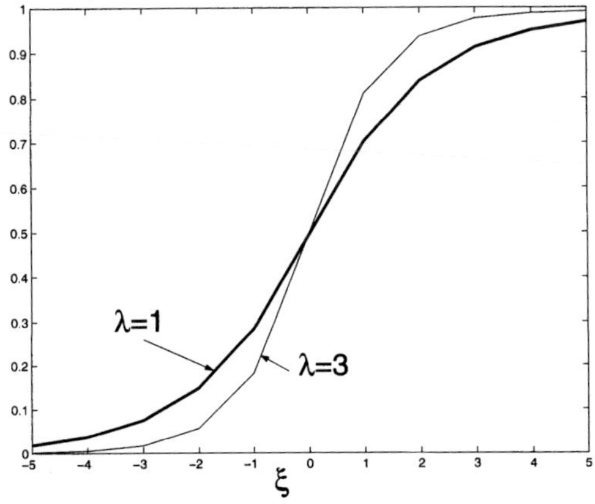

Figure 4.5. The distribution of $I_X(H)$ defined in Fig. 4.4

Note: The increments $B(T_{i+1}) - B(T_i)$ of the Brownian motion are independent Gaussian variables with mean zero and variance $T_{i+1} - T_i$ conditional on the jump times T_i. Hence, $I_X(H)$ is a zero mean Gaussian variable with variance $(T_2 - T_1) + 4\,(T_3 - T_2)$ conditional on the jump times. The formula for $P(I_X(H) \leq \xi)$ follows from the distribution $P(T > \eta) = \exp(-\lambda\,\eta)$ of the time T between consecutive jumps of N. This distribution depends only on η and λ since N has stationary independent increments. ▲

Generally, it is difficult to establish from the definition whether a process is a semimartingale. However, relatively simple criteria are available for identifying semimartingales. Before giving some of these criteria we give a definition. A process is said to have paths of **finite variation on compacts** if almost all its samples are of bounded variation on each compact interval of $[0, \infty)$. Recall that in Section 3.9.3 we have defined the variation of a real-valued deterministic function. Now we state some criteria for identifying semimartingales.

- If X is an adapted process with càdlàg paths of finite variation on compacts, then X is a semimartingale ([147], Theorem 7, p. 47).
- If X is a square integrable martingale with càdlàg paths, then X is a semimartingale ([147], Theorem 8, p. 47).
- If X is a locally square integrable local martingale with càdlàg paths, then X is a semimartingale ([147], Corollary 1, p. 47).
- If X is a local martingale with continuous paths, then X is a semimartingale ([147], Corollary 2, p. 47).

Note: Local martingales were defined in Section 3.11. A (local) martingale X is **locally square integrable** if there exists an increasing sequence T_n, $n \geq 1$, of stopping times such that $\lim_{n\to\infty} T_n = \infty$ a.s. and such that for every $n \geq 1$ the process X^{T_n} is a square integrable martingale.

These criteria show that a Brownian motion B is a semimartingale because B has continuous paths and martingales are local martingales. A compound Poisson process C is also a semimartingale because C is an adapted process with càdlàg paths of finite variation on any bounded time interval $(0, t]$. We have also seen that C is a martingale if its jumps Y_k are in L_1 and have mean zero. ▲

At the end of Section 3.14.2 we introduced briefly a representation for semimartingales that is very useful in applications. We state again this representation here.

An adapted, càdlàg process X is **decomposable** if there exists a locally square integrable martingale M and an adapted process A with càdlàg sample paths of finite variation on compacts such that $M(0) = A(0) = 0$ and ([147], p. 48)

$$X(t) = X(0) + M(t) + A(t). \tag{4.14}$$

• An adapted, càdlàg process X is a **classical semimartingale** if there exists a local martingale M and an adapted process A with càdlàg sample paths of finite variation on compacts such that $M(0) = A(0) = 0$ and

$$X(t) = X(0) + M(t) + A(t). \tag{4.15}$$

• An adapted, càdlàg process X is a semimartingale if and only if X is a classical semimartingale ([147], Theorem 14, p. 105 and Theorem 22, p. 114).

Note: We have used the representation of Lévy processes in Section 3.14.2 to show that these processes are classical semimartingales and therefore semimartingales. The criteria stated previously in this section show that decomposable processes are classical semimartingales. Similar considerations can be applied to show that other processes are semimartingales. For example, a compound Poisson process $C(t) = \sum_{k=1}^{N(t)} Y_k$ with $Y_1 \in L_1$ is a semimartingale since

$$C(t) = (C(t) - \lambda\, t\, E[Y_1]) + \lambda\, t\, E[Y_1] = \tilde{C}(t) + \lambda\, t\, E[Y_1],$$

the compensated compound Poisson process $\tilde{C}(t) = C(t) - \lambda\, t\, E[Y_1]$ is a martingale, and $\lambda\, t\, E[Y_1]$ is adapted with continuous samples of finite variation on compacts (Eq. 4.15). ▲

Example 4.9: The square of a Brownian motion B, that is, the process $X(t) = B(t)^2$, $t \geq 0$, is a semimartingale. ◇

Note: The process X can be given in the form $X(t) = M(t)+A(t)$, where $M(t) = B(t)^2 - t$ is a square integrable martingale, $A(t) = t$ has continuous samples of finite variation on compacts, and $M(0) = A(0) = 0$. Hence, X is a semimartingale.

We have seen that B^2 is a submartingale. That B^2 is a semimartingale also results from the fact that submartingales are semimartingales. Let Y be a submartingale. Then Y is a semimartingale since $Y(t) = (Y(t) - E[Y(t)]) + E[Y(t)]$, $Y(t) - E[Y(t)]$ is a martingale, and $E[Y(t)]$ is an increasing function so that it is of finite variation on compacts. ▲

4.4.2 Simple predictable integrands

The mapping in Eq. 4.12 gives the random variable $I_X(H) \in L_0$. We extend this mapping to define a stochastic process $J_X(H)(t)$, $t \geq 0$, called stochastic integral, and derive properties of this process.

If $H \in \mathcal{S}$, $X \in \mathcal{D}$, and $0 = T_0 \leq T_1 \leq \cdots \leq T_{n+1} < \infty$ are stopping times, the **stochastic integral of H with respect to X on** $[0, t]$ is

$$J_X(H)(t) = H_0\, X(0) + \sum_{i=1}^{n} H_i \left[X(t \wedge T_{i+1}) - X(t \wedge T_i)\right], \quad t \geq 0. \tag{4.16}$$

Alternative notations for the stochastic integral $J_X(H)(t)$ defined by Eq. 4.16 are $\int_0^t H(s)\, dX(s)$, $\int_0^t H\, dX$, and $(H \cdot X)(t)$. If the time $t = 0$ is excluded from the definition in Eq. 4.16, the resulting stochastic integral, denoted by $\int_{0+}^{t} H\, dX$, is

$$\int_{0+}^{t} H(s)\, dX(s) = \sum_{i=1}^{n} H_i \left[X(t \wedge T_{i+1}) - X(t \wedge T_i)\right]$$

so that (Eq. 4.16)

$$\int_{0}^{t} H(s)\, dX(s) = H_0\, X(0) + \int_{0+}^{t} H(s)\, dX(s). \qquad (4.17)$$

Figure 4.6 illustrates the definition of $J_X(H)(\cdot)$ in Eq. 4.16 for $t \in [0, 8]$.

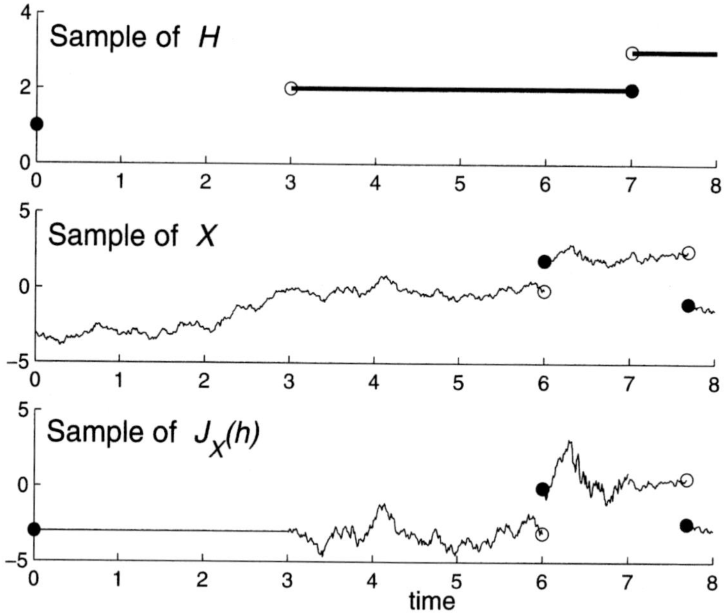

Figure 4.6. A sample of $H \in \mathcal{S}$, a sample of X, and the corresponding sample of $J_X(H)(\cdot)$ in Eq. 4.16

The sample of the simple predictable process has $H_0 = 1$, $H_1 = 2$, $H_2 = 3$, $T_0 = 0$, $T_1 = 3$, and $T_2 = 7$. The sample of the càdlàg process X is a sample of a Brownian motion starting at $B(0) = -3$ that has been modified to have jumps of 2 and -3 at times $t = 6$ and $t = 7.7$, respectively. The corresponding sample of $J_X(H)(\cdot)$ in Fig. 4.6 shows that (1) only the jumps of the integrator result in jumps of the stochastic integral and (2) the stochastic integral has càdlàg sample paths. The remainder of this section gives additional properties of the stochastic integral in Eq. 4.16.

The mappings in Eqs. 4.12 and 4.16 are related by

$$J_X(H)(t) = I_{X^t}(H), \quad \text{where } X^t(\cdot) = X(t \wedge \cdot). \tag{4.18}$$

If $H, G \in \mathcal{S}$, $X_1, X_2 \in \mathcal{D}$, and α, β are some constants, then

$$\begin{aligned} J_X(\alpha\, H + \beta\, G)(t) &= \alpha\, J_X(H)(t) + \beta\, J_X(G)(t), \\ J_{\alpha\, X_1 + \beta\, X_2}(H)(t) &= \alpha\, J_{X_1}(H)(t) + \beta\, J_{X_2}(H)(t). \end{aligned} \tag{4.19}$$

If $0 \le s \le t$, $H \in \mathcal{S}$, and $X \in \mathcal{D}$, then

$$\int_0^t H(u)\, dX(u) = \int_0^s H(u)\, dX(u) + \int_{s+}^t H(u)\, dX(u), \tag{4.20}$$

where $\int_{s+}^t H(u)\, dX(u) = \int_0^t 1_{(s,t]}(u)\, H(u)\, dX(u)$.

Proof: The properties in Eqs. 4.18 and 4.19 follow from the definitions of $I_X(H)$ and $J_X(H)$, the linearity of these integrals with respect to their integrands and integrators, and the fact that $\mathcal{S}$ and $\mathcal{D}$ are vector spaces. For the property in Eq. 4.20 take $0 \le s \le t$ and note that (Eq. 4.19)

$$\begin{aligned} \int_0^t H\, dX &= H_0\, X(0) + \int_0^t \left[1_{(0,s]}(u) + 1_{(s,t]}(u)\right] H(u)\, dX(u) \\ &= H_0\, X(0) + \int_0^t 1_{(0,s]}(u)\, H(u)\, dX(u) + \int_0^t 1_{(s,t]}(u)\, H(u)\, dX(u) \\ &= H_0\, X(0) + \int_{0+}^s H\, dX + \int_{s+}^t H\, dX = \int_0^s H\, dX + \int_{s+}^t H\, dX \end{aligned}$$

as stated. ■

4.4.3 Adapted caglàd integrands

The class of integrands used to define the stochastic integral $J_X(H)$ in Eq. 4.16 is too restrictive for the type of stochastic integrals encountered in many stochastic problems. In this section we extend the definition of $J_X(H)$ to include integrands $H \in \mathcal{L}$ and integrators $X \in \mathcal{D}$. The extension requires a definition of convergence in the collection $\mathcal{S}$ of simple predictable processes $\mathcal{S}$, and is particularly useful for the case in which $X \in \mathcal{D}$ is a semimartingale.

A sequence of processes $H^n \in \mathcal{S}$ is said to converge to a process H **uniformly on compacts in probability**, and denote it by $H^n \xrightarrow{\text{ucp}} H$, if for each $t > 0$

$$\sup_{0 \le s \le t} |H^n(s) - H(s)| \xrightarrow{\text{pr}} 0, \quad \text{as } n \to \infty. \tag{4.21}$$

Note: The ucp convergence over a time interval $[0, t]$ implies the convergence in probability at an arbitrary time $\tau \in [0, t]$ since

$$\{\omega : |H^n(\tau, \omega) - H(\tau, \omega)| > \varepsilon\} \subseteq \{\omega : \sup_{0 \le s \le t} |H^n(s, \omega) - H(s, \omega)| > \varepsilon\}$$

for any $\varepsilon > 0$ and $\tau \in [0, t]$, $P(\sup_{0 \le s \le t} |H^n(s) - H(s)|) \to 0$ as $n \to \infty$, and probability is an increasing set function. ▲

The following two facts, given without proof, yield the desired generalization of the stochastic integral in Eq. 4.16.

1. The space $\mathcal{S}$ of simple predictable processes is dense in the space $\mathcal{L}$ of adapted, càglàd processes under the ucp convergence ([147], Theorem 10, p. 49).

2. If X is a semimartingale, the mapping $J_X : \mathcal{S} \to \mathcal{D}$ is continuous under the ucp convergence in both $\mathcal{S}$ and $\mathcal{D}$ ([147], Theorem 11, p. 50). The continuous linear mapping $J_X : \mathcal{L} \to \mathcal{D}$ extending $J_X : \mathcal{S} \to \mathcal{D}$ is called the **stochastic integral** of $H \in \mathcal{L}$ with respect to a semimartingale process X.

Note: These results show that any process $H \in \mathcal{L}$ can be approximated by a simple predictable process in the sense of the ucp convergence and suggest that most of the properties of the stochastic integral in Eq. 4.16 hold also for its extension $J_X : \mathcal{L} \to \mathcal{D}$.

The extended definition of the stochastic integral includes the integrals $\int_0^t B\,dB$ and $\int_0^t N_-\,dN$ in Examples 4.1 and 4.6 because their integrands, B and N_-, are in $\mathcal{L}$ and their integrators are semimartingales. ▲

4.4.4 Properties of stochastic integrals

We give here some essential properties of the stochastic integral $H \cdot X$ defined in Section 4.4.3, where $H \in \mathcal{L}$ and X is a semimartingale unless stated otherwise.

The **jump process** $\Delta(H \cdot X)(s)$ of the stochastic integral is indistinguishable from $H(s)\,\Delta X(s)$ ([147], Theorem 13, p. 53), that is, the sample paths of these two processes coincide with probability 1.

Note: Recall that $\Delta Y(t) = Y(t) - Y(t-)$ is the jump process of a càdlàg process Y. Our illustration of the definition of the stochastic integral in Fig. 4.6 shows that the jumps of the stochastic integral are equal to $H(T_i)\,\Delta X(T_i)$, where T_i, i=1,2,..., denote the jump times of the integrator, in agreement with the above property. ▲

If X has paths of finite variation on compacts, the stochastic integral $H \cdot X$ is indistinguishable from the corresponding path by path Riemann-Stieltjes integral ([147], Theorem 17, p. 54).

Proof: If H is a simple predictable process, the path by path Riemann-Stieltjes integral $\int_0^t H(s, \omega)\,dX(s, \omega)$ and the stochastic integral in Eq. 4.16 coincide. This property of the stochastic integral has been illustrated in Example 4.6 showing that the Itô integral $\int_0^t N_- \, dN$ and the path by path Riemann-Stieltjes integral $\int_0^t N(s-, \omega)\, dN(s, \omega)$ coincide.

If $H \in \mathcal{L}$, there exists a sequence H^n of processes in $\mathcal{S}$ converging to H in ucp so that we can find a subsequence H^{n_k} of H^n that converges a.s. to H (Section 2.13). The integrals $H^{n_k} \cdot X$ can be calculated as path by path Riemann-Stieltjes integrals, that is,

$$(H^{n_k} \cdot X)(t, \omega) = H_0^{n_k}(\omega)\, X(0, \omega) + \sum_i H_i^{n_k}(\omega) \left[X(t \wedge T_{i+1}, \omega) - X(t \wedge T_i, \omega)\right].$$

The statement in the theorem results by taking the limit as $k \to \infty$ and using properties of H^{n_k} and $H^{n_k} \cdot X$. ■

Preservation. The process $Y = H \cdot X$ is a semimartingale. Moreover, if G is a process in $\mathcal{L}$, then

$$G \cdot Y = G \cdot (H \cdot X) = (G\,H) \cdot X, \tag{4.22}$$

where $G \cdot Y$ denotes the stochastic integral of G with respect to Y, and $G\,H$ is the product of G and H ([147], Theorem 19, p. 55). The property in Eq. 4.22 is referred to as **associativity**.

Preservation. If $H \in \mathcal{L}$ is as stated at the beginning of this section and if

- X is a locally square integrable local martingale, then $H \cdot X$ is a locally square integrable local martingale ([147], Theorem 20, p. 56);
- X is a local martingale, then $H \cdot X$ is a local martingale ([147], Theorem 17, p. 106).

Note: We only show that, if H is a bounded simple predictable process and X is a square integrable martingale, then $H \cdot X$ is a square integrable martingale.

Assume $X(0) = 0$ without loss of generality since $X(\cdot) - X(0)$ is a martingale that starts at zero. The stochastic integral (Eq. 4.16)

$$(H \cdot X)(t) = \sum_{i=1}^{n} H_i \left[X(T_{i+1} \wedge t) - X(T_i \wedge t)\right]$$

is adapted. If H is bounded by $a > 0$, then $(H \cdot X)(t)$ is in L_2 for each t since

$$E[((H \cdot X)(t))^2] = E\left[\left(\sum_{i=1}^{n} H_i \left[X(T_{i+1} \wedge t) - X(T_i \wedge t)\right]\right)^2\right]$$
$$\leq a^2\, E\left[\sum_{i=1}^{n} \left(X(T_{i+1} \wedge t)^2 - X(T_i \wedge t)^2\right)\right]$$
$$= a^2 \left(E[X(T_{n+1} \wedge t)^2] - E[X(T_1 \wedge t)^2]\right)$$

as X is a square integrable martingale. The above calculations have used the martingale properties of X. It remains to show that $H \cdot X$ has the martingale property. For $s \leq t$ the integral $(H \cdot X)(t)$ is the sum of $(H \cdot X)(s)$ and terms $H_i\,\left[X(T_{i+1} \wedge t) - X(T_i \wedge t)\right]$ for $T_i > s$. The conditional expectation of these terms,

$$\begin{aligned} &E\left[H_i\,\left(X(T_{i+1} \wedge t) - X(T_i \wedge t)\right) \mid \mathcal{F}_s\right] \\ &= E\left\{E\left[H_i\,\left[X(T_{i+1} \wedge t) - X(T_i \wedge t)\right] \mid \mathcal{F}_{T_i}\right] \mid \mathcal{F}_s\right\} \\ &= E\left\{H_i\,E\left[X(T_{i+1} \wedge t) - X(T_i \wedge t) \mid \mathcal{F}_{T_i}\right] \mid \mathcal{F}_s\right\} \end{aligned}$$

is zero since $E\left[X(T_{i+1} \wedge t) - X(T_i \wedge t) \mid \mathcal{F}_{T_i}\right] = 0$ by the optional sampling theorem (Section 3.11.2) so that $H \cdot X$ has the property $E\left[(H \cdot X)(t) \mid \mathcal{F}_s\right] = (H \cdot X)(s)$.

Stochastic integrals with Brownian motion integrator are of the type considered here. For example, $\int_0^t B(s)\,dB(s) = \left(B(t)^2 - t\right)/2$ is a martingale that is also square integrable (Example 4.6) and $\int_0^t H(s)\,dB(s)$, $H \in \mathcal{L}$, is a square integrable martingale if $E\left[\int_0^t H(s)^2\,ds\right] < \infty$, as we will see later in this chapter (Eq. 4.36). ▲

We state without proof an additional property of the stochastic integral but first, we need to introduce a new concept. A sequence of stopping times $\sigma_n = (0 = T_0^{(n)} \le T_1^{(n)} \le \cdots \le T_{m_n}^{(n)})$ is said to **tend to the identity** if (1) $\lim_{n\to\infty} \sup_k T_k^{(n)} = \infty$ a.s. and (2) $\sup_k |T_{k+1}^{(n)} - T_k^{(n)}|$ converges a.s. to zero as $n \to \infty$.

If X is a semimartingale as stated at the beginning of this section, Y is a process in $\mathcal{D}$ or $\mathcal{L}$, and σ_n is a sequence of random partitions tending to the identity, then

$$\begin{aligned} \int_{0+}^{\cdot} Y_{\sigma_n}(s)\,dX(s) &= \sum_i Y(T_i^{(n)})(X(T_{i+1}^{(n)} \wedge \cdot) - X(T_i^{(n)} \wedge \cdot) \\ &\overset{\text{ucp}}{\longrightarrow} \int_{0+}^{\cdot} Y_-(s)\,dX(s), \qquad n \to \infty, \end{aligned} \tag{4.23}$$

where $Y_-(s) = Y(s-) = \lim_{s\uparrow t} Y(s)$, $Y_-(0) = 0$, and $Y_{\sigma_n}(t) = Y(0)\,1_{\{0\}} + \sum_i Y(T_i^{(n)})\,1_{(T_{i-1}^{(n)}, T_i^{(n)}]}$ ([147], Theorem 21, p. 57).

Note: This property holds also for a deterministic sequence of partitions. Numerical calculations in this book use almost exclusively deterministic rather than random partitions of the integration interval. There are some situations in which it is convenient to use random partitions ([115], Section 9.5). ▲

Example 4.10: Let N be a Poisson process with intensity $\lambda > 0$ and jump times $T_1, T_2, \ldots$, $M(t) = N(t) - \lambda\,t$ denote the associated compensated Poisson process, $G(t) = 1_{[0,T_1)}(t)$, and $H(t) = 1_{(0,T_1]}(t)$. The stochastic integral $H \cdot M$ is a martingale. However, $G \cdot M$ given by Eq. 4.16 with (G, M) in place of (H, X) is not a martingale. ◇

Proof: We have shown that M is a martingale so that it is also a semimartingale (Section 4.4.1). Note also that $G \in \mathcal{D}$ and $H \in \mathcal{L}$. The process in Eq. 4.16 with (H, X) replaced by (G, M) is

$$\int_0^t G(s)\,dM(s) = \int_0^t G(s)\,dN(s) - \lambda \int_0^t G(s)\,ds = -\lambda\,(t \wedge T_1)$$

so that $G \cdot M$ is not a martingale. On the other hand, Eq. 4.16 with (H, X) replaced by (H, M) yields

$$\int_0^t H(s)\, dM(s) = \int_0^t H(s)\, dN(s) - \lambda \int_0^t G(s)\, ds = N(t \wedge T_1) - \lambda\,(t \wedge T_1).$$

That $H \cdot X$ is a martingale follows from

$$E[(H \cdot M)(t) \mid \mathcal{F}_s] = E\left[N(t \wedge T_1) - \lambda\,(t \wedge T_1) \mid \mathcal{F}_s\right]$$
$$= N(s \wedge t \wedge T_1) - \lambda\,(s \wedge t \wedge T_1) = N(s \wedge T_1) - \lambda\,(s \wedge T_1) = (H \cdot M)(s)$$

for $0 \leq s \leq t$, where the second equality is given by the optional sampling theorem (Section 3.11.2) applied to the martingale $N(t) - \lambda t$. This example suggests that the integrand must have càglàd sample paths to assure that the stochastic integral preserves the martingale property of its integrator. ■

Example 4.11: Consider the stochastic integral $J(t) = \int_0^t C_-\, dB$, where C is a compound Poisson process (Eq. 4.13) and B denotes a Brownian motion that is independent of C. If the jumps Y_k of C are in L_2, then

$$J(t) = \int_0^t C(s-)\, dB(s) = \sum_{i=1}^{\infty} H_i\, [B(T_{i+1} \wedge t) - B(T_i \wedge t)]$$

is a square integrable martingale with mean zero and

$$E[J(t)\, J(s)] = \sum_{i=1}^{\infty} E[H_i^2]\, E[T_{i+1} \wedge t \wedge s - T_i \wedge t \wedge s],$$

where T_i denote the jump times of C and $H_i = \sum_{k=1}^{i} Y_k$. ◇

Proof: That J is a square integrable martingale follows from a preservation property of the stochastic integral. The first two moments of J are

$$E[J(t)] = \sum_{i=1}^{\infty} E[H_i]\, E[B(T_{i+1} \wedge t) - B(T_i \wedge t)] = 0 \quad \text{and}$$

$$E[J(t)\, J(s)] = \sum_{i,j=1}^{\infty} E[H_i\, H_j\, (B(T_{i+1} \wedge t) - B(T_i \wedge t))\, (B(T_{j+1} \wedge s) - B(T_j \wedge s))]$$
$$= \sum_{i=1}^{\infty} E[H_i^2]\, E[(B(T_{i+1} \wedge t \wedge s) - B(T_i \wedge t \wedge s))^2]$$
$$= \sum_{i=1}^{\infty} E[H_i^2]\, E[T_{i+1} \wedge t \wedge s - T_i \wedge t \wedge s],$$

where the last two equalities follow from properties of the increments of the Brownian motion B and the independence between C and B. ■

4.5 Quadratic variation and covariation

In this section, X and Y are semimartingales with the property $X(0-) = Y(0-) = 0$. We define the quadratic variation and covariation processes for semimartingales, give properties of these processes that are useful for applications, and illustrate how the quadratic variation and covariation processes can be used to calculate stochastic integrals. The variation and covariation processes are essential for establishing the Itô formula (Section 4.6). In Section 3.11.4 we have defined variation and covariation processes for martingales.

4.5.1 Definition

The **quadratic variation** process of X, denoted by $[X, X]$ or $[X]$, is defined by

$$[X, X](t) = X(t)^2 - 2\int_0^t X(s-)\, dX(s). \tag{4.24}$$

Note: The stochastic integral $\int_0^t X(s-)\, dX(s)$ is defined since X is a semimartingale so that X_- is in $\mathcal{L}$, where $X_-(s) = X(s-) = \lim_{u \uparrow s} X(u)$. ▲

The **path by path continuous part** $[X, X]^c$ **of** $[X, X]$ is defined by

$$\begin{aligned}[X, X](t) &= [X, X]^c(t) + X(0)^2 + \sum_{0<s\le t} (\Delta X(s))^2 \\ &= [X, X]^c(t) + \sum_{0\le s\le t} (\Delta X(s))^2.\end{aligned} \tag{4.25}$$

Note: It can be shown that any semimartingale X has a unique continuous local martingale part X^c and that $[X^c, X^c] = [X, X]^c$ ([147], p. 63). For example, X^c for the semimartingale $X = B + C$ is $X^c = B$ so that $[X, X]^c(t) = t$, $[X, X](t) = t + \sum_{k=1}^{N(t)} Y_k^2$, and $[X, X](t) - [X, X]^c(t) = \sum_{k=1}^{N(t)} Y_k^2 = \sum_{0\le s\le t}(\Delta C(s))^2$, where B and C denote a Brownian motion and a compound Poisson process, respectively, and N, Y_k are defined by Eq. 4.13. The process $[X, X](t) - [X, X]^c(t)$ is given by the sum of the squares of all jumps of C in $[0, t]$. This result holds for any semimartingale X (Eq. 4.25) because semimartingales can have only jump discontinuities and the number of jumps is at most countable.

We also note that the increment $\Delta X(0) = X(0) - X(0-)$ of X at $t = 0$ is $X(0)$ since $X(0-) = 0$ by assumption so that $[X, X](0) = X(0)^2$ and $[X, X]^c(0) = 0$. ▲

A semimartingale X with $[X, X]^c(t) = 0$ is said to be a **quadratic pure jump** semimartingale. The quadratic variation of a quadratic pure jump semimartingale X is $[X, X](t) = \sum_{0\le s\le t}(\Delta X(s))^2$ (Eq. 4.25).

The **quadratic covariation** or the **bracket** process of X and Y is

$$[X, Y](t) = X(t)\,Y(t) - \int_0^t X(s-)\,dY(s) - \int_0^t Y(s-)\,dX(s). \tag{4.26}$$

Note: The definition of $[X, Y]$ in Eq. 4.26 is meaningful since X, Y are semimartingales so that $X_-, Y_- \in \mathcal{L}$ and the above stochastic integrals are well defined. The quadratic covariation of X and X coincides with the quadratic variation of X in Eq. 4.24. ▲

4.5.2 Properties

Polarization identity.

$$[X, Y] = \frac{1}{2}\left([X+Y, X+Y] - [X, X] - [Y, Y]\right). \tag{4.27}$$

Proof: Since the collection of semimartingales defines a linear space, the quadratic variation process $[X + Y, X + Y]$ is defined. By properties of the stochastic integral and Eqs. 4.24-4.26 we have $[X + Y, X + Y] = [X, X] + [Y, Y] + 2\,[X, Y]$, which yields the polarization identity. ■

Integration by parts.

$$\int_{0+}^t X(s-)\,dY(s) = X(t)\,Y(t) - \int_{0+}^t Y(s-)\,dX(s) - [X, Y](t). \tag{4.28}$$

Note: This formula is a direct consequence of the definition in Eq. 4.26. If $[X, Y] = 0$, Eq. 4.28 coincides with the integration by parts formula in the classical calculus. ▲

The process $[X, X]$ in Eq. 4.24 (1) is adapted with càdlàg, increasing samples, (2) can be obtained for each $t > 0$ and $s \in [0, t]$ from

$$X(0)^2 + \sum_{k=1}^{m_n} \left(X(t_k^{(n)} \wedge s) - X(t_{k-1}^{(n)} \wedge s)\right)^2 \xrightarrow{\text{ucp}} [X, X](s), \tag{4.29}$$

where $p_n = (0 = t_0^{(n)} < t_1^{(n)} < \cdots < t_{m_n}^{(n)} = t)$ is a sequence of partitions with $\Delta(p_n) \to 0$ as $n \to \infty$, and (3) has the properties $[X, X](0) = X(0)^2$ and $\Delta[X, X](t) = [X, X](t) - [X, X](t-) = (\Delta X(t))^2$ ([147], Theorem 22, p. 59).

Proof: We have $S_1^{(n)} = \sum_{k=1}^{m_n} \left(X(t_k^{(n)})^2 - X(t_{k-1}^{(n)})^2\right) = X(t)^2 - X(0)^2$ and

$$S_2^{(n)} = \sum_{k=1}^{m_n} X(t_{k-1}^{(n)})\left(X(t_k^{(n)}) - X(t_{k-1}^{(n)})\right) \xrightarrow{\text{ucp}} \int_{0+}^t X(s-)\,dX(s) = \int_0^t X(s-)\,dX(s)$$

as $n \to \infty$, where we used $\int_0^t X(s-)\,dX(s) = X(0-)\,X(0) + \int_{0+}^t X(s-)\,dX(s)$ and $X(0-) = 0$. The identity $(b-a)^2 = b^2 - a^2 - 2\,a(b-a)$ with $a = X(t_{k-1}^{(n)}))$ and $b = X(t_k^{(n)})$ gives

$$\sum_{k=1}^{m_n} \left(X(t_k^{(n)}) - X(t_{k-1}^{(n)})\right)^2 = S_1^{(n)} - 2\,S_2^{(n)}$$

after summation over k. Hence, $S_1^{(n)} - 2\,S_2^{(n)}$ converges in ucp to

$$\left(X(t)^2 - X(0)^2\right) - 2\int_0^t X(s-)\,dX(s), \quad n \to \infty,$$

as stated (Eq. 4.29).

The quadratic variation process $[X, X]$ increases with t (Eq. 4.29) and has adapted, càdlàg samples (Eq. 4.24). It remains to show that X has the properties in (3). The equality $X(0)^2 = [X, X](0)$ follows directly from the convergence of the series in Eq. 4.29 to $[X, X](t)$. The second equality in property (3) can be obtained from the observation that

$$\begin{aligned}(\Delta X(t))^2 &= (X(t) - X(t-))^2 = X(t)^2 + X(t-)^2 - 2\,X(t)\,X(t-)\\ &= X(t)^2 - X(t-)^2 - 2\,X(t-)\,\Delta X(t) = \Delta(X^2)(t) - 2\,X(t-)\,\Delta X(t) = \Delta[X, X](t)\end{aligned}$$

by the first property in Section 4.4.4 and Eq. 4.24. ■

Example 4.12: Let $C(t) = \sum_{k=1}^{N(t)} Y_k$ be a compound Poisson process. Then $\Delta[C, C](t) = Y_{N(t)}^2\, 1_{\{\Delta N(t)=1\}}$, where $\Delta N(t) = N(t) - N(t-)$. ◇

Proof: If C has a jump at a time $t \geq 0$, then $\Delta C(t) = Y_{N(t)}\, 1_{\{\Delta N(t)=1\}}$. The expression of $\Delta[C, C]$ results from property (3) following Eq. 4.29. ■

- The quadratic variation process $[X, X]$ in Eq. 4.24 is a semimartingale.
- The quadratic covariation process $[X, Y]$ in Eq. 4.26 is a semimartingale.

Proof: Since $[X, X]$ is an adapted process with càdlàg, increasing samples by the above property (Eq. 4.29), the process $[X, X]$ is a semimartingale (Section 4.4.1). The polarization identity and the fact that the collection of martingales is a linear space show that the quadratic covariation process $[X, Y]$ is also a semimartingale.

We also note that the product of two semimartingales is a semimartingale since stochastic integrals and quadratic covariation processes are semimartingales (Eq. 4.28). ■

If X has a.s. continuous sample paths of finite variation on compacts, then $[X, X](t) = X(0)^2$.

Proof: For the sequence of partition p_n in Eq. 4.29

$$\sum_{k=1}^{m_n} \left(X(t_k^{(n)}) - X(t_{k-1}^{(n)})\right)^2 \leq \sup_k \left|X(t_k^{(n)}) - X(t_{k-1}^{(n)})\right| \sum_{k=1}^{m_n} \left|X(t_k^{(n)}) - X(t_{k-1}^{(n)})\right| \to 0$$

as $n \to \infty$ since $\sup_k |X(t_k^{(n)}) - X(t_{k-1}^{(n)})| \to 0$ and $\sum_{k=1}^{m_n} |X(t_k^{(n)}) - X(t_{k-1}^{(n)})| < \infty$ by the continuity and the finite variation of the samples of X, respectively. ■

For each $t > 0$ and partitions $p_n = (0 = t_0^{(n)} < t_1^{(n)} < \cdots < t_{m_n}^{(n)} = t)$ such that $\Delta(p_n) \to 0$ as $n \to \infty$ we have

$$X(0)\,Y(0) + \sum_{k=1}^{m_n} \left(X(t_k^{(n)} \wedge s) - X(t_{k-1}^{(n)} \wedge s) \right) \left(Y(t_k^{(n)} \wedge s) - Y(t_{k-1}^{(n)} \wedge s) \right)$$
$$\xrightarrow{\text{ucp}} [X, Y](s), \quad s \in [0, t], \tag{4.30}$$

$[X, Y](0) = X(0)\,Y(0)$ and $\Delta[X, Y](t) = \Delta X(t)\,\Delta Y(t)$ ([147], Theorem 23, p. 61).

If X is an adapted process with càdlàg paths of finite variation on compacts, then X is a quadratic pure jump semimartingale ([147], Theorem 26, p. 63).

Proof: Note that X is not assumed to be a semimartingale. However, X with the stated properties is a semimartingale (Section 4.4.1) so that Eq. 4.24 applies and gives

$$X(t)^2 = 2 \int_0^t X(s-)\, dX(s) + [X, X](t).$$

Because $X_- \in \mathcal{L}$ and X is of finite variation on compacts, the stochastic integral $X_- \cdot X$ is indistinguishable from the path by path Riemann-Stieltjes integral (Section 4.4.4). Note also that, for each ω,

$$X(t)^2 - X(0)^2 = \sum_{k=1}^{m_n} \left(X(t_k^{(n)})^2 - X(t_{k-1}^{(n)})^2 \right) = \sum_{k=1}^{m_n} \left[X(t_{k-1}^{(n)}) \left(X(t_k^{(n)}) - X(t_{k-1}^{(n)}) \right) \right]$$
$$+ \sum_{k=1}^{m_n} \left[X(t_k^{(n)}) \left(X(t_k^{(n)}) - X(t_{k-1}^{(n)}) \right) \right],$$

where $p_n = (0 = t_0^{(n)} < \cdots < t_{m_n}^{(n)} = t)$ is a partition of $[0, t]$ such that $\Delta(p_n) \to 0$ as $n \to \infty$. The first summation on the right side of the above equation converges in ucp to $\int X_- \, dX$ as $n \to \infty$. The second summation on the right side of the above equation is

$$\sum_{k=1}^{m_n} X(t_{k-1}^{(n)}) \left(X(t_k^{(n)}) - X(t_{k-1}^{(n)}) \right) + \sum_{k=1}^{m_n} \left(X(t_k^{(n)}) - X(t_{k-1}^{(n)}) \right)^2$$

so that it converges in ucp to $\int_0^t X_- \, dX + \sum_{0 < s \le t} (\Delta X_s)^2$ as $n \to \infty$. Hence,

$$X(t)^2 - X(0)^2 = 2 \int_0^t X(s-)\, dX(s) + \sum_{0 < s \le t} (\Delta X(s))^2 \quad \text{or}$$

$$X(t)^2 = 2 \int_0^t X(s-)\, dX(s) + \sum_{0 \le s \le t} (\Delta X(s))^2,$$

which implies $[X, X](t) = \sum_{0 \le s \le t} (\Delta X(s))^2$ (Eq. 4.24). ■

If X is a quadratic pure jump semimartingale and Y is an arbitrary semimartingale, then

$$[X, Y](t) = X(0)\, Y(0) + \sum_{0<s\leq t} \Delta X(s)\, \Delta Y(s). \tag{4.31}$$

Note: The Kunita-Watanabe inequality discussed in the following section (Eq. 4.32) gives $|[X, Y](t)| \leq ([X, X](t)\, [Y, Y](t))^{1/2}$ for $H = K = 1$ so that $[X, X]^c(t) = 0$ implies $[X, Y]^c(t) = 0$. Hence, $[X, Y](t)$ is equal to $\sum_{0\leq s\leq t} \Delta X(s)\, \Delta Y(s)$ by Eq. 4.30.

This result shows that, for example, the quadratic covariation of a compound Poisson process and a Brownian motion is zero. ▲

Example 4.13: Consider the compound Poisson process C in Eq. 4.13. This process is a quadratic pure jump semimartingale with quadratic variation $[C, C](t) = \sum_{k=1}^{N(t)} Y_k^2$. If the jumps of C are one a.s., the compound Poisson process C coincides with a Poisson process N so that $[N, N](t) = \sum_{k=1}^{N(t)} 1 = N(t)$. ◇

Proof: The compound Poisson process C is a semimartingale since it is adapted and has càdlàg paths of finite variation on compacts (Section 4.4.1). We have seen in Example 4.12 that $\Delta[C, C](t) = Y_{N(t)}^2$ if C has a jump at time t and $\Delta[C, C](t) = 0$ otherwise. Hence, the quadratic variation of C in a time interval $(0, t]$ is $[C, C](t) = \sum_{k=1}^{N(t)} Y_k^2$. ■

Example 4.14: We have calculated the stochastic integral $N_- \cdot N$ and found

$$\int_0^t N(s-)\, dN(s) = (N(t)^2 - N(t))/2,$$

in agreement with the definition of the quadratic covariation process of N. ◇

Proof: We have shown that the quadratic variation of a Poisson process N is $[N, N](t) = N(t)$. This result and $[N, N](t) = N(t)^2 - 2\int_0^t N(s-)\, dN(s)$ (Eq. 4.24) yield the above expression for $\int_0^t N_-\, dN$. ■

Example 4.15: If X is a process with continuous paths of finite variation on compacts, then $\int_0^t X(s-)\, dX(s) = (X(t)^2 - X(0)^2)/2$ and $[X, X](t) = X(0)^2$. ◇

Note: We have $[X, X](t) = X(0)^2$ by one of the properties in this section so that $\int_0^t X_-\, dX$ results from Eq. 4.24. For example, take $X(t) = A\cos(\nu\, t)$, where A is a real-valued random variable and $\nu > 0$ is a real number. Then

$$\begin{aligned}\int_0^t X(s-)\, dX(s) &= \int_0^t X(s)\, dX(s) = \int_0^t (A\cos(\nu\, s))\, d(A\cos(\nu\, s)) \\ &= (A^2/2)[(\cos(\nu\, t))^2 - 1] = X(t)^2/2 - A^2/2\end{aligned}$$

so that $[X, X](t) = X(t)^2 - 2\left(X(t)^2/2 - A^2/2\right) = A^2 = X(0)^2$. ▲

Example 4.16: The quadratic covariation $[B, C]$ of a Brownian motion B and the compound Poisson process C is zero. The process $[B + C, B + C]$ is equal to the sum of the quadratic variations of B and C, that is,

$$[B + C, B + C](t) = [B, B](t) + [C, C](t) = t + \sum_{k=1}^{N(t)} Y_k^2.$$

Figure 4.7 shows a hypothetical sample path of C and $[B + C, B + C]$. The linear

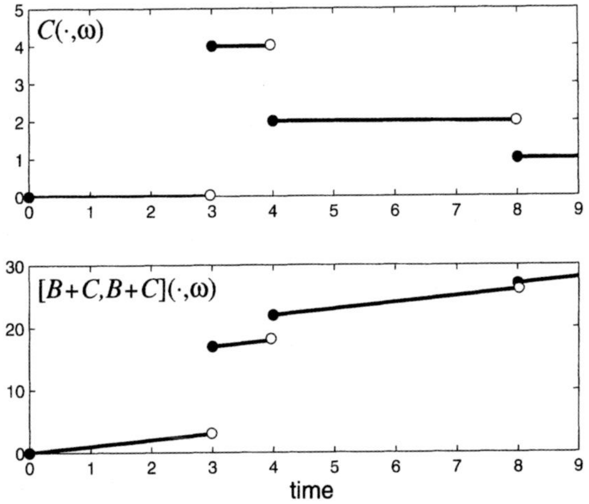

Figure 4.7. Hypothetical samples of C and $B + C$

variation of $[B, B]$ in time is interrupted by jumps caused by the quadratic pure jump semimartingale C. ◇

Proof: The polarization identity (Eq. 4.27) with (B, C) in place of (X, Y) gives $[B, C] = ([B + C, B + C] - [B, B] - [C, C])/2$. We have already found the expressions for the quadratic variation processes $[B, B]$ and $[C, C]$. The quadratic covariation process $[B, C]$ is zero by Eq. 4.31 and properties of (B, C). These observations give the stated expression for $[B + C, B + C]$. ■

Example 4.17: Let X be the Lévy process with characteristic function

$$E[\exp(\sqrt{-1}\, u\, X(t))] = \exp(-t\, \psi(u)),$$

where

$$\psi(u) = \sqrt{-1}\, a\, u + \frac{\sigma^2 u^2}{2} + \int_{\mathbb{R}\setminus\{0\}} \left(1 - e^{\sqrt{-1}\, u\, y} + \sqrt{-1}\, u\, y\, 1_{\{|y|<1\}}\right) \pi(dy),$$

$\pi(dy) = |y|^{-(\alpha+1)}\, dy$, and $\alpha \in (0, 2)$ (Section 3.14.2). Figure 4.8 shows samples of the quadratic variation process $[X, X]$ for $a = 0.1$, $\sigma = 1$, and several values of

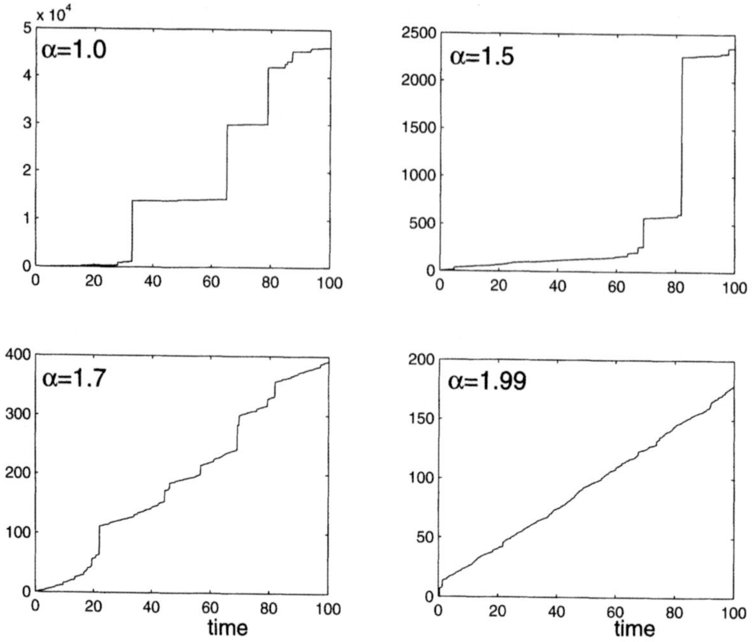

Figure 4.8. Samples of the quadratic variation of a Lévy process with $\alpha = 1, 1.5, 1.7$, and 1.99

α. For small values of α, $[X, X]$ resembles the quadratic variation of a compound Poisson process. For large values of α, $[X, X]$ increases almost linearly in time as the quadratic variation of a Brownian motion. $\diamond$

Note: The generation of samples of X is based on the algorithm in Section 3.14.2 using the approximation $X \simeq X^{(1)} + X^{(2)} + X^{(3,\varepsilon)}$, where $(X^{(1)}, X^{(2)}, X^{(3,\varepsilon)})$ are mutually independent, $X^{(1)}(t) = \sigma\, B(t) - a\, t$, B is a Brownian motion, and $X^{(2)}$ and $X^{(3,\varepsilon)}$ are compound Poisson processes including jumps of X with magnitude in $[1, \infty)$ and $(\varepsilon, 1)$, respectively, and $\varepsilon \in (0, 1)$. The plots in Fig. 4.8 are for $\varepsilon = 0.1$. ▲

4.5.3 Stochastic integrals and covariation processes

We give some useful inequalities and formulas involving stochastic integrals whose integrators are quadratic variation and covariation processes. Also, we present a simple formula for calculating the second moment of stochastic integrals with square integrable martingale integrators.

Kunita-Watanabe inequality. If X, Y are semimartingales and H, K are measurable processes, then a.s. ([147], Theorem 25, p. 61))

$$\int_0^\infty |H(s)|\,|K(s)|\,|d[X,Y](s)| \le \left(\int_0^\infty H(s)^2\, d[X,X](s)\right)^{1/2} \left(\int_0^\infty K(s)^2\, d[Y,Y](s)\right)^{1/2}. \tag{4.32}$$

Note: The integrals in Eq. 4.32 are defined since $[X,X]$, $[Y,Y]$ are càdlàg, increasing, adapted processes and $[X,Y]$ is a process with samples of finite variation on compacts. ▲

If (1) X, Y are semimartingales, (2) H, K are measurable processes, and (3) $1/p + 1/q = 1$, then

$$E\left[\int_0^\infty |H(s)|\,|K(s)|\,|d[X,Y](s)|\right] \le \left\| \left(\int_0^\infty H(s)^2\, d[X,X](s)\right)^{1/2} \right\|_{L_p} \left\| \left(\int_0^\infty K(s)^2\, d[Y,Y](s)\right)^{1/2} \right\|_{L_q}, \tag{4.33}$$

where $\| Z \|_{L_p} = (E[Z^p])^{1/p}$ denotes the p-norm of $Z \in L_p$.

Proof: Set $Z_1 = \left(\int_0^\infty H(s)^2\, d[X,X](s)\right)^{1/2}$ and $Z_2 = \left(\int_0^\infty K(s)^2\, d[Y,Y](s)\right)^{1/2}$. The Kunita-Watanabe inequality gives

$$E\left[\int_0^\infty |H(s)|\,|K(s)|\,|d[X,Y](s)|\right] \le E\left[Z_1\, Z_2\right],$$

by averaging. This result and Hölder's inequality,

$$|E[Z_1\, Z_2]| \le E[|Z_1\, Z_2|] \le (E[|Z_1|^p])^{1/p}\, (E[|Z_2|^q])^{1/q},$$

yield Eq. 4.33. ■

If X, Y are semimartingales and $H, K \in \mathcal{L}$, then

$$[H \cdot X, K \cdot Y](t) = \int_0^t H(s)\, K(s)\, d[X,Y](s). \tag{4.34}$$

Note: The quadratic covariation $[H \cdot X, K \cdot Y]$ is defined because stochastic integrals are semimartingales. We need to show that $[H \cdot X, Y](t) = \int_0^t H(s)\, d[X,Y](s)$ holds for $H \in \mathcal{S}$ and that this equality can be extended to processes $H \in \mathcal{L}$. A proof of this result can be found in [147] (Theorem 29, p. 68), and uses the continuity of the stochastic integral and the fact that $\mathcal{S}$ is dense in $\mathcal{L}$. The above equality and the symmetry of $[\cdot,\cdot]$ give $[H\cdot X, K\cdot Y](t) = \int_0^t H(s)\,d[X, K\cdot Y](s)$ and $[X, K\cdot Y](s) = \int_0^s K(u)\,d[X,Y](u)$ so that Eq. 4.34 results since $d[X, K\cdot Y](s) = K(s)\,d[X,Y](s)$. ▲

If X, Y are semimartingales and $Z \in \mathcal{D}$, then

$$\sum_{k=1}^{m_n} Z(t_{k-1}^{(n)})\left(X(t_k^{(n)}) - X(t_{k-1}^{(n)})\right)\left(Y(t_k^{(n)}) - Y(t_{k-1}^{(n)})\right)$$
$$\stackrel{\text{ucp}}{\longrightarrow} \int_0^t Z(s-)\,d[X,Y](s) \quad \text{as } n\to\infty, \tag{4.35}$$

where $p_n = (0 = t_0^{(n)} \le t_1^{(n)} \le \cdots \le t_{m_n}^{(n)} = t)$ is a sequence of partitions of $[0,t]$ such that $\Delta(p_n)\to 0$ as $n\to\infty$ ([147], Theorem 30, p. 69).

Proof: We have

$$\begin{aligned} Z_-\cdot[X,Y] &= Z_-\cdot(XY) - Z_-\cdot(X_-\cdot Y) - Z_-\cdot(Y_-\cdot X)\\ &= Z_-\cdot(XY) - (Z_-X_-)\cdot Y - (Z_-Y_-)\cdot X\\ &= Z_-\cdot(XY) - (ZX)_-\cdot Y - (ZY)_-\cdot X, \end{aligned}$$

by the definition of the quadratic covariation (Eq. 4.26) and the associativity of the stochastic integral (Eq. 4.22), where $Z_-(t) = Z(t-)$ and $H\cdot V$ denotes $\int H\,dV$. The sequence,

$$\sum_{k=1}^{m_n} Z(t_{k-1}^{(n)})\Big[\left(X(t_k^{(n)})\,Y(t_k^{(n)}) - X(t_{k-1}^{(n)})\,Y(t_{k-1}^{(n)})\right)$$
$$-X(t_{k-1}^{(n)})\left(Y(t_k^{(n)}) - Y(t_{k-1}^{(n)})\right) - Y(t_{k-1}^{(n)})\left(X(t_k^{(n)}) - X(t_{k-1}^{(n)})\right)\Big],$$

in Eq. 4.35 converges in ucp to $Z_-\cdot(XY) - (ZX)_-\cdot Y - (ZY)_-\cdot X$ as $n\to\infty$. This completes the proof since the above limit is equal to $Z_-\cdot[X,Y]$. ■

If M is a square integrable martingale with $M(0) = 0$ and H is an adapted process with càglàd paths such that $E\left[\int_0^t H(s)^2\,d[M,M](s)\right] < \infty$ for each $t\ge 0$, then

$$E\left[\left(\int_0^t H(s)\,dM(s)\right)^2\right] = E\left[\int_0^t H(s)^2 d[M,M](s)\right]. \tag{4.36}$$

Proof: This property is very useful in applications since it provides a procedure for calculating the second moment of stochastic integrals with square integrable martingale integrators. For example, if M is a Brownian motion B, then Eq. 4.36 becomes

$$E\left[\left(\int_0^t H(s)\,dB(s)\right)^2\right] = E\left[\int_0^t H(s)^2 ds\right] \tag{4.37}$$

since $[B, B](t) = t$. The result in Eq. 4.37 is referred to as the **Itô isometry** ([135], Lemma 3.1.5, p. 26).

The expectation on the right side of Eq. 4.36 exists by hypothesis. The expectation on the left side of this equation is defined because the stochastic integral $H \cdot M$ is a square integrable martingale by a preservation property of the stochastic integral (Section 4.4.4). This integral is zero at $t = 0$ by the definition of M. The assumption $M(0) = 0$ and Eq. 4.24 give

$$[H \cdot M, H \cdot M](t) = (H \cdot M)(t)^2 - 2 \int_0^t (H \cdot M)(s-)\, d(H \cdot M)(s).$$

The expectation of the integral in this equation is zero since $H \cdot M$ is a square integrable martingale so that

$$E\left[[H \cdot M, H \cdot M](t)\right] = E\left[(H \cdot M)(t)^2\right].$$

We also have (Eq. 4.34)

$$[H \cdot M, H \cdot M](t) = \int_0^t H(s)^2\, d[M, M](s)$$

so that $E\left[(H \cdot M)(t)^2\right] = E\left[\int_0^t H(s)^2\, d[M, M](s)\right]$. ■

Example 4.18: The second moment of the Itô integral $\int_0^t B\, dB$ is

$$E\left[\left(\int_0^t B(s)\, dB(s)\right)^2\right] = E\left[\int_0^t B(s)^2\, ds\right] = t^2/2,$$

where B is a Brownian motion. ◇

Proof: The isometry in Eq. 4.37 and the Fubini theorem (Section 2.8) give

$$E\left[\int_0^t B(s)^2\, ds\right] = \int_0^t E[B(s)^2]\, ds = \int_0^t s\, ds = t^2/2.$$

An alternative solution is possible in this case since

$$E[(B(t)^2 - t)^2]/4 = E[B(t)^4 - 2t\, B(t)^2 + t^2]/4 = (3t^2 - 2t^2 + t^2)/4 = t^2/2$$

and $\int_0^t B(s)\, dB(s) = (B(t)^2 - t)/2$ (Example 4.1). ■

4.6 Itô's formula

The Itô formula extends the change of variable formula of the classical calculus to stochastic integrals with adapted, càglàd integrands and semimartingale integrators. We will use Itô's formula to develop Monte Carlo simulation techniques (Chapter 5), obtain the local solution of a class of deterministic partial differential equations (Chapter 6), and solve various problems defined by differential equations with random coefficients and/or input (Chapters 7, 8, and 9). This section states and proves the Itô formula, defines the Fisk-Stratonovich integral, and demonstrates by examples the use of Itô's formula and the relationship between the Itô and Fisk-Stratonovich integrals.

4.6.1 One-dimensional case

The classical change of variables formula,

$$g(h(t)) - g(h(0)) = \int_{h(0)}^{h(t)} g'(u)\,du = \int_0^t g'(h(s))\,dh(s), \tag{4.38}$$

gives the increment of a deterministic real-valued function $t \mapsto g(h(t))$ in an interval $[0, t]$. The differential form of this formula is

$$\frac{d}{dt}[g(h(t))] = g'(h(t))\,h'(t) \quad \text{or} \quad d[g(h(t))] = g'(h(t))\,dh(t), \tag{4.39}$$

where the first derivatives g' and h' of functions g and h, respectively, are assumed to exist. The Itô formula extends the rules of the classical calculus in Eqs. 4.38 and 4.39 to the case in which the deterministic function h is replaced with a semimartingale X.

The change of variables formula in Eq. 4.38 gives $\int_0^t B(s)\,dB(s) = \frac{1}{2}\,B(t)^2$ for $g(y) = y^2/2$ and h replaced by a Brownian motion process B. This result is in disagreement with our previous calculations showing that the Itô integral $\int_0^t B(s)\,dB(s)$ is equal to $B(t)^2/2 - t/2$ (Eq. 4.3). On the other hand, the Stratonovich integral $\int_0^t B(s) \circ dB(s) = B(t)^2/2$ (Eq. 4.5) and the classical calculus give the same result.

Itô's formula. If X is a semimartingale and $g \in C^2(\mathbb{R})$, then (1) $g(X)$ is a semimartingale and (2) the **integral form** of the Itô formula,

$$g(X(t)) - g(X(0)) = \int_{0+}^t g'(X(s-))\,dX(s) + \frac{1}{2}\int_{0+}^t g''(X(s-))\,d[X, X]^c(s)$$
$$+ \sum_{0<s\leq t} [g(X(s)) - g(X(s-)) - g'(X(s-))\,\Delta X(s)], \tag{4.40}$$

is valid a.s. at each $t \geq 0$ ([147], Theorem 32, p. 71).

Note: The notation $g \in C^2(\mathbb{R})$ means that the function $g : \mathbb{R} \to \mathbb{R}$ has continuous second order derivative and g'/g'' denote the first/second derivatives of g. The last two terms on the right side of Eq. 4.40 are not present in the classical calculus (Eqs. 4.38 and 4.39).

The Itô formula can also be given in the form

$$g(X(t)) - g(X(0)) = \int_{0+}^t g'(X(s-))\,dX(s) + \frac{1}{2}\int_{0+}^t g''(X(s-))\,d[X, X](s)$$
$$+ \sum_{0<s\leq t} \left[g(X(s)) - g(X(s-)) - g'(X(s-))\,\Delta X(s) - \frac{1}{2} g''(X(s-))\,(\Delta X(s))^2 \right], \tag{4.41}$$

by the relationship between $[X, X]$ and $[X, X]^c$ (Eq. 4.25). ▲

Itô's formula. If X is a continuous semimartingale and $g \in C(\mathbb{R}^2)$, then (1) $g(X)$ is a continuous semimartingale and (2) the **differential and integral forms** of the Itô formula are

$$dg(X(t)) = g'(X(t))\,dX(t) + \frac{1}{2}g''(X(t))\,d[X, X](t) \quad \text{and}$$

$$g(X(t)) - g(X(0)) = \int_0^t g'(X(s))\,dX(s) + \frac{1}{2}\int_0^t g''(X(s))\,d[X, X](s). \tag{4.42}$$

Proof: The proof of the Itô formula in Eq. 4.40 is presented in detail since we will use this formula extensively. The proof, adapted from [147] (Theorem 32, p. 71), has two parts. First, it is assumed that X is a continuous semimartingale. This result is then extended to the case in which X is an arbitrary semimartingale. A heuristic proof of the Itô formula can be found in [117] (Theorem 6.7.1, p. 184).

The following result is used in the proof. If $g \in C^2(\mathbb{R})$ is defined on a bounded interval, the Taylor formula gives

$$g(x + h) - g(x) = h\,g'(x) + \frac{h^2}{2}\,g''(x) + r(x, h), \quad h \in \mathbb{R}, \tag{4.43}$$

where $|r(x, h)| \leq h^2\,\alpha(|h|)$, $\alpha : [0, \infty) \to [0, \infty)$ is increasing, and $\lim_{u \downarrow 0} \alpha(u) = 0$. Two cases are considered.

Case 1: X is a continuous semimartingale. We have $[X, X]$ and $[X, X]^c$ in the Itô formula for X continuous (Eq. 4.42) and X arbitrary (Eq. 4.40), respectively, because $\Delta X(t) = 0$ at all times $t \geq 0$ if X has continuous samples so that $[X, X](t) = [X, X]^c(t)$ in this case.

Since X is continuous and $g \in C(\mathbb{R}^2)$, the process $g(X)$ has continuous samples. If Eq. 4.42 holds, $g(X)$ is a semimartingale since (1) the processes $g'(X)$ and $g''(X)$ are adapted as memoryless transformations of the adapted process X, (2) $g'(X)$ and $g''(X)$ have càdlàg samples by the properties of g and X, for example,

$$\lim_{s \downarrow t} g'(X(s)) = g'(\lim_{s \downarrow t} X(s)) = g'(X(t)) \text{ and } \lim_{s \uparrow t} g'(X(s)) = g'(\lim_{s \uparrow t} X(s)) = g'(X(t-)),$$

(3) the integrals $\int_0^t g'(X(s-))\,dX(s)$ and $\int_0^t g''(X(s-))\,d[X, X]^c(s)$ are semimartingales by a preservation property of the stochastic integral (Section 4.4.4), and (4) sums of semimartingales are semimartingales.

Without loss of generality it can be assumed that $X(0) = 0$. If $X(0) \neq 0$, we can consider the process $X(t) - X(0)$ that starts at zero. The Brownian motion process is an example of a process considered in this part of the proof. The Taylor formula cannot be applied to $g(X)$ because X may not take values in a bounded interval. This difficulty is removed by stopping X when it leaves for the first time an interval $(-a, a)$, $0 < a < \infty$, and then letting $a \to \infty$. Therefore, we can assume that X is bounded and apply Taylor's formula to $g(X)$. Fix $t > 0$ and consider a sequence of partitions $p_n = (0 = t_0^{(n)} \leq t_1^{(n)} \leq$

$\cdots \le t_{m_n}^{(n)} = t$) of $[0, t]$ such that $\Delta(p_n) \to 0$ as $n \to \infty$. Then

$$g(X(t)) - g(X(0)) = \sum_{k=1}^{m_n} \left(g(X(t_k^{(n)})) - g(X(t_{k-1}^{(n)})) \right) = S_1^{(n)} + S_2^{(n)} + S_3^{(n)}, \quad (4.44)$$

where

$$S_1^{(n)} = \sum_{k=1}^{m_n} g'(X(t_{k-1}^{(n)})) \left(X(t_k^{(n)}) - X(t_{k-1}^{(n)}) \right)$$
$$S_2^{(n)} = \frac{1}{2} \sum_{k=1}^{m_n} g''(X(t_{k-1}^{(n)})) \left(X(t_k^{(n)}) - X(t_{k-1}^{(n)}) \right)^2$$
$$S_3^{(n)} = \sum_{k=1}^{m_n} r(X(t_{k-1}^{(n)}), X(t_k^{(n)}) - X(t_{k-1}^{(n)}))$$

correspond to the three terms of the Taylor formula in Eq. 4.43 applied to the intervals of p_n. The sums $S_1^{(n)}$ and $S_2^{(n)}$ converge in ucp to the Itô integrals $\int_0^t g'(X(s-))\, dX(s)$ and $(1/2) \int_0^t g''(X(s-))\, d[X, X](s)$ as $n \to \infty$, respectively (Eqs. 4.23 and 4.35). The absolute value of $S_3^{(n)}$ can be bounded by

$$\left| S_3^{(n)} \right| = \left| \sum_{k=1}^{m_n} r(X(t_{k-1}^{(n)}), X(t_k^{(n)}) - X(t_{k-1}^{(n)})) \right|$$
$$\le \max_{1 \le k \le m_n} \alpha(|X(t_k^{(n)}) - X(t_{k-1}^{(n)})|) \sum_{k=1}^{m_n} (X(t_k^{(n)}) - X(t_{k-1}^{(n)}))^2$$

for each n. Because $\sum_k \left(X(t_k^{(n)}) - X(t_{k-1}^{(n)}) \right)^2 \xrightarrow{\text{ucp}} [X, X](t)$ (Eq. 4.29), this sequence also converges to $[X, X](t)$ in probability. We also have $\max_k \alpha(|X(t_k^{(n)}) - X(t_{k-1}^{(n)})|) \to 0$ as $n \to \infty$ since $\lim_{u \downarrow 0} \alpha(u) = 0$ and the function $s \mapsto X(s, \omega)$ is continuous on $[0, t]$ for almost all ω and, hence, uniformly continuous in this time interval so that $\max_k |X(t_k^{(n)}) - X(t_{k-1}^{(n)})| \to 0$ a.s. as $n \to \infty$.

In summary, we have shown that for each $t \ge 0$ the sequences $S_1^{(n)}$, $S_2^{(n)}$, and $S_3^{(n)}$ converge in probability to $\int_0^t g'(X(s-))\, dX(s)$, $(1/2) \int_0^t g''(X(s-))\, d[X, X](s)$, and zero, respectively, as $n \to \infty$. Hence, we have

$$\lim_{n \to \infty} P\left(\left| S_1^{(n)} + S_2^{(n)} + S_3^{(n)} - \int_0^t g'(X(s-))\, dX(s) - \frac{1}{2} \int_0^t g''(X(s-))\, d[X, X](s) \right| > \varepsilon \right) = 0$$

for any $\varepsilon > 0$ so that the equality

$$g(X(t)) - g(X(0)) = \int_0^t g'(X(s-))\, dX(s) + \frac{1}{2} \int_0^t g''(X(s-))\, d[X, X](s) \quad (4.45)$$

holds a.s. for each t. Note that $X(s-) = X(s)$ for each $s \in [0, t]$ since X is assumed to be continuous. Because the processes on both sides of Eq. 4.45 are continuous, this equality holds for all values of t. If $X(0)$ is not zero, the Itô formula of Eq. 4.45 becomes

$$g(X(t)) - g(X(0)) = \int_0^t g'(X(s))\, dX(s) + \frac{1}{2}\int_0^t g''(X(s))\, d[X, X](s). \tag{4.46}$$

Case 2: X is a general semimartingale. Consider an arbitrary time $t > 0$ and note that $\sum_{0<s\leq t}(\Delta X(s))^2 \leq [X, X](t)$ and $[X, X](t) < \infty$ a.s. because $[X, X]$ is an increasing process with càdlàg paths (Eqs. 4.25 and 4.29). Given $\varepsilon > 0$ and $t > 0$, denote by V a collection of jumps of X in $[0, t]$ such that $\sum_{s\in V}(\Delta X(s))^2 \leq \varepsilon^2$ and such that the set U of jumps in $[0, t]$ that are not in V is finite. Then Eq. 4.44 gives

$$\begin{aligned} g(X(t)) - g(X(0)) &= \sum_{k=1}^{m_n}\left(g(X(t_k^{(n)})) - g(X(t_{k-1}^{(n)})\right) \\ &= \sum_{k,U}\left(g(X(t_k^{(n)})) - g(X(t_{k-1}^{(n)})\right) + \sum_{k,V}\left(g(X(t_k^{(n)})) - g(X(t_{k-1}^{(n)})\right), \end{aligned} \tag{4.47}$$

where $\sum_{k,U}$ corresponds to the intervals of the partition p_n of $[0, t]$ including jumps in U and $\sum_{k,V} = \sum_k - \sum_{k,U}$. Since X has càdlàg sample paths, it has at most a countable number of jumps in $[0, t]$ and, for a sufficiently fine partition p_n, the intervals $(t_{k-1}^{(n)}, t_k^{(n)}]$ contain at most one jump of X in U so that

$$\sum_{k,U}\left[g(X(t_k^{(n)})) - g(X(t_{k-1}^{(n)}))\right] \to \sum_{s\in U}[g(X(s)) - g(X(s-))], \quad \text{as } n \to \infty,$$

for almost all samples of X.

The Taylor formula in Eq. 4.43 can be applied to each term of the summation $\sum_{k,V} = \sum_k - \sum_{k,U}$ and gives

$$\begin{aligned} &\sum_{k,V}\left(g(X(t_k^{(n)})) - g(X(t_{k-1}^{(n)})\right) \\ &= \sum_k\left[g'(X(t_{k-1}^{(n)}))\left(X(t_k^{(n)}) - X(t_{k-1}^{(n)})\right) + \frac{1}{2}g''(X(t_{k-1}^{(n)}))\left(X(t_k^{(n)}) - X(t_{k-1}^{(n)})\right)^2\right] \\ &- \sum_{k,U}\left[g'(X(t_{k-1}^{(n)}))\left(X(t_k^{(n)}) - X(t_{k-1}^{(n)})\right) + \frac{1}{2}g''(X(t_{k-1}^{(n)}))\left(X(t_k^{(n)}) - X(t_{k-1}^{(n)})\right)^2\right] \\ &+ \sum_{k,V} r(X(t_{k-1}^{(n)}), X(t_k^{(n)}) - X(t_{k-1}^{(n)})). \end{aligned}$$

As in the previous case, the first term on the right side of the above equation converges in ucp to $\int_{0+}^t g'(X(s-))\, dX(s) + (1/2)\int_{0+}^t g''(X(s-))\, d[X, X](s)$. The second term converges to

$$-\sum_{s\in U}\left[g'(X(s-))\,\Delta X(s) + \frac{1}{2}g''(X(s-))\,(\Delta X(s))^2\right] \quad \text{as } n \to \infty.$$

It remains to show that, as $\varepsilon \downarrow 0$, $\sum_{s \in U} (g(X(s)) - g(X(s-)))$ and the above summation converge to $\sum_{0<s\leq t} [g(X(s)) - g(X(s-))]$ and

$$-\sum_{0<s\leq t} \left[g'(X(s-)) \, \Delta X(s) + \frac{1}{2} g''(X(s-)) \, (\Delta X(s))^2 \right],$$

respectively, and $\sum_{k,V} r(X(t_{k-1}^{(n)}), X(t_k^{(n)}) - X(t_{k-1}^{(n)}))$ approaches zero. Details on this part of the proof can be found in [147] (pp. 73-74). Hence, the limit of

$$g(X(t)) - g(X(0)) = \sum_{k=1}^{m_n} \left(g(X(t_k^{(n)})) - g(X(t_{k-1}^{(n)})) \right)$$

as $n \to \infty$ and $\varepsilon \downarrow 0$ is

$$\int_{0+}^{t} g'(X(s-)) \, dX(s) + \frac{1}{2} \int_{0+}^{t} g''(X(s-)) \, d[X, X](s)$$
$$+ \sum_{0<s\leq t} \left[g(X(s)) - g(X(s-)) - g'(X(s-)) \, \Delta X(s) - \frac{1}{2} g''(X(s-)) \, (\Delta X(s))^2 \right],$$

which is the right side of Eq. 4.40 since (Eq. 4.25)

$$\int_{0+}^{t} g''(X(s-)) \, d[X, X](s) - \frac{1}{2} \sum_{0<s\leq t} g''(X(s-)) \, (\Delta X(s))^2$$
$$= \int_{0+}^{t} g''(X(s-)) \, d[X, X]^c(s).$$

It remains to show that $g(X)$ is a semimartingale. Considerations as in the previous case show that $g(X)$ is adapted and has càdlàg samples. The two integrals on the right side of Eq. 4.40 are semimartingales by a preservation property of the stochastic integral (Section 4.4.4). The last term in Eq. 4.40 can be split in two summations involving jumps ΔX of X with magnitude larger and smaller than one. The summation including jumps $|\Delta X(s)| > 1$ is a process of finite variation on compacts since X has càdlàg samples, so that it is a semimartingale (Section 4.4.1). The summation including jumps $|\Delta X(s)| \leq 1$ can be written in the form

$$\sum_{0<s\leq t}^{*} [g(X(s)) - g(X(s-)) - g'(X(s-)) \, \Delta X(s)] = \tfrac{1}{2} \sum_{0<s\leq t}^{*} g''(\Theta(s)) \, (\Delta X(s))^2,$$

where $\sum_{0<s\leq t}^{*}$ means $\sum_{0<s\leq t, |\Delta X(s)|\leq 1}$ and $|\Theta(s)| \leq |\Delta X(s)| \leq 1$ a.s. Since $g \in C^2(\mathbb{R})$, there is a constant $a > 0$ such that $|g''(\Theta(s))| \leq a$ and

$$(1/2) \sum_{0<s\leq t}^{*} g''(\Theta(s)) \, (\Delta X(s))^2 \leq (a/2) \sum_{0<s\leq t}^{*} (\Delta X(s))^2,$$

which is bounded since X is a càdlàg process. Hence, $\sum_{0<s\leq t}^{*} [g(X(s)) - g(X(s-)) - g'(X(s-)) \, \Delta X(s)]$ is a semimartingale (Section 4.4.1). Because we have shown that Eq. 4.40 holds and the right side of this equation is a semimartingale as a sum of semimartingales, $g(X)$ is also a semimartingale. ■

The collection of semimartingales has some notable closure properties in the sense that processes obtained by performing some operations on semimartingales are semimartingales. Let X and Y be semimartingales and α, β be some constants. We have seen that (1) $\alpha X + \beta Y$, (2) $\int X_- \, dY$, (3) $[X, X]$ and $[X, Y]$, and (4) $X Y$ are semimartingales. The Itô formula gives another closure property for semimartingales. This formula shows that any memoryless transformation of a semimartingale is a semimartingale, provided that the transformation has continuous second order derivatives. For example, the processes $\sin(X)$, $\cos(X)$, and e^X are semimartingales. In contrast, memoryless transformations of martingales may not be martingales. For example, the square of a Brownian motion is not a martingale.

The Itô formula is applied in the following examples to derive some properties relevant for the analysis of some of the stochastic problems discussed in the following chapters of the book.

Example 4.19: Consider the mapping $B \mapsto g(B) = B^n$ of a Brownian motion, where $n \geq 2$ is an integer. Then

$$B(t)^n = n \int_0^t B(s)^{n-1} \, dB(s) + \frac{n(n-1)}{2} \int_0^t B(s)^{n-2} \, ds \tag{4.48}$$

is an $\mathcal{F}_t$-semimartingale, where $\mathcal{F}_t$ is the natural filtration of B. The formula in Eq. 4.48 is consistent with Eq. 4.3, provides an alternative method for calculating the integral $\int_0^t B \, dB$, and shows that this integral is a martingale since it is equal to $(B(t)^2 - t)/2$. ◇

Proof: The Itô formula in Eq. 4.40 (1) can be applied to the mapping $B \mapsto B^n$ since this mapping is infinitely differentiable and B is a continuous square integrable martingale so that it is a semimartingale, (2) shows that B^n is a semimartingale, and (3) gives a recurrence formula for calculating the stochastic integral $\int_0^t B(s)^q \, dB(s)$, where $q \geq 1$ is an integer (Eq. 4.48).

That $g(B) = B^2$ is a semimartingale also results from the representation $B(t)^2 = A(t) + M(t)$, where $A(t) = t$ is an adapted continuous process with $A(0) = 0$ and paths of finite variation on compacts and $M(t) = B(t)^2 - t$ is a square integrable martingale starting at zero. Hence, the process B^2 is decomposable so that it is a semimartingale (Section 4.4.1).

The equality in Eq. 4.48 also shows that the Itô integral $\int_0^t B(s)^{n-1} \, dB(s)$ can be calculated path by path from $(1/n)\, B(t)^n - ((n-1)/2) \int_0^t B(s)^{n-2} \, ds$ and samples of the Brownian motion B. ■

Example 4.20: The process N^2 is a semimartingale and

$$N(t)^2 = 2 \int_{0+}^t N(s-) \, dN(s) + N(t), \tag{4.49}$$

where N is a Poisson process with intensity $\lambda > 0$ (Eq. 4.10). ◇

Proof: That N^2 is a semimartingale follows from the Itô formula. This formula applied to the function $g(N) = N^2$ gives (Eq. 4.40)

$$N(t)^2 - N(0)^2 = \int_{0+}^{t} 2\,N(s-)\,dN(s) + \frac{1}{2}\int_{0+}^{t} 2\,d[N,N]^c(s) + \sum_{0<s\le t}\left[N(s)^2 - N(s-)^2 - 2\,N(s-)\,\Delta N(s)\right].$$

We have seen that N is a quadratic pure jump semimartingale so that $[N,N]^c(t) = 0$ for all $t \ge 0$. We also have

$$\sum_{0<s\le t}\left[N(s)^2 - N(s-)^2 - 2\,N(s-)\,\Delta N(s)\right] = \sum_{i=1}^{N(t)}\left[N(T_i)^2 - N(T_{i-1})^2 - 2\,N(T_{i-1})\right]$$

$$= \sum_{i=1}^{N(t)}\left[i^2 - (i-1)^2 - 2\,(i-1)\right] = N(t),$$

where T_i, $i = 1, 2, \ldots$, denote the jump times of N. The above results and the Itô formula yield Eq. 4.49 since $N(0) = 0$. ■

Example 4.21: Let C be the compound Poisson process defined by Eq. 4.13 with jumps given by a sequence Y_k of iid real-valued random variables. The process C^2 is a semimartingale and

$$C(t)^2 = 2\int_{0+}^{t} C(s-)\,dC(s) + \sum_{i=1}^{N(t)} Y_k^2. \tag{4.50}$$

This formula provides an alternative way for calculating the stochastic integral $\int C_-\,dC$. ◇

Proof: According to Itô's formula, C^2 is a semimartingale and (Eq. 4.40)

$$C(t)^2 - C(0)^2 = \int_{0+}^{t} 2\,C(s-)\,dC(s) + \frac{1}{2}\int_{0+}^{t} 2\,d[C,C]^c(s) + \sum_{0<s\le t}\left[C(s)^2 - C(s-)^2 - 2\,C(s-)\,\Delta C(s)\right]. \tag{4.51}$$

Because C is a quadratic pure jump semimartingale, we have $[C,C]^c(t) = 0$, $t \ge 0$. The summation on the right side of the above equation is

$$\sum_{0<s\le t}\left[C(s)^2 - C(s-)^2 - 2\,C(s-)\,\Delta C(s)\right]$$

$$= \sum_{i=1}^{N(t)}\left[C(T_i)^2 - C(T_{i-1})^2 - 2\,C(T_{i-1})\,Y_i\right]$$

$$= \sum_{i=1}^{N(t)}\left[\left(\sum_{k=1}^{i} Y_k\right)^2 - \left(\sum_{k=1}^{i-1} Y_k\right)^2 - 2\left(\sum_{k=1}^{i-1} Y_k\right) Y_i\right] = \sum_{i=1}^{N(t)} Y_i^2,$$

and constitutes a compound Poisson process with jumps Y_i^2 occurring at the jump times T_i of C. These considerations and $C(0) = 0$ give Eq. 4.50. ■

Example 4.22: Let C be the compound Poisson process in Eq. 4.13 and g a real-valued function with continuous second order derivative. An alternative form of the Itô formula in Eq. 4.40 is

$$g(C(t)) - g(C(0)) = \int_{0+}^{t} \int_{\mathbb{R}} [g(C(s-) + y) - g(C(s-))] \, \mathcal{M}(ds, dy), \quad (4.52)$$

where $\mathcal{M}(ds, dy)$ denotes a random measure giving the number of jumps of C in the range $(y, y + dy]$ during the time interval $(s, s + ds]$. ◇

Proof: If $g \in C^2(\mathbb{R})$, Eq. 4.40 gives

$$g(C(t)) - g(C(0)) = \int_{0+}^{t} g'(C(s-)) \, dC(s) + \frac{1}{2} \int_{0+}^{t} g''(C(s-)) \, d[C, C]^c(s)$$
$$+ \sum_{0<s\leq t} \left[g(C(s)) - g(C(s-)) - g'(C(s-)) \, \Delta C(s)\right] = \sum_{0<s\leq t} [g(C(s)) - g(C(s-))]$$

since $\int_{0+}^{t} g'(C(s-)) \, dC(s) = \sum_{0<s\leq t} g'(C(s-)) \, \Delta C(s)$ and $[C, C]^c = 0$. We also have

$$\int_{0+}^{t} \int_{\mathbb{R}} [g(C(s-) + y) - g(C(s-))] \, \mathcal{M}(ds, dy) = \sum_{0<s\leq t} [g(C(s)) - g(C(s-))]$$

since the C has piece-wise constant samples with a finite number of jumps in $(0, t]$ a.s. ■

Example 4.23: The $\mathbb{R}^2$-valued process

$$\boldsymbol{X}(t) = (X_1(t) = \cos(B(t)), X_2(t) = \sin(B(t))), \quad t \geq 0, \quad (4.53)$$

takes on values on the unit circle centered at the origin of $\mathbb{R}^2$ and is called the **Brownian motion on the unit circle** ([135], Example 7.5.5, p. 121). This process is the solution of the stochastic differential equation

$$dX_1(t) = \frac{1}{2} X_1(t) \, dt - X_2(t) \, dB(t),$$
$$dX_2(t) = \frac{1}{2} X_2(t) \, dt + X_1(t) \, dB(t). \quad (4.54)$$

Stochastic differential equations are discussed in Section 4.7. Figure 4.9 shows a sample path of $\boldsymbol{X}$. ◇

Note: The differential form of the Itô formula in Eq. 4.42 applied to the coordinates of $\boldsymbol{X}$ gives

$$d(\cos(B(t)) = -\sin(B(t)) \, dB(t) - \frac{1}{2} \cos(B(t)) \, dt = \frac{1}{2} X_1(t) \, dt - X_2(t) \, dB(t),$$
$$d(\sin(B(t)) = \cos(B(t)) \, dB(t) - \frac{1}{2} \sin(B(t)) \, dt = \frac{1}{2} X_2(t) \, dt + X_1(t) \, dB(t)$$

since $d[B, B]^c(t) = d[B, B](t) = dt$. ▲

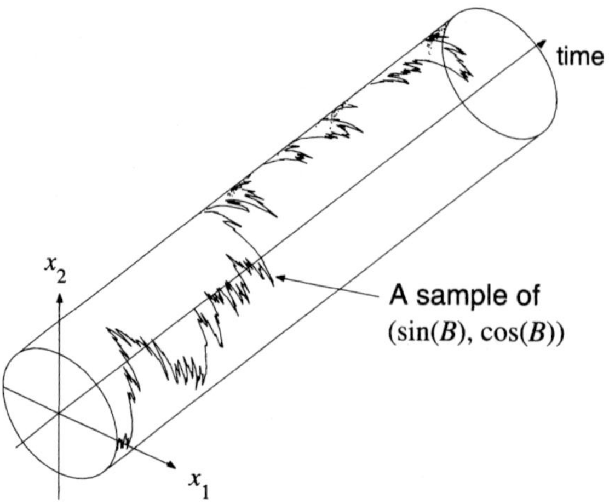

Figure 4.9. A sample path of the Brownian motion on the unit circle

Example 4.24: Let Y and Z be real-valued processes with càdlàg samples adapted to the natural filtration of a Brownian motion B. Let $X(t) = A(t) + M(t)$ be another process, where $A(t) = \int_0^t Y(s-)\, ds$ and $M(t) = \int_0^t Z(s-)\, dB(s)$. Then X is a semimartingale, and

$$g(X(t)) - g(X(0)) = \int_0^t g'(X(s-))\, dX(s) + \frac{1}{2}\int_0^t g''(X(s-))\, d[X, X]^c(s)$$

for $g \in C(\mathbb{R}^2)$, where $[X, X]^c(t) = [X, X](t) = [M, M](t) = \int_0^t Z(s-)^2\, ds$. ◇

Proof: Since A and M are semimartingales (Section 4.4.4), X is also a semimartingale. The expression of $g(X(t)) - g(X(0))$ is given by the Itô formula in Eq. 4.40.

The quadratic variation of X can be calculated from $[X, X] = [M + A, M + A] = [M, M] + [A, A] + 2\,[M, A]$. We have (Eq. 4.34)

$$[M, M](t) = [Z \cdot B, Z \cdot B] = \int_0^t Z(s-)^2\, d[B, B](s) = \int_0^t Z(s-)^2\, ds.$$

The quadratic variation of A is $[A, A](t) = 0$, $t \geq 0$, since

$$\sum_{k=1}^{m_n} \left(A(t_k^{(n)}) - A(t_{k-1}^{(n)})\right)^2 \leq \sup_k |A(t_k^{(n)}) - A(t_{k-1}^{(n)})| \sum_{k=1}^{m_n} |A(t_k^{(n)}) - A(t_{k-1}^{(n)})|$$

converges to zero for partitions p_n of $[0, t]$ with mesh $\Delta(p_n) \to 0$ as $n \to \infty$. Let $i(t) = t$ denote the identity function. Then $A(t) = \int_0^t Y(s-)\, di(s)$ so that the quadratic covariation of M and A is $[M, A] = \int_0^t Y(s-)\, Z(s-)\, d[B, i](s)$, $t \geq 0$, and

$$\sum_{k=1}^{m_n} (B(t_k^{(n)}) - B(t_{k-1}^{(n)}))\,(t_k^{(n)} - t_{k-1}^{(n)}) \leq \sup_k |B(t_k^{(n)}) - B(t_{k-1}^{(n)})| \sum_{k=1}^{m_n} |t_k^{(n)} - t_{k-1}^{(n)}|,$$

converges to zero since $\sum_{k=1}^{m_n} |t_k^{(n)} - t_{k-1}^{(n)}| = t$ and $\sup_k |B(t_k^{(n)}) - B(t_{k-1}^{(n)})| \to 0$ a.s. as $n \to \infty$ by the continuity of B. Hence, the quadratic variation process of X is $[X, X] = [M, M]$ as stated. ■

4.6.2 Multi-dimensional case

Itô's formula. If $\boldsymbol{X}$ is a vector of d semimartingales and $g : \mathbb{R}^d \to \mathbb{R}$ has continuous second order partial derivatives, then (1) $g(\boldsymbol{X})$ is a semimartingale and (2) the **integral form** of the Itô formula is ([147], Theorem 33, p. 74)

$$g(\boldsymbol{X}(t)) - g(\boldsymbol{X}(0)) = \sum_{i=1}^{d} \int_{0+}^{t} \frac{\partial g}{\partial x_i}(\boldsymbol{X}(s-))\, dX_i(s) + \frac{1}{2} \sum_{i,j=1}^{d} \int_{0+}^{t} \frac{\partial^2 g}{\partial x_i\, \partial x_j}(\boldsymbol{X}(s-))\, d[X_i, X_j]^c(s) + \sum_{0<s\le t} \left[g(\boldsymbol{X}(s)) - g(\boldsymbol{X}(s-)) - \sum_{i=1}^{d} \frac{\partial g}{\partial x_i}(\boldsymbol{X}(s-))\, \Delta X_i(s) \right]. \quad (4.55)$$

If the coordinates of $\boldsymbol{X}$ are continuous semimartingales, the **differential and integral forms** of the Itô formula are

$$dg(\boldsymbol{X}(t)) = \sum_{i=1}^{d} \frac{\partial g}{\partial x_i}(\boldsymbol{X}(t))\, dX_i(t) + \frac{1}{2} \sum_{i,j}^{d} \frac{\partial^2 g}{\partial x_i\, \partial x_j}(\boldsymbol{X}(t))\, d[X_i, X_j](t) \quad \text{and}$$

$$g(\boldsymbol{X}(t)) - g(\boldsymbol{X}(0)) = \sum_{i=1}^{d} \int_{0}^{t} \frac{\partial g}{\partial x_i}(\boldsymbol{X}(s))\, dX_i(s) + \frac{1}{2} \sum_{i,j=1}^{d} \int_{0}^{t} \frac{\partial^2 g}{\partial x_i\, \partial x_j}(\boldsymbol{X}(s))\, d[X_i, X_j](s). \quad (4.56)$$

Note: If the coordinates of $\boldsymbol{X}$ are continuous semimartingales, then $[X_i, X_j]^c = [X_i, X_j]$ and $\boldsymbol{X}(s-) = \boldsymbol{X}(s)$ so that the last summation in Eq. 4.55 vanishes.

The differential form of the Itô formula is used in applications to develop evolution equations for moments and other properties of $\boldsymbol{X}$. ▲

Example 4.25: Let $\boldsymbol{B}$ be a Brownian motion in $\mathbb{R}^d$ and let $\boldsymbol{X}(t) = \boldsymbol{g}(\boldsymbol{B}(t)) = \boldsymbol{B}(t)/ \parallel \boldsymbol{B}(t) \parallel$, $t \ge 0$, be the **Brownian motion on the unit sphere** $S_d(\boldsymbol{0}, 1)$ in $\mathbb{R}^d$ centered at the origin of this space, where $\boldsymbol{g} : \mathbb{R}^d \setminus \{0\} \to S_d(\boldsymbol{0}, 1)$ and $\parallel \boldsymbol{x} \parallel = (\sum_{i=1}^{d} x_i^2)^{1/2}$ denotes the Euclidean norm.

The coordinates of X satisfy the differential form

$$dX_p(t) = -\frac{d-1}{2}\frac{B_p(t)}{\| B(t) \|^3}\,dt + \sum_{i=1}^{d} \frac{\delta_{ip} - B_i(t)\,B_p(t)}{\| B(t) \|^3}\,dB_i(t)$$

$$= -\frac{d-1}{2}\frac{X_p(t)}{\| B(t) \|}\,dt + \sum_{i=1}^{d} \frac{\delta_{ip} - X_i(t)\,X_p(t)}{\| B(t) \|}\,dB_i(t)$$

for each $p = 1, \ldots, d$. These forms are not stochastic differential equations for X of the type defined later in this chapter (Eq. 4.69) since the increments $dX(t)$ of this process depend on $X(t)$, $B(t)$, and the increments $dB(t)$ of B rather than just on $X(t)$ and $dB(t)$. The process X can be modified to satisfy a stochastic differential by a random time change ([135], Example 8.5.8, p. 149) or by state augmentation. For example, we can consider the evolution of the process (X, Y), where Y is an $\mathbb{R}^d$-valued process defined by $dY(t) = dB(t)$. ◇

Proof: The differential form of the Itô formula in Eq. 4.56 gives

$$dX_p(t) = \sum_{i=1}^{d} \frac{\partial g_p(B(t))}{\partial x_i}\,dB_i(t) + \frac{1}{2}\sum_{i,j=1}^{d} \frac{\partial^2 g_p(B(t))}{\partial x_i\,\partial x_j}\,d[B_i, B_j](t)$$

for $p = 1, \ldots, d$. The polarization identity (Eq. 4.27) gives

$$d[B_i, B_j](t) = \frac{1}{2}\,d\left([B_i + B_j, B_i + B_j] - [B_i, B_i] - [B_j, B_j]\right)(t),$$

which is $(1/2)(2\,dt - dt - dt) = 0$ for $i \neq j$ since the processes B_i are independent Brownian motions. Hence, the double summation in the above expression of $dX_p(t)$ becomes $\sum_{k=1}^{d}\left(\partial^2 g_p(B(t))/\partial x_i^2\right)\,dt$. ■

Example 4.26: Consider a real-valued function $g : \mathbb{R}^d \to \mathbb{R}$ with continuous and bounded first and second order derivatives. Let B be an $\mathbb{R}^d$-valued Brownian motion starting at $x = (x_1, \ldots, x_d) \in \mathbb{R}^d$ whose coordinates B_i are independent Brownian motions such that $B_i(0) = x_i$, $i = 1, \ldots, d$. Denote by E^x the expectation operator for $B(0) = x$. The limit,

$$\mathcal{A}[g(x)] = \lim_{t \downarrow 0} \frac{E^x\,[g(B(t))] - g(x)}{t} = \frac{1}{2}\sum_{i=1}^{d} \frac{\partial^2 g(x)}{\partial x_i^2}, \qquad (4.57)$$

called the **generator** of B, exists and is proportional with the Laplace operator $\Delta = \sum_{i=1}^{d} \partial^2/\partial x_i^2$. This result provides a link between a class of stochastic processes and some deterministic partial differential equations (Chapter 6). ◇

Proof: The Itô formula in Eq. 4.56 gives

$$g(B(t)) - g(B(0)) = \sum_{i=1}^{d} \int_0^t \frac{\partial g(B(s))}{\partial x_i}\,dB_i(s) + \frac{1}{2}\sum_{i=1}^{d} \int_0^t \frac{\partial^2 g(B(s))}{\partial x_i^2}\,ds$$

since $d[B_i, B_j](s) = \delta_{ij}\, ds$, where $\delta_{ij} = 1$ for for $i = j$ and $\delta_{ij} = 0$ for $i \neq j$.

The integrals in the first summation on the right side of the above equation are $\mathcal{F}_t = \sigma(\boldsymbol{B}(s) : 0 \leq s \leq t)$-martingales starting at zero. Hence, their expectation is zero. The integrals in the second summation on the right side of this equation can be defined as Riemann integrals and can be approximated by $t\, \partial^2 g(\boldsymbol{B}(\theta(\omega)\, t, \omega))/\partial x_i^2$, $\theta(\omega) \in (0, 1)$, for almost all ω's. The limit of these integrals scaled by t as $t \downarrow 0$ is deterministic and equal to $\partial^2 g(\boldsymbol{x})/\partial x_i^2$. These observations yield Eq. 4.57. ■

Example 4.27: A stochastic process X is a standard Brownian motion if and only if X is a continuous local martingale with $X(0) = 0$ and $[X, X](t) = t$. ◇

Proof: Necessity follows from properties of the Brownian motion. For sufficiency, define the function

$$Z(t) = g(X(t), t) = \exp\left(\sqrt{-1}\, u\, X(t) + \frac{1}{2} u^2 t\right), \quad u \in \mathbb{R},$$

that has continuous second and first derivatives relative to x and t, respectively, for each $u \in \mathbb{R}$. The Itô formula applied to $Z(t) = g(X(t), t) = \exp\left(\sqrt{-1}\, u\, X(t) + t\, u^2/2\right)$ gives

$$\begin{aligned}
Z(t) &= Z(0) + \int_0^t \left(\frac{\partial g(X(s), s)}{\partial x}\, dX(s) + \frac{\partial g(X(s), s)}{\partial s}\, ds\right) \\
&\quad + \frac{1}{2} \int_0^t \frac{\partial^2 g(X(s), s)}{\partial x^2}\, d[X, X](s) \\
&= 1 + \sqrt{-1}\, u \int_0^t Z(s)\, dX(s) + \frac{u^2}{2} \int_0^t Z(s)\, ds - \frac{u^2}{2} \int_0^t Z(s)\, d[X, X](s) \\
&= 1 + \sqrt{-1}\, u \int_0^t Z(s)\, dX(s).
\end{aligned}$$

By a preservation property of the stochastic integral, Z is a complex-valued local martingale so that the stopped process Z^t is a martingale. Since Z has the martingale property, we have $E[Z(t) \mid \mathcal{F}_s] = Z(s)$, $t \geq s$, which implies $E\left[e^{\sqrt{-1}\, u (X(t) - X(s))} \mid \mathcal{F}_s\right] = e^{-u^2 (t-s)/2}$ for each $u \in \mathbb{R}$ so that $X(t) - X(s)$ is independent of $\mathcal{F}_s$ and is normally distributed with mean zero and variance $t - s$. In summary, X starts at zero, is $\mathcal{F}_t$-adapted, has continuous samples, and has stationary independent Gaussian increments with mean zero and variance $t - s$. Hence, X is a Brownian motion. ■

4.6.3 Fisk-Stratonovich's integral

Recall that the symbol " $\circ$ " inserted between the integrand and the integrator has been used to denote Stratonovich integrals. We preserve this notation throughout the book. Let X and Y be semimartingales.

The **Fisk-Stratonovich** or **Stratonovich integral** of Y with respect to X is

$$\int_0^t Y(s-) \circ dX(s) = \int_0^t Y(s-)\, dX(s) + \frac{1}{2} [Y, X]^c(t). \tag{4.58}$$

Integration by parts. If at least one of the semimartingales X and Y is continuous, then ([147], Corollary, p. 76)

$$X(t)\,Y(t) - X(0)\,Y(0) = \int_{0+}^{t} X(s-) \circ dY(s) + \int_{0+}^{t} Y(s-) \circ dX(s). \quad (4.59)$$

Proof: If at least one of the semimartingales X and Y is continuous, we have $[X, Y](t) = [X, Y]^c(t) + X(0)\,Y(0)$ so that (Eq. 4.28)

$$X(t)\,Y(t) = \int_{0+}^{t} X(s-)\,dY(s) + \int_{0+}^{t} Y(s-)\,dX(s) + [X, Y]^c(t) + X(0)\,Y(0).$$

This observation, Eq. 4.58, and the equality $[X, Y] = [Y, X]$ give the integration by parts formula based on Stratonovich integrals. Eq. 4.59 coincides with the integration by parts formula in the classical calculus. ■

It is possible under some conditions to develop change of variable formulas similar to the Itô formulas based on Fisk-Stratonovich integrals.

If the third order derivative of a function $g : \mathbb{R} \to \mathbb{R}$ is continuous, then ([147], Theorem 34, p. 75)

$$\begin{aligned} g(X(t)) - g(X(0)) &= \int_{0+}^{t} g'(X(s-)) \circ dX(s) \\ &+ \sum_{0<s\leq t} \left[g(X(s)) - g(X(s-)) - g'(X(s-))\,\Delta X(s)\right]. \end{aligned} \quad (4.60)$$

The **Itô** and **Stratonovich integrals** of $g'(X)$ with respect to X are related by ([147], Theorem 34, p. 75)

$$\begin{aligned} \int_{0+}^{t} g'(X(s-)) \circ dX(s) &= \int_{0+}^{t} g'(X(s-))\,dX(s) \\ &+ \frac{1}{2} \int_{0+}^{t} g''(X(s-))\,d[X, X]^c(s). \end{aligned} \quad (4.61)$$

Proof: If Eq. 4.61 is valid, the Itô formula in Eq. 4.40 gives Eq. 4.60. Hence, it is sufficient to show that (Eq. 4.58)

$$[g'(X), X]^c(t) = \int_{0+}^{t} g''(X(s-))\,d[X, X]^c(s).$$

If $g \in C^3(\mathbb{R})$, the Itô formula can be applied to the semimartingale $g'(X)$ and gives

$$g'(X(t)) - g'(X(0)) = \int_{0+}^{t} g''(X(s-))\, dX(s) + \frac{1}{2}\int_{0+}^{t} g'''(X(s-))\, d[X, X]^c(s)$$
$$+ \sum_{0<s\leq t} \left[g'(X(s)) - g'(X(s-)) - g''(X(s-))\, \Delta X(s)\right].$$

so that

$$[g'(X), X]^c = [g''(X_-) \cdot X, X]^c + \frac{1}{2}[g'''(X_-) \cdot [X, X]^c, X]^c.$$

The terms on the right side of the above equation are (Eq. 4.34)

$$[g''(X_-) \cdot X, X]^c(t) = \int_{0+}^{t} g''(X(s-))\, d[X, X]^c(s) \quad \text{and}$$
$$[g'''(X_-) \cdot [X, X]^c, X](t) = \int_{0+}^{t} g'''(X(s-))\, d[[X, X]^c, X](s)$$
$$= \sum_{0<s\leq t} g'''(X(s-))\, \Delta[X, X]^c(s)\, \Delta X(s).$$

Since $\Delta[X, X](t) = (\Delta X(t))^2$ by Eq. 4.29, we have

$$\sum_{0<s\leq t} g'''(X(s-))\, \Delta[X, X]^c(s)\, \Delta X(s) = \left(\sum_{0<s\leq t} g'''(X(s-))\, (\Delta X(t))^3\right)^c = 0$$

so that $[g'(X), X]^c$ is as stated. ■

It can be shown that the change of variable formula in Eq. 4.60 holds also for functions $g \in C^2(\mathbb{R})$ ([147], Theorem 20, p. 222). The generalization of this change of variable formula to the case of an $\mathbb{R}^d$-valued semimartingale is also possible.

If $\boldsymbol{X}$ is a vector consisting of d semimartingales and $g : \mathbb{R}^d \to \mathbb{R}$ has second order continuous partial derivatives, then (1) $g(\boldsymbol{X})$ is a semimartingale and (2) the **change of variable** formula is ([147], Theorem 21, p. 222)

$$g(\boldsymbol{X}(t)) - g(\boldsymbol{X}(0)) = \sum_{i=1}^{d} \int_{0+}^{t} \frac{\partial g}{\partial x_i}(\boldsymbol{X}(s-)) \circ dX_i(s)$$
$$+ \sum_{0<s\leq t} \left[g(\boldsymbol{X}(s)) - g(\boldsymbol{X}(s-)) - \sum_{i=1}^{d} \frac{\partial g}{\partial x_i}(\boldsymbol{X}(s-))\, \Delta X_i(s)\right]. \quad (4.62)$$

Note: If the coordinates of $\boldsymbol{X}$ are continuous semimartingales, then the change of variable formula in Eq. 4.62 becomes

$$g(\boldsymbol{X}(t)) - g(\boldsymbol{X}(0)) = \sum_{i=1}^{d} \int_{0+}^{t} \frac{\partial g}{\partial x_i}(\boldsymbol{X}(s)) \circ dX_i(s), \quad (4.63)$$

and coincides with the change of variable formula of the classical calculus. ▲

The Itô and the Fisk-Stratonovich integrals in Eqs. 4.55 and 4.62 are related by

$$\sum_{i=1}^{d} \int_{0+}^{t} \frac{\partial g}{\partial x_i}(X(s-)) \circ dX_i(s) = \sum_{i=1}^{d} \int_{0+}^{t} \frac{\partial g}{\partial x_i}(X(s-))\, dX_i(s)$$
$$+ \frac{1}{2} \sum_{i,j=1}^{d} \int_{0+}^{t} \frac{\partial^2 g}{\partial x_i\, \partial x_j}(X(s-))\, d[X_i, X_j]^c(s). \tag{4.64}$$

Note: Generally, the evolution of the state X of a physical system is defined by a differential equation driven by a non-white (colored) noise so that X satisfies a Stratonovich stochastic differential equation ([102], Section 5.4.2). The relationship in Eq. 4.64 and the definition of the Fisk-Stratonovich integral in Section 4.6.3 can be used to express the solution X of a stochastic problem in functions of Riemann-Stieltjes and Itô integrals (Section 4.7.12). This representation of X is particularly useful for calculations because Itô integrals are martingales or semimartingales. ▲

Example 4.28: Let B be a Brownian motion and $h : \mathbb{R} \to \mathbb{R}$ be a function with a continuous second order derivative. Then

$$\int_0^t h(B(s)) \circ dB(s) = \int_0^t h(B(s))\, dB(s) + \frac{1}{2} \int_0^t h'(B(s))\, ds \tag{4.65}$$

gives the relationship between the Stratonovich and the Itô integrals of $h(B)$ with respect to B. ◇

Proof: The integrals $\int_0^t h(B(s)) \circ dB(s)$ and $\int_0^t h(B(s))\, dB(s)$ are defined by the properties of h and B. Eq. 4.64 with $d = 1$ and $h = g'$ gives Eq. 4.65 since $d[B, B]^c(t) = d[B, B](t) = dt$.

For example, if $h(x) = \exp(x)$, we have

$$\int_0^t e^{B(s)} \circ dB(s) = \int_0^t e^{B(s)}\, dB(s) + \frac{1}{2} \int_0^t e^{B(s)}\, ds \tag{4.66}$$

so that $[e^B, B](t) = \int_0^t e^{B(s)}\, ds$ (Eq. 4.58). For $h(x) = x$, we have

$$\int_0^t B(s) \circ dB(s) = \int_0^t B(s)\, dB(s) + \frac{1}{2} \int_0^t ds = \int_0^t B(s)\, dB(s) + \frac{t}{2},$$

in agreement with Examples 4.1 and 4.2. ■

Example 4.29: Let $X = N$ be a Poisson counting process. Then

$$\int_0^t h(N(s-)) \circ dN(s) = \int_0^t h(N(s-))\, dN(s) \tag{4.67}$$

for $h \in C^2(\mathbb{R})$. ◇

Proof: The coincidence of the Itô and the Stratonovich integrals with integrand depending on N and integrator N results from Eq. 4.64 with $d = 1$ because N is a quadratic pure jump semimartingale so that $[N, N]^c = 0$. The same result can be obtained from Eq. 4.58 since $[h(N), N]^c = 0$. The equality in Eq. 4.67 remains valid if N is replaced by a compound Poisson process. ■

4.7 Stochastic differential equations

It is common in applications to characterize the current state of a physical system by an $\mathbb{R}^d$-valued function of time $\boldsymbol{x}(t)$, $t \geq 0$, called the **state vector**. For example, the coordinates of $\boldsymbol{x}$ may denote the displacements and velocities at the degrees of freedom of a discrete dynamical system or at the nodes of a finite difference representation of a continuum. Generally, the behavior of a physical system subjected to an input $\boldsymbol{w}(t)$, $t \geq 0$, is specified by a differential equation of the form

$$\dot{\boldsymbol{x}}(t) = \boldsymbol{a}(\boldsymbol{x}(t), t) + \boldsymbol{b}(\boldsymbol{x}(t), t)\, \boldsymbol{w}(t), \quad t \geq 0, \tag{4.68}$$

giving the rate of change of the state vector, where the functions $\boldsymbol{a}$ and $\boldsymbol{b}$ depend on the system properties. Classical analysis is based on the assumptions that the system properties and the input are perfectly known and deterministic. The evolution of the state vector $\boldsymbol{x}$ is given by Riemann-Stieltjes integrals with integrands depending on the system properties and integrators depending on input and time.

In this section, we generalize Eq. 4.68 by assuming that the input is a stochastic process. The properties of the system are still assumed to be deterministic and perfectly known so that the coefficients of Eq. 4.68 are deterministic, known functions of the state vector and time. Because the input is random, the state vector is a stochastic process. Let $\boldsymbol{X}$ denote the solution of Eq. 4.68 with $\boldsymbol{w}$ replaced by a stochastic process $\boldsymbol{W}$. It is common in applications to assume that $\boldsymbol{W}$ is a white noise process for the following reasons ([175], Chapter 8).

1. In many cases the input has a much shorter memory than the system.

2. A broad class of correlated or colored inputs can be modeled by the output of linear filters to white noise so that the augmented vector consisting of $\boldsymbol{X}$ and the state of the filter defining the colored input satisfies a differential equation driven by white noise.

3. The first two moments of the state $\boldsymbol{X}$ of a linear system driven by white noise can be obtained simply provided that these moments exist. The system defined by Eq. 4.68 is said to be **linear** if $\boldsymbol{a}(\boldsymbol{x}(t), t) = \boldsymbol{\alpha}(t)\, \boldsymbol{x}(t)$ and $\boldsymbol{b}(\boldsymbol{x}(t), t) = \boldsymbol{\beta}(t)\, \boldsymbol{x}(t)$, where the matrices $\boldsymbol{\alpha}$ and $\boldsymbol{\beta}$ are functions of time. We will adopt in our discussion a more restrictive definition for linear systems, which is common in random vibration studies. We say that the system in Eq. 4.68 is linear if $\boldsymbol{a}(\boldsymbol{x}(t), t) = \boldsymbol{\alpha}(t)\, \boldsymbol{x}(t)$ and $\boldsymbol{b}(\boldsymbol{x}(t), t) = \boldsymbol{\beta}(t)$.

Table 4.1. Noise input processes

INPUT NOISE	DEFINITION	
	Increment of	*Formal*
Gaussian white noise	Brownian motion, B	$dB(t)/dt$
Poisson white noise	Compound Poisson, C	$dC(t)/dt$
Lévy noise	Lévy, L	$dL(t)/dt$
Semimartingale noise	Semimartingale, S	$dS(t)/dt$

Table 4.1 summarizes the types of white noise processes used in this book. If the jumps of the compound Poisson process C are in L_2, the second moment properties of the Poisson and the Gaussian white noise processes have the same functional form and these processes can be made equal in the second moment sense. The Lévy white noise includes both the Gaussian and the Poisson white noise processes (Section 3.14.2). The semimartingale noise is considered at the end of this section. The use of the white noise model results in a significant simplification in calculations, for example, linear filters driven by white noise ([175], Section 5.2) but can entail significant technical difficulties. For example, the Gaussian white noise is frequently interpreted in applications as the derivative of the Brownian motion, a stochastic process with a.s. non-differentiable sample paths! In our discussion, we will deal with increments of the Brownian motion, compound Poisson, Lévy, and semimartingale processes rather than the corresponding white noise processes. Accordingly, the evolution of the state vector $\boldsymbol{X}$ is defined by the

Stochastic differential equation

$$d\boldsymbol{X}(t) = \boldsymbol{a}(\boldsymbol{X}(t-), t)\, dt + \boldsymbol{b}(\boldsymbol{X}(t-), t)\, d\boldsymbol{Y}(t), \quad t \geq 0, \tag{4.69}$$

whose meaning is given by the

Stochastic integral equation

$$\boldsymbol{X}(t) = \boldsymbol{X}(0) + \int_0^t \boldsymbol{a}(\boldsymbol{X}(s-), s)\, ds + \int_0^t \boldsymbol{b}(\boldsymbol{X}(s-), s)\, d\boldsymbol{Y}(s), \quad t \geq 0, \tag{4.70}$$

where $\boldsymbol{a}$, $\boldsymbol{b}$ are $(d, 1)$, (d, d')-dimensional matrices whose entries are real-valued Borel measurable functions, $\boldsymbol{Y}$ is a vector in $\mathbb{R}^{d'}$ consisting of d' real-valued semimartingales, and the state $\boldsymbol{X}$ is an $\mathbb{R}^d$-valued stochastic process. The input $\boldsymbol{Y}$ includes the Brownian motion, Poisson, compound Poisson, and Lévy processes. If $\boldsymbol{Y}$ is a Brownian motion $\boldsymbol{B}$, the state $\boldsymbol{X}$ is called a **diffusion** or **Itô** process and the matrices $\boldsymbol{a}$ and $\boldsymbol{b}\,\boldsymbol{b}^T$ are referred to as **drift** and **diffusion** coefficients, respectively. The first and second integrals in Eq. 4.70 are Riemann-Stieltjes and stochastic or Itô integrals, respectively.

Two types of solutions are considered for Eqs. 4.69-4.70, strong and weak solutions. If the initial value $\boldsymbol{X}(0)$ of these equations is deterministic, a **strong solution** is a stochastic process $\boldsymbol{X}(t)$, $t \geq 0$, satisfying the following conditions ([131], p. 137).

1. $\boldsymbol{X}$ is adapted to the natural filtration $\mathcal{F}_t^{\boldsymbol{Y}} = \sigma(\boldsymbol{Y}(s), 0 \leq s \leq t)$ of $\boldsymbol{Y}$, that is, $\boldsymbol{X}(t)$ is a function of $\boldsymbol{Y}(s)$, $s \leq t$, at each time $t \geq 0$.

2. $\boldsymbol{X}$ is a function of the sample paths of $\boldsymbol{Y}$ and the coefficients $\boldsymbol{a}$ and $\boldsymbol{b}$.

3. The Riemann-Stieltjes and Itô integrals in Eq. 4.70 are well defined at all times $t \geq 0$.

If $\boldsymbol{X}(0)$ is random, $\mathcal{F}_t^{\boldsymbol{Y}}$ needs to be extended to $\mathcal{F}_t = \sigma\left(\boldsymbol{X}(0), \mathcal{F}_t^{\boldsymbol{Y}}\right)$. A version of the input $\boldsymbol{Y}$ needs to be specified to construct a strong solution of Eq. 4.70. If we replace one version of $\boldsymbol{Y}$ by another version of this process, a different strong solution may result because these versions of $\boldsymbol{Y}$ can have different sample properties. We construct approximations of the strong solution in Monte Carlo studies because paths of $\boldsymbol{X}$ are calculated from paths of $\boldsymbol{Y}$ by integrating Eq. 4.70. The path behavior is not essential for the weak solution. A **weak solution** of Eqs. 4.69-4.70 is a pair of adapted processes $(\tilde{\boldsymbol{Y}}, \tilde{\boldsymbol{X}})$ defined on a filtered probability space $(\Omega, \mathcal{H}, (\mathcal{H}_t)_{t \geq 0}, P)$ such that $\tilde{\boldsymbol{Y}}$ is a version of $\boldsymbol{Y}$ and the pair $(\tilde{\boldsymbol{Y}}, \tilde{\boldsymbol{X}})$ satisfies Eq. 4.70. The weak solution is completely defined by the initial condition, the functions $\boldsymbol{a}$ and $\boldsymbol{b}$, and the finite dimensional distributions of $\boldsymbol{Y}$. The particular version of the input does not have to be specified. The above definitions show that a strong solution is a weak solution but the converse is not generally true.

The uniqueness of the solution of Eqs. 4.69-4.70 is generally defined in two ways. A strong solution is said to be **unique in the strong** or **pathwise sense** if two different solutions of these equations have the same sample paths except on a subset of Ω of measure zero. The strong uniqueness requires that the set of solutions of Eqs. 4.69-4.70 consists of indistinguishable processes. On the other hand, two solutions, weak or strong, of these equations are **unique in the weak sense** if they have the same finite dimensional distributions. Hence, the collection of weak solutions of Eqs. 4.69-4.70 consists of processes that are versions of each other ([42], Section 10.4, [131], Section 3.2.1, [135], Section 5.3).

Example 4.30: Consider the stochastic differential equation

$$dX(t) = \zeta(X(t))\, dB(t), \quad t \geq 0,$$

where $X(0) = 0$, B is a Brownian motion starting at zero, and $\zeta(x) = -1, 0$, and 1 for $x < 0$, $x = 0$, and $x > 0$, respectively. It can be shown that this equation has a weak solution ([42], Section 7.3). The solution is unique in the weak sense ([42], pp. 248-249) but it is not unique in the strong sense. ◇

Proof: Let $X(\cdot,\omega)$ be a solution of the above stochastic differential equation corresponding to a sample $B(\cdot,\omega)$ of B, that is,

$$X(t,\omega) - X(0,\omega) = \int_0^t \zeta(X(s,\omega))\,dB(s,\omega) \quad \text{so that}$$
$$\tilde{X}(t,\omega) - \tilde{X}(0,\omega) = \int_0^t \zeta(\tilde{X}(s,\omega))\,dB(s,\omega),$$

where $\tilde{X} = -X$. If $X(0) = 0$ and $X(\cdot,\omega)$ is a solution corresponding to a sample $B(\cdot,\omega)$ of B, then $\tilde{X}(\cdot,\omega)$ is also a solution corresponding to the same sample of B. The solutions X and $\tilde{X}$ have the same probability law but their sample paths differ. ■

4.7.1 Brownian motion input

Let $\boldsymbol{Y}$ in Eqs. 4.69 and 4.70 be an $\mathbb{R}^{d'}$-valued Brownian motion process $\boldsymbol{B} = (B_1, \ldots, B_{d'})$, where B_i, $i = 1, \ldots, d'$, are independent real-valued Brownian motions. The meaning of Eq. 4.69 with $\boldsymbol{B}$ in place of $\boldsymbol{Y}$ is given by the stochastic integral equation

$$\boldsymbol{X}(t) = \boldsymbol{X}(0) + \int_0^t \boldsymbol{a}(\boldsymbol{X}(s), s)\,ds + \int_0^t \boldsymbol{b}(\boldsymbol{X}(s), s)\,d\boldsymbol{B}(s), \qquad (4.71)$$

where $\int_0^t \boldsymbol{a}(\boldsymbol{X}(s), s)\,ds$ and $\int_0^t \boldsymbol{b}(\boldsymbol{X}(s), s)\,d\boldsymbol{B}(s)$ are Riemann and Itô integrals, respectively.

Note: Generally, $\int_0^t \boldsymbol{b}(\boldsymbol{X}(s), s)\,d\boldsymbol{B}(s)$ cannot be defined as a path by path Riemann-Stieltjes integral because almost all sample paths of the Brownian motion process are of unbounded variation. We have seen that there is a relatively simple relationship between the Itô and Stratonovich integrals (Eq. 4.58) so that the solution of Eq. 4.71 can also be expressed in terms of Riemann-Stieltjes and Stratonovich integrals. The process $\boldsymbol{X}(s-)$ in Eqs. 4.69 and 4.70 is replaced by $\boldsymbol{X}(s)$ in Eq. 4.71 since diffusion processes have continuous samples, as we will see later in Section 4.7.1.1.

Note that diffusion processes are Markov, that is, the solution $\boldsymbol{X}$ of Eq. 4.71 is a Markov process. Let $t_0 \in (0, t)$ and $\boldsymbol{X}(t_0) = \boldsymbol{x}$. Because the increments of $\boldsymbol{B}$ are independent of the past, future states,

$$\boldsymbol{X}(t) = \boldsymbol{x} + \int_{t_0}^t \boldsymbol{a}(\boldsymbol{X}(s), s)\,ds + \int_{t_0}^t \boldsymbol{b}(\boldsymbol{X}(s), s)\,d\boldsymbol{B}(s), \quad t \geq t_0,$$

of $\boldsymbol{X}$ are independent of the past states $\boldsymbol{X}(u)$, $u < t_0$, of this process conditional on its present state $\boldsymbol{X}(t_0) = \boldsymbol{x}$. The converse is not true. For example, the compound Poisson process is Markov but is not a diffusion process.

We also note that the solution of Eq. 4.71 has the **strong Markov property**, that is, if T is an $\mathcal{F}_t$-stopping time, the process $X(T + \tau)$, $\tau \geq 0$, depends only on $X(T)$, where $\mathcal{F}_t$ denotes the filtration generated by the natural filtration of the Brownian motion $\boldsymbol{B}$ and the initial state $\boldsymbol{X}(0)$. ▲

This section gives conditions for the existence and uniqueness of the solution of Eq. 4.71, outlines some essential properties of the solution, illustrates some theoretical concepts by examples, and demonstrates how Itô's formula can be used to develop differential equations for the moments, the density, and the characteristic function of $\boldsymbol{X}$. In applications Eq. 4.71 is frequently given in the form

$$\dot{\boldsymbol{X}}(t) = \boldsymbol{a}(\boldsymbol{X}(t), t) + \boldsymbol{b}(\boldsymbol{X}(t), t)\, \boldsymbol{W}_g(t), \quad t \geq 0, \tag{4.72}$$

where $\boldsymbol{W}_g$, called **Gaussian white noise**, is interpreted as the formal derivative of the Brownian motion process $\boldsymbol{B}$ ([175], Section 5.2.1).

Example 4.31: Let X be the solution of $dX(t) = -\alpha\, X(t)\, dt + \beta\, dB(t)$, $t \geq 0$, where $\alpha > 0$ and β are some constants, $X(0) \sim N(\mu(0), \gamma(0))$, B is a Brownian motion, and $X(0)$ is independent of B. This real-valued diffusion process, called the **Ornstein-Uhlenbeck** process, is a special case of Eq. 4.71 with $d = d' = 1$, drift $-\alpha\, x$, and diffusion β^2. The theorems in the following section guarantee the existence and the uniqueness of the solution X.

Samples of the strong solution of this equation can be calculated from samples $B(\cdot, \omega)$ of B and the recurrence formula

$$X(t + \Delta t, \omega) = X(t, \omega)\, e^{-\alpha\, \Delta t} + \beta \int_t^{t+\Delta t} e^{-\alpha\,(t+\Delta t - s)}\, dB(s, \omega)$$

giving X at the end of a time interval $[t, t + \Delta t]$ from its value at the beginning of the interval and the input in $[t, t + \Delta t]$. The weak solution of the Ornstein-Uhlenbeck equation is given by a pair of a Brownian motion $\tilde{B}$ and a Gaussian process $\tilde{X}$ with mean $\mu(t) = E[X(t)] = \mu(0)\, e^{-\alpha t}$, variance $\gamma(t) = E[\tilde{X}(t)^2] = \gamma(0)\, e^{-2\alpha t} + \beta^2/(2\,\alpha)\left(1 - e^{-2\alpha t}\right)$, and covariance $c(t, s) = E[\tilde{X}(t)\, \tilde{X}(s)] = \gamma(s \wedge t)\, e^{-\alpha\, |t-s|}$. ◇

Proof: The recurrence formula for X shows that the Ornstein-Uhlenbeck process is Gaussian so that the second moment properties of X define its finite dimensional distributions. We establish here differential equations for some of the second moment properties of X. Complete differential equations for the second moment properties of the state vector of a linear system driven by white noise are in Section 7.2.1.1.

The mean equation can be obtained by averaging the defining equation of X. The stochastic integral equation

$$X(t) = X(s) - \alpha \int_s^t X(u)\, du + \beta \int_s^t dB(u), \quad t \geq s,$$

of X multiplied by $X(s)$ and averaged gives

$$E[X(t)\, X(s)] = E[X(s)^2] - \alpha\, E\left[\int_s^t X(u)\, X(s)\, du\right] + \beta\, E\left[\int_s^t X(s)\, dB(u)\right]$$

or $r(t, s) = r(s, s) - \alpha \int_s^t r(u, s)\, du$ by using the Fubini theorem, the independence of $X(s)$ from future increments of the Brownian motion, and $E[dB(u)] = 0$, where $r(t, s) = E[X(t)\, X(s)]$. This equation gives $\partial r(t, s)/\partial t = -\alpha\, r(t, s)$ by differentiation with respect to t so that $r(t, s) = r(s, s)\, e^{-\alpha\,(t-s)}$. ■

4.7.1.1 Existence and uniqueness of a solution

We need some definitions to state the main results of this section. Let $\boldsymbol{a}(\boldsymbol{x})$ and $\boldsymbol{b}(\boldsymbol{x})$ be $(d, 1)$ and (d, d') matrices whose entries are functions of only $\boldsymbol{x} \in \mathbb{R}^d$. We say that $\boldsymbol{a}$ and $\boldsymbol{b}$ satisfy the **uniform Lipschitz** conditions if there exist a constant $c > 0$ such that

$$\begin{aligned} &\| \boldsymbol{a}(\boldsymbol{x}_1) - \boldsymbol{a}(\boldsymbol{x}_2) \| \leq c \ \| \boldsymbol{x}_1 - \boldsymbol{x}_2 \| \quad \text{and} \\ &\| \boldsymbol{b}(\boldsymbol{x}_1) - \boldsymbol{b}(\boldsymbol{x}_2) \|_m \leq c \ \| \boldsymbol{x}_1 - \boldsymbol{x}_2 \|, \quad \boldsymbol{x}_1, \boldsymbol{x}_2 \in \mathbb{R}^d, \end{aligned} \tag{4.73}$$

where $\| \boldsymbol{\xi} \| = \left(\sum_{i=1}^{d} \xi_i^2 \right)^{1/2}$ and $\| \boldsymbol{b} \|_m = \left(\sum_{i=1}^{d} \sum_{j=1}^{d'} b_{ij}^2 \right)^{1/2}$ denote the usual Euclidean norm and a matrix norm, respectively.

The matrices $\boldsymbol{a}$ and $\boldsymbol{b}$ are said to be **locally Lipschitz** if, for each $\alpha > 0$, there is a constant $c_\alpha > 0$ such that

$$\begin{aligned} &\| \boldsymbol{a}(\boldsymbol{x}_1) - \boldsymbol{a}(\boldsymbol{x}_2) \| \leq c_\alpha \ \| \boldsymbol{x}_1 - \boldsymbol{x}_2 \| \quad \text{and} \\ &\| \boldsymbol{b}(\boldsymbol{x}_1) - \boldsymbol{b}(\boldsymbol{x}_2) \|_m \leq c_\alpha \ \| \boldsymbol{x}_1 - \boldsymbol{x}_2 \| \end{aligned} \tag{4.74}$$

for $\boldsymbol{x}_1, \boldsymbol{x}_2 \in \mathbb{R}^d$ satisfying the condition $\| \boldsymbol{x}_1 \|, \| \boldsymbol{x}_2 \| < \alpha$.

The matrices $\boldsymbol{a}$ and $\boldsymbol{b}$ satisfy the **growth condition** if there is $k > 0$ such that

$$\boldsymbol{x} \cdot \boldsymbol{a}(\boldsymbol{x}) \leq k\,(1 + \| \boldsymbol{x} \|^2) \quad \text{and} \quad \| \boldsymbol{b}(\boldsymbol{x}) \|_m^2 \leq k\,(1 + \| \boldsymbol{x} \|^2), \tag{4.75}$$

where $\boldsymbol{x} \cdot \boldsymbol{a}(\boldsymbol{x}) = \sum_{i=1}^{d} x_i\, a_i(\boldsymbol{x})$. The condition $\boldsymbol{x} \cdot \boldsymbol{a}(\boldsymbol{x}) \leq k\,(1 + \| \boldsymbol{x} \|^2)$ is weaker than the typically stated growth condition, $\| \boldsymbol{a}(\boldsymbol{x}) \|^2 \leq k'\,(1 + \| \boldsymbol{x} \|^2)$, because $(\boldsymbol{x} - \boldsymbol{a}(\boldsymbol{x})) \cdot (\boldsymbol{x} - \boldsymbol{a}(\boldsymbol{x})) \geq 0$ so that

$$\boldsymbol{x} \cdot \boldsymbol{a}(\boldsymbol{x}) \leq \frac{1}{2} \left(\| \boldsymbol{x} \|^2 + \| \boldsymbol{a}(\boldsymbol{x}) \|^2 \right) \leq \frac{k' + 1}{2} \left(1 + \| \boldsymbol{x} \|^2 \right).$$

We present two theorems giving **sufficient conditions for the existence and uniqueness** of the solution of Eq. 4.71 with drift and diffusion coefficients that are not explicit functions of time. Alternative versions of these theorems dealing with drift and diffusion coefficients that may depend explicitly on time are available ([115], Section 4.5, [117], Theorem 7.1.1, p. 195, [135], Theorem 5.2.1, p. 66, [165], Section 4.2). We also note that the theorems in this section can be applied to general diffusion processes by considering an augmented state vector $\tilde{\boldsymbol{X}}(t) = (\tilde{\boldsymbol{X}}_1(t) = \boldsymbol{X}(t), \tilde{X}_2(t) = t) \in \mathbb{R}^{d+1}$, where $\tilde{\boldsymbol{X}}_1 = \boldsymbol{X}$ is defined by Eq. 4.71 and $\tilde{X}_2$ is the solution of the deterministic differential equation $d\,\tilde{X}_2(t) = dt$ with the initial condition $X_2(0) = 0$.

If (1) the drift and diffusion coefficients do not depend on time explicitly, are bounded functions, and satisfy the uniform Lipschitz conditions in Eq. 4.73, (2) $\boldsymbol{B}$ is a Brownian motion martingale on a filtered probability space $(\Omega, \mathcal{F}, (\mathcal{F}_t)_{t\geq 0}, P)$, (3) $\boldsymbol{B}(0) = \mathbf{0}$, and (4) $\boldsymbol{X}(0)$ is $\mathcal{F}_0$-measurable, then

- There exists a strong solution $\boldsymbol{X}$ of Eq. 4.71 unique in the strong sense,
- $\boldsymbol{X}$ is a $\mathcal{B}([0, \infty)) \times \mathcal{F}$-measurable, adapted, and continuous process, and
- The law of $\boldsymbol{X}$ is uniquely determined by the drift and diffusion coefficients and the laws of $\boldsymbol{B}$ and $\boldsymbol{X}(0)$ ([42], Theorem 10.5, p. 228).

Proof: A Brownian motion $\boldsymbol{B}$ is a $\sigma(\boldsymbol{B}(s), 0 \leq s \leq t)$-martingale. The strong uniqueness of the solution means that if $\boldsymbol{X}^{(1)}$ and $\boldsymbol{X}^{(2)}$ are two t-continuous solutions of Eq. 4.71, then $\boldsymbol{X}^{(1)}(t, \omega) = \boldsymbol{X}^{(2)}(t, \omega)$ for all $t \geq 0$ a.s.

If some of the conditions of the above theorem are violated, it does not mean that Eq. 4.71 does not have a solution or that its solution is not unique. The theorem gives only **sufficient conditions** for the existence and uniqueness of the solution of Eq. 4.71. We outline the plan of the proof of this theorem. Technicalities involved in this proof can be found in [42] (Chapter 10). Let

$$\boldsymbol{I}(t, \boldsymbol{X}) = \boldsymbol{X}(0) + \boldsymbol{A}(t) + \boldsymbol{M}(t) \tag{4.76}$$

denote the right side of Eq. 4.71 with drift and diffusion coefficients depending only on the state $\boldsymbol{X}$, where

$$\boldsymbol{A}(t) = \int_0^t \boldsymbol{a}(\boldsymbol{X}(s))\, ds \quad \text{and} \quad \boldsymbol{M}(t) = \int_0^t \boldsymbol{b}(\boldsymbol{X}(s))\, d\boldsymbol{B}(s). \tag{4.77}$$

Generally, the processes $\boldsymbol{I}(\cdot, \boldsymbol{X})$ and $\boldsymbol{X}(\cdot)$ differ. This theorem gives conditions under which these processes are indistinguishable. Let $(\Omega, \mathcal{F}, (\mathcal{F}_t)_{t\geq 0}, P)$ be a filtered probability space with $\mathcal{F}_t = \sigma(\boldsymbol{X}(0), \boldsymbol{B}(s) : 0 \leq s \leq t)$ defined by the initial condition $\boldsymbol{X}(0)$ and the history of the Brownian motion $\boldsymbol{B}$ up to time $t \geq 0$. The Brownian motion $\boldsymbol{B}$ is $\mathcal{F}_t$-adapted by the definition of this filtration. We present now the essential steps of the proof.

(1) *If $\boldsymbol{X}$ is $\mathcal{F}_t$-adapted and $\mathcal{B}([0, \infty)) \times \mathcal{F}$-measurable, then $\boldsymbol{I}(t, \boldsymbol{X})$ is a continuous semimartingale.* Since $\boldsymbol{X}$ is $\mathcal{B}([0, \infty)) \times \mathcal{F}$-measurable and $\boldsymbol{b}$ is continuous, the entries of $\boldsymbol{b}(\boldsymbol{X})$ are also $\mathcal{B}([0, \infty)) \times \mathcal{F}$-measurable. Since the diffusion coefficients are bounded, $\boldsymbol{M}$ is in L_2. These observations and a preservation property of the stochastic integral show that $\boldsymbol{M}$ is a square integrable martingale. The martingale $\boldsymbol{M}$ starts at zero and is continuous. The integral giving $\boldsymbol{A}$ can be defined path by path because the drift coefficients are bounded functions. Since $\boldsymbol{X} \in \mathcal{B}([0, \infty)) \times \mathcal{F}$ and $\boldsymbol{a}$ is continuous, the entries of $\boldsymbol{a}(\boldsymbol{X})$ are $\mathcal{B}([0, \infty)) \times \mathcal{F}$-measurable. Also, $\boldsymbol{A}$ is an adapted continuous process, which starts at zero and is of finite variation on compacts. Hence, $\boldsymbol{A}$ is also a semimartingale. We conclude that $\boldsymbol{I}(\cdot, \boldsymbol{X})$ is a semimartingale with continuous samples.

(2) *Sensitivity of $\boldsymbol{I}(\cdot, \boldsymbol{X})$ to $\boldsymbol{X}$.* The inequality ([42], Lemma 10.1, p. 222)

$$E\left[\sup_{0\leq s\leq t} \| \boldsymbol{I}(s, \boldsymbol{Y}) - \boldsymbol{I}(s, \boldsymbol{Z}) \|^2\right] \leq 3\, E[\| \boldsymbol{Y}(0) - \boldsymbol{Z}(0) \|^2] + c_\tau\, E\left[\int_0^t \| \boldsymbol{Y}(s) - \boldsymbol{Z}(s) \|^2\, ds\right]$$

holds for each $\tau > 0$, $t \in [0, \tau]$, and $\mathcal{B}([0, \infty)) \times \mathcal{F}$-measurable, $\mathcal{F}_t$-adapted processes $\boldsymbol{Y}$ and $\boldsymbol{Z}$ such that $\boldsymbol{Y}(s) - \boldsymbol{Z}(s) \in L_2$, where $c_\tau > 0$ is a constant. The distance between $\boldsymbol{I}(\cdot, \boldsymbol{Y})$ and $\boldsymbol{I}(\cdot, \boldsymbol{Z})$ in $[0, t]$ depends on differences between the initial values and the time histories of $\boldsymbol{Y}$ and $\boldsymbol{Z}$.

(3) *The uniqueness of the solution of* Eq. 4.71. We show that there is at most one $\mathcal{B}([0, \infty)) \times \mathcal{F}$-measurable adapted solution of this equation ([42], Theorem 10.3, p. 224). Let $\boldsymbol{Y}$ and $\boldsymbol{Z}$ be two solutions of Eq. 4.71 so that $\boldsymbol{Y}(0) = \boldsymbol{Z}(0) = \boldsymbol{X}(0)$, $\boldsymbol{Y}(t) = \boldsymbol{I}(t, \boldsymbol{Y})$, and $\boldsymbol{Z}(t) = \boldsymbol{I}(t, \boldsymbol{Z})$ for all $t \geq 0$. For a fixed $\tau > 0$, we have

$$E[\| \boldsymbol{Y}(t) - \boldsymbol{Z}(t) \|^2] \leq c_\tau \, E\left[\int_0^t \| \boldsymbol{Y}(s) - \boldsymbol{Z}(s) \|^2 \, ds\right]$$
$$= c_\tau \int_0^t E\left[\| \boldsymbol{Y}(s) - \boldsymbol{Z}(s) \|^2\right] ds$$

for $t \in [0, \tau]$ by the above inequality and the Fubini theorem. We need one additional fact to establish the uniqueness of the solution. Consider two Lebesgue integrable functions f and g in $[0, \tau]$ for some $\tau \in (0, \infty)$. If there is a constant $\zeta > 0$ such that $f(t) \leq g(t) + \zeta \int_0^t f(s)\, ds$ for all $t \in [0, \tau]$, then ([42], Lemma 10.2, p. 224)

$$f(t) \leq g(t) + \zeta \int_0^t e^{\zeta (t-s)} \, g(s) \, ds.$$

The latter inequality with $f(t) = E[\| \boldsymbol{Y}(t) - \boldsymbol{Z}(t) \|^2]$, $g(t) = 0$, and $\zeta = c_\tau$ gives $E[\| \boldsymbol{Y}(t) - \boldsymbol{Z}(t) \|^2] = 0$ so that $P(\boldsymbol{Y}(t) = \boldsymbol{Z}(t)) = 1$ for each $t \in [0, \tau]$ and, since τ is arbitrary, this result holds for each $t \geq 0$. The processes $\boldsymbol{Y}$ and $\boldsymbol{X}$ are indistinguishable since they are continuous (Section 3.8).

(4) *The existence of the solution of* Eq. 4.71. The proof is based on the construction of a convergent sequence of processes defined by

$$\boldsymbol{X}^{(0)}(t) = \boldsymbol{X}(0),$$
$$\boldsymbol{X}^{(n)}(t) = \boldsymbol{I}(t, \boldsymbol{X}^{(n-1)}), \quad n = 1, 2, \ldots,$$

for all $t \geq 0$. The properties of the $\boldsymbol{I}$ in Eq. 4.76 imply that the processes $\boldsymbol{X}^{(n)}$ defined by the above recurrence formula are $\mathcal{F}_t$-adapted and continuous, and their law is uniquely determined by $\boldsymbol{X}(0)$ and $\boldsymbol{B}$. Moreover, the sequence of processes $\boldsymbol{X}^{(n)}$ converges to a limit $\boldsymbol{X}$ a.s. and uniformly in $[0, \tau]$ and this property holds on any bounded time interval since τ is arbitrary. The limit $\boldsymbol{X}$ has the same properties as the processes $\boldsymbol{X}^{(n)}$, so that $\boldsymbol{X}$ is an adapted continuous process whose probability law is determined by $\boldsymbol{X}(0)$ and $\boldsymbol{B}$. The process $\boldsymbol{X}$ is the strong solution of Eq. 4.71 since it is constructed from a specified version of the Brownian motion and some initial conditions. Hence, the processes $\boldsymbol{I}(\cdot, \boldsymbol{X})$ and $\boldsymbol{X}$ are indistinguishable in $[0, \infty)$. ■

> If (1) $\boldsymbol{X}(0)$ and $\boldsymbol{B}$ satisfy the conditions in the previous existence and uniqueness theorem and (2) the matrices $\boldsymbol{a}$ and $\boldsymbol{b}$ are locally Lipschitz continuous functions (Eq. 4.74) and satisfy some growth conditions (Eq. 4.75), then there is a unique, adapted, and continuous solution $\boldsymbol{X}$ of Eq. 4.71 whose law is uniquely determined by the laws of $\boldsymbol{X}(0)$ and $\boldsymbol{B}$ ([42], Theorem 10.6, p. 229).

Example 4.32: Consider the deterministic differential equation $dx(t) = x(t)^2\,dt$, $t \geq 0$, starting at $x(0) = 1$. The drift coefficient $a(x) = x^2$ does not satisfy the growth condition in Eq. 4.75 for all $x \in \mathbb{R}$. Yet, this equation has the unique solution $x(t) = 1/(1-t)$ in $[0, 1)$. The solution $x(t) = 1/(1-t)$ becomes unbounded as $t \uparrow 1$, so that the solution does not exist at all times. The behavior of $x(t)$ suggests that the growth condition is needed to assure the existence of a solution for all times t. ◇

Example 4.33: Let x be the solution of $dx(t) = 3\,x(t)^{2/3}\,dt$, $t \geq 0$, with the initial condition $x(0) = 0$. The drift $a(x) = 3\,x^{2/3}$ of this equation does not satisfy the Lipschitz condition in Eq. 4.73 at $x = 0$. This equation has a solution but its solution is not unique. For example, any member of the infinite collection of functions $x_a(t)$ defined by $x_a(t) = (t-a)^3$ for $t > a > 0$ and $x_a(t) = 0$ for $t \leq a$ is a solution. This observation suggests that the Lipschitz conditions are needed to have a unique solution. ◇

Example 4.34: The stochastic differential equation

$$dX(t) = c\,X(t)\,dt + \sigma\,X(t)\,dB(t), \qquad t \in [0, \tau], \tag{4.78}$$

has the unique solution

$$X(t) = X(0)\,\exp[(c - \sigma^2/2)\,t + \sigma\,B(t)], \tag{4.79}$$

called the **geometric Brownian motion** process. ◇

Proof: That the above stochastic differential equation has a unique solution follows from the second existence and uniqueness theorem in Section 4.7.1.1 since (1) the coefficients $a(x) = c\,x$ and $b(x) = \sigma\,x$ of this equation satisfy the uniform Lipschitz conditions in Eq. 4.73 and (2) the growth conditions in Eq. 4.75.

The first existence and uniqueness theorem in Section 4.7.1.1 cannot be applied directly because the functions $a(x)$, $b(x)$ are not bounded. However, we can modify the original problem so that its coefficients satisfy the requirements of this theorem. Let X be the solution of the stochastic differential equation

$$dX(t) = a(X(t))\,dt + b(X(t))\,dB(t),$$

where the coefficients a, b satisfy the uniform Lipschitz conditions but may not be bounded. For $\xi > 0$ define the function

$$[x]_\xi = \begin{cases} -\xi & \text{if } x < -\xi, \\ x & \text{if } -\xi \leq x \leq \xi, \\ \xi & \text{if } x > \xi, \end{cases}$$

and consider the stochastic differential equation

$$dX^{(\xi)}(t) = a\left(\left[X^{(\xi)}(t)\right]_\xi\right) dt + a\left(\left[X^{(\xi)}(t)\right]_\xi\right) dB(t).$$

Since the functions $a([\cdot]_\xi)$ and $b([\cdot]_\xi)$ satisfy the conditions of the first existence and uniqueness theorem in Section 4.7.1.1, the solution $X^{(\xi)}$ of the above equation exists and is unique in the strong sense for each $\xi > 0$. Hence, Eq. 4.78 has a unique solution in the strong sense.

We now show that

$$X(t) = g(B(t), t) = X(0)\,\exp[(c - \sigma^2/2)\,t + \sigma\,B(t)]$$

solves Eq. 4.78. Let $i(t) = t$ denote the identity function. The multi-dimensional Itô formula applied to the mapping $(B, i) \mapsto g(B, i)$ gives (Eq. 4.56)

$$\begin{aligned} X(t) - X(0) = g(B(t), t) - g(B(0), 0) &= \int_0^t \frac{\partial g(B(s), s)}{\partial u}\, dB(s) \\ + \int_0^t \frac{\partial g(B(s), s)}{\partial s}\, ds &+ \frac{1}{2}\int_0^t \frac{\partial^2 g(B(s), s)}{\partial u^2}\, d[B, B](s) \\ = \int_0^t \sigma\, X(s)\, dB(s) + \int_0^t (c - \sigma^2/2)\, X(s)\, ds &+ \frac{1}{2}\int_0^t \sigma^2\, X(s)\, ds, \end{aligned}$$

where $\partial g/\partial u$ and $\partial^2 g/\partial u^2$ denote partial derivatives of g with respect to its first argument. Hence, the process X in Eq. 4.79 is the solution of Eq. 4.78. ■

4.7.1.2 Properties of diffusion processes

We have seen that the solution X of Eq. 4.71 is a semimartingale process that is also Markov. The largest difference during a bounded time interval $[0, \tau]$, $\tau > 0$, between two solutions of Eq. 4.71 corresponding to different initial conditions can be bounded by a function depending on τ and the difference between the two initial conditions. We restate these results and illustrate them with simple examples. Examples are also used to show that memoryless monotonic transformations of diffusion processes are diffusion processes. A useful result for applications, the Wong-Zakai theorem, is stated at the end of the section.

Suppose that the drift and diffusion coefficients in Eq. 4.71 satisfy the conditions of the first theorem of existence and uniqueness in Section 4.7.1.1. If $\boldsymbol{X}$ and $\boldsymbol{Z}$ are solutions of Eq. 4.71 corresponding to the initial conditions $\boldsymbol{X}(0)$ and $\boldsymbol{Z}(0)$, respectively, such that $\boldsymbol{X}(0), \boldsymbol{Z}(0) \in \mathcal{F}_0$ and $\boldsymbol{X}(0) - \boldsymbol{Z}(0) \in L_2$, then

$$E\left[\sup_{0 \le t \le \tau} \| \boldsymbol{X}(t) - \boldsymbol{Z}(t) \|^2\right] \le 3\, E\left[\| \boldsymbol{X}(0) - \boldsymbol{Z}(0) \|^2\right] e^{c_\tau \tau}, \quad \tau \ge 0, \tag{4.80}$$

where $c_\tau > 0$ is a constant depending on τ ([42], Theorem 10.5, p. 228).

Example 4.35: Let $\boldsymbol{X}$ be the solution of Eq. 4.71 with $\boldsymbol{a}(\boldsymbol{x}, t) = \boldsymbol{a}\,\boldsymbol{x}$ and $\boldsymbol{b}(\boldsymbol{x}, t) = \boldsymbol{b}$, where $\boldsymbol{a}$ and $\boldsymbol{b}$ are constant matrices. The difference between the solutions $\boldsymbol{X}^{(i)}(t)$ of this equation corresponding to the initial conditions $\boldsymbol{X}^{(i)}(0)$, $i = 1, 2$, is $\boldsymbol{X}^{(1)}(t) - \boldsymbol{X}^{(2)}(t) = \boldsymbol{\phi}(t)\,(\boldsymbol{X}^{(1)}(0) - \boldsymbol{X}^{(2)}(0))$, where the Green function $\boldsymbol{\phi}(t)$

gives the solution X at time $t \geq 0$ for a unit initial condition. For example, $X^{(1)}(t) - X^{(2)}(t) = (X^{(1)}(0) - X^{(2)}(0))\, e^{-\alpha t}$ for the Ornstein-Uhlenbeck process X defined by $dX(t) = -\alpha\, X(t)\, dt + \beta\, dB(t)$, $t \geq 0$, where $\alpha > 0$ and β are some constants. Hence, $\sup_{0 \leq s \leq t} (X^{(1)}(s) - X^{(2)}(s))^2 = (X^{(1)}(0) - X^{(2)}(0))^2$ for this process. ◇

Note: The bound in Eq. 4.80 depends on an undetermined constant c_τ and its definition requires that $X^{(1)}(0)$, $X^{(2)}(0)$ be in L_2. Because c_τ is not known, the bound only provides qualitative information on the evolution in time of the difference between $X^{(1)}$ and $X^{(2)}$.

For the Ornstein-Uhlenbeck process X, Eq. 4.80 gives

$$E\left[\sup_{0 \leq s \leq t} \left(X^{(1)}(s) - X^{(2)}(s)\right)^2\right] \leq 3\, E\left[\left(X^{(1)}(0) - X^{(2)}(0))^2\right)\right] e^{c_\tau\, \tau}.$$

Similar bounds can be obtained for any linear system. Hence, the bound in Eq. 4.80 is not useful for these systems. ▲

Suppose that the drift/diffusion coefficients and the Brownian motion in Eq. 4.71 satisfy the conditions of the first existence and uniqueness theorem in Section 4.7.1.1. Then the solution X of this equation is a semimartingale with the representation

$$X(t) = X(0) + A(t) + M(t), \tag{4.81}$$

where A and M are defined by Eq. 4.77, A is an adapted process with samples of finite variation on compacts, M is an $\mathcal{F}_t$-square integrable martingale, $A(0) = 0$, and $M(0) = 0$ ([42], Theorem 10.5, p. 228).

Note: The stated properties of A and M have been established in the proof of the existence and uniqueness of the solution of Eq. 4.71 (Section 4.7.1.1). We calculate here just the first two moments of a coordinate

$$M_i(t) = \sum_{k=1}^{d'} \int_0^t b_{ik}(X(u))\, dB_k(u) = \sum_{k=1}^{d'} (b_{ik}(X) \cdot B_k)(t)$$

of M. The mean of M_i is zero because $b_{ik}(X) \cdot B_k$ is a martingale starting at zero. The second moment of M_i is

$$\begin{aligned} E[M_i(t)^2] &= \sum_{k,l=1}^{d'} E\left[(b_{ik}(X) \cdot B_k)\,(b_{il}(X) \cdot B_l)\right] \\ &= \sum_{k,l=1}^{d'} E\left[\int_0^t b_{ik}(X(u))\, b_{il}(X(u))\, d[B_k, B_l](u)\right] \\ &= \sum_{k=1}^{d'} E\left[\int_0^t b_{ik}(X(u))^2\, du\right] = \sum_{k=1}^{d'} \int_0^t E[b_{ik}(X(u))^2]\, du \end{aligned}$$

by the definition of M_i, the linearity of the expectation operator, a property of the stochastic integral (Eq. 4.34), and Fubini's theorem. The expectation of M_i is finite since the entries of $\boldsymbol{b}$ are assumed to be bounded. ▲

Example 4.36: Let $d = d' = 1$, $\boldsymbol{a}(\boldsymbol{x}) = a(x)$, and $\boldsymbol{b}(\boldsymbol{x}) = b(x)$ in Eq. 4.71. If the conditions of the first existence and uniqueness theorem in Section 4.7.1.1 are satisfied, then

$$X(t) = X(0) + \int_0^t a(X(s))\, ds + \int_0^t b(X(s))\, dB(s)$$

has a unique solution and

$$g(X(t)) - g(X(0)) = \int_0^t g'(X(s))\, dX(s) + \frac{1}{2} \int_0^t g''(X(s))\, b(X(s))^2\, ds$$

for any function g with continuous second order derivative. ◇

Proof: We have seen that X is a semimartingale so that the Itô formula can be applied to $g(X)$. Because X is continuous, we can use the Itô formula in Eq. 4.42. This formula depends on the quadratic variation process $[X, X]$ of $X(t) = Z + A(t) + M(t)$, where $Z = X(0)$, $A(t) = \int_0^t a(X(s))\, ds$, and $M(t) = \int_0^t b(X(s))\, dB(s)$. We have

$$\begin{aligned}[X, X] &= [Z, Z] + [Z, A] + [Z, M] + [A, Z] + [A, A] \\ &\quad + [A, M] + [M, Z] + [M, A] + [M, M],\end{aligned}$$

by the linearity of quadratic covariation (Eq. 4.26). All quadratic covariations with arguments Z and A or M are zero because Z is a constant process and $A(0) = M(0) = 0$. The quadratic variation of A and the quadratic covariation of A and M are also zero. Hence, we have (Eq. 4.34)

$$[X, X] = Z^2 + [M, M] = Z^2 + \int_0^t b(X(s))^2\, d[B, B](s) = Z^2 + \int_0^t b(X(s))^2\, ds$$

so that $d[X, X](t) = b(X(t))^2\, dt$ since Z^2 is a constant process. ■

Example 4.37: Let X be the diffusion process in Example 4.36 and $g : \mathbb{R} \to \mathbb{R}$ be an increasing function with continuous second order derivative. The memoryless transformation $Y(t) = g(X(t))$ of X is a diffusion process defined by $dY(t) = a_Y(Y(t))\, dt + b_Y(Y(t))\, dB(t)$, where

$$a_Y(y) = g'(x)\, a(x) + \frac{1}{2} g''(x)\, b(x)^2, \quad b_Y(y) = g'(x)\, b(x), \quad \text{and } x = g^{-1}(y).$$

Figure 4.10 shows a sample path of X and the corresponding sample of $Y(t) = g(X(t)) = X(t)^3$, respectively, for $a(x) = -x$ and $b(x) = 1$. The difference between the samples X and Y is significant. Memoryless transformations can be used to generate diffusion processes of specified marginal distribution (Section 3.6.6 in this book, [79], Section 3.1). For example, $Y(t) = F^{-1}(\Phi(\sqrt{2}\, X(t)))$ is a diffusion process with marginal distribution F, where Φ denotes the distribution of the standard Gaussian variable. ◇

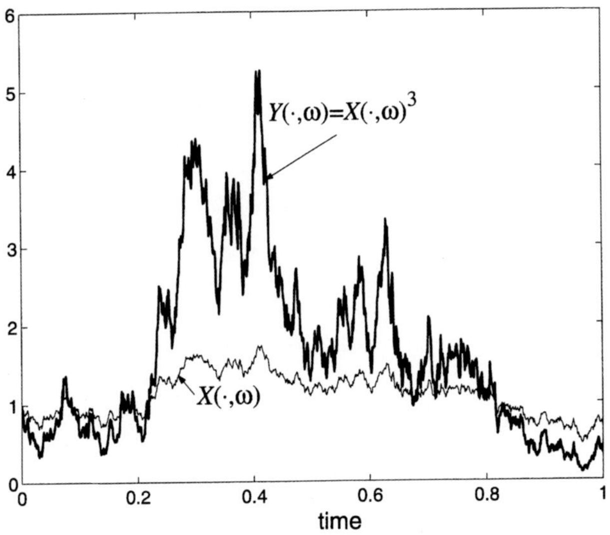

Figure 4.10. Sample paths of X defined by $dX(t) = -X(t)\,dt + dB(t)$ and its memoryless transformation $Y(t) = X(t)^3$

Proof: The Itô formula in Eq. 4.45 applied to $Y = g(X)$ gives

$$Y(t) - Y(0) = \int_0^t g'(X(s))\,[a(X(s))\,ds + b(X(s))\,dB(s)] + \frac{1}{2}\int_0^t g''(X(s))\,b(X(s))^2\,ds$$

and has the differential form

$$dY(t) = \left[g'(X(t))\,a(X(t)) + \frac{1}{2}\,g''(X(t))\,b(X(t))^2\right] dt + g'(X(t))\,b(X(t))\,dB(t).$$

Because g is an increasing function, $x = g^{-1}(y)$ exists so that Y is a diffusion process defined by the stochastic differential equation

$$dY(t) = \left[g'(g^{-1}(Y(t)))\,a(g^{-1}(Y(t))) + \frac{1}{2}\,g''(g^{-1}(Y(t)))\,b(g^{-1}(Y(t)))^2\right] dt + g'(g^{-1}(Y(t)))\,b(g^{-1}(Y(t)))\,dB(t)$$

with the stated drift and diffusion coefficients.

For example, the mapping $x \mapsto y = g(x) = x^3$ has the inverse $x = g^{-1}(y) = |y|^{1/3}\,\mathrm{sign}(y)$, where $\mathrm{sign}(y) = -1, 0,$ and 1 for $y < 0$, $y = 0$, and $y > 0$. The stochastic differential equation of Y for $a(x) = -x$ and $b(x) = 1$ is

$$dY(t) = 3\,(-Y(t) + |Y(t)|^{1/3}\,\mathrm{sign}(Y(t)))\,dt + 3\,|Y(t)|^{2/3}\,dB(t).$$

Sample paths of Y can be obtained from samples of X and $Y = X^3$ or can be generated directly from the above stochastic differential equation of Y. ■

Example 4.38: Let X be the diffusion process in Example 4.36 and $h : \mathbb{R} \to \mathbb{R}$ be a function with continuous second order derivative. Then,

$$\int_0^t h(X(s)) \circ dX(s) = \int_0^t h(X(s))\, dX(s) + \frac{1}{2} \int_0^t h'(X(s))\, b(X(s))^2\, ds,$$

where $\int_0^t h(X(s)) \circ dX(s)$ and $\int_0^t h(X(s))\, dX(s)$ denote Stratonovich and Itô integrals, respectively. ◇

Proof: We have seen that X is a semimartingale with continuous sample paths and that $d[X,X]^c(t) = d[X,X](t) = b(X(t))^2\, dt$. The above relationship between the Itô and Stratonovich integrals follows from Eq. 4.61 with g' replaced by h. ■

Example 4.39: Let X be the solution of the Itô integral equation

$$X(t) = X(0) + \int_0^t a(X(s), s)\, ds + \int_0^t b(X(s), s)\, dB(s). \tag{4.82}$$

This process also satisfies the Stratonovich integral equation

$$X(t) = X(0) + \int_0^t \tilde{a}(X(s), s)\, ds + \int_0^t b(X(s), s) \circ dB(s), \tag{4.83}$$

where $\tilde{a}(x,t) = a(x,t) - (1/2)\, b(x,t)\, [\partial b(x,t)/\partial x]$. It is assumed that $b(x,t)$ has continuous second and first order partial derivatives with respect to x and t, respectively. ◇

Proof: Let $U(s) = b(X(s), s)$. The Itô and Stratonovich integrals of U with respect to B are related by (Eq. 4.58)

$$\int_0^t U(s) \circ dB(s) = \int_0^t U(s)\, dB(s) + \frac{1}{2} [U, B]^c(t).$$

The Itô formula applied to $U(t) = b(X(t), t)$ gives

$$dU(t) = \left[a(X(t), t) \frac{\partial U}{\partial x} + \frac{\partial U}{\partial t} + \frac{1}{2} b(X(t), t)^2 \frac{\partial^2 U}{\partial x^2} \right] dt + b(X(t), t) \frac{\partial U}{\partial x}\, dB(t)$$

so that

$$d[U, B](t) = \left[b(X(t), t) \frac{\partial U}{\partial x}\, dB(t), dB(t) \right] = b(X(t), t) \frac{\partial U}{\partial x}\, dt.$$

Since $[U, B] = [U, B]^c$, the relationship between the Itô and Stratonovich integrals becomes

$$\int_0^t U(s) \circ dB(s) = \int_0^t U(s)\, dB(s) + \frac{1}{2} \int_0^t b(X(s), s) \frac{\partial U(s)}{\partial x}\, ds, \tag{4.84}$$

which yields the expression of $\tilde{a}$ in Eq. 4.83. ■

It is tempting in numerical calculations to (1) approximate B in Eq. 4.82 by a sequence of processes $B^{(n)}$ converging in some sense to B and (2) approximate X by the limit of the sequence of processes $X^{(n)}$ defined by Eq. 4.82 with $B^{(n)}$ in place of B. It turns out that this approximation can be incorrect. Let $p_n = (0 = t_0^{(n)} < t_1^{(n)} < \cdots < t_{m_n}^{(n)} = t)$ be a partition of $[0, t]$, $t > 0$, such that $\Delta(p_n) \to 0$ as $n \to \infty$. Define $B^{(n)}$ to be a linear interpolation function between the values of B at the points of p_n. The process $B^{(n)}$ has continuous samples that approach the samples of B as $n \to \infty$ but it differs from B in an essential way. The Brownian motion is a martingale while $B^{(n)}$ is not. Moreover, $B^{(n)}$ is of finite variation on compacts so that the integral $\int_0^t b(X^{(n)}(s), s)\, dB^{(n)}(s)$ can be calculated as a path by path Riemann-Stieltjes integral.

Wong-Zakai theorem. If $B^{(n)}$ is a sequence of processes with continuous samples of finite variation on compacts and piece-wise continuous derivatives that converges to B a.s. and uniformly, then the solution $X^{(n)}$ of Eq. 4.82 with $B^{(n)}$ in place of B converges a.s. and uniformly to the solution of the Itô equation ([102], Section 5.4.2)

$$X(t) = X(0) + \int_0^t \tilde{a}^*(X(s), s)\, ds + \int_0^t b(X(s), s)\, dB(s),$$

$$\text{where } \tilde{a}^*(x, t) = a(x, t) + \frac{1}{2}\, b(x, t)\, \frac{\partial b(x, t)}{\partial x}. \tag{4.85}$$

Note: The Stratonovich version of Eq. 4.85 is (Eqs. 4.82 and 4.83)

$$\begin{aligned} X(t) &= X(0) + \int_0^t \left[\tilde{a}^*(X(s), s) - \frac{1}{2} b(x, t)\, \frac{\partial b(x, t)}{\partial x}\right] ds + \int_0^t b(X(s), s) \circ dB(s) \\ &= X(0) + \int_0^t a(X(s), s)\, ds + \int_0^t b(X(s), s) \circ dB(s), \end{aligned}$$

showing that the approximation of B in an Itô equation by a colored noise results in the interpretation of this equation as a Stratonovich stochastic integral equation. ▲

4.7.1.3 Moments and other properties of diffusion processes

Examples are presented to illustrate how the Itô formula can be used to derive differential equations for moments, densities, and characteristic functions of $\boldsymbol{X}$. These examples are extended significantly in Chapter 7. We also show how the Stratonovich version of an Itô stochastic differential equation can be used to find properties of $\boldsymbol{X}$.

Example 4.40: Let X be a real-valued diffusion process representing the unique solution of a stochastic differential equation with drift and diffusion coefficients $a(x)$ and $b(x)^2$. If a and b are polynomials, we can develop ordinary differential equations for the moments $\mu(q; t) = E[X(t)^q]$, $q = 1, 2, \ldots$, of X provided these moments exist and are finite. Generally, the resulting moment equations form an infinite hierarchy so that they cannot be solved exactly. ◇

Proof: If $q \geq 0$ is an integer and $g(x) = x^q$, the Itô formula can be applied to the process $g(X)$ since X is a semimartingale and $g \in \mathcal{C}^2(\mathbb{R})$. The expectation of this formula gives

$$\begin{aligned} dE[X(t)^q] &= q\, E[X(t)^{q-1}\, dX(t)] + \frac{q\,(q-1)}{2} E[X(t)^{q-2}\, b(X(t))^2]\, dt \\ &= q\, E[X(t)^{q-1}\, a(X(t))]\, dt + \frac{q\,(q-1)}{2} E[X(t)^{q-2}\, b(X(t))^2]\, dt. \end{aligned}$$

This equation becomes a differential equation for the moments of $X(t)$ if the drift and diffusion coefficients are polynomials. If a and b are polynomials of degree at most one, the moment equations are closed so that they can be solved. ■

Example 4.41: Let X be an Ornstein-Uhlenbeck process defined by $dX(t) = -\alpha\, X(t)\, dt + \beta\, dB(t)$, where B is a Brownian motion and $\alpha > 0$ and β are some constants. The moments $\mu(q;t) = E[X(t)^q]$ of $X(t)$ satisfy the ordinary differential equation

$$\dot{\mu}(q;t) = -\alpha\, q\, \mu(q;t) + \frac{\beta^2\, q(q-1)}{2} \mu(q-2;t), \quad q = 1, 2, \ldots, \tag{4.86}$$

with the convention $\mu(u;t) = 0$ for any integer $u \leq -1$, where $\dot{\mu}(q;t) = d\mu(q;t)/dt$. ◇

Proof: The drift and diffusion coefficients of X satisfy the uniform Lipschitz and the growth conditions so that there exists a unique, adapted, and continuous solution if $X(0) \in \mathcal{F}_0$. The differential form of the Itô formula is (Example 4.36)

$$dg(X(t)) = g'(X(t))\,(-\alpha\, X(t)\, dt + \beta\, dB(t)) + \frac{\beta^2}{2} g''(X(t))\, dt,$$

so that

$$\frac{d}{dt} E[g(X(t))] = E\left[-\alpha\, g'(X(t))\, X(t) + \frac{\beta^2}{2} g''(X(t))\right].$$

Eq. 4.86 results from the above equation with $g(x) = x^q$. ■

Example 4.42: Let X be the Ornstein-Uhlenbeck process in Example 4.41. The characteristic function $\varphi(u;t) = E[\exp(\sqrt{-1}\, u\, X(t))$, $u \in \mathbb{R}$, of $X(t)$ satisfies the partial differential equation

$$\frac{\partial \varphi}{\partial t} = -\alpha\, u\, \frac{\partial \varphi}{\partial u} - \frac{\beta^2\, u^2}{2} \varphi. \tag{4.87}$$

The stationary solution $\varphi_s(u) = \lim_{t\to\infty} \varphi(u;t) = \exp\left(-\beta^2\, u^2/(4\,\alpha)\right)$ of this equation satisfies an ordinary differential equation obtained from Eq. 4.87 by setting $\partial\varphi/\partial t = 0$. Hence, the stationary value of an Ornstein-Uhlenbeck process at any time t is a Gaussian variable with mean zero and variance $\beta^2/(2\,\alpha)$, a known result ([175], Example 5.6, p. 179). ◇

Proof: The partial differential equation of the characteristic function given by Eq. 4.87 can be obtained by applying Itô's formula to $g(X) = \exp(\sqrt{-1}\, u\, X)$, $u \in \mathbb{R}$. Because g is a complex-valued function, the Itô formula previously developed cannot be applied directly. However, this formula is linear in g so that we can apply it to the real and imaginary parts of g separately and add the corresponding contributions.

This extended version of the Itô formula gives

$$\frac{\partial}{\partial t}\varphi(u;t) = E\left[-\alpha\, X(t)\,\sqrt{-1}\,u\, e^{\sqrt{-1}\,u\,X(t)} + \frac{\beta^2}{2}(\sqrt{-1}\,u)^2\, e^{\sqrt{-1}\,u\,X(t)}\right]$$

when applied to $g(X) = \exp(\sqrt{-1}\,u\,X)$. The result of Eq. 4.87 follows since $\partial^k\varphi/\partial u^k = E\left[\left(\sqrt{-1}\,X(t)\right)^k \exp(\sqrt{-1}\,u\,X(t))\right]$, where $k \geq 1$ is an integer. ■

Example 4.43: Let X be the solution of the stochastic differential equation

$$dX(t) = a(X(t),t)\,dt + b(X(t),t)\,dB(t), \quad t \geq t_0, \tag{4.88}$$

starting at $X(t_0) = x_0$. The density $f(x;t \mid x_0;t_0)$ of the conditional variable $X(t) \mid (X(t_0) = x_0)$, referred to as the transition density of X, satisfies the **Fokker-Planck** equation

$$\frac{\partial f}{\partial t} = -\frac{\partial}{\partial x}(a\,f) + \frac{1}{2}\frac{\partial^2}{\partial x^2}(b^2\,f) \tag{4.89}$$

with the initial condition $f(x;t_0 \mid x_0;t_0) = \delta(x - x_0)$ and the boundary conditions $\lim_{|x|\to\infty} a(x,t)\,f(x;t \mid x_0;t_0) = 0$, $\lim_{|x|\to\infty} b(x,t)^2\,f(x;t \mid x_0;t_0) = 0$, and $\lim_{|x|\to\infty} /\partial[b(x,t)^2\,f(x;t \mid x_0;t_0)]/\partial x = 0$. Because X is a Markov process, the density of $X(t_0)$ and the transition density $f(x;t \mid x_0;t_0)$ define all the finite dimensional densities of this process.

The solution of the Fokker-Planck equation is completely defined by the coefficients of Eq. 4.88 and does not depend on the particular version of B. Hence, the Fokker-Planck equation can only deliver the weak solution for Eq. 4.88. We also note that, if a process Y is such that the density of $Y(t) \mid Y(t_0) = y_0$ satisfies Eq. 4.89, it does not mean that Y is a diffusion process. ◇

Proof: The expectation of the Itô formula (Eq. 4.42) gives

$$\frac{d}{dt}E[g] = E\left[a\,\frac{\partial g}{\partial x} + \frac{b^2}{2}\frac{\partial^2 g}{\partial x^2}\right]$$

for any real-valued function $x \mapsto g(x)$ with continuous second order derivative or

$$\frac{d}{dt}\int_{\mathbb{R}} g\, f(x;t \mid x_0;t_0)\,dx = \int_{\mathbb{R}}\left[a\,\frac{\partial g}{\partial x} + \frac{b^2}{2}\frac{\partial^2 g}{\partial x^2}\right] f(x;t \mid x_;,t_0)\,dx.$$

Integrating the last equation by parts and using the stated boundary condition, we find

$$\int_{\mathbb{R}} g\,\frac{\partial f}{\partial t}\,dx = \int_{\mathbb{R}}\left[-\frac{\partial}{\partial x}(a\,f) + \frac{1}{2}\frac{\partial^2}{\partial x^2}(b^2\,f)\right] g\,dx.$$

The Fokker-Planck equation results since g is an arbitrary function. ■

Example 4.44: Let $g(x,t)$ be a function with continuous second and first order partial derivatives in x and t, respectively. The Stratonovich calculus obeys the **classical chain rule of differentiation**, that is,

$$d^\circ g(X(t),t) = \frac{\partial g(X(t),t)}{\partial x}\, d^\circ X(t) + \frac{\partial g(X(t),t)}{\partial t}\, dt, \tag{4.90}$$

where X is defined by Eqs. 4.82-4.83 and d° is used to indicate Stratonovich calculus. ◇

Proof: Let X be a diffusion process defined by Eq. 4.82. The Itô formula applied to $Y(t) = g(X(t),t)$ gives

$$\begin{aligned} Y(t) - Y(0) &= \int_0^t \left(\frac{\partial g}{\partial x}\, dX(s) + \frac{\partial g}{\partial s}\, ds \right) + \frac{1}{2}\int_0^t \frac{\partial^2 g}{\partial x^2}\, b^2\, d[B,B](s) \\ &= \int_0^t \left(a\,\frac{\partial g}{\partial x} + \frac{\partial g}{\partial s} + \frac{1}{2} b^2 \frac{\partial^2 g}{\partial x^2} \right) ds + \int_0^t b\,\frac{\partial g}{\partial x}\, dB(s), \end{aligned}$$

where $g(x,t)$ has continuous second and first order partial derivatives in x and t, respectively.

The relationship between the Itô and Stratonovich integrals given by Eq. 4.84 with U replaced by $b\,(\partial g/\partial x)$ yields

$$\int_0^t b\,\frac{\partial g}{\partial x} \circ dB = \int_0^t b\,\frac{\partial g}{\partial x}\, dB + \frac{1}{2}\int_0^t \left(\frac{\partial b}{\partial x}\frac{\partial g}{\partial x} + b\,\frac{\partial^2 g}{\partial x^2} \right) b\, ds$$

so that

$$Y(t) - Y(0) = \int_0^t \left(a\,\frac{\partial g}{\partial x} + \frac{\partial g}{\partial s} - \frac{1}{2}\, b\,\frac{\partial b}{\partial x}\frac{\partial g}{\partial x} \right) ds + \int_0^t b\,\frac{\partial g}{\partial x} \circ dB$$

or

$$Y(t) - Y(0) = \int_0^t \left(\frac{\partial g}{\partial s} + \tilde{a}\,\frac{\partial g}{\partial x} \right) ds + \int_0^t b\,\frac{\partial g}{\partial x} \circ dB,$$

where $\tilde{a} = a - (1/2)\, b\, \partial b/\partial x$ (Example 4.39). Since we have $d^\circ X(t) = \tilde{a}\, dt + b \circ dB(t)$, the differential form of the above equation is $d^\circ Y(t) = (\partial g/\partial x)\, d^\circ X(t) + (\partial g/\partial t)\, dt$ as stated. ■

Example 4.45: The Stratonovich calculus can be used to find the solution of some Itô differential equations. For example, the solution of the Itô equation $dX(t) = a(X(t))\, dt + b(X(t))\, dB(t)$, $t \geq 0$, with $a(x) = h(x)\, h'(x)/2$, and $b(x) = h(x)$, is $X(t) = g^{-1}(g(X(0)) + B(t))$, where h is a differentiable function and g is the primitive of h^{-1}. ◇

Proof: The Stratonovich integral representation of the diffusion process X is $X(t) = X(0) + \int_0^t h(X(s)) \circ dB(s)$ since $\tilde{a} = a - (1/2)\, b\, \partial b/\partial x = 0$ (Eq. 4.83). The differential form of this equation is $d^\circ X(t) = h(X(t)) \circ dB(t)$. The rules of classical calculus apply and give

$$\int_0^t \frac{d^\circ X(s)}{h(X(s))} = \int_0^t dB(s) = B(t)$$

so that $g(X(t)) - g(X(0)) = B(t)$ ([131], Section 3.2.2). ■

Example 4.46: The solution of the Itô stochastic differential equation $dX(t) = a(t)\,X(t)\,dt + b(t)\,X(t)\,dB(t)$ is

$$X(t) = X(0)\,\exp\left[\int_0^t (a(s) - 0.5\,b(s)^2)\,ds + \int_0^t b(s)\,dB(s)\right].$$

If $a(t) = a$ and $b(t) = b$ are constant, then

$$X(t) = X(0)\,\exp\left[(a - 0.5\,b^2)\,t + b\,B(t)\right]$$

is the geometric Brownian motion (Example 4.34). ◇

Proof: The Itô formula applied to $Y(t) = \ln(X(t))$ gives

$$Y(t) - Y(0) = \int_0^t (a(s) - 0.5\,b(s)^2)\,ds + \int_0^t b(s)\,dB(s),$$

which yields the stated result. The process X is strictly positive a.s. on any bounded time interval by the functional form of its stochastic differential equation and properties of the Brownian motion so that the mapping $x \mapsto \ln(x)$ is in $\mathcal{C}^2(\mathbb{R})$. ■

4.7.2 Semimartingale input

This section considers stochastic differential equations driven by semimartingale noise, that is, $\boldsymbol{X}$ is the solution of Eqs. 4.69 and 4.70, where $\boldsymbol{Y}$ is a semimartingale $\boldsymbol{S}$. The first and second integrals in Eq. 4.70 are Riemann-Stieltjes and stochastic integrals, respectively.

We present without proof conditions for the **existence** and **uniqueness** of the solution of Eq. 4.70 with $\boldsymbol{Y}$ being a semimartingale process $\boldsymbol{S}$ and illustrate these results by examples.

If S is a semimartingale with $S(0) = 0$, $X(0)$ is finite and $\mathcal{F}_0$-measurable, and the function $G : [0, \infty) \times \Omega \times \mathbb{R} \to \mathbb{R}$ is such that (1) the process $(t, \omega) \to G(t, \omega, x)$ is in $\mathcal{L}$ for a fixed x and (2) $|G(t, \omega, x) - G(t, \omega, x')| \leq K(\omega)|x - x'|$ for each (t, ω), where K is a finite-valued random variable, then the stochastic integral equation

$$X(t) = X(0) + \int_0^t G(s, \cdot, X(s-))\,dS(s) \tag{4.91}$$

has a unique solution that is a semimartingale ([147], Theorem 6, p. 194).

Note: The requirement that $G(\cdot, \cdot, x)$ be an adapted, càglàd process for each $x \in \mathbb{R}$ guarantees that the stochastic integral in Eq. 4.91 is defined. The requirement that $G(t, \omega, \cdot)$ satisfy a Lipschitz condition for each (t, ω) resembles the conditions in Section 4.7.1.1 for the existence and uniqueness of the solution of differential equations driven by Gaussian white noise.

The theorem in Eq. 4.91 can be stated for a system of stochastic integral equations involving a finite number of input semimartingales ([147], p. 194). ▲

Example 4.47: Let $dX(t) = b(X(t-), t)\, dS(t)$ be a stochastic differential equation with the integral form

$$X(t) = X(0) + \int_0^t b(X(s-), s)\, dS(s),$$

where S is a semimartingale. If $X(0) \in \mathcal{F}_0$, $t \mapsto b(x, t)$ is continuous for each x, and $x \mapsto b(x, t)$ satisfies a uniform Lipschitz condition for each $t \geq 0$, then the above equation has a unique solution that is a semimartingale.

If S is a compound Poisson process C, we have

$$X(t) = X(0) + \sum_{k=1}^{N(t)} b(X(T_k-), T_k)\, Y_k = X(0) + \int_0^t b(X(s-), s)\, dC(s),$$

where $(T_1, T_2, \ldots)$ and $(Y_1, Y_2, \ldots)$ denote the jump times and the jumps of C, respectively. ◇

Proof: The existence and uniqueness of the solution X follows from the above theorem. The second condition on G in Eq. 4.91 becomes the uniform Lipschitz condition $|b(t, x) - b(t, x')| \leq c\, |x - x'|$ for each $t \geq 0$ and a constant $c > 0$ since b is a deterministic function.

If S is a compound Poison process C, the sample paths of X are constant between consecutive jumps of C and have jumps

$$\Delta X(T_k) = X(T_k) - X(T_k-) = b(X(T_k-), T_k)\, \Delta C(T_k) = b(X(T_k-), T_k)\, Y_k$$

at the jump times of C so that $X(t) = X(T_k-) + b(X(T_k-), T_k)\, Y_k$, $t \in [T_k, T_{k+1})$. This recurrence formula gives the stated expression of X. ▲

Example 4.48: Let S be a semimartingale and X be the solution of $dX(t) = a(X(t-), t)\, dt + b(X(t-), t)\, dS(t)$. If $X(0) \in \mathcal{F}_0$ and a, b are such that (1) the functions $t \mapsto a(x, t), b(x, t)$ are continuous for each x and (2) the functions $x \mapsto a(x, t), b(x, t)$ satisfy uniform Lipschitz conditions for each $t \geq 0$, then X is the unique solution of the stochastic integral equation

$$X(t) = X(0) + \int_0^t a(X(s-), s)\, ds + \int_0^t b(X(s-), s)\, dS(s)$$

and is a semimartingale. ◇

Note: We have mentioned that the statement of theorem guaranteeing the existence and uniqueness of the solution of Eq. 4.91 can be extended to a vector input and/or output. Consider the alternative form,

$$X(t) = X(0) + \int_0^t [a(X(s-), s) \;\; b(X(s-), s)] \begin{bmatrix} ds \\ dS(s) \end{bmatrix},$$

of the integral equation defining X, that extends Eq. 4.91 to the case in which the input is a vector of semimartingales. Since a and b satisfy the conditions for the existence and uniqueness of the solution of Eq. 4.91, X is the a unique solution of the above stochastic differential equation and is a semimartingale since $(s, S(s))$ is a semimartingale. ▲

We need to introduce some notation and definitions to state the following conditions for the existence and uniqueness of the solution of stochastic differential equations driven by semimartingale noise. Let $\mathcal{D}^d$ denote the class of $\mathbb{R}^d$-valued stochastic processes whose coordinates are adapted processes with càdlàg sample paths, that is, are processes in $\mathcal{D}$. An operator $G : \mathcal{D}^d \to \mathcal{D}$ is said to be **functional Lipschitz** if for any processes $\boldsymbol{X}, \boldsymbol{Y} \in \mathcal{D}^d$ we have (1) for any stopping time T, $\boldsymbol{X}^{T-} = \boldsymbol{Y}^{T-}$ implies $G(\boldsymbol{X})^{T-} = G(\boldsymbol{Y})^{T-}$ and (2) there is an increasing finite process $K(t)$, $t \geq 0$, such that $|G(\boldsymbol{X})(t) - G(\boldsymbol{Y})(t)| \leq K(t) \parallel \boldsymbol{X}(t) - \boldsymbol{Y}(t) \parallel$ a.s. for each $t \geq 0$ ([147], p. 195).

If $\boldsymbol{S} = (S_1, \ldots, S_{d'})$ is a vector of semimartingales, $\boldsymbol{S}(0) = \boldsymbol{0}$, $J_i \in \mathcal{D}$, $i = 1, \ldots, d$, and the operators $G_i^j : \mathcal{D}^d \to \mathcal{D}$, $i = 1, \ldots, d$, $j = 1, \ldots, d'$, are functional Lipschitz, then the system of stochastic integral equations

$$X_i(t) = J_i(t) + \sum_{j=1}^{d'} \int_0^t G_i^j(\boldsymbol{X})(s-)\, dS_j(s), \quad i = 1, \ldots, d, \tag{4.92}$$

has a unique solution in $\mathcal{D}^d$. If the processes J_i are semimartingales, then X_i, $i = 1, \ldots, d$, are also semimartingales ([147], Theorem 7, p. 197).

Note: The solution of Eq. 4.92 is an $\mathbb{R}^d$-valued process $\boldsymbol{X} = (X_1, \ldots, X_d)$. This equation extends Eq. 4.91 since it considers vector input/output processes and $\boldsymbol{X}$ depends on some processes J_i. That the integrands in Eq. 4.92 do not depend explicitly on time is not a restriction since the state $\boldsymbol{X}$ can be augmented to include time as a coordinate. ▲

If conditions for the existence and uniqueness for the solution of Eq. 4.92 hold and $X_i^{(0)} \in \mathcal{D}$, $i = 1, \ldots, d$, then $\boldsymbol{X}^{(k)}$, defined by recurrence formula

$$X_i^{(k+1)}(t) = J_i(t) + \sum_{j=1}^{d'} \int_0^t G_i^j(\boldsymbol{X}^{(k)})(s-)\, dS_j(s), \quad i = 1, \ldots, d, \tag{4.93}$$

converges in ucp to the solution $\boldsymbol{X}$ of Eq. 4.92 ([147], Theorem 8, p. 199).

Example 4.49: Let X be the solution of the stochastic differential equation in Example 4.48, where $S = C$ is a compound Poisson process and a, b are not explicit functions of time. This equation has a unique solution under the conditions stated in Example 4.48.

We construct here the solution X based on elementary arguments. If there is a function h with inverse h^{-1} such that $h'(x) = 1/a(x)$, then

$$X(t) = h^{-1}\left(h(X(T_k)) + t - T_k\right), \quad t \in [T_k, T_{k+1}),$$

where $X(T_k) = X(T_k-) + b(X(T_k-))\, Y_k$, T_k are the jump times of C, and Y_k denote the jumps of this process. ◇

Proof: We have already established the relationship between values of X immediately before and after a jump of C. If $t \in [T_k, T_{k+1})$, then X is the solution of the deterministic differential equation $\dot{X}(t) = a(X(t))$ with the random initial condition $X(T_k)$ so that $\int_{X(T_k)}^{X(t)} dx/a(x) = \int_{T_k}^{t} ds$ and $h(X(t)) - h(X(T_k)) = t - T_k$ for $t \in [T_k, T_{k-1})$.

Note also that if $a(\cdot)$ is such that X is a continuous differentiable function during the time intervals $[T_k, T_{k+1})$, then X is a pure jump semimartingale. Under this assumption $[X, X]$ is zero in each time interval $[T_k, T_{k+1})$. The only contribution to the quadratic variation process $[X, X]$ is provided by the jumps $\Delta X(T_k)$ of X so that $[X, X]^c = 0$ and $[X, X](t) = \sum_{k=1}^{N(t)} (\Delta X(T_k))^2$. ■

Example 4.50: Let X be the solution of $dX(t) = X(t-)\,dC(t)$, $t \geq 0$, where C is a compound Poisson process. The Itô and Stratonovich interpretations of this equation,

$$X(t) = X(0) + \int_0^t X(s-)\,dC(s) \quad \text{and} \quad X(t) = X(0) + \int_0^t X(s-) \circ dC(s),$$

coincide and $X(t) = X(0) \prod_{k=1}^{N(t)} (1 + Y_k)$, where $(T_1, T_2, \ldots)$ and $(Y_1, Y_2, \ldots)$ denote the jump times and the jumps of C, respectively. ◇

Proof: It can be shown that ([147], Theorem 23, p. 224)

$$X(t) = X(0)\, e^{C(t)} \prod_{0<s\leq t} (1 + \Delta C(s))\, e^{-\Delta C(s)} = X(0) \prod_{k=1}^{N(t)} (1 + Y_k).$$

This solution can also be obtained by elementary arguments since $dX(t) = 0$ for $t \in [T_k, T_{k-1})$ and each $k \geq 1$. For example, we have $X(t) = X(T_1-) + \Delta X(T_1) = X(T_1-) + X(T_1-)\,\Delta C(T_1)$ for $t \in [T_1, T_2)$ so that

$$X(t) = \begin{cases} X(0) & \text{for } t \in [0, T_1), \\ X(0)\,(1 + Y_1) & \text{for } t \in [T_1, T_2), \\ X(0)\,(1 + Y_1)\,(1 + Y_2) & \text{for } t \in [T_2, T_3). \end{cases}$$

The relationship,

$$\int_0^t X(s-) \circ dC(s) = \int_0^t X(s-)\,dC(s) + \frac{1}{2}\,[X, C]^c(t),$$

between the Itô and Stratonovich integrals (Eq. 4.58) shows that these integrals coincide since X and C are pure jump semimartingales so that $[X, C]^c = 0$. ■

Example 4.51: Let X be the process in Example 4.49 and $g \in C^2(\mathbb{R})$ denote a real-valued function. The change of the process $g(X)$ in a time interval can be obtained from the Itô formula in Eq. 4.40. A more familiar version of this result is ([172], Theorem 4.2.2, p. 199)

$$\begin{aligned} g(X(t)) - g(X(0)) &= \int_0^t a(X(s-))\, g'(X(s-))\, ds \\ &+ \int_0^t \int_{\mathbb{R}} [g(X(s-) + y\, b(X(s-))) - g(X(s-))]\, \mathcal{M}(ds, dy), \end{aligned} \tag{4.94}$$

where $\mathcal{M}$ is a Poisson random measure on $[0, \infty) \times \mathbb{R}$ defining a compound Poisson process $C(t) = \sum_{k=1}^{N(t)} Y_k$, $E[\mathcal{M}(ds, dy)] = (\lambda\, ds)\, dF(y)$, $\lambda > 0$ is the intensity of the Poisson process N, and F denotes the distribution of the iid random variables Y_k (Section 3.12). $\diamond$

Proof: The Itô formula in Eq. 4.40 gives

$$g(X(t)) - g(X(0)) = \int_{0+}^{t} g'(X(s-))\, dX(s) + \frac{1}{2}\int_{0+}^{t} g''(X(s-))\, d[X, X]^c(s) + \sum_{0<s\le t} [g(X(s) - g(X(s-)) - g'(X(s-))\, \Delta X(s)].$$

Because X is a quadratic pure jump semimartingale the above integral with integrator $[X, X]^c$ is zero so that

$$g(X(t)) - g(X(0)) = \int_{0+}^{t} g'(X(s-))\, [a(X(s-))\, ds + b(X(s-))\, dC(s)] + \sum_{0<s\le t} [g(X(s) - g(X(s-)) - g'(X(s-))\, \Delta X(s)].$$

Because $\Delta X(s) = b(X(s-))\Delta C(s)$, we have

$$\int_{0+}^{t} g'(X(s-))\, b(X(s-))\, dC(s) = \sum_{0<s\le t} g'(X(s-))\, b(X(s-))\, \Delta C(s) = \sum_{0<s\le t} g'(X(s-))\, \Delta X(s)$$

so that

$$g(X(t)) - g(X(0)) = \int_{0+}^{t} g'(X(s-))\, a(X(s-))\, ds + \sum_{0<s\le t} [g(X(s) - g(X(s-))].$$

The result in Eq. 4.94 follows from the above equation and the integral form,

$$\int_0^t \int_{\mathbb{R}} [g(X(s-) + y\, b(X(s-)) - g(X(s-))]\, \mathcal{M}(ds, dy),$$

of $\sum_{0<s\le t}[g(X(s)) - g(X(s-))]$. ■

4.7.3 Numerical Solutions

Explicit solutions of stochastic differential equations are only available in a few cases ([115], Section 4.4). Numerical methods are needed for solving most stochastic differential equations encountered in applications. Our discussion on numerical methods is limited to real-valued diffusion processes defined by the Itô differential equation

$$dX(s) = a(X(s))\, ds + b(X(s))\, dB(s), \quad s \in [0, t], \tag{4.95}$$

where a, b are such that this equation has a unique strong solution (Section 4.7.1.1).

An extensive discussion on numerical solutions for stochastic differential equations driven by Brownian motion can be found in [115]. Essentials on numerical solutions of Eq. 4.95 are in [131] (Section 3.4).

4.7.3.1 Definitions

Let p_n be a sequence of partitions of $[0, t]$ with points $0 = t_0^{(n)} < t_1^{(n)} < \cdots < t_{m_n}^{(n)} = t$ such that $\Delta(p_n) \to 0$ as $n \to \infty$. We take $m_n = n$ and $t_k^{(n)} - t_{k-1}^{(n)} = t/n$ so that the partition mesh is $\Delta(p_n) = t/n$. In this section we use the notation $\Delta(p_n) = \delta_n = t/n$ for convenience.

A **numerical solution** $X^{(n)}$ of Eq. 4.95 is a stochastic process that approximates the solution X of this equation. The values of $X^{(n)}$ are calculated only at the times $t_k^{(n)} = k\,t/n$, $k = 0, 1, \ldots, n$, that is, at the points of p_n. Any interpolation method can be used to define $X^{(n)}$ at all times in $[0, t]$. For example, the following definition of $X^{(n)}$ is based on linear interpolation.

Numerical solution based on linear interpolation. If $s \in [t_{k-1}^{(n)}, t_k^{(n)}]$,

$$X^{(n)}(s) = X^{(n)}(t_{k-1}^{(n)}) + \frac{X^{(n)}(t_k^{(n)}) - X^{(n)}(t_{k-1}^{(n)})}{t_k^{(n)} - t_{k-1}^{(n)}}\,(s - t_{k-1}^{(n)}). \quad (4.96)$$

Note: The approximating solution $X^{(n)}(s)$ at a time $s \in [t_{k-1}^{(n)}, t_k^{(n)}]$ depends linearly on the values of $X^{(n)}$ at the two ends of this interval. The samples of $X^{(n)}$ are continuous samples but are not differentiable at the times $t_k^{(n)}$ defining the partition p_n. ▲

To assess the accuracy of an approximating solution, we need to (1) define a measure for the difference between $X^{(n)}$ and X, called the error of $X^{(n)}$, (2) determine the dependence of the error of $X^{(n)}$ on the partition mesh $\Delta(p_n) = \delta_n$, and (3) develop upper bounds on this error.

Error functions. For a partition p_n of $[0, t]$, we define the error functions

$$e_s(p_n, t) = E\left[\left|X(t) - X^{(n)}(t)\right|\right] \quad \text{and}$$
$$e_w(p_n, t) = \left|E\left[X(t)\right] - E\left[X^{(n)}(t)\right]\right|. \quad (4.97)$$

Note: The expectation $E\left[\sup_{0 \le s \le t} |X(s) - X^{(n)}(s)|\right]$ provides a useful measure for the accuracy of the approximation $X^{(n)}$ over the entire time interval $[0, t]$, but it is difficult to evaluate. The error functions in Eq. 4.97 are calculated at the end of the time interval $[0, t]$ for simplicity.

We also note that $e_s(p_n, t)$ measures the difference between the sample values of the random variables $X(t)$ and $X^{(n)}(t)$ while $e_w(p_n, t)$ gives the difference between the expected values of these random variables. ▲

We give now some definitions characterizing the dependence of the approximating solution $X^{(n)}$ on the mesh δ_n of p_n and time. The first definitions are based on the error functions in Eq. 4.97. The last definition involves properties of the drift and diffusion coefficients of Eq. 4.95.

If $e_s(p_n, t) \to 0$ as $n \to \infty$, we say that $X^{(n)}$ **converges strongly at time** t **to** X. The approximation $X^{(n)}$ **converges strongly with order** $\gamma > 0$ **at time** t **to** X if there exist some constants $c > 0$ and $\bar{\delta} > 0$ such that $e_s(p_n, t) \leq c\,(\delta_n)^\gamma$ for each $\delta_n \leq \bar{\delta}$ and c does not depend on δ_n. Weaker versions of these definitions result by replacing $e_s(p_n, t)$ with $e_w(p_n, t)$. We say that $X^{(n)}$ **converges weakly at time** t **to** X if $e_w(p_n, t) \to 0$ as $n \to \infty$. Similarly, $X^{(n)}$ **converges weakly with order** $\gamma > 0$ **at time** t **to** X if there exist some constants $c > 0$ and $\bar{\delta} > 0$ such that $e_w(p_n, t) \leq c\,(\delta_n)^\gamma$ for each $\delta_n \leq \bar{\delta}$ and c does not depend on δ_n.

Let $(\mathcal{F}_t)_{t\geq 0}$ be the natural filtration of X defined by Eq. 4.95. A discrete time approximation $X^{(n)}$ is **strongly consistent** if there exists a positive function $\zeta(u)$ such that $\lim_{u\downarrow 0} \zeta(u) = 0$,

$$E\left[\left(E[(X^{(n)}(t_{k+1}^{(n)}) - X^{(n)}(t_k^{(n)}))/\delta_n \mid \mathcal{F}_{t_k^{(n)}}] - a(X^{(n)}(t_k^{(n)}))\right)^2\right] \leq \zeta(\delta_n)$$

and

$$E\left[\left(X^{(n)}(t_{k+1}^{(n)}) - X^{(n)}(t_k^{(n)}) - E\left[X^{(n)}(t_{k+1}^{(n)}) - X^{(n)}(t_k^{(n)}) \mid \mathcal{F}_{t_k^{(n)}}\right]\right.\right.$$
$$\left.\left. - b(X^{(n)}(t_k^{(n)}))\,\Delta B_k^{(n)}\right)^2 / \delta_n\right] \leq \zeta(\delta_n).$$

Note: This definition requires that the conditional expectation of the increment of $X^{(n)}$ in the time interval $[t_k^{(n)}, t_{k+1}^{(n)}]$ divided by δ_n converges in m.s. to the drift of X and that the variance of the difference of the random parts of the approximate and exact solutions converges in m.s. to zero as $\delta_n \to 0$. An extensive discussion on this topic can be found in [115] (Chapter 9). ▲

4.7.3.2 Euler and Milstein numerical solutions

Euler numerical solution.

$$X^{(n)}(t_k^{(n)}) - X^{(n)}(t_{k-1}^{(n)}) = a(X(t_{k-1}^{(n)}))\,\Delta t_k^{(n)} + b(X(t_{k-1}^{(n)}))\,\Delta B_k^{(n)}, \qquad (4.98)$$

where $\Delta B_k^{(n)} = B(t_k^{(n)}) - B(t_{k-1}^{(n)})$, $\Delta t_k^{(n)} = t_k^{(n)} - t_{k-1}^{(n)} = \delta_n$, and $k = 1, \ldots, n$.

Proof: The increment of the solution X in Eq. 4.95 during a time interval $[t_{k-1}^{(n)}, t_k^{(n)}]$ is given by the recurrence formula

$$X(t_k^{(n)}) - X(t_{k-1}^{(n)}) = \int_{t_{k-1}^{(n)}}^{t_k^{(n)}} a(X(s))\,ds + \int_{t_{k-1}^{(n)}}^{t_k^{(n)}} b(X(s))\,dB(s). \tag{4.99}$$

The Euler numerical solution $X^{(n)}$ results from this equation by approximating the functions $a(X(s))$ and $b(X(s))$, $s \in [t_{k-1}^{(n)}, t_k^{(n)}]$, by their values $a(X(t_{k-1}^{(n)}))$ and $b(X(t_{k-1}^{(n)}))$ at the left end of this time interval. Any interpolation method can be used to define $X^{(n)}$ at all times in $[0, t]$, for example, Eq. 4.96. ■

The Euler numerical solution is the simplest scheme for solving stochastic differential equations. This method is frequently used in the subsequent chapters of the book to generate samples of diffusion and other processes defined by stochastic differential equations.

The Euler numerical solution converges strongly with order $\gamma = 0.5$ and weakly with order $\gamma = 1$ under relatively mild conditions ([115], Theorem 9.6.2, p. 324, [131], p. 162).

Note: An extensive discussion on the convergence of the Euler approximation can be found in [115] (Chapters 8 and 9). For example, if $X(0) \in L_2$ is $\mathcal{F}_0$-measurable and a, b in Eq. 4.95 are measurable functions in L_2 satisfying some Lipschitz and growth conditions (Eqs. 4.73 and 4.75), then a strongly consistent approximation $X^{(n)}$ converges strongly to X. Moreover, we have $E\left[|X^{(n)}(t) - X(t)|\right] \le \rho\,(\delta_n + \zeta(\delta_n))^{1/2}$, where ρ is a positive constant and ζ is the function in the definition of a strongly consistent approximation, that is, $\zeta > 0$ and $\lim_{u \downarrow 0} \zeta(u) = 0$ ([115], Theorem 9.6.2, p. 324). The function ζ is zero for the Euler approximation so that the above upper bound becomes $\rho\,\delta_n^{1/2}$. Hence, the Euler approximation converges strongly with order $\gamma = 0.5$. ▲

Example 4.52: Consider the differential equation $\dot{x}(s) = a(x(s), s)$, $s \in [0, t]$, starting at $x(0) = x_0$. If (1) the drift a is uniform Lipschitz in x, that is, there exists $\beta > 0$ such that $|a(x, s) - a(y, s)| \le \beta |x - y|$ for $x, y \in \mathbb{R}$ and $s \in [0, t]$, (2) the functions $\partial a/\partial t$ and $\partial a/\partial x$ are continuous, and (3) all solutions x are bounded in $[0, t]$, then the error $e_k^{(n)} = x(t_k^{(n)}) - x_k^{(n)}$ satisfies the inequality

$$|e_k^{(n)}| \le \frac{\alpha\,\delta_n}{2\beta}\left(e^{k\,\beta\,\delta_n} - 1\right),$$

where $\alpha = \max|\partial a/\partial t| + \max|a\,\partial a/\partial x|$ and $x_k^{(n)} = x^{(n)}(t_k^{(n)})$ denotes the Euler approximation at time $t_k^{(n)}$.

Figure 4.11 shows time histories of $|e_k^{(n)}|$ and of the above upper bound on the absolute value of the difference between the exact and the Euler numerical solutions for $a(x, t) = -5\,x$, $x(0) = 1$, $\alpha = 25$, $\beta = 5$, and two partitions p_n of $[0, 1]$ with $n = 10$ and 100. The bound on $|e_k^{(n)}|$ is too wide to provide any meaningful information on the accuracy of the Euler numerical solutions. Note that the plots for $n = 10$ and $n = 100$ have different scales. ◇

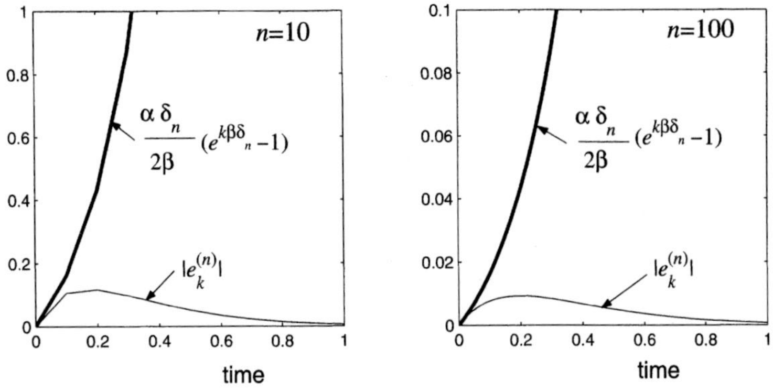

Figure 4.11. Upper bound and error of the Euler numerical solution

Proof: The difference between the Taylor expansion,

$$x(t_{k+1}^{(n)}) = x(t_k^{(n)}) + \dot{x}(t_k^{(n)})\,\delta_n + \frac{1}{2}\,\ddot{x}(\theta_k^{(n)})\,\delta_n^2, \quad \theta_k^{(n)} \in [t_k^{(n)}, t_{k+1}^{(n)}],$$

and the Euler approximation,

$$x_{k+1}^{(n)} = x_k^{(n)} + a(x_k^{(n)}, t_k^{(n)})\,\delta_n,$$

of the solution of $\dot{x}(t) = a(x,t)$ in the interval $[t_k^{(n)}, t_{k+1}^{(n)}]$ is

$$e_{k+1}^{(n)} = e_k^{(n)} + \left[a(x(t_k^{(n)}), t_k^{(n)}) - a(x_k^{(n)}, t_k^{(n)})\right]\delta_n + \frac{1}{2}\,\ddot{x}(\theta_k^{(n)})\,\delta_n^2.$$

The absolute value of the square bracket in the above equation is smaller than $\beta\,|x(t_k^{(n)}) - x_k^{(n)})| = \beta\,|e_k^{(n)}|$ by assumption. We also have $\ddot{x} = da(x,t)/dt = \partial a/\partial t + a\,\partial a/\partial x$ so that there exists $\alpha > 0$ such that $|\ddot{x}(\theta_k^{(n)})| \le \alpha$. The above recurrence formula for $e_k^{(n)}$ and the bounds give

$$|e_{k+1}^{(n)}| \le (1+\beta\,\delta_n)\,|e_k^{(n)}| + \frac{1}{2}\,\alpha\,\delta_n^2 \quad \text{so that } |e_{k+1}^{(n)}| \quad \le \frac{(1+\beta\,\delta_n)^{k+1} - 1}{\beta\,\delta_n}\,\frac{\alpha\,\delta_n^2}{2},$$

since $e_0^{(n)} = 0$. The latter inequality and $(1+c)^n \le e^{n\,c}$, $c > 0$, deliver the stated bound on the error with (c, n) replaced by $(\beta\,\delta_n, k+1)$.

The drift of the differential equation considered in this example is $a(x,t) = -5\,x$ so that $x(t) = e^{-5\,t}$ for $x(0) = 1$. The difference $|a(x,t) - a(y,t)| = 5|x - y|$ so that we can take $\beta = 5$. Because $\ddot{x} = da(x,t)/dt = \partial a/\partial t + a\,\partial a/\partial x$, $\partial a/\partial t = 0$, and $a\,\partial a/\partial x = (-5\,x)(-5) = 25\,x$, we can take $\alpha = 25$. ■

Milstein numerical solution.

$$X^{(n)}(t_k^{(n)}) - X^{(n)}(t_{k-1}^{(n)}) = a(X(t_{k-1}^{(n)}))\,\Delta t_k^{(n)} + b(X(t_{k-1}^{(n)}))\,\Delta B_k^{(n)} + \tilde{R}_{k,1}^{(n)}, \tag{4.100}$$

$$\text{where } \tilde{R}_{k,1}^{(n)} = \frac{1}{2}\,b'(X(t_{k-1}^{(n)}))\,b(X(t_{k-1}^{(n)}))\left[(\Delta B_k^{(n)})^2 - \Delta t_k^{(n)}\right]. \tag{4.101}$$

Proof: Consider the increment of X in Eq. 4.99 during a time interval $[t_{k-1}^{(n)}, t_k^{(n)}]$. Instead of approximating the functions $a(X(s))$ and $b(X(s))$, $s \in [t_{k-1}^{(n)}, t_k^{(n)}]$, by their values $a(X(t_{k-1}^{(n)}))$ and $b(X(t_{k-1}^{(n)}))$ at the left end of this time interval (Eq. 4.98), we use the Itô formula to represent the integrands a and b in $[t_{k-1}^{(n)}, t_k^{(n)}]$. The resulting expression is used to establish the Milstein approximation.

The Itô formula applied to the functions $a(X(s))$ and $b(X(s))$ for $s \in [t_{k-1}^{(n)}, t_k^{(n)}]$ gives

$$a(X(s)) - a(X(t_{k-1}^{(n)})) = \int_{t_{k-1}^{(n)}}^{s} (a'\,a + \frac{1}{2}a''\,b^2)\,du + \int_{t_{k-1}^{(n)}}^{s} a'\,b\,dB(u),$$

$$b(X(s)) - b(X(t_{k-1}^{(n)})) = \int_{t_{k-1}^{(n)}}^{s} (b'\,a + \frac{1}{2}b''\,b^2)\,du + \int_{t_{k-1}^{(n)}}^{s} b'\,b\,dB(u),$$

so that Eq. 4.99 becomes

$$\begin{aligned} X(t_k^{(n)}) &- X(t_{k-1}^{(n)}) \\ &= \int_{t_{k-1}^{(n)}}^{t_k^{(n)}} \left[a(X(t_{k-1}^{(n)})) + \int_{t_{k-1}^{(n)}}^{s} (a'\,a + \frac{1}{2}a''\,b^2)\,du + \int_{t_{k-1}^{(n)}}^{s} a'\,b\,dB(u)\right] ds \\ &+ \int_{t_{k-1}^{(n)}}^{t_k^{(n)}} \left[b(X(t_{k-1}^{(n)})) + \int_{t_{k-1}^{(n)}}^{s} (b'\,a + \frac{1}{2}b''\,b^2)\,du + \int_{t_{k-1}^{(n)}}^{s} b'\,b\,dB(u)\right] dB(s) \end{aligned}$$

or

$$X(t_k^{(n)}) - X(t_{k-1}^{(n)}) = a(X(t_{k-1}^{(n)}))\,\Delta t_k^{(n)} + b(X(t_{k-1}^{(n)}))\,\Delta B_k^{(n)} + R_k^{(n)},$$

where $R_k^{(n)} = R_{k,1}^{(n)} + R_{k,2}^{(n)}$ represents the correction relative to the Euler approximation and

$$\begin{aligned} R_{k,1}^{(n)} &= \int_{t_{k-1}^{(n)}}^{t_k^{(n)}} \left[\int_{t_{k-1}^{(n)}}^{s} b'\,b\,dB(u)\right] dB(s) \\ &\simeq \tilde{R}_{k,1}^{(n)} = b'(X(t_{k-1}^{(n)}))\,b(X(t_{k-1}^{(n)})) \int_{t_{k-1}^{(n)}}^{t_k^{(n)}} \left[\int_{t_{k-1}^{(n)}}^{s} dB(u)\right] dB(s) \\ &= \frac{1}{2}\,b'(X(t_{k-1}^{(n)}))\,b(X(t_{k-1}^{(n)}))\left[(\Delta B_k^{(n)})^2 - \Delta t_k^{(n)}\right]. \end{aligned}$$

The arguments of the functions a and b were not written in some of the previous formulas for simplicity. The final expression of $\tilde{R}_{k,1}^{(n)}$ above is valid since (Eq. 4.3)

$$\int_{t_{k-1}^{(n)}}^{t_k^{(n)}} B(s)\,dB(s) = \frac{1}{2}\left(B(t_k^{(n)})^2 - t_k^{(n)}\right) - \frac{1}{2}\left(B(t_{k-1}^{(n)})^2 - t_{k-1}^{(n)}\right)$$

so that

$$\begin{aligned}
\int_{t_{k-1}^{(n)}}^{t_k^{(n)}} \left[\int_{t_{k-1}^{(n)}}^{s} dB(u)\right] dB(s) &= \int_{t_{k-1}^{(n)}}^{t_k^{(n)}} B(s)\,dB(s) - \int_{t_{k-1}^{(n)}}^{t_k^{(n)}} B(t_{k-1}^{(n)})\,dB(s) \\
&= \frac{1}{2}\left(B(t_k^{(n)})^2 - t_k^{(n)}\right) - \frac{1}{2}\left(B(t_{k-1}^{(n)})^2 - t_{k-1}^{(n)}\right) - B(t_{k-1}^{(n)})\left(B(t_k^{(n)}) - B(t_{k-1}^{(n)})\right) \\
&= \frac{1}{2}\left[\left(B(t_k^{(n)})^2 - B(t_{k-1}^{(n)})\right)^2 - (t_k^{(n)} - t_{k-1}^{(n)})\right] = \frac{1}{2}\left[(\Delta B_k^{(n)})^2 - \Delta t_k^{(n)}\right].
\end{aligned}$$

The Milstein approximation results by neglecting the component $R_{k,2}^{(n)}$ of $R_k^{(n)}$ and replacing $R_{k,1}^{(n)}$ by $\tilde{R}_{k,1}^{(n)}$. ■

If the drift and diffusion coefficients of X satisfy some type of Lipschitz and growth conditions, then the Milstein approximation converges strongly with order $\gamma = 1$ ([115], Theorem 10.3.5, p. 350).

Note: This property shows that the Milstein approximation is superior to the Euler approximation. However, the difference between the Milstein and the Euler approximate solutions for Eq. 4.95 corresponding to the same partition of the time interval $[0, t]$ may not be significant (Example 4.53). ▲

Example 4.53: Consider the stochastic differential equation

$$dX(s) = c\,X(s)\,ds + \sigma\,X(s)\,dB(s), \quad \in [0, t],$$

with the initial condition $X(0) = x_0$, where c and σ are some constants. The geometric Brownian motion $X(t) = x_0\,e^{(c-0.5\,\sigma^2)\,t+\sigma\,B(t)}$ is the solution of this equation (Example 4.34).

Figure 4.12 shows with dotted and solid lines samples of the exact solution and the corresponding samples of $X^{(n)}$ in the time interval $[0, 1]$ obtained by the Euler and Milstein formulas for $c = 0.1$, $\sigma = 2$, and two partitions p_n of $[0, 1]$ with $n = 50$ and $n = 150$. The samples of X and $X^{(n)}$ in the figure correspond to the same sample $B(\cdot, \omega)$ of B. The Milstein approximation is slightly superior to the Euler approximation, an expected result since these approximations converge strongly with order $\gamma = 1$ and $\gamma = 0.5$, respectively. ◇

Note: The Euler and the Milstein formulas for the geometric Brownian motion are

$$\begin{aligned}
X_{k+1}^{(n)} &= (1 + c/n)\,X_k^{(n)} + \sigma\,X_k^{(n)}\,\Delta B_k^{(n)}, \quad \text{and} \\
X_{k+1}^{(n)} &= (1 + c/n)\,X_k^{(n)} + \sigma\,X_k^{(n)}\,\Delta B_k^{(n)} + (1/2)\,\sigma^2\,X_k^{(n)}\,[(\Delta B_k^{(n)})^2 - 1/n],
\end{aligned}$$

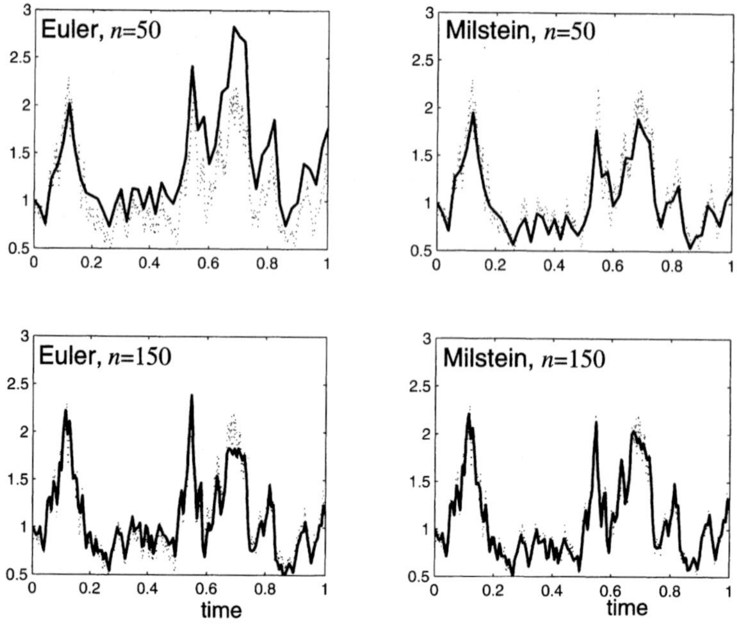

Figure 4.12. Samples of the Euler and Milstein numerical solutions for a geometric Brownian motion process

respectively, with the notations $X_k^{(n)} = X^{(n)}(t_k^{(n)})$ and $\Delta B_k^{(n)} = B((k+1)/n) - B(k/n)$ for the increment of the Brownian motion in $[k/n, (k+1)/n]$.

Estimates of the error function $e_s(p_n, t)$ in Eq. 4.97 must be based on samples of the exact and approximate solutions corresponding to the same sample of the Brownian motion, that is,

$$\hat{e}_s(p_n, t) = \frac{1}{n_s} \sum_{\omega=1}^{n_s} \left[\left| X(t, \omega) - X^{(n)}(t, \omega) \right| \right],$$

where $X(\cdot, \omega)$ and $X^{(n)}(\cdot, \omega)$ are calculated from the same sample $B(\cdot, \omega)$ of B and $n_s \geq 1$ is an integer denoting the number of samples used in estimation. The samples of X and $X^{(n)}$ needed to estimate the error $e_w(p_n, t)$ do not have to correspond to the same samples of the Brownian motion process B. ▲

The difference between the Euler and the Milstein numerical solutions in Eqs. 4.98, 4.100, and 4.101 is the correction term $\tilde{R}_{k,1}^{(n)}$ that depends on the diffusion part of Eq. 4.95. A major source of numerical error and instability relates to the crude approximation of the drift part of Eq. 4.95 by both methods. Methods developed for the numerical solution of deterministic differential equations can provide superior approximations to the solution for Eq. 4.95. Consider a deterministic differential equation

$$\dot{x}(s) = a(x(s), s), \quad s \in [0, t], \tag{4.102}$$

assumed to have a unique solution in $[0, t]$. For this equation the Euler and Milstein approximations coincide. We present an alternative to these numerical solutions, which provides a superior approximation to the solution of Eq. 4.102.

Runge-Kutta numerical solution.

$$x_{k+1}^{(n)} = x_k^{(n)} + \frac{1}{6}\left[\alpha_{k,1} + 2\,\alpha_{k,2} + 2\,\alpha_{k,3} + \alpha_{k,4}\right]\delta_n \tag{4.103}$$

where $x_k^{(n)} = x^{(n)}(t_k^{(n)})$, $\alpha_{k,1} = a(x_k^{(n)}, t_k^{(n)})$, $\alpha_{k,2} = a(x_k^{(n)} + \alpha_{k,1}\,\delta_n/2, t_k^{(n)} + \delta_n/2)$, $\alpha_{k,3} = a(x_k^{(n)} + \alpha_{k,2}\,\delta_n/2, t_k^{(n)} + \delta_n/2)$, and $\alpha_{k,4} = a(x_k^{(n)} + \alpha_{k,3}\,\delta_n, t_{k+1}^{(n)})$.

Note: Additional information on the Runge-Kutta and related numerical methods for solving deterministic differential equations can be found in [115] (Section 8.2). ▲

Example 4.54: Let X be defined by Eq. 4.95 with $a(x) = \rho\, x + \beta\, x^3$ and $b(x) = \sigma$. The Euler and Runge-Kutta numerical solutions for this equation are

$$X_{k+1}^{(n)} = X_k^{(n)} + \left(\rho\, X_k^{(n)} + \beta\left(X_k^{(n)}\right)^3\right)\delta_n + \sigma\,\Delta B_k^{(n)} \quad \text{and}$$

$$X_{k+1}^{(n)} = X_k^{(n)} + \frac{1}{6}\left[A_{k,1} + 2\,A_{k,2} + 2\,A_{k,3} + A_{k,4}\right]\delta_n + \sigma\,\Delta B_k^{(n)} \tag{4.104}$$

where $X_k^{(n)} = X^{(n)}(t_k^{(n)})$, $\Delta B_k^{(n)} = B(t_{k+1}^{(n)}) - B(t_k^{(n)})$, and the coefficients A have the definitions of the coefficients α in Eq. 4.103 with $X_k^{(n)}$ in place of $x_k^{(n)}$.

Figure 4.13 shows samples of the Euler and Runge-Kutta (R-K) solutions for $\rho = 1$, $\beta = -1$, $n = 250$, and $n = 10,000$ corresponding to the same sample of B. The Euler numerical solution diverges if the time step is relatively large ($n = 250$, $\delta_n = 100/250$). On the other hand, the solution given by the Runge-Kutta method is bounded for $n = 250$. Moreover, estimates of moments and other properties of X from samples of $X^{(n)}$ generated by the Runge-Kutta method with $n = 250$ are accurate. The solutions of the Euler and the Runge-Kutta methods are satisfactory and nearly coincide for $\delta_n = 100/10,000$. ◇

Note: The selection of the time step δ_n for a numerical solution $X^{(n)}$ must be such that the solution does not diverge and provides sufficient information on the samples and/or other properties of X. For example, $\delta_n = 100/250$ is inadequate for the Euler method because the corresponding numerical solution $X^{(n)}$ diverges, but seems to be satisfactory if the analysis is based on the Runge-Kutta method. ▲

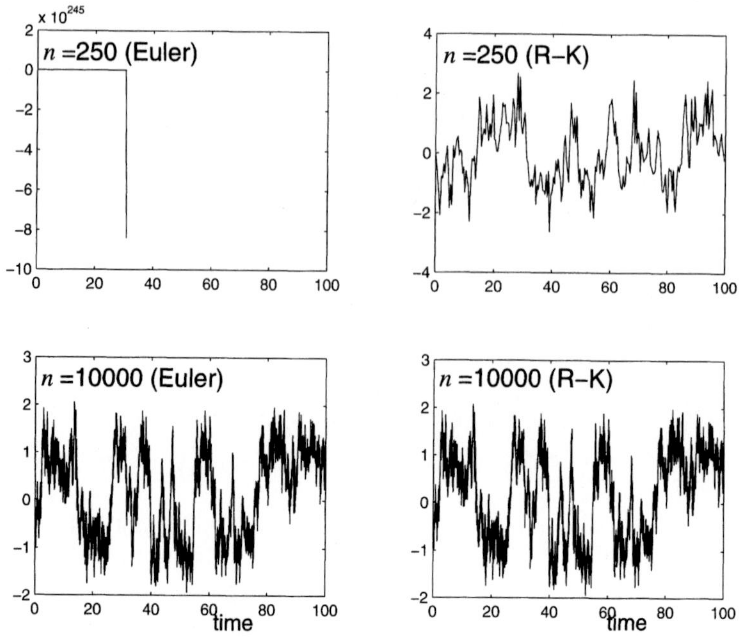

Figure 4.13. Samples of the Euler and Runge-Kutta (R-K) solutions

4.8 Problems

4.1: Show that the stochastic integral $J_X(H)$ in Eq. 4.16 is an adapted càdlàg process.

4.2: Prove the equality $\int_0^t s\,dB(s) = t\,B(t) - \int_0^t B(s)\,ds$, where B is a Brownian motion.

4.3: Prove the associativity and the preservation properties of the stochastic integral (Section 4.4.4) for the special case of simple predictable integrands.

4.4: Show that the jump process $\Delta(H \cdot X)$ is indistinguishable from $H(\Delta X)$, where $H \in \mathcal{L}$ and X is a semimartingale (Section 4.4.4).

4.5: Show that C^2 is a semimartingale, where C is the compound Poisson process in Eq. 4.13.

4.6: Check whether the processes

$$X(t) = B(t) - 4\,t \quad \text{and} \quad X(t) = t^2\,B(t) - 2\int_0^t s\,B(s)\,ds$$

are $\mathcal{F}_t$-martingales, where $\mathcal{F}_t$ denotes the natural filtration of a Brownian motion process B.

4.7: Show that $X(t) = \int_0^t h(s)\, dB(s)$, $t \in [0, \tau]$, $\tau > 0$, is a Gaussian process with mean zero and covariance function $E[X(s)\, X(t)] = \int_0^{s \wedge t} h(u)^2\, du$, where $h : [0, \tau] \to \mathbb{R}$ is a continuous function.

4.8: Use Itô's formula to show that $\int_0^t B\, dB$ is a martingale, where B is a Brownian motion.

4.9: Suppose that X is a real-valued diffusion process. Show that $Y(t) = e^{X(t)}$, $t \geq 0$, is also a diffusion process.

4.10: Write the Itô and the Stratonovich differential equations satisfied by the process $Y(t) = e^{B(t)}$, $t \geq 0$, where B is a Brownian motion.

4.11: Transform the following Stratonovich and Itô differential equations into Itô and Stratonovich differential equations, respectively, that is, the Stratonovich equations

$$dX(t) = a\, X(t)\, dt + b\, X(t) \circ dB(t),$$
$$dX(t) = \sin(X(t))\, \cos(X(t))\, dt + (t^2 + \cos(X(t)) \circ dB(t)$$

and the Itô equations

$$dX(t) = a\, X(t)\, dt + b\, X(t)\, dB(t),$$
$$dX(t) = 2\, \exp(-X(t))\, dt + X(t)^2\, dB(t),$$

where a and b are some constants.

4.12: Show that $X(t) = X(0)\, e^{-\alpha t} + \beta \int_0^t e^{-\alpha(t-s)}\, dB(s)$ is the solution of the Ornstein-Uhlenbeck process in Example 4.41.

4.13: Find the solution of

$$dX(t) = a\, X(t)\, dt + X(t) \left(\sum_{k=1}^{d} b_k\, dB_k(t) \right), \quad X(0) > 0,$$

where B_k are independent Brownian motions and a, b_k are constants.

4.14: Let $\boldsymbol{X}(t) = (X_1(t) = B(t)^2, X_2(t) = e^{B(t)})$ be an $\mathbb{R}^2$-valued process, where B denotes a Brownian motion. Find the stochastic differential equation satisfied by $\boldsymbol{X}$. Is $\boldsymbol{X}$ a diffusion process?

4.15: Let X be a stochastic process defined by the Stratonovich differential equation $d^{\circ}X(t) = a\, X(t)\, dt + b\, X(t) \circ dB(t)$, where a and b are constants. Find the Itô version of this equation.

4.16: Find the solution of $dX(t) = a\, X(t)\, dt + b\, X(t)\, dC(t)$, $X(0) > 0$, where C is the compound Poisson process in Eq. 4.13 and a, b are constants.

Chapter 5

Monte Carlo Simulation

5.1 Introduction

The Monte Carlo simulation method is used extensively in applied science and engineering to solve a variety of stochastic problems. Solutions for this type of problems by Monte Carlo simulation are presented in the following four chapters of the book. We have already used the Monte Carlo method in previous chapters to illustrate theoretical concepts on random variables, conditional expectation, stochastic processes, and stochastic integrals.

The solution of a stochastic problem by the Monte Carlo method involves three steps. First, independent samples of the random parameters and functions in the definition of the stochastic problem need to be generated. Second, the resulting deterministic problems corresponding to the samples generated in the previous step have to be solved. Third, the collection of deterministic solutions needs to be analyzed statistically to estimate properties of the solution of the stochastic problem. The first step of the Monte Carlo solution is the focus of this chapter. We give methods for generating samples of random variables, stochastic processes, and random fields. Improved Monte Carlo simulation methods for estimating system reliability are also presented.

Most algorithms for generating samples of random variables, stochastic processes, and random fields consist of transformations with and without memory of random variables uniformly distributed in $(0, 1)$ and Gaussian variables with mean zero and variance one, denoted by $U(0, 1)$ and $N(0, 1)$, respectively. Most computer software packages include functions producing samples of these random variables. For example, the MATLAB functions **rand** and **randn** generate independent samples of $U(0, 1)$ and $N(0, 1)$.

5.2 Random variables

Let X be an $\mathbb{R}^d$-valued random variable defined on a probability space $(\Omega, \mathcal{F}, P)$. Our objective is to generate independent samples of X.

5.2.1 Gaussian variables

Let $X \sim N(\mu, \gamma)$ be a Gaussian vector in $\mathbb{R}^d$, where μ and γ denote the mean and covariance matrices of X.

Case 1 ($d = 1$). If $X \sim N(\mu, \gamma)$, the scaled and translated MATLAB function $\mu + \sqrt{\gamma}$ **randn** can be used to generate independent samples of X. Some of the algorithms used to generate samples of a Gaussian variable are based on memoryless transformations of some random variables [27, 79, 155]. For example, Gaussian variables can be related simply to uniformly distributed random variables (Eq. 5.1).

Monte Carlo algorithm. The generation of independent samples of a random variable $X \sim N(0, 1)$ involves two steps:

1. Generate samples of the independent random variables $U_1, U_2 \sim U(0, 1)$.

2. Calculate two independent samples of X from

$$Z_1 = \sqrt{-2\,\ln(U_1)}\,\cos(2\pi\,U_2) \quad \text{and} \quad Z_2 = \sqrt{-2\,\ln(U_1)}\,\sin(2\pi\,U_2). \quad (5.1)$$

Proof: If Z_1 and Z_2 are independent copies of $N(0, 1)$, their density is $f(z_1, z_2) = \exp[-(z_1^2 + z_2^2)/2]/(2\pi)$. The density of the random variables (R, Θ) defined by the mapping $Z_1 = R\,\cos(\Theta)$ and $Z_2 = R\,\sin(\Theta)$ is $f_{R,\Theta}(r, \theta) = r\,\exp(-r^2/2)/(2\pi)$. Hence, R has the density $f_R(r) = r\,\exp(-r^2/2)$, Θ is uniformly distributed on $(0, 2\pi]$, and the variables (R, Θ) are independent. Samples of R can be generated from samples of $U_1 \sim U(0, 1)$ and the representation $R \stackrel{d}{=} \sqrt{-2\,\ln(U_1)}$ derived from $F_R(R) = U_1$ and the fact that U_1 and $1 - U_1$ have the same distribution, where F_R denotes the distribution of R. This method for generating samples of random variables is discussed further in Eq. 5.4. The projection of R on the coordinates (z_1, z_2) gives Eq. 5.1. ■

Case 2 ($d > 1$). The generation of independent samples of a Gaussian vector $X \sim N(\mu, \gamma)$ can be based on the MATLAB function **randn** or Eq. 5.1 and the Cholesky decomposition in Eqs. 5.2 and 5.3. The vectors X and

$$Z = \mu + \beta\, G \sim N(\mu, \gamma), \quad (5.2)$$

have the same distribution, where the coordinates of $G \in \mathbb{R}^d$ are independent $N(0, 1)$ and the non-zero entries of the lower triangular matrix β are

$$\beta_{ij} = \frac{\gamma_{ij} - \sum_{r=1}^{j-1} \beta_{ir}\,\beta_{jr}}{\left[\gamma_{jj} - \sum_{r=1}^{j-1} \beta_{jr}^2\right]^{1/2}}, \quad 1 \le j \le i \le d, \text{ with } \sum_{r=1}^{0} \beta_{ir}\,\beta_{jr} = 0. \quad (5.3)$$

Monte Carlo algorithm. The generation of samples of $\boldsymbol{X} \sim N(\boldsymbol{\mu}, \boldsymbol{\gamma})$ involves two steps:

1. Generate samples of $\boldsymbol{G}$.

2. Calculate the corresponding samples of $\boldsymbol{X}$ from Eq. 5.2 with $\boldsymbol{\beta}$ in Eq. 5.3.

Note: A sample of the column vector $\boldsymbol{G}$ can be obtained from the MATLAB function **randn**$(d, 1)$. A proof of the validity of Eqs. 5.2 and 5.3 can be found in [155] (Section 3.5.3). An alternative method for generating samples of $\boldsymbol{X}$ is in Problem 5.1. ▲

5.2.2 Non-Gaussian variables

Let $\boldsymbol{X}$ be an arbitrary $\mathbb{R}^d$-valued random variable with dependent components and distribution F.

Case 1 ($d = 1$). The generation of samples of an arbitrary non-Gaussian random variable can also be based on the MATLAB function **rand**.

Monte Carlo algorithm. The generation of samples of X with distribution F involves two steps:

1. Generate independent samples of $U(0, 1)$.

2. Calculate the corresponding samples of X from

$$Z = F^{-1}(U(0, 1)). \tag{5.4}$$

Proof: The distribution of $Z = F^{-1}(U(0, 1))$ is

$$P(Z \le z) = P(F^{-1}(U(0, 1)) \le z) = P(U(0, 1) \le F(z)) = F(z).$$

Samples of X can be calculated from Eq. 5.4 and independent samples of $U(0, 1)$. For example, n independent samples of an exponential random variable with mean $1/\lambda$, $\lambda > 0$, are given by $-\ln(1 - \textbf{rand}(n, 1))/\lambda \stackrel{d}{=} -\ln(\textbf{rand}(n, 1))/\lambda$. ■

The Monte Carlo simulation algorithm based on Eq. 5.4, referred to as the **inverse transform method**, is general but becomes inefficient if F^{-1} has to be calculated numerically. For example, the algorithm in Eq. 5.1, rather than $\Phi^{-1}(U(0, 1))$, is used in Monte Carlo studies since the inverse Φ^{-1} of the distribution Φ of $N(0, 1)$ is not available in closed form.

Example 5.1: Let $X \sim S_\alpha(\sigma, \beta, \mu)$ be an α-stable random variable with scale $\sigma > 0$, skewness β, and location μ, where $|\beta| \le 1$ and $\alpha \in (0, 2]$. If $\beta = 0$ and $\mu = 0$, X has a symmetric density about zero and its samples for $\sigma = 1$ can be generated from

$$Z = \frac{\sin(\alpha V)}{[\cos(V)]^{1/\alpha}} \left[\frac{\cos((1 - \alpha) V)}{W}\right]^{(1-\alpha)/\alpha}, \tag{5.5}$$

where V is uniformly distributed in $(-\pi/2, \pi/2)$ and W follows an exponential distribution with mean one. The random variables V and W are independent. The above definition of Z becomes

$$Z = \frac{\sin(V)}{\cos(V)} = \tan(V), \qquad \text{for } \alpha = 1 \ (X \sim \text{Cauchy}),$$
$$Z = 2\sqrt{W}\,\sin(V), \qquad \text{for } \alpha = 2 \ (X \sim N(0, 2)). \tag{5.6}$$

If $X \sim S_\alpha(\sigma, 0, \mu)$, samples of X can be obtained from $\mu + \sigma\, Z$ with Z in Eq. 5.5 or 5.6. If $X \sim S_\alpha(\sigma, \beta, \mu)$, samples of X can be calculated from $\mu + \sigma\, Z$ and samples of $Z \sim S_\alpha(1, \beta, 0)$, where

$$Z \stackrel{d}{=} \begin{cases} \frac{\sin(\alpha\,(V-v_0))}{(\cos(V))^{1/\alpha}} \left[\frac{\cos(V-\alpha\,(V-v_0))}{W}\right]^{(1-\alpha)/\alpha}, & \alpha \neq 1, \\ \frac{2}{\pi}\left[\left(\frac{\pi}{2}+\beta\, V\right)\tan(V) - \beta\,\ln\left(\frac{(\pi/2)\,W\,\cos(V)}{\pi/2+\beta\,V}\right)\right], & \alpha = 1, \end{cases} \tag{5.7}$$

with V and W as in Eq. 5.5, $v_0 = -\pi\,\beta\, h(\alpha)/(2\,\alpha)$, and $h(\alpha) = 1 - |1-\alpha|$ ([79], Section 4.8.1, [162], Section 1.7). ◇

Example 5.2: Let $X \sim S_1(\sigma, 0, 0)$, $\sigma > 0$, be a Cauchy variable with density $f(x) = \sigma/[\pi\,(x^2+\sigma^2)]$ and characteristic function $\varphi(u) = e^{-\sigma\,|u|}$. Figure 5.1 shows $n = 100$ independent samples of X and a histogram of X based on 10,000

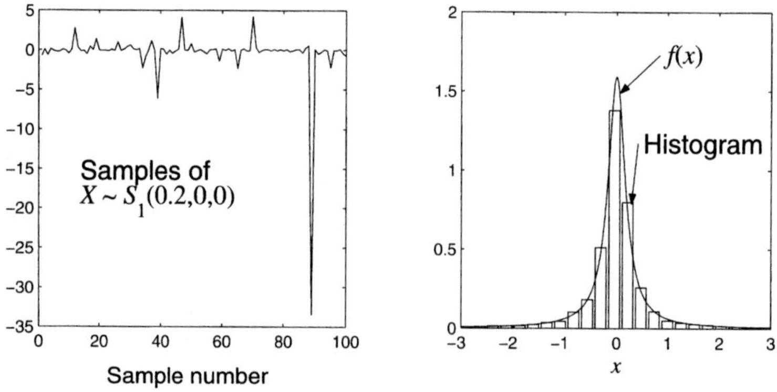

Figure 5.1. Samples of a Cauchy variable $X \sim S_1(0.2, 0, 0)$ and a histogram of X based on 10,000 independent samples

samples of this variable, respectively, for $\sigma = 0.2$. The figure also shows the density f of X. The samples of X have been obtained from $\sigma\, Z$ and samples of Z in Eq. 5.6 for $\alpha = 1$. ◇

Example 5.3: Let N be a Poisson random variable with probability law $P(N = n) = (\lambda^n/n!)\, e^{-\lambda}$, $n = 0, 1, \ldots$, and intensity $\lambda > 0$. It is possible to generate samples of N by the inverse method in Eq. 5.4. However, the approach is inefficient if the probability that N takes relatively large values is not negligible. If $E[N] = \lambda \geq 10$, it is convenient to generate samples of N from

$$Z = \max\left\{0, \left[\lambda + G\,\lambda^{1/2} - 1/2\right]\right\} \tag{5.8}$$

where $[x]$ denotes the integer part of x and $G \sim N(0, 1)$ ([155], Section 3.7.2).

Note: The approximation of N by the random variable Z in Eq. 5.8 is acceptable for large values of λ because the distribution of the random variable $(N - \lambda)/\sqrt{\lambda}$ converges to the distribution of $N(0, 1)$ as λ increases indefinitely. ▲

Case 2 ($d > 1$). Let F_1, $F_{k|k-1,\ldots,1}$, $k = 2, \ldots, d$, and F denote the distributions of the coordinate X_1 of $\boldsymbol{X} \in \mathbb{R}^d$, the conditional random variable $X_k \mid (X_{k-1}, \ldots, X_1)$, and $\boldsymbol{X} = (X_1, \ldots, X_d)$, respectively.

Let $\boldsymbol{Z} = (Z_1, \ldots, Z_d)$ be an $\mathbb{R}^d$-valued random variable defined by

$$\begin{aligned} F_1(Z_1) &= U_1, \\ F_{k|k-1,\ldots,1}(Z_k) &= U_k, \quad k = 2, \ldots, d, \end{aligned} \tag{5.9}$$

where $U_1, \ldots, U_d$ are independent $U(0, 1)$ random variables. Then, $\boldsymbol{X}$ and $\boldsymbol{Z}$ are equal in distribution.

Proof: The random vector $\boldsymbol{Z}$ has the same distribution as $\boldsymbol{X}$ by definition (Eq. 5.9). The representation of $\boldsymbol{Z}$ in Eq. 5.9 constitutes an extension of the inverse transform method in Eq. 5.4 to random vectors. ■

Monte Carlo algorithm. The generation of samples of $\boldsymbol{X}$ involves two steps:

1. Generate independent samples of $U(0, 1)$ and construct samples of $\boldsymbol{U} = (U_1, \ldots, U_d)$.

2. Calculate the corresponding samples of $\boldsymbol{X}$ from Eq. 5.9.

Note: Let $(u_1, \ldots, u_d)$ be a sample of $(U_1, \ldots, U_d)$ and $z = (z_1, \ldots, z_d)$ be the corresponding sample of $\boldsymbol{Z}$ obtained from Eq. 5.9. The first coordinate of z is $z_1 = F_1^{-1}(u_1)$. The other coordinates of z can be calculated from $F_{k|k-1,\ldots,1}(z_k) = u_k$ for increasing values of $k \geq 2$, where the vector $(z_{k-1}, \ldots, z_1)$ has already been calculated in a previous step. ▲

Example 5.4: Let $\boldsymbol{X} = (X_1, X_2)$ be a non-Gaussian vector with $X_1 \sim N(\mu, \sigma^2)$ and $X_2 \mid (X_1 = x_1) \sim N(x_1, \delta^2)$. The density of $\boldsymbol{X}$ is

$$f(x_1, x_2) = \frac{1}{\sigma\,\delta}\,\phi\left(\frac{x_1 - \mu}{\sigma}\right)\,\phi\left(\frac{x_2 - x_1}{\delta}\right).$$

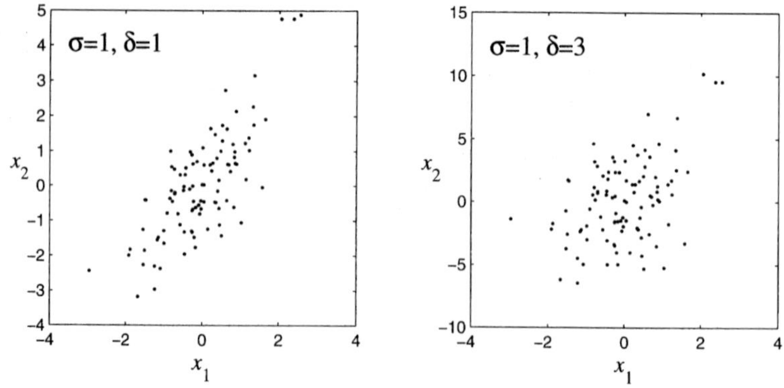

Figure 5.2. Samples $\boldsymbol{X} = (X_1, X_2)$, where $X_1 \sim N(\mu, \sigma^2)$ and $X_2 \mid (X_1 = x_1) \sim N(x_1, \delta^2)$ for $(\sigma = 1, \delta = 1)$ and $(\sigma = 1, \delta = 3)$

The mapping in Eq. 5.9 becomes $Z_1 = \mu + \sigma\, \Phi^{-1}(U_1)$ and $Z_2 \mid (Z_1 = z_1) = z_1 + \delta\, \Phi^{-1}(U_2)$, where U_1 and U_2 are independent copies of $U(0,1)$. Figure 5.2 shows 100 independent samples of $\boldsymbol{X}$ generated by the algorithm in Eq. 5.9 for $\sigma = 1$, $\delta = 1$, and $\delta = 3$. The dependence between X_1 and X_2 is stronger for smaller values of δ. ◇

Example 5.5: Let $\boldsymbol{X}$ be an α-stable vector in $\mathbb{R}^d$ defined by the value of α, location vector $\boldsymbol{\mu} \in \mathbb{R}^d$, and a finite measure γ on the unit sphere S_d centered at the origin of $\mathbb{R}^d$. Define a discrete approximation $\gamma^{(n)}(\boldsymbol{\sigma}) = \sum_{r=1}^n \gamma_r\, \delta(\boldsymbol{\sigma}_r - \boldsymbol{\sigma})$ of γ, where $(A_1, \ldots, A_n)$ is a partition of S_d, $\gamma_r = \gamma(A_r)$, and $\boldsymbol{\sigma}_r \in A_r$ are arbitrary points, $r = 1, \ldots, n$. Samples of the random vector $\boldsymbol{X}$ can be generated from

$$\boldsymbol{Z}^{(n)} = \begin{cases} \sum_{r=1}^n \gamma_r^{1/\alpha}\, V_r\, \boldsymbol{\sigma}_r + \boldsymbol{\mu}, & \alpha \neq 1, \\ \sum_{r=1}^n \gamma_r \left(V_r + \frac{2}{\pi}\, \ln(\gamma_r)\right) \boldsymbol{\sigma}_r + \boldsymbol{\mu}, & \alpha = 1, \end{cases} \tag{5.10}$$

where V_r are independent copies of $S_\alpha(1,1,0)$. ◇

Note: The characteristic function $\varphi(\boldsymbol{u}) = E\left[e^{\sqrt{-1}\,\boldsymbol{u}^T \boldsymbol{X}}\right]$ of $\boldsymbol{X}$ is

$$\begin{array}{ll} \exp\left(a - \int_{S_d} |b|^\alpha\, \gamma(d\boldsymbol{\sigma}) + \sqrt{-1}\, \tan\left(\frac{\pi\,\alpha}{2}\right) \int_{S_d} |b|^\alpha\, \mathrm{sign}(b)\, \gamma(d\boldsymbol{\sigma})\right), & \alpha \neq 1, 2 \\ \exp\left(a - \int_{S_d} |b|\, \gamma(d\boldsymbol{\sigma}) - \sqrt{-1}\, \frac{2}{\pi} \int_{S_d} b\, \ln(|b|)\, \gamma(d\boldsymbol{\sigma})\right), & \alpha = 1 \\ \exp\left(a - \int_{S_d} |b|^2\, \gamma(d\boldsymbol{\sigma})\right), & \alpha = 2 \end{array}$$

where $a = \sqrt{-1}\, \boldsymbol{u}^T\, \boldsymbol{\mu}$ and $b = \boldsymbol{u}^T\, \boldsymbol{\sigma}$. It can be shown that $\boldsymbol{Z}^{(n)}$ converges in distribution to $\boldsymbol{X}$ if $\max_{1 \le r \le d} \gamma(A_r) \to 0$ as $n \to \infty$ ([79], Section 3.5.3). ▲

Example 5.6: Let $X \in \mathbb{R}^d$ be a translation vector, that is, $X = g(Y)$, where Y is an $\mathbb{R}^d$-valued Gaussian vector with mean zero, covariance matrix $\rho = \{\rho_{i,j} = E[Y_i Y_j]\}$ such that $\rho_{i,i} = 1, i = 1, \ldots, d$. Suppose that the memoryless mapping from Y to X is given by $X_i = F_i^{-1}(\Phi(Y_i)) = g_i(Y_i), i = 1, \ldots, d$, where F_i are absolutely continuous distribution functions.

Samples of X can be generated from (1) samples of Y and the definition of X or (2) samples of $U(0, 1)$, the distribution of X, and the algorithm in Eq. 5.9. The latter approach is less efficient for translation random vectors. ◇

Note: Let $y_i = g_i^{-1}(x_i)$ and $y = (y_1, \ldots, y_d)$. The distribution,

$$P(X_1 \leq x_1, \ldots, X_d \leq x_d) = P(Y_1 \leq y_1, \ldots, Y_d \leq y_d) = \Phi_d(y; \rho),$$

called **multivariate translation distribution**, is the input to the Monte Carlo simulation algorithm based on Eq. 5.9, where $\Phi_d(\cdot; \rho)$ denotes the joint distribution function of the Gaussian vector $Y \sim N(\mathbf{0}, \rho)$. ▲

5.3 Stochastic processes and random fields

Properties of stochastic processes and random fields have been discussed in Chapter 3. In this section we present methods for generating sample paths of these random functions. Let $X(t)$ be an $\mathbb{R}^d$-valued stochastic process defined on a probability space $(\Omega, \mathcal{F}, P)$, where the argument $t \in \mathbb{R}$ is viewed as time. We also consider $\mathbb{R}^d$-valued random fields $X(\mathbf{t})$ defined on the same probability space, but $\mathbf{t} \in \mathbb{R}^{d'}$ denotes a space argument.

5.3.1 Stationary Gaussian processes and fields

If all finite dimensional distributions of a random function, that is, a stochastic process $X(t)$ or a random field $X(\mathbf{t})$, are Gaussian, this function is said to be a Gaussian process or field. Our objective is to generate independent samples of a Gaussian function X with mean zero and a prescribed covariance function. If the mean of X is not zero, we generate samples of $X - E[X]$ and add the mean function $E[X]$ to these samples.

5.3.1.1 Spectral representation. Stochastic processes

The spectral representation theorem shows that a weakly stationary process can be approximated by a superposition of harmonics (Sections 3.9.4.1 and 3.9.4.2). This theorem can be used to develop models for generating sample paths of stationary Gaussian processes, which consist of a superposition of harmonics with (1) fixed frequencies and random amplitudes or (2) random frequencies and amplitudes.

Let $X(t)$, $t \in \mathbb{R}$, be a stationary Gaussian process with mean zero, covariance function $c(\tau) = E[X(t + \tau) X(t)^T]$, and spectral density $s(\nu) = \{s_{kl}(\nu)\}$,

$k, l = 1, \ldots, d$. Our objectives are to (1) construct a sequence of processes $\boldsymbol{X}^{(n)}$, $n = 1, 2, \ldots$, such that $\boldsymbol{X}^{(n)}$ depends on a finite number of random variables for each n and becomes a version of $\boldsymbol{X}$ as $n \to \infty$ and (2) develop algorithms for generating samples of $\boldsymbol{X}^{(n)}$.

Fixed frequencies. We first introduce some notation and definitions. Let $\nu^* > 0$ be a cutoff frequency and $p_n = (0 = \alpha_0 < \alpha_1 < \cdots < \alpha_n = \nu^*)$ be a partition of the frequency range $[0, \nu^*]$. Denote by $\Delta\nu_r = \alpha_r - \alpha_{r-1}$ and $\nu_r = (\alpha_{r-1} + \alpha_r)/2$, $r = 1, \ldots, n$, the length and the midpoint of the frequency intervals defined by p_n. It is common to take equal frequency intervals in which case we have $\Delta\nu_r = \nu^*/n$ and $\nu_r = (r - 1/2)\, \nu^*/n$. Consider also the $\mathbb{R}^d$-valued Gaussian variables $\boldsymbol{A}_r$, $\boldsymbol{B}_r$ with mean zero and second moments

$$
\begin{aligned}
E[A_{r,k}\, A_{p,l}] = E[B_{r,k}\, B_{p,l}] &= \delta_{rp} \int_{\alpha_{r-1}}^{\alpha_r} g_{k,l}(\nu)\, d\nu \simeq \delta_{rp}\, g_{k,l}(\nu_r)\, \Delta\nu_r, \\
E[A_{r,k}\, B_{p,l}] = -E[B_{r,k}\, A_{p,l}] &= \delta_{rp} \int_{\alpha_{r-1}}^{\alpha_r} h_{k,l}(\nu)\, d\nu \simeq \delta_{rp}\, h_{k,l}(\nu_r)\, \Delta\nu_r,
\end{aligned}
\tag{5.11}
$$

where $\boldsymbol{g}(\nu) = \boldsymbol{s}(\nu) + \boldsymbol{s}(-\nu)$ and $\boldsymbol{h}(\nu) = -\sqrt{-1}\,(\boldsymbol{s}(\nu) - \boldsymbol{s}(-\nu))$ (Section 3.9.4.2). The approximate equalities in Eq. 5.11 can be used for small values of $\Delta\nu_r$.

If the frequency band of the process $\boldsymbol{X}$ is not bounded, a cutoff frequency $\nu^* > 0$ has to be selected to apply the Monte Carlo simulation algorithm considered in this section. The cutoff frequency $\nu^* > 0$ has to be such that most of the energy of the process $\boldsymbol{X}$ corresponds to harmonics with frequencies in $[0, \nu^*]$.

We now define a sequence of processes $\boldsymbol{X}^{(n)}$ and show that $\boldsymbol{X}^{(n)}$ is approximately a version of $\boldsymbol{X}$ for sufficiently large values of n. This property justifies the use of samples of $\boldsymbol{X}^{(n)}$ as substitutes for samples of $\boldsymbol{X}$.

If $\Delta(p_n) \to 0$ and $\nu^* \to \infty$, then

$$
\boldsymbol{X}^{(n)}(t) = \sum_{r=1}^{n} [\boldsymbol{A}_r \cos(\nu_r\, t) + \boldsymbol{B}_r \sin(\nu_r\, t)]
\tag{5.12}
$$

becomes a version of the stationary Gaussian process $\boldsymbol{X}$.

Proof: Recall that $\Delta(p_n) = \max_{1 \le r \le n} (\alpha_r - \alpha_{r-1})$ denotes the mesh of p_n. For any partition of the frequency range $[0, \nu^*]$, $\boldsymbol{X}^{(n)}$ is a process with mean zero and covariance function with entries

$$
E\left[\left(\boldsymbol{X}^{(n)}(t)\, \boldsymbol{X}^{(n)}(s)^T\right)_{ij}\right] = \sum_{r=1}^{n} \left\{ \left[\int_{\alpha_{r-1}}^{\alpha_r} g_{i,j}(\nu)\, d\nu\right] \cos(\nu_r\, (t-s)) \right.
$$
$$
\left. - \left[\int_{\alpha_{r-1}}^{\alpha_r} h_{i,j}(\nu)\, d\nu\right] \sin(\nu_r\, (t-s)) \right\}, \quad i, j = 1, \ldots, d.
$$

Hence, $X^{(n)}$ is weakly stationary for each n. This process is Gaussian because it depends linearly on the Gaussian variables A_r and B_r. If $d = 1$, then $g_{11} = g$ and $h_{11} = 0$ so that the covariance function of $X^{(n)}$ is

$$E\left[X^{(n)}(t)\, X^{(n)}(s)\right] = \sum_{r=1}^{n} \left[\int_{\alpha_{r-1}}^{\alpha_r} g(\nu)\, d\nu\right] \cos(\nu_r\,(t-s)).$$

For a fixed ν^*, we have

$$\lim_{n\to\infty} E\left[\left(X^{(n)}(t)\, X^{(n)}(s)^T\right)_{ij}\right]$$
$$= \int_0^{\nu^*} \left[g_{i,j}(\nu)\cos(\nu\,(t-s)) - h_{i,j}(\nu)\sin(\nu\,(t-s))\right] d\nu$$

so that the covariance function of $X^{(n)}$ converges to the covariance function of X if in addition to the limit as $n \to \infty$ we also let $\nu^* \to \infty$ (Section 3.9.4.2). Hence, $X^{(n)}$ becomes a version of X since X and $X^{(n)}$ are Gaussian processes.

The model $X^{(n)}$ has two properties that may be undesirable in some applications. The sample paths of this process are (1) periodic with period $2\pi/\nu_1$ if $\Delta\nu_r = \nu^*/n$ and $\nu_r = (r - 1/2)\,\Delta\nu_r$ and (2) infinitely differentiable even if the samples of X are not differentiable. Hence, samples longer than $2\pi/\nu_1$ provide the same information as samples of length $2\pi/\nu_1$, and estimates of, for example, mean crossing rates of X from samples of $X^{(n)}$ are incorrect if X is not differentiable (Section 3.10.1). ■

Monte Carlo algorithm. The generation of samples of X in a time interval $[0, \tau]$, $\tau > 0$, involves three steps:

1. Select a cutoff frequency $\nu^* > 0$, partition the frequency band $[0, \nu^*]$ in n intervals, and select frequencies ν_r, $r = 1, \ldots, n$, in these intervals.

2. Generate samples of the Gaussian random variables A_r and B_r, $r = 1, \ldots, n$, with mean zero and second moments in Eq. 5.11 using the algorithm in Section 5.2.1.

3. Calculate the corresponding samples of $X^{(n)}$ in $[0, \tau]$ from Eq. 5.12.

Example 5.7: Let X be an $\mathbb{R}^2$-valued stationary Gaussian process with spectral density

$$s_{k,l}(\nu) = (1-\rho)\,\delta_{kl}\, s_k(\nu) + \rho\, s_Z(\nu) \quad k, l = 1, 2,$$

where $s_k(\nu) = 1/(2\,\bar\nu_k)\, 1_{[-\bar\nu_k, \bar\nu_k]}(\nu)$, $s_Z(\nu) = 1/(2\,\bar\nu)\, 1_{[-\bar\nu, \bar\nu]}(\nu)$, and $0 < \bar\nu_k, \bar\nu < \infty$. Figure 5.3 shows sample paths of the coordinates of X for $\bar\nu_k = 25$, $\bar\nu = 5$, two values of ρ, and $n = 100$ harmonics (Eq. 5.12). The frequency content of X depends strongly on ρ. The coordinates X_1 and X_2 of X are nearly in phase for values of ρ close to unity. ◇

Note: Set $\nu^* = \max(\bar\nu_1, \bar\nu_2, \bar\nu)$, $\Delta\nu_r = \nu^*/n$, and $\nu_r = (r - 1/2)\,\Delta\nu_r$ for $r = 1, \ldots, n$. The covariance matrix of the random amplitudes A_r and B_r can be calculated from Eq. 5.11. The samples in Fig. 5.3 have been generated by the model in Eq. 5.12. ▲

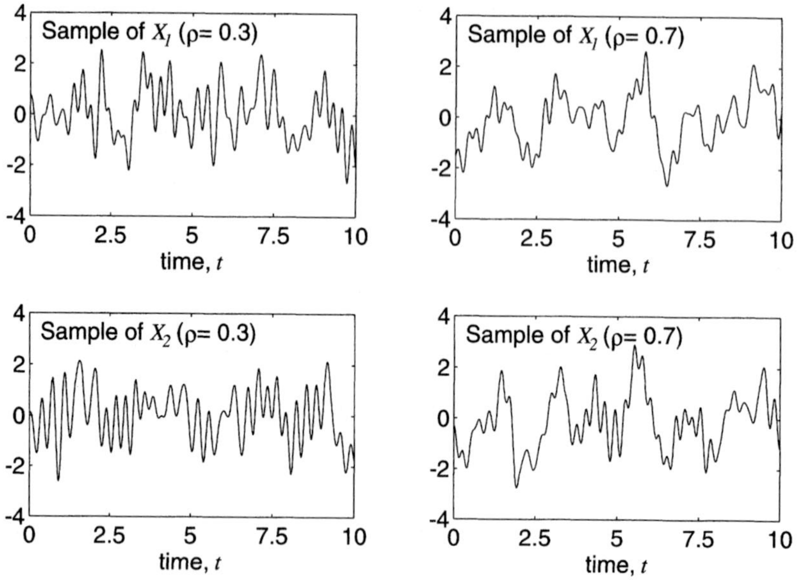

Figure 5.3. Samples of the coordinates X_1 and X_2 of $\boldsymbol{X}$ for $\rho = 0.3$ and $\rho = 0.7$ (adapted from [79], Fig. 4.4)

Random frequencies. We limit our discussion to real-valued processes for simplicity. Our objective is to develop a Monte Carlo simulation algorithm for generating samples of a stationary Gaussian process $\{X(t) \in \mathbb{R}\}$ with mean zero, covariance function $c(\tau) = E[X(t+\tau)\, X(t)]$, and one-sided spectral density g. This algorithm is based on a model consisting of a superposition of harmonics with both random amplitudes and frequencies, in contrast to the model in Eq. 5.12 that has random amplitudes but fixed frequencies.

Let $C_r(\nu)$, $r = 1, 2$, be independent copies of a compound Poisson process $C(\nu) = \sum_{k=1}^{N(\nu)} Y_k$, where N is a Poisson process with intensity $\lambda > 0$ and Y_k are iid random variables with $E[Y_1] = 0$ and $E[Y_1^2] = \mu_2 < \infty$. Denote by $(T_1, T_2, \ldots)$ the random times of the jumps $(Y_1, Y_2, \ldots)$ of C. We continue to call T_k, $k = 1, 2, \ldots$, jump times although they are frequencies. As for the previous algorithm (Eq. 5.12) we select a cutoff frequency $\nu^* > 0$.

If $\zeta(\nu)^2 = g(\nu)/(\lambda\, \mu_2)$, $\lambda \to \infty$, and $\nu^* \to \infty$, then

$$X^{(\lambda)}(t) = \int_0^{\nu^*} \zeta(\nu)\, [\cos(\nu\, t)\, dC_1(\nu) + \sin(\nu\, t)\, dC_2(\nu)]$$
$$= \sum_{k=1}^{N_1(\nu^*)} \zeta(T_{1,k})\, Y_{1,k} \cos\left(T_{1,k}\, t\right) + \sum_{l=1}^{N_2(\nu^*)} \zeta(T_{2,l})\, Y_{2,l} \sin\left(T_{2,l}\, t\right) \tag{5.13}$$

becomes a version of the Gaussian process X [87].

Proof: The parameter $\lambda > 0$ indexing the sequence of processes approximating X has a similar role as $n = 1, 2, \ldots$ in Eq. 5.12. The first equality in Eq. 5.13 is a statement of the spectral representation for an approximation of X obtained by neglecting the power of this process beyond a cutoff frequency $\nu^* > 0$. The second equality in Eq. 5.13 holds since $N_r(\nu^*) < \infty$ a.s. for $\nu^* < \infty$ and the integrators C_r, $r = 1, 2$, have piece-wise constant samples.

The process $X^{(\lambda)}$ has mean zero and covariance function

$$E[X^{(\lambda)}(t)\, X^{(\lambda)}(s)] = \lambda\, \mu_2 \int_0^{\nu^*} \zeta(\nu)^2\, [\cos(\nu\, t)\, \cos(\nu\, s) + \sin(\nu\, t)\, \sin(\nu\, s)]\, d\nu$$
$$= \int_0^{\nu^*} g(\nu)\, \cos(\nu\, (t - s))\, d\nu,$$

where the last equality holds because $\lambda\, \mu_2\, \zeta(\nu)^2 = g(\nu)$ by hypothesis. Hence, the covariance function of $X^{(\lambda)}$ converges to the covariance function of X as $\nu^* \to \infty$, that is, X and $\lim_{\nu^* \to \infty} X^{(\lambda)}$ are equal in the second moment sense for each $\lambda > 0$. It can also be shown that $\tilde{X}^{(\lambda)}(t) = X^{(\lambda)}(t)/\mathrm{Var}[X^{(\lambda)}(t)]$ converges in distribution to $N(0, 1)$ ([79], Example 3.12, p. 83). The approach in [79] can be used to show that the distribution of a vector with coordinates $\tilde{X}^{(\lambda)}(t_k)$ becomes Gaussian as $\lambda \to \infty$, where t_k are arbitrary times. In summary, we showed that (1) the second moments of $X^{(\lambda)}$ approach the second moments of X as $\nu^* \to \infty$ for any $\lambda > 0$ and (2) $X^{(\lambda)}$ converges to a Gaussian process as $\lambda \to \infty$ for any $\nu^* > 0$. Hence, $X^{(\lambda)}$ becomes a version of X as λ and ν^* increase indefinitely. ■

The models $X^{(n)}$ and $X^{(\lambda)}$ in Eqs. 5.12 and 5.13 have similar features but differ in several notable ways. Both models consist of a superposition of a finite number of harmonics with random amplitudes. However, the frequencies of these harmonics are fixed for $X^{(n)}$ but are random for $X^{(\lambda)}$. The frequencies of constituent harmonics of $X^{(\lambda)}$ coincide with the jump times of the compound Poisson processes C_1 and C_2. Both processes, $X^{(n)}$ and $X^{(\lambda)}$, are weakly stationary for any n, λ, and ν^*. However, $X^{(n)}$ is a Gaussian process for any n and ν^* while $X^{(\lambda)}$ is Gaussian asymptotically as $\lambda \to \infty$ for any value of ν^*.

Monte Carlo algorithm. The generation of samples of X in a time interval $[0, \tau]$, $\tau > 0$, involves three steps:

1. Select a cutoff frequency $\nu^* > 0$, an intensity $\lambda > 0$ for the Poisson processes N_r, and a distribution for the random variable $Y_1 \in L_2$ such that $\zeta(\nu)^2 = g(\nu)/(\lambda\, \mu_2)$.

2. Generate samples of the jump times $(T_{r,1}, \ldots, T_{r,N_r(\nu^*)})$ and the jumps $(Y_{r,1}, \ldots, Y_{r,N_r(\nu^*)})$ of the compound Poisson processes C_r, $r = 1, 2$.

3. Calculate the corresponding samples of $X^{(\lambda)}$ from the second equality in Eq. 5.13.

Note: The methods in Section 5.2.1 can be used to generate samples of the jumps of the compound Poisson processes C_r. The jump times of these processes can be generated in at least two ways. One approach is based on the representation $T_{r,i} = \sum_{k=1}^{i} W_{r,i}$ of the jump times of C_r, where $W_{r,i}$ are iid exponential random variables with mean $1/\lambda$. The other approach can first generate samples of the Poisson variables $N_r(\nu^*)$ (Example 5.3) and then place $N_r(\nu^*)$ independent uniformly distributed points in $[0, \nu^*]$. ▲

Example 5.8: Let X be a stationary Gaussian process with mean zero, covariance function $c(\tau) = E[X(t+\tau)\, X(t)] = \sin(\nu^* \tau)/(\nu^* \tau)$, and one-sided spectral density $g(\nu) = (1/\nu^*)\, 1_{[0,\nu^*]}(\nu)$.

Figure 5.4 shows a sample of $X^{(\lambda)}$, the covariance function $c(\tau)$ of X, and

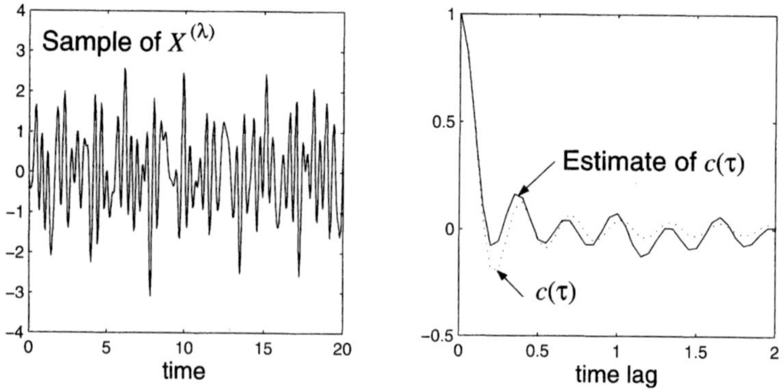

Figure 5.4. A sample of $X^{(\lambda)}$, the covariance function $c(\tau)$ of X, and an estimate of $c(\tau)$ from 500 samples of $X^{(\lambda)}$

an estimate of $c(\tau)$ based on $n_s = 500$ samples of $X^{(\lambda)}$ for $\nu^* = 20$, $\lambda\, \nu^* = 10$, and $Y_{r,p}$ independent $N(0, 1)$ random variables. The figure also shows that $X^{(\lambda)}$ does not have periodic samples and that the estimate of $c(\tau)$ is satisfactory although the samples of $X^{(\lambda)}$ have on average only $\lambda\, \nu^* = 10$ harmonics. In contrast, the samples of the model in Eq. 5.12 with $n = 10$ equally spaced frequencies in $[0, 20]$ are periodic with period 2π. ◇

5.3.1.2 Spectral representation. Random fields

We have seen in Section 3.9.4.3 that the spectral representation theorem applies also to weakly stationary or homogeneous random fields. We apply this theorem as in the previous section to develop models for generating samples of stationary Gaussian fields consisting of a superposition of harmonics with (1) fixed frequencies and random amplitudes and (2) random frequencies and amplitudes.

We discuss only the case of real-valued random fields to simplify the presentation. Considerations in Section 3.9.4.3 can be used to extend results in this section to vector random fields. Let $X(\boldsymbol{t})$, $\boldsymbol{t} \in \mathbb{R}^{d'}$, be a real-valued, homogeneous Gaussian field with mean zero, covariance function $c(\boldsymbol{\tau}) = E[X(\boldsymbol{t}+\boldsymbol{\tau})\, X(\boldsymbol{t})]$, and spectral density $s(\boldsymbol{\nu})$, where $\boldsymbol{t}, \boldsymbol{\tau} \in \mathbb{R}^{d'}$ are spatial coordinates and $\boldsymbol{\nu} \in \mathbb{R}^{d'}$ is a frequency. Our objectives are as in the previous section to (1) construct a sequence of approximations of X whose limit is a version of this random field and (2) develop algorithms for generating samples of X in a bounded subset S of $\mathbb{R}^{d'}$.

Fixed frequencies. Let D be a bounded subset of $\mathbb{R}^{d'}$, $(D_1, \ldots, D_n)$ be a partition of D, and $\boldsymbol{\nu}_r$ denote interior points of D_r. It is common to (1) take D to be a bounded rectangle centered at the origin of $\mathbb{R}^{d'}$, that is, $D = \times_{k=1}^{d'}[-\nu_k^*, \nu_k^*]$, $\nu_k^* > 0$, (2) partition D in rectangles D_r, $r = 1, \ldots, d'$, defined by a grid with step $\Delta\nu_k = \nu_k^*/n_k$, $k = 1, \ldots, d'$, where $n_k \geq 1$ are integers, and (3) choose $\boldsymbol{\nu}_r$ at the center of D_r. Denote by v_r the Lebesgue measure of D_r, that is, the volume of this set, and define a collection of independent Gaussian variables, A_r and B_r, $r = 1, \ldots, n$, with mean zero and variance

$$\gamma_r = E[A_r^2] = E[B_r^2] = \int_{D_r} s(\boldsymbol{\nu})\, d\boldsymbol{\nu}. \tag{5.14}$$

The following equation defines a sequence of random fields $X^{(n)}$, $n = 1, 2, \ldots$, that can be used to approximate a homogeneous Gaussian field X with specified second moment properties.

If $\max_{1 \leq r \leq n} v_r \to 0$ as $n \to \infty$ and $\min_{1 \leq k \leq d'} \nu_k^* \to \infty$, then

$$X^{(n)}(\boldsymbol{t}) = \sum_{r=1}^{n} [A_r \cos(\boldsymbol{\nu}_r \cdot \boldsymbol{t}) + B_r \sin(\boldsymbol{\nu}_r \cdot \boldsymbol{t})], \tag{5.15}$$

becomes a version of X, where $\boldsymbol{\nu}_r \cdot \boldsymbol{t} = \sum_{k=1}^{d'} \nu_{r,k}\, t_k$, $\boldsymbol{\nu}_r = (\nu_{r,1}, \ldots, \nu_{r,d'})$, and $\boldsymbol{t} = (t_1, \ldots, t_{d'})$.

Proof: The random field $X^{(n)}$ is Gaussian with mean zero for any D and partition of it.

This field is homogeneous since

$$E[X^{(n)}(t)\,X^{(n)}(s)]$$
$$= \sum_{r,p=1}^{n} E\left[(A_r\,\cos(\boldsymbol{\nu}_r\cdot\boldsymbol{t})+B_r\,\sin(\boldsymbol{\nu}_r\cdot\boldsymbol{t}))\left(A_p\,\cos(\boldsymbol{\nu}_p\cdot\boldsymbol{s})+B_p\,\sin(\boldsymbol{\nu}_p\cdot\boldsymbol{s})\right)\right]$$
$$= \sum_{r=1}^{n}\gamma_r\,(\cos(\boldsymbol{\nu}_r\cdot\boldsymbol{t})\,\cos(\boldsymbol{\nu}_r\cdot\boldsymbol{s})+\sin(\boldsymbol{\nu}_r\cdot\boldsymbol{t})\,\sin(\boldsymbol{\nu}_r\cdot\boldsymbol{s})) = \sum_{r=1}^{n}\gamma_r\,\cos(\boldsymbol{\nu}_r\cdot(\boldsymbol{t}-\boldsymbol{s})).$$

Considerations as in Eq. 5.12 show that the covariance function of $X^{(n)}$ approaches the covariance function of X as the partition of D is refined and D is increased to $\mathbb{R}^{d'}$. Because $X^{(n)}$ is a Gaussian field for any D and partition of it, $X^{(n)}$ becomes in the limit a version of X. This properties justifies the use of samples of $X^{(n)}$ as a substitute for samples of X.

We note that the second moments γ_r of the random variables A_r and B_r in Eq. 5.14 must be consistent with the properties of the spectral density of real-valued random fields (Section 3.9.4.3). ■

Monte Carlo algorithm. The generation of samples of X in a subset S of $\mathbb{R}^{d'}$ involves three steps:

1. Select cutoff frequencies $\nu_k^* > 0$, $k = 1, \ldots, d'$, partition $D = \times_{k=1}^{d'}[-\nu_k^*, \nu_k^*]$ in n rectangles, and select frequencies $\boldsymbol{\nu}_r$, $r = 1, \ldots, n$, in these rectangles.

2. Generate samples of the independent Gaussian variables A_r, B_r with mean zero and variance in Eq. 5.14.

3. Calculate the corresponding samples of $X^{(n)}(\boldsymbol{t})$, $\boldsymbol{t} \in S$, from Eq. 5.15.

Example 5.9: Let X be a real-valued stationary Gaussian field defined on $\mathbb{R}^{d'}$ with mean zero and spectral density $s(\boldsymbol{\nu}) = \sum_{r=1}^{n}\gamma_r\,\delta(\boldsymbol{\nu}-\boldsymbol{\nu}_r)$ such that $s(\boldsymbol{\nu}) = s(-\boldsymbol{\nu})$ and $\boldsymbol{\nu}_r \in \mathbb{R}^{d'}$ (Section 3.9.4.3). The field X has the representation

$$X(\boldsymbol{t}) = \sum_{r=1}^{n}\left[A_r\,\cos(\boldsymbol{\nu}_r\cdot\boldsymbol{t})+B_r\,\sin(\boldsymbol{\nu}_r\cdot\boldsymbol{t})\right],$$

where A_r and B_r are independent Gaussian variables with mean zero and variance $E[A_r^2] = E[B_r^2] = \gamma_r$. This representation of X and the model of this process in Eq. 5.15 have the same functional form. Figure 5.5 shows two sample of X for $d' = 2$, $n = 6$, $\boldsymbol{\nu}_1 = (1, 2)$, $\boldsymbol{\nu}_2 = (2, 1)$, $\boldsymbol{\nu}_3 = (2, 2)$, $\boldsymbol{\nu}_4 = -\boldsymbol{\nu}_1$, $\boldsymbol{\nu}_5 = -\boldsymbol{\nu}_2$, and $\boldsymbol{\nu}_6 = -\boldsymbol{\nu}_3$. The left sample is for $\gamma_r = 1$, $r = 1, \ldots, 6$. The right sample corresponds to $\gamma_1 = \gamma_4 = 1$ and $\gamma_r = 0.01$ for $r \neq 1, 4$, and has a dominant wave with frequency $\boldsymbol{\nu}_1 = (1, 2)$. ◇

Note: If $d' = 2$, we can think of $\cos(\boldsymbol{\nu}_r\cdot\boldsymbol{t}) = \cos(\nu_{r,1}\,t_1+\nu_{r,2}\,t_2)$ as a wave with length $2\pi/\sqrt{\nu_{r,1}^2+\nu_{r,2}^2}$ traveling in the direction $\theta_r = \tan^{-1}(\nu_{r,2}/\nu_{r,1})$. The wave length is

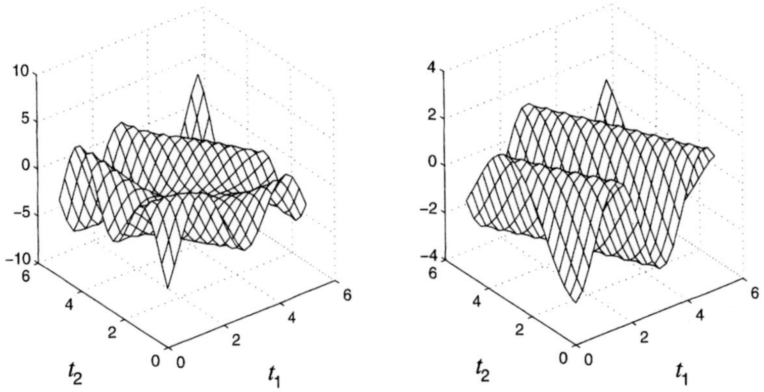

Figure 5.5. Samples of X for two discrete spectral densities

given by the distance between the zeros of $\cos(\boldsymbol{\nu}_r \cdot \boldsymbol{t})$, that is, the lines $\nu_{r,1}\, t_1 + \nu_{r,2}\, t_2 = (2\,p-1)\,\pi/2$ in $\mathbb{R}^2$, where p is an integer. ▲

Random frequencies. We construct a sequence of random fields $X^{(\lambda)}$, $\lambda > 0$, consisting of a superposition of waves with random amplitude and random frequency or length, which becomes in the limit a version of X. A similar construction has been used for the processes in Eq. 5.13.

Let $C(\boldsymbol{\nu}) = \sum_{k=1}^{N(D)} Y_k\, 1_{D(\boldsymbol{T}_k)}(\boldsymbol{\nu})$ be a compound Poisson process, where $D(\boldsymbol{T}) = D \cap \left(\times_{j=1}^{d'}[T_j, \infty)\right)$, $\boldsymbol{T} = (T_1, \ldots, T_{d'})$, D is a bounded Borel subset of $\mathbb{R}^{d'}$, $(\nu_1, \ldots, \nu_{d'})$ are the coordinates of $\boldsymbol{\nu}$, Y_k denote iid random variables with mean zero and variance $\mu_2 = E[Y_1^2] < \infty$, N is a Poisson process defined on $\mathbb{R}^{d'}$ with intensity $\lambda > 0$ and random points $\boldsymbol{T}_k \in \mathbb{R}^{d'}$. We will discuss the generation of point processes in Section 5.3.3.3. Let

$$C_r(\boldsymbol{\nu}) = \sum_{k=1}^{N_r(D)} Y_{r,k}\, 1_{D(\boldsymbol{T}_{r,k})}(\boldsymbol{\nu}), \quad r = 1, 2, \tag{5.16}$$

be two compound Poisson processes, where N_r and $Y_{r,k}$ are independent copies of N and Y_k, respectively, and the random variables $\boldsymbol{T}_{r,k}$ have the same meaning as $\boldsymbol{T}_k$ in the definition of the compound Poisson process C.

If $\zeta(\boldsymbol{\nu})^2 = s(\boldsymbol{\nu})/(\lambda\,\mu_2)$, $\lambda \to \infty$, and D approaches $\mathbb{R}^{d'}$, then

$$X^{(\lambda)}(\boldsymbol{t}) = \int_D \zeta(\boldsymbol{\nu})\,[\cos(\boldsymbol{\nu}\cdot\boldsymbol{t})\,dC_1(\boldsymbol{\nu}) - \sin(\boldsymbol{\nu}\cdot\boldsymbol{t})\,dC_2(\boldsymbol{\nu})]$$
$$= \sum_{k=1}^{N_1(D)} \zeta(\boldsymbol{T}_{1,k})\,Y_{1,k}\cos\left(\boldsymbol{T}_{1,k}\cdot\boldsymbol{t}\right) - \sum_{l=1}^{N_2(D)} \zeta(\boldsymbol{T}_{2,l})\,Y_{2,l}\sin\left(\boldsymbol{T}_{2,l}\cdot\boldsymbol{t}\right) \quad (5.17)$$

becomes a version of the homogeneous Gaussian field X.

Proof: A real-valued random field X can be represented by (Section 3.9.4.3)

$$X(\boldsymbol{t}) = \int_{\mathbb{R}^{d'}} \zeta(\boldsymbol{\nu})\left[\cos(\boldsymbol{\nu}\cdot\boldsymbol{t})\,dC_1(\boldsymbol{\nu}) - \sin(\boldsymbol{\nu}\cdot\boldsymbol{t})\,dC_2(\boldsymbol{\nu})\right],$$

where the processes C_p, $p = 1, 2$, are independent copies of C in Eq. 5.16. This representation suggests the definition of the approximating sequence of fields in Eq. 5.17.

The compound Poisson process C has mean zero and covariance function

$$\begin{aligned}
E[C(\boldsymbol{\nu})\,C(\boldsymbol{\nu}')] &= E\left[\sum_{k,l=1}^{N(D)} Y_k\,Y_l\,1_{D(\boldsymbol{T}_k)}(\boldsymbol{\nu})\,1_{D(\boldsymbol{T}_l)}(\boldsymbol{\nu}')\right] \\
&= E\left\{E\left[\sum_{k,l=1}^{N(D)} Y_k\,Y_l\,1_{D(\boldsymbol{T}_k)}(\boldsymbol{\nu})\,1_{D(\boldsymbol{T}_l)}(\boldsymbol{\nu}') \mid N(D)\right]\right\} \\
&= E\left\{\sum_{k,l=1}^{N(D)} E\left[Y_k\,Y_l\,1_{D(\boldsymbol{T}_k)}(\boldsymbol{\nu})\,1_{D(\boldsymbol{T}_l)}(\boldsymbol{\nu}') \mid N(D)\right]\right\} \\
&= \mu_2\,\lambda\,v_D\,E\left[1_{D(\boldsymbol{T}_1)}(\boldsymbol{\nu})\,1_{D(\boldsymbol{T}_1)}(\boldsymbol{\nu}')\right],
\end{aligned}$$

where $\boldsymbol{\nu}, \boldsymbol{\nu}' \in D$, $D(\boldsymbol{T}_k) = D \cap \left(\times_{j=1}^{d'}[T_{k,j}, \infty)\right)$, $T_{k,j}$ denotes the coordinate j of $\boldsymbol{T}_k$, and $\boldsymbol{T}_1$ is uniformly distributed in D. The above derivations used the fact that the Poisson points $\boldsymbol{T}_k$ are uniformly distributed in D conditional on $N(D)$. If $D = \times_{j=1}^{d'}[-\nu_j^*, \nu_j^*]$, then $E\left[1_{D(\boldsymbol{T}_1)}(\boldsymbol{\nu})\,1_{D(\boldsymbol{T}_1)}(\boldsymbol{\nu}')\right] = \prod_{j=1}^{d'}(\nu_j \wedge \nu_j' + \nu_j^*)/(2\,\nu_j^*)$ since the coordinates $T_{1,j}$ of $\boldsymbol{T}_1$ are uniformly distributed in $[-\nu_j^*, \nu_j^*]$.

The definition of $X^{(\lambda)}$ in Eq. 5.17 gives $E\left[X^{(\lambda)}(t)\right] = 0$ and

$$E[X^{(\lambda)}(\boldsymbol{t})\,X^{(\lambda)}(\boldsymbol{s})] = \lambda\,\mu_2 \int_D \zeta(\boldsymbol{\nu})^2\cos(\boldsymbol{\nu}\cdot(\boldsymbol{t}-\boldsymbol{s}))\,d\boldsymbol{\nu}.$$

The above expression for the covariance function of $X^{(\lambda)}$ results from

$$
\begin{aligned}
E[X^{(\lambda)}(\boldsymbol{t})\, X^{(\lambda)}(\boldsymbol{s})] &= E\left[\sum_{k,l=1}^{N_1(D)} \zeta(\boldsymbol{T}_{1,k})\,\zeta(\boldsymbol{T}_{1,l})\, Y_{1,k}\, Y_{1,l} \cos\left(\boldsymbol{T}_{1,k}\cdot \boldsymbol{t}\right) \cos\left(\boldsymbol{T}_{1,l}\cdot \boldsymbol{s}\right)\right] \\
&+ E\left[\sum_{k,l=1}^{N_2(D)} \zeta(\boldsymbol{T}_{2,k})\,\zeta(\boldsymbol{T}_{2,l})\, Y_{2,k}\, Y_{2,l} \sin\left(\boldsymbol{T}_{2,k}\cdot \boldsymbol{t}\right) \sin\left(\boldsymbol{T}_{2,l}\cdot \boldsymbol{s}\right)\right] \\
&= E\left[\sum_{k=1}^{N_1(D)} \zeta(\boldsymbol{T}_{1,k})^2\, Y_{1,k}^2 \cos\left(\boldsymbol{T}_{1,k}\cdot \boldsymbol{t}\right) \cos\left(\boldsymbol{T}_{1,k}\cdot \boldsymbol{t}\right)\right] \\
&+ E\left[\sum_{k=1}^{N_2(D)} \zeta(\boldsymbol{T}_{2,k})^2\, Y_{2,k}^2 \sin\left(\boldsymbol{T}_{2,k}\cdot \boldsymbol{t}\right) \sin\left(\boldsymbol{T}_{2,k}\cdot \boldsymbol{t}\right)\right],
\end{aligned}
$$

the fact that the random variables $\boldsymbol{T}_{r,k}$, $k = 1, \ldots, N_r(D)$, $r = 1, 2$, are independent and uniformly distributed in D conditional on $N_r(D)$, the properties of the random variables $Y_{r,k}$, and $E[N_r(D)] = \lambda\, v_D$, where $v_D = \int_D d\boldsymbol{v}$ denotes the volume of D.

Hence, $X^{(\lambda)}$ is a weakly stationary random field for any $\lambda > 0$ and $D \subset \mathbb{R}^{d'}$. Moreover, $X^{(\lambda)}$ becomes equal to X in the second moment sense if D is increased to $\mathbb{R}^{d'}$. Also, $X^{(\lambda)}$ converges to a Gaussian field as $\lambda \to \infty$ for any D. We conclude that $X^{(\lambda)}$ becomes a version of X if $\lambda \to \infty$ and D is increased to $\mathbb{R}^{d'}$. ■

The models in Eqs. 5.15 and 5.17 have features and limitations similar to those of the models in Eqs. 5.12 and 5.13.

Monte Carlo simulation. The generation of samples of X in a bounded subset S of $\mathbb{R}^{d'}$ involves three steps:

1. Select cutoff frequencies ν_k^*, $k = 1, \ldots, d'$, defining $D = \times_{k=1}^{d'}[-\nu_k^*, \nu_k^*]$, an intensity $\lambda > 0$ for the Poisson processes N_r, and a distribution for $Y_1 \in L_2$ such that $\zeta(\boldsymbol{\nu})^2 = s(\boldsymbol{\nu})/(\lambda\, \mu_2)$.

2. Generate samples of the Poisson points $(\boldsymbol{T}_{r,1}, \ldots, \boldsymbol{T}_{r,N_r(D)})$ and the jumps $(Y_{r,1}, \ldots, Y_{r,N_r(D)})$ of the compound Poisson processes C_r, $r = 1, 2$.

3. Calculate the corresponding samples of $X^{(\lambda)}(\boldsymbol{t})$, $\boldsymbol{t} \in S$, from the second equality in Eq. 5.17.

Note: The generation of Poisson points in a subset of $\mathbb{R}^{d'}$ is discussed in Section 5.3.3.3. The methods described in Section 5.2 can be used to generate samples of the random variable $(Y_{r,1}, \ldots, Y_{r,N_r(D)})$. ▲

Example 5.10: Let X be a zero mean, homogeneous Gaussian field defined on

$\mathbb{R}^2$ with spectral density and covariance functions

$$s(\boldsymbol{\nu}) = \frac{1}{2\pi\sqrt{1-\rho^2}} \exp\left[-\frac{\nu_1^2 - 2\rho\,\nu_1\,\nu_2 + \nu_2^2}{2(1-\rho^2)}\right], \quad \boldsymbol{\nu} \in \mathbb{R}^2,$$

$$c(\boldsymbol{\tau}) = E[X(\boldsymbol{t})\,X(\boldsymbol{t}+\boldsymbol{\tau})] = \exp\left[-\frac{\tau_1^2 + 2\rho\,\tau_1\,\tau_2 + \tau_2^2}{2}\right], \quad \boldsymbol{\tau} \in \mathbb{R}^2,$$

where $|\rho| < 1$. The spectral density of X is shown in Fig. 5.6 for $\rho = 0.7$.

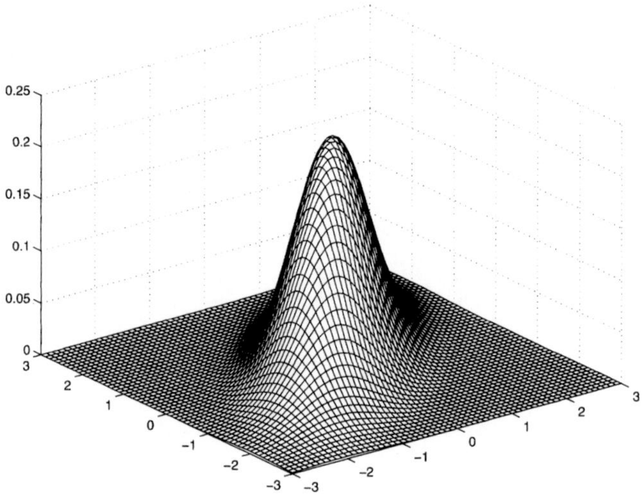

Figure 5.6. Spectral density of X for $\rho = 0.7$

Figure 5.7 shows four samples of X for $\boldsymbol{t} \in S = [0, 10] \times [0, 10]$, $\lambda = 0.5$, Y_1 a Gaussian random variable with mean zero and variance one, $D = [-3, 3] \times [-3, 3]$, and $\zeta(\boldsymbol{\nu})^2 = s(\boldsymbol{\nu})/\lambda$. The random field $X^{(\lambda)}$ has approximately the same second moment properties as X for any λ and $\zeta(\boldsymbol{\nu})$ satisfying the above condition. The model $X^{(\lambda)}$ in Eq. 5.17 has been used to generate the samples shown in this figure. The average number of Poisson points in D is $\lambda\, v_D = 18$, where v_D denotes the volume of D. ◇

5.3.1.3 Sampling theorem. Stochastic processes

We show that a sampling theorem for deterministic functions can be extended to define a sequence of processes $\boldsymbol{X}^{(n)}$, $n = 1, 2, \ldots$, such that each $\boldsymbol{X}^{(n)}$ depends on a finite number of random variables and the limit of this sequence as $n \to \infty$ is a version of a specified stationary Gaussian process $\boldsymbol{X}$. The following equation states the sampling theorem for deterministic functions.

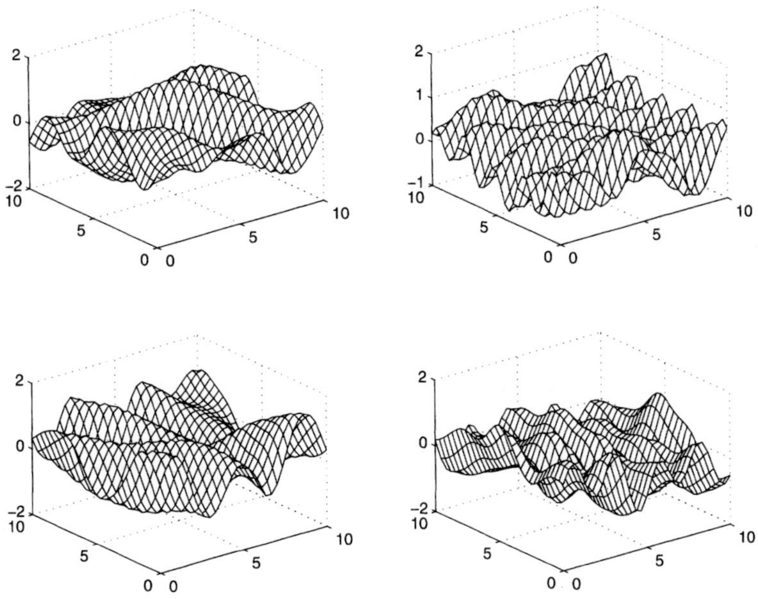

Figure 5.7. Four samples of $X^{(\lambda)}$ approximating X

Let $[-\nu^*, \nu^*]$, $0 < \nu^* < \infty$, be the frequency band of a deterministic function $x(t)$, $t \in \mathbb{R}$. Then ([29], Sections 5-4, 5-5)

$$x(t) = \lim_{n \to \infty} \sum_{k=-n}^{n} x(k\,t^*)\,\alpha_k(t;t^*), \quad \text{where}$$

$$\alpha_k(t;t^*) = \frac{\sin(\pi\,(t/t^* - k))}{\pi\,(t/t^* - k)} \quad \text{and} \quad t^* = \pi/\nu^*. \tag{5.18}$$

Note: It is sufficient to know the values $x(k\,t^*)$, $k \in \mathbb{Z}$, of x to reconstruct the entire signal. The function α_k is 1 if $t/t^* = k$, is zero if t/t^* is an integer different from k, and decreases rapidly to zero as $|t/t^* - k|$ increases. ▲

Let $\boldsymbol{X}(t)$, $t \in \mathbb{R}$, be an $\mathbb{R}^d$-valued stationary Gaussian process with mean zero, covariance function $\boldsymbol{c}(\tau) = E[\boldsymbol{X}(t+\tau)\,\boldsymbol{X}(t)^T]$, and spectral density $\boldsymbol{s}(\nu) = \{s_{kl}(\nu)\}$, $k, l = 1, \ldots, d$. It is assumed that the coordinates X_i of $\boldsymbol{X}$ have a bounded frequency range $[-\nu_i^*, \nu_i^*]$, $0 < \nu_i^* < \infty$. If this condition is not satisfied, we approximate $\boldsymbol{s}$ by a spectral density $\tilde{\boldsymbol{s}}$ that is equal to $\boldsymbol{s}$ in $\times_{i=1}^{d}[-\nu_i^*, \nu_i^*]$ and is zero outside this rectangle, and develop a representation for the process with the spectral density $\tilde{\boldsymbol{s}}$. We give first a version of the sampling theorem in Eq. 5.18 for real-valued processes ($d = 1$) and then extend this result to vector processes ($d > 1$).

If X is a real-valued stationary Gaussian process with zero spectral density outside a frequency band $[-\nu^*, \nu^*]$, $0 < \nu^* < \infty$, then

$$X^{(n)}(t) = \sum_{k=n_t-n}^{n_t+n+1} X(k\,t^*)\,\alpha_k(t; t^*), \tag{5.19}$$

becomes a version of X as $n \to \infty$, where n_t denotes the integer part of t/t^* and $n \geq 1$ is an integer.

Proof: We call nodes the times $k\,t^*$ at which the process is sampled, where $\nu = \nu_1$, $\nu^* = \nu_1^*$, $t^* = \pi/\nu^*$, and $k \in \mathbb{Z}$. The integer $n \geq 1$ is half the size of the window centered on the current cell $[n_t\,t^*, (n_t + 1)\,t^*]$, that is, the cell containing the time argument t (Fig. 5.8). The model $X^{(n)}$ in Eq. 5.19 depends linearly on the values of X at $2\,(n + 1)$ nodes and coincides with X at the nodes included in its window.

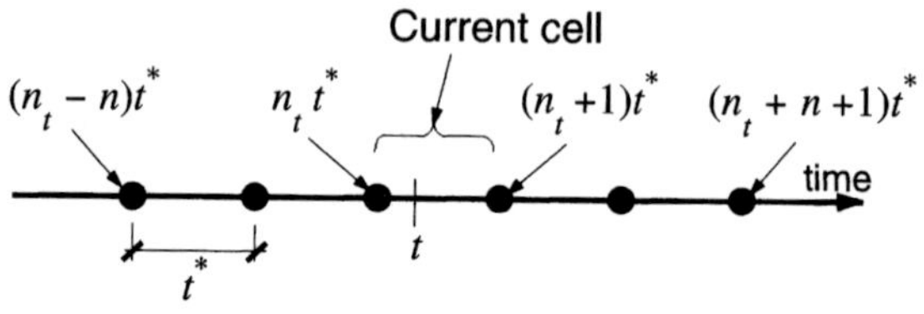

Figure 5.8. Moving window for $X^{(n)}$ with $n = 2$

We have (Eq. 5.18)

$$X(t, \omega) = \lim_{n\to\infty} \sum_{k=-n}^{n} X(k\,t^*, \omega)\,\alpha_k(t; t^*)$$

for almost all ω's since the samples of X consists of a superposition of harmonics with frequencies in the range $[-\nu^*, \nu^*]$. The representation $\lim_{n\to\infty} X^{(n)}$ cannot be used to generate samples of X because it depends on an infinite number of random variables, the random variables $X(k\,t^*)$, $k \in \mathbb{Z}$.

Consider the model in Eq. 5.19. This stochastic process is Gaussian with mean zero and covariance function

$$E[X^{(n)}(t)\,X^{(n)}(s)] = \sum_{k=n_t-n}^{n_t+n+1} \alpha_k(t; t^*) \left(\sum_{l=n_s-n}^{n_s+n+1} c((k-l)\,t^*)\,\alpha_l(s; t^*) \right),$$

where c denotes the covariance function of X. Hence, $X^{(n)}$ is not stationary for finite values of n because $E[X^{(n)}(t)\,X^{(n)}(s)]$ depends explicitly on the times t and s. However, $E[X^{(n)}(t)\,X^{(n)}(s)]$ converges to $c(t - s)$ as $n \to \infty$ This follows from the sampling theorem in Eq. 5.18 applied twice to the covariance function $c(\cdot)$ in the above equation. We can apply this theorem to the covariance function of X because the frequency band of its Fourier transform is $[-\nu^*, \nu^*]$ by hypothesis. Therefore, $X^{(n)}$ becomes a version of X as the window size is increased to infinity. This property justifies the use of samples of $X^{(n)}$ as a substitute for samples of X provided that n is sufficiently large. Numerical results show that $X^{(n)}$ with $n \simeq 10$ approximates satisfactorily X [78, 88]. ■

The model $X^{(n)}$ in Eq. 5.19 can be used to develop an efficient Monte Carlo simulation algorithm for generating samples of stationary Gaussian processes.

Monte Carlo algorithm. Suppose that a sample $X^{(n)}(\cdot, \omega)$ of $X^{(n)}$ has been generated up to a time $(n_t + 1)\, t^*$. The extension of this sample in the next cell $[(n_t + 1)\, t^*, (n_t + 2)\, t^*]$ involves two steps:

1. Generate a sample of the conditional random variable

$$\begin{aligned}
&\hat{\hat{X}}((n_t + n + 2)\, t^*) \\
&\quad = X((n_t + n + 2)\, t^*) \mid \big[X((n_t + n + 1)\, t^*) = X((n_t + n + 1)\, t^*, \omega), \\
&\qquad X((n_t + n)\, t^*) = X((n_t + n)\, t^*, \omega), \ldots, \\
&\qquad X((n_t - n)\, t^*) = X((n_t - n)\, t^*, \omega)\big]. \qquad (5.20)
\end{aligned}$$

2. Calculate the corresponding sample of $X^{(n)}$ in $[(n_t + 1)\, t^*, (n_t + 2)\, t^*]$ from Eq. 5.19.

Note: The real-valued random variable $\hat{\hat{X}}((n_t + n + 2)\, t^*)$ is Gaussian with known mean and variance (Section 2.11.5). Because X is stationary and we consider in Eq. 5.20 only values of X at the past $2\,n+2$ nodes, the second moment properties of $\hat{\hat{X}}((n_t+n+2)\, t^*)$ do not change in time so that they have to be calculated once. The algorithms in Section 5.2.1 can be used to generate a sample of $\hat{\hat{X}}((n_t + n + 2)\, t^*)$.

The extension of the sample of $X^{(n)}$ beyond the time $(n_t+1)\, t^*$ should use a sample of the conditional variable

$$\begin{aligned}
&\hat{X}((n_t + n + 2)\, t^*) \\
&\quad = X((n_t + n + 2)\, t^*) \mid \big[X((n_t + n + 1)\, t^*) = X((n_t + n + 1)\, t^*, \omega), \\
&\qquad X((n_t + n)\, t^*) = X((n_t + n)\, t^*, \omega), \ldots\big],
\end{aligned}$$

which accounts for the entire time history, rather than a sample of $\hat{\hat{X}}((n_t + n + 2)\, t^*)$. However, this exact formulation is impractical since the properties of $\hat{X}((n_t + n + 2)\, t^*)$ have to be recalculated at each node and depend on a vector of increasing length as time progresses. Moreover, the improved accuracy of this formulation may be insignificant because of the reduced influence of values of X at nodes far away from the cell containing the current time. Numerical results support this statement for processes that do not exhibit long range dependence. A process is said to have long range dependence if its covariance function is such that $c(\tau) \sim \tau^{-\beta}$ for $\tau \to \infty$, where $\beta \in (0, 1)$ is a constant [78, 88].

The algorithm in Eq. 5.20 can be applied to narrowband processes, that is, processes whose spectral density is zero outside a small vicinity of a central frequency $\nu_0 > 0$. However, the Monte Carlo simulation algorithm based on Eq. 5.20 can be inefficient if the central frequency ν_0 is large. A modified version of this algorithm developed for narrowband processes can be found in [78]. ▲

If $\boldsymbol{X}$ is an $\mathbb{R}^d$-valued stationary Gaussian process whose coordinates X_i have a bounded frequency range $[-\nu_i^*, \nu_i^*]$, $\nu_i^* > 0$, $i = 1, \ldots, d$, then the process $\boldsymbol{X}^{(\boldsymbol{n})}$ with coordinates

$$X_i^{(n_i)} = \sum_{k=n_{i,t}-n_i}^{n_{i,t}+n_i+1} X_i(k\, t_i^*)\, \alpha_k(t; t_i^*) \tag{5.21}$$

becomes a version of $\boldsymbol{X}$ as $n_i \to \infty$, $i = 1, \ldots, d$, where $t_i^* = \pi/\nu_i^*$, $n_i \geq 1$ are integers, $n_{i,t}$ is the integer part of t/t_i^*, and $\boldsymbol{n} = (n_1, \ldots, n_d)$ [88].

Note: The process $\boldsymbol{X}^{(\boldsymbol{n})}$ with coordinates in Eq. 5.21 is Gaussian with mean zero for any values of n_i. Straightforward calculations show that $\boldsymbol{X}^{(\boldsymbol{n})}$ becomes a version of $\boldsymbol{X}$ as $\min_{1 \leq i \leq d} n_i \to \infty$ so that we can use samples of $\boldsymbol{X}^{(\boldsymbol{n})}$ as a substitute for samples of $\boldsymbol{X}$ provided that n_i are sufficiently large. ▲

Example 5.11: Consider the $\mathbb{R}^2$-valued process $\boldsymbol{X}$ with coordinates

$$X_i(t) = \sqrt{1-\rho}\, Y_i(t) + \sqrt{\rho}\, Y(t), \quad 0 < \rho < 1, \quad i = 1, 2,$$

where Y_i and Y are independent band-limited Gaussian white noise processes with mean zero and spectral densities $s_{Y_i}(\nu) = (1/(2\,\bar{\nu}_i))\, 1_{[-\bar{\nu}_i, \bar{\nu}_i]}(\nu)$ and $s_Y(\nu) = (1/(2\,\bar{\nu}_0))\, 1_{[-\bar{\nu}_0, \bar{\nu}_0]}(\nu)$, respectively. The spectral density $\boldsymbol{s}$ of $\boldsymbol{X}$ has the entries $s_{i,i}(\nu) = (1-\rho)\, s_{Y_i}(\nu) + \rho\, s_Y(\nu)$, $i = 1, 2$, and $s_{1,2}(\nu) = s_{2,1}(\nu) = \rho\, s_Y(\nu)$. Figure 5.9 shows the exact and approximate covariance function $E[X_1(t+\tau)\, X_1(t)]$ for t at a cell midpoint, lag times $\tau \geq 0$, $\bar{\nu}_1 = 2\pi$, $\bar{\nu}_2 = 0.4\pi$, and $\bar{\nu} = 0.2\pi$ so that $\nu_1^* = 2\pi$, $\nu_2^* = 0.4\pi$, $t_1^* = 0.5$, and $t_2^* = 2.5$. The approximate covariance functions for $\rho = 0$ and $\rho = 0.5$ corresponding to $n_1 = 10$ nearly coincide with the exact covariance functions of X_1 for these values of ρ. ◇

Note: If the spacing $t_i^* = t^*$ is the same for all the coordinates of $\boldsymbol{X}^{(\boldsymbol{n})}$ in Eq. 5.21, we have $n_{i,t} = n_t$. Suppose that $n_i = n$ and $n_{i,t} = n_t$, and that a sample of $\boldsymbol{X}$ has been generated up to a time $(n_t + 1)\, t^*$. The extension of this sample to the cell $[(n_t + 1)\, t^*, (n_t + 2)\, t^*]$ requires a sample of $\boldsymbol{X}((n_t + n + 2)\, t^*)$.

The simulation algorithm is less simple if the nodes in the representation of the coordinates of $\boldsymbol{X}^{(\boldsymbol{n})}$ in Eq. 5.21 are not equally spaced. Details on a simulation algorithm for this case can be found in [88]. ▲

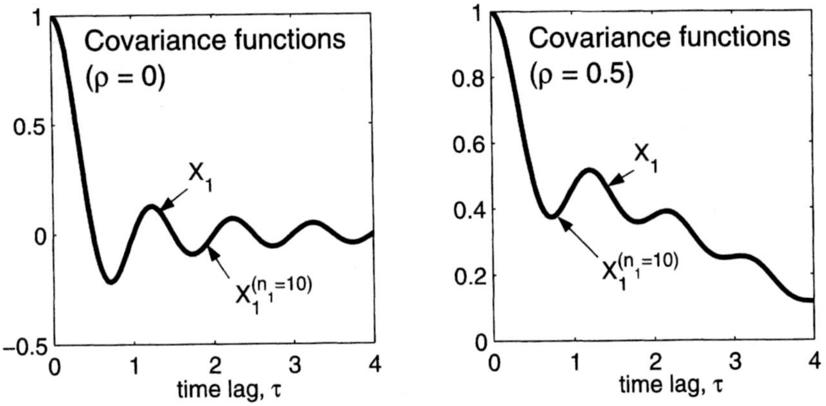

Figure 5.9. Covariance functions of X_1 and $X_1^{(n_1)}$

5.3.1.4 Sampling theorem. Random fields

Let $X(\boldsymbol{t})$, $\boldsymbol{t} \in \mathbb{R}^{d'}$, be a real-valued, homogeneous, Gaussian random field with mean zero, covariance function $c(\boldsymbol{\tau}) = E[X(\boldsymbol{t}+\boldsymbol{\tau})\, X(\boldsymbol{t})]$, $\boldsymbol{\tau} \in \mathbb{R}^{d'}$, and spectral density $s(\boldsymbol{\nu})$, $\boldsymbol{\nu} \in \mathbb{R}^{d'}$, $0 < \nu_\sigma^* < \infty$. It is assumed that the ordinates of the spectral density are zero outside a rectangle $D = \times_{\sigma=1}^{d'}[-\nu_\sigma^*, \nu_\sigma^*]$. If this condition is not satisfied, we can truncate the spectral density of X, develop a representation of X for this version of the field, and then let the cutoff frequencies ν_σ^* increase indefinitely. Results in this section can be extended simply to vector random fields [88].

If X is a real-valued, homogeneous Gaussian field with zero spectral density outside $D = \times_{\sigma=1}^{d'}[-\nu_\sigma^*, \nu_\sigma^*]$, $\sigma = 1, \ldots, d'$, then

$$X^{(\boldsymbol{n})}(\boldsymbol{t}) = \sum_{k_1 = n_{t_1} - n_1}^{n_{t_1}+n_1+1} \cdots \sum_{k_{d'} = n_{t_{d'}} - n_{d'}}^{n_{t_{d'}}+n_{d'}+1} X(k_1\, t_1^*, \ldots, k_{d'}\, t_{d'}^*) \prod_{\sigma=1}^{d'} \alpha_{k_\sigma}(t_\sigma; t_\sigma^*) \tag{5.22}$$

becomes a version of X as $n_\sigma \to \infty$, $\sigma = 1, \ldots, d'$, where $t_\sigma^* = \pi/\nu_\sigma^*$, n_{t_σ} is the integer part of t_σ/t_σ^*, $n_\sigma \geq 1$ are integers, and $\boldsymbol{n} = (n_1, \ldots, n_{d'})$.

Note: The random field $X^{(\boldsymbol{n})}$ is Gaussian with mean zero for any window width n_σ. The field is not homogeneous since $E[X^{(\boldsymbol{n})}(\boldsymbol{t})\, X^{(\boldsymbol{n})}(\boldsymbol{s})]$ depends explicitly on $\boldsymbol{t}$ and $\boldsymbol{s}$. The covariance function of $X^{(\boldsymbol{n})}$ approaches the covariance function of X as $n_\sigma \to \infty$, $\sigma = 1, \ldots, d'$, so that $X^{(\boldsymbol{n})}$ becomes a version of X for a window with infinite width. This property justifies the use of samples of $X^{(\boldsymbol{n})}$ as a substitute for samples of X.

The representation of X in Eq. 5.22 is for $\boldsymbol{t} \in \times_{\sigma=1}^{d'}[n_\sigma\, t_\sigma^*, (n_\sigma+1)\, t_\sigma^*]$ and depends on values of X at nodes around this cell. Suppose that a sample of $X^{(\boldsymbol{n})}$ has been generated for values of t in this cell. To extend this sample to a neighboring cell, it is necessary to generate new values of X conditional on the previously generated values of this field. An algorithm for generating samples of X based on the model in Eq. 5.22 is given in [88]. ▲

Example 5.12: Let $X(\boldsymbol{t})$, $\boldsymbol{t} \in \mathbb{R}^2$, be a real-valued, homogeneous, Gaussian field with mean zero, spectral density $s(\boldsymbol{\nu}) = 1/(4\,\nu_1^*\,\nu_2^*)$ in $[-\nu_1^*, \nu_1^*] \times [-\nu_2^*, \nu_2^*]$ and zero outside this rectangle, and covariance $c(\boldsymbol{\tau}) = E[X(\boldsymbol{t}+\boldsymbol{\tau})\,X(\boldsymbol{t})] = \prod_{\sigma=1}^{2} \sin(\nu_\sigma^*\,\tau_\sigma)/(\nu_\sigma^*\,\tau_\sigma)$. The covariance function of $X^{(\boldsymbol{n})}$ in Eq. 5.22 is

$$E[X^{(\boldsymbol{n})}(\boldsymbol{t})\,X^{(\boldsymbol{n})}(\boldsymbol{s})] = \sum_{k_1=n_{t_1}-n_1}^{n_{t_1}+n_1+1} \sum_{k_2=n_{t_2}-n_2}^{n_{t_2}+n_2+1} \sum_{l_1=n_{s_1}-n_1}^{n_{s_1}+n_1+1} \sum_{l_2=n_{s_2}-n_2}^{n_{2_2}+n_2+1}$$

$$c((k_1-l_1)\,t_1^*, (k_2-l_2)\,t_2^*) \prod_{u=1}^{2} \alpha_{k_u}(t_u; t_u^*) \prod_{v=1}^{2} \alpha_{l_v}(s_v; t_v^*),$$

so that $X^{(\boldsymbol{n})}$ is not stationary for a finite window width.

Numerical results in [88] show that the covariance function of $X^{(\boldsymbol{n})}$ approaches rapidly the covariance function of X. Satisfactory approximations are reported in [88] for $n_1 = n_2 \simeq 10$. ◇

5.3.2 Non-stationary Gaussian processes and fields

Let $\boldsymbol{X}(t)$, $t \in [0, \tau]$, be an $\mathbb{R}^d$-valued stochastic process and $\boldsymbol{X}(\boldsymbol{t})$, $\boldsymbol{t} \in S$, denote an $\mathbb{R}^d$-valued random field, where $\tau \in (0, \infty)$ and S is a bounded subset of $\mathbb{R}^{d'}$. It is assumed that $\boldsymbol{X}(t)$ and $\boldsymbol{X}(\boldsymbol{t})$ are non-stationary Gaussian functions with mean zero. Their covariance functions, $c_{i,j}(t,s) = E[X_i(t)\,X_j(s)]$ and $c_{i,j}(\boldsymbol{t},\boldsymbol{s}) = E[X_i(\boldsymbol{t})\,X_j(\boldsymbol{s})]$, depend on two time arguments $t, s \in [0, \tau]$ for stochastic processes and two space arguments $\boldsymbol{t}, \boldsymbol{s} \in S$ for random fields. Our objective is to generate independent samples of these Gaussian functions. If the mean of a Gaussian function $\boldsymbol{X}$ is not zero, we generate samples of $\boldsymbol{X} - E[\boldsymbol{X}]$ and then add the mean function $E[\boldsymbol{X}]$ to these samples.

5.3.2.1 Linear differential equations

The Monte Carlo simulation algorithm in this section applies only to Gaussian processes. Let $\boldsymbol{X}(t)$, $t \geq 0$, be a Gaussian process with mean zero and covariance function $\boldsymbol{c}(t,s) = E[\boldsymbol{X}(t)\,\boldsymbol{X}(s)^T]$. Suppose that $\boldsymbol{X}$ is defined by the linear stochastic differential equation

$$\dot{\boldsymbol{X}}(t) = \boldsymbol{a}(t)\,\boldsymbol{X}(t) + \boldsymbol{b}(t)\,\boldsymbol{Y}(t), \quad t \geq 0, \tag{5.23}$$

where $\boldsymbol{Y} \in \mathbb{R}^{d'}$ is a stationary Gaussian process with mean zero and a prescribed covariance function.

Note: It is assumed that the (d, d) and (d, d') matrices $\boldsymbol{a}$ and $\boldsymbol{b}$ are such that Eq. 5.23 has a unique solution. If $\boldsymbol{Y}$ is a Gaussian white noise, then $\boldsymbol{X}$ is called a **filtered Gaussian process**.

The solution of Eq. 5.23 is

$$\boldsymbol{X}(t) = \boldsymbol{\theta}(t, 0)\, \boldsymbol{X}(0) + \int_0^t \boldsymbol{\theta}(t, s)\, \boldsymbol{Y}(s)\, ds = \boldsymbol{\theta}(t, 0)\, \boldsymbol{X}(0) + \int_{\mathbb{R}} \tilde{\boldsymbol{\theta}}(t, s)\, d\boldsymbol{Z}(\nu),$$

where $\tilde{\boldsymbol{\theta}}(t, s) = \int_0^t \boldsymbol{\theta}(t, s)\, e^{\sqrt{-1}\,\nu s}\, ds$, $\boldsymbol{\theta}(t, s)$ denotes the Green function for Eq. 5.23, $\boldsymbol{Y}(t) = \int_{\mathbb{R}} e^{\sqrt{-1}\,\nu t}\, d\boldsymbol{Z}(\nu)$, and $\boldsymbol{Z}$ is a process with stationary, orthogonal increments (Section 3.9.4). The second moment properties of $\boldsymbol{X}$ in Eq. 5.23 can be calculated by the methods in Section 7.2.1.2. ▲

Monte Carlo algorithm. The generation of samples of a Gaussian process $\boldsymbol{X}$ with specified second moment properties involves two steps:

1. Define the coefficients $\boldsymbol{a}$, $\boldsymbol{b}$ and the properties of the process $\boldsymbol{Y}$ in Eq. 5.23.

2. Generate samples of $\boldsymbol{Y}$ by any of the methods in Section 5.3.1 and integrate Eq. 5.23 to obtain samples of $\boldsymbol{X}$.

Note: The determination of the coefficients $\boldsymbol{a}$ and $\boldsymbol{b}$ of Eq. 5.23 is difficult in a general setting. Relatively simple results are available in special cases, for example, $\boldsymbol{Y}$ is a stationary Gaussian process with mean zero and the matrices $\boldsymbol{a}$, $\boldsymbol{b}$ are constant ([30], Theorem 1, p. 106).

A finite difference approximation of Eq. 5.23 can be used to obtain samples of $\boldsymbol{X}$ from samples of $\boldsymbol{Y}$ (Section 4.7.3). For example, we can use the forward finite difference scheme

$$\boldsymbol{X}(t + \Delta t) = [\boldsymbol{i} + \boldsymbol{a}(t)\, \Delta t]\; \boldsymbol{X}(t) + \boldsymbol{b}(t)\, \Delta \boldsymbol{Y}(t),$$

where $\boldsymbol{i}$ denotes the (d, d) identity matrix, $\Delta t > 0$ is the time step, and $\Delta \boldsymbol{Y}(t) = \boldsymbol{Y}(t + \Delta t) - \boldsymbol{Y}(t)$. ▲

Example 5.13: Let X be a real-valued stationary Gaussian process defined by the stochastic differential equation $dX(t) = -\rho\, X(t)\, dt + \sqrt{2\,\rho}\, dB(t)$ for $t \geq 0$, where $\rho > 0$ and B denotes a Brownian motion. The above equation admits a stationary solution X_s, where X_s is a zero-mean Gaussian process with covariance function $c(\tau) = E[X(t+\tau)\, X(t)] = e^{-\rho\,|\tau|}$. If the initial state $X(0)$ has the same marginal distribution as X_s, that is, $X(0) \sim N(0, 1)$, and is independent of the driving noise, then X is a version of X_s.

The recurrence formula $X_{k+1} = (1 - \rho\, \Delta t)\, X_k + \sqrt{2\,\rho}\, W_k$, $k = 0, 1, \ldots$, can be used to generate samples of X, where $\Delta t > 0$ denotes the time step, $X_k = X(k\, \Delta t)$, and W_k are independent copies of $N(0, \Delta t)$. ◇

Proof: The solution of the above stochastic differential equation is

$$X(t) = X(0)\, e^{-\rho\, t} + \sqrt{2\,\rho} \int_0^t e^{-\rho\,(t-s)}\, dB(s).$$

If $E[X(0)] = 0$ and $E[X(0)^2] = 1$, then $E[X(t)] = 0$ and

$$\begin{aligned} E[X(t)\, X(s)] &= e^{-\rho\,(t+s)} + 2\,\rho \int_0^t du \int_0^s dv\, e^{-\rho\,(t-u)}\, e^{-\rho\,(s-v)}\, E[dB(u)\, dB(v)] \\ &= e^{-\rho\,(t+s)} + 2\,\rho \int_0^{t\wedge s} e^{-\rho\,(t-u)}\, e^{-\rho\,(s-u)}\, du = e^{-\rho\,|t-s|} \end{aligned}$$

since the increments $dB(t) \sim N(0, dt)$ of B are independent. The above second moment properties of X can be obtained directly from the mean and covariance equations (Section 7.2.1.2). If $X(0)$ is not a zero mean, unit variance random variable, then X is not a weakly stationary process. ■

5.3.2.2 Fourier series. Stochastic processes

Let $\boldsymbol{X}(t)$, $t \in [0, \tau]$, be an $\mathbb{R}^d$-valued non-stationary Gaussian process with mean zero and covariance functions $c_{i,j}(t, s) = E[X_i(t)\, X_j(s)]$ for $i, j = 1, \ldots, d$ and $t, s \in [0, \tau]$.

Our objective is to develop an algorithm for generating samples of $\boldsymbol{X}$. A family of processes $\boldsymbol{X}^{(\boldsymbol{n})}$ defined by partial sums of the Fourier series of $\boldsymbol{X}$ is used to approximate this process. We show that the family of processes $\boldsymbol{X}^{(\boldsymbol{n})}$ can be used to generate samples of $\boldsymbol{X}$ in $[0, \tau]$ if the Fourier series of the covariance functions $c_{i,j}$ converge to $c_{i,j}$ in the rectangle $R = [0, \tau] \times [0, \tau]$, $\tau \in (0, \infty)$.

We introduce some notation and definitions prior to presenting a Monte Carlo algorithm for generating samples of $\boldsymbol{X}$. Let $\nu_1 = 2\pi/\tau$ and $\nu_k = k\,\nu_1$, where $k = 1, \ldots, n_i$, $n_i \geq 1$ are integers, $i = 1, \ldots, d$, and $\boldsymbol{n} = (n_1, \ldots, n_d)$. Consider the weighted averages

$$\begin{aligned} A_{i,k} &= \frac{2}{\tau} \int_0^\tau X_i(u)\, \cos(\nu_k\, u)\, du, \quad k = 0, 1, \ldots, n_i, \\ B_{i,k} &= \frac{2}{\tau} \int_0^\tau X_i(u)\, \sin(\nu_k\, u)\, du, \quad k = 1, \ldots, n_i, \end{aligned} \tag{5.24}$$

of X_i in $[0, \tau]$. The following equation defines the sequence of processes $\boldsymbol{X}^{(\boldsymbol{n})}$ used to approximate $\boldsymbol{X}$ and generate samples of this process.

If the Fourier series of $c_{i,j}$ converge to the covariance functions $c_{i,j}$ of $\boldsymbol{X}$ at every interior point of the rectangle $R = [0, \tau] \times [0, \tau]$, then the process $\boldsymbol{X}^{(\boldsymbol{n})}$ with coordinates

$$X_i^{(n_i)}(t) = \frac{A_{i,0}}{2} + \sum_{k=1}^{n_i} \left[A_{i,k}\, \cos(\nu_k\, t) + B_{i,k}\, \sin(\nu_k\, t)\right], \quad i = 1, \ldots, d, \tag{5.25}$$

becomes a version of the Gaussian process $\boldsymbol{X}$ as $n_i \to \infty$, $i = 1, \ldots, d$.

Proof: The convergence of the Fourier series of $c_{i,j}$ to $c_{i,j}$ in R can be proved under various conditions. For example, if $c_{i,j}$ is continuous and its first order partial derivatives are bounded in R, then the Fourier series of $c_{i,j}$ converges to $c_{i,j}$ at every interior point of R in the vicinity of which the mixed derivative $\partial^2 c_{i,j}/\partial t\,\partial s$ exists. This property holds in the special case in which $c_{i,j}$ and its partial derivatives $\partial c_{i,j}/\partial t$, $\partial c_{i,j}/\partial s$, and $\partial^2 c_{i,j}/\partial t\,\partial s$ are continuous in R ([183], Chapter 7).

The Fourier series of $c_{i,j}$ in R is

$$\tilde{c}_{i,j}(t,s) = \sum_{k,l=0}^{\infty} \lambda_{k,l} \Big[a_{kl}^{(i,j)} \cos(\nu_k t)\cos(\nu_l s) + b_{kl}^{(i,j)} \cos(\nu_k t)\sin(\nu_l s) + c_{kl}^{(i,j)} \sin(\nu_k t)\cos(\nu_l s) + d_{kl}^{(i,j)} \sin(\nu_k t)\sin(\nu_l s)\Big]$$

where

$$a_{k,l}^{(i,j)} = \frac{4}{\tau^2}\int_0^\tau du \int_0^\tau dv\, c_{i,j}(u,v)\cos(\nu_k u)\cos(\nu_l v),$$
$$b_{k,l}^{(i,j)} = c_{l,k}^{(i,j)} = \frac{4}{\tau^2}\int_0^\tau du \int_0^\tau dv\, c_{i,j}(u,v)\cos(\nu_k u)\sin(\nu_l v) \quad \text{and}$$
$$d_{k,l}^{(i,j)} = \frac{4}{\tau^2}\int_0^\tau du \int_0^\tau dv\, c_{i,j}(u,v)\sin(\nu_k u)\sin(\nu_l v),$$

for $i, j = 1, \ldots, d$, $k, l = 0, 1, \ldots$, $\nu_0 = 0$, $\lambda_{k,l} = 1/4$ for $k = l = 0$, $\lambda_{k,l} = 1/2$ for $k > 0, l = 0$ or $k = 0, l > 0$, and $\lambda_{k,l} = 1$ for $k, l > 0$ ([183], Section 7.2).

The covariance functions $c_{i,j}^{(n_i,n_j)}(t,s) = E\left[X_i^{(n_i)}(t)\,X_j^{(n_j)}(s)\right]$ represent partial sums of the Fourier series $\tilde{c}_{i,j}(t,s)$ since $a_{k,l}^{(i,j)} = E[A_{i,k}\,A_{j,l}]$, $b_{k,l}^{(i,j)} = E[A_{i,k}\,B_{j,l}]$, $c_{k,l}^{(i,j)} = E[B_{i,k}\,A_{j,l}]$, and $d_{k,l}^{(i,j)} = E[B_{i,k}\,B_{j,l}]$. Hence, $c_{i,j}^{(n_i,n_j)}$ converges to $c_{i,j}$ in R as $n_i, n_j \to \infty$ because the Fourier series of $c_{i,j}$ is convergent in R by hypothesis. Hence, $\boldsymbol{X}^{(\boldsymbol{n})}$ becomes equal to $\boldsymbol{X}$ in the second moment sense as $n_1, \ldots, n_d \to \infty$. Because $\boldsymbol{X}^{(\boldsymbol{n})}$ and $\boldsymbol{X}$ are Gaussian processes, $\boldsymbol{X}^{(\boldsymbol{n})}$ becomes a version of $\boldsymbol{X}$ as more and more terms are used in Eq. 5.25. This property justifies the use of samples of $\boldsymbol{X}^{(\boldsymbol{n})}$ as a substitute for samples of $\boldsymbol{X}$ if all n_i's are sufficiently large.

We also note that the coefficients $A_{i,k}$, $B_{i,k}$ exist in m.s. since $E[|A_{i,k}\,B_{j,l}|] \le \left(E[A_{i,k}^2]\,E[B_{j,l}^2]\right)^{1/2}$ and, for example,

$$E[A_{i,k}^2] = \frac{4}{\tau^2}\int_0^\tau du \int_0^\tau dv\, c_{i,i}(u,v)\cos(\nu_k u)\cos(\nu_k v) \le \frac{4}{\tau^2}\int_0^\tau du \int_0^\tau dv\, |c_{i,i}(u,v)|,$$

and the last integral is bounded. ■

Monte Carlo algorithm. The generation of samples of $\boldsymbol{X}$ in a time interval $[0, \tau]$ involves two steps:

1. Generate samples of the correlated Gaussian random variables $(A_{i,k}, B_{i,k'})$ for $i = 1, \ldots, d$, $k = 0, 1, \ldots, n_i$, and $k' = 1, \ldots, n_i$.

2. Calculate the corresponding samples of $\boldsymbol{X}^{(\boldsymbol{n})}$ in $[0, \tau]$ from Eq. 5.25.

Note: The methods in Section 5.2.1 can be used to generate samples of the correlated Gaussian variables $(A_{i,k}, B_{i,k'})$ defined by Eq. 5.25. ▲

Example 5.14: Let X be an Ornstein-Uhlenbeck process defined by the stochastic differential equation $dX(t) = -\rho\, X(t)\, dt + \sqrt{2\,\rho}\, dB(t)$, where $t \geq 0$, B is a Brownian motion, and $\rho > 0$. The mean and correlation functions of X are $\mu(t) = E[X(t)] = x_0\, e^{-\rho\, t}$ and $r(t, s) = E[X(t)\, X(s)] = \rho\, e^{-\rho\,(t+s)} + e^{2\,\rho\, \min(t,s) - \rho\,(t+s)}$ for $X(0) = x_0$. The coefficients of the family of approximations $X^{(n)}$ of X in Eqs. 5.24 and 5.25 with $d = 1$ and $n_1 = n$ can be obtained in closed form [77].

Figure 5.10 shows with heavy and thin lines the evolution in time of the

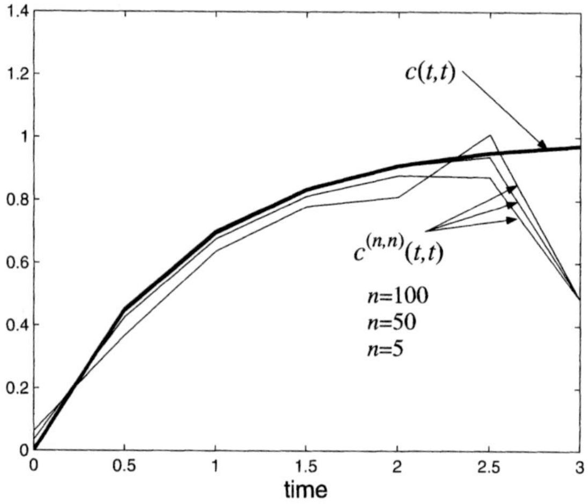

Figure 5.10. Time evolution of exact and approximate variance functions, $c(t, t)$ and $c^{(n,n)}(t, t)$ for $x_0 = 0$, $\rho = 0.6$, and $\tau = 3$

exact and approximate variance functions, $c(t, t)$ and $c^{(n,n)}(t, t)$, for $x_0 = 0$, $\rho = 0.6$, $\tau = 3$, and several values of n. The approximate variance functions $c^{(n,n)}(t, t)$ correspond to a function c^* defined on $[-\tau, \tau] \times [-\tau, \tau]$ and extended periodically to the entire plane $\mathbb{R}^2$. This function is such that $c^*(t, s) = c(t, s)$ for $(t, s) \in R = [0, \tau] \times [0, \tau]$, $c^*(t, s) = c(-t, -s)$ for $(t, s) \in [-\tau, 0] \times [-\tau, 0]$, and $c^*(t, s) = 0$ for $(t, s) \in [0, \tau] \times [-\tau, 0] \cup [-\tau, 0] \times [0, \tau]$. The approximate covariance function $c^{(n,n)}(\tau, \tau)$ approaches $c(\tau, \tau)/2$ as n increases since the periodic extension c^* of c has a jump discontinuity on the boundary of $[-\tau, \tau] \times [-\tau, \tau]$. The difference between $c^{(n,n)}(\tau, \tau)$ and $c(\tau, \tau)$ can be eliminated by using a periodic function c^{**} that coincides with c in $[0, \tau] \times [0, \tau]$ and is continuous on the boundaries of $[0, \tau] \times [0, \tau]$. $\diamond$

Note: The Fourier series of the covariance function $c(t, s) = r(t, s) - \mu(t)\, \mu(s)$ of X is convergent and converges to $c(t, s)$ since this function is continuous in $[0, \tau] \times [0, \tau]$ and its right and left partial derivatives are continuous almost everywhere in this set. ■

5.3.2.3 Fourier series. Random fields

We restrict our discussion to real-valued random fields defined on $\mathbb{R}^2$, that is, $d = 1$ and $d' = 2$. The extension to the general case is straightforward. Our objective is to generate samples of an inhomogeneous Gaussian field X with mean zero and covariance function $c(\boldsymbol{t}, \boldsymbol{s}) = E[X(\boldsymbol{t})\, X(\boldsymbol{s})]$ in a bounded subset of $\mathbb{R}^2$. We first introduce some notation and definitions. It can be shown that the functions $e^{\sqrt{-1}\,(k_1 t_1 + k_2 t_2)}$, $k_1, k_2 = 0, 1, \ldots$, define an orthogonal system on $R = [-\pi, \pi] \times [-\pi, \pi]$ in $\mathbb{R}^2$. An alternative representation of these functions is $\cos(k_1 t_1)\, \cos(k_2 t_2)$, $\cos(k_1 t_1)\, \sin(k_2 t_2)$, $\sin(k_1 t_1)\, \cos(k_2 t_2)$, and $\sin(k_1 t_1)\, \sin(k_2 t_2)$. Similarly, the family of functions $e^{\sqrt{-1}\,(k_1 t_1 + k_2 t_2 + k_3 t_3 + k_4 t_4)}$, $k_1, k_2, k_3, k_4 = 0, 1, \ldots$, defines an orthogonal system on $R^* = R \times R$. An alternative representation of this system is $\prod_{r=1}^{4} g_r(k_r\, t_r)$, where $g_1\, g_2\, g_3\, g_4$ are distinct products of two sine and two cosine functions with the arguments of the functions g_r. The Fourier series of a real-valued periodic function defined on R or R^* converges absolutely and uniformly if its second partial derivatives exist and are piece-wise continuous ([43], Section II.5).

Consider the weighted averages

$$A_0 = \frac{1}{(2\pi)^2} \int_R X(\boldsymbol{t})\, d\boldsymbol{t},$$
$$A_{k_1,k_2} = \frac{1}{\pi^2} \int_R X(\boldsymbol{t})\, \cos(k_1 t_1)\, \cos(k_2 t_2)\, d\boldsymbol{t},$$

of X in R. Let also B_{k_1,k_2}, C_{k_1,k_2}, and D_{k_1,k_2} be averages of X over R weighted by the functions $\cos(k_1 t_1)\, \sin(k_2 t_2)$, $\sin(k_1 t_1)\, \cos(k_2 t_2)$, and $\sin(k_1 t_1)\, \sin(k_2 t_2)$, respectively. The coefficients A_0, A_{k_1,k_2}, B_{k_1,k_2}, C_{k_1,k_2}, and D_{k_1,k_2} are correlated Gaussian variables.

If the Fourier series of c converges to the covariance function c of X at every interior point of R^*, then

$$\begin{aligned} X^{(n)}(\boldsymbol{t}) = A_0 + \sum_{k_1,k_2=1}^{n} \big[& A_{k_1,k_2} \cos(k_1 t_1)\, \cos(k_2 t_2) \\ & + B_{k_1,k_2} \cos(k_1 t_1)\, \sin(k_2 t_2) + C_{k_1,k_2} \sin(k_1 t_1)\, \cos(k_2 t_2) \\ & + D_{k_1,k_2} \sin(k_1 t_1)\, \sin(k_2 t_2) \big] \end{aligned} \tag{5.26}$$

becomes a version of the Gaussian field X as $n \to \infty$.

Proof: The Fourier series of $c(\boldsymbol{t}, \boldsymbol{s})$ is

$$\tilde{c}(\boldsymbol{t}, \boldsymbol{s}) = a_0 + \sum_{k_1,k_2,k_3,k_4=1}^{\infty} a_{k_1,k_2,k_3,k_4}\, g_1(k_1 t_1)\, g_2(k_2 t_2)\, g_3(k_3 s_1)\, g_4(k_4 s_2),$$

where $\boldsymbol{t} = (t_1, t_2)$, $\boldsymbol{s} = (s_1, s_2)$ and

$$a_0 = \frac{1}{(2\pi)^4} \int_{R^*} c(\boldsymbol{t}, \boldsymbol{s})\, d\boldsymbol{t}\, d\boldsymbol{s},$$

$$a_{k_1,k_2,k_3,k_4} = \frac{1}{\pi^4} \int_{R^*} c(\boldsymbol{t}, \boldsymbol{s})\, g_1(k_1 t_1)\, g_2(k_2 t_2)\, g_3(k_3 s_1)\, g_4(k_4 s_2)\, d\boldsymbol{t}\, d\boldsymbol{s}$$

with the appropriate expressions for the functions g_r. The covariance function of $X^{(n)}$ is equal to the partial sum of the Fourier series $\tilde{c}$. If the Fourier series of the covariance function of X is convergent in R^*, then the covariance function of $X^{(n)}$ converges to the covariance function of X as $n \to \infty$. Because X and $X^{(n)}$ are Gaussian, $X^{(n)}$ becomes a version of X as $n \to \infty$. Conditions for the convergence of Fourier series can be found, for example, in [43] (Section II.5) and [183] (Chapters 1 and 7). ■

Monte Carlo algorithm. The generation of samples of X in a bounded rectangle R involves two steps:

1. Generate samples of the correlated Gaussian variables A_0 and A_{k_1,k_2}, B_{k_1,k_2}, C_{k_1,k_2}, and D_{k_1,k_2}, $k_1, k_2 = 1, \ldots, n$ (Section 5.2.1).

2. Calculate the corresponding samples of $X^{(n)}$ from Eq. 5.26.

The above Monte Carlo simulation algorithm is useful in many applications involving functions that vary randomly in space. We will use this algorithm in Chapters 8 and 9 to generate samples of the spatial variation of the atomic lattice orientation and other material properties.

5.3.3 Non-Gaussian processes and fields

We consider non-Gaussian functions defined by transformations with and without memory of Gaussian functions as well as point and related processes. These processes have been discussed in Chapter 3. Our objective is to develop Monte Carlo algorithms for generating samples of these classes of non-Gaussian random functions.

5.3.3.1 Memoryless transformations

Let $\boldsymbol{Y}(\boldsymbol{t})$, $\boldsymbol{t} \in \mathbb{R}^{d'}$, be an $\mathbb{R}^d$-valued stationary Gaussian process ($d' = 1$) or homogeneous Gaussian field ($d' \geq 1$). The argument of $\boldsymbol{Y}$ is viewed as time for stochastic processes and space for random fields. The coordinates of $\boldsymbol{Y}$ have mean zero, variance one, and covariance functions $\rho_{i,j}(\boldsymbol{\tau}) = E[Y_i(\boldsymbol{t} + \boldsymbol{\tau})\, Y_j(\boldsymbol{t})]$, $i, j = 1, \ldots, d$. Consider also a continuous mapping $\boldsymbol{g} : \mathbb{R}^d \to \mathbb{R}^d$. The non-Gaussian random function

$$\boldsymbol{X}(\boldsymbol{t}) = \boldsymbol{g}(\boldsymbol{Y}(\boldsymbol{t})), \tag{5.27}$$

is called a **translation** stochastic process or random field (Section 3.6.6). It is common to define the coordinates of $\boldsymbol{X}$ by the transformations $X_i(\boldsymbol{t}) = g_i(Y_i(\boldsymbol{t}))$,

$i = 1, \ldots, d$, depending only on the corresponding coordinates of $\boldsymbol{Y}$. For example, we may take $g_i = F_i^{-1} \circ \Phi$, where F_i, $i = 1, \ldots, d$, are some distributions and Φ denotes the distribution of $N(0, 1)$ ([79], Section 3.1). The random functions X_i have the marginal distribution F_i and correlation functions

$$\xi_{i,j}(\boldsymbol{\tau}) = E[X_i(\boldsymbol{t} + \boldsymbol{\tau}) X_j(\boldsymbol{t})] = \int_{\mathbb{R}^2} g_i(u) g_j(v) \phi(u, v; \rho_{i,j}(\boldsymbol{\tau})) \, du \, dv, \quad (5.28)$$

where $\phi(\cdot, \cdot; \rho)$ denotes the density of an $\mathbb{R}^2$-valued Gaussian variable with mean zero, variance one, and correlation coefficient ρ. If $d = 1$, the finite dimensional densities of $\boldsymbol{X}$ are

$$f_n(x_1, \ldots, x_n; \boldsymbol{t}_1, \ldots, \boldsymbol{t}_n) = \phi(y_1, \ldots, y_n; \boldsymbol{\rho}) \prod_{i=1}^{n} \frac{f(x_i)}{\phi(y_i)}, \quad (5.29)$$

where $\boldsymbol{Y} = (Y(\boldsymbol{t}_1), \ldots, Y(\boldsymbol{t}_n))$ is a Gaussian vector with mean zero and covariance matrix $\boldsymbol{\rho} = \{\rho(\boldsymbol{t}_i - \boldsymbol{t}_j) = E[Y(\boldsymbol{t}_i) Y(\boldsymbol{t}_j)]\}$, $x_i \in \mathbb{R}$, $\boldsymbol{t}_i \in \mathbb{R}^{d'}$, $i = 1, \ldots, n$, $\phi(y_1, \ldots, y_n; \boldsymbol{\rho})$ denotes the density of $\boldsymbol{Y}$, and $y_i = \Phi^{-1}(F(x_i))$. Similar results can be found for $\mathbb{R}^d$-valued random functions with $d > 1$.

The translation processes and fields are stationary, can follow any marginal distribution, and their correlation function is completely defined by the marginal distributions F_i and the covariance functions $\rho_{i,j}$ of $\boldsymbol{Y}$. Generally, the marginal distributions F_i and the correlation functions $\xi_{i,j}$ are available in applications, for example, they may be estimated from records of $\boldsymbol{X}$. Hence, the covariance functions $\rho_{i,j}$ defining the underlying Gaussian function $\boldsymbol{Y}$ have to be calculated from Eq. 5.28. It turns out that the functions $(F_i, \xi_{i,j})$ cannot be arbitrary. They must be such that the solutions $\rho_{i,j}$ of Eq. 5.28 are covariance functions ([79], Section 3.1.1). If this condition is not satisfied, there is no translation model with the specified properties $(F_i, \xi_{i,j})$.

Monte Carlo algorithm. The generation of samples of $\boldsymbol{X}$ involves two steps:

1. Generate samples of $\boldsymbol{Y}$ by any of the methods in Section 5.3.

2. Calculate the corresponding samples of $\boldsymbol{X}$ from Eq. 5.27.

Example 5.15: Suppose that X is a real-valued translation process with mean zero, covariance function $E[X(t + \tau) X(t)] = e^{-\alpha|\tau|}(3 + 2e^{-2\alpha|\tau|})/5$, $\alpha > 0$, and marginal distribution $F(x) = \Phi\left(|x|^{1/3} \operatorname{sign}(x)\right)$. This process can be defined by the memoryless transformation $X(t) = Y(t)^3$, where Y is a stationary Gaussian process with mean zero and covariance function $\rho(\tau) = e^{-\alpha|\tau|}$ ([79], Section 3.1.1).

Figure 5.11 shows a sample of Y and the corresponding sample of X for $\alpha = 1$. The sample of Y has been obtained by the spectral representation method in Eq. 5.12 with $\nu^* = 50$ and $n = 500$. ◇

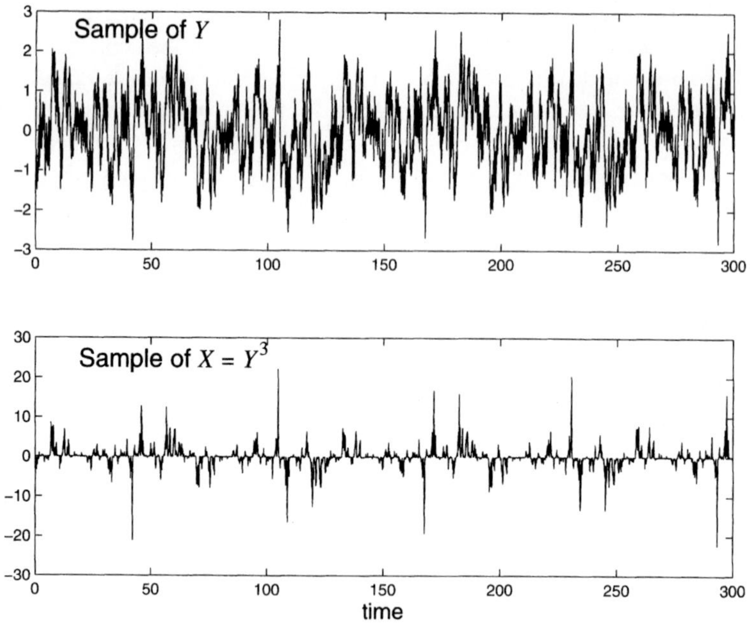

Figure 5.11. Samples of a Gaussian process Y with an exponential covariance function and of its translation image $X = Y^3$

Example 5.16: Let $Y(\boldsymbol{t}) = (a_1\, Y_1(t_1) + a_1\, Y_2(t_2)))\, (a_1^2+a_2^2)^{-1/2}$ be a real-valued Gaussian field, where $\boldsymbol{t} = (t_1, t_2) \in \mathbb{R}^2$, a_k are constants, and Y_k are stationary Gaussian fields with mean zero, variance one, and spectral densities that are zero outside the frequency ranges $[-\nu_k^*, \nu_k^*]$, $k = 1, 2$, and take constant non-zero values in these ranges.

Figure 5.12 shows a sample of $Y(\boldsymbol{t})$ for $\boldsymbol{t} \in [0, 5] \times [0, 5]$ and the translation image of this samples defined by the memoryless transformation $X(\boldsymbol{t}) = Y(\boldsymbol{t})^3$. Results are for $a_1 = a_2 = 1$, $\nu_1^* = 10$, and $\nu_2^* = 5$. The samples of Y_1 and Y_2 have been obtained by the algorithm in Eq. 5.22 with $n_1 = n_2 = 5$. $\diamond$

The class of translation processes in Eq. 5.27 can be extended to the class of mixtures of translation processes (Section 3.6.7). The non-Gaussian processes in this class can match not only correlation and marginal distributions as translation processes, but also higher order correlation functions. This property is illustrated by the following example.

Example 5.17: Let $X_k(t) = \left(e^{Y_k(t)} - e^{1/2}\right) / \sqrt{e^2 - e}$, $k = 1, 2$, be lognormal translation processes, where Y_1 and Y_2 denote two independent stationary Gaus-

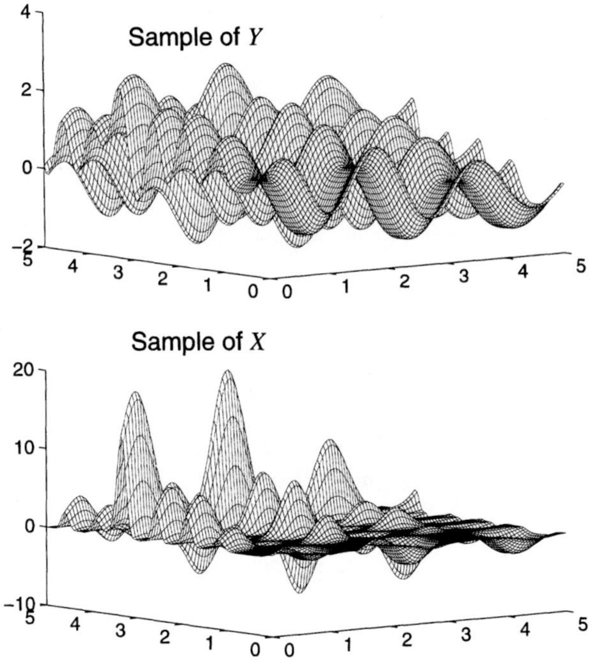

Figure 5.12. Samples of a Gaussian field Y and of its translation image $X = Y^3$ (adapted from [79], Fig. 3.7)

sian processes with mean zero and covariance functions

$$\rho_1(\alpha) = e^{-\lambda\,|\alpha|}, \quad \lambda > 0,$$

$$\rho_2(\alpha) = \frac{\sin\left((\nu_b - \nu_a)\,\alpha/2\right)\,\cos\left((\nu_b + \nu_a)\,\alpha/2\right)}{(\nu_b - \nu_a)\,\alpha/2}, \quad 0 \le \nu_a < \nu_b.$$

The marginal distribution and density of the processes X_k are

$$F(x) = \Phi(\log(x\,\sqrt{e-1} + 1) + 1/2)),$$

$$f(x) = \frac{\sqrt{e-1}}{x\,\sqrt{e-1}+1}\,\phi(\log(x\,\sqrt{e-1} + 1) + 1/2) \tag{5.30}$$

for $x > -e/\sqrt{e^2 - e}$. The first and second order correlation functions of the processes X_k are

$$\xi_k(\alpha) = E\,[X_k(t)\,X_k(t+\alpha)] = \frac{e^{\rho_k(\alpha)} - 1}{e-1},$$

$$\zeta_k(\alpha, \sigma) = E\,[X_k(t)\,X_k(t+\alpha)\,X_k(t+\sigma)] = \frac{e^{\rho_k(\alpha)+\rho_k(\sigma)+\rho_k(\alpha-\sigma)}}{(e-1)^{3/2}}. \tag{5.31}$$

Let X be a mixture of the translation processes X_1 and X_2 so that the finite dimensional distributions, F_n and $F_n^{(k)}$, of any order $n \geq 1$ of X and X_k, respectively, are related by $F_n = \sum_{k=1}^{2} p_k \, F_n^{(k)}$, where $p_k \geq 0$ and $\sum_{k=1}^{2} p_k = 1$ (Section 3.6.7). The Monte Carlo algorithm in Section 5.3.1.3 based on the sampling theorem can be applied to generate samples of X. However, the generation is less simple in this case since the distribution of the conditional random variable $\hat{\hat{X}}((n_t + n + 2)\, t^*)$ in Eq. 5.20 cannot be obtained analytically.

Let $f^{(2n+2)}$ and $f^{(2n+3)}$ be the joint density functions of

$$(X((n_t - n)\,\tau), \ldots, X((n_t + n + 1)\,\tau)) \quad \text{and}$$
$$(X((n_t - n)\,\tau), \ldots, X((n_t + n + 1)\,\tau), X((n_t + n + 2)\,\tau)),$$

respectively. Let $(X((n_t - n)\,\tau, \omega) = z_1, \ldots, X((n_t + n + 1)\,\tau, \omega) = z_{2n+2})$ be a sample of the first vector. The density of the conditional vector in Eq. 5.20 is $\hat{f}(x \mid z) = f^{(2n+3)}(z, x)/f^{(2n+2)}(z)$, where $z = (z_1, \ldots, z_{2n+2})$.

Figure 5.13 shows five samples of X for $\lambda = 1/5$, $\nu_a = 7$, and $\nu_b = 10$ corresponding to $p_1 = 0$, 1/4, 1/2, 3/4, and 1. The samples have been generated by the Monte Carlo algorithm in Eqs. 5.20 and 5.21 using a nodal spacing $t^* = 0.1$ and a window size $n = 5$. The samples illustrate the dependence of the correlation structure of X on the correlation functions of the constituent processes X_k and the weights p_k of these processes in the definition of X. Figure 5.14 shows the first order correlation function of X for $p_1 = 1/2$. Consider also a translation process X_T with the marginal distribution F in Eq. 5.30 and the covariance function in Fig. 5.14. Figure 5.15 shows a three dimensional view and contour lines of the difference $\zeta - \zeta_T$ between the second order correlation functions of X and X_T. The figure suggests that the translation process X_T may be inadequate to model X if the second order correlation function of this process needs to be represented accurately in addition to its marginal distribution and first order correlation functions. ◇

Note: There is no simple and efficient way to generate samples from $\hat{f}(\cdot \mid z)$ since the vector z changes in time. The following algorithm has been used for the calculations in the figures. Let u be a sample of a random variable uniformly distributed in (0, 1) and denote by (a, b) the range of $\hat{f}(\cdot \mid z)$ used for numerical calculations. If $u \leq 0.5$, integrate the conditional density from the left to find x such that $\int_a^x \hat{f}(\alpha \mid z)\, d\alpha = u$. If $u > 0.5$, integrate from the right to find x satisfying the condition $\int_x^b \hat{f}(\alpha \mid z)\, d\alpha = 1 - u$. The solution x is a sample of $\hat{f}(\cdot \mid z)$. ▲

5.3.3.2 Transformations with memory

If the current value of a stochastic process $\boldsymbol{X}$ is obtained from the past history of another process $\boldsymbol{Y}$, the mapping from $\boldsymbol{Y}$ to $\boldsymbol{X}$ is said to have memory. Transformations with memory can be defined by differential equations with input $\boldsymbol{Y}$ and output $\boldsymbol{X}$. These transformations cannot be used to define random fields because, unlike time, space does not have a natural flow.

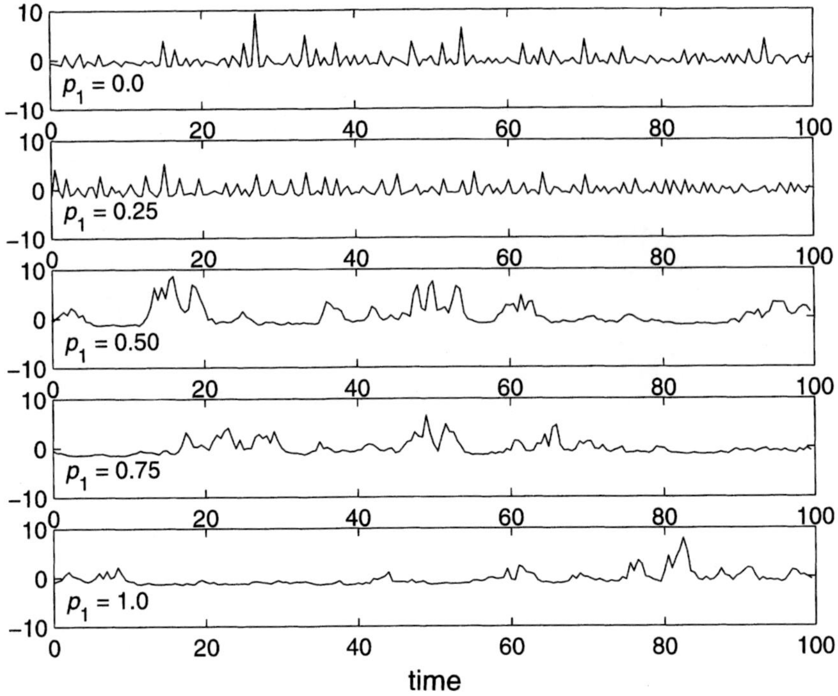

Figure 5.13. Five samples of X(t) corresponding to $p_1 = 0$, 1/4, 1/2, 3/4, and 1

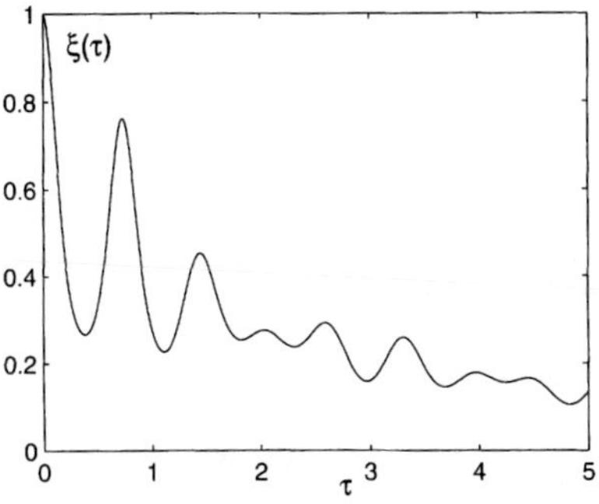

Figure 5.14. First order correlation function ξ of X for $p_1 = 1/2$

Our objective is to generate samples of $\boldsymbol{X}$. It is assumed that (1) $\boldsymbol{X}$ is an $\mathbb{R}^d$-valued process defined by a differential equation with random input $\boldsymbol{Y}$, (2) the defining differential equation for $\boldsymbol{X}$ is known, and (3) $\boldsymbol{Y}$ is Gaussian, Poisson, Lévy, or semimartingale noise.

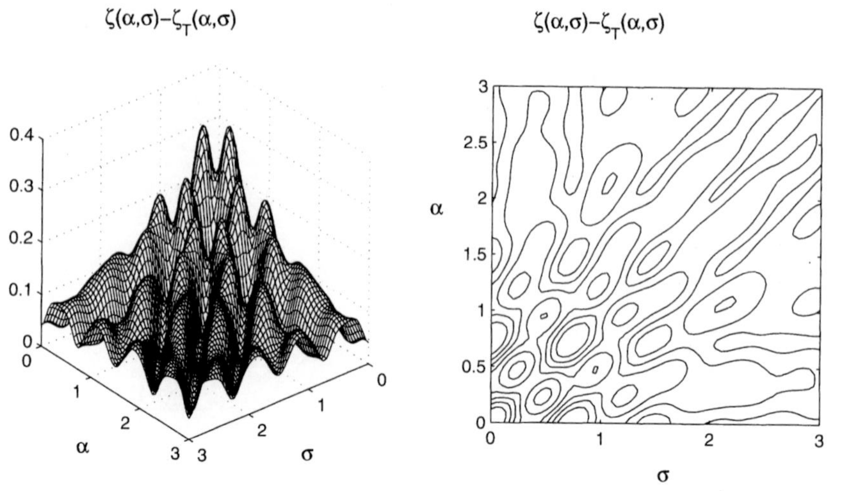

Figure 5.15. Difference $\zeta - \zeta_T$ of the second order correlation functions of the processes X and X_T

Monte Carlo algorithm. The generation of samples of $\boldsymbol{X}$ involves two steps:

1. Generate samples of the driving noise $\boldsymbol{Y}$.

2. Integrate the differential equation of $\boldsymbol{X}$ driven by samples of $\boldsymbol{Y}$ to obtain samples of this process.

Note: The numerical methods in Section 4.7.3 can be applied to the differential equation defining $\boldsymbol{X}$ to calculate samples of $\boldsymbol{X}$ from samples of $\boldsymbol{Y}$. This section considers only the first step of the above Monte Carlo algorithm, that is, the generation of samples of $\boldsymbol{Y}$, assumed to be a Gaussian, Poisson, or Lévy noise. ▲

Gaussian white noise. Let B be a real-valued Brownian motion. The increments of B in non-overlapping time intervals are independent Gaussian variables with mean zero and variance equal to the duration of these time intervals, that is, the successive time increments $\Delta B(t) = B(t+\Delta t) - B(t)$, $\Delta t > 0$, of B are independent $N(0, \Delta t)$ variables. If the driving noise is an $\mathbb{R}^{d'}$-valued Brownian motion $\boldsymbol{B}$ with independent coordinates, then its increments $\Delta \boldsymbol{B}(t) = \boldsymbol{B}(t+\Delta t) - \boldsymbol{B}(t)$ are independent $\mathbb{R}^{d'}$-valued Gaussian variables with mean zero and covariance matrix $\boldsymbol{i}\,\Delta t$, where $\boldsymbol{i}$ denotes the (d', d') identity matrix.

Samples of $\boldsymbol{B}$ can be generated by the recurrence formula

$$\boldsymbol{B}(t_{k+1}) = \boldsymbol{B}(t_k) + \sqrt{\Delta t}\ \mathbf{randn}(d', 1), \tag{5.32}$$

where $\Delta t > 0$ is a time increment, $t_k = k\,\Delta t$, $k = 0, 1, \ldots$, and $\boldsymbol{B}(t_0) = \mathbf{0}$.

Note: The time increment Δt should correspond to the time step used to integrate the differential equation for $\boldsymbol{X}$. The algorithm in Eq. 5.32 is based on properties of the Brownian motion process. ▲

Poisson white noise. Let $C(t) = \sum_{k=1}^{N(t)} Y_k$ be a compound Poisson process, where N denotes a Poisson process with intensity $\lambda > 0$ and Y_k are independent copies of a random variable Y_1. The process C has stationary independent increments. If $\lambda\,\Delta t \ll 1$, the increment of C in $(t, t+\Delta t]$ is either one of the variables Y_k or zero, with probabilities $1 - e^{-\lambda\,\Delta t} \simeq \lambda\,\Delta t$ or $e^{-\lambda\,\Delta t} \simeq 1 - \lambda\,\Delta t$, respectively. If $Y_1 \in L_2$, the mean and covariance function of C are $E[C(t)] = \lambda\,t\,E[Y_1]$ and $E[(C(t) - E[C(t)])(C(s) - E[C(s)])] = \lambda\,\min(t,s)\,E[Y_1^2]$. If $E[Y_1] = 0$ and $\lambda\,E[Y_1^2] = 1$, C and the Brownian motion B are equal in the second moment sense. However, the samples of these processes differ significantly. The Brownian motion has continuous samples a.s. while the samples of C have jump discontinuities.

Samples of C in a time interval $[0, \tau]$, $\tau > 0$, can be generated in two steps:

1. Generate samples of the jump times T_k and the jumps Y_k of C.

2. Calculate samples of C corresponding to the samples of T_k and Y_k.

Note: The number of jumps of C in $(0, \tau]$ is given by a Poisson random variable $N(\lambda\,\tau)$ with intensity $\lambda\,\tau$ (Example 5.3). The jump times of C are uniformly distributed in $(0, \tau]$ conditional on the number of jumps. The generation of $N(\lambda\,\tau)$ independent samples of Y_1 can be based on the methods in Sections 5.2.1 and 5.2.2. Additional information on the generation of samples of C can be found in Sections 5.3.1.1 and 5.3.1.2.

The above approach can be extended to $\mathbb{R}^{d'}$-valued compound Poisson processes $\boldsymbol{C}(t) = \sum_{k=1}^{N(t)} \boldsymbol{Y}_k$, where $\boldsymbol{Y}_k$ are iid $\mathbb{R}^{d'}$-valued random variables. ▲

Lévy white noise. The general form of a Lévy process L is given by the Lévy decomposition in Section 3.14.2. The Lévy process includes the Brownian motion and the compound Poisson processes so that the Gaussian and Poisson white noise processes are special cases of the Lévy noise. We have already generated samples of L using the representation of this process in Section 3.14.2. We do not restate this Monte Carlo algorithm. Instead, we give a Monte Carlo algorithm for generating samples of an α-stable process L_α, that is, a Lévy process whose increments $\Delta L_\alpha(t) = L_\alpha(t+\Delta t) - L_\alpha(t)$ are independent $S_\alpha((\Delta t)^{1/\alpha}, 0, 0)$ random variables (Example 5.1).

Samples of L_α can be generated by the recurrence formula

$$L_\alpha(t_{k+1}) = L_\alpha(t_k) + (\Delta t)^{1/\alpha}\,Z_k, \tag{5.33}$$

where $t_k = k\,\Delta t$, $k = 0, 1, \ldots$, $L_\alpha(0) = 0$, and Z_k are iid $\sim S_\alpha(1, 0, 0)$.

Example 5.18: Let $L_\alpha(t)$, $t \in [0, \tau]$, be an α-stable Lévy process. Samples of this process for $\alpha = 1.0, 1.5, 1.7$, and 1.99 have been generated in Section 3.14.2 based on a representation of L_α as a sum of compound Poisson processes with small and large jumps. Figure 5.16 shows samples of L_α for the same values of

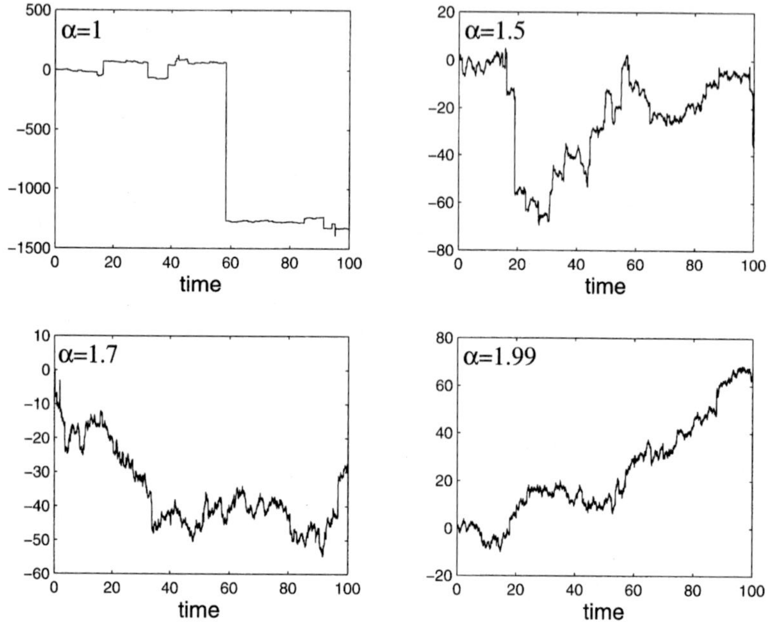

Figure 5.16. Samples of a Lévy process

α generated by the algorithm in Eq. 5.33. The size of the jumps of L_α decreases with α. The sample of L_α for $\alpha = 1.99$ resembles the sample of a Brownian motion process. ◇

Semimartingale noise. We have seen that any semimartingale admits a decomposition in an adapted process with càdlàg samples of finite variation on compacts and a local martingale (Section 4.4.1). The generation of samples of a process with finite variation on compacts does not pose difficulties. An example of such a process is the compound Poisson process. We focus here on the generation of the martingale component of a semimartingale. Most of the martingales used in this book are defined by (1) sums of two stochastic integrals with Brownian motion and compensated Poisson processes integrators or (2) memoryless transformations of Brownian motions. Samples of these semimartingales can be calculated from samples of Brownian motion and compound Poisson processes.

Example 5.19: We have seen that a Brownian motion B and a compound Poisson process C can have the same second moment properties. Another example of a process that is equal to B in the second moment sense is the process M^* with increments $dM^*(t) = B(t)\,dB(t)/\sqrt{t}$, $t > 0$. Figure 5.17 shows several samples of B and the corresponding samples of M^*. The sample properties of these two

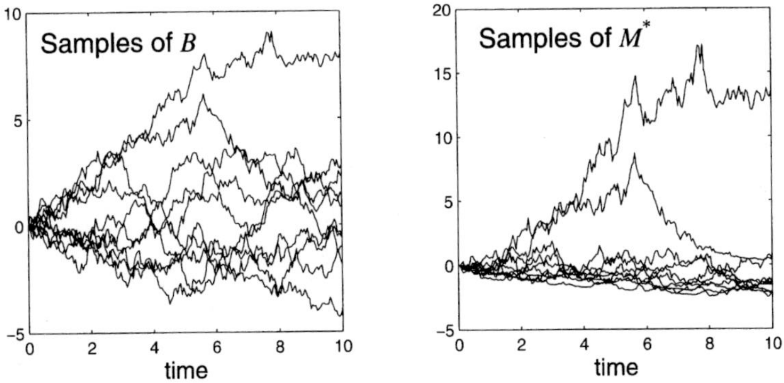

Figure 5.17. Samples of a Brownian motion and the corresponding samples of M^*

processes differ significantly although their increments have the same first two moments and are not correlated.

It is common in the engineering literature to define white noise in the second moment sense. This example demonstrates one more time the need for a precise definition of the white noise process that has to go beyond second moment properties. ◇

Proof: Consider the martingale $M(t) = B(t)^2 - t$. Denote by $dM^*(t) = B(t)\,dB(t)/\sqrt{t}$, $t > 0$, the increments $dM(t) = 2\,B(t)\,dB(t)$ of M scaled by $2\sqrt{t}$. The first two moments of the increments of M^* are $E[dM^*(t)] = 0$ and $E[dM^*(t)\,dM^*(s)] = \delta(t-s)\,dt$ so that M^* and B are equal in the second moment sense. However, while the increments of B are independent, those of M^* are not.

An algorithm for generating samples of B was discussed previously in this section. The recurrence formula $M^*(t+dt) = M^*(t) + dM^*(t)$ can also be used to produce samples of M^*. ■

5.3.3.3 Point and related processes

We have already used point processes in Examples 5.8 and 5.10 to produce samples of stationary Gaussian processes and fields. This section provides algorithms for generating samples of (1) two point processes, the binomial and the Poisson processes, (2) a compound Poisson processes, and (3) the Voronoi tessellation.

Binomial process. Consider a bounded Borel set B in $\mathbb{R}^{d'}$ with Lebesgue measure v_B, that is, v_B is the volume of B. A random point $\boldsymbol{T} \in B$ is said to be **uniformly**

distributed in B if $\boldsymbol{T}$ belongs to $B' \subset B$, $B' \in \mathcal{B}(\mathbb{R}^{d'})$, with the probability $p_{B'} = P(\boldsymbol{T} \in B') = v_{B'}/v_B$.

Let $A_i^{(n)}$, $i = 1, \ldots, n$, be a partition of B in Borel sets and denote by $v_i^{(n)}$ the volume of $A_i^{(n)}$. The probability that a uniformly distributed point in B falls in $A_i^{(n)}$ is $p_i^{(n)} = v_i^{(n)}/v_B$. Consider k independent uniformly distributed points in B, and let $k_i \geq 0$ denote the number of these points falling in $A_i^{(n)}$, so that we have $\sum_{i=1}^{n} k_i = k$. The probability that k_i points are in $A_i^{(n)}$, $i = 1, \ldots, n$, referred to as the multinomial probability, is

$$\frac{k!}{k_1! \cdots k_n!} \prod_{i=1}^{n} \left(\frac{v_i^{(n)}}{v_B} \right)^{k_i}. \tag{5.34}$$

The special case of Eq. 5.34 for $n = 2$ and $k_2 = k - k_1$ gives the probability that $k_1 \geq 0$ out of k independent uniformly distributed points in B fall in $A_1^{(n)}$, that is, the **binomial distribution**. If the number of points k in B and the probability p_{A_1} that a uniformly distributed point in B falls in A_1 tend to infinity and zero, respectively, such that $k\, p_{A_1}$ is a constant $\lambda\, v_B$, the binomial distribution converges to a Poisson distribution with intensity λ.

Monte Carlo algorithm. The generation of n independent uniformly distributed points in a Borel set $B \subset R^{d'}$ involves two steps:

1. Select a rectangle $R \subset \mathbb{R}^{d'}$ including B.

2. Generate independent uniformly distributed points in R until n of these points fall in B. The points in B are uniformly distributed in this set.

Note: The MATLAB function **rand** can be used to generate independent, uniformly distributed samples in the unit cube $[0, 1]^{d'}$. The columns of **rand**(d', n) give the coordinates of n such samples. Uniformly distributed samples in R can be obtained by translating and scaling the entries of **rand**(d', n). Alternative methods can be developed to produce uniformly distributed samples in B ([180], pp. 38-40). ▲

Poisson process. A stationary or homogeneous Poisson process can be viewed as a random measure

$$N : \left(\mathbb{R}^{d'}, \mathcal{B}(\mathbb{R}^{d'})\right) \longrightarrow (\mathbb{Z}_+, \mathcal{K}), \tag{5.35}$$

where the σ-field $\mathcal{K}$ consists of all the parts of $\mathbb{Z}_+ = \{0, 1, \ldots\}$ and $N(B)$ gives the number of points in a bounded Borel set $B \in \mathcal{B}(\mathbb{R}^{d'})$. This random measure is defined by two properties: (1) $N(B)$ is a Poisson variable with intensity $\lambda\, v_B$, where v_B is the volume of B and $\lambda > 0$ is a constant and (2) the random variables $N(B)$ and $N(B')$ are independent for $B, B' \in \mathcal{B}(\mathbb{R}^{d'})$ and $B \cap B' = \emptyset$. These properties of N imply

$$P\left(N(B_1) = k_1, \ldots, N(B_m) = k_m\right) = \prod_{i=1}^{m} \frac{(\lambda\, v_i)^{k_i}}{k_i!} \exp\left(-\sum_{i=1}^{m} \lambda\, v_i\right), \tag{5.36}$$

where $B_i \in \mathcal{B}(\mathbb{R}^{d'})$, $i = 1, \ldots, m$, are disjoint subsets with volume v_i. We also note that (*a*) N is stationary and isotropic, that is, the process is invariant to translation and rotation, (*b*) the conditional process $N \mid (N(B) = n)$ is binomial, and (*c*) $E[N(B)] = \lambda\, v_B$ ([180], Section 2.4.1).

Consider a generalized Poisson process $\mathcal{N}$ characterized by a point dependent intensity $\tilde{\lambda}(\boldsymbol{x})$, $\boldsymbol{x} \in \mathbb{R}^{d'}$, and properties resembling N. For example, the probability of the event $\{\mathcal{N}(B) = n\}$ is $(\lambda_B\, v_B)^n \exp(-\lambda_B\, v_B)/n!$ for $n = 0, 1, \ldots$, and $B \in \mathcal{B}(\mathbb{R}^{d'})$ so that $E[\mathcal{N}(B)] = \lambda_B\, v_B$, where $\lambda_B\, v_B = \int_B \tilde{\lambda}(\boldsymbol{x})\, d\boldsymbol{x}$. Also, the random variables $\mathcal{N}(B)$ and $\mathcal{N}(B')$ are independent for $B, B' \in \mathcal{B}(\mathbb{R}^{d'})$ and $B \cap B' = \emptyset$.

Monte Carlo algorithm. The generation of samples of N in B involves two steps:

1. Generate a sample of the Poisson random variable $N(B)$.

2. Place $N(B)$ independent, uniformly distributed points in B.

Monte Carlo algorithm. The generation of samples of $\mathcal{N}$ in B involves two steps:

1. Apply the above algorithm to generate samples of N^* in B, where N^* is a Poisson process with intensity $\lambda^* = \max_{\boldsymbol{x} \in B} \tilde{\lambda}(\boldsymbol{x})$.

2. Modify the resulting points of N^* according to the following thinning rule. A point $\boldsymbol{x} \in B$ of N^* is eliminated with probability $1 - \tilde{\lambda}(\boldsymbol{x})/\lambda^*$. This modification is performed independently of the other points of this process.

Proof: The expectation $n_B = E\left[\sum_{i=1}^{N^*(B)} \tilde{\lambda}(\boldsymbol{T}_i)/\lambda^*\right]$ gives the average number of points retained from the $N^*(B)$ points in B since a Poisson point $\boldsymbol{T}_i$ is kept with probability $\tilde{\lambda}(\boldsymbol{T}_i)/\lambda^*$. We have

$$n_B = E\left\{E\left[\sum_{i=1}^{N^*(B)} \tilde{\lambda}(\boldsymbol{T}_i)/\lambda^* \mid N^*(B)\right]\right\} = \frac{1}{\lambda^*}\, E\left\{N^*(B)\, E\left[\tilde{\lambda}(\boldsymbol{T}_i)\right]\right\}$$
$$= \frac{\lambda_B}{\lambda^*}\, E[N^*(B)] = \lambda_B\, v_B$$

since the points $\boldsymbol{T}_i$ are mutually independent and uniformly distributed in B conditional on $N^*(B)$, $\lambda_B = (1/v_B) \int_B \tilde{\lambda}(\boldsymbol{x})\, d\boldsymbol{x} = E[\tilde{\lambda}(\boldsymbol{T}_i)]$, and $E[N^*(B)] = \lambda^*\, v_B$. Hence, the expected number of points kept from $N^*(B)$ is equal to $E[\mathcal{N}] = \lambda_B\, v_B$. ■

Example 5.20: Let N be a Poisson process with intensity $\lambda > 0$ defined on $\mathbb{R}^2$ and $B = [0, 1] \times [0, 1]$ be a Borel set in this space. Consider also a generalized Poisson process $\mathcal{N}$ with intensity $\tilde{\lambda}(\boldsymbol{x}) = 6\, \lambda\, x_1^2\, x_2$ defined on the same space. The processes N and $\mathcal{N}$ have the same average number of points in B. Figure 5.18 shows samples of N and $\mathcal{N}$ in B, respectively, for $\lambda = 100$. The points of N are uniformly spread in B while the points of $\mathcal{N}$ are concentrated in the upper right corner of B where $\tilde{\lambda}$ takes on larger values. ◇

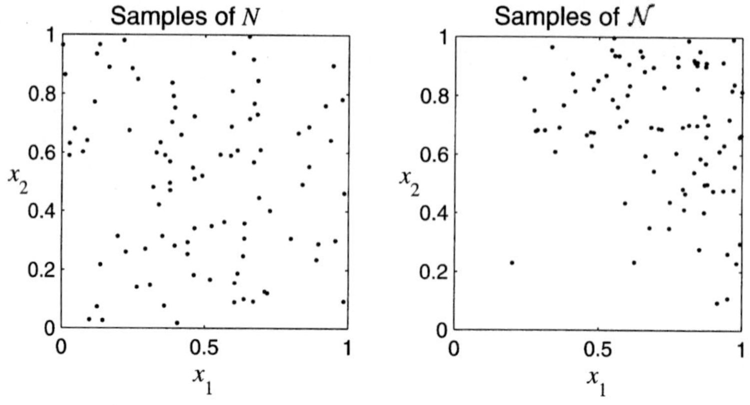

Figure 5.18. Samples of the Poison processes N and $\mathcal{N}$

Note: We have $v_B = 1$, $\lambda^* = 6\,\lambda$, $\lambda_B\, v_B = \lambda_B = \lambda$, and $E[\mathcal{N}] = \lambda_B\, v_B = \lambda$. ▲

Compound Poisson processes in space. Consider the real-valued compound Poisson process $C(\boldsymbol{v}) = \sum_{k=1}^{N(B)} Y_k\, 1_{B(\boldsymbol{T}_k)}(\boldsymbol{v})$ in Eq. 5.16, where $B \subset \mathbb{R}^{d'}$ is a bounded Borel set, $N(B)$ denotes a Poisson variable with intensity $\lambda\, v_B$, $\lambda > 0$, v_B is the Lebesgue measure of B, $\boldsymbol{T} = (T_1, \ldots, T_{d'})$ are points of N in B, $B(\boldsymbol{T}) = B \cap \left(\times_{j=1}^{d'} [T_j, \infty)\right)$, and Y_k denote independent copies of a random variable Y_1 with finite first two moments.

Monte Carlo algorithm. The generation of samples of C in B involves two steps:

1. Generate a sample of the Poisson random variable $N(B)$, $N(B)$ independent, uniformly distributed points $\boldsymbol{T}_k$ in B, and $N(B)$ independent samples of Y_1.

2. Calculate the corresponding samples of C from its definition and samples of $\boldsymbol{T}_k$ and Y_k.

Voronoi tessellation. The Voronoi tessellation is used in forestry, astrophysics, biology, material science, geology, physics, ecology, and other fields of applied science and engineering ([180], Section 10.2).

Let $\mathcal{P}$ denote the class of open convex non-empty polygons $\boldsymbol{p}$ in $\mathbb{R}^2$. A subset of $\mathcal{P}$ is said to be a **planar tessellation** if its elements $\boldsymbol{p}$ are disjoint, the union of their closure fills the plane, and the set $\{\boldsymbol{p} : \boldsymbol{p} \cap B \neq \emptyset\}$ is finite for any bounded $B \in \mathcal{B}(\mathbb{R}^2)$. A **random tessellation** is a random variable with values in the class of tessellations.

Let N be a Poisson process defined on $\mathbb{R}^{d'}$ with intensity $\lambda > 0$. Almost all points $\boldsymbol{x} \in \mathbb{R}^{d'}$ have a nearest point $\boldsymbol{T_x}$ in the set $\{\boldsymbol{T}_1, \boldsymbol{T}_2, \ldots\}$ of points of N.

The open convex polytopes defining the **Voronoi tessellation** are

$$\boldsymbol{p}_k = \{\boldsymbol{x} \in \mathbb{R}^{d'} : \| \boldsymbol{x} - \boldsymbol{T}_k \| < \| \boldsymbol{x} - \boldsymbol{T}_l \|, l \neq k\}, \quad k = 1, 2, \ldots, \tag{5.37}$$

that is, the polytope k consists of the points in $\mathbb{R}^{d'}$ that are closest to $\boldsymbol{T}_k$.

Monte Carlo algorithm. The generation of a sample of a Voronoi tessellation in a bounded Borel set B of $\mathbb{R}^{d'}$ involves two steps:

1. Generate a sample of a Poisson process N in B, that is, a sample of the vector $(\boldsymbol{T}_1, \ldots, \boldsymbol{T}_{N(B)})$.

2. Construct the edges of the polytopes defining the Voronoi tessellation, that is, the hyperplanes that pass through the midpoint of the segments connecting adjacent points of $N(B)$ and are orthogonal to these segments.

Note: The Voronoi tessellation can be generalized in various directions, for example, the Poisson process N may not be stationary and/or the hyperplanes defining the polytope edges may not intersect the segment between adjacent Poisson points at its midpoint. ▲

Example 5.21: Consider the inhomogeneous Poisson process $\mathcal{N}$ in Example 5.20. Figure 5.19 shows samples of Voronoi tessellations in $B = [0, 1] \times [0, 1]$ corre-

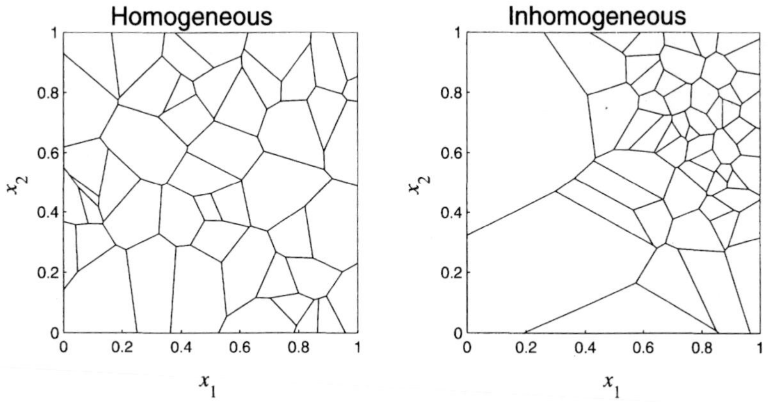

Figure 5.19. Samples of Voronoi tessellations corresponding to the homogeneous and inhomogeneous Poisson processes in Example 5.20

sponding to the samples of N and $\mathcal{N}$ in Fig. 5.18. The sample of the Voronoi tessellation obtained from the sample of $\mathcal{N}$ shows that it is possible to generate microstructures with grains of a wide range of sizes and shapes. ◇

5.4 Improved Monte Carlo simulation

The main limitation of the Monte Carlo method is the computation time, which can be excessive. The time required to obtain a sample of the state $\boldsymbol{X}$ of a physical system and the number of samples needed to estimate properties of $\boldsymbol{X}$ determine the computation time. Software packages developed for the integration of deterministic differential equations are used to calculate samples of $\boldsymbol{X}$ from input samples. Generally, (1) marginal reductions of the computation time can be obtained by modifying existing packages for integrating deterministic differential equations since most of these packages are efficient and (2) a large number of output samples may be needed in a broad range of applications. For example, suppose that the probability $P_s(\tau)$ that $\boldsymbol{X}$ does not leave a safe set D during a time interval $[0, \tau]$, that is, the system reliability in $[0, \tau]$, needs to be calculated. If $P_s(\tau)$ is of the order $1 - 10^{-5}$, we need at least 10^6 samples of $\boldsymbol{X}$ to estimate this probability.

It is shown that a stochastic problem can be modified by changing the time and/or the measure such that the Monte Carlo solution of the modified problem can be based on fewer samples of $\boldsymbol{X}$ and the calculation of each sample of this process may even require less computation time. We use numerical examples to assess the usefulness of changing the time and/or the measure in Monte Carlo studies.

5.4.1 Time change

Suppose that $\boldsymbol{X} \in \mathbb{R}^d$ is the solution of the linear stochastic differential equation

$$d\boldsymbol{X}(t) = \boldsymbol{a}(t)\,\boldsymbol{X}(t)\,dt + \boldsymbol{b}(t)\,d\boldsymbol{B}(t), \quad t \geq 0, \tag{5.38}$$

driven by a Brownian motion $\boldsymbol{B} \in \mathbb{R}^{d'}$, where $\boldsymbol{a}$ and $\boldsymbol{b}$ are such that this equation has a unique solution (Section 4.7.1.1). Our objective is to estimate properties of the solution $\boldsymbol{X}$ of this equation.

Let $\boldsymbol{\theta}$ and $\boldsymbol{X}(0)$ be the Green function and the initial state, respectively, for Eq. 5.38. The solution of this equation is $\boldsymbol{X}(t) = \boldsymbol{\theta}(t, 0)\,\boldsymbol{X}(0) + \boldsymbol{M}(t)$, where

$$\boldsymbol{M}(t) = \int_0^t \boldsymbol{\theta}(t, s)\,\boldsymbol{b}(s)\,d\boldsymbol{B}(s), \quad t \geq 0, \tag{5.39}$$

is a martingale under some conditions. For example, $\boldsymbol{M}$ is a square integrable martingale if $\boldsymbol{\theta}\,\boldsymbol{b}$ is square integrable in $(0, t)$. We can generate samples of $\boldsymbol{X}$ from (1) numerical solutions of Eq. 5.38 with $\boldsymbol{B}$ replaced by its samples or (2) samples of $\boldsymbol{M}$ added to the function $\boldsymbol{\theta}(t, 0)\,\boldsymbol{X}(0)$. If $\boldsymbol{X}(0)$ is random, we need to use samples of $\boldsymbol{X}(0)$ in $\boldsymbol{\theta}(t, 0)\,\boldsymbol{X}(0)$. The latter approach is used in this section to generate samples of $\boldsymbol{X}$ in conjunction with the time change theorem for martingales. Our discussion is limited to a single driving noise, that is, $\boldsymbol{B} = B$ is a real-valued Brownian motion.

Time change theorem. If $M(t)$, $t \geq 0$, is a continuous local martingale such that $M(0) = 0$ and $\lim_{t\to\infty}[M, M](t) = \infty$ a.s., then

$$M(t) = \tilde{B}([M, M](t)) \quad \text{a.s. for } t \geq 0, \tag{5.40}$$

where $\tilde{B}$ is a Brownian motion ([147], Theorem 41, p. 81). We refer to t and $T(t) = \inf\{s > 0 : [M, M](s) > t\}$ as the **original** and the **new clocks** or **times**, respectively. If T is deterministic, it is denoted by τ.

Note: We only give here a heuristic justification of the time change theorem in a simplified setting. Let $(\Omega, \mathcal{F}, P)$ be a probability space with the filtration $\mathcal{F}_t = \sigma(B(s) : 0 \leq s \leq t)$, $t \geq 0$, where B is a Brownian motion process. Let $M(t) = \int_0^t A(s)\, dB(s)$ and assume that A is an $\mathcal{F}_t$-adapted real-valued process that is independent of B and has the property $\lim_{t\to\infty} \int_0^t A(s)^2\, ds = \infty$ a.s. Conditional on A, the increments $A(s)\, dB(s)$ of M are independent and follow a Gaussian distribution with mean zero and variance $A(s)^2\, ds$. These increments have the same distribution as a Brownian motion $\tilde{B}$ running in a different clock T defined locally by $dT(s) = A(s)^2\, ds$ so that $T(t) = \int_0^t A(s)^2\, ds$ and $\lim_{n\to\infty} T(t) = \infty$ a.s. by hypothesis. If $A(\cdot)$ is a deterministic function $a(\cdot)$, the new clock $T(t) = \tau(t) = \int_0^t a(s)^2\, ds$ is a deterministic increasing function.

We note that the time change in Eq. 5.40 modifies the diffusion coefficient since the martingale in Eq. 5.39 giving the noise contribution to the solution of Eq. 5.38 becomes a Brownian motion in the new clock. We will see that the change of measure considered in the following section causes a change of the drift coefficient. ▲

Monte Carlo algorithm. The generation of samples of $\boldsymbol{X}$ involves two steps:

1. Generate samples $\tilde{B}$ in the new clock and map them into the original clock to find samples of M (Eq. 5.40).

2. Calculate the corresponding samples of $\boldsymbol{X}$ from $\boldsymbol{X}(t) = \boldsymbol{\theta}(t, 0)\, \boldsymbol{X}(0) + \boldsymbol{M}(t)$, where samples of $\boldsymbol{X}(0)$ need to be used for random initial conditions.

Note: The following examples show that this Monte Carlo algorithm is preferable to algorithms based on a direct integration of Eq. 5.38 if the relationship between the new and original clocks is deterministic. If this condition is not satisfied, the use of the time change theorem in Monte Carlo simulation seems to be impractical. ▲

Example 5.22: Let X be defined by $\ddot{X}(t) + 2\,\zeta\,\nu\,\dot{X}(t) + \nu^2\, X(t) = \beta\, W(t)$, $t \geq 0$, where $\zeta \in (0, 1)$, $\nu > 0$, $W(t)$ stands for the formal derivative of a Brownian motion B, β is a constant, and $(X(0), \dot{X}(0)) = (X_0, \dot{X}_0)$. We have

$$X(t) = e^{-\zeta\,\nu\,t}\left[X_0\,\theta_{11}(t) + \dot{X}_0\,\theta_{12}(t) + M(t)\right], \quad \text{where}$$
$$M(t) = \beta \int_0^t e^{\zeta\,\nu\,s}\,\theta_{12}(t - s)\, dB(s),$$

$\theta_{11}(t) = \cos(\tilde{\nu}\,t) + \zeta\,\nu\,\sin(\tilde{\nu}\,t)/\tilde{\nu}$, $\theta_{12}(t) = \sin(\tilde{\nu}\,t)/\tilde{\nu}$, and $\tilde{\nu} = \nu\sqrt{1 - \zeta^2}$. The time change theorem gives the representation

$$X(t) = e^{-\zeta \nu t} \left[X_0 \theta_{11}(t) + \dot{X}_0 \theta_{12}(t) + \tilde{B}(\tau) \right], \quad \text{where}$$

$$\tau(t) = \left(\frac{\beta}{\tilde{\nu}}\right)^2 \left[\frac{(1-\zeta^2)\, e^{2\zeta \nu t} - 1}{4 \zeta \nu} + \frac{\zeta \cos(2\tilde{\nu} t) - \sqrt{1-\zeta^2} \sin(2 \tilde{\nu} t)}{4 \nu} \right]$$

denotes the new time. The mapping from the original to the new times is deterministic in this case.

Figure 5.20 shows the relationship between the original and the new times

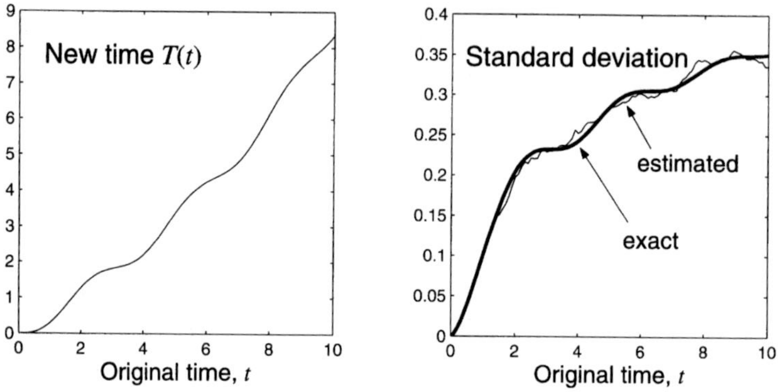

Figure 5.20. The time change and the evolution of the exact and the estimated standard deviation of X

and the evolution of the exact and estimated standard deviation of X for $\nu = 1$, $\zeta = 0.05$, and $\beta = 0.2$. The estimated standard deviation of X is based on 200 samples of this process generated in the new clock. ◇

Note: The state $\boldsymbol{X} = (X_1 = X, X_2 = \dot{X})$ is the solution of Eq. 5.38 with $d = 2$, $d' = 1$, $(\boldsymbol{a}\,\boldsymbol{X})_1 = X_2$, $(\boldsymbol{a}\,\boldsymbol{X})_2 = -\nu^2 X_1 - 2\zeta \nu X_2$, $b_1 = 0$, and $b_2 = \beta$. The process X represents the displacement of a linear oscillator with damping ratio ζ and natural frequency ν subjected to white noise.

The process M is a square integrable, continuous martingale. The new clock, $\tau(t) = [M, M](t) = \beta^2 \int_0^t e^{2\zeta \nu s} \theta_{12}(t-s)^2\, ds$, is deterministic and approaches infinity as $t \to \infty$. The time change theorem states that $M(t) = \tilde{B}(\tau)$ a.s. The samples of X used to estimate the standard deviation of this process were obtained in two steps. First, samples of $\tilde{B}$ have been generated. Second, the corresponding samples of X have been calculated from the samples of $\tilde{B}$ and the relationship between M and $\tilde{B}$. The method is more efficient than the classical Monte Carlo algorithms, which produce samples of X by integrating the differential equation for this process driven by samples of W. ▲

Example 5.23: Consider the martingale $M(t) = B(t)^2 - t$, $t \geq 0$, where B is a Brownian motion. The quadratic variation of this martingale, $T(t) = [M, M](t) =$

$4 \int_0^t B(s)^2\, ds$, converges to infinity a.s. as $t \to \infty$ so that the time change theorem can be applied. Because the new time T is random, the mapping from $\tilde{B}$ to M is complex and its use in Monte Carlo simulation may be impractical. Figure 5.21 shows 20 samples of T obtained from samples of B in the original clock and the

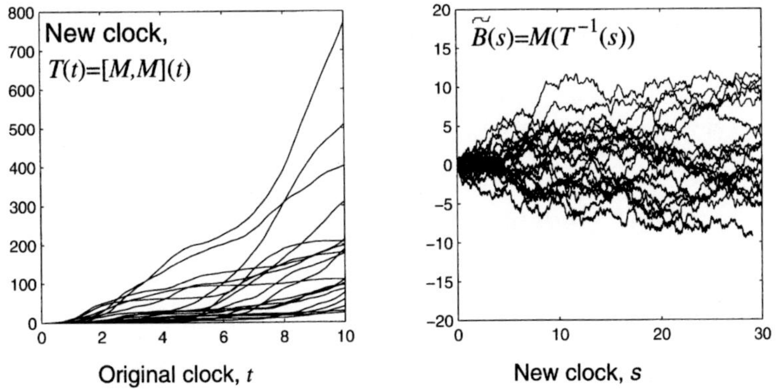

Figure 5.21. Samples of the new clock T and the Brownian motion $\tilde{B}$

corresponding samples of $\tilde{B}(s) = M(T^{-1}(s))$. The process $\tilde{B}$ is a Brownian motion in the new clock according to the time change theorem. $\diamond$

Note: Because $M(t) = 2 \int_0^t B(s)\, dB(s)$, we have (Section 4.5.3)

$$T(t) = [M, M](t) = 4 \left[\int_0^t B(s)\, dB(s), \int_0^t B(s)\, dB(s) \right] = 4 \int_0^t B(s)^2\, ds$$

so that T is a random process with increasing samples. Define the sequence of stopping times $T_1 = \inf\{t > 1 : B(t) = 0\}$, $T_2 = \inf\{t > T_1 + 1 : B(t) = 0\}, \ldots$, and note that (1) $\int_0^\infty B(s)^2\, ds = X_1 + X_2 + \cdots$, where $X_1 = \int_0^{T_1} B(s)^2\, ds$, $X_2 = \int_{T_1}^{T_2} B(s)^2\, ds, \ldots$, (2) the random variables $X_1, X_2, \ldots$ are independent and identically distributed, and (3) the expectation of X_1 is

$$E[X_1] = E\left[\int_0^{T_1} B(s)^2\, ds \right] \geq E\left[\int_0^1 B(s)^2\, ds \right] = \int_0^1 E[B(s)^2]\, ds = 1/2.$$

We have $\int_0^t B(s)^2\, ds \to \infty$ a.s. as $t \to \infty$ because $\sum_{k=1}^n X_i$ converges a.s. to infinity as $n \to \infty$ (Section 2.14 in this book, [150], Proposition 7.2.3, p. 563) and its limit is $\int_0^\infty B(s)^2\, ds$.

The time change theorem states that $M(t) = \tilde{B}([M, M](t))$ a.s., $t \geq 0$. The samples of $\tilde{B}$ in the new clock can be obtained from the recurrence formula

$$\tilde{B}(s + \Delta s, \omega) = \tilde{B}(s, \omega) + \left(M(T^{-1}(s + \Delta s, \omega), \omega) - M(T^{-1}(s, \omega), \omega) \right), \quad s \geq 0,$$

where $\Delta s > 0$ is a time increment in the new clock. ▲

Example 5.24: Let N be a Poisson process with jump times T_k, $k = 1, 2, \ldots$, and intensity $\lambda > 0$. The compensated Poisson process $\tilde{N}(t) = N(t) - \lambda t$ is a martingale with quadratic variation $[\tilde{N}, \tilde{N}](t) = [N, N](t) = N(t)$ (Section 3.12). However, the process $\tilde{B}(s) = \tilde{N}([\tilde{N}, \tilde{N}]^{-1}(s))$, $s \geq 0$, in Eq. 5.40 is not a Brownian motion since $\tilde{N}$ is not a continuous martingale as required by the time change theorem in Eq. 5.40. ◇

Proof: Let $\tilde{B}_k$ denote values of a $\tilde{B}$ at time $k\,\Delta s$, where $k \geq 0$ is an integer and $\Delta s = 1$ is a time unit in the new clock. The definition of $\tilde{B}_k$ in Eq. 5.40 gives $\tilde{B}_k = \tilde{N}(N^{-1}(k)) = k - \lambda\, T_k$ so that $\tilde{B}_k - \tilde{B}_{k-1} = 1 - \lambda\, Z$, where Z is an exponential random variable with mean $1/\lambda$. Hence, $\tilde{B}$ cannot be a Brownian motion since its increments are not Gaussian variables. ■

5.4.2 Measure change

Consider a probability space $(\Omega, \mathcal{F}, P)$ and suppose that the Monte Carlo solution of a stochastic problem defined on this space requires a large number of output samples. Our objective is to define a new probability measure Q on the measurable space $(\Omega, \mathcal{F})$ such that the Monte Carlo solution of the original stochastic problem solved in the probability space $(\Omega, \mathcal{F}, Q)$ requires fewer samples. This Monte Carlo solution is referred to as the **importance sampling** method.

5.4.2.1 Time invariant problems

Let P and Q be two probability measures on a measurable space $(\Omega, \mathcal{F})$. Recall that P is said to be absolutely continuous with respect to Q if $Q(A) = 0$ implies $P(A) = 0$, $A \in \mathcal{F}$, and we denote this property by $P \ll Q$. The measures P and Q are said to be equivalent if $P \ll Q$ and $Q \ll P$. We indicate that two measures are equivalent by the notation $P \sim Q$ (Section 2.9). The Radon-Nikodym theorem states that if $P \ll Q$ there exists a positive measurable function $g = dP/dQ : (\Omega, \mathcal{F}) \to ([0, \infty), \mathcal{B}([0, \infty)))$, called the Radon-Nikodym derivative, such that $P(A) = \int_A g(\omega)\, Q(d\omega)$, $A \in \mathcal{F}$ (Section 2.9). The examples in this section illustrate how this theorem can be used to improve the efficiency of Monte Carlo simulation.

Example 5.25: Let $(\Omega, \mathcal{F}, P)$ be a probability space and X denote a real-valued random variable defined on this space. Our objective is to estimate the probability $P(X > x)$ by Monte Carlo simulation. The estimates of $P(X > x)$ by direct Monte Carlo and importance sampling simulation are

$$\hat{p}_{MC}(x) = \frac{1}{n} \sum_{i=1}^{n} 1_{(x,\infty)}(x_i) \quad \text{and} \quad \hat{p}_{IS}(x) = \frac{1}{n} \sum_{i=1}^{n} 1_{(x,\infty)}(z_i)\, \frac{f(z_i)}{q(z_i)},$$

respectively, where x_i and z_i are independent samples generated from the densities f and q, respectively, where $f(\xi) = dP(X \leq \xi)/d\xi$, $q(\xi) = dQ(X \leq \xi)/d\xi$, and Q is a measure on $(\Omega, \mathcal{F})$ such that $P \ll Q$.

Let $X = \exp(Y)$, $Y \sim N(1, (0.2)^2)$. The exact probability $P(X > x)$ is $0.3712, 0.0211, 0.1603 \times 10^{-3}, 0.3293 \times 10^{-4}$, and 0.7061×10^{-5} for $x = 3, 5$, 8, 9, and 10, respectively. The corresponding estimates $\hat{p}_{MC}(x)$ based on 10,000 samples are 0.3766, 0.0235, 0.2×10^{-3}, 0, and 0. The estimates $\hat{p}_{IS}(x)$ based on the same number of samples are 0.3733, 0.0212, 0.1668×10^{-3}, 0.3304×10^{-4}, and 0.7084×10^{-5} for $x = 3, 5, 8, 9$, and 10, where the density $q(z) = \phi((z - x)/\sigma)/\sigma$ corresponds to a Gaussian variable with mean x and variance σ^2. While the importance sampling provides satisfactory estimates of $P(X > x)$ for levels up to $x = 10$, the estimates $\hat{p}_{\text{MC}}(x)$ are inaccurate for $x \geq 8$. ◇

Note: The required probability is

$$P(X > x) = \int_{\mathbb{R}} 1_{(x,\infty)}(\xi)\, f(\xi)\, d\xi = E_P\left[1_{(x,\infty)}(X)\right],$$

where E_P denotes the expectation operator under P. The estimate $\hat{p}_{MC}(x)$ approximates this expectation. We also have

$$P(X > x) = \int_{\mathbb{R}} \left[1_{(x,\infty)}(\xi)\, \frac{f(\xi)}{q(\xi)}\right] q(\xi)\, d\xi = E_Q\left[1_{(x,\infty)}(X)\, \frac{f(X)}{q(X)}\right],$$

where q is the density of X under Q. This density has been selected such that 50% of its samples exceed x and the ratio f/q is bounded in $\mathbb{R}$. ▲

Example 5.26: In reliability studies we need to calculate the probabilities $P_s = P(X \in D)$ and $P_f = P(X \in D^c)$, called reliability and probability of failure, where $D \subset \mathbb{R}^d$ is a safe set. The probability of failure is

$$P_f = \int_{\mathbb{R}^d} 1_{D^c}(\boldsymbol{x})\, f(\boldsymbol{x})\, d\boldsymbol{x} = E_P[1_{D^c}(\boldsymbol{X})] \quad \text{or}$$

$$P_f = \int_{\mathbb{R}^d} \left[1_{D^c}(\boldsymbol{x})\, \frac{f(\boldsymbol{x})}{q(\boldsymbol{x})}\right] q(\boldsymbol{x})\, d\boldsymbol{x} = E_Q\left[1_{D^c}(\boldsymbol{X})\, \frac{f(\boldsymbol{X})}{q(\boldsymbol{X})}\right],$$

where f and q are the densities of $\boldsymbol{X}$ under the probability measures P and Q, respectively. The measure P defines the original reliability problem. The measure Q is selected to recast the original reliability problem and is such that $P \ll Q$. The estimates of P_f by direct Monte Carlo and importance sampling are denoted by $\hat{p}_{f,MC}$ and $\hat{p}_{f,IS}$, respectively (Example 5.25).

Let $\boldsymbol{X}$ have independent $N(0, 1)$ coordinates and D be a sphere of radius $r > 0$ centered at the origin of $\mathbb{R}^d$, that is, $D = \{\boldsymbol{x} \in \mathbb{R}^d : \| \boldsymbol{x} \| \leq r\}$. Table 5.1 lists the exact and estimated probabilities of failure for $d = 10$, and several values of r. The estimates of P_f are based on 10,000 samples. The density q corresponds to an $\mathbb{R}^d$-valued Gaussian variable with mean $(r, 0, \ldots, 0)$ and covariance matrix $\sigma^2 \boldsymbol{i}$, where $\sigma > 0$ is a constant and $\boldsymbol{i}$ denotes the identity matrix. The direct

Table 5.1. Exact and estimated failure probabilities ($d = 10$)

r	σ	P_f	$\hat{p}_{f,IS}$	$\hat{p}_{f,MC}$
5	0.5	0.053	0.0	0.0053
	1		0.0009	
	2		0.0053	
	3		0.0050	
6	0.5	0.8414×10^{-4}	0.0001×10^{-4}	0.0
	1		0.1028×10^{-4}	
	2		0.5697×10^{-4}	
	3		1.1580×10^{-4}	
	4		1.1350×10^{-4}	
7	0.5	0.4073×10^{-6}	0.0	0.0
	1		0.0016×10^{-6}	
	2		0.1223×10^{-6}	
	3		0.6035×10^{-6}	
	4		0.4042×10^{-6}	

Monte Carlo method is inaccurate for relatively large values of r, that is, small probabilities of failure. The success of the importance sampling method depends on the density q. For example, $\hat{p}_{f,IS}$ is in error for $d = 10$ and $\sigma = 0.5$ but becomes accurate if σ is increased to 3 or 4. ◇

Note: The exact reliability,

$$1 - P_f = P\left(\sum_{i=1}^{d} X_i^2 \leq r^2\right) = \frac{1}{\Gamma(d/2)} \int_0^{r^2/2} \xi^{d/2-1} e^{\xi} \, d\xi,$$

is an incomplete gamma function, where $\Gamma(\cdot)$ denotes the gamma function. The densities of $\boldsymbol{X}$ under the measures P and Q are

$$f(\boldsymbol{x}) = (2\pi)^{-d/2} \exp\left[-\frac{1}{2}\sum_{i=1}^{d} x_i^2\right] \quad \text{and}$$

$$q(\boldsymbol{z}) = [(2\pi)\,\sigma^2]^{-d/2} \exp\left\{-\frac{1}{2\sigma^2}\left[(z_1 - r)^2 + \sum_{i=2}^{d} z_i^2\right]\right\}.$$

The estimate of P_f by the importance sampling method is

$$\hat{p}_{f,IS} = \frac{1}{n}\sum_{k=1}^{n} 1_{D^c}(\boldsymbol{z}_k)\,\frac{f(\boldsymbol{z}_k)}{q(\boldsymbol{z}_k)},$$

where $\boldsymbol{z}_k$ are independent samples of $\boldsymbol{X}$ under Q. ▲

5.4.2.2 Time variant problems

The Girsanov theorem can be used to define a new measure Q that is absolutely continuous with respect to the original measure P and provides the framework for a more efficient Monte Carlo solution of the original stochastic problem. We state the Girsanov theorem without proof and illustrate its use in Monte Carlo simulation by examples.

Let $(\Omega, \mathcal{F}, (\mathcal{F}_t)_{t\geq 0}, P)$ be a filtered probability space and let $\boldsymbol{B}$ be an $\mathbb{R}^{d'}$-valued Brownian motion on this space. Let $\boldsymbol{X}$ be an $\mathbb{R}^d$-valued diffusion process defined by

$$\boldsymbol{X}(t) = \boldsymbol{x} + \int_0^t \boldsymbol{a}(\boldsymbol{X}(s), s)\, ds + \int_0^t \boldsymbol{b}(\boldsymbol{X}(s), s)\, \boldsymbol{B}(s), \tag{5.41}$$

where $\boldsymbol{X}(0) = \boldsymbol{x}$ denotes the initial state and the coefficients $\boldsymbol{a}, \boldsymbol{b}$ are such that the solution of this equation exists and is unique (Section 4.7.1.1).

We need some definitions to state the Girsanov theorem. Let

$$\boldsymbol{g}(\boldsymbol{x}, t) = \boldsymbol{a}(\boldsymbol{x}, t) + \boldsymbol{b}(\boldsymbol{x}, t)\, \boldsymbol{\gamma}(\boldsymbol{x}, t), \tag{5.42}$$

where $\boldsymbol{\gamma}(\boldsymbol{X}(t), t) \in \mathbb{R}^{d'}$ is a bounded, measurable function that is $\mathcal{F}_t$-adapted with càglàd samples. Define an $\mathbb{R}^d$-valued stochastic process $\boldsymbol{Z}$ by

$$\boldsymbol{Z}(t) = \boldsymbol{Z}(0) + \int_0^t \boldsymbol{g}(\boldsymbol{Z}(s), s)\, ds + \int_0^t \boldsymbol{b}(\boldsymbol{Z}(s), s)\, d\tilde{\boldsymbol{B}}(s), \quad t \in [0, \tau], \tag{5.43}$$

that starts as $\boldsymbol{X}$ at $\boldsymbol{Z}(0) = \boldsymbol{X}(0) = \boldsymbol{x}$, where $\tau > 0$ and

$$\tilde{\boldsymbol{B}}(t) = \boldsymbol{B}(t) - \int_0^t \boldsymbol{\gamma}(\boldsymbol{X}(s), s)\, ds. \tag{5.44}$$

Define also a new probability measure Q by

$$\left(\frac{dQ}{dP}\right)(t) = \exp\left[\int_0^t \boldsymbol{\gamma}(\boldsymbol{X}(s), s)^T\, d\boldsymbol{B}(s) - \frac{1}{2}\int_0^t \sum_{r=1}^{d'} \gamma_r(\boldsymbol{X}(s), s)^2\, ds\right], \tag{5.45}$$

where $t \in [0, \tau]$.

Girsanov's theorem. This theorem gives the following properties that are essential for Monte Carlo simulation ([147], pp. 108-114):

1. The process $\tilde{\boldsymbol{B}}$ in Eq. 5.44 is a Brownian motion under Q.

2. The solution $\boldsymbol{X}$ of Eq. 5.41 under P satisfies also Eq. 5.43 under Q.

Note: The stochastic equations for $\boldsymbol{X}$ and $\boldsymbol{Z}$ in Eqs. 5.41 and 5.43 (1) have the same diffusion coefficients but different drift coefficients and (2) are driven by Brownian motions under the measures P and Q, respectively. The drift of $\boldsymbol{Z}$ can be obtained from the drift of $\boldsymbol{X}$ and Eq. 5.42. The sample paths of $\boldsymbol{X}$ under P and $\boldsymbol{Z}$ under Q differ. However, the processes $\boldsymbol{X}$ and $\boldsymbol{Z}$ have the same probability law. ▲

The Girsanov theorem can be used to improve the efficiency and accuracy of the direct Monte Carlo method. For example, let X denote the state of a system that fails if X leaves a safe set D in a time interval $[0, \tau]$. Suppose that our objective is to estimate the probability of failure $P_f(\tau)$ of this system by Monte Carlo simulation. The classical Monte Carlo method generates n samples of X, count the number of samples n_f of X that leave D at least once in $[0, \tau]$, and estimate $P_f(\tau)$ by n_f/n. This solution can be inefficient in applications because (1) the estimates of $P_f(\tau)$ have to be based on a very large number of samples of X for highly reliable systems, that is, systems with $P_f(\tau) \ll 1$, and (2) the computation time for obtaining a single sample of X can be significant for realistic systems. The Girsanov theorem provides a procedure for modifying the drift of X if considered under a new probability Q. If a significant number of samples X under Q cause system failure, then the number n of samples needed to estimate $P_f(\tau)$ can be reduced significantly relative to the classical Monte Carlo method based on the definition of X under the original probability measure P. The following algorithm outlines the estimation of $P_f(\tau)$ based on samples of X under Q.

Monte Carlo algorithm. The estimation of the failure probability $P_f(\tau)$ involves three steps:

1. Select the function γ in Eq. 5.42 such that the fraction of samples of X exiting D at least once during $[0, \tau]$ is relatively large.

2. Generate samples $z_i(t)$ of X under Q and of the Radon-Nikodym derivative dQ/dP, $i = 1, \ldots, n$, for $t \in [0, \tau]$.

3. Estimate the failure probability $P_f(\tau)$ by (Girsanov's theorem)

$$\hat{p}_{f,IS}(\tau) = \frac{1}{n} \sum_{i=1}^{n} 1_{\{z_i(t) \in D,\ 0 \le t \le \tau\}^c} \left(\frac{dP}{dQ}\right)_i (\tau). \tag{5.46}$$

Note: The success of the Monte Carlo simulation algorithm in Eq. 5.46 depends essentially on the choice of the function γ in Eq. 5.42. Unfortunately, there exists no simple method for selecting γ. The following two examples are presented to clarify the statement of the Girsanov theorem. Estimates of $P_f(\tau)$ based on this theorem are given in Examples 5.29 and 5.30. These examples also present some practical ideas for selecting an adequate expression for γ in Eq. 5.42. ▲

Example 5.27: Let $(\Omega, \mathcal{F}, (\mathcal{F}_t)_{t \ge 0}, P)$ be a filtered probability space and X be a process on this space defined by $dX(t) = a\,dt + b\,dB(t)$, $t \in [0, \tau]$, where $\tau > 0$, $X(0) = x$, B is a Brownian motion under P, and a, b are constants. Let Z be another process on the measurable space $(\Omega, \mathcal{F})$ satisfying the equation $dZ(t) = b\,d\tilde{B}(t)$, $t \in [0, \tau]$, where $Z(0) = x$ and $\tilde{B}$ is a Brownian motion with respect to a probability measure Q defined by $(dQ/dP)(t) = \exp[-(a/b)\,B(t) - (a^2/(2b^2))\,t]$. Then X and Z have the same law. The expectation of $X(t)$ and the probability of the event $A = \{X(t) \le \xi\}$, $\xi \in \mathbb{R}$, are $x + a\,t$ and $\Phi((\xi - x -$

$a\,t)/b\,\sqrt{t})$, respectively, and can be calculated from X under P or from Z, that is, X under the probability measure Q. ◇

Proof: Take $\gamma = -a/b$ so that $g = 0$ (Eq. 5.42) and (Eq. 5.45),

$$\left(\frac{dQ}{dP}\right)(t) = \exp\left[\int_0^t (-a/b)\,dB(s) - \frac{1}{2}\int_0^t (-a/b)^2\,ds\right]$$
$$= \exp\left[-(a/b)\,\tilde{B}(t) + a^2\,t/(2\,b^2)\right]$$

since $\tilde{B}(t) = B(t) - \int_0^t (-a/b)\,ds = B(t) + a\,t/b$ (Eq. 5.44) and $Z(t) = x + b\,\tilde{B}(t)$ (Eq. 5.43).

We have $E_P[X(t)] = E_P[x + a\,t + b\,B(t)] = x + a\,t$ under P. This expectation is also

$$E_Q\left[Z(t)\,\frac{dP}{dQ}(t)\right] = E_Q[(x + b\,\tilde{B}(t))\,\exp(a\,\tilde{B}(t)/b - a^2\,t/(2\,b^2))]$$
$$= e^{-a^2\,t/(2\,b^2)}\left(x\,E_Q[\exp(a\,\tilde{B}(t)/b)] + b\,E_Q[\tilde{B}(t)\,\exp(a\,\tilde{B}(t)/b)]\right) = x + a\,t$$

since $\tilde{B}(t) \stackrel{d}{=} \sqrt{t}\,G$, $G \sim N(0,1)$, under Q, and $E[\exp(u\,G)] = \exp(u^2/2)$, $u \in \mathbb{R}$. Similar considerations can be used to calculate higher order moments of $X(t)$.

The probability of $A = \{X(t) \le \xi\}$ is as stated since $X(t) \sim N(x + a\,t, b^2\,t)$ under P. We also have

$$P(A) = \int_A \left(\frac{dP}{dQ}\right)(t)\,dQ(t) = e^{-a^2\,t/(2\,b^2)}\int_A \exp(a\,\tilde{B}(t)/b)\,dQ(t)$$
$$= e^{-a^2\,t/(2\,b^2)}\int_{(-\infty,(\xi-x)/b]} e^{a\,y/b}\,\phi(y/\sqrt{t})/\sqrt{t}\,dy = \Phi\left(\frac{\xi - x - a\,t}{b\,\sqrt{t}}\right),$$

where Φ denotes the distribution of $N(0,1)$. ■

Example 5.28: Let X be defined by $dX(t) = a\,X(t)\,dt + b\,X(t)\,dB(t)$ under a probability P. Let Z be the solution of $dZ(t) = b\,Z(t)\,d\tilde{B}(t)$ under a probability Q given by $(dQ/dP)(t) = \exp(-(a/b)\,\tilde{B}(t) + a^2\,t/(2\,b^2))$. Then X and Z have the same law. ◇

Proof: The Radon-Nikodym derivative in Eq. 5.45 with $\gamma = -a/b$ gives the above expression of dQ/dP. The process Z is a geometric Brownian motion equal to $Z(t) = x\,\exp(-0.5\,b^2\,t + b\,\tilde{B}(t))$ for $Z(0) = x$. ■

Example 5.29: Let X be the process in Example 5.27 with $X(0) = x = 0$. Our objective is to estimate $P_f(\tau) = P(\max_{0\le t\le\tau} X(t) > x_{cr})$. If X is a performance index and x_{cr} denotes a critical level for a system, then $P_f(\tau)$ is the probability of failure in $[0, \tau]$.

Let Z be a process defined by $dZ(t) = (a + b\,\gamma)\,dt + b\,d\tilde{B}(t)$, where $\tilde{B}$ is a Brownian motion under a probability measure Q defined by Eq. 5.45 and

γ is a constant. The estimates of $P_f(\tau)$ delivered by the classical Monte Carlo simulation and the importance sampling methods are

$$\hat{p}_{f,MC}(\tau) = \frac{1}{n} \sum_{i=1}^{n} 1_{(x_{cr},\infty)} \left(\max_{0 \le t \le \tau} \{x_i(t)\} \right) \quad \text{and}$$

$$\hat{p}_{f,IS}(\tau) = \frac{1}{n} \sum_{i=1}^{n} 1_{(x_{cr},\infty)} \left(\max_{0 \le t \le \tau} \{z_i(t)\} \right) \left(\frac{dP}{dQ} \right)_i (\tau),$$

respectively, where x_i and z_i denote independent samples of X and Z. If $a = 2$, $b = 2$, and $x_{cr} = 6$, the exact value of $P_f(\tau)$ is 0.1316×10^{-11} and 0.0098 for $\tau = 0.5$ and $\tau = 1.5$, respectively. The estimate $\hat{p}_{f,MC}(\tau)$ is zero for sample sizes smaller than 1,000. On the other hand, the estimates $\hat{p}_{f,IS}(\tau)$ based on 500 samples and Z defined by $dZ(t) = (a + b\,\gamma)\,dt + b\,d\tilde{B}(t)$ are 0.1311×10^{-11}, 0.1460×10^{-11}, 0.1023×10^{-11}, and 0.0919×10^{-11} for $\gamma = 6, 8, 10$, and 12 if $\tau = 0.5$ and 0.0099, 0.0092, 0.0085, and 0.0077 for $\gamma = 1, 1.5, 2.0$, and 2.5 if $\tau = 1.5$. The latter estimates become 0.0079, 0.0088, 0.0084, and 0.0085 for $\gamma = 1, 1.5, 2.0$, and 2.5 if based on 1,000 independent samples of Z. ◇

Proof: The exact probability of failure is ([111], Section 15.4)

$$P_f(\tau) = \int_0^\tau \frac{x_{cr} - x}{b\,(2\pi\,t^3)^{1/2}} \exp\left[-\frac{1}{2\,b^2\,t} (x_{cr} - x - a\,t)^2 \right] dt.$$

Because the probability that the samples of X reach x_{cr} in $[0, \tau]$ is very low, a large number of samples of X is needed to estimate $P_f(\tau)$ by $\hat{p}_{f,MC}$, for example, about $10/10^{-11} = 10^{12}$ samples for $\tau = 0.5$. This number of samples can be reduced significantly if we modify the drift coefficient. In this example we have changed the drift such that it reaches x_{cr} at time τ, that is, $(a + b\,\gamma)\,\tau = x_{cr}$ so that $\gamma = (x_{cr}/\tau - a)/b$. For $x_{cr} = 6$, $a = 2$, and $b = 1$, we have $\gamma = 10$ and $\gamma = 2$ for $\tau = 0.5$ and $\tau = 1.5$, respectively. Samples $(dQ/dP)_i$ of the Radon-Nikodym derivative dQ/dP can be calculated from

$$\left(\frac{dQ}{dP} \right)(t) = \exp[\gamma\,\tilde{B}(t) + \gamma^2\,t/2],$$

and samples of $\tilde{B}$. Alternative forms for γ can be found in [181]. ■

Example 5.30: Let X be an Ornstein-Uhlenbeck process defined by

$$dX(t) = -\alpha\,X(t)\,dt + \sqrt{2\,\alpha}\,dB(t), \quad t \ge 0,$$

for $X(0) = x$ and $\alpha > 0$. Let $\hat{p}_{f,MC}(\tau)$ and $\hat{p}_{f,IS}(\tau)$ be the estimates of $P_f(\tau)$ in Example 5.29, where Z is defined by

$$dZ(t) = \left(-\alpha\,Z(t) + \gamma\,\sqrt{2\,\alpha} \right) dt + \sqrt{2\,\alpha}\,d\tilde{B}(t), \quad t \ge 0,$$

$Z(0) = x$, and $\tilde{B}$ is a Brownian motion under Q.

If $\alpha = 1$, $x_{cr} = 2.5$, and $\tau = 1$, then $\hat{p}_{f,MC}(\tau)$ is zero for sample sizes $n = 500$, 1,000, 5,000, and 10,000, and 0.5×10^{-4} for $n = 100,000$ samples. The corresponding values of $\hat{p}_{f,IS}(\tau)$ with $\gamma = 4$ are 0.2340×10^{-4}, 0.3384×10^{-4}, 0.3618×10^{-4}, 0.3402×10^{-4}, and 0.3981×10^{-4} for $n = 500$, 1,000, 5,000, 10,000, and 100,000 samples, respectively.

Figure 5.22 shows ten samples of X under the measure P and the corresponding samples of this process under the measure Q. Because of the drift correction obtained by Girsanov's theorem, the samples of X under Q are more likely to exceed x_{cr}. This property of the samples of X under Q results in superior estimates of $P_f(\tau)$. ◇

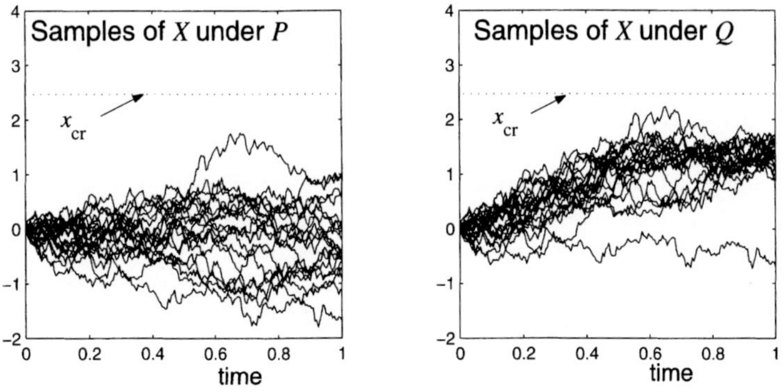

Figure 5.22. Ten samples of X under P and under Q

Proof: As in the previous example we have modified the drift such that it reaches x_{cr} at time τ. This condition gives $\gamma = (x_{cr} - x\, e^{-\alpha\,\tau})/[\sqrt{2/\alpha}\,(1 - e^{-\alpha\,\tau})]$ so that $\gamma = 4.47$ for the values considered. The Radon-Nikodym derivative $(dQ/dP)(t)$ is equal to $\exp[\gamma\,\tilde{B}(t) + \gamma^2\,t/2]$.

The measure change used in this example has been established by elementary considerations and is not optimal in the sense that the number of samples of X under the above measure Q that exceed x_{cr} in $[0, \tau]$ is relatively small (Fig. 5.22). Alternative measures Q can be selected such that approximately 50% of the samples of X under Q exceed x_{cr} in $[0, \tau]$. The use of such measures would further increase the efficiency of the Monte Carlo algorithm for estimating $P_f(\tau)$. ■

5.5 Problems

5.1: Let $\boldsymbol{X} \sim N(\boldsymbol{\mu}, \boldsymbol{\gamma})$ be a Gaussian vector in $\mathbb{R}^d$. Develop an algorithm for generating samples of $\boldsymbol{X}$ using the fact that conditional Gaussian vectors are Gaussian vectors.

5.2: Prove Eqs. 5.5 and 5.6.

5.3: Use Eq. 5.9 to develop a Monte Carlo algorithm for generating samples of the translation vector considered in Example 5.6.

5.4: Complete calculations in the text giving the covariance function of $\boldsymbol{X}^{(n)}$ in Eq. 5.12.

5.5: Complete the details of the proof showing that $\boldsymbol{X}^{(n)}$ in Eq. 5.12 becomes a version of $\boldsymbol{X}$ as $\Delta(p_n) \to 0$ and $\nu^* \to \infty$.

5.6: Complete the details of the proof showing that $X^{(\lambda)}$ in Eq. 5.13 becomes equal to X in the second moment sense as $\nu^* \to \infty$ and that $X^{(\lambda)}$ becomes a Gaussian process as $\lambda \to \infty$.

5.7: Show that the second moment properties of the compound Poisson processes C_p, $p = 1, 2$, used to define the process $X^{(\lambda)}$ in Eq. 5.17 are as stated. Find the second moment properties of the increments $dC_p(\boldsymbol{\nu})$ of these processes.

5.8: Suppose that X is a stationary Gaussian process with mean zero and one-sided spectral density $g(\nu) = (1/\nu^*)\, 1_{[0,\nu^*]}(\nu)$, $\nu \geq 0$, where $\nu^* > 0$ is a constant. Calculate the mean square error $e = E[(X(t) - X^{(n)}(t))]$ of the approximate representation $X^{(n)}$ of X given by Eqs. 5.12 and 5.19.

5.9: Complete the details of the proof showing that $X^{(\boldsymbol{n})}$ in Eq. 5.22 becomes a version of X as $n_\sigma \to \infty$, $\sigma = 1, \ldots, d'$.

5.10: Redo Example 5.14 with an alternative periodic extension of the covariance function of X such that $c^{(n,n)}(\tau, \tau)$ is approximately equal to $c(\tau, \tau)$.

5.11: Prove Eq. 5.29. Find also the finite dimensional density of $\boldsymbol{X}$ for $d > 1$.

5.12: Prove the convergence of a binomial distribution to a Poisson distribution under the conditions stated following Eq. 5.34.

5.13: Show that the points of $N \mid N(B)$ are uniformly distributed in B, where N is a Poisson random measure defined on $\mathbb{R}^{d'}$ and $B \in \mathcal{B}(\mathbb{R}^{d'})$ is bounded.

5.14: Repeat the analysis in Example 5.22 for the case in which X is an Ornstein-Uhlenbeck process.

5.15: Calculate moments of order 3 and higher of X in Example 5.27 using the definition of this process under the probability measure Q.

5.16: Find the density of the random variable $T = \inf\{t \geq 0 : X(t) = x\}$, $x > 0$, by using the change of measure in Example 5.27 and the distribution of the first time a Brownian motion B starting at $B(0) = 0$ reaches x (Section 3.4).

5.17: Extend the analysis in Example 5.30 to a process X defined by the stochastic differential equation $dX(t) = (\alpha\, X(t) + \beta\, X(t)^3)\, dt + \sigma\, dB(t)$, where α, β, σ are some constants and B denotes a Brownian motion.

Chapter 6

Deterministic Systems and Input

6.1 Introduction

The current state of a system is commonly given by the solution of a deterministic differential, algebraic, or integral equation. For example, the functions giving the displacements of the points of a solid satisfy partial differential equations obtained from equilibrium conditions, kinematic constraints, and material constitutive laws. Generally, it is not possible to solve analytically the equations defining the system state. Numerical methods are needed for their solution. Most numerical methods are **global**, that is, they determine the solution everywhere or at a large number of points in a system even if the solution is needed at a single point.

This chapter develops alternative methods for solving deterministic differential, algebraic, and integral equations that have two common features.

(1) The methods are **local**, that is, they give directly the value of the solution of a differential or integral equation at an arbitrary point in the set on which the solution is defined, rather than extracting its value from the field solution. For algebraic equations, the methods give the value of a particular unknown directly.

(2) The methods employ **Monte Carlo simulation.** Local solutions can be expressed as averages of functions of some stochastic processes that can be estimated by Monte Carlo simulation algorithms. A MATLAB function for finding the local solution of a partial differential equation is given in Section 6.2.1.3 for illustration.

We consider the following classes of differential, algebraic, and integral equations and develop methods for finding their local solutions.

Differential equation (Sections 6.2 and 6.3):

$$\frac{\partial u(\boldsymbol{x},t)}{\partial t} = \sum_{i=1}^{d} \alpha_i(\boldsymbol{x}) \frac{\partial u(\boldsymbol{x},t)}{\partial x_i} + \frac{1}{2} \sum_{i,j=1}^{d} \beta_{ij}(\boldsymbol{x}) \frac{\partial^2 u(\boldsymbol{x},t)}{\partial x_i \, \partial x_j}$$
$$+ q(\boldsymbol{x},t)\, u(\boldsymbol{x},t) + p(\boldsymbol{x},t), \quad (\boldsymbol{x},t) \in D \times (0,\infty), \tag{6.1}$$

where D is an open subset of $\mathbb{R}^d$, α_i, β_{ij} are real-valued functions defined on $D \subset \mathbb{R}^d$, $d \geq 1$ is an integer, and q, p denote real-valued functions defined on $D\times[0,\infty)$. The solution $u : D\times(0,\infty) \to \mathbb{R}$ of Eq. 6.1 depends on boundary and initial conditions. This equation is meaningful if the first order partial derivative and the second order partial derivatives of u with respect to t and $\boldsymbol{x}$, respectively, are continuous. If the functions q and p depend on only $\boldsymbol{x}$ and $\partial u/\partial t = 0$, Eq. 6.1 is said to be **time-invariant**.

In Section 6.4 we consider a class of partial differential equations that define linear elasticity problems and differ from Eq. 6.1. The local solutions developed in Sections 6.2 and 6.3 for Eq. 6.1 cannot be applied to elasticity problems. An alternative solution based on Neumann series representations will be used to solve locally some elasticity problems.

Algebraic equation (Section 6.5):

$$(\boldsymbol{i} - \lambda\, \boldsymbol{g})\boldsymbol{u} = \boldsymbol{\phi}, \tag{6.2}$$

where $\boldsymbol{i}$ is the (m,m) identity matrix, $\boldsymbol{g}$ is an (m,m) matrix, $\boldsymbol{\phi}$ and $\boldsymbol{u}$ denote m-dimensional vectors, and λ is a parameter. It is assumed that the entries of $\boldsymbol{g}$ and $\boldsymbol{\phi}$ are real-valued. If $\boldsymbol{\phi} \neq \boldsymbol{0}$, λ has a fixed value, and $\det(\boldsymbol{i} - \lambda\, \boldsymbol{g}) \neq 0$, Eq. 6.2 has a unique solution. If $\boldsymbol{\phi} = \boldsymbol{0}$, Eq. 6.2 defines an eigenvalue problem for matrix $\boldsymbol{g}$.

Integral equation (Section 6.6):

$$u(\boldsymbol{x}) = \phi(\boldsymbol{x}) + \lambda \int_D g(\boldsymbol{x},\boldsymbol{y})\, u(\boldsymbol{y})\, d\boldsymbol{y}, \quad \boldsymbol{x}, \boldsymbol{y} \in D, \tag{6.3}$$

where D is an open subset of $\mathbb{R}^d$ and λ is a constant. If $\phi = 0$, this Fredholm integral equation defines an eigenvalue problem for the kernel g.

The emphasis in this chapter is on local solutions for Eq. 6.1. Three methods for obtaining local solutions are considered, the **random walk method** (RWM), the **sphere walk method** (SWM), and the **boundary walk method** (BWM). Of these three local methods, the RWM is discussed in detail. The method is a direct consequence of properties of diffusion processes and the Itô formula. The RWM, SWM, and RWM provide an alternative to traditional computation methods, for example, the finite element, boundary element, and finite difference methods, used currently to obtain numerical solutions of partial differential equations.

Some possible limitations of the traditional computational methods are: (1) the computer codes used for solution are relatively complex and may require extensive preprocessing to define a particular problem in a specified format, (2) the numerical algorithms may become unstable in some cases, (3) the errors caused by the discretization of the domain of integration and the numerical integration methods used in analysis cannot be bounded, and (4) the field solution must be calculated even if the solution is needed at a single point or a small collection of points in D. In contrast, numerical algorithms based on the random walk method are simple to program, always stable, accurate, local, and ideal for parallel computation. However, the local solutions are less general than those using the traditional computational methods.

6.2 Random walk method

We consider the local solution of Eq. 6.1 with the **initial condition**

$$u(\boldsymbol{x}, 0) = \eta(\boldsymbol{x}), \quad \boldsymbol{x} \in D, \tag{6.4}$$

giving the values of function u in D at the initial time $t = 0$. The boundary conditions for Eq. 6.1 specify the values of u and/or its derivatives on the boundary ∂D of D for $t > 0$. If derivatives of u are specified on a subset ∂D_n of ∂D and values of u are given on $\partial D_d = \partial D \setminus \partial D_n$, the problem is said to have **mixed boundary conditions**. The subsets ∂D_d and ∂D_n are referred to as **Dirichlet** and **Neumann boundaries**. The conditions that the solution must satisfy on the boundaries ∂D_d and ∂D_n are called **Dirichlet** and **Neumann conditions**, respectively.

The partial differential equation in Eq. 6.1 has numerous applications in science and engineering. For example:

1. The **beam** equation in bending

$$\frac{d^2 u(x)}{dx^2} + p(x) = 0, \quad x \in D = (0, l), \tag{6.5}$$

giving the displacement function u of a beam of span l is a special case of the time-invariant version of Eq. 6.1 with $d = 1$, $u(\boldsymbol{x}, t) = u(x)$, $\alpha_1 = 0$, $\beta_{11}(x) = 2$, $q = 0$, and $p(x, t) = p(x)$.

2. The **Poisson** equation

$$\Delta u(\boldsymbol{x}) + p(\boldsymbol{x}) = 0, \quad \boldsymbol{x} \in D, \tag{6.6}$$

corresponds to the time-invariant solution $u(\boldsymbol{x}, t) = u(\boldsymbol{x})$ of Eq. 6.1 with $\alpha_i(\boldsymbol{x}) = 0$, $\beta_{ij}(\boldsymbol{x}) = 2\,\delta_{ij}$, $i, j = 1, \ldots, d$, $q = 0$, and $p(\boldsymbol{x}, t) = p(\boldsymbol{x})$, where $\Delta = \sum_{i=1}^{d} \partial^2/\partial x_i^2$ is the Laplace operator, $\delta_{ij} = 1$ for $i = j$, and $\delta_{ij} = 0$ for $i \neq j$. If $p = 0$, Eq. 6.6 becomes $\Delta u(\boldsymbol{x}) = 0$, and is referred to as the **Laplace** equation.

3. The **heat, transport**, and **diffusion** equations for heterogeneous media are given by Eq. 6.1 with $q = 0$, where p has the meaning of flux. The heat equation for a homogeneous medium with diffusion coefficient $\alpha > 0$ is

$$\frac{\partial u(\boldsymbol{x}, t)}{\partial t} = \alpha^2 \, \Delta u(\boldsymbol{x}, t) + p(\boldsymbol{x}, t), \tag{6.7}$$

and is given by Eq. 6.1 with $\alpha_i(\boldsymbol{x}) = 0$, $\beta_{ij}(\boldsymbol{x}) = 2\,\alpha^2\,\delta_{ij}$, and $q = 0$.

4. The **Schrödinger** equation for a heterogeneous medium is

$$\sum_{i=1}^{d} \alpha_i(\boldsymbol{x}) \frac{\partial u(\boldsymbol{x})}{\partial x_i} + \frac{1}{2} \sum_{i,j=1}^{d} \beta_{ij}(\boldsymbol{x}) \frac{\partial^2 u(\boldsymbol{x})}{\partial x_i \, \partial x_j} + q(\boldsymbol{x})\, u(\boldsymbol{x}) + p(\boldsymbol{x}) = 0, \tag{6.8}$$

and corresponds to the time-invariant solution $u(\boldsymbol{x}, t) = u(\boldsymbol{x})$ of Eq. 6.1 for $q(\boldsymbol{x}, t) = q(\boldsymbol{x})$ and $p(\boldsymbol{x}, t) = p(\boldsymbol{x})$. The Schrödinger equation for a homogeneous medium is

$$\frac{1}{2}\,\Delta u(\boldsymbol{x}) + q(\boldsymbol{x})\, u(\boldsymbol{x}) + p(\boldsymbol{x}) = 0, \tag{6.9}$$

and can be obtained from Eq. 6.8 by setting $\alpha_i(\boldsymbol{x}) = 0$ and $\beta_{ij}(\boldsymbol{x}) = \delta_{ij}$. The above Schrödinger equations are referred to as inhomogeneous and homogeneous if $p \neq 0$ and $p = 0$, respectively.

6.2.1 Dirichlet boundary conditions ($q = 0$)

Let D be an open subset of $\mathbb{R}^d$ and let ξ be a specified real-valued function defined on the boundary ∂D of D. Denote by u the solution of Eq. 6.1 with $q = 0$, the initial condition in Eq. 6.4, and the **Dirichlet boundary condition**

$$u(\boldsymbol{x}, t) = \xi(\boldsymbol{x}, t), \quad \boldsymbol{x} \in \partial D, \quad t > 0. \tag{6.10}$$

Note: The boundary condition for the time-invariant version of Eq. 6.1 is $u(\boldsymbol{x}) = \xi(\boldsymbol{x})$, $\boldsymbol{x} \in \partial D$. The real-valued function ξ giving the boundary values of u in this case differs from ξ in Eq. 6.10. We use the same notation for simplicity. ▲

6.2.1.1 Local solution

We establish the local solution of Eq. 6.1 with the initial and boundary conditions in Eqs. 6.4 and 6.10 under the following assumptions.

1. The entries of the matrix $\boldsymbol{\beta}(\boldsymbol{x}) = \{\beta_{ij}(\boldsymbol{x})\}$ are real-valued functions and $\boldsymbol{\beta}$ is a symmetric positive definite matrix for each $\boldsymbol{x} \in \mathbb{R}^d$, that is, the eigenvalues of $\boldsymbol{\beta}(\boldsymbol{x})$ are strictly positive for all $\boldsymbol{x} \in D$. Hence, there exists a matrix $\boldsymbol{b}$ such that $\boldsymbol{\beta}(\boldsymbol{x}) = \boldsymbol{b}(\boldsymbol{x})\,\boldsymbol{b}(\boldsymbol{x})^T$ for each $\boldsymbol{x} \in D$.

Note: There are several methods for calculating a matrix $\boldsymbol{b}$ satisfying the condition $\boldsymbol{\beta}(\boldsymbol{x}) = \boldsymbol{b}(\boldsymbol{x})\,\boldsymbol{b}(\boldsymbol{x})^T$ for a fixed $\boldsymbol{x} \in \mathbb{R}^d$. The Cholesky factorization provides such a representation for $\boldsymbol{\beta}$ (Section 5.2.1 in this book, [155], Section 3.5.3). An alternative solution is given by the equality $\boldsymbol{\varphi}(\boldsymbol{x})^T\,\boldsymbol{\beta}(\boldsymbol{x})\,\boldsymbol{\varphi}(\boldsymbol{x}) = \boldsymbol{\lambda}(\boldsymbol{x})$, where $\boldsymbol{\varphi}(\boldsymbol{x})$ is a (d, d) matrix whose columns are the eigenvectors of $\boldsymbol{\beta}(\boldsymbol{x})$ and $\boldsymbol{\lambda}(\boldsymbol{x})$ is a diagonal matrix whose non-zero entries are the eigenvalues of $\boldsymbol{\beta}(\boldsymbol{x})$. If the eigenvalues of $\boldsymbol{\beta}(\boldsymbol{x})$ are distinct, then we can take $\boldsymbol{b}(\boldsymbol{x}) = \boldsymbol{\varphi}(\boldsymbol{x})\,\boldsymbol{\mu}(\boldsymbol{x})$, where $\boldsymbol{\mu}(\boldsymbol{x})$ is a diagonal matrix with $\mu_{ii}(\boldsymbol{x}) = \sqrt{\lambda_{ii}(\boldsymbol{x})}$. ▲

2. Set $a_i(\boldsymbol{x}) = \alpha_i(\boldsymbol{x})$, $i = 1, \ldots, d$, and $\boldsymbol{b}$ such that $\boldsymbol{\beta}(\boldsymbol{x}) = \boldsymbol{b}(\boldsymbol{x})\,\boldsymbol{b}(\boldsymbol{x})^T$ for all $\boldsymbol{x} \in \mathbb{R}^d$. Let $\boldsymbol{B}$ be an $\mathbb{R}^d$-valued Brownian motion and let $\tilde{\boldsymbol{X}} = (\boldsymbol{X}, X_{d+1})$ be an $\mathbb{R}^{d+1}$-valued diffusion process defined by the stochastic differential equation

$$\begin{aligned} d\boldsymbol{X}(s) &= \boldsymbol{a}(\boldsymbol{X}(s))\,ds + \boldsymbol{b}(\boldsymbol{X}(s))\,d\boldsymbol{B}(s), \\ dX_{d+1}(s) &= -ds. \end{aligned} \tag{6.11}$$

It is assumed that the drift and diffusion coefficients of $\tilde{\boldsymbol{X}}$ satisfy the conditions in Section 4.7.1.1 so that the solution $\tilde{\boldsymbol{X}}$ of Eq. 6.11 exists and is unique. These requirements impose additional conditions on the coefficients of Eq. 6.1. We also note that $\tilde{\boldsymbol{X}}$ is a semimartingale with continuous samples.

3. For the time-invariant version of Eq. 6.1, let $\boldsymbol{X}$ be defined by Eq. 6.11 with $\boldsymbol{X}(0) = \boldsymbol{x} \in D$ and let $\mathcal{F}_t = \sigma(\boldsymbol{B}(s) : 0 \le s \le t)$ denote the natural filtration of $\boldsymbol{B}$. If D is a Borel set, then

$$T = \inf\{s > 0 : \boldsymbol{X}(s) \notin D\} \tag{6.12}$$

is an $\mathcal{F}_t$-stopping time. It is assumed that T corresponding to $\boldsymbol{X}$ starting at $\boldsymbol{x} \in D$ has a finite expectation, that is, $E^{\boldsymbol{x}}[T] < \infty$.

Proof: Since D is an open subset of $\mathbb{R}^d$, it is a Borel set. For all $t \ge 0$ and $s \in [0, t)$ rational the event $\{T < t\} = \cup_{0 \le s < t}\{\boldsymbol{X}(s) \notin D\}$ is in $\mathcal{F}_t$ since $\{\boldsymbol{X}(s) \notin D\} \in \mathcal{F}_s \subset \mathcal{F}_t$. Hence, T is an $\mathcal{F}_t$-stopping time (Section 3.4).

The assumption that T has a finite expectation is not restrictive in many cases. For example, the average time $\boldsymbol{B}$ needs to leave a spherical domain is finite (Section 8.5.1.1). Also, we will see in Section 7.3.1.3 that differential equations can be obtained for the expectation of T from the Fokker-Planck-Kolmogorov equations. The structure of these differential equations can be used to establish conditions under which $E^{\boldsymbol{x}}[T]$ is finite. ■

4. For the general version of Eq. 6.1, let $\tilde{\boldsymbol{X}}$ be defined by Eq. 6.11 with $\boldsymbol{X}(0) = \boldsymbol{x} \in D$ and $X_{d+1}(0) = t > 0$. Define also the set $D_t = D \times (0, t)$ for any $t > 0$. If D is a Borel set, then

$$\tilde{T} = \inf\{s > 0 : \tilde{X}(s) \notin D_t\} \tag{6.13}$$

is an $\mathcal{F}_t = \sigma(B(s) : 0 \leq s \leq t)$-stopping time. Note that $\tilde{T}$ cannot exceed t.

Proof: Since D is an open subset of $\mathbb{R}^d$, $D_t = D \times (0, t) \subset \mathbb{R}^{d+1}$ is a Borel set. Also, the event $\{\tilde{T} < \tilde{t}\} = \cup_{0 \leq s < \tilde{t}} \{\tilde{X}(s) \notin D_t\}$ is in $\mathcal{F}_{\tilde{t}}$ for all $\tilde{t} \in (0, t)$ and $s \in [0, \tilde{t})$ rational since $\{\tilde{X}(s) \notin D_t\} \in \mathcal{F}_s \subset \mathcal{F}_{\tilde{t}}$. Hence, $\tilde{T}$ is an $\mathcal{F}_{\tilde{t}}$-stopping time (Section 3.4). ■

5. The functions η and ξ are continuous in their arguments, the function p is Hölder continuous in D, the boundaries of D are regular, and the partial derivatives of u in Eq. 6.1 are bounded in D_t.

Note: Consider Eq. 6.1 with $q = 0$. This equation can be written in the form

$$p = \frac{\partial u}{\partial t} - \sum_i \alpha_i \frac{\partial u}{\partial x_i} - \frac{1}{2} \sum_{i,j} \beta_{ij} \frac{\partial^2 u}{\partial x_i \, \partial x_j}$$

showing that p must be continuous in its arguments. It turns out that p has to satisfy a stronger condition for u to have continuous partial derivatives of order 1 in t and order 2 in x. The function p must be Hölder continuous in D, that is, there exist constants $c, \alpha \in (0, \infty)$ such that $|p(x, t) - p(x', t)| \leq c \, \| x - x' \|^\alpha$, where $x, x' \in D$ ([58], p. 133).

Let X be a diffusion process starting at $x \in \partial D$. We say that x is a **regular** point of D if the first time T, when X exits D, is zero a.s., that is, $P(T = 0) = 1$. A boundary point that does not satisfy this condition is said to be an **irregular** point. The boundary point $\{0\}$ of $D = \{x \in \mathbb{R}^2, \| x \| < 1\} \setminus \{0\}$ is an irregular point with respect to a Brownian motion B in $\mathbb{R}^2$ because B will not hit $\{0\}$ in a finite time even if it starts at this point ([58], p. 99). Useful considerations and examples on regular and irregular boundary points can be found in [58] (Section 4.4) and [134] (Section 9.2). That D may have irregular points does not have significant practical implications on the local solution of Eq. 6.1 because (1) D can be approximated by a regular subset of $\mathbb{R}^d$ if it has irregular points and (2) almost no path of the diffusion processes used to estimate the local solution of Eq. 6.1 will ever hit the irregular points of D, provided that D has a finite number of irregular points ([41], pp. 49-50). ▲

If the conditions **1, 2, 4** and **5** are satisfied, then the **local solution** of Eq. 6.1 with the initial and boundary conditions in Eqs. 6.4 and 6.10 is

$$u(x, t) = E^{(x,t)}[u(\tilde{X}(\tilde{T}))] + E^{(x,t)}\left[\int_0^{\tilde{T}} p(\tilde{X}(s)) \, ds\right]$$
$$= E^{(x,t)}[\eta(X(t)) \mid \tilde{T} = t] \, P(\tilde{T} = t) + E^{(x,t)}[\xi(\tilde{X}(t - \tilde{T})) \mid \tilde{T} < t] \, P(\tilde{T} < t)$$
$$+ E^{(x,t)}\left[\int_0^{\tilde{T}} p(\tilde{X}(s)) \, ds\right]. \tag{6.14}$$

Proof: The superscripts of the expectation operator $E^{(\boldsymbol{x},t)}$ indicate that $\tilde{\boldsymbol{X}}(0) = (\boldsymbol{x}, t)$, $\boldsymbol{x} \in D$ and $t > 0$. The boundary value of u depends on the exit point of $\tilde{\boldsymbol{X}}$. This function is equal to $\eta(\boldsymbol{X}(t))$ if $\tilde{T} = t$ (Eq. 6.4) and $\xi(\tilde{\boldsymbol{X}}(t - \tilde{T}))$ if $\tilde{T} < t$ (Eq. 6.10).

The generator of the diffusion process $\tilde{\boldsymbol{X}}$ is defined by the limit

$$\mathcal{A}[g(\boldsymbol{x}, t)] = \lim_{s \downarrow 0} \frac{E^{(\boldsymbol{x},t)}[g(\tilde{\boldsymbol{X}}(s))] - g(\boldsymbol{x}, t)}{s}, \tag{6.15}$$

where g is a real-valued function defined on $\mathbb{R}^{d+1}$ with continuous second order partial derivatives in $\boldsymbol{x} \in \mathbb{R}^d$ and continuous first order partial derivative in $s \geq 0$. Denote by $\partial g/\partial \tilde{x}_i$ and $\partial^2 g/\partial \tilde{x}_i\, \partial \tilde{x}_j$ the partial derivatives of g with respect to the coordinates $\tilde{x}_i$ and $\tilde{x}_i, \tilde{x}_j$ of $\tilde{\boldsymbol{x}} = (\boldsymbol{x}, x_{d+1})$ so that $\tilde{x}_i = x_i$ for $i = 1, \ldots, d, d+1$. We have seen that $\tilde{\boldsymbol{X}}$ is a continuous semimartingale (Section 4.7.1.1) so that we can apply the multi-dimensional Itô formula (Section 4.6.2) to $g(\tilde{\boldsymbol{X}}(s))$. This formula gives

$$E^{(\boldsymbol{x},t)}[g(\tilde{\boldsymbol{X}}(s))] - g(\tilde{\boldsymbol{X}}(0)) = \sum_{i=1}^{d+1} E^{(\boldsymbol{x},t)} \left[\int_0^s \frac{\partial g(\tilde{\boldsymbol{X}}(\sigma))}{\partial \tilde{x}_i}\, d\tilde{X}_i(\sigma) \right]$$
$$+ \frac{1}{2} \sum_{i,j=1}^{d} E^{(\boldsymbol{x},t)} \left[\int_0^s \left(\boldsymbol{b}(\boldsymbol{X}(\sigma))\, \boldsymbol{b}(\boldsymbol{X}(\sigma))^T \right)_{ij} \frac{\partial^2 g(\tilde{\boldsymbol{X}}(\sigma))}{\partial \tilde{x}_i \partial \tilde{x}_j}\, d\sigma \right]$$

by averaging. The processes $\int_0^s [\partial g(\tilde{\boldsymbol{X}}(\sigma))/\partial \tilde{x}_i]\, b_{ij}(\tilde{\boldsymbol{X}}(\sigma))\, dB_j(\sigma)$, $i, j = 1, \ldots, d$, are martingales starting at zero since the functions $\partial g(\tilde{\boldsymbol{X}}(\sigma))/\partial \tilde{x}_i$ are bounded in $[0, s]$ by the continuity of $\partial g/\partial \tilde{x}_i$ and $\tilde{\boldsymbol{X}}$, b_{ij} are bounded functions by hypothesis, and $\tilde{\boldsymbol{X}}$ is adapted to the natural filtration of $\boldsymbol{B}$. Hence, the above equation becomes

$$E^{(\boldsymbol{x},t)}[g(\tilde{\boldsymbol{X}}(s))] - g(\boldsymbol{x}, t) = E^{(\boldsymbol{x},t)} \left[\int_0^s \left(\sum_{i=1}^{d} \frac{\partial g(\tilde{\boldsymbol{X}}(\sigma))}{\partial x_i}\, a_i(\boldsymbol{X}(\sigma)) - \frac{\partial g(\tilde{\boldsymbol{X}}(\sigma))}{\partial x_{d+1}} \right) d\sigma \right.$$
$$\left. + \frac{1}{2} \sum_{i,j=1}^{d} \int_0^s \left(\boldsymbol{b}(\boldsymbol{X}(\sigma)\, \boldsymbol{b}(\boldsymbol{X}(\sigma))^T \right)_{ij} \frac{\partial^2 g(\tilde{\boldsymbol{X}}(\sigma))}{\partial x_i \partial x_j}\, d\sigma \right].$$

For a small $s > 0$, the sum of integrals under the expectation on the right side of the above equation divided by s can be approximated by

$$\sum_{i=1}^{d} \frac{\partial g(\tilde{\boldsymbol{X}}(\theta(\omega)\, s, \omega))}{\partial x_i}\, a_i(\boldsymbol{X}(\theta(\omega)\, s, \omega)) - \frac{\partial g(\tilde{\boldsymbol{X}}(\theta(\omega)\, s, \omega))}{\partial x_{d+1}}$$
$$+ \frac{1}{2} \sum_{i,j=1}^{d} \left(\boldsymbol{b}(\boldsymbol{X}(\theta(\omega)\, s, \omega))\, \boldsymbol{b}(\boldsymbol{X}(\theta(\omega)\, s, \omega))^T \right)_{ij} \frac{\partial^2 g(\tilde{\boldsymbol{X}}(\theta(\omega)\, s, \omega))}{\partial x_i \partial x_j}$$

for each $\omega \in \Omega$, where $\theta(\omega) \in [0, 1]$. Since the drift and diffusion coefficients of $\tilde{\boldsymbol{X}}$ are bounded functions, $\tilde{\boldsymbol{X}}$ has continuous sample paths, and g has continuous partial derivatives of order 1 and order 2 with respect to $\tilde{\boldsymbol{x}}$ and $\boldsymbol{x}$, respectively, the limit of the above function as $s \downarrow 0$ is deterministic so that $\mathcal{A}$ in Eq. 6.15 is

$$\mathcal{A} = -\frac{\partial}{\partial x_{d+1}} + \sum_{i=1}^{d} a_i \frac{\partial}{\partial x_i} + \frac{1}{2} \sum_{i,j=1}^{d} (\boldsymbol{b}\, \boldsymbol{b}^T)_{ij} \frac{\partial^2}{\partial x_i\, \partial x_j}. \tag{6.16}$$

The generator of $\tilde{X}$ coincides with the differential operator of Eq. 6.1 for the drift and diffusion coefficients of $\tilde{X}$ in Eq. 6.11. We have performed similar calculations in Section 4.6.2 to show that the generator of a Brownian motion is proportional to the Laplace operator. The expectation of the Itô formula applied to the function $g(\tilde{X}(s))$ can be given in the form (Eq. 6.16)

$$E^{(\boldsymbol{x},t)}[g(\tilde{\boldsymbol{X}}(s))] - g(\boldsymbol{x},t) = E^{(\boldsymbol{x},t)}\left[\int_0^s \mathcal{A}[g(\tilde{\boldsymbol{X}}(\sigma))]\,d\sigma\right].$$

We have seen that any semimartingale $\tilde{X}$ has the representation $\tilde{X}(t) = \tilde{X}(0) + \tilde{A}(t) + \tilde{M}(t)$, where $\tilde{A}$ is an adapted process with càdlàg samples of finite variation on compacts and $\tilde{M}$ is a local martingale such that $\tilde{A}(0) = \tilde{M}(0) = \mathbf{0}$ (Section 4.4.1). Let $\tilde{X}^{\tilde{T}}(t) = \tilde{X}(0) + (\tilde{A} + \tilde{M})^{\tilde{T}}(t) = \tilde{X}(0) + \tilde{A}^{\tilde{T}}(t) + \tilde{M}^{\tilde{T}}(t)$ be the process $\tilde{X}$ stopped at $\tilde{T}$. Since $\tilde{A}^{\tilde{T}}$ and $\tilde{M}^{\tilde{T}}$ have the same properties as $\tilde{A}$ and $\tilde{M}$, respectively, then $\tilde{X}^{\tilde{T}}$ is a semimartingale so that Eq. 6.16 can be applied to the function $g(\tilde{X}^{\tilde{T}})$, and gives

$$E^{(\boldsymbol{x},t)}[g(\tilde{\boldsymbol{X}}^{\tilde{T}}(s))] - g(\boldsymbol{x},t) = E^{(\boldsymbol{x},t)}\left[\int_0^s \mathcal{A}[g(\tilde{\boldsymbol{X}}^{\tilde{T}}(\sigma))]\,d\sigma\right].$$

Also, $g(\tilde{X}^{\tilde{T}}(s)) = g(\tilde{X}(\tilde{T} \wedge s))$, $\int_0^s \mathcal{A}[g(\tilde{X}^{\tilde{T}}(\sigma))]\,d\sigma = \int_0^{\tilde{T}\wedge s} \mathcal{A}[g(\tilde{X}(\sigma))]\,d\sigma$, and $\tilde{T} \leq t$, so that the above equation gives

$$g(\boldsymbol{x},t) = E^{(\boldsymbol{x},t)}[g(\tilde{\boldsymbol{X}}(\tilde{T}))] - E^{(\boldsymbol{x},t)}\left[\int_0^{\tilde{T}} \mathcal{A}[g(\tilde{\boldsymbol{X}}(\sigma))]\,d\sigma\right] \tag{6.17}$$

for $s \geq t$. Eq. 6.17 is referred to as the Dynkin formula ([135], Theorem 7.4.1, p. 118).

If $\tilde{X}$ is defined by Eq. 6.11 and the solution u of Eq. 6.1 with the initial and boundary conditions in Eqs. 6.4 and 6.10 is used in place of g, then Eq. 6.17 gives Eq. 6.14 since $\tilde{X}(s) \in D_t$ for $s < \tilde{T}$ and $\tilde{X}(\tilde{T})$ is on the boundary of D_t so that $\mathcal{A}[u(\tilde{X}(s))] = -p(\tilde{X}(s))$ (Eq. 6.1) and

$$u(\tilde{\boldsymbol{X}}(\tilde{T})) = \begin{cases} \xi(\tilde{\boldsymbol{X}}(t-\tilde{T})), & \text{if } \tilde{T} < t, \\ \eta(\boldsymbol{X}(t)), & \text{if } \tilde{T} = t. \end{cases}$$

We can replace g with u in Eq. 6.17 since the solution u has a continuous first order derivative in t and continuous second order partial derivatives in $\boldsymbol{x}$. Additional technical considerations related to the above proof can be found in [42] (Chapter 6), [58] (Chapter 4), and [135] (Chapter 9). ■

If the conditions **1, 2, 3** and **5** are satisfied, the **local solution** of the time-invariant version of Eq. 6.1 with the boundary condition $u(\boldsymbol{x}) = \xi(\boldsymbol{x})$, $\boldsymbol{x} \in \partial D$, is

$$u(\boldsymbol{x}) = E^{\boldsymbol{x}}[\xi(\boldsymbol{X}(T))] + E^{\boldsymbol{x}}\left[\int_0^T p(\boldsymbol{X}(\sigma))\,d\sigma\right]. \tag{6.18}$$

Proof: The expectation of the Itô formula applied to a real-valued function $g \in \mathcal{C}^2(\mathbb{R}^d)$

and the $\mathbb{R}^d$-valued diffusion process $\boldsymbol{X}$ in Eq. 6.11 is

$$E^{\boldsymbol{x}}[g(\boldsymbol{X}(s))] - g(\boldsymbol{x}) = E^{\boldsymbol{x}}\left[\sum_{i=1}^{d}\int_0^s \frac{\partial g(\boldsymbol{X}(\sigma))}{\partial x_i}\, a_i(\boldsymbol{X}(\sigma))\,d\sigma \right.$$
$$\left. + \frac{1}{2}\sum_{i,j=1}^{d}\int_0^s \left(\boldsymbol{b}(\boldsymbol{X}(\sigma)\,\boldsymbol{b}(\boldsymbol{X}(\sigma))^T\right)_{ij}\frac{\partial^2 g(\boldsymbol{X}(\sigma))}{\partial x_i \partial x_j}\,d\sigma\right].$$

Arguments similar to those used to derive Eq. 6.14 show that the generator of $\boldsymbol{X}$ is

$$\mathcal{A} = \sum_{i=1}^{d} a_i(\boldsymbol{x})\,\frac{\partial}{\partial x_i} + \frac{1}{2}\sum_{i,j=1}^{d}(\boldsymbol{b}(\boldsymbol{x})\,\boldsymbol{b}(\boldsymbol{x})^T)_{ij}\,\frac{\partial^2}{\partial x_i\,\partial x_j},$$

and the Itô formula applied to $g(\boldsymbol{X})$ gives

$$E^{\boldsymbol{x}}[g(\boldsymbol{X}(s))] - g(\boldsymbol{x}) = E^{\boldsymbol{x}}\left[\int_0^s \mathcal{A}[g(\boldsymbol{X}(\sigma))]\,d\sigma\right].$$

The local solution in Eq. 6.18 results by substituting in the above equation T and u for s and g, respectively, since $\int_0^T \mathcal{A}[u(\boldsymbol{X}(\sigma))]\,d\sigma = -\int_0^T p(\boldsymbol{X}(\sigma))\,d\sigma$.

It remains to show that we can use T in place of s in the Itô formula. Let $n \geq 1$ be an integer and let $T_n = T \wedge n$ be a sequence of stopping times. The Itô formula can be applied to the function $u(\boldsymbol{X}^{T_n})$ for each $n \geq 1$, and gives

$$E^{\boldsymbol{x}}\left[u(\boldsymbol{X}^{T_n}(s))\right] - u(\boldsymbol{x}) = E^{\boldsymbol{x}}\left[\int_0^s \sum_{j=1}^{d} h_j(\boldsymbol{X}^{T_n}(\sigma))\,dB_j(\sigma)\right]$$
$$+ E^{\boldsymbol{x}}\left[\int_0^s \mathcal{A}[u(\boldsymbol{X}^{T_n}(\sigma))]\,d\sigma\right],$$

where $h_j(\boldsymbol{X}^{T_n}(\sigma)) = \sum_{i=1}^{d}[\partial u(\boldsymbol{X}^{T_n}(\sigma))/\partial x_i]\,b_{ij}(\boldsymbol{X}^{T_n}(\sigma))$. Since the stochastic integral

$$\int_0^{T_n} h_j(\boldsymbol{X}(\sigma))\,dB_j(\sigma) = \int_0^n 1_{\{\sigma<T\}}\,h_j(\boldsymbol{X}(\sigma))\,dB_j(\sigma)$$

is a martingale starting at zero, its expectation is zero. If h_j is a bounded Borel function in D, then there is a constant $a > 0$ such that $|h_i(\boldsymbol{x})| \leq a$ for all $\boldsymbol{x} \in D$ and

$$E^{\boldsymbol{x}}\left[\left(\int_0^{T_n} h_j(\boldsymbol{X}(\sigma))\,dB_j(\sigma)\right)^2\right] = E^{\boldsymbol{x}}\left[\int_0^{T_n} h_j(\boldsymbol{X}(\sigma))^2\,d\sigma\right] \leq a^2\,E^{\boldsymbol{x}}[T].$$

Hence, the family $\int_0^{T_n} h_j(\boldsymbol{X}(\sigma))\,dB_j(\sigma)$, $n = 1, 2, \ldots$, is uniformly integrable ([135], Appendix C) so that

$$0 = \lim_{n\to\infty} E^{\boldsymbol{x}}\left[\int_0^{T_n} h_j(\boldsymbol{X}(\sigma))\,dB_j(\sigma)\right] = E^{\boldsymbol{x}}\left[\lim_{n\to\infty}\int_0^{T_n} h_j(\boldsymbol{X}(\sigma))\,dB_j(\sigma)\right]$$
$$= E^{\boldsymbol{x}}\left[\int_0^{T} h_j(\boldsymbol{X}(\sigma))\,dB_j(\sigma)\right].$$

The functions h_j are bounded Borel functions in D by the properties of the diffusion coefficients of $\boldsymbol{X}$ and of the solution u of the time-invariant version of Eq. 6.1. ■

Example 6.1: Let D be a simply connected open bounded subset of $\mathbb{R}^d$ and let u be the solution of the Laplace equation $\Delta u(\boldsymbol{x}) = 0$, $\boldsymbol{x} \in D$, with the boundary condition $u(\boldsymbol{x}) = \xi(\boldsymbol{x})$, $\boldsymbol{x} \in \partial D$. Denote by $T = \inf\{t > 0 : \boldsymbol{B}(t) \notin D\}$ the first time when an $\mathbb{R}^d$-valued Brownian motion $\boldsymbol{B}$ starting at $\boldsymbol{x} \in D$ leaves D. The local solution is $u(\boldsymbol{x}) = E^{\boldsymbol{x}}[\xi(\boldsymbol{B}(T))]$ for each $\boldsymbol{x} \in D$. ◇

Proof: The process $\boldsymbol{B}^T$ is a square integrable martingale since T is a stopping time and $\boldsymbol{B}$ is a square integrable martingale. The Itô formula applied to the function $u(\boldsymbol{B}^T)$ gives

$$\begin{aligned} u(\boldsymbol{B}^T(t)) &= u(\boldsymbol{x}) + \sum_{i=1}^{d} \int_0^t \frac{\partial u(\boldsymbol{B}^T(s))}{\partial x_i} \, dB_i^T(s) + \frac{1}{2} \int_0^t \Delta u(\boldsymbol{B}^T(s))\, ds \\ &= u(\boldsymbol{x}) + \sum_{i=1}^{d} \int_0^t \frac{\partial u(\boldsymbol{B}^T(s))}{\partial x_i} \, dB_i^T(s), \end{aligned}$$

where the latter equality holds since $\Delta u(\boldsymbol{B}^T(s)) = 0$ for all $s \geq 0$. The stochastic integrals $M_i(t) = \int_0^t [\partial u(\boldsymbol{B}^T(s))/\partial x_i]\, dB_i^T(s)$, $t \geq 0$, are $\mathcal{F}_t$-measurable, have finite expectation since $E[M_i(t)^2] = E\left[\int_0^t (\partial u(\boldsymbol{B}(s))/\partial x_i)^2 \, ds\right]$ and the functions $\partial u/\partial x_i$ are bounded in D, and

$$E[M_i(t) \mid \mathcal{F}_s] = E\left[M_i(s) + \int_s^t \frac{\partial u(\boldsymbol{B}^T(\xi))}{\partial x_i} \, dB_i^T(\xi) \mid \mathcal{F}_s\right] = M_i(s), \quad s \leq t.$$

Hence, the processes M_i are martingales. Then $u(\boldsymbol{B}^T(t)) = u(\boldsymbol{x}) + \sum_{i=1}^d M_i(t)$ is a martingale with expectation $E^{\boldsymbol{x}}[u(\boldsymbol{B}^T(t))] = u(\boldsymbol{x})$ so that $E^{\boldsymbol{x}}[u(\boldsymbol{B}(T))] = u(\boldsymbol{x})$ by letting $t \to \infty$. This result and the specified boundary conditions yield $u(\boldsymbol{x}) = E^{\boldsymbol{x}}[\xi(\boldsymbol{B}(T))]$. ■

Example 6.2: The RWM can solve locally elliptic and parabolic partial differential equations. The method cannot be used to solve hyperbolic partial differential equations. ◇

Proof: Let $\boldsymbol{X}$ be the diffusion process in Eq. 6.11. The generator of this process is

$$\mathcal{A} = \sum_{i=1}^{d} a_i \frac{\partial}{\partial x_i} + \frac{1}{2} \sum_{i,j=1}^{d} (\boldsymbol{b}\,\boldsymbol{b}^T)_{ij} \frac{\partial^2}{\partial x_i \, \partial x_j}.$$

Since $\boldsymbol{b}\,\boldsymbol{b}^T$ is a real-valued symmetric matrix, its eigenvalues cannot be negative so that the RWM cannot be applied to solve parabolic partial differential equations ([73], Chapter 26).

For example, consider the time-invariant version of Eq. 6.1 with $d = 2$ and $\alpha_i(\boldsymbol{x}) = 0$ and let $\boldsymbol{X}$ be an $\mathbb{R}^2$-valued diffusion process defined by the stochastic differential equation

$d\boldsymbol{X}(t) = \boldsymbol{b}\, d\boldsymbol{B}(t)$, $t \geq 0$, where $\boldsymbol{B}$ is a Brownian motion in $\mathbb{R}^2$ and $\boldsymbol{b}$ denotes a (2, 2) real-valued matrix. The generator of $\boldsymbol{X}$ is (Eq. 6.16)

$$\mathcal{A} = \frac{1}{2} \sum_{i,j=1}^{2} (\boldsymbol{b}\,\boldsymbol{b}^T)_{ij} \frac{\partial^2}{\partial x_i\, \partial x_j}$$
$$= \frac{1}{2} \left[(b_{11}^2 + b_{12}^2) \frac{\partial^2}{\partial x_1^2} + 2\,(b_{11}\, b_{21} + b_{12}\, b_{22}) \frac{\partial^2}{\partial x_1\, \partial x_2} + (b_{21}^2 + b_{22}^2) \frac{\partial^2}{\partial x_2^2} \right].$$

Since

$$(b_{11}\, b_{21} + b_{12}\, b_{22})^2 - (b_{11}^2 + b_{12}^2)(b_{21}^2 + b_{22}^2) = -(b_{11}\, b_{22} - b_{12}\, b_{21})^2 \leq 0,$$

the differential operator defined by the generator of $\boldsymbol{X}$ cannot match the operator of a hyperbolic equation ([73], Chapter 26). ■

6.2.1.2 Monte Carlo algorithm

The RWM in Eqs. 6.14 and 6.18 gives the value of the solution u of Eq. 6.1 at an arbitrary point $\boldsymbol{x} \in D$ and time $t > 0$ as a sum of expectations depending on the sample paths of $\tilde{\boldsymbol{X}}$. Generally, it is not possible to find these expectations analytically, but they can be estimated from samples of $\tilde{\boldsymbol{X}}$ generated by Monte Carlo simulation.

Figure 6.1 shows a set D_t for $d = 2$ and two sample paths of $\tilde{\boldsymbol{X}}$ that exit D_t at the times $\tilde{T} = t$ and $\tilde{T} < t$. The sample paths of $\tilde{\boldsymbol{X}}$ can be divided in two classes, sample paths exiting D_t through D at $\tilde{T} = t$ and sample paths leaving D_t

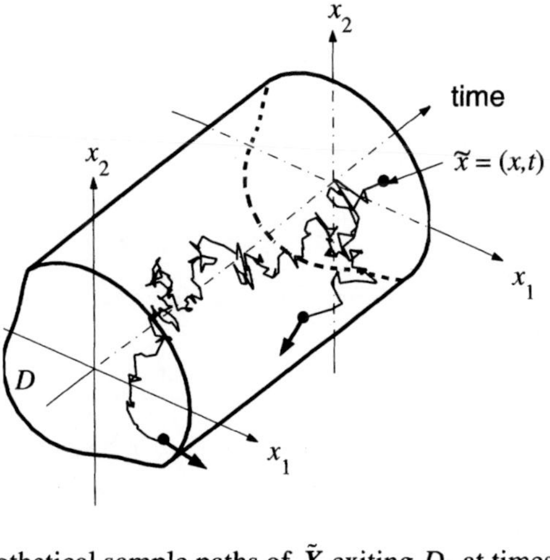

Figure 6.1. Hypothetical sample paths of $\tilde{\boldsymbol{X}}$ exiting D_t at times $\tilde{T} = t$ and $\tilde{T} < t$ for $d = 2$

through ∂D at an earlier time $\tilde{T} < t$. The values of $u(\tilde{X}(\tilde{T}))$ at these exit points are $\eta(X(t))$ and $\xi(\tilde{X}(t - \tilde{T}))$, respectively.

Suppose that n_s independent samples of $\tilde{X}$ have been generated. Denote by n'_s and n''_s the number of sample paths of $\tilde{X}$ that exit D_t through D and ∂D, respectively. We have $n'_s \geq 0$, $n''_s \geq 0$, and $n_s = n'_s + n''_s$.

The local solution in Eq. 6.14 can be estimated by

$$\hat{u}(x, t) = \frac{n'_s}{n_s} \left[\frac{1}{n'_s} \sum_{\omega'=1}^{n'_s} \eta(X(t, \omega')) \right] + \frac{n''_s}{n_s} \left[\frac{1}{n''_s} \sum_{\omega''=1}^{n''_s} \xi(\tilde{X}(t - \tilde{T}(\omega''), \omega'')) \right] + \frac{1}{n_s} \sum_{\omega=1}^{n_s} \int_0^{\tilde{T}(\omega)} p(\tilde{X}(s, \omega))\, ds. \tag{6.19}$$

Note: There is no reason to continue the calculation of a sample of $\tilde{X}$ after this sample has exited D_t. Hence, we need to generate samples of the process $\tilde{X}$ stopped at $\tilde{T}$.

The terms in the square brackets of Eq. 6.19 are estimates of the expectations $E^{(x,t)}[u(\tilde{X}(\tilde{T})) \mid \tilde{T} = t]$ and $E^{(x,t)}[u(\tilde{X}(\tilde{T})) \mid \tilde{T} < t]$ and the ratios n'_s/n_s and n''_s/n_s weighting these terms approximate the probabilities $P(\tilde{T} = t)$ and $P(\tilde{T} < t)$, respectively. Hence, the sum of the first two terms in Eq. 6.19 is an estimate for $E^{(x,t)}[u(\tilde{X}(\tilde{T}))]$. The expectation $E^{(x,t)}\left[\int_0^{\tilde{T}} p(\tilde{X}(s))\, ds\right]$ in Eq. 6.14 is approximated by the last term in Eq. 6.19.

The accuracy of the local solution given by Eq. 6.19 depends on (1) the sample size $n_s = n'_s + n''_s$, (2) the time step used to generate sample paths of the diffusion process $\tilde{X}$, and (3) the accuracy if the recurrence formula used for generating sample paths of $\tilde{X}$. The algorithms in Section 4.7.3 can be used to generate samples of $\tilde{X}$. ▲

The local solution in Eq. 6.18 can be estimated by

$$\hat{u}(x) = \frac{1}{n_s} \sum_{\omega=1}^{n_s} \xi(X(T(\omega), \omega)) + \frac{1}{n_s} \sum_{\omega=1}^{n_s} \int_0^{T(\omega)} p(X(s, \omega))\, ds, \tag{6.20}$$

where $X(\cdot, \omega)$ and $T(\omega)$, $\omega = 1, \ldots, n_s$, are samples of X and T, respectively.

Example 6.3: Lct u be the solution of the partial differential equation

$$x_1^2 \frac{\partial^2 u}{\partial x_1^2} + x_2^2 \frac{\partial^2 u}{\partial x_2^2} = 2\, x_1^2\, x_2^2$$

defined in $D = (\varepsilon, 1) \times (\varepsilon, 1)$, where $\varepsilon \in (0, 1)$. The values of the solution u on the boundary ∂D of D are given by the values of function $x_1^2 x_2^2/2$ on ∂D. The local solution is

$$u(x) = E^x\,[u(X(T))] - E^x \left[\int_0^T X_1(s)^2\, X_2(s)^2\, ds \right],$$

where $x \in D$, the coordinates of X are defined by the stochastic differential equations $dX_i(t) = X_i(t)\, dB_i(t)$, B_1 and B_2 are independent Brownian motions, and $T = \inf\{t > 0 : X(t) \notin D\}$. Estimates of $u(x)$ at many points in D for $\varepsilon = 0.01$ calculated from $n_s = 5,000$ samples of X generated with a time step $\Delta t = 0.0005$ are in error by less than 3%. ◇

Proof: The Itô formula applied to $X \mapsto u(X)$ gives

$$E^{x}\left[u(X(t))\right] - u(x) = \frac{1}{2}\, E^{x}\left[\int_0^t \left(X_1(s)^2\, \frac{\partial^2 u(X(s))}{\partial x_1^2} + X_2(s)^2\, \frac{\partial^2 u(X(s))}{\partial x_2^2}\right) ds\right]$$

by expectation. The stated local solution results by substituting T for t and using the defining equation of u.

It is simple to verify that the solution of the above partial differential equation is $u(x_1, x_2) = x_1^2\, x_2^2/2$. This exact solution has been used to assess the accuracy of the estimates delivered by the local solution. ■

6.2.1.3 The Laplace equation

Let D be an open bounded subset in $\mathbb{R}^d$ with boundaries ∂D_r, $r = 1, \ldots, m$, and let ξ_r be functions defined on these boundaries. If $m = 1$, then D is a simply connected subset of $\mathbb{R}^d$. Otherwise, D is multiply connected.

Let u be the solution of the Laplace equation

$$\begin{aligned} \Delta u(x) &= 0, \quad x \in D, \\ u(x) &= \xi_r(x), \quad x \in \partial D_r,\ r = 1, \ldots, m. \end{aligned} \tag{6.21}$$

Denote by B an $\mathbb{R}^d$-valued Brownian motion starting at $x \in D$. The samples of B will exit D through one of the boundaries ∂D_r of D. Figure 6.2 shows two samples of a Brownian motion B in $\mathbb{R}^2$ and their exit points from a multiply connected set D. We define as in Eq. 6.13 a stopping time T as the first time B with $B(0) = x \in D$ exits D, that is,

$$T = \inf\{t > 0 : B(t) \notin D\}. \tag{6.22}$$

The **local solution** of the Laplace equation at an arbitrary point $x \in D$ is

$$u(x) = \sum_{r=1}^{m} E^{x}\left[\xi_r(B(T)) \mid B(T) \in \partial D_r\right] P(B(T) \in \partial D_r) \tag{6.23}$$

Note: The local solution in Eq. 6.23 is a special case of Eq. 6.18 for $\alpha_i(x) = 0$, $\beta_{ij}(x) = 2\,\delta_{ij}$, and $q = p = 0$. The Brownian motion B is adequate for solution because its

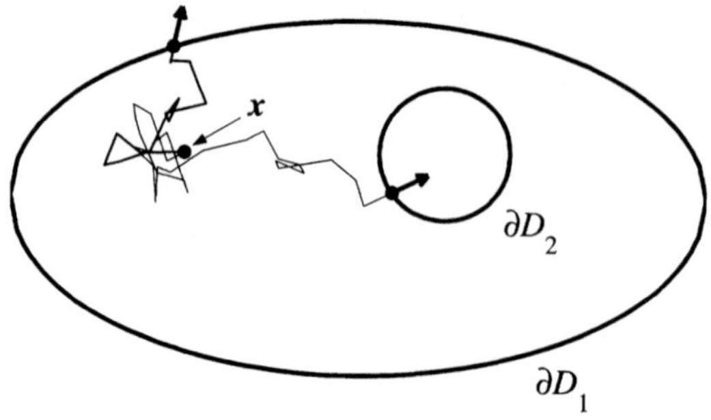

Figure 6.2. Local solution of the Laplace equation for a multiply connected domain D in $\mathbb{R}^2$

generator is proportional to the Laplace operator. If D is simply connected, then $m = 1$ in Eq. 6.23, $\partial D_1 = \partial D$, $E^{\boldsymbol{x}}[\xi_r(\boldsymbol{B}(T)) \mid \boldsymbol{B}(T) \in \partial D_1] = E^{\boldsymbol{x}}[\xi_1(\boldsymbol{B}(T))]$, and $P(\boldsymbol{B}(T) \in \partial D_1) = 1$.

If the boundary conditions are $\xi_r(\boldsymbol{x}) = k_r, \boldsymbol{x} \in \partial D_r$, then Eq. 6.23 gives

$$u(\boldsymbol{x}) = E^{\boldsymbol{x}}[u(\boldsymbol{B}(T))] = \sum_{r=1}^{m} k_r\, p_r(\boldsymbol{x}) \quad \boldsymbol{x} \in D, \tag{6.24}$$

where $p_r(\boldsymbol{x}) = P(\boldsymbol{B}(T) \in \partial D_r)$ [82, 84]. ▲

Example 6.4: Let $u(\boldsymbol{x})$ be the steady-state temperature at an arbitrary point $\boldsymbol{x} = (x_1, x_2)$ of an eccentric annulus

$$D = \{(x_1, x_2) : x_1^2 + x_2^2 - 1 < 0, (x_1 - 1/4)^2 + x_2^2 - (1/4)^2 > 0\}$$

in $\mathbb{R}^2$ with boundaries $\partial D_1 = \{(x_1, x_2) : x_1^2 + x_2^2 - 1 = 0\}$ and $\partial D_2 = \{(x_1, x_2) : (x_1 - 1/4)^2 + x_2^2 - (1/4)^2 = 0\}$ kept at the constant temperatures 50 and 0, respectively. The local solution by the RWM is $u(\boldsymbol{x}) = 50\, p_1(\boldsymbol{x})$ for any point $\boldsymbol{x} \in D$ (Eq. 6.24). Because $p_1(\boldsymbol{x})$ cannot be found analytically, it has been calculated from Eq. 6.20 and samples of $\boldsymbol{B}$ starting at $\boldsymbol{x}$. The largest error recorded at $\boldsymbol{x} = (0.7, 0), (0.9, 0), (0, 0.25), (0, 0.5)$, and $(0, 0.75)$ was found to be 2.79% for $n_s = 1,000$ independent samples of $\boldsymbol{B}$ with time steps $\Delta t = 0.001$ or 0.0001. Smaller time steps were used at points $\boldsymbol{x}$ close to the boundary of D. The error can be reduced by decreasing the time step and/or increasing the sample size.

Figure 6.3 shows the exact and RWM solutions along a circle of radius $r = 3/4$ centered at the origin, where the angle $\theta \in [0, 2\pi)$ measured from the coordinate x_1 gives the location on this circle. The RWM solutions with ($\Delta t = 0.01$, $n_s = 100$) and ($\Delta t = 0.0001$, $n_s = 5000$) are shown with circles and stars, respectively. The accuracy of the RWM method improves as the time step decreases and the sample size increases [38]. ◇

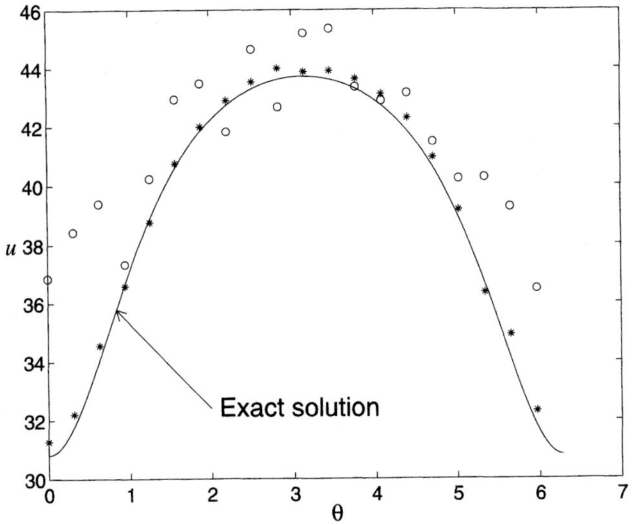

Figure 6.3. Solution along a circle of radius $r = 3/4$ centered at the origin

Note: The temperature u is the solution of $\Delta u = 0$ with the boundary conditions $u(\boldsymbol{x}) = 50$ for $\boldsymbol{x} \in \partial D_1$ and $u(\boldsymbol{x}) = 0$ for $\boldsymbol{x} \in \partial D_2$. The exact solution is ([73], Example 16.3, p. 296)

$$u(\boldsymbol{x}) = 50 \left(1 - \frac{\log(\zeta(\boldsymbol{x})^2 + \rho(\boldsymbol{x})^2)}{2 \log(a)}\right), \quad \text{where } a = 2 + \sqrt{3},$$

$$\zeta(\boldsymbol{x}) = \frac{(a\,x_1 - 1)(x_1 - a) + a\,x_2^2}{(a\,x_1 - 1)^2 + (a\,x_2)^2}, \quad \text{and } \rho(\boldsymbol{x}) = \frac{(a^2 - 1)\,x_2}{(a\,x_1 - 1)^2 + (a\,x_2)^2}.$$

This exact result was used to evaluate the accuracy of the local solution. ▲

Example 6.5: Let $D = \{\boldsymbol{x} \in \mathbb{R}^2 : (x_1/a_1)^2 + (x_2/a_2)^2 < 1\}$ be an ellipse, where $a_1, a_2 > 0$ are some constants. Let u be the solution of the Laplace equation $\Delta u(\boldsymbol{x}) = 0$, $\boldsymbol{x} \in D$, such that $u(\boldsymbol{x}) = \xi(\boldsymbol{x})$, $\boldsymbol{x} \in \partial D$, where ξ is a specified function and $\partial D = \{\boldsymbol{x} \in \mathbb{R}^2 : (x_1/a_1)^2 + (x_2/a_2)^2 = 1\}$ denotes the boundary of D. The local solution of this equation is $u(\boldsymbol{x}) = E^{\boldsymbol{x}}[\xi(\boldsymbol{B}(T))]$ at any point $\boldsymbol{x} \in D$, where $\boldsymbol{B}$ denotes an $\mathbb{R}^2$-valued Brownian motion starting at $\boldsymbol{x}$ and $T = \inf\{t \geq 0 : \boldsymbol{B}(t) \notin D\}$ is the first time $\boldsymbol{B}$ exits D.

The following MATLAB function can be used to estimate $u(\boldsymbol{x})$ (Eq. 6.19).

```
function uest =laplace(a1,a2,'ξ',dt,ns,nseed)
dts=sqrt(dt);
randn('seed',nseed)
for ks=1:ns,
      xbr=x; xbr1=abs(xbr(1)); xbr2=abs(xbr(2));
      xx=(xbr1/a1)^2 + (xbr2/a2)^2;
      while xx < 1
         xbr=xbr+randn(2,1)*dts;xbr1=abs(xbr(1)); xbr2=abs(xbr(2));
         xx=(xbr1/a1)^2 + (xbr2/a2)^2;
      end,
      u(ks)=ξ(xbr1,xbr2); end,
uest=mean(u);
```

The output "uest" is the estimated value of u at $\boldsymbol{x} \in D$ based on "ns" independent samples of $\boldsymbol{B}$ generated with a time step "dt". ◇

Note: The input to this MATLAB function consists of the dimensions a_1 and a_2 of D, the point $\boldsymbol{x} \in D$ at which the local solution is to be determined, the seed "nseed" and the time step "dt" used to generate "ns" independent samples of $\boldsymbol{B}$, and the function ξ giving the values of u on ∂D. ▲

Example 6.6: Let u be the solution of the partial differential equation

$$\sum_{i=1}^{2} \alpha_i(\boldsymbol{x}) \frac{\partial u(\boldsymbol{x})}{\partial x_i} + \frac{1}{2} \sum_{i,j=1}^{2} \beta_{ij}(\boldsymbol{x}) \frac{\partial^2 u(\boldsymbol{x})}{\partial x_i \partial x_j} = 0, \quad \boldsymbol{x} \in D \subset \mathbb{R}^2,$$

with the boundary conditions $u(\boldsymbol{x}) = \xi(\boldsymbol{x})$, $\boldsymbol{x} \in \partial D$. If $\boldsymbol{\beta}(\boldsymbol{x}) = \{\beta_{ij}(\boldsymbol{x})\}$ admits the decomposition $\boldsymbol{\beta}(\boldsymbol{x}) = \boldsymbol{b}(\boldsymbol{x})\, \boldsymbol{b}(\boldsymbol{x})^T$ for each $\boldsymbol{x} \in D$, then the local solution is $u(\boldsymbol{x}) = E^{\boldsymbol{x}}[u(\boldsymbol{X}(T))]$, where $\boldsymbol{X}$ is an $\mathbb{R}^2$-valued diffusion process with drift and diffusion coefficients $a_i(\boldsymbol{x}) = \alpha_i(\boldsymbol{x})$ and $b_{ij}(\boldsymbol{x})$ and T denotes the first time $\boldsymbol{X}$ starting at $\boldsymbol{x} \in D$ exits D (Eq. 6.14). It is assumed that $E^{\boldsymbol{x}}[T] < \infty$.

Generally, the expectation $E^{\boldsymbol{x}}[u(\boldsymbol{X}(T))]$ cannot be obtained analytically but it can be estimated from samples of $\boldsymbol{X}$ generated by Monte Carlo simulation. Numerical results were obtained for $D = (-2, 2) \times (-1, 1)$, boundary conditions $\xi(\boldsymbol{x}) = 1$ for $\boldsymbol{x} \in (-2, 2) \times \{-1\}$ and zero on the other boundaries of D, and the coefficients $\alpha_1(\boldsymbol{x}) = 1$, $\alpha_2(\boldsymbol{x}) = 2$, $\beta_{11}(\boldsymbol{x}) = 1.25$, $\beta_{12}(\boldsymbol{x}) = \beta_{21}(\boldsymbol{x}) = 1.5$, and $\beta_{22}(\boldsymbol{x}) = 4.25$ of the partial differential equation of u. The local approximations of u based on $n_s = 1,000$ samples of $\boldsymbol{X}$ and a time step of $\Delta t = 0.001$ are satisfactory. For example, these values are 0.2720, 0.2650, and 0.2222 at $\boldsymbol{x} = (0.5, 0)$, $(1.0, 0)$, and $(1.5, 0)$. They differ from the finite element solution by less than 5%. ◇

Note: The diffusion process $\boldsymbol{X}$ needed to estimate $u(\boldsymbol{x})$ at an arbitrary point $\boldsymbol{x} \in D$ is defined by the stochastic differential equation

$$\begin{cases} dX_1(t) = dt + dB_1(t) + (1/2)\, dB_2(t), \\ dX_2(t) = 2\, dt + (1/2)\, dB_1(t) + 2\, dB_2(t), \end{cases}$$

where B_1 and B_2 are independent Brownian motions. A finite element solution has been used to assess the accuracy of the local solution because the exact solution is not known in this case. ▲

6.2.1.4 The Poisson equation

Let D be an open bounded subset in $\mathbb{R}^d$ with boundaries $\partial D_r, r = 1, \ldots, m$, and let ξ_r be functions defined on these boundaries. Consider an $\mathbb{R}^d$-valued Brownian motion $\boldsymbol{B}$ starting at $\boldsymbol{x} \in D$ and let T be the stopping time in Eq. 6.22.

Let u be the solution of the Poisson equation

$$\begin{aligned} \Delta u(\boldsymbol{x}) &= -p(\boldsymbol{x}), \quad \boldsymbol{x} \in D, \\ u(\boldsymbol{x}) &= \xi_r(\boldsymbol{x}), \quad \boldsymbol{x} \in \partial D_r, r = 1, \ldots, m. \end{aligned} \tag{6.25}$$

The **local solution** of a Poisson equation at an arbitrary point $\boldsymbol{x} \in D$ is

$$\begin{aligned} u(\boldsymbol{x}) &= E^{\boldsymbol{x}}[u(\boldsymbol{B}(T))] - E^{\boldsymbol{x}}\left[\int_0^T \mathcal{A}[u(\boldsymbol{B}(s))]\,ds\right] \\ &= \sum_{r=1}^m E^{\boldsymbol{x}}[\xi_r(\boldsymbol{B}(T)) \mid \boldsymbol{B}(T) \in \partial D_r]\, p_r(\boldsymbol{x}) + \frac{1}{2} E^{\boldsymbol{x}}\left[\int_0^T p(\boldsymbol{B}(s))\,ds\right]. \end{aligned} \tag{6.26}$$

Note: The function $p_r(\boldsymbol{x}) = P(\boldsymbol{B}(T) \in \partial D_r)$ is the probability that $\boldsymbol{B}$ starting at $\boldsymbol{x} \in D$ exits D for the first time through $\partial D_r, r = 1, \ldots, m$ (Eq. 6.23). The operator $\mathcal{A} = (1/2)\,\Delta$ is the generator of the Brownian motion $\boldsymbol{B}$. Generally, the probabilities and expectations in Eq. 6.26 cannot be obtained analytically but can be estimated from samples of $\boldsymbol{B}$. ▲

Example 6.7: The Prandtl stress function for a bar in torsion is the solution of $\Delta\, u(\boldsymbol{x}) = -2\,G\,\beta, \quad \boldsymbol{x} \in D \subset \mathbb{R}^2$, where G is the modulus of elasticity in shear and β denotes the angle of twist. If D is a multiply connected domain with the external boundary ∂D_1 and the interior boundaries ∂D_r, $r = 2, \ldots, m$, delineating the cavities of the beam cross section, the Prandtl function satisfies the conditions $u(\boldsymbol{x}) = k_r$ for $\boldsymbol{x} \in \partial D_r$, $r = 1, \ldots, m$, where k_r are constants. One of the constants k_r is arbitrary and can be set to zero, for example $k_1 = 0$ ([23], Section 7-6).

The local solution for the Prandtl stress function is

$$\begin{aligned} u(\boldsymbol{x}) &= E^{\boldsymbol{x}}[u(\boldsymbol{B}(T))] - \frac{1}{2} E^{\boldsymbol{x}}\left[\int_0^T \Delta\, u(\boldsymbol{B}(s))\,ds\right] \\ &= \sum_{r=2}^m p_r(\boldsymbol{x})\, k_r + G\,\beta\, E^{\boldsymbol{x}}[T] \end{aligned}$$

with the notation in Eq. 6.26. If the Brownian motion $\boldsymbol{B}$ starts at a point on the boundary of D, for example, $\boldsymbol{x} \in \partial D_r$, and D is regular, then $p_r(\boldsymbol{x}) = 1$, $p_q(\boldsymbol{x}) = 0$ for $q \neq r$, and $E^{\boldsymbol{x}}[T] = 0$ so that $u(\boldsymbol{x}) = k_r$, that is, the local solution satisfies the boundary conditions exactly [82, 84]. ◇

Note: The constant β can be calculated from the condition that the applied twist is twice the volume of the Prandtl function over D ([23], Section 7-6). Therefore, the global solution is needed to find β. However, the value of β may not be necessary in the design phase selecting the shape of D from several possible alternatives. Because local solutions help detect regions of stress concentration in D, they provide a useful tool for the selection of adequate shapes for D, that is, shapes which do not cause large localized stresses.

If D is a simply connected set, then $p_r(\boldsymbol{x}) = 0$ for $r = 2, \ldots, m$ so that the local solution becomes $u(\boldsymbol{x}) = G\,\beta\,E^{\boldsymbol{x}}[T]$ ▲

Example 6.8: Suppose D is a simply connected circular cross section of radius $r > 0$. Let T be the first time an $\mathbb{R}^2$-valued Brownian motion $\boldsymbol{B}$ starting at $\boldsymbol{x} \in D$ leaves D. The Prandtl function given by the local solution in Eq. 6.26 is

$$u(\boldsymbol{x}) = G\,\beta\,E^{\boldsymbol{x}}[T] = \frac{G\,\beta(r^2 - x_1^2 - x_2^2)}{2}, \quad \boldsymbol{x} \in D,$$

and coincides with the expression of this function in solid mechanics books ([23], Section 7-4). ◇

Note: Consider a sphere of radius $r > 0$ centered at the origin of $\mathbb{R}^d$ and an $\mathbb{R}^d$-valued Brownian motion $\boldsymbol{B}$ starting at $\boldsymbol{x}$, $\| \boldsymbol{x} \| < r$. Let T be the first time that $\boldsymbol{B}$ starting at $\boldsymbol{x}$ exits the sphere. Then $E^{\boldsymbol{x}}[T] = \left(r^2 - \| \boldsymbol{x} \|^2\right)/d$ (Section 8.5.1.1 in this book, [135], Example 7.4.2, p. 119). ▲

Example 6.9: Estimates of the Prandtl stress function $u(\boldsymbol{x})$ in Example 6.7 at $\boldsymbol{x} = (2, 1), (3, 1), (4, 1), (2, 2)$, and $(3, 2)$ for an elliptical cross section $D = \{\boldsymbol{x} : (x_1/a_1)^2 + (x_2/a_2)^2 < 1\}$ with $a_1 = 5$ and $a_2 = 3$ have been obtained by Monte Carlo simulation for $G\,\beta = 1/2$. For $n_s = 100$ samples of $\boldsymbol{B}$ generated with a time step $\Delta t = 0.001$ the largest error of the estimated Prandtl function is nearly 33%. The error decreases to 2% if the sample size is increased to $n_s = 1,000$. ◇

Note: The exact expression,

$$u(\boldsymbol{x}) = -\frac{a_1^2\,a_2^2\,G\,\beta}{2(a_1^2 + a_2^2)}\left(\frac{x_1^2}{a_1^2} + \frac{x_2^2}{a_2^2} - 1\right),$$

of the Prandtl stress function ([23], Section 7-4) was used to assess the accuracy of the local solution. ▲

Example 6.10: Consider a hollow circular shaft with inner radius $a > 0$ and outer radius $b > a$ and let T be the first time an $\mathbb{R}^2$-valued Brownian motion $\boldsymbol{B}$ starting at $\boldsymbol{x} \in D$ leaves $D = \{\boldsymbol{x} \in \mathbb{R}^2 : a <\| \boldsymbol{x} \|< b\}$. The local solution is

$$u(\boldsymbol{x}) = p_2(\boldsymbol{x})\, k_2 + G\, \beta\, E^{\boldsymbol{x}}[T]$$

for $u(\boldsymbol{x}) = k_1 = 0$ and $u(\boldsymbol{x}) = k_2 = 0.5\, G\, \beta(b^2 - a^2)$ on the exterior and the interior boundaries, $\partial D_1 = \{\boldsymbol{x} :\| \boldsymbol{x} \|= b\}$ and $\partial D_2 = \{\boldsymbol{x} :\| \boldsymbol{x} \|= a\}$, respectively (Eqs. 6.14 and 6.26).

No analytical results are available for $E^{\boldsymbol{x}}[T]$ and $p_2(\boldsymbol{x})$. The Monte Carlo simulation method was therefore used. Numerical results have been obtained for $a = 1$, $b = 3$, and $G\,\beta = 2$. The largest errors of the estimated values of the Prandtl function at $\boldsymbol{x} = (1.5, 0)$, $(2, 0)$, and $(2.5, 0)$ are 21% for $n_s = 100$ and 1.35% for $n_s = 1,000$ independent samples of $\boldsymbol{B}$ generated with a time step of $\Delta t = 0.001$. ◇

Note: The Prandtl stress function is $u(\boldsymbol{x}) = -\frac{1}{2} G\, \beta(\boldsymbol{x}^T \boldsymbol{x} - b^2)$ for $a^2 < \boldsymbol{x}^T \boldsymbol{x} < b^2$ ([23], Section 7-9). This expression of u has been used to assess the accuracy of the local solution. ▲

6.2.1.5 Heat and transport equations

Let D be an open bounded subset in $\mathbb{R}^d$, $\alpha > 0$ denote a constant thermal diffusion coefficient, and $u(\boldsymbol{x}, t)$ be the temperature at $\boldsymbol{x} \in D$ and time $t > 0$.

Let u be the solution of the heat equation

$$\frac{\partial u(\boldsymbol{x}, t)}{\partial t} = \alpha^2 \sum_{i=1}^{d} \frac{\partial^2 u(\boldsymbol{x}, t)}{\partial x_i^2} \tag{6.27}$$

with specified initial and boundary conditions giving the functions $u(\boldsymbol{x}, 0)$ for $\boldsymbol{x} \in D$ and $u(\boldsymbol{x}, t)$ for $\boldsymbol{x} \in \partial D$ and $t > 0$, respectively.

Consider an $\mathbb{R}^d$-valued Brownian motion $\boldsymbol{B}$ starting at $\boldsymbol{x} \in D$ and the stopping time $\tilde{T}$ in Eq. 6.13, where $\tilde{\boldsymbol{X}}$ is an $\mathbb{R}^{d+1}$-valued diffusion process starting at $(\tilde{X}_1(0), \ldots, \tilde{X}_d(0)) = \boldsymbol{x} \in D$ and $\tilde{X}_{d+1}(0) = t > 0$. The process $\tilde{\boldsymbol{X}}(s)$, $s \geq 0$, is defined by the following stochastic differential equation.

$$\begin{aligned} d\tilde{X}_i(s) &= \sqrt{2}\,\alpha\, dB_i(s), \quad i = 1, \ldots, d, \\ d\tilde{X}_{d+1}(s) &= -ds. \end{aligned} \tag{6.28}$$

The **local solution** of Eq. 6.27 is

$$u(\boldsymbol{x}, t) = E^{(\boldsymbol{x},t)}\left[u(\tilde{X}(\tilde{T}))\right]. \tag{6.29}$$

Proof: The generator of $\tilde{X}$ is $\mathcal{A} = -\partial/\partial x_{d+1} + \alpha^2\,\Delta$ so that Eq. 6.29 giving the local solution of Eq. 6.27 is a special case of Eq. 6.14 for $\alpha_i = 0$, $\beta_{ij} = 2\,\alpha^2\,\delta_{ij}$, $q = 0$, and $p = 0$. The **steady-state solution** $u_s(\boldsymbol{x}) = \lim_{t\to\infty} u(\boldsymbol{x}, t)$ satisfies the Laplace equation $\Delta u_s = 0$ since $\partial u_s/\partial t = 0$.

The local solution in Eq. 6.29 can be extended to random heterogeneous media. This extension is used in Section 8.5.1 to find effective conductivity coefficients for random heterogeneous media. ■

Example 6.11: Consider a rod of length $l > 0$ with initial temperature $u(x, 0) = 100$, $x \in D = (0, l)$, and boundary temperatures $u(0, t) = u(l, t) = 0$ for $t > 0$. The temperature at an arbitrary time $t > 0$ and location $x \in (0, l)$ is $u(x, t) = 100\,p_3(x, t)$, where $p_1(x, t)$, $p_2(x, t)$, and $p_3(x, t)$ are the probabilities that the $\mathbb{R}^2$-valued diffusion process $\tilde{X} = (X_1, X_2)$ defined by

$$\begin{cases} dX_1(s) &= \sqrt{2}\,\alpha\,dB(s), \quad X_1(0) = x \in (0, l), \\ dX_2(s) &= -ds, \quad X_2(0) = t > 0, \end{cases}$$

exits $D_t = (0, l) \times (0, t)$ for the first time through the boundaries $\{0\} \times (0, t)$, $\{l\} \times (0, t)$, and $(0, l) \times \{0\}$, respectively. The probabilities $p_r(x, t)$ satisfy the condition $\sum_{r=1}^{3} p_r(x, t) = 1$. Estimates of $u(x, t)$ for $\alpha^2 = 1/2$, $l = 1$, $t \in (0, 2)$, $x = 0.1, 0.3, 0.5, 0.7$, and 0.9, $n_s = 1{,}000$ samples, and a time step of $\Delta t = 0.001$ are in error by less than 3%. Figure 6.4 shows three samples of $\tilde{X}$ starting at $(x = 0.25, t = 0.3)$ that exit D_t at times $\tilde{T} < t$ and $\tilde{T} = t$. ◇

Note: The generator of $\tilde{X} = (X_1, X_2)$ is $\mathcal{A} = \alpha^2\,\frac{\partial^2}{\partial x_1^2} - \frac{\partial}{\partial x_2}$ so that (Eq. 6.14)

$$E^{(x,t)}\left[u(\tilde{X}(\tilde{T}))\right] = u(x, t) + E^{(x,t)}\left[\int_0^{\tilde{T}} \mathcal{A}[u(\tilde{X}(s))]\,ds\right] = u(x, t),$$

where $\tilde{T}$ is the first time that $\tilde{X}$ starting at (x, t) exits $D_t = (0, l) \times (0, t)$, $x \in (0, l)$, and $t > 0$. The last equality in the above equation holds since u is the solution of the heat equation, $\tilde{X}(s) \in D_t$ for $s \in (0, \tilde{T})$, and $\mathcal{A}$ coincides with the differential operator of Eq. 6.27 for $d = 1$.

The exact solution is

$$u(x, t) = \sum_{n=1,3,\ldots}^{\infty} \frac{400}{n\,\pi}\,\sin\frac{n\,\pi\,x}{l}\,\exp\{-(n\,\pi\,\alpha/l)^2\,t\}$$

for $x \in (0, l)$ and $t \geq 0$ ([73], Example 26.2, pp. 528-531). ▲

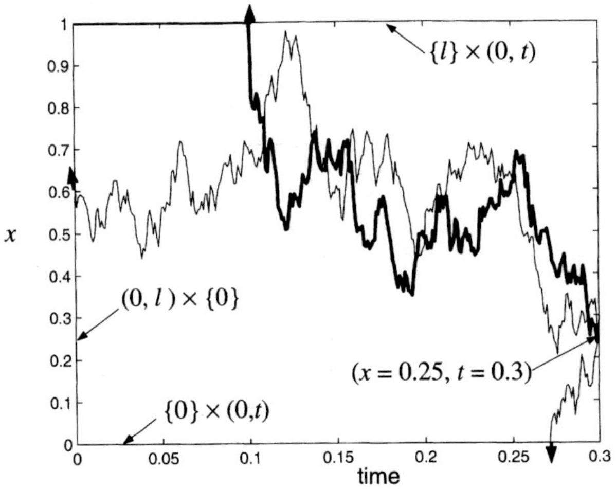

Figure 6.4. Samples of $\tilde{X}$ with generator $\alpha^2\,(\partial^2/\partial x_1^2) - \partial/\partial x_2$

Example 6.12: Consider the mass transport equation

$$\frac{\partial u}{\partial t} = -c\,\frac{\partial u}{\partial x} + d\,\frac{\partial^2 u}{\partial x^2}$$

defined in $D = \{x : x > 0\}$ for $t > 0$ with the initial and boundary conditions $u(x, 0) = 0$ and $u(0, t) = u_0$, $t > 0$, respectively. The local solution of this equation at (x, t) is $u(x, t) = u_0\, P(\tilde{T} < t)$, where $D_t = D \times (0, t)$, $\tilde{T} = \inf\{s > 0 : \tilde{X}(s) \notin D_t\}$, $\tilde{X} = (X_1, X_2)$ is defined by

$$\begin{cases} dX_1(s) &= -c\,ds + (2d)^{1/2}\,dB(s), \\ dX_2(s) &= -ds, \end{cases}$$

and B denotes a Brownian motion. Figure 6.5 shows two samples of $\tilde{X}$ starting at $(x = 0.4, t = 0.2)$ and exiting D_t at times $\tilde{T} < t$ and $\tilde{T} = t$.

Estimates of $u(x, t)$ calculated at several locations x and times t are in error by less than 2% for $c = 1$, $d = 1$, $u_0 = 1$, a time step $\Delta t = 0.001$, and $n_s = 500$ samples. The error can be reduced by increasing the sample size. For example, the error decreases to approximately 1% if $n_s = 1,000$ samples are used in calculations [86]. ◇

Note: The generator $\mathcal{A} = d\,(\partial^2/\partial x_1^2) - c\,(\partial/\partial x_1) - \partial/\partial x_2$ of $\tilde{X}$, Eq. 6.14, and the boundary and initial conditions yield the stated local solution.

The exact solution of the mass transport equation giving the concentration of pollutant at an arbitrary point x and time $t > 0$ is

$$u(x, t) = \frac{u_0}{2}\left[\operatorname{erfc}\left(\frac{x - c\,t}{2\sqrt{d\,t}}\right) + \exp(c\,x/d)\operatorname{erfc}\left(\frac{x + c\,t}{2\sqrt{d\,t}}\right)\right],$$

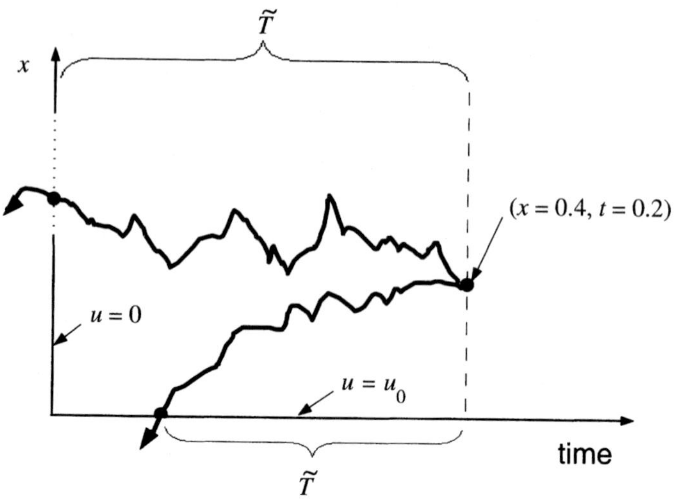

Figure 6.5. Samples of $\tilde{X}$ with generator $\mathcal{A} = d\,(\partial^2/\partial x_1^2) - c\,(\partial/\partial x_1) - \partial/\partial x_2$ for $c = 1$ and $d = 1$

where $\mathrm{erfc}(x) = (2/\sqrt{\pi}) \int_0^\infty e^{-\xi^2}\, d\xi$ is the complementary error function ([195], Example Problem 3.1, p. 65). ▲

6.2.2 Dirichlet boundary conditions ($q \neq 0$)

Our objective is the development of a local solution for Eq. 6.1 with the initial and boundary conditions given by Eqs. 6.4 and 6.10. The method is based on the Itô formula and the Feynman-Kac functional.

6.2.2.1 The Feynman-Kac functional

Let $\boldsymbol{X}$ be an $\mathbb{R}^d$-valued diffusion process defined by the stochastic differential equation

$$d\boldsymbol{X}(s) = \boldsymbol{a}(\boldsymbol{X}(s))\, ds + \boldsymbol{b}(\boldsymbol{X}(s))\, d\boldsymbol{B}(s), \quad s \geq 0, \tag{6.30}$$

starting at $\boldsymbol{X}(0) = \boldsymbol{x}$, where $\boldsymbol{B}$ is a Brownian motion in $\mathbb{R}^d$ and the entries of the $(d, 1)$, (d, d) matrices $\boldsymbol{a}, \boldsymbol{b}$ do not depend explicitly on time. We have seen in the proof of Eq. 6.18 that the generator of $\boldsymbol{X}$ is

$$\mathcal{A} = \sum_{i=1}^{d} a_i(\boldsymbol{x}) \frac{\partial}{\partial x_i} + \frac{1}{2} \sum_{i,j=1}^{d} (\boldsymbol{b}(\boldsymbol{x})\, \boldsymbol{b}(\boldsymbol{x})^T)_{ij} \frac{\partial^2}{\partial x_i\, \partial x_j}. \tag{6.31}$$

Let $\rho, \zeta : \mathbb{R}^d \to \mathbb{R}$ be two functions such that (1) ρ has a compact support and continuous second order partial derivatives, (2) ζ is a bounded Borel measurable function, and (3) the drift and diffusion coefficients in Eq. 6.30 are such that this equation has a unique solution $\boldsymbol{X}$ that belongs a.s. to the support of ρ.

The **Feynman-Kac functional**

$$v(\boldsymbol{x}, t) = E^{\boldsymbol{x}}\left[\rho(\boldsymbol{X}(t))\, e^{\int_0^t \zeta(\boldsymbol{X}(u))\, du}\right] \tag{6.32}$$

is the solution of the partial differential equation

$$\frac{\partial v}{\partial t} = \mathcal{A}[v] + \zeta\, v \tag{6.33}$$

with $v(\boldsymbol{x}, 0) = \rho(\boldsymbol{x})$ and some boundary conditions.

Proof: The derivation of the Feynman-Kac formula involves two steps. First, consider the $\mathbb{R}^2$-valued process (Y, Z) defined by

$$\begin{cases} Y(t) & = \rho(\boldsymbol{X}(t)), \\ Z(t) & = \exp\left(\int_0^t \zeta(\boldsymbol{X}(u))\, du\right), \end{cases}$$

$Y(0) = \rho(\boldsymbol{X}(0)) = \rho(\boldsymbol{x})$, and $Z(0) = 1$. The expectation of the product of these two processes is $v(\boldsymbol{x}, t) = E^{\boldsymbol{x}}[Y(t)\, Z(t)]$ in Eq. 6.32. The Itô formula applied to the above functions of $\boldsymbol{X}$ defining the processes Y and Z gives

$$\begin{cases} dY(t) & = \mathcal{A}[\rho(\boldsymbol{X}(t))]\, dt + \sum_{j=1}^d \left(\sum_{i=1}^d b_{ij}(\boldsymbol{X}(t)) \frac{\partial \rho(\boldsymbol{X}(t))}{\partial x_i}\right) dB_j(t), \\ dZ(t) & = \zeta(\boldsymbol{X}(t))\, Z(t)\, dt. \end{cases}$$

Second, the product $Y(t)\, Z(t)$ can be given in the form

$$\begin{aligned} Y(t)\, Z(t) = \rho(\boldsymbol{X}(t)) &\left[\exp\left(\int_0^h \zeta(\boldsymbol{X}(u))\, du\right) - 1\right] \exp\left(\int_h^t \zeta(\boldsymbol{X}(u))\, du\right) \\ &+ \rho(\boldsymbol{X}(t)) \exp\left(\int_h^t \zeta(\boldsymbol{X}(u))\, du\right) \end{aligned}$$

for $0 < h < t$ so that we have

$$\begin{aligned} v(\boldsymbol{x}, t) = E^{\boldsymbol{x}} &\left\{\left[\exp\left(\int_0^h \zeta(\boldsymbol{X}(u))\, du\right) - 1\right]\right. \\ &\left.\times E^{\boldsymbol{X}(h)}\left[\rho(\boldsymbol{X}(t-h)) \exp\left(\int_0^{t-h} \zeta(\boldsymbol{X}(u))\, du\right)\right]\right\} \\ &+ E^{\boldsymbol{x}}\left\{E^{\boldsymbol{X}(h)}\left[\rho(\boldsymbol{X}(t-h) \exp\left(\int_0^{t-h} \zeta(\boldsymbol{X}(u))\, du\right)\right]\right\} \end{aligned}$$

by averaging, or

$$\begin{aligned} v(\boldsymbol{x}, t) &= E^{\boldsymbol{x}} \left\{ \left[\exp\left(\int_0^h \zeta(\boldsymbol{X}(u))\, du \right) - 1 \right] v(\boldsymbol{X}(h), t - h) \right\} \\ &\quad + E^{\boldsymbol{x}} \left[v(\boldsymbol{X}(h), t - h) \right] \\ &= E^{\boldsymbol{x}} \left[\exp\left(\int_0^h \zeta(\boldsymbol{X}(u))\, du \right) v(\boldsymbol{X}(h), t - h) \right], \end{aligned}$$

so that, for a small $h > 0$, $v(\boldsymbol{x}, t)$ can be approximated by

$$v(\boldsymbol{x}, t) \simeq (1 + h\,\zeta(\boldsymbol{x}) + o(h))\, E^{\boldsymbol{x}} [v(\boldsymbol{X}(h), t - h)].$$

We now apply the Itô formula to the function $v(\boldsymbol{X}(\sigma), t - h)$ of $\boldsymbol{X}(\sigma)$ for $\sigma \in (0, h)$. The expectation of this formula is

$$E^{\boldsymbol{x}} [v(\boldsymbol{X}(h), t - h)] = v(\boldsymbol{x}, t - h) + E^{\boldsymbol{x}} \left[\int_0^h \mathcal{A} [v(\boldsymbol{X}(\sigma), t - h)]\, d\sigma \right].$$

Since h is small, we have

$$E^{\boldsymbol{x}} \left[\int_0^h \mathcal{A} [v(\boldsymbol{X}(\sigma), t - h)]\, d\sigma \right] \simeq h\, \mathcal{A} [v(\boldsymbol{x}, t - h)] + o(h)$$

so that

$$E^{\boldsymbol{x}} [v(\boldsymbol{X}(h), t - h)] \simeq v(\boldsymbol{x}, t - h) + h\, \mathcal{A} [v(\boldsymbol{x}, t - h)] + o(h).$$

The approximation $v(\boldsymbol{x}, t) \simeq (1 + h\,\zeta(\boldsymbol{x}) + o(h))\, E^{\boldsymbol{x}} [v(\boldsymbol{X}(h), t - h)]$ of v established previously and the above approximate expression of $E^{\boldsymbol{x}} [v(\boldsymbol{X}(h), t - h)]$ give

$$v(\boldsymbol{x}, t) \simeq v(\boldsymbol{x}, t - h) + h\,\zeta(\boldsymbol{x})\, v(\boldsymbol{x}, t - h) + h\, \mathcal{A} [v(x, t - h)] + o(h),$$

which yields Eq. 6.33 by moving $v(\boldsymbol{x}, t - h)$ to the left side of this equality, dividing by h, and taking the limit as $h \to 0$.

It can also be shown that a bounded function $w(\boldsymbol{x}, t)$ satisfying Eq. 6.33 coincides with $v(\boldsymbol{x}, t)$ provided that it has the same initial and boundary conditions as v ([134], Theorem 8.2.1, p. 135). ■

Example 6.13: Let Γ denote the time a Brownian motion B starting at zero spends in the positive half line up to a time $t > 0$. The distribution and density of Γ are

$$F(\tau) = \frac{2}{\pi} \arcsin\left(\sqrt{\frac{\tau}{t}} \right) \quad \text{and} \quad f(\tau) = \frac{1}{\pi \sqrt{\tau\,(t - \tau)}}, \qquad \tau \in (0, t).$$

This property of Γ is referred to as the **arcsine law**. ◇

Proof: Take $\rho = 1$, $d = 1$, $\zeta(x) = -\beta\, 1_{\{x > 0\}}$, and $X = B$ in Eq. 6.32, where $\beta > 0$ is arbitrary. Hence, $\int_0^t \zeta(X(u))\, du$ is equal to $-\beta \int_0^t 1_{\{B(u) > 0\}}\, du$, and is proportional to the time B spends in $(0, \infty)$ during $[0, t]$. The corresponding Feynman-Kac functional is the solution of the partial differential equation

$$\frac{\partial v(x,t)}{\partial t} = \begin{cases} -\beta\, v(x,t) + \frac{1}{2}\frac{\partial^2 v(x,t)}{\partial x^2}, & x > 0, \quad t > 0, \\ \frac{1}{2}\frac{\partial^2 v(x,t)}{\partial x^2}, & x \le 0, \quad t > 0. \end{cases}$$

The function v must be bounded and satisfies the continuity conditions $v(0+, t) = v(0-, t)$ and $\partial v(0+, t)/\partial x = \partial v(0-, t)/\partial x$ at each $t > 0$. It can be shown that

$$v(0,t) = \int_0^t \frac{1}{\pi\sqrt{\tau\,(t-\tau)}}\, e^{-\beta\,\tau}\, d\tau = \int_0^t e^{-\beta\,\tau}\, f(\tau)\, d\tau$$

so that the density of Γ is as asserted ([111], pp. 224-226). ■

6.2.2.2 The inhomogeneous Schrödinger equation

Let $\tilde{\boldsymbol{X}} = (\boldsymbol{X}, X_{d+1})$ be the $\mathbb{R}^{d+1}$-valued diffusion process in Eq. 6.11 and let $\tilde{T}$ be the first time $\tilde{\boldsymbol{X}}$ starting at $(\boldsymbol{x}, t)$ exits $D_t = D \times (0, t)$, where $\boldsymbol{x} \in D$ and $t > 0$ (Eq. 6.13). It is assumed that the solution u of Eq. 6.1 and the function q in this equation are such that Eqs. 6.32 and 6.33 can be applied with $\tilde{\boldsymbol{X}}$, u, and q in place of $\boldsymbol{X}$, ρ, and ζ.

The **local solution** of Eq. 6.1 is

$$u(\boldsymbol{x}, t) = E^{(\boldsymbol{x},t)}\left[u(\tilde{\boldsymbol{X}}(\tilde{T}))\, \exp\left(\int_0^{\tilde{T}} q(\tilde{\boldsymbol{X}}(\sigma))\, d\sigma\right)\right] + E^{(\boldsymbol{x},t)}\left[\int_0^{\tilde{T}} p(\tilde{\boldsymbol{X}}(s))\, \exp\left(\int_0^{s} q(\tilde{\boldsymbol{X}}(\sigma))\, d\sigma\right) ds\right]. \tag{6.34}$$

Proof: We first note that the local solution in Eq. 6.14 is a special case of Eq. 6.34 corresponding to $q = 0$.

Consider an $\mathbb{R}^{d+2}$-valued diffusion process $(\tilde{\boldsymbol{X}}, X_{d+2}) = (\boldsymbol{X}, X_{d+1}, X_{d+2} = Z)$ defined by the stochastic differential equation

$$\begin{cases} d\boldsymbol{X}(s) & = \boldsymbol{a}(\boldsymbol{X}(s))\, ds + \boldsymbol{b}(\boldsymbol{X}(s))\, d\boldsymbol{B}(s), \\ dX_{d+1}(s) & = -ds, \\ dZ(s) & = q(\boldsymbol{X}(s), s)\, Z(s)\, ds, \end{cases}$$

for $s \ge 0$ and the initial condition $\boldsymbol{X}(0) = \boldsymbol{x} \in D$, $X_{d+1}(0) = t > 0$, and $Z(0) = 1$. The generator of the diffusion process $(\tilde{\boldsymbol{X}}, Z) = (\boldsymbol{X}, X_{d+1}, X_{d+2} = Z)$ is

$$\mathcal{A}^* = -\frac{\partial}{\partial x_{d+1}} + \sum_{i=1}^{d} a_i\, \frac{\partial}{\partial x_i} + q\, z\, \frac{\partial}{\partial x_{d+2}} + \frac{1}{2}\sum_{i,j=1}^{d} (\boldsymbol{b}\,\boldsymbol{b}^T)_{ij}\, \frac{\partial^2}{\partial x_i\, \partial x_j},$$

and can be applied to an arbitrary function $g(\boldsymbol{x}, s, z)$ with continuous first order partial derivatives in s and z and continuous second order partial derivatives in $\boldsymbol{x}$. For the function $g(\boldsymbol{x}, x_{d+1}, x_{d+2} = z) = \psi(\boldsymbol{x}, x_{d+1})\, z$, we have

$$\mathcal{A}^*[g(\boldsymbol{x}, s, z)] = z \left[-\frac{\partial}{\partial x_{d+1}} + \sum_{i=1}^{d} a_i(\boldsymbol{x}) \frac{\partial}{\partial x_i} + q(\boldsymbol{x}, s) \right.$$
$$\left. + \frac{1}{2} \sum_{i,j=1}^{d} (\boldsymbol{b}(\boldsymbol{x})\, \boldsymbol{b}(\boldsymbol{x})^T)_{ij} \frac{\partial^2}{\partial x_i\, \partial x_j} \right] \psi(\boldsymbol{x}, x_{d+1}).$$

For $g(\tilde{\boldsymbol{X}}, Z) = \psi(\tilde{\boldsymbol{X}})\, Z$ and the solution u of Eq. 6.1 in place of ψ, the Itô formula gives

$$E^{(\boldsymbol{x},t)}[u(\tilde{\boldsymbol{X}}(\tilde{T}))\, Z(\tilde{T})] - u(\tilde{\boldsymbol{X}}(0))\, Z(0) = E^{(\boldsymbol{x},t)} \left[\int_0^{\tilde{T}} \mathcal{A}^*[u(\tilde{\boldsymbol{X}}(s))\, Z(s)]\, ds \right]$$

by averaging. The above formula yields Eq. 6.34 since $(\tilde{\boldsymbol{X}}, Z)$ starts at $(\boldsymbol{x}, t, 1)$ so that $u(\tilde{\boldsymbol{X}}(0))\, Z(0) = u(\boldsymbol{x}, t)$, and $\tilde{\boldsymbol{X}}(s) \in D_t$ for $s \in (0, \tilde{T})$ implies

$$\int_0^{\tilde{T}} \mathcal{A}^*\, [u(\tilde{\boldsymbol{X}}(s))\, Z(s)]\, ds = - \int_0^{\tilde{T}} p(\boldsymbol{X}(s), s)\, Z(s)\, ds$$

for any $(\boldsymbol{x}, t)$ in $D \times (0, \infty)$ (Eq. 6.1). Additional technical considerations on the local solution of the Schrödinger equation can be found in [42] (Section 6.4). ■

Generally, the expectations in Eq. 6.34 cannot be obtained analytically but can be estimated from sample paths of $\tilde{\boldsymbol{X}}$ generated by Monte Carlo simulation. The estimation of these expectations involves a notable computational effort since they involve integrals depending on the entire history of $\tilde{\boldsymbol{X}}$ in D_t. A simpler case is considered in the following equation.

If $q(\boldsymbol{x}) = q$ is a constant, the local solution in Eq. 6.34 becomes

$$u(\boldsymbol{x}, t) = E^{(\boldsymbol{x},t)} \left[u(\tilde{\boldsymbol{X}}(\tilde{T}))\, e^{q\, \tilde{T}} \right] + E^{(\boldsymbol{x},t)} \left[\int_0^{\tilde{T}} p(\tilde{\boldsymbol{X}}(s))\, e^{q\, s}\, ds \right]. \tag{6.35}$$

Example 6.14: If $q(\boldsymbol{x}) = q$ is constant, $p(\boldsymbol{x}) = 0$, $\alpha_i = 0$, and $\beta_{ij} = 2\, \delta_{ij}$ in Eq. 6.8, the local solution of this special case of the Schrödinger equation is

$$u(\boldsymbol{x}) = E^{\boldsymbol{x}} \left[\xi(\sqrt{2}\, \boldsymbol{B}(T))\, e^{q\, T} \right], \quad \boldsymbol{x} \in D,$$

for the boundary condition $u(\boldsymbol{x}) = \xi(\boldsymbol{x})$, where $\boldsymbol{B}$ is a Brownian motion in $\mathbb{R}^d$ starting at $\boldsymbol{x} \in D$ and $T = \inf\{t > 0 : \sqrt{2}\, \boldsymbol{B}(t) \notin D\}$ is a stopping time. The value of u can be estimated at $\boldsymbol{x} \in D$ from n_s samples of T and $\boldsymbol{B}(T)$ by

$$\hat{u}(\boldsymbol{x}) = \frac{1}{n_s} \sum_{\omega=1}^{n_s} \xi \left(\sqrt{2}\, \boldsymbol{B}(T(\omega), \omega) \right) e^{q\, T(\omega)}.$$

Numerical results obtained for $D = \{\boldsymbol{x} \in \mathbb{R}^2 : x_1^2 + x_2^2 < 4\}$, $q = -\sum_{i=1}^{2} a_i^2$, and $\xi(\boldsymbol{x}) = \exp(a_0 + a_1 x_1 + a_2 x_2)$ are in error by less then 1.5% for $a_0 = 0.5$, $a_1 = 0.3$, $a_2 = 0.3$, and $n_s = 500$ samples of $\boldsymbol{B}$ generated with a time step $\Delta t = 0.001$. ◇

Note: The errors were calculated with respect to the exact solution $u(\boldsymbol{x}) = \exp(a_0 + a_1 x_1 + a_2 x_2)$. That this function satisfies the Schrödinger equation results by elementary calculations. ▲

Example 6.15: The local solution of the one-dimensional Schrödinger equation $(1/2)\, u''(x) + q\, u(x) = 0$, $x \in D = (0, l)$, with the boundary conditions $u(0) = \alpha$ and $u(l) = \beta$ is $u(x) = E^x\left[\xi(B(T))\, e^{q\,T}\right]$, where $\xi(B(T)) = \alpha$ or β if $B(T) = 0$ or $B(T) = l$, respectively. The local solution can be estimated by (Eq. 6.35)

$$\hat{u}(x) = \frac{1}{n_s}\left[\alpha \sum_{\omega'=1}^{n'_s} e^{q\,T(\omega')} + \beta \sum_{\omega''=1}^{n''_s} e^{q\,T(\omega'')}\right] = c'(x)\,\alpha + c''(x)\,\beta,$$

where $x \in (0, 1)$, n'_s and n''_s denote the number of samples of a Brownian motion B starting at $x \in D$ that exit $D = (0, l)$ through the left and the right ends of d, $n_s = n'_s + n''_s$ denotes the sample size, and $c'(x)$, $c''(x)$ giving the weights of the boundary values $u(0) = \alpha$, $u(l) = \beta$ result from the above expression of $\hat{u}(x)$. The largest value, $e_{\max}$, of the error $e = |u(x) - \hat{u}(x)|/|u(x)|$ at $x = k/10$, $k = 1, \ldots, 9$, is smaller than 5% for $l = 1$, $\alpha = 1$, $\beta = 2$, $q = 1$, $\Delta t = 0.001$, and $n_s = 500$. The error can be reduced by increasing the sample size and/or reducing the time step Δt used to generate Brownian motion samples. For example, $e_{\max}$ is less than 1.5% for $n_s = 1,000$ and $\Delta t = 0.0005$ [85]. ◇

Note: The exact solution of this Schrödinger equation used to asses the accuracy of the estimator $\hat{u}$ is

$$u(x) = \alpha\,\cos(\sqrt{2\,q}\,x) + \frac{\beta - \alpha\,\cos(\sqrt{2\,q}\,l)}{\sin(\sqrt{2\,q}\,l)}\,\sin(\sqrt{2\,q}\,x).$$

The estimator $\hat{u}$ of u is less accurate when D and/or q is large. This unsatisfactory performance is caused by the dependence of $\hat{u}(x)$ on $\exp(q\,T)$, a random variable that can have a heavy tail. To clarify this statement, consider the special case $\alpha = \beta = 1$, $q = 1$, $l = 2\,a$, and $x = a$ in which $\hat{u}(x)$ is equal to the expected value of $\exp(T)$. Estimates of $E[T]$ based on 1,000 sample paths of B generated with a time step $\Delta t = 0.001$ are 1.0097 for $a = 1$ and 4.0946 for $a = 2$. The corresponding estimates of coefficient of variation v_T of T are 0.7554 and 0.7663, respectively. The estimates of the coefficient of variation of $\exp(T)$ are 1.7506 for $a = 1$ and 31.6228 for $a = 2$. The large coefficients of variation are the cause of unstable estimates for the expectation of $\exp(T)$. In fact, $E[\exp(T)]$ may not be bounded. For example, if T is an exponential random variable with expectation $1/\lambda$ and $\lambda \in (0, 1]$, then $E[\exp(T)] = +\infty$ so that it will not be possible to obtain stable estimates of the mean of $\exp(T)$ from samples of T ([79], Section 2.2.6).

The performance of the estimator $\hat{u}$ can be improved by dividing D in sufficiently small parts for which this estimator is accurate. Suppose that $D = (0, l)$ is divided in $m+1$

equal intervals and let $\alpha_k = u(k\,l/(m+1))$ be the unknown value of u at the division point $x_k = k\,l/(m+1)$, $k = 1, \ldots, m$. The relationships $\alpha_k = c'_k\,\alpha_{k-1} + c''_k\,\alpha_{k+1}$, between the values of u at the division points can be obtained from the functional form of the estimate of $\hat{u}$ applied in the intervals $((k-1)l/(m+1), (k+1)l/(m+1))$. The unknown values of u at the division points can be obtained from these relationships and the boundary conditions giving two additional equations, $\alpha_0 = \alpha$ and $\alpha_{m+1} = \beta$. For example, let $\alpha = 1$, $\beta = 2$, $q = 1$, $l = 2$, and $m = 3$. The relationships between the values of u at the division points are $\alpha_k = (\alpha_{k-1} + \alpha_{k+1})/2$, $k = 1, 2, 3$, by the symmetry of the Brownian motion. These equations and the boundary conditions give $\alpha_1 = 1.25$, $\alpha_2 = 1.50$, and $\alpha_3 = 1.75$. Suppose that $u(0.3)$ needs to be estimated. The estimate of $u(0.3)$ can be obtained by this approach for the domain $(0, 0.5)$ and the boundary conditions $u(0) = 1$, $u(0.5) = 1.25$. The error of this estimate is under 1.2% for $n_s = 500$ and $\Delta t = 0.0005$. ▲

6.2.2.3 The homogeneous Schrödinger equation

Let u be the solution of Eq. 6.8 or 6.9, where $q(\boldsymbol{x}) = \lambda$ is a parameter and $p(\boldsymbol{x}) = 0$. Our objective is to find a local solution for these equations, that is, to find the lowest eigenvalue for Eqs. 6.8 and 6.9. We use Eq. 6.9 to illustrate the construction of the local solution.

Let (λ, u) be the solution of the eigenvalue problem

$$-\frac{1}{2}\,\Delta u(\boldsymbol{x}) = \lambda\, u(\boldsymbol{x}), \quad \boldsymbol{x} \in D,$$
$$u(\boldsymbol{x}) = 0, \quad \boldsymbol{x} \in \partial D. \tag{6.36}$$

The **lowest eigenvalue** of the operator in Eq. 6.36 is

$$\lambda_0 = \sup\{\rho : E^{\boldsymbol{x}}[\exp(\rho\, T)] < \infty, \ \text{ for all } \boldsymbol{x} \in D\}, \tag{6.37}$$

where $T = \inf\{t \geq 0 : \boldsymbol{B}(t) \notin D\}$ and $\boldsymbol{B}(0) = \boldsymbol{x} \in D$.

Note: The trivial solution $u \equiv 0$ always satisfies Eq. 6.36. A constant λ is said to be an eigenvalue of Eq. 6.36 if there is a function u that is not zero everywhere in D and the pair (λ, u) satisfies Eq. 6.36. If D is smooth, the operator $-(1/2)\,\Delta$ has a countable number of distinct eigenvalues $0 < \lambda_0 < \lambda_1 < \cdots$ such that $\lambda_n \to \infty$ as $n \to \infty$ ([43], Volume 1, Section V.5). Also, the eigenvalues of the operator $-(1/2)\,\Delta$ are strictly positive as it can be found from the equality $\langle (1/2)\,\Delta u, u\rangle = -\lambda\,\langle u, u\rangle$ using integration by parts and the fact that u is an eigenfunction, where $\langle u, v\rangle = \int_D u(\boldsymbol{x})\,v(\boldsymbol{x})\,d\boldsymbol{x}$.

Let $\tilde{\boldsymbol{X}} = (\boldsymbol{X} = \boldsymbol{B}, X_{d+1})$ be an $\mathbb{R}^{d+1}$-valued diffusion process, where $d X_{d+1}(t) = dt$, $\boldsymbol{X}(0) = \boldsymbol{x} \in D$, and $X_{d+1}(0) = 0$. The generator of $\tilde{\boldsymbol{X}}$ is $\mathcal{A} = \partial/\partial x_{d+1} + (1/2)\,\Delta = \partial/\partial t + (1/2)\,\Delta$, where $\Delta = \sum_{i=1}^{d} \partial^2/\partial x_i^2$. Suppose that u_ρ is a solution of Eq. 6.36 for $\lambda = \rho > 0$. The Itô formula applied to the function $\exp(\rho\, X_{d+1}(t))\, u_\rho(\boldsymbol{X}(t)) =$

$\exp(\rho\, t)\, u_\rho(\boldsymbol{B}(t))$ gives

$$
\begin{aligned}
u_\rho(\boldsymbol{x}) &= E^{\boldsymbol{x}}\left[e^{\rho\, T}\, u_\rho(\boldsymbol{B}(T))\right] - E^{\boldsymbol{x}}\left[\int_0^T \mathcal{A}[e^{\rho\, s}\, u_\rho(\boldsymbol{B}(s))]\, ds\right] \\
&= E^{\boldsymbol{x}}\left[e^{\rho\, T}\, u_\rho(\boldsymbol{B}(T))\right] - E^{\boldsymbol{x}}\left[\int_0^T e^{\rho\, s}\, (\rho + (1/2)\,\Delta)\, u_\rho(\boldsymbol{B}(s))\, ds\right], \quad \boldsymbol{x} \in D.
\end{aligned}
$$

If $E^{\boldsymbol{x}}[e^{\rho\, T}]$ is finite, then $u_\rho(\boldsymbol{x}) = 0$ for all $\boldsymbol{x} \in D$ since $u_\rho(\boldsymbol{B}(T)) = 0$ by the boundary condition in Eq. 6.36, $\boldsymbol{B}(s) \in D$ for $s \in (0, T)$, and $u_\rho(\boldsymbol{B}(s))$ satisfies Eq. 6.36 for $\lambda = \rho$ by assumption. In this case ρ cannot be an eigenvalue of Eq. 6.36 so that λ_0 must be at least $\sup\{\rho : E^{\boldsymbol{x}}[e^{\rho\, T}] < \infty, \forall \boldsymbol{x} \in D\}$. It can be shown that the lowest eigenvalue of the operator of Eq. 6.36 is given by ([57], Section 8.8, pp. 263-270)

$$\lambda_0 = \sup\{\rho : E^{\boldsymbol{x}}\left[e^{\rho\, T}\right] < \infty, \forall \boldsymbol{x} \in D\}.$$

The use of Eq. 6.37 is impractical because it requires us to estimate the expectation $E^{\boldsymbol{x}}[\exp(\rho\, T)]$ for many values of $\boldsymbol{x} \in D$ and values of ρ. A more efficient Monte Carlo method for estimating the lowest eigenvalue of integral and differential operators is discussed in Section 6.6.2. ▲

6.2.3 Mixed boundary conditions

The RWM in Eqs. 6.14 and 6.18 has been developed to solve locally Eq. 6.1 with Dirichlet boundary conditions. If the unknown function u is not specified everywhere on the boundary ∂D of D, Eqs. 6.14 and 6.18 cannot deliver the local solution of Eq. 6.1.

Example 6.16: Let u be the solution of $u''(x) = -m(x)/\chi$, $x \in D = (0, 1)$, with the boundary conditions $u'(0) = 0$ and $u(1) = 0$, where $m(\cdot)$ is a specified continuous function and $\chi > 0$ is a constant. The local solution in Eq. 6.18 with $d = 1$, $D = (0, 1)$, and $X = B$ gives

$$
\begin{aligned}
u(x) &= E^x[u(B(T))] - E^x\left[\int_0^T \mathcal{A}\,[u(B(s))]\, ds\right] \\
&= E^x[u(B(T))] + \frac{1}{2\,\chi} E^x\left[\int_0^T m(B(s))\, ds\right],
\end{aligned}
$$

where B is a Brownian motion and $T = \inf\{t \geq 0 : B(t) \notin D\}$ is the first time B starting at $x \in D$ exits D. We cannot calculate the value of $u(x)$ from the above equation since B can exit D through both its ends so that $u(B(T))$ is either $u(0)$ or $u(1)$ and $u(0)$ is not known. Hence, the expectation $E^x[u(B(T))]$, and therefore the value $u(x)$ of u, cannot be calculated. ◇

We will see that mixed boundary value problems can be solved locally by using Brownian motion and diffusion processes, but these processes have to be reflected toward the interior of D when they reach ∂D at a point where Neumann conditions are prescribed. Also, we will have to extend the classical Itô formula to handle functions of reflected Brownian motion and diffusion processes.

6.2.3.1 Brownian motion reflected at zero

Let B be a Brownian motion. The absolute value of B, that is, the process $|B|$, is called the **Brownian motion reflected at zero**. We (1) state without proof Tanaka's formula providing a representation of the Brownian motion reflected at zero, (2) prove a first extension of Tanaka's formula for the process $g(|B|)$, $g \in C^2(\mathbb{R})$, and (3) apply the extended version of Tanaka's formula to solve locally some mixed boundary value problems.

The **local time** process is

$$L(t) = \lim_{\varepsilon \downarrow 0} \frac{1}{2\varepsilon} \int_0^t 1_{(-\varepsilon,\varepsilon)}(B(s))\, ds, \tag{6.38}$$

where the limit exists in L_2 and a.s. ([42], Chapter 7).

Note: The local time L (1) is adapted to the natural filtration of B, (2) has continuous increasing samples starting at zero, so that it is of finite variation on compacts, and (3) can increase a.s. only when B is zero, that is, $\int_0^\infty 1_{\{t: B(t) \neq 0\}} dL(t) = 0$ a.s. ([42], Theorem 7.6, p. 150). An extensive discussion on the local time process can be found in [42] (Chapters 7 and 8). ▲

Example 6.17: The expectation of the local time process L exists and is equal to $E[L(t)] = \sqrt{2\,t/\pi}$ for all $t \geq 0$. ◇

Proof: The integral $\int_0^t 1_{(-\varepsilon,\varepsilon)}(B(s))\, ds$ in the definition of the local time process is bounded by t for any $\varepsilon > 0$ so that

$$E\left[\int_0^t 1_{(-\varepsilon,\varepsilon)}(B(s))\, ds\right] = \int_0^t E\left[1_{(-\varepsilon,\varepsilon)}(B(s))\right] ds = \int_0^t \left[2\,\Phi(\varepsilon/\sqrt{s}) - 1\right] ds$$

by Fubini's theorem and properties of the Brownian motion. Hence,

$$E\left[\int_0^t 1_{(-\varepsilon,\varepsilon)}(B(s))\, ds\right] \simeq \int_0^t \left(\frac{2\,\varepsilon}{\sqrt{2\,\pi\, s}} + o(\varepsilon)\right) ds = \frac{4\,\varepsilon\,\sqrt{t}}{\sqrt{2\,\pi}} + o(\varepsilon), \quad \varepsilon \downarrow 0,$$

and $\left(4\,\varepsilon\,\sqrt{t}/\sqrt{2\,\pi} + o(\varepsilon)\right)/(2\,\varepsilon)$ converges to $\sqrt{2\,t/\pi}$ as $\varepsilon \downarrow 0$ so that the expectation of $L(t)$ is as stated.

Tanaka's formula given later in this section (Eq. 6.39) provides an alternative way for calculating $E[L(t)]$. According to this formula we have $E[|B(t)|] - E[|B(0)|] = E[\hat{B}(t)] + E[L(t)]$, where $\hat{B}$ is a Brownian motion. For $B(0) = 0$ we have $E[L(t)] = E[|B(t)|]$, which gives the stated expectation for $L(t)$ since $B(t) \sim N(0, t)$. ■

Example 6.18: Let X be a real-valued mean square differentiable stationary Gaussian process with mean zero and unit variance. The expectation of the local time of X is $E[L(t)] = t/\sqrt{2\,\pi}$ for all $t \geq 0$, where L is given by Eq. 6.38 with X in place of B.

We note that the expectations of the local time process $L(t)$, $t \geq 0$, for a Brownian and an m.s. differentiable Gaussian process with mean zero and variance 1 are proportional to $t^{1/2}$ and t, respectively. ◇

Proof: The mean rate at which X crosses zero is $\dot{\sigma}/\pi$, where $\dot{\sigma}^2 = E[\dot{X}(t)^2]$ (Section 3.10.1). Let $N(t)$ be the number of zero-crossings of X during a time interval $[0, t]$ and let D_i be the largest time interval including the time T_i of the zero-crossing i of X such that $X(s) \in (-\varepsilon, \varepsilon)$ for all $s \in D_i$. The approximate length of D_i is $2\,\varepsilon/\dot{X}(T_i)$ and has the expectation $2\,\varepsilon\,\sqrt{\pi/2}/\dot{\sigma}$ since the density of $\dot{X}$ at the time of a zero-crossing of X is $f(z) = (z/\dot{\sigma}^2)\exp(-z^2/(2\,\dot{\sigma}^2))$, $z \geq 0$.

The expected value of the total residence $D(t) = \sum_{i=1}^{N(t)} D_i$ of X in $(-\varepsilon, \varepsilon)$ during a time interval $(0, t)$ is

$$E[D(t)] = E\left\{E\left[\sum_{i=1}^{N(t)} D_i \mid N(t)\right]\right\} = E[N(t)]\,E\left[\frac{2\,\varepsilon}{\dot{X}(T_i)}\right] = \frac{\dot{\sigma}\,t}{\pi}\,\frac{2\,\varepsilon\,\sqrt{\pi/2}}{\dot{\sigma}}$$

so that $E[D(t)]/(2\,\varepsilon) = t/\sqrt{2\pi}$. Note that the expectation of the local time is a fixed fraction of t for stationary mean square differentiable Gaussian processes. ■

The following equation provides a representation of the Brownian motion reflected at zero, that is based on the local time process in Eq. 6.38. The representation is very useful for both theoretical developments and applications.

Tanaka's formula. If B is a Brownian motion, then

$$\begin{aligned} |B(t)| - |B(0)| &= \int_0^t \operatorname{sign}(B(s))\,dB(s) + \lim_{\varepsilon \downarrow 0} \frac{1}{2\varepsilon} \int_0^t 1_{(-\varepsilon,\varepsilon)}(B(s))\,ds \\ &= \hat{B}(t) + L(t) \quad \text{a.s.,} \end{aligned} \tag{6.39}$$

where the limit is in L_2 and $\hat{B}$ is a Brownian motion ([42], Section 7.3).

Note: The process $\hat{B}(t) = |B(0)| + \int_0^t \operatorname{sign}(B(s))\,dB(s)$ is a square integrable martingale with continuous samples and quadratic variation (Section 4.5.3)

$$[\hat{B}, \hat{B}](t) = \int_0^t (\operatorname{sign}(B(s)))^2\, d[B, B](s) = t \quad \text{a.s.,}$$

where $\operatorname{sign}(x)$ is -1, 0, and 1 for $x < 0$, $x = 0$, and $x > 0$, respectively. Hence, $\hat{B}$ is a Brownian motion (Section 4.6.2). The second integral on the right side of Eq. 6.39 is the local time process $L(t)$ in Eq. 6.38. Figure 6.6 shows a few sample paths of B and the corresponding samples of the Brownian motion process $\hat{B}(t) = \int_0^t \operatorname{sign}(B(s))\,dB(s)$ for $B(0) = 0$.

The Tanaka formula also shows that $|B|$ is a semimartingale since $\hat{B}$ is a square integrable martingale and L is a continuous increasing process that is adapted to the natural filtration of B (Section 4.4.1). ▲

Let h_ε be a real-valued function defined by $h_\varepsilon(x) = |x|$ for $|x| \geq \varepsilon$ and $h_\varepsilon(x) = \varepsilon/2 + x^2/(2\,\varepsilon)$ for $|x| \leq \varepsilon$, where $\varepsilon > 0$ is arbitrary (Fig. 6.7).

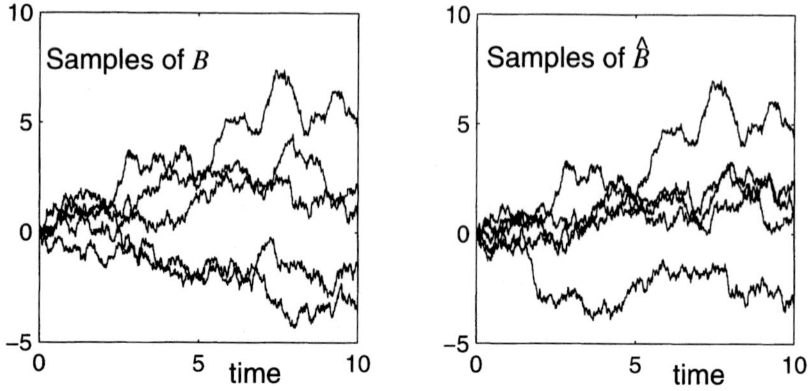

Figure 6.6. Five sample paths of a Brownian motion B and the corresponding samples of $\hat{B}$ for $B(0) = 0$

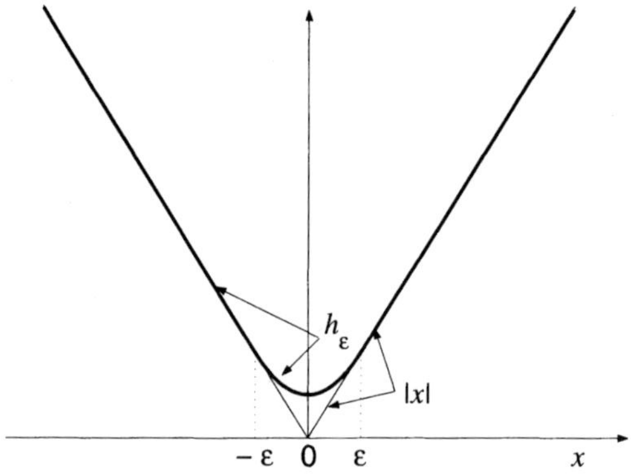

Figure 6.7. The approximation $h_\varepsilon(x)$ of $|x|$

Extended local time process. If $g \in \mathcal{C}^2(\mathbb{R})$,

$$\tilde{L}(t) = \int_0^t g'(|B(s)|)\, dL(s) = g'(0)\, L(t). \tag{6.40}$$

Note: If g is the identity function, then the process $\tilde{L}$ becomes the local time process L defined by Eq. 6.38. ▲

First extension of Tanaka's formula. If $g \in \mathcal{C}^2(\mathbb{R})$, then

$$g(|B(t)|) - g(|B(0)|)$$
$$= \int_0^t g'(|B(s)|)\,\text{sign}(B(s))\,dB(s) + \frac{1}{2}\int_0^t g''(|B(s)|)\,ds + \tilde{L}(t), \qquad (6.41)$$

where $\tilde{L}(t) = g'(0)\,L(t)$ (Eq. 6.40).

Proof: The classical Itô formula cannot be applied to $g(X) = g(|B|)$ since the mapping $B \mapsto g(|B|)$ is not differentiable. Formal calculations using Itô's formula yield

$$g(|B(t)|) - g(|B(0)|) = \int_0^t g'(|B(s)|)\,\text{sign}(B(s))\,dB(s) + \frac{1}{2}\int_0^t g''(|B(s)|)\,ds + \int_0^t g'(|B(s)|)\,\delta(B(s))\,ds$$

since

$$dg(|x|)/dx = g'(|x|)\,d|x|/dx = g'(|x|)\,\text{sign}(x) \quad \text{and}$$
$$d^2g(|x|)/dx^2 = g''(|x|)\,(\text{sign}(x))^2 + g'(|x|)\,d\,\text{sign}(x)/dx = g''(|x|) + 2\,g'(|x|)\,\delta(x),$$

where δ is the Dirac delta function. We show that the above equation holds if the integral $\int_0^t g'(|B(s)|)\,\delta(B(s))\,ds$ is the process $\tilde{L}$ in Eq. 6.40. The formula in Eq. 6.41 can be obtained from Tanaka's formula or by direct calculations. If g is the identity function, Eq. 6.41 coincides with Tanaka's formula.

Tanaka's formula: Because g has a continuous second derivative and the $\mathbb{R}^2$-valued process $(\hat{B}, L)$ is a semimartingale, the classical Itô formula can be applied to $g(X) = g(\hat{B} + L)$ and gives

$$\begin{aligned} g(|B(t)|) - g(|B(0)|) &= \int_0^t g'(|B(s)|)\,d(\hat{B} + L)(s) + \frac{1}{2}\int_0^t g''(|B(s)|)\,ds \\ &= \int_0^t g'(|B(s)|)\,d\hat{B}(s) + \frac{1}{2}\int_0^t g''(|B(s)|)\,ds + \int_0^t g'(|B(s)|)\,dL(s) \\ &= \int_0^t g'(|B(s)|)\,d\hat{B}(s) + \frac{1}{2}\int_0^t g''(|B(s)|)\,ds + g'(0)\,L(t), \end{aligned}$$

that is, the result in Eq. 6.41.

Direct calculations: The proof of Eq. 6.41 involves three steps. First, a sequence of approximations $h_{n,\varepsilon}(x)$ converging to $|x|$ is constructed, where $h_{n,\varepsilon} \in \mathcal{C}^2(\mathbb{R})$ for each integer $n \geq 1$ and $\varepsilon > 0$. Second, the Itô formula is applied to the sequence of approximations $g(h_{n,\varepsilon}(B))$ of $g(|B|)$. Third, it is shown that the resulting sequence of equations giving the increment of $g(h_{n,\varepsilon}(B))$ in a time interval $[0, t]$ converges to Eq. 6.41 as $n \to \infty$ and $\varepsilon \downarrow 0$. Similar steps are used in [42] (Chapter 7) to prove Tanaka's formula.

1. Let h_ε be the function in Fig. 6.7 defined for some $\varepsilon > 0$. The first derivative of h_ε is $h'_\varepsilon(x) = -1, x/\varepsilon$, and 1 for $x \leq \varepsilon$, $x \in [-\varepsilon, \varepsilon]$, and $x \geq \varepsilon$, respectively. The function h_ε is not twice differentiable at $x = \pm\varepsilon$, and we set h''_ε zero at $x = \pm\varepsilon$. With this definition the second derivative of h_ε is $h''_\varepsilon(x) = (1/\varepsilon)\,1_{(-\varepsilon,\varepsilon)}(x)$.

Consider the sequence of functions

$$h_{n,\varepsilon}(x) = \int_{-\infty}^{\infty} h_\varepsilon(x-y)\,\varphi_n(y)\,dy = \int_{-\infty}^{\infty} h_\varepsilon(z)\,\varphi_n(x-z)\,dz,$$

where the functions $\varphi_n \in C^\infty(\mathbb{R})$ have compact support shrinking to $\{0\}$ as $n \to \infty$, for example, $\varphi_n(y) = n\,\varphi(n\,y)$ and $\varphi(y) = c\,\exp(-(1-y^2)^{-1})$ for $|y| < 1$ and $\varphi(y) = 0$ for $|y| \geq 1$, where $c > 0$ is a constant such that $\int_{-1}^{1} \varphi(y)\,dy = 1$. The functions $h_{n,\varepsilon}$ are infinitely differentiable and are such that (1) $h_{n,\varepsilon} \to h_\varepsilon$, $h'_{n,\varepsilon} \to h'_\varepsilon$ uniformly in $\mathbb{R}$ as $n \to \infty$ and (2) $h''_{n,\varepsilon} \to h''_\varepsilon$ pointwise as $n \to \infty$ except at $x = \pm\varepsilon$.

Consider first the functions $h_{n,\varepsilon}$. Since

$$|h_{n,\varepsilon}(x) - h_\varepsilon(x)| \leq \int_{-\infty}^{\infty} |h_\varepsilon(x-y) - h_\varepsilon(x)|\,\varphi_n(y)\,dy, \quad x \in \mathbb{R},$$

it is sufficient to show that for $\eta > 0$ arbitrary there is $n(\eta)$ such that $\int_{-\infty}^{\infty} |h_\varepsilon(x-y) - h_\varepsilon(x)|\,\varphi_n(y)\,dy < \eta$ for all $n > n(\eta)$ and for all $x \in \mathbb{R}$. The function h_ε is uniformly continuous so that there is a $\delta > 0$ for which $|h_\varepsilon(u) - h_\varepsilon(v)| < \eta$ if $|u - v| < \delta$. Divide the domain of integration of $\int_{-\infty}^{\infty} |h_\varepsilon(x-y) - h_\varepsilon(x)|\,\varphi_n(y)\,dy$ in two intervals, $(-\delta, \delta)$ and $(-\delta, \delta)^c$. The integral on $(-\delta, \delta)$ is smaller than η for all n. The integral on $(-\delta, \delta)^c$ is equal to zero for n large enough so that there is $n(\eta)$ such that $|h_{n,\varepsilon}(x) - h_\varepsilon(x)|$ is smaller than η for all $n > n(\eta)$. The same arguments can be used to prove the uniform convergence of $h'_{n,\varepsilon}$ to h'_ε. That $h''_{n,\varepsilon}$ converges to h''_ε pointwise for $x \neq \pm\varepsilon$ results by similar considerations. At $x = \varepsilon$, the difference $h''_{n,\varepsilon}(\varepsilon) - h''_\varepsilon(\varepsilon)$ is $\int_{-\infty}^{\infty} \left(h''_\varepsilon(\varepsilon - y) - h''_\varepsilon(\varepsilon)\right)\varphi_n(y)\,dy = (1/\varepsilon)\int_0^{2\varepsilon} \varphi_n(y)\,dy$ and approaches $1/(2\varepsilon) \neq 0$ as $n \to \infty$. Hence, $h''_{n,\varepsilon}$ does not converge to h''_ε at $x = \varepsilon$ and at $x = -\varepsilon$.

2. The Itô formula applied to $B(t) \mapsto (g \circ h_{n,\varepsilon})(B(t)) = g(h_{n,\varepsilon}(B(t)))$ gives

$$\begin{aligned} g(h_{n,\varepsilon}(B(t))) - g(h_{n,\varepsilon}(B(0))) &= \int_0^t g'(h_{n,\varepsilon}(B(s)))\,h'_{n,\varepsilon}(B(s))\,dB(s) \\ &+ \frac{1}{2}\int_0^t g''(h_{n,\varepsilon}(B(s)))\,(h'_{n,\varepsilon}(B(s)))^2\,ds \\ &+ \frac{1}{2}\int_0^t g'(h_{n,\varepsilon}(B(s)))\,h''_{n,\varepsilon}(B(s))\,ds \quad \text{a.s.} \end{aligned} \tag{6.42}$$

for each $t \geq 0$. The expectation,

$$\begin{aligned} E^x[g(h_{n,\varepsilon}(B(t)))] - g(h_{n,\varepsilon}(x)) &= \frac{1}{2}E^x\left[\int_0^t g''(h_{n,\varepsilon}(B(s)))\,(h'_{n,\varepsilon}(B(s)))^2\,ds\right] \\ &+ \frac{1}{2}E^x\left[\int_0^t g'(h_{n,\varepsilon}(B(s)))\,h''_{n,\varepsilon}(B(s))\,ds\right], \end{aligned}$$

of the above equation with $B(0) = x$ can be used to estimate the local solution of mixed boundary value problems by Monte Carlo simulation, as illustrated later in this section. The stochastic integral $\int_0^t g'(h_{n,\varepsilon}(B(s)))\,h'_{n,\varepsilon}(B(s))\,dB(s)$ is a martingale starting at zero so that its expectation is zero.

The first extension of the Tanaka formula in Eq. 6.41 results by taking the limits of Eq. 6.42 as $n \to \infty$ and $\varepsilon \downarrow 0$. Before performing these operations we make the following observation. Consider the sequence $T_q = \inf\{t \geq 0 : |B(t)| \geq q\}$, $q = 1, 2, \ldots$, of stopping times and let B^{T_q} be the Brownian motion B stopped at T_q. The process B^{T_q} is bounded by q. If Eq. 6.41 with B^{T_q} in place of B is valid for each q, then this equation is valid. Hence, we can assume that B takes values in a compact set. This assumption is made in the following step.

3(a). For each $t \geq 0$ and $g \in \mathcal{C}^2(\mathbb{R})$, we have almost surely

$$g(h_\varepsilon(B(t))) - g(h_\varepsilon(B(0))) = \int_0^t g'(h_\varepsilon(B(s)))\, h'_\varepsilon(B(s))\, dB(s)$$
$$+ \frac{1}{2}\int_0^t g''(h_\varepsilon(B(s)))\,(h'_\varepsilon(B(s)))^2\, ds + \frac{1}{2}\int_0^t g'(h_\varepsilon(B(s)))\, h''_\varepsilon(B(s))\, ds. \tag{6.43}$$

Consider first the left side of Eq. 6.42. By the definitions and the properties of $h_{n,\varepsilon}$ and h_ε the sequence of functions $g \circ h_{n,\varepsilon}$ converges to $g \circ h_\varepsilon$ uniformly in $\mathbb{R}$ as $n \to \infty$ so that $(g \circ h_{n,\varepsilon})(B)$ converges to $(g \circ h_\varepsilon)(B)$ in L_2 and a.s. Hence, the left side of Eq. 6.42 converges to the left side of Eq. 6.43 in L_2 and a.s. for each $t \geq 0$ as $n \to \infty$.

Consider the first integral on the right side of Eq. 6.42. For each $t \geq 0$ the sequence of functions $1_{[0,t]}\,(g' \circ g_{n,\varepsilon})\,(g'_{n,\varepsilon})(B)$ converges to $1_{[0,t]}\,(g' \circ h_\varepsilon)\,(h'_\varepsilon)(B)$ uniformly on $[0, \infty) \times \Omega$ as $n \to \infty$. This observation and the equality (Section 4.5.3)

$$E\left[\left(\int_0^t g'(h_{n,\varepsilon}(B(s)))\, h'_{n,\varepsilon}(B(s))\, dB(s) - \int_0^t g'(h_\varepsilon(B(s)))\, h'_\varepsilon(B(s))\, dB(s)\right)^2\right]$$
$$= \int_0^t E\left[\left(g'(h_{n,\varepsilon}(B(s)))\, h'_{n,\varepsilon}(B(s)) - g'(h_\varepsilon(B(s)))\, h'_\varepsilon(B(s))\right)^2\right] ds,$$

imply the L_2 convergence of $\int_0^t g'(h_{n,\varepsilon}(B(s)))\, h'_{n,\varepsilon}(B(s))\, dB(s)$ to the stochastic integral $\int_0^t g'(h_\varepsilon(B(s)))\, h'_\varepsilon(B(s))\, dB(s)$ as $n \to \infty$. The a.s. convergence of this sequence of integrals results by bounded convergence.

Consider the second integral on the right side of Eq. 6.42. The sequence of functions $(g'' \circ h_{n,\varepsilon})\,(h'_{n,\varepsilon})$ converges to $(g'' \circ h_\varepsilon)\,(h'_\varepsilon)$ uniformly in $\mathbb{R}$ as $n \to \infty$ so that $(g'' \circ h_{n,\varepsilon})\,(h'_{n,\varepsilon})(B)$ converges to $(g'' \circ h_\varepsilon)\,(h'_\varepsilon)(B)$ in L_2. Because the functions $(g'' \circ h_{n,\varepsilon})\,(h'_{n,\varepsilon})(B)$ are bounded a.s. on $[0, t]$, $\int_0^t g''(h_{n,\varepsilon}(B(s)))\,(h'_{n,\varepsilon}(B(s)))^2\, ds$ converges to $\int_0^t g''(h_\varepsilon(B(s)))\,(h'_\varepsilon(B(s)))^2\, ds$ in L_2 and a.s. as $n \to \infty$ by bounded convergence.

Consider now the last integral in Eq. 6.42. The integrand $g'(h_{n,\varepsilon}(B(s)))\, h''_{n,\varepsilon}(B(s))$ converges to $g'(h_\varepsilon(B(s)))\,(h''_\varepsilon(B(s)))$ as $n \to \infty$ for $\omega \in \Omega \setminus N(s)$ and a fixed $s \geq 0$, where $N(s) = \{\omega : B(s, \omega) = \pm\varepsilon\} \subset \Omega$ and $P(N(s)) = 0$. Hence,

$$\lim_{n\to\infty} g'(h_{n,\varepsilon}(B(s)))\, h''_{n,\varepsilon}(B(s)) = g'(h_\varepsilon(B(s)))\, h''_\varepsilon(B(s))$$

holds a.s. for each fixed $s \geq 0$. Let $N = \{(s, \omega) \in [0, t] \times \Omega : B(s, \omega) = \pm\varepsilon\}$ be a subset of $[0, t] \times \Omega$ where the preceding equality does not hold. The set N is $\lambda \times P$-measurable, where λ denotes the Lebesgue measure. This set has measure zero since $\int_N (\lambda \times P)(ds, d\omega) = \int_{[0,t]} \lambda(ds) \int_{N(s)} P(d\omega)$ (Fubini's theorem) and $\int_{N(s)} P(d\omega) = 0$ for each $s \in [0, t]$. Therefore,

$$\lim_{n\to\infty} g'(h_{n,\varepsilon}(B(s)))\, h''_{n,\varepsilon}(B(s)) = g'(h_\varepsilon(B(s)))\, h''_\varepsilon(B(s))$$

is true for λ-almost all $s \geq 0$, almost surely. Hence, the third integral on the right side of Eq. 6.42 converges to $\int_0^t g'(h_\varepsilon(B(s)))\, h''_\varepsilon(B(s))\, ds$ in L_2 and a.s. for each $t \geq 0$ as $n \to \infty$ by bounded convergence (Section 2.5.2.2). This completes the proof of Eq. 6.43.

3(b). For each $t \geq 0$ we have a.s.

$$g(|B(t)|) - g(|B(0)|) = \int_0^t g'(|B(s)|)\,\mathrm{sign}(B(s))\,dB(s)$$
$$+ \frac{1}{2}\int_0^t g''(|B(s)|)\,ds + \lim_{\varepsilon \downarrow 0} \frac{1}{2\varepsilon}\int_0^t g'(h_\varepsilon(B(s)))\,1_{(-\varepsilon,\varepsilon)}(B(s))\,ds, \tag{6.44}$$

where the above limit is in L_2 and is equal to $g'(0)\,L(t)$.

Consider first the left side of Eqs. 6.43 and 6.44. The function $(g \circ h_\varepsilon)(B(t))$ converges to $g(|B(t)|)$ in L_2 and a.s. as $\varepsilon \downarrow 0$ since these functions differ only in $(-\varepsilon, \varepsilon)$, their difference is bounded in this interval so that $E\left[(g(|B(t)|) - g(h_\varepsilon)(B(t)))^2\right]$ is smaller than a finite constant times the probability $P(|B(t)| \leq \varepsilon)$, and this probability approaches zero as $\varepsilon \downarrow 0$.

Consider now the integrals on the right side of Eqs. 6.43 and 6.44. The second moment of the difference

$$I(t) = \int_0^t g'(h_\varepsilon(B(s)))\,h'_\varepsilon(B(s))\,dB(s) - \int_0^t g'(|B(s)|)\ \mathrm{sign}(B(s))\,dB(s)$$

of the first integrals on the right side of Eqs. 6.43 and 6.44 is (Section 4.5.3)

$$E[I(t)^2] = \int_0^t E\left[\left(g'(h_\varepsilon(B(s)))h'_\varepsilon(B(s)) - g'(|B(s)|)\,\mathrm{sign}(B(s))\right)^2\right] ds.$$

The above integrand is bounded by

$$2\,E\,[(g'(h_\varepsilon(B(s))) - g'(|B(s)|))^2 (h'_\varepsilon(B(s)))^2]$$
$$+ 2E[g'(|B(s)|)^2 (h'_\varepsilon(B(s)) - \mathrm{sign}(B(s)))^2]$$

and this bound converges to zero as $\varepsilon \downarrow 0$ since $(h'_\varepsilon(x)))^2 \leq 1$, $|h'_\varepsilon(x) - \mathrm{sign}(x)| \leq 1_{(-\varepsilon,\varepsilon)}(x)$, and g' is continuous.

The expectation of the square of the difference,

$$J(t) = \int_0^t g''(h_\varepsilon(B(s)))\,(h'_\varepsilon(B(s)))^2\,ds - \int_0^t g''(|B(s)|))\,ds,$$

of the second integrals on the right side of Eqs. 6.43 and 6.44 is

$$E[J(t)^2] = E\left[\left(\int_0^t (g''(h_\varepsilon(B(s))) - g''(|B(s)|))\,(h'_\varepsilon(B(s)))^2\,ds\right.\right.$$
$$\left.\left. + \int_0^t g''(|B(s)|)\,(h'_\varepsilon(B(s))^2 - 1)\,ds\right)^2\right] = E\left[(J_1(t) + J_2(t))^2\right]$$
$$\leq 2\,E\left[J_1(t)^2\right] + 2\,E\left[J_2(t)^2\right],$$

where J_1 and J_2 denote the first and second integrals, respectively, in the above equation.

Also, we have

$$|J_1(t)| \le \int_0^t |g''(h_\varepsilon(B(s))) - g''(|B(s)|)| \, (h'_\varepsilon(B(s)))^2 \, ds$$
$$\le \int_0^t |g''(h_\varepsilon(B(s))) - g''(|B(s)|)| \, 1_{(-\varepsilon,\varepsilon)}(B(s)) \, ds \le c_1 \int_0^t 1_{(-\varepsilon,\varepsilon)}(B(s)) \, ds,$$
$$|J_2(t)| \le \int_0^t |g''(|B(s)|)| \, |h'_\varepsilon(B(s))^2 - 1| \, ds \le c_2 \int_0^t 1_{(-\varepsilon,\varepsilon)}(B(s)) \, ds,$$

where the constants $c_1, c_2 > 0$ exist since the functions g'' and h_ε are continuous and the Brownian motion is restricted to the interval $(-\varepsilon, \varepsilon)$. Hence,

$$E[J_i(t)^2] \le c_i^2 \, E\left[\left(\int_0^t 1_{(-\varepsilon,\varepsilon)}(B(s)) \, ds\right)^2\right], \quad i = 1, 2,$$

and

$$E\left[\left(\int_0^t 1_{(-\varepsilon,\varepsilon)}(B(s)) \, ds\right)^2\right] = t^2 \int_\Omega \left(\frac{1}{t} \int_0^t 1_{(-\varepsilon,\varepsilon)}(B(s)) \, ds\right)^2 dP$$
$$\le t^2 \int_\Omega \left(\frac{1}{t} \int_0^t 1_{(-\varepsilon,\varepsilon)}(B(s)) \, ds\right) dP = t \int_0^t \left(\int_\Omega 1_{(-\varepsilon,\varepsilon)}(B(s)) \, dP\right) ds$$
$$= t \int_0^t P(|B(s)| < \varepsilon) \, ds = t \int_0^t \left(\Phi(\varepsilon/\sqrt{s}) - \Phi(-\varepsilon/\sqrt{s})\right) ds \to 0, \quad \varepsilon \to 0,$$

proving the convergence of $\int_0^t g''(h_\varepsilon(B(s))) \, (h'_\varepsilon(B(s)))^2 \, ds$ to $\int_0^t g''(|B(s)|) \, ds$ in L_2 as $\varepsilon \downarrow 0$. Hence, the limit in Eq. 6.44 must also exist in L_2 and the equality in Eq. 6.41 holds a.s. for each fixed $t \ge 0$. That $\lim_{\varepsilon \downarrow 0} (1/2\,\varepsilon) \int_0^t g'(h_\varepsilon(B(s))) \, 1_{(-\varepsilon,\varepsilon)}(B(s)) \, ds$ is equal to $g'(0) \, L(t)$ follows from the observations that $\int_0^t 1_{(-\varepsilon,\varepsilon)}(B(s)) \, ds$ converges to zero as $\varepsilon \downarrow 0$ and

$$\left| \lim_{\varepsilon \downarrow 0} \frac{1}{2\,\varepsilon} \int_0^t g'(h_\varepsilon(B(s))) \, 1_{(-\varepsilon,\varepsilon)}(B(s)) \, ds - g'(0) \, L(t) \right|$$
$$= \left| \lim_{\varepsilon \downarrow 0} \frac{1}{2\,\varepsilon} \int_0^t \left[g'(h_\varepsilon(B(s))) - g'(0)\right] 1_{(-\varepsilon,\varepsilon)}(B(s)) \, ds \right|$$
$$\le \lim_{\varepsilon \downarrow 0} \frac{1}{2\,\varepsilon} \int_0^t \left|g'(h_\varepsilon(B(s))) - g'(0)\right| 1_{(-\varepsilon,\varepsilon)}(B(s)) \, ds$$
$$\le \lim_{\varepsilon \downarrow 0} \frac{c\,\varepsilon}{2\,\varepsilon} \int_0^t 1_{(-\varepsilon,\varepsilon)}(B(s)) \, ds,$$

where $c > 0$ and ε bound $|g''(x)|$ and $h_\varepsilon(x)$, respectively, for $x \in (-\varepsilon, \varepsilon)$. ■

The expectation of Eq. 6.41 for $|B(0)| = x$ is

$$E^x[g(|B(t)|)] - g(x) = \frac{1}{2} E^x\left[\int_0^t g''(|B(s)|) \, ds\right] + g'(0) \, E^x[L(t)]. \quad (6.45)$$

Note: This formula can be used to find the local solution for mixed boundary value problems. For example, if g is the solution of a partial differential equation defined in $(0, 1)$ and g' is prescribed at $x = 0$, the last term in the above equation captures the contribution of this boundary condition to the local solution. ▲

The process $\tilde{L}$ in Eq. 6.40 can be approximated by

$$\tilde{L}(t, \omega) \simeq \frac{g'(0)}{2\,\varepsilon} \int_0^t 1_{(-\varepsilon,\varepsilon)}(B(s, \omega)\, ds,$$

for a small $\varepsilon > 0$, that is, by the time B spends in the range $(-\varepsilon, \varepsilon)$ during the time interval $(0, t)$ scaled by $g'(0)/(2\,\varepsilon)$. The approximation has been used in the following two examples to find local solutions by Monte Carlo simulation.

Example 6.19: Let u be the solution of $u''(x) = -m(x)/\chi$, $x \in (0, 1)$ (Example 6.16). The function u gives the displacement in a beam with unit span and stiffness $\chi > 0$, where m is the bending moment in the beam. It is assumed that the beam supports a uniform load of intensity $q > 0$. Three boundary conditions are considered.

(i) Fixed left end and free right end implying $u(0) = 0$, $u'(0) = 0$, and $-m(x)/\chi = a\,(1 - x)^2$, where $a = q/(2\,\chi)$.

(ii) Fixed left end and simply supported right end implying $u(0) = 0$, $u'(0) = 0$, $u(1) = 0$, and $-m(x)/\chi = a\,(x^2 - x) + \beta\,(x - 1)$, where $\beta = m_0/\chi$ and m_0 denotes the unknown bending moment at the left end of the beam.

(iii) Simply supported ends with a rotational spring of stiffness k at the left end implying $u(0) = 0$, $u(1) = 0$, and $-m(x)/\chi = a\,(x^2 - x) - (k\,u'(0)/\chi)\,(x - 1)$.

Let $X = |B|$ be a Brownian motion reflected at zero, starting at $x \in (0, 1)$. Let $T = \inf\{t > 0 : |B(t)| \geq 1\}$ denote the first time $X = |B|$ exits $(0, 1)$. The local solution is (Eq. 6.45, [82]),

$$u(x) = E^x[u(X(T))] + E^x\left[\int_0^T \frac{m(X(s))}{2\,\chi}\, ds\right] - u'(0)\, E^x\,[L(T)]\,.$$

This equation specialized for the above boundary conditions gives the following local solutions.

(i) The local solution cannot be applied directly since $E^x[u(X(T))] = u_{\max}$ is not known. The boundary condition $u(0) = 0$ and the local solution at $x = 0$ yield $u_{\max} = \frac{a}{2} E^0\left[\int_0^T (1 - X(s))^2\, ds\right]$. Once $u_{\max}$ has been calculated, we can apply the local solution to find the value of u at an arbitrary $x \in (0, 1)$. The exact value of the tip displacement is $u_{\max} = 0.25\,a$. The estimated value of $u_{\max}$ based on $n_s = 1,000$ samples of X generated with a time step $\Delta t = 0.001$ is in error by 0.24%. Note that the value of $u_{\max}$ can also be used to solve locally the Dirichlet boundary value problem for $u(0) = 0$ and $u(1) = u_{\max}$.

(ii) The local solution,

$$u(x) = \frac{1}{2} E^x \left[\int_0^T \left(a\,(X(s) - X(s)^2) + \frac{m_0}{\chi}\,(1 - X(s)) \right) ds \right],$$

depends on the unknown moment m_0. We can find m_0 from the local solution applied at $x = 0$ and the boundary condition $u(0) = 0$. The estimate of m_0 can be used in the above equation to find $u(x)$ at any $x \in (0, 1)$. The estimated value of m_0 from $n_s = 1,000$ samples of X with a time step of $\Delta t = 0.001$ is in error by 0.16%.

(iii) The local solution is

$$u(x) = -u'(0)\, E^x[L(T)] + \frac{1}{2} E^x \left[\int_0^T \left(\frac{0.5\,a}{\chi}\,(X(s) - X(s)^2) + k\,u'(0)\,(1 - X(s)) \right) ds \right],$$

since $u(X(T)) = u(1) = 0$. The Monte Carlo estimate of $E^x[L(T)]$ based on $n_s = 500$ samples and a time step $\Delta t = 0.001$ is 1.0248. From the condition $u(0) = 0$, the resulting estimate of the moment m_0 is 0.0494. The error in the estimated value of m_0 is 1.2%. The value of $u'(0)$ and the above expression of the displacement function u can be used to estimate the value of this function at any $x \in (0, 1)$. ◇

Proof: Let $X = |B| = \hat{B} + L$ be a Brownian motion reflected at zero, where $\hat{B}$ is a Brownian motion and L denotes the local time process in Eq. 6.38 and $X(0) = x$. The Itô formula in Eq. 6.41 gives

$$u(X(t)) - u(X(0)) = \int_0^t u'(X(s))\, d\hat{B}(s) + u'(0)\, L(t) + \frac{1}{2} \int_0^t u''(X(s))\, ds$$

for $g = u$, where $u'(x) = du(x)/dx$ and $u''(x) = d^2u(x)/dx^2$. The expectation of the above equation in the time interval $(0, T)$ is

$$E^x[u(X(T))] - u(x) = u'(0)\, E^x\,[L(T)] + \frac{1}{2}\, E^x \left[\int_0^T u''(X(s))\, ds \right].$$

Because u is the solution of $u''(X(s)) = -m(X(s))/\chi$ for $s < T$, the above equation yields the stated local solution [82]. ■

Example 6.20: Let u be the solution of $(1/2)\,u''(x) + q\,u(x) = 0$, $x \in D = (0, l)$, with the Dirichlet and Neumann boundary conditions $u'(0) = \alpha$ and $u(l) = \beta$, where α, β are some constants. The local solution of this Schrödinger equation is

$$u(x) = \beta\, E^x \left[e^{q\,T} \right] - \alpha\, E^x \left[\int_0^T e^{q\,s}\, dL(s) \right],$$

where $T = \inf\{t > 0 : |B(t)| \notin D\}$, $B(0) = x \in D$.

The estimates of u given by the local solution are in error by less than 2% for $\alpha = 1$, $\beta = 2$, $q = 1$, $l = 1$, and $n_s = 500$ samples of $\boldsymbol{X}$ generated with a time step of $\Delta t = 0.0005$. ◇

Proof: Let X be an $\mathbb{R}^2$-valued process with coordinates $X_1 = |B|$ and X_2 defined by the differential equation $d X_2(t) = q\, X_2(t)\, dt$ with $X_2(0) = 1$ (Section 6.2.2.1). The Itô formula applied to the function $u(|B(t)|)\, X_2(t) = u(\hat{B}(t) + L(t))\, X_2(t)$ gives

$$u(|B(t)|)\, X_2(t) - u(x) = \int_0^t u'(\hat{B}(s) + L(s))\, X_2(s)\, d(\hat{B}(s) + L(s))$$
$$+ \int_0^t u(\hat{B}(s) + L(s))\, d X_2(s) + \frac{1}{2} \int_0^t u''(\hat{B}(s) + L(s))\, X_2(s)\, ds$$

so that

$$E^x \left[u(|B(T)|)\, X_2(T)\right] - u(x) = E^x \left[\int_0^T u'(|B(s)|)\, X_2(s)\, dL(s)\right]$$
$$+ q\, E^x \left[\int_0^T u(|B(s)|)\, X_2(s)\, ds\right] + \frac{1}{2} E^x \left[\int_0^T u''(|B(s)|)\, X_2(s)\, ds\right]$$

by considering the time interval $[0, T]$, using the definition of X_2, and averaging. Since $|B(s)|$ is in D for $s < T$, then $(1/2)\, u''(|B(s)|) + q\, u(|B(s)|) = 0$ so that

$$E^x \left[u(|B(T)|\, X_2(T))\right] - u(x) = u'(0)\, E^x \left[\int_0^T X_2(s)\, dL(s)\right].$$

This formula yields the stated expression of $u(x)$ since $u(|B(T)|) = \beta$ and $X_2(s) = e^{q\, s}$.

The exact solution of the Schrödinger equation is

$$u(x) = \frac{\beta - (\alpha/\sqrt{2\, q}) \sin(\sqrt{2\, q}\, l)}{\cos(\sqrt{2\, q}\, l)} \cos(\sqrt{2\, q}\, x) + \frac{\alpha}{\sqrt{2\, q}} \sin(\sqrt{2\, q}\, x)$$

for $q > 0$. This solution has been used to calculate the error of the local solution. ■

Example 6.21: Let X be a diffusion process defined by the stochastic differential equation $dX(t) = a(X(t))\, dt + b(X(t))\, dB(t)$, $t \geq 0$, where B is a Brownian motion. It is assumed that the coefficients a, b are bounded and satisfy the uniform Lipschitz conditions so that the above equation has a unique solution (Section 4.7.1.1). The absolute value of X, referred to as X **reflected at zero**, is given by

$$|X(t)| - |X(0)| = \int_0^t \operatorname{sign}(X(s))\, dX(s) + \lim_{\varepsilon \downarrow 0} \frac{1}{2\, \varepsilon} \int_0^t b(X(s))^2\, 1_{(-\varepsilon,\varepsilon)}(X(s))\, ds$$
$$= \hat{X}(t) + b(0)^2\, L^{(x)}(t),$$

where $L^{(x)}(t) = \lim_{\varepsilon \downarrow 0} 1/(2\, \varepsilon) \int_0^t 1_{(-\varepsilon,\varepsilon)}(X(s))\, ds$ and the limit is in L_2. If $a = 0$ and $b = 1$, X is a Brownian motion and the above equation coincides with Tanaka's formula (Eq. 6.39). ◇

Proof: Let $h_{n,\varepsilon}$ and h_ε be the functions used to prove Eq. 6.41, where $\varepsilon > 0$ and $n \geq 1$ is an integer. The classical Itô formula applied to function $h_{n,\varepsilon}(X)$ gives

$$h_{n,\varepsilon}(X(t)) - h_{n,\varepsilon}(X(0)) = \int_0^t h'_{n,\varepsilon}(X(s))\, dX(s) + \frac{1}{2}\int_0^t h''_{n,\varepsilon}(X(s))\, b(X(s))^2\, ds.$$

The limit of this equation as $n \to \infty$ and $\varepsilon \downarrow 0$ yields the stated representation of $|X(t)| - |X(0)|$ based on arguments similar to those used to prove Eq. 6.41. The second integral on the right side of the above equation is equal to $b(0)^2\, L^{(x)}(t)$ since it converges to $(1/2)\int_0^t h''_\varepsilon(X(s))\, b(X(s))^2\, ds = (1/2\,\varepsilon)\int_0^t b(X(s))^2\, 1_{(-\varepsilon,\varepsilon)}(X(s))\, ds$ as $n \to \infty$. The proof can be completed by considerations of the type used to prove Eq. 6.44. ■

6.2.3.2 Brownian motion reflected at two thresholds

Let a and $b > a$ be some constants and let r be a periodic function with period $2\,(b-a)$ defined by $r(x) = |x-a|$ for $|x-a| \leq b-a$. Figure 6.8 shows this function in $[-3, 3]$ for $a = 0$ and $b = 1$. Let B be a Brownian motion.

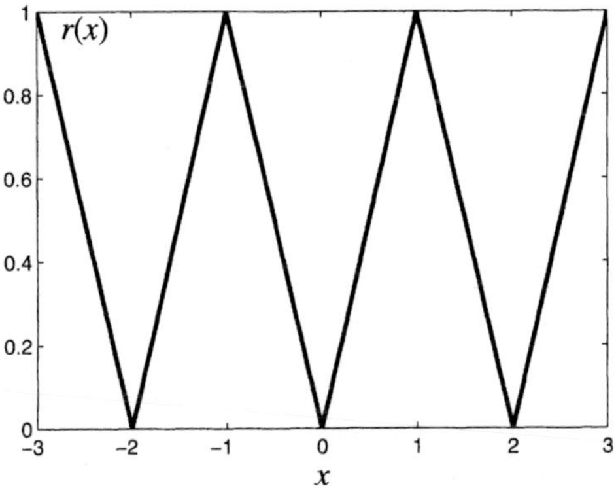

Figure 6.8. Mapping for Brownian motion reflected at $a = 0$ and $b = 1$

The process $r(B)$ has the range $[0, b-a]$, and is referred to as **Brownian motion reflected at two thresholds**. Figure 6.9 shows a sample of a Brownian motion B and the corresponding sample of $r(B)$ for $a = 0$ and $b = 1$.

Let $x_k(a) = a + 2\,k\,(b-a)$ and $x_k(b) = b + 2\,k\,(b-a)$, $k \in \mathbb{Z}$, be the collection of points on the real line where $r(x)$ is equal to 0 and $b-a$, respectively. For $\varepsilon \in (0, (b-a)/2)$ define the intervals $I_{k,\varepsilon}(a) = \{x : |x - x_k(a)| \leq \varepsilon\}$ and $I_{k,\varepsilon}(b) = \{x : |x - x_k(b)| \leq \varepsilon\}$ centered on $x_k(a)$ and $x_k(b)$, respectively, and set $I(a) = \cup_{k=-\infty}^{\infty}(x_k(a), x_k(b))$ and $I(b) = \cup_{k=-\infty}^{\infty}(x_k(b), x_{k+1}(a))$.

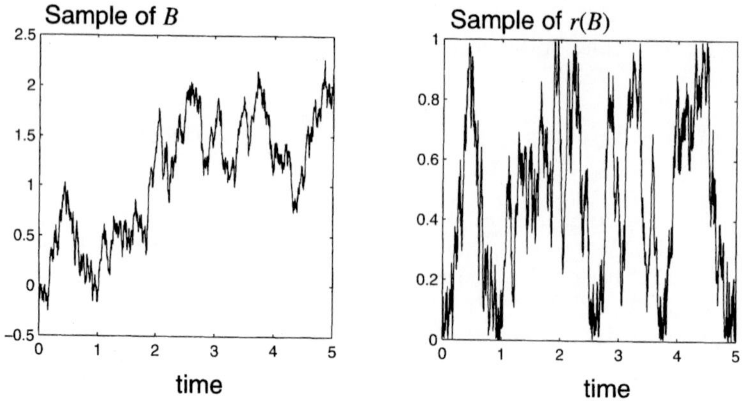

Figure 6.9. A sample of B and the corresponding sample of $r(B)$ for $a = 0$ and $b = 1$

Second extension of Tanaka's formula. If $g \in \mathcal{C}^2(\mathbb{R})$, then

$$g(r(B(t))) - g(r(B(0))) = \int_0^t g'(r(B(s)))\,\left(1_{I(a)} - 1_{I(b)}\right)(B(s))\,dB(s)$$
$$+ \sum_{k=-\infty}^{\infty} \left[g'(0)\,L(t; x_k(a)) - g'(b-a)\,L(t; x_k(b))\right) + \frac{1}{2}\int_0^t g''(r(B(s)))\,ds, \tag{6.46}$$

where $L(t; x_k(a)) = \lim_{\varepsilon \downarrow 0}(1/2\,\varepsilon)\int_0^t 1_{I_k(a)}(B(s))\,ds$ and the expression of $L(t; x_k(b))$ is given by $L(t; x_k(a))$ with $I_k(a)$ in place of $I_k(a)$.

Proof: The Tanaka formulas in Eqs. 6.39 and 6.41 cannot be used to prove Eq. 6.46 since they deal with Brownian motion processes reflected at a single threshold. Therefore, we have to use direct calculations to establish the second extension of Tanaka's formula.

We only prove that $r(B)$ satisfies the condition,

$$r(B(t)) - r(B(0)) = \int_0^t \left(1_{I(a)} - 1_{I(b)}\right)(B(s))\,dB(s)$$
$$+ \sum_{k=-\infty}^{\infty} \left(L(t; x_k(a)) - L(t; x_k(b))\right), \quad a < b, \tag{6.47}$$

that is, a special case of Eq. 6.46 for g equal to the identity function. The proof involves three steps (Eq. 6.41). First, a sequence of smooth approximations $h_{n,\varepsilon}(B)$ converging to $r(B)$ is developed. Second, the Itô formula is applied to the function $h_{n,\varepsilon}(B)$. Third, the limit of the resulting formula is calculated for $n \to \infty$ and $\varepsilon \downarrow 0$.

1. Let $\varepsilon > 0$ and define

$$h_\varepsilon(x) = \begin{cases} r(x), & x \notin \left(\cup_k I_{k,\varepsilon}(a)\right) \cup \left(\cup_l I_{l,\varepsilon}(b)\right), \\ \frac{\varepsilon}{2} + \frac{(x-x_k(a))^2}{2\varepsilon}, & x \in I_{k,\varepsilon}(a), \quad k \in \mathbb{Z}, \\ b - a - \frac{\varepsilon}{2} - \frac{(x-x_k(b))^2}{2\varepsilon}, & x \in I_{k,\varepsilon}(b), \quad k \in \mathbb{Z}. \end{cases}$$

Let $h_{n,\varepsilon}(x) = \int_{\mathbb{R}} h_\varepsilon(x-y)\,\varphi_n(y)\,dy$ be a sequence of approximations of h_ε, where the functions $\varphi_n \in C^\infty(\mathbb{R})$ have compact support shrinking to $\{0\}$ as $n \to \infty$. The functions $h_{n,\varepsilon}$ are infinitely differentiable for each n and $\varepsilon > 0$. Note that the sequence of functions $h_{n,\varepsilon}$ used in the proofs of Eqs. 6.41 and 6.46-6.47 share the same notation but have different definitions.

For every $\varepsilon > 0$, the functions $h_{n,\varepsilon}$ and $h'_{n,\varepsilon}$ converge uniformly to h_ε and h'_ε in $\mathbb{R}$ as $n \to \infty$, respectively, and $h''_{n,\varepsilon}$ converges pointwise to h''_ε in $\mathbb{R}$ as $n \to \infty$ except at $x = x_k(a) \pm \varepsilon$ and $x = x_k(b) \pm \varepsilon$ for $k \in \mathbb{Z}$. The function h''_ε is not defined at $x = x_k(a) \pm \varepsilon$, $x = x_k(b) \pm \varepsilon$ and is set to zero at these points (Section 6.2.3.1).

2. The Itô formula applied to the function $h_{n,\varepsilon}(B(t))$ yields

$$h_{n,\varepsilon}(B(t)) - h_{n,\varepsilon}(B(0)) = \int_0^t h'_{n,\varepsilon}(B(s))\,dB(s) + \frac{1}{2}\int_0^t h''_{n,\varepsilon}(B(s))\,ds.$$

3. For each $t \geq 0$ we have almost surely

$$\begin{aligned} &h_\varepsilon(B(t)) - h_\varepsilon(B(0)) \\ &= \int_0^t h'_\varepsilon(B(s))\,dB(s) + \frac{1}{2\,\varepsilon}\int_0^t \sum_{k=-\infty}^{\infty} \left(1_{I_{k,\varepsilon}(a)} - 1_{I_{k,\varepsilon}(b)}\right)(B(s))\,ds \\ &= \int_0^t h'_\varepsilon(B(s))\,dB(s) + \sum_{k=-\infty}^{\infty} \frac{1}{2\,\varepsilon}\int_0^t \left(1_{I_{k,\varepsilon}(a)} - 1_{I_{k,\varepsilon}(b)}\right)(B(s))\,ds \end{aligned} \tag{6.48}$$

as $n \to \infty$, since $h''_\varepsilon(x) = \sum_{k=-\infty}^{\infty}\left(1_{I_{k,\varepsilon}(a)} - 1_{I_{k,\varepsilon}(b)}\right)(x)$. The last equality in the above equation holds since the sequence

$$\frac{1}{2\,\varepsilon}\int_0^t \sum_{k=-m}^{m} 1_{I_{k,\varepsilon}(a)}(B(s))\,ds = \sum_{k=-m}^{m} \frac{1}{2\,\varepsilon}\int_0^t 1_{I_{k,\varepsilon}(a)}(B(s))\,ds$$

is positive, increasing, and bounded by $t/(2\,\varepsilon)$ for each integer $m \geq 1$. The L_2 limit of the above equation as $\varepsilon \downarrow 0$ yields Eq. 6.47 for each $t \geq 0$. Similar arguments can be used to establish Eq. 6.46.

The Tanaka formula in Eq. 6.39 is a special case of Eq. 6.47 for $a = 0$ and $b \to \infty$. Under these conditions (1) the subsets $I(a)$ and $I(b)$ of $\mathbb{R}$ approach $(0, \infty)$ and $(-\infty, 0)$, respectively, so that $1_{I(a)}(x) - 1_{I(b)}(x) = \text{sign}(x)$, (2) the summation on the right side of Eq. 6.47 degenerates in a single term corresponding to $I_{k,0}(\varepsilon) = 1_{(-\varepsilon,\varepsilon)}$, and (3) the function $r(x)$ becomes $|x|$. ■

Example 6.22: Consider a steady-state transport problem defined in $D = (0, a) \times (0, b)$, $b \gg a > 0$. The pollutant concentration u satisfies the partial differential equation

$$-c_1 \frac{\partial u}{\partial x_1} + d_1 \frac{\partial^2 u}{\partial x_1^2} + d_2 \frac{\partial^2 u}{\partial x_2^2} = 0$$

with the boundary conditions $u(x_1, 0) = u_0$ for $x_1 \in (0, \zeta\, a)$, $0 < \zeta < 1$, $u(x_1, 0) = 0$ for $x_1 \in (\zeta\, a, a)$, $u(x_1, b) = \zeta\, u_0$, and $\partial u(\boldsymbol{x})/\partial x_1 = 0$ at $x_1 = 0$, $x_1 = a$.

The local solution is $u(\boldsymbol{x}) = E^{\boldsymbol{x}}[u(\tilde{\boldsymbol{X}}(T))]$, where $\tilde{\boldsymbol{X}} = (r(X_1), X_2)$, $\boldsymbol{X} = (X_1, X_2)$ is a diffusion process defined by the stochastic differential equation

$$\begin{cases} dX_1(t) &= -c_1\, dt + (2d_1)^{1/2}\, dB_1(t), \\ dX_2(t) &= (2d_2)^{1/2}\, dB_2(t), \end{cases}$$

with the initial condition $\boldsymbol{X}(0) = \boldsymbol{x} \in D$, B_i, $i = 1, 2$, are independent Brownian motions, and $T = \inf\{t > 0 : \tilde{\boldsymbol{X}}(t) \notin D\}$. The process $\tilde{\boldsymbol{X}}$ can exit D only through the boundaries $(0, a) \times \{0\}$ and $(0, a) \times \{b\}$ since its samples are reflected at the boundaries $x_1 = 0$ and $x_1 = a$.

Numerical results have been obtained for $c_1 = 1$, $d_1 = 1$, $d_2 = 1$, $a = 1$, $\zeta = 0.3$, $b = 20$, and a large number of points in D. The resulting estimates of $u(\boldsymbol{x})$ given by the local solution based on $n_s = 1,000$ samples of $\tilde{\boldsymbol{X}}$ generated with a time step of $\Delta t = 0.005$ are in error by approximately 1%. ◇

Proof: The generator

$$\mathcal{A} = -c_1 \frac{\partial}{\partial x_1} + d_1 \frac{\partial^2}{\partial x_1^2} + d_2 \frac{\partial^2}{\partial x_2^2}$$

of $\boldsymbol{X}$ matches the differential operator of the transport equation. However, $\boldsymbol{X}$ cannot be used to solve the posed boundary value problem since u is not known on the boundaries $x_1 = 0$ and $x_1 = a$. We modify the process X_1 such that it is reflected at these boundaries of D. The extended Tanaka formula in Eq. 6.46 cannot be applied since X_1 is a Brownian motion with drift. We have to derive another extension of the Tanaka formula to solve this transport problem.

Let $\tilde{\boldsymbol{X}} = (r(X_1), X_2)$ be a new process, where r is a periodic function with period $2\,a$ defined by $r(\xi) = |\xi|$ for $\xi \in [-a, a]$ and $\boldsymbol{X} = (X_1, X_2)$ denotes the diffusion process defined above. The process $\tilde{\boldsymbol{X}}$ can exit D only through the boundaries $x_2 = 0$ and $x_2 = b$ because its first coordinate takes values in the range $(0, a)$ so that $T = \inf\{t \geq 0 : \tilde{\boldsymbol{X}}(t) \notin D\} = \inf\{t \geq 0 : X_2(t) \notin (0, b)\}$ is the first time that $\tilde{\boldsymbol{X}}$ exits $D = (0, a) \times (0, b)$.

Consider the sequence of approximations $u(h_{n,\varepsilon}(X_1), X_2)$ of $u(\tilde{\boldsymbol{X}})$, where $h_{n,\varepsilon}$ is the function used in the proof of Eq. 6.46 with a, b replaced by 0, a. Accordingly, the intervals $I(a)$ and $I(b)$ in this proof become $I(0)$ and $I(a)$, respectively. Since $u(\xi_1, \xi_2)$ has continuous second order partial derivatives with respect to $\xi_1 = h_{n,\varepsilon}(x_1)$ and $\xi_2 = x_2$, and $\xi_1 = h_{n,\varepsilon}(x_1)$ is infinitely differentiable with respect to x_1 for each n and $\varepsilon > 0$, the Itô formula can be applied to $(X_1, X_2) \mapsto u(h_{n,\varepsilon}(X_1), X_2)$ and gives

$$\begin{aligned} &u(h_{n,\varepsilon}(X_1(t)), X_2(t)) - u(h_{n,\varepsilon}(X_1(0)), X_2(0)) \\ &= \int_0^t \left[-c_1 \frac{\partial u}{\partial \xi_1} h'_{n,\varepsilon} + d_1 \left(\frac{\partial^2 u}{\partial \xi_1^2} (h'_{n,\varepsilon})^2 + \frac{\partial u}{\partial \xi_1} h''_{n,\varepsilon} \right) + d_2 \frac{\partial^2 u}{\partial \xi_2^2} \right] ds \\ &+ \int_0^t \left(\sqrt{2\, d_1} \frac{\partial u}{\partial \xi_1} h'_{n,\varepsilon}\, dB_1 + \sqrt{2\, d_2} \frac{\partial u}{\partial \xi_2}\, dB_2 \right). \end{aligned}$$

The limit of the above equation as $n \to \infty$ is

$$
\begin{aligned}
&u(h_\varepsilon(X_1(t)), X_2(t)) - u(h_\varepsilon(X_1(0)), X_2(0)) \\
&= \int_0^t \left[-c_1 \frac{\partial u}{\partial \xi_1} h'_\varepsilon + d_1 \left(\frac{\partial^2 u}{\partial \xi_1^2} (h'_\varepsilon)^2 + \frac{\partial u}{\partial \xi_1} h''_\varepsilon \right) + d_2 \frac{\partial^2 u}{\partial \xi_2^2} \right] ds \\
&+ \int_0^t \left(\sqrt{2\,d_1}\, \frac{\partial u}{\partial \xi_1} h'_\varepsilon \, dB_1 + \sqrt{2\,d_2}\, \frac{\partial u}{\partial \xi_2} \, dB_2 \right),
\end{aligned}
$$

where the arguments of functions u and h_ε are $(h_\varepsilon(X_1(s)), X_2(s))$ and $X_1(s)$, respectively.

The limit of the above equation as $\varepsilon \downarrow 0$ provides a further extension of the Tanaka formula, which can be used to develop local solutions for transport equations. This approach has not been used here. The local solution and the numerical results reported in these examples have been obtained from the expectation of the above equation with ε in the range $(0, 0.1)$.

The exact steady-state concentration distribution in $(0, a) \times (0, \infty)$ is ([194], Example Problem 3.2, p. 68)

$$
u(\boldsymbol{x}) = \zeta\, u_0 + \sum_{n=1}^{\infty} \frac{2 u_0}{n \pi} \sin(n\,\pi\,\zeta)\, \cos\left(\frac{n\,\pi\,x_1}{a}\right) \exp\left[0.5\,(c_1/d_2 - j_n)\, x_2\right],
$$

where $j_n = [(c_1/d_2)^2 + (2\,n\,\pi\,d_1/(a\,d_2))^2]^{1/2}$. The exact solution shows that the concentration approaches a constant value $\zeta\, u_0$ as $x_2 \to \infty$. This solution is used to assess the accuracy of the local solution calculated in $D = (0, a) \times (0, b)$ with the additional boundary condition $u(x_1, b) = \zeta\, u_0$. ■

6.2.3.3 Brownian motion in the first orthant of $\mathbb{R}^2$

The definition of a reflected Brownian motion in spaces of dimension 2 and higher requires us to specify the subsets of the boundary ∂D of D where the Brownian motion is reflected and the direction of reflection at each point of these subsets. Let $\boldsymbol{X}$ be a Brownian motion in the first orthant of $\mathbb{R}^2$, that is, $\boldsymbol{X}(t)$ is in the closure of $D = \{\boldsymbol{x} \in \mathbb{R}^2 : x_1 > 0, x_2 > 0\}$ at each $t \geq 0$. Three cases are examined: (1) $\boldsymbol{X}$ has independent coordinates and its reflections are orthogonal to the boundaries of D, (2) $\boldsymbol{X}$ has independent coordinates and its reflections are at an arbitrary angle relative to the boundaries of D, and (3) $\boldsymbol{X}$ has dependent coordinates and its reflections are at an arbitrary angle relative to the boundaries of D.

Case 1: $X_i = |B_i| = \hat{B}_i + L_i$, $i = 1, 2$, (Eq. 6.39), $g \in \mathcal{C}^2(\mathbb{R}^2)$, and

$$
g(\boldsymbol{X}(t)) - g(\boldsymbol{X}(0)) = \sum_{i=1}^{2} \int_0^t \frac{\partial g(\boldsymbol{X}(s))}{\partial x_i} (d\hat{B}_i(s) + dL_i(s)) + \frac{1}{2} \int_0^t \Delta g(\boldsymbol{X}(s))\, ds. \tag{6.49}
$$

Proof: Let (B_1, B_2) be two independent Brownian motions. The $\mathbb{R}^2$-valued process $\boldsymbol{X} = (|B_1|, |B_2|)$ lives in the first orthant of the plane and its reflections are orthogonal to the boundaries of D. For example, if $\boldsymbol{X}$ is on the boundary $x_1 = 0$, then L_1 has a positive increment causing an increase of X_1 in the positive direction of coordinate x_1.

The Itô formula can be applied to $g(\boldsymbol{X})$ in conjunction with Tanaka's formula providing a representation for $|B_i|$ and gives Eq. 6.49. Direct calculations based on arguments similar to those used to derive Eqs. 6.41 and 6.46 give

$$\begin{aligned} g(\boldsymbol{X}(t)) - g(\boldsymbol{X}(0)) &= \sum_{i=1}^{2} \int_0^t \frac{\partial g(\boldsymbol{X}(s))}{\partial x_i} \operatorname{sign}(B_i(s))\, dB_i(s) \\ &+ \frac{1}{2} \int_0^t \Delta g(\boldsymbol{X}(s))\, ds + \sum_{i=1}^{2} \tilde{L}_i(t) \end{aligned}$$

for each $t \geq 0$, where $\tilde{L}_i(t) = (\partial g(0)/\partial x_i)\, L_i(t)$. ■

Case 2: Let d_{ij}, $i, j = 1, 2$, be a constant matrix with $d_{11}, d_{22} > 0$. Then

$$\boldsymbol{X}(t) = \begin{bmatrix} \hat{B}_1(t) \\ \hat{B}_2(t) \end{bmatrix} + \begin{bmatrix} d_{11} & d_{12} \\ d_{21} & d_{22} \end{bmatrix} \begin{bmatrix} L_1(t) \\ L_2(t) \end{bmatrix} \tag{6.50}$$

lives in the first orthant D of $\mathbb{R}^2$, where B_1 and B_2 are independent Brownian motion processes and $|B_i| = \hat{B}_i + L_i$, $i = 1, 2$ (Eq. 6.39).

Proof: The coordinates of $\boldsymbol{X}$ are independent Brownian motions in D because $\hat{B}_1$ and $\hat{B}_2$ are independent Brownian motions and the local time processes L_1 and L_2 are zero if $\boldsymbol{X}(t) \in D$. The classical Itô formula can be applied directly to functions $g(\boldsymbol{X}(t))$, $g \in \mathcal{C}^2(\mathbb{R}^2)$, by the definition of $\boldsymbol{X}$.

The process $\boldsymbol{X}$ takes values in the first orthant of $\mathbb{R}^2$ because d_{11} and d_{22} are strictly positive. For example, suppose that $\boldsymbol{X}(t) = (0, x_2)$, $x_2 > 0$, at a time t. The increment $dL_1(t)$ of the local time L_1 causes a change of $\boldsymbol{X}$ from $(0, x_2)$ to $(d_{11}\, dL_1(t), x_2 + d_{21}\, dL_1(t)) \in D$. Similar considerations apply for the other coordinate. Additional information and examples can be found in [42] (Chapter 8). ■

Case 3: $X_i = |B_i^*|$, $i = 1, 2$, $\boldsymbol{X} = (X_1, X_2)$, $\boldsymbol{B}^*(t) = \boldsymbol{b}\,\boldsymbol{B}(t)$, $\boldsymbol{c} = \boldsymbol{b}\,\boldsymbol{b}^T$, $g \in \mathcal{C}^2(\mathbb{R})$, and

$$g(\boldsymbol{X}(t)) - g(\boldsymbol{X}(0)) = \sum_{i=1}^{2} \int_0^t g_{,i}(\boldsymbol{X}(s))\,\mathrm{sign}(B_i^*(s))\,dB_i^*(s)$$
$$+\frac{1}{2}\int_0^t \big[c_{11}\,g_{,11}(\boldsymbol{X}(s)) + c_{22}\,g_{,22}(\boldsymbol{X}(s))$$
$$+2\,c_{12}\,g_{,12}(\boldsymbol{X}(s))\,\mathrm{sign}(B_1^*(s))\,\mathrm{sign}(B_2^*(s))\big]\,ds$$
$$+\sum_{i=1}^{2} c_{ii} \lim_{\varepsilon \downarrow 0} \frac{1}{2\varepsilon} \int_0^t g_{,i}(\boldsymbol{X}(s))\,1_{(-\varepsilon,\varepsilon)}(B_i^*(s))\,ds. \quad (6.51)$$

Note: The processes $|B_i^*|$ can be approximated by the sequence of processes $h_{n_i,\varepsilon}(B_i^*)$, $n_i = 1, 2, \ldots$, for $i = 1, 2$ and $\varepsilon > 0$. The Itô formula applied to the mapping $(B_1^*, B_2^*) \mapsto g(h_{n_1,\varepsilon}(B_1^*), h_{n_2,\varepsilon}(B_2^*))$ gives

$$g(h_{n_1,\varepsilon}(B_1^*(t)), h_{n_2,\varepsilon}(B_2^*(t))) - g(h_{n_1,\varepsilon}(B_1^*(0)), h_{n_2,\varepsilon}(B_2^*(0)))$$
$$= \sum_{i=1}^{2} \int_0^t g_{,i}\,h'_{n_i,\varepsilon}\,dB_i^*(s) + \frac{1}{2}\sum_{i=1}^{2}\int_0^t c_{ii}\,g_{,i}\,h''_{n_i,\varepsilon}\,ds$$
$$+\frac{1}{2}\int_0^t \left[c_{11}\,g_{,11}\,(h'_{n_1,\varepsilon})^2 + 2c_{12}\,g_{,12}\,h'_{n_1,\varepsilon}\,h'_{n_2,\varepsilon} + c_{22}\,g_{,22}\,(h'_{n_2,\varepsilon})^2\right] ds,$$

where the arguments are omitted, $g_{,i} = \partial g/\partial \xi_i$, $g_{,ij} = \partial^2 g/(\partial \xi_i\,\partial \xi_j)$, and $\xi_i = h_{n_i,\varepsilon}(x_i)$, $i = 1, 2$. The formula in Eq. 6.51 results by taking the limit of the above expression as $n_i \to \infty$ and $\varepsilon_i \downarrow 0$, $i = 1, 2$.

The formula in Eq. 6.51 can be used to find the solution of the Laplace equation defined on a wedge $D^* = \{\boldsymbol{x} \in \mathbb{R}^2 : x_1 > 0, x_2 > a x_1, a > 0\}$ in $\mathbb{R}^2$ by the change of variables $y_1 = x_1$, $y_2 = x_2 - a\,x_1$. In the new coordinates D^* is mapped into the first orthant D of $\mathbb{R}^2$ and $\boldsymbol{B}$ becomes a Gaussian process $\boldsymbol{B}^* = (B_1^* = B_1, B_2^* = B_2 - a\,B_1)$ with dependent coordinates. The condition that $\boldsymbol{B}$ is in D^* is equivalent with the condition that $\boldsymbol{B}^*$ is in D. The solution of the two problems corresponding to D and D^* coincide if the reflections at the Neumann boundaries of the two problems are consistent. An alternative solution for Neumann boundary value problems defined in a wedge set can be found in [42] (pp. 174-177). ▲

Example 6.23: Consider the partial differential equation

$$\frac{\partial^2 u}{\partial x_1^2} + 3\frac{\partial^2 u}{\partial x_2^2} = -16, \qquad \boldsymbol{x} \in D = (0, 1) \times (0, 1),$$

corresponding to the time invariant version of Eq. 6.1 with $d = 2$, $\alpha_i(\boldsymbol{x}) = 0$, $\beta_{11}(\boldsymbol{x}) = 2$, $\beta_{12}(\boldsymbol{x}) = \beta_{21}(\boldsymbol{x}) = 0$, $\beta_{22}(\boldsymbol{x}) = 6$, $q(\boldsymbol{x}) = 0$, and $p(\boldsymbol{x}) = 16$. The solution satisfies the mixed boundary conditions $\partial u/\partial x_1 = 0$ on $\{0\} \times (0, 1)$,

$u = 0$ on $\{1\} \times (0, 1)$, $\partial u/\partial x_2 = 0$ on $(0, 1) \times \{0\}$, and $u = -3.5349x_1^2 + 0.5161x_1 + 3.0441$ on $(0, 1) \times \{1\}$.

The local solution of this equation is

$$u(\boldsymbol{x}) = 8E^{\boldsymbol{x}}[T] + E^{\boldsymbol{x}}[u(\boldsymbol{X}(T))], \quad \boldsymbol{x} \in D,$$

where the $\mathbb{R}^2$-valued process $\boldsymbol{X}$ has the coordinates $X_1 = |B_1| = \hat{B}_1 + L_1$ and $X_2 = \sqrt{3}\,|B_2| = \sqrt{3}\,(\hat{B}_2 + L_2)$ (Eq. 6.39), B_i are independent Brownian motions, $i = 1, 2$, and $T = \inf\{t \geq 0 : \boldsymbol{X}(t) \notin D\}$ is the first time $\boldsymbol{X}$ exits D. The local solution is the sum of two expectations that can be estimated from samples of $\boldsymbol{X}$ and T.

Estimated values of the local solution at $\boldsymbol{x} = (0.25, 0.25)$, (0.5,0.5), and (0.75,0.75) based on $n_s = 500$ samples of $\boldsymbol{X}$ generated with a time step $\Delta t = 0.001$ were found to be 3.2571, 4.1808, and 1.8325. The corresponding values of u obtained by the finite difference method are 3.258, 4.195, and 1.801. ◇

Proof: A slight generalization of the Itô formula in Eq. 6.49 gives

$$E^{\boldsymbol{x}}[u(\boldsymbol{X}(T))] - u(\boldsymbol{x}) = \sum_{i=1}^{2} c_i E^{\boldsymbol{x}}\left[\int_0^T \frac{\partial u(\boldsymbol{X}(s))}{\partial x_i}(d\hat{B}_i(s) + dL_i(s))\right]$$
$$+ \frac{1}{2}E^{\boldsymbol{x}}\left[\int_0^T \left(\frac{\partial^2 u(\boldsymbol{X}(s))}{\partial x_1^2} + 3\frac{\partial^2 u(\boldsymbol{X}(s))}{\partial x_2^2}\right) ds\right],$$

where $c_1 = 1$ and $c_2 = \sqrt{3}$. If u is a solution, the above equation gives the stated result because (1) the partial derivatives $\partial u/\partial x_i$ are zero on the Neumann boundaries and the stochastic integrals $\int [\partial u/\partial x_i]\, d\hat{B}$ are martingales starting at zero so that the first two expectations on the right side of the above equation are zero and (2) $\partial^2 u(\boldsymbol{X}(s))/\partial x_1^2 + 3\,(\partial^2 u(\boldsymbol{X}(s))/\partial x_2^2) = -16$ for $s \in (0, T)$. ■

6.2.3.4 General case

There are no simple extensions of the Tanaka and Itô formulas for an arbitrary $\mathbb{R}^d$-valued diffusion process reflected at a boundary of a subset D of $\mathbb{R}^d$ (Method 1). However, heuristic approximations can be developed for finding the local solution of general Neumann boundary value problems (Method 2).

Method 1: Let u be the solution of a partial differential equation defined in a rectangle $D = \times_{i=1}^{d}(a_i, b_i)$ of $\mathbb{R}^d$, $d \geq 1$. Let $\boldsymbol{X}$ be a diffusion process whose generator coincides with the operator of this differential equation. Suppose that the equation has mixed boundary conditions and that the partial derivative $\partial u/\partial x_i$ is specified on the boundary $x_i = a_i$. To incorporate the contribution of the Neumann boundary conditions to the solution, we need to reflect one of the coordinates of $\boldsymbol{X}$ when the process reaches this boundary of D, for example, the coordinate X_1 if $\boldsymbol{X}$ becomes equal to $x_1 = a_1$ and $x_1 = a_1$ is a Neumann boundary. The method in Examples 6.19 and 6.20 can be used to find the local solution of this class of problems. However, the approach cannot be extended to solve locally general mixed boundary value problems characterized by less simple Neumann boundary conditions and arbitrary subsets D in $\mathbb{R}^d$.

Method 2: Consider the time-invariant version of Eq. 6.1 defined on an open bounded subset D of $\mathbb{R}^d$ with the mixed boundary conditions $u(\boldsymbol{x}) = \xi_d(\boldsymbol{x})$, $\boldsymbol{x} \in \partial D_d$, and $\nabla u(\boldsymbol{x}) \cdot \boldsymbol{c}(\boldsymbol{x}) = \xi_n(\boldsymbol{x})$, $\boldsymbol{x} \in \partial D_n$, where ξ_d and ξ_n are prescribed functions, ∂D_d and ∂D_n are subsets of ∂D such that $\partial D = \partial D_d \cup \partial D_n$ and $\partial D_d \cap \partial D_n = \emptyset$, $\nabla = (\partial/\partial x_1, \ldots, \partial/\partial x_d)$, and $\boldsymbol{c}$ denotes a specified $\mathbb{R}^d$-valued function. Our objective is to develop a heuristic algorithm for calculating the local solution of this partial differential equation.

Let $\boldsymbol{x} \in D$ be a point at which we would like to calculate the local solution and let $\boldsymbol{X}$ be an $\mathbb{R}^d$-valued diffusion process whose generator coincides with the differential operator of Eq. 6.1. The process $\boldsymbol{X}$ is started at $\boldsymbol{x} \in D$ if we seek the solution of Eq. 6.1 at this point. The sample paths of $\boldsymbol{X}$ can be divided in two classes:

1. *Sample paths that reach ∂D for the first time at $\boldsymbol{y} \in \partial D_d$.* The contribution of these sample paths of $\boldsymbol{X}$ to the local solution is given by results for Dirichlet boundary value problems since the classical Itô formula is valid in D and $u(\boldsymbol{y})$ is known.

2. *Sample paths that reach ∂D for the first time at $\boldsymbol{y} \in \partial D_n$.* The solution for Dirichlet boundary value problems cannot be applied since u is not known on ∂D_n. The contribution of these samples of $\boldsymbol{X}$ to the local solution can be calculated by the following heuristic algorithm:

- Apply the classical Itô formula between $\boldsymbol{x}$ and $\boldsymbol{y} \in \partial D_n$ to derive a relationship between the values of u at $\boldsymbol{x} \in D$ and $\boldsymbol{y} \in \partial D_n$.

- Reflect $\boldsymbol{X}$, that is, shift $\boldsymbol{y} \in \partial D_n$ to a point $\boldsymbol{x}' \in D$ by a selected amount in a direction consistent with the Neumann boundary conditions at $\boldsymbol{y}$.

- Restart $\boldsymbol{X}$ at $\boldsymbol{x}' \in D$ and find the point where the restarted process reaches ∂D for the first time. If this point is on ∂D_d the sample of $\boldsymbol{X}$ is stopped because u is known on ∂D_d. Otherwise, the previous step is repeated till the restarted process reaches ∂D_d.

The above algorithm giving the contribution of the Neumann boundaries to the local solution is illustrated in detail by the following example. In this algorithm the shift $\| \boldsymbol{y} - \boldsymbol{x} \|$ from $\boldsymbol{y} \in \partial D$ to $\boldsymbol{x} \in D$ must be small but remains unspecified.

Example 6.24: Consider the local solution of the partial differential equation

$$\frac{\partial^2 u}{\partial x_1^2} + 3\,\frac{\partial^2 u}{\partial x_2^2} = -16, \qquad \boldsymbol{x} = (x_1, x_2) \in D = (0,1) \times (0,1),$$

with the mixed boundary conditions $\partial u/\partial x_1 = \beta(x_2)$ on $\{0\} \times (0,1)$, $u = 0$ on $\{1\} \times (0,1)$, $\partial u/\partial x_2 = \alpha(x_1)$ on $(0,1) \times \{0\}$, and $u = \gamma(x_1)$ on $(0,1) \times$

$\{1\}$, where α, β, and γ are prescribed functions. The generator of the diffusion process $\boldsymbol{X} = (X_1 = B_1, X_2 = \sqrt{3}\, B_2)$ coincides with the operator of the above partial differential equation, where B_1 and B_2 are independent Brownian motion processes.

Let $\boldsymbol{X}(\cdot, \omega)$ be a sample of $\boldsymbol{X}$ that reaches ∂D for the first time at $\boldsymbol{Y}(\omega) \in \partial D_d$. This sample is stopped at $\boldsymbol{Y}(\omega)$ because $u(\boldsymbol{Y}(\omega))$ is known. The contribution of the sample $\boldsymbol{X}(\cdot, \omega)$ to the local solution is $u(\boldsymbol{x}, \omega) = \xi(\boldsymbol{Y}(\omega)) + 8\, T(\omega)$, where $T(\omega)$ is the time it takes this sample of $\boldsymbol{X}$ to reach ∂D_d (Example 6.23). For example, $\xi_d(\boldsymbol{Y}(\omega))$ is equal to zero or $\gamma(\boldsymbol{Y}(\omega))$ if $\boldsymbol{Y}(\omega)$ belongs to $\{1\} \times (0, 1)$ or $(0, 1) \times \{1\}$, respectively.

Let $\boldsymbol{X}(\cdot, \omega)$ be a sample of $\boldsymbol{X}$ that reaches ∂D for the first time at $\boldsymbol{Y}^{(1)}(\omega) \in \partial D_n$ (Fig. 6.10). This sample cannot be stopped because u is not known at

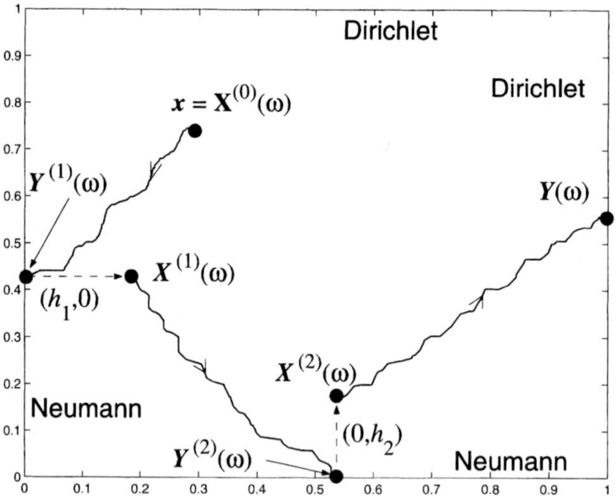

Figure 6.10. A hypothetical sample path of $\boldsymbol{X}$ with two reflections

$\boldsymbol{Y}^{(1)}(\omega)$. The point $\boldsymbol{Y}^{(1)}(\omega)$ is reflected to $\boldsymbol{X}^{(1)}(\omega) = \boldsymbol{Y}^{(1)}(\omega) + (h_1, 0) \in D$, $h_1 > 0$, and the sample of $\boldsymbol{X}$ is restarted at $\boldsymbol{X}^{(1)}(\omega)$. Suppose that the restarted sample reaches ∂D for the first time at $\boldsymbol{Y}^{(2)}(\omega) \in \partial D_n$ (Fig. 6.10). Because u is not known at $\boldsymbol{Y}^{(2)}(\omega)$, we reflect the sample to $\boldsymbol{X}^{(2)}(\omega) = \boldsymbol{Y}^{(2)}(\omega) + (0, h_2) \in D$, $h_2 > 0$. Suppose, as illustrated in Fig. 6.10, that the sample of $\boldsymbol{X}$ restarted at $\boldsymbol{X}^{(2)}(\omega)$ reaches ∂D at $\boldsymbol{Y}(\omega) \in \partial D_d$. This sample of $\boldsymbol{X}$ can now be stopped because $u(\boldsymbol{Y}(\omega))$ is known. Let $T^{(i)}(\omega)$ and $T(\omega)$ be the times the sample ω of $\boldsymbol{X}$ takes to travel from $\boldsymbol{X}^{(i-1)}(\omega)$ to $\boldsymbol{Y}^{(i)}(\omega)$, $i = 1, 2$, and from $\boldsymbol{X}^{(2)}(\omega)$ to $\boldsymbol{Y}(\omega)$, respectively, where $\boldsymbol{X}^{(0)}(\omega) = \boldsymbol{x} \in D$. It is assumed that reflections from $\boldsymbol{Y}^{(i)}(\omega)$ to $\boldsymbol{X}^{(i)}(\omega)$ take no time. The contribution of this sample of $\boldsymbol{X}$ to the local solution is (Fig. 6.10)

$$u(\boldsymbol{x}, \omega) \simeq \xi_d(\boldsymbol{Y}(\omega)) + 8\,[T^{(1)}(\omega) + T^{(2)}(\omega) + T(\omega)]$$
$$- \beta(Y_2^{(1)}(\omega))\,h_1 - \alpha(Y_1^{(2)}(\omega))\,h_2$$

where $\boldsymbol{Y}^{(i)} = \left(Y_1^{(i)}, Y_2^{(i)}\right)$ and $-\beta(Y_2^{(1)}(\omega))\,h_1 - \alpha(Y_1^{(2)}(\omega))\,h_2$ represents the contribution of the Neumann boundary to the local solution. ◇

Note: Let $\boldsymbol{Y}^{(i)}(\omega) \in \partial D_n$ and let $\boldsymbol{X}^{(i)}(\omega) \in D$ be the reflection of $\boldsymbol{Y}^{(i)}(\omega)$. The first order Taylor expansion of u at $\boldsymbol{X}^{(i)}(\omega)$ gives

$$u(\boldsymbol{X}^{(i)}(\omega)) = u(\boldsymbol{Y}^{(i)}(\omega) + (\boldsymbol{X}^{(i)}(\omega) - \boldsymbol{Y}^{(i)}(\omega)))$$
$$\simeq u(\boldsymbol{Y}^{(i)}(\omega)) + \nabla u(\boldsymbol{Y}^{(i)}(\omega)) \cdot (\boldsymbol{X}^{(i)}(\omega) - \boldsymbol{Y}^{(i)}(\omega))$$

for $\| \boldsymbol{X}^{(i)}(\omega) - \boldsymbol{Y}^{(i)}(\omega) \|$ small, or

$$u(\boldsymbol{Y}^{(i)}(\omega) \simeq u(\boldsymbol{X}^{(i)}(\omega)) - \nabla u(\boldsymbol{Y}^{(i)}(\omega)) \cdot (\boldsymbol{X}^{(i)}(\omega) - \boldsymbol{Y}^{(i)}(\omega)).$$

The approximate contribution to the local solution of the sample of $\boldsymbol{X}$ shown in Fig. 6.10 can be calculated from the following considerations. First, the contribution of sample ω in the time interval $[0, T^{(1)}(\omega)]$ to the local solution at $\boldsymbol{x}$ is

$$u(\boldsymbol{x}, \omega) = u(\boldsymbol{Y}^{(1)}(\omega)) + 8\,T^{(1)}(\omega).$$

The first order Taylor expansion applied to $u(\boldsymbol{X}^{(1)}(\omega))$ gives

$$u(\boldsymbol{Y}^{(1)}(\omega)) \simeq u(\boldsymbol{X}^{(1)}(\omega)) - \beta(Y_2^{(1)}(\omega))\,h_1$$

since $\boldsymbol{X}^{(1)}(\omega) - \boldsymbol{Y}^{(1)}(\omega) = (h_1, 0)$, $h_1 > 0$, so that

$$\nabla u(\boldsymbol{Y}^{(1)}(\omega)) \cdot (\boldsymbol{X}^{(1)}(\omega) - \boldsymbol{Y}^{(1)}(\omega)) = \beta(Y_2^{(1)}(\omega))\,h_1$$

according to the boundary condition. Second, the contribution during the time interval $[T^{(1)}(\omega), T^{(2)}(\omega)]$ of the process restarted at $\boldsymbol{X}^{(1)}(\omega)$ to the local solution is

$$u(\boldsymbol{X}^{(1)}(\omega)) = u(\boldsymbol{Y}^{(2)}(\omega)) + 8\,T^{(2)}(\omega).$$

The first order Taylor expansion applied to $u(\boldsymbol{X}^{(2)}(\omega))$ gives

$$u(\boldsymbol{Y}^{(2)}(\omega)) \simeq u(\boldsymbol{X}^{(2)}(\omega)) - \alpha(Y_1^{(2)}(\omega))\,h_2$$

for a shift $h_2 > 0$ in the positive direction of coordinate x_2. Third, the contribution during the time interval $[T^{(2)}(\omega), T(\omega)]$ of the process restarted at $\boldsymbol{X}^{(2)}(\omega)$ to the local solution is

$$u(\boldsymbol{X}^{(2)}(\omega)) = \xi_d(\boldsymbol{Y}(\omega)) + 8\,T(\omega).$$

The stated result follows from the above equations. ▲

The approach illustrated in Fig. 6.10 can be applied to arbitrary Neumann conditions. For example, suppose that the boundary condition at an arbitrary point $\boldsymbol{x} \in \partial D_n$ is $\nabla u(\boldsymbol{x}) \cdot \boldsymbol{c}(\boldsymbol{x}) = \xi_n(\boldsymbol{x})$ and that a sample of $\boldsymbol{X}$ has reached a point $\boldsymbol{Y}^{(i)}(\omega) \in \partial D_n$. Following the approach in Fig. 6.10, this point is reflected to $\boldsymbol{X}^{(i)}(\omega) \in D$ such that $\boldsymbol{X}^{(i)}(\omega) - \boldsymbol{Y}^{(i)}(\omega) = h\,\boldsymbol{c}(\boldsymbol{Y}^{(i)})$, where $h > 0$ is a specified constant. The sample values of u at $\boldsymbol{X}^{(i)}(\omega)$ and $u(\boldsymbol{Y}^{(i)}(\omega))$ can be related approximately by

$$u(\boldsymbol{Y}^{(i)}(\omega)) \simeq u(\boldsymbol{X}^{(i)}(\omega)) - h\,\nabla u(\boldsymbol{Y}^{(i)}(\omega)) \cdot \boldsymbol{c}(\boldsymbol{Y}^{(i)}),$$

where $\nabla u(\boldsymbol{Y}^{(i)}(\omega)) \cdot \boldsymbol{c}(\boldsymbol{Y}^{(i)}(\omega))) = \xi_n(\boldsymbol{Y}^{(i)}(\omega))$ by the prescribed Neumann boundary condition.

6.3 Sphere walk method

The RWM examined in Section 6.2 and the SWM considered here are similar in the sense that both methods (1) deliver local solutions for partial differential equations and (2) are based on processes that originate at a point $\boldsymbol{x} \in D$ at which the solution of a partial differential equation has to be calculated and exit D after a random time T. However, the stochastic processes used by these methods are different. The RWM is based on Brownian motion and diffusion processes while the SWM uses spherical processes.

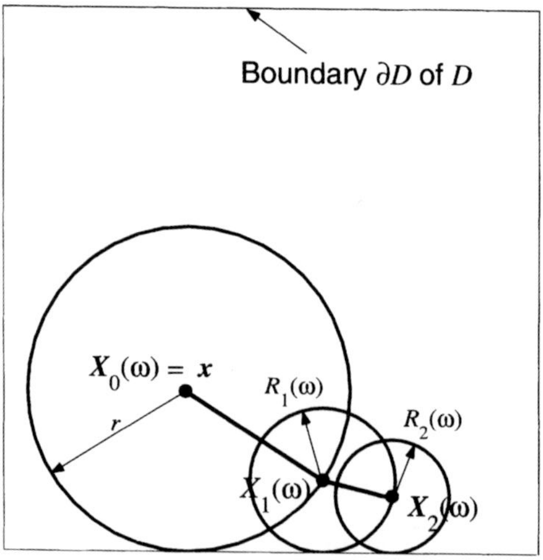

Figure 6.11. Definition of the spherical process

Figure 6.11 illustrates the construction of the **spherical process**. The process consists of a sequence of open spheres included in D whose centers approach the boundary ∂D of D. The point $\boldsymbol{x} \in D$ where the local solution has to be calculated is the center of the first sphere. This sphere has radius $r = \min_{\boldsymbol{y} \in \partial D} \| \boldsymbol{x} - \boldsymbol{y} \|$ so that it is tangent to ∂D. The center $\boldsymbol{X}_1$ of the second sphere is a random point on the surface of the first sphere and its radius R_1 is such that it is tangent to ∂D. The spherical process in Fig. 6.11 is stopped when the center of a sphere in this sequence enters a small vicinity of ∂D for the first time ([160], Section 1.2.1).

The local solution given by the SWM is based on the Green function, the mean value property, and the Monte Carlo simulation method. The SWM can be applied to find the local solution of partial differential equations of the type given by Eq. 6.1 with Dirichlet and mixed boundary conditions. Efficient local solutions by the SWM have been reported only for the Laplace and Poisson equations [200].

6.3.1 The Green function

Consider the partial differential equation

$$\mathcal{L}\,[u(\boldsymbol{x})] = -p(\boldsymbol{x}), \quad \boldsymbol{x} \in D, \tag{6.52}$$

where $\mathcal{L}$ is a linear partial differential operator, for example, the operator of the time-invariant version of Eq. 6.1, D denotes an open bounded subset of $\mathbb{R}^d$, and p is a specified real-valued function defined on $\mathbb{R}^d$. Let $g(\boldsymbol{x}, \boldsymbol{y})$ be the solution of Eq. 6.52 at $\boldsymbol{x}$ corresponding to a unit action applied at a "source point" $\boldsymbol{y} \in D$, that is, the solution of $\mathcal{L}\,[g(\boldsymbol{x}, \boldsymbol{y})] = -\delta(\boldsymbol{x} - \boldsymbol{y})$, $\boldsymbol{x}, \boldsymbol{y} \in D$, with $g(\boldsymbol{x}, \boldsymbol{y}) = 0$ for $\boldsymbol{x} \in \partial D$. The function $g(\boldsymbol{x}, \boldsymbol{y})$ is the **Green function** for $\mathcal{L}$ and can be used to calculate the solution u of Eq. 6.52 at an arbitrary point $\boldsymbol{x} \in D$ ([43], Section V.14).

If u in Eq. 6.52 satisfies the boundary condition $u(\boldsymbol{x}) = 0$, $\boldsymbol{x} \in \partial D$, then

$$u(\boldsymbol{x}) = \int_D g(\boldsymbol{x}, \boldsymbol{y})\, p(\boldsymbol{y})\, d\boldsymbol{y}. \tag{6.53}$$

If u in Eq. 6.52 satisfies the boundary condition $u(\boldsymbol{x}) = \xi(\boldsymbol{x})$, $\boldsymbol{x} \in \partial D$, then

$$u(\boldsymbol{x}) = \tilde{\xi}(\boldsymbol{x}) + \int_D g(\boldsymbol{x}, \boldsymbol{y}) \left(p(\boldsymbol{y}) + \mathcal{L}\,[\tilde{\xi}(\boldsymbol{y})]\right) d\boldsymbol{y}, \tag{6.54}$$

where $\tilde{\xi}$ denotes a continuous extension of ξ to $D \cup \partial D$.

Note: The Dirichlet boundary conditions in Eqs. 6.53 and 6.54 are referred to as homogeneous and inhomogeneous, respectively. The solution in Eq. 6.53 holds since $\mathcal{L}$ is a linear operator. The formula in Eq. 6.54 is valid since $v = u - \tilde{\xi}$ satisfies the partial differential equation

$$\mathcal{L}[v] = -p(\boldsymbol{x}) - \mathcal{L}[\tilde{\xi}(\boldsymbol{x})]$$

with $v(\boldsymbol{x}) = 0$ for $\boldsymbol{x} \in \partial D$ so that $v(\boldsymbol{x}) = \int_D g(\boldsymbol{x}, \boldsymbol{y}) \left(p(\boldsymbol{y}) + \mathcal{L}[\tilde{\xi}(\boldsymbol{y})]\right) d\boldsymbol{y}$ by Eq. 6.53, which gives Eq. 6.54. Useful information on Green's functions can be found in [43] (Volume 1, Chapter V), [73] (Section 22.5), and [128]. ▲

Example 6.25: Let u be defined in D by $\mathcal{L}[u(\boldsymbol{x})] = -q(\boldsymbol{x})\,u(\boldsymbol{x})$, $\boldsymbol{x} \in D$, with the boundary condition $u(\boldsymbol{x}) = 0$, $\boldsymbol{x} \in \partial D$. If $\mathcal{L}$ is a linear differential operator and q is integrable in D, then u also satisfies the integral equation $u(\boldsymbol{x}) = \int_D g(\boldsymbol{x}, \boldsymbol{y})\,q(\boldsymbol{y})\,u(\boldsymbol{y})\,d\boldsymbol{y}$, where g denotes the Green function of $\mathcal{L}$. ◇

Note: The formula in Eq. 6.53 with $q\,u$ in place of p gives an integral equation for u. If $g(\boldsymbol{x}, \boldsymbol{y})$ is positive for all $\boldsymbol{x}, \boldsymbol{y} \in D$, the integral equation for u can be given in the form $u(\boldsymbol{x}) = c(\boldsymbol{x})\,E[q(\boldsymbol{Y})\,u(\boldsymbol{Y})]$, where $c(\boldsymbol{x}) = \int_D g(\boldsymbol{x}, \boldsymbol{y})\,d\boldsymbol{y}$ and $\boldsymbol{Y}$ is a random vector with density $\tilde{g}(\boldsymbol{x}, \cdot) = g(\boldsymbol{x}, \cdot)/c(\boldsymbol{x})$. ▲

Example 6.26: Consider the Laplace operator $\Delta = \sum_{i=1}^{d} \partial^2/\partial x_i^2$ in $\mathbb{R}^d$, where $d \geq 2$ is an integer. The Green function of this operator is $g(\boldsymbol{x}, \boldsymbol{y}) = \gamma(r) + w(\boldsymbol{x}, \boldsymbol{y})$, where w is regular, that is, it satisfies the equation $\Delta w = 0$ and has continuous second order partial derivatives, $r = \| \boldsymbol{x} - \boldsymbol{y} \|$,

$$\gamma(r) = \begin{cases} \frac{1}{2\pi} \log(1/r), & d = 2, \\ \frac{1}{(d-2)\,a_d}\, r^{d-2}, & d \geq 3, \end{cases}$$

and $a_d = 2\,\pi^{d/2}/\Gamma(d/2)$ is the surface area of the unit sphere in $\mathbb{R}^d$ ([128], Section 1.2). The Green functions have various singularities depending on the functional form of the operator $\mathcal{L}$ and the dimension d. For example, the singularities of the Green function for the Laplace operator in $\mathbb{R}^2$ and $\mathbb{R}^3$ are given by $\log(r)$ and r^{-1}, respectively. ◇

Note: The Green function can be calculated as the limit as $\varepsilon \downarrow 0$ of a sequence of solutions g_ε of $\Delta g_\varepsilon(\boldsymbol{x}, \boldsymbol{y}) = -p_\varepsilon(\boldsymbol{x}, \boldsymbol{y})$, $\boldsymbol{x}, \boldsymbol{y} \in D$, with the homogeneous boundary conditions $g_\varepsilon(\boldsymbol{x}, \boldsymbol{y}) = 0$, where $p_\varepsilon(\cdot, \boldsymbol{y})$ is zero outside the disc $D_\varepsilon(\boldsymbol{y}) = \{\boldsymbol{x} : \| \boldsymbol{x} - \boldsymbol{y} \| < \varepsilon\}$ of radius ε centered at $\boldsymbol{y} \in D$ and satisfies the conditions $\int_{D_\varepsilon(\boldsymbol{y})} p_\varepsilon(\boldsymbol{x}, \boldsymbol{y})\,d\boldsymbol{x} = 1$ and $D_\varepsilon(\boldsymbol{y}) \subset D$ for each $\varepsilon > 0$. The limit of $g_\varepsilon(\cdot, \cdot)$ as $\varepsilon \downarrow 0$ yields the Green function. A comprehensive review on Green's function can be found in [128]. ▲

6.3.2 Mean value property

Let u be the solution of

$$\mathcal{L}[u(\boldsymbol{x})] = 0, \quad \boldsymbol{x} \in D, \tag{6.55}$$

where D is an open bounded subset of $\mathbb{R}^d$ and $\mathcal{L}$ is the differential operator in Eq. 6.52. Consider a sphere $S(\boldsymbol{x}, r) = \{\boldsymbol{y} \in D : \| \boldsymbol{x} - \boldsymbol{y} \| < r\}$ with radius $r > 0$, center $\boldsymbol{x} \in D$, and boundary $\partial S(\boldsymbol{x}, r) = \{\boldsymbol{y} \in D : \| \boldsymbol{x} - \boldsymbol{y} \| = r\}$ whose closure $S(\boldsymbol{x}, r) \cup \partial S(\boldsymbol{x}, r)$ is in D. The area of $\partial S(\boldsymbol{x}, r)$ is $a_d(r) = 2\,\pi^{d/2}\,r^{d-1}/\Gamma(d/2)$. Let $d\sigma(\boldsymbol{y})$ denote an infinitesimal element on $\partial S(\boldsymbol{x}, r)$ so that $(1/a_d(r)) \int_{\partial S(\boldsymbol{x},r)} d\sigma(\boldsymbol{y}) = 1$.

The **mean value theorem** states that ([44], Section IV.3)

$$u(\boldsymbol{x}) = \frac{1}{a_d(r)} \int_{\partial S(\boldsymbol{x},r)} k(\boldsymbol{x}, \boldsymbol{y})\, u(\boldsymbol{y})\, d\sigma(\boldsymbol{y}), \tag{6.56}$$

where the kernel k depends on $\mathcal{L}$.

Note: The formula in Eq. 6.56 states that the value of u at the center of $S(\boldsymbol{x}, r)$ is equal to the average of the values of this function on $\partial S(\boldsymbol{x}, r)$ weighted by the kernel k of $\mathcal{L}$. We will see that the kernel k is an essential ingredient of the SWM. Efficient algorithms for finding the local solution of partial differential equations by the SWM are possible only if k has a known analytical expression.

Two methods can be used to find the kernel of a differential equation. The kernel of $\mathcal{L}$ can be obtained from Eq. 6.52 using integration by parts and the Green formulas ([44], Volume 2, Chapter IV). There are few differential operators for which the kernel k in Eq. 6.56 has a simple form, for example,

$$k(\boldsymbol{x}, \boldsymbol{y}) = \begin{cases} 1, & \text{for } \Delta, \\ \dfrac{\|\boldsymbol{x}-\boldsymbol{y}\|\, c^{1/2}}{\sinh(\|\boldsymbol{x}-\boldsymbol{y}\|\, c^{1/2})}, & \text{for } \Delta - c, \quad c > 0. \end{cases}$$

Other examples can be found in [128, 160].

An alternative method for finding the kernel k of $\mathcal{L}$ can be based on the RWM in Section 6.2. Let u be such that $\mathcal{L}[u] = 0$ in $S(\boldsymbol{x}, r)$. If there is a diffusion process $\boldsymbol{X}$ with generator $\mathcal{L}$ that starts at $\boldsymbol{x}$, then

$$E^{\boldsymbol{x}}\left[u(\boldsymbol{X}(T))\right] - u(\boldsymbol{x}) = E^{\boldsymbol{x}}\left[\int_0^T \mathcal{L}[\boldsymbol{X}(s)]\right] = 0,$$

where $T = \inf\{t > 0 : \boldsymbol{X}(t) \notin S(\boldsymbol{x}, r)\}$ (Eq. 6.14). Hence,

$$u(\boldsymbol{x}) = E^{\boldsymbol{x}}\left[u(\boldsymbol{X}(T))\right] = \int_{\partial S(\boldsymbol{x},r)} u(\boldsymbol{y})\, f_{\boldsymbol{X}(T)|\boldsymbol{X}(0)}(\boldsymbol{y} \mid \boldsymbol{x})\, d\sigma(\boldsymbol{y}),$$

where $f_{\boldsymbol{X}(T)|\boldsymbol{X}(0)}$ is the density of $\boldsymbol{X}(T)$ conditional on $\boldsymbol{X}(0) = \boldsymbol{x}$. If the kernel k is positive for all $\boldsymbol{x}, \boldsymbol{y} \in D$, the above representation of u and Eq. 6.56 imply

$$f_{\boldsymbol{X}(T)|\boldsymbol{X}(0)}(\boldsymbol{y} \mid \boldsymbol{x}) = k(\boldsymbol{x}, \boldsymbol{y})/a_d(r).$$

Under this condition the kernel k of $\mathcal{L}$ can be approximated by a histogram of the conditional random variable $\boldsymbol{X}(T) \mid (\boldsymbol{X}(0) = \boldsymbol{x})$. ▲

Example 6.27: The local solution of the Laplace equation in Eq. 6.23 for $D = S(\boldsymbol{x}, r)$ is $u(\boldsymbol{x}) = E^{\boldsymbol{x}}\left[u(\boldsymbol{B}(T))\right]$, where $\boldsymbol{B}$ denotes an $\mathbb{R}^d$-valued Brownian motion starting at $\boldsymbol{B}(0) = \boldsymbol{x}$ and $T = \inf\{t > 0 : \boldsymbol{B}(t) \notin S(\boldsymbol{x}, r)\}$. Because $\boldsymbol{B}(T)$ is uniformly distributed on $\partial S(\boldsymbol{x}, r)$, we have

$$u(\boldsymbol{x}) = \frac{1}{a_d(r)} \int_{\partial S(\boldsymbol{x},r)} u(\boldsymbol{y})\, d\sigma(\boldsymbol{y})$$

so that $k = 1$ for the Laplace operator since $a_d(r)$ is the area of $\partial S(\boldsymbol{x}, r)$. ◇

If $\boldsymbol{Y}$ is a uniformly distributed random variable on $\partial S(\boldsymbol{x}, r)$, then

$$u(\boldsymbol{x}) = E^{\boldsymbol{x}}\left[k(\boldsymbol{x}, \boldsymbol{Y})\, u(\boldsymbol{Y})\right], \tag{6.57}$$

$$\frac{\partial u(\boldsymbol{x})}{\partial x_i} = E^{\boldsymbol{x}}\left[\frac{\partial k(\boldsymbol{x}, \boldsymbol{Y})}{\partial x_i}\, u(\boldsymbol{Y})\right], \quad i = 1, \ldots, d. \tag{6.58}$$

Note: The expression of u in Eq. 6.56 implies Eq. 6.57 since $\boldsymbol{Y}$ is a uniformly distributed $\partial S(\boldsymbol{x}, r)$-valued random variable, and Eq. 6.58 follows from Eq. 6.57 by differentiation with respect to x_i. The notation $E^{\boldsymbol{x}}$ is used to indicate that the expectation is performed on a sphere centered at $\boldsymbol{x}$.

The relationships in Eqs. 6.57 and 6.58 state that u and its partial derivatives at an arbitrary point $\boldsymbol{x} \in D$ are equal to averages of the values of u on $\partial S(\boldsymbol{x}, r)$ weighted by the kernel k and by its partial derivatives $\partial k/\partial x_i$, respectively. Hence, we can calculate $u(\boldsymbol{x})$ and $\partial u(\boldsymbol{x})/\partial x_i$ at $\boldsymbol{x}$ from the values of u on $\partial S(\boldsymbol{x}, r)$ provided that the kernel k is known.

If $k \geq 0$ on $\partial S(\boldsymbol{x}, r)$, the mean value property in Eq. 6.57 can also be given in the form $u(\boldsymbol{x}) = \alpha_d(\boldsymbol{x}, r)\, E^{\boldsymbol{x}}[u(\mathbf{Z})]$, where $\mathbf{Z}$ is an $\partial S(\boldsymbol{x}, r)$-valued random variable with the probability density function

$$\tilde{k}(\boldsymbol{x}, \boldsymbol{y}) = k(\boldsymbol{x}, \boldsymbol{y})/\alpha_d(\boldsymbol{x}, r) \tag{6.59}$$

where $\alpha_d(\boldsymbol{x}, r) = \int_{\partial S(\boldsymbol{x}, r)} k(\boldsymbol{x}, \boldsymbol{\xi})\, d\sigma(d\boldsymbol{\xi})$. ▲

Let $\boldsymbol{Y}(\omega)$, $\omega = 1, \ldots, n_s$, be n_s uniformly distributed and independent points on $\partial S(\boldsymbol{x}, r)$. Assume that u, k, and $\partial k/\partial x_i$ are known on $\partial S(\boldsymbol{x}, r)$.

The estimates of u and $\partial u/\partial x_i$ at $\boldsymbol{x} \in D$ are

$$\hat{u}(\boldsymbol{x}) = \frac{1}{n_s} \sum_{\omega=1}^{n_s} k(\boldsymbol{x}, \boldsymbol{Y}(\omega))\, u(\boldsymbol{Y}(\omega)) \quad \text{and}$$

$$\frac{\partial \hat{u}(\boldsymbol{x})}{\partial x_i} = \frac{1}{n_s} \sum_{\omega=1}^{n_s} \frac{\partial k(\boldsymbol{x}, \boldsymbol{Y}(\omega))}{\partial x_i}\, u(\boldsymbol{Y}(\omega)). \tag{6.60}$$

Note: Let G_i, $i = 1, \ldots, d$, be independent copies of $N(0, 1)$ and define the random variable $U = \left(\sum_{i=1}^{d} G_i^2\right)^{1/2}$. The vector $(G_1/U, \ldots, G_d/U)$ is uniformly distributed on $\partial S(\mathbf{0}, 1)$ in $\mathbb{R}^d$ ([155], Procedure N-4, p. 89). This observation can be used to generate independent samples of $\boldsymbol{Y}$. ▲

In applications the function u and its derivatives are specified on the boundary ∂D of D, which generally is not a sphere. Hence, Eqs. 6.57 and 6.58 cannot be applied directly. The following two sections use the spherical process in Fig. 6.11 to find the local solution for Dirichlet and mixed boundary value problems.

6.3.3 Dirichlet boundary conditions

Let u be the solution of Eq. 6.55 defined in an open bounded subset D of $\mathbb{R}^d$ with the boundary conditions $u(\boldsymbol{x}) = \xi(\boldsymbol{x})$, $\boldsymbol{x} \in \partial D$. Let $\boldsymbol{X}(\omega) = \{\boldsymbol{X}_0(\omega) = \boldsymbol{x}, \boldsymbol{X}_1(\omega), \ldots, \boldsymbol{X}_{N(\omega)}(\omega)\}$, $\omega = 1, \ldots, n_s$, be n_s independent samples of a spherical process starting at $\boldsymbol{x} \in D$, where $N(\omega)$ denotes the number of steps until a sample $\boldsymbol{X}(\omega)$ of $\boldsymbol{X}$ reaches for the first time a small vicinity of ∂D, that is, the first time that the distance between $\boldsymbol{X}_{N(\omega)}(\omega)$ and ∂D is smaller than a specified $\varepsilon > 0$ (Figs. 6.11 and 6.12). This condition defines the number of steps N after which $\boldsymbol{X}$ is stopped. The radii of the spheres centered at $\boldsymbol{x}, \boldsymbol{X}_1, \ldots, \boldsymbol{X}_N$ are $r, R_1, \ldots, R_N$.

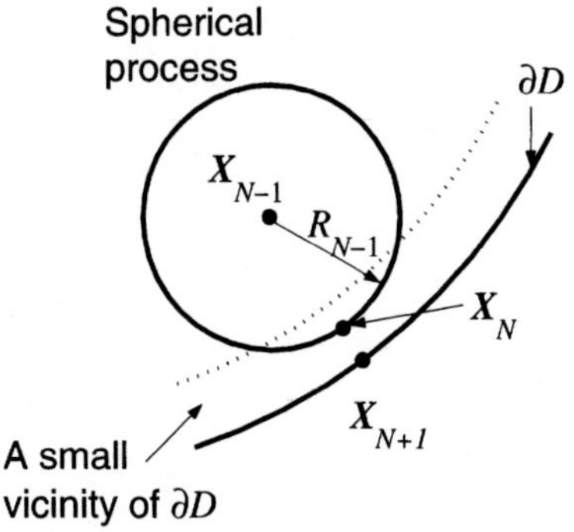

Figure 6.12. Stopping rule for spherical processes

The points $\boldsymbol{X}_i$ are uniformly distributed on $\partial S(\boldsymbol{X}_{i-1}, R_{i-1})$ (Eq. 6.57). Let $\boldsymbol{X}_{N+1}$ be the point of ∂D that is the closest to $\boldsymbol{X}_N$ (Fig. 6.12). By the definition of N, we have $\| \boldsymbol{X}_N - \boldsymbol{X}_{N+1} \| < \varepsilon$.

The **local solution** of Eq. 6.55 is

$$u(\boldsymbol{x}) = E^{\boldsymbol{x}}\left[\left(\prod_{i=1}^{N+1} k(\boldsymbol{X}_{i-1}, \boldsymbol{X}_i)\right) \xi(\boldsymbol{X}_{N+1})\right]. \tag{6.61}$$

Proof: Suppose that $N(\omega) = 2$ (Fig. 6.11). The use of Eq. 6.57 along the sample $(\boldsymbol{X}_0(\omega), \boldsymbol{X}_1(\omega), \boldsymbol{X}_2(\omega), \boldsymbol{X}_3(\omega))$ of $\boldsymbol{X}$ gives

$$\begin{aligned} u(\boldsymbol{X}_2(\omega)) &= k(\boldsymbol{X}_2(\omega), \boldsymbol{X}_3(\omega))\, u(\boldsymbol{X}_3(\omega)), \\ u(\boldsymbol{X}_1(\omega)) &= k(\boldsymbol{X}_1(\omega), \boldsymbol{X}_2(\omega))\, u(\boldsymbol{X}_2(\omega)), \quad \text{and} \\ u(\boldsymbol{x}, \omega) &= k(\boldsymbol{x}, \boldsymbol{X}_1(\omega))\, u(\boldsymbol{X}_1(\omega)), \end{aligned}$$

so that $u(\boldsymbol{x}, \omega) = \prod_{i=1}^{3} k(\boldsymbol{X}_{i-1}(\omega), \boldsymbol{X}_i(\omega))\, \xi(\boldsymbol{X}_3(\omega))$ since the $\boldsymbol{X}_3(\omega)$ is on the boundary ∂D of D so that $u(\boldsymbol{X}_3(\omega)) = \xi(\boldsymbol{X}_3(\omega))$. The equality $u(\boldsymbol{X}_N) = k(\boldsymbol{X}_N, \boldsymbol{X}_{N+1})\, \xi(\boldsymbol{X}_{N+1})$

is approximate since the spherical process is stopped before reaching the boundary of D. The approximation is likely to be satisfactory since u is a smooth function.

We have seen that samples of a Brownian motion (Example 6.27) or properties of Gaussian variables (Eq. 6.60) can be used to generate uniformly distributed points on $\partial S(X_i, R_i)$. We will see in Section 8.5.1.1 that the exit time for a Brownian motion from a sphere is finite a.s. This property guarantees that the number of steps N in Eq. 6.61 that a spherical process takes to reach the boundary of D is finite a.s. ■

The local solution in Eq. 6.61 can be estimated by

$$\hat{u}(\boldsymbol{x}) = \frac{1}{n_s} \sum_{\omega=1}^{n_s} u(\boldsymbol{x}, \omega), \quad \text{where}$$

$$u(\boldsymbol{x}, \omega) = \left(\prod_{i=1}^{N(\omega)+1} k(\boldsymbol{X}_{i-1}(\omega), \boldsymbol{X}_i(\omega)) \right) \xi(\boldsymbol{X}_{N(\omega)+1}(\omega)). \tag{6.62}$$

Note: The calculation of the estimate $\hat{u}$ of u can be inefficient since the radii $R_i(\omega)$ of the spheres $S(\boldsymbol{X}_i(\omega), R_i(\omega))$, $i = 0, 1, \ldots, N(\omega)$, are the solutions of nonlinear optimization problems. This difficulty can be reduced by using specialized algorithms for calculating $R_i(\omega)$ based on simplified representations of ∂D [200]. ▲

Example 6.28: Let u be a harmonic function on an open bounded set $D \subset \mathbb{R}^d$, that is, $\Delta u(\boldsymbol{x}) = 0$, that satisfies the Dirichlet boundary conditions $u(\boldsymbol{x}) = \xi(\boldsymbol{x})$, $\boldsymbol{x} \in \partial D$. The local solution at an arbitrary point $\boldsymbol{x} \in D$ can be estimated by $\hat{u}(\boldsymbol{x}) = (1/n_s) \sum_{\omega=1}^{n_s} \xi(\boldsymbol{X}_{N(\omega)+1}(\omega))$ (Eq. 6.62). ◇

Note: The above local solution by the spherical process method follows from Eq. 6.62 since the kernel of Δ is $k = 1$. We also note that the local solution $u(\boldsymbol{x}) = E^{\boldsymbol{x}}[\xi(\boldsymbol{X}_{N+1})]$ delivered by the SWM coincides with the local solution by the RWM in Eq. 6.23. The coincidence is of no surprise since the points $\boldsymbol{X}_0 = \boldsymbol{x}, \boldsymbol{X}_1(\omega), \ldots$ are on the sample paths of a Brownian motion starting at $\boldsymbol{x}$. ▲

Example 6.29: Let u be the solution of Eq. 6.55 with

$$\mathcal{L} = \sum_{i=1}^{2} a_i \frac{\partial}{\partial x_i} + \frac{1}{2} \sum_{i,j=1}^{2} d_{ij} \frac{\partial^2}{\partial x_i \, \partial x_j}, \qquad \boldsymbol{x} \in D = (-2, 2) \times (-1, 1),$$

and the Dirichlet boundary conditions $u(\boldsymbol{x}) = 1$ for $\boldsymbol{x} \in (-2, 2) \times \{-1\}$ and $u(\boldsymbol{x}) = 0$ on the other boundaries of D, where a_i and d_{ij} are some constants.

Two versions of the SWM can be implemented depending on the sign of the kernel k of $\mathcal{L}$. If k does not have a constant sign in D, the local solution must be based on Eq. 6.57. If k has a constant sign, for example, $k \geq 0$ in D, the local solution can be based on Eq. 6.59 and a modified spherical process whose points are not uniformly distributed on the boundary of the spheres $S(\boldsymbol{X}_i, R_i)$.

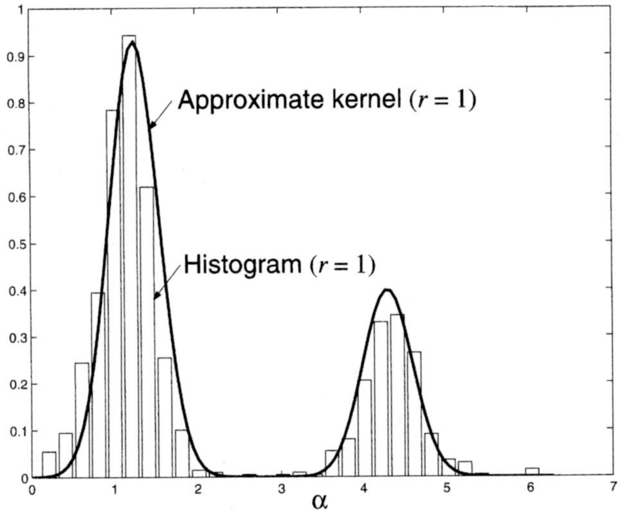

Figure 6.13. Histogram and approximation of $\tilde{k}$ for $r = 1$

Figure 6.13 shows a histogram of $X(T) \mid X(0)$ and an analytical approximation of $\tilde{k}$ in Eq. 6.59 for a sphere of radius $r = 1$ and the coefficients $a_1 = 1$, $a_2 = 2$, $d_{11} = 1$, $d_{12} = d_{21} = 0.5$, and $d_{22} = 2$ of $\mathcal{L}$. Because the coefficients of $\mathcal{L}$ are constants, the kernel of $\mathcal{L}$ does not depend on the location of the center $\boldsymbol{x}$ of the sphere $S(\boldsymbol{x}, r)$. The parameter $\alpha \in [0, 2\pi)$ defines the position on $\partial S(\boldsymbol{x}, r)$. Estimates of the local solution u based on $n_s = 1,000$ samples of the modified spherical process and the analytical approximation of $\tilde{k}$ in Fig. 6.13 are in error by less than 3% relative to a finite element solution. ◇

Note: Let $\boldsymbol{X}$ be an $\mathbb{R}^2$-valued diffusion process defined by the stochastic differential equation $d\boldsymbol{X}(t) = \boldsymbol{a}\, dt + \boldsymbol{b}\, d\boldsymbol{B}(t)$, $t \geq 0$, where $\boldsymbol{a}, \boldsymbol{b}$ denote (2, 1), (2, 2) constant matrices and $\boldsymbol{B}$ is an $\mathbb{R}^2$-valued Brownian motion. If $\boldsymbol{a} = (a_1, a_2)$ and $\boldsymbol{b}\,\boldsymbol{b}^T = \boldsymbol{d} = \{d_{ij}\}$, the generator of $\boldsymbol{X}$ coincides with $\mathcal{L}$. It was found that the function

$$h(\alpha, r) = \sum_{i=1}^{2} \frac{p_i(r)}{\sigma_i}\, \phi\left(\frac{\alpha - \mu_i}{\sigma_i}\right)$$

with $\mu_1 = 1.25$, $\mu_2 = 3.4$, $\sigma_1 = \sigma_2 = 0.3$, and $p_2(r) = 1 - p_1(r)$ approximates $\tilde{k}$ satisfactorily, where $p_1(r) = 0.6; 0.7; 0.8; 0.9$ for $r = 0.5; 1.0; 1.5; 2.0$, and $\phi(\rho) = (2\pi)^{-1/2} \exp(-0.5\,\rho^2)$. The histograms in Fig. 6.13 are based on $n_s = 1,000$ samples of $\boldsymbol{X}$ generated with a time step of 0.001. ▲

We have seen that the spherical process used to define the local solution in Eq. 6.61 reaches a small vicinity of the boundary ∂D of D in a finite number of steps. However, the product of kernels $k(\boldsymbol{X}_{i-1}, \boldsymbol{X}_i)$ in Eq. 6.61 may not converge as the number of steps N increases, in which case the local solution given by this equation cannot be used, as demonstrated by the following example.

Example 6.30: Consider a thin rectangular plate in $D = (0, a) \times (0, b), a, b > 0$, in $\mathbb{R}^2$. The plate is subjected to a uniform traction at the boundaries $\{0\} \times (0, b)$ and $\{a\} \times (0, b)$. The state of stress at a point $\boldsymbol{x} \in D$ in the plate is given by the vector $\boldsymbol{\tau}(\boldsymbol{x}) = (\tau_{11}(\boldsymbol{x}), \tau_{22}(\boldsymbol{x}), \tau_{12}(\boldsymbol{x}))$. There is a mean value theorem for $\boldsymbol{\tau}$ similar to Eq. 6.57 stating that

$$\boldsymbol{\tau}(\boldsymbol{x}) = E^{\boldsymbol{x}} \left[\boldsymbol{a}(\Theta)\, \boldsymbol{\tau}(\Theta)\right],$$

where $\Theta \sim U(0, 2\pi)$ gives the location on the unit circle centered at $\boldsymbol{x}$, $\boldsymbol{a}(\Theta) = \boldsymbol{a}_1(\Theta)^T \boldsymbol{a}_2(\Theta)$,

$$\begin{aligned} \boldsymbol{a}_1(\Theta) &= [4\cos^2(\Theta) - 1, 4\sin^2(\Theta) - 1, 2\sin(2\Theta)], \\ \boldsymbol{a}_2(\Theta) &= [\cos^2(\Theta), \sin^2(\Theta), \sin(2\Theta)], \end{aligned}$$

and $\boldsymbol{\tau}$ is a column vector ([51], Eq. 5.22).

The local solution in Eq. 6.61 depends on the product $\boldsymbol{B}_n = \prod_{i=1}^n \boldsymbol{a}(\Theta_i)$ of random matrices. Let $\Lambda(n)^2$ be the largest eigenvalue of $\boldsymbol{B}_n^T \boldsymbol{B}_n$. We will see in Section 8.8.2 that $\Lambda(n)$ converges to a deterministic value λ as $n \to \infty$ under some conditions and gives the rate of change of $\boldsymbol{\tau}$ for large values of n. If $\lambda > 0$, the algorithm in Eq. 6.61 diverges so that the SWM cannot be applied. Extensive Monte Carlo simulations have shown that the samples $\Lambda(n, \omega)$ of $\Lambda(n)$ are positive and they seem to converge to a deterministic value $\lambda > 0$ as n increases for all ω's. For example, $\Lambda(10{,}000, \omega) \simeq 0.015$ for all the generated samples suggesting that the SWM diverges. ◇

6.3.4 Mixed boundary conditions

Let $u : \mathbb{R}^d \to \mathbb{R}$ be the solution of Eq. 6.55 defined in an open bounded subset D of $\mathbb{R}^d$ with Dirichlet and Neumann boundary conditions on ∂D_d and ∂D_n, respectively, where $\partial D_d \cup \partial D_n = \partial D$ and $\partial D_d \cap \partial D_n = \emptyset$. For example, consider the Laplace equation

$$\begin{cases} \Delta u(\boldsymbol{x}) = 0, & \boldsymbol{x} \in D, \\ u(\boldsymbol{x}) = \xi_d(\boldsymbol{x}), & \boldsymbol{x} \in \partial D_d, \\ \nabla u(\boldsymbol{x}) \cdot \boldsymbol{c}(\boldsymbol{x}) = \xi_n(\boldsymbol{x}), & \boldsymbol{x} \in \partial D_n, \end{cases} \tag{6.63}$$

where $\boldsymbol{c}$ and ξ_d, ξ_n are $\mathbb{R}^d$- and $\mathbb{R}$-valued prescribed functions, respectively.

The SWM in the previous section cannot solve mixed boundary value problems since u is not known on ∂D_n. We present here an extended version of the SWM for the local solution of Eq. 6.63. Alternative versions of the SWM as well as numerical examples can be found in [200] (Sections 3.3 and 5.3).

Suppose we need to calculate the local solution of Eq. 6.63 at a point $\boldsymbol{x} \in D$ and that the spherical process in Fig. 6.12 has reached a point $\boldsymbol{X}_N$ in a small vicinity of ∂D. If $\boldsymbol{X}_{N+1} \in \partial D_d$, then we can calculate $u(\boldsymbol{X}_N) = k(\boldsymbol{X}_N, \boldsymbol{X}_{N+1})\,\xi(\boldsymbol{X}_{N+1})$ and determine the contribution to $u(\boldsymbol{x})$ from Eq. 6.61. If $\boldsymbol{X}_{N+1} \in \partial D_n$, then $u(\boldsymbol{X}_{N+1})$ is not known. Method 2 in Section 6.2.3.4 can be used for solution. Accordingly, we reflect $\boldsymbol{X}_{N+1}$ in the direction of $\boldsymbol{c}(\boldsymbol{X}_{N+1})$ to $\boldsymbol{Y}_0 \in D$ at a distance $\rho > 0$ from $\boldsymbol{X}_{N+1}$. We have

$$\begin{aligned} u(\boldsymbol{Y}_0) &= u(\boldsymbol{X}_{N+1} + (\boldsymbol{Y}_0 - \boldsymbol{X}_{N+1})) \\ &\simeq u(\boldsymbol{X}_{N+1}) + \nabla u(\boldsymbol{X}_{N+1}) \cdot (\boldsymbol{Y}_0 - \boldsymbol{X}_{N+1}) \\ &= u(\boldsymbol{X}_{N+1}) + \rho\,\xi_n(\boldsymbol{X}_{N+1}) \end{aligned}$$

since $\boldsymbol{Y}_0 - \boldsymbol{X}_{N+1} = \rho\,\boldsymbol{c}(\boldsymbol{X}_{N+1})$. Hence, the unknown value of u at $\boldsymbol{X}_{N+1}$ can be related to $\rho\,\xi_n(\boldsymbol{X}_{N+1})$, a known value, and $u(\boldsymbol{Y}_0)$. Now we restart the spherical process from $\boldsymbol{Y}_0$. Suppose that after M steps the restarted spherical process is at $\boldsymbol{Y}_M$ in a small vicinity of ∂D and that $\boldsymbol{Y}_{M+1} \in \partial D_d$. By Eq. 6.61 $u(\boldsymbol{Y}_0)$ can be related to $u(\boldsymbol{Y}_{M+1}) = \xi_d(\boldsymbol{Y}_{M+1})$, which is known. If $\boldsymbol{Y}_{M+1} \in \partial D_n$, then $\boldsymbol{Y}_{M+1}$ has to be reflected in D and the previous algorithm needs to be repeated until a Dirichlet boundary is reached. Hence, the value of $u(\boldsymbol{x})$ can be calculated for any sample of the spherical process.

6.4 Boundary walk method

The differential equations of linear elasticity do not have the form in Eq. 6.1 so that the RWM cannot be used to solve elasticity problems. We have seen in Example 6.30 that the SWM may give unreliable results because it involves infinite products of random matrices, which may diverge. Some preliminary results suggest that the boundary walk method (BWM) can be used to solve linear elasticity problems [118]. The BWM is based on a representation of the local solution by an infinite series whose terms are boundary integrals. We present essentials of the BWM and illustrate it by a simple example.

Let D be an open bounded subset of $\mathbb{R}^d$, $d \geq 1$, with boundary ∂D on which tractions or displacements are prescribed. Our objective is to find the solution $\boldsymbol{u}(\boldsymbol{x})$ of a linear elasticity problem at an arbitrary point $\boldsymbol{x} \in D$, where $\boldsymbol{u}$ is a scalar or vector function.

To establish the local solution for a linear elasticity problem, we need to introduce some notation and definitions. Let $\boldsymbol{k}(\boldsymbol{x}, \boldsymbol{y})$ be a (d, d) matrix whose entries are real-valued functions defined on $\mathbb{R}^{2d}$. Define the operators

$$\mathcal{K}^{(i)}[\boldsymbol{h}(\boldsymbol{x})] = \int_{\partial D} \boldsymbol{k}^{(i)}(\boldsymbol{x}, \boldsymbol{y})\,\boldsymbol{h}(\boldsymbol{y})\,d\sigma(\boldsymbol{y}), \quad i = 1, 2, \ldots, \tag{6.64}$$

for $\boldsymbol{x} \in \partial D$, where

$$\boldsymbol{k}^{(i)}(\boldsymbol{x}, \boldsymbol{y}) = \int_{\partial D} \boldsymbol{k}(\boldsymbol{x}, \boldsymbol{\xi})\,\boldsymbol{k}^{(i-1)}(\boldsymbol{\xi}, \boldsymbol{y})\,d\sigma(\boldsymbol{\xi}), \quad i = 2, 3, \ldots, \tag{6.65}$$

and $\boldsymbol{k}^{(0)}(\boldsymbol{\xi}, \boldsymbol{y}) = 1$. It can be shown that the equations of linear elasticity can be given in the form

$$\boldsymbol{\mu}(\boldsymbol{x}) = \boldsymbol{g}(\boldsymbol{x}) + \int_{\partial D} \boldsymbol{k}(\boldsymbol{x}, \boldsymbol{y})\, \boldsymbol{\mu}(\boldsymbol{y})\, d\sigma(\boldsymbol{y}) = \boldsymbol{g}(\boldsymbol{x}) + \mathcal{K}^{(1)}[\boldsymbol{\mu}(\boldsymbol{x})], \quad \boldsymbol{x} \in \partial D, \tag{6.66}$$

where $\boldsymbol{\mu}$ and $\boldsymbol{g}$ are unknown and known functions on ∂D, respectively. The function $\boldsymbol{\mu}$ can be used to obtain the solution $\boldsymbol{u}$ for a linear elasticity problem in D from [118]

$$\boldsymbol{u}(\boldsymbol{\xi}) = \int_{\partial D} \boldsymbol{r}(\boldsymbol{\xi}, \boldsymbol{y})\, \boldsymbol{\mu}(\boldsymbol{y})\, d\sigma(\boldsymbol{y}) = \mathcal{R}[\boldsymbol{\mu}(\boldsymbol{\xi})], \quad \boldsymbol{x} \in D, \tag{6.67}$$

where $\mathcal{R}$ denotes an integral operator and the kernel $\boldsymbol{r}(\boldsymbol{x}, \boldsymbol{y})$ is a (d, d) matrix with real-valued entries defined on $\mathbb{R}^{2d}$. The kernels $\boldsymbol{k}(\boldsymbol{x}, \boldsymbol{y})$ and $\boldsymbol{r}(\boldsymbol{x}, \boldsymbol{y})$ are closely related [118].

If $\sup_{\boldsymbol{x} \in \partial D} \int_{\partial D} \| \boldsymbol{k}(\boldsymbol{x}, \boldsymbol{y}) \| \, d\sigma(\boldsymbol{y}) < 1$, the **local solution** for a linear elasticity problem is

$$\boldsymbol{u}(\boldsymbol{x}) = \mathcal{R}[\boldsymbol{g}(\boldsymbol{x})] + \sum_{i=1}^{\infty} \mathcal{R}\,\mathcal{K}^{(i)}[\boldsymbol{g}(\boldsymbol{x})], \quad \boldsymbol{x} \in D. \tag{6.68}$$

Note: Let $\mathcal{I}$ denote the identity operator. Under the above condition, the Neumann series of the operator $\mathcal{I} - \mathcal{K}^{(1)}$ is absolutely and uniformly convergent, so that the solution of $\left(\mathcal{I} - \mathcal{K}^{(1)}\right)[\boldsymbol{\mu}(\boldsymbol{\xi})] = \boldsymbol{g}(\boldsymbol{\xi})$, that is, the solution of Eq. 6.66 is (Section 8.3.1.4)

$$\boldsymbol{\mu}(\boldsymbol{\xi}) = g(\boldsymbol{\xi}) + \sum_{i=1}^{\infty} \mathcal{K}^{(i)}[g(\boldsymbol{\xi})], \quad \boldsymbol{\xi} \in \partial D.$$

The above series representation of $\boldsymbol{\mu}$ and Eq. 6.67 yield Eq. 6.68. Generally, the condition under which Eq. 6.68 holds is not satisfied for elasticity problems. However, it is possible to modify the series representation of the local solution such that it becomes convergent, as demonstrated later in this section (Example 6.31). ▲

Suppose that the representation of the local solution in Eq. 6.68 is valid. This representation cannot be used for numerical calculation since it has an infinite number of terms. We have to approximate the local solution $\boldsymbol{u}(\boldsymbol{x})$, $\boldsymbol{x} \in D$, by the first $m < \infty$ terms in Eq. 6.68, that is,

$$\boldsymbol{u}(\boldsymbol{x}) \simeq \tilde{\boldsymbol{u}}(\boldsymbol{x}) = \mathcal{R}[\boldsymbol{g}(\boldsymbol{x})] + \sum_{i=1}^{m} \mathcal{R}\,\mathcal{K}^{i}[\boldsymbol{g}(\boldsymbol{x})]. \tag{6.69}$$

No probabilistic concepts have been used to establish Eqs. 6.68 and 6.69. We will use Monte Carlo simulation to calculate the terms of $\tilde{\boldsymbol{u}}(\boldsymbol{x})$ in Eq. 6.69. A random walk on ∂D, referred to as the **boundary walk method** (BWM), is used to evaluate $\tilde{\boldsymbol{u}}(\boldsymbol{x})$ in Eq. 6.69.

Let $\boldsymbol{Y}_0, \boldsymbol{Y}_1, \ldots, \boldsymbol{Y}_i, \ldots$ be a ∂D-valued Markov chain, where f_0 and $f_{i|i-1}$ denote the densities of $\boldsymbol{Y}_0$ and $\boldsymbol{Y}_i \mid \boldsymbol{Y}_{i-1}$, $i = 1, 2, \ldots$, respectively. Define also an $\mathbb{R}^d$-valued process with states $\boldsymbol{Z}_0, \boldsymbol{Z}_1, \ldots, \boldsymbol{Z}_i, \ldots$, where

$$\begin{aligned} \boldsymbol{Z}_i &= \boldsymbol{q}_i\, \boldsymbol{g}(\boldsymbol{Y}_i), \quad i = 0, 1, \ldots, \\ \boldsymbol{q}_i &= \frac{\boldsymbol{q}_{i-1}\, \boldsymbol{k}(\boldsymbol{Y}_{i-1}, \boldsymbol{Y}_i)}{f_{i|i-1}(\boldsymbol{Y}_i \mid \boldsymbol{Y}_{i-1})}, \quad i = 1, 2, \ldots, \quad \text{and} \\ \boldsymbol{q}_0 &= \frac{\boldsymbol{r}(\boldsymbol{x}, \boldsymbol{Y}_0)}{f_0(\boldsymbol{Y}_0)}. \end{aligned} \tag{6.70}$$

The states $\boldsymbol{Z}_0, \boldsymbol{Z}_1, \ldots$, and $\boldsymbol{Z}_i$ depend on $\boldsymbol{Y}_0$, $(\boldsymbol{Y}_0, \boldsymbol{Y}_1), \ldots$, and $(\boldsymbol{Y}_0, \ldots, \boldsymbol{Y}_i)$, respectively. The joint density $f_{0,1,\ldots,i}$ of $(\boldsymbol{Y}_0, \ldots, \boldsymbol{Y}_i)$ is $f_{0,1,\ldots,i}(\boldsymbol{y}_0, \boldsymbol{y}_1, \ldots, \boldsymbol{y}_i) = f_0(\boldsymbol{y}_0) \prod_{j=1}^{i} f_{j|j-1}(\boldsymbol{y}_j \mid \boldsymbol{y}_{j-1})$ for $j \geq 1$.

If $f_{0,1,\ldots,i}(\boldsymbol{y}_0, \boldsymbol{y}_1, \ldots, \boldsymbol{y}_i) = 0$ implies that the entries of the matrix $\boldsymbol{r}(\boldsymbol{x}, \boldsymbol{y}_0) \prod_{j=1}^{i} \boldsymbol{k}(\boldsymbol{y}_{j-1}, \boldsymbol{y}_j)\, \boldsymbol{g}(\boldsymbol{y}_i)$ are zero for all $\boldsymbol{x} \in D$, all boundary points $\boldsymbol{y}_0, \boldsymbol{y}_1, \ldots, \boldsymbol{y}_i \in \partial D$, and $i \geq 0$, then

$$\mathcal{RK}^{(i)}[\boldsymbol{g}(\boldsymbol{x})] = E\left[\boldsymbol{Z}_i\right], \quad i \geq 0. \tag{6.71}$$

Proof: We use the definitions in Eqs. 6.66, 6.67, 6.70, and 6.70 to prove Eq. 6.71. Note that

$$\mathcal{R}[\boldsymbol{g}(\boldsymbol{x})] = \int_{\partial D} \boldsymbol{r}(\boldsymbol{x}, \boldsymbol{y})\, \boldsymbol{g}(\boldsymbol{y})\, d\sigma(\boldsymbol{y}) = \int_{\partial D} \left[\frac{\boldsymbol{r}(\boldsymbol{x}, \boldsymbol{y})}{f_0(\boldsymbol{y})}\, \boldsymbol{g}(\boldsymbol{y})\right] f_0(\boldsymbol{y})\, d\sigma(\boldsymbol{y}) = E\left[\boldsymbol{Z}_0\right]$$

provided that $f_0(\boldsymbol{y}) = 0$ implies that the entries of $\boldsymbol{r}(\boldsymbol{x}, \boldsymbol{y})\, \boldsymbol{g}(\boldsymbol{y})$ are zero for all $\boldsymbol{x} \in D$ and $\boldsymbol{y} \in \partial D$. The second term in Eq. 6.69 is

$$\begin{aligned} \mathcal{RK}^{(1)}[\boldsymbol{g}(\boldsymbol{x})] &= \int_{\partial D \times \partial D} \boldsymbol{r}(\boldsymbol{x}, \boldsymbol{y}_0)\, \boldsymbol{k}(\boldsymbol{y}_0, \boldsymbol{y}_1)\, \boldsymbol{g}(\boldsymbol{y}_1)\, d\sigma(\boldsymbol{y}_0)\, d\sigma(\boldsymbol{y}_1) \\ &= \int_{\partial D \times \partial D} \left[\frac{\boldsymbol{r}(\boldsymbol{x}, \boldsymbol{y}_0)}{f_0(\boldsymbol{y}_0)}\, \frac{\boldsymbol{k}(\boldsymbol{y}_0, \boldsymbol{y}_1)}{f_{1|0}(\boldsymbol{y}_1 \mid \boldsymbol{y}_0)}\, \boldsymbol{g}(\boldsymbol{y}_1)\right] f_{0,1}(\boldsymbol{y}_0, \boldsymbol{y}_1)\, d\sigma(\boldsymbol{y}_0)\, d\sigma(\boldsymbol{y}_1) = E\left[\boldsymbol{Z}_1\right] \end{aligned}$$

since $f_{0,1}(\boldsymbol{y}_0, \boldsymbol{y}_1) = 0$ implies the entries of $\boldsymbol{r}(\boldsymbol{x}, \boldsymbol{y}_0)\, \boldsymbol{k}(\boldsymbol{y}_0, \boldsymbol{y}_1)\, \boldsymbol{g}(\boldsymbol{y}_1)$ are zero for all $\boldsymbol{x} \in D$ and $\boldsymbol{y}_0, \boldsymbol{y}_1 \in \partial D$. ■

Let $(\boldsymbol{Y}_0(\omega), \boldsymbol{Y}_1(\omega), \ldots)$, $\omega = 1, \ldots, n_s$, be n_s independent samples of the Markov process $(\boldsymbol{Y}_0, \boldsymbol{Y}_1, \ldots)$ and let $(\boldsymbol{Z}_0(\omega), \boldsymbol{Z}_1(\omega), \ldots)$ denote the corresponding samples of $(\boldsymbol{Z}_0, \boldsymbol{Z}_1, \ldots)$.

The terms of the local solution in Eq. 6.68 can be approximated by the following estimates of the expectation of the random variables $\boldsymbol{Z}_i$, that is,

$$\mathcal{RK}^{(i)}[\boldsymbol{g}(\boldsymbol{x})] \simeq \frac{1}{n_s} \sum_{\omega=1}^{n_s} \boldsymbol{Z}_i(\omega), \quad i \geq 0. \tag{6.72}$$

Example 6.31: Let u be the solution of $\Delta u(\boldsymbol{x}) = 0$, $\boldsymbol{x} \in D$, with the boundary conditions $u(\boldsymbol{x}) = \xi(\boldsymbol{x})$, $\boldsymbol{x} \in \partial D$, where D is an open, convex, bounded subset in $\mathbb{R}^2$ and ξ is a prescribed function. The local solution of this equation is

$$\begin{aligned} u(\boldsymbol{x}) &= \frac{1}{2}\,\mathcal{R}[g(\boldsymbol{x})] + \frac{1}{2}\sum_{i=1}^{\infty}\left(\mathcal{R}\mathcal{K}^{(i-1)}[g(\boldsymbol{x})] + \mathcal{R}\mathcal{K}^{(i)}[g(\boldsymbol{x})]\right) \\ &= \frac{1}{2}\,E[\mathbf{Z}_0] + \frac{1}{2}\sum_{i=1}^{\infty}\left(E[\mathbf{Z}_{i-1}] + E[\mathbf{Z}_i]\right). \end{aligned}$$

The local solution u has been calculated for $D = \{\boldsymbol{x} \in \mathbb{R}^2, \| \boldsymbol{x} \| < 1\}$ and $\xi(\boldsymbol{x}) = 100\cos(\theta)$, where $x_1 = \cos(\theta)$, $x_2 = \sin(\theta)$, and $\theta \in [0, 2\pi)$. Numerical results on a circle of radius 1/2 centered at zero are in error by less than 1% when based on the approximation $\tilde{u}(\boldsymbol{x}) = (1/2)\,E[\mathbf{Z}_0] + (1/2)\,(E[\mathbf{Z}_0] + E[\mathbf{Z}_1]) + (1/2)\,(E[\mathbf{Z}_1] + E[\mathbf{Z}_2])$, where the expectations $E[\mathbf{Z}_i]$ in the expression of $\tilde{u}$ have been calculated from 20,000 independent samples of the boundary walk. ◇

Proof: The initial and transition densities $f_0(\boldsymbol{y}_0) = r(\boldsymbol{x}, \boldsymbol{y}_0)$ and $f_{i|i-1}(\boldsymbol{y}_i \mid \boldsymbol{y}_{i-1}) = |k(\boldsymbol{y}_{i-1}, \boldsymbol{y}_i)|$, $i = 1, 2, \ldots$, have been used in calculations, where

$$\begin{aligned} r(\boldsymbol{x}, \boldsymbol{y}_0) &= \cos(\varphi_{\boldsymbol{x},\boldsymbol{y}_0})/(2\pi\, r_0), \\ k(\boldsymbol{y}_{i-1}, \boldsymbol{y}_i) &= -\cos(\varphi_{\boldsymbol{y}_{i-1},\boldsymbol{y}_i})/(\pi\, r_{i-1,i}), \end{aligned}$$

$r_0 = \| \boldsymbol{x} - \boldsymbol{y}_0 \|$, $r_{i-1,i} = \| \boldsymbol{y}_{i-1} - \boldsymbol{y}_i \|$, $\varphi_{\boldsymbol{x},\boldsymbol{y}_0}$ denotes the angle between the interior normal at $\boldsymbol{y}_0 \in \partial D$ and the ray from $\boldsymbol{y}_0 \in \partial D$ to $\boldsymbol{x} \in D$, and $\varphi_{\boldsymbol{y}_{i-1},\boldsymbol{y}_i}$ is the angle between the interior normal at $\boldsymbol{y}_i$ and the ray from $\boldsymbol{y}_i \in \partial D$ to $\boldsymbol{y}_{i-1} \in \partial D$.

The series $\sum_{i=0}^{\infty} E[Z_i]$ cannot be used to calculate $u(\boldsymbol{x})$ since it is not convergent. Consider the series $v(\boldsymbol{x}) = \sum_{i=0}^{\infty} \lambda^i\, E[Z_i]$. This series has a single pole at $\lambda = -1$ and coincides with the series expansion of $u(\boldsymbol{x})$ for $\lambda = 1$. The idea is to eliminate the pole of $v(\boldsymbol{x})$ and then use the resulting series to calculate the local solution $u(\boldsymbol{x})$.

Note that

$$(1+\lambda)\,v(\boldsymbol{x}) = (1+\lambda)\sum_{i=0}^{\infty}\lambda^i\, E[Z_i] = E[Z_0] + \sum_{i=1}^{\infty}\lambda^i\,\left(E[Z_{i-1}] + E[Z_i]\right)$$

or

$$v(\boldsymbol{x}) = \frac{1}{1+\lambda}\,E[\mathbf{Z}_0] + \frac{1}{1+\lambda}\sum_{i=1}^{\infty}\lambda^i\,\left(E[Z_{i-1}] + E[Z_i]\right).$$

Since the above series is convergent and $u(\boldsymbol{x})$ is equal to $v(\boldsymbol{x})$ for $\lambda = 1$, $u(\boldsymbol{x})$ has the stated representation. ■

6.5 Algebraic equations

Consider the algebraic equation (Eq. 6.2)

$$(\boldsymbol{i} - \lambda\, \boldsymbol{g})\boldsymbol{u} = \boldsymbol{\phi}, \tag{6.73}$$

where $\boldsymbol{i}$ is the (m, m) identity matrix, $\boldsymbol{g}$ is an (m, m) matrix, $\boldsymbol{\phi}$ and $\boldsymbol{u}$ denote m-dimensional vectors, and λ is a parameter. The entries of $\boldsymbol{g}$ and $\boldsymbol{\phi}$ are real numbers. If $\boldsymbol{\phi} \neq \mathbf{0}$ and $\det(\boldsymbol{i} - \lambda\, \boldsymbol{g}) \neq 0$ for a specified value of λ, then Eq. 6.73 has a unique solution. If $\boldsymbol{\phi} = \mathbf{0}$, Eq. 6.73 becomes

$$(\boldsymbol{i} - \lambda\, \boldsymbol{g})\boldsymbol{u} = 0 \tag{6.74}$$

and defines an eigenvalue problem. We refer to Eqs. 6.73 and 6.74 as inhomogeneous and homogeneous, respectively.

Our objectives are to develop methods for finding directly the value of a particular coordinate of the solution $\boldsymbol{u}$ of Eq. 6.73 and the dominant eigenvalue/eigenvector of Eq. 6.74, that is, to develop methods for the local solution of Eqs. 6.73 and 6.74.

6.5.1 Inhomogeneous equations

Consider the algebraic equation

$$\boldsymbol{u} = \boldsymbol{\phi} + \boldsymbol{g}\, \boldsymbol{u} \tag{6.75}$$

defined by Eq. 6.73 with $\lambda = 1$. This choice of λ is not restrictive. The objective is to find the value of a coordinate u_i of $\boldsymbol{u}$ directly rather than extracting its value from the global solution of Eq. 6.75.

Let X be a discrete time Markov process with discrete states $\{1, \ldots, m+1\}$, where $m + 1$ is an absorbing state. The process X starts at a state $i \in \{1, \ldots, m\}$ and exits $\{1, \ldots, m\}$ or, equivalently, enters the absorbing state $m + 1$, for the first time in T transitions, where

$$T = \inf\{r > 0 : X(r) = m + 1, X(0) = i \in \{1, \ldots, m\}\}. \tag{6.76}$$

Let $\boldsymbol{p} = \{p_{i,j}\}$, $i, j = 1, \ldots, m$, be a matrix such that $p_{i,j} \geq 0$, $p_i = 1 - \sum_{j=1}^{m} p_{i,j} \geq 0$, $i = 1, \ldots, m$, and at least one of the p_i's is strictly positive. Let $\boldsymbol{v} = \{v_{i,j}\}$ be an (m, m) matrix with entries

$$v_{i,j} = \begin{cases} g_{i,j}/p_{i,j}, & \text{if } p_{i,j} > 0, \\ 0, & \text{if } p_{i,j} = 0, \end{cases} \tag{6.77}$$

where $g_{i,j}$ are the entries of $\boldsymbol{g}$ in Eq. 6.75.

Let $V = (V(0), V(1), \ldots)$ be a process defined by

$$\begin{aligned} V(k+1) &= V(k)\, v_{X(k),X(k+1)}, \quad k = 0, 1, \ldots, \\ V(0) &= 1. \end{aligned} \tag{6.78}$$

Note: The $(m+1, m+1)$ matrix $\tilde{\boldsymbol{p}}$ with the first m rows and columns given by $\boldsymbol{p}$, $p_{i,m+1} = p_i$ for $i = 1, \ldots, m$, $\tilde{p}_{m+1,j} = 0$ for $j = 1, \ldots, m$, and $\tilde{p}_{m+1,m+1} = 1$ is a transition probability matrix that is used to define the Markov chain X with the discrete states $\{1, \ldots, m+1\}$ and a unit time step, where $m+1$ is an absorbing state. The time T in Eq. 6.76 gives the number of transitions till X starting at $X(0) = i \in \{1, \ldots, m\}$ enters the absorbing state $m+1$ for the first time.

If all probabilities p_i were zero, the Markov chain X would not reach the absorbing state $m+1$ a.s. so that T would be unbounded. Because at least one of the probabilities p_i is strictly positive, X reaches the absorbing state $m+1$ in a finite number of transitions a.s. (Eq. 6.80). ▲

If $\det(\boldsymbol{i} - \boldsymbol{g}) \neq 0$ and $\| \boldsymbol{g} \|_m < 1$, the **local solution** of Eq. 6.75 is

$$u_i = E[V(T-1)\, \phi_{i_{T-1}} / p_{i_{T-1}} \mid X(0) = i]. \tag{6.79}$$

Proof: The above matrix norm $\| \cdot \|_m$ is $\| \boldsymbol{g} \|_m = \max_{\|\boldsymbol{\xi}\|=1} \| \boldsymbol{g}\, \boldsymbol{\xi} \|$, where $\| \cdot \|$ is the Euclidian norm.

Let $S = T - 1$ and $i, i_1, \ldots, i_{S(\omega)}, i_{T(\omega)}$ be a sample path $X(\cdot, \omega)$ of X, which starts at state $i \in \{1, \ldots, m\}$ and reaches the absorbing state $m+1$ in $T(\omega)$ transitions. The expectation in Eq. 6.79 is

$$E[V(S)\, \phi_{i_S} / p_{i_S} \mid X(0) = i] = \sum_{\omega} \left(V(S(\omega))\, \phi_{i_{S(\omega)}} / p_{i_{S(\omega)}} \right) P(\omega),$$

where the sum is performed over all sample paths ω of X,

$$P(\omega) = \begin{cases} p_{i,i_1}\, p_{i_1,i_2} \cdots p_{i_{S(\omega)-1}, i_{S(\omega)}}\, p_{i_{S(\omega)}} & \text{for } S(\omega) > 0, \\ p_{i_0} = p_i & \text{for } S(\omega) = 0 \end{cases}$$

is the probability of sample path ω, and $V(S(\omega)) = v_{i,i_1}\, v_{i_1,i_2} \cdots v_{i_{S(\omega)-1}, i_{S(\omega)}}$ denotes the value of V just prior to X entering the absorbing state $m+1$ calculated along a sample path ω of X. An alternative form of Eq. 6.79 is

$$u_i = \sum_{\omega} \left(v_{i,i_1}\, v_{i_1,i_2} \cdots v_{i_{S(\omega)-1}, i_{S(\omega)}}\, \phi_{i_{S(\omega)}} / p_{i_{S(\omega)}} \right) P(\omega)$$

or, equivalently,

$$\begin{aligned} u_i &= \sum_{k=0}^{\infty} \sum_{i_1=1}^{m} \cdots \sum_{i_k=1}^{m} g_{i,i_1}\, g_{i_1,i_2} \cdots g_{i_{k-1},i_k}\, \phi_{i_k} \\ &= \phi_i + (\boldsymbol{g}\, \boldsymbol{\phi})_i + (\boldsymbol{g}^2\, \boldsymbol{\phi})_i + \cdots, \end{aligned}$$

where the first value of k is zero since, if X enters the absorbing state at the first transition, $T = 1$ and $S = 0$. The resulting expression of u_i represents the entry i of the Neumann series expansion $\sum_{s=0}^{\infty} \boldsymbol{g}^s\, \boldsymbol{\phi}$, which is convergent since $\| \boldsymbol{g} \|_m < 1$ by hypothesis. Since $(\boldsymbol{i} - \boldsymbol{g})^{-1} \boldsymbol{\phi} = \sum_{s=0}^{\infty} \boldsymbol{g}^s\, \boldsymbol{\phi}$, u_i in Eq. 6.79 solves locally Eq. 6.75. ■

If at least one of the probabilities p_i is not zero, the average number of transitions of X to the absorbing state $m+1$ is finite and equal to

$$\boldsymbol{w} = (\boldsymbol{i} - \boldsymbol{p})^{-1}\,\mathbf{1}, \tag{6.80}$$

where $\mathbf{1} \in \mathbb{R}^m$ has unit entries and $w_i = E[T \mid X(0) = i]$, $i = 1, \ldots, m$.

Proof: Let ρ be a real-valued function defined on $\{1, \ldots, m\}$ and

$$w_i = E\left[\sum_{k=0}^{S} \rho(X(k)) \mid X(0) = i\right]$$

for $i \in \{1, \ldots, m\}$ and $S = T - 1$. It can be shown that $w_i = \rho(i) + \sum_{j=1}^{m} p_{i,j}\, w_j$ ([150], pp. 108-110) or $\boldsymbol{w} = \boldsymbol{\rho} + \boldsymbol{p}\,\boldsymbol{w}$. If $\rho(i) = 1$ for all $i \in \{1, \ldots, m\}$, then

$$w_i = E[S + 1 \mid X(0) = i] = E[T \mid X(0) = i]$$

is the expected number of transitions that X, starting at $X(0) = i$, takes to enter the absorbing state $m+1$ and $\boldsymbol{w} = (\boldsymbol{i} - \boldsymbol{p})^{-1}\,\mathbf{1}$ as stated. For example, let $m = 3$ and

$$\boldsymbol{p} = \begin{bmatrix} 0.2 & 0.4 & 0.0 \\ 0.4 & 0.2 & 0.2 \\ 0.0 & 0.5 & 0.2 \end{bmatrix}.$$

The expected values of the time to absorption are $(3.09, 3.68, 3.55)$ for the initial states $(1, 2, 3)$ of X. ■

The efficiency and accuracy of the local solution in Eq. 6.79 depends on matrices $\boldsymbol{g}$ and $\boldsymbol{p}$. We have no control on $\boldsymbol{g}$ but the selection of $\boldsymbol{p}$ is rather arbitrary provided it defines a Markov chain X with a reachable absorbing state. There are no simple rules for selecting an optimal transition probability matrix $\boldsymbol{p}$. Some suggestions for the selection of matrix $\boldsymbol{p}$ can be found in [94] (Section 7.1).

Let $i \mapsto X(1,\omega) \mapsto \cdots \mapsto X(T(\omega)-1, \omega) \mapsto X(T(\omega), \omega)$ be a sample of X and let $X(T(\omega) - 1, \omega) = X(S(\omega), \omega)$ denote the last value of X in $\{1, \ldots, m\}$ prior to its exit from this set. Suppose that n_s independent samples of X have been generated by using the transition probability matrix $\tilde{\boldsymbol{p}}$ defined following Eq. 6.78. The values of V along a sample ω of X can be calculated from Eq. 6.78.

The **local solution** in Eq. 6.79 can be estimated by

$$\hat{u}_i = \frac{1}{n_s} \sum_{\omega=1}^{n_s} V(T(\omega) - 1, \omega)\, \frac{\phi_{i_{T(\omega)-1}}}{p_{i_{T(\omega)-1}}}. \tag{6.81}$$

Example 6.32: Let u be defined by Eq. 6.75 with $m = 1$. The exact solution of this equation is $u = \phi/(1 - g)$. The associated Markov process X has two states and the transition probability matrix

$$\tilde{\boldsymbol{p}} = \begin{bmatrix} p & 1-p \\ 0 & 1 \end{bmatrix}.$$

If $|g| < 1$, Eq. 6.79 gives $u = \sum_{k=0}^{\infty} g^k \phi = \phi \sum_{k=0}^{\infty} g^k$, which coincides with the exact solution $\phi/(1-g)$ since the Neumann series $\sum_{k=0}^{\infty} g^k$ is convergent with sum $1/(1-g)$. ◇

Example 6.33: Suppose that all probabilities $p_i = 1 - \sum_{j=1}^{m} p_{i,j}$ are equal to $p \in (0, 1)$. Then the average number of transitions to absorption for the Markov chain X with transition probabilities $p_{i,j}$ is $w_i = 1/p$, for each $i \in \{1, \ldots, m\}$. ◇

Proof: The solution $w_i = 1/p$, $i \in \{1, \ldots, m\}$, can be obtained directly by noting that X is absorbed with probability p or takes values in $\{1, \ldots, m\}$ with probability $1 - p$ at each transition. Hence, T is the time to the first success for a sequence of independent Bernoulli trials with probability of success p. Hence, T is independent of the initial value of X and has the mean and variance $E[T] = 1/p$ and $\mathrm{Var}[T] = (1-p)/p^2$, respectively.

Consider the Neumann series $\boldsymbol{w} = \mathbf{1} + \sum_{s=1}^{\infty} \boldsymbol{p}^s \mathbf{1}$ corresponding to the solution $\boldsymbol{w} = (\boldsymbol{i} - \boldsymbol{p})^{-1} \mathbf{1}$. Because the terms $\boldsymbol{p}^s \mathbf{1}$ of $\sum_{s=1}^{\infty} \boldsymbol{p}^s \mathbf{1}$ are equal to $(1-p)^s \mathbf{1}$, this series is convergent and all coordinates of $\boldsymbol{w}$ are equal to $\sum_{s=0}^{\infty} (1-p)^s = 1/p$. For example, the entries of $\boldsymbol{p}\,\mathbf{1}$ are $\sum_{j=1}^{m} p_{i,j} = 1 - p$ so that $\boldsymbol{p}^2 \mathbf{1} = \boldsymbol{p}\,(\boldsymbol{p}\,\mathbf{1}) = (1-p)\,\boldsymbol{p}\,\mathbf{1}$. ■

Example 6.34: Let $\boldsymbol{u}$ be the solution of Eq. 6.75 with $m = 3$, $\phi_1 = 1$, $\phi_2 = 2$, $\phi_3 = 3$, and

$$\boldsymbol{g} = \begin{bmatrix} 0.2 & 0.4 & 0.0 \\ 0.4 & 0.2 & 0.2 \\ 0.0 & 0.5 & 0.2 \end{bmatrix}.$$

Suppose we take $\boldsymbol{p}$ equal to

$$\boldsymbol{p}_1 = \begin{bmatrix} 0.3 & 0.3 & 0.3 \\ 0.2 & 0.2 & 0.5 \\ 0.5 & 0.2 & 0.1 \end{bmatrix}$$

or $\boldsymbol{p}_2$ with all entries equal to 0.1. The average number of transitions of X to absorption are $(7.75, 7.63, 7.11)$ and $(1.43, 1.43, 1.43)$ under $\boldsymbol{p}_1$ and $\boldsymbol{p}_2$, respectively, for the initial states $(1, 2, 3)$ (Eq. 6.80). The ratio, $((7.75 + 7.63 + 7.11)/3))/1.43 = 5.24$, shows that the computation times for finding the local solution based on 2,000 and 10,000 samples of X defined by $\boldsymbol{p}_1$ and $\boldsymbol{p}_2$, respectively, are similar.

The exact solution is $\boldsymbol{u} = (4.68, 6.84, 8.03)$. The algorithm in Eq. 6.81 can be applied since $\| \boldsymbol{g} \|_m < 1$. The estimates of $\boldsymbol{u}$ are $(4.23, 6.65, 8.01)$ and $(3.90, 5.78, 7.14)$ for the transition probability matrices $\boldsymbol{p}_1$ and $\boldsymbol{p}_2$ based on 2,000 and 10,000 samples, respectively. The errors of these estimates are $(-9.62, -2.78, -0.25)\%$ for $\boldsymbol{p}_1$ and $(-16.67, -12.35, -11.08)\%$ for $\boldsymbol{p}_2$. The probability matrix with smaller absorbing probabilities yields a superior local solution in this case for approximately the same amount of calculations. ◇

Note: Experience provides some guidelines for selecting the probabilities $p_{i,j}$. If the probabilities p_i are "too large", the average number of transitions of X to absorption is very small and the estimates of the local solutions can be unstable and less accurate, as demonstrated by this example. ▲

Example 6.35: Consider the Poisson equation $\Delta u(\boldsymbol{x}) + p(\boldsymbol{x}) = 0$ in Eq. 6.6 defined on an open bounded set $D \subset \mathbb{R}^2$ with the Dirichlet boundary condition $u(\boldsymbol{x}) = \xi(\boldsymbol{x})$, $\boldsymbol{x} \in \partial D$. The local solution of this problem by the RWM is in Section 6.2.1.4. We present here an alternative local solution based on the finite difference approximation,

$$u_{i,j} = \frac{1}{4}(u_{i+1,j} + u_{i-1,j} + u_{i,j+1} + u_{i,j-1}) + \frac{a^2}{4}\,p(\boldsymbol{x}_{i,j}), \tag{6.82}$$

of this Poisson equation, where $\boldsymbol{x}_{i,j}$ denotes the coordinate of the node (i, j) of the finite difference mesh in D, a denotes the step of the finite difference mesh, and $u_{i,j} = u(\boldsymbol{x}_{i,j})$.

The method in this section can be applied to solve locally the linear system defined by Eq. 6.82. However, we present an alternative local solution based on n_s walkers that start at a node $\boldsymbol{x}$ of the finite difference mesh and can move to a node just right, left, up, or down from $\boldsymbol{x}$ with equal probability. A walker k travels through the nodes $\boldsymbol{x}_s^{(k)}$, $s = 1, \ldots, m_k$, of the finite difference mesh and exits D at $\boldsymbol{x}_{m_k}^{(k)} \in \partial D$, where m_k denotes the number of transitions walker k takes to reach the boundary ∂D of D. For simplicity, we change notation and identify the nodes of the finite difference mesh by a single subscript. The local solution can be estimated by

$$\hat{u}(\boldsymbol{x}) = \frac{1}{n_s}\sum_{k=1}^{n_s}\left[\xi(\boldsymbol{x}_{m_k}^{(k)}) + \frac{a^2}{4}\sum_{s=1}^{m_k} p(\boldsymbol{x}_s^{(k)})\right].$$

The result provides an approximation for the local solution in Eq. 6.26. ◇

Proof: Note that the walkers paths define a random walk in $\mathbb{R}^2$. The first term in the expression of $\hat{u}$, that is, the sum $(1/n_s)\sum_{k=1}^{n_s}\xi(\boldsymbol{x}_{m_k}^{(k)})$, is an estimate of the expectation $E^{\boldsymbol{x}}[u(\boldsymbol{B}(T))]$ in Eq. 6.26. The connection between the second term in the expression of $\hat{u}$ and the integral $\int_0^T p(\boldsymbol{B}(s))\,ds$ in Eq. 6.26 is less obvious.

To show that $E^{\boldsymbol{x}}\left[\int_0^T p(\boldsymbol{B}(s))\,ds\right]$ is approximated by the second term in the expression of $\hat{u}(\boldsymbol{x})$, we develop an alternative interpretation of $\int_0^T p(\boldsymbol{B}(s))\,ds$. Consider a circle $S(\boldsymbol{x}, a)$ of radius $a > 0$ centered at $\boldsymbol{x} \in D$, where a is the step of the finite difference mesh. Let ΔT_1 be the time a Brownian motion $\boldsymbol{B}$ starting at $\boldsymbol{x}$ needs to reach the boundary of $S(\boldsymbol{x}, a)$ for the first time. Denote by $\boldsymbol{Z}_1$ the exit point of $\boldsymbol{B}$ from $S(\boldsymbol{x}, a)$. Consider another circle $S(\boldsymbol{Z}_1, a)$ and a Brownian motion starting at $\boldsymbol{Z}_1$. Let ΔT_2 be the time it takes this Brownian motion to reach for the first time the boundary of $S(\boldsymbol{Z}_1, a)$ and so on. We stop this construction at step M if $S(\boldsymbol{Z}_M, a)$ is the first circle that is not included in D. The process $\boldsymbol{x} = \boldsymbol{Z}_0 \mapsto \boldsymbol{Z}_1 \mapsto \cdots \mapsto \boldsymbol{Z}_M$ is similar to the spherical process in Fig. 6.11. Because the radius a is small, the last point of the process $\boldsymbol{Z} = (\boldsymbol{x} = \boldsymbol{Z}_0, \boldsymbol{Z}_1, \ldots, \boldsymbol{Z}_M)$ is close to the boundary ∂D of D so that $\int_0^T p(\boldsymbol{B}(s))\,ds$

can be approximated by $\sum_{r=1}^{M} p(\mathbf{Z}_{r-1})\,\Delta T_r$. Also,

$$E\left[\sum_{r=1}^{M} p(\mathbf{Z}_{r-1})\,\Delta T_r\right] = E\left[\sum_{r=1}^{M-1} p(\mathbf{Z}_{r-1})\,\Delta T_r\right] + E\left\{E\left[p(\mathbf{Z}_{M-1})\,\Delta T_M \mid \mathbf{Z}_{M-1}\right]\right\}$$

$$= E\left[\sum_{r=1}^{M-1} p(\mathbf{Z}_{r-1})\,\Delta T_r\right] + \frac{a^2}{2}\,E\left[p(\mathbf{Z}_{M-1})\right]$$

so that $E\left[\sum_{r=1}^{M} p(\mathbf{Z}_{r-1})\,\Delta T_r\right] = (a^2/2)\,E\left[\sum_{r=1}^{M} p(\mathbf{Z}_{r-1})\right]$, where we used $E[\Delta T_r] = a^2/2$ (Section 8.5.1.1). These considerations show that the second term in the expression of $\hat{u}$ constitutes a Monte Carlo estimate for $(1/2)\,E\left[\sum_{r=1}^{M} p(\mathbf{Z}_{r-1})\right]\,(a^2/2)$. ■

Example 6.36: Consider the partial differential equation in the previous example with $p(\boldsymbol{x}) = 0$ and $D = (0, 1) \times (0, 1)$. The coordinate of node (i, j) of the finite difference mesh is $\boldsymbol{x}_{i,j} = ((i-1)\,a, (j-1)\,a)$, $i, j = 1, \ldots, n+1$, where n denotes the number of equal intervals in $(0, 1)$ and $a = 1/n$. The estimate of the solution u of the Laplace equation $\Delta u(\boldsymbol{x}) = 0$ in D at an arbitrary node (i, j) of the finite difference mesh is $\hat{u}_{i,j} = (1/n_s)\sum_{k=1}^{n_s} \xi(\boldsymbol{x}_{m_k}^{(k)})$ (Example 6.35), where ξ is the boundary value of u and $\boldsymbol{x}_{m_k}^{(k)}$ denotes the terminal point of the sample paths k of the random walk.

If ξ is equal to 100 on $\{1\} \times (0, 1)$ and zero on the other boundaries of D, the above estimate of u gives $\hat{u}_{i,j} = 100\,n_s'/n_s$, where n_s' denotes the number of samples of the random walk that exit D through boundary $\{1\} \times (0, 1)$. The estimates of u calculated at a relatively large number of points in D are in error by less than 3% for a finite difference mesh with $a = 1/1,000$ and $n_s = 1,000$ samples. ◇

Note: The exact solution of the Laplace equation with the domain and Dirichlet boundary conditions considered in this numerical example is

$$u(\boldsymbol{x}) = \frac{400}{\pi} \sum_{n=1,3,\ldots}^{\infty} \frac{\sinh(n\,\pi\,x_1)\,\sin(n\,\pi\,x_2)}{\sinh(n\,\pi)}$$

and can be found in [73] (Example 26.13, pp. 550-551). ▲

Example 6.37: Let $\boldsymbol{u}$ be the solution of Eq. 6.75. Suppose that the condition $\|\,\boldsymbol{g}\,\|_m < 1$ is not satisfied so that the local solution in Eq. 6.79 cannot be used. Let $\boldsymbol{a} = \boldsymbol{i} - \boldsymbol{g}$, μ be an eigenvalue of $\boldsymbol{a}$, $\bar{\mu} > \max\{|\mu|\}$, $c = (\bar{\mu})^{-2}$, $\boldsymbol{\phi}^* = c\,\boldsymbol{a}\,\boldsymbol{\phi}$, and $\boldsymbol{g}^* = \boldsymbol{i} - c\,\boldsymbol{a}^2$. Then $\boldsymbol{u}$ is also the solution $\boldsymbol{u} = \boldsymbol{\phi}^* + \boldsymbol{g}^*\,\boldsymbol{u}$ and $\|\,\boldsymbol{g}^*\,\|_m < 1$. Hence, Eq. 6.79 can be used to solve $\boldsymbol{u} = \boldsymbol{\phi}^* + \boldsymbol{g}^*\,\boldsymbol{u}$ locally. ◇

Proof: That $\boldsymbol{u}$ is the solution of $\boldsymbol{u} = \boldsymbol{\phi}^* + \boldsymbol{g}^*\,\boldsymbol{u}$ results by straightforward calculations using the definitions of $\boldsymbol{\phi}^*$ and $\boldsymbol{g}^*$.

We can assume $g = g^T$ without loss of generality since, if this condition is not satisfied, we can solve $a^T\, a\, u = a^T\, \phi$, and this equation can be set in the form of Eq. 6.75 with g symmetric. Also, we have $a^2\, v = \mu\, a\, v = \mu^2\, v$ so that the eigenvalues $\mu^* = c\,\mu^2$ of matrix $c\, a^2$ are positive and strictly smaller than one. The eigenvalues λ^* of g^* are the solution of

$$0 = \det(g^* - \lambda^*\, i) = \det((1 - \lambda^*)\, i - c\, a^2)$$

so that they are equal to $1 - \mu^*$. Because $\mu^* \in (0, 1)$ so are the eigenvalues λ^*. Hence, the Monte Carlo solution is convergent for the modified problem [93]. ■

6.5.2 Homogeneous equations

Consider the eigenvalue problem

$$g\, u = \mu\, u \tag{6.83}$$

obtained from Eq. 6.73 with $\phi = 0$ and $\mu = 1/\lambda$, where g is assumed to be a real-valued symmetric matrix with simple positive eigenvalues. Our objective is to find the largest eigenvalue of g and the corresponding eigenvector, referred to as the dominant eigenvalue and eigenvector, respectively.

Example 6.38: Consider the eigenvalue problem in Eq. 6.83. Suppose that the eigenvalues μ_k of g satisfy the condition $\mu_1 > \mu_k$ for all $k > 1$. Let $y^{(r)} \in \mathbb{R}^m$, $r = 0, 1, \ldots$, be a sequence of vectors generated by the recurrence formula $y^{(r)} = g\, y^{(r-1)}$, where $y^{(0)}$ does not exclude the dominant eigenvector of g, that is, $y^{(0)} = \sum_{i=1}^m \beta_i\, u_i$, where u_i denote the eigenvectors of g and $\beta_1 \neq 0$. Then $y^{(r)}/\mu_1^r$ and $y_i^{(r+1)}/y_i^{(r)}$ converge to $\beta_1\, u_1$ and μ_1, respectively, as $r \to \infty$, where $y_i^{(r)}$ denotes the coordinate i of $y^{(r)}$. ◇

Proof: We have

$$y^{(r)} = g\, y^{(r-1)} = g^2\, y^{(r-2)} = \cdots = g^r\, y^{(0)} = \sum_{i=1}^m \beta_i\, g^r\, u_i = \sum_{i=1}^m \beta_i\, \mu_i^r\, u_i$$

so that

$$y^{(r)} = \mu_1^r \left[\beta_1\, u_1 + \sum_{i=2}^m \beta_i \left(\frac{\mu_i}{\mu_1}\right)^r u_i \right].$$

Hence, the scaled vector $y^{(r)}/\mu_1^r$ converges to $\beta_1\, u_1$ as $r \to \infty$ since $\mu_i/\mu_1 < 1$, and the ratio $y_i^{(r+1)}/y_i^{(r)}$ approaches μ_1 as $r \to \infty$. ■

Let $p = \{p_{i,j}\}$ be an (m, m) matrix such that $p_{i,j} \geq 0$ and $\sum_{j=1}^m p_{i,j} = 1$ and let X be a Markov chain with states $\{1, \ldots, m\}$ defined by the transition probability matrix p. Let $a = (a_1, \ldots, a_m) \in \mathbb{R}^m$ and $\pi_i > 0$, $i = 1, \ldots, m$, such that $\sum_{i=1}^m \pi_i = 1$. Define a discrete random variable Y taking the values $\alpha_i = a_i/\pi_i$ with probabilities π_i, $i = 1, \ldots, m$, and let V be the process defined by the recurrence formula in Eq. 6.78, but starting at $V(0) = \alpha_i$ rather than 1. Consider also the recurrence formula

$$\boldsymbol{g}\,\boldsymbol{x}^{(r-1)} = \rho_r\,\boldsymbol{x}^{(r)}, \quad r = 1, 2, \ldots, \tag{6.84}$$

starting at $\boldsymbol{x}^{(0)} \in \mathbb{R}^m$ such that $\boldsymbol{x}^{(0)}$ does not exclude the dominant eigenvector of $\boldsymbol{g}$. Set $\rho_r = \| \boldsymbol{g}\,\boldsymbol{x}^{(r-1)} \|$ in Eq. 6.84 so that the vectors $\boldsymbol{x}^{(1)}, \boldsymbol{x}^{(2)}, \ldots$ have unit norm. Let $v_{i,j} = g_{i,j}/p_{i,j}$ if $p_{i,j} > 0$ and $v_{i,j} = 0$ if $p_{i,j} = 0$.

If the eigenvalues μ_k of $\boldsymbol{g} = \boldsymbol{g}^T$ are such that $\mu = \mu_1 > \mu_k$ for $k > 1$, then

$$\mu \simeq \left(\frac{e(q')}{e(q)}\right)^{1/(q'-q)} \quad \text{for } q' > q \text{ and large } q, \text{ where} \tag{6.85}$$

$$e(q) = \sum_{j=1}^{m} E[V(q) \mid X(0) = Y, X(q) = j] = \rho_1 \ldots \rho_q \sum_{j=1}^{m} (\boldsymbol{x}^{(q)})_j. \tag{6.86}$$

Proof: Let $\omega_{i,j}$ be a sample path of X starting at state i and reaching a state j for the first time after q transitions. The value of V along this sample is $\alpha_i\, v_{i,i_1}\, v_{i_1,i_2} \cdots v_{i_{q-1},j}$ so that the expectation of V along all samples $\omega_{i,j}$ of X is

$$\begin{aligned} E[V(q) \mid X(0) = i, X(q) = j] &= \sum_{i_1,\ldots,i_{q-1} \in \{1,\ldots,m\}} (\alpha_i\, v_{i,i_1}\, v_{i_1,i_2} \cdots v_{i_{q-1},j})\,(p_{i,i_1}\, p_{i_1,i_2} \cdots p_{i_{q-1},j}) \\ &= \sum_{i_1,\ldots,i_{q-1} \in \{1,\ldots,m\}} \alpha_i\, g_{i,i_1}\, g_{i_1,i_2} \cdots g_{i_{q-1},j} = \alpha_i\, (\boldsymbol{g}^q)_{i,j}\,. \end{aligned}$$

If the initial value of X is Y, the above expectation becomes

$$E[V(q) \mid X(0) = Y, X(q) = j] = \sum_{i=1}^{m} (\boldsymbol{g}^q)_{i,j}\, a_i = (\boldsymbol{g}^q\, \boldsymbol{a})_j\,.$$

For $\boldsymbol{x}^{(0)} = \boldsymbol{a}$ we have $\boldsymbol{g}^r\,\boldsymbol{a} = \rho_1 \ldots \rho_{r-1}\,\rho_r\,\boldsymbol{x}^{(r)}$ (Eq. 6.84) so that

$$E[V(q) \mid X(0) = Y, X(q) = j] = \rho_1 \ldots \rho_q\,(\boldsymbol{x}^{(q)})_j,$$

and summation with respect to j gives Eq. 6.86.

The dominant eigenvalue in Eq. 6.85 results from (Eq. 6.86)

$$\frac{e(q')}{e(q)} = \rho_{q+1} \cdots \rho_{q'}\,\frac{\sum_{j=1}^{m} (\boldsymbol{x}^{(q')})_j}{\sum_{j=1}^{m} (\boldsymbol{x}^{(q)})_j} \simeq \mu^{q'-q}, \quad q' > q,$$

where the approximate equality holds for large values of q since the vectors $\boldsymbol{x}^{(q)}$ and $\boldsymbol{x}^{(q')}$ nearly coincide with the dominant eigenvector and the numbers $\rho_{q+1}, \ldots, \rho_{q'}$ are approximately equal to μ (Example 6.38). ■

If the conditions under which Eqs. 6.85 and 6.86 hold, the coordinate u_j of the dominant eigenvector of Eq. 6.83 can be approximated by

$$u_j \simeq \frac{1-\mu}{\mu^{q+1}(1-\mu^{q'-q})} \sum_{r=q+1}^{q'} E[V(r) \mid X(0) = Y, X(r) = j]. \tag{6.87}$$

Proof: Suppose that q is sufficiently large so that the expectation $E[V(r) \mid X(0) = Y, X(r) = j]$ can be approximated by $\mu^r u_j$. For $q' > q$ we have

$$\sum_{r=q+1}^{q'} E[V(r) \mid X(0) = Y, X(r) = j] \simeq \sum_{r=q+1}^{q'} \mu^r u_j = \mu^{q+1} \frac{1-\mu^{q'-q}}{1-\mu} u_j$$

showing that Eq. 6.87 holds. ■

Example 6.39: The dominant eigenvalue of the matrix

$$\boldsymbol{g} = \begin{bmatrix} 0.2 & 0.4 & 0.1 \\ 0.4 & 0.2 & 0.2 \\ 0.1 & 0.5 & 0.2 \end{bmatrix}$$

is $\mu = 0.7653$. The estimate of this eigenvalue in Eq. 6.85 varies between 0.7 and 0.7663 for $q' = 11, 12, \ldots, 20$, $q = 10$, the transition probability matrix

$$\boldsymbol{p} = \begin{bmatrix} 0.3 & 0.3 & 0.4 \\ 0.2 & 0.3 & 0.5 \\ 0.5 & 0.2 & 0.3 \end{bmatrix},$$

probabilities $\pi_i = 1/3$, and $\boldsymbol{a} = (100, 100, 100)$. The values of $e(q)$ and $e(q')$ have been estimated from $n_s = 1,000$ samples. ◇

Note: The expectations in Eqs. 6.86 and 6.87 have been estimated from samples of X generated from the transition probability matrix $\boldsymbol{p}$ defining this process and the probability $(\pi_1, \ldots, \pi_m)$ of its initial state. ▲

Example 6.40: The dominant eigenvalue and eigenvector of the matrix

$$\boldsymbol{g} = \begin{bmatrix} 0.6 & 0.2 & 0.1 \\ -0.2 & 0.7 & -0.1 \\ 0.1 & -0.1 & 0.7 \end{bmatrix}$$

are, respectively, 0.9388 and $(0.5441, -0.6681, 0.5075)$. The dominant eigenvalue and eigenvector of $\boldsymbol{g}$ given by the recurrence formula in Eq. 6.84 with $\boldsymbol{x}^{(0)} = (1, 1, 1)/\sqrt{3}$ are 0.9377 and $(0.5346, -0.6470, 0.5437)$ after 10 iterations, and 0.9388 and $(0.5441, -0.6681, 0.5075)$ after 30 iterations. These estimates are based on the transition probability matrix and the probability of $X(0)$ in the previous example. ◇

6.6 Integral equations

We consider the inhomogeneous and homogeneous Fredholm equations of the second kind defined by (Eq. 6.3)

$$u(\boldsymbol{x}) = \phi(\boldsymbol{x}) + \lambda \int_D g(\boldsymbol{x}, \boldsymbol{y})\, u(\boldsymbol{y})\, d\boldsymbol{y}, \quad \boldsymbol{x}, \boldsymbol{y} \in D, \quad \text{and} \tag{6.88}$$

$$u(\boldsymbol{x}) = \lambda \int_D g(\boldsymbol{x}, \boldsymbol{y})\, u(\boldsymbol{y})\, d\boldsymbol{y}, \quad \boldsymbol{x}, \boldsymbol{y} \in D, \tag{6.89}$$

respectively, where D is an open bounded set in $\mathbb{R}^d$, λ is a parameter, and ϕ and g are real-valued functions defined on D and $D \times D$, respectively. These equations are frequently encountered in science and engineering ([37], Section 5.3). Our objectives are to find local solutions for Eqs. 6.88 and 6.89, that is, the value of u in Eq. 6.88 for a specified value of λ at an arbitrary point $\boldsymbol{x} \in D$ and the lowest eigenvalue and eigenfunction of Eq. 6.89, respectively.

The methods of the previous section can be applied to solve locally Eqs. 6.88 and 6.89 if they are approximated by algebraic equations, as demonstrated by the following three examples. Alternative local solutions for Eqs. 6.88 and 6.89 are discussed in the remainder of this section.

Example 6.41: Let D_i, $i = 1, \ldots, m$, be a partition of D, that is, $D = \cup_{i=1}^m D_i$ and $D_i \cap D_j = \emptyset$, $i \neq j$. If the partition is sufficiently fine and the functions ϕ, g, and u are smooth, the solution of Eq. 6.88 can be approximated by the solution $\tilde{u}$ of the algebraic equations

$$\tilde{u}(\boldsymbol{x}_i) = \phi(\boldsymbol{x}_i) + \lambda \sum_{j=1}^m g(\boldsymbol{x}_i, \boldsymbol{x}_j)\, \tilde{u}(\boldsymbol{x}_j)\, v_j,$$

where $\boldsymbol{x}_i \in D_i$ and v_i is the volume D_i. The above equations defining $\tilde{u}$ has the form of Eq. 6.73 in which the entries of the matrices $\boldsymbol{g}$, $\boldsymbol{\phi}$, and $\boldsymbol{u}$ are $g(\boldsymbol{x}_i, \boldsymbol{x}_j)\, v_j$, $\phi(\boldsymbol{x}_i)$, and $\tilde{u}(\boldsymbol{x}_i)$. Similar arguments show that Eq. 6.89 can be cast in the form of Eq. 6.74. ◇

Example 6.42: Suppose that the kernel of Eqs. 6.88 and 6.89 is degenerate, that is, $g(\boldsymbol{x}, \boldsymbol{y}) = \sum_{s=1}^m g_{1,s}(\boldsymbol{x})\, g_{2,s}(\boldsymbol{y})$, $\boldsymbol{x}, \boldsymbol{y} \in D$, and that the functions $g_{1,s}$ are linearly independent. Then the solution of Eq. 6.88 is

$$u(\boldsymbol{x}) = \phi(\boldsymbol{x}) + \lambda \sum_{s=1}^m g_{1,s}(\boldsymbol{x})\, h_s,$$

in which the coefficients $h_s = \int_D g_{2,s}(\boldsymbol{y})\, u(\boldsymbol{y})\, d\boldsymbol{y}$ are given by the linear system of algebraic equations

$$h_s = a_s + \lambda \sum_{t=1}^m h_t\, b_{ts}, \quad s = 1, \ldots, m,$$

where $a_s = \int_D g_{2,s}(y)\,\phi(y)\,dy$ and $b_{ts} = \int_D g_{1,t}(y)\,g_{2,s}(y)\,dy$ (Eq. 6.73). If $a_s = 0, s = 1, \ldots, m$, the above equations define an algebraic eigenvalue problem for the matrix $\{b_{ts}\}$. ◇

Note: If u has the above representation, Eq. 6.88 becomes

$$\phi(x) + \lambda \sum_{s=1}^{m} g_{1,s}(x)\,h_s = \phi(x) + \lambda \sum_{s=1}^{m} g_{1,s}(x) \left(a_s + \lambda \sum_{t=1}^{m} h_t\, b_{ts} \right),$$

or

$$\sum_{s=1}^{m} g_{1,s}(x) \left[h_s - a_s - \lambda \sum_{t=1}^{r} h_t\, b_{st} \right] = 0, \quad x \in D.$$

The coefficients of the above linear form must be zero since the functions $g_{1,s}$ are assumed to be linearly independent. This condition yields the defining equations for h_s. ▲

Example 6.43: Let $\chi_s, s = 1, 2, \ldots$, be a complete collection of real-valued functions defined on D so that the representations

$$\phi(x) = \sum_s \alpha_s\, \chi_s(x) \quad \text{and} \quad u(x) = \sum_s \beta_s\, \chi_s(x)$$

hold. The coefficients α_s can be calculated since ϕ is a known function. If the series in the representations of ϕ and u are absolutely and uniformly convergent, the unknown coefficients β_s are the solutions of the algebraic system

$$\beta_s - \alpha_s - \lambda \sum_q \beta_q\, \gamma_{sr} = 0, \quad s = 1, 2, \ldots,$$

with an infinite number of equations, where γ_{sr} are some coefficients. ◇

Note: Because $\int_D g(x, y)\,\chi_s(y)\,dy$ is an element of the space spanned by the functions χ_r, it has the representation $\sum_q \gamma_{sq}\,\chi_q(x)$. The coefficients γ_{sq} can be calculated since $\int_D g(x, y)\,\chi_s(y)\,dy$ is a known function.

If the integral in Eq. 6.88 can be performed term by term, we have

$$\sum_s \beta_s\, \chi_s(x) = \sum_s \alpha_s\, \chi_s(x) + \lambda \sum_s \beta_s \sum_q \gamma_{sq}\, \chi_q(x).$$

Hence,

$$\sum_s \left[\beta_s - \alpha_s - \lambda \sum_t \beta_t\, \gamma_{ts} \right] \chi_s(x) = 0$$

so that the square brackets must be zero. The resulting conditions for β_s constitute a system with an infinite number of linear equations. The solution technique for this system resembles the local method for solving Eq. 6.73 ([94], p. 89). ▲

6.6.1 Inhomogeneous equations

Our objective is to find the value of the solution u of Eq. 6.88 at an arbitrary point $\mathbf{x} \in D$, that is, to find the local solution of this equation.

Let $u_1, u_2, \ldots$ be a sequence of functions defined by

$$u_k(\mathbf{x}) = \int_D g_k(\mathbf{x}, \mathbf{y})\, \phi(\mathbf{y})\, d\mathbf{y}, \quad \text{where}$$

$$g_{k+1}(\mathbf{x}, \mathbf{y}) = \int_D g_1(\mathbf{x}, \mathbf{z})\, g_k(\mathbf{z}, \mathbf{y})\, d\mathbf{z}, \quad k = 1, 2, \ldots, \text{ and } g_1 = g. \tag{6.90}$$

If $|\phi(\mathbf{x})| \leq c_1$, $|g(\mathbf{x}, \mathbf{y})| \leq c_2$ for all $\mathbf{x}, \mathbf{y} \in D$, $0 < c_1, c_2 < \infty$, $v = \int_D d\mathbf{x} < \infty$, and $|\lambda|\, c_2\, v < 1$, then the **local solution** of Eq. 6.88 is

$$u(\mathbf{x}) = \phi(\mathbf{x}) + \sum_{k=1}^{\infty} \lambda^k\, u_k(\mathbf{x}), \quad \mathbf{x} \in D. \tag{6.91}$$

Proof: The Neumann series in Eq. 6.91 introduced in Eq. 6.88 gives

$$\phi(\mathbf{x}) + \sum_{k=1}^{\infty} \lambda^k\, u_k(\mathbf{x}) = \phi(\mathbf{x}) + \lambda \int_D g(\mathbf{x}, \mathbf{y})\, \phi(\mathbf{y})\, d\mathbf{y} + \sum_{k=1}^{\infty} \lambda^{k+1} \int_D g(\mathbf{x}, \mathbf{y})\, u_k(\mathbf{y})\, d\mathbf{y}$$

under the assumption that the integral $\int_D g(\mathbf{x}, \mathbf{y})\, u(\mathbf{y})\, d\mathbf{y}$ with u in Eq. 6.91 can be performed term by term. These calculations are formal since we have not yet imposed any conditions on the series in Eq. 6.91. The functions u_k in Eq. 6.90 result by equating the terms of the above equation that have λ at the same power. For example, the first two functions of the series expansion in Eq. 6.91 are (Eq. 6.90)

$$u_1(\mathbf{x}) = \int_D g(\mathbf{x}, \mathbf{y})\, \phi(\mathbf{y})\, d\mathbf{y}$$

$$u_2(\mathbf{x}) = \int_D g(\mathbf{x}, \mathbf{y})\, u_1(\mathbf{y})\, d\mathbf{y} = \int_D g(\mathbf{x}, \mathbf{y}) \left(\int_D g(\mathbf{y}, \mathbf{z})\, \phi(\mathbf{z})\, d\mathbf{z} \right) d\mathbf{y}$$

$$= \int_D g_2(\mathbf{x}, \mathbf{z})\, \phi(\mathbf{z})\, d\mathbf{z}.$$

Because the inequalities $|\phi(\mathbf{x})| \leq c_1$ and $|g(\mathbf{x}, \mathbf{y})| \leq c_2$ are assumed to hold for some constants $c_1, c_2 > 0$ and all $\mathbf{x}, \mathbf{y} \in D$, $|g_2(\mathbf{x}, \mathbf{y})| \leq \int_D |g(\mathbf{x}, \mathbf{z})|\, |g(\mathbf{z}, \mathbf{y})|\, d\mathbf{z} \leq v\, c_2^2$, and, generally, $|g_k(\mathbf{x}, \mathbf{y})| \leq (v\, c_2)^k / v$ for all $k = 1, 2, \ldots$, the absolute value of u in

Eq. 6.91 is dominated by

$$|u(\boldsymbol{x})| \leq |\phi(\boldsymbol{x})| + \sum_{k=1}^{\infty} |\lambda|^k \int_D |g_k(\boldsymbol{x}, \boldsymbol{y})| \, |\phi(\boldsymbol{y})| \, d\boldsymbol{y}$$
$$\leq c_1 + c_1 \sum_{k=1}^{\infty} (|\lambda| \, c_2 \, v)^k.$$

If the condition $|\lambda| \, c_2 \, v < 1$ is satisfied, then $\sum_k (|\lambda| \, c_2 \, v)^k$ is convergent implying that the Neumann series in Eq. 6.91 is absolutely and uniformly convergent so that the term by term integration performed above is valid. Moreover, the value of $u(\boldsymbol{x})$ at an arbitrary point $\boldsymbol{x} \in D$ can be approximated by the first $\bar{n} < \infty$ terms of the Neumann series in Eq. 6.91.

The method in this section for solving locally Eq. 6.88 relates to the BWM developed in Section 6.4 for solving locally linear elasticity problems. Both methods are based on Neumann series representations of the local solutions and the terms of these series are calculated by Monte Carlo simulation (Eqs. 6.71 and 6.92). ■

If the Neumann series of Eq. 6.91 is absolutely and uniformly convergent, and the kernel g is positive, then

$$u_k(\boldsymbol{x}) = a_k(\boldsymbol{x}) \, E\left[\phi\left(\boldsymbol{Y}^{(k)}(\boldsymbol{x})\right)\right], \tag{6.92}$$

where $\boldsymbol{Y}^{(k)}(\boldsymbol{x})$ is a D-valued random variable with probability density $\tilde{g}_k(\boldsymbol{x}, \cdot) = g_k(\boldsymbol{x}, \cdot)/a_k(\boldsymbol{x})$ and $a_k(\boldsymbol{x}) = \int_D g_k(\boldsymbol{x}, \boldsymbol{y}) \, d\boldsymbol{y}$.

Note: The functions $\tilde{g}_k(\boldsymbol{x}, \cdot)$ are proper density functions with support D because they are positive and the volume under their graph is unity. The definition of u_k in Eq. 6.90 has also the form $u_k(\boldsymbol{x}) = a_k(\boldsymbol{x}) \int_D \tilde{g}_k(\boldsymbol{x}, \boldsymbol{y}) \, \phi(\boldsymbol{y}) \, d\boldsymbol{y}$ showing that the integral is the expectation of $\phi(\boldsymbol{Y}^{(k)}(\boldsymbol{x}))$. The recurrence formula given by Eq. 6.90 shows that, if g is positive, so are the kernels g_k. Generally, the integrals in Eq. 6.90 must be calculated numerically.

If the kernel g is not positive on $D \times D$, $u_k(\boldsymbol{x})$ can be interpreted as the expectation of the random variable $g(\boldsymbol{x}, \boldsymbol{Z}^{(k)}) \, \phi(\boldsymbol{Z}^{(k)}) \, v$, where $\boldsymbol{Z}^{(k)}$ is uniformly distributed in D and $v = \int_D d\boldsymbol{x}$ is the volume of D. ▲

The local solution in the preceding section (Eq. 6.79) can be extended to find local solutions for Eq. 6.88 ([48] and [95], p. 90). Let $\boldsymbol{X} = (\boldsymbol{X}(0) = \boldsymbol{x}, \boldsymbol{X}(1), \ldots)$ be a discrete time, continuous state Markov process starting at the point $\boldsymbol{x} \in D$ where the local solution needs to be determined. Let $f(\cdot \mid \boldsymbol{y})$ denote the density of the conditional random variable $\boldsymbol{X}(i) \mid (\boldsymbol{X}(i-1) = \boldsymbol{y}), i \geq 1$. This density, referred to as transition density, needs to be specified. Given a transition density $f(\cdot \mid \cdot)$, the probability that X starting at $\boldsymbol{x} \in D$ exits D in one transition is $p(\boldsymbol{x}) = \int_{D^c} f(\boldsymbol{\xi} \mid \boldsymbol{x}) \, d\boldsymbol{\xi}$. The following two equations define a stopping time T giving the number of transitions that $\boldsymbol{X}$ starting at $\boldsymbol{x} \in D$ takes to exit D for the first time and a discrete time process $V = (V(0), V(1), \ldots)$. It is assumed that the transition density $f(\cdot \mid \boldsymbol{x})$ is such that $T < \infty$ a.s. for all $\boldsymbol{x} \in D$.

$$T = \inf\{r > 0 : X(r) \notin D, X(0) = x \in D\}. \quad (6.93)$$

$$V(k+1) = \lambda\, V(k)\, \frac{g(X(k), X(k+1))}{f(X(k+1)) \mid X(k))} \quad k = 0, 1, \ldots, \quad (6.94)$$

where V starts at V(0) =1.

If the Neumann series in Eq. 6.91 is absolutely and uniformly convergent, the **local solution** of Eq. 6.88 at $x \in D$ is

$$u(x) = E\left[V(T-1)\, \phi(X(T-1))/p(X(T-1)) \mid X(0) = x \in D\right]. \quad (6.95)$$

Proof: Let $x = x_0 \mapsto x_1 \mapsto \cdots x_k \mapsto x_{k+1}$ with $x_0, \ldots, x_k \in D$ and $x_{k+1} \in D^c$ be a sample path of X that starts at $x \in D$ and exits D for the first time in $k+1$ transitions. Note that $X(T-1)$ and $X(T)$ are the last and the first values of X in D and D^c, respectively (Eq. 6.93). The expectation of $V(T-1)\, \phi(X(T-1))/p(X(T-1))$ conditional on $X(0) = x$ and $T = k+1$, $k = 0, 1, \ldots$, is $\phi(x)/p(x)$ for $k = 0$ and

$$\begin{aligned}&\int_{D^k} V(k)\, \frac{\phi(x_k)}{p(x_k)}\, f(x_1 \mid x_0)\, f(x_2 \mid x_1) \cdots f(x_k \mid x_{k-1})\, dx_1 \cdots dx_k \\ &= \lambda^k \int_{D^k} g(x_0, x_1)\, g(x_1, x_2) \cdots g(x_{k-1}, x_k)\, \frac{\phi(x_k)}{p(x_k)}\, dx_1 \cdots dx_k\end{aligned}$$

for $k = 1, 2, \ldots$, where $D^k = \times_{j=1}^k D$ so that

$$\begin{aligned}&E\left[V(T-1)\phi(X(T-1))/p(X(T-1)) \mid X(0) = x \in D\right] = \phi(x) \\ &+ \lambda \int_D g(x, x_1)\, \phi(x_1)\, dx_1 + \lambda^2 \int_{D\times D} g(x, x_1)\, g(x_1, x_2)\, \phi(x_2)\, dx_1\, dx_2 + \cdots \\ &= \phi(x) + \lambda\, u_1(x) + \lambda^2\, u_2(x) + \cdots\end{aligned}$$

with the notation in Eq. 6.90. Hence, the expectation in Eq. 6.95 coincides with the local solution in Eq. 6.91. Because the Neumann series in Eq. 6.91 is absolutely and uniformly convergent by hypothesis, Eq. 6.95 gives the local solution of Eq. 6.88. ■

The local solution in Eq. 6.95 can be estimated from samples of X and V. Let $x \mapsto X(1, \omega) \mapsto \cdots \mapsto X(T(\omega) - 1, \omega) \mapsto X(T(\omega), \omega)$, $\omega = 1, \ldots, n_s$, be n_s independent samples of X starting at $x \in D$.

The local solution in Eq. 6.95 can be estimated by

$$\hat{u}(x) = \frac{1}{n_s} \sum_{\omega=1}^{n_s} V(T(\omega) - 1, \omega)\, \frac{\phi(X(T(\omega) - 1, \omega))}{\tilde{f}(X(T(\omega) - 1, \omega))}. \quad (6.96)$$

Note: Let $X(r, \omega) = x_r \in D$ be the value of a sample $X(\cdot, \omega)$ of X after $r \geq 0$ transitions. The value of this sample at time $r+1$ can be generated from the transition density $f(\cdot \mid x_r)$. The sample $X(\cdot, \omega)$ is stopped when it leaves D for the first time. Samples of V can be calculated from Eq. 6.94 and samples of X. ▲

Example 6.44: Consider the differential equation

$$\frac{d^4 u(x)}{dx^4} + \chi(x)\, u(x) = q(x), \quad x \in D = (0, l),$$

with the boundary conditions $u(0) = u(l) = 0$ and $u''(0) = u''(l) = 0$, where $\chi > 0$ and q are specified functions. The integral form of the above differential equation is

$$\begin{aligned} u(\xi) &= \int_0^1 g^*(\xi, \eta)\, [q(\eta) - \chi(\eta)\, u(\eta)]\, d\eta \\ &= \phi(\xi) - \int_0^1 g(\xi, \eta)\, u(\eta)\, d\eta, \end{aligned}$$

where g^* denotes the Green function of the differential operator d^4/dx^4, $\xi = x/l$, $\eta = y/l$, $\phi(\xi) = \int_0^1 g^*(\xi, \eta)\, q(\eta)\, d\eta$ and $g(\xi, \eta) = g^*(\xi, \eta)\, \chi(\eta)$. Because $\max_{\xi, \eta \in (0,1)} |g(\xi, \eta)| < 1$, the local solution in Eq. 6.95 is valid and has been used to calculate the local solution.

Numerical results have been obtained for $l = 1$, $\chi = 1$, $q = 1$, and X defined by $X(k+1) \mid (X(k) = \xi) \sim N(\xi, \sigma^2)$, where $\sigma > 0$ is a specified parameter. These results are in error by less than 2% for σ in the range $(0.2, 0.6)$ and $n_s = 500$ samples. ◇

Note: The Green function of the differential operator d^4/dx^4 is

$$g^*(x, y) = \begin{cases} (l^3/6)(1-\eta)\,\xi\,[1 - \xi^2 - (1-\eta)^2], & \xi \leq \eta, \\ (l^3/6)(1-\xi)\,\eta\,[1 - \eta^2 - (1-\xi)^2], & \xi \geq \eta, \end{cases}$$

where $\xi = x/l$, $\eta = y/l \in (0, 1)$. This function gives the value of u at x caused by a unit action applied at y.

The solution of the above differential equation is the displacement function for a simply supported beam with unit stiffness and span l that is supported on an elastic foundation of stiffness χ and is subjected to a load q. ▲

6.6.2 Homogeneous equations

Suppose that the real-valued kernel g in Eq. 6.89 is symmetric, that is, $g(\boldsymbol{x}, \boldsymbol{y}) = g(\boldsymbol{y}, \boldsymbol{x})$ for all $\boldsymbol{x}, \boldsymbol{y} \in D$, and is square integrable. Also, assume that the eigenvalues of Eq. 6.89 are simple. Our objective is to find the lowest eigenvalue and the corresponding eigenfunction. Before discussing local solutions for Eq. 6.89, we will review briefly some essential properties of the eivenvalues/eigenfunctions of this equation. Let λ_k and u_k, $k = 0, 1, \ldots$, be the eigenvalues and eigenfunctions of Eq. 6.89. The eigenfunctions cannot be zero everywhere in D, $u_k \in L_2$, and the pair (λ_k, u_k), $k = 0, 1, \ldots$, satisfies the equation

$$u_k(\boldsymbol{x}) = \lambda_k \int_D g(\boldsymbol{x}, \boldsymbol{y})\, u_k(\boldsymbol{y})\, d\boldsymbol{y}, \quad \boldsymbol{x}, \boldsymbol{y} \in D. \tag{6.97}$$

- The eigenvalues are real-valued.
- The eigenfunctions corresponding to distinct eigenvalues are orthogonal and linearly independent.

Proof: Let $v, w : D \to \mathbb{R}$ be square integrable functions. The inner product, $\langle v, w \rangle = \int_D v(\boldsymbol{x})\, w(\boldsymbol{x})^*\, d\boldsymbol{x}$, induces the norm $\| v \| = \langle v, v \rangle^{1/2}$, where z^* denotes the complex conjugate of $z \in \mathbb{C}$. This definition is meaningful by the Cauchy-Schwarz inequality since the functions v, w are square integrable in D.

Suppose that there exists a complex-valued solution (λ, u) of the eigenvalue problem in Eq. 6.89. The complex conjugate,

$$u(\boldsymbol{x})^* = \lambda^* \int_D g(\boldsymbol{x}, \boldsymbol{y})\, u(\boldsymbol{y})^*\, d\boldsymbol{y},$$

of Eq. 6.89 multiplied by $u(\boldsymbol{x})$ and integrated with respect to $\boldsymbol{x}$ over D yields

$$\int_D u(\boldsymbol{x})\, u(\boldsymbol{x})^*\, d\boldsymbol{x} = \lambda^* \int_D d\boldsymbol{x}\, u(\boldsymbol{x}) \int_D g(\boldsymbol{x}, \boldsymbol{y})\, u(\boldsymbol{y})^*\, d\boldsymbol{y}.$$

We also have

$$\int_D u(\boldsymbol{x})\, u(\boldsymbol{x})^*\, d\boldsymbol{x} = \lambda \int_D d\boldsymbol{x}\, u(\boldsymbol{x})^* \int_D g(\boldsymbol{x}, \boldsymbol{y})\, u(\boldsymbol{y})\, d\boldsymbol{y}$$

by multiplying Eq. 6.89 with $u(\boldsymbol{x})^*$ and integrating the result with respect to $\boldsymbol{x}$ over D. The above two equations imply $\lambda = \lambda^*$ since g is symmetric and u is an eigenfunction, so that the inner product $\langle u, u \rangle$ is real and non-zero.

Let u' and u'' be two eigenfunctions corresponding to the eigenvalues λ' and $\lambda'' \neq \lambda'$, respectively. The inner product of these eigenfunctions,

$$\begin{aligned} \langle u', u'' \rangle &= \int_D u'(\boldsymbol{x})\, u''(\boldsymbol{x})^*\, d\boldsymbol{x} = \lambda' \int_{D \times D} g(\boldsymbol{x}, \boldsymbol{y})\, u'(\boldsymbol{y})\, u''(\boldsymbol{x})^*\, d\boldsymbol{x}\, d\boldsymbol{y} \\ &= \lambda' \int_D \left[\int_D g(\boldsymbol{x}, \boldsymbol{y})\, u''(\boldsymbol{x})^*\, d\boldsymbol{x} \right] u'(\boldsymbol{y})\, d\boldsymbol{y} = \frac{\lambda'}{\lambda''} \langle u', u'' \rangle, \end{aligned}$$

shows that $\langle u', u'' \rangle$ must be zero since $\lambda'' \neq \lambda'$ by hypothesis.

Let $u_1, \ldots, u_n$ be eigenfunctions corresponding to distinct eigenvalues of Eq. 6.89 and suppose that there exist some coefficients $c_1, \ldots, c_n$ which are not all zero such that $\sum_{k=1}^n c_k\, u_k(\boldsymbol{x}) = 0$. The inner product of $\sum_k c_k\, u_k$ with u_i gives $c_i \langle u_i, u_i \rangle = 0$ so that all c_i must be zero for all $i = 1, \ldots, n$ in contradiction with our assumption. ■

Let $u^{(0)} : D \to \mathbb{R}$ be an initial trial function for the lowest eigenfunction u_0 of Eq. 6.89 and construct a sequence of functions $u^{(i)} : D \to \mathbb{R}$, $i = 1, 2, \ldots$, by the recurrence formula

$$u^{(i+1)}(\boldsymbol{x}) = \int_D g(\boldsymbol{x}, \boldsymbol{y})\, \bar{u}^{(i)}(\boldsymbol{y})\, d\boldsymbol{y}, \quad i = 0, 1, \ldots, \tag{6.98}$$

where $\bar{u}^{(i)}(\boldsymbol{y}) = u^{(i)}(\boldsymbol{y}) / \parallel u^{(i)} \parallel$.

If (1) the kernel g in Eq. 6.89 is real-valued and symmetric, (2) the initial trial function $u^{(0)}$ does not exclude u_0, and (3) $0 < \lambda_0 < \lambda_k$ for $k \geq 1$, then

$$\lambda_0 \simeq \frac{\parallel u^{(i)} \parallel}{\langle u^{(i+1)}, u^{(i)} \rangle} \quad \text{and } u_0(\boldsymbol{x}) \text{ is proportional to } u^{(i)}(\boldsymbol{x}) \text{ for } i \text{ large.} \tag{6.99}$$

Proof: Because the collection of eigenfunctions of Eq. 6.89 is complete, we can represent the initial trial function by

$$u^{(0)}(\boldsymbol{x}) = \sum_{k=0}^{\infty} c_k^{(0)}\, u_k(\boldsymbol{x}), \quad \boldsymbol{x} \in D.$$

Suppose that $c_0^{(0)} \neq 0$ so that the eigenfunction u_0 is not excluded from $u^{(0)}$. The proof of Eq. 6.99 involves three steps. First, we note that

$$u^{(i)}(\boldsymbol{x}) = \sum_{k=0}^{\infty} \frac{c_k^{(i)}}{\lambda_k^i}\, u_k(\boldsymbol{x}), \quad i = 1, 2, \ldots,$$

where $c_k^{(i)} = c_k^{(i-1)} / \parallel u^{(i-1)} \parallel$. For example, (Eqs. 6.97 and 6.98)

$$u^{(1)}(\boldsymbol{x}) = \int_D g(\boldsymbol{x}, \boldsymbol{y})\, \bar{u}^{(0)}(\boldsymbol{y})\, d\boldsymbol{y} = \frac{1}{\parallel u^{(0)} \parallel} \sum_{k=0}^{\infty} c_k^{(0)} \int_D g(\boldsymbol{x}, \boldsymbol{y})\, u_k(\boldsymbol{y})\, d\boldsymbol{y}$$

$$= \frac{1}{\parallel u^{(0)} \parallel} \sum_{k=0}^{\infty} \frac{c_k^{(0)}}{\lambda_k}\, u_k(\boldsymbol{x}) = \sum_{k=0}^{\infty} \frac{c_k^{(1)}}{\lambda_k}\, u_k(\boldsymbol{x}),$$

where $c_k^{(1)} = c_k^{(0)} / \parallel u^{(0)} \parallel$. Because $u^{(1)}$ has the same functional form as $u^{(0)}$, we have $u^{(2)}(\boldsymbol{x}) = \sum_k (c_k^{(2)} / \lambda_k^2)\, u_k(\boldsymbol{x})$. The above representation of $u^{(i)}$ follows by induction.

Second, the first term of the series giving $u^{(i)}$ is dominant for a sufficiently large value of i since $0 < \lambda_0 < \lambda_k$ for all $k \geq 1$ so that $\lambda_0^i\, u^{(i)}(\boldsymbol{x}) / c_0^{(i)}$ converges to $u_0(\boldsymbol{x})$ as $i \to \infty$.

Third, consider two successive approximations,

$$u^{(i)} \simeq \frac{c_0^{(i)}}{\lambda_0^i}\, u_0 \quad \text{and} \quad u^{(i+1)} \simeq \frac{c_0^{(i+1)}}{\lambda_0^{i+1}}\, u_0 = \frac{c_0^{(i)}}{\parallel u^{(i)} \parallel\, \lambda_0^{i+1}}\, u_0.$$

Their inner product,

$$\langle u^{(i+1)}, u^{(i)} \rangle \simeq \frac{(c_0^{(i)})^2}{\lambda_0^{2i+1} \parallel u^{(i)} \parallel} \langle u_0, u_0 \rangle,$$

divided by $\parallel u^{(i)} \parallel^2 \simeq (c_0^{(i)})^2 \langle u_0, u_0 \rangle / \lambda_0^{2i}$ gives the approximate expression of λ_0 in Eq. 6.99. The above equations also show that u_0 becomes proportional to $u^{(i)}$ for a sufficiently large i. ■

If the conditions in Eq. 6.99 are satisfied and g is a positive function, then

$$u^{(i+1)}(\boldsymbol{x}) = a(\boldsymbol{x})\, E[\bar{u}^{(i)}(\boldsymbol{Y}(\boldsymbol{x}))], \qquad (6.100)$$

where $a(\boldsymbol{x}) = \int_D g(\boldsymbol{x}, \boldsymbol{y})\, d\boldsymbol{y}$ and $\boldsymbol{Y}(\boldsymbol{x})$ denotes an $\mathbb{R}^d$-valued random variable with the probability density function $\tilde{g}(\boldsymbol{x}, \cdot) = g(\boldsymbol{x}, \cdot)/a(\boldsymbol{x})$.

The functions in Eqs. 6.98 and 6.100 can be estimated from n_s independent samples of $\boldsymbol{Y}(\boldsymbol{x})$ (Eq. 6.100) by

$$\hat{u}^{(i+1)}(\boldsymbol{x}) = a(\boldsymbol{x}) \left[\frac{1}{n_s} \sum_{\omega=1}^{n_s} \bar{u}^{(i)}(\boldsymbol{Y}(\boldsymbol{x}, \omega)) \right]. \qquad (6.101)$$

Example 6.45: Let u be the solution of the ordinary differential equation

$$u''(x) + \lambda\, u(x) = 0, \quad x \in D = (-1, 1), \quad \lambda > 0,$$

with the boundary conditions $u(\pm 1) = 0$. The integral form of this equation is

$$u(x) = \lambda \int_{-1}^{1} g(x, y)\, u(y)\, dy$$

where g is the Green function of the differential operator d^2/dx^2. This equation is a simple version of the Schrödinger equation, which can also be used to analyze the stability of a simply supported beam subjected to an axial load.

Figure 6.14 shows the lowest eigenfunction and eigenvalue of the above differential equation. The approximate eigenfunction calculated from Eq. 6.101 with $i = 100$ iterations and $n_s = 500$ samples at each iteration is practically indistinguishable from the exact solution $u_0(x) = \cos(\pi\, x/2)$. In fact, a very good approximation of u_0 results in a few iterations. However, the estimate of λ_0 in Eq. 6.99 becomes stable after a larger number of iterations, as demonstrated by its variation with the number of iterations. ◇

Note: The Green function of the differential operator d^2/dx^2 is the solution of the differential equation $g''(x, y) = \delta(x - y)$, $x, y \in D = (-1, 1)$, with the boundary conditions

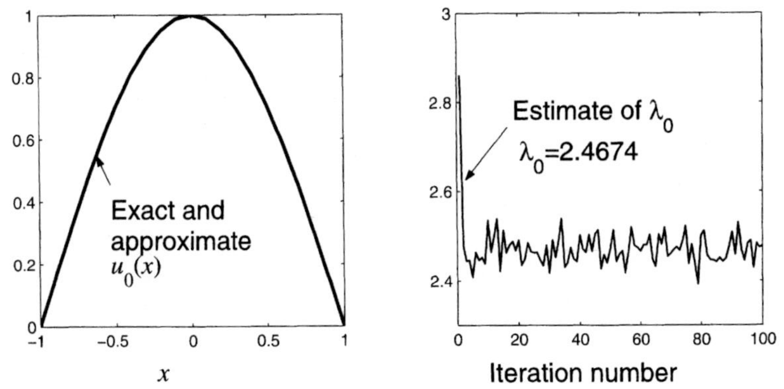

Figure 6.14. Exact/estimated eigenfunction u_0 and the variation of the estimate of λ_0 with the number of iteration

$g(\pm 1, y) = 0$, where the primes denote derivatives with respect to x and $\delta(\cdot)$ is the Dirac delta function. Elementary calculations give

$$g(x, y) = \begin{cases} (1-y)(1+x)/2, & x \le y, \\ (1+y)(1-x)/2, & x \ge y. \end{cases}$$

The Green function g is positive, symmetric, and square integrable in D so that Eq. 6.101 can be applied. The density of $Y(x)$ is $\tilde{g} = g/a(x)$, where $a(x) = \int_{-1}^{1} g(x, y)\, dy = (1 - x^2)/2$.

The solution of $u'' + \lambda\, u = 0$ is an even function since, if $u(x)$ satisfies this equation, so does $u(-x)$. Hence, we can replace D by $D^* = (0, 1)$ and solve $u'' + \lambda\, u = 0$ in D^* with the boundary conditions $u(1) = 0$ and $u'(0) = 0$. The general solution of this equation,

$$u(x) = \alpha \cos(\sqrt{\lambda}\, x) + \beta \sin(\sqrt{\lambda}\, x), \quad x \in D^*,$$

where α and β are some constants. Since the function u must satisfy the boundary conditions and cannot be zero everywhere in D^*, we have

$$u_k(x) = \cos(\sqrt{\lambda_k}\, x), \quad \text{where } \lambda_k = (2k+1)^2\, (\pi/2)^2.$$

The parameters λ_k and the functions u_k, $k = 0, 1, \ldots$, are the eigenvalues and the eigenfunctions of $u'' + \lambda\, u = 0$ with the specified boundary conditions. ▲

6.7 Problems

6.1: Develop an algorithm for solving locally Eq. 6.21 for the case in which ξ_r are random fields defined on the boundaries ∂D_r of D.

6.2: Extend the algorithm in Problem 6.1 to solve locally Eq. 6.25 with p and ξ random fields defined on D and ∂D_r, respectively.

6.3: Solve the Schrödinger equation $(1/2)\,u''(x) + q(x)\,u(x)$, $0 < x < l$, with the Dirichlet boundary conditions $u(0) = \alpha$ and $u(0) = \beta$ for two cases: (1) q is a continuous function and (2) q is a random field with continuous samples.

6.4: Show that the eigenvalues of Eq. 6.36 with $d = 1$ and $D = (0, 1)$ must be strictly positive.

6.5: Let $g(|B(t)|)$ and $g(h_{n,\varepsilon}(B(t)))$ be the processes given by Eq. 6.41 and Eq. 6.42, respectively. Evaluate numerically the difference between these two processes in a time interval $[0, 1]$ for several values of n and $\varepsilon > 0$, $g(x) = x^2$.

6.6: Write a MATLAB function generating samples of $(1/\varepsilon)\int_0^t 1_{(-\varepsilon,\varepsilon)}(B(s))\,ds$ for $t \geq 0$ and some $\varepsilon > 0$. Calculate statistics of this process approximating the local time process L in Eq. 6.38 for decreasing values of ε.

6.7: Complete the details of the proof given in the text for the second extension of Tanaka's formula in Eq. 6.46.

6.8: Generalize Eq. 6.41 to find a formula for the increment $g(X(t)) - g(X(0))$, where $g \in C^2(\mathbb{R})$ and X denotes an arbitrary diffusion process. Apply your result to two diffusion processes, an Ornstein-Uhlenbeck process and a geometric Brownian motion process.

6.9: Prove Eq. 6.51.

6.10: Apply the algorithm described in Fig. 6.10 to solve locally the partial differential equation in Example 6.22.

6.11: Use the SWM to solve locally the Laplace equation $\Delta u(\boldsymbol{x}) = 0$, $\boldsymbol{x} \in D = (-2, 2) \times (-1, 1)$ with the Dirichlet boundary conditions $u(\boldsymbol{x}) = 1$ for $\boldsymbol{x} \in (-2, 2) \times \{-1\}$ and $u(\boldsymbol{x}) = 0$ on the other boundaries of D.

6.12: Repeat the analysis of the partial differential equation in Example 6.22 for the same Dirichlet boundary conditions on $(0, a) \times \{0\}$ and $(0, a) \times \{b\}$ but different Neumann boundary conditions. Use in your calculations $\partial u(\boldsymbol{x})/\partial x_1 = \alpha$ on $\{0\} \times (0, b)$ and $\partial u(\boldsymbol{x})/\partial x_1 = \beta$ on $\{a\} \times (0, b)$, where α and β are some constants.

6.13: Show that the solution of Eq. 6.75 is $\boldsymbol{u} = (\boldsymbol{i} - \boldsymbol{g})^{-1}\,\boldsymbol{\phi} = \left(\sum_{r=0}^{\infty} \boldsymbol{g}^r\right)\boldsymbol{\phi}$ if there exists $\gamma \in (0, 1)$ such that $\| \boldsymbol{g}\,\boldsymbol{x} \| < \gamma \; \| \boldsymbol{x} \|$ for all $\boldsymbol{x} \in \mathbb{R}^m$, where $\boldsymbol{g}^0 = \boldsymbol{i}$ denotes the identity matrix.

6.14: Apply Eq. 6.79 to solve locally the problem in Example 6.34 by using transition probability matrices other than $\boldsymbol{p}_1$ and $\boldsymbol{p}_2$ in this example. Comment on the dependence of the solution accuracy on the transition probability matrix. Extend your analysis to the case in which $\boldsymbol{\phi}$ in Eq. 6.75 is random.

6.15: Apply Eq. 6.85 and 6.87 to find the dominant eigenvalue and eigenvector for some real-valued symmetric matrices of your choice.

6.16: Use Eq. 6.95 to solve locally the differential equation in Example 6.44. Consider for $X(k+1) \mid X(k)$ a different probability law than the one used in the text.

Chapter 7

Deterministic Systems and Stochastic Input

7.1 Introduction

This chapter examines stochastic problems defined by

$$\mathcal{D}[\mathcal{X}(\boldsymbol{x}, t)] = \boldsymbol{Y}(\boldsymbol{x}, t), \quad t \geq 0, \quad \boldsymbol{x} \in D \subset \mathbb{R}^n, \tag{7.1}$$

where $n \geq 1$ is an integer, $\boldsymbol{x}$ and t denote space and time parameters, respectively, $\mathcal{D}$ can be an algebraic, differential, or integral operator with deterministic coefficients that may or may not depend on time, $\boldsymbol{Y}$ is the random input, and $\mathcal{X}$ denotes the output. It is common in applications to concentrate on values of $\mathcal{X}$ at a finite number of points $\boldsymbol{x}_k \in D$ rather then all points of D. Systems described by the evolution in time of $\mathcal{X}$ at a finite number of points and at all points in D are called **discrete** and **continuous**, respectively. The focus of this chapter is on discrete systems since they are used extensively in applications and are simpler to analyze than continuous systems. The vector $\boldsymbol{X}(t) \in \mathbb{R}^d$ collecting the processes $\mathcal{X}(\boldsymbol{x}_k, t)$, called the **state vector**, defines the evolution of a discrete system. The mapping from $\boldsymbol{Y}$ to $\boldsymbol{X}$ can be with or without memory. An extensive discussion on memoryless transformations of random processes can be found in [79] (Chapters 3, 4, and 5) and is not presented here.

Our objective is to determine the probability law of $\boldsymbol{X}$ from the probability law of $\boldsymbol{Y}$, the defining equation of $\boldsymbol{X}$ (Eq. 7.1), and the initial and/or boundary conditions for this equation. This objective cannot be achieved for all types of operators and inputs in Eq. 7.1. Therefore, in some cases, we will derive only second moment properties of $\boldsymbol{X}$. The methods needed to characterize $\boldsymbol{X}$ depend on the properties of the input, the definition of $\mathcal{D}$, and the required output statistics. For example, if $\mathcal{D}$ is a linear operator and $\boldsymbol{Y}$ is Gaussian, then $\boldsymbol{X}$ is a Gaussian process so that the first two moments of this process define its probability law.

On the other hand, the second moment properties of X are insufficient to define its probability law if Y is not Gaussian or if Y is Gaussian but $\mathcal{D}$ is a nonlinear operator. Linear and nonlinear operators are examined separately. It is also shown that some of the methods developed for properties of the solution of initial value problems can be extended to a class of boundary value problems.

Table 7.1. System, input, solutions

EQ	Brownian motion	Semimartingale, DM	Semimartingale, SAM
LINEAR	• Mean/corr. • Linear random vibration (formal calculus)	• Mean/corr. if $M \in L_2$ • PDE for φ if A, H, and K are deterministic • HOM for special cases	• PDE for φ in special cases • HOM if GWN/PWN input
NONLINEAR	• HOM if drift/ diffusion are poly(X) • PDE for φ if drift/diffusion are poly(X) • Fokker-Planck-Kolmogorov eqs • Approximations	• Mean/corr. if a, b are poly(X) • PDE for φ in special cases	• HOM if drift/diffusion are poly(X, S) • PDE for φ in special cases

• **System:**

$$dX(t) = a(X(t-), t)\,dt + b(X(t-), t)\,dS(t)$$

$$\dot{X}(t) = a(X(t-), t) + b(X(t-), t)\,p(S(t-))$$

• **Input:**

$$dS(t) = dA(t) + H(t)\,dB(t) + K(t)\,\tilde{C}(t) = dA(t) + dM(t)$$

Table 7.1 lists the type of problem considered and results obtained in the following two sections. The notations *mean/corr, HOM*, and *PDE for* φ mean that differential equations are obtained for the mean/correlation, the higher order moments, and the characteristic function φ of X. Qualifiers such as Z = *solution of an SDE with GWN/PWN* or H = *poly(*B*)* mean that Z satisfies a stochastic differential equation driven by Gaussian and/or Poisson white noise processes or H is a polynomial of a Brownian motion B. The qualifiers *DM* and *SAM* in the table heading refer to direct and state augmentation methods used in the analysis. The defining equations for the state X and the driving noise S are given at the bottom of the table.

The analysis is based on the Itô formula for semimartingales. We give this formula here for convenience. Let $\boldsymbol{Y}$ be an $\mathbb{R}^m$-valued semimartingale and let $g : \mathbb{R}^m \to \mathbb{R}$ be a function with continuous second order partial derivatives. Then $g(\boldsymbol{Y})$ is a semimartingale and (Section 4.6.2)

$$
\begin{aligned}
g(\boldsymbol{Y}(t)) - g(\boldsymbol{Y}(0)) &= \sum_{i=1}^{d} \int_{0+}^{t} \frac{\partial g}{\partial y_i}(\boldsymbol{Y}(s-))\, dY_i(s) \\
&+ \frac{1}{2} \sum_{i,j=1}^{d} \int_{0+}^{t} \frac{\partial^2 g}{\partial y_i\, \partial y_j}(\boldsymbol{Y}(s-))\, d[Y_i, Y_j]^c(s) \\
&+ \sum_{0<s\le t} \left[g(\boldsymbol{Y}(s)) - g(\boldsymbol{Y}(s-)) - \sum_{i=1}^{d} \frac{\partial g}{\partial y_i}(\boldsymbol{Y}(s-))\, \Delta Y_i(s) \right]. \qquad (7.2)
\end{aligned}
$$

If $\boldsymbol{Y}$ has continuous samples, Eq. 7.2 becomes

$$
\begin{aligned}
g(\boldsymbol{Y}(t)) - g(\boldsymbol{Y}(0)) &= \sum_{i=1}^{d} \int_{0}^{t} \frac{\partial g}{\partial y_i}(\boldsymbol{Y}(s))\, dY_i(s) \\
&+ \frac{1}{2} \sum_{i,j=1}^{d} \int_{0}^{t} \frac{\partial^2 g}{\partial y_i\, \partial y_j}(\boldsymbol{Y}(s))\, d[Y_i, Y_j](s). \qquad (7.3)
\end{aligned}
$$

If $\boldsymbol{Y}$ is a pure jump semimartingale, Eq. 7.2 becomes

$$
\begin{aligned}
g(\boldsymbol{Y}(t)) - g(\boldsymbol{Y}(0)) &= \sum_{i=1}^{d} \int_{0+}^{t} \frac{\partial g}{\partial y_i}(\boldsymbol{Y}(s-))\, dY_i(s) \\
&+ \sum_{0<s\le t} \left[g(\boldsymbol{Y}(s)) - g(\boldsymbol{Y}(s-)) - \sum_{i=1}^{d} \frac{\partial g}{\partial y_i}(\boldsymbol{Y}(s-))\, \Delta Y_i(s) \right]. \qquad (7.4)
\end{aligned}
$$

Note: If $\boldsymbol{Y}$ has continuous samples, then $\boldsymbol{Y}(s-) = \boldsymbol{Y}(s)$, $\Delta Y_i(s) = 0$, and $[Y_i, Y_j]^c = [Y_i, Y_j]$ so that Eq. 7.2 becomes Eq. 7.3. If $\boldsymbol{Y}$ is a quadratic pure jump semimartingale, then $[\boldsymbol{Y}, \boldsymbol{Y}]^c = 0$ (Section 4.5.2), and Eq. 7.2 becomes Eq. 7.4. The formula in Eq. 7.4 applies if, for example, $\boldsymbol{Y}$ is a compound Poisson process. ▲

7.2 Linear systems

For a discrete linear system, Eq. 7.1 is an initial value problem that can be given in the form

$$\dot{\boldsymbol{X}}(t) = \boldsymbol{a}(t)\,\boldsymbol{X}(t) + \boldsymbol{b}(t)\,\boldsymbol{Y}(t), \quad t \geq 0, \tag{7.5}$$

where the state $\boldsymbol{X}$ is an $\mathbb{R}^d$-valued stochastic process, the input $\boldsymbol{Y}$ takes values in $\mathbb{R}^{d'}$, $\boldsymbol{a}$ denotes a (d, d)-matrix, and $\boldsymbol{b}$ is a (d, d')-matrix. The solution of this differential equation is

$$\boldsymbol{X}(t) = \boldsymbol{\theta}(t, 0)\,\boldsymbol{X}(0) + \int_0^t \boldsymbol{\theta}(t, s)\,\boldsymbol{b}(s)\,\boldsymbol{Y}(s)\,ds, \tag{7.6}$$

where the Green function $\boldsymbol{\theta}$, also called the unit impulse response function in dynamics, satisfies the equation ([175], Section 5.1.1)

$$\frac{\partial \boldsymbol{\theta}(t, s)}{\partial t} = \boldsymbol{a}(t)\,\boldsymbol{\theta}(t, s), \quad t \geq s \geq 0, \tag{7.7}$$

and $\boldsymbol{\theta}(s, s)$ is the identity matrix.

Example 7.1: Suppose that Eq. 7.1 is

$$\ddot{X}(t) + \alpha(t)\,\dot{X}(t) + \beta(t)\,X(t) = Y(t), \quad t \geq 0,$$

and set $X_1 = X$ and $X_2 = \dot{X}$. The $\mathbb{R}^2$-valued process $\boldsymbol{X} = (X_1, X_2)$ is the solution of

$$\frac{d}{dt}\begin{bmatrix} X_1(t) \\ X_2(t) \end{bmatrix} = \begin{bmatrix} 0 & 1 \\ -\beta(t) & -\alpha(t) \end{bmatrix}\begin{bmatrix} X_1(t) \\ X_2(t) \end{bmatrix} + \begin{bmatrix} 0 \\ 1 \end{bmatrix} Y(t),$$

that is, it satisfies a differential equation of the type given by Eq. 7.5. ◇

Our objective is to develop differential equations for moments and other probabilistic properties of $\boldsymbol{X}$. If $\boldsymbol{Y}$ is a Gaussian process, the probability law of $\boldsymbol{X}$ is defined completely by its first two moments since Eq. 7.5 is linear. If $\boldsymbol{Y}$ is not Gaussian, $\boldsymbol{X}$ is generally non-Gaussian so that moments of order higher than two and/or other statistics are needed to characterize its probability law ([79], Section 5.2).

7.2.1 Brownian motion input

If $\boldsymbol{Y}$ in Eq. 7.5 is a Gaussian white noise, the state $\boldsymbol{X}$ is a diffusion process defined by the stochastic differential equation

$$d\boldsymbol{X}(t) = \boldsymbol{a}(t)\,\boldsymbol{X}(t)\,dt + \boldsymbol{b}(t)\,d\boldsymbol{B}(t), \quad t \geq 0, \tag{7.8}$$

where the coordinates of $\boldsymbol{B} \in \mathbb{R}^{d'}$ are independent Brownian motions. The solution of this differential equation is (Eq. 7.6)

$$\boldsymbol{X}(t) = \boldsymbol{\theta}(t,0)\,\boldsymbol{X}(0) + \int_0^t \boldsymbol{\theta}(t,s)\,\boldsymbol{b}(s)\,d\boldsymbol{B}(s). \tag{7.9}$$

In Eq. 7.8 we can assume without loss of generality that the input is white noise. If the input to this equation is colored, it can be approximated by the output $\boldsymbol{Y}$ of a linear filter driven by white noise ([30], Section 2.17). The augmented vector $\boldsymbol{Z} = (\boldsymbol{X}, \boldsymbol{Y})$ satisfies a differential equation of the same form as Eq. 7.8.

Note: If the coefficients $\boldsymbol{a}$ and $\boldsymbol{b}$ of Eq. 7.8 satisfy the conditions in Section 4.7.1.1, then the equation has a unique solution. If the functions $\boldsymbol{\theta}$ and $\boldsymbol{b}$ are bounded, the integral in Eq. 7.9 with Brownian motion integrator is a $\sigma(\boldsymbol{B}(s), 0 \le s \le t)$-square integrable martingale. We have also seen that $\boldsymbol{X}$ is a semimartingale (Section 4.7.1.1). ▲

Example 7.2: Let X be the process in Example 7.1, where Y is the solution of $dY(t) = \rho(t)\,Y(t)\,dt + \zeta(t)\,dB(t)$, $t \ge 0$, and B denotes a Brownian motion. The augmented state vector $\boldsymbol{Z} = (X_1, X_2, Y)$ is the solution of

$$d\begin{bmatrix} X_1(t) \\ X_2(t) \\ Y(t) \end{bmatrix} = \begin{bmatrix} 0 & 1 & 0 \\ -\beta(t) & -\alpha(t) & 1 \\ 0 & 0 & \rho(t) \end{bmatrix} \begin{bmatrix} X_1(t) \\ X_2(t) \\ Y(t) \end{bmatrix} dt + \begin{bmatrix} 0 \\ 0 \\ \zeta(t) \end{bmatrix} dB(t)$$

so that $\boldsymbol{Z}$ is a diffusion process (Eq. 7.8). ◇

7.2.1.1 Mean and correlation equations

Our objective is to calculate the second moment properties of $\boldsymbol{X}$, that is, the **mean function** $\boldsymbol{\mu}(t) = E[\boldsymbol{X}(t)]$ and the **correlation function** $\boldsymbol{r}(t,s) = E[\boldsymbol{X}(t)\,\boldsymbol{X}(s)^T]$ or the **covariance function** $\boldsymbol{c}(t,s) = E[(\boldsymbol{X}(t) - \boldsymbol{\mu}(t))\,(\boldsymbol{X}(s) - \boldsymbol{\mu}(s))^T]$ of this process for deterministic and random initial conditions. If $\boldsymbol{X}(0)$ is random, it is assumed to be independent of the driving noise. The state $\boldsymbol{X}$ is a Gaussian process if $\boldsymbol{X}(0)$ is a constant or a Gaussian vector. Otherwise, the probability of $\boldsymbol{X}$ can be calculated from the second moment properties of the conditional Gaussian process $\boldsymbol{X}(t) \mid \boldsymbol{X}(0)$ and the distribution of $\boldsymbol{X}(0)$.

The **mean and correlation functions** of $\boldsymbol{X}$ in Eq. 7.8 satisfy the differential equations

$$\begin{aligned}
&\dot{\boldsymbol{\mu}}(t) = \boldsymbol{a}(t)\,\boldsymbol{\mu}(t), \quad t \ge 0,\\
&\dot{\boldsymbol{r}}(t,t) = \boldsymbol{a}(t)\,\boldsymbol{r}(t,t) + \boldsymbol{r}(t,t)\,\boldsymbol{a}(t)^T + \boldsymbol{b}(t)\,\boldsymbol{b}(t)^T, \quad t \ge 0, \quad \text{and}\\
&\frac{\partial \boldsymbol{r}(t,s)}{\partial t} = \boldsymbol{a}(t)\,\boldsymbol{r}(t,s), \quad t > s \ge 0,
\end{aligned} \tag{7.10}$$

with $\boldsymbol{\mu}(0) = E[\boldsymbol{X}(0)]$ and $\boldsymbol{r}(0,0) = E[\boldsymbol{X}(0)\,\boldsymbol{X}(0)^T]$. The covariance function $\boldsymbol{c}$ of $\boldsymbol{X}$ satisfies the same differential equation as $\boldsymbol{r}$.

Proof: The mean equation results by averaging Eq. 7.8.

The integral form of the Itô formula for continuous semimartingales (Section 4.6.2 and Eq. 7.3) applied to function $\boldsymbol{X}(t) \mapsto X_p(t)\, X_q(t)$, $p, q = 1, \ldots, d$, gives

$$X_p(t)\, X_q(t) - X_p(0)\, X_q(0) = \sum_{k=1}^{d} \int_0^t \left(\delta_{pk}\, X_q(s) + X_p(s)\, \delta_{qk}\right) dX_k(s)$$
$$+ \frac{1}{2} \sum_{k,l=1}^{d} \int_0^t \left(\delta_{pk}\, \delta_{ql} + \delta_{pl}\, \delta_{qk}\right) d[X_k, X_l](s) \quad \text{or}$$

$$X_p(t)\, X_q(t) - X_p(0)\, X_q(0) = \int_0^t X_q(s)\, dX_p(s) + \int_0^t X_p(s)\, dX_q(s)$$
$$+ \frac{1}{2} \int_0^t \left(d[X_p, X_q](s) + d[X_q, X_p](s)\right).$$

The average of the left side of the above equation is $r_{pq}(t,t) - r_{pq}(0,0)$ so that its derivative with respect to t is $\dot{r}_{pq}(t,t)$. The average of the first term on the right side of this equation is

$$E\left[\int_0^t X_q(s)\, dX_p(s)\right] = E\left[\int_0^t X_q(s) \left(\sum_u a_{pu}(s)\, X_u(s)\, ds + \sum_v b_{pv}(s)\, dB_v(s)\right)\right]$$
$$= \sum_u E\left[\int_0^t a_{pu}(s)\, X_q(s)\, X_u(s)\, ds\right] + \sum_v E\left[\int_0^t b_{pv}(s)\, X_q(s)\, dB_v(s)\right]$$
$$= \sum_u \int_0^t a_{pu}(s)\, r_{qu}(s,s)\, ds,$$

so that its derivative with respect to time is $\sum_u a_{pu}(t)\, r_{qu}(t,t)$. In the same way, the average of $\int_0^t X_p(s)\, dX_q(s)$ can be obtained. The quadratic covariation process $[X_p, X_q]$ is (Section 4.5)

$$[X_p, X_q](t) = \sum_{v=1}^{d'} \int_0^t b_{pv}(s)\, b_{qv}(s)\, ds,$$

with time derivative $\sum_{v=1}^{d'} b_{pv}(t)\, b_{qv}(t)$. The expectation of the last term on the right side of the Itô formula is $\sum_{v=1}^{d'} \int_0^t b_{pv}(s)\, b_{qv}(s)\, ds$. Hence, the Itô formula gives by expectation and differentiation

$$\dot{r}_{pq}(t,t) = \sum_u a_{pu}(t)\, r_{qu}(t,t) + \sum_u a_{qu}(t)\, r_{pu}(t,t) + \sum_{v=1}^{d'} b_{pv}(t)\, b_{qv}(t).$$

The matrix form of this equation is Eq. 7.10.

The Itô formula applied to mapping $\boldsymbol{X}(t) \mapsto X_p(t)$ in the time interval $[s, t]$ gives

$$X_p(t) - X_p(s) = \sum_{k=1}^{d} \int_s^t \delta_{pk}\, dX_k(\sigma) = \int_s^t dX_p(\sigma) \quad \text{for } t \geq s.$$

Note that the above equation can also be obtained by using classical analysis. We have

$$X_p(t)\, X_q(s) - X_p(s)\, X_q(s) = \int_s^t X_q(s)\, dX_p(\sigma)$$
$$= \sum_{u=1}^{d} \int_s^t a_{pu}(\sigma)\, X_u(\sigma)\, X_q(s)\, d\sigma + \sum_{v=1}^{d'} \int_s^t b_{pv}(\sigma)\, X_q(s)\, dB_v(\sigma)$$

by multiplying the above equation with $X_q(s)$. The average of the above equation is

$$r_{pq}(t, s) - r_{pq}(s, s) = \sum_{u=1}^{d} \int_s^t a_{pu}(\sigma)\, r_{uq}(\sigma, s)\, d\sigma$$

since

$$E[X_q(s)\, dB_v(\sigma)] = E\left\{E[X_q(s)\, dB_v(\sigma) \mid \mathcal{F}_s]\right\} = E\left\{X_q(s)\, E[dB_v(\sigma) \mid \mathcal{F}_s]\right\} = 0$$

where $\mathcal{F}_s = \sigma\{\boldsymbol{B}(u), 0 \le u \le s\}$ is the natural filtration of $\boldsymbol{B}$. The differentiation of the above equation with respect to t yields the last formula in Eq. 7.10.

The calculation of $\boldsymbol{r}$ in Eq. 7.10 involves two steps. First, the function $\boldsymbol{r}(t, t), t \ge 0$, needs to be obtained for the specified initial condition $\boldsymbol{r}(0, 0) = E[\boldsymbol{X}(0)\, \boldsymbol{X}(0)^T]$. Second, the last differential equation in Eq. 7.10 has to be integrated for $t \ge s$ and known $\boldsymbol{r}(s, s)$.

That $\boldsymbol{c}$ satisfies the same differential equations as $\boldsymbol{r}$ results from the relationship

$$\boldsymbol{r}(t, s) = \boldsymbol{c}(t, s) + \boldsymbol{\mu}(t)\, \boldsymbol{\mu}(s)^T$$

and the mean equation. For example, the second formula in Eq. 7.10 is

$$\frac{d}{dt}\left(\boldsymbol{c}(t, t) + \boldsymbol{\mu}(t)\, \boldsymbol{\mu}(t)^T\right) = \boldsymbol{a}(t)\left(\boldsymbol{c}(t, t) + \boldsymbol{\mu}(t)\, \boldsymbol{\mu}(t)^T\right)$$
$$+ \left(\boldsymbol{c}(t, t) + \boldsymbol{\mu}(t)\, \boldsymbol{\mu}(t)^T\right) \boldsymbol{a}(t)^T + \boldsymbol{b}(t)\, \boldsymbol{b}(t)^T$$

showing that $\boldsymbol{c}$ and $\boldsymbol{r}$ are the solutions of the same differential equations since

$$\frac{d}{dt}\left(\boldsymbol{c}(t, t) + \boldsymbol{\mu}(t)\, \boldsymbol{\mu}(t)^T\right) = \dot{\boldsymbol{c}}(t, t) + \dot{\boldsymbol{\mu}}(t)\, \boldsymbol{\mu}(t)^T + \boldsymbol{\mu}(t)\, \dot{\boldsymbol{\mu}}(t)^T$$

and $\dot{\boldsymbol{\mu}}(t) = \boldsymbol{a}(t)\, \boldsymbol{\mu}(t)$. ■

It is useful for applications to consider a noise $\boldsymbol{B}_b$ with increments

$$d\boldsymbol{B}_b(t) = \boldsymbol{\mu}_b(t)\, dt + \boldsymbol{\zeta}_b(t)\, d\boldsymbol{B}(t), \tag{7.11}$$

where the $(d', 1)$ and (d', d') matrices $\boldsymbol{\mu}_b$ and $\boldsymbol{\zeta}_b$ are interpreted as the mean and the intensity of the driving white noise, respectively. The coordinates of the noise $\boldsymbol{B}_b$ are correlated.

If $\boldsymbol{X}$ is the solution of Eq. 7.8 with $\boldsymbol{B}$ replaced by $\boldsymbol{B}_b$ in Eq. 7.11, the **mean and correlation functions** of $\boldsymbol{X}$ are given by

$$\begin{aligned}
\dot{\boldsymbol{\mu}}(t) &= \boldsymbol{a}(t)\,\boldsymbol{\mu}(t) + \boldsymbol{b}(t)\,\boldsymbol{\mu}_b(t), \quad t \geq 0, \\
\dot{\boldsymbol{r}}(t,t) &= \boldsymbol{a}(t)\,\boldsymbol{r}(t,t) + \boldsymbol{r}(t,t)\,\boldsymbol{a}(t)^T + \boldsymbol{b}(t)\,\boldsymbol{\zeta}_b(t)\,\boldsymbol{\zeta}_b(t)^T\,\boldsymbol{b}(t)^T \\
&\quad + \boldsymbol{b}(t)\,\boldsymbol{\mu}_b(t)\,\boldsymbol{\mu}(t)^T + \boldsymbol{\mu}(t)\,\boldsymbol{\mu}_b(t)^T\,\boldsymbol{b}(t)^T, \quad t \geq 0, \\
\frac{\partial \boldsymbol{r}(t,s)}{\partial t} &= \boldsymbol{a}(t)\,\boldsymbol{r}(t,s) + \boldsymbol{b}(t)\,\boldsymbol{\mu}_b(t)\,\boldsymbol{\mu}(s)^T, \quad t > s \geq 0,
\end{aligned} \tag{7.12}$$

with $\boldsymbol{\mu}(0) = E[\boldsymbol{X}(0)]$ and $\boldsymbol{r}(0,0) = E[\boldsymbol{X}(0)\,\boldsymbol{X}(0)^T]$. The last two formulas hold with $\boldsymbol{c}$ in place of $\boldsymbol{r}$ and $\boldsymbol{\mu}_b = \boldsymbol{0}$.

Proof: The mean equation results by averaging the defining equation of $\boldsymbol{X}$. The difference between the equations for $\boldsymbol{X}$ and $\boldsymbol{\mu}$ gives

$$d\tilde{\boldsymbol{X}}(t) = \boldsymbol{a}(t)\,\tilde{\boldsymbol{X}}(t)\,dt + \boldsymbol{b}(t)\,\boldsymbol{\zeta}_b(t)\,d\boldsymbol{B}(t),$$

where $\tilde{\boldsymbol{X}} = \boldsymbol{X} - \boldsymbol{\mu}$. The covariance function of $\tilde{\boldsymbol{X}}$ satisfies the last two equations in Eq. 7.10 with $\boldsymbol{b}\,\boldsymbol{\zeta}_b$ in place of $\boldsymbol{b}$. These equations and the relationship between the correlation and the covariance functions give the conditions satisfied by the correlation function of $\boldsymbol{X}$. If $\boldsymbol{\mu}_b = \boldsymbol{0}$ and $\boldsymbol{\zeta}_b = \boldsymbol{i}$ is the identity matrix, Eq. 7.12 coincides with Eq. 7.10. ■

Example 7.3: Let $X \in \mathbb{R}$ be an Ornstein-Uhlenbeck process defined by $dX(t) = -\alpha\,X(t)\,dt + \beta\,dB(t)$, $t \geq 0$, where $\alpha > 0$ and β are constants. If $X(0) \in L_2$ is independent of B, the mean and correlation functions of X are

$$\begin{aligned}
\mu(t) &= E[X(0)]\,e^{-\alpha t}, \quad t \geq 0, \\
r(t,t) &= E[X(0)^2]\,e^{-2\alpha t} + \frac{\beta^2}{2\alpha}\left(1 - e^{-2\alpha t}\right), \quad t \geq 0, \\
r(t,s) &= r(s,s)\,e^{-\alpha(t-s)}, \quad t \geq s \geq 0.
\end{aligned}$$

The probability law of X can be obtained from the properties of the Gaussian process $X(t) \mid (X(0) = \xi)$ whose first two moments are given by the above equations with (ξ, ξ^2) in place of $(E[X(0)], E[X(0)^2])$.

The differential equations for the correlation function of X become

$$\begin{aligned}
\dot{r}(t,t) &= -2\alpha\,r(t,t) + \beta^2\,\zeta_b(t)^2 + 2\,\beta\,\mu_b(t)\,\mu(t), \quad t \geq 0, \\
\frac{\partial r(t,s)}{\partial t} &= -\alpha\,r(t,s) + \beta\,\mu_b(t)\,\mu(s), \quad t \geq s,
\end{aligned}$$

if $dB(t)$ is replaced by $dB_b(t) = \mu_b(t)\,dt + \zeta_b(t)\,dB(t)$ (Eq. 7.11). ◇

Proof: The second moment properties of X satisfy the differential equations $\dot{\mu}(t) = -\alpha\,\mu(t)$, $\dot{r}(t,t) = -2\alpha\,r(t,t) + \beta^2$, and $\partial r(t,s)/\partial t = -\alpha\,r(t,s)$ (Eq. 7.10). To solve these equations, the mean and variance of $X(0)$ need to be specified. The differential equations for the mean and correlation functions of X driven by the white noise B_b are given by Eq. 7.12. ■

7.2.1.2 Linear random vibration

One of the main objectives of linear random vibration studies is the determination of the second moment properties of the state $\boldsymbol{X}$ of a linear dynamic system driven by white noise. The state $\boldsymbol{X}$ is the solution of Eq. 7.5 with $\boldsymbol{Y} = \boldsymbol{W}$, where $\boldsymbol{W}$ is a white noise process specified by

$$E[\boldsymbol{W}(t)] = \boldsymbol{\mu}_w(t),$$
$$E[(\boldsymbol{W}(t) - \boldsymbol{\mu}_w(t))(\boldsymbol{W}(s) - \boldsymbol{\mu}_w(s))^T] = \boldsymbol{q}_w(t)\,\delta(t-s). \tag{7.13}$$

The functions $\boldsymbol{\mu}_w(t)$ and $\boldsymbol{q}_w(t)$ denote the mean and the intensity of $\boldsymbol{W}$ at time t, respectively, and $\delta(\cdot)$ is the Dirac delta function. It is assumed that there is no correlation between the noise and the initial state $\boldsymbol{X}(0)$.

The classical **mean** and **correlation/covariance** equations in linear random vibration coincide with Eq. 7.12 for $\boldsymbol{\mu}_b = \boldsymbol{\mu}_w$ and $\boldsymbol{\zeta}_b \boldsymbol{\zeta}_b^T = \boldsymbol{q}_w$. However, the derivation of these equations in linear random vibration is based on a heuristic definition of white noise (Eq. 7.13) and on formal calculations ([175], Chapter 5).

Example 7.4: Let X be the solution of the differential equation

$$\dot{X}(t) = -\alpha\, X(t) + \beta\, W(t), \quad t \geq 0, \tag{7.14}$$

where $X(0) = 0$, $\alpha > 0$ and β are constants, and W is a white noise with $\mu_w = 0$ and $q_w = 1$ (Eq. 7.13). Denote the solutions of Eq. 7.14 by X_G and X_P if $W(t)$ is a Gaussian white noise $W_G(t) = dB(t)/dt$ and a Poisson white noise $W_P(t) = dC(t)/dt$, respectively. Suppose that W_G and W_P have the same first two moments as W so that X_G and X_P are equal in the second moment sense. The classical linear random vibration can only deliver the second moment properties of the state so that it cannot distinguish between the processes X_G and X_P. This is a severe limitation since the samples of these processes can differ significantly. Figure 7.1 shows a sample of X_G and three samples of X_P for $C(t) = \sum_{k=1}^{N(t)} Y_k$, where N is a Poisson process with intensity $\lambda = 5$, 10, and 100, and Y_k are independent copies of a Gaussian variable Y_1 with mean zero and variance $E[Y_1^2]$ such that $\lambda\, E[Y_1^2] = 1$. The qualitative difference between the samples of X_P and X_G decreases as λ increases. ◇

Note: Recall that the mean and correlation functions of C are $E[C(t)] = \lambda\, E[Y_1] = 0$ and $E[C(t)\, C(s)] = \lambda\, E[Y_1^2] \min(t, s)$ (Section 3.12). Formal calculations of the type used in the classical linear random vibration show that the mean and the covariance functions of the Poisson white noise $W_P(t) = dC(t)/dt$ are zero and $\lambda\, E[Y_1^2]\, \delta(t-s)$, respectively. If we set $\lambda\, E[Y_1^2] = 1$, the white noise processes W_G and W_P are equal in the second moment sense. ▲

In some applications it is of interest to (1) establish conditions under which $\boldsymbol{X}$ becomes a weakly stationary process $\boldsymbol{X}_s$ and (2) determine the second moment

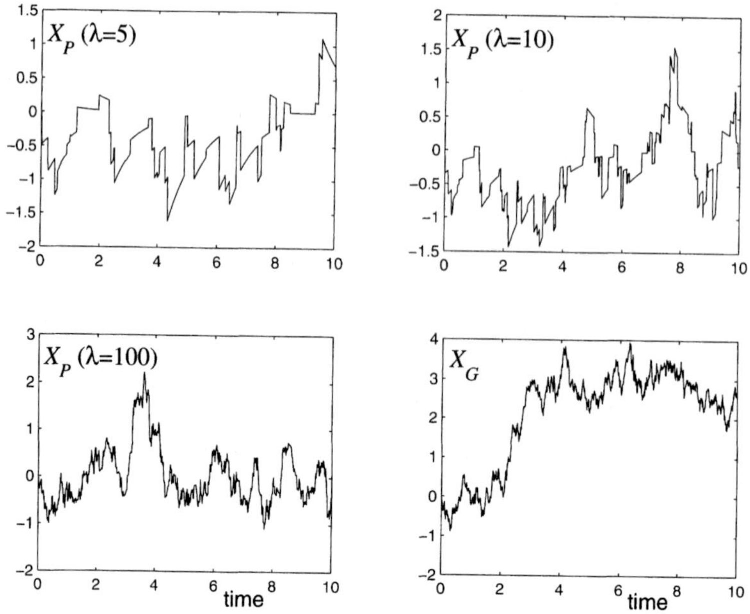

Figure 7.1. A sample of X_G and three samples of X_P for $\lambda = 5$, 10, and 100 such that $\lambda\, E[Y_1^2] = 1$

properties of $\boldsymbol{X}_s$. Let $\boldsymbol{\mu}_s$, $\boldsymbol{r}_s$, and $\boldsymbol{c}_s$ denote the mean, correlation, and covariance functions of $\boldsymbol{X}_s$ provided they exist.

If (1) the coefficients of a linear system are **time-invariant** and the driving noise is weakly stationary, that is, $\boldsymbol{a}(t) = \boldsymbol{a}$, $\boldsymbol{b}(t) = \boldsymbol{b}$, $\boldsymbol{\mu}_w(t) = \boldsymbol{\mu}_w$, and $\boldsymbol{q}_w(t) = \boldsymbol{q}_w$, and (2) $\boldsymbol{\mu}(t)$ and $\boldsymbol{r}(t, t)$ in Eq. 7.12 become time-invariant as $t \to \infty$, then $\boldsymbol{X}_s$ exists and its mean and covariance functions are given by

$$\begin{aligned}
&\boldsymbol{a}\,\boldsymbol{\mu}_s + \boldsymbol{b}\,\boldsymbol{\mu}_w = \boldsymbol{0}, \\
&\boldsymbol{a}\,\boldsymbol{c}_s + \boldsymbol{c}_s\,\boldsymbol{a}^T + \boldsymbol{b}\,\boldsymbol{q}_w\,\boldsymbol{b}^T = \boldsymbol{0}, \\
&\dot{\boldsymbol{c}}_s(\tau) = \boldsymbol{a}\,\boldsymbol{c}_s(\tau), \quad \tau \geq 0,
\end{aligned} \tag{7.15}$$

where $\boldsymbol{c}_s$ and $\boldsymbol{c}_s(\tau)$ are shorthand notations for $\boldsymbol{c}_s(t, t)$ and $\boldsymbol{c}_s(t + \tau, t)$, respectively.

Proof: If $\boldsymbol{X}$ is weakly stationary, the above equations follow from Eq. 7.12 since $\boldsymbol{\mu}(t)$ and $\boldsymbol{c}(t, t)$ do not vary in time so that their time derivatives are zero and $\boldsymbol{c}(t, s)$ depends on only the time lag $\tau = t - s$ rather than the times t and s.

The differential equations for $\boldsymbol{\mu}(t)$ and the entries $c_{ij}(t, t)$, $i \geq j$, of $\boldsymbol{c}(t, t)$ have the form $\dot{\boldsymbol{\xi}}(t) = \boldsymbol{\alpha}\,\boldsymbol{\xi}(t) + \boldsymbol{\beta}$, $t \geq 0$, where $\boldsymbol{\alpha}$ and $\boldsymbol{\beta}$ are constant matrices. The matrix $\boldsymbol{\alpha}$

is equal to $\boldsymbol{a}$ for $\boldsymbol{\mu}(t)$. Hence, $\boldsymbol{\mu}_s = \lim_{t\to\infty} \boldsymbol{\mu}(t)$ exists and is finite if the eigenvalues of $\boldsymbol{a}$ have negative real part. Similarly, $\boldsymbol{c}_s = \lim_{t\to\infty} \boldsymbol{c}(t,t)$ exists and is finite if the eigenvalues of the corresponding matrix $\boldsymbol{\alpha}$ have negative real part ([166], Section 4.1.3, [175], Section 5.1.1).

Suppose that $\boldsymbol{\alpha}$ has distinct eigenvalues λ_i with negative real part and let $\boldsymbol{\xi}_s = \sum_i d_i\, \boldsymbol{v}_i$ be the stationary solution of $\dot{\boldsymbol{\xi}}(t) = \boldsymbol{\alpha}\, \boldsymbol{\xi}(t) + \boldsymbol{\beta}$, where $\boldsymbol{v}_i$ denote the eigenvectors of $\boldsymbol{\alpha}$. Then $\boldsymbol{\xi}(t) = \sum_i \left(c_i\, e^{\lambda_i t} + d_i\right) \boldsymbol{v}_i$, where c_i are some constants. If $\boldsymbol{\xi}(0) = \sum_i d_i'\, \boldsymbol{v}_i$, then $\boldsymbol{\xi}(t) = \sum_i \left((d_i' - d_i)\, e^{\lambda_i t} + d_i\right) \boldsymbol{v}_i$. If $\boldsymbol{\xi}(0) = \boldsymbol{\xi}_s$, then $d_i' - d_i = 0$ for all i's so that the general solution is $\boldsymbol{\xi}(t) = \sum_i d_i\, \boldsymbol{v}_i = \boldsymbol{\xi}_s$, as stated. ■

Example 7.5: Consider a linear dynamic system with n degrees of freedom, mass matrix $\boldsymbol{m}$, damping matrix $\boldsymbol{\zeta}$, and stiffness matrix $\boldsymbol{k}$ that is subjected to a random forcing function $\boldsymbol{Y}$. Let $\boldsymbol{X}$ be an $\mathbb{R}^n$-valued process collecting the system displacements at its degrees of freedom. This process is defined by ([175], Section 5.2.1.1)

$$\boldsymbol{m}\, \ddot{\boldsymbol{X}}(t) + \boldsymbol{\zeta}\, \dot{\boldsymbol{X}}(t) + \boldsymbol{k}\, \boldsymbol{X}(t) = \boldsymbol{Y}(t), \quad t \geq 0. \tag{7.16}$$

Let ν_i and $\boldsymbol{\phi}_i$, $i = 1, \ldots, n$, be the eigenvalues and eigenvectors of the eigenvalue problem $\nu^2\, \boldsymbol{m}\, \boldsymbol{\phi} = \boldsymbol{k}\, \boldsymbol{\phi}$. Suppose that (1) the damping matrix is $\boldsymbol{\zeta} = \alpha\, \boldsymbol{m} + \beta\, \boldsymbol{k}$, where α, β are some constants, and (2) the eigenvalues ν_i^2 are distinct. Then $\boldsymbol{X}$ has the representation

$$\boldsymbol{X}(t) = \sum_{i=1}^{n} \boldsymbol{\phi}_i\, Q_i(t), \quad t \geq 0, \tag{7.17}$$

where

$$\ddot{Q}_i(t) + 2\,\zeta_i\, \nu_i\, \dot{Q}_i(t) + \nu_i^2\, Q_i(t) = \left(\boldsymbol{\Phi}^T\, \boldsymbol{Y}(t)\right)_i / \tilde{m}_i, \quad i = 1, \ldots, n, \tag{7.18}$$

the columns of the (n, n) matrix $\boldsymbol{\Phi}$ are the eigenvectors $\boldsymbol{\phi}_i$, $\tilde{m}_i$ are the non-zero entries of the diagonal matrix $\boldsymbol{\Phi}^T\, \boldsymbol{m}\, \boldsymbol{\Phi}$, and the modal damping ratios ζ_i are given by $\left(\boldsymbol{\Phi}^T\, \boldsymbol{\zeta}\, \boldsymbol{\Phi}\right)_{ii} = 2\,\zeta_i\, \nu_i\, \tilde{m}_i$. Any system response can be represented in the form

$$\tilde{\boldsymbol{X}}(t) = \sum_{i=1}^{n} \boldsymbol{\rho}\, \boldsymbol{\phi}_i\, Q_i(t), \quad t \geq 0, \tag{7.19}$$

where $\boldsymbol{\rho}$ is a matrix depending on the response considered. For example, the relative displacement between the first two degrees of freedom of the system is $\tilde{X}(t) = \sum_{i=1}^{n} (\phi_{i,2} - \phi_{i,1})\, Q_i(t)$ so that $\boldsymbol{\rho}$ is a $(1, n)$ matrix with $\rho_{11} = -1$, $\rho_{12} = 1$, and $\rho_{1i} = 0$ for $i = 3, \ldots, n$.

The definitions of $\boldsymbol{X}$ and $\tilde{\boldsymbol{X}}$ in Eqs. 7.17 and 7.19 show that the second moment properties of any response depend only on the system characteristics and the second moment properties of the modal coordinates Q_i, $i = 1, \ldots, n$. ◇

Note: The matrices $\boldsymbol{\Phi}^T\, \boldsymbol{m}\, \boldsymbol{\Phi}$ and $\boldsymbol{\Phi}^T\, \boldsymbol{k}\, \boldsymbol{\Phi}$ are diagonal by the definition of the eigenvalue problem. Because $\boldsymbol{\zeta} = \alpha\, \boldsymbol{m} + \beta\, \boldsymbol{k}$, the matrix $\boldsymbol{\Phi}^T\, \boldsymbol{\zeta}\, \boldsymbol{\Phi}$ is also diagonal. The parameters $\tilde{m}_i$, ζ_i, and $\tilde{k}_i = \nu_i^2/\tilde{m}_i$ are called modal mass, damping ratio, and stiffness, respectively.

If the matrix $\boldsymbol{\Phi}^T\, \boldsymbol{\zeta}\, \boldsymbol{\Phi}$ is not diagonal, a similar approach can be used for the state vector $(\boldsymbol{X}, \dot{\boldsymbol{X}})$ rather then $\boldsymbol{X}$. In this case, the system eigenvalues are complex-valued, and a right and a left complex-valued eigenvector corresponds to each eigenvalue ([127], Section 3.7). ▲

Example 7.6: Consider a special case of the previous example with $\boldsymbol{Y}$ replaced by $-\boldsymbol{m}\,\boldsymbol{1}\, W(t)$, where $\boldsymbol{1}$ denotes an $(n, 1)$ matrix with unit entries and W is a scalar white noise with mean zero and intensity $q_w(t) = q_w = \pi\, g_0$. This case corresponds to a building with n-degrees of freedom subjected to seismic ground acceleration modeled as a white noise with a one-sided spectral density of intensity $g_0 > 0$. The stationary displacement vector $\boldsymbol{X}(t)$ has mean zero and covariance $\boldsymbol{\Phi}\, \boldsymbol{\gamma}_q\, \boldsymbol{\Phi}^T$, where $\boldsymbol{\gamma}_q$ denotes the covariance matrix of the n-dimensional vector $\boldsymbol{Q}(t) = (Q_1(t), \ldots, Q_n(t))$. The entries (i, i) of $\boldsymbol{\Phi}\, \boldsymbol{\gamma}_q\, \boldsymbol{\Phi}^T$ are

$$\gamma_{q,ii} = \frac{\pi\, g_0\, r_i^2}{4\, \zeta_i\, \nu_i^3}, \qquad i = 1, \ldots, n, \tag{7.20}$$

where $r_i = (\boldsymbol{\Phi}\, \boldsymbol{m}\, \boldsymbol{1})_i\, / \tilde{m}_i$ are called modal participation factors. The correlation coefficients between the modal coordinates are

$$\begin{aligned}
\rho_{Q_i,Q_j} &= 8\, \lambda\, \sqrt{\lambda\, \mu}\, (\lambda\, \mu + 1)\, \zeta_j^2 / \chi, \\
\rho_{\dot{Q}_i,\dot{Q}_j} &= \rho_{Q_i,Q_j}\, (\lambda + \mu)/(\lambda\, \mu + 1), \\
\rho_{Q_i,\dot{Q}_j} &= -\rho_{\dot{Q}_i,Q_j}\, \lambda - 4\, \lambda\, \sqrt{\lambda\, \mu}\, (1 - \lambda^2)\, \zeta_j / \chi,
\end{aligned} \tag{7.21}$$

where $\chi = 4\, \zeta_j^2\, \lambda\, (\lambda\, \mu + 1)\, (\lambda + \mu) + (1 - \lambda^2)^2$, $\lambda = \nu_i/\nu_j$, and $\mu = \zeta_i/\zeta_j$. The above results can be used to calculate the second moment properties of the response of a linear system subjected to seismic ground acceleration and develop approximations for response maxima ([175], Example 5.10, p. 190). ◇

Proof: An arbitrary pair of modal coordinates, Q_i and Q_j, $i \neq j$, satisfy the differential equations

$$\frac{d}{dt}\begin{bmatrix} Q_i(t) \\ \dot{Q}_i(t) \\ Q_j(t) \\ \dot{Q}_j(t) \end{bmatrix} = \begin{bmatrix} 0 & 1 & 0 & 0 \\ -\nu_i^2 & -2\, \zeta_i\, \nu_i^2 & 0 & 0 \\ 0 & 0 & 0 & 1 \\ 0 & 0 & -\nu_j^2 & -2\, \zeta_j\, \nu_j^2 \end{bmatrix} \begin{bmatrix} Q_i(t) \\ \dot{Q}_i(t) \\ Q_j(t) \\ \dot{Q}_j(t) \end{bmatrix} + \begin{bmatrix} 0 \\ -r_i \\ 0 \\ -r_j \end{bmatrix} W(t).$$

The above equation defines the evolution of the state vector $(Q_i, \dot{Q}_i, Q_j, \dot{Q}_j)$. The second moment properties given by Eqs. 7.20 and 7.21 result from Eq. 7.15 applied to this state vector. ■

Example 7.7: Let $\boldsymbol{X}$ be the solution of Eq. 7.5 with constant coefficients, that is, $\boldsymbol{a}(t) = \boldsymbol{a}$ and $\boldsymbol{b}(t) = \boldsymbol{b}$ so that the Green function $\boldsymbol{\theta}$ in Eq. 7.7 depends only on the time lag. Suppose that (1) the system is causal, that is, $\boldsymbol{\theta}(\tau) = 0$ for $\tau < 0$, (2) $\boldsymbol{Y}$ is a weakly stationary process with mean zero, correlation function $\boldsymbol{r}_y$, and spectral density $\boldsymbol{s}_y$, and (3) Eq. 7.5 admits a weakly stationary solution. The spectral density of $\boldsymbol{X}$ is

$$\boldsymbol{s}(\nu) = \boldsymbol{h}(\nu)^* \, \boldsymbol{s}_y(\nu) \, \boldsymbol{h}(\nu)^T, \tag{7.22}$$

where z^* denotes the complex conjugate of $z \in \mathbb{C}$ and

$$\boldsymbol{h}(\nu) = \int_{\mathbb{R}} \boldsymbol{\theta}(u) \, \boldsymbol{b} \, e^{-\sqrt{-1}\,\nu u} \, du \tag{7.23}$$

is the system transfer function. ◇

Proof: We have

$$\boldsymbol{X}(t) = \int_{-\infty}^{\infty} \boldsymbol{\theta}(t-\sigma) \, \boldsymbol{b} \, \boldsymbol{Y}(\sigma) \, d\sigma = \int_{-\infty}^{\infty} \boldsymbol{\theta}(u) \, \boldsymbol{b} \, \boldsymbol{Y}(t-u) \, du,$$

since the system is causal and time invariant. The starting time $-\infty$ is needed for stationary response ([175], Section 5.2.2). The second equality in the above equation follows from the change of variable $u = t - \sigma$. Because $\boldsymbol{Y}$ has mean zero, the expectation of $\boldsymbol{X}$ is zero at all times. If the mean of $\boldsymbol{Y}$ is a constant $\boldsymbol{\mu}_y \neq \boldsymbol{0}$, then (Section 3.9.3.3)

$$E[\boldsymbol{X}(t)] = E\left[\int_{-\infty}^{\infty} \boldsymbol{\theta}(u) \, \boldsymbol{b} \, \boldsymbol{Y}(t-u) \, du\right] = \int_{-\infty}^{\infty} \boldsymbol{\theta}(u) \, \boldsymbol{b} \, \boldsymbol{\mu}_y \, du$$

is a constant.

The correlation function of $\boldsymbol{X}$,

$$\begin{aligned} E[\boldsymbol{X}(t) \, \boldsymbol{X}(s)^T] &= \int_{-\infty}^{\infty} \int_{-\infty}^{\infty} \boldsymbol{\theta}(u) \, \boldsymbol{b} \, E[\boldsymbol{Y}(t-u) \, \boldsymbol{Y}(s-v)^T] \, \boldsymbol{b}^T \, \boldsymbol{\theta}(v)^T \, du \, dv \\ &= \int_{-\infty}^{\infty} \int_{-\infty}^{\infty} \boldsymbol{\theta}(u) \, \boldsymbol{b} \, \boldsymbol{r}_y(t-s-u+v) \, \boldsymbol{b}^T \, \boldsymbol{\theta}(v)^T \, du \, dv = \boldsymbol{r}(t-s), \end{aligned}$$

depends on the time lag $\tau = t - s$ rather than the times t and s, so that this process is weakly stationary. The Fourier transform of the above equation is (Section 3.7.2.4)

$$\begin{aligned} \boldsymbol{s}(\nu) &= \frac{1}{2\pi} \int_{\mathbb{R}} e^{-\sqrt{-1}\,\tau\,\nu} \, \boldsymbol{r}(\tau) \, d\tau = \frac{1}{2\pi} \int_{\mathbb{R}} e^{(-\sqrt{-1}\,(\tau-u+v)+(u-v))\,\nu} \, \boldsymbol{r}(\tau) \, d\tau \\ &= \int_{-\infty}^{\infty} \int_{-\infty}^{\infty} e^{-\sqrt{-1}\,(u-v)\,\nu} \, \boldsymbol{\theta}(u) \, \boldsymbol{b} \, \boldsymbol{s}_y(\nu) \, \boldsymbol{b}^T \, \boldsymbol{\theta}(v)^T \, du \, dv \\ &= \boldsymbol{h}(\nu) \, \boldsymbol{s}_y(\nu) \, \boldsymbol{h}(\nu)^T, \end{aligned}$$

where $\boldsymbol{s}_y(\nu) = (1/(2\pi)) \int_{\mathbb{R}} e^{-\sqrt{-1}\,(\tau-u+v)\,\nu} \, \boldsymbol{r}_y(\tau-u+v) \, d\tau$ for each (u, v). ■

Example 7.8: Let $\boldsymbol{X}(t)$, $t \geq 0$, be the solution of Eq. 7.5 with $\boldsymbol{a}(t) = \boldsymbol{a}$, $\boldsymbol{b}(t) = \boldsymbol{b}$, $\boldsymbol{X}(0) = \boldsymbol{x}$, and a Gaussian white noise input $\boldsymbol{Y} = \boldsymbol{W}$ with second moment properties in Eq. 7.13. Suppose that Eq. 7.5 has a stationary solution $\boldsymbol{X}_s$ and let $\boldsymbol{\mu}_s$ and $\boldsymbol{c}_s(\cdot)$ denote its mean and covariance functions. The process

$$\hat{\boldsymbol{X}}(t) = \boldsymbol{X}_s(t) \mid (\boldsymbol{X}_s(0) = \boldsymbol{x}), \quad t \geq 0, \tag{7.24}$$

is a version of $\boldsymbol{X}(t)$, $t \geq 0$, defined by Eq. 7.5 with $\boldsymbol{X}(0) = \boldsymbol{x}$. The mean and covariance functions of $\hat{\boldsymbol{X}}$ are

$$\begin{aligned} \hat{\boldsymbol{\mu}}(t) &= \boldsymbol{\mu}_s + \boldsymbol{\theta}(t)\,(\boldsymbol{x} - \boldsymbol{\mu}_s), \\ \hat{\boldsymbol{c}}(t,s) &= \boldsymbol{\theta}(t-s)\,\boldsymbol{c}_s(0) - \boldsymbol{\theta}(t)\,\boldsymbol{c}_s(0)\,\boldsymbol{\theta}(s)^T, \quad t \geq s, \end{aligned} \tag{7.25}$$

where $\boldsymbol{\theta}$ is the Green function defined by Eq. 7.7. ◇

Proof: That $\hat{\boldsymbol{X}}$ is Gaussian with the first two moments in Eq. 7.25 follows from properties of Gaussian vectors (Section 2.11.5). Because $\hat{\boldsymbol{X}}$ and $\boldsymbol{X}$ are Gaussian processes, they are versions if they have the same second moment properties. The stationary solution $\boldsymbol{X}_s$ satisfies the differential equation $\dot{\boldsymbol{X}}_s(t) = \boldsymbol{a}\,\boldsymbol{X}_s(t) + \boldsymbol{b}\,\boldsymbol{Y}(t)$ so that $\dot{\hat{\boldsymbol{X}}}(t) = \boldsymbol{a}\,\hat{\boldsymbol{X}}(t) + \boldsymbol{b}\,\hat{\boldsymbol{Y}}(t)$, $t \geq 0$, results from Eq. 7.5 by conditioning with respect to $\boldsymbol{X}_s(0) = \boldsymbol{x}$, where $\hat{\boldsymbol{Y}}(t) = \boldsymbol{Y}(t) \mid \boldsymbol{X}_s(0)$ and $\hat{\boldsymbol{X}}(0) = \boldsymbol{x}$. Because $\boldsymbol{Y}$ is a white noise process, $\hat{\boldsymbol{Y}}(t) = \hat{\boldsymbol{W}}(t) = \boldsymbol{W}(t)$ so that $\hat{\boldsymbol{X}}$ and $\boldsymbol{X}$ satisfy the same differential equation and initial condition.

The extension to a colored stationary Gaussian input $\boldsymbol{Y}$ is discussed in [76] for the case in which $\boldsymbol{Y}$ is defined by some of the coordinates of the state of a linear filter driven by Gaussian white noise. ■

Example 7.9: Let X be the solution of $\dot{X}(t) = -\alpha\,X(t) + Y(t)$, $t \geq 0$, where $X(0) = x$, Y is the stationary solution of $\dot{Y}(t) = -\beta\,Y(t) + W(t)$, W is a stationary Gaussian white noise with mean μ_w and intensity q_w, $\alpha \neq \beta$, and $\alpha, \beta > 0$. The mean and variance functions of $X(t)$, $t \geq 0$, are

$$\begin{aligned} \mu(t) &= x\,e^{-\alpha t} + \mu_1\left(1 - e^{-\alpha t}\right), \\ c(t,t) &= \frac{q_w}{2\,\beta\,(\alpha+\beta)}\left[\frac{1 - e^{-2\alpha t}}{\alpha} - 2\,\frac{e^{-(\alpha+\beta)t} - e^{-2\alpha t}}{\alpha - \beta}\right] \end{aligned} \tag{7.26}$$

where $\mu_1 = \mu_w/(\alpha\,\beta)$. ◇

Proof: The augmented state vector $(X, Y) \in \mathbb{R}^2$ satisfies the differential equation

$$\frac{d}{dt}\begin{bmatrix} X(t) \\ Y(t) \end{bmatrix} = \begin{bmatrix} -\alpha & 1 \\ 0 & -\beta \end{bmatrix}\begin{bmatrix} X(t) \\ Y(t) \end{bmatrix} + \begin{bmatrix} 0 \\ 1 \end{bmatrix} W(t).$$

This system has a stationary solution (X_s, Y_s) since it has constant coefficients satisfying the conditions in Eq. 7.15. The first two moments of (X_s, Y_s) are $\mu_1 = E[X_s(t)] = \mu_w/(\alpha\,\beta)$, $\mu_2 = E[Y_s(t)] = \mu_w/\beta$, $\gamma_{1,1} = E[\tilde{X}_s(t)^2] = q_w/[2\,\alpha\,\beta\,(\alpha+\beta)]$, $\gamma_{1,2} =$

$\gamma_{2,1} = E[\tilde{X}_s(t)\,\tilde{Y}_s(t)] = q_w/[2\,\beta\,(\alpha+\beta)]$, and $\gamma_{2,2} = E[\tilde{Y}_s(t)^2] = q_w/2\,\beta$, where $\tilde{Z} = Z - E[Z]$ denotes the centered variable Z. The Green function is

$$\boldsymbol{\theta}(\tau) = \begin{bmatrix} e^{-\alpha\,\tau} & \frac{1}{\alpha-\beta}\left(e^{-\beta\,\tau} - e^{-\alpha\,\tau}\right) \\ 0 & e^{-\beta\,\tau} \end{bmatrix}$$

so that the mean and covariance functions of $(\hat{X}(t), \hat{Y}(t)) = (X_s(t), Y_s(t)) \mid (X_s(0) = x, Y_s(0) = y)$ are (Eq. 7.25)

$$\begin{aligned}
\hat{\mu}_1(t) &= \mu_1 + (x-\mu_1)\,e^{-\alpha\,t} + \frac{y-\mu_2}{\alpha-\beta}\left(e^{-\beta\,t} - e^{-\alpha\,t}\right),\\
\hat{\mu}_2(t) &= \mu_2 + (y-\mu_2)\,e^{-\beta\,t},\\
\hat{c}_{1,1}(t,t) &= \frac{q_w}{2\,\beta}\left\{\frac{1}{\alpha\,(\alpha+\beta)} - e^{-\alpha\,t}\left[\frac{1}{\alpha\,(\alpha+\beta)}\,e^{-\alpha\,t} + \frac{1}{\alpha+\beta}\,\frac{e^{-\beta\,t} - e^{-\alpha\,t}}{\alpha-\beta}\right]\right.\\
&\quad \left. - \frac{e^{-\beta\,t} - e^{-\alpha\,t}}{\alpha-\beta}\left[\frac{1}{\alpha+\beta}\,e^{-\alpha\,t} + \frac{e^{-\beta\,t} - e^{-\alpha\,t}}{\alpha-\beta}\right]\right\},\\
\hat{c}_{1,2}(t,t) &= \hat{c}_{2,1}(t,t)\\
&= \frac{q_w}{2\,\beta}\left\{\frac{1}{\alpha+\beta} - \frac{1}{\alpha+\beta}\,e^{-(\alpha+\beta)\,t} - \frac{1}{\alpha-\beta}\,e^{-\beta\,t}\left(e^{-\beta\,t} - e^{-\alpha\,t}\right)\right\},\\
\hat{c}_{2,2}(t,t) &= \frac{q_w}{2\,\beta}\left(1 - e^{-2\,\beta\,t}\right),
\end{aligned}$$

where the subscripts 1 and 2 refer to X and Y, respectively.

If $Y_s(0)$ is a random variable with mean $E[Y_s(0)] = \mu_2$, we have $\mu(t) = E[X_s(t) \mid X_s(0) = x]$ since the first two terms in the expression for $\hat{\mu}_1(t) = E[X_s(t) \mid X_s(0) = x, Y_s(0) = y]$ do not depend on $Y_s(0)$ and the expectation with respect to $Y_s(0)$ of the last term in the expression for $\hat{\mu}_1(t)$ is zero. Similar considerations can be used to calculate the variance of $X_s(t) \mid X_s(0) = x$ [76]. ■

Example 7.10: Let $t \in (0, l)$ in Eq. 7.5 be a space parameter and let $\boldsymbol{Y} = \boldsymbol{W}$ be a white noise with the second moment properties in Eq. 7.13. The solution of this equation satisfies the homogeneous boundary conditions $\sum_{i=1}^{n} \alpha_{pi}\, X_i(0) = 0$ and $\sum_{i=1}^{n} \beta_{qi}\, X_i(l) = 0$, where $p = 1, \ldots, d_1$, $q = 1, \ldots, d - d_1$, $0 \le d_1 \le d$, and $(\alpha_{pi}, \beta_{qi})$ are some constants. Generally, one or more coordinates of $\boldsymbol{X}(0)$ are not known so that Eqs. 7.10 and 7.12 cannot be solved. This is the essential difference between the initial value problem considered previously in this chapter and the boundary value problem in this example.

The mean and covariance function of $\boldsymbol{X}$ are [83]

$$\begin{aligned}
\dot{\boldsymbol{\mu}}(t) &= \boldsymbol{a}(t)\,\boldsymbol{\mu}(t) + \boldsymbol{b}(t)\,\boldsymbol{\mu}_w(t), \\
\dot{\boldsymbol{c}}(t,t) &= \boldsymbol{a}(t)\,\boldsymbol{c}(t,t) + \boldsymbol{c}(t,t)\,\boldsymbol{a}(t)^T + \boldsymbol{b}(t)\,\boldsymbol{q}_w(t)\,\boldsymbol{b}(t)^T \\
&\quad + \boldsymbol{\theta}(t,0)\,\boldsymbol{\gamma}_{w,0}(t)\,\boldsymbol{b}(t)^T + \boldsymbol{b}(t)\,\boldsymbol{\gamma}_{w,0}(t)^T\,\boldsymbol{\theta}(t,0)^T, \\
\boldsymbol{c}(t,s) &= \boldsymbol{\theta}(t,s)\,\boldsymbol{c}(s,s) \\
&\quad + \left[\int_s^t \boldsymbol{\theta}(t,u)\,\boldsymbol{b}(u)\,\boldsymbol{\gamma}_{w,0}(u)^T\,du\right]\boldsymbol{\theta}(s,0)^T, \quad 0 < s < t < l, \qquad (7.27)
\end{aligned}$$

where $\boldsymbol{\gamma}_{w,0}(t) = E[\tilde{\boldsymbol{X}}(0)\,\tilde{\boldsymbol{W}}(t)^T]$ and $\boldsymbol{\theta}$ denotes the Green function in Eq. 7.7. Initial conditions for these equations are given by the second moment properties of $\boldsymbol{X}(0) = \boldsymbol{\zeta}\,\boldsymbol{Z}$, where $\boldsymbol{\zeta}$ is a (d, d) constant matrix depending on $\boldsymbol{\theta}(l, 0)$ and the coefficients $(\alpha_{pi}, \beta_{qi})$ in the boundary conditions and $\boldsymbol{Z} = \int_0^l \boldsymbol{\theta}(l,s)\,\boldsymbol{b}(s)\,\boldsymbol{W}(s)\,ds$ is an $\mathbb{R}^d$-valued random variable with mean and covariance matrices

$$\boldsymbol{\mu}_z = \int_0^l \boldsymbol{\theta}(l,s)\,\boldsymbol{b}(s)\,\boldsymbol{\mu}_w(s)\,ds \quad \text{and}$$

$$\boldsymbol{\gamma}_z = \int_0^l \boldsymbol{\theta}(l,s)\,\boldsymbol{b}(s)\,\boldsymbol{q}_w(s)\,\boldsymbol{b}(s)^T\,\boldsymbol{\theta}(l,s)^T\,ds.$$

The above covariance equations differ in two ways from the covariance equation used in linear random vibration. The equations for $\boldsymbol{c}(t,t)$ and $\boldsymbol{c}(t,s)$ include new terms and $\boldsymbol{c}(0,0)$ is not generally available but can be determined from the relationship between $\boldsymbol{X}(0)$ and $\boldsymbol{X}(l)$. ◇

Proof: To derive Eq. 7.27, we use the formal definition of white noise in Eq. 7.13. The solution of Eq. 7.8 at $t = l$,

$$\boldsymbol{X}(l) = \boldsymbol{\theta}(l,0)\,\boldsymbol{X}(0) + \int_0^l \boldsymbol{\theta}(l,s)\,\boldsymbol{b}(s)\,\boldsymbol{W}(s)\,ds = \boldsymbol{\theta}(l,0)\,\boldsymbol{X}(0) + \boldsymbol{Z},$$

and the boundary conditions give $\boldsymbol{X}(0) = \boldsymbol{\zeta}\,\boldsymbol{Z}$. To see that $\boldsymbol{X}(0) = \boldsymbol{\zeta}\,\boldsymbol{Z}$ holds, consider the linear system of equations defined by the above equation and the boundary conditions for the coordinates of $\boldsymbol{X}(0)$ and $\boldsymbol{X}(l)$, which are assumed to be homogeneous.

The equation for the mean of $\boldsymbol{X}$ results by averaging and differentiation. The centered state vector is

$$\tilde{\boldsymbol{X}}(t) = \boldsymbol{X}(t) - \boldsymbol{\mu}(t) = \boldsymbol{\theta}(t,0)\,\tilde{\boldsymbol{X}}(0) + \int_0^t \boldsymbol{\theta}(t,s)\,\boldsymbol{b}(s)\,\tilde{\boldsymbol{W}}(s)\,ds$$

so that

$$\begin{aligned}
\boldsymbol{c}(t,t) &= E[\tilde{\boldsymbol{X}}(t)\,\tilde{\boldsymbol{X}}(t)^T] \\
&= \boldsymbol{\theta}(t,0)\,\boldsymbol{c}(0,0)\,\boldsymbol{\theta}(t,0)^T + \boldsymbol{\theta}(t,0)\int_0^t \boldsymbol{\gamma}_{w,0}(s)\,\boldsymbol{b}(s)^T\,\boldsymbol{\theta}(t,s)^T\,ds \\
&\quad + \left[\int_0^t \boldsymbol{\theta}(t,s)\,\boldsymbol{b}(s)\,\boldsymbol{\gamma}_{w,0}(s)^T\,ds\right]\boldsymbol{\theta}(t,0)^T \\
&\quad + \int_0^t ds \int_0^t du\,\boldsymbol{\theta}(t,s)\,\boldsymbol{b}(s)\,E[\tilde{\boldsymbol{W}}(s)\,\tilde{\boldsymbol{W}}(u)^T]\,\boldsymbol{b}(u)^T\,\boldsymbol{\theta}(t,u)^T,
\end{aligned}$$

where $\boldsymbol{\gamma}_{w,0}(t) = E[\tilde{\boldsymbol{X}}(0)\, \tilde{\boldsymbol{W}}(t)^T]$. The above double integral degenerates into

$$\int_0^t \boldsymbol{\theta}(t,s)\, \boldsymbol{b}(s)\, \boldsymbol{q}_w(s)\, \boldsymbol{b}(s)^T\, \boldsymbol{\theta}(t,s)^T\, ds$$

by the definition of the driving noise. The derivative of the covariance function of $\boldsymbol{X}(t)$ is

$$\begin{aligned}
\dot{\boldsymbol{c}}(t,t) &= \dot{\boldsymbol{\theta}}(t,0)\, \boldsymbol{c}(0,0)\, \boldsymbol{\theta}(t,0)^T + \boldsymbol{\theta}(t,0)\, \boldsymbol{c}(0,0)\, \dot{\boldsymbol{\theta}}(t,0)^T \\
&+ \dot{\boldsymbol{\theta}}(t,0) \int_0^t \boldsymbol{\gamma}_{w,0}(s)\, \boldsymbol{b}(s)^T\, \boldsymbol{\theta}(t,s)^T\, ds + \boldsymbol{\theta}(t,0) \Big[\boldsymbol{\gamma}_{w,0}(t)\, \boldsymbol{b}(t)^T\, \boldsymbol{\theta}(t,t)^T \\
&+ \int_0^t \boldsymbol{\gamma}_{w,0}(s)\, \boldsymbol{b}(s)^T\, \dot{\boldsymbol{\theta}}(t,s)^T\, ds\Big] + \left[\int_0^t \boldsymbol{\theta}(t,s)\, \boldsymbol{b}(s)\, \boldsymbol{\gamma}_{w,0}(s)^T\, ds\right] \dot{\boldsymbol{\theta}}(t,0)^T \\
&+ \left[\boldsymbol{\theta}(t,t)\, \boldsymbol{b}(t)\, \boldsymbol{\gamma}_{w,0}(t)^T + \int_0^t \dot{\boldsymbol{\theta}}(t,s)\, \boldsymbol{b}(s)\, \boldsymbol{\gamma}_{w,0}(s)^T\, ds\right] \boldsymbol{\theta}(t,0)^T \\
&+ \int_0^t \dot{\boldsymbol{\theta}}(t,s)\, \boldsymbol{b}(s)\, \boldsymbol{q}_w(s)\, \boldsymbol{b}(s)^T\, \boldsymbol{\theta}(t,s)^T\, ds + \int_0^t \boldsymbol{\theta}(t,s)\, \boldsymbol{b}(s)\, \boldsymbol{q}_w(s)\, \boldsymbol{b}(s)^T\, \dot{\boldsymbol{\theta}}(t,s)^T\, ds \\
&+ \boldsymbol{\theta}(t,t)\, \boldsymbol{b}(t)\, \boldsymbol{q}_w(t)\, \boldsymbol{b}(t)^T\, \boldsymbol{\theta}(t,t)^T,
\end{aligned}$$

where $\dot{\boldsymbol{\theta}}(t,s)$ denotes $\partial\boldsymbol{\theta}(t,s)/\partial t$. The differentiation rule $\frac{d}{dt}\int_0^t g(t,s)\,ds = g(t,t) + \int_0^t \frac{\partial}{\partial t} g(t,s)\,ds$ was used in the above calculations. The second formula in Eq. 7.27 follows from the above equation, the definition of the Green function $\boldsymbol{\theta}$, and the expression of $\boldsymbol{c}(t,t)$.

The last formula in Eq. 7.27 follows from

$$\begin{aligned}
\boldsymbol{c}(t,s) &= E\left[\tilde{\boldsymbol{X}}(t)\, \tilde{\boldsymbol{X}}(s)^T\right] = E\left[\boldsymbol{\theta}(t,s)\, \tilde{\boldsymbol{X}}(s) + \int_s^t \boldsymbol{\theta}(t,u)\, \boldsymbol{b}(u)\, \tilde{\boldsymbol{W}}(u)\, du\right] \tilde{\boldsymbol{X}}(s)^T \\
&= \boldsymbol{\theta}(t,s)\, \boldsymbol{c}(s,s) + \int_s^t \boldsymbol{\theta}(t,u)\, \boldsymbol{b}(u)\, E\left[\tilde{\boldsymbol{W}}(u)\, \tilde{\boldsymbol{X}}(s)^T\right] du, \quad t \geq s,
\end{aligned}$$

since

$$\begin{aligned}
E\left[\tilde{\boldsymbol{W}}(u)\, \tilde{\boldsymbol{X}}(s)^T\right] &= E\left[\tilde{\boldsymbol{W}}(u)\, \tilde{\boldsymbol{X}}(0)^T\right] \boldsymbol{\theta}(s,0)^T \\
&+ \int_0^s E\left[\tilde{\boldsymbol{W}}(u)\, \tilde{\boldsymbol{W}}(v)^T\right] \boldsymbol{b}(v)^T\, \boldsymbol{\theta}(s,v)^T\, du,
\end{aligned}$$

and $E\left[\tilde{\boldsymbol{W}}(u)\, \tilde{\boldsymbol{W}}(v)^T\right] = 0$ for $u \neq v$.

The mean and covariance equations in Eq. 7.27 can also be solved as initial value problems for the second moment properties, $\boldsymbol{\mu}(0) = E[\boldsymbol{X}(0)] = \boldsymbol{\zeta}\, \boldsymbol{\mu}_z$ and $\boldsymbol{c}(0,0) = E\left[(\boldsymbol{X}(0) - \boldsymbol{\mu}(0))(\boldsymbol{X}(0) - \boldsymbol{\mu}(0))^T\right] = \boldsymbol{\zeta}\, \boldsymbol{\gamma}_z\, \boldsymbol{\zeta}^T$. ■

Example 7.11: Let V, Φ, M, and Q be the displacement, the rotation, the bending moment, and the shear force in a beam of length l fixed at the left end and simply supported at the right end. The state vector $\boldsymbol{X} = (V, \Phi, M, Q)$ satisfies the differential equations $\dot{V}(t) = \Phi(t)$, $\dot{\Phi}(t) = -\alpha\, M(t)$, $\dot{M}(t) = Q(t)$, and $\dot{Q}(t) = -W(t)$, where $1/\alpha$ is the beam stiffness and W is a distributed white noise load with mean μ_w and intensity q_w.

The mean functions of the coordinates of $\boldsymbol{X}$ are (Eq. 7.27)

$$\mu_1(t) = -\alpha\,\mu_w\,l^4\,(-\frac{1}{24}\xi^4 + \frac{5}{48}\xi^3 - \frac{1}{16}\xi^2),$$
$$\mu_2(t) = -\alpha\,\mu_w\,l^3\,(-\frac{1}{6}\xi^3 + \frac{5}{16}\xi^2 - \frac{1}{8}),$$
$$\mu_3(t) = \mu_w\,l^2\,(-\frac{1}{2}\xi^2 + \frac{5}{8}\xi - \frac{1}{8}),$$
$$\mu_4(t) = \mu_w\,l\,(-\xi + \frac{5}{8}),$$

where $\xi = t/l \in (0, 1)$. Figure 7.2 shows the functions $\tilde{\mu}_1 = \mu_1/(\alpha\,\mu_w\,l^4)$ and $\tilde{\mu}_2 = \mu_2/(\alpha\,\mu_w\,l^3)$, $\tilde{\mu}_3 = \mu_3/(\mu_w\,l^2)$, and $\tilde{\mu}_4 = \mu_4/(\mu_w\,l)$.

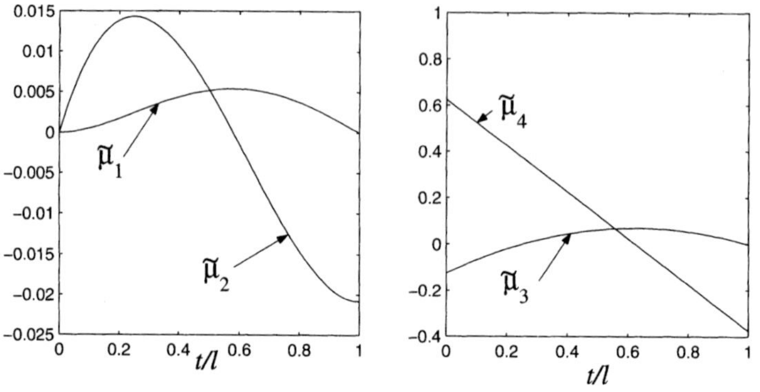

Figure 7.2. Mean function of $\boldsymbol{X}$

The differential equations for the variance functions of $\boldsymbol{X}$ are (Eq. 7.27)

$$\dot{c}_{1,1}(t,t) = 2\,c_{1,2}(t,t),$$
$$\dot{c}_{1,2}(t,t) = c_{2,2}(t,t) - \alpha\,c_{1,3}(t,t),$$
$$\dot{c}_{1,3}(t,t) = c_{2,3}(t,t) + c_{1,4}(t,t),$$
$$\dot{c}_{1,4}(t,t) = c_{2,4}(t,t) + \frac{\alpha}{2}\,t^2\,E\left[\tilde{M}(0)\,\tilde{W}(t)\right] + \frac{\alpha}{6}\,t^3\,E\left[\tilde{Q}(0)\,\tilde{W}(t)\right],$$
$$\dot{c}_{2,2}(t,t) = -2\,\alpha\,c_{2,3}(t,t),$$
$$\dot{c}_{2,3}(t,t) = -\alpha\,c_{3,3}(t,t) + c_{2,4}(t,t),$$
$$\dot{c}_{2,4}(t,t) = -\alpha\,c_{3,4}(t,t) + \alpha\,t\,E\left[\tilde{M}(0)\,\tilde{W}(t)\right] + \frac{\alpha}{2}\,t^2\,E\left[\tilde{Q}(0)\,\tilde{W}(t)\right],$$
$$\dot{c}_{3,3}(t,t) = 2\,c_{3,4}(t,t),$$
$$\dot{c}_{3,4}(t,t) = c_{4,4}(t,t) - E\left[\tilde{M}(0)\,\tilde{W}(t)\right] - t\,E\left[\tilde{Q}(0)\,\tilde{W}(t)\right],$$
$$\dot{c}_{4,4}(t,t) = q_w - 2\,E\left[\tilde{Q}(0)\,\tilde{W}(t)\right].$$

Figure 7.3 shows the scaled variance functions

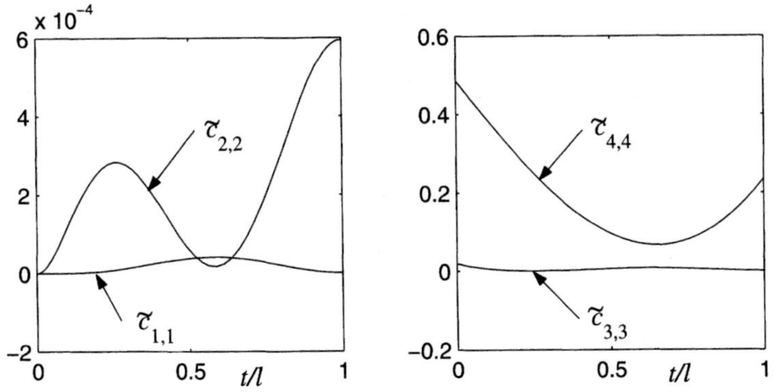

Figure 7.3. Variance function of $\boldsymbol{X}$

$$\tilde{c}_{1,1}(t,t) = c_{1,1}(t,t)/(\alpha^2\,\gamma_w\,l^7), \quad \tilde{c}_{2,2}(t) = c_{2,2}(t)/(\alpha^2\,\gamma_w\,l^5)\},$$
$$\tilde{c}_{3,3}(t) = c_{3,3}(t)/(\gamma_w\,l^3), \quad \text{and } \tilde{c}_{4,4}(t) = c_{4,4}(t)/(\gamma_w\,l)\}.$$

The variance functions are consistent with the boundary conditions. For example, $c_{1,1}(0,0) = c_{1,1}(l,l) = 0$ since the displacement is zero at the beam ends. ◇

Proof: The state vector $\boldsymbol{X}$ satisfies Eq. 7.5 with $\boldsymbol{Y} = \boldsymbol{W}$ in Eq. 7.13, and the boundary conditions $V(0) = 0$, $\Phi(0) = 0$, $V(l) = 0$, and $M(l) = 0$. The non-zero entries of matrices $\boldsymbol{a}(t)$ and $\boldsymbol{b}(t)$ in Eq. 7.5 are $a_{12} = 1$, $a_{23} = -\alpha$, $a_{14} = 1$, and $b_4 = -1$. The non-zero entries of the Green function $\boldsymbol{\theta}(t,0)$ are $\theta_{11} = 1$, $\theta_{12} = t$, $\theta_{13} = -\alpha\,t^2/2$, $\theta_{14} = -\alpha\,t^3/6$, $\theta_{22} = 1$, $\theta_{23} = -\alpha\,t$, $\theta_{24} = -\alpha\,t^2/2$, $\theta_{33} = 1$, $\theta_{34} = t$, and $\theta_{44} = 1$ so that the coordinates of $\boldsymbol{X}(0)$ and $\boldsymbol{X}(l)$ are related by [83]

$$V(l) = V(0) + l\,\Phi(0) - \frac{\alpha\,l^2}{2}\,M(0) - \frac{\alpha\,l^3}{6}\,Q(0) + \frac{\alpha}{6}\int_0^l (l-s)^3\,W(s)\,ds,$$

$$\Phi(l) = \Phi(0) - \alpha\,l\,M(0) - \frac{\alpha l^2}{2}\,Q(0) + \frac{\alpha}{2}\int_0^l (l-s)^2\,W(s)\,ds,$$

$$M(l) = M(0) + l\,Q(0) - \int_0^l (l-s)\,W(s)\,ds,$$

$$Q(l) = Q(0) - \int_0^l W(s)\,ds.$$

The last two equalities constitute global equilibrium conditions for the beam. The first and third equalities together with the boundary conditions $V(0) = 0$, $\Phi(0) = 0$, $V(l) = 0$, and $M(l) = 0$ can be used to find the unknown entries in $\boldsymbol{X}(0)$. The bending moment and the shear force at the left end of the beam are $M(0) = \frac{1}{2}(3\,Y_1 - Y_2)$ and $Q(0) = \frac{3}{2l}(Y_2 - Y_1)$, where $Y_1 = \frac{1}{3\,l^2}\int_0^l (l-s)^3\,W(s)\,ds$ and $Y_2 = \int_0^l (l-s)\,W(s)\,ds$ so that $c_{3,3}(0,0) = 2\,q_w\,l^3/105$, $c_{3,4}(0,0) = -3\,q_w\,l^2/35$, and $c_{4,4}(0,0) = 17\,q_w\,l/35$. The covariances

$c_{i,j}(0,0)$ are zero for $i = 1, 2$ and $j = 1, \ldots, 4$ because $V(0) = 0$ and $\Phi(0) = 0$. The expressions of $M(0)$ and $Q(0)$ can also be used to calculate the expectations

$$E\left[\tilde{M}(0)\,\tilde{W}(t)\right] = \frac{q_w\,l}{2}\,(1-\xi)\left[(1-\xi)^2 - 1\right],$$
$$E\left[\tilde{Q}(0)\,\tilde{W}(t)\right] = \frac{3\,q_w}{2}\,(1-\xi)\left[1 - \frac{1}{3}(1-\xi)^2\right],$$

giving the non-zero entries in $\gamma_{w,0}(t)$. ■

Example 7.12: Let X be the solution of

$$\ddot{X}(t) + \beta\,\dot{X}(t) + \nu_0^2\,X(t) - \int_{-\infty}^{\infty} k(t-s)\,X(s)\,ds = Y(t),$$

where $\beta, \nu_0 > 0$ are some constants, the kernel k is such that $k(u) = 0$ for $u < 0$ and $\int_{-\infty}^{\infty} |k(u)|\,du < \infty$, and Y denotes a weakly stationary process with mean zero and spectral density s_y. If the above equation admits a weakly stationary solution X_s, then this solution has mean zero and spectral density

$$s_x(\nu) = \frac{s_y(\nu)}{\left|-\nu^2 + \sqrt{-1}\,\beta\,\nu + \nu_0^2 - h(\nu)\right|^2},$$

where $h(\nu) = \int_{-\infty}^{\infty} k(u)\,e^{-\sqrt{-1}\,\nu\,u}\,du$. The process X describes the torsional motion of an airplane wing in a turbulent wind and Y models the effect of the buffeting forces on the wing. ◇

Proof: We first note that the moment equations developed in this section cannot be applied to find the second moment properties of X defined by the above equation.

Let $X_s(t) = \int_{-\infty}^{\infty} e^{\sqrt{-1}\,\nu\,t}\,dZ(\nu)$ be the spectral representation of X_s, where Z is a process with $E[dZ(\nu)] = 0$ and $E[dZ(\nu)\,dZ(\nu')^*] = s_x(\nu)\,\delta(\nu-\nu')\,d\nu$ (Section 3.9.4.1). The second moment of the increments of Z depend on the spectral density s_x of X, which is not known. The representation of X_s gives

$$\int_{-\infty}^{\infty} k(t-\sigma)\,X_s(\sigma)\,d\sigma = \int_{-\infty}^{\infty}\left[\int_{-\infty}^{\infty} k(u)\,e^{-\sqrt{-1}\,\nu\,u}\,du\right] e^{\sqrt{-1}\,\nu\,t}\,dZ(\nu)$$
$$= \int_{-\infty}^{\infty} h(\nu)\,e^{\sqrt{-1}\,\nu\,t}\,dZ(\nu)$$

by the change of variables $u = t - \sigma$ and the definition of h. Also, $h(t)$ is bounded for each $t \geq 0$ since $|h(\nu)| \leq \int_{-\infty}^{\infty} |k(u)|\,du$, which is finite. The defining equation of X and the above equality give

$$\int_{-\infty}^{\infty}\left[-\nu^2 + \sqrt{-1}\,\beta\,\nu + \nu_0^2 - h(\nu)\right] e^{\sqrt{-1}\,\nu\,t}\,dZ(\nu) = Y(t).$$

Let $\tilde{X}$ denote the left side of the above equation. Then, $\tilde{X}$ and Y must have the same second moment properties. The process $\tilde{X}$ has mean zero consistently with Y. The second moments of $\tilde{X}$ and Y are

$$E[\tilde{X}(t)^2] = \int_{-\infty}^{\infty} \left| -\nu^2 + \sqrt{-1}\,\beta\,\nu + \nu_0^2 - h(\nu) \right|^2 s_x(\nu)\,d\nu,$$
$$E[Y(t)^2] = \int_{-\infty}^{\infty} s_y(\nu)\,d\nu,$$

so that

$$\int_{-\infty}^{\infty} \left[\left| -\nu^2 + \sqrt{-1}\,\beta\,\nu + \nu_0^2 - h(\nu) \right|^2 s_x(\nu) - s_y(\nu) \right] d\nu = 0,$$

which gives the stated expression of s_x. ■

7.2.2 Semimartingale input

Let $\boldsymbol{S}$ be an $\mathbb{R}^{d'}$-valued semimartingale defined on a filtered probability space $(\Omega, \mathcal{F}, (\mathcal{F}_t)_{t\geq 0}, P)$. Denote by $\boldsymbol{X}$ the solution of Eq. 7.5 with $\boldsymbol{Y}$ being the formal derivative of $\boldsymbol{S}$ with respect to time or a polynomial of $\boldsymbol{S}$, that is,

$$d\boldsymbol{X}(t) = \boldsymbol{a}(t)\,\boldsymbol{X}(t-)\,dt + \boldsymbol{b}(t)\,d\boldsymbol{S}(t), \quad t \geq 0, \tag{7.28}$$
$$\dot{\boldsymbol{X}}(t) = \boldsymbol{a}(t)\,\boldsymbol{X}(t-) + \boldsymbol{b}(t)\,\boldsymbol{p}(\boldsymbol{S}(t-), t), \quad t \geq 0, \tag{7.29}$$

where the coordinates of the $\mathbb{R}^{d_p}$-valued process $\boldsymbol{p}$ are polynomials of $\boldsymbol{S}$ with coefficients that may depend on time.

The semimartingale noise $\boldsymbol{S}$ is represented by

$$\boldsymbol{S}(t) = \boldsymbol{S}(0) + \boldsymbol{A}(t) + \boldsymbol{M}(t), \quad t \geq 0, \tag{7.30}$$

where $\boldsymbol{S}(0) \in \mathcal{F}_0$, $\boldsymbol{A}$ is an $\mathcal{F}_t$-adapted process with càdlàg samples of finite variation on each compact of $[0, \infty)$, $\boldsymbol{A}(0) = \boldsymbol{0}$, $\boldsymbol{M}$ is an $\mathcal{F}_t$-local martingale, and $\boldsymbol{M}(0) = \boldsymbol{0}$.

Note: We have seen that a real-valued process S defined on a probability space $(\Omega, \mathcal{F}, P)$ endowed with a right continuous filtration $(\mathcal{F}_t)_{t\geq 0}$ is a semimartingale if and only if it is a classical semimartingale, that is, it has the representation

$$S(t) = S(0) + A(t) + M(t), \quad t \geq 0, \tag{7.31}$$

where $S(0)$, A, and M have the properties in Eq. 7.30 (Section 4.4.1 in this book, [147], Theorem 14, p. 105).

The matrices $\boldsymbol{b}$ in Eqs. 7.28 and 7.29 have the dimensions (d, d') and (d, d_p), respectively. We use the same notation for simplicity. It is assumed that $\boldsymbol{a}$ and $\boldsymbol{b}$ are such that the solutions of Eqs. 7.28 and 7.29 exist and are unique (Sections 4.7.1.1 and 4.7.2). ▲

The martingale component of S in Eq. 7.30 is assumed to admit the representation

$$dM(t) = H(t)\, dB(t) + K(t)\, d\tilde{C}(t), \tag{7.32}$$

where the entries of the (d', d_b) and (d', d_c) matrices H and K are processes in $\mathcal{L}$, the coordinates of B and $\tilde{C}$ are independent Brownian motion and compensated compound Poisson processes, respectively, the jumps of the coordinates of $\tilde{C}$ have finite mean, B and $\tilde{C}$ are independent of each other, H is independent of $(K, \tilde{C})$, and K is independent of (H, B).

Note: Recall that the processes in $\mathcal{L}$ are $\mathcal{F}_t$-adapted and have càglàd paths (Section 4.4.1). The stochastic integrals defined in Section 4.4.3 are for integrands in $\mathcal{L}$ and semimartingale integrators. We also note that the processes defined by Eqs. 7.30 and 7.32 do not include all semimartingales. For example, a general Lévy process (Section 3.14.2) does not admit the representation in these equations.

Since $M(0) = 0$, we have (Eq. 7.32)

$$M(t) = \int_{0+}^{t} H(s)\, dB(s) + \int_{0+}^{t} K(s)\, d\tilde{C}(s).$$

The integrals $\int_{0+}^{t} H_{iu}(s)\, dB_u(s)$ and $\int_{0+}^{t} K_{iv}(s)\, d\tilde{C}_v(s)$, $i = 1, \ldots, d'$, $u = 1, \ldots, d_b$, and $v = 1, \ldots, d_c$, in the above definition of M are local martingales by a preservation property of the stochastic integral (Section 4.4.4). Hence, the coordinates M_i, $i = 1, \ldots, d'$, of M are local martingales as sums of local martingales.

If in addition $\tilde{C} \in L_2$ and the conditions

$$E\left[\int_{0+}^{t} H_{iu}(s)^2\, ds\right] < \infty \quad \text{and}$$
$$E\left[\int_{0+}^{t} K_{iv}(s)^2\, d[\tilde{C}, \tilde{C}](s)\right] < \infty \tag{7.33}$$

are satisfied at each time $t \geq 0$ for all indices $i = 1, \ldots, d'$, $u = 1, \ldots, d_b$, and $v = 1, \ldots, d_c$, then M_i, $i = 1, \ldots, d'$, are square integrable martingales. The second moment of $M_i(t) = \sum_u H_{iu} \cdot B_u(t) + \sum_v K_{iv} \cdot \tilde{C}_v(t)$ is $E[M_i(t)^2] = \sum_u E\left[(H_{iu} \cdot B_u(t))^2\right] + E\left[\left(K_{iv} \cdot \tilde{C}_v(t)\right)^2\right]$ by the properties of H, K, B, and $\tilde{C}$. We have seen in Section 4.5.3 that $E\left[\left(\int_{0+}^{t} H_{iu}(s)\, dB_u(s)\right)^2\right] = E\left[\int_{0+}^{t} H_{iu}(s)^2\, ds\right]$ and $E\left[\left(\int_{0+}^{t} K_{iv}(s)\, d\tilde{C}_v(s)\right)^2\right] = E\left[\int_{0+}^{t} K_{iv}(s)^2\, d[\tilde{C}_v, \tilde{C}_v](s)\right]$. Hence, M_i is a square integrable martingale under the above conditions.

Let $\mathcal{G}_t$, $t \geq 0$, be the filtration generated by B and $\tilde{C}$. Generally, $\mathcal{G}_t$ is smaller than the filtration $\mathcal{F}_t$ considered at the beginning of this section. If the processes H and K are $\mathcal{G}_t$-adapted and have càglàd paths, then M is an $\mathcal{G}_t$-local martingale. If in addition the process A in the representation of S (Eq. 7.30) is also $\mathcal{G}_t$-adapted, then S is an $\mathcal{G}_t$-semimartingale. For example, A can be a deterministic function with continuous samples of finite variation on compacts and the processes H and K can be memoryless transformations of B and $\tilde{C}_-$, respectively. ▲

Example 7.13: Let B be a Brownian motion and $C = \sum_{k=1}^{N(t)} Y_k$ be a compound Poisson process that is independent of B, where N denotes a Poisson process with intensity $\lambda > 0$ and Y_k are iid random variables such that $E[Y_1] = 0$. Then the process M defined by $dM(t) = 3\,(B(t)^2 - t)\,dB(t) + 2\,C(t-)\,dC(t)$ with $B(0) = C(0) = 0$ is a martingale. If $E[Y_1^2] < \infty$, M is a square integrable martingale. ◇

Proof: Let $\mathcal{G}_t = \sigma(B(s), C(s), 0 \le s \le t)$, $t \ge 0$, be the filtration generated by B and C. That M is a $\mathcal{G}_t$-local martingale follows from a preservation property of the stochastic integral since $3\,(B(t)^2 - t), 2\,C(t-)$ are $\mathcal{G}_t$-adapted processes with càglàd sample paths and B, C are martingales (Section 4.4.4). The process M is a martingale since $E\left[\int_{0+}^t 3\,(B(s)^2 - s)\,dB(s)\right] = 0$ and $E\left[\int_{0+}^t 2\,C(s-)\,dC(s)\right] = 0$ for all $t \ge 0$.

The process $\int_{0+}^t 3\,(B(s)^2 - s)\,dB(s)$ is a square integrable martingale since its square has the expectation $9\,E\left[\int_{0+}^t (B(s)^4 + s^2 - 2\,s\,B(s)^2)\,ds\right] = 9\int_{0+}^t (3\,s^2 + s^2 - 2\,s\,s)\,ds = 6\,t^3$. The square of $\int_{0+}^t 2\,C(s-)\,dC(s)$ is $4\sum_{k=2}^{N(t)} \left(\sum_{i=1}^{k-1} Y_i\right)^2 Y_k^2$ so that its expectation is finite provided that $E[Y_1^2] < \infty$ since C has a.s. a finite number of jumps in $(0, t]$. Hence, $\int_{0+}^t 2\,C(s-)\,dC(s)$ is also a square integrable martingale. Since this process is independent of $\int_{0+}^t 3\,(B(s)^2 - s)\,dB(s)$, M is a square integrable martingale. ■

Let

$$\begin{aligned} \boldsymbol{X}_1(t) &= \int_0^t \boldsymbol{\theta}(t,\sigma)\,\boldsymbol{b}(\sigma)\,d\boldsymbol{A}(\sigma) \quad \text{and} \\ \boldsymbol{X}_2(t) &= \int_0^t \boldsymbol{\theta}(t,\sigma)\,\boldsymbol{b}(\sigma)\,d\boldsymbol{M}(\sigma) \end{aligned} \tag{7.34}$$

be two processes corresponding to the components $\boldsymbol{A}$ and $\boldsymbol{M}$ of $\boldsymbol{S}$ (Eq. 7.30).

The solutions of Eqs. 7.28 and 7.29 are, respectively,

$$\boldsymbol{X}(t) = \boldsymbol{\theta}(t,0)\,\boldsymbol{X}(0) + \boldsymbol{X}_1(t) + \boldsymbol{X}_2(t), \quad t \ge 0, \quad \text{and} \tag{7.35}$$

$$\boldsymbol{X}(t) = \boldsymbol{\theta}(t,0)\,\boldsymbol{X}(0) + \int_0^t \boldsymbol{\theta}(t,\sigma)\,\boldsymbol{b}(\sigma)\,\boldsymbol{p}(\boldsymbol{S}(\sigma-),\sigma)\,d\sigma, \quad t \ge 0, \tag{7.36}$$

where $\boldsymbol{\theta}$ is defined by Eq. 7.7.

Our objective is to find the first two moments and/or other probabilistic properties of the $\mathbb{R}^d$-valued processes $\boldsymbol{X}$ in Eqs. 7.35 and 7.36 provided they exist. The analysis is based on the Itô formula for semimartingales applied to functions of $\boldsymbol{X}$ or to functions of an augmented state vector $\boldsymbol{Z}$, which includes $\boldsymbol{X}$. We refer to these two approaches as the **direct** and the **state augmentation** methods, respectively (Table 7.1). The first method is used in the next two sections to find properties of $\boldsymbol{X}$ in Eq. 7.28 for square integrable and general martingale inputs. Section 7.7.2.3 applies the state augmentation method to solve Eqs. 7.28 and 7.29. Numerical examples are used to illustrate the application of these methods.

7.2.2.1 Direct method. Square integrable martingales

Let $\boldsymbol{X}$ be the solution of Eq. 7.28 with $\boldsymbol{S}$ in Eq. 7.30 and $\boldsymbol{M}$ in Eq. 7.32, where $\tilde{C}_q(t) = C_q(t) - \lambda_q\, t\, E[Y_{q,1}]$,

$$C_q(t) = \sum_{k=1}^{N_q(t)} Y_{q,k}, \quad t \geq 0, \tag{7.37}$$

N_q are Poisson processes with intensities $\lambda_q > 0$, and $Y_{q,k}$ are independent copies of $Y_{q,1}$, $q = 1, \ldots, d_c$. It is assumed that $\boldsymbol{C}$ is in L_2, the conditions in Eq. 7.33 are satisfied, and the coefficients $\boldsymbol{a}, \boldsymbol{b}$ in Eq. 7.28 are such that this equation has a unique solution.

We now derive differential equations for the mean function $\boldsymbol{\mu}(t) = E[\boldsymbol{X}(t)]$ and the correlation function $\boldsymbol{r}(t,s) = E[\boldsymbol{X}(t)\,\boldsymbol{X}(s)^T]$ of $\boldsymbol{X}$.

If $\boldsymbol{A}$ in Eq. 7.30 is a deterministic function $\boldsymbol{\alpha}$ with continuous first order derivative, then

$$\begin{aligned}
\dot{\boldsymbol{\mu}}(t) &= \boldsymbol{a}(t)\,\boldsymbol{\mu}(t) + \boldsymbol{b}(t)\,\dot{\boldsymbol{\alpha}}(t), \\
\dot{\boldsymbol{r}}(t,t) &= \boldsymbol{a}(t)\,\boldsymbol{r}(t,t) + \boldsymbol{r}(t,t)\,\boldsymbol{a}(t)^T \\
&\quad + \boldsymbol{\mu}(t)\,\dot{\boldsymbol{\alpha}}(t)^T\,\boldsymbol{b}(t)^T + \boldsymbol{b}(t)\,\dot{\boldsymbol{\alpha}}(t)\,\boldsymbol{\mu}(t)^T \\
&\quad + E[\tilde{\boldsymbol{H}}(t)\,\tilde{\boldsymbol{H}}(t)^T] + E[\tilde{\boldsymbol{K}}(t)\,\boldsymbol{e}\,\boldsymbol{e}^T\,\tilde{\boldsymbol{K}}(t)^T], \\
\frac{\partial \boldsymbol{r}(t,s)}{\partial t} &= \boldsymbol{a}(t)\,\boldsymbol{r}(t,s) + \boldsymbol{b}(t)\,\dot{\boldsymbol{\alpha}}(t)\,\boldsymbol{\mu}(t)^T, \quad t \geq s,
\end{aligned} \tag{7.38}$$

where $\tilde{\boldsymbol{H}}(t) = \boldsymbol{b}(t)\,\boldsymbol{H}(t)$, $\tilde{\boldsymbol{K}}(t) = \boldsymbol{b}(t)\,\boldsymbol{K}(t)$, and $\boldsymbol{e}$ is a (d_c, d_c) matrix with non-zero entries $e_{uu} = \sqrt{\lambda_u\, E[Y_{u,1}^2]}$, $u = 1, \ldots, d_c$.

Proof: The assumption that $\boldsymbol{A}$ is deterministic restricts the class of acceptable inputs. However, this assumption is adequate for the type of stochastic problems considered in the book, as demonstrated by the examples in this and following chapters.

The average of Eq. 7.35 gives

$$\boldsymbol{\mu}(t) = \boldsymbol{\theta}(t,0)\,\boldsymbol{\mu}(0) + \int_0^t \boldsymbol{\theta}(t,\sigma)\,\boldsymbol{b}(\sigma)\,d\boldsymbol{\alpha}(\sigma)$$

since $E[d\boldsymbol{M}(t)] = \boldsymbol{0}$. The derivative with respect to time of this equation yields the above differential equation for $\boldsymbol{\mu}$ (Eq. 7.7).

The Itô formula in Eq. 7.2 applied to $\boldsymbol{X}(t) \mapsto X_p(t)\,X_q(t)$ gives

$$X_p(t)\,X_q(t) - X_p(0)\,X_q(0)$$
$$= \sum_{i=1}^{d} \int_{0+}^{t} \left(\delta_{pi}\,X_q(s-) + X_p(s-)\,\delta_{qi}\right) \left[\sum_{j=1}^{d} a_{ij}(s)\,X_j(s-)\,ds\right.$$
$$\left. + \sum_{u=1}^{d'} b_{iu}(s)\,\left(d\alpha_u(s) + d M_u(s)\right)\right]$$
$$+ \frac{1}{2} \sum_{i,j=1}^{d} \int_{0+}^{t} \left(\delta_{pi}\,\delta_{qj} + \delta_{pj}\,\delta_{qi}\right) \sum_{u,v=1}^{d'} \tilde{H}_{iu}(s)\,\tilde{H}_{jv}(s)\,d[B_u, B_v](s)$$
$$+ \sum_{0<s\leq t} \left[X_p(s)\,X_q(s) - X_p(s-)\,X_q(s-) - \sum_{i=1}^{d} \left(\delta_{pi}\,X_q(s-) + X_p(s-)\,\delta_{qi}\right)\,\Delta X_i(s)\right],$$

where $\Delta X_i(s) = X_i(s) - X_i(s-)$ is the jump of X_i at time s. The differentiation with respect to time of the expected value of the above equation gives the second formula in Eq. 7.38. Except for the last term on the right side of the above equation, the calculations are straightforward and are not presented. The last term takes the simpler form

$$\sum_{0<s\leq t} [(X_p(s-) + \Delta X_p(s))\,(X_q(s-) + \Delta X_q(s))$$
$$- \sum_{i=1}^{d} (\delta_{pi}\,X_q(s-) + X_p(s-)\,\delta_{qi})\,\Delta X_i(s)] = \sum_{0<s\leq t} \Delta X_p(s)\Delta X_q(s),$$

which is zero at a time s unless both X_p and X_q have a jump at this time. The jumps of X_p are $\Delta X_p(s) = \sum_{v=1}^{d_c} \tilde{K}_{pv}(s)\,\Delta\tilde{C}_v(s)$. Two distinct coordinates of $\tilde{\boldsymbol{C}}$ have a jump at the same time with zero probability since $\tilde{\boldsymbol{C}}$ has independent coordinates so that $\Delta X_p(s)\Delta X_q(s) = \sum_{v=1}^{d_c} \tilde{K}_{pv}(s)\,\tilde{K}_{qv}(s)\,\left(\Delta\tilde{C}_v(s)\right)^2$.

Consider a small time interval $(s, s + \Delta s]$, that will be reduced to a point by taking the limit $\Delta s \downarrow 0$. If $\lambda_v\,\Delta s \ll 1$, a component $\tilde{C}_v$ of $\tilde{\boldsymbol{C}}$ has at least one jump during the time interval $(s, s + \Delta s]$ with probability $1 - e^{\lambda_v\,\Delta s} \simeq \lambda_v\,\Delta s$. We have

$$E\left[\tilde{K}_{pv}(s)\,\tilde{K}_{qv}(s)\,\left(\Delta\tilde{C}_v(s)\right)^2\right] = \left(\lambda_v\,E[Y_{v,1}^2]\,\Delta s\right)\,E\left[\tilde{K}_{pv}(s)\,\tilde{K}_{qv}(s)\right]$$

by conditioning with respect to $\mathcal{F}_s$ since $\Delta\tilde{C}_v(s)$ is independent of $\mathcal{F}_s$ and the random variables $\tilde{K}_{pv}(s)$, $\tilde{K}_{qv}(s)$ are $\mathcal{F}_s$-measurable.

The last formula in Eq. 7.38 is obtained by multiplying the equality

$$X_p(t) - X_p(s) = \int_{s+}^{t} dX_p(u), \quad t \geq s,$$

with $X_q(s)$, calculating the expectation of the resulting equation, and differentiating this equation with respect to t. ■

If $\boldsymbol{H} = \boldsymbol{h}$ and $\boldsymbol{K} = \boldsymbol{k}$ in Eq. 7.38 are also deterministic functions, the state $\boldsymbol{X}$ is driven by Gaussian and Poisson white noise processes. In this case, Eq. 7.38 coincides with Eq. 7.12 if $\dot{\boldsymbol{\alpha}} = \boldsymbol{\mu}_b$ and the functions $\boldsymbol{h}$ and $\boldsymbol{k}$ are such that the combined intensity of the Gaussian and Poisson noise processes has the same intensity as $\boldsymbol{B}_b$ in Eq. 7.11, that is, $\boldsymbol{\zeta}_b(t)\,\boldsymbol{\zeta}_b(t)^T$ is equal to

$$
\begin{aligned}
&E[(d\boldsymbol{S}(t) - d\boldsymbol{\alpha}(t))(d\boldsymbol{S}(t) - d\boldsymbol{\alpha}(t))^T]/dt \\
&\qquad = \left[\boldsymbol{h}(t)\,\boldsymbol{h}(t)^T + \boldsymbol{k}(t)\,\boldsymbol{e}\,(\boldsymbol{k}(t)\,\boldsymbol{e})^T\right],
\end{aligned}
$$

where $\boldsymbol{e}$ is defined in Eq. 7.38 and $E[d\boldsymbol{S}(t)] = d\boldsymbol{\alpha}(t)$. Hence, the formal second moment calculus used in classical linear random vibration theory can be applied to find the second moment properties of $\boldsymbol{X}$ under the above conditions.

Example 7.14: Let X be a real-valued process defined by the stochastic differential equation $dX(t) = a\,X(t-)\,dt + b\,dS(t)$, where $dS(t) = dA(t) + dM(t)$ and $dM(t) = H(t)\,dB(t) + K(t)\,d\tilde{C}(t)$ with the notation of Eq. 7.32. The first two moments of X satisfy the differential equations

$$
\begin{aligned}
\dot{\mu}(1;t) &= a\,\mu(1;t) + b\,E[\dot{A}(t)], \\
\dot{\mu}(2;t) &= 2\,a\,\mu(2;t) + 2\,b\,E[X(t-)\,\dot{A}(t)] \\
&\quad + b^2\,E[H(t)^2] + \lambda\,b^2\,E[Y_1^2]\,E[K(t)^2],
\end{aligned}
$$

where $\mu(p;t) = E[X(t-)^p] = E[X(t)^p]$. The expectations $E[X(t-)^p]$ and $E[X(t)^p]$ coincide since $P(X(t) = X(t-)) = 1$ at a fixed t.

If $A(t) = 0$ and $H(t) = K(t) = 1$, X is the solution of a stochastic differential equation driven by Brownian and Poisson white noise, and its first two moments satisfy the differential equations $\dot{\mu}(1;t) = a\,\mu(1;t)$ and $\dot{\mu}(2;t) = 2\,a\,\mu(2;t) + b^2\,(1 + \lambda\,E[Y_1^2])$. These moment equations are in agreement with our previous calculations (Section 4.7.1.3). ◇

Proof: The formulas in Eq. 7.38 do not apply since A is a random process. The above equations involving moments of X have been obtained by Itô's formula applied to the function $X(t)^p$. Recall the notation $\mu(t) = \mu(1;t) = E[X(t)]$, $r(t,t) = \mu(2;t) = E[X(t)^2]$, $\tilde{H}(t) = b\,H(t)$, $\tilde{K}(t) = b\,K(t)$, and $e = \sqrt{\lambda\,E[Y_1^2]}$. ■

Example 7.15: Let X be defined by $dX(t) = a\,X(t)\,dt + b\,dS(t)$, where $dS(t) = dM(t)$ and $M(t) = B(t)^3 - 3\,t\,B(t)$. The differential equations,

$$
\begin{aligned}
\dot{\mu}(t) &= a\,\mu(t), \quad t \geq 0, \\
\dot{r}(t,t) &= 2\,a\,r(t,t) + 18\,b^2\,t^2, \quad t \geq 0, \\
\frac{\partial r(t,s)}{\partial t} &= a\,r(t,s), \quad t \geq s \geq 0,
\end{aligned}
$$

can be solved to find the second moment properties of X. ◇

Note: The differential form of Itô's formula applied to the function $B(t) \mapsto B(t)^n$ for an integer $n \geq 1$ gives

$$d(B(t)^n) = n\, B(t)^{n-1}\, dB(t) + \frac{n(n-1)}{2}\, B(t)^{n-2}\, dt$$

so that $dM(t) = 3\,(B(t)^2 - t)\, dB(t)$. Hence, $H(t) = 3\,(B(t)^2 - t)$, $E[\tilde{H}(t)^2] = E[(b\, H(t))^2] = 18\, b^2\, t^2$, and $K(t) = 0$. The above moment equations result from Eq. 7.38. ▲

The Itô formula in Eq. 7.2 can be applied to develop differential equations for higher order moments of $\boldsymbol{X}$ and other probabilistic properties. However, the determination of some coefficients of these equations can be cumbersome. Some computational difficulties involved in the determination of higher order moments of $\boldsymbol{X}$ are illustrated by the following example.

Example 7.16: Let X be defined by $dX(t) = a(t)\, X(t)\, dt + b(t)\, dS(t)$, where $dS(t) = dM(t)$ and $dM(t) = H(t)\, dB(t)$. The moments $\mu(p; t) = E[X(t)^p]$ are the solutions of

$$\dot{\mu}(p; t) = p\, a(t)\, \mu(p; t) + \frac{p(p-1)}{2}\, b(t)^2\, E\left[X(t)^{p-2}\, H(t)^2\right].$$

The expectation $E\left[X(t)^{p-2}\, H(t)^2\right]$ is difficult to calculate for $p \geq 3$ because of the complex dependence between X and M. ◇

Proof: Itô's formula applied to the function $X(t) \mapsto X(t)^p$ gives

$$X(t)^p - X(0)^p = \int_0^t p\, X(s)^{p-1}\, dX(s) + \frac{1}{2} \int_0^t p(p-1)\, X(s)^{p-2}\, d[X, X](s),$$

which yields the stated moment equation by averaging and differentiation with respect to time since $d[X, X](s) = b(s)^2\, H(s)^2\, dt$.

For $X(0) = 0$, the state is $X(t) = \int_0^t \theta(t, s)\, b(s)\, H(s)\, dB(s)$ so that

$$E\left[X(t)^{p-2}\, H(t)^2\right] = \int_{[0,t]^q} E\left[\left(\prod_{k=1}^{q} H(s_k)\, dB(s_k)\right) H(t)^2\right] \left(\prod_{k=1}^{q} \theta(t, s_k)\, b(s_k)\right) ds_k,$$

where $q = p - 2$ and θ is the Green function for X. ■

7.2.2.2 Direct method. General martingales

Let $\boldsymbol{X}$ be the solution of Eq. 7.28 with $\boldsymbol{S}$ defined by Eqs. 7.30 and 7.32. We only require that the jumps of the coordinates of $\tilde{\boldsymbol{C}}$ are in L_1. The conditions in Eq. 7.33 need not be satisfied so that the martingale component of $\boldsymbol{S}$ is not generally square integrable and the expectations $E[\tilde{\boldsymbol{H}}(t)\, \tilde{\boldsymbol{H}}(t)^T]$ and

$E[\tilde{\boldsymbol{K}}(t)\,\boldsymbol{e}\,\boldsymbol{e}^T\,\tilde{\boldsymbol{K}}(t)^T]$ in Eq. 7.38 may not exist. We attempt here to develop a differential equation for the characteristic function $\varphi(\boldsymbol{u};t) = E\left[e^{\sqrt{-1}\,\boldsymbol{u}^T\,\boldsymbol{X}(t-)}\right]$, $\boldsymbol{u} \in \mathbb{R}^d$, of $\boldsymbol{X}(t)$ because this function is always defined.

If $\boldsymbol{A}$ in Eq. 7.30 is a deterministic function $\boldsymbol{\alpha}$ with continuous first order derivative, then

$$\frac{\partial \varphi}{\partial t} = \sum_{k,l=1}^{d} u_k\, a_{kl}(t)\, \frac{\partial \varphi}{\partial u_l} + \sqrt{-1} \sum_{k=1}^{d} u_k \sum_{r=1}^{d'} b_{kr}(t)\, \dot{\alpha}_r(t)\, \varphi$$
$$- \frac{1}{2} \sum_{k,l=1}^{d} u_k\, u_l \sum_{p=1}^{d_b} E\left[\tilde{H}_{kp}(t)\, \tilde{H}_{lp}(t)\, \exp\left(\sqrt{-1}\,\boldsymbol{u}^T\, \boldsymbol{X}(t-)\right)\right]$$
$$+ \sum_{q=1}^{d_c} \lambda_q \left\{ E\left[\exp\left(\sqrt{-1}\left[\boldsymbol{u}^T\, \boldsymbol{X}(t-) + \sum_{q=1}^{d_c} u_k\, \tilde{K}_{kq}(t)\, Y_{q,1}\right]\right)\right] - \varphi \right\}. \tag{7.39}$$

Proof: The arguments $\boldsymbol{u}$ and t of the characteristic function $\varphi(\boldsymbol{u};t)$ are not shown in the above equation for simplicity.

The Itô formula in Eq. 7.2 applied to $\boldsymbol{X}(t) \mapsto \exp\left(\sqrt{-1}\,\boldsymbol{u}^T\,\boldsymbol{X}(t)\right)$ gives

$$e^{\sqrt{-1}\,\boldsymbol{u}^T\,\boldsymbol{X}(t)} - e^{\sqrt{-1}\,\boldsymbol{u}^T\,\boldsymbol{X}(0)} = \sum_{k=1}^{d} \int_{0+}^{t} \left(\sqrt{-1}\, u_k\, e^{\sqrt{-1}\,\boldsymbol{u}^T\,\boldsymbol{X}(s-)}\, dX_k(s)\right)$$
$$- \frac{1}{2} \sum_{k,l=1}^{d} \int_{0+}^{t} u_k\, u_l\, e^{\sqrt{-1}\,\boldsymbol{u}^T\,\boldsymbol{X}(s-)}\, d[X_k, X_l]^c(s)$$
$$+ \sum_{0<s\le t} \left[e^{\sqrt{-1}\,\boldsymbol{u}^T\,\boldsymbol{X}(s)} - e^{\sqrt{-1}\,\boldsymbol{u}^T\,\boldsymbol{X}(s-)} - \sum_{k=1}^{d} \sqrt{-1}\, u_k\, e^{\sqrt{-1}\,\boldsymbol{u}^T\,\boldsymbol{X}(s-)}\, \Delta X_k(s) \right], \tag{7.40}$$

where

$$dX_k(s) = \sum_{l=1}^{d} a_{kl}(s)\, X_l(s-)\, ds + \sum_{r=1}^{d'} b_{kr}(s)\, d\alpha_r(s)$$
$$+ \sum_{p=1}^{d_b} \tilde{H}_{kp}(s)\, dB_p(s) + \sum_{q=1}^{d_c} \tilde{K}_{kq}(s)\, d\tilde{C}_q(s),$$

$$d[X_k, X_l]^c(s) = \sum_{p,p'=1}^{d_b} \tilde{H}_{kp}(s)\, \tilde{H}_{lp'}(s)\, d[B_p, B_{p'}](s) = \sum_{p=1}^{d_b} \tilde{H}_{kp}(s)\, \tilde{H}_{lp}(s)\, ds,$$

and $\Delta X_k(s) = X_k(s) - X_k(s-) = \sum_{q=1}^{d_c} \tilde{K}_{kq}(s)\,\Delta\tilde{C}_q(s)$. The expectation of the left side of Eq. 7.40 is $\varphi(\boldsymbol{u};t) - \varphi(\boldsymbol{u};0)$, and its derivative with respect to time is $\partial\varphi(\boldsymbol{u};t)/\partial t$.

Consider now the first term on the right side of Eq. 7.40. There is no contribution from the Brownian component of the driving noise. The terms corresponding to the compensated compound Poisson process of the driving noise and the last term including the jumps $\Delta X_k(s)$ cancel. The only contribution of the first term comes from the drift of $\boldsymbol{X}$ and the component $\boldsymbol{\alpha}$ of the driving noise. The differentiation with respect to time of the expectation of these contributions and properties of the characteristic function yield the first two terms in Eq. 7.39.

The expectation of the second term on the right side of Eq. 7.40 and the expression for $[X_k, X_l]^c$ give the third term in Eq. 7.39.

The expectation of $e^{\sqrt{-1}\,\boldsymbol{u}^T \boldsymbol{X}(s)} - e^{\sqrt{-1}\,\boldsymbol{u}^T \boldsymbol{X}(s-)}$ in the last term of Eq. 7.40 can be obtained from $\boldsymbol{X}(s) = \boldsymbol{X}(s-) + \Delta\boldsymbol{X}(s)$, the expression of the jumps of $\boldsymbol{X}$, and properties of $\tilde{\boldsymbol{C}}$. If $\lambda_q\,\Delta t \ll 1$, the probability that a coordinate $\tilde{C}_q$ of $\tilde{\boldsymbol{C}}$ has at least a jump in $(t, t+\Delta t]$ is approximately $\lambda_q\,\Delta t$ and the corresponding jump of X_k is $\tilde{K}_{kq}(t)\,\Delta\tilde{C}_q(t)$, where $\Delta\tilde{C}_q(t)$ is equal to $Y_{q,k}$ for some k. The events D_q, $q = 1, \ldots, d_c$, that $\tilde{C}_q$ jumps in $(t, t+\Delta t]$ and $\tilde{C}_r$, $r \neq q$, have no jumps in this time interval are disjoint so that

$$\sum_{q=1}^{d_c} E\left[e^{\sqrt{-1}\,\boldsymbol{u}^T \boldsymbol{X}(t-)}\, e^{\sqrt{-1}\,\sum_{k=1}^{d} u_k \tilde{K}_{kq}(t)\, Y_{q,1}} - e^{\sqrt{-1}\,\boldsymbol{u}^T \boldsymbol{X}(t-)} \mid D_q\right] P(D_q)$$

$$= \Delta t \sum_{q=1}^{d_c} \lambda_q \left(E\left[e^{\sqrt{-1}\,\boldsymbol{u}^T \boldsymbol{X}(t-)}\, e^{\sqrt{-1}\,\sum_{k=1}^{d} u_k \tilde{K}_{kq}(t)\, Y_{q,1}}\right] - \varphi(\boldsymbol{u};t)\right)$$

since $P(D_q) = \left(1 - e^{-\lambda_q\,\Delta t}\right)\prod_{r=1, r\neq q}^{d_c} e^{-\lambda_r\,\Delta t} \simeq \lambda_q\,\Delta t$ provided that $\lambda_q\,\Delta t \ll 1$. The probability of the events corresponding to jumps of two or more processes $\tilde{C}_q$ in $(t, t+\Delta t]$ are of order $(\Delta t)^2$ or higher so that they will not contribute to the partial differential equation for φ. These considerations yield the last term in Eq. 7.39. ■

Generally, Eq. 7.39 is not a partial differential equation for the characteristic function of $\boldsymbol{X}$ since some of the expectations in this equation cannot be expressed in terms of derivatives of φ. The condition in Eq. 7.39 becomes a partial differential equation for φ if, for example, $\boldsymbol{H}$ and $\boldsymbol{K}$ are deterministic functions, in which case $\boldsymbol{X}$ is the state of a linear system driven by Gaussian and Poisson white noise (Table 7.1). This special case has been considered in the previous section under the assumption $Y_{q,1} \in L_2$, $q = 1, \ldots, d_c$.

Example 7.17: Let X be the solution of

$$dX(t) = a(t)\,X(t-)\,dt + b_1(t)\,dB(t) + b_2(t)\,d\tilde{C}(t), \quad t \geq 0,$$

where a, b_1, and b_2 are deterministic functions of time. The input consists of Brownian motion B and a compensated compound Poisson process $\tilde{C}$. The characteristic function of X satisfies the partial differential equation

$$\frac{\partial\varphi}{\partial t} = u\,a(t)\,\frac{\partial\varphi}{\partial u} - \frac{u^2\,b_1(t)^2}{2}\,\varphi + \lambda\left(E\left[e^{\sqrt{-1}\,u\,b_2(t)\,Y_1}\right] - 1\right)\varphi,$$

where the arguments of the characteristic function are not shown. ◇

Proof: Apply Eq. 7.39 for $d = d' = 1$, $\alpha(t) = 0$, $H(t) = b_1(t)$, and $K(t) = b_2(t)$. A special case of the above equation corresponding to $b_2(t) = 0$ has been discussed in Section 4.7.1.3. ■

Example 7.18: If the coefficients b_1, b_2 in the previous example are replaced by some processes $H, K \in \mathcal{L}$ (Eq. 7.32), the characteristic function of X satisfies the condition (Eq. 7.39)

$$\frac{\partial \varphi}{\partial t} = u\, a(t)\, \frac{\partial \varphi}{\partial u} - \frac{u^2}{2}\, E\left[\tilde{H}(t)^2\, e^{\sqrt{-1}\, u\, X(t-)}\right] + \lambda \left(E\left[e^{\sqrt{-1}\, u\, \left(X(t-) + \tilde{K}(t)\, Y_1\right)}\right] - \varphi\right).$$

Because the expectations on the right side of this equation cannot be expressed as partial derivatives of φ, it is not possible to obtain a partial differential equation for the characteristic function of X. ◇

Example 7.19: Suppose that $\tilde{H}(t) = 3\, b\, (B(t)^2 - t)$ and $\tilde{K}(t) = 0$ in Example 7.18 so that X is the solution of

$$dX(t) = a\, X(t)\, dt + 3\, b\, (B(t)^2 - t)\, dB(t),$$

where a, b are some constants and B denotes a Brownian motion. The characteristic function of X satisfies the condition (Eq. 7.39)

$$\frac{\partial \varphi}{\partial t} = u\, a\, \frac{\partial \varphi}{\partial u} - \frac{9\, b^2}{2}\, u^2\, E\left[\left(B(t)^2 - t\right)^2\, e^{\sqrt{-1}\, u\, X(t)}\right].$$

This is not a partial differential equation for φ since the complex dependence between $B(t)$ and $X(t)$ prevents the representation of the above expectation as a function of a partial derivative of φ. ◇

Example 7.20: Let X be the solution of $dX(t) = -\rho\, X(t-)\, dt + dL_\alpha(t)$, $t \geq 0$, where $X(0) = 0$, $\rho > 0$, $\alpha \in (1, 2)$, and L_α is an α-stable process with $E[\exp(\sqrt{-1}\, u\, dL_\alpha(t))] = \exp(-|u|^\alpha\, dt)$ (Sections 2.10.3.3 and 3.14.2). The characteristic function $\varphi(u; t) = E[\exp(\sqrt{-1}\, u\, X(t))]$ of $X(t)$ is the solution of the partial differential equation

$$\frac{\partial \varphi}{\partial t} = -\rho\, u\, \frac{\partial \varphi}{\partial u} - |u|^\alpha\, \varphi$$

with the initial condition $\varphi(u; 0) = 1$. ◇

Proof: Note that the driving noise L_α does not admit the representation in Eqs. 7.30 and 7.32 so that Eq. 7.39 cannot be applied. The Itô formula in Eq. 7.3 applied to the function $X(t) \mapsto e^{\sqrt{-1}\,u\,X(t)}$ gives

$$e^{\sqrt{-1}\,u\,X(t)} - 1 = \int_{0+}^{t} \sqrt{-1}\,u\,e^{\sqrt{-1}\,u\,X(s-)}\,(-\rho\,X(s-)\,ds + dL_\alpha(s))$$
$$+ \sum_{0<s\le t} \left[e^{\sqrt{-1}\,u\,X(s)} - e^{\sqrt{-1}\,u\,X(s-)} - \sqrt{-1}\,u\,e^{\sqrt{-1}\,u\,X(s-)}\,\Delta X(s)\right]$$
$$= -\rho\,\sqrt{-1}\,u \int_{0+}^{t} X(s-)\,e^{\sqrt{-1}\,u\,X(s-)}\,ds + \sum_{0<s\le t} \left[e^{\sqrt{-1}\,u\,X(s)} - e^{\sqrt{-1}\,u\,X(s-)}\right]$$

since $[X, X]^c(t) = 0$ and $\int_{0+}^{t} e^{\sqrt{-1}\,u\,X(s-)}\,dL_\alpha(s) = \sum_{0<s\le t} e^{\sqrt{-1}\,u\,X(s-)}\,\Delta X(s)$, so that

$$\varphi(u;t) - 1 = E\left\{-\sqrt{-1}\,\rho\,u \int_{0+}^{t} e^{\sqrt{-1}\,u\,X(s-)}\,X(s-)\,ds\right.$$
$$\left.+ \sum_{0<s\le t} e^{\sqrt{-1}\,u\,X(s-)}\left(e^{\sqrt{-1}\,u\,\Delta L_\alpha(s)} - 1\right)\right\}.$$

Errors can result if we calculate the above expectation term by term since the expectations of the individual terms may not be finite, as we will see in Example 7.53. However, we will continue with formal calculations and perform the expectation term by term. The expectation of the derivative with respect to time of the first term is

$$-\rho\,\sqrt{-1}\,u\,E\left[X(s-)\,e^{\sqrt{-1}\,u\,X(s-)}\right] = -\rho\,u\,\frac{\partial \varphi}{\partial u}.$$

The expectation of the increment of the second term in a small time interval $(t, t+\Delta t]$ can be approximated by

$$\sum_{t<s\le t+\Delta t} E\left[e^{\sqrt{-1}\,u\,X(s-)}\left(e^{\sqrt{-1}\,u\,\Delta L_\alpha(s)} - 1\right)\right] \simeq \varphi(u;t)\left(e^{-|u|^\alpha\,\Delta t} - 1\right)$$
$$\simeq \varphi(u;t)\left(-|u|^\alpha\,\Delta t\right)$$

as $\Delta t \downarrow 0$. Since the time derivative of the above term is $\varphi(u;t)\,(-|u|^\alpha)$, our formal calculations suggest that φ satisfies the stated partial differential equation.

We also note that the solution $X(t) = \int_0^t e^{-\rho\,(t-s)}\,dL_\alpha(s)$ is an α-stable variable $S_\alpha(\sigma, \beta, \mu)$ with $\sigma = \left((1 - e^{-\rho\,\alpha\,t})/(\rho\,\alpha)\right)^{1/\alpha}$, $\beta = 0$, and $\mu = 0$ (Section 2.10.3.2 in this book, [79], Example 3.27, p. 112). The characteristic function of $X(t)$ is

$$\varphi(u;t) = E\left[e^{\sqrt{-1}\,u\,X(t)}\right] = \exp\left(-|u|^\alpha\,\frac{1 - e^{-\rho\,\alpha\,t}}{\rho\,\alpha}\right)$$

so that

$$\frac{\partial \varphi}{\partial t} = -e^{-\rho\,\alpha\,t}\,|u|^\alpha\,\varphi, \quad \frac{\partial \varphi}{\partial u} = -\alpha\,\frac{1 - e^{-\rho\,\alpha\,t}}{\rho\,\alpha}\,\mathrm{sign}(u)\,|u|^{\alpha-1}\,\varphi,$$

and $\partial\varphi/\partial t + \rho\,u\,\partial\varphi/\partial u = -|u|^\alpha\,\varphi$ showing that the above formal calculations yield a correct result in this case. ■

7.2.2.3 State augmentation method

The state augmentation method applies if it is possible to define a vector process $\boldsymbol{Z}$ that includes the state $\boldsymbol{X}$ and satisfies a stochastic differential equation driven by Gaussian and/or Poisson white noise. Generally, the differential equations defining the augmented state $\boldsymbol{Z}$ are nonlinear so that the methods in the second part of this chapter dealing with nonlinear systems can be used to find properties of $\boldsymbol{Z}$.

It is possible to write conditions for various properties of the augmented vector $\boldsymbol{Z}$, for example, higher order moments and the characteristic or marginal density function of $\boldsymbol{Z}$. The resulting conditions may or may not be differential equations for these properties of $\boldsymbol{Z}$ depending on the noise and system properties. Instead of presenting general conditions for properties of $\boldsymbol{Z}$, we give several examples illustrating the use of the state augmentation method for two special cases of semimartingale noise (Eqs. 7.43 and 7.44). The last two examples in this section examine the response of linear systems to polynomials of filtered Gaussian and Poisson processes, that is, processes defined as the output of linear filters to Gaussian and Poisson white noise, respectively.

We now outline the essentials of the **state augmentation** method for a state vector $\boldsymbol{X}$ defined by

$$d\boldsymbol{X}(t) = \boldsymbol{a}(\boldsymbol{X}(t-), t)\, dt + \boldsymbol{b}(\boldsymbol{X}(t-), t)\, d\boldsymbol{S}(t), \quad t \geq 0, \tag{7.41}$$

$$\dot{\boldsymbol{X}}(t) = \boldsymbol{a}(\boldsymbol{X}(t-), t) + \boldsymbol{b}(\boldsymbol{X}(t-), t)\, \boldsymbol{p}(\boldsymbol{S}(t-), t), \quad t \geq 0, \tag{7.42}$$

rather than Eqs. 7.28 and 7.29. This definition of $\boldsymbol{X}$ includes the class of nonlinear systems, which will be considered in Section 7.3.2.3. If $\boldsymbol{a}$ is linear in $\boldsymbol{X}$ and $\boldsymbol{b}$ does not depend on $\boldsymbol{X}$, the above equations coincide with Eqs. 7.28 and 7.29. It is assumed that the matrices $\boldsymbol{a}, \boldsymbol{b}$ are such that Eqs. 7.41 and 7.42 have a unique solution (Sections 4.7.1.1 and 4.7.2). Let $\boldsymbol{S}$ be the semimartingale process in Eqs. 7.30 and 7.32. We consider two special cases for $\boldsymbol{S}$ corresponding to the following definitions of the processes $\boldsymbol{A}$, $\boldsymbol{H}$, and $\boldsymbol{K}$.

Case 1. Let $\boldsymbol{\alpha}$, $\boldsymbol{h}$, and $\boldsymbol{k}$ be some deterministic functions and define

$$\boldsymbol{A}(t) = \boldsymbol{\alpha}(t), \quad \boldsymbol{H}(t) = \boldsymbol{h}(\boldsymbol{B}(t), t), \quad \text{and } \boldsymbol{K}(t) = \boldsymbol{k}(\tilde{\boldsymbol{C}}(t-), t). \tag{7.43}$$

Case 2. Let $\boldsymbol{\beta}$, $\boldsymbol{h}$, and $\boldsymbol{k}$ be some deterministic functions and define

$$d\boldsymbol{A}(t) = \boldsymbol{\beta}(\boldsymbol{S}(t-), t)\, dt, \quad \boldsymbol{H}(t) = \boldsymbol{h}(\boldsymbol{S}(t-), t), \quad \text{and } \boldsymbol{K}(t) = \boldsymbol{k}(\boldsymbol{S}(t-), t). \tag{7.44}$$

State augmentation formulation. If $\boldsymbol{X}$ satisfies Eq. 7.41 and $\boldsymbol{S}$ is defined by Eqs. 7.30, 7.32, and 7.43, then $\boldsymbol{Z} = (\boldsymbol{X}, \boldsymbol{U} = \boldsymbol{B}, \boldsymbol{V} = \tilde{\boldsymbol{C}})$ is the solution of

$$\begin{cases} d\boldsymbol{X}(t) & = \boldsymbol{a}(\boldsymbol{X}(t-), t)\, dt + \boldsymbol{b}(\boldsymbol{X}(t-), t)\, [d\boldsymbol{\alpha}(t) \\ & \quad + \boldsymbol{h}(\boldsymbol{U}(t), t)\, d\boldsymbol{B}(t) + \boldsymbol{k}(\boldsymbol{V}(t-), t)\, d\tilde{\boldsymbol{C}}(t)], \\ d\boldsymbol{U}(t) & = d\boldsymbol{B}(t), \\ d\boldsymbol{V}(t) & = d\tilde{\boldsymbol{C}}(t). \end{cases} \tag{7.45}$$

Note: The augmented vector $\boldsymbol{Z}$ in Eq. 7.45 satisfies a stochastic differential equation driven by Gaussian and Poisson white noise. Generally, this equation is nonlinear even if the differential equation for $\boldsymbol{X}$ is linear.

If $\boldsymbol{S}$ corresponds to the special case in Eq. 7.44, then the augmented vector $\boldsymbol{Z} = (\boldsymbol{X}, \boldsymbol{S})$ is the solution of the stochastic differential equation

$$\begin{cases} d\boldsymbol{X}(t) & = \boldsymbol{a}(\boldsymbol{X}(t-), t)\, dt + \boldsymbol{b}(\boldsymbol{X}(t-), t)\, [d\boldsymbol{\beta}(\boldsymbol{S}(t-), t) \\ & \quad + \boldsymbol{h}(\boldsymbol{S}(t-), t)\, d\boldsymbol{B}(t) + \boldsymbol{k}(\boldsymbol{S}(t-), t)\, d\tilde{\boldsymbol{C}}(t)], \\ d\boldsymbol{S}(t) & = \boldsymbol{\beta}(\boldsymbol{S}(t-), t)\, dt + \boldsymbol{h}(\boldsymbol{S}(t-), t)\, d\boldsymbol{B}(t) + \boldsymbol{k}(\boldsymbol{S}(t-), t)\, d\tilde{\boldsymbol{C}}(t) \end{cases} \tag{7.46}$$

driven by Gaussian and Poisson white noise. ▲

State augmentation formulation. If $\boldsymbol{X}$ satisfies Eq. 7.42 and $\boldsymbol{S}$ is defined by Eqs. 7.30, 7.32, and 7.44, then $\boldsymbol{Z} = (\boldsymbol{X}, \boldsymbol{S})$ is the solution of

$$\begin{cases} d\boldsymbol{X}(t) & = \boldsymbol{a}(\boldsymbol{X}(t-), t)\, dt + \boldsymbol{b}(\boldsymbol{X}(t-), t)\, \boldsymbol{p}(\boldsymbol{S}(t-), t)\, dt, \\ d\boldsymbol{S}(t) & = \boldsymbol{\beta}(\boldsymbol{S}(t-), t)\, dt + \boldsymbol{h}(\boldsymbol{S}(t-), t)\, d\boldsymbol{B}(t) + \boldsymbol{k}(\boldsymbol{S}(t-), t)\, d\tilde{\boldsymbol{C}}(t). \end{cases} \tag{7.47}$$

Example 7.21: Let X be the solution of $dX(t) = a\,X(t)\,dt + b\,dS(t)$, where $S(t) = M(t) = B(t)^3 - 3\,t\,B(t)$ and B is a Brownian motion. The moments $\mu(p, q; t) = E[X(t)^p\, Y(t)^q]$ of the augmented vector $\boldsymbol{Z} = (X, Y = B)$ satisfy the differential equations

$$\begin{aligned} \dot{\mu}(p, q; t) = p\,a\,\mu(p, q; t) &+ \frac{9\,b^2}{2}\, p\,(p-1)\, \Big[\mu(p-2, q+4; t) \\ &-2\,t\,\mu(p-2, q+2; t) + t^2\,\mu(p-2, q; t)\Big] \\ &+ 3\,b\,p\,q\,[\mu(p-1, q+1; t) - t\,\mu(p-1, q-1; t)] \\ &+ \frac{1}{2}\, q\,(q-1)\,\mu(p, q-2; t), \end{aligned}$$

where $p, q \geq 0$ are integers and $\mu(u, v; t) = 0$ if at least one of its arguments u, v is strictly negative. The equations of order $p+q$ contain the moments $\mu(p-2, q+4; t)$ of order $(p-2)+(q+4) = p+q+2$. It may seem that we cannot calculate the moments $\mu(p, q; t)$ of $\boldsymbol{Z}$. However, the moments $\mu(p-2, q+4; t)$ can be obtained sequentially for increasing values of p since the moments $\mu(0, q; t)$ depend only on S, so that they can be calculated from the defining equation of this process. We say that the above infinite hierarchy of equations is closed. ◇

Proof: The augmented vector $\mathbf{Z}$ is a diffusion process defined by the stochastic differential equation

$$\begin{cases} dX(t) &= a\,X(t)\,dt + 3\,b\,(Y(t)^2 - t)\,dB(t), \\ dY(t) &= dB(t), \end{cases}$$

since $dS(t) = 3\,(B(t)^2 - t)\,dB(t)$. The Itô formula applied to function $\mathbf{Z} = (X, Y) \mapsto X^p\,Y^q$ gives the above moment equations by averaging and differentiation. This example illustrates the state augmentation method in Eq. 7.45. ■

Example 7.22: Let X be the process in Example 7.21. We have found that the direct method cannot deliver a partial differential equation for the characteristic function of X (Example 7.19). However, we have

$$\frac{\partial \varphi}{\partial t} = a\,u\,\frac{\partial \varphi}{\partial u} - \frac{9\,b^2\,u^2}{2}\left[\frac{\partial^4 \varphi}{\partial v^4} + 2\,t\,\frac{\partial^2 \varphi}{\partial v^2} + t^2\,\varphi\right]$$
$$+ 3\,b\,u\,v\left(\frac{\partial^2 \varphi}{\partial v^2} + t\,\varphi\right) - \frac{v^2}{2}\,\varphi,$$

where $\varphi(u, v; t) = E\left[e^{\sqrt{-1}\,(u\,X(t) + v\,Y(t))}\right]$ is the characteristic function of the augmented state vector $\mathbf{Z} = (X, Y = B)$. ◇

Proof: We have seen in the previous example that $\mathbf{Z} = (X, Y)$ is a diffusion process. Let $\chi(u, v; t) = e^{\sqrt{-1}\,(u\,X(t) + v\,Y(t))}$ be a differentiable function of $\mathbf{Z}(t)$. The Itô formula in Eq. 7.3 applied to the mapping $(X, Y) \mapsto \chi(u, v; t)$ gives

$$\chi(u, v; t) - \chi(u, v; 0) = \int_0^t \sqrt{-1}\,u\,\chi(u, v; s)\,dX(s) + \int_0^t \sqrt{-1}\,v\,\chi(u, v; s)\,dY(s)$$
$$+ \frac{1}{2}\int_0^t \Big((\sqrt{-1}\,u)^2\,\chi(u, v; s)\,d[X, X](s) + 2\,(\sqrt{-1}\,u)\,(\sqrt{-1}\,v)\,\chi(u, v; s)\,d[X, Y](s)$$
$$+ (\sqrt{-1}\,v)^2\,\chi(u, v; s)\,d[Y, Y](s)\Big),$$

where $d[X, X](s)$, $d[X, Y](s)$, and $d[Y, Y]$ are equal to $9\,b^2\,(Y(s)^2 - s)^2\,ds$, $3\,b\,(Y(s)^2 - s)\,ds$, and ds, respectively. By averaging the above equation and differentiating the result with respect to time, we obtain the stated partial differential equation of φ. The characteristic function φ must satisfy the condition $\varphi(0, 0; t) = 1$ (Section 2.11.3). Also, φ has to satisfy the initial condition $\varphi(u, v; 0) = \varphi_0(u)$ for $B(0) = 0$, as considered here, where φ_0 is the characteristic function of $X(0)$. Boundary conditions needed to solve the partial differential equation for φ are considered later in this chapter (Eq. 7.62).

The formula in Eq. 7.39 can also be used to obtain the above partial differential equation for the characteristic function of $\mathbf{Z}$. ■

The previous examples have considered stochastic problems of the type in Eq. 7.45. In the remainder of this section it is assumed that the state $\boldsymbol{X}$ is contained in a vector $\boldsymbol{Z} = (\boldsymbol{X}, \boldsymbol{S})$ satisfying Eq. 7.47. Our objective is to calculate moments

$$\mu(p_1, \ldots, p_d, q_1, \ldots, q_{d'}; t) = E\left[\prod_{i=1}^{d} X_i(t)^{p_i} \prod_{k=1}^{d'} S_k(t)^{q_k}\right] \tag{7.48}$$

of order $p+q$ of $\mathbf{Z}$, where $p_i, q_k \geq 0$ are integers, $p = \sum_{i=1}^{d} p_i$, and $q = \sum_{k=1}^{d'} q_k$. The convention $\mu(p_1, \ldots, p_d, q_1, \ldots, q_{d'}; t) = 0$ is used if at least one of the p_i's or q_k's is strictly negative. If the state vector $\mathbf{Z}$ becomes a stationary process $\mathbf{Z}_s$ as $t \to \infty$, the moments of $\mathbf{Z}_s$ are time invariant and are denoted by $\mu(p_1, \ldots, p_d, q_1, \ldots, q_{d'})$, that is, the last argument indicating the dependence of time is eliminated.

If $\boldsymbol{Z} = (\boldsymbol{X}, \boldsymbol{S})$ is defined by Eq. 7.47, then

$$\begin{aligned}
&\dot{\mu}(p_1, \ldots, p_d, q_1, \ldots, q_{d'}; t) \\
&= \sum_{j=1}^{d} p_j\, E\left[\prod_{i=1}^{d} X_i(t-)^{p_i} \prod_{k=1}^{d'} S_k(t-)^{q_k}\, X_j(t-)^{-1}\, dX_j(t)\right] \\
&+ \sum_{l=1}^{d'} q_l\, E\left[\prod_{i=1}^{d} X_i(t-)^{p_i} \prod_{k=1}^{d'} S_k(t-)^{q_k}\, S_l(t-)^{-1}\, dS_l(t)\right] \\
&+ \frac{1}{2} \sum_{u,v=1}^{d'} q_u\, q_v\, (\boldsymbol{h}\,\boldsymbol{h}^T)_{uv}\, E\left[\prod_{i=1}^{d} X_i(t-)^{p_i} \prod_{k=1}^{d'} S_k(t-)^{q_k}\, S_u(t-)^{-1}\, S_v(t-)^{-1}\right] \\
&+ \frac{d}{dt} \sum_{0<s\leq t} E\left[\prod_{i=1}^{d} X_i(s)^{p_i} \prod_{k=1}^{d'} S_k(s)^{q_k} - \prod_{i=1}^{d} X_i(s-)^{p_i} \prod_{k=1}^{d'} S_k(s-)^{q_k}\right. \\
&\left. - \sum_{l=1}^{d'} q_l \prod_{i=1}^{d} X_i(s-)^{p_i} \prod_{k=1}^{d'} S_k(s-)^{q_k}\, S_l(s-)^{-1}\, \Delta S_l(s)\right],
\end{aligned} \tag{7.49}$$

provided that the above expectations are finite.

Note: The Itô formula in Eq. 7.2 applied to the function

$$(\boldsymbol{X}(t), \boldsymbol{S}(t)) \mapsto \prod_{i=1}^{d} X_i(t)^{p_i} \prod_{k=1}^{d'} S_k(t)^{q_k}$$

gives the above condition. If the coefficients of the differential equations defining $\mathbf{Z}$ are polynomials of $\boldsymbol{X}$ and $\boldsymbol{S}$, Eq. 7.49 is a differential equation for the moments of the augmented state vector $\mathbf{Z}$. Generally, these equations form an infinite hierarchy. However, there are numerous special cases of Eq. 7.47 that are relevant in applications and the resultant moment equations are closed, as demonstrated by the following examples. ▲

Example 7.23: Let X be defined by

$$\ddot{X}(t) + 2\,\zeta\,\nu\,\dot{X}(t) + \nu^2\,X(t) = \sum_{l=0}^{n} a_l(t)\,S(t)^l$$

where $dS(t) = -\alpha\,S(t)\,dt + \sigma\,\sqrt{2\alpha}\,dB(t)$, $\alpha > 0$, σ, $\zeta > 0$, and $\nu > 0$ are some constants, B denotes a Brownian motion, $n \geq 1$ is an integer, and a_l are continuous functions. The process X represents the displacement of a linear oscillator with natural frequency ν and damping ratio ζ driven by a polynomial of the Ornstein-Uhlenbeck process S.

The moments $\mu(p, q, r; t) = E\left[X(t)^p\,\dot{X}(t)^q\,S(t)^r\right]$ of order $s = p+q+r$ of $\mathbf{Z} = (X, \dot{X}, S)$ satisfy the ordinary differential equation

$$\begin{aligned}\dot{\mu}(p, q, r; t) &= p\,\mu(p-1, q+1, r; t) - q\,\nu^2\,\mu(p+1, q-1, r; t)\\ &\quad - (q\,\beta + r\,\alpha)\,\mu(p, q, r; t) + q\sum_{l=0}^{n} a_l(t)\,\mu(p, q-1, r+l; t)\\ &\quad + r(r-1)\,\alpha\,\sigma^2\,\mu(p, q, r-2; t)\end{aligned}$$

at each time t, where $\beta = 2\,\zeta\,\nu$ and $\mu(p, q, r; t) = 0$ if at least one of the arguments p, q, or r of $\mu(p, q, r; t)$ is strictly negative. If $(X, \dot{X}, S)$ becomes stationary as $t \to \infty$, the stationary moments of this vector process can be obtained from a system of algebraic equations derived from the above moment equation by setting $\dot{\mu}(p, q, r; t) = 0$.

Figure 7.4 shows the dependence on damping $\beta = 2\,\zeta\,\nu$ of the coefficients

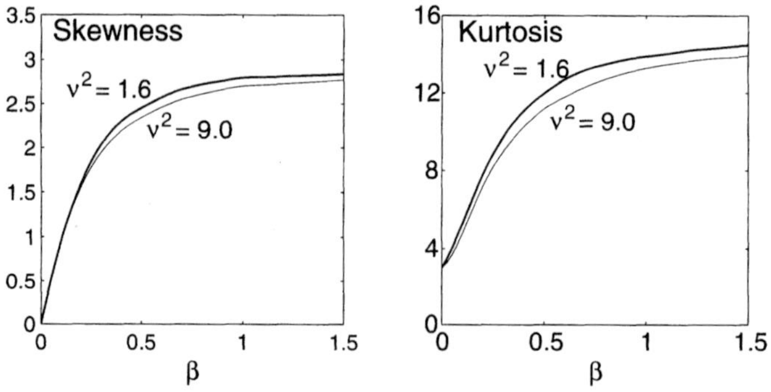

Figure 7.4. Coefficients of skewness and kurtosis of the stationary displacement X of a simple linear oscillator driven by the square of an Ornstein-Uhlenbeck process

of skewness and kurtosis, γ_3 and γ_4, of the stationary process X for $\nu^2 = 1.6$, $\nu^2 = 9.0$, and the input $\sum_{l=0}^{n} a_l(t)\,S(t)^l$ with $n = 2$, $a_0(t) = a_1(t) = 0$,

$a_2(t) = 1$, $\alpha = 0.12$, and $\sigma = 1$. The skewness and the kurtosis coefficients are increasing functions of β. If the linear system has no damping ($\beta = 0$), these coefficients are zero and 3, respectively, that is, the values corresponding to Gaussian variables. This result is consistent with a theorem stating roughly that the output of a linear filter with infinite memory to a random input becomes Gaussian as time increases indefinitely ([79], Section 5.2). The linear filter in this example has infinite memory for $\beta = 0$. ◇

Proof: The state vector $\mathbf{Z} = (X, \dot{X}, S)$ is a diffusion process defined by

$$d\begin{bmatrix} X(t) \\ \dot{X}(t) \\ S(t) \end{bmatrix} = \begin{bmatrix} \dot{X}(t) \\ -\nu^2 X(t) - \beta \dot{X}(t) + \sum_{l=0}^{n} a_l(t)\, S(t)^l \\ -\alpha\, S(t) \end{bmatrix} dt + \begin{bmatrix} 0 \\ 0 \\ \sigma\sqrt{2\alpha} \end{bmatrix} dB(t).$$

The Itô formula applied to $(X(t), \dot{X}(t), S(t)) \mapsto X(t)^p\, \dot{X}(t)^q\, S(t)^r$ gives the above differential equation for the moments of $(X, \dot{X}, S)$ following averaging and differentiation.

To see that the moment equations are closed, consider the collection of equations,

$$\frac{d}{dt}\begin{bmatrix} \mu(s,0,r;t) \\ \mu(s-1,1,r;t) \\ \vdots \\ \mu(1,s-1,r;t) \\ \mu(0,s,r;t) \end{bmatrix} = \boldsymbol{c} \begin{bmatrix} \mu(s,0,r;t) \\ \mu(s-1,1,r;t) \\ \vdots \\ \mu(1,s-1,r;t) \\ \mu(0,s,r;t) \end{bmatrix}$$

$$+ \begin{bmatrix} 0 \\ \sum_l a_l(t)\, \mu(s-1,0,r+l;t) \\ \vdots \\ (s-1)\sum_l a_l(t)\, \mu(1,s-2,r+l;t) \\ s\sum_l a_l(b)\, \mu(0,s-1,r+l;t) \end{bmatrix} + r(r-1)\,\alpha\,\sigma^2 \begin{bmatrix} \mu(s,0,r-2;t) \\ \mu(s-1,1,r-2;t) \\ \vdots \\ \mu(1,s-1,r-2;t) \\ \mu(0,s,r-2;t) \end{bmatrix},$$

for a fixed value of $s = p + q$, where $\boldsymbol{c}$ is a time-invariant matrix that can be constructed from the moment equations. The differential equations for the moments of $\mathbf{Z}$ with $s = 1$ can be used sequentially for increasing values of r to find $\mu(1,0,r;t)$ and $\mu(0,1,r;t)$ for any value of r. These moments are needed to solve the moment equations for $\mathbf{Z}$ with $s = 2$. These calculations can be continued to obtain moments of any order of the state vector $\mathbf{Z} = (X, \dot{X}, S)$. ■

Example 7.24: Let X be the solution of $\dot{X}(t) = -\rho\, X(t) + g(S(t))$, where g is an increasing function with continuous second order derivative and S denotes the Ornstein-Uhlenbeck process in Example 7.23. The augmented vector $\mathbf{Z} = (X, Y = g(S))$ is a diffusion process defined by

$$\begin{cases} dX(t) = (-\rho\, X(t) + Y(t))\, dt, \\ dY(t) = -a_y(Y(t))\, dt + b_y(Y(t))\, dB(t), \end{cases}$$

where $a_y(y) = \alpha\,\sigma^2\, g''(z) - \alpha\, z\, g'(z)$ and $b_y(z)^2 = 2\,\alpha\,\sigma^2 (g'(z))^2$, $z = g^{-1}(y)$, are the drift and the diffusion coefficients of Y, respectively. ◇

Proof: Because g is an increasing function, it has an inverse so that it is possible to derive a stochastic differential equation for Y from the Itô formula applied to $Y = g(S)$ since g^{-1} exists and $S = g^{-1}(Y)$. Hence, (X, Y) is an $\mathbb{R}^2$-valued diffusion process. ■

Example 7.25: Let X be the process in Example 7.23. The characteristic function of the augmented state vector $\boldsymbol{Z} = (Z_1 = X, Z_2 = \dot{X}, Z_3 = S)$ satisfies the partial differential equation

$$\frac{\partial \varphi}{\partial t} = -\alpha\,\sigma^2\,w^2\,\varphi - \nu^2\,v\,\frac{\partial \varphi}{\partial u} + (u - \beta\,v)\,\frac{\partial \varphi}{\partial v} - \alpha\,w\,\frac{\partial \varphi}{\partial w} + v \sum_{l=0}^{n} a_l(t)\,(-1)^l\,(\sqrt{-1})^{l+1}\,\frac{\partial^l \varphi}{\partial w^l},$$

where $\varphi(u, v, w; t) = E[\exp(\sqrt{-1}\,(u\,Z_1(t) + v\,Z_2(t) + w\,Z_3(t)))]$ denotes the characteristic function of $\boldsymbol{Z}(t)$ and $\beta = 2\,\zeta\,\nu$. ◇

Proof: Apply the Itô formula in Eq. 7.3 to function $\boldsymbol{Z}(t) \mapsto \exp(\sqrt{-1}\,(u\,Z_1(t) + v\,Z_2(t) + w\,Z_3(t)))$. The derivative with respect to time of the expected value of this formula gives the equation for φ. ■

The state vector $\boldsymbol{Z} = (\boldsymbol{X}, \boldsymbol{S})$ in the following two examples is defined by

$$d\begin{bmatrix} \boldsymbol{X}(t) \\ \boldsymbol{S}(t) \end{bmatrix} = \begin{bmatrix} \boldsymbol{a}(t)\,\boldsymbol{X}(t-) + \boldsymbol{p}(\boldsymbol{S}(t-), t) \\ \boldsymbol{\alpha}(t)\,\boldsymbol{S}(t-) \end{bmatrix} dt + \int_{\mathbb{R}^{d'}} \begin{bmatrix} \boldsymbol{0} \\ \boldsymbol{y} \end{bmatrix} \mathcal{M}(dt, d\boldsymbol{y}), \tag{7.50}$$

where $\mathcal{M}$ is a Poisson measure (Section 3.12) and $\boldsymbol{p}$ is a polynomial in $\boldsymbol{S}$ with coefficients that may depend on time. We note that the driving compound Poisson process $\boldsymbol{C}(t) = \int_{0+}^{t} \int_{\mathbb{R}^{d'}} \boldsymbol{y}\,\mathcal{M}(ds, d\boldsymbol{y})$ is not compensated and differs from the process in Eq. 7.37.

Note: The coordinates of $\boldsymbol{p}$ are

$$p_i(\boldsymbol{S}(t), t) = \sum_{\boldsymbol{d}_i} \zeta_{i,\boldsymbol{d}_i}(t) \prod_{r=1}^{d'} S_r(t)^{d_{i,r}}, \tag{7.51}$$

where $\boldsymbol{d}_i = (d_{i,1}, \ldots, d_{i,d'})$, $d_{i,r} \geq 0$ denote integers, and $\zeta_{i,\boldsymbol{d}}(t)$ are time dependent coefficients. The compound Poisson process driving this equation has the representation

$$\boldsymbol{C}(t) = \sum_{k=1}^{N(t)} \boldsymbol{Y}_k = \int_{0+}^{t} \boldsymbol{y}\,\mathcal{M}(dt, d\boldsymbol{y}), \tag{7.52}$$

where N is a Poisson process of intensity $\lambda > 0$, $\boldsymbol{Y}_k$ are independent copies of an $\mathbb{R}^{d'}$-valued random variable $\boldsymbol{Y}_1$ with distribution F and finite moments, and $\mathcal{M}(dt, d\boldsymbol{y})$ is a Poisson random measure with expectation $E[\mathcal{M}(dt, d\boldsymbol{y})] = \lambda\,dt\,dF(\boldsymbol{y})$. The moments $\mu(q_1, \ldots, q_n; t) = E\left[\prod_{i=1}^{n} Z_i(t)^{q_i}\right]$ of $\boldsymbol{Z} = (\boldsymbol{X}, \boldsymbol{S})$ are defined as in Eq. 7.48, where $q_i \geq 0$ are integers, $q = \sum_{i=1}^{n} q_i$, and $n = d + d'$. ▲

If $\boldsymbol{Z}$ is the solution of Eq. 7.50, then

$$\begin{aligned}
&\dot{\mu}(q_1, \ldots, q_n; t) \\
&= \sum_{k,l=1}^{d} q_k\, a_{kl}(t)\, \mu(q_1, \ldots, \hat{q}_k, \ldots, \hat{q}_l, \ldots, q_d, \ldots, q_n; t) \\
&+ \sum_{k=1}^{d} \sum_{\boldsymbol{d}_k} q_k\, \zeta_{k,\boldsymbol{d}_k}\, \mu(q_1, \ldots, q_k - 1, \ldots, q_d, q_{d+1} + d_{k,1}, \ldots, q_n + d_{k,d'}; t) \\
&+ \sum_{k,l=d+1}^{n} q_k\, \alpha_{k-d,l-d}\, \mu(q_1, \ldots, q_d, \ldots, \hat{q}_k, \ldots, \hat{q}_l, \ldots, q_n; t) \\
&+ \lambda\, E\left[\prod_{i=1}^{d} X_i(t-)^{q_i} \int_{\mathbb{R}^{d'}} \prod_{j=1}^{d'} \sum_{r_i=0}^{q_i} \frac{q_i!}{r_i!(q_i - r_i)!}\, S_j(t-)^{r_i}\, y_j^{q_i - r_i}\, dF(\boldsymbol{y}) \right] \\
&- \lambda\, \mu(q_1, \ldots, q_n; t),
\end{aligned} \tag{7.53}$$

where $\hat{q}_k = q_k - 1$, $\hat{q}_l = q_l + 1$ if $k \neq l$ and $\hat{q}_k = q_k$, $\hat{q}_l = q_l$ if $k = l$.

Proof: The augmented vector $\boldsymbol{Z} = (\boldsymbol{X}, \boldsymbol{S}) \in \mathbb{R}^n$, $n = d + d'$, is the solution of

$$d\boldsymbol{Z}(t) = \boldsymbol{m}(\boldsymbol{Z}(t-), t)\, dt + \int_{\mathbb{R}^{d'}} \boldsymbol{c}(\boldsymbol{Z}(t-), t, \boldsymbol{y})\, \mathcal{M}(dt, d\boldsymbol{y}), \quad t \geq 0,$$

where matrices $\boldsymbol{m}(z, t)$ and $\boldsymbol{c}(z, t, \boldsymbol{y})$ are such that the above equation has a unique solution (Section 4.7.2). The differential form of the Itô formula applied to $g(\boldsymbol{Z}(t)) = \prod_{i=1}^{n} Z_i(t)^{q_i}$ gives (Eq. 7.2)

$$\begin{aligned}
dg(\boldsymbol{Z}(t)) = &\left\{ \sum_{k=1}^{d} \left[\sum_{l=1}^{d} a_{kl}(t) X_l(t-) + p_k(\boldsymbol{S}(t-), t) \right] q_k\, g_{,k}(\boldsymbol{Z}(t)) \right. \\
&\left. + \sum_{k=d+1}^{n} \left[\sum_{l=d+1}^{n} \alpha_{k-d,l-d}(t)\, S_{l-d}(t-) \right] q_k\, g_{,k}(\boldsymbol{Z}(t-)) \right\} dt \\
&+ \int_{\mathbb{R}^{d'}} \left[\prod_{i=1}^{d} Z_i(t-)^{q_i} \prod_{i=d+1}^{n} (Z_i(t-) + y_{i-d})^{q_i} - \prod_{i=1}^{n} Z_i(t-)^{q_i} \right] \mathcal{M}(dt, d\boldsymbol{y}),
\end{aligned}$$

where $g_{,k}(z) = \partial g(z)/\partial z_k$. The expectation of this equation is

$$
\begin{aligned}
d\mu(q_1, \ldots, q_n; t) \\
= \Bigg\{ \sum_{k,l=1}^{d} q_k\, a_{kl}(t)\, \mu(q_1, \ldots, \hat{q}_k, \ldots, \hat{q}_l, \ldots, q_d, \ldots, q_n; t) \\
+ \sum_{k=1}^{d} \sum_{\boldsymbol{d}_k} q_k\, \zeta_{k,\boldsymbol{d}_k}\, \mu(q_1, \ldots, q_k - 1, \ldots, q_d, q_{d+1} + d_{k,1}, \ldots, q_n + d_{k,d'}; t) \\
+ \sum_{k,l=d+1}^{n} q_k\, \alpha_{k-d,l-d}\, \mu(q_1, \ldots, q_d, \ldots, \hat{q}_k, \ldots, \hat{q}_l, \ldots, q_n; t) \Bigg\}\, dt \\
+ \lambda\, dt\, E \Bigg\{ \prod_{i=1}^{d} X_i(t)^{q_i} \int_{\mathbb{R}^{d'}} \prod_{j=1}^{d'} \sum_{r_i=0}^{q_i} \frac{q_i!}{r_i!(q_i - r_i)!}\, S_j(t)^{r_i}\, y_j^{q_i - r_i}\, dF(\boldsymbol{y}) \\
- \int_{\mathbb{R}^{d'}} \prod_{i=1}^{n} Z_i(t)^{q_i}\, dF(\boldsymbol{y}) \Bigg\},
\end{aligned}
$$

with the convention that moments $\mu(q_1, \ldots, q_n; t)$ with at least one strictly negative argument are zero.

An algorithm can be developed for solving exactly Eq. 7.53 following the approach in Example 7.23. The moments $\mu(0, \ldots, 0, r_{d+1}, \ldots, r_n)$ can be calculated exactly since $\boldsymbol{S}$ is a filtered Poisson process ([79], Section 3.3). If the coefficients of the differential equation defining $\boldsymbol{Z}$ are time invariant and this process becomes stationary as $t \to \infty$, the moments $\mu(q_1, \ldots, q_n; t) = \mu(q_1, \ldots, q_n)$ do not depend on time so that they satisfy algebraic equations obtained from Eq. 7.53 by setting $\dot{\mu}(q_1, \ldots, q_n; t) = 0$. ■

Example 7.26: Let $\boldsymbol{Z} = (X, S)$ be the solution of

$$
d\begin{bmatrix} X(t) \\ S(t) \end{bmatrix} = \begin{bmatrix} a\, X(t-) + S(t-)^l \\ \alpha\, S(t-) \end{bmatrix} dt + \int_{\mathbb{R}} \begin{bmatrix} 0 \\ y \end{bmatrix} \mathcal{M}(dt, dy)
$$

with the notation in Eq. 7.50. If $E[Y_1^k]$ exists and is finite for $k \geq 1$, the moments $\mu(p, q; t) = E[X(t)^p\, S(t)^q]$ of $\boldsymbol{Z} = (X, S)$ satisfy the differential equation

$$
\begin{aligned}
\dot{\mu}(p, q; t) &= (p\, a + q\, \alpha)\, \mu(p, q; t) + p\, \mu(p - 1, q + l; t) \\
&+ \lambda \sum_{k=1}^{q} \frac{q!}{k!(q - k)!} \mu(p, q - k; t)\, E[Y_1^k].
\end{aligned}
$$

Figure 7.5 shows the first four moments $\mu(p, 0; t)$, $p = 1, \ldots, 4$, of X calculated from the above moment equations and their Monte Carlo estimates for $\boldsymbol{Z}(0) = \boldsymbol{0}$, $a = -2\pi$, $\alpha = -1$, $l = 2$, $\lambda = 2$, and $Y_1 \sim N(0, 5)$. The dependence of the stationary skewness and kurtosis coefficients of X on the parameter a is illustrated in Fig. 7.6 for $\lambda = 2$ and $\lambda = 200$. The stationary skewness and kurtosis

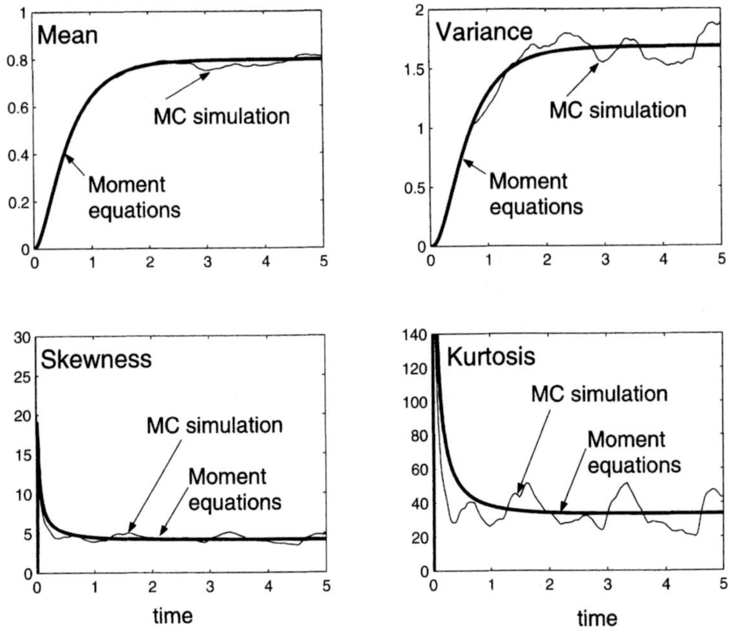

Figure 7.5. The mean, variance, coefficient of skewness, and coefficient of kurtosis functions of X

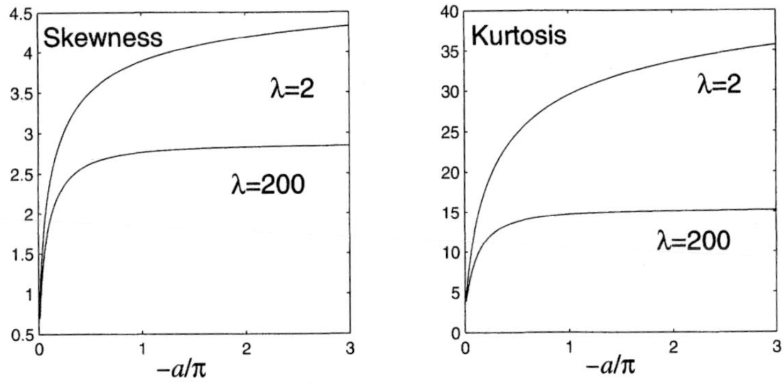

Figure 7.6. Stationary coefficients of skewness and kurtosis of X

coefficients of X differ significantly from their Gaussian counterparts $\gamma_{g,3} = 0$ and $\gamma_{g,4} = 3$ indicating that X is a non-Gaussian process. The skewness and kurtosis coefficients of X approach $\gamma_{g,3}$ and $\gamma_{g,4}$ as $a \to 0$, consistent with a result in ([79], Section 5.2). The stationary skewness and kurtosis coefficients of X also approach $\gamma_{g,3}$ and $\gamma_{g,4}$ if λ increases indefinitely and a is kept constant since the input becomes Gaussian as $\lambda \to \infty$ under adequate scaling. $\diamond$

Proof: The expectation in Eq. 7.53 for the mapping $\mathbf{Z} \mapsto X^p\, S^q$ is

$$\begin{aligned} E\left[X(t-)^p\, \left((S(t-)+Y_1)^q - S(t-)^q\right)\right] &= \sum_{k=1}^{q} \frac{q!}{k!(q-k)!}\, E\left[X(t-)^p\, S(t-)^{q-k}\, Y_1^k\right] \\ &= \sum_{k=1}^{q} \frac{q!}{k!(q-k)!}\, \mu(p, q-k; t)\, E[Y_1^k] \end{aligned}$$

and involves moments of order smaller than $p+q$. We also note that the moments $\mu(0, q; t)$ can be calculated in advance from the characteristic function

$$\varphi(u;t) = E\left[e^{\sqrt{-1}\,u\,S(t)}\right] = \exp\left\{-\lambda\, t\left[1 - \frac{1}{t}\int_0^t E\left[\exp(\sqrt{-1}\,u\,Y_1\, e^{\alpha\,(t-s)})\right] ds\right]\right\}$$

or the cumulants

$$\chi_q(t) = \lambda \int_0^t E\left[\left(Y_1\, e^{\alpha(t-s)}\right)^q\right] ds = \frac{\lambda\, E[Y_1^q]}{q\,\alpha}\, [\exp(q\,\alpha\, t) - 1]$$

of $S(t)$ ([79], Section 3.3).

Consider now the set of moment equations corresponding to increasing values of $s = p+q = 1, 2, \ldots$, that is, the equations

$$\begin{aligned} \dot{\mu}(1,0;t) &= a\,\mu(1,0;t) + \mu(0,l;t), \\ \dot{\mu}(0,1;t) &= \alpha\,\mu(0,1;t) + \lambda\, E[Y_1]\,\mu(0,0;t), \end{aligned}$$

for $s = 1$,

$$\begin{aligned} \dot{\mu}(2,0;t) &= 2\,a\,\mu(2,0;t) + 2\,\mu(1,l;t), \\ \dot{\mu}(1,1;t) &= (a+\alpha)\,\mu(1,1;t) + \mu(0,l+1;t) + \lambda\, E[Y_1]\,\mu(1,0;t), \\ \dot{\mu}(0,2;t) &= 2\,\alpha\,\mu(0,2;t) + \lambda\,(2\,E[Y_1]\,\mu(0,1;t) + E[Y_1^2]\,\mu(0,0;t)), \end{aligned}$$

for $s = 2$, and

$$\begin{aligned} \dot{\mu}(3,0;t) &= 3\,a\,\mu(3,0;t) + 3\,\mu(2,l;t), \\ \dot{\mu}(2,1;t) &= (2\,a+\alpha)\,\mu(2,1;t) + 2\,\mu(1,l+1;t) + \lambda\, E[Y_1]\,\mu(2,0;t), \\ \dot{\mu}(1,2;t) &= (a+2\,\alpha)\,\mu(1,2;t) + \mu(0,l+2;t) + \lambda\,(2\,\mu(1,1;t)\, E[Y_1] \\ &\quad + \mu(1,0;t)\, E[Y_1^2]), \\ \dot{\mu}(0,3;t) &= 3\,\alpha\,\mu(0,3;t) + \lambda\left(3\,\mu(0,2;t)\, E[Y_1] + 3\,\mu(0,1;t)\, E[Y_1^2]\right. \\ &\quad \left. + \mu(0,0;t)\, E[Y_1^3]\right), \end{aligned}$$

for $s = 3$. The moments of order $s = 1$ can be calculated since $\mu(0, q; t)$ are known for all values of q. The moment equations for $s = 2$ involve the unknown moments $\mu(1, l; t)$. These moments can be determined recursively from the moment equations

$$\dot{\mu}(1,q;t) = (a+q\,\alpha)\,\mu(1,q;t) + \mu(0,q+l;t) + \lambda \sum_{k=1}^{q} \frac{q!}{k!(q-k)!}\mu(1,q-k)\, E[Y_1^k]$$

by increasing q. The moment equations of order $s = 3$ involve the unknown moments $\mu(2, l; t)$ that can be determined recursively from

$$\dot{\mu}(2, q; t) = (2\,a + q\,\alpha)\,\mu(2, q; t) + 2\,\mu(1, q + l; t) + \lambda \sum_{k=1}^{q} \frac{q!}{k!(q-k)!}\mu(2, q-k)\,E[Y_1^k],$$

starting with $q = 1$. ■

Example 7.27: Let $\mathbf{Z} = (X, S)$ be the process on Example 7.26. The characteristic function of the augmented state vector $\mathbf{Z}$ satisfies the partial differential equation

$$\frac{\partial \varphi}{\partial t} = u\,a\,\frac{\partial \varphi}{\partial u} + (-1)^l\,(\sqrt{-1})^{l+1}\,u\,\frac{\partial^l \varphi}{\partial v^l} + \alpha\,v\,\frac{\partial \varphi}{\partial v} + \lambda\,\varphi\,[\varphi_y(v) - 1],$$

where $\varphi(u, v; t) = E[\exp(\sqrt{-1}\,(u\,X(t) + v\,S(t)))]$ and the characteristic function of Y_1 is denoted by $\varphi_y(v) = E[\exp(\sqrt{-1}\,v\,Y_1)]$. ◇

Proof: Apply Itô's formula in Eq. 7.4 to $(X(t), S(t)) \mapsto \exp(\sqrt{-1}\,(u\,X(t) + v\,S(t)))$. We discuss here only the contribution of the last term in Eq. 7.4. The jumps of the coordinates of $\mathbf{Z}$ at a time $\sigma > 0$ are $\Delta X(\sigma) = 0$ and $\Delta S(\sigma) = \Delta C(\sigma)$, where $C(t) = \int_{0+}^{t} y\,\mathcal{M}(d\sigma, dy)$. We have

$$\begin{aligned} & E\left[e^{\sqrt{-1}\,(u\,X(t)+v\,S(t))} - e^{\sqrt{-1}\,(u\,X(t-)+v\,S(t-))}\right] \\ &= E\left[e^{\sqrt{-1}\,(u\,X(t-)+v\,S(t-))}\left(e^{\sqrt{-1}\,v\,\Delta C(t)} - 1\right)\right] \\ &= E\left[e^{\sqrt{-1}\,(u\,X(t-)+v\,S(t-))}\right]\left(E[e^{\sqrt{-1}\,v\,\Delta C(t)}] - 1\right) \end{aligned}$$

where the last equality holds since $\mathbf{Z}(t-)$ and $\Delta C(t)$ are independent.

Note also that the characteristic function $\tilde{\varphi}(v) = E[e^{\sqrt{-1}\,v\,S(t)}]$ of $S(t)$ satisfies the partial differential equation

$$\frac{\partial \tilde{\varphi}}{\partial t} = \alpha\,v\,\frac{\partial \tilde{\varphi}}{\partial u} + \lambda\,\tilde{\varphi}\,[\varphi_y(v) - 1],$$

which can be obtained from the above partial differential equation of φ for $u = 0$. ■

Example 7.28: Let X be the displacement of a linear oscillator with damping ratio $\zeta > 0$ and frequency $\nu > 0$ satisfying the equation

$$\ddot{X}(t) + 2\,\zeta\,\nu\,\dot{X}(t) + \nu^2\,X(t) = S(t)^l,$$

where S is a filtered Poisson process defined by

$$dS(t) = \alpha\,S(t)\,dt + dC(t),$$

C denotes a compound Poisson process, and $l > 0$ is an integer. The moments $\mu(p,q,r;t) = E[X(t)^p \dot{X}(t)^q S(t)^r]$ of $\mathbf{Z} = (X, \dot{X}, S)$ are given by

$$\begin{aligned}\dot{\mu}(p,q,r;t) &= (-2\zeta\nu q + \alpha r)\,\mu(p,q,r;t) + p\,\mu(p-1,q+1,r;t)\\ &\quad - \nu^2 q\,\mu(p+1,q-1,r;t) + q\,\mu(p,q-1,r+l;t)\\ &\quad + \lambda\sum_{k=1}^{r}\frac{r!}{k!(r-k)!}\,\mu(p,q,r-k;t)\,E[Y_1^k].\end{aligned}$$

Figure 7.7 shows the first four moments of X calculated from the above moment equations and their Monte Carlo estimates for $\mathbf{Z}(0) = \mathbf{0}$, $\nu = 2\pi$, $\zeta = 0.05$, $\alpha = -1$, $l = 2$, $\lambda = 2$, and $Y_1 \sim N(0, 12)$. The dependence of the stationary

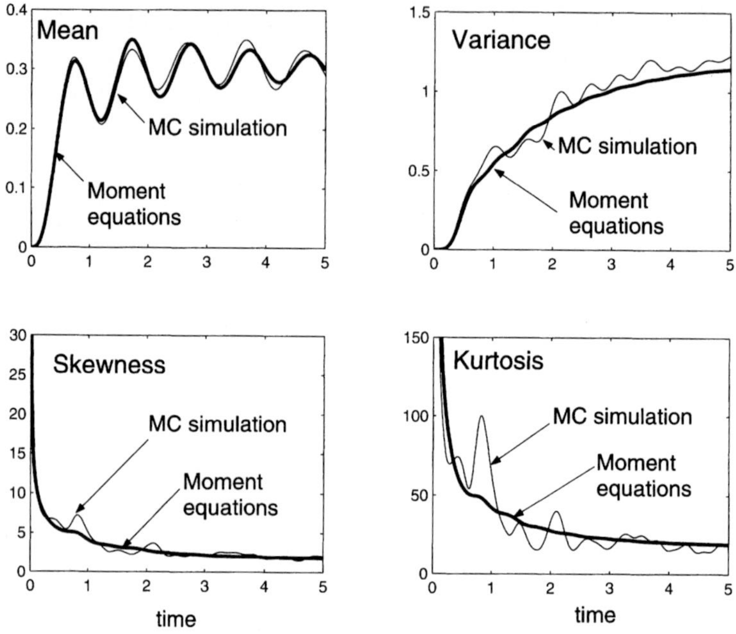

Figure 7.7. The mean, variance, coefficient of skewness, and coefficient of kurtosis functions

skewness and kurtosis coefficients of X on ζ is illustrated in Fig. 7.8 and Fig. 7.9 for $\lambda = 2$ and $\lambda = 200$, respectively, and $\nu = 2\pi, \pi$, and $\pi/2$. These coefficients approach $\gamma_{g,3} = 0$ and $\gamma_{g,4} = 3$, that is, the values corresponding to Gaussian variables, as $\zeta \to 0$. The stationary skewness and kurtosis coefficients of X differ significantly from $\gamma_{g,3}$ and $\gamma_{g,4}$ for other values of ζ indicating that X is a non-Gaussian process. The differences between the skewness and kurtosis coefficients of the stationary displacement X and $\gamma_{g,3}$ and $\gamma_{g,4}$ decrease as λ increases, an expected finding since the input approaches a Gaussian process as $\lambda \to \infty$. ◇

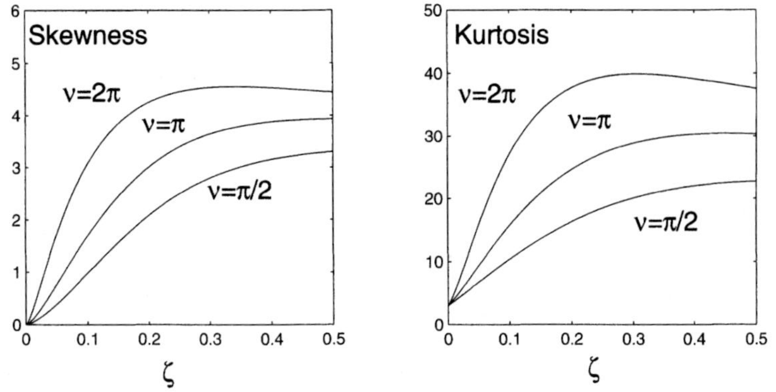

Figure 7.8. Stationary coefficients of skewness and kurtosis of X for $\lambda = 2$

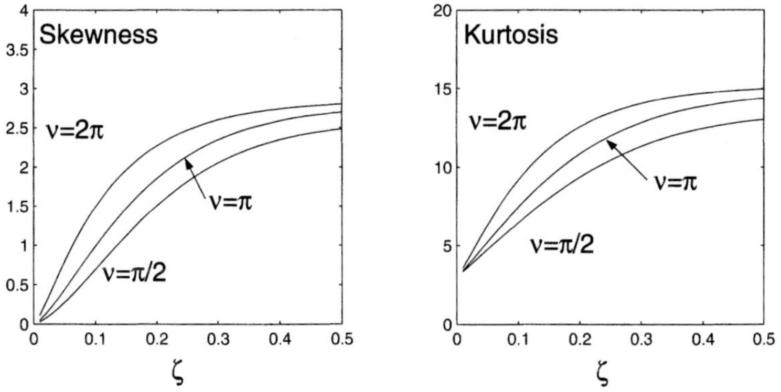

Figure 7.9. Stationary coefficients of skewness and kurtosis of X for $\lambda = 200$

Proof: The above moment equations have been obtained from Eq. 7.53 applied to the augmented vector $\boldsymbol{Z} = (X, \dot{X}, S)$. The resulting system of equations for the moments of $\boldsymbol{Z}$ is closed so that we can calculate moments of any order of $\boldsymbol{Z}$. ■

7.3 Nonlinear systems

If the matrix $\boldsymbol{a}$ in Eq. 7.5 is a nonlinear function of $\boldsymbol{X}$ or both $\boldsymbol{a}$ and $\boldsymbol{b}$ depend on the state vector, the system is **nonlinear**. The matrices $\boldsymbol{a}$ and $\boldsymbol{b}$ may also depend on time explicitly. The general form of a nonlinear system is

$$\dot{\boldsymbol{X}}(t) = \boldsymbol{a}(\boldsymbol{X}(t), t) + \boldsymbol{b}(\boldsymbol{X}(t), t)\, \boldsymbol{Y}(t), \quad t \geq 0, \tag{7.54}$$

where the driving noise $\boldsymbol{Y}$ is an $\mathbb{R}^{d'}$-valued process and the entries of the $(d, 1)$ and (d, d') matrices $\boldsymbol{a}$ and $\boldsymbol{b}$ are real-valued functions of state and time. Systems defined by Eq. 7.54 in which $\boldsymbol{a}$ and $\boldsymbol{b}$ are linear functions of $\boldsymbol{X}$ are viewed as nonlinear. This definition of a nonlinear system is not general ([131], Section 3.3).

We have defined a linear system by Eq. 7.5 consistently with the theory of linear random vibration [175]. Moreover, there are relatively simple and general methods for finding properties of the state $\boldsymbol{X}$ of Eq. 7.5, as we have seen in Section 7.2. For example, the probability law of $\boldsymbol{X}$ is completely defined by its second moment properties if the input to Eq. 7.5 is a Gaussian process. In contrast, there is no general method for finding the probability law and other properties of $\boldsymbol{X}$ defined by Eq. 7.54. Some examples of nonlinear systems for which it is possible to find analytically properties of the state defined by Eq. 7.54 are in Section 7.3.1.4.

The state vector $\boldsymbol{X}$ in Eq. 7.54 can define the evolution of a dynamic system, material structure, damage state in a system, portfolio value, and other processes (Section 7.4). The determination of probabilistic properties of $\boldsymbol{X}$ is the main objective of **nonlinear random vibration**, a branch of stochastic mechanics and dynamics analyzing the response of nonlinear dynamic systems subjected to random input [175]. Two cases are distinguished in nonlinear random vibration depending on the functional form of $\boldsymbol{b}$. If $\boldsymbol{b}$ does not depend on the state $\boldsymbol{X}$, that is, $\boldsymbol{b}(\boldsymbol{X}(t), t) = \boldsymbol{b}(t)$, the nonlinear system is said to have **additive noise**. Otherwise, we say that the system has **multiplicative noise**. The behavior of systems with additive and multiplicative noise can differ significantly.

Example 7.29: Let X be the solution of the stochastic differential equation

$$dX(t) = (\beta\, X(t) - X(t)^3)\, dt + \sigma\, b(X(t))\, dB(t), \quad t \geq 0,$$

where $\beta = \alpha + \sigma^2/2$, α and σ are constants, $b(x) = 1$ or $b(x) = x$, and B denotes a Brownian motion. The noise is additive if $b(x) = 1$ and multiplicative if b is a function of x. Figure 7.10 shows samples of X for $b(x) = 1$, $b(x) = x$, $\alpha = \pm 1$, and $\sigma = 1$. The samples of X corresponding to multiplicative noise converge to zero for $\alpha = -1$ and oscillate about a constant value for $\alpha = 1$. The samples of X corresponding to additive noise have a very different behavior. For $\alpha = -1$, X oscillates about zero. For $\alpha = 1$, X oscillates for some time about $\sqrt{1.5}$ then jumps and continues to oscillate about $-\sqrt{1.5}$. ◇

Note: We will come back to this example in our discussion on the stability of the solution of dynamic systems with multiplicative random noise, referred to as **stochastic stability** (Section 8.7). ▲

Our objectives in this section are as in the first part of the chapter. We attempt to find properties of $\boldsymbol{X}$ satisfying Eq. 7.54 driven by various noise processes $\boldsymbol{Y}$. Because $\boldsymbol{X}$ is the solution of a nonlinear differential equation, it is not a Gaussian process even if $\boldsymbol{Y}$ is Gaussian. Most available results are for inputs $\boldsymbol{Y}$ that can be modeled by Gaussian white noise, but even for these inputs, probabilistic properties of $\boldsymbol{X}$ are not generally available in analytical form. Numerical methods, Monte Carlo simulation, and heuristic assumptions are used in most applications to characterize $\boldsymbol{X}$.

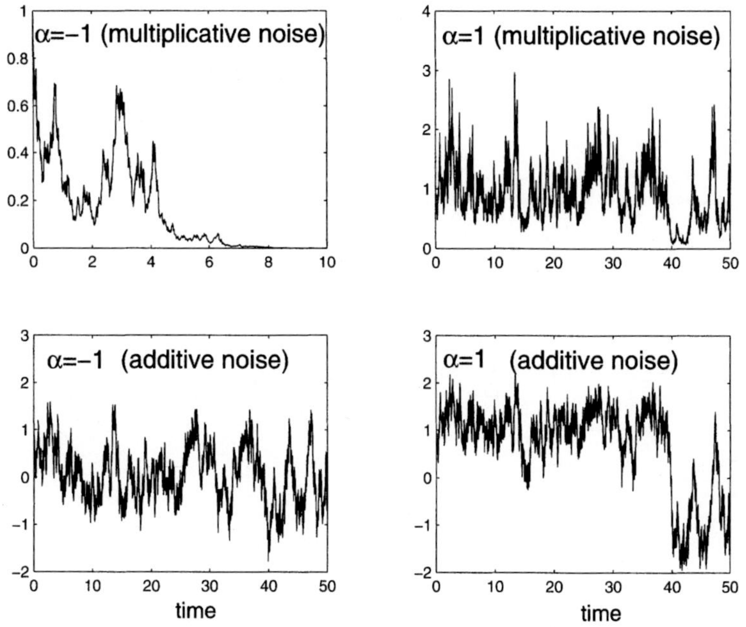

Figure 7.10. Samples of X for multiplicative and additive noise, $\alpha = \pm 1$, and $\sigma = 1$

7.3.1 Brownian motion input

Let $\boldsymbol{Y}$ in Eq. 7.54 be a Gaussian white noise defined formally by the derivative $d\boldsymbol{B}/dt$ of a Brownian motion $\boldsymbol{B}$ with respect to time. Then $\boldsymbol{X}$ is a diffusion process satisfying the stochastic differential equation

$$d\boldsymbol{X}(t) = \boldsymbol{a}(\boldsymbol{X}(t), t)\, dt + \boldsymbol{b}(\boldsymbol{X}(t), t)\, d\boldsymbol{B}(t), \quad t \geq 0. \tag{7.55}$$

Conditions for the existence and uniqueness of the solution of Eq. 7.55 are discussed in Section 4.7.1. The coefficients $\boldsymbol{a}$ and $\boldsymbol{b}\,\boldsymbol{b}^T$ in Eq. 7.55 are called **drift** and **diffusion**, respectively.

We develop differential equations for moments, densities, and characteristic functions of $\boldsymbol{X}$ and discuss the solution of these equations. Solutions of Eq. 7.55 by Monte Carlo simulation and various approximations are also presented.

7.3.1.1 Moment equations

Let $g(\boldsymbol{X}(t)) = \prod_{i=1}^{d} X_i(t)^{q_i}$ and denote by

$$\mu(q_1, \ldots, q_d; t) = E\left[g(\boldsymbol{X}(t))\right] = E\left[\prod_{i=1}^{d} X_i(t)^{q_i}\right], \tag{7.56}$$

the moments of order $q = \sum_{i=1}^{d} q_i$ of $X(t)$, where $q_i \geq 0$ are integers and X is the solution of Eq. 7.55. The convention $\mu(q_1, \ldots, q_d; t) = 0$ is used if at least one of the q_i's is strictly negative. If X becomes a stationary process X_s as $t \to \infty$, the moments of X_s are time invariant, and we denote these moments by $\mu(q_1, \ldots, q_d)$.

The moments of X defined by Eq. 7.55 satisfy the condition

$$\dot{\mu}(q_1, \ldots, q_d; t) = \sum_{i=1}^{d} E\left[a_i(X(t), t) \frac{\partial g(X(t))}{\partial X_i(t)}\right] + \frac{1}{2} \sum_{i,j=1}^{d} E\left[\left(b(X(t))\, b(X(t))^T\right)_{ij} \frac{\partial^2 g(X(t))}{\partial X_i(t)\, \partial X_j(t)}\right]. \tag{7.57}$$

If the state X becomes a stationary process X_s as $t \to \infty$, then

$$\sum_{i=1}^{d} E\left[a_i(X_s(t), t) \frac{\partial g(X_s(t))}{\partial X_{s,i}(t)}\right] + \frac{1}{2} \sum_{i,j=1}^{d} E\left[\left(b(X_s(t))\, b(X_s(t))^T\right)_{ij} \frac{\partial^2 g(X_s(t))}{\partial X_{s,i}(t)\, \partial X_j(t)}\right] = 0, \tag{7.58}$$

where $X_{s,i}$ is the coordinate i of X_s.

Proof: Apply Itô's formula in Eq. 7.3 to function $X(t) \mapsto g(X(t))$. This formula can be used since $g \in C^2(\mathbb{R}^d)$ and X is a semimartingale with continuous samples (Section 4.7.1.1), and gives

$$dg(X(t)) = \sum_{i=1}^{d} \frac{\partial g(X(t))}{\partial X_i(t)} \left(a_i(X(t), t)\, dt + \sum_{k=1}^{d'} b_{ik}(X(t), t)\, dB_k(t) \right) + \frac{1}{2} \sum_{i,j=1}^{d} \sum_{k,l=1}^{d'} \frac{\partial^2 g(X(t))}{\partial X_i(t)\, \partial X_j(t)} b_{ik}(X(t), t)\, b_{jl}(X(t), t)\, d[B_k, B_l](t).$$

The second term, $dM(t) = (\partial g(X(t))/\partial X_i(t))\, b_{ik}(X(t))\, dB_k(t)$, on the right side of the above equation is a local martingale by a preservation property of the stochastic integral so that $E[dM(t)] = 0$ since $M(0) = 0$. The expectation of the above equation divided by dt gives Eq. 7.57 because $d[B_k, B_l](t) = \delta_{kl}\, dt$.

If $X = X_s$, its moments are time invariant so that $\dot{\mu}(q_1, \ldots, q_d; t) = 0$ and Eq. 7.57 yields Eq. 7.58. ■

The condition in Eq. 7.57 cannot be used to calculate moments of X since the expectations on its right side are not moments of X. If the drift and diffusion coefficients are polynomials of the state vector, these expectations become moments of $\boldsymbol{X}$. Under these conditions, Eqs. 7.57 and 7.58 are differential and algebraic equations for the moments of the state vector, respectively, and are referred to as **moment equations**.

Example 7.30: Let X be defined by

$$dX(t) = a(X(t))\,dt + b(X(t))\,dB(t), \quad t \geq 0,$$

where the drift and diffusion are such that the solution X exists and is unique. The moments of $X(t)$ satisfy the condition (Eq. 7.57)

$$\dot{\mu}(q;t) = q\,E\left[X(t)^{q-1}\,a(X(t))\right] + \frac{q\,(q-1)}{2}\,E\left[X(t)^{q-2}\,b(X(t))^2\right].$$

If the drift and diffusion coefficients are polynomials of X, this condition becomes a differential equation for the moments of the state X. ◇

Example 7.31: Suppose that the drift and diffusion coefficients in Example 7.30 are $a(x) = \beta\,x - x^3$ and $b(x) = \sigma$. The moments of X satisfy the differential equation

$$\dot{\mu}(q;t) = q\,\beta\,\mu(q;t) - q\,\mu(q+2;t) + \frac{q\,(q-1)\,\sigma^2}{2}\,\mu(q-2;t).$$

It is not possible to calculate the moments of X exactly since the above moment equations form an **infinite hierarchy**. The calculation of a moment of any order of $X(t)$ involves the solution of an infinite set of equations simultaneously. We present in Section 7.3.1.5 some heuristic methods, called **closure methods**, used in physics and nonlinear random vibration theory to solve infinite hierarchies of moment equations ([175], Section 6.1.1). If the moment equations do not form an infinite hierarchy, we say that the moment equations are **closed**. In this case it is possible to calculate exactly moments of any order of $\boldsymbol{X}$. ◇

Note: The moment equations up to order $q \leq 3$ are

$$\begin{aligned}
\dot{\mu}(1;t) &= \beta\,\mu(1;t) - \mu(3;t),\\
\dot{\mu}(2;t) &= 2\,\beta\,\mu(2;t) - 2\,\mu(4;t) + \sigma^2,\\
\dot{\mu}(3;t) &= 3\,\beta\,\mu(3;t) - 3\,\mu(5;t) + 3\,\sigma^2\,\mu(1;t).
\end{aligned}$$

These moment equations involve five unknowns, the moments $\mu(1;t)$, $\mu(2;t)$, $\mu(3;t)$, $\mu(4;t)$, and $\mu(5;t)$. Hence, it is not possible to find the first three moments of $X(t)$. The situation persists for any value of q.

The stochastic differential equation for X has a unique solution because its coefficients satisfy local Lipschitz and growth conditions (Section 4.7.1.1). The theory of deterministic ordinary differential equations can be used to assess the existence and uniqueness of the solution of the moment equations. ▲

Linear systems driven by polynomials of filtered Gaussian and Poisson processes are examples of nonlinear systems for which the moment equations are closed (Section 7.2.2.3). The following example presents another nonlinear system characterized by closed moment equations.

Example 7.32: Let X be defined by the stochastic differential equation $dX(t) = -\alpha\, X(t)\, dt + \sigma\, X(t)\, dB(t),\ t \geq 0$, where α, σ are constants and B denotes a Brownian motion. We have

$$\mu(q; t) = E[X(t)^q] = \mu(q; 0)\, \exp\left[\left(-\alpha + \frac{(q-1)\,\sigma^2}{2}\right) q\, t\right]$$

for any integer $q \geq 0$. The moment of order q of X approaches zero and $\pm\infty$ as $t \to \infty$ if $\alpha > \alpha^*(q)$ and $\alpha < \alpha^*(q)$, respectively, where $\alpha^*(q) = \sigma^2\,(q-1)/2$. If $\alpha = \alpha^*(q)$, then $\mu(q; t) = \mu(q; 0)$ is time invariant. ◇

Note: The moment equation is

$$\dot{\mu}(q; t) = -\alpha\, q\, \mu(q; t) + \frac{q\,(q-1)\,\sigma^2}{2}\, \mu(q; t),$$

and has the stated solution. If $-\alpha + (q-1)\,\sigma^2/2 < 0$, or equivalently, $\alpha > \alpha^*(q)$, then $\mu(q; t)$ decreases to zero as $t \to \infty$. The moment $\mu(q; t)$ is time invariant if $\alpha = \alpha^*(q)$. If $\alpha < \alpha^*(q)$, $\mu(q; t)$ converges to $\pm\infty$ as $t \to \infty$ depending on the sign of $\mu(q; 0)$.

We also note that X is a geometric Brownian motion process with solution (Section 4.7.1.3)

$$X(t) = X(0)\, \exp\left[\left(-(\alpha + \sigma^2/2)\, t + \sigma\, B(t)\right)\right].$$

If $X(0)$ and B are independent, then

$$E\left[X(t)^q\right] = E\left[X(0)^q\right]\, E\left[e^{q\,\left(-(\alpha+\sigma^2/2)\, t + \sigma\, B(t)\right)}\right],$$

which coincides with $\mu(q; t)$ since $E[e^{u\, G}] = e^{u^2/2}$ for $G \sim N(0, 1)$. ▲

7.3.1.2 Differential equation for characteristic function

Let $\boldsymbol{X}$ be the solution of Eq. 7.55 corresponding to a deterministic or random initial condition $\boldsymbol{X}(0)$ that needs to be specified. The expectation,

$$\varphi(\boldsymbol{u}; t) = E\left[e^{\sqrt{-1}\,\boldsymbol{u}^T\, \boldsymbol{X}(t)}\right], \quad \boldsymbol{u} \in \mathbb{R}^d, \tag{7.59}$$

of $g(\boldsymbol{X}(t)) = \exp\left(\sqrt{-1}\, \boldsymbol{u}^T\, \boldsymbol{X}(t)\right)$ is the characteristic function of the $\mathbb{R}^d$-valued random variable $\boldsymbol{X}(t)$.

The characteristic function of $\boldsymbol{X}$ satisfies the condition

$$\frac{\partial \varphi(\boldsymbol{u}; t)}{\partial t} = \sqrt{-1} \sum_{i=1}^{d} u_i \, E\left[e^{\sqrt{-1}\,\boldsymbol{u}^T \boldsymbol{X}(t)}\, a_i(\boldsymbol{X}(t), t)\right] - \frac{1}{2} \sum_{i,j=1}^{d} u_i \, u_j \, E\left[e^{\sqrt{-1}\,\boldsymbol{u}^T \boldsymbol{X}(t)} \left(\boldsymbol{b}(\boldsymbol{X}(t), t)\, \boldsymbol{b}(\boldsymbol{X}(t), t)^T\right)_{ij}\right]. \tag{7.60}$$

If the state $\boldsymbol{X}$ becomes a stationary process $\boldsymbol{X}_s$ as $t \to \infty$, then

$$\sqrt{-1} \sum_{i=1}^{d} u_i \, E\left[e^{\sqrt{-1}\,\boldsymbol{u}^T \boldsymbol{X}_s(t)}\, a_i(\boldsymbol{X}_s(t), t)\right] - \frac{1}{2} \sum_{i,j=1}^{d} u_i \, u_j \, E\left[e^{\sqrt{-1}\,\boldsymbol{u}^T \boldsymbol{X}_s(t)} \left(\boldsymbol{b}(\boldsymbol{X}_s(t), t)\, \boldsymbol{b}(\boldsymbol{X}_s(t), t)^T\right)_{ij}\right] = 0. \tag{7.61}$$

Proof: Apply Eq. 7.3 to the mapping $\boldsymbol{X}(t) \mapsto e^{\sqrt{-1}\,\boldsymbol{u}^T \boldsymbol{X}(t)}$. We can use this version of the Itô formula since $\boldsymbol{X}$ is a continuous semimartingale. The expectation of the Itô formula scaled by dt yields Eq. 7.60 since $\int_0^t e^{\sqrt{-1}\,\boldsymbol{u}^T \boldsymbol{X}(s)}\, b_{ik}(\boldsymbol{X}(s))\, dB_k(s)$ is a local martingale starting at zero and $d[B_k, B_l](t) = \delta_{kl}\, dt$.

Let $q_i \geq 1$ be integers and $q = \sum_{i=1}^{d} q_i$. If the entries of the matrices $\boldsymbol{a}$ and $\boldsymbol{b}$ are polynomials of $\boldsymbol{X}$ and the partial derivatives,

$$\frac{\partial^q \varphi(\boldsymbol{u}; t)}{\partial u_1^{q_1} \cdots \partial u_d^{q_d}} = \left(\sqrt{-1}\right)^q \, E\left[\prod_{i=1}^{d} X_i(t)^{q_i}\, e^{\sqrt{-1}\,\boldsymbol{u}^T \boldsymbol{X}(t)}\right],$$

of φ exist, the right sides of Eqs. 7.60 and 7.61 can be expressed as partial derivatives of φ with respect to the coordinates of $\boldsymbol{u}$.

The characteristic function of the stationary solution $\boldsymbol{X}_s$ satisfies the condition in Eq. 7.60 but it is time invariant so that Eq. 7.60 becomes Eq. 7.61. ■

We have seen in the previous section that moment equations can be obtained for a diffusion process $\boldsymbol{X}$ only if its drift and diffusion coefficients are polynomials of its coordinates. The same condition is needed in the current setting. If the drift and diffusion coefficients of $\boldsymbol{X}$ are polynomials of this process, then Eqs. 7.60 and 7.61 become partial differential equations for the characteristic function of $\boldsymbol{X}$. In this case, it is possible to find the characteristic function of the state if initial and boundary conditions are specified for Eq. 7.60 and boundary conditions are given for Eq. 7.61. The initial condition for Eq. 7.60 is $\varphi(\boldsymbol{u}; 0) = e^{\sqrt{-1}\,\boldsymbol{u}^T \boldsymbol{x}_0}$ and $\varphi(\boldsymbol{u}; 0) = E\left[e^{\sqrt{-1}\,\boldsymbol{u}^T \boldsymbol{X}(0)}\right]$ if $\boldsymbol{X}(0) = \boldsymbol{x}_0$ and $\boldsymbol{X}(0)$ is an $\mathbb{R}^d$-valued random variable, respectively. The boundary conditions for Eqs. 7.60 and 7.61 are discussed later in this section.

Example 7.33: The displacement X of an oscillator with cubic stiffness, called the Duffing oscillator, satisfies the equation

$$\ddot{X}(t) + 2\,\zeta\,\nu\,\dot{X}(t) + \nu^2\left(X(t) + \alpha\,X(t)^3\right) = W(t),$$

where $\zeta \in (0, 1)$ is the damping ratio, ν denotes the initial frequency, α is a real constant, and W is a Gaussian white noise with mean zero and one-sided spectral density of intensity $g_0 > 0$. The characteristic function of X satisfies the partial differential equation

$$\frac{\partial \varphi}{\partial t} = u_1\,\frac{\partial \varphi}{\partial u_2} - \nu^2\,u_2\,\frac{\partial \varphi}{\partial u_1} + \nu^2\,\alpha\,u_2\,\frac{\partial^3 \varphi}{\partial u_1^3} - 2\,\zeta\,\nu\,u_2\,\frac{\partial \varphi}{\partial u_2} - \frac{\pi\,g_0}{2}\,u_2^2\,\varphi$$

with initial condition given by the distribution of the state X at $t = 0$. ◇

Proof: The vector process $(X_1 = X, X_2 = \dot{X})$ satisfies the differential equation

$$\begin{cases} dX_1(t) &= X_2(t)\,dt, \\ dX_2(t) &= -\nu^2\left(X_1(t) + \alpha\,X_1(t)^3\right)dt - 2\,\zeta\,\nu\,X_2(t)\,dt + \sqrt{\pi\,g_0}\,dB(t), \end{cases}$$

where B is a Brownian motion. The partial differential equation for the characteristic function of (X_1, X_2) is given by Eq. 7.60. This equation must satisfy the boundary conditions $\varphi(0, 0; t) = 1$ and φ, $\partial^k \varphi/\partial u_1^k$, $\partial \varphi/\partial u_2 \to 0$ as $u_1^2 + u_2^2 \to \infty$ for $k = 1, 2, 3$ since the random variables $X_1(t)$ and $X_2(t)$ have finite moments (Eq. 7.62). ■

Boundary conditions. Suppose that the drift and diffusion coefficients in Eq. 7.55 are polynomials of $\boldsymbol{X}$ so that Eqs. 7.60 and 7.61 are partial differential equations for the characteristic function $\varphi(\cdot; t)$ of $\boldsymbol{X}(t)$. Initial and boundary conditions need to be specified to solve these equations. The initial conditions result directly from the properties of the state at time $t = 0$. The boundary conditions can be obtained from the following properties of the characteristic function.

- $\varphi(\boldsymbol{0}; t) = 1$,
- $|\varphi(\boldsymbol{u}; t)| \leq 1$,
- If $\boldsymbol{X}(t)$ has a density, then $\varphi(\boldsymbol{u}; t) \to 0$ as $\|\,\boldsymbol{u}\,\| \to \infty$,
- If $\boldsymbol{X}(t)$ has a density and finite moments of order q, then

$$\frac{\partial^q \varphi(\boldsymbol{u}; t)}{\partial u_1^{q_1} \cdots u_d^{q_d}} \to 0 \text{ as } \|\,\boldsymbol{u}\,\| \to \infty,$$

where $q_i \geq 0$ are integers and $\sum_{i=1}^{d} q_i = q$. (7.62)

Note: The first two conditions result from the definition of the characteristic function.

The last two properties follow from the fact that, if g is an integrable function and $\gamma(u) = \int_{\mathbb{R}} e^{\sqrt{-1}\,u\,x}\, g(x)\,dx$, then $\gamma(u) \to 0$ as $|u| \to \infty$ ([62], Lemma 3, p. 513). This fact implies $\varphi(\boldsymbol{u}; t) \to 0$ as $\| \boldsymbol{u} \| \to \infty$ since the density f of $\boldsymbol{X}(t)$ is assumed to be integrable. If $\boldsymbol{X}(t)$ has finite moments of order q, then $x_1^{q_1} \cdots x_d^{q_d}\, f(\boldsymbol{x})$ is integrable so that the partial derivatives of order $q = \sum_{k=1}^{d} q_k$ of φ converge to zero as $\| \boldsymbol{u} \| \to \infty$. ▲

7.3.1.3 Fokker-Planck-Kolmogorov equations

The density of the solution $\boldsymbol{X}$ of Eq. 7.55 satisfies two partial differential equations, the **Fokker-Planck** or the **forward Kolmogorov** equation and the **backward Kolmogorov** equation. These equations can be derived from (1) the relationship between the characteristic and density functions and Eq. 7.60, (2) the Itô formula (Section 4.7.1.3), or (3) the Chapman-Kolmogorov formula (Examples 7.38 and 7.40).

Let $f(\boldsymbol{x}; t \mid \boldsymbol{x}_0; 0)$ be the density of $\boldsymbol{X}(t) \mid (\boldsymbol{X}(0) = \boldsymbol{x}_0)$, $t \geq 0$, where $\boldsymbol{X}(0) = \boldsymbol{x}_0$ is a deterministic initial state. If the starting time is $s > 0$ rather then zero, the density of $\boldsymbol{X}(t) \mid (\boldsymbol{X}(s) = \boldsymbol{x}_0)$ is denoted by $f(\boldsymbol{x}; t \mid \boldsymbol{x}_0; s)$, $0 < s \leq t$. The shorthand notation $f(\boldsymbol{x}; t)$ will also be used for $f(\boldsymbol{x}; t \mid \boldsymbol{x}_0; 0)$ provided there is no ambiguity. If the initial conditions are random, that is, $\boldsymbol{X}(0)$ is an $\mathbb{R}^d$-valued random variable, the density of $\boldsymbol{X}(t)$ can be calculated from $f(\boldsymbol{x}; t \mid \boldsymbol{x}_0; 0)$ and the density of $\boldsymbol{X}(0)$. It is assumed that $f(\boldsymbol{x}; t \mid \boldsymbol{x}_0; 0)$ satisfies the conditions

$$\begin{aligned}
&\lim_{|x_i| \to \infty} [a_i(\boldsymbol{x}, t)\, f(\boldsymbol{x}; t \mid \boldsymbol{x}_0; 0)] = 0, \quad i = 1, \ldots, d, \\
&\lim_{|x_i| \to \infty} (\boldsymbol{b}(\boldsymbol{x}, t)\, \boldsymbol{b}(\boldsymbol{x}, t)^T)_{ij}\, f(\boldsymbol{x}; t \mid \boldsymbol{x}_0; 0) = 0, \quad i, j = 1, \ldots, d, \quad \text{and} \\
&\lim_{|x_i| \to \infty} \partial[(\boldsymbol{b}(\boldsymbol{x}, t)\, \boldsymbol{b}(\boldsymbol{x}, t)^T)_{ij}\, f(\boldsymbol{x}; t \mid \boldsymbol{x}_0; 0)]/\partial x_i = 0, \quad i, j = 1, \ldots, d.
\end{aligned} \tag{7.63}$$

If $f(\boldsymbol{x}; t \mid \boldsymbol{x}_0; 0)$ satisfies the conditions in Eq. 7.63, then f is the solution of the **Fokker-Planck** equation

$$\begin{aligned}
\frac{\partial f(\boldsymbol{x}; t \mid \boldsymbol{x}_0; 0)}{\partial t} &= -\sum_{i=1}^{d} \frac{\partial}{\partial x_i} [a_i(\boldsymbol{x}, t)\, f(\boldsymbol{x}; t \mid \boldsymbol{x}_0; 0)] \\
&+ \frac{1}{2} \sum_{i,j=1}^{d} \frac{\partial^2}{\partial x_i\, \partial x_j} \left[\left(\boldsymbol{b}(\boldsymbol{x}, t)\, \boldsymbol{b}(\boldsymbol{x}, t)^T \right)_{ij} f(\boldsymbol{x}; t \mid \boldsymbol{x}_0; 0) \right].
\end{aligned} \tag{7.64}$$

If the state X becomes a stationary process X_s with density f_s as $t \to \infty$ and the conditions in Eq. 7.63 are satisfied, then f_s is the solution of the **stationary Fokker-Planck** equation

$$-\sum_{i=1}^{d} \frac{\partial}{\partial x_i} [a_i(\boldsymbol{x}, t)\, f_s(\boldsymbol{x})] + \frac{1}{2} \sum_{i,j=1}^{d} \frac{\partial^2}{\partial x_i\, \partial x_j} \left[\left(\boldsymbol{b}(\boldsymbol{x}, t)\, \boldsymbol{b}(\boldsymbol{x}, t)^T \right)_{ij} f_s(\boldsymbol{x}) \right] = 0. \tag{7.65}$$

Proof: We need to show that Eq. 7.60 multiplied by $e^{-\sqrt{-1}\,\boldsymbol{u}^T \boldsymbol{x}}/(2\pi)^d$ and integrated with respect to $\boldsymbol{u}$ over $\mathbb{R}^d$ gives Eq. 7.64. Alternatively, we can show that Eq. 7.64 multiplied by $e^{\sqrt{-1}\,\boldsymbol{u}^T \boldsymbol{x}}$ and integrated over $\mathbb{R}^d$ with respect to $\boldsymbol{x}$ gives Eq. 7.60. We follow the latter approach. The left side of Eq. 7.64 gives

$$\int_{\mathbb{R}^d} \frac{\partial f(\boldsymbol{x}; t)}{\partial t} e^{\sqrt{-1}\,\boldsymbol{u}^T \boldsymbol{x}}\, d\boldsymbol{x} = \frac{\partial}{\partial t} \int_{\mathbb{R}^d} e^{\sqrt{-1}\,\boldsymbol{u}^T \boldsymbol{x}} f(\boldsymbol{x}; t)\, d\boldsymbol{x} = \frac{\partial \varphi(\boldsymbol{u}; t)}{\partial t}.$$

A term of the first summation on the right side of Eq. 7.64 multiplied by $e^{\sqrt{-1}\,\boldsymbol{u}^T \boldsymbol{x}}$ and integrated over $\mathbb{R}^d$ with respect to $\boldsymbol{x}$ can be written as

$$\int_{\mathbb{R}^{d-1}} \prod_{j=1, j \neq i}^{d} \left(e^{\sqrt{-1} u_j x_j}\, dx_j \right) \left\{ \int_{\mathbb{R}} \frac{\partial}{\partial x_i} [a_i(\boldsymbol{x}, t)\, f(\boldsymbol{x}; t)]\, e^{\sqrt{-1} u_i x_i}\, dx_i \right\}.$$

The integral in brackets becomes

$$\begin{aligned}
&a_i(\boldsymbol{x}, t)\, f(\boldsymbol{x}; t)\, e^{\sqrt{-1} u_i x_i} \,|_{x_i = \pm\infty} - \sqrt{-1}\, u_i \int_{\mathbb{R}} a_i(\boldsymbol{x}, t)\, f(\boldsymbol{x}; t)\, e^{\sqrt{-1} u_i x_i}\, dx_i \\
&= -\sqrt{-1}\, u_i \int_{\mathbb{R}} a_i(\boldsymbol{x}, t)\, f(\boldsymbol{x}; t)\, e^{\sqrt{-1} u_i x_i}\, dx_i
\end{aligned}$$

using integration by parts and the first set of conditions in Eq. 7.63. Hence, the term we considered becomes $-\sqrt{-1}\, u_i\, E\left[a_i(\boldsymbol{X}(t), t)\, e^{\sqrt{-1}\,\boldsymbol{u}^T \boldsymbol{X}(t)}\right]$.

A term of the second summation on the right side of Eq. 7.64 can be written as

$$\begin{aligned}
\eta_{ij} = &\int_{\mathbb{R}^{d-2}} \prod_{k=1, k \neq i,j}^{d} \left(e^{\sqrt{-1} u_k x_k}\, dx_k \right) \\
&\times \left\{ \int_{\mathbb{R}^2} \frac{\partial^2}{\partial x_i\, \partial x_j} \left[\left(\boldsymbol{b}(\boldsymbol{x}, t)\, \boldsymbol{b}(\boldsymbol{x}, t)^T \right)_{ij} f(\boldsymbol{x}; t) \right] e^{\sqrt{-1}(u_i x_i + u_j x_j)}\, dx_i\, dx_j \right\}
\end{aligned}$$

following multiplication with $e^{\sqrt{-1}\,\boldsymbol{u}^T \boldsymbol{x}}$ and integration over $\mathbb{R}^d$ with respect to $\boldsymbol{x}$. The double integral in the brackets has also the form

$$\xi_{ij} = \int_{\mathbb{R}} e^{\sqrt{-1} u_j x_j}\, dx_j \int_{\mathbb{R}} \frac{\partial}{\partial x_i} \left\{ \frac{\partial}{\partial x_j} \left[\left(\boldsymbol{b}(\boldsymbol{x}, t)\, \boldsymbol{b}(\boldsymbol{x}, t)^T \right)_{ij} f(\boldsymbol{x}; t) \right] \right\} e^{\sqrt{-1} u_i x_i}\, dx_i.$$

Integration by parts of the inner integral in the expression of ξ_{ij} gives

$$\frac{\partial}{\partial x_j}\left[\left(b(x,t)\,b(x,t)^T\right)_{ij} f(x;t)\right] e^{\sqrt{-1}\,u_i\,x_i}\,|_{x_i=\pm\infty}$$
$$-\sqrt{-1}\,u_i \int_{\mathbb{R}} \frac{\partial}{\partial x_j}\left[\left(b(x,t)\,b(x,t)^T\right)_{ij} f(x;t)\right] e^{\sqrt{-1}\,u_i\,x_i}\,dx_i$$

so that (Eq. 7.63)

$$\xi_{ij} = -\sqrt{-1}\,u_i \int_{\mathbb{R}^2} e^{\sqrt{-1}(u_i\,x_i+u_j\,x_j)} \frac{\partial}{\partial x_j}\left[\left(b(x,t)\,b(x,t)^T\right)_{ij} f(x;t)\right] dx_i\,dx_j$$
$$= -\sqrt{-1}\,u_i \int_{\mathbb{R}} e^{\sqrt{-1}\,u_i\,x_i}\,dx_i \int_{\mathbb{R}} \frac{\partial}{\partial x_j}\left[\left(b(x,t)\,b(x,t)^T\right)_{ij} f(x;t)\right] e^{\sqrt{-1}\,u_j\,x_j}\,dx_j.$$

The above inner integral gives

$$\left(b(x,t)\,b(x,t)^T\right)_{ij} f(x;t)\,e^{\sqrt{-1}\,u_j\,x_j}\,|_{x_j=\pm\infty}$$
$$-\sqrt{-1}\,u_j \int_{\mathbb{R}} \left(b(x,t)\,b(x,t)^T\right)_{ij} f(x;t)\,e^{\sqrt{-1}\,u_j\,x_j}\,dx_j$$

upon integration by parts, so that (Eq. 7.63)

$$\xi_{ij} = -u_i\,u_j \int_{\mathbb{R}^2} e^{\sqrt{-1}\,(u_i\,x_i+u_j\,x_j)} \left(b(x,t)\,b(x,t)^T\right)_{ij} f(x;t)\,dx_i\,dx_j$$

and

$$\eta_{ij} = -u_i\,u_j \int_{\mathbb{R}^d} e^{\sqrt{-1}\,u^T x} \left(b(x,t)\,b(x,t)^T\right)_{ij} f(x;t)\,dx$$
$$= -u_i\,u_j\,E\left[e^{\sqrt{-1}\,u^T X(t)} \left(b(X(t),t)\,b(X(t),t)^T\right)_{ij}\right].$$

The proof is completed by assembling the above results.

Generally, it is not possible to find analytical solutions for Eqs. 7.64 and 7.65 (Section 7.3.1.4) so that the solution of these equations has to be obtained numerically in most applications. The available numerical solutions of the Fokker-Planck equations (1) replace the domain of definition of these equations with an open bounded subset D of $\mathbb{R}^d$ and (2) impose the boundary conditions $f(x;t \mid x_0;0) = 0$ and $\partial f(x;t \mid x_0;0)/\partial x_i = 0$ for $x \in \partial D$, where ∂D denotes the boundary of D. The solution of the resulting problem can be solved by the finite difference or finite element methods [179]. ■

Example 7.34: Let X be defined by $dX(t) = -\rho\,X(t-)\,dt + dS(t)$ for $\rho > 0$, $t \geq 0$, and $X(0) = x$, where S is a Gaussian (GWN), Poisson (PWN), or Lévy (LWN) white noise. The GWN, PWN, and LWN are interpreted as the formal derivative of the Brownian motion, compound Poisson, and α-stable processes, respectively. The increments of the Lévy white noise $dS(t) = dL_\alpha(t)$ are α-stable random variables with the characteristic function $E\left[e^{\sqrt{-1}\,u\,dL_\alpha(t)}\right] = e^{-|u|^\alpha\,dt}$, $u \in \mathbb{R}$.

The characteristic function $\varphi(u;t) = E[e^{\sqrt{-1}\,u\,X(t)}]$ of X(t) satisfies the partial differential equations

$$\frac{\partial\varphi}{\partial t} = -\rho\,u\,\frac{\partial\varphi}{\partial u} - \frac{u^2}{2}\,\varphi \qquad \text{for GWN,}$$
$$\frac{\partial\varphi}{\partial t} = -\rho\,u\,\frac{\partial\varphi}{\partial u} + \lambda\left(E\left[e^{\sqrt{-1}\,u\,Y_1}\right] - 1\right)\varphi \qquad \text{for PWN,}$$
$$\frac{\partial\varphi}{\partial t} = -\rho\,u\,\frac{\partial\varphi}{\partial u} - |u|^{\alpha}\,\varphi \qquad \text{for LWN,}$$

where $\lambda > 0$ and Y_1 are parameters of the compound Poisson process and $\alpha \in (0,2)$ defines the Lévy white noise. The density, $f(x;t \mid x_0;0)$, of the conditional process $X(t) \mid (X(0) = x_0)$ satisfies the partial differential equations

$$\frac{\partial f}{\partial t} = \rho\,\frac{\partial(x\,f)}{\partial x} + \frac{1}{2}\,\frac{\partial^2 f}{\partial x^2} \qquad \text{for GWN,}$$
$$\frac{\partial f}{\partial t} = \rho\,\frac{\partial(x\,f)}{\partial x} + \lambda\sum_{k=1}^{\infty}\frac{(-1)^k\,E[Y_1^k]}{k!}\,\frac{\partial^k f}{\partial x^k} \qquad \text{for PWN.}$$

The first formula is a Fokker-Planck equation given by Eq. 7.64. The second formula is not a classical Fokker-Planck equation, and we refer to it as a **generalized** or **extended Fokker-Planck** equation. The generalized Fokker-Planck equation is valid if Y_1 has finite moments of any order and some additional conditions stated in the following comments are satisfied. It is not possible to derive a partial differential equation for f under Lévy white noise with $\alpha \in (0,2)$. ◇

Proof: The derivation of the partial differential equations for the characteristic and density functions of X under Poisson white noise is outlined here. The differential equation for the characteristic function of X driven by a Lévy white noise is in Example 7.20. The corresponding equation for a Gaussian noise results from Eq. 7.60.

Let $C(t) = \sum_{k=1}^{N(t)} Y_k$ be a compound Poisson process, where N is a Poisson process with intensity $\lambda > 0$ and Y_k are independent copies of a random variable Y_1. The Itô formula in Eq. 7.4 applied to $e^{\sqrt{-1}\,u\,X(t)}$ gives

$$e^{\sqrt{-1}\,u\,X(t)} - 1 = -\sqrt{-1}\,\rho\,u\int_{0+}^{t} X(s-)\,e^{\sqrt{-1}\,u\,X(s-)}\,ds + \sum_{0<s\le t}\left[e^{\sqrt{-1}\,u\,X(s)} - e^{\sqrt{-1}\,u\,X(s-)}\right]$$

so that we have

$$\varphi(u;t) - 1 = -\rho\,u\int_{0+}^{t}\frac{\partial\varphi(u;s)}{\partial u}\,ds + \sum_{0<s\le t}\varphi(u;s)\,E\left[e^{\sqrt{-1}\,u\,\Delta C(s)} - 1\right]$$

by averaging. The last term has this form since

$$X(s) = X(s-) + \Delta X(s) = X(s-) + \Delta C(s),$$

$X(s-)$ is independent of future jumps $\Delta C(s) = C(s) - C(s-)$ of C, and $\Delta C(s)$ is either Y_k for some k or zero in any small time interval $\Delta s > 0$ so that the random variable $e^{\sqrt{-1}\,u\,\Delta C(s)} - 1$ is $e^{\sqrt{-1}\,u\,Y_k} - 1$ with probability $\lambda\,\Delta s$ and zero with probability $1 - \lambda\,\Delta s$ as $\Delta s \downarrow 0$. The partial differential equation for φ results by differentiation.

The Fourier transform of the left side of the partial differential equation for φ is $\partial f/\partial t$. The first term on the right side of the Fokker-Planck equation is given by the corresponding term of the equation for φ by using integration by parts, the boundary conditions, and the equality

$$\frac{\partial^k f(x;t)}{\partial x^k} = \frac{1}{2\pi}\int_{\mathbb{R}} (-\sqrt{-1}\,u)^k\, e^{-\sqrt{-1}\,u\,x}\,\varphi(u;t)\,du,$$

which is valid for any integer $k \geq 1$. The Fourier transform of $\varphi(u;t)\,E[e^{\sqrt{-1}\,u\,Y_1}]$ is

$$\begin{aligned}
&\frac{1}{2\pi}\int_{\mathbb{R}} e^{-\sqrt{-1}\,u\,x}\,\varphi(u;t)\,E[e^{\sqrt{-1}\,u\,Y_1}]\,du \\
&= \frac{1}{2\pi}\int_{\mathbb{R}} e^{-\sqrt{-1}\,u\,x}\,\varphi(u;t)\,E\left[\sum_{k=0}^{\infty}\frac{(\sqrt{-1}\,u\,Y_1)^k}{k!}\right]du \\
&= \sum_{k=0}^{\infty}\frac{E[Y_1^k]}{k!}\,\frac{1}{2\pi}\int_{\mathbb{R}} e^{-\sqrt{-1}\,u\,x}\,\varphi(u;t)\,(\sqrt{-1}\,u)^k\,du = \sum_{k=1}^{\infty}\frac{(-1)^k\,E[Y_1^k]}{k!}\,\frac{\partial^k f}{\partial x^k},
\end{aligned}$$

where it is assumed that Y_1 has finite moments of any order,

$$E[e^{\sqrt{-1}\,u\,Y_1}]\,du = \sum_{k=0}^{\infty} E\left[\frac{(\sqrt{-1}\,u\,Y_1)^k}{k!}\right],$$

and the above integration can be performed term by term. The last equality follows from the relationships between the density and the characteristic functions given above. ■

The Fokker-Planck equations (Eqs. 7.64 and 7.65) are deterministic, linear partial differential equations of the second order defined for all $t \geq 0$ and $\boldsymbol{x} \in \mathbb{R}^d$. Conditions for the existence and uniqueness of the solution of these equations can be found in [44] (Chapter IV). Comprehensive discussions on the Fokker-Planck equations are in [70, 174].

In reliability studies and other applications we are interested in some global properties of the state $\boldsymbol{X}$ defined by Eq. 7.55, for example, the distribution of the stopping time $T = \inf\{t > 0 : \boldsymbol{X}(t) \notin D\}$ corresponding to an open subset D of $\mathbb{R}^d$ and deterministic initial state $\boldsymbol{X}(0) = \boldsymbol{x}_0 \in D$. Let $\boldsymbol{X}^T(t) = \boldsymbol{X}(t \wedge T)$ denote the process $\boldsymbol{X}$ stopped at T. The samples of $\boldsymbol{X}^T$ and $\boldsymbol{X}$ coincide for times $t \leq T$ but differ for $t > T$. The samples of $\boldsymbol{X}^T$ are constant and equal to $\boldsymbol{X}(T) \in \partial D$ at times $t \geq T$. If $\boldsymbol{X}$ is the state of a physical system and D denotes a safe set for this system, the probability of the event $\{\boldsymbol{X}^T(t) \in D\}$ gives the fraction of samples of $\boldsymbol{X}$ that never left D in $[0, t]$, so that

$$P_s(t) = P(T > t) = P\left(\boldsymbol{X}^T(t) \in D\right), \quad t \geq 0, \tag{7.66}$$

is the system reliability in $[0, t]$. We will see that the probability $P_s(t)$ in Eq. 7.66 can be calculated from the solution of the Fokker-Planck equation for the density of $\boldsymbol{X}$ with appropriate boundary conditions. We review now some boundary conditions that are relevant for reliability studies and other applications. Extensive discussions on the type and classification of boundary conditions for diffusion processes and the Fokker-Planck equation can be found elsewhere ([70], Chapter 5, [111], Section 15.6).

Boundary conditions. Let $p_D(t) = \int_D f(\boldsymbol{x}; t \mid \boldsymbol{x}_0; 0)\, d\boldsymbol{x}$ be the probability that the solution $\boldsymbol{X}$ of Eq. 7.55 belongs to an open subset D of $\mathbb{R}^d$ at a time $t \geq 0$, where $\boldsymbol{X}(0) = \boldsymbol{x}_0 \in D$ and $f(\boldsymbol{x}; t \mid \boldsymbol{x}_0; 0)$ denotes the density of the $\mathbb{R}^d$-valued random variable $\boldsymbol{X}(t) \mid (\boldsymbol{X}(0) = \boldsymbol{x}_0)$. The evolution of $f(\boldsymbol{x}; t \mid \boldsymbol{x}_0; 0)$ and p_D in time is given by the following two equations.

$$\frac{\partial f}{\partial t} = -\sum_{i=1}^{d} \frac{\partial \lambda_i(\boldsymbol{x}; t)}{\partial x_i}, \quad \text{where} \tag{7.67}$$

$$\lambda_i(\boldsymbol{x}; t) = a_i\, f - \frac{1}{2} \sum_{j=1}^{d} \frac{\partial}{\partial x_j}[(\boldsymbol{b}\,\boldsymbol{b}^T)_{ij}\, f], \quad i = 1, \ldots, d, \tag{7.68}$$

are the coordinates of a vector $\boldsymbol{\lambda}(\boldsymbol{x}; t) \in \mathbb{R}^d$, called the **probability current**.

$$\frac{\partial p_D(t)}{\partial t} = -\sum_{i=1}^{d} \int_D \frac{\partial \lambda_i(\boldsymbol{x}; t)}{\partial x_i}\, d\boldsymbol{x} = -\sum_{i=1}^{d} \int_{\partial D} \lambda_i(\boldsymbol{x}; t)\, n_i(\boldsymbol{x})\, d\sigma(\boldsymbol{x}), \tag{7.69}$$

where $\boldsymbol{n}(\boldsymbol{x}) = (n_1(\boldsymbol{x}), \ldots, n_d(\boldsymbol{x}))$ is the exterior normal at $\boldsymbol{x} \in \partial D$.

Note: We note that Eq. 7.67 is an alternative form of the Fokker-Planck equation (Eq. 7.64). The rate of change of $p_D(t)$ is

$$\begin{aligned}\frac{\partial p_D(t)}{\partial t} &= \int_D \frac{\partial f(\boldsymbol{x}, t \mid \boldsymbol{x}_0, 0)}{\partial t}\, d\boldsymbol{x} \\ &= -\sum_{i=1}^{d} \int_D \frac{\partial \lambda_i(\boldsymbol{x}, t)}{\partial x_i}\, d\boldsymbol{x} = -\sum_{i=1}^{d} \int_{\partial D} \lambda_i(\boldsymbol{x}, t)\, n_i(\boldsymbol{x})\, d\sigma(\boldsymbol{x}),\end{aligned}$$

where the last equality in the above equation follows from the Gauss or the divergence theorem ([68], p. 116). The function

$$-\sum_{i=1}^{d} \lambda_i(\boldsymbol{x}, t)\, n_i(\boldsymbol{x})\, d\sigma(\boldsymbol{x}) = -\boldsymbol{\lambda}(\boldsymbol{x}, t) \cdot \boldsymbol{n}(\boldsymbol{x})\, d\sigma(\boldsymbol{x})$$

gives the **probability flow** from D to D^c through an infinitesimal surface element $d\sigma(\boldsymbol{x})$ of ∂D. Hence, $-\int_{\partial D'} \boldsymbol{\lambda}(\boldsymbol{x}, t) \cdot \boldsymbol{n}(\boldsymbol{x})\, d\sigma(\boldsymbol{x})$ represents the probability flow from D to D^c through a subset $\partial D'$ of ∂D.

For numerical calculations, if D is not bounded, we have to approximate it by a bounded subset $\tilde{D} \subset D$. For example, if $D = \{x \in \mathbb{R}^d : x_1 < a\}$, $a \in \mathbb{R}$, we can take $\tilde{D} = \{x \in \mathbb{R}^d : x_1 \in (-\alpha_1, a), x_i \in (-\alpha_i, \alpha_i), i = 2, \ldots, d\}$ and solve Eq. 7.69 with $\tilde{D}$ in place of D, where $\alpha_i > 0$ are some constants. This approximation is adequate for large values of α_i since f and its first order partial derivatives vanish as $\| x \| \to \infty$ so that the contribution of the boundaries of $\tilde{D}$ to the surface integral in Eq. 7.69 is virtually limited to the hyperplane $x_1 = a$ [179]. ▲

Let T be the stopping time in Eq. 7.66 corresponding to a solution X of Eq. 7.55 with $X(0) = x_0 \in D$ and an open subset D of $\mathbb{R}^d$. We define two types of boundaries depending on the properties of T.

- If $T = \infty$ a.s., ∂D is an **inaccessible boundary**. An inaccessible boundary is **natural** or **attracting** if X never reaches ∂D or $\lim_{t\to\infty} X(t) \in \partial D$, respectively.
- If $T < \infty$ a.s., ∂D is an **accessible boundary**. An accessible boundary is **reflecting** or **absorbing** if $\boldsymbol{\lambda}(x; t) \cdot n(x) = 0$, $x \in \partial D$, or $f(x; t \mid x_0; 0) = 0$, $x \in \partial D$, respectively.

Note: The above definitions can be applied to subsets of the boundary ∂D of D. Distinct subsets of ∂D can be accessible or inaccessible boundaries.

The probability $P\left(X^T(t) \in D^c\right)$ is zero at $t = 0$ since $X(0) = x_0 \in D$. This probability is equal to $P(T \leq t)$ (Eq. 7.66) and increases in time since more and more samples of X reach the boundary of D. The density of $X^T(t)$ in D is the solution of Eq. 7.64 with the absorbing boundary condition $f(x; t \mid x_0; 0) = 0$ for $x \in \partial D$. This boundary condition eliminates the samples of X once they reach ∂D so that $P(T > t)$ is equal to $p_D(t)$ in Eq. 7.69.

Absorbing boundary conditions are used in reliability analysis to characterize system performance (Section 7.4.3). We have used reflecting boundaries in Section 6.2.3 to find local solutions for deterministic partial differential equations with Neumann boundary conditions. ▲

Example 7.35: Let X be a geometric Brownian motion defined by $dX(t) = c\,X(t)\,dt + \sigma\,X(t)\,dB(t)$, $t \geq 0$, where c, σ are constants, B denotes a Brownian motion, and $X(0) > 0$. If $c - \sigma^2/2 \geq 0$, the boundary $x = 0$ is natural. Otherwise, $x = 0$ is an attracting boundary. ◇

Proof: We have seen that $X(t) = X(0)\,e^{(c-\sigma^2/2)\,t+\sigma\,B(t)}$ (Example 7.32) so that $X(t) \geq 0$ at all times since $X(0) > 0$ by hypothesis. The process can be zero if and only if $(c - \sigma^2/2)\,t + \sigma\,B(t)$ becomes $-\infty$. The function $(c - \sigma^2/2)\,t + \sigma\,B(t)$ is bounded a.s. for t finite and

$$\lim_{t\to\infty} \frac{1}{t}[(c - \sigma^2/2)\,t + \sigma\,B(t)] = (c - \sigma^2/2) + \lim_{t\to\infty}[\sigma\,B(t)/t] = (c - \sigma^2/2)$$

since $\lim_{t\to\infty} B(t)/t = 0$ a.s. ([150], Section 6.4). If $c - \sigma^2/2 < 0$, then $\lim_{t\to\infty} X(t) = 0$ a.s. so that $x = 0$ is an inaccessible attracting boundary. Otherwise, the boundary $x = 0$ cannot ever be reached so that $x = 0$ is an inaccessible natural boundary. ■

Example 7.36: Let B be a Brownian motion. Then

$$f(x;t \mid y;0) = \frac{2}{\sqrt{t}}\,\phi\left(\frac{x}{\sqrt{t}}\right)$$

for $B(0) = 0$ and a reflecting boundary at $x = 0$. ◇

Proof: The density $f(x;t \mid y;0)$ satisfies the Fokker-Planck equation

$$\frac{\partial f}{\partial t} = \frac{1}{2}\frac{\partial^2 f}{\partial x^2}$$

with the boundary condition $\partial f(x;t \mid 0;0)/\partial x = 0$ at $x = 0$, $t > 0$, since $\lambda(x;t) = -(1/2)\,\partial f(x;t \mid 0;0)/\partial x$. The density f can be calculated from the above Fokker-Planck equation with the stated boundary condition and the initial condition $f(x;0 \mid 0;0) = 1$ for $x = 0$ and zero otherwise.

We follow here a different approach using the observation that $B \mid (B(0) = 0)$ reflected at $x = 0$ is equal in distribution with $|B| \mid (B(0) = 0)$. Because $B(t) \mid (B(0) = 0)$ and $\sqrt{t}\,N(0,1)$ have the same distribution, we have for $\xi > 0$

$$P(|B(t)| \le \xi \mid (B(0) = 0)) = P(-\xi < \sqrt{t}\,N(0,1) \le \xi) = \Phi\left(\frac{\xi}{\sqrt{t}}\right) - \Phi\left(\frac{-\xi}{\sqrt{t}}\right),$$

so that the corresponding density is $f(x;t \mid 0;0) = (d/d\xi)P(|B| \le \xi \mid B(0) = y)$ is as stated ([154], Section 6.3). ■

Example 7.37: Let $dX(t) = (\alpha\,X(t) + \beta\,X(t)^3)\,dt + \sigma\,dB(t)$ be a stochastic differential equation defining a real-valued diffusion process X starting at $X(0) = x_0 \in \mathbb{R}$, where α, β, σ are some constants.

Figure 7.11 shows the evolution in time of the density $f(x;t \mid x_0;0)$ of $X(t) \mid (X(0) = x_0)$ for $t \in [0, 0.125]$, $\alpha = \beta = -1$, $\sigma = 1$, and $x_0 = 0$. This density has been obtained by a finite difference solution of the Fokker-Planck equation in the interval $[-2, 2]$ with zero boundary conditions, time step $\Delta t = 0.00625$, and space step $\Delta x = 0.1$. The plot shows that $f(x;t \mid x_0;0)$ is practically time invariant toward the end of the time interval. We also can see that the selected integration interval $[-2, 2]$ is adequate since the entire probability mass of $X(t)$ seems to be contained in this interval. ◇

Note: The density $f(x;t \mid x_0;0)$ of $X(t) \mid (X(0) = x_0)$ is the solution of the Fokker-Planck equation

$$\frac{\partial f}{\partial t} = -\frac{\partial}{\partial x}((\alpha\,x + \beta\,x^3)\,f) + \frac{\sigma^2}{2}\frac{\partial^2 f}{\partial x^2}$$

with $f(-2,t \mid x_0;0) = f(2,t \mid x_0;0) = 0$.

Let $\Delta x > 0$ and $\Delta t > 0$ be the space and time steps defining the nodes $x_i = -2 + i\,\Delta x$, $i = 0, 1, \ldots$, and $t_j = j\,\Delta t$, $j = 0, 1, \ldots$, of the finite difference mesh. The finite difference approximation of the Fokker-Planck equation at node (i, j) is

$$\frac{f_{i,j+1} - f_{i,j}}{\Delta t} = -(\alpha + 3\,\beta\,x_i^2)\,f_{i,j} - (\alpha\,x_i + \beta\,x_i^3)\,\frac{f_{i+1,j} - f_{i-1,j}}{2\,\Delta x} + \frac{\sigma^2}{2}\,\frac{f_{i+1,j} - 2\,f_{i,j} + f_{i-1,j}}{(\Delta x)^2},$$

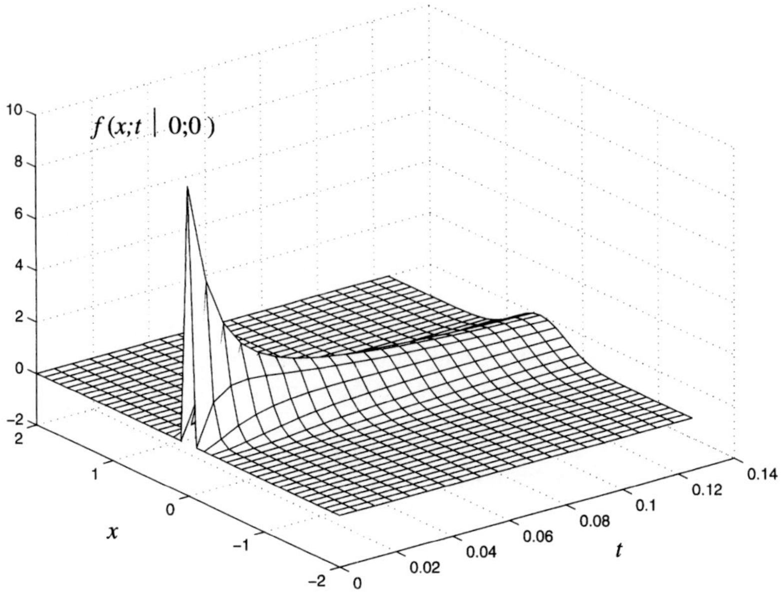

Figure 7.11. The evolution of the density of $X(t) \mid X(0) = x_0$ for $\alpha = \beta = -1$, $\sigma = 1$, and $x_0 = 0$

where $f_{i,j} = f(x_i; t_j \mid x_0; 0)$. The above linear system of equations has been used to propagate the vector $\boldsymbol{f}^{(j)} = (f_{1,j}, f_{2,j}, \ldots)$ in time. ▲

Example 7.38: Let X be a diffusion process defined by the stochastic differential equation $dX(t) = a(X(t))\,dt + b(X(t))\,dB(t)$ and $f(x; t \mid x_0; 0)$ denote the density of $X(t) \mid (X(0) = x_0)$. The Markov property of X can be used to derive the Fokker-Planck equation for the density function $f(x; t \mid x_0; 0)$ under the conditions in Eq. 7.63. ◇

Proof: Let $g \in \mathcal{C}(\mathbb{R}^2)$ be an arbitrary real-valued function and consider the integral $\eta = \int_{\mathbb{R}} \left[f(x; t+h \mid x_0; 0) - f(x; t \mid x_0; 0) \right] g(x)\,dx$, $h > 0$. Because X is a Markov process, the Chapman-Kolmogorov equation can be used and gives (Section 3.6.3)

$$f(x; t+h \mid x_0; 0) = \int_{\mathbb{R}} f(x; t+h \mid \xi; t)\, f(\xi; t \mid x_0; 0)\,d\xi$$

so that

$$\begin{aligned}\eta &= \int_{\mathbb{R}} \left[\int_{\mathbb{R}} f(x; t+h \mid \xi; t)\, f(\xi; t \mid x_0; 0)\,d\xi - f(x; t \mid x_0; 0) \right] g(x)\,dx \\ &= \int_{\mathbb{R}} \left[\int_{\mathbb{R}} f(\xi; t+h \mid x; t)\, g(\xi)\,d\xi - g(x) \right] f(x; t \mid x_0; 0)\,dx\end{aligned}$$

and

$$\eta/h = \int_{\mathbb{R}} \left[\int_{\mathbb{R}} f(\xi; t+h \mid x; t) \, \frac{g(\xi) - g(x)}{h} \, d\xi \right] f(x; t \mid x_0; 0) \, dx.$$

The Fokker-Planck equation results by taking the limit as $h \downarrow 0$ of the above expression with the approximation

$$\frac{g(\xi) - g(x)}{h} = g'(x)\,(\xi - x) + \frac{1}{2}\, g''(x)\,(\xi - x)^2 + o((\xi - x)^2), \quad h \downarrow 0,$$

and using integration by parts and the fact that g is an arbitrary function. ■

The density $f(\boldsymbol{x}; t \mid \boldsymbol{\xi}; s)$ of the conditional vector $\boldsymbol{X}(t) \mid (\boldsymbol{X}(s) = \boldsymbol{\xi})$, $0 < s < t$, satisfies, in addition to the Fokker-Planck equation, another partial differential equation, called the backward Kolmogorov equation.

If $\boldsymbol{X}$ is the solution of Eq. 7.55, then $f(\boldsymbol{x}; t \mid \boldsymbol{y}; s)$, $0 < s < t$, satisfies the **backward Kolmogorov** equation

$$\frac{\partial f(\boldsymbol{x}; t \mid \boldsymbol{y}; s)}{\partial s} = -\sum_{i=1}^{d} a_i(\boldsymbol{y}, s) \, \frac{\partial f(\boldsymbol{x}; t \mid \boldsymbol{y}; s)}{\partial y_i} - \frac{1}{2} \sum_{i,j=1}^{d} \left(\boldsymbol{b}(\boldsymbol{y}, s)\, \boldsymbol{b}(\boldsymbol{y}, s)^T \right)_{ij} \frac{\partial^2 f(\boldsymbol{x}; t \mid \boldsymbol{y}; s)}{\partial y_i \, \partial y_j}. \tag{7.70}$$

Proof: Let $u(\boldsymbol{\xi}, \sigma) = E^{(\boldsymbol{\xi}, \sigma)}[g(\boldsymbol{X}(t))]$, where σ is an arbitrary point in $(0, t)$, the superscripts of $E^{(\boldsymbol{\xi}, \sigma)}$ indicate the condition $\boldsymbol{X}(\sigma) = \boldsymbol{\xi}$, and $g \in C^2(\mathbb{R}^d)$. Take $s, h > 0$ such that $0 < s - h < s < t$ and $\boldsymbol{X}(s-h) = \boldsymbol{y}$. We have

$$\lim_{h \downarrow 0} \frac{E^{(\boldsymbol{y}, s-h)}[g(\boldsymbol{X}(s))] - g(\boldsymbol{y})}{h} = \lim_{h \downarrow 0} \frac{E^{(\boldsymbol{y}, s-h)}[g(\boldsymbol{X}(s)) - g(\boldsymbol{y})]}{h}$$

$$= \lim_{h \downarrow 0} \frac{1}{h} E^{(\boldsymbol{y}, s-h)} \left[\int_{s-h}^{s} \left(\sum_{i=1}^{d} \frac{\partial g(\boldsymbol{X}(\sigma))}{\partial x_i} \, dX_i(\sigma) + \frac{1}{2} \sum_{i,j=1}^{d} \left(\boldsymbol{b}(\boldsymbol{X}(\sigma), \sigma)\, \boldsymbol{b}(\boldsymbol{X}(\sigma), \sigma)^T \right)_{ij} \frac{\partial^2 g(\boldsymbol{X}(\sigma))}{\partial x_i \, \partial x_j} \, d\sigma \right) \right]$$

$$= \sum_{i=1}^{d} a_i(\boldsymbol{y}, s) \, \frac{\partial g(\boldsymbol{y})}{\partial y_i} + \frac{1}{2} \sum_{i,j=1}^{d} \left(\boldsymbol{b}(\boldsymbol{y}, s)\, \boldsymbol{b}(\boldsymbol{y}, s)^T \right)_{ij} \frac{\partial^2 g(\boldsymbol{y})}{\partial y_i \, \partial y_j} = \mathcal{A}[g(\boldsymbol{y})],$$

where the second equality in the above equation holds by Itô's formula and $\mathcal{A}$ is the generator of the diffusion process $\boldsymbol{X}$ (Section 6.2.1.1). The definition of the function u gives

$$u(\boldsymbol{y}, s-h) = E^{(\boldsymbol{y}, s-h)}[g(\boldsymbol{X}(t))] = E^{(\boldsymbol{y}, s-h)} \left\{ E^{(\boldsymbol{X}(s), s)}[g(\boldsymbol{X}(t))] \right\}$$

$$= E^{(\boldsymbol{y}, s-h)}[u(\boldsymbol{X}(s), s)]$$

so that

$$\frac{\partial u(\boldsymbol{y}, s)}{\partial s} = \lim_{h \downarrow 0} \frac{u(\boldsymbol{y}, s) - u(\boldsymbol{y}, s-h)}{h}$$
$$= -\lim_{h \downarrow 0} \frac{E^{(\boldsymbol{y}, s-h)}[u(\boldsymbol{X}(s), s)] - u(\boldsymbol{y}, s)}{h} = -\mathcal{A}[u(\boldsymbol{y}, s)]$$

or $\partial u(\boldsymbol{y}, s)/\partial s + \mathcal{A}[u(\boldsymbol{y}, s)] = 0$. The functions $\partial u/\partial s$ and $\mathcal{A}[u]$ are

$$\frac{\partial u(\boldsymbol{y}, s)}{\partial s} = \int_{\mathbb{R}^d} g(\boldsymbol{x}) \frac{\partial f(\boldsymbol{x}; t \mid \boldsymbol{y}; s)}{\partial s} d\boldsymbol{x} \quad \text{and}$$
$$\mathcal{A}[u(\boldsymbol{y}, s)] = \int_{\mathbb{R}^d} g(\boldsymbol{x}) \mathcal{A}[f(\boldsymbol{x}; t \mid \boldsymbol{y}; s)] d\boldsymbol{x},$$

where $f(\boldsymbol{x}; t \mid \boldsymbol{y}; s)$ denotes the density of $\boldsymbol{X}(t) \mid (\boldsymbol{X}(s) = \boldsymbol{y})$. The last equality in the above equation holds since $\mathcal{A}$ does not depend on the variable of integration. Because g is arbitrary, we have

$$\frac{\partial f(\boldsymbol{x}; t \mid \boldsymbol{y}; s)}{\partial s} + \mathcal{A}[f(\boldsymbol{x}; t \mid \boldsymbol{y}; s)] = 0.$$

If the drift and diffusion coefficients of $\boldsymbol{X}$ do not depend explicitly on time, then the transition density of $\boldsymbol{X}$ depends only on the time lag, that is, $f(\boldsymbol{x}; t \mid \boldsymbol{y}; s) = f(\boldsymbol{x}; \tau \mid \boldsymbol{y}; 0)$, where $\tau = t - s$ is the time lag. Because $\partial f/\partial s = -\partial f/\partial \tau$, the above equation becomes $\partial f/\partial \tau = \mathcal{A}[f]$. This version of the backward Kolmogorov equation is particularly useful for calculating moments of the first exit time T in Eq. 7.66 as illustrated by the following example and [175] (Section 7.4.2). ■

Example 7.39: Let X be defined by $dX(t) = a(X(t))\,dt + b(X(t))\,dB(t)$. The mean μ_T of the first exit time $T = \inf\{t \geq 0 : X(t) \notin (\alpha, \beta)\}$ for $X(0) = x_0 \in D = (\alpha, \beta)$ is the solution of

$$-1 = a(x_0) \frac{d\mu_T}{dx_0} + \frac{1}{2} b(x_0)^2 \frac{d^2 \mu_T}{dx_0^2}$$

with the boundary conditions $\mu_T(\alpha) = \mu_T(\beta) = 0$. ◇

Proof: Let $f(x, t \mid \boldsymbol{x}_0; 0)$ be the solution of the Fokker-Planck equation for X with the boundary conditions $f(\alpha, t \mid \boldsymbol{x}_0; 0) = f(\beta, t \mid \boldsymbol{x}_0; 0) = 0$. The function $p_D(t) = \int_\alpha^\beta f(x; t \mid \boldsymbol{x}_0; 0)\,dx$ gives the fraction of samples of X starting at $x_0 \in D$ that have not reached the boundaries of D in $[0, t]$, that is, the probability $P(T > t)$. Integrating the backward Kolmogorov equation with respect to x over (α, β), we find

$$\frac{\partial p_D}{\partial \tau} = a(x_0) \frac{\partial p_D}{\partial x_0} + \frac{1}{2} b(x_0)^2 \frac{\partial^2 p_D}{\partial x_0^2}$$

since the drift and diffusion coefficients of X do not depend explicitly on time. The integral of the above equation over the time interval $[0, \infty)$ gives an equation for the mean of T since $\int_0^\infty (\partial p_D/\partial \tau)\,d\tau = p_D(\infty) - p_D(0) = -1$ and $\int_0^\infty p_D(\tau)\,d\tau = \mu_T$. ■

Example 7.40: The Chapman-Kolmogorov equation can be used to derive the backward Kolmogorov equation in Eq. 7.70.

Proof: Let X be a diffusion process defined by Eq. 7.55 with $d = d' = 1$ and let $0 < s < s + h < t$, where $h > 0$. Consider the density $f(x; t \mid z; s + h)$ of $X(t) \mid (X(s + h) = z)$ and view it as a function of z. The second order Taylor expansion of this function about $z = \xi$ is

$$f(x; t \mid z; s+h) = f(x; t \mid \xi; s+h) + \frac{\partial f(x; t \mid z; s+h)}{\partial z} \Big|_{z=\xi} (z - \xi) + \frac{1}{2} \frac{\partial^2 f(x; t \mid z; s+h)}{\partial z^2} \Big|_{z=\xi} (z-\xi)^2 + o((z-\xi)^2).$$

This representation of $f(x; t \mid z; s + h)$ and the Chapman-Kolmogorov equation,

$$f(x; t \mid \xi; s) = \int_{\mathbb{R}} f(x; t \mid z; s+h)\, f(z; s+h \mid \xi; s)\, dz,$$

give

$$\begin{aligned} & f(x; t \mid \xi; s) \\ & = f(x; t \mid \xi; s+h) + \frac{\partial f(x; t \mid z; s+h)}{\partial z} \Big|_{z=\xi} \int_{\mathbb{R}} (z-\xi)\, f(z; s+h \mid \xi; s)\, dz \\ & + \frac{1}{2} \frac{\partial^2 f(x; t \mid z; s)}{\partial z^2} \Big|_{z=\xi} \int_{\mathbb{R}} (z-\xi)^2 f(z; s+h \mid \xi; s)\, dz \\ & + \int_{\mathbb{R}} o((z-\xi)^2)\, f(z; s+h \mid \xi; s)\, dz \end{aligned}$$

or

$$\begin{aligned} & \frac{f(x; t \mid \xi; s) - f(x; t \mid \xi; s+h)}{h} \\ & = \frac{\partial f(x; t \mid z; s+h)}{\partial z} \Big|_{z=\xi} \frac{1}{h} \int_{\mathbb{R}} (z-\xi)\, f(z; s+h \mid \xi; s)\, dz \\ & + \frac{1}{2} \frac{\partial^2 f(x; t \mid z; s+h)}{\partial z^2} \Big|_{z=\xi} \frac{1}{h} \int_{\mathbb{R}} (z-\xi)^2 f(z; s+h \mid \xi; s)\, dz \\ & + \frac{1}{h} \int_{\mathbb{R}} o((z-\xi)^2)\, f(z; s+h \mid \xi; s)\, dz. \end{aligned}$$

The above equation yields the backward Kolmogorov equation by taking the limit as $h \downarrow 0$ and noting that the last two integrals in this equation converge to the drift and diffusion coefficients of X ([111], Section 15.2). ■

7.3.1.4 Exact solutions

Generally, it is not possible to find analytical solutions for the state $\boldsymbol{X}$ in Eq. 7.55 or the Fokker-Planck and backward Kolmogorov equations (Eqs. 7.64, 7.65, and 7.70). We have obtained in Section 4.7.1.2 analytical solutions for the state of some simple diffusion equations. Additional examples can be found in [115] (Section 4.4). This section gives examples of nonlinear systems for which the Fokker-Planck equation admits analytical solutions. Other examples are available in [67].

Example 7.41: Let X be a diffusion process defined by (Example 7.30)

$$dX(t) = a(X(t))\,dt + b(X(t))\,dB(t).$$

If the conditions in Eq. 7.63 are satisfied and the drift and diffusion coefficients are such that X becomes a stationary process X_s as $t \to \infty$, then the density of X_s is

$$f_s(x) = \frac{c}{b(x)^2} \exp\left[2 \int^x \frac{a(\sigma)}{b(\sigma)^2}\,d\sigma\right],$$

where $c > 0$ is a constant and $\int^x \alpha(\sigma)\,d\sigma$ denotes the primitive of α. ◇

Proof: We have (Eq. 7.65)

$$\frac{d}{dx}\left(a(x)\,f_s(x)\right) - \frac{1}{2}\frac{d^2}{dx^2}\left(b(x)^2\,f_s(x)\right) = 0$$

so that $a(x)\,f_s(x) - (1/2)\left(b(x)^2\,f_s(x)\right)' = -c_1$ or $h'(x) - 2\,\xi(x)\,h(x) = -2\,c_1$, where $h = b^2\,f_s$, $\xi = a/b^2$, and c_1 is a constant, which must be zero by Eq. 7.63. Therefore, we have $h'(x) - 2\,\xi(x)\,h(x) = 0$, which yields the stated expression for f_s. ■

Example 7.42: Let X be the displacement of a simple oscillator defined by

$$\ddot{X}(t) + \rho\,\dot{X}(t) + u'(X(t)) = W(t),$$

where $\rho > 0$ is the viscous damping coefficient, u denotes a potential, $u'(x) = du(x)/dx$, and W is a stationary Gaussian white noise process with mean zero and one-sided spectral density $g(\nu) = g_0 > 0$ for all $\nu \geq 0$. The total mechanical energy of the system is $h(X, \dot{X}) = u(X) + \dot{X}^2/2$. Denote the partial derivatives of this function by $h_{,k} = \partial h/\partial x_k$, where $X_1 = X$ and $X_2 = \dot{X}$. The stationary density of $(X, \dot{X})$ is

$$f_s(x, \dot{x}) = c\,\exp\left[-\frac{2\,\rho}{\pi\,g_0}\,h(x, \dot{x})\right],$$

where $c > 0$ is a constant. ◇

Proof: The diffusion process $\boldsymbol{X} = (X_1 = X, X_2 = \dot{X})$ is the solution of

$$\begin{cases} dX_1(t) &= h_{,2}(\boldsymbol{X}(t))\,dt, \\ dX_2(t) &= -\left[h_{,1}(\boldsymbol{X}(t)) + \rho\,h_{,2}(\boldsymbol{X}(t))\right]dt, + \sqrt{\pi\,g_0}\,dB(t), \end{cases}$$

where B is a Brownian motion. The stationary Fokker-Planck equation for the density f_s of the stationary solution X_s is (Eq. 7.65)

$$-\frac{\partial}{\partial x_1}\left(h_{,2}\, f_s\right) - \frac{\partial}{\partial x_2}\left(-(h_{,1} + \rho\, h_{,2})\, f_s\right) + \frac{\pi\, g_0}{2}\,\frac{\partial^2 f_s}{\partial x_2^2} = 0.$$

Suppose that there exists a function g such that $f_s(\boldsymbol{x}) = g(h(\boldsymbol{x}))$. Then the above Fokker-Planck equation becomes

$$\rho\left[(h_{,2})^2\, g'(h) + g(h)\right] + \frac{\pi\, g_0}{2}\left[(h_{,2})^2\, g''(h) + g'(h)\right] = 0$$

since $f_{s,1} = g'(h)\, h_{,1}$, $f_{s,2} = g'(h)\, h_{,2}$, $f_{s,22} = (h_{,2})^2\, g''(h) + g'(h)$, $h_{,12} = 0$, and $h_{,22} = 1$. Hence, f_s viewed as a function of h is the solution of

$$\rho\,\frac{\partial}{\partial x_2}\left[h_{,2}\, f_s\right] + \frac{\pi\, g_0}{2}\,\frac{\partial^2 f_s}{\partial x_2^2} = 0$$

so that $\rho\, h_{,2}\, f_s + (\pi\, g_0/2)\,(\partial f_s/\partial x_2) = l(x_1)$, where l is an arbitrary function of x_1. Because the left side of this equation approaches zero as $|x_2| \to \infty$ for each x_1 (Eq. 7.63), $l(x_1)$ must be zero. We have $\rho\, h_{,2}\, f_s + (\pi\, g_0/2)\, h_{,2}\, g' = 0$ or $\rho\, g + (\pi\, g_0/2)\, g' = 0$ following simplification with $h_{,2} = x_2$ so that $dg/g = -(2\,\rho/(\pi\, g_0))\, dh$. The solution of this equation is the stated stationary density f_s. ■

7.3.1.5 Nonlinear random vibration

The main objective of most random vibration studies is the determination of properties of the response or state $\boldsymbol{X}$ of a dynamic system defined by Eq. 7.55, for example, moments, marginal densities, mean crossing rates, and first passage times. We have already seen in the previous section that analytical solutions of Eqs. 7.55, 7.64, and 7.70 are possible only in a few cases. These difficulties justify the extensive use of numerical methods, Monte Carlo simulation, and approximations for solving random vibration problems. We discuss here approximate solutions of nonlinear random vibration problems, which are used extensively in applications.

Fokker-Planck equation. The previous section shows that analytical solutions of the Fokker-Planck and backward Kolmogorov equations (Eqs. 7.64 and 7.70) are possible only in special cases. Numerical solutions of the Fokker-Planck and backward Kolmogorov equations are limited to state vectors $\boldsymbol{X}$ with dimensions $d \leq 3$ or 4 and are based on finite difference and finite element methods [179, 196].

Path integral method. Let $\boldsymbol{X}$ be defined by Eq. 7.55 and suppose that the coefficients $\boldsymbol{a}$ and $\boldsymbol{b}$ are such that this equation has a unique solution. Our objective is to find the global solution of Eq. 7.55, that is, the probability law of $\boldsymbol{X}$ over a time interval $[0, t]$, $t \geq 0$.

The path integral method constructs the global solution of Eq. 7.55 by linking local solutions of this equation, that is, solutions over relatively small time intervals $[t_{k-1}, t_k]$, where $p = (0 = t_0 < t_1 < \cdots < t_n)$ is a partition of $[0, t]$ with mesh $\Delta(p) = \max_{1 \leq k \leq n}(t_k - t_{k-1})$. Two properties of $\boldsymbol{X}$ are used in the path integral method. First, $\boldsymbol{X}$ is a Markov process so that its finite dimensional distributions can be obtained from the density of the initial state $\boldsymbol{X}(0)$ and the density $f(\cdot; t \mid \boldsymbol{y}; s)$ of the conditional vector $\boldsymbol{X}(t) \mid (\boldsymbol{X}(s) = \boldsymbol{y})$, $0 \leq s \leq t$, referred to as the transition density (Section 3.6.3). Second, if the time step $t_k - t_{k-1}$ is sufficiently small, the vector $\boldsymbol{X}(t_k) \mid (\boldsymbol{X}(t_{k-1}) = \boldsymbol{y})$ is approximately Gaussian with mean $\boldsymbol{y} + \boldsymbol{a}(\boldsymbol{y}, t_{k-1})\,(t_k - t_{k-1})$ and covariance matrix $\boldsymbol{b}(\boldsymbol{y}, t_{k-1})\,\boldsymbol{b}(\boldsymbol{y}, t_{k-1})^T\,(t_k - t_{k-1})$.

For the numerical implementation of the path integral method both time and space need to be partitioned. Let C_i, $i = 1, \ldots, n$, be a partition of the set $D \subset \mathbb{R}^d$ in which $\boldsymbol{X}$ takes values. If $\boldsymbol{X}(t_{k-1}) \in C_i$, the probability of the event $\{\boldsymbol{X}(t_k) \in C_j\}$ can be approximated by $\int_{C_j} f(\boldsymbol{x}; t_k \mid \boldsymbol{y}_i; t_{k-1})\,d\boldsymbol{x}$, where $f(\cdot; t_k \mid \boldsymbol{y}; t_{k-1})$ denotes the density of $\boldsymbol{X}(t_k) \mid (\boldsymbol{X}(t_{k-1}) = \boldsymbol{y})$ and $\boldsymbol{y}_i \in C_i$ may be selected at the center of C_i. These considerations show that the evolution of $\boldsymbol{X}$ can be approximated by a Markov chain with known transition probability matrix. If the drift and diffusion coefficients of $\boldsymbol{X}$ do not depend explicitly on time, the transition probability matrix has to be calculated once. Otherwise, the value of this matrix needs to be updated at each time step. Theoretical considerations on the path integral method and examples illustrating its application can be found [132].

Moment closure. Generally, the moment equations cannot be solved exactly because they form an infinite hierarchy (Example 7.31). Several heuristic solutions have been advanced in nonlinear random vibration theory and physics to close the moment equations. Most of these solutions are based on postulated relationships between moments of the state vector $\boldsymbol{X}$ providing additional equations, which are used to close the infinite hierarchy of moment equations [22, 175]. For example:

- **Central moment closure** postulates that all central moments of order $q = \sum_{i=1}^{d} q_i > q_0$ of $\boldsymbol{X}$ are zero, where q_0 is called the closure level ([175], Section 6.1.1).

- **Gaussian closure** assumes that the moments of $\boldsymbol{X}$ of order $q \geq 3$ have the same properties as the corresponding moments of Gaussian vectors, that is, all odd and even central moments of $\boldsymbol{X}$ of order $q \geq 3$ are assumed to be zero and be related in the same way as for Gaussian variables, respectively ([175], Section 6.1.1).

- **Cumulant closure** postulates that the cumulants of $\boldsymbol{X}$ of order higher than a closure level q_0 are zero. The cumulant closure with $q_0 = 2$ and the Gaussian closure approximations give the same answer ([175], Section 6.1.1).

- **Consistent closure** postulates a functional form,

$$\tilde{f}(\boldsymbol{x}; t) = \sum_{k=1}^{n} p_k(t)\, f_k(\boldsymbol{x}),$$

for the density of $\boldsymbol{X}(t)$, where $f_k(\boldsymbol{x})$ are fixed densities, and $p_k(t)$ are weights such that $p_k(t) \geq 0$ and $\sum_{k=1}^{n} p_k(t) = 1$ for all $t \geq 0$. If $\boldsymbol{X}(t)$ is stationary, the weights p_k are time-invariant. Let $\boldsymbol{X} = X$ be a real-valued stationary diffusion process and let $e_i(p_1, \ldots, p_n)$ be the expression of a moment equation of order i in which the moments of X have been obtained from the density $\tilde{f}$. Define an objective function $e(p_1, \ldots, p_n) = \sum_{i=1}^{q_0} e_i(p_1, \ldots, p_n)^2$, where q_0 denotes a selected closure level. The optimal functions p_k minimize the objective function, that is, they are the solutions of the system of equations $\partial e / \partial p_k = 0$, $k = 1, \ldots, n$ [74].

There is no justification for the relationships between moments of $\boldsymbol{X}$ postulated by the central moment, Gaussian, and cumulant closure methods. Moreover, the closure methods can deliver incorrect results, for example, negative even-order moments [97]. Yet, these methods are used in large scale applications since they are relatively simple and provide useful results in some cases [22, 105].

Example 7.43: Let X be the solution of

$$dX(t) = -[X(t) + \alpha\, X(t)^3]\, dt + \sqrt{\pi\, g_0}\, dB(t),$$

where α and $g_0 > 0$ are some constants. The moments of X satisfy the infinite hierarchy of equations

$$\dot{\mu}(p; t) = -p\, \mu(p; t) - \alpha\, p\, \mu(p+2; t) + \frac{\pi\, g_0}{2}\, p\,(p-1)\, \mu(p-2; t).$$

If X is a stationary process, the Gaussian closure method gives the nonlinear algebraic equations

$$\begin{cases} \mu(1) - \alpha\,(3\,\mu(1)\,\mu(2) - 2\,\mu(1)^3) = 0, \\ -2\,\mu(2) - 2\,\alpha\,[3\,(\mu(2) - \mu(1)^2)^2 + 6\,\mu(1)^2\,\mu(2) - 5\,\mu(1)^4] + \pi\, g_0 = 0, \end{cases}$$

for the first two moments of X. Approximate/exact values of the second moment $\mu(2)$ are 0.2773/0.2896, 0.3333/0.3451, 0.4415/0.4443, and 0.4927/0.4927 for $\alpha = 1, 0.5, 0.1$, and 0.01, respectively, and $\pi\, g_0/2 = 1/2$. The approximations in this case are satisfactory ([175], Example 6.7, p. 228). ◇

Proof: The moment equations are given by Eq. 7.57. Let $\tilde{\mu}(p; t) = E[(X(t) - \mu(1; t))^p]$ denote the central moment of order p of $X(t)$. The Gaussian closure requires that the moments of $X(t)$ of order three and higher behave as moments of Gaussian variables, for example, $\tilde{\mu}(3; t) = 0$ and $\tilde{\mu}(4; t) = 3\,\tilde{\mu}(2; t)^2$ for the third and fourth order moments, so that

$$\mu(3; t) = 3\,\mu(1; t)\,\mu(2; t) - 2\,\mu(1; t)^3,$$

$$\mu(4; t) = 3\,[\mu(2; t) - \mu(1; t)^2]^2 + 6\,\mu(1; t)^2\,\mu(2; t) - 5\,\mu(1; t)^4.$$

These conditions and the moment equations for $p = 1$ and $p = 2$ are used in the Gaussian closure procedure to calculate the first four moments of X. If X is stationary, we need to solve the above coupled nonlinear algebraic equations to find the first two moments of the state. ■

Example 7.44: Let X be the displacement of the Duffing oscillator in Example 7.33 with $\alpha > 0$. The moments $\mu(p, q; t) = E[X(t)^p \dot{X}(t)^q]$ satisfy the differential equation

$$\begin{aligned}\dot{\mu}(p, q; t) &= p\,\mu(p-1, q+1; t) - \nu^2 q\,\mu(p+1, q-1; t)\\ &\quad - \alpha\,\nu^2 q\,\mu(p+3, q-1; t)\\ &\quad - 2\,\zeta\,\nu\,q\,\mu(p, q; t) + \frac{\pi\,g_0}{2}\,q\,(q-1)\,\mu(p, q-2; t)\end{aligned}$$

with the usual convention that $\mu(p, q; t) = 0$ if at least one of the arguments (p, q) is strictly negative. These moment equations form an infinite hierarchy that cannot be solved exactly.

For a closure level $q_0 = 2$, the central moment and cumulant closure give the relationship

$$\begin{aligned}\mu(3, 0; t) &= 3\,\mu(1, 0; t)\,\mu(2, 0; t) - 2\,\mu(1, 0; t)^3,\\ \mu(4, 0; t) &= 3\,\mu(2, 0; t)^2 - 2\,\mu(1, 0; t)^4,\\ \mu(3, 1; t) &= -2\,\mu(1, 0; t)^3\,\mu(0, 1; t) + 3\,\mu(2, 0; t)\,\mu(1, 1; t),\end{aligned}$$

which can be used to close the moment equations of order $p + q = 1$ and 2. The corresponding approximate stationary solutions are $\mu(1, 0) = 0$, $\mu(0, 1) = 0$, $\mu(2, 0) = \frac{\nu^2}{6\alpha}\left(\sqrt{1 + \frac{3\,\alpha\,\pi\,g_0}{\zeta\,\nu^5}} - 1\right)$, $\mu(1, 1) = 0$, and $\mu(0, 2) = \frac{\pi\,g_0}{4\,\zeta\,\nu}$. If $\alpha = 0$, the Duffing oscillator becomes a linear system so that the variance of the stationary displacement is $(\pi\,g_0)/(4\,\zeta\,\nu^3)$. ◇

Note: The above moment equations have been obtained from Eq. 7.57. They can also be derived by applying Itô's formula to the mapping $(X(t), \dot{X}(t)) \mapsto X(t)^p\,\dot{X}(t)^q$. ■

Equivalent linearization. Let $\boldsymbol{X}$ be the solution of Eq. 7.55 with $\boldsymbol{b}(\boldsymbol{x}, t) = \boldsymbol{b}(t)$. Denote by $\boldsymbol{X}_l$ the solution of the linear stochastic differential equation

$$d\boldsymbol{X}_l(t) = \tilde{\boldsymbol{a}}(t)\,\boldsymbol{X}_l(t)\,dt + \boldsymbol{b}(t)\,d\boldsymbol{B}(t), \quad t \geq 0, \tag{7.71}$$

driven by the noise in Eq. 7.55. The objective of the equivalent linearization method is to find the matrix $\tilde{\boldsymbol{a}}$ in Eq. 7.71 such that $\boldsymbol{X}_l$ is a satisfactory approximation for $\boldsymbol{X}$. Accordingly, the entries of $\tilde{\boldsymbol{a}}$ are selected to minimize the m.s. difference between the drift coefficients of the exact and approximate differential equations, that is, the mean square error [153, 175]

$$e = E\left[(\boldsymbol{a}(\boldsymbol{X}(t), t) - \tilde{\boldsymbol{a}}(t)\,\boldsymbol{X}_l(t))^T\,(\boldsymbol{a}(\boldsymbol{X}(t), t) - \tilde{\boldsymbol{a}}(t)\,\boldsymbol{X}_l(t))\right]. \tag{7.72}$$

This seems to be a reasonable criterion for finding the parameters of the approximating equation. The problem is that we cannot calculate the mean square error in Eq. 7.72 since the probability law of $\boldsymbol{X}$ is not known. If the law of $\boldsymbol{X}$ were known, there would be no need for constructing an approximation. To overcome this difficulty, we consider the following objective function.

Objective function. If Eq. 7.55 has additive noise and $\boldsymbol{X}_l$ is defined by Eq. 7.71, then $\tilde{\boldsymbol{a}}$ is selected to minimize the m.s. error

$$\tilde{e} = E\left[(\boldsymbol{a}(\boldsymbol{X}_l(t), t) - \tilde{\boldsymbol{a}}(t)\,\boldsymbol{X}_l(t))^T\,(\boldsymbol{a}(\boldsymbol{X}_l(t), t) - \tilde{\boldsymbol{a}}(t)\,\boldsymbol{X}_l(t))\right]. \quad (7.73)$$

Note: The objective function $\tilde{e}$ in Eq. 7.73 depends only on the probability law of the Gaussian process $\boldsymbol{X}_l$ defined completely by its first two moments, which can be calculated from Eq. 7.10. Because the moments of $\boldsymbol{X}_l$ depend on the entries of $\tilde{\boldsymbol{a}}$, the objective function $\tilde{e}$ in Eq. 7.73 is a function of $\tilde{\boldsymbol{a}}$. The optimal matrix $\tilde{\boldsymbol{a}}$ minimizes $\tilde{e}$. The entries of the optimal matrix $\tilde{\boldsymbol{a}}$ satisfy a coupled system of nonlinear algebraic equations.

There are several reasons to question the accuracy of the approximating solution $\boldsymbol{X}_l$ delivered by the equivalent linearization method: (1) there is little justification for using $\tilde{e}$ in place of e, (2) the approximation $\boldsymbol{X}_l$ is a Gaussian process while $\boldsymbol{X}$ is not, and (3) the frequency contents of $\boldsymbol{X}_l$ and $\boldsymbol{X}$ may differ significantly. ▲

Despite all these limitations that can cause significant errors in reliability studies when using $\boldsymbol{X}_l$ as an approximation for $\boldsymbol{X}$, the equivalent linearization method is used extensively in engineering applications. Numerical results and theoretical considerations suggest that the method gives satisfactory approximations for the first two moments of the state vector at a fixed time [177, 153]. Theoretical considerations on the equivalent linearization method can be found in [19, 177, 153]. The equivalent linearization method has been extended to (1) accommodate Poisson rather than Gaussian noise [80] and (2) use nonlinear systems with known solutions to approximate the properties of the state of nonlinear systems whose solution is not known [175].

Example 7.45: Let X be the displacement of a Duffing oscillator defined by

$$\ddot{X}(t) + 2\,\beta\,\dot{X}(t) + \nu^2\left(X(t) + \alpha\,X(t)^3\right) = W(t)$$

where $\beta > 0$ denotes the damping ratio, $\alpha > 0$ provides a measure of nonlinearity, $\nu > 0$ is the frequency of the associated linear oscillator ($\alpha = 0$), and W is a stationary Gaussian white noise with mean zero and one-sided spectral intensity $g_0 > 0$. The displacement X_l of the equivalent linear system is the solution of

$$\ddot{X}_l(t) + 2\,\beta\,\dot{X}_l(t) + \nu_{\text{eq}}^2\,X_l(t) = W(t),$$

where ν_{eq} is the only unknown coefficient. If the response is stationary, the natural frequency of the equivalent linear oscillator is

$$\nu_{\text{eq}} = \frac{\nu^2}{2}\left(1 + \sqrt{1 + \frac{3\,\alpha\,\pi\,g_0}{\beta\,\nu^2}}\right).$$

The frequency ν_{eq} is equal to the frequency of the associated linear oscillator ($\alpha = 0$) corrected by a factor depending on the magnitude of α. ◇

Proof: The linear differential equation for X_l is constructed by replacing the nonlinear restoring force $\nu^2\,(X(t) + \alpha\,X(t)^3)$ by the linear restoring force $\nu_{\text{eq}}^2\,X_l(t)$. The solution of $\partial \tilde{e}/\partial \nu_{\text{eq}}^2 = 0$ under the assumption that X_l is a stationary process is

$$\nu_{\text{eq}}^2 = \nu^2 \left(1 + \alpha\,\frac{E[X_l(t)^4]}{E[X_l(t)^2]}\right) = \nu^2 \left(1 + \alpha\,\frac{3\,\pi\,g_0}{4\,\beta\,\nu_{\text{eq}}^2}\right),$$

where $\tilde{e} = E\left[\left(\nu^2\,(X_l(t) + \alpha\,X_l(t)^3) - \nu_{\text{eq}}^2\,X_l(t)\right)^2\right]$. The last equality in the expression of ν_{eq}^2 holds because X_l is a Gaussian process and $E[X_l(t)^2] = \pi\,g_0/(4\,\beta\,\nu_{\text{eq}}^2)$ for a linear oscillator. The above nonlinear algebraic equation for ν_{eq}^2 can be solved exactly in this case. Generally, the equations satisfied by the unknown parameters of an equivalent linear system have to be solved numerically. ■

Perturbation method. Suppose that the linear and nonlinear parts of the matrices $\boldsymbol{a}$ and $\boldsymbol{b}$ in Eq. 7.55 are of order one and ε, $0 < \varepsilon \ll 1$, respectively, that is, $\boldsymbol{a} = \boldsymbol{a}_l + \varepsilon\,\boldsymbol{a}_n$ and $\boldsymbol{b} = \boldsymbol{b}_l + \varepsilon\,\boldsymbol{b}_n$, where the entries of matrices $\boldsymbol{a}$, $\boldsymbol{a}_l$, $\boldsymbol{a}_n$, $\boldsymbol{b}$, $\boldsymbol{b}_l$, and $\boldsymbol{b}_n$ are of order one. Then the defining equation for $\boldsymbol{X}$ is

$$\begin{aligned} d\boldsymbol{X}(t) &= (\boldsymbol{a}_l(\boldsymbol{X}(t), t) + \varepsilon\,\boldsymbol{a}_n(\boldsymbol{X}(t), t))\,dt \\ &\quad + (\boldsymbol{b}_l(\boldsymbol{X}(t), t) + \varepsilon\,\boldsymbol{b}_n(\boldsymbol{X}(t), t))\,d\boldsymbol{B}(t). \end{aligned} \tag{7.74}$$

The solution is sought as a power series in ε, that is,

$$\boldsymbol{X}(t) = \boldsymbol{X}^{(0)}(t) + \varepsilon\,\boldsymbol{X}^{(1)}(t) + \varepsilon^2\,\boldsymbol{X}^{(2)}(t) + \cdots, \tag{7.75}$$

and is usually approximated by $\boldsymbol{X}(t) \simeq \boldsymbol{X}^{(0)}(t) + \varepsilon\,\boldsymbol{X}^{(1)}(t)$, referred to as the **first order perturbation** solution.

The processes $\boldsymbol{X}^{(0)}$ and $\boldsymbol{X}^{(1)}$ in the first order perturbation solution satisfy the differential equations

$$\begin{aligned} d\boldsymbol{X}^{(0)}(t) &= \boldsymbol{a}_l(\boldsymbol{X}^{(0)}(t), t)\,dt + \boldsymbol{b}_l(\boldsymbol{X}^{(0)}(t), t)\,d\boldsymbol{B}(t), \\ d\boldsymbol{X}^{(1)}(t) &= \boldsymbol{a}_l(\boldsymbol{X}^{(1)}(t), t)\,dt + \boldsymbol{b}_l(\boldsymbol{X}^{(1)}(t), t)\,d\boldsymbol{B}(t) \\ &\quad + \boldsymbol{a}_n(\boldsymbol{X}^{(0)}(t), t)\,dt + \boldsymbol{b}_n(\boldsymbol{X}^{(0)}(t), t)\,d\boldsymbol{B}(t). \end{aligned} \tag{7.76}$$

Proof: The perturbation solution and the definition of $\boldsymbol{X}$ (Eqs. 7.74 and 7.75) give

$$\begin{aligned} &d\left(\boldsymbol{X}^{(0)}(t) + \varepsilon\,\boldsymbol{X}^{(1)}(t) + \cdots\right) \\ &= \left[\boldsymbol{a}_l\left(\boldsymbol{X}^{(0)}(t) + \varepsilon\,\boldsymbol{X}^{(1)}(t) + \cdots, t\right) + \varepsilon\,\boldsymbol{a}_n\left(\boldsymbol{X}^{(0)}(t) + \varepsilon\,\boldsymbol{X}^{(1)}(t) + \cdots, t\right)\right] dt \\ &+ \left[\boldsymbol{b}_l\left(\boldsymbol{X}^{(0)}(t) + \varepsilon\,\boldsymbol{X}^{(1)}(t) + \cdots, t\right) + \varepsilon\,\boldsymbol{b}_n\left(\boldsymbol{X}^{(0)}(t) + \varepsilon\,\boldsymbol{X}^{(1)}(t) + \cdots, t\right)\right] d\boldsymbol{B}(t). \end{aligned}$$

Because $\boldsymbol{a}_l$ and $\boldsymbol{b}_l$ are linear functions of $\boldsymbol{X}$, we have

$$\begin{aligned}\boldsymbol{a}_l(\boldsymbol{X}(t),t) &= \boldsymbol{a}_l(\boldsymbol{X}^{(0)}(t),t) + \varepsilon\,\boldsymbol{a}_l(\boldsymbol{X}^{(1)}(t),t) + \cdots,\\ \boldsymbol{b}_l(\boldsymbol{X}(t),t) &= \boldsymbol{b}_l(\boldsymbol{X}^{(0)}(t),t) + \varepsilon\,\boldsymbol{b}_l(\boldsymbol{X}^{(1)}(t),t) + \cdots.\end{aligned}$$

First order Taylor approximations of $\boldsymbol{a}_n$ and $\boldsymbol{b}_n$ about $\boldsymbol{X}^{(0)}(t)$ are

$$\begin{aligned}\boldsymbol{a}_n(\boldsymbol{X}(t),t) &= \boldsymbol{a}_n(\boldsymbol{X}^{(0)}(t),t) + \sum_{i=1}^{d} \frac{\partial \boldsymbol{a}_n(\boldsymbol{X}^{(0)},t)}{\partial x_i}\left(\varepsilon\,X_i^{(1)}(t) + \varepsilon^2\,X_i^{(2)}(t) + \cdots\right),\\ \boldsymbol{b}_n(\boldsymbol{X}(t),t) &= \boldsymbol{b}_n(\boldsymbol{X}^{(0)}(t),t) + \sum_{i=1}^{d} \frac{\partial \boldsymbol{b}_n(\boldsymbol{X}^{(0)},t)}{\partial x_i}\left(\varepsilon\,X_i^{(1)}(t) + \varepsilon^2\,X_i^{(2)}(t) + \cdots\right).\end{aligned}$$

The above equations and approximations yield Eq. 7.76 by equating terms of the same order of magnitude. ■

Additional considerations on the perturbation method are in Chapter 8. We only state here that the method applies for small values of ε, that is, in the context of this chapter, for nearly linear dynamic systems. However, in some cases, the method can give incorrect results even for small values of ε (Example 7.47).

Example 7.46: Let X be the displacement of the Duffing oscillator in Example 7.33 in which $\alpha = \varepsilon > 0$ is the only small parameter in the defining equation for X. The components of the first order perturbation of X are defined by

$$\begin{aligned}\ddot{X}^{(0)}(t) + 2\,\zeta\,\nu\,\dot{X}^{(0)} + \nu^2\,X^{(0)}(t) &= W(t),\\ \ddot{X}^{(1)}(t) + 2\,\zeta\,\nu\,\dot{X}^{(1)} + \nu^2\,X^{(1)}(t) &= -\nu^2\,X^{(0)}(t)^3,\end{aligned}$$

so that $X^{(0)}$ is a Gaussian process but $X^{(1)}$ is not. The process $X^{(1)}$ is the solution of a linear system driven by a polynomial of a filtered Gaussian process (Example 7.23). The stationary mean and variance of X obtained from the first order perturbation solution are zero and $\mathrm{Var}[X(t)] \simeq \sigma_0^2\left(1 - 3\,\varepsilon\,\sigma_0^2\right) + o(\varepsilon)$, where $\sigma_0^2 = \pi\,g_0/(4\,\zeta\,\nu^3)$ is the variance of the displacement of the associated linear system ($\varepsilon = 0$), that is, the process $X^{(0)}$. ◇

Proof: The differential equations for $X^{(0)}$ and $X^{(1)}$ follow from Eq. 7.76.

The stationary processes $X^{(0)}$ and $X^{(1)}$ have mean zero. The variance of the first order perturbation solution is

$$\begin{aligned}\mathrm{Var}[X(t)] &\simeq E[(X^{(0)}(t) + \varepsilon\,X^{(1)}(t))^2]\\ &= E[X^{(0)}(t)^2] + 2\,\varepsilon\,E[X^{(0)}(t)\,X^{(1)}(t)] + \varepsilon^2\,E[X^{(1)}(t)^2].\end{aligned}$$

The approach in Example 7.23 or direct calculations using the solutions of $X^{(0)}$ and $X^{(1)}$ can be used to find the approximate variance of X based on the first order perturbation solution ([175], Example 6.11, p. 237). ■

Example 7.47: Suppose that the damping ratio ζ of the Duffing oscillator in the previous example is also a small parameter, that is, $\zeta = \varepsilon > 0$. The perturbation method fails because the variance of the perturbation solutions increases indefinitely in time while the density of the stationary displacement X is (Example 7.42)

$$f_s(x) = c \exp\left[-\frac{1}{2\sigma_0^2}\left(x^2 + \frac{\varepsilon}{2}x^4\right)\right], \quad \varepsilon > 0,$$

where $c > 0$ is a constant. ◇

Proof: The first two terms of the perturbation solution are defined by (Eq. 7.76)

$$\ddot{X}^{(0)}(t) + \nu^2 X^{(0)}(t) = W(t),$$
$$\ddot{X}^{(1)}(t) + \nu^2 X^{(1)}(t) = -2\nu \dot{X}^{(0)}(t) - \nu^2 X^{(0)}(t)^3,$$

so that the variance of the processes $X^{(0)}$ and $X^{(1)}$ approaches infinity as $t \to \infty$ since the linear oscillators defining these processes have no damping. ■

Stochastic averaging. There are many applications in which the state X varies slowly in time, that is, $\dot{X}(t) \sim O(\varepsilon)$, where $\varepsilon > 0$ is a small parameter ([122], Section 4.7). It is shown that approximate equations can be developed for the evolution of X by averaging the differential equations developed for some memoryless transformations of this process. We illustrate this approach, referred to as the method of averaging, with two examples. A theorem justifying the validity of the method of averaging is stated.

Example 7.48: Let x be the displacement of a nonlinear oscillator defined by

$$\ddot{x}(t) + \nu^2 x(t) = \varepsilon\, g(x(t), \dot{x}(t), t),$$

where $\nu > 0$ is a constant, g is an integrable function, and $\varepsilon > 0$ is a small parameter. The solution of this equation can be approximated by $\tilde{x}(t) = \tilde{a}(t) \cos(\tilde{\psi}(t))$, where $\tilde{\psi}(t) = \nu t + \tilde{\phi}(t)$,

$$\dot{\tilde{a}}(t) = -\frac{\varepsilon}{2\pi\nu}\int_0^{2\pi} g(\tilde{a}\cos(\tilde{\psi}), -\tilde{a}\nu\sin(\tilde{\psi}), t)\sin(\tilde{\psi})\, d\tilde{\psi},$$
$$\dot{\tilde{\phi}}(t) = -\frac{\varepsilon}{2\pi\nu\tilde{a}}\int_0^{2\pi} g(\tilde{a}\cos(\tilde{\psi}), -\tilde{a}\nu\sin(\tilde{\psi}), t)\cos(\tilde{\psi})\, d\tilde{\psi},$$

and $\tilde{a}$ and $\tilde{\phi}$ represent averages of the amplitude and phase of $x(t)$ over a window of size $2\pi/\nu$. ◇

Proof: The change of variables,

$$x(t) = a(t)\cos(\psi(t)) \quad \text{and} \quad \dot{x}(t) = -\nu\, a(t)\sin(\psi(t)),$$

and the defining equation for x give

$$\dot{a}(t) = -\frac{\varepsilon}{\nu}\, g(a\,\cos(\psi), -a\,\nu\,\sin(\psi), t)\,\sin(\psi),$$

$$\dot{\phi}(t) = -\frac{\varepsilon}{\nu\,a}\, g(a\,\cos(\psi), -a\,\nu\,\sin(\psi), t)\,\cos(\psi)$$

for the amplitude a and the phase ϕ of x, where $\psi(t) = \nu\,t + \phi(t)$. The differential equations for a and ϕ are exact. They show that (1) a and ϕ vary slowly in time since their rate of change is of order ε and (2) the right side of the above equations are nearly periodic functions of time with period $2\,\pi/\nu$. The approximate equations for the amplitude and phase result by averaging the above equations over a time interval $[0, 2\,\pi/\nu]$ under the assumption that $a(t)$ and $\phi(t)$ do not change during this time interval, so that these functions can be approximated by their average values $\tilde{a}(t)$ and $\tilde{\phi}(t)$ in $[0, 2\,\pi/\nu]$. The average value of ψ is $\tilde{\psi}(t) = \nu\,t + \tilde{\phi}(t)$. An alternative version of this approach, referred to as the Krylov-Bogoliubov-Mitropolsky averaging method, can be used to derive more accurate approximations for x ([175], Section 6.3). ■

Example 7.49: Let X be a real-valued stochastic process defined by

$$\ddot{X}(t) + \varepsilon^2\,h(X(t), \dot{X}(t)) + \nu^2\,X(t) = \varepsilon\,Y(t),$$

where h is a nonlinear function, $\nu, \varepsilon > 0$ are some constants, and Y is a stationary, broadband Gaussian process with mean zero, correlation function $r(\tau) = E[Y(t)\,Y(t+\tau)]$, spectral density $s(\cdot)$, and correlation time τ_c. The correlation time provides information on the time lag beyond which the dependence between values of Y is weak. The correlation time can be calculated, for example, from the condition $r(\tau_c)/r(0) = e^{-1}$. Consider the change of variable

$$X(t) = A(t)\,\cos(\Psi(t)) \quad \text{and} \quad \dot{X}(t) = -\nu\,A(t)\,\sin(\Psi(t)),$$

where $\Psi(t) = \nu\,t + \Phi(t)$. If ε is a small parameter, the evolution of the amplitude A and phase Φ can be approximated by

$$d\tilde{A}(t) = \varepsilon^2\left[\frac{1}{\nu}\,h_s(\tilde{A}(t)) + \frac{\pi\,s(\nu)}{2\,\nu^2\,\tilde{A}(t)}\right]dt + \varepsilon\,\frac{\sqrt{\pi\,s(\nu)}}{\nu}\,dB_1(t)$$

$$d\tilde{\Phi}(t) = \varepsilon^2\,\frac{1}{\nu\,\tilde{A}(t)}\,h_c(\tilde{A}(t))\,dt + \varepsilon\,\frac{\sqrt{\pi\,s(\nu)}}{\nu\,\tilde{A}(t)}\,dB_2(t),$$

where

$$h_s(a) = \frac{1}{2\pi}\int_0^{2\pi} h(a\,\cos(\psi), -\nu\,a\,\sin(\psi))\,\sin(\psi)\,d\psi,$$

$$h_c(a) = \frac{1}{2\pi}\int_0^{2\pi} h(a\,\cos(\psi), -\nu\,a\,\sin(\psi))\,\cos(\psi)\,d\psi,$$

and B_1, B_2 denote independent Brownian motions [152]. ◇

Proof: The change of variables $(X, \dot{X}) \mapsto (A, \Psi)$ applied to the defining equation for X gives

$$\dot{A}(t) = \frac{\varepsilon^2}{\nu}\, h(A\,\cos(\Psi), -\nu\, A\,\sin(\Psi))\,\sin(\Psi) - \frac{\varepsilon}{\nu}\, Y(t)\,\sin(\Psi),$$
$$\dot{\Phi}(t) = \frac{\varepsilon^2}{\nu\, A}\, h(A\,\cos(\Psi), -\nu\, A\,\sin(\Psi))\,\cos(\Psi) - \frac{\varepsilon}{\nu\, A}\, Y(t)\,\cos(\Psi).$$

The temporal average over a time interval of duration $2\pi/\nu$ of the terms in the above equations that do not include the input Y is

$$\dot{\tilde{A}} = \frac{\varepsilon^2}{\nu}\, h_s(\tilde{A}) - \frac{\varepsilon}{\nu}\, Y(t)\,\sin(\tilde{\Psi}),$$
$$\dot{\tilde{\Phi}} = \frac{\varepsilon^2}{\nu\,\tilde{A}}\, h_c(\tilde{A}) - \frac{\varepsilon}{\nu\,\tilde{A}}\, Y(t)\,\cos(\tilde{\Psi}),$$

where $\tilde{\Psi}(t) = \nu\, t + \tilde{\Phi}(t)$. We have performed a similar temporal average in the previous example.

The next step of the stochastic averaging method is to approximate the random input. It is assumed that the correlation time τ_c of the driving noise Y is such that there exists a time $\Delta t > 0$ with the properties (1) $\tilde{A}$ and $\tilde{\Phi}$ are nearly constant in $[t - \Delta t, t]$ for any $t \geq 0$ and (2) $\Delta t \gg \tau_c$. Conditional on $\tilde{A}(t - \Delta t) = a$ and $\tilde{\Phi}(t - \Delta t) = \phi$, we have

$$Y_1(t) = Y(t)\,\sin(\nu\, t + \tilde{\Phi}(t)) \simeq Y(t)\,[\sin(\nu\, t + \phi) + \cos(\nu\, t + \phi)\,\Delta\Phi(t)]\,,$$

since

$$\sin(\nu\, t + \tilde{\Phi}(t)) = \sin(\nu\, t + \phi + \Delta\tilde{\Phi}(t))$$
$$= \sin(\nu\, t + \phi) + \cos(\nu\, t + \phi)\,\Delta\tilde{\Phi}(t) + O((\Delta\tilde{\Phi}(t))^2),$$

where $\Delta\tilde{\Phi}(t) = (\tilde{\Phi}(t) - \tilde{\Phi}(t - \Delta t)) \mid (\tilde{A}(t - \Delta t) = a, \tilde{\Phi}(t - \Delta t) = \phi)$. We need to determine the second moment properties of the processes $Y_1(t) = Y(t)\,\sin(\nu\, t + \tilde{\Phi}(t))$ and $Y_2(t) = Y(t)\,\cos(\nu\, t + \tilde{\Phi}(t))$ driving the amplitude and phase equations. The expectation $E[Y_1(t)] \simeq \cos(\nu\, t + \phi)\, E[Y(t)\,\Delta\tilde{\Phi}(t)]$ can be calculated from the evolution equation for $\tilde{\Phi}$ giving

$$\Delta\tilde{\Phi}(t) \simeq \frac{\varepsilon^2}{\nu\, a}\, h_c(a) - \frac{\varepsilon}{\nu\, a}\int_{t-\Delta t}^{t} Y(\sigma)\,\cos(\nu\,\sigma + \phi)\, d\sigma$$

so that

$$E[Y_1(t)] = -\frac{\varepsilon}{\nu\, a}\int_{t-\Delta t}^{t} E[Y(t)\, Y(\sigma)]\,\cos(\nu\, t + \phi)\,\cos(\nu\,\sigma + \phi)\, d\sigma$$
$$\simeq -\frac{\varepsilon}{\nu\, a}\int_{-\infty}^{0} r(\tau)\,\cos(\nu\, t + \phi)\,\cos(\nu\,(t + \tau) + \phi)\, d\tau$$
$$= -\frac{\varepsilon}{2\,\nu\, a}\int_{-\infty}^{0} r(\tau)\,[\cos(\nu\,\tau) + \cos(\nu\,(2\, t + \tau) + 2\,\phi)]\, d\tau$$
$$= -\frac{\varepsilon}{2\,\nu\, a}\int_{-\infty}^{0} r(\tau)\,\cos(\nu\,\tau)\, d\tau + \text{ oscillatory terms.}$$

The second equality in the above equation holds approximately since the correlation function is nearly zero for time lags larger than $\tau_c \ll \Delta t$. The oscillatory terms vanish by temporal averaging over a time interval $[0, 2\pi/\nu]$ so that

$$\begin{aligned} E[Y_1(t)] &\simeq -\frac{\varepsilon}{2\nu a} \int_{-\infty}^{0} r(\tau) \cos(\nu \tau) \, d\tau \\ &= -\frac{\varepsilon}{4\nu a} \int_{-\infty}^{\infty} r(\tau) \cos(\nu \tau) \, d\tau = -\frac{\pi \varepsilon}{2\nu a} s(\nu). \end{aligned}$$

The correlation and covariance functions of Y_1 are

$$\begin{aligned} r_1(\tau) &= E[Y_1(t) Y_1(t+\tau)] \simeq E[Y(t) Y(t+\tau) \sin(\nu t + \phi) \sin(\nu (t+\tau) + \phi)] \\ &= \frac{1}{2} r(\tau) [\cos(\nu \tau) - \cos(\nu (2t + \tau) + 2\phi)] \\ &= \frac{1}{2} r(\tau) \cos(\nu \tau) + \text{ oscillatory terms} \end{aligned}$$

and $c_1(\tau) \simeq r_1(\tau) - \left(\frac{\pi \varepsilon}{2\nu a} s(\nu)\right)^2 = r_1(\tau) + O(\varepsilon^2)$, respectively. Because r_1 and c_1 are proportional to r, Y_1 is a broadband process that can be approximated by a white noise with intensity

$$q_1 = \frac{1}{2} \int_{-\infty}^{\infty} r(\tau) \cos(\nu \tau) \, d\tau + \text{ oscillatory terms} \simeq \pi \, s(\nu),$$

where the last equality results by neglecting the oscillatory terms. Hence, Y_1 can be approximated by

$$Y_1(t) \simeq -\frac{\pi \varepsilon}{2\nu a} s(\nu) - \sqrt{\pi \, s(\nu)} \, W_1(t),$$

where W_1 is a stationary Gaussian white noise with mean zero. Similar considerations can be used to show that $Y_2(t) \simeq -\sqrt{\pi \, s(\nu)} \, W_2(t)$, where W_2 is a zero-mean Gaussian white noise. The above approximate representations of Y_1 and Y_2 introduced in the differential equations for $(\tilde{A}, \tilde{\Phi})$ give the stochastic differential equations for the approximate amplitude and phase of X. ■

It turns out that the approach used in the previous two examples to derive simplified differential equations for the state of a system driven by random noise is valid under some conditions. Consider an $\mathbb{R}^d$-valued stochastic process defined by the differential equation

$$\dot{\boldsymbol{X}}(t) = \varepsilon^2 \boldsymbol{\alpha}(\boldsymbol{X}(t), t) + \varepsilon \, \boldsymbol{\beta}(\boldsymbol{X}(t), t, \boldsymbol{Y}(t)), \tag{7.77}$$

where the $\mathbb{R}^d$-valued functions $\boldsymbol{\alpha}$, $\boldsymbol{\beta}$ satisfy a sequence of conditions given by the **Stratonovich-Khas'minskii** theorem [112], $\varepsilon > 0$ denotes a small parameter, and $\boldsymbol{Y}$ is an $\mathbb{R}^{d'}$-valued stationary Gaussian process whose coordinates are broadband processes with mean zero.

If $\boldsymbol{\alpha}$ and $\boldsymbol{\beta}$ in Eq. 7.77 satisfy the conditions in [112], then the solution $\boldsymbol{X}$ of this equation converges in the weak sense to the solution of

$$d\tilde{\boldsymbol{X}}(t) = \varepsilon^2\, \boldsymbol{a}(\tilde{\boldsymbol{X}}(t))\, dt + \varepsilon\, \boldsymbol{b}(\tilde{\boldsymbol{X}}(t))\, d\boldsymbol{B}(t) \tag{7.78}$$

as $\varepsilon \downarrow 0$, where $\boldsymbol{B}$ an $\mathbb{R}^{d'}$-valued Brownian motion and

$$\begin{aligned} \boldsymbol{a} &= T^{\mathrm{av}}\left\{ E[\boldsymbol{\alpha}] + \int_{-\infty}^{0} E\left[\left(\frac{\partial \boldsymbol{\beta}}{\partial \boldsymbol{X}}\right)_t (\boldsymbol{\beta})_{t+\tau} \right] d\tau \right\}, \\ \boldsymbol{b}\,\boldsymbol{b}^T &= T^{\mathrm{av}}\left\{ \int_{-\infty}^{\infty} E\left[\boldsymbol{\beta}_t\, \boldsymbol{\beta}_{t+\tau}^T \right] \right\} dt, \\ T^{\mathrm{av}}\{\cdot\} &= \lim_{u\to\infty} \frac{1}{u} \int_t^{t+u} \{\cdot\}\, ds. \end{aligned} \tag{7.79}$$

Note: The approximate solution of $\boldsymbol{X}$ in Eq. 7.78 has the same functional form as the stochastic differential equation for $(\tilde{A}, \tilde{\Phi})$ in Example 7.49. The conditions that the matrices $\boldsymbol{\alpha}$ and $\boldsymbol{\beta}$ in Eq. 7.77 need to satisfy and the type of weak convergence of $\boldsymbol{X}$ to $\tilde{\boldsymbol{X}}$ are defined in [112], and are not stated here.

The average operator T^{av} is applied over all explicit time arguments. If the functions $\boldsymbol{\alpha}$ and $\boldsymbol{\beta}$ are periodic with period τ_0, this operator becomes

$$T_0^{\mathrm{av}}\{\cdot\} = \frac{1}{\tau_0} \int_t^{t+\tau_0} \{\cdot\}\, ds.$$

The symbol $\partial \boldsymbol{\beta}/\partial \boldsymbol{X}$ denotes a (d, d) matrix with rows $(\partial \beta_k/\partial X_1, \ldots, \partial \beta_k/\partial X_d)$, $k = 1, \ldots, d$. The entries b_{ij} of $\boldsymbol{b}$ can be calculated from the expressions of $\boldsymbol{b}\,\boldsymbol{b}^T$ given by Eq. 7.79. ▲

Example 7.50: The Stratonovich-Khas'minskii theorem applied to the differential equation in Example 7.49 yields the stochastic differential equations for the approximate amplitude and phase of the state X obtained by direct calculations in this example [152]. ◇

Proof: The exact differential equations for $\boldsymbol{X} = (A, \Phi)$ in Example 7.49 show that

$$\begin{aligned} \alpha_1 &= \frac{1}{\nu}\, h(A\cos(\Psi), -\nu\, A \sin(\Psi)) \sin(\Psi), \\ \alpha_2 &= \frac{1}{\nu\, A}\, h(A\cos(\Psi), -\nu\, A \sin(\Psi)) \cos(\Psi), \\ \beta_1 &= -\frac{1}{\nu}\, Y \sin(\Psi), \\ \beta_2 &= -\frac{1}{\nu\, A}\, Y \cos(\Psi), \end{aligned}$$

with the notation in Eq. 7.77, where $\Psi(t) = \nu t + \Phi(t)$. The first coordinate of the drift of the stochastic differential equation for the approximate amplitude and phase of $\boldsymbol{X}$ is

(Eqs. 7.78 and 7.79)

$$a_1 = T^{\mathrm{av}}\, E\left\{\frac{1}{\nu}\, h(A\,\cos(\Psi), -\nu\, A\,\sin(\Psi))\,\sin(\Psi)\right\}$$
$$+\, T^{\mathrm{av}}\left\{\int_{-\infty}^{0} E\left[\left(\frac{\partial\beta_1}{\partial X_1}\right)_t (\beta_1)_{t+\tau} + \left(\frac{\partial\beta_1}{\partial X_2}\right)_t (\beta_2)_{t+\tau}\right] d\tau\right\}.$$

Because $\partial\beta_1/\partial X_1 = \partial\beta_1/\partial A = 0$ and $\partial\beta_1/\partial X_2 = \partial\beta_1/\partial\Phi = -Y\,\cos(\Psi)/\nu$, the second term on the right side of the above equation is

$$\frac{1}{\nu^2\, A}\, T^{\mathrm{av}}\left\{\int_{-\infty}^{0} E\left[Y(t)\, Y(t+\tau)\,\cos(\Psi(t))\,\cos(\Psi(t+\tau))\right] d\tau\right\}$$
$$= \frac{1}{\nu^2\, A}\, T^{\mathrm{av}}\left\{\int_{-\infty}^{0} r(\tau)\,\cos(\Psi(t))\,\cos(\Psi(t+\tau))\, d\tau\right\} = \frac{\pi\, s(\nu)}{2\,\nu^2\, A},$$

where the last equality results by time averaging, as performed in Example 7.49 to find $E[Y_1(t)]$. Hence, we have $a_1 = h_s(A)/\nu + \pi\, s(\nu)/(2\,\nu^2\, A)$. The first and second components of a_1, that is, $h_s(A)/\nu$ and $\pi\, s(\nu)/(2\,\nu^2\, A)$, correspond to the time and stochastic averaging (Example 7.49). Similar calculations can be used to find the remaining drift and diffusion coefficients. For example, the entry (1, 1) of $\boldsymbol{b}\,\boldsymbol{b}^T$ is

$$b_{11}^2 + b_{12}^2 = \frac{1}{\nu^2}\, T^{\mathrm{av}}\left\{\int_{-\infty}^{\infty} r(\tau)\,\sin(\nu\, t + \phi)\,\sin(\nu\,(t+\tau) + \phi)\, d\tau\right\}.$$

The expression under the integral coincides with the approximate correlation function of the process Y_1 in Example 7.49. ■

Example 7.51: Let X be the solution of

$$\ddot{X}(t) + 2\,\varepsilon^2\,\nu\,\dot{X}(t) + \nu^2\, X(t) = \varepsilon\, Y(t),$$

with the same notation as in Examples 7.49 and 7.50. The approximate amplitude and phase of X are the solutions of

$$d\tilde{A}(t) = -\varepsilon^2\left(\nu\,\tilde{A}(t) - \frac{\pi\, s(\nu)}{2\,\nu^2\,\tilde{A}(t)}\right) dt + \varepsilon\,\frac{\sqrt{\pi\, s(\nu)}}{\nu}\, dB_1(t),$$
$$d\tilde{\Phi}(t) = \varepsilon\,\frac{\sqrt{\pi\, s(\nu)}}{\nu\,\tilde{A}(t)}\, dB_2(t).$$

An alternative form of the amplitude equation is

$$dR(t) = -\varepsilon^2\,\nu\left(R(t) - \frac{1}{2\,R(t)}\right) dt + \varepsilon\,\sqrt{\nu}\, dB_1(t),$$

where $R = \tilde{A}/(\sigma\,\sqrt{2})$, where $\sigma^2 = \pi\, s(\nu)/(2\,\nu^3)$. The approximate amplitude equations in Example 7.49 and this example are not coupled with the phase so that they can be solved independently. ◇

Note: The above stochastic differential equations for $\tilde{A}$ and $\tilde{\Phi}$ can be obtained by following the approach in Example 7.49 or using the formulation in Eqs. 7.78 and 7.79. The equations for $\tilde{A}$ and $\tilde{\Phi}$ were originally derived in [8]. ▲

Monte Carlo simulation. Monte Carlo simulation is the most general and conceptually simple method for estimating properties of the state vector $\boldsymbol{X}$. The method involves (1) the generation of samples of the driving noise $\boldsymbol{B}$ in Eq. 7.55, (2) numerical solutions of Eq. 7.55 to obtain samples of $\boldsymbol{X}$ (Sections 4.7.3 and 5.3.3.2), and (3) statistical techniques for estimating the required properties of $\boldsymbol{X}$.

The only limitation of the Monte Carlo method is the computation time, which can be excessive. For example, the estimation of the reliability of very safe dynamic systems requires the generation of many samples of $\boldsymbol{X}$ since the probability that the design conditions are violated is extremely low. The following example presents another practical situation in which Monte Carlo simulation can be inefficient.

Example 7.52: The process $X = x\, e^{-t/2+B(t)}$, called geometric Brownian motion, is the solution of $dX(t) = X(t)\, dB(t)$, $t \geq 0$, with the initial condition $X(0) = x$, where B denotes a Brownian motion (Example 7.35). Suppose we want to estimate the moment of order q of $X(t)$ from samples of this process generated by Monte Carlo simulation. Let $\hat{\mu}(q;t)$ be an estimator of $E[X(t)^q]$ defined as the arithmetic average of the random variables $X_i(t)^p$, where $X_i(t)$, $i = 1, \ldots$, are independent copies of $X(t)$. The number of samples of X required such that the coefficient of variation of $\hat{\mu}(q;t)$ has a specified value v_{req} is $n_{\text{req}} = \left(e^{q^2 t} - 1\right)/v_{\text{req}}^2$. This number can be very large, for example, $n_{\text{req}} = 1,192,000$ samples are needed so that $v_{\text{req}} = 0.05$ for $t = 1$ and $q = 4$. ◇

Proof: We have

$$E[X(t)^q] = E\left[x^q\, e^{q\,(-t/2+B(t))}\right] = x^q\, e^{-q\,t/2}\, E\left[e^{q\,B(t)}\right] = x^q\, e^{(q^2-q)\,t/2},$$

where the last equality holds because $q\,B(t) \sim N(0, q^2\,t)$. This result can also be obtained by Itô's formula applied to the function $X(t)^q$.

Let $\hat{\mu}(q;t) = \frac{1}{n}\sum_{i=1}^{n} X_i(t)^q$ be an estimator of $E[X(t)^q]$, where $X_i(t)$ are independent copies of $X(t)$. This estimator is unbiased with variance

$$\text{Var}[\hat{\mu}(q;t)] = \left(E[X(t)^{2q}] - \left(E[X(t)^q]\right)^2\right)/n = x^{2q}\left[e^{(2q^2-q)\,t} - e^{(q^2-q)\,t}\right]/n$$

and coefficient of variation $\text{C.o.v.}[\hat{\mu}(q;t)] = \sqrt{(e^{q^2 t} - 1)/n}$. The required number of samples n_{req} results from the condition $\text{C.o.v.}[\hat{\mu}(q;t)] = v_{\text{req}}$. ■

7.3.2 Semimartingale input

Let X be the solution of Eq. 7.54 in which the input Y is a semimartingale noise or a polynomial of a semimartingale process, that is, X is the solution of (Eqs. 7.41 and 7.42)

$$dX(t) = a(X(t-), t)\, dt + b(X(t-), t)\, dS(t), \quad t \geq 0, \quad \text{or} \tag{7.80}$$

$$\dot{X}(t) = a(X(t-), t) + b(X(t-), t)\, p(S(t-), t), \quad t \geq 0, \tag{7.81}$$

where S is an $\mathbb{R}^{d'}$-valued semimartingale defined by Eqs. 7.30 and 7.32 and the coordinates of the $\mathbb{R}^{d_p}$-valued process p are polynomials of S (Eq. 7.29). Polynomials of semimartingales are semimartingales since sums of semimartingales are semimartingales and so are smooth memoryless transformations of semimartingales (Itô's formula, Section 4.6.2).

As in the previous sections, our objective is to develop differential equations for properties of the state vector X. The analysis is based on Itô's formula for semimartingales (Eq. 7.2). This formula can be applied to functions of the state X in Eqs. 7.80 and 7.81 or functions of an augmented state vector Z including X. We have referred to these two ways of using the Itô formula as the direct and the state augmentation methods, respectively.

7.3.2.1 Direct method. Square integrable martingales

We have seen that the moment equations for the state X of a nonlinear system driven by Gaussian white noise generally form an infinite hierarchy so that the moments of X cannot be calculated exactly. This situation does not change if the Brownian motion in Eq. 7.55 is replaced by a semimartingale noise.

Let S be defined by Eqs. 7.30 and 7.32 with $A = 0$ and $H = 0$, that is, $dS(t) = K(t)\, d\tilde{C}(t)$. It is assumed that $K \in \mathcal{L}$ satisfies a condition in Eq. 7.33. We derive differential equations satisfied by the mean and correlation functions, $\mu(t) = E[X(t)]$ and $r_{pq}(t,t) = E[X_p(t)\, X_q(t)]$, of X in Eq. 7.80, and use the notation $\tilde{K}(s) = b(X(s-), s)\, K(s)$.

If X is defined by Eq. 7.80 with $dS(t) = K(t)\, d\tilde{C}(t)$, then

$$\begin{aligned} \dot{\mu}(t) &= E[a(X(t-), t)], \\ \dot{r}_{pq}(t,t) &= E[a_p(X(t-), t)\, X_q(t-)] + E[a_q(X(t-), t)\, X_p(t-)] \\ &\quad + \sum_{\alpha=1}^{d_c} E\left[\left(\Delta \tilde{C}_\alpha(s)\right)^2\right] E\left[\tilde{K}_{p\alpha}(t)\, \tilde{K}_{q\alpha}(t)\right], \end{aligned} \tag{7.82}$$

provided that the above expectations exist and are finite.

Proof: The expectation of Eq. 7.80 is

$$dE[dX(t)] = E\left[a(X(t-), t)\right] dt + E\left[b(X(t-), t)\, K(t)\, d\tilde{C}(t)\right].$$

Because the increment $d\tilde{C}(t)$ of $\tilde{C}$ is independent of $b(X(t-), t)\, K(t)$ and $E\left[d\tilde{C}(t)\right] = \mathbf{0}$, the second expectation on the right side of the above equation is zero. These observations yield the first formula in Eq. 7.82.

The Itô formula applied to the function $X(t) \mapsto X_p(t)\, X_q(t)$ gives

$$X_p(t)\, X_q(t) - X_p(0)\, X_q(0) = \sum_{i=1}^{d} \int_{0+}^{t} \left[\delta_{pi}\, X_q(s-) + X_p(s-)\, \delta_{qi}\right] dX_i(s)$$

$$+ \sum_{0<s\le t} \left[X_p(s)\, X_q(s) - X_p(s-)\, X_q(s-) - \sum_{i=1}^{d} \left[\delta_{pi}\, X_q(s-) + X_p(s-)\, \delta_{qi}\right] \Delta X_i(s) \right].$$

The equality $X_p(s) = X_p(s-) + \Delta X_p(s)$ and previous calculations (Eq. 7.38) show that the last term on the right side of the above equation is $\sum_{0<s\le t} \Delta X_p(s)\, \Delta X_q(s)$. The average of the resulting equation is

$$r_{pq}(t,t) - r_{pq}(0,0) = E\left[\int_{0+}^{t} a_p(X(s-), s)\, X_q(s-)\, ds \right.$$

$$\left. + \int_{0+}^{t} a_q(X(s-), s)\, X_p(s-)\, ds + \sum_{0<s\le t} \Delta X_p(s)\, \Delta X_q(s)\right].$$

If the above expectations exist and are finite, the equalities,

$$E\left[\sum_{0<s\le t} \Delta X_p(s)\, \Delta X_q(s)\right] = E\left[\sum_{0<s\le t} \sum_{\alpha,\beta=1}^{d_c} \tilde{K}_{p\alpha}(s)\, \tilde{K}_{q\beta}(s)\, \Delta\tilde{C}_\alpha(s)\, \Delta\tilde{C}_\beta(s)\right]$$

$$= \sum_{0<s\le t} \sum_{\alpha=1}^{d_c} E\left[\tilde{K}_{p\alpha}(s)\, \tilde{K}_{q\alpha}(s)\right] E\left[\left(\Delta\tilde{C}_\alpha(s)\right)^2\right],$$

yield Eq. 7.82 by differentiation with respect to time. ■

Generally, Eq. 7.82 is not a differential equation for the moments of X but it becomes such an equation if a and b are polynomials of X. Usually, the collection of equations for moments of X forms an infinite hierarchy so that these equations cannot be solved exactly. Approximations, for example, closure methods, can be used to find the second moment properties of X.

7.3.2.2 Direct method. General martingales

Suppose that the input is $dS(t) = H(t)\, dB(t) + K(t)\, \tilde{C}(t)$, that is, S is given by Eqs. 7.30 and 7.32 with $A = \mathbf{0}$. The state X may not have moments so that we attempt to derive a differential equation for the characteristic function $\varphi(u; t) = E\left[\exp(\sqrt{-1}\, u^T X(t))\right]$, $u \in \mathbb{R}^d$, of $X(t)$. The notation $\tilde{H}(t) = b(X(t-), t)\, H(t)$ and $\tilde{K}(t) = b(X(t-), t)\, K(t)$ used in the following equation differ from $\tilde{H}$ and $\tilde{K}$ in Eq. 7.38 since b in the current setting depends on X.

The characteristic function of X satisfies the condition

$$\frac{\partial \varphi(\boldsymbol{u};t)}{\partial t} = \sqrt{-1}\sum_{i=1}^{d} u_i\, E\left[e^{\sqrt{-1}\,\boldsymbol{u}^T \boldsymbol{X}(t-)}\, a_i(\boldsymbol{X}(t-),t)\right]$$

$$-\frac{1}{2}\sum_{i,j}^{d} u_i\, u_j\, E\left[e^{\sqrt{-1}\,\boldsymbol{u}^T \boldsymbol{X}(t-)}\left(\tilde{\boldsymbol{H}}(t)\,\tilde{\boldsymbol{H}}(t)^T\right)_{ij}\right]$$

$$+\sum_{\alpha=1}^{d_c}\lambda_\alpha\left\{E\left[e^{\sqrt{-1}\,\boldsymbol{u}^T \boldsymbol{X}(t-)}\, e^{\sqrt{-1}\sum_{i=1}^{d} u_i \sum_{\alpha=1}^{d_c}\tilde{K}_{i\alpha}(t)\,\Delta\tilde{C}_\alpha(t)}\right]-\varphi(\boldsymbol{u};t)\right\}. \tag{7.83}$$

Proof: The increment of function $\boldsymbol{X}(t)\mapsto e^{\sqrt{-1}\,\boldsymbol{u}^T \boldsymbol{X}(t)}$, $\boldsymbol{u}\in\mathbb{R}^d$, in $[0,t]$ is

$$e^{\sqrt{-1}\,\boldsymbol{u}^T \boldsymbol{X}(t)} - e^{\sqrt{-1}\,\boldsymbol{u}^T \boldsymbol{X}(0)} = \sum_{i=1}^{d}\int_{0+}^{t}\sqrt{-1}\,u_i\, e^{\sqrt{-1}\,\boldsymbol{u}^T \boldsymbol{X}(s-)}\, dX_i(s)$$

$$-\frac{1}{2}\sum_{i,j=1}^{d}\int_{0+}^{t} u_i\, u_j\, e^{\sqrt{-1}\,\boldsymbol{u}^T \boldsymbol{X}(s-)}\left[\sum_{k=1}^{d_b}\tilde{H}_{ik}(s)\,dB_k(s),\sum_{l=1}^{d_b}\tilde{H}_{jl}(s)\,dB_l(s)\right]$$

$$+\sum_{0<s\le t}\left[e^{\sqrt{-1}\,\boldsymbol{u}^T \boldsymbol{X}(s)} - e^{\sqrt{-1}\,\boldsymbol{u}^T \boldsymbol{X}(s-)} - \sum_{i=1}^{d}\sqrt{-1}\,u_i\, e^{\sqrt{-1}\,\boldsymbol{u}^T \boldsymbol{X}(s-)}\,\Delta X_i(s)\right],$$

where

$$dX_i(s) = a_i(\boldsymbol{X}(s-),s)\,ds + \sum_{k=1}^{d_b}\tilde{H}_{ik}(s)\,dB_k(s) + \sum_{\alpha=1}^{d_c}\tilde{K}_{i\alpha}(s)\,d\tilde{C}_\alpha(s)$$

and $\Delta X_i(s) = \sum_{\alpha=1}^{d_c}\tilde{K}_{i\alpha}(s)\,\Delta\tilde{C}_\alpha(s)$. The above Itô formula, the assumption that all expectations in Eq. 7.83 exist and are finite, and considerations similar to those used to derive Eq. 7.39 give Eq. 7.83 by differentiation with respect to time. ■

The condition in Eq. 7.83 is not a partial differential equation for φ because the expectations in this equation cannot be expressed as functions of φ and its partial derivatives for arbitrary $\boldsymbol{a}$, $\boldsymbol{b}$, $\tilde{\boldsymbol{H}}$, and $\tilde{\boldsymbol{K}}$. This condition becomes a partial differential equation for φ under rather restrictive conditions. More general systems can be analyzed by the state augmentation method discussed in the next section.

Example 7.53: Let X be the solution of the stochastic differential equation

$$dX(t) = -(\rho\, X(t) + \zeta\, X(t)^3)\,dt + \sigma\, dL_\alpha(t), \quad t\ge 0,$$

where $\rho > 0$, ζ, and σ are some constants and L_α is an α-stable process. The marginal distribution of X has a heavy tail, that is,

$$P(X(t) > x) \sim c\, x^{-\beta} \quad \text{as } x \to \infty,$$

where $c, \beta > 0$ are some constants. Estimates of β based on a sample of X of length 1,000,000 generated with a time step $\Delta t = 0.001$ are in the range $(2.5, 5)$ for $\rho = 1$, $\zeta = 1$, $\sigma = 0.2$, and $\alpha = 1$.

Formal calculations show that the characteristic function $\varphi(\cdot; t)$ of $X(t)$ is the solution of

$$\frac{\partial \varphi}{\partial t} = u \left(-\rho \frac{\partial \varphi}{\partial u} + \zeta \frac{\partial^3 \varphi}{\partial u^3} \right) - \sigma^\alpha |u|^\alpha \varphi.$$

This equation cannot explain the heavy tail of the distribution of X.

We have seen that the characteristic function $\varphi(u; t) = E[e^{\sqrt{-1}\, u\, X(t)}]$ of X is the solution of the partial differential equation (Example 7.20)

$$\frac{\partial \varphi}{\partial t} = -\rho\, u \frac{\partial \varphi}{\partial u} - |u|^\alpha \varphi.$$

for $\zeta = 0$ and $\sigma = 1$. This equation is a special case of the above partial differential equation of φ. ◇

Proof: Note that the process L_α in this example does not belong to the class of inputs $\mathcal{S}$ in Eq. 7.83 so that this equation cannot be used. The Itô formula applied to the function $X(t) \mapsto e^{\sqrt{-1}\, u\, X(t)}$ gives

$$\begin{aligned} e^{\sqrt{-1}\, u\, X(t)} - e^{\sqrt{-1}\, u\, X(0)} &= -\sqrt{-1}\, u \int_{0+}^{t} \left(\rho\, X(s-) + \zeta\, X(s-)^3 \right) e^{\sqrt{-1}\, u\, X(s-)}\, ds \\ &+ \sum_{0<s\le t} \left[e^{\sqrt{-1}\, u\, X(s-)} \left(e^{\sqrt{-1}\, u\, \Delta X(s)} - 1 \right) \right]. \end{aligned}$$

The expectation of the terms on the left side of the above equation gives φ at times t and $t = 0$. However, the expectation of the right side of this equation cannot be calculated term by term since individual terms may not have finite expectations. Hence, it is not possible to find a partial differential equation for φ. If we calculate the expectation term by term, the result of these formal calculations is

$$\begin{aligned} \varphi(u; t) - \varphi(u; 0) &= -\sqrt{-1}\, u \int_{0+}^{t} E\left[\left(\rho\, X(s-) + \zeta\, X(s-)^3 \right) e^{\sqrt{-1}\, u\, X(s-)} \right] ds \\ &+ \sum_{0<s\le t} E\left[e^{\sqrt{-1}\, u\, X(s-)} \left(e^{\sqrt{-1}\, u\, \Delta X(s)} - 1 \right) \right]. \end{aligned}$$

Differentiation with respect to time and considerations as in Example 7.20 yield the above equation for the characteristic function.

Let $X_i = X(i\, \Delta t)$, $i = 1, \ldots, n$, be a sample of X generated by Monte Carlo simulation with a time step $\Delta t > 0$. Let $X_{(1)} \ge \cdots \ge X_{(n)}$ be the order statistics derived from the sample $X_1, \dots, X_n$. For example, $X_{(1)} = \max_{1 \le i \le n} X_i$ and $X_{(n)} = \min_{1 \le i \le n} X_i$ are the largest and the smallest observations in the record of X. It can be shown that the Hill estimator [100]

$$H_{k,n} = \frac{1}{k} \sum_{j=1}^{k} \log\left(\frac{X_{(j)}}{X_{(k+1)}}\right)$$

converges in probability to $1/\beta$ as $k, n \to \infty$ such that $k/n \to 0$ [100]. The range (2.5,5) of the estimated values of β correspond to ratios $k/n = 0.0001$, 0,0005, 0.001, 0.005, and 0.01 and $n = 1,000,000$.

The parameter β gives the rate of decay of the tail of the distribution of $X(t)$. We also note that the asymptotic expression of $P(X(t) > x)$ gives

$$\log\left(P(X(t) > x)\right) \sim \log(c) - \beta \log(x) \quad \text{as } x \to \infty$$

so that the graph $(\log(x), \log(P(X(t) > x)))$ is a straight line with slope $-\beta$. It is common to refer to this behavior of the tail of the distribution of X as **power law** [15]. ■

7.3.2.3 State augmentation method

The state augmentation method can be applied to both evolution equations for $\boldsymbol{X}$, that is, Eqs. 7.80 and 7.81. The analysis of the two evolution equations is based on the same formulas but the augmented state vector used in analysis may differ. Details of the state augmentation method are given in Section 7.2.2.3 and are not repeated here. We only present an example illustrating the application of the method.

Example 7.54: Let X be the solution of

$$dX(t) = a(X(t))\, dt + b(X(t))\, dS(t)$$

where $dS(t) = 3\left(B(t)^2 - t\right) dB(t)$. The moments $\mu(p, q; t) = E[X(t)^p\, Y(t)^q]$ of order $p + q$ of $\boldsymbol{Z} = (X, Y = B)$ satisfy the condition

$$\begin{aligned}
&\dot{\mu}(p, q; t) = p\, E[X(t)^{p-1}\, Y(t)^q\, a(X(t))] \\
&+ (9\, p\, (p-1)/2)\, E[X(t)^{p-2}\, Y(t)^q\, b(X(t))^2\, (Y(t)^2 - t)^2] \\
&+ 3\, p q\, E[X(t)^{p-1}\, Y(t)^{q-1}\, b(X(t))\, (Y(t)^2 - t)] + (q\, (q-1)/2)\, \mu(p, q-2; t).
\end{aligned}$$

If a and b are polynomials of X, this equation becomes a differential equation for the moments of $\boldsymbol{Z} = (X, Y)$. However, the resulting moment equations are not closed.

The characteristic function $\varphi(\boldsymbol{u}; t)$ of $\boldsymbol{Z}(t)$ satisfies the differential form

$$\begin{aligned}
\frac{\partial \varphi}{\partial t} &= \sqrt{-1}\, u_1\, E\left[e^{\sqrt{-1}\, \boldsymbol{u}^T \boldsymbol{Z}(t)}\, a(X(t))\right] \\
&- (9/2)\, u_1^2\, E\left[e^{\sqrt{-1}\, \boldsymbol{u}^T \boldsymbol{Z}(t)}\, b(X(t))^2\, (Y(t)^2 - t)^2\right] \\
&- 3\, u_1\, u_2\, E\left[e^{\sqrt{-1}\, \boldsymbol{u}^T \boldsymbol{Z}(t)}\, b(X(t))\, (Y(t)^2 - t)\right] - (1/2)\, u_2^2\, \varphi,
\end{aligned}$$

which becomes a partial differential equation for φ if the coefficients a and b are polynomials of X. ◇

Proof: The augmented vector $\mathbf{Z}$ is the solution of

$$\begin{cases} dX(t) &= a(X(t))\,dt + 3\,b(X(t))\,(Y(t)^2 - t)\,dB(t), \\ dY(t) &= dB(t). \end{cases}$$

The differential equations for the moments and the characteristic function of $\mathbf{Z}$ result by applying the Itô formula to the functions $X(t)^p\,Y(t)^q$ and $e^{\sqrt{-1}\,\boldsymbol{u}^T\,\mathbf{Z}(t)}$, respectively, taking the average, and differentiating the result with respect to time. ■

7.4 Applications

The developments in the first part of this chapter are used to illustrate solutions of some stochastic problems, which can be defined by Eq. 7.1. We present elementary criteria for model selection (Section 7.4.1), demonstrate how stochastic differential equations and diffusion processes can be used to describe the evolution of some properties of random heterogeneous materials (Section 7.4.2), discuss methods for reliability analysis based on the crossing theory for stochastic processes, Fokker-Planck equation, probability bounds, and Monte Carlo simulation (Section 7.4.3), introduce elements of finance and establish the Black-Scholes formula (Section 7.4.4), and give essentials of linear estimation and Kalman-Bucy filtering (Section 7.4.5).

7.4.1 Models

In many cases, the laws defining the evolution in time of the state $\boldsymbol{X}$ of a system are either partially understood or too complex to be derived from basic principles, for example, climatic changes, wave and wind forces, stock prices, seismic ground acceleration, and many other phenomena. A common approach in these cases is to select a probabilistic model for $\boldsymbol{X}$.

The selection of a model for $\boldsymbol{X}$ is based on the **available information**, which consists of (1) one finite record or a set of finite records of $\boldsymbol{X}$ and (2) physical properties of this process, for example, if $\boldsymbol{X}(t)$ represents the strength of a fiber at location t, it must be positive. Because of the limited information the model selection problem does not have a unique solution. Generally, there are many **competing models**, that is, models that are consistent with the available information. The selection of a model for $\boldsymbol{X}$ involves the following two steps.

1. Specify a finite collection of competing models $\mathcal{M}_k$, $k = 1, 2, \ldots, m$. There are no theorems defining the models $\mathcal{M}_k$. Simplicity and use should guide the selection of the collection of competing models. Nevertheless, in applications, familiarity with a particular model plays a significant role in the construction of the collection of competing models.

2. Find the model $\mathcal{M}_{\text{opt}} = \mathcal{M}_{k_0}$, $k_0 \in \{1, \ldots, m\}$, that is optimal in some sense. In Section 9.8.2 we present a Bayesian framework for finding $\mathcal{M}_{\text{opt}}$.

We use examples in this section to illustrate the classical approach for model selection and demonstrate the practical importance of accounting for use in the solution of the model selection problem.

7.4.1.1 Earth climate

It has been shown that earth climate is strongly correlated with calcium concentration in ice deposits [199]. This dependence suggests that calcium concentration records obtained from ice-core samples can be used as a proxy for the evolution of earth climate in time. Figure 7.12 shows an 80,000 year record of the

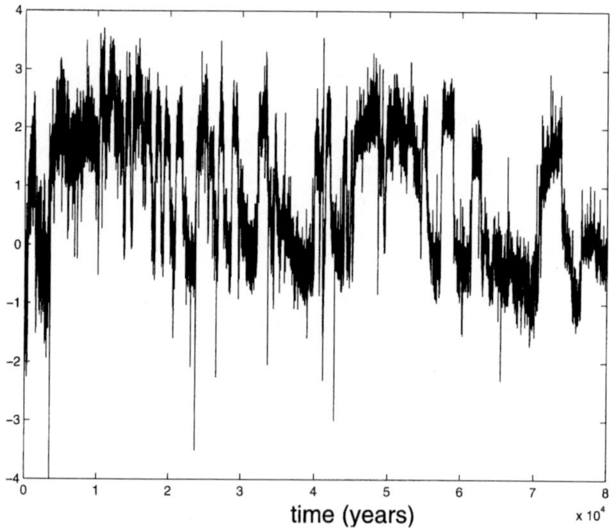

Figure 7.12. Logarithm of calcium concentration in an ice-core record

logarithm of calcium concentration X obtained from an ice-core sample.

The sample in Fig. 7.12 shows that X has two equilibrium points corresponding to the interstadial and the full glacial states. The typical residence of X in a state is between 1,000 and 2,000 years. These observations suggest that we model the evolution of X by a stochastic differential equation whose drift corresponds to a potential with two wells.

A model of X. The process X defined by

$$dX(t) = -u'(X(t))\,dt + \sigma_y\,dY(t) + \sigma_l\,dL_\alpha(t),$$
$$dY(t) = -Y(t)\,dt + \sqrt{1+Y(t)^2}\,dB(t) \tag{7.84}$$

has been proposed to model the evolution of earth climate, where u is a two-well potential, B denotes a Brownian motion, L_α is an α-stable process, and σ_y, σ_l are some constants [55].

Note: According to [55], Eq. 7.84 is a competing model since it is consistent with (1) the record in Fig. 7.12 for $\alpha = 1.75$ and $\sigma_y/\sigma_l = 3$ and (2) current climate dynamics theories.

Note that the differential equation of X can be given in the form

$$dX(t) = -u'(X(t))\,dt + dS(t),$$

where S is defined by

$$dS(t) = \sigma_y\left[-Y(t)\,dt + \sqrt{1+Y(t)^2}\,dB(t)\right] + \sigma_l\,dL_\alpha(t),$$

so that it is a semimartingale. Hence, properties of X can be obtained by the methods presented previously in this chapter. ▲

The selection of the stochastic process in Eq. 7.84 essentially follows the classical approach in modeling. First, a functional form has been postulated for the logarithm of the calcium concentration X. Second, the record in Fig. 7.12 has been used to estimate the parameters of the postulated model. Details of the method used to estimate the parameters of the processes in Eq. 7.84 are in [55]. The potential in Eq. 7.84 has been obtained from the marginal histogram of X and the relationship between the marginal distribution and the potential (Example 7.41). The noise properties have been obtained from the statistical analysis of the record in Fig. 7.12 corrected by the estimated drift of X. The analysis has shown that the driving noise has a component of the α-stable type with $\alpha = 1.75$ [55].

In most applications the collection $\mathcal{M}_k$, $k = 1, \ldots, m$, of competing models consists of a single model $\mathcal{M}$ with a specified functional form but unknown parameters. Therefore, the model selection problem is reduced to the estimation of the unknown parameters of $\mathcal{M}$. This classical solution of the model selection problem has been used in [55] to establish the model $\mathcal{M}$ in Eq. 7.84.

7.4.1.2 Non-Gaussian input

Non-Gaussian processes such as (1) memoryless nonlinear transformations of Gaussian processes, referred to as translation processes (2) Gaussian processes scaled by random parameters and other conditional Gaussian processes, and (3) processes defined by stochastic differential equations, for example, diffusion and filtered Poisson processes, have been used in many fields of applied science and engineering ([79], Chapter 2).

Translation processes can have any marginal distribution F but their correlation function r cannot be chosen independently of F, and their finite dimensional distributions F_n, $n \geq 2$, are uniquely defined by F and r (Section 3.6.6 in this book, [79], Section 3.1.1). Conditional Gaussian processes can be useful in some particular applications but they cannot be calibrated to a specified marginal distribution and correlation function ([79], Section 3.4). Diffusion processes can match the marginal distribution of any signal provided that its correlation is exponential [122]. Filtered Poisson processes can match the correlation function of any time series but they can only approximate its marginal distribution ([79], Section 3.3).

Translation processes have attractive features for applications. However, they cannot be viewed as universal non-Gaussian models. In addition to restrictions on their correlation and finite dimensional distribution functions, these models are not recommended in some cases for other reasons. For example, suppose that our objective is to find properties of the state $\boldsymbol{X}$ of a system driven by a non-Gaussian process Y. The representation of Y by a translation model Y_T causes difficulties since there is no theory for finding properties of $\boldsymbol{X}$ driven by Y_T. The output of a linear filter to Poisson white noise or to a polynomial of a class of Gaussian processes is a preferable model for Y since the augmented system of equations consisting of the defining differential equations for Y and $\boldsymbol{X}$ can be solved by the methods discussed in this chapter. We consider the latter model of Y in this section.

Suppose that the non-Gaussian input Y to a system with state $\boldsymbol{X}$ can be modeled by the output of the linear filter in Eq. 7.85 to a polynomial $\sum_{l=1}^{n} S^l$ of an Ornstein-Uhlenbeck process S. Properties of $\boldsymbol{X}$ can be obtained by the state augmentation method discussed earlier in this chapter, the representation of Y, and the evolution equation for $\boldsymbol{X}$. The selection of a model for Y is by no means trivial and may require many iterations.

Model of input Y.

$$\ddot{Y}(t) + \beta\,\dot{Y}(t) + \nu^2\,Y(t) = \sum_{l=0}^{n} S(t)^l,$$

$$dS(t) = -\alpha\,S(t)\,dt + \sigma\,\sqrt{2\,\alpha}\,dB(t), \tag{7.85}$$

where $\alpha > 0$, σ are some constants and B denotes a Brownian motion.

Note: We have shown in Section 7.2.2.3 that the moment equations for the diffusion process $\boldsymbol{Z} = (Y, \dot{Y}, S)$ are closed so that the moments of any order of the $\mathbb{R}^3$-valued random variable $\boldsymbol{Z}(t)$ can be calculated exactly for any $t \geq 0$. Let

$$d\boldsymbol{Z}(t) = \boldsymbol{a}(\boldsymbol{Z}(t), t)\,dt + \boldsymbol{b}\,dB(t)$$

be the defining differential equation for $\boldsymbol{Z}$ (Example 7.23).

The correlation function $r_{pq}(t,s) = E[Z_p(t)\, Z_q(s)]$ of $\mathbf{Z}$ satisfies the condition

$$\frac{\partial r_{pq}(t,s)}{\partial t} = E\left[a_p(\mathbf{Z}(t),t)\, Z_q(s)\right], \quad t \geq s,$$

since $Z_p(t) - Z_p(s) = \int_s^t \left[a_p(\mathbf{Z}(u),u)\, du + b_p\, dB(u)\right]$, $p = 1, 2, 3$, so that

$$Z_p(t)\, Z_q(s) - Z_p(s)\, Z_q(s) = \int_s^t a_p(\mathbf{Z}(u),u)\, Z_q(s)\, du + \int_s^t b_p\, Z_q(s)\, dB(u), \quad t \geq s,$$

which gives the above differential equation for $r_{pq}(t,s)$ by averaging and differentiating with respect to t.

The following example shows that the correlation functions r_{pq} can be calculated from the above equation. Hence, we can calculate exactly the correlation function of the non-Gaussian process Y in Eq. 7.85 and moments of any order of $Y(t)$, $t \geq 0$. ▲

Example 7.55: Suppose that $\sum_{l=0}^n S(t)^l = S(t)^2$, $\beta = 0.25$, $\nu^2 = 1.6$, $\alpha = 0.12$, and $\sigma = 1$ in Eq. 7.85. The stationary skewness and kurtosis coefficients of Y are $\gamma_3 \simeq 1.8$ and $\gamma_4 \simeq 8$, respectively. Hence, if we need to analyze the response of a system to an input with $\gamma_3 \simeq 1.8$ and $\gamma_4 \simeq 8$, the process Y can be used to model this input. Figure 7.13 shows some of the correlation functions of the stationary

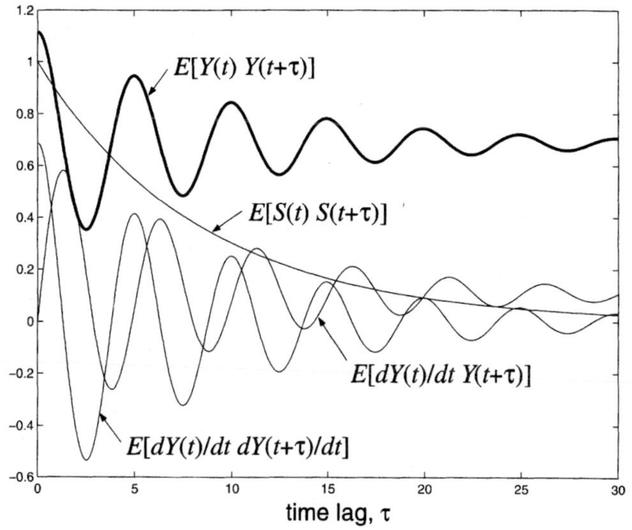

Figure 7.13. Some of the correlation functions of $\mathbf{Z}$

$\mathbb{R}^3$-valued process $\mathbf{Z} = (Y, \dot{Y}, S)$. The properties of the linear filter defining Y control to a great extent the correlation function of this process if $\sum_{l=0}^n S^l$ is a broadband process. ◇

Proof: The moments $\mu(p,q,r;s)$ are given in Example 7.23. The differential equations giving the correlation functions $r_{pq}(t,s) = E[Z_p(t)\, Z_q(s)]$ involve only moments of $\mathbf{Z}$

since the drift coefficients $a_p(\boldsymbol{Z}(t), t)$ are polynomials of $\boldsymbol{Z}$. However, some of these moments are of order higher than 2 so that they are not correlations. The moments, $E\left[Z_3(t)^2\, Z_1(s)\right]$, $E\left[Z_3(t)^2\, Z_2(s)\right]$, and $E\left[Z_3(t)^2\, Z_3(s)\right]$, $t \geq s$, in the differential equations for the correlation function of $\boldsymbol{Z}$ are not known but can be calculated simply. For example,

$$E\left[Z_3(t)^2\, Z_1(s)\right] = E\left[(Z_3(s) + G)^2\, Z_1(s)\right] = \mu(1, 0, 2; s) + E[G^2]\, \mu(1, 0, 0; s),$$

where G is a zero-mean Gaussian variable with variance $\sigma^2\left(1 - e^{-2\alpha(t-s)}\right)$, which is independent of $Z_1(s)$ and $Z_3(s)$. That this representation is valid follows from the solution,

$$Z_3(t) = Z_3(s) + \int_s^t e^{-\alpha\,(t-u)}\, \sigma\, \sqrt{2\,\alpha}\, dB(u) = Z_3(s) + G,$$

of the last coordinate of $\boldsymbol{Z}$. ■

Consider a system with state $\boldsymbol{X}$ and let Y and Y_T be two models of the input to this system, which are consistent with the available information. The model Y is defined by a differential equation of the type in Eq. 7.85 and the model Y_T is a translation process. The construction of the model of Y exceeds by far the effort require to define Y_T. However, once Y has been obtained, properties of the state $\boldsymbol{X}$ corresponding to Y can be established by the methods in this chapter. On the other hand, there are no practical methods for calculating properties of $\boldsymbol{X}$ driven by Y_T.

7.4.2 Materials science

The evolution in time of stiffness, atomic lattice orientation, and other properties of random heterogeneous media subjected to persistent actions is discussed extensively in Chapter 8. In this section we show that some global measures of these properties satisfy differential equations with deterministic coefficients driven by Gaussian white noise (Section 7.3.1).

Consider a random heterogeneous medium in a bounded subset D of $\mathbb{R}^q$, $q = 1, 2, 3$, so that the properties of this material can be modeled by a random field defined on a probability space $(\Omega, \mathcal{F}, P)$. The material is subjected to an external deterministic strain $\gamma_e(t)$, $t \geq 0$, with a constant rate $\dot{\gamma}_e(t) = \dot{\gamma}_e > 0$. Denote by $\Gamma(\boldsymbol{\xi}, t)$ an $\mathbb{R}$-valued continuous function of the strain tensor at location $\boldsymbol{\xi} \in D$ and time t induced by $\gamma_e(t)$. Because the medium is random and heterogeneous, $\Gamma : D \times [0, \infty) \times \Omega \to \mathbb{R}$ is a random field defined on the probability space $(\Omega, \mathcal{F}, P)$ that depends on time and space arguments. A sample $\Gamma(\cdot, t, \omega)$ of Γ is a real-valued function depending on the strain field in D at time t for a material specimen ω. Because strains are bounded in D, the moments of any order of Γ exist and are finite, so that we can write

$$\Gamma(\boldsymbol{\xi}, t) = E[\Gamma(\boldsymbol{\xi}, t)] + \tilde{\Gamma}(\boldsymbol{\xi}, t) = \gamma(t) + \tilde{\Gamma}(\boldsymbol{\xi}, t). \tag{7.86}$$

It is assumed that Γ is ergodic in $\boldsymbol{\xi}$ and that D is sufficiently large with respect to the correlation length of Γ so that the approximation,

$$E[\Gamma(\boldsymbol{\xi}, t)] = \gamma(t) \simeq \frac{1}{v_D} \int_D \Gamma(\boldsymbol{\xi}, t, \omega)\, d\boldsymbol{\xi}, \tag{7.87}$$

holds for all $t \geq 0$ and almost all ω's, where v_D denotes the volume of D. Hence, the ensemble average $E[\Gamma(\boldsymbol{\xi}, t)] = \gamma(t)$ can be obtained from the spatial average of an arbitrary sample of Γ at any time t and can be interpreted as a strain measure at time t induced by the external action γ_e in a homogeneous material.

Let $\mathcal{X}(\boldsymbol{\xi}, t)$ be an internal variable characterizing a material property at location $\boldsymbol{\xi} \in D$ and time $t \geq 0$, for example, one of the Euler angles of the atomic lattice orientation in a polycrystal. Because material properties and internal strains vary randomly in space, $\mathcal{X}$ is a random function defined on the same probability space as Γ. Our objective is to derive an evolution equation for $\mathcal{X}$. We proceed under the assumptions that (1) the evolution of material properties is governed only by plastic straining and (2) the internal variable $\mathcal{X}$ takes values on the real line. Generally, $\mathcal{X}$ is an $\mathbb{R}^d$-valued random function but the assumption that $\mathcal{X}$ is real-valued is satisfactory in many cases [14, 92].

If the material were homogeneous, then $\mathcal{X}(\boldsymbol{\xi}, t) = X_h(t)$ would have the same value everywhere in D and its evolution would be given by a deterministic differential equation of the type $\dot{X}_h(t) = a(X_h(t))\, \dot{\gamma}(t)$. The function a is problem dependent, for example, $a(x)$ is proportional with $2\cos(x) - c$ if x denotes the atomic lattice orientation of a planar crystal with two slip systems, where c is a constant (Section 8.6.2.1).

If the medium is random and heterogeneous, the internal variable X characterizing the evolution of its properties can be modeled by a diffusion process satisfying the stochastic differential equation

$$dX(\zeta) = a(X(\zeta))\, d\zeta + q_{w,0}\, b(X(\zeta))\, dB(\zeta), \tag{7.88}$$

where ζ is a new time defined by $d\zeta(t) = \dot{\gamma}(t)\, dt$, B denotes a Brownian motion, $q_{w,0}$ is a constant, and $b(\cdot)$ is a function of the state X.

Proof: Because the medium is random and heterogeneous, the internal variable $\mathcal{X}$ is a random function depending on time and space. The evolution of $\mathcal{X}$ is characterized by the change of its probability law in time. However, the determination of the evolution of the probability law of $\mathcal{X}$ is a very difficult task. A simplified description of this evolution has been proposed in [92]. Let C be a collection of points $\boldsymbol{\xi}$ in D that are sufficiently far apart so that the correlation between strain processes at distinct points in C can be neglected, that is, the processes $\tilde{\Gamma}(\boldsymbol{\xi}, \cdot)$ and $\tilde{\Gamma}(\boldsymbol{\xi}', \cdot)$ in Eq. 7.86 can be assumed to be uncorrelated, where $\boldsymbol{\xi}, \boldsymbol{\xi}' \in C$ and $\boldsymbol{\xi} \neq \boldsymbol{\xi}'$. The collection of functions of time $\dot{\tilde{\Gamma}}(\boldsymbol{\xi}, \cdot)$, $\boldsymbol{\xi} \in C$, can be viewed as samples of a stochastic process $\dot{\gamma}(\cdot) + Y(\cdot)$, where $\dot{\gamma}$ is the spatial average of the strain rates in C and the noise Y captures the sample to sample variation of these strain rates. Hence, the evolution of the internal variable X in a random heterogeneous medium can be described by

$$\dot{X}(t) = a(X(t))\,\dot{\gamma}(t) + a(X(t))\,Y(t),$$

that is, the evolution equation $\dot{X}_h(t) = a(X_h(t))\,\dot{\gamma}(t)$ for a homogeneous medium in which the strain rate $\dot{\gamma}(t)$ is replaced with $\dot{\gamma}(t) + Y(t)$. The equation of X becomes

$$\frac{dX}{d\zeta} = a(X) + a(X)\,\frac{Y}{\dot{\gamma}},$$

in the new "time" ζ defined by $d\zeta(t) = \dot{\gamma}(t)\,dt$. Because $\dot{\gamma}_e(t) > 0$ by hypothesis, we have $\dot{\gamma}(t) > 0$ so that the new time is well defined. It is assumed in [14] that $W = Y/\dot{\gamma}$ in Eq. 7.88 is a Gaussian white noise W with mean zero and intensity

$$q_w(\zeta)^2 = \frac{E[\tilde{\Gamma}(\boldsymbol{\xi}, \zeta)^2]}{(\dot{\gamma})^2}\,\zeta_{\text{corr}},$$

where $\zeta_{\text{corr}} = \dot{\gamma}\,t_{\text{corr}}$ and t_{corr} are correlation times. The expectation $E[\tilde{\Gamma}(\boldsymbol{\xi}, \zeta)^2]$ is performed with respect to the space parameter, and can be related to properties of the random medium [14]. Generally, q_w may depend on X in which case it can be given in the form $q_w(\zeta)^2 = q_{w,0}^2\,b(X(\zeta))^2$ [14]. ■

The methods in Section 7.3.1 dealing with nonlinear systems driven by Brownian motion can be used to calculate properties of the diffusion process X in Eq. 7.88. The Fokker-Planck equation (Section 7.3.1.3) can be used to assess changes in the properties of a material subjected to a sustained action.

The density $f(x; \zeta \mid x_0; \zeta_0)$ of the conditional process $X(\zeta) \mid (X(\zeta_0) = x_0)$ is the solution of the Fokker-Planck equation (Eq. 7.64)

$$\frac{\partial f}{\partial \zeta} = -\frac{\partial}{\partial x}\,(a(x)\,f) + \frac{q_{w,0}^2}{2}\,\frac{\partial^2}{\partial x^2}\left(b(x)^2\,f\right). \tag{7.89}$$

Note: The dependence of the density $f(x; \zeta \mid x_0; \zeta_0)$ on ζ can be used to characterize the change in time of some material properties, referred to as the **structural evolution**. If $f(x; \zeta \mid x_0; \zeta_0)$ converges to a stationary density $f_s(x) = \lim_{\zeta\to\infty} f(x; \zeta \mid x_0; \zeta_0)$, we say that the **structural evolution emerges in a pattern** (Section 8.6). The stationary density f_s is the solution of

$$\frac{d}{dx}\,(a(x)\,f_s(x)) - \frac{q_{w,0}^2}{2}\,\frac{d^2}{dx^2}\left(b(x)^2\,f_s(x)\right) = 0$$

so that $f_s(x) = (c/b(x)^2)\,\exp\left[2\int^x a(\sigma)/(q_{w,0}^2\,b(\sigma)^2)\,d\sigma\right]$, where $c > 0$ is a scaling constant (Example 7.41).

We also note that $f(x; \zeta \mid x_0; \zeta_0)\,dx$ can be interpreted as the probability that the state of a randomly selected point $\boldsymbol{\xi}$ in D belongs to an infinitesimal range $(x, x + dx]$ if the medium in D is subjected to an average strain ζ and the initial state is $X(\zeta_0) = x_0$. The stationary density f_s has a similar meaning. If the initial conditions are random, this probability is given by the integral $dx \int f(x; \zeta \mid x_0; \zeta_0)\,f_0(x_0)\,dx_0$, where f_0 is the density of $X(\zeta_0)$. ▲

Example 7.56: Consider a polycrystal consisting of brick-type grains of identical dimensions characterized by 12 slip systems with random orientation specified by their Euler angles (Section 8.6.2). The polycrystal is subjected to the deterministic strain history ε_{11} in the direction of coordinate x_1 (Fig. 7.14, left plot). The right

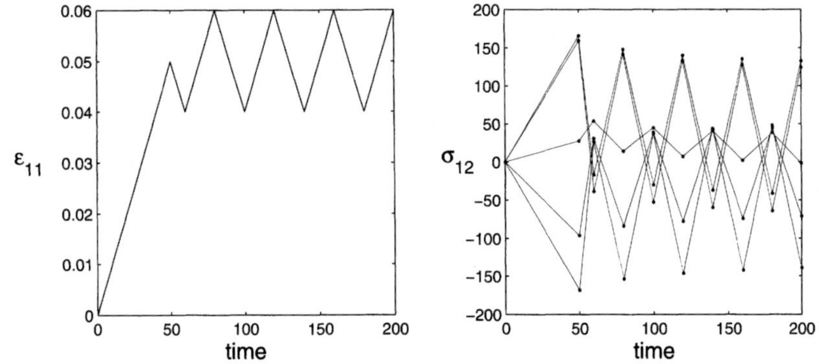

Figure 7.14. Applied strain history and corresponding stress histories in five grains

plot in Fig. 7.14 shows the corresponding shear stress histories in the (x_1, x_2) plane in five different grains.

The stress histories in Fig. 7.14 represent samples of the random function $\Gamma(\boldsymbol{\xi}, t)$ in Eq. 7.86, recorded at five different locations $\boldsymbol{\xi}$ in the polycrystal. By following the arguments used to define the process X in Eq. 7.88, the stress samples in Fig. 7.14 can be used to define a stochastic process for the shear stress in the polycrystals. The differences between these stress samples is caused by the uncertainty in the material properties, since the input is deterministic.

Note: The stress analysis is based on the Taylor hypothesis, assuming that there is no interaction between grains. Crystal plasticity theory has been used to calculate stress histories in individual grains. Details on this topic are in Section 8.6.2.2. ▲

Example 7.57: Let X be the atomic lattice orientation in a single crystal with two slip systems subjected to a random shear strain rate $\dot{\gamma} + Y = 1 + q\, W$, where q is a constant and W is a Gaussian white noise. The atomic lattice orientation process is the solution of (Section 8.6.2.1 in this book, [12], [14])

$$\dot{X}(\zeta) = (2\,\cos(X(\zeta)) - 1)\;(1 + q\,W(\zeta))\,, \quad \zeta \geq 0.$$

Figure 7.15 shows the evolution of the density $f(x; \zeta)$ for an initial state $X(0) \sim U([0, 2\pi])$ and two values of the noise intensity q. The evolution of $f(x; \zeta)$ reaches a steady configuration that depends on the noise intensity. ◇

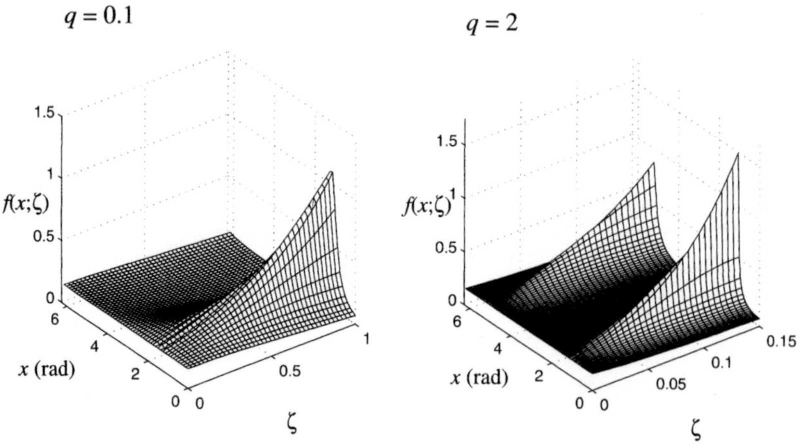

Figure 7.15. Density $f(x;\zeta)$ for $X(0) \sim U([0, 2\pi])$ and two noise intensities

Note: Generalities on crystal plasticity are in Section 8.6.2. The internal variable X in this example is equal to twice the atomic lattice orientation. The general evolution equation for a single crystal is in Section 8.6.2.1.

Because we assume that the crystal has two slip systems with deterministic properties, the material does not have random properties. The internal variable X is a random process since the initial condition and applied strain are uncertain. ▲

7.4.3 Reliability analysis

Engineering systems are designed for specified functions and level of performance. For example, a bridge has to be sufficiently strong to support its own weight and traffic loads without excessive deformations and/or vibrations over a specified time interval. Cars and planes are designed to carry passengers comfortably and safely. Computers controlling air traffic, nuclear power plants, or bank accounts need to function according to specified performance criteria. The height of a flood protection system must be equal to a water level that is exceeded on the average, for example, at most once in a hundred years. It is not possible to manufacture a system that fulfills some design conditions with certainty because of the randomness in both system properties and actions. However, we can calculate the probability that these conditions are met during a specified time interval and modify a design such that this probability is sufficiently large.

Let $\boldsymbol{X}(t)$ be an $\mathbb{R}^d$-valued process defining the state of a system at time $t \geq 0$ and denote by $\tau > 0$ the projected life of the system, called **design life**. The coordinates of $\boldsymbol{X}$ may denote displacements and velocities at some points of a bridge, stresses at critical points of an aircraft, the current earth temperature, or the value of a stock. The design life τ can vary significantly depending on the system. Common values of τ are 50 years for a bridge, 20 years for an aircraft, thousands of years for earth temperature, and weeks or months for a stock price. The **reliability** in τ is the probability

$$P_s(\tau) = P(\boldsymbol{X}(t) \in D, 0 \le t \le \tau), \tag{7.90}$$

where D is an open subset of $\mathbb{R}^d$, called the **safe set**, and results from the design conditions. Our objective is to calculate the reliability $P_s(\tau)$ or the **probability of failure** $P_f(\tau) = 1 - P_s(\tau)$ in τ.

Example 7.58: Let X and $\dot{X}$ be the displacement and velocity of a linear or nonlinear oscillator that is subjected to a random input. Denote by $\boldsymbol{X} = (X_1 = X, X_2 = \dot{X})$ the state vector of this system. The safe set is

$$D = \{\boldsymbol{x} \in \mathbb{R}^2 : |x_1| < a\}, \quad a > 0, \quad \text{or}$$
$$D = \{\boldsymbol{x} \in \mathbb{R}^2 : x_1^2 + x_2^2 < e\}, \quad e > 0,$$

if the design objective is to control the displacement or the total energy of the oscillator, respectively. ◇

Generally, the reliability of an engineering system is found by using Monte Carlo simulation, crossing theory for stochastic processes, or numerical solution of Fokker-Planck-Kolmogorov equations. The following sections illustrate the use of these methods.

7.4.3.1 Crossing theory

Crossing properties of stochastic processes were discussed in Section 3.10.1. We have shown that the frequency at which a state vector $\boldsymbol{X}$ crosses the boundary ∂D of the safe set D from D to D^c can be used to bound and approximate $P_s(\tau)$. A ∂D-crossing of $\boldsymbol{X}$ from D to D^c is called a **D-outcrossing**.

Let $N(D; \tau)$ be a random variable giving the number of D-outcrossings of $\boldsymbol{X}$ during a time interval $[0, \tau]$. The random variable $N(D; \tau)$ takes values in $\{0, 1, \ldots\}$. It is assumed that $\boldsymbol{X}$ is such that $E[N(D; \tau)] < \infty$. If $\boldsymbol{X}$ is a stationary process, then $E[N(D; \tau)] = \tau\, E[N(D; 1)]$ so that $\lambda(D) = E[N(D; 1)]$ represents the average number of D-outcrossings per unit of time, and is called the **mean D-outcrossing rate** of $\boldsymbol{X}$. The corresponding values of $E[N(D; 1)]$ for the x-upcrossings and x-downcrossings of a real-valued process are denoted by $\lambda(x)^+$ and $\lambda(x)^-$, respectively (Section 3.10.1).

Let $A = \{\boldsymbol{X}(0) \in D\}$ and $B = \{N(D; \tau) = 0\}$. A common approximation for $P_s(\tau)$ in reliability studies is

$$P_s(\tau) \simeq P(A)\, P(B) \simeq P(\boldsymbol{X}(0) \in D)\, \exp\left(-E[N(D; \tau)]\right). \tag{7.91}$$

Note: The first approximate equality is based on the assumption that the events A and B are independent. The second approximate equality assumes that the D-outcrossings of $\boldsymbol{X}$ follow a Poisson process. There is no justification for the first assumption but numerical results support it. The second assumption is correct asymptotically as the size of D and the reference period τ increase indefinitely ([45], Section 12.2). ▲

The probability of failure $P_f(\tau)$ takes values in $[P_{f,l}(\tau), P_{f,u}(\tau)]$, where

$$P_{f,u}(\tau) = \min\left(1, 1 - P(\boldsymbol{X}(0) \in D) + E[N(D;\tau)]\right),$$
$$P_{f,l}(\tau) = (\boldsymbol{q}^T \boldsymbol{p})^2/(\boldsymbol{q}^T \boldsymbol{\pi}\, \boldsymbol{q}), \quad \boldsymbol{q} \in \mathbb{R}^m, \tag{7.92}$$

$0 \le t_1 < \cdots < t_m \le \tau$, $A_i = \{\boldsymbol{X}(t_i) \notin D\}$, $p_i = P(A_i)$, $\pi_{ij} = P(A_i \cap A_j)$, $\boldsymbol{p} = (p_1, \ldots, p_m)$, and $\boldsymbol{\pi} = \{\pi_{ij}\}$, $i, j = 1, \ldots, m$.

Proof: Let $A = \{\boldsymbol{X}(0) \in D\}$ and $B = \{N(D;\tau) = 0\}$. Then $P_s(\tau) = P(A \cap B)$ so that

$$P_f(\tau) = P((A \cap B)^c) = P(A^c \cup B^c) \le P(A^c) + P(B^c),$$

where

$$P(A^c) = P(\boldsymbol{X}(0) \in D^c) = 1 - P(\boldsymbol{X}(0) \in D),$$
$$P(B^c) = \sum_{k=1}^{\infty} P(N(D;\tau) = k) \le \sum_{k=1}^{\infty} k\, P(N(D;\tau) = k) = E[N(D;\tau)].$$

These results give the upper bound $P_{f,u}(\tau)$.

For the lower bound $P_{f,l}$, note that $\{\boldsymbol{X}(t) \in D, 0 \le t \le \tau\} \subset \cap_{i=1}^m A_i^c$ for any distinct times $t_i \in [0,\tau]$, $i = 1, \ldots, m$, so that $P_s(\tau) \le P\left(\cap_{i=1}^m A_i^c\right)$ and

$$P_f(\tau) \ge P_{f,l}(\tau) = P\left(\cup_{i=1}^m A_i\right) = P(\cup_{i=1}^m \{\boldsymbol{X}(t_i) \in D^c\}).$$

If $m = 1$, then $P_{f,l}(\tau) = P(\boldsymbol{X}(t) \in D^c)$ with the optimal value $P(\boldsymbol{X}(t^*) \in D^c)$, where $t^* \in [0,\tau]$ maximizes $P(\boldsymbol{X}(t) \in D^c)$. If $\boldsymbol{X}$ is stationary, the probability $P(\boldsymbol{X}(t) \in D^c)$ is time invariant so that t can be any time in $[0,\tau]$. If $m > 1$ is fixed, an optimization algorithm can be used to find the times $(t_1^*, \ldots, t_m^*)$ that maximize $P_{f,l}(\tau)$ in Eq. 7.92. Consider some fixed distinct times $(t_1, \ldots, t_m)$ and define a random variable $Z = \sum_{i=1}^m q_i\, 1_{A_i}$, where q_i's are arbitrary constants. The random variable Z is zero on the subset $\left(\cup_{i=1}^m A_i\right)^c$ of Ω and has the properties

$$P(Z \ne 0) \le P\left(\cup_{i=1}^m A_i\right),$$
$$E[Z] = \sum_{i=1}^m q_i\, E\left[1_{A_i}\right] = \sum_{i=1}^m q_i\, P(A_i) = \boldsymbol{q}^T \boldsymbol{p}, \quad \text{and}$$
$$E[Z^2] = \sum_{i,j=1}^m q_i\, q_j\, E\left[1_{A_i} 1_{A_j}\right] = \sum_{i,j=1}^m q_i\, q_j\, P(A_i \cap A_j) = \boldsymbol{q}^T \boldsymbol{\pi}\, \boldsymbol{q},$$

where $\boldsymbol{q} = (q_1, \ldots, q_m)$ and $\boldsymbol{p}$ are viewed as column vectors. Because

$$E\left[\left(x\, Z - 1_{\{Z \ne 0\}}\right)^2\right] = x^2\, E[Z^2] - 2x\, E[Z] + P(Z \ne 0) \ge 0, \quad \forall x \in \mathbb{R},$$

we have $(E[Z])^2 - E[Z^2]\, P(Z \neq 0) \leq 0$ or $P(Z \neq 0) \geq (E[Z])^2 / E[Z^2]$ which gives

$$P_f(\tau) \geq P(\cup_{i=1}^m A_i) \geq P(Z \neq 0) \geq \frac{(E[Z])^2}{E[Z^2]} = \frac{(\boldsymbol{q}^T\, \boldsymbol{p})^2}{\boldsymbol{q}^T\, \boldsymbol{\pi}\, \boldsymbol{q}}$$

for any $\boldsymbol{q} \in \mathbb{R}^m$. Tighter lower bounds can be obtained by selecting $\boldsymbol{q}$ such that it maximizes $(\boldsymbol{q}^T\, \boldsymbol{p})^2/(\boldsymbol{q}^T\, \boldsymbol{\pi}\, \boldsymbol{q})$ [170].

Alternative lower bounds on the failure probability can be obtained. For example, we have shown that the probability

$$\tilde{P}_{f,l}(\tau) = P(A_1) + \sum_{i=2}^{m} \max\left(0,\, P(A_i) - \sum_{j=1}^{i-1} P(A_i \cap A_j)\right)$$

is a lower bound on $P_{f,l}(\tau)$ (Section 2.2.3). ■

Example 7.59: Consider an elasto-plastic oscillator with a small damping ratio ζ, elastic frequency $\nu > 0$, and yield displacement $u > 0$. The oscillator is subjected to a broadband, stationary, Gaussian input with mean zero. Let T^{el} denote the random time between consecutive plastic excursions of the oscillator displacement X, that is, the time between consecutive $(-u, u)$-outcrossings of X. If $u \to \infty$, then $T^{\text{el}} \to \infty$ and X approaches a stationary, narrowband Gaussian process X^{el} with mean zero as time increases (Section 7.2.1.2). It is assumed that (1) the input is scaled such that X^{el} has unit-variance and (2) $u < \infty$ is such that T^{el} is much larger than the duration of the transient displacement following a plastic excursion. Hence, X can be approximated by X^{el} in T^{el} and the plastic excursions are rare events.

Let $N(\tau)$ and Δ_i be the number of plastic excursions in $[0, \tau]$ and the plastic deformation at the end of plastic excursion $i = 1, \ldots, N(\tau)$, respectively. The reliability in τ can be defined by $P_s(\tau) = P\left(\sum_{i=1}^{N(\tau)} |\Delta_i| \leq \delta_{\text{cr}}\right)$, where δ_{cr} is a specified limit value. The definition of $P_s(\tau)$ is meaningful since $\sum_{i=1}^{N(\tau)} |\Delta_i|$ is proportional to the energy dissipated by plastic deformation in $[0, \tau]$ so that it can be used as a damage measure [54]. It is assumed that Δ_i are iid random variables that are independent of the counting process N so that, if the laws of the random variables Δ_1 and $N(\tau)$ are known, we can calculate the reliability $P_s(\tau)$ from the characteristic function,

$$\begin{aligned} E\left[e^{\sqrt{-1}\, u \sum_{i=1}^{N(\tau)} |\Delta_i|}\right] &= E\left\{E\left[e^{\sqrt{-1}\, u \sum_{i=1}^{N(\tau)} |\Delta_i|} \mid N(\tau)\right]\right\} \\ &= \sum_{n=0}^{\infty} \left(\varphi_{|\Delta_1|}(u)\right)^n\, P(N(\tau) = n), \end{aligned}$$

of $\sum_{i=1}^{N(\tau)} |\Delta_i|$, where $\varphi_{|\Delta_1|}$ denotes the characteristic function of $|\Delta_1|$.

Methods of various complexity and accuracy have been proposed for calculating the laws of Δ_1 and $N(\tau)$ [54]. One of the simplest methods for finding the distribution of Δ_1 uses (1) the equivalence between the kinetic energy of the oscillator mass at the initiation of a plastic excursion and the energy dissipated by plastic deformation Δ_1 and (2) the distribution of $\dot{X}^{el}$ at the time of a $\pm u$-crossing of X^{el}, which can be obtained from the properties of this process [189]. The assumption that $N(\tau)$ is a Poisson random variable with intensity λ given by the mean $(-u, u)$-outcrossing of X^{el} is not satisfactory since X^{el} is a narrowband process so that its $(-u, u)$-outcrossings occur in clusters and therefore they cannot be independent. This difficulty can be overcome by replacing N with another counting process $\hat{N}$, where $\hat{N}(\tau)$ gives the number of u-upcrossings in $[0, \tau]$ of the envelope process $R(t) = [X^{el}(t)^2 + \hat{X}^{el}(t)^2]^{1/2}$, where $\hat{X}^{el}$ denotes the Hilbert transform of X^{el} (Section 3.9.4.1). The envelope process (1) is larger than $|X^{el}|$ at all times, (2) is equal to $|X^{el}|$ at the zero-crossing times of $\hat{X}^{el}$, (3) has a single excursion above u for each cluster of $(-u, u)$-outcrossings of X^{el}, and (4) may have excursions outside $(-u, u)$ during which X^{el} does not exit $(-u, u)$. The excursions of R which are not associated with $\pm u$-crossings of X^{el} are called empty excursions, and are not considered in the calculation of $P_s(\tau)$. The process $\hat{N}$ can be approximated by a homogeneous Poisson process with intensity

$$\lambda_R(u) = \delta \sqrt{\lambda_2/(2\pi)}\, u\, e^{-u^2/2},$$

where $\delta = \sqrt{1 - \lambda_1^2/\lambda_2}$, $\lambda_k = \int_0^\infty \nu^k\, g(\nu)\, d\nu$, and g denotes the one-sided spectral density of X^{el} ([45], Sections 11.7 and 11.8).

To complete the analysis we need to estimate the number of empty envelope crossings and modify $\hat{N}$ so that only the qualified envelope excursions, that is, the envelope crossings that are accompanied by crossings of X^{el}, will be counted. The marked process $\hat{N}$ counting only the qualified envelope crossings has the intensity $r\,\lambda_R(u)$, where r denotes the long run fraction of qualified envelope excursions, and can be calculated, for example, from

$$r = \frac{1}{\pi\,\delta\,u}\left(1 - e^{-\pi\,\delta\,u}\right) \quad \text{or}$$

$$r = 1 - 2\int_0^u \phi(\sigma)\left[1 - \sqrt{2\pi}\,\frac{\Phi(\gamma\,\pi\,(u^2 - \sigma^2)/u) - 1/2}{\gamma\,\pi\,(u^2 - \sigma^2)/u}\right],$$

where $\gamma = \delta/\sqrt{1 - \delta^2}$. Comparisons of the fraction of empty envelopes $1 - r$ in the above expressions with simulation results show that the first expression becomes less satisfactory as the spectral bandwidth δ of X increases [54]. ◇

Note: The first expression of r results from the distribution $P(R(t) > u) = e^{-u^2/2}$ of R, the expression of the mean crossing rates $\lambda_R(u)$ and $\lambda(u)$ of R and X^{el}, respectively, and the following approximations and assumptions.

Let T be the random duration of an excursion of R above u. The long run average of T is $P(R(t) > u)/\lambda_R = 1/(u\,\pi\,\sqrt{\lambda_2})$ since $\tau\, P(R(t) > u)$ approximates the average time R is above u during a time interval $[0, \tau]$, provided that τ is large, and $\lambda_R\,\tau$ gives the average number of u-upcrossings of R in $[0, \tau]$. The average interarrival time between consecutive zero crossings of X^{el} is $1/\lambda(0) = \pi/\sqrt{\lambda_2}$. If $T \geq 1/\lambda(0)$, it is assumed that corresponding envelope excursions are qualified. If $T < 1/\lambda(0)$, empty envelope crossings are possible, and the probability of an empty envelope crossing can be approximated by $[1/\lambda(0) - T]/(1/\lambda(0))$ or $1 - T\sqrt{\lambda_2}/\pi$ by assuming that T is randomly located on the time axis and the zeros of X are multiples of $1/\lambda(0)$. The first expression of r above results under the assumption that T is an exponentially distributed random variable [189]. The second expression of r is based on rather complex arguments regarding properties of the envelope process [54]. These considerations also allow us to derive the probability of empty envelope excursions. ▲

7.4.3.2 First passage time

Let $T = \inf\{t \geq 0 : \boldsymbol{X}(t) \notin D\}$ be a stopping time giving the first time when the state $\boldsymbol{X}$ starting at $\boldsymbol{X}(0) \in D$ leaves the safe set D. The reliability in $[0, \tau]$ is (Eqs. 7.66 and 7.90)

$$P_s(\tau) = P(T > \tau) = 1 - F_T(\tau), \tag{7.93}$$

where F_T denotes the distribution of T. We discuss the calculation of F_T by Monte Carlo simulation, Durbin's formula, and Fokker-Planck's equation.

Monte Carlo simulation. In Section 5.4.2 the Monte Carlo simulation method was used to estimate the probability that the state of a physical system does not leave a safe set during a specified time interval. Here we estimate the probability that a structural system performs according to some design specifications during an earthquake. The coordinates of the state vector $\boldsymbol{X}$ are displacements and velocities at a finite number of points of a structural system. The safe set D results from conditions on $\boldsymbol{X}$. The input is the seismic ground acceleration, which can be modeled by a process $A(t) = \xi(t)\, A_s(t)$, $t \in [0, \tau]$, where $\tau > 0$ is the duration of the seismic event, $A_s(t)$ is a stationary Gaussian process with mean zero and spectral density $s(\nu)$, and $\xi(t)$ denotes a deterministic modulation function.

Statistical analysis of a large number of seismic ground acceleration records and plate tectonics considerations have been used to develop expressions for the spectral density $s(\nu)$, modulation function $\xi(t)$, and earthquake duration τ of the seismic ground acceleration process $A(t)$ at many sites in the US. It has been shown that $s(\nu)$, $\xi(t)$, and τ are completely defined by the moment magnitude M of the seismic sources in the proximity of a site and the site-source distance R [96, 139].

Let $P_f(m, r)$ denote the probability that a structural system subjected to the ground acceleration process A corresponding to a seismic source with $(M = m, R = r)$ does not satisfy some specified performance requirements, that is,

$$P_f(m, r) = 1 - P(\boldsymbol{X}(t) \in D, 0 \leq t \leq \tau \mid M = m, R = r).$$

The plot of $P_f(m, r)$ versus all possible values (m, r) of (M, R) is referred to as the **fragility surface**. The unconditional probability of failure

$$P_f = \int\int P_f(m,r)\, f_{M,R}(m,r)\, dm\, dr$$

can be used to assess the seismic performance of the system under consideration, where $f_{M,R}$ denotes the density of (M, R). Estimates of this density for the central and eastern United States can be found in [64].

Example 7.60: A steel water tank with dimensions $20.3 \times 16.8 \times 4.8$ ft is located at the top of a 20 story steel frame building. The tank is supported at its corners by four vertical legs of length 6.83 ft. It is assumed that (1) the steel frames of the building remain linear and elastic during the seismic event but the tank legs may yield, (2) the building and the tank oscillate in a vertical plane containing the direction of the seismic ground motion, (3) there is no feedback from the tank to the building, (4) the tank can be modeled as a rigid body, and (5) the design condition for the tank requires that the largest deformation of its legs be smaller than a critical value d_{cr}.

Let A_{bt} be the absolute acceleration process at the connection point between the building and the tank, so that this process represents the seismic input to the tank. The calculation of fragility surfaces by Monte Carlo simulation involves three steps. First, a pair of values (m, r) needs to be selected for (M, R) and n_s samples of A_{bt} corresponding to these values of (M, R) have to be generated. Second, the dynamic response of the tank has to be calculated for each sample of A_{bt}. Third, the probability $P_f(m, r)$ has to be estimated by, for example, the ratio n_s^*/n_s, where n_s^* denotes the number of tank response samples that exceed a critical displacement d_{cr}. Figure 7.16 shows fragility surfaces for the tank corresponding to several values of d_{cr} and two different designs, which are sketched in the figure. The fragility surfaces in Fig. 7.16 corresponding to the same critical value d_{cr} differ significantly for the two designs illustrated in this figure. Fragility surfaces of the type shown in Fig. 7.16 and information on the cost of various design alternatives can be used in earthquake engineering to select an optimal repair strategy for an existing system and/or develop efficient new seismic designs. ◇

Note: The building's steel frame can be modeled by a system with $r = 20$ translation degrees of freedom. If the system has classical modes of vibration, the displacement at the structure-tank connection point is $Y(t) = \sum_{i=1}^{r} \phi_i\, Q_i(t)$, where the constants ϕ_i can be obtained from the modal shapes of the building and the functions Q_i denote modal coordinates. The modal coordinates satisfy the differential equations

$$\ddot{Q}_i(t) + 2\,\zeta_i\,\nu_i\,\dot{Q}_i(t) + \nu_i^2\,Q_i(t) = -\gamma_i\,A(t), \quad t \geq 0,$$

with zero initial conditions, where ζ_i and ν_i denote modal damping ratios and frequencies, respectively, and γ_i are modal participation factors (Examples 7.5 and 7.6). We have

$$Q_i(t) = \int_0^t \theta_i(t-s)\,[-\gamma_i\,A(s)]\,ds, \quad t \geq 0,$$

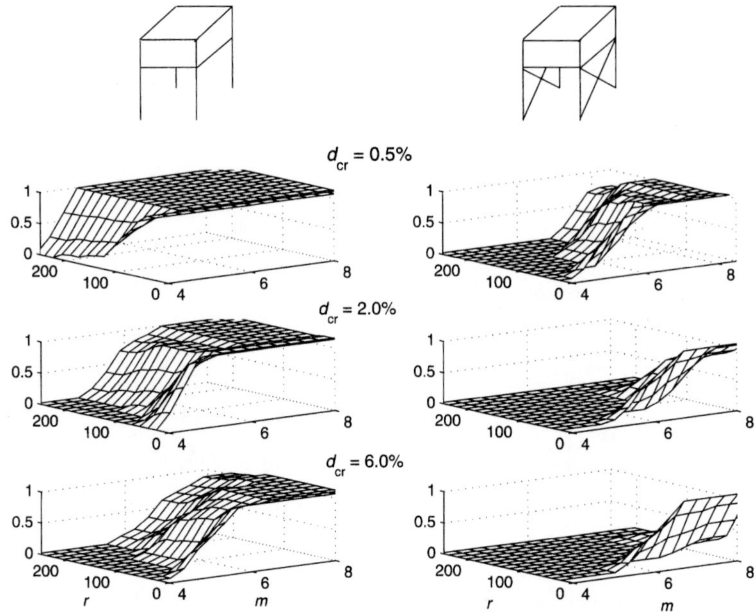

Figure 7.16. Fragility surfaces for several values of d_{cr}

where θ_i is the unit impulse response function for mode i (Example 7.5). Let

$$A(t) = \xi(t) \sum_{k=1}^{n} \sigma_k \left[A_k \cos(\nu_k t) + B_k \sin(\nu_k t)\right]$$

be an approximate representation of the seismic ground acceleration for Monte Carlo simulation, where $\sigma_k, \nu_k > 0$ are constants and A_k, B_k are independent $N(0, 1)$ variables (Section 5.3.1.1). Then

$$Q_i(t) = \sum_{k=1}^{n} \sigma_k \left[A_k\, g_{c,ik}(t) + B_k\, g_{s,ik}(t)\right] \quad \text{and}$$

$$Y(t) = \sum_{k=1}^{n} \sigma_k \left[A_k \left(\sum_{i=1}^{r} \phi_i\, g_{c,ik}(t)\right) + B_k \left(\sum_{i=1}^{r} \phi_i\, g_{s,ik}(t)\right)\right]$$

where

$$g_{c,ik}(t) = -\gamma_i \int_0^t \theta_i(t-s)\, \xi(s) \cos(\nu_k s)\, ds,$$

$$g_{s,ik}(t) = -\gamma_i \int_0^t \theta_i(t-s)\, \xi(s) \sin(\nu_k s)\, ds.$$

The absolute acceleration at the structure-tank connection point is $A_{bt}(t) = A(t) + \ddot{Y}(t)$.

The stress -strain relationship for the tank legs considered in Fig. 7.16 is linear from (0,0) to (0.0012, 36 ksi), constant from (0.0012, 36 ksi) to (0.1356, 36 ksi), linear from (0.1356, 36 ksi) to (0.585, 60 ksi), and constant for larger strains, where the first and second numbers in parentheses denote strains and stresses, respectively. ▲

Durbin's formula. We have given in Section 3.10.2 formulas for the densities of T and $(T, \dot{X}(T))$, where X is a real-valued process, $T = \inf\{t > 0 : X(t) \geq u\}$, and $u \in \mathbb{R}$. These formulas are repeated in the following equation.

The joint density of the random variables $(T, \dot{X}(T))$ and the density of T are

$$f_{T,\dot{X}(T)}(t, z) = E\left[1_{\{X(s)<u, 0\leq s<t\}}\, \dot{X}(t)^+ \mid X(t) = u, \dot{X}(t) = z\right] f_{X(t),\dot{X}(t)}(u, z),$$
$$f_T(t) = E\left[1_{\{X(s)<u, 0\leq s<t\}}\, \dot{X}(t)^+ \mid X(t) = u\right] f_{X(t)}(u). \qquad (7.94)$$

We show that Eq. 7.94 can be used to bound the density of T. Consider n distinct times $0 < t_1 < \cdots < t_n < t$ in $[0, t]$ and the events $A = \{X(t) < u < X(t + \Delta t)\}$ and $B_n = \cap_{i=1}^n \{X(t_i) < u\}$, where $\Delta t > 0$ is a small time interval. Then f_T can be bounded by the functions in Eq. 7.95. The bounds on f_T in this equation have been used to develop computer codes for calculating the density of T approximately [159].

The density of T satisfies the inequalities

$$f_T(t) \geq \lambda(u; t)^+ - \int_0^t E\left[1_{\{T_1=s\}}\, \dot{X}(s)^+\, \dot{X}(t)^+ \mid X(s) = X(t) = u\right] f_{X(t),X(s)}(u, u)\, ds,$$
$$f_T(t) \leq E\left[1_{B_n}\, \dot{X}(t)^+ \mid X(t) = u\right] f_{X(t)}(u) \leq \lambda(u; t)^+. \qquad (7.95)$$

Proof: The upper bound on f_T results from Eq. 7.94 since $1_{\{X(s)<u, 0\leq s<t\}} \leq 1_{B_n} \leq 1$.

Let T_k, $k = 1, 2, \ldots$, denote the time of the k'th u-upcrossing of X. The probability that this upcrossing occurs in $[t, t+\Delta t]$ can be approximated by $f_{T_k}(t)\,\Delta t$, where f_{T_k} is the density of T_k. Hence, the mean u-upcrossing rate $\lambda(u; t)^+$ of X at time t is $\sum_{k=1}^\infty f_{T_k}(t)$ so that

$$f_{T_1}(t) = \lambda(u; t)^+ - \sum_{k=2}^\infty f_{T_k}(t) = \lambda(u; t)^+ - \sum_{k=2}^\infty \int_0^t f_{T_1,T_k}(s, t)\, ds.$$

We also have

$$\sum_{k=2}^{\infty} \int_0^t f_{T_1,T_k}(s,t)\, ds$$

$$= \sum_{k=2}^{\infty} \int_0^t E\left[1_{\{T_1=s\}}\, 1_{\{T_k=t\}}\, \dot{X}(s)^+\, \dot{X}(t)^+ \mid X(s) = X(t) = u\right] f_{X(s),X(t)}(u,u)\, ds$$

$$= \int_0^t E\left[1_{\{T_1=s\}}\, \dot{X}(s)^+ \left(\sum_{k=2}^{\infty} 1_{\{T_k=t\}}\, \dot{X}(t)^+\right) \mid X(s) = X(t) = u\right] f_{X(s),X(t)}(u,u)\, ds$$

$$\leq \int_0^t E\left[1_{\{T_1=s\}}\, \dot{X}(s)^+\, \dot{X}(t)^+ \mid X(s) = X(t) = u\right] f_{X(s),X(t)}(u,u)\, ds$$

since $\sum_{k=2}^{\infty} 1_{\{T_k=t\}}\, \dot{X}(t)^+ = (1 - 1_{\{T_1=t\}})\, \dot{X}(t)^+ \leq \dot{X}(t)^+$. The expression of f_{T_1} and the above inequality give the lower bound on f_T in Eq. 7.95. ■

Fokker-Planck equation. In Section 7.3.1.3 we have established the boundary conditions that the Fokker-Planck equation must satisfy to deliver the distribution and/or moments of the stopping time T in Eq. 7.93. Generally, the Fokker-Planck equation cannot be solved analytically. Numerical solutions have been obtained for low dimension state vectors ($d \leq 3$). The following examples illustrate how the distribution of T can be obtained from the solution of the Fokker-Planck equation.

Example 7.61: Let $X = B$ be a Brownian motion and let

$$P_f(t) = 1 - P\left(\sup_{0 \leq s \leq t} B(s) < a\right)$$

be the probability of failure for a safe set $D = (-\infty, a)$, $a > 0$. Figure 7.17 shows the evolution in time of the exact and approximate probability of failure $P_f(t)$ for $a = 1$ and $X(0) = 0$. The approximate solutions correspond to Monte Carlo simulation and a finite difference solution of the Fokker-Planck equation for the density of $X(t) \mid (X(0) = 0)$ with an absorbing boundary at $x = a$. ◇

Note: Let $T = \inf\{t > 0 : X(t) \notin (-\infty, a)\}$. The distribution of this stopping time is

$$P(T \leq t) = 2\left[1 - \Phi\left(a/\sqrt{t}\right)\right],$$

where Φ denotes the distribution of $N(0,1)$ ([154], Section 6.2). By the definitions of T and $P_f(t)$ we have $P(T \leq t) = P_f(t)$.

The Monte Carlo estimate of $P_f(t)$ is given by the fraction of samples of X that have left $D = (-\infty, a)$ at least once in $[0, t]$.

A finite difference representation of the type in Example 7.37 has been used to solve the Fokker-Planck equation $\partial f/\partial t = (1/2)\, \partial^2 f/\partial x^2$ with the initial and boundary conditions $f(x; 0 \mid 0; 0) = \delta(x)$ and $f(a; t \mid 0; 0) = 0$, $t > 0$, respectively. The probability of failure has been calculated from $P_f(t) = 1 - \int_{-\infty}^{a} f(x; t \mid 0; 0)\, dx$. ▲

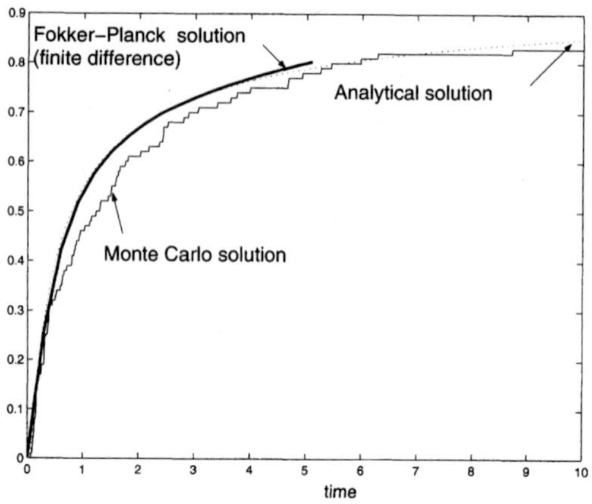

Figure 7.17. Evolution of the exact and approximate probability of failure $P_f(t)$ for $X = B$ and a safe set $(-\infty, 1)$

Example 7.62: Let X be the solution of $dX(t) = (\zeta\, X(t) - X(t)^3)\, dt + \sigma\, dB(t)$, $t \geq 0$, where ζ, σ are constants and B is a Brownian motion. Consider a safe set $D = (-a, a)$ and an initial state $X(0) = x_0 \in D$, where $a > 0$ is a constant. Figure 7.18 shows the evolution of the probability of failure $P_f(t)$ for $\zeta = \pm 1$,

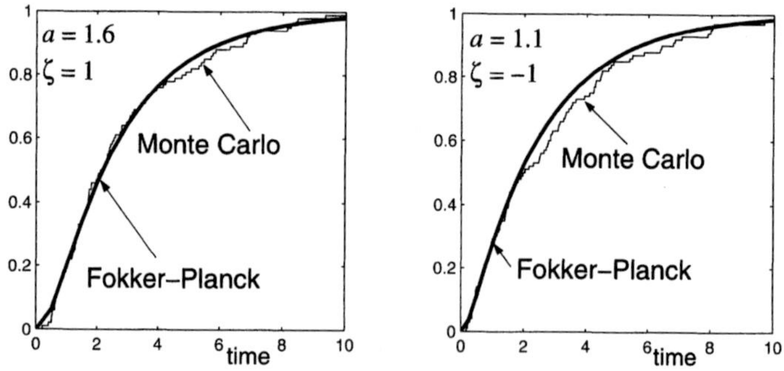

Figure 7.18. Probability of failure $P_f(t)$) for X defined by $dX(t) = (\pm X(t) - X(t)^3)\, dt + dB(t)$ and a safe set $D = (-a, a)$

$\sigma = 1$, $x_0 = 0$, $a = 1.6$, and $a = 1.1$. The finite difference and the Monte Carlo simulation methods have been used for calculations. ◇

Note: The initial and boundary conditions for the finite difference solution of the Fokker-Planck equation are $f(x; 0 \mid x_0; 0) = \delta(x - x_0)$ and $f(x; t \mid x_0; 0) = 0$ for $t > 0$, respectively. The condition $f(x; t \mid x_0; 0) = 0$, $t > 0$, defines an absorbing boundary for the Fokker-Planck equation associated with the diffusion process X. ▲

Example 7.63: Let X be the solution of

$$\ddot{X}(t) + 2\,\varepsilon^2\,\nu_0\,\dot{X}(t) + \nu_0^2\,X(t) = \varepsilon\,Y(t), \quad t > 0,$$

where $\zeta, \nu, \varepsilon > 0$ are some constants and Y is a stationary Gaussian process with mean zero and a broadband spectral density $s(\nu)$. The envelope $A(t) = \left(X(t)^2 + \dot{X}(t)^2/\nu_0^2\right)^{1/2}$ of X can be approximated by a diffusion process (Examples 7.48 and 7.51). Let $P_f(t) = 1 - P(\max_{0 \le s \le t} |X(t)| < a)$ be the probability of failure for the state X in the above equation. An approximation of this probability of failure is $\tilde{P}_f(t) = 1 - P(\max_{0 \le s \le t} A(t) < a)$.

Figure 7.19 shows Monte Carlo estimates of $P_f(t)$ and $\tilde{P}_f(t)$ obtained from

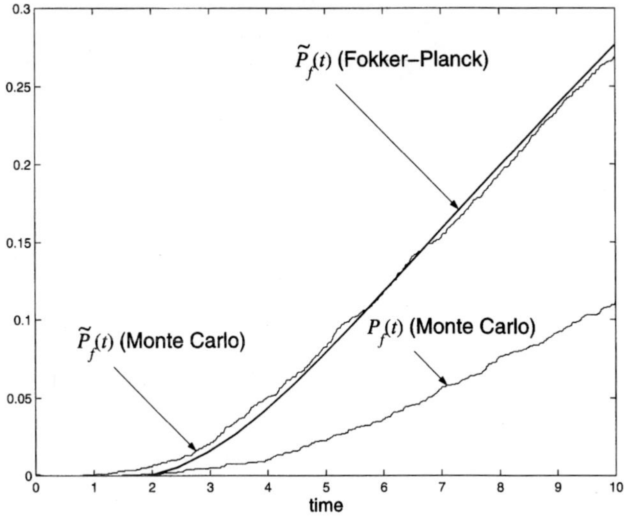

Figure 7.19. Evolution of probabilities P_f and $\tilde{P}_f$ in time

1,000 samples of X and A, respectively. The figure also shows the probability $\tilde{P}_f(t)$ obtained from the finite difference solution of the Fokker-Planck equation for the density of A. Numerical results are for $\varepsilon = 0.283$, $\nu = 1$, $s(\nu_0) = 3.978$, and $a = 5$. That $\tilde{P}_f(t)$ is larger than $P_f(t)$ is consistent with our discussion in Example 7.59 on the relationship between crossings of A and X.

Figure 7.20 shows the evolution of the joint density of $(X, \dot{X})$ in time for the case in which the driving noise is $Y = W$, where W is a zero-mean stationary Gaussian white noise with spectral density $s_w(\nu) = s(\nu_0)$, $\nu \in \mathbb{R}$. The results in this figure have been obtained by a finite difference solution of the Fokker-Planck equation for the density $(X, \dot{X})$ [179]. ◇

Note: Let $\boldsymbol{X} = (X_1 = X, X_2 = \dot{X})$ be the state of the above oscillator driven by W rather than Y. Hence, $\boldsymbol{X}$ is an $\mathbb{R}^2$-valued diffusion process. The density $f(\boldsymbol{x}; t \mid \boldsymbol{x}_0; 0)$ of

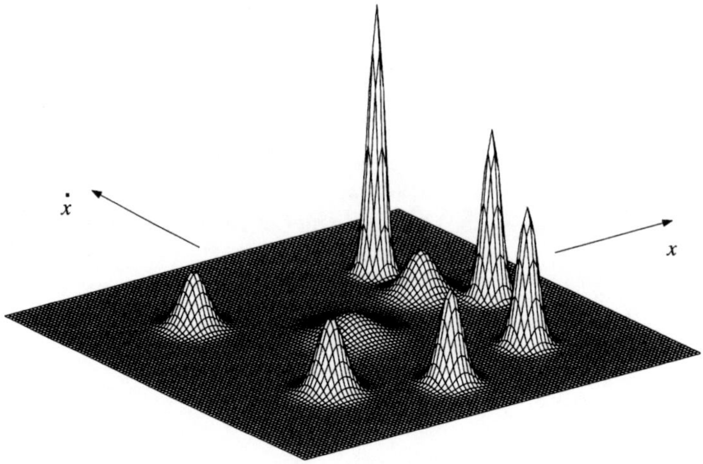

Figure 7.20. Evolution of joint density of $(X, \dot{X})$ in time (adapted from [179])

$\boldsymbol{X}(t) \mid (\boldsymbol{X}(0) = \boldsymbol{x}_0)$ satisfies the Fokker-Planck equation

$$\frac{\partial f}{\partial t} = -\frac{\partial (x_2 f)}{\partial x_1} + \frac{\partial [(v_0^2 x_1 + 2\varepsilon^2 v_0 x_2) f]}{\partial x_2} + \pi s(v_0) \frac{\partial^2 f}{\partial x_2^2}.$$

The probability of failure of this oscillator for the safe set $D = (-a, a) \times \mathbb{R}$ is given by $P_f(t) = 1 - \int_{-a}^{a} \int_{-\infty}^{\infty} f(\boldsymbol{x}; t \mid \boldsymbol{x}_0; 0)\, dx_1\, dx_2$, where $f(\boldsymbol{x}; t \mid \boldsymbol{x}_0; 0)$ is the solution of the above Fokker-Planck equation with the initial and boundary conditions $f(\boldsymbol{x}; 0 \mid \boldsymbol{x}_0; 0) = \delta(\boldsymbol{x} - \boldsymbol{x}_0)$ and $f(\pm a, x_2; t \mid \boldsymbol{x}_0; 0) = 0$ for $t > 0$, $x_2 \in \mathbb{R}$, and $\boldsymbol{x}_0 \in D$. ▲

7.4.4 Finance

The discussion in this section is based on Chapter 4 in [131]. Useful references on this topic are [60], [126], and [151] (Section 10.16).

Let $X(t)$ denote the price of a stock at time t. The relative return in a small time interval $[t, t + \Delta t]$, $\Delta t > 0$, can be modeled by

$$\frac{X(t + \Delta t) - X(t)}{X(t)} = c\,\Delta t + \sigma \text{ “noise”},$$

where $c > 0$ and $\sigma > 0$ are the **mean return rate** and **volatility**, respectively. Bond, money market, savings account, and other similar assets are viewed as riskless so that their evolution is given by a deterministic differential equation.

The following models are considered for the stock and bond prices and for the value of a portfolio.

Stock price: $dX(t) = c\,X(t)\,dt + \sigma\,X(t)\,dB(t)$.

Bond price: $d\beta(t) = r\,\beta(t)\,dt$.

Portfolio value for trading strategy $(U_s(t), U_b(t))$**:**

$$V(t) = U_s(t)\,X(t) + U_b(t)\,\beta(t).$$

Note: Because we use Gaussian white noise to model market volatility, the value of a stock at time t, $X(t) = X(0)\,e^{(c-0.5\,\sigma^2)\,t+\sigma\,B(t)}$, is a geometric Brownian motion. The evolution of the price of a bond is given by $\beta(t) = \beta(0)\,e^{r\,t}$, where $r > 0$ denotes the **return rate**. Alternative models have been proposed for the evolution of stock prices, for example, stochastic differential equations driven by Lévy or Poisson white noise and the ARCH and GARCH processes ([126], Chapter 10).

The pair $(U_s(t), U_b(t))$ is called a **trading strategy** and gives the number of shares held in stocks and bonds at time t, respectively. The processes U_s and U_b are assumed to be $\sigma(B(s), 0 \le s \le t)$-adapted. The above expression for V gives the **value of a portfolio** at time t corresponding to the trading strategy $(U_s(t), U_b(t))$. The processes U_s and U_b can take positive and negative values. A negative value of $U_s(t)$ means that stocks are sold at time t, that is, **short sale**. A negative value of U_b means that money was borrowed at the bond interest rate. ▲

Options are contracts agreed at time $t = 0$ that entitle one to buy one share of stock for a specified price k, called **strike** or **exercise** price, at or prior to a time $t = \tau$. If an option can be exercised at time $t = \tau$ or at any time $t \le \tau$, it is called a **European call option** or an **American call option**, respectively. The holder of a call option does not have to exercise it. An option to sell a stock at a specified price k at or until a time of maturity $t = \tau$ is called a **put**. European and American put options are defined in the same way as the corresponding call options. The purchaser of a European call option can make a profit of $\max(0, X(\tau) - k)$ by buying the stock at the strike price k and selling immediately at the market value $X(\tau)$. The purchaser of a European put makes a profit $\max(0, k - X(\tau))$. Hence, the European call and put options with strike price k are the random variables $\max(0, X(\tau) - k)$ and $\max(0, k - X(\tau))$, respectively. There are many other types of options, for example, Asian and Russian options [60].

The following assumptions are used in our discussion:

- The options are European calls, there is no cost for transactions, and riskless assets have a constant rate of return.
- The **market is rational.**
- The **self-financing** condition $dV(t) = U_s(t)\,dX(t) + U_b(t)\,d\beta(t)$ holds.

Note: In a rational market there is no opportunity for **arbitrage**, that is, riskless strategies guaranteeing a profit do not exist. The self-financing condition implies that changes in a portfolio's value are caused solely by changes in the prices of stocks and bonds. There is no infusion of new capital in or consumption of wealth from a portfolio. ▲

Consider an investor starting a portfolio, that is, buying a number of stocks and bonds at time $t = 0$. The invested capital and the initial balance between stocks and bonds is a personal decision. The price of riskless assets is not in question. However, the price the investor is willing to pay for a stock at time $t = 0$ is in question because the time evolution of stock prices is random. Our objectives are to find (1) a rational price $X(0)$ for a stock at time $t = 0$ and (2) a self-financing strategy yielding a payoff $\max(0, X(\tau) - k)$, that is, the payoff corresponding to a European call option for this stock. If these objectives are satisfied we say that the investor is **hedged**. The Black-Scholes formula provides a self-financing strategy yielding the desired payoff. Let $u(X(t), \tau - t) = V(t)$ be the value of a portfolio V at time t. Because we require that V at maturity be $\max(0, X(\tau) - k)$, the function u must satisfy the terminal condition for $V(\tau) = u(X(\tau), 0) = (X(\tau) - k)^+ = \max(0, X(\tau) - k)$.

The **Black-Scholes formula** states that the rational price for a European call option at time $t = 0$ and exercise price k is $V(0)$, where

$$V(t) = u(X(t), \tau - t) = X(t)\,\Phi\left(g(X(t), \tau - t)\right) - k\,e^{-r(\tau - t)}\,\Phi\left(h(X(t), \tau - t)\right), \tag{7.96}$$

$$g(x, t) = \frac{\ln(x/k) + (r + \sigma^2/2)\,t}{\sigma\,t^{1/2}}, \qquad h(x, t) = g(x, t) - \sigma\,t^{1/2}, \tag{7.97}$$

and Φ denotes the distribution of $N(0, 1)$.

The process $V(t)$ in Eqs. 7.96 and 7.97 is the value of a portfolio at time t defined by the self-financing strategy

$$U_s(t) = \frac{\partial u(X(t), \tau - t)}{\partial x}, \qquad U_b(t) = \frac{u(X(t), \tau - t) - U_s(t)\,X(t)}{\beta(t)}. \tag{7.98}$$

Proof: Consider the $\mathbb{R}^2$-valued diffusion process $X = (X_1, X_2)$ defined by $X_1 = X$ and $dX_2(t) = -dt$ with the initial condition $X_2(0) = \tau$. The Itô formula applied to the function $V(t) = u(X_1(t), X_2(t)) = u(X(t), \tau - t)$ gives (Eq. 7.3)

$$V(t) - V(0) = \int_0^t \left[c\,X(s)\,u_{,1}(X(s), \tau - s) - u_{,2}(X(s), \tau - s)\right] ds + \frac{\sigma^2}{2}\int_0^t X(s)^2\,u_{,11}(X(s), \tau - s)\,ds + \int_0^t \sigma\,X(s)\,u_{,1}(X(s), \tau - s)\,dB(s),$$

where $u_{,1}$ and $u_{,11}$ denote the first and second partial derivatives of u with respect to x_1 and $u_{,2}$ is the partial derivative of u relative to x_2.

The self-financing strategy and the portfolio value

$$\begin{aligned} dV(t) &= U_s(t)\,dX(t) + U_b(t)\,d\beta(t) \\ &= \left[c\,U_s(t)\,X(t) + r\,U_b(t)\,\beta(t)\right] dt + \sigma\,U_s(t)\,X(t)\,dB(t) \quad \text{and} \\ U_b(t) &= \frac{V(t) - U_s(t)\,X(t)}{\beta(t)} \end{aligned}$$

imply

$$dV(t) = [(c-r)\,U_s(t)\,X(t) + r\,V(t)]\,dt + \sigma\,U_s(t)\,X(t)\,dB(t)$$

so that

$$V(t) - V(0) = \int_0^t [(c-r)\,U_s(s)\,X(s) + r\,V(s)]\,ds + \sigma \int_0^t U_s(s)\,X(s)\,dB(s).$$

Because the above two expressions for $V(t) - V(0)$ must coincide, we have

$$\begin{aligned} &(c-r)\,U_s(t)\,X(t) + r\,u(X(t), \tau - t) \\ &\quad = c\,X(t)\,u_{,1}(X(t), \tau - t) - u_{,2}(X(t), \tau - t) + \frac{\sigma^2}{2}\,X(t)^2\,u_{,11}(X(t), \tau - t), \\ &U_s(t) = u_{,1}(X(t), \tau - t). \end{aligned}$$

The first equation, with U_s in the second equation above, gives

$$u_{,2}(X(t), \tau-t) = \frac{\sigma^2}{2}\,X(t)^2\,u_{,11}(X(t), \tau-t) + r\,X(t)\,u_{,1}(X(t), \tau-t) - r\,u(X(t), \tau-t).$$

This identity provides a partial differential equation,

$$u_{,2}(x, s) = \frac{\sigma^2}{2}\,x^2\,u_{,11}(x, s) + r\,x\,u_{,1}(x, s) - r\,u(x, s)$$

for function u defined for $x > 0$ and $s \in [0, \tau]$ with the deterministic terminal condition $u(x, 0) = (x - k)^+$. The solution of this differential equation is

$$u(x, t) = x\,\Phi(g(x, t)) - k\,e^{-r\,t}\,\Phi(h(x, t))$$

with the notation in Eq. 7.97. ■

Example 7.64: Let $c = 1$, $\sigma = 0.5$, and $r = 0.5$ be the mean return rate, volatility, and return rate, respectively. Suppose that (1) current stock and bond prices are $X(0) = 6$ and $\beta(0) = 6$ and (2) a European call option is available at a strike price k at time $\tau = 1$. The question is whether the current offering of $X(0) = 6$ is acceptable.

Figure 7.21 shows ten samples of the self-financing strategies U_s and U_b in Eq. 7.98 and the portfolio value V in Eq. 7.96 for $k = 10$ and ten samples of V for $k = 20$. These samples suggest that it is unlikely to make any profit at these strike prices. Estimates of the probability p_{prf} of making a profit and the expected profit e_{prf} are $p_{\text{prf}} = 0.03$ and $e_{\text{prf}} = 2.2$ for $k = 10$ and $p_{\text{prf}} \simeq 0$ and $e_{\text{prf}} = 2.1$ for $k = 20$. If $k = 5$, the estimates are $p_{\text{prf}} = 0.55$ and $e_{\text{prf}} = 7.03$, so that we have a better then 50% chance to profit. The above estimates have been calculated from 500 samples. ◇

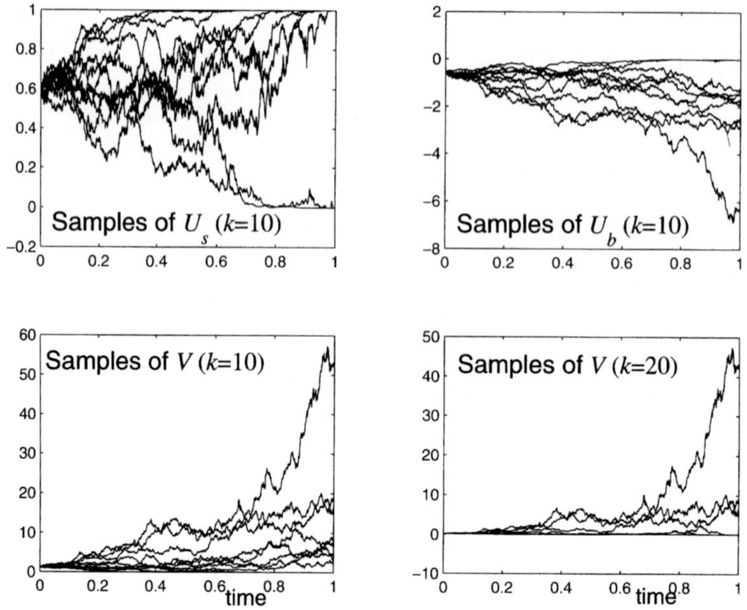

Figure 7.21. Samples of trading strategies, stock price, and portfolio value

Example 7.65: Let $\tilde{X}(t) = e^{-rt}\, X(t)$ and $\tilde{V}(t) = e^{-rt}\, V(t)$ be the discounted stock price and portfolio value at a time $t \in [0, \tau]$, where X and V are defined on a probability space $(\Omega, \mathcal{F}, P)$ with a filtration $\mathcal{F}_t = \sigma(B(s), 0 \le s \le t)$ and B is the Brownian motion in the defining equation for X. The processes $\tilde{X}$ and $\tilde{V}$ satisfy the differential equations $d\tilde{X}(t) = \sigma\, \tilde{X}(t)\, d\tilde{B}(t)$ and $d\tilde{V}(t) = U_s(t)\, d\tilde{X}(t)$, where $\tilde{B}$ is a Brownian motion under a probability measure Q defined by $(dQ/dP)(t) = \exp[-(c/\sigma)\, B(t) - (c^2/2\,\sigma)\, t]$. The process $\tilde{V}$ is an $\mathcal{F}_t$-martingale and

$$V(t) = E_Q\left[e^{-r\,(\tau - t)}\, h(X(\tau)) \mid \mathcal{F}_t\right],$$

where $h(X(\tau)) = (X(\tau) - k)^+$ for a European call option and E_Q denotes expectation with respect to probability measure Q. The above expression for $V(t)$ gives the Black-Scholes formula. ◇

Proof: Itô's formula and the self-financing condition give

$$d\tilde{X}(t) = -r\, e^{-rt}\, X(t) + e^{-rt}\, dX(t) = \sigma\, \tilde{X}(t)\, [(c - \sigma)\, dt + \sigma\, dB(t)],$$
$$d\tilde{V}(t) = U_s(t)\, \left(-r\, e^{-rt}\, X(t) + e^{-rt}\, dX(t)\right) = U_s(t)\, d\tilde{X}(t).$$

That $\tilde{B}(t) = (c - r)\, t/\sigma + B(t)$ is a Brownian motion under Q was shown in Section 5.4.2.2. The process

$$\tilde{V}(t) = V(0) + \sigma \int_0^t U_s(s)\, \tilde{X}(s)\, d\tilde{B}(s)$$

is an $\mathcal{F}_t$-martingale under probability Q since it is defined by a stochastic integral whose integrator is a Brownian motion under Q and the process $U_s\, X$ is $\mathcal{F}_t$-adapted and has continuous samples. The martingale property of $\tilde{V}$ implies $\tilde{V}(t) = E_Q[\tilde{V}(\tau) \mid \mathcal{F}_t]$ for $t \in [0, \tau]$ so that

$$\tilde{V}(t) = e^{-r\,t}\, V(t) = E_Q\left[e^{-r\,\tau}\, h(X(\tau)) \mid \mathcal{F}_t\right],$$

as stated.

The above expression of $\tilde{V}(t)$ and the relationships

$$\tilde{X}(\tau) = \tilde{X}(t)\, e^{-(\sigma^2/2)\,(\tau-t)+\sigma\,(\tilde{B}(\tau)-\tilde{B}(t))},$$
$$X(\tau) = X(t)\, e^{(r-\sigma^2/2)\,(\tau-t)+\sigma\,(\tilde{B}(\tau)-\tilde{B}(t))},$$

give

$$V(t) = E_Q\left[e^{-r\,\theta}\, h(X(\tau)) \mid \mathcal{F}_t\right] = E_Q\left[e^{-r\,\theta}\, h\left(X(t)\, e^{(r-\sigma^2/2)\,\theta+\sigma\,(\tilde{B}(\tau)-\tilde{B}(t))}\right) \mid \mathcal{F}_t\right]$$

with the notation $\theta = \tau - t$. Because $X(t) \in \mathcal{F}_t$ and $\tilde{B}(\tau) - \tilde{B}(t)$ is independent of $\mathcal{F}_t$, we have $V(t) = \xi(X(t), t)$, where

$$\xi(x, t) = e^{-r\,\theta} \int_{\mathbb{R}} h\left(x\, e^{(r-\sigma^2/2)\,\theta+\sigma\,\sqrt{\theta}\,z}\right)\, \phi(z)\, dz.$$

This function coincides with u in Eqs. 7.96 and 7.97 for a European call option ([131], Section 4.2.2). ■

7.4.5 Estimation

The evolution of the state $\boldsymbol{X}$ of many physical systems can only be modeled approximately because of complexity and/or limited understanding. Generally, the resulting approximate models are satisfactory for short term predictions of future values of $\boldsymbol{X}$ but cannot be used for long term predictions of a system state. Measurements of the actual state of a system can be used to update $\boldsymbol{X}$ and reduce errors caused by approximations in modeling. The Kalman-Bucy filtering approach is an efficient tool for propagating the state $\boldsymbol{X}$ based on its model and measurements. We present the essentials of this approach starting with time invariant estimation problems and then extending these results to time dependent linear problems with discrete and continuous time.

7.4.5.1 Time invariant problems

Let $\boldsymbol{X}$ be an unknown $\mathbb{R}^d$-valued random variable and let $\boldsymbol{Z} \in \mathbb{R}^{d'}$ be a measurement of $\boldsymbol{X}$. This measurement may not include all the coordinates of $\boldsymbol{X}$ and may be altered by noise. Our objective is to find an estimator $\hat{\boldsymbol{X}}$ of $\boldsymbol{X}$ that is optimal in some sense and is based on the observation vector $\boldsymbol{Z}$. The following formulas use the notations $\boldsymbol{\mu}_x = E[\boldsymbol{X}]$, $\boldsymbol{\mu}_z = E[\boldsymbol{Z}]$, and $\boldsymbol{\gamma}_{uv}$ for the covariance matrix of a random vector $(\boldsymbol{U}, \boldsymbol{V})$.

• The **best mean square (m.s.) estimator** of X given Z is the conditional expectation $\hat{X} = g(Z) = E\left[X \mid \mathcal{F}^Z\right] = E[X \mid Z]$.

• The **best m.s. linear estimator** of X given Z is

$$\hat{X} = \mu_x + \gamma_{xz}\,\gamma_{zz}^{-1}\,(Z - \mu_z), \tag{7.99}$$

and has the **error covariance matrix**

$$\gamma = E\left[(X - \hat{X})\,(X - \hat{X})^T\right] = \gamma_{xx} - \gamma_{xz}\,\gamma_{zz}^{-1}\,\gamma_{zx}. \tag{7.100}$$

Proof: Let $\hat{X} = g(Z)$ be an estimator of X. The m.s. error of this estimator is

$$\begin{aligned}
&E\left[(X - g(Z))^T\,(X - g(Z))\right] \\
&= \int_{\mathbb{R}^{d'}} \left[\int_{\mathbb{R}^d} (x - g(z))^T\,(x - g(z))\, f_{x|z}(x \mid z)\, dx\right] f_z(z)\, dz \\
&= \int_{\mathbb{R}^{d'}} E\left[(X - g(Z))^T\,(X - g(Z)) \mid Z = z\right] f_z(z)\, dz
\end{aligned}$$

where X, $g(Z)$, and Z are column vectors, $f_{x|z}$ is the density of $X \mid Z$, and f_z denotes the density of Z. The expectation in the above integral,

$$E[X^T\,X \mid Z = z] - 2\,g(z)^T\,E[X \mid Z = z] + g(z)^T\,g(z),$$

is positive, and is minimized by $g(z) = E[X \mid Z = z]$ for each $z \in \mathbb{R}^{d'}$. Hence, $\hat{X} = E[X \mid Z]$ is the best m.s. estimator, and its error $X - \hat{X} = X - E[X \mid Z]$ is orthogonal to the σ-field $\mathcal{F}^Z$ generated by Z (Section 2.17.2).

Assume that the first two moments of (X, Z) exist and are finite. Consider the linear estimator $\hat{X} = a\,Z + b$, where a and b are unknown matrices. The requirements that (1) $\hat{X}$ is unbiased, that is, $E[\hat{X}] = E[X] = \mu_x$, and (2) the estimation error $X - \hat{X}$ is orthogonal on $\mathcal{F}^Z$ give Eqs. 7.99 and 7.100. We note that the error covariance matrix can be calculated from the second moment properties of (X, Z) prior to taking any observations. If X and Z are not correlated, that is, $\gamma_{xz} = 0$, then $\hat{X} = \mu_x$ is the best m.s. linear estimator of X and its error covariance matrix is $\gamma = \gamma_{xx}$. In this case, Z provides no information on X. ■

Example 7.66: Suppose that $Z = h\,X + V$ is the observation vector, where h is a known deterministic matrix and V is an $\mathbb{R}^{d'}$-valued observation noise with mean μ_v and covariance matrix γ_{vv}. It is assumed that X and V are not correlated. The best m.s. linear estimator of X and its error covariance matrix are

$$\begin{aligned}
\hat{X} &= \mu_x + \gamma_{xx}\,h^T\,(h\,\gamma_{xx}\,h^T + \gamma_{vv})^{-1}\,(Z - \mu_z), \\
\gamma &= \gamma_{xx} - \gamma_{xx}\,h^T\,(h\,\gamma_{xx}\,h^T + \gamma_{vv})^{-1}\,h\,\gamma_{xx}.
\end{aligned} \tag{7.101}$$

If $d = d' = 1$, $h = 1$, and $\mu_x = \mu_z = 0$, then

$$\hat{X} = \frac{1}{1 + \gamma_{vv}/\gamma_{xx}}\,Z \quad \text{and} \quad \gamma = \gamma_{xx}\left(1 - \frac{1}{1 + \gamma_{vv}/\gamma_{xx}}\right) \le \gamma_{xx}.$$

The quality of the estimation depends on the noise to signal ratio γ_{vv}/γ_{xx}. The error covariance matrix is zero for perfect measurements ($\gamma_{vv} = 0$) and approaches γ_{xx} as $\gamma_{vv}/\gamma_{xx} \to \infty$. ◇

Proof: The above results follow by elementary calculations from the expressions for the estimator $\hat{X}$ in Eq. 7.99 and of its error covariance matrix. ■

7.4.5.2 Time dependent problems. Discrete time

Let $\boldsymbol{X}(t) \in \mathbb{R}^d$, $t = 0, 1, \ldots$, be the state of a system that is observed at each time step. The defining equations for $\boldsymbol{X}$ and the observation vector $\boldsymbol{Z} \in \mathbb{R}^{d'}$ are

$$\begin{aligned} \boldsymbol{X}(t+1) &= \boldsymbol{a}(t)\,\boldsymbol{X}(t) + \boldsymbol{b}(t)\,\boldsymbol{W}(t), \\ \boldsymbol{Z}(t) &= \boldsymbol{h}(t)\,\boldsymbol{X}(t) + \boldsymbol{V}(t), \end{aligned} \tag{7.102}$$

where $E[\boldsymbol{X}(0)] = \boldsymbol{0}$, $E[\boldsymbol{W}(t)] = E[\boldsymbol{V}(t)] = \boldsymbol{0}$ for $t \geq 0$, $E[\boldsymbol{W}(t)\,\boldsymbol{W}(s)^T] = \boldsymbol{q}(t)\,\delta_{ts}$, $E[\boldsymbol{V}(t)\,\boldsymbol{V}(s)^T] = \boldsymbol{r}(t)\,\delta_{ts}$, $E[\boldsymbol{X}(0)\,\boldsymbol{X}(0)^T] = \boldsymbol{\psi}$, $E[\boldsymbol{W}(t)\,\boldsymbol{V}(s)^T] = \boldsymbol{0}$ for $t, s \geq 0$, there is no correlation between the random process $\boldsymbol{V}$, $\boldsymbol{W}$ and the random variable $\boldsymbol{X}(0)$, and $\boldsymbol{a}$, $\boldsymbol{b}$, $\boldsymbol{h}$ are specified deterministic matrices. The assumption that $\boldsymbol{X}(0)$, $\boldsymbol{V}$, and $\boldsymbol{W}$ have mean zero is not restrictive. If this assumption is not valid, the original equation for $\boldsymbol{X}$ can be replaced with an equation for the state $\boldsymbol{X}(t) - E[\boldsymbol{X}(t)]$, which has mean zero.

Our objectives are to determine (1) the best m.s. linear estimator

$$\hat{\boldsymbol{X}}(t \mid s) = E\left[\boldsymbol{X}(t) \mid \boldsymbol{Z}(1), \ldots, \boldsymbol{Z}(s)\right] \tag{7.103}$$

of $\boldsymbol{X}(t)$ given the observation vector $(\boldsymbol{Z}(1), \ldots, \boldsymbol{Z}(s))$ and (2) the error covariance matrix,

$$\boldsymbol{\gamma}(t \mid s) = E[(\boldsymbol{X}(t) - \hat{\boldsymbol{X}}(t \mid s))\,(\boldsymbol{X}(t) - \hat{\boldsymbol{X}}(t \mid s))^T], \tag{7.104}$$

for $\hat{\boldsymbol{X}}(t \mid s)$. The solution of this problem is referred to as **filtering**, **prediction**, and **smoothing** if $t = s$, $t > s$, and $t < s$, respectively. The definitions in Eqs. 7.103 and 7.104 imply $\hat{\boldsymbol{X}}(0 \mid 0) = E[\boldsymbol{X}(0)] = \boldsymbol{0}$ and $\boldsymbol{\gamma}(0 \mid 0) = \boldsymbol{\psi}$.

We note that $\hat{\boldsymbol{X}}(t \mid s)$ and $\boldsymbol{\gamma}(t \mid s)$ can be obtained from Eqs. 7.99 and 7.100. However, the approach is inefficient for a large observation vector $(\boldsymbol{Z}(1), \ldots, \boldsymbol{Z}(s))$. The following formulation provides a procedure for calculating $\hat{\boldsymbol{X}}(t \mid s)$ and $\boldsymbol{\gamma}(t \mid s)$ efficiently. We give equations for the one-step predictor $\hat{\boldsymbol{X}}(t \mid t-1)$, the estimator $\hat{\boldsymbol{X}}(t \mid t)$, and their error covariance matrices.

The **best m.s. linear one-step predictor** and the **best m.s. estimator** of $\boldsymbol{X}$ satisfy the equations

$$\begin{aligned}\hat{\boldsymbol{X}}(t \mid t-1) &= \boldsymbol{a}(t-1)\,\hat{\boldsymbol{X}}(t-1 \mid t-1),\\ \hat{\boldsymbol{X}}(t \mid t) &= \hat{\boldsymbol{X}}(t \mid t-1) + \boldsymbol{k}(t)\left(\boldsymbol{Z}(t) - \boldsymbol{h}(t)\,\hat{\boldsymbol{X}}(t \mid t-1)\right),\end{aligned} \tag{7.105}$$

where $\hat{\boldsymbol{X}}(0 \mid 0) = E[\boldsymbol{X}(0)] = \boldsymbol{0}$, $\boldsymbol{\gamma}(0 \mid 0) = \boldsymbol{\psi}$,

$$\begin{aligned}\boldsymbol{\gamma}(t \mid t-1) &= \boldsymbol{a}(t-1)\,\boldsymbol{\gamma}(t-1 \mid t-1)\,\boldsymbol{a}(t-1)^T\\ &\quad + \boldsymbol{b}(t-1)\,\boldsymbol{q}(t-1)\,\boldsymbol{b}(t-1)^T,\\ \boldsymbol{\gamma}(t \mid t) &= (\boldsymbol{i} - \boldsymbol{k}(t)\,\boldsymbol{h}(t))\,\boldsymbol{\gamma}(t \mid t-1), \quad \text{and}\\ \boldsymbol{k}(t) &= \boldsymbol{\gamma}(t \mid t-1)\,\boldsymbol{h}(t)^T\left(\boldsymbol{h}(t)\,\boldsymbol{\gamma}(t \mid t-1)\,\boldsymbol{h}(t)^T + \boldsymbol{r}(t)\right)^{-1}.\end{aligned} \tag{7.106}$$

Proof: The first two formulas in Eq. 7.105 are based on the state propagation equation and the estimation theory. The formulas in Eq. 7.106 are the **Kalman-Bucy** equations.

The first formula in Eq. 7.105 results from Eq. 7.102 applied in the time interval $[t-1, t]$, that gives $\boldsymbol{a}(t-1)\,\hat{\boldsymbol{X}}(t-1 \mid t-1) + \boldsymbol{b}(t-1)\,\boldsymbol{W}(t-1)$ for the initial condition $\hat{\boldsymbol{X}}(t-1 \mid t-1)$. Hence, the best m.s. estimator of $\boldsymbol{X}(t)$ is $\hat{\boldsymbol{X}}(t \mid t-1) = \boldsymbol{a}(t-1)\,\hat{\boldsymbol{X}}(t-1 \mid t-1)$ since $\boldsymbol{W}(t-1)$ and $\boldsymbol{X}(t-1)$ are uncorrelated and $\boldsymbol{W}$ has mean zero. If we assume that the noise processes $\boldsymbol{V}$ and $\boldsymbol{W}$ and the random variable $\boldsymbol{X}(0)$ are mutually independent, then

$$\begin{aligned}\hat{\boldsymbol{X}}(t \mid t-1) &= E[\boldsymbol{X}(t) \mid \boldsymbol{Z}(1), \ldots, \boldsymbol{Z}(t-1)]\\ &= E[\boldsymbol{a}(t-1)\,\boldsymbol{X}(t-1) + \boldsymbol{b}(t-1)\,\boldsymbol{W}(t-1) \mid \boldsymbol{Z}(1), \ldots, \boldsymbol{Z}(t-1)]\\ &= \boldsymbol{a}(t-1)\,\hat{\boldsymbol{X}}(t-1 \mid t-1)\end{aligned}$$

since $\boldsymbol{W}(t)$ is independent of $\mathcal{F}_t = \sigma(\boldsymbol{Z}(1), \ldots, \boldsymbol{Z}(t-1))$ so that

$$E[\boldsymbol{b}(t)\,\boldsymbol{W}(t) \mid \boldsymbol{Z}(1), \ldots, \boldsymbol{Z}(t-1)] = \boldsymbol{b}(t)\,E[\boldsymbol{W}(t)] = \boldsymbol{0}.$$

As expected, we obtained the first formula in Eq. 7.105.

The expression for $\hat{\boldsymbol{X}}(t \mid t)$ is based on the following property. Let $\boldsymbol{X} \in \mathbb{R}^d$ and $\boldsymbol{Z}_i \in \mathbb{R}^{d'}$, $i = 1, 2$, be random vectors in L_2 with mean zero. Denote the best m.s. linear estimator of $\boldsymbol{X}$ given an observation vector $\boldsymbol{U} \in \mathbb{R}^{d'}$ by $\hat{\boldsymbol{X}}\,|_{\boldsymbol{U}}$. Straightforward calculations show that

$$\hat{\boldsymbol{X}}\,|_{(\boldsymbol{Z}_1, \boldsymbol{Z}_2)} = \hat{\boldsymbol{X}}\,|_{\boldsymbol{Z}_1} + \hat{\boldsymbol{X}}\,|_{\boldsymbol{N}_2} = \hat{\boldsymbol{X}}\,|_{\boldsymbol{Z}_1} + \boldsymbol{\gamma}_{x,n_2}\,\boldsymbol{\gamma}_{n_2,n_2}^{-1}\left(\boldsymbol{Z}_2 - \boldsymbol{\gamma}_{z_2,z_1}\,\boldsymbol{\gamma}_{z_1,z_1}^{-1}\,\boldsymbol{Z}_1\right),$$

where $\boldsymbol{N}_2 = \boldsymbol{Z}_2 - \hat{\boldsymbol{Z}}_2\,|_{\boldsymbol{Z}_1} = \boldsymbol{Z}_2 - \boldsymbol{\gamma}_{z_2,z_1}\,\boldsymbol{\gamma}_{z_1,z_1}^{-1}\,\boldsymbol{Z}_1$, called the **innovation** vector, includes only information in $\boldsymbol{Z}_2$ that is not in $\boldsymbol{Z}_1$ ([166], Section 6.2) and, for example, $\boldsymbol{\gamma}_{z_2,z_1}$ denotes the covariance matrix of $(\boldsymbol{Z}_2, \boldsymbol{Z}_1)$. The error covariance matrix of $\hat{\boldsymbol{X}}\,|_{(\boldsymbol{Z}_1, \boldsymbol{Z}_2)}$ is the error covariance matrix of $\hat{X} \mid_{Z_1}$ less a positive definite matrix of the type in Eq. 7.100 corresponding to the information content in N_2, that is,

$$\gamma_{\hat{X}|_{(Z_1,Z_2)}} = \gamma_{\hat{X}|_{Z_1}} - \gamma_{x,n_2}\, \gamma_{n_2,n_2}^{-1}\, \gamma_{n_2,x}.$$

The formulas in Eqs. 7.105 and 7.106 result by replacing Z_1, Z_2, and $\hat{Z}_2 \mid_{Z_1}$ in the above equations by $(Z(1), \ldots, Z(t-1))$, $Z(t)$, and $\hat{X}(t \mid t-1)$, respectively. ■

To find $X(t \mid t-1)$, $X(t \mid t)$, and their properties, we need to perform the following operations. First, we need to solve the Kalman-Bucy equations for the initial condition $\gamma(0 \mid 0) = \psi$. The formulas in Eq. 7.106 give $\gamma(1 \mid 0)$, $k(1)$, and $\gamma(1 \mid 1)$ for $t = 1$. Now we can apply the Kalman-Bucy equations for $t = 2$ to find $\gamma(2 \mid 1)$, $k(2)$, and $\gamma(2 \mid 2)$ and so on. Note that the covariance matrices $\gamma(t \mid t-1)$ and $\gamma(t \mid t)$ and the gain $k(t)$ can be calculated independently of the observations. Second, calculate $\hat{X}(t \mid t-1)$ and $\hat{X}(t \mid t)$ from Eq. 7.105 for the initial condition $\hat{X}(0 \mid 0) = \mathbf{0}$ and the given observation vector $(Z(1), Z(2), \ldots)$.

Example 7.67: Let $d = d' = 1$, $b = h = 1$, and $|a| < 1$, $q > 0$, $r > 0$ be some constants (Eq. 7.102). The error covariance matrix of $\hat{X}(t \mid t)$ is (Eq. 7.106)

$$\gamma(t \mid t) = \left(1 - \frac{a^2\,\gamma(t-1 \mid t-1) + q}{a^2\,\gamma(t-1 \mid t-1) + q + r}\right)\left(a^2\,\gamma(t-1 \mid t-1) + q\right).$$

The error covariance matrix $\gamma(t \mid t)$ is zero for perfect measurements ($r = 0$) and approaches $a^2\,\gamma(t-1 \mid t-1) + q$ as $r \to \infty$. The measurements do not help in the latter case since they are polluted by a large noise. ◇

7.4.5.3 Time dependent problems. Continuous time

Let $X(t)$, $t \geq 0$, be an $\mathbb{R}^d$-valued stochastic process in L_2 that is observed continuously. The defining equations for X and the observation process Z are

$$\begin{aligned} \dot{X}(t) &= a(t)\, X(t) + b(t)\, W(t), \\ Z(t) &= h(t)\, X(t) + V(t), \end{aligned} \tag{7.107}$$

where $X(0)$, $W(t)$, and $V(s)$ are uncorrelated with mean zero for $t, s \geq 0$, $E[X(0)\, X(0)^T] = \psi$, $E[W(t)\, W(s)^T] = q(t)\, \delta(t-s)$, and $E[V(t)\, V(s)^T] = r(t)\, \delta(t-s)$. The processes W and V are white noise processes defined heuristically by their first two moments (Eq. 7.13).

Our objectives are to find (1) the best m.s. linear estimator $\hat{X}(t) = \hat{X}(t \mid t)$ of $X(t)$ given the observations $(Z(s), 0 \leq s \leq t)$ and (2) the error covariance matrix

$$\gamma(t) = \gamma(t \mid t) = E[(X(t) - \hat{X}(t))\,(X(t) - \hat{X}(t))^T] \tag{7.108}$$

of $\hat{X}(t)$. We give equations for the evolution of $\hat{X}(t)$ and the propagation of the error covariance matrix $\gamma(t)$.

The **best m.s. linear estimator** of X is the solution of

$$\frac{d}{dt}\hat{X}(t) = a(t)\,\hat{X}(t) + k(t)\,N(t), \tag{7.109}$$

where $\hat{X}(0) = \mathbf{0}$, the **innovation** N is a zero-mean white noise process with $E[N(t)\,N(s)^T] = r(t)\,\delta(t-s)$, and γ satisfies the **Riccati equation**

$$\dot{\gamma}(t) = a(t)\,\gamma(t) + \gamma(t)\,a(t)^T + b(t)\,q(t)\,b(t)^T - \gamma(t)\,h(t)^T\,r(t)^{-1}\,h(t)\,\gamma(t)$$
$$\text{where } k(t) = \gamma(t)\,h(t)\,r(t)^{-1} \text{ and } \gamma(0) = \psi. \tag{7.110}$$

Note: We have seen that the best m.s. linear estimator $\hat{X}(t \mid t)$ of the state $X(t)$ of a discrete time system depends on observations linearly. We consider a similar estimator for the state X in Eq. 7.107, that is, we take

$$\hat{X}(t) = \hat{X}(t \mid t) = \int_0^t \rho(t,s)\,Z(s)\,ds,$$

where the matrix $\rho(t,s)$ will be determined from the condition that it minimizes the covariance of the error $X(t) - \hat{X}(t)$. The above definition of $\hat{X}$ implies $\hat{X}(0) = \mathbf{0}$ so that $\gamma(0) = \psi$ (Eq. 7.108).

Under the assumption that W in Eq. 7.107 is a Gaussian white noise, it can be shown that ρ satisfies the Wiener-Hopf equation ([33], Chapter 4, [135], Chapter 6))

$$E[X(t)\,Z(s)^T] = \int_0^t \rho(t,u)\,E[Z(u)\,Z(s)^T]\,du, \quad s \le t.$$

Because $E\left[\hat{X}(t)\,Z(s)^T\right] = \int_0^t \rho(t,u)\,E[Z(u)\,Z(s)^T]\,du$ by the above definition of $\hat{X}$, the Wiener-Hopf equation implies $E\left[\left(X(t) - \hat{X}(t)\right)\,Z(s)^T\right] = \mathbf{0}$ for all $s \le t$ so that $\hat{X}(t) = E\,[X(t) \mid \mathcal{F}_t]$, where $\mathcal{F}_t = \sigma(Z(s), 0 \le s \le t)$.

A heuristic proof of Eqs. 7.109 and 7.110 can be found in [166] (Section 7.2). The proof is based on the finite difference approximation

$$X(t+\Delta t) \simeq (i + a(t)\,\Delta t)\,X(t) + b(t)\,\Delta W(t)$$

of Eq. 7.107, where $\Delta t > 0$ is a time step, i denotes the identity matrix, and $\Delta W(t) = W(t+\Delta t) - W(t)$. This approximation has the functional form of the defining equation for X in Eq. 7.102 so that Eqs. 7.105 and 7.106 can be applied to find the evolution of $\hat{X}(t \mid t)$. The limit of the resulting equations as $\Delta t \to 0$ gives Eqs. 7.109 and 7.110. Heuristic arguments need to be invoked to perform this limit. ▲

To find $\hat{X}(t)$ and its properties, we need to perform the following operations. First, the Riccati equation has to be solved for the initial condition $\gamma(0) = \psi$. The solution of this equation and the second formula in Eq. 7.110 yield the time evolution of the gain k. Second, Eq. 7.109 has to be solved for the initial condition $\hat{X}(0) = \mathbf{0}$. The methods in Section 7.2.1.2 can be used to find the second moment properties of $\hat{\boldsymbol{X}}$. If the state $\boldsymbol{X}$ is observed at some discrete times, the procedure in the previous section can be applied in conjunction with the recurrence formula (Eq. 7.6)

$$\boldsymbol{X}(t) = \boldsymbol{\theta}(t, s)\, \boldsymbol{X}(s) + \int_s^t \boldsymbol{\theta}(t, u)\, \boldsymbol{b}(u)\, \boldsymbol{W}(u)\, du, \quad t > s.$$

Example 7.68: Let X be a real-valued process observed at the discrete times $t = n\,\Delta t$, where $n \geq 1$ is an integer and $\Delta t > 0$ denotes the time step. The observation equation is $Z(n\,\Delta t) = X(n\,\Delta t) + V(n)$, where $E[V(n)] = 0$, $E[V(n)\,V(m)] = r\,\delta_{nm}$, and $r > 0$. The state is defined by $\dot{X}(t) = W(t)$ with $X(0) = 0$, $t \geq 0$, where W is a white noise with $E[W(t)] = 0$, $E[W(t)\,W(s)] = \delta(t - s)$, and $E[W(t)\,V(s)] = 0$ for all $s, t \geq 0$. The error covariance matrices of $\hat{X}(n \mid n-1)$ and $\hat{X}(n \mid n) = E[X(n\,\Delta) \mid Z(\Delta t), \ldots, Z(n\,\Delta t)]$ are

$$\gamma(n \mid n-1) = \gamma(n-1 \mid n-1) + \Delta t,$$
$$\gamma(n \mid n) = \frac{r\,\gamma(n \mid n-1)}{r + \gamma(n \mid n-1)}.$$

If the observation noise is much smaller than the driving noise ($\Delta t \gg r$), then $\gamma(n \mid n) \simeq r$ and $\gamma(n \mid n-1) \simeq \Delta t$. Hence, the prediction uncertainty is approximately equal to the variance of the driving noise in a time step Δt. This uncertainty is reduced in the filtering phase to the observation variance. ◇

Proof: The discrete version of the state equation is $X(n\,\Delta t) = X((n-1)\,\Delta t) + \tilde{W}(n)$, where $E[\tilde{W}(n)] = 0$ and $E[\tilde{W}(n)\,\tilde{W}(m)] = \Delta t\,\delta_{nm}$. The formulas in Eq. 7.106 give the above recurrence equations for $\gamma(n \mid n-1)$ and $\gamma(n \mid n)$. Since $\gamma(0 \mid 0) = 0$, we have $\gamma(1 \mid 0) = \Delta t$ and $\gamma(1 \mid 1) = r\,\Delta t/(r + \Delta t) \simeq r$, where the approximate equality holds for $\Delta t \gg r$. Similar approximations hold for $\gamma(n \mid n-1)$ and $\gamma(n \mid n)$. ■

Example 7.69: Let $\boldsymbol{X}$ be the state in Eq. 7.5, where $\boldsymbol{Y} = \boldsymbol{W}$ and $\boldsymbol{W}$ is the white noise in Eq. 7.13 assumed to have mean zero, that is, $\boldsymbol{\mu}_w(t) = \boldsymbol{0}$. The mean and covariance matrices of the initial state are $\boldsymbol{\mu}(0)$ and $\boldsymbol{c}(0,0)$, respectively. Suppose that we observe the state $\boldsymbol{X}$ through an imperfect device giving the readings $\boldsymbol{Z}(t) = \boldsymbol{h}(t)\,\boldsymbol{X}(t) + \boldsymbol{V}(t)$ (Eq. 7.107), where $\boldsymbol{h}$ is a (d, d) matrix and $\boldsymbol{V}$ is a white noise with $E[\boldsymbol{V}(t)] = \boldsymbol{0}$ and $E[\boldsymbol{V}(t)\,\boldsymbol{V}(s)^T] = \boldsymbol{q}_v(t)\,\delta(t-s)$ that is uncorrelated with $\boldsymbol{X}$ and $\boldsymbol{W}$. The best m.s. linear estimator of the state $\boldsymbol{X}$ is the solution of

$$\dot{\hat{\boldsymbol{X}}}(t) = \boldsymbol{a}(t)\,\hat{\boldsymbol{X}}(t) + \boldsymbol{k}(t)\left(\boldsymbol{Z}(t) - \boldsymbol{h}(t)\,\hat{\boldsymbol{X}}(t)\right), \tag{7.111}$$

where $\boldsymbol{k}(t) = \hat{\boldsymbol{c}}(t, t)\,\boldsymbol{h}(t)^T\,\boldsymbol{q}_v(t)^{-1}$ and $\hat{\boldsymbol{c}}$ is the covariance function of $\hat{\boldsymbol{X}}$. This result is consistent with Eq. 7.109. ◇

Proof: Define $\hat{\boldsymbol{X}}$ by $\dot{\hat{\boldsymbol{X}}}(t) = \boldsymbol{\alpha}(t)\,\hat{\boldsymbol{X}}(t) + \boldsymbol{k}(t)\,\boldsymbol{Z}(t)$ and impose two conditions: (1) the estimator $\hat{\boldsymbol{X}}$ of $\boldsymbol{X}$ is unbiased, that is, $E[\hat{\boldsymbol{X}}(t)] = E[\boldsymbol{X}(t)]$ for all $t \geq 0$ and (2) the error

$\tilde{X}(t) = X(t) - \hat{X}(t)$ is minimized in the m.s. sense, for example, we may require that the trace of $E[\tilde{X}(t)\,\tilde{X}(t)^T]$ be minimized.

The means μ and $\hat{\mu}$ of X and $\hat{X}$ satisfy the equations

$$\dot{\mu}(t) = a(t)\,\mu(t)) \quad \text{and}$$
$$\dot{\hat{\mu}}(t) = \alpha(t)\,\hat{\mu}(t) + k(t)\,\mu_z(t) = \alpha(t)\,\hat{\mu}(t) + k(t)\,h(t)\,\mu(t),$$

where the last equality holds since $\mu_z(t) = E[Z(t)] = h(t)\,\mu(t)$. To satisfy the first condition, that is, that $\hat{X}$ is an unbiased estimator of X, we must have $\alpha(t) + k(t)\,h(t) = a(t)$ and $\hat{\mu}(0) = \mu(0)$. Accordingly, $\hat{X}$ is the solution of

$$\dot{\hat{X}}(t) = a(t)\,\hat{X}(t) + k(t)\,\left(Z(t) - h(t)\,\hat{X}(t)\right)$$

with $\hat{X}(0) = \mu(0)$. The difference between the differential equations for X and $\hat{X}$ yields

$$\dot{\tilde{X}}(t) = [a(t) - k(t)\,h(t)]\,\tilde{X}(t) + b(t)\,W(t) - k(t)\,V(t).$$

The evolution of $\tilde{\mu}(t) = E[\tilde{X}(t)]$ and $\tilde{c}(t,t) = E\left[\left(\tilde{X}(t) - \tilde{\mu}(t)\right)\left(\tilde{X}(t) - \tilde{\mu}(t)\right)^T\right]$ can be calculated from results in Section 7.2.1 since $b(t)\,W(t) - k(t)\,V(t)$ is a white noise process with mean zero and intensity $b(t)\,q_w(t)\,b(t)^T + k(t)\,q_v(t)\,k(t)^T$. The initial condition for the differential equation of $\tilde{c}$ is $\tilde{c}(0,0) = c(0,0)$ since $\hat{X}(0) = \mu(0)$ so that $\tilde{X}(0) - \tilde{\mu}(0) = X(0) - \mu(0)$.

The second condition requiring the minimization of the error $\tilde{X}(t) = X(t) - \hat{X}(t)$ is used to find the optimal gain, that is, the function k that minimizes $\tilde{c}(t,t)$ at each time $t \geq 0$. These calculations can be found in [129] (pp. 265-267). ∎

7.5 Problems

7.1: Derive the mean, correlation, and covariance equations in Eq. 7.12.

7.2: Find the second moment properties of the stochastic process $X(t)$, $t \geq 0$, in Example 7.5 for $\zeta \neq \alpha\,m + \beta\,k$.

7.3: Extend the approach in Example 7.8 to the case in which the driving noise Y is a colored stationary Gaussian process given by some of the coordinates of the state of a linear time-invariant filter driven by Gaussian white noise.

7.4: Let X be the solution of Eq. 7.5 with Y being the white noise W in Eq. 7.13. Suppose that the drift, diffusion, and noise parameters are time invariant, that is, $a(t) = a$, $b(t) = b$, $\mu_w(t) = \mu_w$, and $q_w(t) = q_w$, and that the drift coefficients are such that Eq. 7.5 has a stationary solution X_s with second moment properties given by Eq. 7.15. Show that the second moment properties of X are given by Eq. 7.15 if $X(0)$ is random and follows the marginal distribution of X_s.

7.5: Find the first four moments of the martingale $M(t) = B(t)^3 - 3\,t\,B(t)$, $t \geq 0$, where B is a Brownian motion.

7.6: Let $C(t) = \sum_{k=1}^{N(t)} Y_k$ be a compound Poisson process, where N is a Poisson process with intensity $\lambda > 0$ and the random variables Y_k are independent copies of Y_1 with mean zero and finite moments. Let $M(t) = C(t)^2 - C_2(t)$, where $C_2 = \sum_{k=1}^{N(t)} Y_k^2$. Calcutate the first two moments of M. Is M a martingale with respect to the natural filtration of C?

7.7: Find the first four moments of a process X defined by the stochastic differential equation $dX(t) = -\alpha\, X(t)\, dt + \beta\, dM(t)$ for $t \geq 0$ and $X(0) = 0$, where $\alpha > 0$, β are some constants, $M(t) = B(t)^3 - 3\,t\,B(t)$, and B is a Brownian motion.

7.8: Complete the details of the proof of Eq. 7.38.

7.9: Let $\boldsymbol{X}$ be the solution of Eq. 7.28, where $\boldsymbol{S} = \boldsymbol{M}$ and $\boldsymbol{M}$ is a continuous square integrable martingale. Find whether it is possible to develop differential equations for the moments $\mu(q_1, \ldots, q_d) = E\left[\prod_{k=1}^{d} X_k(t)^{q_k}\right]$ of $\boldsymbol{X}$.

7.10: Let X be the solution of the stochastic differential equation

$$dX(t) = -\alpha\, X(t)\, dt + dM(t), \quad \alpha > 0, \quad t \geq 0,$$

where $dM(t) = \chi(M(t))\, dB(t)$ and χ is a continuous function. Find whether the moment equations corresponding to the augmented vector (X, M) are closed for $\chi(\xi) = \xi$ and $\chi(\xi) = \xi^2$.

7.11: Complete the details of the proof of Eq. 7.39.

7.12: Write a differential equation involving moments of the augmented state vector $\boldsymbol{Z} = (\boldsymbol{X}, \boldsymbol{S})$ in Eq. 7.47. Outline conditions under which these equations can be used to find the moments of $\boldsymbol{Z}$.

7.13: Repeat the calculations in Example 7.23 for $l = 3$.

7.14: Prove the result in Example 7.25.

7.15: Find the solution for X defined in Example 7.32 in which B is replaced by a compound Poisson process. Develop moment equations for X.

7.16: Let X be the process defined in Example 7.33. Develop moment equations for $(X, \dot{X})$. Are these moment equations closed?

7.17: Let $f(x) = (1/h)\, 1_{[x_0 - h/2, x_0 + h/2]}(x)$, $h > 0$, be a density function. Find the characteristic function $\varphi(u)$ of this density. Comment on the behavior of $\varphi(u)$ as $|u| \to \infty$.

7.18: Show that $\lim_{t\to\infty} B(t)/t = 0$ a.s.

7.19: Derive the partial differential equation for the characteristic function of the state vector $(X, \dot{X}, S)$ defined in Example 7.28 and specify the initial and boundary conditions needed to solve this equation.

7.20: Derive the Fokker-Planck equation for the density of X defined by $dX(t) = a(X(t))\,dt + b(X(t))\,dB(t)$, $t \geq 0$, starting at $X(0) = x_0$ by using Itô's formula. Assume that the conditions in Eq. 7.63 are satisfied.

7.21: Find the density of the geometric Brownian motion in Example 7.36 by solving the Fokker-Planck equation numerically.

7.22: Find the stationary density of the Ornstein-Uhlenbeck process X defined by $dX(t) = -\alpha\, X(t)\, dt + \sqrt{2\,\alpha}\, dB(t)$, where $\alpha > 0$ is a constant.

7.23: Complete the details of the proof in Example 7.38.

7.24: Complete the details of the proof in Example 7.39.

7.25: Take $u'(x) = \nu_0^2\, x\,(1 + \varepsilon\, x^2)$ in Example 7.42, where ε is a constant. Show that the stationary density of the state vector $(X_1 = X, X_2 = \dot{X})$ is

$$f_s(x_1, x_2) = c\,\sqrt{2\,\pi}\,\dot{\sigma}_0\, \exp\left[-\frac{1}{2\,\sigma_0^2}\left(x_1^2 + \frac{1}{2}\,\varepsilon\, x^4\right)\right]$$
$$\times\, \frac{1}{\sqrt{2\,\pi}\,\dot{\sigma}_0}\, \exp\left[-\frac{x_2^2}{2\,\dot{\sigma}_0^2}\right],$$

where $\sigma_0 = \pi\, g_0/(2\,\rho\,\nu_0^2)$, $\dot{\sigma}_0^2 = \nu_0^2\,\sigma_0^2$, and $c > 0$ is a constant.

7.26: Complete the calculations in Example 7.49.

7.27: Prove the statement in Example 7.51.

7.28: Develop an alternative model of the ice-core data in Fig. 7.12.

7.29: Develop the differential equations for the correlation functions $r_{pq}(t, s)$ of the augmented state $\mathbf{Z} = (X, \dot{X}, S)$ in Section 7.4.1.2 for both the transient and stationary solutions. Confirm the numerical results in Fig. 7.13.

7.30: Define a reliability problem and calculate the bounds in Eq. 7.92 on the probability of failure $P_f(\tau)$. Find also Monte Carlo estimates of $P_f(\tau)$.

7.31: Find the solution of the problem in Example 7.68 for the case in which X is an Ornstein-Uhlenbeck process.

Chapter 8

Stochastic Systems and Deterministic Input

8.1 Introduction

This chapter examines algebraic, differential, and integral equations with random coefficients. The boundary and/or initial conditions required for the solution of differential and integral equations can be deterministic or random. The input is deterministic for all types of equations. We refer to these problems as **stochastic systems with deterministic input**.

Our objective is to find probabilistic properties of the solution of stochastic systems subjected to deterministic inputs. Two classes of methods are available for solving this type of stochastic problems. The methods in the first class constitute direct extensions of some of the results in Chapters 3 and 6, for example, the local solution for a class of differential equations with random coefficients (Section 8.2), the crossing solution for the random eigenvalue problem (Section 8.3.2.6), and some of the homogenization methods for random heterogeneous media (Section 8.5.1). These methods involve advanced probabilistic concepts and in many cases deliver detailed properties of solutions. The methods in the second class are largely based on relatively simple probabilistic concepts on random variables, processes, and fields. The main objective of these methods is the calculation of the second moment properties of the solution of algebraic, differential, and integral equations with random coefficients.

We (1) examine partial differential equations with the functional form in Chapter 6 (Section 8.2), inhomogeneous and homogeneous algebraic equations (Sections 8.3.1 and 8.3.2), and inhomogeneous and homogeneous differential and integral equations (Sections 8.4.1 and 8.4.2), (2) calculate effective or macroscopic constants for random heterogeneous materials (Sections 8.5.1 and 8.5.2), (3) study the evolution of material properties and the possibility of pattern formation in the context of elasticity and crystal plasticity (Sections 8.6.1 and 8.6.2), (4) define the stochastic stability problem and illustrate its solution by some simple examples, and (5) examine localization phenomena in heterogeneous soil layers and nearly periodic dynamic systems (Section 8.8.1 and 8.8.2).

8.2 Local solutions

Consider the partial differential equations defined in Section 6.1 and suppose that they have random coefficients, boundary conditions, and/or initial conditions. Assume that the local solutions developed in Chapter 6 hold for almost all samples of the random parameters in these equations. The local solutions are random variables that can be characterized by estimates of their mean, covariance, and other properties.

A three step procedure can be applied to find the local solution for partial differential equations with random coefficients, boundary conditions, and/or initial conditions. First, the local solutions in Sections 6.2, 6.3, and 6.4 need to be calculated for a sample ω' of the random parameters in the definition of these equations. Second, the first step has to be repeated for a collection of samples $\{\omega'\}$. Third, statistics of the local solution need to be inferred from results in the previous step. We illustrate this procedure by a simple example.

Example 8.1: Let U be the solution of the partial differential equation

$$\sum_{i=1}^{2} A_i(\boldsymbol{x}) \frac{\partial U(\boldsymbol{x})}{\partial x_i} + \frac{1}{2} \sum_{i,j=1}^{2} H_{ij}(\boldsymbol{x}) \frac{\partial^2 U(\boldsymbol{x})}{\partial x_i \, \partial x_j} = 0, \quad \boldsymbol{x} \in D,$$

with the boundary condition $U(\boldsymbol{x}) = \Xi(\boldsymbol{x})$, $\boldsymbol{x} \in \partial D$, where D is an open set in $\mathbb{R}^2$, and (A_i, H_{ij}) and Ξ are real-valued random fields defined on D and ∂D, respectively. It is assumed that (1) the matrix $\boldsymbol{H} = \{H_{ij}\}$ admits the representation $\boldsymbol{H}(\boldsymbol{x}) = \boldsymbol{C}(\boldsymbol{x}) \, \boldsymbol{C}(\boldsymbol{x})^T$ at each $\boldsymbol{x} \in D$ and (2) the random fields A_i, H_{ij}, C_{ij}, and Ξ are such that the local solution in Section 6.2 can be applied for almost all samples of these fields.

Let $A_i(\cdot, \omega')$, $C_{ij}(\cdot, \omega')$, and $\Xi(\cdot, \omega')$ be samples of the random fields A_i, C_{ij}, and Ξ, respectively, and define for each ω' an $\mathbb{R}^2$-valued diffusion process by the stochastic differential equation

$$d\boldsymbol{X}(t; \omega') = \boldsymbol{A}(\boldsymbol{X}(t; \omega'), \omega') \, dt + \boldsymbol{C}(\boldsymbol{X}(t; \omega'), \omega') \, d\boldsymbol{B}(t, \omega'),$$

where $\boldsymbol{B}$ is an $\mathbb{R}^2$-valued Brownian motion. For a sample ω' of A_i, C_{ij}, and Ξ, the local solution is

$$U(\boldsymbol{x}, \omega') = E^{\boldsymbol{x}}[\Xi(\boldsymbol{X}(T(\omega'), \omega'), \omega')], \quad \boldsymbol{x} \in D,$$

where $T(\omega') = \inf\{t > 0 : \boldsymbol{X}(t, \omega') \notin D\}$ denotes the first time $\boldsymbol{X}(\cdot; \omega')$ starting at $\boldsymbol{x} \in D$ exits D and the above expectation is with respect to the random variable

$X(T(\omega'), \omega')$ for a fixed ω' (Section 6.2.1). Statistics of the local solution $U(x)$ can be obtained from its samples $U(x, \omega')$.

Numerical results have been obtained for $D = (-2, 2) \times (-1, 1)$,

$$A_i(x), C_{ii}(x) \sim U(i - a_i, i + a_i)$$

with $|a_i| < i$, $i = 1, 2$, $C_{12}(x) \sim U(1/2 - a_3, 1/2 + a_3)$ with $|a_3| < 1/2$, $\Xi(x) \sim U(1 - a_1, i + a_1)$ on the side $(-2, 2) \times \{-1\}$ of D, and $\Xi(x) = 0$ on all other boundaries of D, where $a_1 = 0.4$, $a_2 = 1$, and $a_3 = 0.2$, and the random variables A_i, C_{ij}, and Ξ are independent of each other. Figure 8.1 shows a histogram of $U((0.5, 0))$ based on 100 independent samples of A_i, C_{ij}, and Ξ.

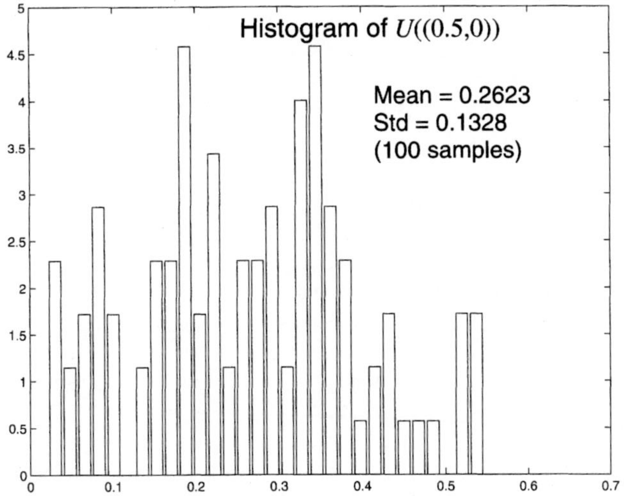

Figure 8.1. Histogram of the local solution $U(x)$ at $x = (0.5, 0)$

The solutions $U((0.5, 0), \omega')$ for each ω' are based on 1,000 independent samples of X generated with a time step of $\Delta t = 0.001$. For $a_i = 0$, $i = 1, 2, 3$, the local solution is deterministic, and its estimate at $x = (0.5, 0)$ for the same sample size is 0.2731. ◇

8.3 Algebraic equations

Consider the algebraic equation

$$(A - \lambda\, i)\, X = q, \tag{8.1}$$

where i denotes the (n, n) identity matrix, q is an n-dimensional column vector with real coordinates, A is an (n, n) matrix whose entries are real-valued random variables defined on a probability space $(\Omega, \mathcal{F}, P)$, and λ is a parameter. We refer to Eq. 8.1 with $\boldsymbol{q} \neq \mathbf{0}$ and $\boldsymbol{q} = \mathbf{0}$ as inhomogeneous and homogeneous equations, respectively. Our objective is to find probabilistic properties of the solution $\boldsymbol{X}$ for the inhomogeneous version of Eq. 8.2 with a fixed value of λ and of the eigenvalues and eigenvectors of the homogeneous version of this equation. The solution $\boldsymbol{X}$ is an n-dimensional column vector.

The local solutions in Sections 6.5.1 and 6.5.2 can be extended to solve Eq. 8.1 locally by following the approach in Section 8.2. This method of solution is not considered here.

8.3.1 Inhomogeneous equations

Let $\boldsymbol{Y} = \boldsymbol{A} - \lambda\,\boldsymbol{i}$ for a fixed value of λ such that $\boldsymbol{Y}^{-1}$ exists a.s. Then Eq. 8.1 becomes

$$\boldsymbol{Y}\,\boldsymbol{X} = \boldsymbol{q}, \quad \boldsymbol{q} \in \mathbb{R}^n. \tag{8.2}$$

Note: Because $\boldsymbol{X} = \boldsymbol{Y}^{-1}\,\boldsymbol{q}$ and the mapping $\boldsymbol{Y} \mapsto \boldsymbol{Y}^{-1}\,\boldsymbol{q}$ is measurable, $\boldsymbol{X}$ is a random variable on $(\Omega, \mathcal{F}, P)$. The uncertainty in $\boldsymbol{Y}$ can have many sources, for example, geometrical imperfections for mechanical systems, limited understanding of a phenomenon, and random errors related to the definition of the boundary conditions.

The conditions in Eq. 8.2 are encountered frequently in applications [173]. For example, the finite difference and finite element formulations of differential and integral equations defining problems in solid mechanics, physics, economics, and other fields yield conditions of the type in Eq. 8.2. ▲

Our objective is to find the probability law of $\boldsymbol{X}$ from its definition in Eq. 8.2 and the probability law of $\boldsymbol{Y}$. We will be able to achieve this objective in only a few cases since (1) the inverse $\boldsymbol{Y}^{-1}$ of a random matrix $\boldsymbol{Y}$ cannot generally be found analytically and (2) the determination of the probability law of $\boldsymbol{X}$ from $\boldsymbol{X} = \boldsymbol{Y}^{-1}\,\boldsymbol{q}$ is usually impractical even if $\boldsymbol{Y}^{-1}$ is known. Most of the methods in this section focus on the approximate calculation of the first two moments of $\boldsymbol{X}$ under the assumption $\boldsymbol{Y} \in L_2$.

8.3.1.1 Monte Carlo simulation method

The Monte Carlo simulation method provides the most general solution for Eq. 8.2 in the sense that it can deliver an estimate of the probability law of $\boldsymbol{X}$. The essential limitation of the method is the computation time, that can be excessive if the solution of the associated deterministic problem (Eq. 8.3) is computationally intensive and/or the required sample size n_s is large. Another potential limitation of Monte Carlo simulation relates to the available information on $\boldsymbol{Y}$. For example, if only the first two moments of $\boldsymbol{Y}$ are available, the use of Monte Carlo simulation requires us to postulate the probability law of this random vector.

1. Generate n_s independent samples $\boldsymbol{y}_k$, $k = 1, \ldots, n_s$, of $\boldsymbol{Y}$ and calculate

$$\boldsymbol{x}_k = \boldsymbol{y}_k^{-1}\, \boldsymbol{q}, \quad k = 1, \ldots, n_s. \tag{8.3}$$

2. Estimate statistics of $\boldsymbol{X}$, for example, the mean and covariance matrices of $\boldsymbol{X}$, $\boldsymbol{\mu}_x$ and $\boldsymbol{\gamma}_x$, can be approximated by

$$\hat{\boldsymbol{\mu}}_x = \frac{1}{n_s}\sum_{k=1}^{n_s} \boldsymbol{x}_k \quad \text{and} \quad \hat{\boldsymbol{\gamma}}_x = \frac{1}{n_s}\sum_{k=1}^{n_s} (\boldsymbol{x}_k - \hat{\boldsymbol{\mu}}_x)\,(\boldsymbol{x}_k - \hat{\boldsymbol{\mu}}_x)^T. \tag{8.4}$$

Note: The algorithms in Section. 5.2 can be used to generate samples of $\boldsymbol{Y}$. The Monte Carlo simulation method is not restricted to linear equations. The method can be applied to nonlinear equations in which case the determination of the samples $\boldsymbol{x}_k$ of $\boldsymbol{X}$ requires solution of a nonlinear algebraic equation rather than Eq. 8.3. ▲

Example 8.2: Let $\boldsymbol{X} = (X_1, X_2)$ be the solution of

$$\begin{bmatrix} \cos(\Theta_1) & \cos(\Theta_2) \\ \sin(\Theta_1) & \sin(\Theta_2) \end{bmatrix} \begin{bmatrix} X_1 \\ X_2 \end{bmatrix} = \begin{bmatrix} 0 \\ q \end{bmatrix},$$

where $\Theta_1 = \tan^{-1}(Z_2/(Z_1 + a))$, $\Theta_2 = \tan^{-1}((a - Z_2)/(Z_1 + a))$, Z_1 and Z_2 are independent random variables uniformly distributed in $(-\varepsilon\, a, \varepsilon\, a)$, and $0 \le \varepsilon \le 1/3$. The random vector $\boldsymbol{X}$ gives the forces in the two-member truss in Fig. 8.2 under a deterministic load q. Because of geometric imperfections, the

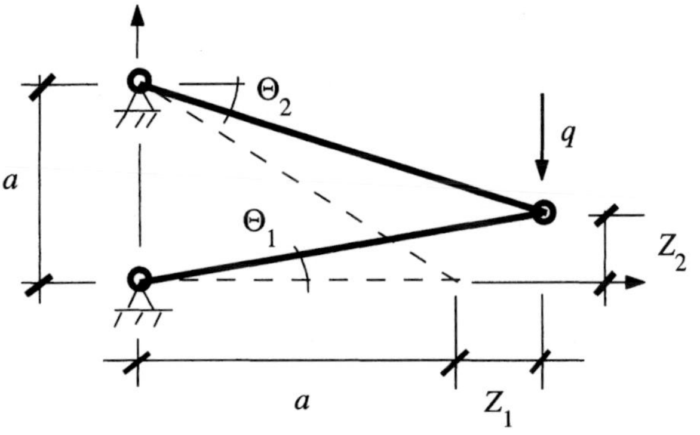

Figure 8.2. A two-member truss with random geometric imperfections

position of the joint defined by the intersection of the two bars is random with the coordinates $(a + Z_1, Z_2)$.

Figure 8.3 shows histograms of X_1 and X_2 obtained from $n_s = 1,000$ truss samples for $a = 1$, $\varepsilon = 0.3$, and $q = 1$. The estimated mean, standard deviation,

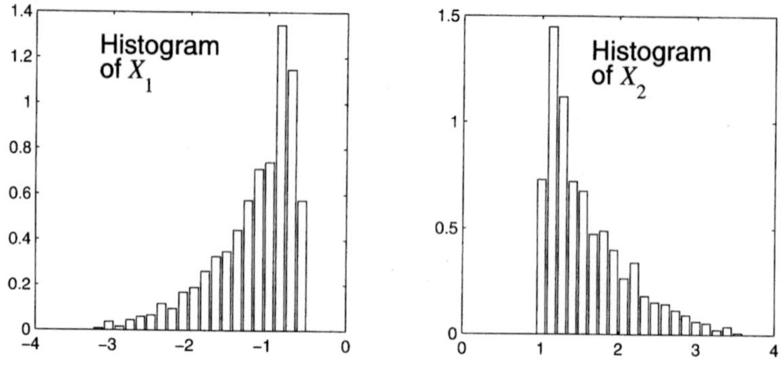

Figure 8.3. Histograms of X_1 and X_2

skewness, and kurtosis are $-1.16, 0.52, 1.23$, and 4.17 for X_1 and $1.58, 0.52, 1.18$, and 3.94 for X_2. If $\varepsilon = 0$, that is, there is no uncertainty in the truss geometry, the member forces are $X_1 = -1$ and $X_2 = \sqrt{2}$. ◇

Note: The defining equations for (X_1, X_2) result from the equilibrium conditions, where Θ_1 and Θ_2 are the random orientations of the bars (Fig. 8.2). The entries of matrix $\boldsymbol{Y}$ in Eq. 8.2 are $Y_{11} = \cos(\Theta_1)$, $Y_{12} = \cos(\Theta_2)$, $Y_{21} = \sin(\Theta_1)$, and $Y_{22} = \sin(\Theta_2)$. The Monte Carlo simulation was based on Eqs. 8.3 and 8.4. First, n_s independent samples of the joint coordinates $(a + Z_1, Z_2)$ were generated and the corresponding samples of the angles (Θ_1, Θ_2) and member forces (X_1, X_2) were calculated. Second, the resulting member forces were used to estimate moments and develop histograms for (X_1, X_2). ▲

8.3.1.2 Taylor series method

Suppose that all entries of $\boldsymbol{Y}$ in Eq. 8.2 are measurable functions of an $\mathbb{R}^m$-valued random variable $\boldsymbol{Z} \in L_2$ with mean $\boldsymbol{\mu}_z = \{\mu_{z,u}\}$ and covariance $\boldsymbol{\gamma}_z = \{\gamma_{z,uv}\}$, $u, v = 1, \ldots, m$. Hence, the solution $\boldsymbol{X}$ can also be viewed as a function $\boldsymbol{Z} \mapsto \boldsymbol{X}(\boldsymbol{Z})$. It is assumed that this function has continuous second order partial derivatives in $\mathbb{R}^m$. The following two equations give the first order Taylor approximation of $\boldsymbol{X}$ and the first two moments of this approximation.

1. Approximation of $\boldsymbol{X}$:

$$\boldsymbol{X} \simeq \boldsymbol{X}(\boldsymbol{\mu}_z) + \sum_{u=1}^{m} \frac{\partial \boldsymbol{X}(\boldsymbol{\mu}_z)}{\partial z_u} (Z_u - \mu_{z,u}). \tag{8.5}$$

2. Approximate second moment properties of $\boldsymbol{X}$:

$$\boldsymbol{\mu}_x \simeq \boldsymbol{X}(\boldsymbol{\mu}_z) \quad \text{and} \quad \boldsymbol{\gamma}_x \simeq \sum_{u,v=1}^{m} \frac{\partial \boldsymbol{X}(\boldsymbol{\mu}_z)}{\partial z_u} \frac{\partial \boldsymbol{X}(\boldsymbol{\mu}_z)}{\partial z_v}^T \gamma_{z,uv}. \tag{8.6}$$

Note: The approximation of the solution $\boldsymbol{X}$ in Eq. 8.5 is referred to as the **linear** or **first order Taylor** approximation. We also note that the approximate second moment properties of $\boldsymbol{X}$ in Eq. 8.6 depend only on the first two moments of $\mathbf{Z}$.

The Taylor expansion is

$$\boldsymbol{X}(\mathbf{Z}) = \boldsymbol{X}(\mathbf{z}^{(0)}) + \sum_{u=1}^{m} \frac{\partial \boldsymbol{X}(\mathbf{z}^{(0)})}{\partial z_u} \left(Z_u - z_u^{(0)}\right) + \boldsymbol{R}_2(\mathbf{z}^{(0)}),$$

where

$$\boldsymbol{R}_2(\mathbf{z}^{(0)}) = \frac{1}{2} \sum_{u,v=1}^{m} \frac{\partial^2 \boldsymbol{X}(\mathbf{Z}^*)}{\partial z_u \, \partial z_v} \left(Z_u - z_u^{(0)}\right) \left(Z_v - z_v^{(0)}\right)$$

and $\mathbf{Z}^* = \theta \, \mathbf{z}^{(0)} + (1-\theta) \, \mathbf{Z}$, $\theta \in (0,1)$ ([149], Remark 2, p. 453, [26],Theorem 21.1, p. 142). The approximation in Eq. 8.5 is given by the above Taylor expansion for $\mathbf{z}^{(0)} = \boldsymbol{\mu}_z$ without the remainder $\boldsymbol{R}_2$.

The error of the linear approximation in Eq. 8.5 can be bounded by using the expression of $\boldsymbol{R}_2$. This bound is rarely used in applications since it requires the calculation of the second order partial derivatives $\partial^2 \boldsymbol{X}/\partial z_u \, \partial z_v$, $u, v = 1, 2, \ldots, m$, which is a time consuming task for large vectors $\mathbf{Z}$. ▲

The approximate solutions in Eqs. 8.5 and 8.6 depend on the values of $\boldsymbol{X}$ and its gradient for $\mathbf{Z}$ equal to its mean value, which are given by the following two equations.

$$\boldsymbol{X}(\boldsymbol{\mu}_z) = \boldsymbol{Y}(\boldsymbol{\mu}_z)^{-1} \boldsymbol{q} \quad \text{and} \tag{8.7}$$

$$\frac{\partial \boldsymbol{X}(\boldsymbol{\mu}_z)}{\partial z_u} = \boldsymbol{Y}(\boldsymbol{\mu}_z)^{-1} \left(-\frac{\partial \boldsymbol{Y}(\boldsymbol{\mu}_z)}{\partial z_u} \boldsymbol{X}(\boldsymbol{\mu}_z)\right). \tag{8.8}$$

Proof: The solution of Eq. 8.7 corresponds to Eq. 8.2 with $\mathbf{Z}$ replaced by its mean value and involves the inversion of a deterministic matrix $\boldsymbol{Y}(\boldsymbol{\mu}_z)$. The derivative of Eq. 8.2 with respect to a coordinate of $\mathbf{Z}$ is

$$\frac{\partial \boldsymbol{Y}(\mathbf{Z})}{\partial z_u} \boldsymbol{X}(\mathbf{Z}) + \boldsymbol{Y}(\mathbf{Z}) \frac{\partial \boldsymbol{X}(\mathbf{Z})}{\partial z_u} = \frac{\partial \boldsymbol{q}}{\partial z_u}$$

and gives Eq. 8.8 by setting $\boldsymbol{z} = \boldsymbol{\mu}_z$ since $\partial \boldsymbol{q}/\partial z_u = 0$. The partial derivatives $\partial \boldsymbol{X}(\boldsymbol{\mu}_z)/\partial z_u$ are called **sensitivity factors** because they give the rate of change of $\boldsymbol{X}$ with respect to perturbations in the coordinates of $\boldsymbol{Z}$ about $\boldsymbol{\mu}_z$. If both the sensitivity factor with respect to a coordinate Z_k of $\boldsymbol{Z}$ and the uncertainty in this coordinate are small, then Z_k can be set equal to its mean value as a first approximation. ■

Example 8.3: Let X be the solution of Eq. 8.2 with $m = n = 1$. Suppose that Z takes values in a bounded interval (a, b) and that the function $Z \mapsto X(Z)$ has a continuous second order derivative. The error of the first order approximation, $\tilde{X}(Z) = X(\mu_z) + X'(\mu_z)(Z - \mu_z)$, can be bounded by

$$|X(Z(\omega)) - \tilde{X}(Z(\omega))| \leq \max_{a<\xi<b} |X''(\xi)| \, (Z(\omega) - \mu_z)^2$$

for almost all $\omega \in \Omega$. The calculation of this bound is impractical for realistic applications since X'' is not known and its numerical calculation can be prohibitive. Moreover, the resulting bound may be too wide to be informative. ◇

Example 8.4: Consider the algebraic equation in Example 8.2 with $a = 1, q = 1$, and $\varepsilon = 0.3$. The approximate mean and standard deviation given by Eq. 8.6 are -1 (-14%) and 0.6 (14.81%) for X_1 and $\sqrt{2}$ (-10.65%) and 0.49 (-6.42%) for X_2. The numbers in parentheses give differences between these approximations and the Monte Carlo simulation in Example 8.2, and can be viewed as errors since the Monte Carlo solution is based on a large sample. The errors of the approximations in Eq. 8.6 decrease with the uncertainty in $\boldsymbol{Z}$ and vanish if there is no uncertainty in the truss geometry ($\varepsilon = 0$). ◇

Note: The approximate mean of $\boldsymbol{X}$ given by Eq. 8.7 can be obtained from the equilibrium conditions for $Z_1 = 0$ and $Z_2 = 0$. The gradients of the coordinates of $\boldsymbol{X}$ relative to $\boldsymbol{Z}$ are given by Eq. 8.8. The partial derivatives $\partial \cos(\Theta_u)/\partial Z_v$ and $\partial \sin(\Theta_u)/\partial Z_v$, $u, v = 1, 2$, have to be calculated to find the entries of matrices $\partial \boldsymbol{Y}(\boldsymbol{Z})/\partial z_u$, for example, the element (1, 1) of this matrix is

$$\frac{\partial \cos(\Theta_1)}{\partial z_2} = -(a + Z_1)\, Z_2 \, [(a + Z_1^2) + Z_2^2]^{-3/2}$$

for $u = 2$. The entries (1,1), (1,2), (2,1), and (2,2) of matrices $\partial \boldsymbol{Y}(\boldsymbol{\mu}_z)/\partial z_u$ are 0, $1/\sqrt{2}-1$, 0, and -1 for $u = 1$ and 0, 0, 1, and $-1/\sqrt{2}$ for $u = 2$. The derivatives $\partial \boldsymbol{X}(\boldsymbol{\mu}_z)/\partial z_u$ in Eq. 8.8 are $(2, -2)$ for $u = 1$ and $(2.83, 2)$ for $u = 2$. The second formula in Eq. 8.6 gives an approximation for the covariance matrix of $\boldsymbol{X}$.

Figure 8.4 shows the functions $\boldsymbol{Z} \mapsto X_u(\boldsymbol{Z})$, $u = 1, 2$. The approximate coordinate of $\boldsymbol{X}$ in Eq. 8.5 are hyperplanes π_u tangent to the graphs of these functions at $(\boldsymbol{\mu}_z, x_u(\boldsymbol{\mu}_z))$, $u = 1, 2$. The error of the approximate covariance in Eq. 8.6 depends on (1) the differences between the hyperplanes π_u and the functions $\boldsymbol{Z} \mapsto X_u(\boldsymbol{Z})$ and (2) the uncertainty in $\boldsymbol{Z}$. For example, the approximate second moments of $\boldsymbol{X}$ are likely to be inaccurate if the above differences are significant and/or the variance of the coordinates of $\boldsymbol{Z}$ is relatively large. ▲

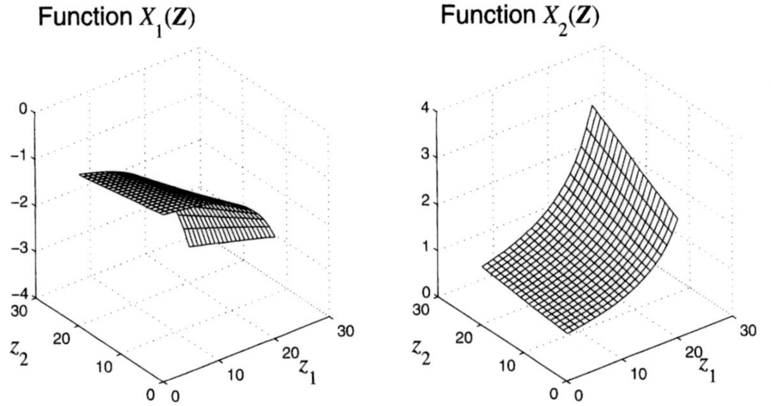

Figure 8.4. The mapping $\boldsymbol{Z} \mapsto X_u(\boldsymbol{Z})$ for a two-member truss

The approximations of the second moment properties of $\boldsymbol{X}$ in Eqs. 8.5-8.8 can be (*a*) improved by retaining additional terms of the Taylor series representing the function $\boldsymbol{Z} \mapsto \boldsymbol{X}(\boldsymbol{Z})$ and (*b*) generalized to solve nonlinear algebraic equations. We examine briefly these extensions.

Higher order approximations. If the function $\boldsymbol{Z} \mapsto \boldsymbol{X}(\boldsymbol{Z})$ has continuous third order partial derivatives, then $\boldsymbol{X}$ can be approximated by the quadratic form

$$\boldsymbol{X}(\boldsymbol{Z}) \simeq \boldsymbol{X}(\boldsymbol{\mu}_z) + \sum_{u=1}^{m} \frac{\partial \boldsymbol{X}(\boldsymbol{\mu}_z)}{\partial z_u}(Z_u - \mu_{z,u}) + \frac{1}{2} \sum_{u,v=1}^{m} \frac{\partial^2 \boldsymbol{X}(\boldsymbol{\mu}_z)}{\partial z_u \, \partial z_v}(Z_u - \mu_{z,u})\,(Z_v - \mu_{z,v}), \tag{8.9}$$

of $\boldsymbol{Z}$. The calculation of the second moment properties of $\boldsymbol{X}$ based on this approximation requires knowledge of the first four moments of $\boldsymbol{Z}$. For example, the mean of $\boldsymbol{X}$ can be approximated by

$$\boldsymbol{\mu}_x \simeq \boldsymbol{X}(\boldsymbol{\mu}_z) + \frac{1}{2} \sum_{u,v=1}^{m} \frac{\partial^2 \boldsymbol{X}(\boldsymbol{\mu}_z)}{\partial z_u \, \partial z_v} \gamma_{z,uv}.$$

The above second order partial derivatives of $\boldsymbol{X}$ can be calculated from the second partial derivatives of Eq. 8.2.

Nonlinear equations. Suppose that $\boldsymbol{X}$ is the solution of

$$\boldsymbol{Y}\,\boldsymbol{X} + \boldsymbol{N} = \boldsymbol{q}, \tag{8.10}$$

where $\boldsymbol{Y}$ is as in Eq. 8.2, $\boldsymbol{N}$ denotes an $(n, 1)$ matrix whose entries are nonlinear real-valued functions of $(\boldsymbol{Z}, \boldsymbol{X})$, and the $\mathbb{R}^m$-valued random variable $\boldsymbol{Z}$ collects the distinct random parameters in this equation. Suppose that X is approximated by Eq. 8.5 so that we need to find $X(\mu_z)$ and $\partial X(\mu_z)/\partial z_u$, $u = 1, \ldots, m$, to apply Eq. 8.6. The solution of Eq. 8.10 with Z replaced by its mean value gives $X(\mu_z)$. The derivative of Eq. 8.10 with respect to a coordinate Z_u of Z is

$$\frac{\partial Y}{\partial z_u} X + Y \frac{\partial X}{\partial z_u} + \frac{\partial N}{\partial z_u} + \sum_{k=1}^{n} \frac{\partial N}{\partial x_k} \frac{\partial X_k}{\partial z_u} = 0.$$

The gradients of X can be obtained from the above equation since the functions $Z \mapsto Y(Z)$ and $(Z, X) \mapsto N(Z, X)$ are known so that the partial derivatives $\partial Y/\partial z_u$, $\partial N/\partial z_u$, and $\partial N/\partial x_k$ can be calculated.

8.3.1.3 Perturbation method

Suppose that the random matrix Y in Eq. 8.2 has the representation

$$Y = a + \varepsilon \tilde{R}, \tag{8.11}$$

where a is a deterministic matrix, ε denotes a small parameter, and $\tilde{R}$ is the random part of Y. It is common to set $a = E[Y]$ so that $\varepsilon \tilde{R} = Y - a$, $E[\tilde{R}] = 0$, and $\tilde{\gamma}_{ij,kl} = E[\tilde{R}_{ij} \tilde{R}_{kl}]$. The approximate perturbation solution of Eq. 8.2 with Y in Eq. 8.11 and its second moment properties are given by the following equations.

1. Approximation of X:

$$X = X_0 + \varepsilon X_1 + \varepsilon^2 X_2 + O(\varepsilon^3), \quad \text{where}$$
$$X_r = -a^{-1} \tilde{R} X_{r-1}, \quad r = 1, 2, \ldots, \text{and } X_0 = a^{-1} q. \tag{8.12}$$

2. Approximate second moment properties of X:

$$E[X] = a^{-1} q + \varepsilon^2 a^{-1} E[\tilde{R} a^{-1} \tilde{R}] a^{-1} q + O(\varepsilon^3),$$
$$E[X X^T] = E[X_0 X_0^T] + \varepsilon^2 E[X_0 X_2^T + X_1 X_1^T + X_2 X_0^T] + O(\varepsilon^3). \tag{8.13}$$

Proof: The above perturbation solution is regular because the random part of Y is small and the solution X does not differ significantly from the solution $x = a^{-1} q$ of Eq. 8.2 with $\varepsilon = 0$ (Example 8.5 in this section, [101], Section 1.2).

The representation of X in Eq. 8.12 and Eq. 8.2 give

$$(a + \varepsilon \tilde{R})(X_0 + \varepsilon X_1 + \varepsilon^2 X_2 + \cdots) = q \quad \text{or}$$
$$(a X_0 - q) + \varepsilon (a X_1 + \tilde{R} X_0) + \varepsilon^2 (a X_2 + \tilde{R} X_1) + \cdots = 0,$$

that is, a power series that must be zero for all values of ε. We require that the coefficients of all powers of ε be zero according to a fundamental theorem of perturbation theory ([171],

p. 12). This requirement gives

$$\begin{aligned} \boldsymbol{a}\,\boldsymbol{X}_0 &= \boldsymbol{q} \quad \text{(order 1)},\\ \boldsymbol{a}\,\boldsymbol{X}_1 &= -\tilde{\boldsymbol{R}}\,\boldsymbol{X}_0 \quad (\text{order } \varepsilon),\\ \boldsymbol{a}\,\boldsymbol{X}_2 &= -\tilde{\boldsymbol{R}}\,\boldsymbol{X}_1 \quad (\text{order } \varepsilon^2) \end{aligned}$$

for the terms of order ε^r, $r = 0, 1, 2$. The **solution of order** $p \geq 1$ of $\boldsymbol{X}$ is

$$\boldsymbol{X}^{(p)} = \sum_{k=0}^{p} (-1)^k\,\varepsilon^k \left(\boldsymbol{a}^{-1}\,\tilde{\boldsymbol{R}}\right)^k \boldsymbol{a}^{-1}\,\boldsymbol{q},$$

where $\left(\boldsymbol{a}^{-1}\,\tilde{\boldsymbol{R}}\right)^k$ is the identity matrix for $k = 0$. Note that the equations for $\boldsymbol{X}_0, \boldsymbol{X}_1, \ldots$ have the same deterministic operator.

The moments in Eq. 8.13 show that it is not possible to approximate $E[\boldsymbol{X}\,\boldsymbol{X}^T]$ to the order ε^2 by using the first order approximation of $\boldsymbol{X}$. The first order approximation of $\boldsymbol{X}$ delivers only one of the three terms of order ε^2, the term $E[\boldsymbol{X}_1\,\boldsymbol{X}_1^T]$. The other two terms cannot be obtained from the first order approximation. ■

Generally, it is difficult to assess whether a perturbation series, $\boldsymbol{X}_0 + \varepsilon\,\boldsymbol{X}_1 + \varepsilon^2\,\boldsymbol{X}_2 + \cdots$, is convergent since its general term is rarely known even for simple problems (Example 8.5). The representation given by Eq. 8.12 can be generalized in at least two directions, as shown by the following two results.

1. The perturbation method can be used to find the solution of Eq. 8.2 when the random matrix $\boldsymbol{Y}$ has a more general form than in Eq. 8.11, for example, $\boldsymbol{Y} = \boldsymbol{a} + \varepsilon\,\tilde{\boldsymbol{R}}_1 + \varepsilon^2\,\tilde{\boldsymbol{R}}_2 + O(\varepsilon^3)$.

2. The perturbation method can also be applied to determine the solution of non-linear problems defined by

$$\boldsymbol{Y}\,\boldsymbol{X} + \varepsilon\,\tilde{\boldsymbol{N}} = \boldsymbol{q}, \tag{8.14}$$

where $\boldsymbol{Y}$ is given by Eq. 8.11 and the entries of the $(n, 1)$ matrix $\tilde{\boldsymbol{N}}$ are real-valued nonlinear functions of $\boldsymbol{X}$ and may be random. The first three terms of the perturbation solution of this equation can be calculated from

$$\begin{aligned} \boldsymbol{a}\,\boldsymbol{X}_0 &= \boldsymbol{q},\\ \boldsymbol{a}\,\boldsymbol{X}_1 &= -\tilde{\boldsymbol{R}}\,\boldsymbol{X}_0 - \tilde{\boldsymbol{N}}(\boldsymbol{X}_0),\\ \boldsymbol{a}\,\boldsymbol{X}_2 &= \tilde{\boldsymbol{R}}\,\boldsymbol{X}_1 - \boldsymbol{N}^*(\boldsymbol{X}_0)^T\,\boldsymbol{X}_1, \end{aligned} \tag{8.15}$$

where $\boldsymbol{N}^*(\boldsymbol{X}_0)$ is an (n, n) matrix with columns $\partial\tilde{\boldsymbol{N}}/\partial x_i$ evaluated at $\boldsymbol{X}_0$.

Proof: The perturbation series in Eq. 8.12 and Eq. 8.14 give

$$(\boldsymbol{a} + \varepsilon\,\tilde{\boldsymbol{R}})(\boldsymbol{X}_0 + \varepsilon\,\boldsymbol{X}_1 + \cdots) + \varepsilon\,\tilde{\boldsymbol{N}}(\boldsymbol{X}_0 + \varepsilon\,\boldsymbol{X}_1 + \cdots) = \boldsymbol{q},$$

which can be approximated by

$$(\boldsymbol{a} + \varepsilon\,\tilde{\boldsymbol{R}})(\boldsymbol{X}_0 + \varepsilon\,\boldsymbol{X}_1 + \cdots) + \varepsilon\,\tilde{\boldsymbol{N}}(\boldsymbol{X}_0) + \varepsilon \sum_{i=1}^{n} \frac{\partial\tilde{\boldsymbol{N}}(\boldsymbol{X}_0)}{\partial x_i} \left(\varepsilon\,X_{1,i} + \cdots\right) = \boldsymbol{q},$$

where $X_{1,i}$ are the coordinates of $\boldsymbol{X}_1$. ■

Example 8.5: Let x be the solution of the deterministic nonlinear algebraic equation $x + \varepsilon\, x^3 = 1$, where ε is a small parameter. The approximation of the real root of this equation to the order ε^3 is $x \simeq x^{(3)} = 1 - \varepsilon + 3\,\varepsilon^2 - 12\,\varepsilon^3$.

Figure 8.5 shows the exact root and the perturbation solutions of the first

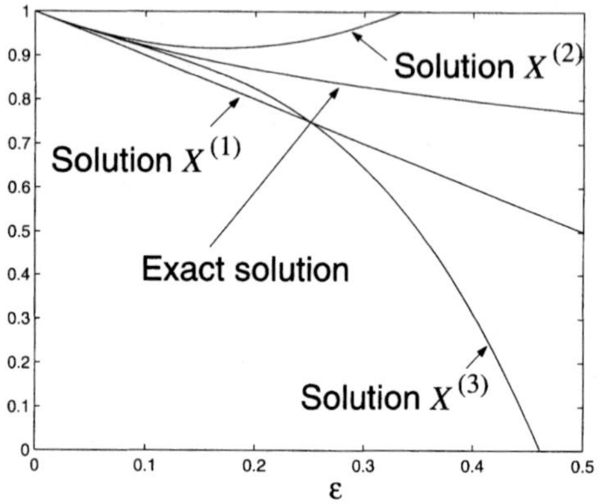

Figure 8.5. Exact and perturbation solutions

three orders. For small values of ε the perturbation solutions are satisfactory, and their accuracy improves slightly with the order of the approximation. For large values of ε the perturbation solutions are inaccurate. ◇

Note: The power series $x = \sum_{k=0,1,\ldots} \varepsilon^k\, x_k$ and the equation $x + \varepsilon\, x^3 = 1$ give

$$(x_0 + \varepsilon\, x_1 + \varepsilon^2\, x_2 + \cdots) + \varepsilon\,(x_0 + \varepsilon\, x_1 + \varepsilon^2\, x_2 + \cdots)^3 = 1 \quad \text{or}$$

$$x_0 + \varepsilon\, x_1 + \varepsilon^2\, x_2 + \varepsilon^3\, x_3 + \cdots + \varepsilon\,[x_0^3 + 3\,\varepsilon\, x_0^2\, x_1 + 3\,\varepsilon^2\,(x_0^2\, x_2 + x_0\, x_1^2) + \cdots] = 1$$

so that we have $x_0 = 1$, $x_1 + x_0^3 = 0$, $x_2 + 3\,x_0^2\, x_1 = 0$, and $x_3 + 3\,(x_0^2\, x_2 + x_0\, x_1^2) = 0$ by equating terms of the same order of magnitude. These equations give $x_0 = 1$, $x_1 = -x_0^3 = -1$, $x_2 = -3\,x_0^2\, x_1 = 3$, and $x_3 = -3\,(x_0^2\, x_2 + x_0\, x_1^2) = -12$.

We note that the algebraic equation in this example provides an illustration for a **singular** perturbation problem since the limit point $\varepsilon = 0$ differs in an essential way from the limit $\varepsilon \to 0$. The equations with $\varepsilon = 0$ and $\varepsilon > 0$ have a single root and three roots, respectively. Problems that are not singular are said to be **regular** ([101], Section 1.2). ▲

Example 8.6: Consider the algebraic equation in Example 8.2 and represent the random matrix $\boldsymbol{Y}$ as the sum of a deterministic part $\boldsymbol{a} = E[\boldsymbol{Y}]$ and a random part $\varepsilon\,\tilde{\boldsymbol{R}} = \boldsymbol{Y} - \boldsymbol{a}$. For $\varepsilon = 0.3$ the approximate means of X_1 and X_2 are -1.0130 (1.30%) and 1.4245 (0.73%), and the approximate standard deviations of X_1 and X_2 are 0.3825 (-26.81%) and 0.3831 (-26.82%), respectively. These approximate results are based on Eq. 8.13. The numbers in parentheses give errors relative to the Monte Carlo solution in Example 8.2. ◇

Note: The expectations $E[Y]$ and $E[\tilde{R}_{ij}\ \tilde{R}_{kl}]$ needed for solution are difficult to find analytically because of the complex dependence of $\tilde{\boldsymbol{R}}$ on the random variables specifying the random geometry of the truss. These averages were estimated from $n_s = 1,000$ samples of $\boldsymbol{Y}$ generated by Monte Carlo simulation. For example, the estimates of the entries $(1, 1)$, $(1, 2)$, $(2, 1)$, and $(2, 2)$ of $E[\boldsymbol{Y}]$ are 0.9848, 0.7003, 0.0025, and 0.7038, respectively. ▲

8.3.1.4 Neumann series method

The solution $\boldsymbol{X} = \boldsymbol{Y}^{-1}\,\boldsymbol{q}$ of Eq. 8.2 requires the inversion of an arbitrary random matrix, and there are no efficient algorithms for finding $\boldsymbol{Y}^{-1}$. The Taylor expansion and the perturbation methods calculate approximately probabilistic characteristics of $\boldsymbol{X}$ without evaluating $\boldsymbol{Y}^{-1}$. In contrast, the Neumann series method approximates $\boldsymbol{Y}^{-1}$ and uses this approximation to calculate probabilistic properties of $\boldsymbol{X}$. The Neumann series method is based on the following fact.

If $\boldsymbol{m}$ is an (n, n)-matrix such that

$$\|\,\boldsymbol{m}\,\boldsymbol{x}\,\| \leq \gamma\,\|\,\boldsymbol{x}\,\|, \quad 0 < \gamma < 1, \quad \forall \boldsymbol{x} \in \mathbb{R}^n, \quad \text{then} \tag{8.16}$$

$$(\boldsymbol{i} + \boldsymbol{m})^{-1} = \sum_{r=0}^{\infty} (-1)^r\,\boldsymbol{m}^r, \tag{8.17}$$

where $\boldsymbol{i}$ denotes the (n, n) identity matrix, $\boldsymbol{m}^0 = \boldsymbol{i}$, $\|\,\boldsymbol{\xi}\,\|$ is the Euclidean norm of $\boldsymbol{\xi} \in \mathbb{R}^n$, and the series in Eq. 8.17 is absolutely convergent ([187], Chapter 2).

Proof: Consider the sequence of sums $s_k = \sum_{r=0}^{k} (-1)^r\,\boldsymbol{m}^r$. For $k > l$, we have

$$\begin{aligned}\|\,s_k\,\boldsymbol{x} - s_l\,\boldsymbol{x}\,\| &= \|\sum_{r=l+1}^{k} (-1)^r\,\boldsymbol{m}^r\,\boldsymbol{x}\,\| \leq \sum_{r=l+1}^{k} \|\,\boldsymbol{m}^r\,\boldsymbol{x}\,\| \leq \sum_{r=l+1}^{k} \gamma^r\,\|\,\boldsymbol{x}\,\| \\ &= \frac{\gamma^{l+1}(1-\gamma^{k-l})}{1-\gamma}\,\|\,\boldsymbol{x}\,\| \leq \frac{\gamma^{l+1}}{1-\gamma}\,\|\,\boldsymbol{x}\,\|\end{aligned}$$

by norm properties and Eq. 8.16. Because the upper bound on $\|\,s_k\,\boldsymbol{x} - s_l\,\boldsymbol{x}\,\|$ approaches zero for each $\boldsymbol{x} \in \mathbb{R}^n$ as $l, k \to \infty$ and $\mathbb{R}^n$ is complete ([32], Theorem 3.8, p. 41), $s_k\,\boldsymbol{x}$ is Cauchy in $\mathbb{R}^n$ and has a limit

$$s\,\boldsymbol{x} = \lim_{k\to\infty} s_k\,\boldsymbol{x} = \lim_{k\to\infty} \sum_{r=0}^{k} (-1)^r\,\boldsymbol{m}^r\,\boldsymbol{x} = \sum_{r=0}^{\infty} (-1)^r\,\boldsymbol{m}^r\,\boldsymbol{x}.$$

It is left to show that $s = (\boldsymbol{i} + \boldsymbol{m})^{-1}$, that is, that the equalities $s\,(\boldsymbol{i} + \boldsymbol{m})\,\boldsymbol{x} = (\boldsymbol{i} + \boldsymbol{m})\,s\,\boldsymbol{x} = \boldsymbol{x}$ hold for each $\boldsymbol{x}$. The difference,

$$\|\,s_k\,(\boldsymbol{i} + \boldsymbol{m})\,\boldsymbol{x} - \boldsymbol{x}\,\| = \|\,s_k\,\boldsymbol{x} + s_k\,\boldsymbol{m}\,\boldsymbol{x} - \boldsymbol{x}\,\| = \|\,(-1)^k\,\boldsymbol{m}^{k+1}\,\boldsymbol{x}\,\| \leq \gamma^{k+1}\,\|\,\boldsymbol{x}\,\|,$$

between $s_k\,(\boldsymbol{i} + \boldsymbol{m})\,\boldsymbol{x}$ and $\boldsymbol{x}$ approaches zero as $k \to \infty$ because $\gamma \in (0, 1)$ so that $\lim_{k\to\infty} s_k\,(\boldsymbol{i} + \boldsymbol{m})\,\boldsymbol{x} = \boldsymbol{x}$. Because $s\,\boldsymbol{x} = \lim_{k\to\infty} s_k\,\boldsymbol{x}$ holds for each $\boldsymbol{x} \in \mathbb{R}^n$ and

$(\boldsymbol{i}+\boldsymbol{m})\,\boldsymbol{x} \in \mathbb{R}^n$, we have $\boldsymbol{s}\,(\boldsymbol{i}+\boldsymbol{m})\,\boldsymbol{x} = \boldsymbol{x}$. Similar considerations can be used to show that $(\boldsymbol{i}+\boldsymbol{m})\,\boldsymbol{s}\,\boldsymbol{x} = \boldsymbol{x}$.

We also note that under the condition in Eq. 8.16 the series $\sum_{r=0}^{\infty}(-1)^r\,\boldsymbol{m}^r\,\boldsymbol{x}$ is absolutely convergent since

$$\left\| \frac{1}{2}\sum_{r=0}^{\infty}(-1)^r\,\boldsymbol{m}^r\,\boldsymbol{x} \right\| \le \sum_{r=0}^{\infty} \|\,\boldsymbol{m}^r\,\boldsymbol{x}\,\| \le \sum_{r=0}^{\infty} \gamma^r\ \|\,\boldsymbol{x}\,\| = \frac{1}{1-\gamma}\ \|\,\boldsymbol{x}\,\|,$$

which is finite. ■

Example 8.7: Let $X \in \mathbb{R}$ be the solution of $(a+R)\,X = q$, where a, q are some constants and R is a random variable. If $|R/a| \le \gamma < 1$ a.s., the sequence

$$S_k\,X_0 = \left(\sum_{r=0}^{k}(-1)^r\left(\frac{R}{a}\right)^r\right)(q/a), \quad k = 1, 2, \ldots,$$

converges a.s. to X so that $X = \left(1+\sum_{r=1}^{\infty}(-1)^r\,(R/a)^r\right)(q/a)$ is the solution of $(a+R)\,X = q$. ◇

Proof: An alternative form of $(a+R)\,X = q$ is $(1+a^{-1}\,R)\,X = a^{-1}\,q$ so that

$$X = (1+a^{-1}\,R)^{-1}\,(a^{-1}\,q).$$

Let $S_k(\omega) = 1+\sum_{r=1}^{k}(-1)^r\,a^{-r}\,R(\omega)^r$ be a sample of S_k. Because $|R(\omega)/a| \le \gamma < 1$ for almost all ω's, then

$$|S_k(\omega)| \le 1+\sum_{r=1}^{k}|a^{-r}\,R(\omega)^r| \le 1+\sum_{r=1}^{k}\gamma^r = \frac{1-\gamma^{k+1}}{1-\gamma}$$

is bounded by a geometric series with finite sum since $\gamma < 1$ by hypothesis. Hence, the numerical series $S_k(\omega)$ approaches a limit $S(\omega) = X(\omega)$ as $k \to \infty$ for almost all ω's. ▲

Let $\boldsymbol{X}$ be the solution of Eq. 8.2, where $\boldsymbol{Y} \in L_2$. Denote by $\boldsymbol{a} = E[\boldsymbol{Y}]$ and $\boldsymbol{R} = \boldsymbol{Y} - \boldsymbol{a}$ the deterministic and random parts of $\boldsymbol{Y}$, respectively. The Neumann series solution for Eq. 8.2 and the approximate second moment properties of $\boldsymbol{X}$ can be calculated from the following equations.

1. If there exists a $\gamma \in (0, 1)$ such that $\|\,\boldsymbol{a}^{-1}\,\boldsymbol{R}\,\boldsymbol{x}\,\| \le \gamma\ \|\,\boldsymbol{x}\,\|$ a.s., then $\boldsymbol{X}$ has the Neumann series representation:

$$\boldsymbol{X} = \sum_{r=0}^{\infty}\boldsymbol{X}_r, \quad \text{where } \boldsymbol{X}_0 = \boldsymbol{a}^{-1}\,\boldsymbol{q} \text{ and}$$
$$\boldsymbol{X}_r = -\boldsymbol{a}^{-1}\,\boldsymbol{R}\,\boldsymbol{X}_{r-1}, \quad r = 1, 2, \ldots. \tag{8.18}$$

2. Approximate second moment properties of $\boldsymbol{X}$ based on the first two terms of the Neumann series, that is, $\boldsymbol{X} \simeq \boldsymbol{X}_0 + \boldsymbol{X}_1$:

$$\boldsymbol{\mu}_x = \boldsymbol{a}^{-1}\,\boldsymbol{q} \quad \text{and} \quad \boldsymbol{\gamma}_x = E[(\boldsymbol{a}^{-1}\,\boldsymbol{R}\,\boldsymbol{a}^{-1}\,\boldsymbol{q})(\boldsymbol{a}^{-1}\,\boldsymbol{R}\,\boldsymbol{a}^{-1}\,\boldsymbol{q})^T]. \tag{8.19}$$

Proof: Consider the alternative form,

$$\boldsymbol{X} = (\boldsymbol{a} + \boldsymbol{R})^{-1}\,\boldsymbol{q} = (\boldsymbol{i} + \boldsymbol{a}^{-1}\,\boldsymbol{R})^{-1}\,\boldsymbol{a}^{-1}\,\boldsymbol{q},$$

of Eq. 8.2. If $\| \boldsymbol{a}^{-1}\,\boldsymbol{R}\,\boldsymbol{x} \| \leq \gamma \; \| \boldsymbol{x} \|$ a.s. for $0 < \gamma < 1$ and $\boldsymbol{x} \in \mathbb{R}^n$, then (Eq. 8.17)

$$\boldsymbol{X} = \left(\boldsymbol{i} + \sum_{r=1}^{\infty} (-1)^r \left(\boldsymbol{a}^{-1}\,\boldsymbol{R}\right)^r\right) \boldsymbol{a}^{-1}\,\boldsymbol{q} \quad \text{a.s.}$$

Generally, the first few terms of the above series are used to approximate $\boldsymbol{X}$. For example, the second moment properties of $\boldsymbol{X}$ given by Eq. 8.19 are based on the approximation

$$\boldsymbol{X} \simeq \boldsymbol{X}_0 + \boldsymbol{X}_1 = \boldsymbol{a}^{-1}\,\boldsymbol{q} - \boldsymbol{a}^{-1}\,\boldsymbol{R}\,\boldsymbol{a}^{-1}\,\boldsymbol{q}.$$

Similar expressions can be found for higher order approximations of $\boldsymbol{X}$. ■

Example 8.8: Consider the equation $Y\,X = q$, where $Y = a + R$, R is uniformly distributed in $(-\alpha, \alpha)$, $q = 10$, $a = 2$, and $\alpha = 0.5$. The solution X can be interpreted as the deformation of a linear spring with random stiffness Y under a constant force q.

The mean and variance of X are 5.00 and 0.52 for the approximation $X \simeq [1 - (R/a)]\,(q/a)$. The corresponding moments given by the approximation $X \simeq [1 - (R/a) + (R/a)^2]\,(q/a)$ are 5.10 and 0.57, respectively. The exact mean and variance of X are 5.11 and 0.55. The first order approximation is satisfactory in this case. ◇

Note: The Neumann series solution in Eq. 8.18 is convergent a.s. because there exists a constant $\gamma \in (0, 1)$ such that $|(R/a)\,x| \leq \gamma |x|$ for each $x \in \mathbb{R}$. Accordingly, we have $X = [1 - (R/a) + (R/a)^2 - (R/a)^3 + \cdots]\,(q/a)$.

The distribution of X is $P(q/(a+R) \leq x) = 1 - F_R(q/x - a)$, where F_R denotes the distribution of R. The density of X is equal to $q/(2\,\alpha\,x^2)$ in the range $(q/(a+\alpha), q/(a-\alpha))$ and zero outside this range so that the first two moments of this variable are $E[X] = (0.5\,q/\alpha)\,\ln((a+\alpha)/(a-\alpha))$ and $E[X^2] = q^2/(a^2 - \alpha^2)$. ▲

8.3.1.5 Decomposition method

The solution for Eq. 8.2 is represented by the series

$$\boldsymbol{X} = \sum_{r=0}^{\infty} \boldsymbol{U}_r, \quad \text{where } \boldsymbol{U}_0 = \boldsymbol{a}^{-1}\,\boldsymbol{q} \text{ and}$$

$$\boldsymbol{U}_r = -\boldsymbol{a}^{-1}\,\boldsymbol{R}\,\boldsymbol{U}_{r-1}, \quad r = 1, 2, \ldots. \tag{8.20}$$

Note: Let $\boldsymbol{X} = \sum_{r=0}^{\infty} \lambda^r\,\boldsymbol{U}_r$ be a representation of $\boldsymbol{X}$, where λ is an arbitrary parameter. The equation,

$$\sum_{r=0}^{\infty} \lambda^r\,\boldsymbol{U}_r = \boldsymbol{a}^{-1}\,\boldsymbol{q} - \lambda\,\boldsymbol{a}^{-1}\,\boldsymbol{R} \sum_{r=0}^{\infty} \lambda^r\,\boldsymbol{U}_r,$$

coincides with the representation $\boldsymbol{X} = \boldsymbol{a}^{-1}\,\boldsymbol{q} - \boldsymbol{a}^{-1}\,\boldsymbol{R}\,\boldsymbol{X}$ of Eq. 8.2 with $\boldsymbol{X}$ in Eq. 8.20 for $\lambda = 1$. The formulas in Eq. 8.20 result by setting the coefficients of each power of λ equal to zero. The representation of $\boldsymbol{X}$ in Eq. 8.20 corresponds to $\lambda = 1$. The above manipulations are formal since the series in Eq. 8.20 may not be convergent. Extensive considerations on this method can be found in [3]. We note that the representations of the solution $\boldsymbol{X}$ in the Neumann series method (Eq. 8.18) and the decomposition method (Eq. 8.20) have the same general terms. ▲

8.3.1.6 Equivalent linearization method

Generally, the solution $\boldsymbol{X}$ of Eq. 8.2 is a nonlinear function of the uncertain parameters $\boldsymbol{Z}$ (Fig. 8.4). The objective of the equivalent linearization method is to express $\boldsymbol{X}$ as a linear function of $\boldsymbol{Z}$ that is optimal in some sense. The following two equations give the approximate solution by the equivalent linearization method and the corresponding second moment properties of $\boldsymbol{X}$.

1. Approximate solution $\boldsymbol{X}$:

$$\boldsymbol{X} \simeq \boldsymbol{X}_l = \boldsymbol{\alpha}\,\boldsymbol{Z} + \boldsymbol{\beta}, \tag{8.21}$$

where $\boldsymbol{\alpha}$ and $\boldsymbol{\beta}$ minimize $e = E[\| \boldsymbol{q} - \boldsymbol{Y}\,\boldsymbol{X}_l \|^2]$.

2. Approximate second moment properties of $\boldsymbol{X}$:

$$\boldsymbol{\mu}_x \simeq \boldsymbol{\alpha}\,\boldsymbol{\mu}_z + \boldsymbol{\beta} \quad \text{and} \quad \boldsymbol{\gamma}_x \simeq \boldsymbol{\alpha}\,\boldsymbol{\gamma}_z\,\boldsymbol{\alpha}^T. \tag{8.22}$$

Proof: The unknown coefficients $\boldsymbol{\alpha}$ and $\boldsymbol{\beta}$ should be determined from the condition that the difference between $\boldsymbol{X}$ and $\boldsymbol{X}_l$ is minimized in some sense. However, this difference cannot be calculated directly since the probability law of $\boldsymbol{X}$ is unknown. If $\| \boldsymbol{Y}^{-1}\,\boldsymbol{x} \| \leq \gamma\, \| \boldsymbol{x} \|$ a.s. for all $\boldsymbol{x} \in \mathbb{R}^n$ and a constant $\gamma < \infty$, then

$$\| \boldsymbol{X} - \boldsymbol{X}_l \| = \| \boldsymbol{Y}^{-1}\boldsymbol{q} - \boldsymbol{Y}^{-1}\,\boldsymbol{Y}\,\boldsymbol{X}_l \| \leq \gamma\, \| \boldsymbol{q} - \boldsymbol{Y}\,\boldsymbol{X}_l \| \quad \text{a.s.}$$

so that the use of the objective function $e = E[\| \boldsymbol{q} - \boldsymbol{Y}\,\boldsymbol{X}_l \|^2]$ is meaningful. The entries of the matrices $\boldsymbol{\alpha}$ and $\boldsymbol{\beta}$ are the solutions of the linear system of equations given by $\partial e/\partial \alpha_{ij} = 0$ and $\partial e/\partial \beta_i = 0$, $i, j = 1, 2, \ldots, n$. ■

Example 8.9: Consider the problem in Example 8.8 and assume that Y is a lognormal variable with mean μ and variance σ^2 and $q > 0$ is a constant. Figure 8.6 shows the exact mean and variance of X scaled by μ/q and $(\mu/q)^2$, respectively, as a function of the coefficient of variation $v = \sigma/\mu$. The figure also shows the approximate second moment properties of the solution X given by the equivalent linearization and the first order Taylor expansion. The approximate solutions are accurate for small values of the coefficient of variation but become unsatisfactory for relatively large values of v. ◇

Proof: The equivalent linearization approximates X by $X_l = \alpha\,Y + \beta$ so that the objective function is $e = E[(q - \alpha\,Y^2 - \beta\,Y)^2]$. The conditions $\partial e/\partial \alpha = 0$ and $\partial e/\partial \beta = 0$ give

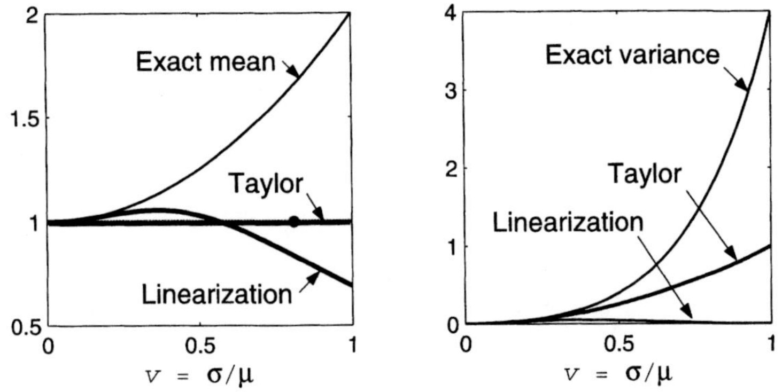

Figure 8.6. Exact and approximate mean and variance of X

$\alpha = -q\,\mu^{-2}\,(1+v^2)^{-4}$ and $\beta = q\,\mu^{-1}\,(2+v^2)\,(1+v^2)^{-2}$, so that the first two moments of X_l are

$$E[X_l] = q\,\mu^{-1}\,(1+v^2)^{-4}\,(1+5\,v^2+4\,v^4+v^6),$$
$$\mathrm{Var}[X_l] = q^2\,\mu^{-2}\,(1+v^2)^{-8}\,v^2.$$

These moments are used as approximations for the corresponding moments of X. ■

8.3.1.7 Iteration method

The iteration method is based on (1) the recurrence formula

$$\boldsymbol{X} = \boldsymbol{a}^{-1}\,\boldsymbol{q} - \boldsymbol{a}^{-1}\,\boldsymbol{R}\,\boldsymbol{X} \tag{8.23}$$

derived from Eq. 8.2, where $\boldsymbol{a} = E[\boldsymbol{Y}]$ and $\boldsymbol{R} = \boldsymbol{Y} - \boldsymbol{a}$, and (2) a heuristic approximation, referred to as the **local independence** hypothesis, stating that the correlation between $\boldsymbol{X}$ and $\boldsymbol{R}$ can be disregarded ([164], p. 18, [173], Section II.8). There is no theoretical justification for this hypothesis. We review briefly this method because of its wide use in some fields [164, 173]. Two versions of the iteration method are presented.

Version 1. Approximate second moment properties of $\boldsymbol{X}$:

$$E[\boldsymbol{X}] \simeq \boldsymbol{a}^{-1}\,\boldsymbol{q},$$
$$E[\boldsymbol{X}\,\boldsymbol{X}^T] \simeq \boldsymbol{a}^{-1}\,\boldsymbol{q}(\boldsymbol{a}^{-1}\,\boldsymbol{q})^T + \boldsymbol{a}^{-1}\,E[\boldsymbol{R}\,\boldsymbol{X}\,\boldsymbol{X}^T\,\boldsymbol{R}^T]\,(\boldsymbol{a}^{-1})^T. \tag{8.24}$$

Proof: The above equations for the first two moments of $\boldsymbol{X}$ can be obtained from the iteration of order zero of Eq. 8.2, that is, Eq. 8.23.

The expectation of Eq. 8.23 and the local independence hypothesis, $E[\boldsymbol{R}\,X] = E[\boldsymbol{R}]\,E[X] = \boldsymbol{0}$, yield the first formula in Eq. 8.24.

The expression of the product of Eq. 8.23 with itself and the local independence hypothesis give the second formula in Eq. 8.24. This formula and the local independence hypothesis can be used to calculate the expectations $E[X_i\,X_j]$. ■

Version 2. Approximate second moment properties of $\boldsymbol{X}$:

$$\begin{aligned} E[X] &\simeq (\boldsymbol{i} - \boldsymbol{a}^{-1}\,E[\boldsymbol{R}\,\boldsymbol{a}^{-1}\,\boldsymbol{R}])^{-1}\,\boldsymbol{a}^{-1}\,\boldsymbol{q}, \\ E[X\,X^T] &\simeq E[(\boldsymbol{a}^{-1}\boldsymbol{q} - \boldsymbol{a}^{-1}\,\boldsymbol{R}\,X)\,(\boldsymbol{a}^{-1}\,\boldsymbol{q} - \boldsymbol{a}^{-1}\,\boldsymbol{R}\,X)^T] \\ &= \boldsymbol{a}^{-1}\left(\boldsymbol{q}\,\boldsymbol{q}^T - \boldsymbol{q}\,E[X^T\,\boldsymbol{R}^T] - E[\boldsymbol{R}\,X]\,\boldsymbol{q}^T \right. \\ &\quad \left. + E[\boldsymbol{R}\,X\,X^T\,\boldsymbol{R}^T]\right)(\boldsymbol{a}^{-1})^T. \end{aligned} \tag{8.25}$$

Proof: The average of Eq. 8.23 involves the unknown expectation $E[\boldsymbol{R}\,X]$. The approach here is to develop an equation for $E[\boldsymbol{R}\,X]$ rather then using the local independence hypothesis in this equation. By multiplying Eq. 8.23 to the left with $\boldsymbol{R}$ and then averaging, we have

$$E[\boldsymbol{R}\,X] = E[\boldsymbol{R}]\,\boldsymbol{a}^{-1}\boldsymbol{q} - E[\boldsymbol{R}\,\boldsymbol{a}^{-1}\,\boldsymbol{R}\,X] = -E[\boldsymbol{R}\,\boldsymbol{a}^{-1}\,\boldsymbol{R}\,X] \simeq -E[\boldsymbol{R}\,\boldsymbol{a}^{-1}\,\boldsymbol{R}]\,E[X],$$

where the last approximate equality results by the local independence hypothesis. This equality and Eq. 8.23 give the first formula in Eq. 8.25. The second formula in Eq. 8.25 results from the hypothesis of independence and the expression of $\boldsymbol{X}$ in Eq. 8.23. This formula and the hypothesis of independence can be used to calculate the expectations $E[X_i\,X_j]$. ■

8.3.2 Homogeneous equations

Let $\boldsymbol{q} = \boldsymbol{0}$ in Eq. 8.1. Then this equation becomes

$$A\,X = \Lambda\,X. \tag{8.26}$$

It is assumed that the matrix $\boldsymbol{a} = E[\boldsymbol{A}]$ is symmetric.

Our objective is to solve the above random eigenvalue problem, that is, to find probabilistic properties of the eigenvalues Λ_i and the eigenvectors $\boldsymbol{X}_i$ of $\boldsymbol{A}$.

Note: It can be shown that the eigenvalues Λ_i are Borel measurable functions of the entries of $\boldsymbol{A}$ ([21], Theorem 2.2, p. 23) so that they are random variables on $(\Omega, \mathcal{F}, P)$. The eigenvectors $\boldsymbol{X}_i$ of $\boldsymbol{A}$ are also random variables on this probability space ([164], Theorems 1.6, p. 37, and Theorems 1.8, p. 43). ▲

We consider eigenvalue problems defined by Eq. 8.26 and by equations of similar form in which $\boldsymbol{A}$ is a linear differential or integral equation with random coefficients. These types of problems are referred to as **discrete** and **continuous random eigenvalue problems**, respectively. This section deals with discrete random eigenvalue problems. Continuous problems are discussed in Section 8.4.2.

The determination of probabilistic properties of Λ_i and $\boldsymbol{X}_i$ can be based on (1) extensions of the local solution in Section 6.5.2, (2) Monte Carlo simulation, (3) exact or approximate expressions for the eigenvalues and eigenvectors of $\boldsymbol{A}$ and bounds on Λ_i, and (4) equations for moments of the eigenvalues and eigenvectors of $\boldsymbol{A}$. The first two classes of methods are not discussed since they constitute direct extensions of previous developments in the book. The methods in the third class are illustrated by exact expressions for Λ_i and $\boldsymbol{X}_i$ and bounds on Λ_i (Section 8.3.2.2), Taylor expansion (Section 8.3.2.3), and perturbation (Sec. 8.3.2.4). The iteration (Section 8.3.2.5) and the crossing theory for stochastic processes (Section 8.3.2.6) belong to methods in the fourth class. The next section reviews essential properties of the eigenvalues and eigenvectors for a deterministic matrix.

8.3.2.1 Deterministic eigenvalue problem

Let $\boldsymbol{a}$ be an (n, n)-symmetric, non-singular matrix with real-valued entries. The roots $\lambda_1, \ldots, \lambda_n$ of the **characteristic equation** $\det(\boldsymbol{a} - \lambda\, \boldsymbol{i}) = 0$ are the **eigenvalues** of $\boldsymbol{a}$, where $\boldsymbol{i}$ denotes the (n, n) identity matrix. These roots may not be distinct. We index the eigenvalues of $\boldsymbol{a}$ in increasing order, that is, $\lambda_1 \leq \lambda_2 \leq \cdots \leq \lambda_n$. The polynomial $\det(\boldsymbol{a} - \lambda\, \boldsymbol{i})$ of λ is called the **characteristic polynomial**. An alternative form of the characteristic equation is

$$\det(\boldsymbol{a} - \lambda\, \boldsymbol{i}) = c_0\, \lambda^n + c_1\, \lambda^{n-1} + \cdots + c_{n-1}\, \lambda + c_n = 0,$$

where $c_k = -(c_{k-1}\, d_1 + c_{k-2}\, d_2 + \cdots + c_1\, d_{k-1} + d_k)/k$, $k = 1, \ldots, n$, $c_0 = 1$, and $d_k = \mathrm{tr}(\boldsymbol{a}^k)$, $k = 1, \ldots, n$ ([143], p. 84). A solution $\boldsymbol{x}$ of $\boldsymbol{a}\, \boldsymbol{x} = \lambda\, \boldsymbol{x}$ is said to be non-trivial if $\boldsymbol{x} \neq \boldsymbol{0}$. A non-trivial solution $\boldsymbol{x}_i$ of $\boldsymbol{a}\, \boldsymbol{x}_i = \lambda_i\, \boldsymbol{x}_i$, $i = 1, \ldots, n$, is called an **eigenvector** of $\boldsymbol{a}$ associated with the eigenvalue λ_i of this matrix.

We summarize essential properties of the eigenvalues λ_i and eigenvectors $\boldsymbol{x}_i$ of $\boldsymbol{a}$ and provide bounds on λ_i and a formula for calculating λ_i.

- The eigenvalues λ_i are real and the eigenvectors $\boldsymbol{x}_i$ are in $\mathbb{R}^n$.
- If $\boldsymbol{a}$ is positive definite, its eigenvalues are positive.
- The eigenvectors corresponding to distinct eigenvalues are orthogonal and linearly independent.
- The eigenvectors are unique in the sense that an eigenvalue that is not multiple can have only one eigenvector.
- If $\boldsymbol{a}$ has a multiple eigenvalue, it is possible to define a set of n linearly independent eigenvectors providing a basis in $\mathbb{R}^n$.

Proof: The equation $\boldsymbol{a}\, \boldsymbol{x}_i = \lambda_i\, \boldsymbol{x}_i$ implies the equalities $[((\boldsymbol{a}\, \boldsymbol{x}_i)^*)^T]\, \boldsymbol{x}_i = \lambda_i^*\, (\boldsymbol{x}_i^*)^T\, \boldsymbol{x}_i$ and $(\boldsymbol{x}_i^*)^T\, (\boldsymbol{a}\, \boldsymbol{x}_i) = \lambda_i\, (\boldsymbol{x}_i^*)^T\, \boldsymbol{x}_i$ so that $(\lambda_i^* - \lambda_i)\, (\boldsymbol{x}_i^*)^T\, \boldsymbol{x}_i = 0$, where $\boldsymbol{x}_i^*$ is the complex conjugate of x_i. Because x_i is an eigenvector, $(x_i^*)^T x_i \neq 0$ so that $\lambda_i^* - \lambda_i = 0$. This property also shows that the eigenvectors of a are elements of $\mathbb{R}^n$.

A matrix a is positive definite if $x^T a x \geq 0$ for all $x \in \mathbb{R}^n$. We have $x_i^T a x_i = \lambda_i x_i^T x_i$ from $a x_i = \lambda_i x_i$ so that $\lambda_i \geq 0$ because $x_i^T a x_i \geq 0$ by hypothesis and x_i is an eigenvector such that $x_i^T x_i > 0$.

Two vectors, x_i and x_j, are said to be **orthogonal** if their inner product $\langle x_i, x_j \rangle = x_i^T x_j$ is zero. The eigenvalue problems $a x_i = \lambda_i x_i$ and $a x_j = \lambda_j x_j$ imply $x_j^T a x_i = \lambda_i x_j^T x_i$ and $(a x_j)^T x_i = \lambda_j x_j^T x_i$ so that $(\lambda_i - \lambda_j) \langle x_i, x_j \rangle = 0$. If $\lambda_i \neq \lambda_j$, then $\langle x_i, x_j \rangle$ must be zero.

Suppose that $\lambda_1 \neq \lambda_2$ and assume that the corresponding eigenvectors are linearly dependent, that is, $\alpha_1 x_1 + \alpha_2 x_2 = \mathbf{0}$ with $\alpha_1, \alpha_2 \neq 0$. We have $\langle x_1, \alpha_1 x_1 + \alpha_2 x_2 \rangle = \alpha_1 \langle x_1, x_1 \rangle = 0$ so that α_1 must be zero in contradiction with our assumption. Hence, x_1 and x_2 are linearly independent. If the eigenvalues of a are distinct, its eigenvectors span the space $\mathbb{R}^n$ and define a basis of this space.

Let $(x_1, \ldots, x_n)$ be n linearly independent eigenvectors of a and suppose that λ_1 has two eigenvectors, x_1 and y_1. We have $y_1 = \sum_{k=1}^n \beta_k x_k$ so that

$$\lambda_1 y_1 = a y_1 = \sum_{k=1}^n \beta_k a x_k = \sum_{k=1}^n \beta_k \lambda_k x_k \quad \text{and} \quad \lambda_1 y_1 = \sum_{k=1}^n \beta_k \lambda_1 x_k.$$

The difference, $\sum_{k=2}^n \beta_k (\lambda_1 - \lambda_k) x_k = \mathbf{0}$, of the above equalities shows that the coefficients $\beta_k (\lambda_1 - \lambda_k)$ must be zero because the eigenvectors are linearly independent so that $\beta_k = 0$ for $k > 1$ since $\lambda_1 \neq \lambda_k$ for $k > 1$. Hence, y_1 must be proportional to x_1.

Suppose that λ_1 is a double root of the characteristic equation and all the other roots of this equation are simple. Hence, the eigenvectors $(x_3, \ldots, x_n)$ corresponding to the distinct eigenvalues $(\lambda_3, \ldots, \lambda_n)$ are linearly independent so that they span a subspace $\mathcal{S}$ of $\mathbb{R}^n$ with dimension $n - 2$. The eigenvectors associated with λ_1 are linearly independent of $(x_3, \ldots, x_n)$ since $\lambda_1 \neq \lambda_i$, $i \geq 3$, so that they are in the subspace $\mathbb{R}^n \setminus \mathcal{S}$. Under some conditions it is possible to select two linearly independent eigenvectors for λ_1 ([193], pp. 7-21). ■

- $\lambda_i = \langle a x_i, x_i \rangle / \langle x_i, x_i \rangle$.
- $\lambda_1 \leq \lambda_R = \langle a x, x \rangle / \langle x, x \rangle \leq \lambda_n$ for $x \in \mathbb{R}^n \setminus \{\mathbf{0}\}$.
- $|\lambda_i - \mu_i| \leq \left(\sum_{i,j=1}^n r_{ij}^2 \right)^{1/2} \leq n \max |r_{ij}|$, where r is an (n, n) symmetric matrix with real-valued entries and μ_i denote the eigenvalues of $a + r$ ([127], pp. 96-99, and [193], pp. 101-103).
- $|\lambda - a_{kk}| \leq r_k = \sum_{i=1, i \neq k}^n |a_{ki}|$ for at least one k ([120], p. 371, [193], p. 71).

Proof: We prove the first two statements. The first statement follows from the inner product of $a x_i = \lambda_i x_i$ with x_i.

Suppose that the eigenvectors of a are normalized such that their norm is $\| x_k \| = \langle x_k, x_k \rangle^{1/2} = 1$. If $x = \sum_{k=1}^n c_k x_k$ is a vector in $\mathbb{R}^n$ with $\| x \| = \langle x, x \rangle^{1/2} = 1$, the Rayleigh quotient is

$$\lambda_R = \frac{\langle a x, x \rangle}{\langle x, x \rangle} = \sum_{k=1}^n c_k^2 \lambda_k = \lambda_1 + \sum_{k=2}^n c_k^2 (\lambda_k - \lambda_1) \geq \lambda_1$$

since $\| \boldsymbol{x} \|^2 = \sum_{k=1}^{n} c_k^2 = 1$ and $\lambda_k \geq \lambda_1$ for $k \geq 2$. Also, $\lambda_R = \sum_{k=1}^{n} c_k^2 \lambda_k \leq \lambda_n \sum_{k=1}^{n} c_k^2 = \lambda_n$. If $c_k = \varepsilon_i \, c_p$ for $k \neq p$,

$$\lambda_R = \lambda_p + c_p^2 \sum_{k=1, k \neq p}^{n} (\lambda_k - \lambda_p) \, \varepsilon_k^2,$$

so that $\lambda_R = \lambda_p + O(\varepsilon^2)$ for $\varepsilon_i = \varepsilon$ and $|\varepsilon| \ll 1$. These properties imply

$$\lambda_1 = \min_{\boldsymbol{x} \in \mathbb{R}^n \setminus \{\boldsymbol{0}\}} \frac{\langle \boldsymbol{a} \, \boldsymbol{x}, \boldsymbol{x} \rangle}{\langle \boldsymbol{x}, \boldsymbol{x} \rangle} \quad \text{and} \quad \lambda_n = \max_{\boldsymbol{x} \in \mathbb{R}^n \setminus \{\boldsymbol{0}\}} \frac{\langle \boldsymbol{a} \, \boldsymbol{x}, \boldsymbol{x} \rangle}{\langle \boldsymbol{x}, \boldsymbol{x} \rangle},$$

and provide bounds on the smallest and largest eigenvalues of $\boldsymbol{a}$. ■

8.3.2.2 Exact expressions and bounds

Suppose that it is possible to express the eigenvalues and the eigenvectors of $\boldsymbol{A}$ as continuous functions of the entries of $\boldsymbol{A}$. Because continuous functions are Borel measurable the corresponding eigenvalues and eigenvectors are random variables on $(\Omega, \mathcal{F}, P)$. Statistics of Λ_i and $\boldsymbol{X}_i$ can be obtained by direct calculations, Monte Carlo simulation, or other techniques using the mapping $\boldsymbol{A} \mapsto (\Lambda_i, \boldsymbol{X}_i)$.

Example 8.10: Let S_{ij}, $i, j = 1, 2$, be a random stress tensor at a point in a thin plate loaded in its own plane. Our objective is to determine the principal stresses and stress directions of the stress tensor S_{ij}, that is, the eigenvalues and eigenvectors of random matrix $\boldsymbol{A} = \{S_{ij}\}$. The eigenvalues and eigenvectors of this matrix are, respectively,

$$\Lambda_{1,2} = \frac{S_{11} + S_{22}}{2} \pm \left[\left(\frac{S_{11} - S_{22}}{2} \right)^2 + S_{12}^2 \right]^{1/2}$$

and $\boldsymbol{X}_i = (1, -(S_{11} - \Lambda_i)/S_{12})$, $i = 1, 2$. Probabilistic properties of Λ_i and $\boldsymbol{X}_i$ can be obtained from the above equations by Monte Carlo simulation or direct calculations. ◇

Note: The characteristic equation of $\boldsymbol{A}$ is $\Lambda^2 - (S_{11} + S_{22}) \, \Lambda + S_{11} \, S_{22} - S_{12}^2 = 0$. The eigenvectors of $\boldsymbol{A}$ can be obtained from any of the two equations of the linear system $(\boldsymbol{A} - \Lambda_i \, \boldsymbol{i}) \, \boldsymbol{X}_i = \boldsymbol{0}$, $i = 1, 2$, since $\det(\boldsymbol{A} - \Lambda_i \, \boldsymbol{i}) = 0$ a.s. ▲

Example 8.11: Consider the random matrix

$$\boldsymbol{A} = \begin{bmatrix} 3 & -1 & 0 \\ -1 & 1 & -1 \\ 0 & -1 & 5 \end{bmatrix} + Z \begin{bmatrix} 1 & -1 & 0 \\ -1 & 1 & 0 \\ 0 & 0 & 0 \end{bmatrix} = \boldsymbol{a} + Z \, \boldsymbol{b},$$

where Z is a random variable. The smallest eigenvalue of $\boldsymbol{A}$ is bounded by $\Lambda_1 \leq \boldsymbol{x}_1^T \, (\boldsymbol{a} + Z \, \boldsymbol{b}) \, \boldsymbol{x}_1 = \boldsymbol{x}_1^T \, \boldsymbol{a} \, \boldsymbol{x} + Z \, \boldsymbol{x}_1^T \, \boldsymbol{b} \, \boldsymbol{x}_1 = 0.03983 + 0.3172 \, Z$ a.s., where $\boldsymbol{x}_1 = [0.9295, -0.3159, 0.1903]^T$, $\| \boldsymbol{x} \| = 1$, denotes the first eigenvector of $\boldsymbol{a}$. The bound follows from the properties of the Rayleigh quotient. ◇

8.3.2.3 Taylor series method

Suppose that the entries of A are measurable functions of an $\mathbb{R}^m$-valued random variable $Z \in L_2$ with mean $\mu_z = \{\mu_{z,u}\}$ and covariance $\gamma_z = \{\gamma_{z,uv}\}$, $u, v = 1, \ldots, m$. Let $a = A(\mu_z)$, $\lambda_i = \Lambda_i(\mu_z)$, $x_i = X_i(\mu_z)$, $a_{,p} = \partial A(\mu_z)/\partial z_p$, $x_{i,p} = \partial X_i(\mu_z)/\partial z_p$, $\lambda_{i,p} = \partial \Lambda_i(\mu_z)/\partial z_p$, and $c_{i,p} = \partial C_i(\mu_z)/\partial z_p$, where C_i are the coefficients of the characteristic polynomial $\det(A - \Lambda\, i)$. It is assumed that the eigenvalues λ_i of a are distinct and that the eigenvectors x_i of this matrix are normalized to have unit norm.

If a is a symmetric matrix with $\det(a) \neq 0$ and distinct eigenvalues λ_i, then

$$\Lambda_i \simeq \lambda_i + \sum_{p=1}^{m} \lambda_{i,p}\,(Z_p - E[Z_p]), \quad \text{where}$$

$$\lambda_{i,p} = -\frac{c_{1,p}\,\lambda_i^{n-1} + \cdots + c_{n-1,p}\,\lambda_i + c_{n,p}}{n\,\lambda_i^{n-1} + (n-1)\,c_1\,\lambda_i^{n-2} + \cdots + c_{n-1}} \quad \text{and} \qquad (8.27)$$

$$X_i \simeq x_i + \sum_{p=1}^{m} x_{i,p}\,(Z_p - E[Z_p]), \quad \text{where } x_{i,p} = \sum_{k=1,k\neq i}^{n} \frac{x_k^T\, a_{,p}\, x_i}{\lambda_i - \lambda_k}\, x_k. \qquad (8.28)$$

Proof: The partial derivative $\lambda_{i,p}$ and $x_{i,p}$ of Λ_i and X_i relative to a coordinate Z_p of Z in Eqs. 8.27 and 8.28 are called **sensitivity factors**. The expressions of Λ_i and X_i in Eqs. 8.27 and 8.28 are referred to as **first order** approximations.

Let $\Lambda^n + C_1\,\Lambda^{n-1} + \cdots + C_{n-1}\,\Lambda + C_n = 0$ be the characteristic equation for A. The differentiation of this equation with respect to Z_p yields

$$\begin{aligned} &n\,\Lambda^{n-1}\,\Lambda_{,p} + C_{1,p}\,\Lambda^{n-1} + (n-1)\,C_1\,\Lambda^{n-2}\,\Lambda_{,p} \\ &\quad + \cdots + C_{n-1,p}\,\Lambda + C_{n-1}\,\Lambda_{,p} + C_{n,p} = 0 \quad \text{or} \\ &(n\,\Lambda^{n-1} + (n-1)\,C_1\,\Lambda^{n-2} + \cdots + C_{n-1})\,\Lambda_{,p} \\ &\quad = -(C_{1,p}\,\Lambda^{n-1} + \cdots + C_{n-1,p}\,\Lambda + C_{n,p}), \end{aligned}$$

where $W_{,p}$ stand for $\partial W/\partial z_p$. The above equation written for $\Lambda = \Lambda_i$ and $Z = \mu_z$ gives $\lambda_{i,p}$ in Eq. 8.27. Similar equations can be obtained for higher order derivatives of the eigenvalues of A.

The differentiation of $A\,X_i = \Lambda_i\,X_i$ with respect to Z_p gives $A_{,p}\,X_i + A\,X_{i,p} = \Lambda_{i,p}\,X_i + \Lambda_i\,X_{i,p}$ or, by setting Z equal to μ_z,

$$(a - \lambda_i\, i)\, x_{i,p} = \lambda_{i,p}\, x_i - a_{,p}\, x_i.$$

The above equation can be used to calculate the coordinates $b_{il}^{(p)}$ of $x_{i,p} = \sum_{l=1}^{n} b_{il}^{(p)}\, x_l$ in the basis defined by the eigenvectors of a by noting that (1) $(a - \lambda_i\, i)\, x_{i,p}$ is orthogonal to $\boldsymbol{x}_i$ since $(\boldsymbol{x}_i^T \boldsymbol{a} - \lambda_i \boldsymbol{x}_i^T)\boldsymbol{x}_{i,p} = 0$ and (2) the product of the above equation with $\boldsymbol{x}_k^T$ to the left yields

$$(\lambda_k - \lambda_i)\, b_{ik}^{(p)} = \lambda_{i,k}\, \delta_{ki} - \boldsymbol{x}_k^T\, \boldsymbol{a}_{,p}\, \boldsymbol{x}_i.$$

The latter equation gives the coordinates $b_{ik}^{(p)}$ of $\boldsymbol{x}_{i,p}$ for $i \neq k$ since $\boldsymbol{a}$ has distinct eigenvalues by assumption. The coordinate $b_{ii}^{(p)}$ remains undetermined and can be taken as zero so that $\boldsymbol{x}_{i,p} = \sum_{k=1,k\neq i}^{n} b_{ik}^{(p)}\, \boldsymbol{x}_k$. The choice $b_{ii}^{(p)} = 0$ is not restrictive since $(\boldsymbol{a} - \lambda_i\, \boldsymbol{i})\,\boldsymbol{x}_{i,p}$ and $\boldsymbol{x}_i$ are orthogonal so that the above results do not change if we add to $\boldsymbol{x}_{i,p}$ a vector proportional to $\boldsymbol{x}_i$. ■

The first order approximations of Λ_i and $\boldsymbol{X}_i$ in Eqs. 8.27 and 8.28 are linear functions of $\boldsymbol{Z}$ so that the second moment properties of $\boldsymbol{Z}$ are sufficient to calculate the first two moments of the eigenvalues and eigenvectors of $\boldsymbol{A}$. For example, the approximate mean and variance of Λ_i are λ_i and $\sum_{p,q=1}^{m} \lambda_{i,p}\, \lambda_{i,q}\, \gamma_{z,pq}$, respectively.

Example 8.12: Consider a (3, 3) random matrix $\boldsymbol{A} = \boldsymbol{a} + \boldsymbol{R}$, where

$$\boldsymbol{a} = \begin{bmatrix} 2.6 & -1.1 & 0.0 \\ -1.1 & 5.2 & -\sqrt{2} \\ 0.0 & -\sqrt{2} & 10 \end{bmatrix},$$

$$\boldsymbol{R} = Z \begin{bmatrix} 0.1 & -0.1 & 0.0 \\ -0.1 & 0.2 & 0.0 \\ 0.0 & 0.0 & 0.0 \end{bmatrix},$$

and Z is a random variable with mean zero ($\mu_z = 0$) and variance σ_z^2. The eigenvalues and eigenvectors of $\boldsymbol{a}$, that is, of $\boldsymbol{A}$ with $Z = 0$, are $\lambda_1 = 2.1647$, $\lambda_2 = 5.2385$, $\lambda_3 = 10.3967$, and

$$\boldsymbol{b} = \begin{bmatrix} 0.9278 & 0.3711 & 0.0381 \\ 0.3671 & -0.8902 & -0.2699 \\ 0.0663 & -0.2646 & 0.9621 \end{bmatrix},$$

where the columns of $\boldsymbol{b}$ are the eigenvectors of $\boldsymbol{a}$. The first order Taylor approximations of the eigenvalues and eigenvectors of $\boldsymbol{A}$ given by Eqs. 8.27 and 8.28 are $\Lambda_1 \simeq 2.1647 + 0.0449\, Z$, $\Lambda_2 \simeq 5.2385 + 0.2383\, Z$, $\Lambda_3 \simeq 10.3967 + 0.0168\, Z$, and

$$\boldsymbol{X} \simeq \boldsymbol{b} + Z \begin{bmatrix} -0.0046 & 0.0118 & 0.0054 \\ 0.0113 & 0.0025 & -0.0105 \\ 0.0024 & -0.0082 & -0.0032 \end{bmatrix},$$

respectively. ◇

Note: The random coefficients of the characteristic polynomial $\det(\boldsymbol{A} - \Lambda\, \boldsymbol{i})$ are $C_0 = 1$, $C_1 = -0.3\, Z - 17.8$, $C_2 = 0.01\, Z^2 + 3.82\, Z + 88.31$, and $C_3 = -0.1\, Z^2 - 8\, Z - 117.9$

so that $C_{0,1} = 0$, $C_{1,1} = -0.3$, $C_{2,1} = 0.02\,Z + 3.82$, and $C_{3,1} = -0.2\,Z - 8$. The sensitivity factors for the eigenvalues can be calculated from (Eq. 8.27)

$$\lambda_{i,1} = -\frac{c_{1,1}\,\lambda_i^2 + c_{2,1}\,\lambda_i + c_{3,1}}{3\,\lambda_i^2 + 2\,c_1\,\lambda_i + c_2}$$

with Z set as zero. The first order approximations of the eigenvectors of $\boldsymbol{A}$ is given by Eq. 8.28. For example, the coordinates of $\boldsymbol{x}_{1,1}$ are

$$b_{12}^{(1)} = \frac{\boldsymbol{x}_2^T\,\boldsymbol{a}_{,1}\,\boldsymbol{x}_1}{\lambda_1 - \lambda_2} = -0.0124 \quad \text{and} \quad b_{13}^{(1)} = \frac{\boldsymbol{x}_3^T\,\boldsymbol{a}_{,1}\,\boldsymbol{x}_1}{\lambda_1 - \lambda_3} = -0.00089$$

so that $\boldsymbol{X}_1 = \boldsymbol{x}_1 + (b_{12}^{(1)}\,\boldsymbol{x}_2 + b_{13}^{(1)}\,\boldsymbol{x}_3)\,Z$. ▲

8.3.2.4 Perturbation method

Suppose that

$$\boldsymbol{A} = \boldsymbol{a} + \varepsilon\,\tilde{\boldsymbol{R}}, \tag{8.29}$$

where $\boldsymbol{a}$ and $\tilde{\boldsymbol{R}}$ are (n, n) matrices with real-valued entries that are deterministic and random, respectively, $\det(\boldsymbol{a}) \neq 0$, $E[\tilde{\boldsymbol{R}}] = \boldsymbol{0}$, and ε is a small parameter. It is assumed that $\boldsymbol{a}$ is symmetric and has distinct eigenvalues. Let λ_i and $\boldsymbol{x}_i$ be the eigenvalues and eigenvectors of $\boldsymbol{a}$. The eigenvectors $\boldsymbol{x}_i$ are scaled to have unit norm. It is assumed that $\boldsymbol{a}$ has distinct eigenvalues.

The first order approximations of the eigenvalues and eigenvectors of $\boldsymbol{A}$ are

$$\Lambda_i \simeq \lambda_i + \varepsilon\,\boldsymbol{x}_i^T\,\tilde{\boldsymbol{R}}\,\boldsymbol{x}_i + O(\varepsilon^2) \quad \text{and} \tag{8.30}$$

$$\boldsymbol{X}_i \simeq \boldsymbol{x}_i + \varepsilon \sum_{k=1,k\neq i}^{n} \frac{\boldsymbol{x}_k^T\,\tilde{\boldsymbol{R}}\,\boldsymbol{x}_i}{\lambda_i - \lambda_k}\,\boldsymbol{x}_k + O(\varepsilon^2). \tag{8.31}$$

Proof: The representations,

$$\Lambda_i = \lambda_i + \varepsilon\,\Lambda_i^{(1)} + \varepsilon^2\,\Lambda_i^{(2)} + \cdots \quad \text{and} \quad \boldsymbol{X}_i = \boldsymbol{x}_i + \varepsilon\,\boldsymbol{X}_i^{(1)} + \varepsilon^2\,\boldsymbol{X}_i^{(2)} + \cdots,$$

of the eigenvalues and eigenvectors on $\boldsymbol{A}$ introduced in $\boldsymbol{A}\,\boldsymbol{X}_i = \Lambda_i\,\boldsymbol{X}_i$ with $\boldsymbol{A}$ in Eq. 8.29 yields

$$(\boldsymbol{a} + \varepsilon\,\tilde{\boldsymbol{R}})(\boldsymbol{x}_i + \varepsilon\,\boldsymbol{X}_i^{(1)} + \cdots) = (\lambda_i + \varepsilon\,\Lambda_i^{(1)} + \cdots)(\boldsymbol{x}_i + \varepsilon\,\boldsymbol{X}_i^{(1)} + \cdots)$$

so that

$$\begin{aligned} \boldsymbol{a}\,\boldsymbol{x}_i &= \lambda_i\,\boldsymbol{x}_i && \text{(order 1)}, \\ \boldsymbol{a}\,\boldsymbol{X}_i^{(1)} + \tilde{\boldsymbol{R}}\,\boldsymbol{x}_i &= \lambda_i\,\boldsymbol{X}_i^{(1)} + \Lambda_i^{(1)}\,\boldsymbol{x}_i && \text{(order } \varepsilon). \end{aligned}$$

The alternative form, $(\boldsymbol{a} - \lambda_i\, \boldsymbol{i})\, \boldsymbol{X}_i^{(1)} = \Lambda_i^{(1)}\, \boldsymbol{x}_i - \tilde{\boldsymbol{R}}\, \boldsymbol{x}_i$, of the above equation for order ε and the orthogonality of the vectors $(\boldsymbol{a} - \lambda_i\, \boldsymbol{i})\, \boldsymbol{X}_i^{(1)}$ and $\boldsymbol{x}_i$ imply $\boldsymbol{x}_i^T \left(\Lambda_i^{(1)}\, \boldsymbol{x}_i - \tilde{\boldsymbol{R}}\, \boldsymbol{x}_i\right) = 0$ or $\Lambda_i^{(1)} = \boldsymbol{x}_i^T\, \tilde{\boldsymbol{R}}\, \boldsymbol{x}_i$ since $\boldsymbol{x}_i^T\, \boldsymbol{x}_i = 1$.

Alternative calculations can be performed to find this result. For example, the equation for order ε multiplied to the left by $\boldsymbol{x}_j^T$ gives

$$\boldsymbol{x}_j^T\, \boldsymbol{a} \sum_{k=1}^{n} B_{ik}\, \boldsymbol{x}_k + \boldsymbol{x}_j^T\, \tilde{\boldsymbol{R}}\, \boldsymbol{x}_i = \lambda_i\, \boldsymbol{x}_j^T \sum_{k=1}^{n} B_{ik}\, \boldsymbol{x}_k + \Lambda_i^{(1)}\, \boldsymbol{x}_j^T\, \boldsymbol{x}_i,$$

where $\boldsymbol{X}_i^{(1)} = \sum_{k=1}^{n} B_{ik}\, \boldsymbol{x}_k$ and B_{ik}, $k = 1, \ldots, n$, denote the projections of $\boldsymbol{X}_i^{(1)}$ on the eigenvectors of $\boldsymbol{a}$. The last equation simplifies to $B_{ij}\,(\lambda_j - \lambda_i) + \boldsymbol{x}_j^T\, \tilde{\boldsymbol{R}}\, \boldsymbol{x}_i = \Lambda_i^{(1)}\, \delta_{ij}$ since $\boldsymbol{x}_j^T\, \boldsymbol{a}\, \boldsymbol{x}_i = \lambda_i\, \delta_{ij}$ and $\boldsymbol{x}_j^T\, \boldsymbol{x}_i = \delta_{ij}$ so that

$$\Lambda_i^{(1)} = \boldsymbol{x}_i^T\, \tilde{\boldsymbol{R}}\, \boldsymbol{x}_i \qquad \text{for } i = j \text{ and}$$

$$B_{ij} = \frac{\boldsymbol{x}_j^T\, \tilde{\boldsymbol{R}}\, \boldsymbol{x}_i}{\lambda_i - \lambda_j} \qquad \text{for } i \neq j.$$

The projection B_{ii} of $\boldsymbol{X}_i^{(1)}$ on $\boldsymbol{x}_i$ remains undetermined. Adding a vector proportional with $\boldsymbol{x}_i$ to $\boldsymbol{X}_i^{(1)} = \sum_{j=1, j \neq i} B_{ij}\, \boldsymbol{x}_j$ does not change the above results so that we take $B_{ii} = 0$, and the resulting first order approximation of $\boldsymbol{X}_i$ is $\boldsymbol{x}_i + \varepsilon\, \boldsymbol{X}_i^{(1)}$. Additional considerations on the determination of $\boldsymbol{X}_i^{(1)}$ and approximations of order 2 and higher of the eigenvalues and eigenvectors of matrix $\boldsymbol{A}$ can be found, for example, in [120] (Chapters. 9 and 11). ■

The first order approximations of Λ_i and $\boldsymbol{X}_i$ in Eqs. 8.30 and 8.31 are linear functions of $\tilde{\boldsymbol{R}}$ so that the second moment properties of this random matrix are sufficient to calculate the first two moments of the eigenvalues and eigenvectors of the random matrix $\boldsymbol{A}$.

Example 8.13: Suppose that the deterministic and stochastic parts, $\boldsymbol{a}$ and $\tilde{\boldsymbol{R}}$, of a (3, 3) random matrix $\boldsymbol{A}$ (Eq. 8.29) are the matrices $\boldsymbol{a}$ and $\boldsymbol{R}$ in Example 8.12. It is assumed that the random variable Z has mean zero and variance of order 1.

The zero order perturbation solution is given by the eigenvalues and the eigenvectors of $\boldsymbol{a}$ (Example 8.12). The first order corrections of the eigenvalues and eigenvectors of $\boldsymbol{A}$ are $\Lambda_1^{(1)} = \boldsymbol{x}_1^T\, \tilde{\boldsymbol{R}}\, \boldsymbol{x}_1 = 0.045971\, Z$, $\Lambda_2^{(1)} = \boldsymbol{x}_2^T\, \tilde{\boldsymbol{R}}\, \boldsymbol{x}_2 = 0.238710\, Z$, $\Lambda_3^{(1)} = \boldsymbol{x}_3^T\, \tilde{\boldsymbol{R}}\, \boldsymbol{x}_3 = 0.015319\, Z$, and

$$\boldsymbol{X}^{(1)} = Z \begin{bmatrix} -0.0046 & 0.0118 & 0.0054 \\ 0.0113 & 0.0025 & -0.0105 \\ 0.0024 & -0.0082 & -0.0032 \end{bmatrix},$$

respectively. The eigenvalues and eigenvectors to the order ε are $\Lambda_i \simeq \lambda_i + \varepsilon\, \Lambda_i^{(1)}$ and $\boldsymbol{X}_i \simeq \boldsymbol{x}_i + \varepsilon\, \boldsymbol{X}_i^{(1)}$. The above expressions for $\boldsymbol{X}^{(i)}$ and Λ_i can be used to calculate approximately moments and other probabilistic properties of the eigenvalues and eigenvectors of $\boldsymbol{A}$. ◇

Note: The first order approximation of the first eigenvector of $\boldsymbol{A}$ is (Eq. 8.31)

$$\boldsymbol{X}_1^{(1)} = \frac{\boldsymbol{x}_2^T \tilde{\boldsymbol{R}} \boldsymbol{x}_1}{\lambda_1 - \lambda_2} \boldsymbol{x}_2 + \frac{\boldsymbol{x}_3^T \tilde{\boldsymbol{R}} \boldsymbol{x}_1}{\lambda_1 - \lambda_3} \boldsymbol{x}_3 = (-0.0124\, \boldsymbol{x}_2 - 0.00089\, \boldsymbol{x}_3)\, Z.$$

The calculations of moments and other probabilistic properties of $\Lambda_i^{(1)}$ and $\boldsymbol{X}_i^{(1)}$ are very simple in this case since they depend linearly on Z. ▲

8.3.2.5 Iteration method

Let $\boldsymbol{A}$ be an (n, n) random symmetric matrix whose entries are real-valued square integrable random variables and let

$$\boldsymbol{A} = \boldsymbol{a} + \boldsymbol{R}, \tag{8.32}$$

where $\boldsymbol{a} = E[\boldsymbol{A}]$ is a symmetric matrix and $\boldsymbol{R} = \boldsymbol{A} - \boldsymbol{a}$. Our objective is to develop approximations for the first two moments of the eigenvalues and eigenvectors of $\boldsymbol{A}$. The analysis is based on a version of the local independence hypothesis (Section 8.3.1.7) stating that the dependence among Λ, $\boldsymbol{X}$, and $\boldsymbol{R}$ can be disregarded. As previously stated, this hypothesis has no justification. We include the iteration method in our discussion because of its wide use in some fields [164, 173].

We present just a few results based on both versions of the iteration method discussed in Section 8.3.1.7. Additional information on this method can be found in [164] (pp. 18-19).

The expectation of the eigenvalues and eigenvectors of $\boldsymbol{A}$ can be approximated by the solution of the eigenvalue problem

$$\boldsymbol{a}\, E[\boldsymbol{X}] \simeq E[\Lambda]\, E[\boldsymbol{X}]. \tag{8.33}$$

Note: The above approximations of the expectations of the eigenvalues and eigenvectors of $\boldsymbol{A}$ coincide with the eigenvalues and eigenvectors of $\boldsymbol{a}$.

The equation, $\boldsymbol{a}\, E[\boldsymbol{X}] + E[\boldsymbol{R}\, \boldsymbol{X}] = E[\Lambda\, \boldsymbol{X}]$, obtained by averaging $(\boldsymbol{a} + \boldsymbol{R})\, \boldsymbol{X} = \Lambda\, \boldsymbol{X}$ cannot be used to calculate the first order moments of Λ and $\boldsymbol{X}$ since $E[\boldsymbol{R}\, \boldsymbol{X}]$ and $E[\Lambda\, \boldsymbol{X}]$ are not known. The approximations

$$E[\boldsymbol{R}\, \boldsymbol{X}] \simeq E[\boldsymbol{R}]\, E[\boldsymbol{X}] = \boldsymbol{0} \quad \text{and} \quad E[\Lambda\, \boldsymbol{X}] \simeq E[\Lambda]\, E[\boldsymbol{X}],$$

based on the hypothesis of local independence, yield Eq. 8.33. ▲

The second moment of the eigenvalues of $\boldsymbol{A}$ can be approximated by

$$E[\Lambda_i^2] \simeq \lambda_i^2 + \boldsymbol{x}_i^T\, E[\boldsymbol{R}\, \boldsymbol{R}]\, \boldsymbol{x}_i. \tag{8.34}$$

Note: Consider the alternative form,

$$X = \Lambda^{-1} (a + R) X = \Lambda^{-1} (a + R) [\Lambda^{-1} (a + R) X]$$
$$= \Lambda^{-2} (a\,a + a\,R + R\,a + R\,R) X,$$

of $A\,X = \Lambda\,X$. The expectation of this equation is

$$E[\Lambda^2 X] = a\,a\,E[X] + a\,E[R\,X] + E[R\,a\,X] + E[R\,R\,X]$$

and depends on the unknown averages $E[\Lambda^2 X]$, $E[R\,X]$, $E[R\,a\,X]$, and $E[R\,R\,X]$. The local independence hypothesis and the last equation give $E[\Lambda^2]\,E[X] \simeq a\,a\,E[X] + E[R\,R]\,E[X]$ or

$$E[\Lambda_i^2]\,x_i^T\,x_i \simeq x_i^T\,a\,a\,x_i + x_i^T\,E[R\,R]\,x_i,$$

which yields Eq. 8.34 since $x_i^T\,x_i = 1$ and $x_i^T\,a\,a\,x_i = \lambda_i^2\,x_i^T\,x_i = \lambda_i^2$. ▲

Example 8.14: Consider the (3, 3) random matrix A in Example 8.13 with $\varepsilon = 1$. The first moments of the eigenvalues and eigenvectors of A coincide with the deterministic solution corresponding to matrix a. We have obtained the same approximation for the mean values of the eigenvalues and eigenvectors by the perturbation method. However, the approximate second moment of the first eigenvalue of A is $E[\Lambda_1^2] \simeq 4.69 + 0.0038\,\sigma_z^2$, and differs from the perturbation solution. ◇

8.3.2.6 Level crossing for stochastic processes

We use the Rice formula discussed in Section 3.10.1 to (1) find the likely values of the eigenvalues of an (n, n) real-valued random matrix A and (2) reduce the uncertainty in the eigenvalues of A if some of its eigenvalues are specified. For example, A may characterize a physical system with uncertain properties whose lower eigenvalues have been measured without errors. Since A is assumed to be symmetric and to have real-valued entries, its eigenvalues are real-valued. The level crossing method discussed here does not seem to be useful for finding properties of the eigenvectors of A.

Let

$$V(\lambda) = \lambda^n + C_1\,\lambda^{n-1} + \cdots + C_{n-1}\,\lambda + C_n = \lambda^n + \alpha(\lambda)\,C, \tag{8.35}$$

be the characteristic polynomial, $\det(A - \lambda\,i)$, of A, where

$$\alpha(\lambda) = [\lambda^{n-1}, \lambda^{n-2}, \ldots, \lambda, 1] \quad \text{and} \quad C = [C_1, C_2, \ldots, C_{n-1}, C_n]^T. \tag{8.36}$$

The derivative $V'(\lambda) = dV(\lambda)/d\lambda$ of V is

$$V'(\lambda) = n\,\lambda^{n-1} + \beta(\lambda)\,C, \quad \text{where} \tag{8.37}$$

$$\beta(\lambda) = [(n-1)\,\lambda^{n-2}, (n-2)\,\lambda^{n-3}, \ldots, 1, 0]. \tag{8.38}$$

The coefficients C_k, $k = 1, \ldots, n$, are measurable functions of the entries of A so that they are random variables on the probability space $(\Omega, \mathcal{F}, P)$ on which A is defined. The zeros of the polynomial $V(\lambda)$ are the eigenvalues of A, and these zeros are on the real line since A is real-valued and symmetric.

Example 8.15: Let $\boldsymbol{A}$ be a (3, 3) symmetric matrix with entries $A_{11} = K_1 + K_2$, $A_{22} = K_2 + K_3$, $A_{33} = K_3$, $A_{12} = -K_2$, $A_{13} = 0$, and $A_{23} = -K_3$, where $\boldsymbol{K} = (K_1, K_2, K_3)$, $\boldsymbol{K} = \exp(\boldsymbol{G})$, and $\boldsymbol{G}$ is an $\mathbb{R}^3$-valued Gaussian variable with mean (1, 1, 1) and covariance matrix $\boldsymbol{\gamma} = \{\rho^{|i-j|}\}$, $i, j = 1, 2, 3$. Figure 8.7 shows three samples of the polynomial $V(\lambda)$ corresponding to three independent

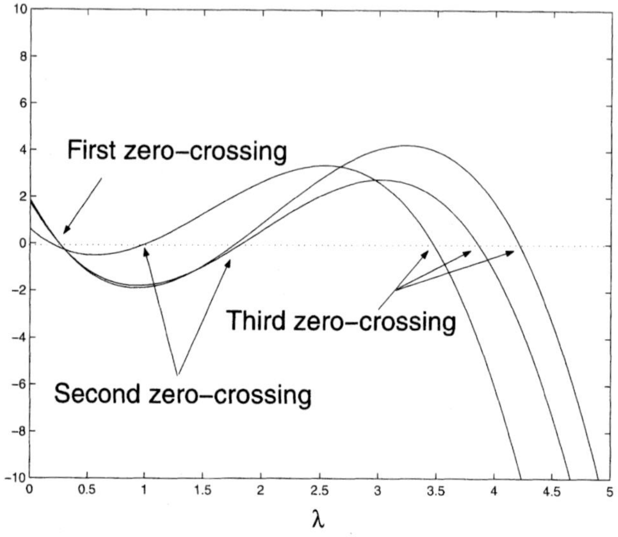

Figure 8.7. Three samples of $V(\lambda)$ for $\rho = 0.3$

samples of $\boldsymbol{A}$ with $\rho = 0.3$. The zero-crossings of stochastic process V give the eigenvalues of the random matrix $\boldsymbol{A}$. ◇

The polynomial $V(\lambda)$, $\lambda \in [0, \infty)$, in Eq. 8.35 is a real-valued stochastic process defined on the real line that has continuous and differentiable sample paths. The process V is not stationary and its finite dimensional distributions are not usually Gaussian. The mean zero-crossing rate $\nu(\lambda)$ of V at each $\lambda \geq 0$ is given by the Rice formula (Section 3.10.1). It is difficult if not impossible in applications to calculate $\nu(\lambda)$ analytically because of the properties of V and the complex dependence of the coefficients $\boldsymbol{C}$ in Eq. 8.35 on $\boldsymbol{A}$. To simplify calculations, we assume that V is a non-stationary Gaussian process V_G, which has the same second moment properties as V. There is no theoretical justification for this assumption, but numerical results in [75] provide a heuristic support for it. Note that the probability law of V_G is completely determined by the second moment properties of V.

Let $\mu(\lambda) = E[V(\lambda)]$ and $\mu'(\lambda) = E[V'(\lambda)]$ be the mean functions of the processes $V^{(0)}(\lambda) = V(\lambda)$ and $V^{(1)}(\lambda) = V'(\lambda)$, respectively, and denote by $\gamma_{pq}(\lambda, \rho) = \text{Cov}[V^{(p)}(\lambda), V^{(q)}(\rho)]$, $p, q = 0, 1$, their covariance functions.

Consider also the functions

$$m(\lambda) = \mu'(\lambda) - \frac{\gamma_{01}(\lambda, \lambda)}{\gamma_{00}(\lambda, \lambda)}\,\mu(\lambda),$$

$$\sigma(\lambda)^2 = \gamma_{11}(\lambda, \lambda) - \frac{\gamma_{01}(\lambda, \lambda)\,\gamma_{10}(\lambda, \lambda)}{\gamma_{00}(\lambda, \lambda)}, \tag{8.39}$$

where $\chi(u) = u\,(2\,\Phi(u) - 1) + 2\,\phi(u)$, $\Phi(u) = \int_{-\infty}^{u} \phi(s)\,ds$ is the distribution of $N(0, 1)$, and $\phi(u) = (2\,\pi)^{-1/2}\,\exp(-0.5\,u^2)$.

If V is assumed to be a Gaussian process V_G, then $\nu(\lambda)$ becomes [75]

$$\nu_G(\lambda) = \frac{\sigma(\lambda)}{\sqrt{\gamma_{00}(\lambda, \lambda)}}\,\chi\left(\frac{m(\lambda)}{\sigma(\lambda)}\right)\,\phi\left(-\frac{\mu(\lambda)}{\sqrt{\gamma_{00}(\lambda, \lambda)}}\right), \quad \lambda \geq 0. \tag{8.40}$$

Proof: By the Rice formula the mean zero-crossing rate of V is

$$\nu(\lambda) = \int_{-\infty}^{\infty} |z|\, f_{V(\lambda), V'(\lambda)}(0, z)\,dz \quad \text{or}$$

$$\nu(\lambda) = \left[\int_{-\infty}^{\infty} |z|\, f_{V'(\lambda)|V(\lambda)}(z \mid 0)\,dz\right] f_{V(\lambda)}(0) = E\left[\left|V'(\lambda)\right| \mid V(\lambda) = 0\right] f_{V(\lambda)}(0),$$

where $f_{V(\lambda), V'(\lambda)}$ and $f_{V'(\lambda)|V(\lambda)}$ denote the densities of $(V(\lambda), V'(\lambda))$ and $V'(\lambda) \mid V(\lambda)$, respectively. The mean rate $\nu(\lambda)$ includes zero-crossings with both positive and negative slopes. The second moment properties of the $\mathbb{R}^2$-valued random variable $(V(\lambda), V'(\lambda))$ are

$$\mu(\lambda) = E[V(\lambda)] = \lambda^n + \boldsymbol{\alpha}(\lambda)\,\boldsymbol{\mu}, \quad \mu'(\lambda) = E[V'(\lambda)] = n\,\lambda^{n-1} + \boldsymbol{\beta}(\lambda)\,\boldsymbol{\mu},$$

$\gamma_{00}(\lambda, \rho) = \boldsymbol{\alpha}(\lambda)\,\boldsymbol{\gamma}\,\boldsymbol{\alpha}(\rho)^T$, $\gamma_{01}(\lambda, \rho) = \boldsymbol{\alpha}(\lambda)\,\boldsymbol{\gamma}\,\boldsymbol{\beta}(\rho)^T$, $\gamma_{10}(\lambda, \rho) = \boldsymbol{\beta}(\lambda)\,\boldsymbol{\gamma}\,\boldsymbol{\alpha}(\rho)^T$, and $\gamma_{11}(\lambda, \rho) = \boldsymbol{\beta}(\lambda)\,\boldsymbol{\gamma}\,\boldsymbol{\beta}(\rho)^T$, where $\boldsymbol{\mu}$ and $\boldsymbol{\gamma}$ denote the mean and covariance matrices of the random vector $\boldsymbol{C}$.

If it is assumed that $\boldsymbol{C}$ is an $\mathbb{R}^n$-valued Gaussian variable, then $(V(\lambda), V'(\lambda))$ is a Gaussian vector, and the mean and variance of the conditional random variable $V'(\lambda) \mid V(\lambda)$ are $m(\lambda)$ and $\sigma(\lambda)^2$ in Eq. 8.39, respectively (Section 2.11.5). The Rice formula applied to the process V_G yields $\nu_G(\lambda)$ in Eq. 8.40.

The second moment properties of $\boldsymbol{C}$ may be difficult to obtain analytically, but can be estimated simply from samples $\boldsymbol{A}$ generated by Monte Carlo simulation. ■

Suppose that the first $r < n$ eigenvalues of a physical system characterized by a random matrix $\boldsymbol{A}$ have been measured without errors on a particular sample of $\boldsymbol{A}$. Our objective is to find properties of the remaining eigenvalues of $\boldsymbol{A}$ conditional on this information. Let $f(\boldsymbol{z} \mid \boldsymbol{y})$ be the joint density of $(V'(\zeta_1), \ldots, V'(\zeta_r))$, $r < n$, conditional on $(V(\zeta_1) = y_1, \ldots, V(\zeta_r) = y_r)$, where $\zeta_1 < \cdots < \zeta_r$. Denote by $\nu(\lambda \mid \boldsymbol{0}, \boldsymbol{z}, \boldsymbol{\zeta})$ the mean zero-crossing rate of V at

$\lambda > \zeta_r$ conditional on V having zero-crossings at ζ_s with slopes z_s, $s = 1, \ldots, r$, where $\boldsymbol{y} = (y_1, \ldots, y_r)$, $\boldsymbol{z} = (z_1, \ldots, z_r)$, and $\boldsymbol{\zeta} = (\zeta_1, \ldots, \zeta_r)$.

The joint density of $V'(\zeta_s)$ at the y_s-crossings of V, $s = 1, \ldots, r$, that is, the density of $(V'(\zeta_1), \ldots, V'(\zeta_r))$ conditional on $(V(\zeta_1) = y_1, \ldots, V(\zeta_r) = y_r)$, is

$$g(\boldsymbol{z} \mid \boldsymbol{y}, \boldsymbol{\zeta}) = \frac{\prod_{s=1}^r |z_s| \, f(\boldsymbol{z} \mid \boldsymbol{y})}{\int_{\mathbb{R}^r} \prod_{s=1}^r |z_s| \, f(\boldsymbol{z} \mid \boldsymbol{y}) \, d\boldsymbol{z}}. \tag{8.41}$$

The mean zero-crossing rate of $V(\lambda) = \det(\boldsymbol{A} - \lambda\, \boldsymbol{i})$ conditional on $\Lambda_1 = \zeta_1 < \cdots < \Lambda_r = \zeta_r$, $r < n$, is

$$\hat{\nu}(\lambda \mid \boldsymbol{\zeta}) = \int_{\mathbb{R}^r} g(\boldsymbol{z} \mid \boldsymbol{0}, \boldsymbol{\zeta}) \, \nu(\lambda \mid \boldsymbol{0}, \boldsymbol{z}, \boldsymbol{\zeta}) \, d\boldsymbol{z}, \quad \lambda > \zeta_r. \tag{8.42}$$

The corresponding mean zero-crossing rate for V_G is denoted by $\hat{\nu}_G(\lambda \mid \boldsymbol{\zeta})$.

Proof: We present a heuristic proof for Eqs. 8.41 and 8.42. Rigorous arguments can be constructed based on properties of Slepian processes ([121], Chapter 10).

Let $\mathcal{C}$ be a set defined by the y_s-crossings of V at ζ_s, $s = 1, \ldots, r$. The set can be approximated by $\cap_{s=1}^r (U_s \cup D_s)$ for small values of $\Delta\zeta_s$, $s = 1, \ldots, r$, where

$$U_s = \{V(\zeta_s) < y_s < V(\zeta_s) + V'(\zeta_s)\, \Delta\zeta_s\} \quad \text{and}$$
$$D_s = \{V(\zeta_s) > y_s > V(\zeta_s) + V'(\zeta_s)\, \Delta\zeta_s\}, \quad s = 1, \ldots, r.$$

Consider also the event $\mathcal{G} = \cap_{t=1}^q \{V(\lambda_t) \in (x_t, x_t + \Delta x_t]\}$, $\Delta x_t > 0$, for some arbitrary values of λ such that $\lambda_t > \zeta_r$. The probability of the simultaneous occurrence of $\mathcal{C}$ and $\mathcal{G}$ can be approximated by

$$P(\mathcal{C} \cap \mathcal{G}) \simeq \int_{\mathbb{R}^r} \left(\prod_{s=1}^r |z_s| \right) f(\boldsymbol{z}, \boldsymbol{y}, \boldsymbol{x}) \, d\boldsymbol{z} \prod_{t=1}^q \Delta x_t,$$

where Δx_t is small and $f(\boldsymbol{z}, \boldsymbol{y}, \boldsymbol{x})$ is the joint density of $(V'(\zeta_s), V(\zeta_s))$, $s = 1, \ldots, r$, and $V(\lambda_t)$, $t = 1, \ldots, q$, so that

$$\begin{aligned}
P(\mathcal{G} \mid \mathcal{C}) &= \prod_{t=1}^q \Delta x_t \, \frac{\int_{\mathbb{R}^r} \prod_{s=1}^r |z_s| \, f(\boldsymbol{z}, \boldsymbol{y}, \boldsymbol{x}) \, d\boldsymbol{z}}{\int_{\mathbb{R}^r} \prod_{s=1}^r |u_s| \, f(\boldsymbol{u}, \boldsymbol{y}) \, d\boldsymbol{u}} \\
&= \prod_{t=1}^q \Delta x_t \int_{\mathbb{R}^r} \frac{\prod_{s=1}^r |z_s| \, f(\boldsymbol{z}, \boldsymbol{y}) \, f(\boldsymbol{x} \mid \boldsymbol{y}, \boldsymbol{z})}{\int_{\mathbb{R}^r} \prod_{s=1}^r |u_s| \, f(\boldsymbol{u}, \boldsymbol{y}) \, d\boldsymbol{u}} \, d\boldsymbol{z} \\
&= \prod_{t=1}^q \Delta x_t \int_{\mathbb{R}^r} g(\boldsymbol{z} \mid \boldsymbol{y}, \boldsymbol{\zeta}) \, f(\boldsymbol{x} \mid \boldsymbol{y}, \boldsymbol{z}),
\end{aligned}$$

where $f(\boldsymbol{x} \mid \boldsymbol{y}, \boldsymbol{z})$ denotes the density of $V(\lambda_t)$, $t = 1, \ldots, q$, conditional on $(V(\zeta_s) = y_s, V'(\zeta_s) = z_s)$, $s = 1, \ldots, r$, and $f(\boldsymbol{x}, \boldsymbol{y})$ denotes the density of $(V(\zeta_s), V'(\zeta_s))$, $s = 1, \ldots, r$. The above result gives Eq. 8.41.

Let $\nu(\lambda \mid \boldsymbol{y}, \boldsymbol{z}, \boldsymbol{\zeta})$ be the mean zero-crossing rate of the conditional stochastic process $V(\lambda) \mid (V(\zeta_s) = y_s, V'(\zeta_s) = z_s, s = 1, \ldots, r)$ for $\lambda > \zeta_r$. The mean zero-crossing rate of $V(\lambda)$, $\lambda > \zeta_r$, conditional on only $V(\zeta_s) = y_s$, $s = 1, \ldots, r$, can be obtained by integrating $\nu(\lambda \mid \boldsymbol{y}, \boldsymbol{z}, \boldsymbol{\zeta})$ weighted by $g(\boldsymbol{z} \mid \boldsymbol{y}, \boldsymbol{\zeta})$ over all possible values of $\boldsymbol{z}$. This mean crossing rate with $\boldsymbol{y} = \boldsymbol{0}$ yields $\hat{\nu}(\lambda \mid \boldsymbol{\zeta})$ in Eq. 8.42.

The determination of the functions $g(\boldsymbol{z} \mid \boldsymbol{y}, \boldsymbol{\zeta})$ and $\nu(\lambda \mid \boldsymbol{0}, \boldsymbol{z})$ is difficult for an arbitrary vector $\boldsymbol{C}$. If $\boldsymbol{C}$ is Gaussian, the conditional zero-crossing rate $\nu(\lambda \mid \boldsymbol{0}, \boldsymbol{z})$, denoted by $\nu_G(\lambda \mid \boldsymbol{0}, \boldsymbol{z})$, can be obtained from Eq. 8.40 with the appropriate parameters and $g(\boldsymbol{z} \mid \boldsymbol{y}, \boldsymbol{\zeta})$ becomes

$$g_G(\boldsymbol{z} \mid \boldsymbol{y}, \boldsymbol{\zeta}) = \frac{\prod_{s=1}^{r} |z_s| \exp\left[-(1/2)\,(\boldsymbol{z} - \boldsymbol{m})^T\, \boldsymbol{\gamma}^{-1}\, (\boldsymbol{z} - \boldsymbol{m})\right]}{\int_{\mathbb{R}^r} \prod_{s=1}^{r} |u_s| \exp\left[-(1/2)\,(\boldsymbol{u} - \boldsymbol{m})^T\, \boldsymbol{\gamma}^{-1}\, (\boldsymbol{u} - \boldsymbol{m})\right] d\boldsymbol{u}},$$

where $\boldsymbol{m}$ and $\boldsymbol{\gamma}$ denote the mean and covariance matrices of the conditional random vector $(V'(\zeta_1), \ldots, V'(\zeta_r)) \mid (V(\zeta_1) = y_1, \ldots, V(\zeta_r) = y_r)$ [75]. ■

The mean zero-crossing rates $\nu(\lambda)$ and $\hat{\nu}(\lambda \mid \boldsymbol{\zeta})$ provide useful information on the eigenvalues Λ_i of $\boldsymbol{A}$. For example, (1) the average number of eigenvalues of $\boldsymbol{A}$ in an interval I of the real line is $\int_I \nu(\lambda)\, d\lambda$ or $\int_I \hat{\nu}(\lambda \mid \boldsymbol{\zeta})\, d\lambda$, (2) the probability that the number of eigenvalues $N(I)$ in I exceeds a specified value $q \geq 0$ is bounded by

$$P(N(I) \geq q) \leq \frac{E[N(I)]}{q} = \frac{\int_I \nu(\lambda)\, d\lambda}{q}, \quad q > 0,$$

and (3) the probability that $\boldsymbol{A}$ has at least one eigenvalue in I can be approximated by $1 - \exp\left(-\int_I \nu(\lambda)\, d\lambda\right)$ or $1 - \exp\left(-\int_I \hat{\nu}(\lambda \mid \boldsymbol{\zeta})\, d\lambda\right)$.

Note: The definitions of ν and $\hat{\nu}$ give the first property. The second property follows from the Chebychev inequality. The last property is based on the heuristic assumption that the zero-crossings of the characteristic polynomial define an inhomogeneous Poisson process with intensity $\nu(\lambda)$ or $\hat{\nu}(\lambda \mid \boldsymbol{\zeta})$. ▲

Example 8.16: Consider a random matrix

$$\boldsymbol{A} = \begin{bmatrix} \frac{g}{l} + \frac{a^2}{m\, l^2}\, K & -\frac{a^2}{m\, l^2}\, K \\ -\frac{a^2}{m\, l^2}\, K & \frac{g}{l} + \frac{a^2}{m\, l^2}\, K \end{bmatrix},$$

where g denotes the constant of gravity, $0 < a < l$ and $m > 0$ are some constants, and $K > 0$ is a random variable. The eigenvalues, $\Lambda_1 = g/l$ and $\Lambda_2 = g/l + 2\,a^2\,K/(m\,l^2)$, of $\boldsymbol{A}$ are the square of the natural frequencies of a mechanical system consisting of two pendulums with length l and mass m coupled by a spring of random stiffness K. The mean zero-crossing rate of the process V can be calculated exactly and is [75]

$$\nu(\lambda) = \delta\left(\lambda - \frac{g}{l}\right) + \frac{m\, l^2}{2\, a^2}\, f_K\left(\frac{m\, l^2}{2\, a^2}\, (\lambda - \frac{g}{l})\right).$$

The process V has a zero-crossing at $\lambda = g/l$, indicating a deterministic eigenvalue. The average number of zero-crossings of V is $\int_0^\infty \nu(\lambda)\, d\lambda = 2$. $\diamond$

Note: The probability density functions of the eigenvalues of A,

$$f_{\Lambda_1}(\lambda) = \delta(\lambda - g/l) \quad \text{and}$$
$$f_{\Lambda_2}(\lambda) = \frac{m\, l^2}{2\, a^2}\, f_K\left(\frac{m\, l^2}{2\, a^2}\,(\lambda - g/l)\right),$$

can be obtained by direct calculations and show that the first eigenvalue of A is deterministic, in agreement with the crossing solution. ■

Example 8.17: Consider a random matrix

$$A = \begin{bmatrix} (K_1 + K_2)/m & (K_2\, l_2 - K_1\, l_1)/m \\ (K_2\, l_2 - K_1\, l_1)/m & (K_1\, l_1^2 + K_2\, l_2^2)/\xi \end{bmatrix},$$

where $m, l_1, l_2, \xi > 0$ are some constants and $K_1, K_2 > 0$ are random variables. This matrix defines the eigenvalue problem for a rigid bar with mass m, length l, and moment of inertia ξ relative to its mass center, which is supported at its ends by two springs with stiffness K_1 and K_2. The mass center is at distances l_1 and l_2 from the two springs so that $l = l_1 + l_2$. Numerical results are given for $m\, g =$ 3220 lb, $l_1 = 4.5$ ft, $l_2 = 5.5$ ft, $\xi = m\, r^2$, $r = 4$ ft, and lognormal variables K_1 and K_2 with $E[K_1] = 2400$ lb/ft, $E[K_2] = 2600$ lb/ft, coefficients of variation $v_1 = v_2 = 0.3$, and correlation coefficient ρ [75]. The lognormal vector (K_1, K_2) represents the image of an $\mathbb{R}^2$-valued Gaussian variable by a nonlinear mapping ([79], pp. 48-49). The matrix A has two eigenvalues and we have $\int_0^\infty \nu(\lambda)\, d\lambda =$ 2.

Figure 8.8 shows the mean zero-crossing rates ν and ν_G of V and V_G, respectively, for $\rho = 0$, 0.5, and 0.9 in the left three plots. The mean zero-crossing rates ν have been obtained by numerical integration. The plots also show the conditional mean zero-crossing rates $\hat{\nu}_G(\lambda \mid \Lambda_1 = 46)$. The mean zero-crossing rates $\hat{\nu}_G(\lambda \mid \Lambda_1 = 46)$ are also shown in the three plots on the right side of Fig. 8.8 together with Monte Carlo (MC) simulation results. The Gaussian assumption is satisfactory for the matrix A in this example. The additional information $\Lambda_1 = 46$ reduces the uncertainty in Λ_2. The reduction in uncertainty increases with the correlation coefficient ρ between K_1 and K_2. $\diamond$

Note: The joint density of $(V(\lambda), V'(\lambda))$ is

$$f_{V(\lambda),V'(\lambda)}(x, z) = \frac{f_{K_1,K_2}(k_1, k_2)}{|(a\,\lambda + c\, k_2)\, b - (b\,\lambda + c\, k_1)|\, a},$$

where $a = -(1/m + l_1^2/\xi)$, $b = -(1/m + l_2^2/\xi)$, and $c = (l_1^2 + l_2^2)/(m\,\xi)$. The arguments k_s, $s = 1, 2$, in the above expression are related to x and z by the mappings $V(\lambda) = x$ and $V'(\lambda) = z$. ▲

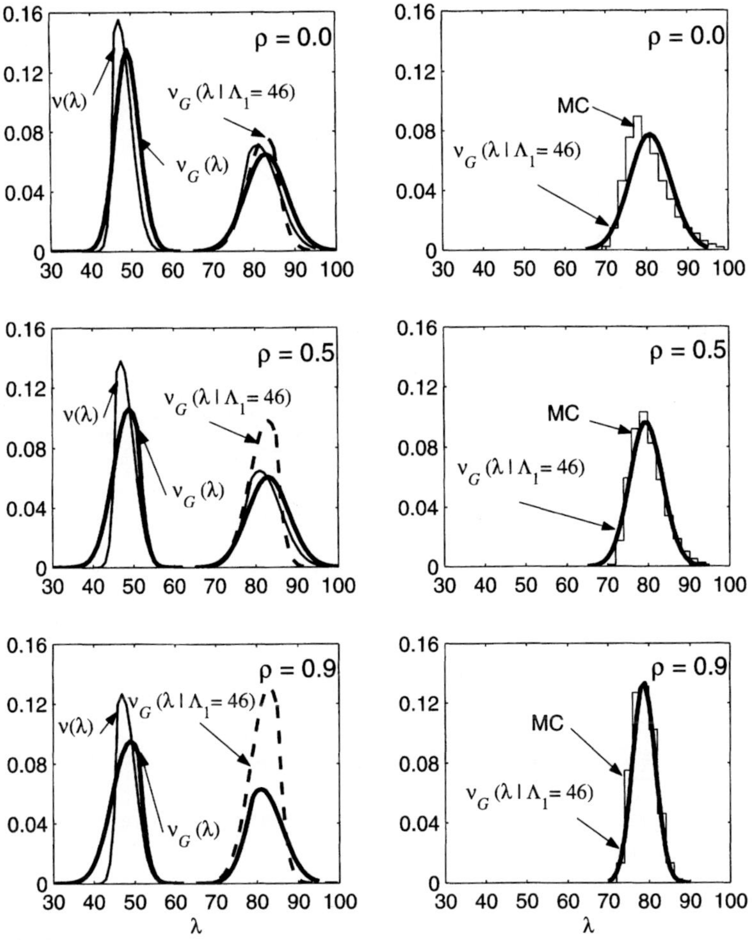

Figure 8.8. Exact and approximate zero-crossing rates

Example 8.18: Consider a system with n identical masses $m > 0$ connected in series by n springs of random stiffnesses $K_1, \ldots, K_n$. The first spring of stiffness K_1 is attached to a fixed support. Figure 8.9 shows the approximate mean zero-crossing rate of V assumed to be a Gaussian process (Eq. 8.40) and a histogram of the eigenvalues of the system obtained by Monte Carlo simulation for $n = 4$, independent Gaussian variables K_i with means $E[K_1] = 30,000$, $E[K_2] = 20,000$, and $E[K_3] = E[K_4] = 10,000$ lb/in and coefficients of variation $v_i = v = 0.15$, and $m = 100$ lb sec^2/in [75]. Results show that the approximate solution given by $\nu_G(\lambda)$ is consistent with simulation results. ◇

Note: The square frequencies of the system coincide with the eigenvalues of the (n, n)

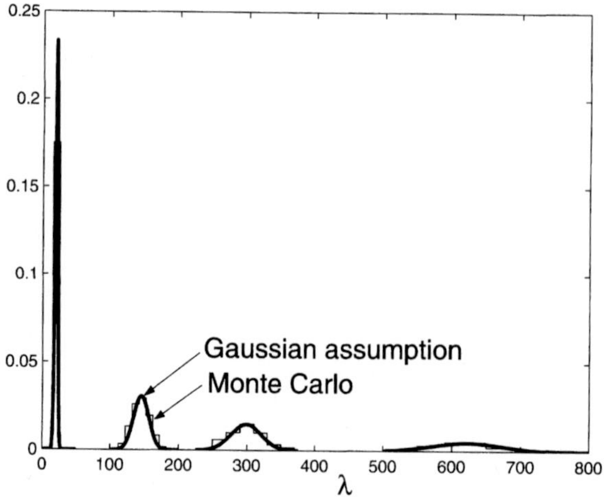

Figure 8.9. Approximate zero-crossing rates based on the Gaussian assumption and Monte Carlo simulation

matrix $\boldsymbol{A}$. The non-zero entries of $\boldsymbol{A}$ are $A_{i,i} = (K_i + K_{i+1})/m$, $i = 1, \ldots, n-1$, $A_{n,n} = K_n/m$, and $A_{i,i+1} = A_{i+1,i} = -K_{i+1}/m$, $i = 1, \ldots, n-1$. ▲

8.4 Differential and integral equations

Let D be an open bounded subset of $\mathbb{R}^d$ and $t \geq 0$ be a time argument. Consider the differential equation

$$(\mathcal{A} - \lambda \mathcal{I})\,[\boldsymbol{U}(\boldsymbol{x}, t)] = \boldsymbol{q}(\boldsymbol{x}, t), \quad \boldsymbol{x} \in D, \; t > 0, \tag{8.43}$$

where $\mathcal{I}$ denotes the identity operator, λ is a parameter, $\boldsymbol{q}$ is an $\mathbb{R}^n$-valued deterministic function, and $\mathcal{A}$ is a differential operator with random coefficients defined on a probability space $(\Omega, \mathcal{F}, P)$. Initial and boundary conditions need to be specified to find the solution $\boldsymbol{U}$ of Eq. 8.43. These conditions can be deterministic or random. Generally, the solution of Eq. 8.43 is an $\mathbb{R}^n$-valued random function defined on $D^* = D \times [0, \infty)$. An alternative form of Eq. 8.43 is the integral equation

$$\boldsymbol{U}(\boldsymbol{x}^*) = \boldsymbol{F}(\boldsymbol{x}^*) + \lambda \int_{D^*} \boldsymbol{K}(\boldsymbol{x}^*, \boldsymbol{\xi}^*)\, \boldsymbol{U}(\boldsymbol{\xi}^*)\, d\boldsymbol{\xi}^*, \tag{8.44}$$

where $\boldsymbol{x}^* = (\boldsymbol{x}, t)$, $\boldsymbol{\xi}^* = (\boldsymbol{\xi}, s)$, $\boldsymbol{x}, \boldsymbol{\xi} \in D \subset \mathbb{R}^d$, $s \in (0, t)$, $t > 0$, $\boldsymbol{F}$ is a random function depending on the initial and boundary conditions, and $\boldsymbol{K}$ denotes a random kernel.

Note: The operator $\mathcal{A}$ in Eq. 8.43 and the generator of diffusion processes, used extensively in Chapter 6 and other parts of the book, share the same notation but are different mathematical objects and should not be confused. The symbol $\mathcal{A}$ in Eq. 8.43 has been used to parallel the notation $\boldsymbol{A}$ in Eq. 8.1. ▲

Example 8.19: Let U be the solution of $U''(x) + \lambda\, Y_1\, U(x) = q(x)$, $x \in D = (0, 1)$, with the boundary conditions $U(0) = Y_2$ and $U'(0) = 0$, where Y_1, Y_2 are real-valued random variables and q is an integrable function. The operator $\mathcal{A}$ in Eq. 8.43 for this differential equation is $\mathcal{A} = d^2/dx^2 + \lambda\,(Y_1 + 1)$. The integral form of the differential equation for U is ([99], p. 226)

$$U(x) = \left[Y_2 - \int_0^1 (x - \xi)\, q(\xi)\, 1_{[0,x]}(\xi)\, d\xi \right] + \lambda\, Y_1 \int_0^1 (\xi - x)\, U(\xi)\, 1_{[0,x]}(\xi)\, d\xi,$$

where the term in square bracket is a random function $F(x)$ and the kernel of the second integral $K(x, \xi) = 1_{[0,x]}(\xi)\,(\xi - x)\, Y_1$ is random (Eq. 8.44). ◇

The problem defined by Eq. 8.43 with $\boldsymbol{q} \neq \boldsymbol{0}$ or Eq. 8.44 with $\boldsymbol{F} \neq \boldsymbol{0}$ is said to be inhomogeneous. If $\boldsymbol{q} = \boldsymbol{0}$ and $\boldsymbol{F} = \boldsymbol{0}$ in these equations, we deal with a homogeneous problem. Our objective is to find probabilistic properties of the solution of the inhomogeneous and homogeneous versions of the above equations.

The local solutions in Sections 6.2, 6.3, 6.4, and 6.5 can be extended to solve locally some forms of Eqs. 8.43 and 8.44 by following the approach in Section 8.2. These extensions are not considered here.

8.4.1 Inhomogeneous equations

Let $\mathcal{D} = \mathcal{A} - \lambda\, \mathcal{I}$ for a fixed λ such that Eq. 8.43, that is,

$$\mathcal{D}\,[\boldsymbol{U}(\boldsymbol{x}, t)] = \boldsymbol{q}(\boldsymbol{x}, t), \quad \boldsymbol{x} \in D \subset \mathbb{R}^d, \quad t \geq 0, \tag{8.45}$$

has a.s. a unique solution.

Note: The solution $\boldsymbol{U}$ of this equation is a random field on $(\Omega, \mathcal{F}, P)$ because it is a measurable function of the random coefficients of $\mathcal{D}$ and of the random initial/boundary conditions for Eq. 8.45, which are defined on this probability space. ▲

It is assumed that (1) all random parameters in Eq. 8.45 have finite second moments, (2) $\mathcal{D}$ is a linear operator, that is, $\mathcal{D}[\alpha_1\, \boldsymbol{U}_1 + \alpha_2\, \boldsymbol{U}_2] = \alpha_1\, \mathcal{D}[\boldsymbol{U}_1] + \alpha_2\, \mathcal{D}[\boldsymbol{U}_2]$, where α_i, $i = 1, 2$, are some constants, and (3) the deterministic part $\mathcal{L} = E[\mathcal{D}]$ of $\mathcal{D}$ has an inverse $\mathcal{L}^{-1}$ that is a linear operator. The random part $\mathcal{R} = \mathcal{D} - \mathcal{L}$ of $\mathcal{D}$ has zero expectation by definition. For example, the random operator in Example 8.19 is $\mathcal{D} = d^2/dx^2 + \lambda\, Y_1$ so that $\mathcal{L} = d^2/dx^2 + \lambda\, E[Y_1]$ and $\mathcal{R} = \lambda\,(Y_1 - E[Y_1])$.

Our objective is to find probabilistic properties of the solution $\boldsymbol{U}$ of Eq. 8.45. We have already mentioned that some of the local solutions in Chapter 6 can be extended to solve a class of differential equations with random coefficients. We also can approximate Eq. 8.45 by an algebraic equation with random coefficients and use any of the methods in Section 8.3.1 to solve the resulting equation. The finite difference, finite element, or other methods can be used to discretize Eq. 8.45.

8.4.1.1 Monte Carlo simulation method

The approach outlined in Sec. 8.3.1.1 for solving Eq. 8.2 can be applied to find probabilistic properties of the solution $\boldsymbol{U}$ of Eq. 8.45. Our comments on features and limitations of Monte Carlo simulation in Section 8.3.1.1 remain valid. The application of the Monte Carlo simulation involves the following two steps.

1. Generate n_s independent samples $\mathcal{D}_k = \mathcal{L} + \mathcal{R}_k$, $k = 1, \ldots, n_s$, of the stochastic operator $\mathcal{D}$ and calculate

$$\mathcal{D}_k[\boldsymbol{u}_k(\boldsymbol{x}, t)] = \boldsymbol{q}(\boldsymbol{x}, t), \quad k = 1, \ldots, n_s. \tag{8.46}$$

2. Estimate statistics of $\boldsymbol{U}$, for example, the mean and covariance functions, $\boldsymbol{\mu}_u$ and $\boldsymbol{c}_u$, of $\boldsymbol{U}$, can be approximated by

$$\begin{aligned}
\hat{\boldsymbol{\mu}}_u(\boldsymbol{x}, t) &= \frac{1}{n_s} \sum_{k=1}^{n_s} \boldsymbol{u}_k(\boldsymbol{x}, t), \\
\hat{\boldsymbol{c}}_u(\boldsymbol{x}_1, \boldsymbol{x}_2, t_1, t_2) &= \frac{1}{n_s} \sum_{k=1}^{n_s} (\boldsymbol{u}_k(\boldsymbol{x}_1, t_1) - \hat{\boldsymbol{\mu}}_u(\boldsymbol{x}_1, t_1)) \\
&\quad \times (\boldsymbol{u}_k(\boldsymbol{x}_2, t_2) - \hat{\boldsymbol{\mu}}_u(\boldsymbol{x}_2, t_2))^T.
\end{aligned} \tag{8.47}$$

Note: The algorithms in Chapter 5 can be used to generate samples of $\mathcal{D}$. The Monte Carlo simulation method is not restricted to linear operators. It can be applied to nonlinear equations, in which case the determination of the samples $\boldsymbol{u}_k$ of $\boldsymbol{U}$ requires solving a nonlinear differential equation rather than Eq. 8.46. ▲

8.4.1.2 Taylor series method

Suppose that $\mathcal{D}$ in Eq. 8.45 and the initial/boundary conditions of this equation depend on $m < \infty$ random parameters collected in an $\mathbb{R}^m$-valued random variable $\boldsymbol{Z} \in L_2$ defined on a probability space $(\Omega, \mathcal{F}, P)$ with mean $\boldsymbol{\mu}_z = \{\mu_{z,u}\}$ and covariance $\boldsymbol{\gamma}_z = \{\gamma_{z,uv}\}$. If a coefficient of $\mathcal{D}$ is, for example, a random field, it can be approximated by a parametric model consisting of a finite sum of deterministic functions weighted by random coefficients. We have used this approximation in Section 5.3 to generate samples of an arbitrary random field.

The solution $\boldsymbol{U}$ of Eq. 8.45 can be viewed as a function not only of $(\boldsymbol{x}, t)$ but also of $\boldsymbol{Z}$, that is, $\boldsymbol{U} = \boldsymbol{U}(\boldsymbol{x}, t; \boldsymbol{Z})$. It is assumed that the mapping $\boldsymbol{Z} \mapsto \boldsymbol{U}$ has continuous second order partial derivatives for each $(\boldsymbol{x}, t)$. Since the mapping $\boldsymbol{Z} \mapsto \boldsymbol{U}$ is measurable, $\boldsymbol{U}$ is a random function on the probability space $(\Omega, \mathcal{F}, P)$.

1. Approximation of $\boldsymbol{U}$:

$$\boldsymbol{U}(\boldsymbol{x}, t; \boldsymbol{Z}) \simeq \boldsymbol{U}(\boldsymbol{x}, t; \boldsymbol{\mu}_z) + \sum_{u=1}^{m} \frac{\partial \boldsymbol{U}(\boldsymbol{x}, t; \boldsymbol{\mu}_z)}{\partial z_u}(Z_u - \mu_{z,u}). \tag{8.48}$$

2. Approximate second moment properties of $\boldsymbol{U}$:

$$\boldsymbol{\mu}_u(\boldsymbol{x}, t) \simeq \boldsymbol{U}(\boldsymbol{x}, t; \boldsymbol{\mu}_z),$$

$$\boldsymbol{c}_u(\boldsymbol{x}_1, \boldsymbol{x}_2, t_1, t_2) \simeq \sum_{u,v=1}^{m} \frac{\partial \boldsymbol{U}(\boldsymbol{x}_1, t_1; \boldsymbol{\mu}_z)}{\partial z_u} \frac{\partial \boldsymbol{U}(\boldsymbol{x}_2, t_2; \boldsymbol{\mu}_z)}{\partial z_v}^T \gamma_{z,uv}. \tag{8.49}$$

Note: The approximate solution for $\boldsymbol{U}$ in Eq. 8.48 is referred to as the **linear** or **first order Taylor** solution. Considerations similar to the discrete case (Eqs. 8.5 and 8.6) can be used to derive Eqs. 8.48 and 8.49. ▲

The functions $\boldsymbol{U}(\boldsymbol{x}, t; \boldsymbol{\mu}_z)$ and $\partial \boldsymbol{U}(\boldsymbol{x}, t; \boldsymbol{\mu}_z)/\partial z_u$ needed to apply Eqs. 8.48 and 8.49 can be calculated from the following two equations. Let $\tilde{\mathcal{L}}$ be the operator $\mathcal{D}$ with $\boldsymbol{\mu}_z$ in place of $\boldsymbol{Z}$.

$$\tilde{\mathcal{L}}[\boldsymbol{U}(\boldsymbol{x}, t; \boldsymbol{\mu}_z)] = \boldsymbol{q}(\boldsymbol{x}, t), \tag{8.50}$$

$$\tilde{\mathcal{L}}\left[\frac{\partial \boldsymbol{U}(\boldsymbol{x}, t; \boldsymbol{\mu}_z)}{\partial z_u}\right] = -\frac{\partial \mathcal{D}(\boldsymbol{x}, t; \boldsymbol{\mu}_z)}{\partial z_u} \boldsymbol{U}(\boldsymbol{x}, t; \boldsymbol{\mu}_z). \tag{8.51}$$

Proof: Generally, $\tilde{\mathcal{L}}$ differs from $\mathcal{L} = E[\mathcal{D}]$. The function $\boldsymbol{U}(\boldsymbol{x}, t; \boldsymbol{\mu}_z)$ in Eq. 8.50 results from Eq. 8.45 with its initial and boundary conditions in which $\boldsymbol{Z}$ is set equal to $\boldsymbol{\mu}_z$.

The partial derivative $\frac{\partial \mathcal{D}}{\partial z_u}[\boldsymbol{U}] + \mathcal{D}\left[\frac{\partial \boldsymbol{U}}{\partial z_u}\right] = \boldsymbol{0}$ of Eq. 8.45 relative to a coordinate Z_u of $\boldsymbol{Z}$ gives Eq. 8.51 by taking $\boldsymbol{Z}$ equal to $\boldsymbol{\mu}_z$ in this equation.

The functions $\boldsymbol{U}$ and $\partial \boldsymbol{U}/\partial z_u$ satisfy deterministic differential equations with the same differential operator, the operator $\tilde{\mathcal{L}}$. The derivatives $\partial \boldsymbol{U}/\partial z_u$ give the rate of change of $\boldsymbol{U}$ with respect to the coordinates of $\boldsymbol{Z}$ and are called **sensitivity factors**. If a sensitivity factor $\partial \boldsymbol{U}/\partial z_k$ is small and the variance of Z_k is not excessive, the effect of the uncertainty in Z_k on the solution is limited so that Z_k may be taken equal to its mean value as a first approximation. ■

The approximate second moment properties of $\boldsymbol{U}$ in Eq. 8.49 can be (a) improved by retaining additional terms of the Taylor series approximating the mapping $\boldsymbol{Z} \mapsto \boldsymbol{U}$ and (b) generalized to solve nonlinear equations. For example, let $\boldsymbol{U}$ be the solution of

$$\mathcal{D}[\boldsymbol{U}(\boldsymbol{x}, t)] + \mathcal{N}[\boldsymbol{U}(\boldsymbol{x}, t)] = \boldsymbol{q}(\boldsymbol{x}, t), \tag{8.52}$$

where $\mathcal{D}$ is given by Eq. 8.45 and $\mathcal{N}$ is an n-dimensional column vector whose entries are nonlinear functions of $\boldsymbol{Z}$ and $\boldsymbol{U}$.

Note: Suppose that the approximation in Eq. 8.48 is augmented with second order terms. In addition to Eqs. 8.50 and 8.51, we need a new equation giving the mixed derivatives, $\partial^2 \boldsymbol{U}(\boldsymbol{x}, t; \boldsymbol{\mu}_z)/\partial z_u \partial z_v$, $u, v = 1, \ldots, m$. This equation can be obtained by differentiating Eq. 8.45 with respect to Z_u and Z_v, $u, v = 1, \ldots, m$, and setting $\boldsymbol{Z}$ equal to its mean value.

The first order approximation of Eq. 8.52 cannot use Eqs. 8.50 and 8.51 if $\boldsymbol{U}$ is defined by Eq. 8.52 rather then Eq. 8.45. The function $\boldsymbol{U}(\boldsymbol{x}, t; \boldsymbol{\mu}_z)$ is the solution of Eq. 8.52 with $\boldsymbol{Z}$ replaced by its mean value. The sensitivity factors for $\boldsymbol{U}$ can be obtained from Eq. 8.52 by differentiation. Since

$$\frac{\partial \mathcal{D}}{\partial z_u}[\boldsymbol{U}] + \mathcal{D}\left[\frac{\partial \boldsymbol{U}}{\partial z_u}\right] + \frac{\partial \mathcal{N}}{\partial z_u}[\boldsymbol{U}] + \sum_{k=1}^{n} \frac{\partial \mathcal{N}}{\partial u_k}[\boldsymbol{U}]\,\frac{\partial U_k}{\partial z_u} = \boldsymbol{0}$$

and the functions $\boldsymbol{Z} \mapsto \mathcal{D}$ and $(\boldsymbol{Z}, \boldsymbol{U}) \mapsto \mathcal{N}$ are known, the derivatives $\partial \mathcal{D}/\partial z_u$, $\partial \mathcal{N}/\partial z_u$, and $\partial \boldsymbol{U}/\partial u_k$ can be calculated, and therefore the gradients $\partial \boldsymbol{U}(\boldsymbol{x}, t; \boldsymbol{\mu}_z)/\partial z_u$ can be obtained from the above equation. ▲

Example 8.20: Let $U(x; \mathbf{Z}) \in \mathbb{R}$, $x \geq 0$, be the solution of the differential equation $U'(x; \mathbf{Z}) + (x\, Z_1 + x^2\, Z_2)\, U(x; \mathbf{Z}) = 0$, where $U(0; \mathbf{Z}) = h(\mathbf{Z})$ and $\mathbf{Z} = (Z_1, Z_2) \in \mathbb{R}^2$ is the vector of uncertain parameters. Hence, the operator in Eq. 8.45 is $\mathcal{D} = d/dx + x\, Z_1 + x^2\, Z_2$.

The first order solution of U is given by Eq. 8.48, where $U(x; \boldsymbol{\mu}_z)$ and $\partial U(x; \boldsymbol{\mu}_z)/\partial z_u$ can be calculated from

$$\tilde{\mathcal{L}}\,[U(x; \boldsymbol{\mu}_z)] = 0 \quad \text{with } U(0; \boldsymbol{\mu}_z) = h(\boldsymbol{\mu}_z) \text{ and}$$
$$\tilde{\mathcal{L}}\,\left[\partial U(x; \boldsymbol{\mu}_z)/\partial z_u\right] = -x^u\, U(x; \boldsymbol{\mu}_z) \quad \text{with } \partial U(0; \boldsymbol{\mu}_z)/\partial z_u = \partial h(\boldsymbol{\mu}_z)/\partial z_u.$$

In this case $\tilde{\mathcal{L}}$ coincides with $\mathcal{L} = E[\mathcal{D}]$ since the coefficients of $\mathcal{D}$ are linear in the coordinates of $\mathbf{Z}$. ◇

Note: The differential equations for $U(x; \boldsymbol{\mu}_z)$ and $\partial U(x, ; \boldsymbol{\mu}_z)/\partial z_u$, $u = 1, 2$, result from Eqs. 8.50 and 8.51. Classical methods for solving deterministic differential equations can be used to find $U(x; \boldsymbol{\mu}_z)$ and $\partial U(x, ; \boldsymbol{\mu}_z)/\partial z_u$. ▲

Example 8.21: Let $U(t)$, $t \geq 0$, be the solution of the nonlinear differential equation $\left(d/dt - \alpha - Z\, U(t)^2\right) U(t) = q(t)$, where $t \geq 0$, α is a constant, Z denotes a random variable, and $U(0) = 0$. The function $U(t; \mu_z)$ and the sensitivity factor $V(t; \mu_z) = \partial U(t; \mu_z)/\partial z$ satisfy the differential equations

$$\dot{U}(t; \mu_z) = \alpha\, U(t; \mu_z) + \mu_z\, U(t; \mu_z)^3 + q(t),$$
$$\dot{V}(t; \mu_z) = \alpha\, V(t; \mu_z) + U(t; \mu_z)^3 + 3\,\mu_z\, U(t; \mu_z)^2\, V(t; \mu_z),$$

with zero initial conditions.

Figure 8.10 shows the evolution of $U(t; \mu_z)$ and $V(t; \mu_z)$ for $\alpha = 1, q(t) =$

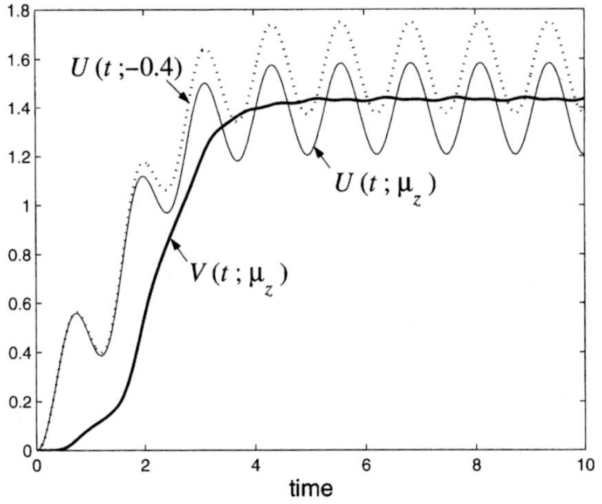

Figure 8.10. Solution and sensitivity factor for $Z = \mu_z$

$\sin(\nu t)$, $\nu = 5$, and $\mu_z = -0.5$. The sensitivity factor increases rapidly in time and reaches a plateau at about $t = 4$, showing that the dependence of the solution on the value of Z increases in time for $t \le 4$ and is nearly constant for $t > 4$. This observation is consistent with the graph of $U(t; -0.4)$, showing a notable difference from $U(t; \mu_z)$ for times $t \ge 4$. ◇

Note: The differential equation defining U is nonlinear. The linear and nonlinear parts of the differential operator are $\mathcal{D}[U] = (d/dt - \alpha)\, U$ and $\mathcal{N}[U] = -Z\, U^3$, respectively (Eq. 8.52).

The equation

$$\dot{V}(t; Z) = \alpha\, V(t; Z) + U(t; Z)^3 + 3\, Z\, U(t; Z)^2\, V(t; Z),$$

for $V(t; Z) = \partial U(t; Z)/\partial z$, results from the equation for U by differentiation with respect to Z, where $\dot{V}(t; Z) = \partial V(t; Z)/\partial t$. The sensitivity factor $V(t; \mu_z)$ is the solution of the above equation with $V(0; Z) = 0$ and Z set equal to its mean value. ▲

8.4.1.3 Perturbation method

Suppose that the random operator $\mathcal{D}$ in Eq. 8.45 has the representation

$$\mathcal{D} = \mathcal{L} + \varepsilon\, \tilde{\mathcal{R}}, \tag{8.53}$$

where $\mathcal{L} = E[\mathcal{D}]$, ε is a small parameter, $E[\tilde{\mathcal{R}}] = 0$, and $\boldsymbol{q}$ is of order 1. The perturbation series solution for Eq. 8.45 with $\mathcal{D}$ in Eq. 8.53 and its second moment properties are given by the following equations.

1. Approximation of $\boldsymbol{U}$:

$$\begin{aligned}
&\boldsymbol{U}(\boldsymbol{x},t) \simeq \boldsymbol{U}_0(\boldsymbol{x},t) + \varepsilon\,\boldsymbol{U}_1(\boldsymbol{x},t) + \varepsilon^2\,\boldsymbol{U}_2(\boldsymbol{x},t) + O(\varepsilon^3), \quad \text{where} \\
&\qquad \mathcal{L}[\boldsymbol{U}_r(\boldsymbol{x},t)] = -\tilde{\mathcal{R}}\,[\boldsymbol{U}_{r-1}(\boldsymbol{x},t)], \quad r = 1, 2, \ldots, \quad \text{and} \\
&\qquad \mathcal{L}[\boldsymbol{U}_0(\boldsymbol{x},t)] = \boldsymbol{q}(\boldsymbol{x},t).
\end{aligned} \tag{8.54}$$

2. Approximate second moment properties of $\boldsymbol{U}$:

$$\begin{aligned}
&E[\boldsymbol{U}(\boldsymbol{x},t)] \simeq \boldsymbol{U}_0(\boldsymbol{x},t) + \varepsilon^2\,E[\boldsymbol{U}_2(\boldsymbol{x},t)] + O(\varepsilon^3), \\
&E[\boldsymbol{U}(\boldsymbol{x}_1,t_1)\,\boldsymbol{U}(\boldsymbol{x}_2,t_2)^T] \simeq \boldsymbol{U}_0(\boldsymbol{x}_1,t_1)\,\boldsymbol{U}_0(\boldsymbol{x}_2,t_2)^T \\
&\quad + \varepsilon^2\,E[\boldsymbol{U}_0(\boldsymbol{x}_1,t_1)\,\boldsymbol{U}_2(\boldsymbol{x}_2,t_2)^T + \boldsymbol{U}_1(\boldsymbol{x}_1,t_1)\,\boldsymbol{U}_1(\boldsymbol{x}_2,t_2)^T \\
&\quad + \boldsymbol{U}_2(\boldsymbol{x}_1,t_1)\,\boldsymbol{U}_0(\boldsymbol{x}_2,t_2)^T] + O(\varepsilon^3).
\end{aligned} \tag{8.55}$$

Proof: The approximation for $\boldsymbol{U}$ based on the first q terms of the power series in Eq. 8.54 is called approximation of **order** q. The **first order** approximation $\boldsymbol{U} \simeq \boldsymbol{U}_0 + \varepsilon\,\boldsymbol{U}_1$ corresponds to $q = 1$. The initial and boundary conditions for the differential equation defining $\boldsymbol{U}_k$, $k = 0, 1, \ldots$, consist of the terms of order ε^k of the corresponding conditions for Eq. 8.45.

The recurrence formulas in Eq. 8.54 giving $\boldsymbol{U}_k$ can be obtained from Eq. 8.45 with $\mathcal{D}$ in Eq. 8.53 and the solution $\boldsymbol{U}$ is represented by the power series in Eq. 8.54. These equations give

$$(\mathcal{L} + \varepsilon\,\tilde{\mathcal{R}})[\boldsymbol{U}_0 + \varepsilon\,\boldsymbol{U}_1 + \varepsilon^2\,\boldsymbol{U}_2 + O(\varepsilon^3)] = \boldsymbol{q}$$

or, by grouping terms of the same order of magnitude,

$$(\mathcal{L}\,[\boldsymbol{U}_0] - \boldsymbol{q}) + \varepsilon\,(\mathcal{L}\,[\boldsymbol{U}_1] + \tilde{\mathcal{R}}\,[\boldsymbol{U}_0]) + \varepsilon^2\,(\mathcal{L}\,[\boldsymbol{U}_2] + \tilde{\mathcal{R}}\,[\boldsymbol{U}_1]) + O(\varepsilon^3) = \boldsymbol{0}.$$

The coefficients of each power of ε must be zero by the fundamental theorem of perturbation theory ([171], p. 12) so that we have

$$\begin{aligned}
&\mathcal{L}\,[\boldsymbol{U}_0] - \boldsymbol{q} = \boldsymbol{0} && \text{(order 1)}, \\
&\mathcal{L}\,[\boldsymbol{U}_1] + \tilde{\mathcal{R}}\,[\boldsymbol{U}_0] = \boldsymbol{0} && \text{(order } \varepsilon), \\
&\mathcal{L}\,[\boldsymbol{U}_2] + \tilde{\mathcal{R}}\,[\boldsymbol{U}_1] = \boldsymbol{0} && \text{(order } \varepsilon^2),
\end{aligned}$$

for the first three terms of the perturbation series. These conditions give Eq. 8.54. The expressions of the first two moments of $\boldsymbol{U}$ depend on the order of the approximation considered for the solution. For example, the approximations of the mean of $\boldsymbol{U}$ to the orders ε and ε^2 are, respectively, $E[\boldsymbol{U}] = \boldsymbol{U}_0 + O(\varepsilon^2)$ and $E[\boldsymbol{U}] = \boldsymbol{U}_0 + \varepsilon^2\,E[\boldsymbol{U}_2] + O(\varepsilon^3)$. Errors may result if the order of the approximating solution is insufficient to deliver all the terms needed for calculating moments for a specified accuracy. For example, the first order approximation for $\boldsymbol{U}$ delivers only one of the three terms of order ε^2 in the expression of $E[\boldsymbol{U}(\boldsymbol{x}_1,t_1)\,\boldsymbol{U}(\boldsymbol{x}_2,t_2)^T]$ (Eq. 8.55). ■

The perturbation series method has notable features: (1) the differential equations delivering the terms $\boldsymbol{U}_k$, $k \geq 0$, have the same deterministic operator

$\mathcal{L}$ but different inputs (Eq. 8.54), (2) the calculation of the terms $\boldsymbol{U}_k$ can be performed sequentially starting with $\boldsymbol{U}_0$, (3) the formulas in Eq. 8.54 can be used to express the solution of Eq. 8.45 in terms of the input $\boldsymbol{q}$, and (4) the method can be applied to solve nonlinear problems. For these problems, $\boldsymbol{U}_0$ is the solution of a nonlinear differential equation, but higher order terms satisfy linear differential equations (Example 8.24).

Example 8.22: Let $\mathcal{D} = d/dt + \beta + \varepsilon\, Y(t)$, $t \geq 0$, in Eq. 8.45 with $d = 1$, where q and $\beta > 0$ are some constant, Y is a stochastic process with mean zero, and $U(0) = 0$. Hence, the deterministic and random parts of $\mathcal{D}$ are $\mathcal{L} = (d/dt) + \beta$ and $\tilde{\mathcal{R}} = Y$, respectively. The first two terms of the perturbation series are

$$U_0(t) = \frac{q}{\beta}\left(1 - e^{-\beta t}\right) \quad \text{and} \quad U_1(t) = -\frac{q}{\beta}\, e^{-\beta t} \int_0^t \left(e^{\beta s} - 1\right) Y(s)\, ds.$$

The approximate mean of U is $E[U(t)] = U_0(t) + O(\varepsilon^2)$. If Y is a white noise process with unit spectral intensity, then

$$E[U_1(t)\, U_1(s)] = \frac{q^2}{\beta^2}\, e^{-\beta\,(t+s)} \left[\frac{1}{2\,\beta}\, e^{2\,\beta\,(t\wedge s)} - \frac{2}{\beta}\, e^{\beta\,(t\wedge s)} + t \wedge s + \frac{3}{2\,\beta}\right]$$

is one of the three terms entering the approximation of order ε^2 of the correlation function of U (Eq. 8.55). ◇

Note: The functions U_0 and U_1 satisfy the differential equations $\mathcal{L}[U_0(t)] = q$ and $\mathcal{L}[U_1(t)] = -Y(t)\, U_0(t)$ with zero initial conditions. The Green function giving the kernel of the integral representation of $\mathcal{L}^{-1}$ is $e^{-\beta\,(t-s)}$ for $t \geq s$. The term of order ε^2 of the correlation function of U is $E[U_0(t)\, U_2(s) + U_1(t)\, U_1(s) + U_2(t)\, U_0(s)]$. ▲

Example 8.23: Let $U(t)$ be the solution of the differential equation in the previous example with $Y(t) = Y$ a random variable uniformly distributed in $(0, 1)$, q a constant, and $U(0) = 0$. Denote by $U_{st} = \lim_{t\to\infty} U(t)$ the stationary solution of this equation. The distribution of this random variable is $P(U_{st} \leq u) = P(Y > (q - u\,\beta)/(\varepsilon\, u))$ and can be approximated by $P(U_{st} \leq u) \simeq P(Y > (\beta^2/(\varepsilon\, q))(q/\beta - u))$ for $U \simeq U_0 + \varepsilon\, U_1$. Figure 8.11 shows with dotted and solid lines the exact and approximate probabilities $P(U_{st} \leq u)$, respectively, for $q = 1$, $\beta = 1$, and $\varepsilon = 0.1, 0.5, 1.0$. As expected, the perturbation solution is accurate for small values of ε. ◇

Proof: The exact solution is

$$U(t) = \int_0^t e^{-(\beta+\varepsilon\, Y)(t-s)}\, q\, ds = \frac{q}{\beta + \varepsilon\, Y}\left[1 - e^{-(\beta+\varepsilon\, Y)\, t}\right]$$

so that $U_{st} = q/(\beta + \varepsilon\, Y)$ and $P(U_{st} \leq u)$ is as stated. The first two terms of the perturbation solution satisfy the equations

$$(d/dt + \beta) U_0(t) = q \qquad \text{(order 1)},$$
$$(d/dt + \beta) U_1(t) = -Y\, U_0(t) \qquad \text{(order } \varepsilon\text{)},$$

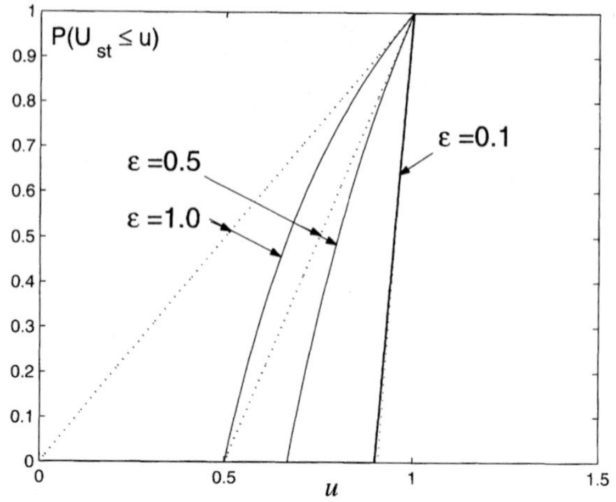

Figure 8.11. Exact and approximate distributions of the stationary solution

so that $U_0(t) = (q/\beta)\left(1 - e^{-\beta t}\right)$ and $U_1(t) = -(q\,Y/\beta^2)\left[1 - (1+\beta\,t)\,e^{-\beta\,t}\right]$ since $U(0) = U_1(0) = 0$. The first order approximation, $U_{st} \simeq q/\beta - \varepsilon\,(q\,Y/\beta^2)$, of the stationary solution yields the stated approximation for $P(U_{st} \le u)$. ■

Example 8.24: Let U be the solution of the nonlinear differential equation $\dot{U}(t) = -U(t)^2 - (U(t) + Y)^2 \sin(\varepsilon\,t)$, $t \ge 0$, starting at $U(0) = 1$, where Y is a random variable taking values in $[0, 1]$ and ε is a small parameter. The terms of the perturbation solution

$$U(t) = U_0(t) + \varepsilon\,U_1(t) + \varepsilon^2\,U_2(t) + O(\varepsilon^3),$$

are given by $\dot{U}_0 = -U_0^2$, $\dot{U}_1 = -2\,U_0\,U_1 - (U_0 + Y)^2\,t$, and $\dot{U}_2 = -U_1^2 - 2\,U_0\,U_2 - 2\,t\,U_1\,(U_0 + Y)$ with the initial conditions $U_0(0) = 1$, $U_1(0) = 0$, and $U_2(0) = 0$. The differential equation for U_0 is nonlinear, but the equations defining the terms U_k, $k \ge 1$, are linear. The approximate expression of U can be used to calculate probabilistic properties of the solution ([17], pp. 56-62). ◇

Note: The function $\sin(\varepsilon\,t)$ needs to be approximated by a polynomial in ε to develop differential equations for the terms U_k, $k \ge 0$, of the perturbation series. The resulting approximation of $\sin(\varepsilon\,t)$ is accurate for small values of $\varepsilon\,t$.

The expressions for U_k, $k = 0, 1, 2$, are

$$U_0(t) = (t+1)^{-1}, \quad U_1(t) = -\frac{t^2}{(t+1)^2}\,(1/2 + Z\,\varphi_1(t) + Z^2\,\varphi_2(t)), \quad \text{and}$$

$$U_2(t) = \frac{1}{(t+1)^2}\left[\sum_{k=1}^{6} \gamma_k\,\frac{t^{k+1}}{k+1} - \frac{(\gamma_0 + \gamma_7)\,t}{t+1} + \gamma_7\,\ln(t+1)\right],$$

where φ_i, $i = 1, 2$, and γ_k, $k = 1, \ldots, 7$, are some deterministic functions of t and Y ([17], pp. 57-58).

Approximate mappings relating the random parameter Y to $U(t)$ can be used to find the marginal distribution of the solution at an arbitrary time t. For example, consider the mapping $Y \mapsto \tilde{U}(t) = h(t, \varepsilon, Y)$ corresponding to the first order approximation $\tilde{U} = U_0 + \varepsilon\, U_1$ for U. This mapping has the inverse ([17], p. 61),

$$Y = \frac{-1}{2\,\varphi_2(t)} \left[\varphi_1(t) - (\varphi_1(t)^2 - 4\,\varphi_2(t)\,\psi(\tilde{U}, \varepsilon))^{1/2}\right],$$

in [0, 1], where $\varphi_1(t) = 2\,t/3 + 1$, $\varphi_2(t) = t^2/4 + 2\,t/3 + 1/2$, and $\psi(u, \varepsilon) = 1/2 - (1 + t - u\,(1+t)^2)/(\varepsilon\, t^2)$, which can be used to calculate approximately the probability law of $\tilde{U}(t)$ ([17], p. 62). ▲

8.4.1.4 Neumann series method

Suppose that Eq. 8.45 has the integral form in Eq. 8.44. We have used such integral representation in Sections 6.3, 6.4 and 6.6.1. A useful discussion on the relations between differential and integral equations can be found in [99] (pp. 225-228).

Example 8.25: Consider the differential equation

$$\frac{d^4 U(x)}{dx^4} + Y(x)\, U(x) = q(x), \quad x \in D = (0, 1)$$

with the boundary conditions $U(0) = U(1) = 0$ and $U''(0) = U''(1) = 0$, where Y is a real-valued random field with strictly positive samples a.s. and q is an integrable function. The solution U is a real-valued random field satisfying the integral equation

$$\begin{aligned} U(x) &= \mathcal{L}^{-1}\,[q(x)] - \left(\mathcal{L}^{-1}\,\mathcal{R}\right)\,[U(x)] \\ &= \int_0^1 g(x, v)\, q(v)\, dv - \int_0^1 g(x, v)\, Y(v)\, U(v)\, dv, \end{aligned}$$

where $\mathcal{L} = d^4/dx^4$, $\mathcal{R} = Y(x)$, and

$$g(x, v) = \begin{cases} (1-v)\,x\,[1 - x^2 - (1-v)^2]/6, & x \le v, \\ (1-x)\,v\,[1 - v^2 - (1-x)^2]/6, & x \ge v \end{cases} \tag{8.56}$$

is the Green function of $\mathcal{L} = d^4/dx^4$. ◇

Note: The solution U can be interpreted as the deflection of a beam on elastic foundation that has unit length, is simply supported at its ends, and is subjected to a load q. The random field Y defines the stiffness of the beam-elastic support ensemble.

The above integral equation defining U can be approximated by

$$U(x_i) \simeq h \sum_{j=1}^{n} g(x_i, x_j)\, q(x_j) - h \sum_{j=1}^{n} g(x_i, x_j)\, Y(x_j)\, U(x_j),$$

where $x_i = i\,h$, $i = 0, 1, \ldots, n$, and $h = 1/n$. The matrix form of the above set of equations is $\boldsymbol{Y}\,\boldsymbol{U} = \boldsymbol{q}$, in which $\boldsymbol{U}$ collects the unknown values $U(x_i)$ of U, the coordinate i of $\boldsymbol{q}$ is $h \sum_{j=1}^{n} g(x_i, x_j)\, q(x_j)$, and $\boldsymbol{Y}$ is a random matrix depending on the values of the field Y at the nodes x_i of the finite difference mesh. The resulting algebraic equation has random coefficients and can be solved by the methods in Section 8.3.1. ▲

Example 8.26: Suppose that $U(t)$, $t \geq 0$, is the solution of the differential equation of $(d/dt + \beta + Y(t))\; U(t) = q(t)$, where $U(0) = Z$ is a random variable, Y denotes a stochastic process with positive sample paths a.s., $\beta > 0$, and q is an integrable function of time. The deterministic and random parts of the differential operator of this equation are $\mathcal{L} = d/dt + \beta$ and $\mathcal{R} = Y(t)$, respectively. The integral form of the differential equation for U is

$$U(t) = F(t) + \int_0^t e^{-\beta\,(t-s)}\, q(s)\, ds - \int_0^t K(t,s)\, U(s)\, ds$$

where $F(t) = Z\,e^{-\beta t}$, the Green function of $\mathcal{L}$ is $g(t,s) = \exp(-\beta\,(t-s))$ for $t \geq s$ and zero otherwise, and $K(t,s) = g(t,s)\,Y(s)$ is a stochastic kernel. ◇

Note: The integral equation for U can be obtained by multiplying the differential equation of U with the Green function of $\mathcal{L}$ and integrating over the time interval $[0, t]$, that is,

$$\int_0^t e^{-\beta\,(t-s)} \left(\frac{d}{ds} + \beta\right) U(s)\, ds = \int_0^t e^{-\beta\,(t-s)}\, q(s)\, ds - \int_0^t e^{-\beta\,(t-s)}\, Y(s)\, U(s)\, ds.$$

Integration by parts of this equation gives the stated integral equation for U.

The integral equation for U can also be obtained by viewing $\dot{U}(t) = -\beta\, U(t) + q(t) - Y(t)\, U(t)$ as a linear system with Green function $g(t,s) = \exp(-\beta\,(t-s))$, $t \geq s$, driven by $q(t) - Y(t)\, U(t)$. ▲

Consider the deterministic Fredholm integral equation of the second kind

$$\varphi(x) = f(x) + \lambda \int_0^1 k(x,y)\, \varphi(y)\, dy, \tag{8.57}$$

defining a real-valued function $\varphi(x)$ for $x \in [0, 1]$, where f is a real-valued function and λ is a constant. The Neumann series solution of Eq. 8.44 is based on the following fact.

If the kernel k is square integrable on $[0, 1] \times [0, 1]$ and $|\lambda| < \| k \|^{-1}$, the solution of Eq. 8.57 is

$$\varphi(x) = f(x) - \lambda \int_0^1 h(x, y; \lambda)\, f(y)\, dy, \quad \text{where} \tag{8.58}$$

$$h(x, y; \lambda) = -\sum_{r=1}^{\infty} \lambda^{r-1}\, k_r(x, y), \quad k_1(x, y) = k(x, y), \text{ and}$$

$$k_r(x, y) = \int_0^1 k(x, z)\, k_{r-1}(z, y)\, dz, \quad k = 2, 3, \ldots. \tag{8.59}$$

Proof: Let $a(x)^2 = \int_0^1 k(x, y)^2\, dy$ and $b(y)^2 = \int_0^1 k(x, y)^2\, dx$. The norm,

$$\| k \|^2 = \int_0^1 \int_0^1 k(x, y)^2\, dx\, dy = \int_0^1 a(x)^2\, dx = \int_0^1 b(y)^2\, dy \le \alpha^2, \quad 0 < \alpha < \infty,$$

is finite by hypothesis so that there exists an α with the above property. We have

$$k_2(x, y)^2 = \left(\int_0^1 k(x, z)\, k(z, y)\, dz \right)^2 \le a(x)^2\, b(y)^2,$$

$$k_3(x, y)^2 = \left(\int_0^1 k(x, z)\, k_2(z, y)\, dz \right)^2 \le a(x)^2 \int_0^1 k_2(z, y)^2\, dz \le \alpha^2\, a(x)^2\, b(y)^2,$$

$$k_4(x, y)^2 = \left(\int_0^1 k(x, z)\, k_3(z, y)\, dz \right)^2 \le a(x)^2 \int_0^1 k_3(z, y)^2\, dz$$

$$\le a(x)^2\, b(y)^2\, \alpha^2 \int_0^1 a(z)^2\, dz \le \alpha^4\, a(x)^2\, b(y)^2,$$

from Eq. 8.59, properties of the norm of k, and the Cauchy-Schwarz inequality. For example,

$$k_3(x, y)^2 \le \int_0^1 k(x, z)^2\, dz \int_0^1 k_2(z, y)^2\, dz$$

by the Cauchy-Schwarz inequality,

$$\int_0^1 k(x, z)^2\, dz = a(x)^2, \quad \text{and}$$

$$\int_0^1 k_2(z, y)^2\, dz \le \int_0^1 a(z)^2\, b(y)^2\, dz = b(y)^2\, \| k \|^2 \le b(y)^2\, \alpha^2$$

so that $k_3(x, y)^2 \leq a(x)^2\, b(y)^2\, \alpha^2$ and $k_{n+2}(x, y)^2 \leq \alpha^{2n}\, a(x)^2\, b(y)^2$ or $|k_{n+2}(x, y)| \leq \alpha^n\, a(x)\, b(y)$, $n = 0, 1, \ldots$, which gives

$$\left|\sum_{r=1}^{\infty} \lambda^r\, k_r(x, y)\right| \leq \sum_{r=1}^{\infty} |\lambda|^r\, |k_r(x, y)| \leq \sum_{r=1}^{\infty} |\lambda|^r\, a(x)\, b(y)\, \alpha^{r-2}$$
$$= \alpha^{-2}\, a(x)\, b(y) \sum_{r=1}^{\infty} (|\lambda|\, \alpha)^r.$$

The geometric series $\sum_{r=1}^{\infty} (|\lambda|\, \alpha)^r$ converges if $|\lambda| < \alpha^{-1}$ or $|\lambda| < \| k \|^{-1}$ since we can take $\alpha = \| k \|$. Hence, the series $\sum_{r=1}^{\infty} \lambda^r\, k_r(x, y)$ is absolutely and uniformly convergent in $[0, 1] \times [0, 1]$ because its absolute value is bounded by a convergent series independent of the arguments (x, y) multiplied by the function $\alpha^{-2}\, a(x)\, b(y)$, which is bounded almost everywhere in $[0, 1] \times [0, 1]$.

If $|\lambda| < \| k \|^{-1}$, the solution of the Fredholm equation is given by Eq. 8.58 since the series in Eq. 8.59 giving the resolvent kernel h is absolutely and uniformly convergent so that it can be integrated term by term, that is,

$$\lambda \int_0^1 k(x, z)\, h(x, y; \lambda)\, dz = -\lambda \int_0^1 k(x, z) \left[\sum_{r=1}^{\infty} \lambda^{r-1}\, k_r(z, y)\right] dz$$
$$= -\int_0^1 k(x, z) \sum_{r=1}^{\infty} \lambda^r\, k_r(z, y)\, dz = -\sum_{r=1}^{\infty} \lambda^r \int_0^1 k(x, z)\, k_r(z, y)\, dz$$
$$= -\sum_{r=1}^{\infty} \lambda^r\, k_{r+1}(x, y) = k(x, y) + h(x, y; \lambda).$$

Hence, we have $\lambda \int_0^1 k(x, z)\, h(z, y; \lambda)\, dz = k(x, y) + h(x, y; \lambda)$.

It remains to show that φ in Eq. 8.58 satisfies Eq. 8.57, that is, that

$$f(x) - \lambda \int_0^1 h(x, y; \lambda)\, f(y)\, dy = f(x)$$
$$+ \lambda \int_0^1 k(x, y) \left[f(y) - \lambda \int_0^1 h(y, z; \lambda)\, f(z)\, dz\right] dy,$$

holds, or

$$\int_0^1 \left[-h(x, y; \lambda) - k(x, y) + \lambda \int_0^1 k(x, z)\, h(z, y; \lambda)\, dz\right] f(y)\, dy = 0.$$

The last condition is satisfied since the above integrand is zero. Hence, φ given by Eqs. 8.58 and 8.59 is the solution of Eq. 8.57. Additional considerations on the Neumann series in Eq. 8.57 can be found in [99] (pp. 266-269), [110] (Chapter 3), and [187] (pp. 49-53).

The proof that Eqs. 8.58 and 8.59 give the solution of Eq. 8.57 can be obtained under other conditions. For example, this result holds if the kernel is bounded in its domain of definition, that is, $|k(x, y)| \leq \beta$, $0 < \beta < \infty$, and $|\lambda| < \beta^{-1}$. Moreover, the requirement $|\lambda| < \beta^{-1}$ can be replaced by $|\lambda| < |\lambda_1|$, where λ_1 denotes the smallest eigenvalue of k ([99], p. 267).

Because the series representation for $h(x, y; \lambda)$ in Eq. 8.59 is absolutely and uniformly convergent almost everywhere in $[0, 1] \times [0, 1]$, the integration in Eq. 8.58 can be performed term by term. Therefore, the Neumann series solution is

$$\varphi(x) = f(x) + \sum_{r=1}^{\infty} \lambda^r \int_0^1 k_r(x, y)\, f(y)\, dy.$$

Numerical calculations based on the Neumann series solution are usually based on the first two or three terms of this series. ■

Example 8.27: Consider the integral equation

$$\boldsymbol{\varphi}(\boldsymbol{x}) = \boldsymbol{f}(\boldsymbol{x}) + \lambda \int_D \boldsymbol{k}(\boldsymbol{x}, \boldsymbol{y})\, \boldsymbol{\varphi}(\boldsymbol{y})\, d\boldsymbol{y}, \quad \boldsymbol{x} \in D, \tag{8.60}$$

where D is an open bounded subset of $\mathbb{R}^d$, $\boldsymbol{\varphi}, \boldsymbol{f} : D \to \mathbb{R}^n$, and $\boldsymbol{k} : D \times D \to \mathbb{R}^{n\times n}$. If (1) $|k_{ij}(\boldsymbol{x}, \boldsymbol{y})| \leq \beta$, $0 < \beta < \infty$, for all $\boldsymbol{x}, \boldsymbol{y} \in D$ and $i, j = 1, \ldots, n$, (2) $\int_D d\boldsymbol{x} = v_D < \infty$, and (3) $|\lambda| < (n\, v_D\, \beta)^{-1}$, then the Neumann series solution of Eq. 8.60 is

$$\boldsymbol{\varphi}(\boldsymbol{x}) = \boldsymbol{f}(\boldsymbol{x}) - \lambda \int_D \boldsymbol{h}(\boldsymbol{x}, \boldsymbol{y}; \lambda)\, \boldsymbol{f}(\boldsymbol{y})\, d\boldsymbol{y}, \tag{8.61}$$

where $\boldsymbol{h}(\boldsymbol{x}, \boldsymbol{y}; \lambda) = -\sum_{r=1}^{\infty} \lambda^{r-1}\, \boldsymbol{k}_r(\boldsymbol{x}, \boldsymbol{y})$ and the kernels $\boldsymbol{k}_r$ can be obtained from $\boldsymbol{k}_1(\boldsymbol{x}, \boldsymbol{y}) = \boldsymbol{k}(\boldsymbol{x}, \boldsymbol{y})$ and $\boldsymbol{k}_r(\boldsymbol{x}, \boldsymbol{y}) = \int_D \boldsymbol{k}(\boldsymbol{x}, \boldsymbol{z})\, \boldsymbol{k}_{r-1}(\boldsymbol{z}, \boldsymbol{y})\, d\boldsymbol{z}$ for $r \geq 2$. ◇

Proof: We have

$$|k_{2,ij}(\boldsymbol{x}, \boldsymbol{y})| \leq \sum_{p=1}^{n} \int_D |k_{1,ip}(\boldsymbol{x}, \boldsymbol{z})|\, |k_{1,pj}(\boldsymbol{z}, \boldsymbol{y})|\, d\boldsymbol{z} \leq n\, v_D\, \beta^2$$

by the definition of $\boldsymbol{k}_r$ and the above requirements on $\boldsymbol{k}$. Similar considerations give $|k_{r,ij}| \leq n^{r-1}\, v_D^{r-1}\, \beta^r$ so that

$$\left| \sum_{r=1}^{\infty} \lambda^r\, k_{r,ij}(\boldsymbol{x}, \boldsymbol{y}) \right| \leq \sum_{r=1}^{\infty} |\lambda|^r\, n^{r-1}\, v_D^{r-1}\, \beta^r$$

showing that $\sum_{r=1}^{\infty} \lambda^r\, k_{r,ij}(\boldsymbol{x}, \boldsymbol{y})$ is absolutely and uniformly convergent in $D \times D$ if the condition $|\lambda| < (n\, v_D\, \beta)^{-1}$ is satisfied. The function $\boldsymbol{h}$ satisfies the equation

$$\lambda \int_D \boldsymbol{k}(\boldsymbol{x}, \boldsymbol{z})\, \boldsymbol{h}(\boldsymbol{z}, \boldsymbol{y}; \lambda)\, d\boldsymbol{z} = \boldsymbol{k}(\boldsymbol{x}, \boldsymbol{y}) + \boldsymbol{h}(\boldsymbol{x}, \boldsymbol{y}; \lambda)$$

since its series representation can be integrated term by term. The above property of $\boldsymbol{h}$ shows that Eq. 8.61 is the solution of Eq. 8.60. ■

The solutions of Eqs. 8.57-8.59 and Eqs. 8.60-8.61 can be extended to obtain a Neumann series solution for Eq. 8.44. There is a notable difference between these equations. The kernels of the integrals in Eqs. 8.57 and 8.60 are deterministic while the kernel of the integral in Eq. 8.44 is stochastic.

If (1) there is $\beta \in (0, \infty)$ such that $|K_{ij}(\boldsymbol{x}^*, \boldsymbol{y}^*)| \leq \beta$ a.s. for all $\boldsymbol{x}^*, \boldsymbol{y}^* \in D^*$ and $i, j = 1, \ldots, n$, (2) $\int_{D^*} d\boldsymbol{x}^* = v_D^* < \infty$, and (3) $|\lambda| < (n\, v_D^*\, \beta)^{-1}$, then the solution of Eq. 8.44 completed with initial and boundary conditions is

$$\boldsymbol{U}(\boldsymbol{x}^*) = \boldsymbol{F}(\boldsymbol{x}^*) - \lambda \int_{D^*} \boldsymbol{H}(\boldsymbol{x}^*, \boldsymbol{y}^*; \lambda)\, \boldsymbol{F}(\boldsymbol{y}^*)\, d\boldsymbol{y}^*, \quad \text{where} \tag{8.62}$$

$$\boldsymbol{H}(\boldsymbol{x}^*, \boldsymbol{y}^*; \lambda) = -\sum_{r=1}^{\infty} \lambda^{r-1}\, \boldsymbol{K}_r(\boldsymbol{x}^*, \boldsymbol{y}^*), \quad \boldsymbol{K}_1(\boldsymbol{x}^*, \boldsymbol{y}^*) = \boldsymbol{K}(\boldsymbol{x}^*, \boldsymbol{y}^*), \text{ and}$$

$$\boldsymbol{K}_r(\boldsymbol{x}^*, \boldsymbol{y}^*) = \int_{D^*} \boldsymbol{K}(\boldsymbol{x}^*, \boldsymbol{z}^*)\, \boldsymbol{K}_{r-1}(\boldsymbol{z}^*, \boldsymbol{y}^*)\, d\boldsymbol{z}^*, \quad r = 2, 3, \ldots. \tag{8.63}$$

Note: The result in Example 8.27 can be applied for almost all samples of the stochastic kernel $\boldsymbol{K}$ since the entries of $\boldsymbol{K}$ are assumed to be bounded almost surely. Hence, the resulting Neumann series is absolutely and uniformly convergent a.s. We also note that the condition $|\lambda| < (n\, v_D^*\, \beta)^{-1}$ implies that the Neumann series is absolutely and uniformly convergent a.s. on any $\tilde{D} \subset D^*$ since $\int_{\tilde{D}} d\boldsymbol{x}^* = \tilde{v} \leq v^*$. ▲

Approximate second moments properties of $\boldsymbol{U}$ can be calculated from a few terms of the Neumann series. For example, the approximate first two moments of $\boldsymbol{U}$ are

$$E[\boldsymbol{U}(\boldsymbol{x}^*)] \simeq E[\boldsymbol{F}(\boldsymbol{x}^*)] + \lambda \int_{D^*} E[\boldsymbol{K}(\boldsymbol{x}^*, \boldsymbol{y}^*)\, \boldsymbol{F}(\boldsymbol{y}^*)]\, d\boldsymbol{y}^*, \tag{8.64}$$

$$\begin{aligned} E[\boldsymbol{U}(\boldsymbol{x}^*)\, \boldsymbol{U}(\boldsymbol{\xi}^*)^T] \simeq\, & E[\boldsymbol{F}(\boldsymbol{x}^*)\, \boldsymbol{F}(\boldsymbol{\xi}^*)^T] \\ & + \lambda \int_{D^*} E[\boldsymbol{F}(\boldsymbol{x}^*)\, \boldsymbol{F}(\boldsymbol{z}^*)^T\, \boldsymbol{K}(\boldsymbol{\xi}^*, \boldsymbol{z}^*)^T]\, d\boldsymbol{z}^* \\ & + \lambda \int_{D^*} E[\boldsymbol{K}(\boldsymbol{x}^*, \boldsymbol{y}^*)\, \boldsymbol{F}(\boldsymbol{y}^*)\, \boldsymbol{F}(\boldsymbol{\xi}^*)^T]\, d\boldsymbol{y}^* \\ & + \lambda^2 \int_{D^* \times D^*} E[\boldsymbol{K}(\boldsymbol{x}^*, \boldsymbol{y}^*)\, \boldsymbol{F}(\boldsymbol{y}^*)\, \boldsymbol{F}(\boldsymbol{z}^*)^T\, \boldsymbol{K}(\boldsymbol{\xi}^*, \boldsymbol{z}^*)^T]\, d\boldsymbol{y}^*\, d\boldsymbol{z}^* \end{aligned} \tag{8.65}$$

for $\boldsymbol{U}(\boldsymbol{x}^*) \simeq \boldsymbol{F}(\boldsymbol{x}^*) + \lambda \int_{D^*} \boldsymbol{K}(\boldsymbol{x}^*, \boldsymbol{y}^*)\, \boldsymbol{F}(\boldsymbol{y}^*)\, d\boldsymbol{y}^*$.

Example 8.28: Let U be the solution of $(d/dt + \beta + Y)\, U(t) = q$, $t \in [0, \tau]$, where Y is a random variable and $U(0) = 0$. The integral form of this equation is

$U(t) = f(t) + \lambda \int_0^\tau K(t,s)\,U(s)\,ds$ where $f(t) = (q/\beta)(1 - e^{-\beta t})$, $K(t,s) = 1_{[0,t]}(s)\,Y\,e^{-\beta(t-s)}$, and $\lambda = -1$ (Example 8.26).

If $|Y| < 1/\tau$ a.s., the Neumann series is absolutely and uniformly convergent in $[0, \tau]$ a.s., and the solution is

$$U(t) = f(t) - \lambda \int_0^\tau H(t,s;\lambda)\,f(s)\,ds,$$

where $H(t,s;\lambda) = -\sum_{r=1}^{\infty} \lambda^{r-1} K_r(t,s)$ and the kernels K_r are $K_1(t,s) = K(t,s)$ and $K_r(t,s) = \int_0^\tau K(t,\sigma)\,K_{r-1}(\sigma,s)\,d\sigma$ for $r \geq 2$. Figure 8.12 shows the variation in time of the exact and approximate mean and standard deviation of

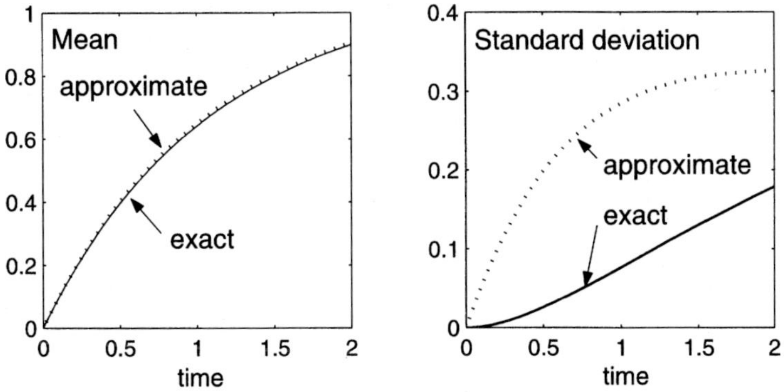

Figure 8.12. Exact and approximate mean and standard deviation functions of U

U for $\beta = 1$, $Y \sim U(-0.5, 0.5)$, and $\tau = 2$. The approximate moments of U are based on the first two terms of the Neumann series solution. The approximate mean of U is accurate but the approximate standard deviation of the solution is unsatisfactory. Superior approximations can be obtained by retaining additional terms of the Neumann series solution, but the calculation becomes tedious. ◇

Note: The absolute value of the kernel K can be bounded by

$$|K(t,s)| = \left| 1_{[0,t]}(s)\,Y\,e^{-\beta(t-s)} \right| \leq |Y|$$

so that the Neumann series is absolutely and uniformly convergent in $[0, \tau]$ if $\tau\,|Y| < 1$ a.s. Because $|Y| < 0.5$ a.s., the Neumann series giving the solution U is convergent for $\tau < 2$. The first two kernels are $K_1 = K$ and

$$\begin{aligned} K_2(t,s) &= Y^2 \int_0^\tau 1_{[0,t]}(\sigma)\,e^{-\beta(t-\sigma)}\,1_{[0,\sigma]}(s)\,e^{-\beta(\sigma-s)}\,d\sigma \\ &= Y^2\,1_{[0,t]}(s)\,(t-s)\,e^{-\beta(t-s)} \end{aligned}$$

so that $U(t) \simeq f(t) - Y\, f_1(t) + Y^2\, f_2(t)$, where

$$f_1(t) = \int_0^\tau K(t,s)\, f(s)\, ds = \frac{q}{\beta^2}\left(1 - (1+\beta\, t)\, e^{-\beta\, t}\right),$$

$$\begin{aligned} f_2(t) &= \int_0^\tau K_2(t,s)\, f(s)\, ds = \frac{q\, t}{\beta^2}\left(1 - (1+\beta\, t)\, e^{-\beta\, t}\right) \\ &- \frac{q}{\beta^3}\left(\beta\, t - 1 + (1 - (\beta\, t)^2/2)\, e^{-\beta\, t}\right). \end{aligned}$$

The approximate solution can be used to calculate moments and other properties of U. The exact solution $U(t) = (q/(q+Y))\left(1 - e^{-(\beta+Y)\, t}\right)$ is a simple function of Y so that its moments can be obtained without difficulty. ▲

8.4.1.5 Other methods

The decomposition and iteration methods are briefly discussed. The equivalent linearization method introduced in Section 8.3.1.6 appears to be impractical, and is not considered. The following two examples illustrate the decomposition and the iteration methods. We have outlined the essentials of these methods in Sections 8.3.1.5 and 8.3.1.7.

Example 8.29: Let $U(x)$, $x \in [0, 1]$, be the solution of the differential equation $(d/dt + \beta + Y(t))\; U(t) = q$ with $U(0) = 0$, where Y is a zero-mean process. The decomposition method is based on a series representation $U(t) = \sum_{r=0}^{\infty} U_r(t)$ of U, where

$$U_0(t) = \mathcal{L}^{-1}[q] = \int_0^t e^{-\beta\,(t-s)}\, q\, ds = \frac{q}{\beta}\,(1 - e^{-\beta\, t}) \quad \text{and}$$

$$\begin{aligned} U_1(t) &= -(\mathcal{L}^{-1}\mathcal{R})[U_0(t)] = -\int_0^t e^{-\beta\,(t-s)}\, Y(s)\, \frac{q}{\beta}\,(1 - e^{-\beta\, s})\, ds \\ &= -\frac{q}{\beta}\, e^{-\beta\, t} \int_0^t (e^{\beta\, s} - 1)\, Y(s)\, ds \end{aligned}$$

are the first two terms of the above decomposition of U, $\mathcal{L} = d/dt + \beta$, and $\mathcal{R} = Y(t)$ ([3], Sections 4.7, 4.10, 5.1, 7.2, and 8.2). The functions U_0 and U_1 coincide with the first two terms of the Neumann series solution for U. ◇

Note: The differential equation satisfied by U is $(\mathcal{L} + \mathcal{R})\,[U(t)] = q$. The integral version of the differential equation for U can be written symbolically as $U(t) = \mathcal{L}^{-1}[q] - \mathcal{L}^{-1}\mathcal{R}[U(t)]$ or

$$\begin{aligned} U(t) &= U_0(t) + U_1(t) + \cdots = \mathcal{L}^{-1}[q] + \mathcal{L}^{-1}\mathcal{R}[U_0(t) + U_1(t) + \cdots] \\ &= \mathcal{L}^{-1}[q] + \mathcal{L}^{-1}\mathcal{R}[U_0(t)] + \mathcal{L}^{-1}\mathcal{R}[U_1(t)] + \cdots, \end{aligned}$$

suggesting the recurrence formulas $U_0 = \mathcal{L}^{-1}[q]$ and $U_k = \mathcal{L}^{-1}\mathcal{R}[U_{k-1}]$ for $k \geq 1$. ▲

Example 8.30: Let $U(t)$, $t \geq 0$, be the solution of the differential equation $(\mathcal{L}+\mathcal{R})\,[U(t)] = q(t)$ in the previous example, where $\mathcal{L}$ and $\mathcal{R}$ are deterministic and random differential operators and q is an integrable function. The mean of U can be calculated approximately from

$$E[U(t)] \simeq \mathcal{L}^{-1}[q(t)] - \mathcal{L}^{-1}[E[\mathcal{R}\,\mathcal{L}^{-1}\,\mathcal{R}]\,E[U(t)]]$$

with $E[U(0)] = 0$, according to the second version of the iteration method (Section 8.3.1.7). ◇

Note: The expectation, $\mathcal{L}[E[U(t)]]+E[\mathcal{R}\,U(t)] = q(t)$, of the differential equation for U cannot be used to calculate the first moment of the solution since $E[\mathcal{R}\,U(t)]$ is not known. An equation can be developed for this unknown expectation by applying the operator $\mathcal{R}$ to the solution $U(t) = \mathcal{L}^{-1}[q(t)] - \mathcal{L}^{-1}\,\mathcal{R}[U(t)]$, which gives

$$\mathcal{R}[U(t)] = \mathcal{R}\,\mathcal{L}^{-1}[q(t)] - \mathcal{R}\,\mathcal{L}^{-1}\,\mathcal{R}[U(t)].$$

The stated mean equation follows from the expectation of the above equation and the approximations,

$$E[\mathcal{R}\,\mathcal{L}^{-1}[q(t)]] \simeq E[\mathcal{R}]\,\mathcal{L}^{-1}[q(t)] = 0,$$
$$E[\mathcal{R}\,\mathcal{L}^{-1}\,\mathcal{R}[U(t)]] \simeq E[\mathcal{R}\,\mathcal{L}^{-1}\,\mathcal{R}]\,E[U(t)],$$

based on the local independence hypothesis. ▲

8.4.2 Homogeneous equations

Let $\boldsymbol{q} = \boldsymbol{0}$ in Eq. 8.43. Then this equation becomes

$$\mathcal{A}[\Phi(\boldsymbol{x})] = \Lambda\,\Phi(\boldsymbol{x}), \quad \boldsymbol{x} \in D \subset \mathbb{R}^d. \tag{8.66}$$

It is assumed that (1) $\mathcal{A}$ is a linear differential operator with time-invariant random coefficients and (2) $\mathcal{L} = E[\mathcal{A}]$ is a linear self-adjoint operator.

Our objective is to solve the random eigenvalue problem defined by Eq. 8.66, that is, to find properties of the **eigenvalues** Λ_i and the **eigenfunctions** Φ_i of $\mathcal{A}$. To complete the definition of the eigenvalue problem in Eq. 8.66, boundary conditions need to be specified for this equation.

Note: It can be shown that the eigenvalues and eigenfunctions of $\mathcal{A}$ are random variables and random fields, respectively, on a probability space $(\Omega, \mathcal{F}, P)$ on which $\mathcal{A}$ is defined ([164], Section 1.2).

More general eigenvalue problems arise in some applications. For example, we may be interested in finding the eigenvalues and eigenfunctions of the homogeneous equation $\mathcal{A}_1[\Phi] = \Lambda\,\mathcal{A}_2[\Phi]$, where $\mathcal{A}_1$ and $\mathcal{A}_2$ are differential or integral operators with random coefficients. ▲

The analysis can be based on (1) extensions of the local solutions in Sections 6.2.2.3 and 6.6.2, (2) Monte Carlo simulation, (3) algebraic eigenvalue problems approximating Eq. 8.66, which can be obtained, for example, by finite difference or finite element representations of Eq. 8.66, (4) exact or approximate expressions for the solution of Eq. 8.66 and bounds on the eigenvalues of $\mathcal{A}$ (Section 8.4.2.2), and (5) moment equations for the eigenvalues and eigenvectors of $\mathcal{A}$ (Sections 8.4.2.3 and 8.4.2.4).

We do not discuss solutions of Eq. 8.66 by Monte Carlo simulation and algebraic eigenvalue problems because these solutions can be implemented on the basis of previous developments in this chapter. Also, an example illustrating the Monte Carlo solution for a deterministic eigenvalue problem has been presented in Section 6.6.2. Numerical comparisons between various methods for solving Eq. 8.66 can be found in [25] (Section VI, pp. 53-71).

8.4.2.1 Deterministic eigenvalue problem

Let $\mathcal{H}$ be the collection of complex-valued square integrable functions defined on a subset D of $\mathbb{R}^d$, that is, $\int_D f(\boldsymbol{x})\, f(\boldsymbol{x})^*\, d\boldsymbol{x} < \infty$ for $f \in \mathcal{H}$. The inner product $\langle f, g\rangle = \int_D f(\boldsymbol{x})\, g(\boldsymbol{x})^*\, d\boldsymbol{x}$, $f, g \in \mathcal{H}$, on $\mathcal{H}$ is finite since $|\langle f, g\rangle| \leq [\langle f, f\rangle\, \langle g, g\rangle]^{1/2}$ by the Cauchy-Schwarz inequality and properties of $\mathcal{H}$. The norm on $\mathcal{H}$ induced by the inner product is $\| f \| = \langle f, f\rangle^{1/2}$, $f \in \mathcal{H}$.

An operator $\mathcal{L} : \mathcal{H} \to \mathcal{H}$ is said to be **linear** and **self-adjoint** if $\mathcal{L}[\alpha\, f + \beta\, g] = \alpha\, \mathcal{L}[f] + \beta\, \mathcal{L}[g]$ and $\langle \mathcal{L}[f], g\rangle = \langle f, \mathcal{L}[g]\rangle$ for $f, g \in \mathcal{H}$ and $\alpha, \beta \in \mathbb{C}$. We assume that $\mathcal{L}$ is a linear and self-adjoint operator. The eigenvalues and eigenfunctions, λ_i and ϕ_i, of $\mathcal{L}$ satisfy the equation

$$\mathcal{L}\,[\phi_i(\boldsymbol{x})] = \lambda_i\, \phi_i(\boldsymbol{x}) \tag{8.67}$$

completed with boundary conditions. Let $\mathcal{U}$ be a vector space consisting of square integrable functions $u : D \to \mathbb{C}$ satisfying the boundary conditions specified for Eq. 8.67 such that $(\mathcal{L} - \lambda\, \mathcal{I})\, u$ is defined for each $u \in \mathcal{U}$. The members of $\mathcal{U}$ are called **admissible functions**. Hence, the eigenfunctions of $\mathcal{L}$ are in $\mathcal{U}$.

Suppose that $\mathcal{L}$ is a linear self-adjoint operator with real-valued coefficients. We summarize some essential properties of the eigenvalues and eigenfunctions of $\mathcal{L}$ defined by Eq. 8.67.

• The eigenvalues and the eigenfunctions of $\mathcal{L}$ are real-valued.

• If $\mathcal{L}$ is positive definite and self-adjoint, then $\mathcal{L}$ possesses a countable number of eigenvalues and all eigenvalues are positive.

• The eigenfunctions corresponding to distinct eigenvalues are orthogonal and linearly independent.

• The eigenfunctions are unique in the sense that an eigenvalue that is not multiple can have only one eigenfunction.

• The lowest eigenvalue $\mathcal{L}$ satisfies the condition $\lambda_1 \leq \langle \mathcal{L}[u], u \rangle$ for all $u \in \mathcal{U}$ with $\| u \| = 1$ so that $\lambda_1 = \min_{u \in \mathcal{U},\, \|u\|=1} \langle \mathcal{L}[u], u \rangle$.

• The eigenfunctions $\{\phi_k\}$ of $\mathcal{L}$ span $\mathcal{U}$ and $u = \sum_{k=1}^{\infty} c_k \, \phi_k$ for any $u \in \mathcal{U}$, where $c_k = \langle u, \phi_k \rangle$ ([133], Chapter V).

Proof: Because $\mathcal{L}$ is self-adjoint, we have $\langle \mathcal{L}[\phi], \phi \rangle = \langle \phi, \mathcal{L}[\phi] \rangle$ so that $\lambda \langle \phi, \phi \rangle = \lambda^* \langle \phi, \phi \rangle$. If ϕ is an eigenfunction, $\| \phi \| \neq 0$ so that $\lambda = \lambda^*$. This property also shows that the eigenfunctions of $\mathcal{L}$ are real-valued functions.

An operator $\mathcal{L}$ is positive definite if $\langle \mathcal{L}[u], u \rangle$ is positive for $u \in \mathcal{U}$ and $\langle \mathcal{L}[u], u \rangle = 0$ if and only if $u \equiv 0$. If $\mathcal{L}$ is positive definite, then $\langle \mathcal{L}[u], u \rangle = \lambda \langle u, u \rangle$ is positive so that $\lambda \geq 0$. It can be shown that the eigenvalues of $\mathcal{L}$ form a countable sequence converging to infinity under relatively mild conditions ([65], Chapter 2 and [133], Chapter V). We index the eigenvalues of $\mathcal{L}$ in increasing order, that is, $\lambda_1 \leq \lambda_2 \leq \cdots$ so that $\lim_{i \to \infty} \lambda_i = \infty$.

We have $\langle \mathcal{L}[\phi_i], \phi_j \rangle = \langle \phi_i, \mathcal{L}[\phi_j] \rangle$ since $\mathcal{L}$ is self-adjoint so that $\lambda_i \langle \phi_i, \phi_j \rangle = \lambda_j \langle \phi_i, \phi_j \rangle$ or $(\lambda_i - \lambda_j) \langle \phi_i, \phi_j \rangle = 0$. If $\lambda_i \neq \lambda_j$, we have $\langle \phi_i, \phi_j \rangle = 0$, that is, the eigenfunctions corresponding to distinct eigenvalues are **orthogonal**. If $\langle \phi_i, \phi_i \rangle = 1$ for all values of i and $\langle \phi_i, \phi_j \rangle = 0$ for $\lambda_i \neq \lambda_j$, $i \neq j$, the eigenfunctions are said to be **orthonormal**.

If $\alpha \, \phi_i + \beta \, \phi_j = 0$ implies $\alpha = \beta = 0$, ϕ_i and ϕ_j are said to be **linearly independent**. Suppose that the orthonormal eigenfunctions ϕ_i and ϕ_j are linearly dependent, that is, there are some non-zero coefficients α, β such that $\alpha \, \phi_i + \beta \, \phi_j = 0$. Then $\langle \alpha \, \phi_i + \beta \, \phi_j, \phi_i \rangle$ is zero, but is also equal to $\alpha \langle \phi_i, \phi_i \rangle$ since ϕ_i and ϕ_j are orthogonal so that we must have $\alpha = 0$. Similar considerations show that $\beta = 0$, in contradiction with our assumption.

Suppose that the first eigenvalue has two eigenfunctions, the functions ϕ_1 and ϕ. Because the set of eigenfunctions ϕ_i is orthonormal and complete, we have $\phi = \sum_i c_i \, \phi_i$ and $c_k = \langle \phi, \phi_k \rangle = 0$ for $k \neq 1$ so that ϕ differs from ϕ_1 by a constant.

Let $u \in \mathcal{U}$ such that $\| u \| = 1$ and $\{\phi_i\}$ be orthonormal eigenfunctions of $\mathcal{L}$. We have $u(x) = \sum_i c_i \, \phi_i(x)$, where c_i are constants, and

$$\langle \mathcal{L}[u], u \rangle = \sum_{i,j} \lambda_i \, c_i \, c_j \, \langle \phi_i, \phi_j \rangle = \sum_i \lambda_i \, c_i^2 = \lambda_1 + \sum_{i \geq 2} (\lambda_i - \lambda_1) \, c_i^2 \geq \lambda_1$$

since $\sum_i c_i^2 = 1$ and $\lambda_i \geq \lambda_1$. The inner product can be calculated term by term because the space spanned by the eigenfunctions of $\mathcal{L}$ is complete.

The last property implies that any function in $\mathcal{U}$ can be represented as a linear form of the eigenfunctions of $\mathcal{L}$. ■

8.4.2.2 Exact expressions and bounds

Denote the deterministic and random parts of the operator $\mathcal{A}$ in Eq. 8.66 by $\mathcal{L}$ and $\mathcal{R}$, respectively. We assume that the random coefficients of $\mathcal{A}$ are square integrable. Set $\mathcal{L} = E[\mathcal{A}]$ and $\mathcal{R} = \mathcal{A} - \mathcal{L}$. Bounds on the eigenvalues of $\mathcal{A}$ can provide useful information on, for example, the dynamic properties of uncertain dynamic systems. Exact expressions for the eigenvalues and eigenfunctions of $\mathcal{A}$ can be most valuable in applications. However, these expressions are available only for very simple problems.

Example 8.31: Consider the eigenvalue problem $d^2\Phi(x)/dx^2 + \Lambda\,\Phi(x) = 0$, $x \in D = (0, 1)$, with the boundary conditions $\Phi(x) = 0$ at $x = 0$ and $d\Phi(x)/dx + Z\,\Phi(x) = 0$ at $x = 1$, where Z is a positive random variable. The lowest eigenvalue of this boundary value problem is the solution of $\tan(\Lambda_1^{1/2}) = -\Lambda_1^{1/2}/Z$ in the interval $[\pi/2, \pi]$ ([25], p. 53). The solution is a function of Z, and its expression can be used to calculate moments and the distribution of Λ_1. ◇

Note: The differential operator for this boundary value problem $\mathcal{A} = \mathcal{L} = d^2/dx^2$ is deterministic. The boundary condition at $x = 1$ is the only source of uncertainty. The set of admissible functions for a sample $Z(\omega)$ of Z consists of real-valued twice differentiable functions $\Phi(\cdot, \omega)$ with support $[0, 1]$ and boundary conditions $\Phi(0, \omega) = 0$ and $d\Phi(1, \omega)/dx + Z(\omega)\,\Phi(1, \omega) = 0$. The solution of this equation is $\Phi(x) = a\,\cos(\Lambda^{1/2}\,x) + b\,\sin(\Lambda^{1/2}\,x)$, where a and b are arbitrary constants. The first and second boundary conditions imply $a = 0$ and $\Lambda^{1/2}\,\cos(\Lambda^{1/2}) + Z\,\sin(\Lambda^{1/2}) = 0$, respectively. ▲

Example 8.32: Consider the boundary value problem in the previous example. The inner product $\langle (d^2\Phi(x)/dx^2), \Phi(x)\rangle$ provides an upper bound on the lowest eigenvalue Λ_1 of this problem so that $E[\Lambda_1] \leq E[\langle (d^2\Phi(x)/dx^2), \Phi(x)\rangle]$, where

$$\Phi(x) = \left(x^2 - \frac{Z+2}{Z+1}\,x\right)\left[\frac{1}{5} - \frac{Z+2}{2\,(Z+1)} + \frac{(Z+2)^2}{3\,(Z+1)^2}\right]^{-1/2}$$

is an admissible function. ◇

Note: The polynomial $\Phi(x) = C_1\,x^2 + C_2\,x + C_3$ is differentiable and satisfies the boundary conditions stated in Example 8.31 for $C_3 = 0$ and $C_2 = -C_1\,(Z+2)/(Z+1)$. Hence, $\Phi(x) = C_1\left[x^2 - (Z+2)\,x/(Z+1)\right]$ is an admissible function. The square of the norm of this function is

$$\| \Phi \|^2 = \int_0^1 \Phi(x)^2\,dx = C_1^2\left[\frac{1}{5} - \frac{Z+2}{2\,(Z+1)} + \frac{(Z+2)^2}{3\,(Z+1)^2}\right].$$

The constant C_1 is selected such that $\| \Phi \| = 1$. ▲

8.4.2.3 Perturbation method

Suppose that

$$\mathcal{A} = \mathcal{L} + \varepsilon\, \tilde{\mathcal{R}}, \tag{8.68}$$

where $\mathcal{L} = E[\mathcal{A}]$, $\varepsilon\, \tilde{\mathcal{R}} = \mathcal{A} - \mathcal{L}$, and ε is a small parameter. It is assumed as previously that the random coefficients of $\mathcal{A}$ are in L_2. Let λ_i and ϕ_i denote the eigenvalues and eigenfunctions of the eigenvalue problem defined by $\mathcal{L}$ with boundary conditions given by the order 1 boundary conditions of the eigenvalue problem for $\mathcal{A}$.

If $\| \phi_i \| = 1$ and λ_i is not a multiple eigenvalue, the first order approximation of the eigenvalues and eigenfunctions of $\mathcal{A}$ are:

$$\Lambda_i \simeq \lambda_i + \varepsilon\, \langle \tilde{\mathcal{R}}[\phi_i], \phi_i \rangle + O(\varepsilon^2) \quad \text{and} \tag{8.69}$$

$$\Phi_i(\boldsymbol{x}) \simeq \phi_i(\boldsymbol{x}) + \varepsilon \sum_{k \neq i} \frac{\langle \tilde{\mathcal{R}}[\phi_i], \phi_k \rangle}{\lambda_i - \lambda_k}\, \phi_k(\boldsymbol{x}) + O(\varepsilon^2). \tag{8.70}$$

Proof: We assume that the eigenvalues and eigenfunctions of $\mathcal{A}$ can be expanded in the convergent power series

$$\Lambda_i = \lambda_i + \varepsilon\, \Lambda_i^{(1)} + \varepsilon^2\, \Lambda_i^{(2)} + \cdots,$$
$$\Phi_i(\boldsymbol{x}) = \phi_i(\boldsymbol{x}) + \varepsilon\, \Phi_i^{(1)} + \varepsilon^2\, \Phi_i^{(2)} + \cdots \tag{8.71}$$

in the small parameter ε. This representation and Eq. 8.66 give

$$(\mathcal{L} + \varepsilon\, \tilde{\mathcal{R}})[\phi_i + \varepsilon\, \Phi_i^{(1)} + \cdots] = (\lambda_i + \varepsilon\, \Lambda_i^{(1)} + \cdots)\, (\phi_i(\boldsymbol{x}) + \varepsilon\, \Phi_i^{(1)} + \cdots)$$

so that $\mathcal{L}[\phi_i] = \lambda_i\, \phi_i$ and $\mathcal{L}[\Phi_i^{(1)}] + \tilde{\mathcal{R}}[\phi_i] = \lambda_i\, \Phi_i^{(1)} + \Lambda_i^{(1)}\, \phi_i$ for order 1 and order ε, respectively. These equations are accompanied by conditions given by the terms of order 1 and ε of the boundary conditions prescribed for the eigenvalue problem of $\mathcal{A}$. The inner product of the equation of order ε and ϕ_j gives

$$\langle \mathcal{L}[\Phi_i^{(1)}], \phi_j \rangle + \langle \tilde{\mathcal{R}}[\phi_i], \phi_j \rangle = \lambda_i\, \langle \Phi_i^{(1)}, \phi_j \rangle + \Lambda_i^{(1)}\, \langle \phi_i, \phi_j \rangle$$

or $B_{ij}\, (\lambda_j - \lambda_i) + \langle \tilde{\mathcal{R}}[\phi_i], \phi_j \rangle = \Lambda_i^{(1)}\, \delta_{ij}$ since the collection of eigenfunctions is complete so that $\Phi_i^{(1)} = \sum_k B_{ik}\, \phi_k$ and $\langle \phi_i, \phi_j \rangle = \delta_{ij}$. This equation yields $\Lambda_i^{(1)} = \langle \tilde{\mathcal{R}}[\phi_i], \phi_i \rangle$ for $i = j$ and $B_{ij} = \langle \tilde{\mathcal{R}}[\phi_i], \phi_j \rangle / (\lambda_i - \lambda_j)$ for $i \neq j$. The projection B_{ii} of $\Phi_i^{(1)}$ on ϕ_i remains undetermined. We take $B_{ii} = 0$ since the above results do not change if we add to $\Phi_i^{(1)}$ a function proportional to ϕ_i. Moreover, the eigenfunction $\Phi_i \simeq \phi_i + \varepsilon\, \Phi_i^{(1)}$ approaches ϕ_i as $\varepsilon \to 0$. ■

Example 8.33: Consider the eigenvalue problem $(1 + \varepsilon\, Z)\, \Delta\Phi(\boldsymbol{x}) = \Lambda\, \Phi(\boldsymbol{x})$, $\boldsymbol{x} \in D \subset \mathbb{R}^d$, where $\Delta = \sum_{r=1}^{d} \partial^2/\partial x_r^2$, Z is a random variable with mean zero and finite second moment, and $\Phi(\boldsymbol{x}) = 0$ on the boundary of D. The approximate eigenvalues and eigenfunctions to the order ε are $\Lambda_i \simeq \lambda_i\, (1 + \varepsilon\, Z)$ and $\Phi_i(\boldsymbol{x}) \simeq \phi_i(\boldsymbol{x})$. ◇

Note: The deterministic and random parts $\mathcal{A}$ are $\mathcal{L} = \Delta$ and $\mathcal{R} = Z\,\Delta$, respectively. The first order correction of the eigenvalue i relative to λ_i is (Eq. 8.69)

$$\Lambda_i^{(1)} = \langle Z\,\Delta[\phi_i], \phi_i\rangle = Z\,\langle\lambda_i\,\phi_i, \phi_i\rangle = Z\,\lambda_i.$$

The coordinates of the first order correction of the eigenfunctions of $(1 + \varepsilon\,Z)\,\Delta$ are (Eq. 8.70)

$$B_{ij} = \frac{\langle Z\,\Delta[\phi_i], \phi_j\rangle}{\lambda_i - \lambda_j} = \frac{Z\,\lambda_i\,\langle\phi_i, \phi_j\rangle}{\lambda_i - \lambda_j} = 0 \quad \text{for } i \neq j,$$

so that $\Phi_i^{(1)}(\boldsymbol{x})$ is zero. The result also follows from the alternative form $\Delta\Phi(\boldsymbol{x}) = (\Lambda/(1 + \varepsilon\,Z))\,\Phi(\boldsymbol{x})$ of the original boundary value problem, showing that the random and deterministic operators, $(1 + \varepsilon\,Z)\,\Delta$ and Δ, have the same eigenfunctions but different eigenvalues. The eigenvalues of the random operator scaled by $(1 + \varepsilon\,Z)$ coincide with the eigenvalues of Δ. ▲

8.4.2.4 Iteration method

Consider the decomposition $\mathcal{A} = \mathcal{L} + \mathcal{R}$ of the random operator $\mathcal{A}$, where $\mathcal{L} = E[\mathcal{A}]$ and $\mathcal{R} = \mathcal{A} - \mathcal{L}$ has mean zero. It is assumed that the random coefficients of $\mathcal{A}$ are in L_2.

Our objective is to find moments of the eigenvalues and eigenfunctions of $\mathcal{A}$. The analysis is based on the local independence hypothesis, which states that the dependence between the eigenvalues, eigenfunctions, and the random part of $\mathcal{A}$ can be ignored. We mention again that there is no theoretical justification for this hypothesis. Results are given only for some of the first two moments of the eigenvalues and eigenfunctions of $\mathcal{A}$. Additional information on this method can be found in [25] (pp. 43-52).

> The expectation of the eigenvalues and eigenfunctions of $\mathcal{A}$ can be approximated by the solution of the eigenvalue problem
>
> $$\mathcal{L}[E[\Phi_i(\boldsymbol{x})]] \simeq E[\Lambda_i]\,E[\Phi_i(\boldsymbol{x})]. \tag{8.72}$$

Note: The expectation, $\mathcal{L}[E[\Phi]] + E[\mathcal{R}[\Phi]] = E[\Lambda\,\Phi]$, of Eq. 8.66 cannot be used to find the mean values of eigenvalues and eigenfunctions of $\mathcal{A}[\Phi] = \Lambda\,\Phi$ since $E[\mathcal{R}[\Phi]]$ and $E[\Lambda\,\Phi]$ are not known. The expectation of Eq. 8.66 and the local independence hypothesis, giving the approximations $E[\mathcal{R}[\Phi]] \simeq E[\mathcal{R}]\,E[\Phi] = 0$ and $E[\Lambda\,\Phi] \simeq E[\Lambda]\,E[\Phi]$, yield Eq. 8.72. ▲

> The second moment of the eigenvalues of $\mathcal{A}$ can be approximated by
>
> $$E[\Lambda_i^2] \simeq \lambda_i^2 + \langle(E[\mathcal{R}\,\mathcal{R}])[\phi_i], \phi_i\rangle. \tag{8.73}$$

Note: The sequence of equalities,

$$\begin{aligned}\Phi &= \Lambda^{-1}\,(\mathcal{L}+\mathcal{R})[\Phi] = \Lambda^{-2}\,(\mathcal{L}+\mathcal{R})\,(\mathcal{L}+\mathcal{R})[\Phi] \\ &= \Lambda^{-2}\,(\mathcal{L}\,\mathcal{L}+\mathcal{L}\,\mathcal{R}+\mathcal{R}\,\mathcal{L}+\mathcal{R}\,\mathcal{R})[\Phi],\end{aligned}$$

gives $E[\Lambda^2]\,E[\Phi] \simeq \mathcal{L}\,\mathcal{L}[E[\Phi]] + E[\mathcal{R}\,\mathcal{R}]\,E[\Phi]$ by averaging and using the local independence hypothesis. The inner product of the last formula for $\Lambda = \Lambda_i$ and $\Phi = \Phi_i$ with ϕ_i yields Eq. 8.73 since $E[\Phi_i] \simeq \phi_i$, $\langle \phi_i, \phi_i \rangle = 1$,

$$E[\Lambda_i^2]\,\langle \phi_i, \phi_i \rangle \simeq \langle \mathcal{L}\,\mathcal{L}[\phi_i], \phi_i \rangle + \langle E[\mathcal{R}\,\mathcal{R}]\,[\phi_i], \phi_i \rangle,$$

and $\mathcal{L}$ is a self-adjoint operator, so that $\langle \mathcal{L}\,\mathcal{L}[\phi_i], \phi \rangle = \langle \mathcal{L}[\phi_i], \mathcal{L}[\phi_i] \rangle = \lambda_i^2\,\langle \phi_i, \phi_i \rangle$. There is no simple formula for the second moment of the eigenfunctions of a random operator $\mathcal{A}$. ▲

Example 8.34: Let (Λ_i, Φ_i) be the solution of the eigenvalue problem

$$(1+Z)\,\Delta\Phi(\boldsymbol{x}) = \Lambda\,\Phi(\boldsymbol{x}),$$

$\boldsymbol{x} \in D$, here Z is a random variable with mean zero and finite second moment. The approximate mean solution is $E[\Lambda_i] \simeq \lambda_i$ and $E[\Phi_i] \simeq \phi_i$, where λ_i and ϕ_i denote the eigenvalues and eigenfunctions of the deterministic part $\mathcal{L} = \Delta$ of $\mathcal{A} = (1+Z)\,\Delta$. The approximate second moment of an eigenvalue of $\mathcal{A}$ is $E[\Lambda_i^2] \simeq \lambda_i^2\,(1 + E[Z^2])$, so that the approximate variance of Λ_i is $\lambda_i^2\,E[Z^2]$. ◇

Note: The inner product $\langle E[\mathcal{R}\,\mathcal{R}]\,[\phi_i], \phi_i \rangle$ is

$$\langle E[Z^2\,\Delta\,\Delta][\phi_i], \phi_i \rangle = E[Z^2]\,\langle \Delta\,\Delta[\phi_i], \phi_i \rangle = E[Z^2]\,\lambda_i^2$$

since Δ is a self-adjoint operator, $\Delta[\phi_i] = \lambda_i\,\phi_i$, and $\langle \phi_i, \phi_i \rangle = 1$. ▲

We conclude here our discussion on methods for solving stochastic systems with deterministic input. The remaining part of this chapter presents stochastic problems selected from various fields of applied science and engineering. The solutions of the problems in the following sections are based on the methods in Chapters 5, 6, 7, and the first part of this chapter.

8.5 Effective material properties

Most materials are heterogeneous at the scale of their constituents, referred to as the **microscopic scale** or **microscale**. For example, concrete at the scale of its aggregates, granular materials at the scale of their particles, metals at the scale of their grains or crystals, and multi-phase media or suspensions at the scale of their inclusions. Moreover, material properties at the microscale tend to fluctuate randomly in space so that they can be modeled by random fields at this scale. Generally, we say that the **microstructure** of a material is known if there exists complete information on the material features at the microscopic level, for example, we know the concentration, the properties, and the geometry of the constituent phases of a multiphase material as well as their spatial distribution.

Our objective is to calculate **bulk**, **global**, or **effective properties**, that is, material properties at the **macroscale** or **laboratory scale**. These properties are some type of averages of the material features at the microscale. Effective properties (1) are measured in most laboratory tests, (2) are used as material constants in continuum theories, for example, the Lamé constants of linear elasticity, and (3) can be obtained in some cases even if the available information on the microstructure is partial, for example, we may know only the volume fraction and mechanical properties of the constituents of a multi-phase media but not their geometry and spatial distribution.

Consider a microstructure and some differential equations describing the state of the microstructure subjected to a specified action. Because material properties are random at the microscopic scale, these equations have random coefficients. The solution of these equations provides detailed information on the state of the microstructure, which may not be needed. We can derive equations for the material state at the macroscopic scale by averaging the solution of the state equations for the microstructure, but this approach is computationally inefficient since it requires finding first the state of the microstructure. Two methods have been developed for establishing state equations at the macroscopic scale, the **homogenization** and **representative volume** methods. The homogenization method considers a family of problems for microstructures with scale $\varepsilon > 0$ and calculates the limit of the solution of these problems as $\varepsilon \downarrow 0$. The method also delivers effective material properties. The representative volume method develops differential equations for the macroscopic state of a material, that is, for an average of the microscopic state over a window much larger than the scale of the microstructure, referred to as representative volume. The macroscopic state equations can be obtained directly, rather than by averaging the microscopic state equations.

It is common to assume that the material properties at the microscopic scale can be modeled by ergodic random fields. Spatial averages of these properties over a representative volume much larger than the microscale are nearly deterministic and practically independent of the particular microstructure sample used for calculations.

Generally, numerical algorithms are needed for calculating effective properties. However, some homogenization problems can be solved analytically, for example, the determination of the effective properties of series and parallel systems. The components of a series and a parallel system carry the same load, the applied load, and experience the same deformation, the system deformation, respectively. A series system fails by its weakest component while a parallel system fails when its last surviving component breaks.

Example 8.35: Consider a series and a parallel system with n linear elastic components of stiffness K_i, $i = 1, \ldots, n$, where K_i are independent copies of a real-valued random variable $K > 0$ with finite mean. These heterogeneous systems can be replaced by homogeneous systems, that is, systems with components of the same stiffness $K_{\text{eff}} = n/\left(\sum_{i=1}^{n} 1/K_i\right)$ and $K_{\text{eff}} = (1/n)\sum_{i=1}^{n} K_i$ for series and parallel systems, respectively. The equivalence between the original heterogeneous systems and their homogenized versions is very strong in this case. The homogenized and the heterogeneous systems have the same force-displacement relationship.

The effective stiffness K_{eff} converges a.s. as $n \to \infty$ to $1/E[1/K]$ and $E[K]$ for series and parallel systems, respectively, by the strong law of large numbers since the mean of K exists and is finite (Section 2.13). ◇

Proof: Let $d > 0$ be an imposed deterministic displacement at the boundaries, that is, the two ends, of a series system. This displacement is generated by a force F which is random because of the uncertainty in the system components. The force in each component is F so that the elongation of component i is a random variable $D_i = F/K_i$. The system deformation, $\sum_{i=1}^{n} D_i = \sum_{i=1}^{n} F/K_i = F\sum_{i=1}^{n} 1/K_i$ must coincide with d so that the overall stiffness of the system is $F/d = 1/\sum_{i=1}^{n} 1/K_i$. Hence, the stiffness of the components of a homogeneous series system with n components and the same global constitutive law is $n/\sum_{i=1}^{n} 1/K_i$. If a deterministic force $f > 0$ is applied at the system boundaries, the elongation of component i is f/K_i so that $f\sum_{i=1}^{n}(1/K_i)$ is the system deformation. The ratio of f to the system deformation gives the system stiffness, which has the same expression as in the deformation controlled experiment.

Similar considerations can be applied to find the effective stiffness for a parallel system. For example, suppose that the system is subjected to a deterministic force $f > 0$ causing a random deformation D. Because the components have the same deformation D, the force carried by component i is $K_i\, D$ so that $f = D\sum_{i=1}^{n} K_i$, showing that the system stiffness is $\sum_{i=1}^{n} K_i$. ■

Example 8.36: Suppose that the components of the series and parallel systems in Example 8.35 are made of Maxwell or Kelvin materials, that is, they are modeled by a spring and a damper in series or a spring and a damper in parallel, respectively. The overall force-displacement relations for a parallel system with Maxwell and Kelvin components are

$$\frac{f}{\Gamma(t)} = \sum_{i=1}^{n} \frac{1}{1/K_i + t/C_i} \quad \text{and} \quad \frac{f}{\Gamma(t)} = \sum_{i=1}^{n} \frac{K_i}{1 - \exp[-(K_i/C_i)\, t]},$$

respectively, where K_i and C_i, $i = 1, \ldots, n$, denote the stiffness and damping parameters of component i, n is the number of components, $t \geq 0$ is the time measured from the application of the force $f > 0$, and $\Gamma(t)$ denotes the system elongation at time t.

The overall force-displacement relations for a series system are

$$\frac{f}{\Gamma(t)} = \frac{1}{\sum_{i=1}^{n} (1/K_i + t/C_i)} \quad \text{and} \quad \frac{f}{\Gamma(t)} = \frac{1}{\sum_{i=1}^{n} \left[1 - \exp(-(K_i/C_i)\, t)\right]/K_i}$$

for Maxwell and Kelvin components, respectively. ◇

Proof: The above results are consistent with findings in Example 8.35. If $C_i \to \infty$ and $C_i \to 0$, the Maxwell and Kelvin materials degenerate into linear springs and the above equations yield the results in Example 8.35.

The elongation $\Gamma(t)$ of Maxwell and Kelvin materials with parameters (K_i, C_i) subjected to a force $x > 0$ are

$$\Gamma(t) = x\left(\frac{1}{K_i} + \frac{t}{C_i}\right) \quad \text{and} \quad \Gamma(t) = \frac{x}{K_i}\left(1 - e^{-(K_i/C_i)\,t}\right),$$

respectively.

Suppose that a parallel system with Maxwell components is loaded by $f > 0$. The deformation $\Gamma(t)$ at time t of all system components is $\Gamma(t) = F_i(t)\,(1/K_i + t/C_i)$, where the force $F_i(t)$ in component $i = 1, \ldots, n$ at time t is unknown. Because $\sum_{i=1}^n F_i(t) = f$ at each t by equilibrium, we have

$$f = \sum_{i=1}^n F_i(t) = \Gamma(t) \sum_{i=1}^n \frac{1}{1/K_i + t/C_i}.$$

The elongation $\Gamma(t)$ can be determined from the above equation for a specified load and time and then used to find internal forces $F_i(t)$. The deformation $\Gamma(t)$ and internal forces $F_i(t)$ are parametric stochastic processes (Section 3.6.5) since they are deterministic functions of time depending on a finite number of random parameters, the component stiffness and damping. A similar approach can be applied to find the global force-displacement relations for other systems. ■

Example 8.37: Consider a rod of length $l > 0$ and random stiffness $K(\xi) > 0$, $\xi \in [0, l]$. The effective stiffness of the rod is $K_{\text{eff}} = 1/\int_0^l d\xi/K(\xi)$. Figure 8.13 shows samples of K_{eff} as functions of the rod length. The samples nearly coincide as the rod length increases. The plots are for the stiffness random field $K(\xi) = a + \Phi(X(\xi))(b - a)$, $\xi \in [0, l]$, $a = 1$, $b = 2$, and a process X defined by $dX(\xi) = -\rho\, X(\xi)\, d\xi + \sqrt{2\,\rho}\, dB(\xi)$, where $\rho = 10$, B is a Brownian motion, and $X(0) \sim N(0, 1)$. ◇

Proof: The determination of the effective stiffness is based on the approach in Example 8.35 for series systems. Suppose that the rod is subjected to a fixed elongation $d > 0$. A random force $F > 0$ needs to be applied at the ends of the rod to generate this elongation. The strain caused by F at $\xi \in [0, l]$ is $\Gamma(\xi) = F/K(\xi)$ so that $d = \int_0^l \Gamma(\xi)\, d\xi$ or $F \int_0^l d\xi/K(\xi) = d$, which gives the rod stiffness. The equivalent homogeneous rod has stiffness $\tilde{K}(l) = l/K_{\text{eff}}$.

We note that the stiffness K is a translation random field (Section 3.6.6) with the properties $K(\xi) \sim U(a, b)$ for all $\xi \in [0, l]$ and scaled covariance function that is approximately equal to $\exp(-\rho\,|\xi_1 - \xi_2|)$, where ξ_1, ξ_2 are some points in $[0, l]$. ■

Consider a multi-phase material. The phase properties and their volume fraction provide useful information on the microstructure. This partial information on the microstructure suffices to calculate effective material properties in some

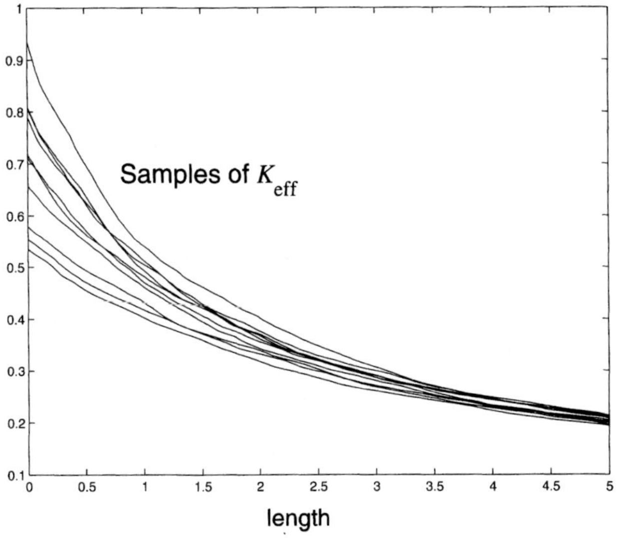

Figure 8.13. Samples of effective stiffness K_{eff}

cases. For example, suppose that the series system in Example 8.35 has two types of components with stiffnesses k_a and k_b and volume fractions ξ_a and ξ_b, respectively, that is, there are $n_a = n\,\xi_a$ and $n_b = n\,\xi_b$ components with stiffness k_a and k_b. The effective stiffness of the components of a homogeneous series system with n components is $k_{\text{eff}} = 1/(\xi_a/k_a + \xi_b/k_b)$. However, the above partial information on the microstructure is insufficient in other cases. For example, consider two microstructures of a material consisting of disconnected inclusions imbedded in a matrix with a 50% inclusion volume fraction. In the first microstructure the inclusions have much higher conductivity than the matrix. In the second microstructure the matrix has much higher conductivity than the inclusions. The microstructures are indistinguishable if the information is limited to phase properties and concentration. However, the effective conductivity of the second microstructure exceeds by far the corresponding property of the first microstructure. In this case we need detailed information on the microstructure to calculate the effective conductivity.

The following two sections illustrate methods for calculating effective properties for random heterogeneous materials. In Section 8.5.1 we estimate the effective conductivity of multi-phase random materials. The analysis is based on an extension of the random walk method (Section 6.2) and smooth approximations of the conductivity field. In Section 8.5.2 we illustrate the determination of the effective material properties for linear elasticity. The analysis is based on properties of random fields, concepts of linear elasticity, and perturbation and Neumann series methods.

8.5.1 Conductivity

There are numerous studies providing approximations of and bounds on macroscopic properties of heterogeneous media, such as electric or thermal conductivity, magnetic permeability, and diffusion coefficients. Most of the available results are for special microstructures, such as two-phase composites with inclusions of simple geometries [184, 192]. A notable exception is the Monte Carlo simulation method in [113, 186] for calculating the effective conductivity of two-phase materials. The method uses samples of Brownian motions having different speeds in different phases and non-zero probabilities of reflection at phase interface. Numerical results by this Monte Carlo simulation method have been reported for two-phase media. The method appears to be impractical for calculating the effective conductivity of general heterogeneous media because it requires the solution of complex first passage problems for $\mathbb{R}^2$- and $\mathbb{R}^3$-valued Brownian motion processes.

In this section we discuss an alternative method for estimating the effective conductivity of random heterogeneous media. The method is general, and can be applied to estimate the effective conductivity of heterogeneous media with an arbitrary number of phases, which can have any geometry. As previously stated, the analysis is based on an extension of the random walk method discussed in Section 6.2.

8.5.1.1 Homogeneous media

Suppose that u is the solution of the Dirichlet boundary value problem

$$\sigma \Delta u(\boldsymbol{x}) = -1, \quad \boldsymbol{x} \in D, \tag{8.74}$$

defined on an open bounded subset D of $\mathbb{R}^d$ with $u(\boldsymbol{x}) = 0$ for $\boldsymbol{x} \in \partial D$, where ∂D denotes the boundary of D, $\sigma > 0$ is a constant, and $\Delta = \sum_{p=1}^{d} \partial^2/\partial x_p^2$. The solution u can be interpreted as the temperature in a homogeneous medium with constant conductivity $\sigma > 0$, which is subjected to a unit flux everywhere in D and zero boundary conditions. Let

$$T = \inf\{t \geq 0 : \boldsymbol{B}(t) \notin D, \boldsymbol{B}(0) = \boldsymbol{x} \in D\} \tag{8.75}$$

be the first time an $\mathbb{R}^d$-valued Brownian motion $\boldsymbol{B}$ starting at $\boldsymbol{x} \in D$ exits D.

The local solution of Eq. 8.74 is

$$u(\boldsymbol{x}) = \frac{1}{2\sigma} E^{\boldsymbol{x}}[T], \quad \boldsymbol{x} \in D. \tag{8.76}$$

Proof: The Itô formula applied to the mapping $\boldsymbol{B} \mapsto g(\boldsymbol{B})$, $g \in C^2(\mathbb{R}^d)$, in the time interval $[0, T]$ gives

$$E^{\boldsymbol{x}}[g(\boldsymbol{B}(T))] - g(\boldsymbol{x}) = E^{\boldsymbol{x}}\left[\int_0^T \mathcal{A}[g(\boldsymbol{B}(s))]\, ds\right]$$

by expectation, where $\mathcal{A} = (1/2)\sum_{p=1}^{d} \partial^2/\partial x_p^2 = (1/2)\,\Delta$ is the generator of $\boldsymbol{B}$ (Section 6.2.1.3). The notation $\mathcal{A}$ used in this section for the generator of $\boldsymbol{B}$ and in the following section for the generator of a diffusion process $\boldsymbol{X}$, has no relation to a similar notation used previously in this chapter for some differential operators (Section 8.4).

The local solution in Eq. 8.76 results by writing the above equation with u in place of g and noting that $\boldsymbol{B}(T) \in \partial D$ so that $u(\boldsymbol{B}(T)) = 0$ and that $\boldsymbol{B}(s) \in D$ for $s < T$ so that $\mathcal{A}[u(\boldsymbol{B}(s))] = -1/(2\sigma)$. ■

Example 8.38: If D in Eq. 8.74 is a sphere $S_d(r) = \{\boldsymbol{\xi} \in \mathbb{R}^d : \| \boldsymbol{\xi} \| < r\}$ of radius $r > 0$ centered at the origin of $\mathbb{R}^d$, then

$$u(\boldsymbol{x}) = \frac{E^{\boldsymbol{x}}[T(r)]}{2\,\sigma} = \frac{r^2 - ||\boldsymbol{x}||^2}{2\,d\,\sigma}, \quad \boldsymbol{x} \in S_d(r),$$

where $T(r)$ is given by Eq. 8.75 with $D = S_d(r)$. Hence, the temperature at the center of the sphere is $u(\boldsymbol{0}) = r^2/(2\,d\,\sigma)$. ◇

Proof: Let $k > 0$ be an integer and let $T_k = \min(k, T(r))$. The expectation of the Itô formula applied to the mapping $\boldsymbol{B} \mapsto \| \boldsymbol{B} \|^2$ in the time interval $(0, T_k)$ gives

$$E^{\boldsymbol{x}}\left[\| \boldsymbol{B}(T_k) \|^2\right] - \| \boldsymbol{x} \|^2 = E^{\boldsymbol{x}}\left[\int_0^{T_k} \frac{1}{2}\,\Delta \| \boldsymbol{B}(s) \|^2\, ds\right]$$

or $E^{\boldsymbol{x}}\left[\| \boldsymbol{B}(T_k) \|^2\right] - \| \boldsymbol{x} \|^2 = E^{\boldsymbol{x}}[T_k]\,d$ since $\Delta \| \boldsymbol{x} \|^2 = 2\,d$. We have

$$E^{\boldsymbol{x}}[T_k] = \frac{E^{\boldsymbol{x}}\left[\| \boldsymbol{B}(T_k) \|^2\right] - \| \boldsymbol{x} \|^2}{d} \le \frac{r^2 - \| \boldsymbol{x} \|^2}{d}$$

so that $T(r) = \lim_{k\to\infty} T_k < \infty$ a.s. and $E^{\boldsymbol{x}}[T(r)] = (r^2 - \| \boldsymbol{x} \|^2)/d$. Additional considerations on this proof and related topics can be found in [135] (Chapter 7 and Example 7.4.2, p. 119). ■

8.5.1.2 Heterogeneous media

Consider a heterogeneous medium in $D \subset \mathbb{R}^d$ characterized by the conductivity field $\sigma(\boldsymbol{x}) > 0$, $\boldsymbol{x} \in D$. Then Eq. 8.74 becomes

$$\nabla \cdot (\sigma(\boldsymbol{x})\,\nabla u(\boldsymbol{x})) = -\gamma(\boldsymbol{x}), \quad \boldsymbol{x} \in D, \quad \text{or}$$

$$\sum_{p=1}^{d} \frac{\partial \sigma(\boldsymbol{x})}{\partial x_p}\frac{\partial u(\boldsymbol{x})}{\partial x_p} + \sigma(\boldsymbol{x})\Delta u(\boldsymbol{x}) = -\gamma(\boldsymbol{x}), \quad \boldsymbol{x} \in D, \tag{8.77}$$

with $u(\boldsymbol{x}) = 0$ for $\boldsymbol{x} \in \partial D$, where $\nabla = (\partial/\partial x_1, \ldots, \partial/\partial x_d)$ and $\gamma(\boldsymbol{x})$ denotes the flux at $\boldsymbol{x} \in D$.

Let $\boldsymbol{X}$ be a diffusion process defined by

$$d\boldsymbol{X}(t) = \boldsymbol{a}(\boldsymbol{X}(t))\,dt + \boldsymbol{b}(\boldsymbol{X}(t))\,d\boldsymbol{B}(t), \tag{8.78}$$

where the entries of the $(d, 1)$ and (d, d) matrices $\boldsymbol{a}$ and $\boldsymbol{b}$ are real-valued functions defined on $\mathbb{R}^d$ and $\boldsymbol{B}$ is an $\mathbb{R}^d$-valued Brownian motion. Define the drift and diffusion coefficients of $\boldsymbol{X}$ by

$$a_p(\boldsymbol{x}) = \frac{\partial \sigma(\boldsymbol{x})}{\partial x_p}, \quad p = 1, \ldots, d, \quad \text{and } \boldsymbol{b}(\boldsymbol{x}) = \sqrt{2\sigma(\boldsymbol{x})}\, \boldsymbol{i}, \tag{8.79}$$

where $\boldsymbol{i}$ is the (d, d) identity matrix. It is assumed that the drift and diffusion coefficients defined by Eq. 8.79 satisfy the conditions in Section 4.7.1.1 so that the solution of Eq. 8.78 exists and is unique. Let

$$T^* = \inf\{t \geq 0 : \boldsymbol{X}(t) \notin D, \boldsymbol{X}(0) = \boldsymbol{x} \in D\} \tag{8.80}$$

denote the first time when $\boldsymbol{X}$ starting at $\boldsymbol{X}(0) = \boldsymbol{x} \in D$ exits D. It is assumed that T^* has a finite expectation (Section 6.2.1.1).

The local solution of Eq. 8.77 is

$$u(\boldsymbol{x}) = E^{\boldsymbol{x}}\left[\int_0^{T^*} \gamma(\boldsymbol{X}(s))\, ds\right]. \tag{8.81}$$

Proof: We first note that, if the flux γ is unity, then the local solution of Eq. 8.77 becomes $u(\boldsymbol{x}) = E^{\boldsymbol{x}}[T^*]$.

Let g be a function with continuous second order partial derivatives and assume $\boldsymbol{X}(0) = \boldsymbol{x}$. The average of the Itô formula applied to the mapping $\boldsymbol{X} \mapsto g(\boldsymbol{X})$, $g \in C^2(\mathbb{R}^d)$, in a time interval $[0, t]$ gives

$$E^{\boldsymbol{x}}[g(\boldsymbol{X}(t))] - g(\boldsymbol{x}) = E^{\boldsymbol{x}}\left[\int_0^t \mathcal{A}[g(\boldsymbol{X}(s))]\, ds\right],$$

where

$$\mathcal{A} = \sum_{p=1}^d a_p(\boldsymbol{x}) \frac{\partial}{\partial x_p} + \frac{1}{2} \sum_{p,q=1}^d (\boldsymbol{b}(\boldsymbol{x})\boldsymbol{b}(\boldsymbol{x})^T)_{pq} \frac{\partial^2}{\partial x_p \partial x_q},$$

is the generator of $\boldsymbol{X}$ (Eqs. 8.78-8.79 and Section 6.2.1.1). The above Itô formula applied to the solution u of Eq. 8.77 in the time interval $(0, T^*)$ yields

$$E^{\boldsymbol{x}}[u(\boldsymbol{X}(T^*))] - u(\boldsymbol{x}) = E^{\boldsymbol{x}}\left[\int_0^{T^*} \mathcal{A}[u(\boldsymbol{X}(s))]\, ds\right],$$

where $\mathcal{A}[u(\boldsymbol{x})] = \sum_{p=1}^d (\partial\sigma(\boldsymbol{x})/\partial x_p)(\partial u(\boldsymbol{x})/\partial x_p) + \sigma(\boldsymbol{x})\Delta u(\boldsymbol{x})$ (Eq. 8.79). The local solution in Eq. 8.81 follows since $\boldsymbol{X}(T^*) \in \partial D$ and $\boldsymbol{X}(s) \in D$ for $s < T^*$, so that $u(\boldsymbol{X}(T^*)) = 0$ and $\mathcal{A}[u(\boldsymbol{X}(s))] = -\gamma(\boldsymbol{X}(s))$.

Generally, the expectation in Eq. 8.81 cannot be calculated analytically. However, this expectation can be estimated from samples of $\boldsymbol{X}$. Methods for generating sample paths of diffusion processes are discussed in Section 5.3.3.2.

The conductivity field in a multi-phase medium has jumps at the interface between its constituents so that the drift and diffusion coefficients given by Eq. 8.79 do not satisfy the uniform Lipschitz conditions. However, we can develop smooth approximations $\tilde{\sigma}$ for the conductivity field σ of a multi-phase material, and use these approximations to calculate the effective conductivity, as illustrated later in Example 8.39. ■

8.5.1.3 Effective conductivity

Consider a deterministic heterogeneous medium with conductivity field $\sigma(\boldsymbol{x})$ and a deterministic homogeneous medium of unknown conductivity σ_{eff}. The two media are contained in two d-dimensional spheres $S_d(r)$ with the same radius $r > 0$. We define σ_{eff} from the condition that the temperatures at the center of the heterogeneous and the homogeneous spheres coincide and referred to it as the **effective conductivity** of the heterogeneous medium ([185], Chapter 18). The effective conductivity depends on the radius $r > 0$ of the sphere.

Let $T^*(r)$ be the stopping time in Eq. 8.80 for $D = S_d(r)$, $\gamma(\boldsymbol{x}) = 1$, the diffusion process $\boldsymbol{X}$ in Eq. 8.78 with coefficients in Eq. 8.79, and initial state $\boldsymbol{X}(0) = \boldsymbol{0}$.

The effective conductivity for a deterministic heterogeneous material contained in a sphere $S_d(r)$, $r > 0$, is

$$\sigma_{\text{eff}}(r) = \frac{r^2}{2\,d\,E^{\boldsymbol{0}}[T^*(r)]}. \tag{8.82}$$

Proof: We have shown in the previous sections that the values of function u at the center of a sphere $S_d(r)$ containing a homogeneous material with conductivity σ_{eff} and a heterogeneous material with conductivity field $\sigma(\boldsymbol{x})$, $\boldsymbol{x} \in S_d(r)$, are $r^2/(2\,d\,\sigma_{\text{eff}})$ and $E^{\boldsymbol{0}}[T^*(r)]$, respectively (Eqs. 8.76 and 8.81 and Example 8.38). The effective conductivity of a heterogeneous medium results from the condition $E^{\boldsymbol{0}}[T^*(r)] = r^2/(2\,d\,\sigma_{\text{eff}}(r))$.

If the conductivity field is periodic, the dependence of σ_{eff} on r becomes weak for values of r exceeding the scale of material periodicity because additional segments of material beyond this value do not bring any new information. ■

Consider now a random heterogeneous material subjected to a unit flux, that is, $\gamma(\boldsymbol{x}) = 1$ in Eq. 8.77, and zero boundary condition. Let $S_d(r)$ be two identical spheres containing a random heterogeneous material with conductivity given by a random field $\Sigma(\boldsymbol{x})$, $\boldsymbol{x} \in S_d(r)$, and a homogeneous material of unknown conductivity Σ_{eff}. Our objective is to find properties of Σ_{eff}. We determine Σ_{eff}, as in Eq. 8.82, from the condition that the temperatures at the center of the heterogeneous and homogeneous spheres coincide. The problem is less simple in this case since the temperature $U(\boldsymbol{x})$ in the heterogeneous sphere is a random field. For a sample $\omega \in \Omega$ of the random heterogeneous material the temperature at the center of $S_d(r)$ is $U(\boldsymbol{0}, \omega) = E^{\boldsymbol{0}}[T^*(r, \omega)]$, where $T^*(r, \omega)$ is given by Eq. 8.80 for $D = S_d(r)$, material sample ω, and $\boldsymbol{X}(0) = \boldsymbol{0}$.

To find Σ_{eff}, we require that the temperature at the center of the sphere $S_d(r)$ containing the homogeneous material matches $U(\mathbf{0}, \omega) = E^{\mathbf{0}}[T^*(r, \omega)]$ for almost all ω's, that is, $r^2/(2\, d\, \Sigma_{\text{eff}}(r, \omega)) = U(\mathbf{0}, \omega)$, which yields (Eq. 8.82)

$$\Sigma_{\text{eff}}(r, \omega) = \frac{r^2}{2\, d\, E^{\mathbf{0}}[T^*(r, \omega)]}, \qquad \omega \in \Omega. \tag{8.83}$$

The resulting effective conductivity is a random variable. The following Monte Carlo algorithm can be used to calculate the effective conductivity Σ_{eff}.

- Select a sphere $S_d(r)$, that is, a radius $r > 0$, and generate n_ω independent samples $\Sigma(\boldsymbol{x}, \omega)$, $\boldsymbol{x} \in S_d(r)$, of the conductivity random field Σ.
- Estimate for each material sample ω the effective conductivity $\Sigma_{\text{eff}}(r, \omega)$ in Eq. 8.83. The estimation of the expectation $E^{\mathbf{0}}[T^*(r, \omega)]$ is based on n_s samples of the diffusion process $\boldsymbol{X}$ in Eqs. 8.78 and 8.79 for each material sample ω.
- Estimate properties of $\Sigma_{\text{eff}}(r)$, for example, $E[\Sigma_{\text{eff}}(r)]$ can be estimated by

$$\hat{\Sigma}_{\text{eff}}(r) = \frac{1}{n_\omega} \sum_{\omega=1}^{n_\omega} \Sigma_{\text{eff}}(r, \omega). \tag{8.84}$$

Note: The conductivity random field can be defined by a translation random field, that is, $\Sigma(\boldsymbol{x}) = F^{-1} \circ \Phi(G(\boldsymbol{x}))$, where F is a specified absolutely continuous distribution, Φ denotes the distribution of $N(0, 1)$, and G is a Gaussian random field with differentiable samples (Sections 3.6.6 and 5.3.3.1 in this book, [79], Section 3.1).

Let $\Sigma(\cdot, \omega)$ be a sample of Σ and let $\boldsymbol{X}^{(\omega)}$ denote the diffusion process $\boldsymbol{X}$ in Eqs. 8.78 and 8.79, in which $\sigma(\cdot)$ is replaced by $\Sigma(\cdot, \omega)$. Let $T^*(r, \omega)$ be the first time $\boldsymbol{X}^{(\omega)}$ starting at $\boldsymbol{x} = \mathbf{0}$ exits $S_d(r)$. The expectation of $T^*(r, \omega)$ can be estimated from n_s independent samples of $\boldsymbol{X}^{(\omega)}$ generated by the methods in Section 5.3.3.2. The effective conductivity corresponding to a material sample ω in $S_d(r)$ is given by Eq. 8.83. An estimate of the expected value of $\Sigma_{\text{eff}}(r)$ associated with $S_d(r)$ can be obtained from Eq. 8.84.

If the radius r of $S_d(r)$ is much larger than the correlation distance of the conductivity field and Σ is an ergodic random field, the dependence of $\Sigma_{\text{eff}}(r, \omega)$ on ω is likely to be insignificant for almost all material samples so that $\Sigma_{\text{eff}}(r)$ can be obtained from a single material sample and is nearly deterministic. ▲

Example 8.39: Consider a random heterogeneous medium in $S_d(r) = S_1(r)$ with two homogeneous phases of conductivity σ_1 for the matrix and $\sigma_2 > \sigma_1$ for inclusions. The conductivity field $\sigma(x)$ of this medium is a piece-wise constant function taking the values σ_1 and σ_2 on intervals of random location and length. Let the intervals (ξ_i, η_i) and (η_i, ξ_{i+1}), $\xi_i < \eta_i < \xi_{i+1}$, give the location in $S_1(r)$ of inclusion i and the matrix to the right end of this inclusion for an arbitrary realization ω of the medium. The spatial distribution of the inclusions is given by a Poisson process N with intensity $\lambda > 0$.

Numerical results have been obtained for $\sigma_1 = 1$, $\sigma_2 = 2$, $r = 10$, and $\lambda = 1$. The variance of the estimator $\hat{\Sigma}_{\text{eff}}(r)$ in Eq. 8.84 can be relatively large for small values of n_ω and n_s so that the resulting estimates of the effective conductivity can be unstable, that is, their values may differ significantly from sample to sample. For example, $\hat{\Sigma}_{\text{eff}}(r)$ is 1.7270 and 1.5812 for $(n_\omega = 10, n_s = 10)$ and $(n_\omega = 50, n_s = 10)$, respectively. The corresponding estimates of the coefficients of variation of $\hat{\Sigma}_{\text{eff}}(r)$ are 8.33% and 4.12%. Stable values of $\hat{\Sigma}_{\text{eff}}(r)$ result for $n_\omega \geq 50$ and $n_s \geq 500$. For $n_\omega = 50$ and $n_s = 500$ the estimated effective conductivity is $\hat{\Sigma}_{\text{eff}}(r)$ is 1.4997, and has a coefficient of variation of 1.6%. ◇

Note: The above numerical results are for a smooth approximation $\tilde{\Sigma} \in C^2(S_1(r))$ of the conductivity random field Σ of the two-phase material. The approximation $\tilde{\Sigma}$ is such that Eq. 8.78 with drift and diffusion coefficients in Eq. 8.79 has a unique solution. ▲

Example 8.40: Let U be the solution of $d\left(\Sigma(x)\,\frac{dU(x)}{dx}\right)/dx = -1$ for $x \in S(r) = (-r, r)$, $r > 0$, and $U(\pm r) = 0$, where the conductivity Σ is a random field with smooth strictly positive samples. The effective conductivity is

$$\Sigma_{\text{eff}} = \frac{r^2}{2}\left[-\int_{-r}^{0}\frac{y\,dy}{\Sigma(y)} + \int_{-r}^{r}\frac{y\,dy}{\Sigma(y)}\int_{-r}^{0}\frac{dy}{\Sigma(y)}\Big/\int_{-r}^{r}\frac{dy}{\Sigma(y)}\right]^{-1}.$$

Suppose that $\Sigma(t) = c_1 + c_2\exp(G(t))$, $t \in S(r)$, where $c_1, c_2 > 0$ are some constants and the values of the Gaussian field G at the locations $t_k = -r + k\,\Delta t$, $\Delta t = r/n$, are given by the autoregressive Gaussian sequence $G_{k+1} = \rho\,G_k + \sqrt{1-\rho^2}\,W_k$, k=0,1,...,2n-1, with G_0 and W_k independent $N(0,1)$ variables.

Figure 8.15 shows estimates of the expectation and the standard deviation

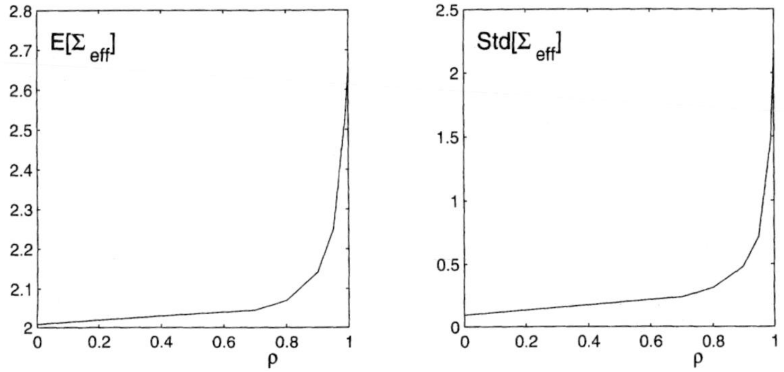

Figure 8.14. Second moment properties of Σ_{eff}

of Σ_{eff} based on 1,000 material samples. Small and large values of ρ correspond physically to rapid and slow spatial fluctuations of conductivity, respectively. These values of ρ can also be viewed as corresponding to large and small material specimens with respect to the microscale fluctuations. If $\rho \simeq 1$, the conductivity is nearly constant in each material specimen so that $\Sigma_{\text{eff}}(\omega) \simeq \alpha + \beta\, e^{G_0}$, and the mean and standard deviation of Σ_{eff} are $1 + e^{1/2} \simeq 2.65$ and $(e^2 - e)^{1/2} \simeq 2.16$, respectively. If $\rho \simeq 0$, the conductivity is nearly a white noise, the dependence of $\Sigma_{\text{eff}}(\omega)$ on ω is weak so that $\text{Std}[\Sigma_{\text{eff}}]$ is small. The relationship between the estimate of $E[\Sigma_{\text{eff}}]$ and ρ indicate a size effect phenomenon for the effective conductivity. ◇

Proof: The solution of the above equation for U is

$$U(x) = -\int_{-r}^{x} \frac{y\,dy}{\Sigma(y)} + \frac{\int_{-r}^{r} \frac{y\,dy}{\Sigma(y)}}{\int_{-r}^{r} \frac{dy}{\Sigma(y)}} \int_{-r}^{x} \frac{dy}{\Sigma(y)}.$$

The conditions in Eqs. 8.82 and 8.83 give the stated expression for Σ_{eff}. This expression and samples of Σ were used to construct the plots in Fig. 8.15. ■

Example 8.41: Consider a random heterogeneous material in an open bounded subset D of $\mathbb{R}^2$ with conductivity $\Sigma(\boldsymbol{t}) = \alpha + \beta \sum_{k=1}^{N(D')} Y_k\, h(\boldsymbol{t} - \boldsymbol{\Gamma}_k)$, where α, β are some constants, N is a Poisson process with intensity $\lambda > 0$, and $D' \supset D$ is such that points of N outside D' have nearly no contribution to Σ. Figure 8.15 shows contour lines for two samples of Σ corresponding to $\alpha = 2$, $\beta = 0.05$,

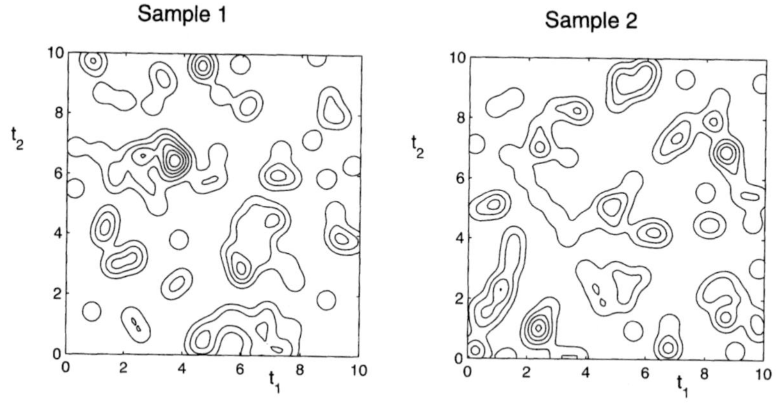

Figure 8.15. Contour lines for two samples of Σ

$Y_k = 1$, and $h(\boldsymbol{t}) = \phi(t_1/0.2)\, \phi(t_2/0.2)$, where ϕ denotes the density of $N(0, 1)$. Figure 8.16 shows the dependence of estimates of the expectation of Σ_{eff} on λ. The estimates of $E\,[\Sigma_{\text{eff}}]$ for each value of λ have been obtained from ten material samples and thirty samples of $\boldsymbol{X}$ for each material sample (Eq. 8.84). The estimates of $E\,[\Sigma_{\text{eff}}]$ nearly coincide with and are much larger than the matrix conductivity α for small and large values of λ, respectively. ◇

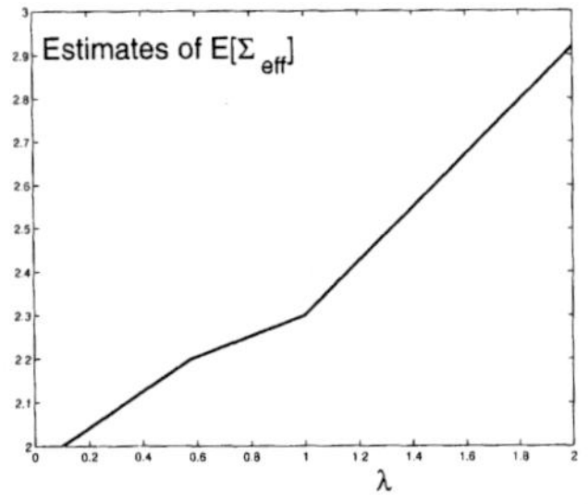

Figure 8.16. Estimates of $E\,[\Sigma_{\text{eff}}]$

Note: The average number of inclusions in D is equal to $\lambda\, v_{D'}$, where $v_{D'}$ denotes the volume of D'. The effective conductivity increases with λ since the inclusions have higher conductivity than the matrix. ▲

8.5.2 Elasticity

Our objective is the determination of effective properties for a class of random heterogeneous materials characterized by properties with a scale of spatial fluctuations much smaller than the overall dimension of material specimens. This type of heterogeneity is common to polycrystalline metals and multi-phase materials, for example, aluminum and glass fiber reinforced plastics.

Consider a random heterogeneous material in an open bounded subset D of $\mathbb{R}^d$ that is in equilibrium under some external actions. Let $S_{ij}(\boldsymbol{x})$, $\Gamma_{ij}(\boldsymbol{x})$, $A_{ijkl}(\boldsymbol{x})$, and $C_{ijkl}(\boldsymbol{x})$ denote, respectively, the strain, stress, stiffness, and compliance tensors at $\boldsymbol{x} \in D$. It is assumed that:

1. The material is linearly elastic with the stress-strain relationship

$$S_{ij}(\boldsymbol{x}) = A_{ijkl}(\boldsymbol{x})\,\Gamma_{kl}(\boldsymbol{x}) \quad \text{and} \quad \Gamma_{ij}(\boldsymbol{x}) = C_{ijkl}(\boldsymbol{x})\,S_{kl}(\boldsymbol{x}) \tag{8.85}$$

where $\boldsymbol{x} \in D$, $i, j, k, l = 1, \ldots, d$, and summation is performed over repeated subscripts.

Note: The matrix form of these equations is

$$\boldsymbol{S}(\boldsymbol{x}) = \boldsymbol{A}(\boldsymbol{x})\,\boldsymbol{\Gamma}(\boldsymbol{x}), \quad \boldsymbol{x} \in D \quad \text{and} \quad \boldsymbol{\Gamma}(\boldsymbol{x}) = \boldsymbol{C}(\boldsymbol{x})\,\boldsymbol{S}(\boldsymbol{x}), \quad \boldsymbol{x} \in D, \tag{8.86}$$

where $\boldsymbol{S} = (S_{11}, S_{22}, S_{33}, S_{12}, S_{23}, S_{31})$ and $\boldsymbol{\Gamma} = (\Gamma_{11}, \Gamma_{22}, \Gamma_{33}, \Gamma_{12}, \Gamma_{23}, \Gamma_{31})$ denote column vectors, and $\boldsymbol{A}$ and $\boldsymbol{C}$ are the corresponding matrices of stiffness and compliance coefficients for three-dimensional elasticity problems ($d = 3$). The fourth order tensor A_{ijkl} has the properties: (a) $A_{ijkl} = A_{ijlk}$ and $A_{ijkl} = A_{jikl}$ by the symmetry of the stress and strain tensors and (b) $A_{ijkl} = A_{klij}$ by the existence of the strain energy ([125], p. 141). The compliance coefficients C_{ijkl} have the similar properties. Hence, the tensors A_{ijkl} and C_{ijkl} in $\mathbb{R}^3$ have 21 distinct entries. ▲

2. The deformations are small so that

$$\Gamma_{ij}(\boldsymbol{x}) = \frac{1}{2}\left(U_{i,j}(\boldsymbol{x}) + U_{j,i}(\boldsymbol{x})\right), \quad \boldsymbol{x} \in D, \tag{8.87}$$

where $U_i(\boldsymbol{x})$ denotes the displacement function at $\boldsymbol{x}$ in the direction of coordinate x_i and $U_{i,j} = \partial U_i/\partial x_j$.

3. The stiffness $\boldsymbol{A}(\boldsymbol{x})$, $\boldsymbol{x} \in D \subset \mathbb{R}^d$, is an $\mathbb{R}^q$-valued random field defined on a probability space $(\Omega, \mathcal{F}, P)$, where the integer $q \geq 1$ gives the number of distinct entries of $\boldsymbol{A}$. The random field $\boldsymbol{A}$ is homogeneous, square integrable, and ergodic. Moreover, D is sufficiently large so that the spatial average of an arbitrary sample of $\boldsymbol{A}$ over D is nearly constant and equal to its ensemble average $E[\boldsymbol{A}(\boldsymbol{x})]$.

The constitutive law (Eq. 8.85), the kinematics relationships (Eq. 8.87), the equilibrium conditions

$$S_{ij,j}(\boldsymbol{x}) + F_i(\boldsymbol{x}) = 0, \quad \boldsymbol{x} \in D, \tag{8.88}$$

where F_i is the coordinate i of the body force vector, and the boundary conditions, for example, traction, displacements, or a mixture of these actions on the boundary ∂D of D, need to be used to find the displacement, strain, and stress fields in D. The equilibrium conditions expressed in terms of displacements are

$$\frac{1}{2}\left[A_{ijkl}(\boldsymbol{x})\,\left(U_{k,l}(\boldsymbol{x}) + U_{l,k}(\boldsymbol{x})\right)\right]_{,j} + F_i(\boldsymbol{x}) = 0$$

or $[A_{ijkl}(\boldsymbol{x})\, U_{k,l}(\boldsymbol{x})]_{,j} + F_i(\boldsymbol{x}) = 0$, so that the displacement field is the solution of a system of partial differential equations with random coefficients. The solution U_i, $i = 1, \ldots, d$, of these equations is a random field with differentiable functions defined on $(\Omega, \mathcal{F}, P)$. Hence, the strains and stresses are also random fields defined on the same probability space (Eqs. 8.85 and 8.87). Note that the local solution in Sections 6.2, 6.3, and 6.4 can be used to solve some elasticity problems for samples $\boldsymbol{A}(\cdot, \omega)$ of $\boldsymbol{A}$.

Let $(S_{ij}(\boldsymbol{x}), \Gamma_{ij}(\boldsymbol{x}), U_i(\boldsymbol{x}))$, $\boldsymbol{x} \in D$, $i, j = 1, \ldots, d$, be the solution of the stochastic boundary value problem defined by Eqs. 8.85-8.88. This solution provides detailed information on the stress, strain, and displacement fields that may be unnecessary. Coarser information provided, for example, by local averages of these fields may suffice in many applications. Denote by

$$\bar{S}_{ij}(\boldsymbol{x}) = \frac{1}{v(\boldsymbol{x})}\int_{D(\boldsymbol{x})} S_{ij}(\boldsymbol{\xi})\, d\boldsymbol{\xi} \quad \text{and} \quad \bar{\Gamma}_{ij}(\boldsymbol{x}) = \frac{1}{v(\boldsymbol{x})}\int_{D(\boldsymbol{x})} \Gamma_{ij}(\boldsymbol{\xi})\, d\boldsymbol{\xi} \tag{8.89}$$

the stress and strain moving average fields, where $v(\boldsymbol{x}) = \int_{D(\boldsymbol{x})} d\boldsymbol{x}$ and $D(\boldsymbol{x})$ is a subset of D centered on $\boldsymbol{x}$ such that $D(\boldsymbol{x})$ is small relative to D but large with respect to the scale of fluctuation of S_{ij} and Γ_{ij}. Also, $D(\boldsymbol{x})$ has the same shape and size for every $\boldsymbol{x}$. A set $D(\boldsymbol{x})$ with this property is referred to as a **representative volume**. Our objective is to find properties of the moving average fields $\bar{S}_{ij}$ and $\bar{\Gamma}_{ij}$. Two options are available. The first option is to derive properties of $\bar{S}_{ij}$ and $\bar{\Gamma}_{ij}$ from Eq. 8.89. The disadvantage of this approach is that it requires knowledge of the probability laws of the random fields $(S_{ij}, \Gamma_{ij}, U_i)$. The second option is to find properties of $\bar{S}_{ij}$ and $\bar{\Gamma}_{ij}$ directly. This approach requires the development of new constitutive laws for the average stress and strain fields, and such laws are difficult to derive in a general setting. For example, Eqs. 8.85 and 8.89 yield

$$\begin{aligned}\bar{S}_{ij}(\boldsymbol{x}) &= \frac{1}{v(\boldsymbol{x})}\int_{D(\boldsymbol{x})} A_{ijkl}(\boldsymbol{\xi})\,\Gamma_{kl}(\boldsymbol{\xi})\,d\boldsymbol{\xi} \\ &= \left(\frac{1}{v(\boldsymbol{x})}\int_{D(\boldsymbol{x})} A_{ijkl}(\boldsymbol{\xi})\,\frac{\Gamma_{kl}(\boldsymbol{\xi})}{\bar{\Gamma}_{kl}(\boldsymbol{x})}\,d\boldsymbol{\xi}\right)\bar{\Gamma}_{kl}(\boldsymbol{x}) = \bar{A}_{ijkl}(\boldsymbol{x})\,\bar{\Gamma}_{kl}(\boldsymbol{x}) \qquad (8.90)\end{aligned}$$

and

$$\begin{aligned}\bar{\Gamma}_{ij}(\boldsymbol{x}) &= \frac{1}{v(\boldsymbol{x})}\int_{D(\boldsymbol{x})} C_{ijkl}(\boldsymbol{\xi})\,S_{kl}(\boldsymbol{\xi})\,d\boldsymbol{\xi} \\ &= \left(\frac{1}{v(\boldsymbol{x})}\int_{D(\boldsymbol{x})} C_{ijkl}(\boldsymbol{\xi})\,\frac{S_{kl}(\boldsymbol{\xi})}{\bar{S}_{kl}(\boldsymbol{x})}\,d\boldsymbol{\xi}\right)\bar{S}_{kl}(\boldsymbol{x}) = \bar{C}_{ijkl}(\boldsymbol{x})\,\bar{S}_{kl}(\boldsymbol{x}), \qquad (8.91)\end{aligned}$$

respectively, provided that $\bar{\Gamma}_{kl}(\boldsymbol{x})$ and $\bar{S}_{kl}(\boldsymbol{x})$ are not zero. The matrix form of the above equations is $\bar{\boldsymbol{S}}(\boldsymbol{x}) = \bar{\boldsymbol{A}}(\boldsymbol{x})\,\bar{\boldsymbol{\Gamma}}(\boldsymbol{x})$ and $\bar{\boldsymbol{\Gamma}}(\boldsymbol{x}) = \bar{\boldsymbol{C}}(\boldsymbol{x})\,\bar{\boldsymbol{S}}(\boldsymbol{x})$, respectively, with notation as in Eq. 8.86. If the average is performed over D the argument $\boldsymbol{x}$ is dropped.

The weighted averages of the stiffness and compliance coefficients defined by Eqs. 8.90 and 8.91 give the constitutive law for the average stresses and strains, where $\bar{A}_{ijkl}$ and $\bar{C}_{ijkl}$ are called the **overall** or **effective stiffness** and **compliance** coefficients, respectively. These definitions of $\bar{A}_{ijkl}$ and $\bar{C}_{ijkl}$ are impractical since they involve the fields S_{ij} and Γ_{ij}. The following sections present approximations of and bounds on the effective stiffness and compliance coefficients, and are based on developments in [137, 194].

8.5.2.1 Displacement controlled experiment. Voigt's average

Consider an arbitrary linear elastic heterogeneous continuum in D with volume $v = \int_D d\boldsymbol{x}$, subjected to the boundary displacements

$$U_i(\boldsymbol{x}) = \bar{\gamma}_{ij}\,x_j, \quad \boldsymbol{x} \in \partial D, \qquad (8.92)$$

where $\bar{\gamma}_{ij} = \bar{\gamma}_{ji}$ are specified constants defining the magnitude of the applied boundary deformation. Let $\bar{S}_{ij}$, $\bar{\Gamma}_{ij}$, and $\bar{\mathcal{U}}$ denote the average over D of S_{ij}, Γ_{ij}, and $\mathcal{U}$, respectively, where $\mathcal{U}$ is the strain energy of the body. It is assumed that the body forces F_i in Eq. 8.88 are zero.

The average strain and the average strain energy are ([194], Section II.B)

$$\bar{\Gamma}_{ij} = \frac{1}{v} \int_D U_{i,j}(\boldsymbol{\xi})\, d\boldsymbol{\xi} = \bar{\gamma}_{ij},$$
$$\bar{\mathcal{U}} = \frac{1}{2\,v} \int_D S_{ij}(\boldsymbol{\xi})\, U_{i,j}(\boldsymbol{\xi})\, d\boldsymbol{\xi} = \frac{1}{2}\, \bar{S}_{ij}\, \bar{\gamma}_{ij}, \tag{8.93}$$

Proof: Let $\boldsymbol{\zeta} : \mathbb{R}^d \to \mathbb{R}^d$ be a differentiable function. The Gauss divergence theorem states

$$\int_D \operatorname{div}(\boldsymbol{\zeta}(\boldsymbol{x}))\, d\boldsymbol{x} = \int_{\partial D} \boldsymbol{\zeta}(\boldsymbol{x}) \cdot \boldsymbol{n}(\boldsymbol{x})\, d\sigma(\boldsymbol{x}) = \int_{\partial D} \zeta_i(\boldsymbol{x})\, n_i(\boldsymbol{x})\, d\sigma(\boldsymbol{x}),$$

where $\operatorname{div}(\boldsymbol{\zeta}(\boldsymbol{x})) = \zeta_{i,i}(\boldsymbol{x})$, $n_i(\boldsymbol{x})$ is the coordinate i of the exterior normal $\boldsymbol{n}(\boldsymbol{x})$ at $\boldsymbol{x} \in \partial D$, and $d\sigma(\boldsymbol{x})$ denotes a surface element on ∂D ([73], Section 9.4).

Let $\boldsymbol{U}^{(i)}(\boldsymbol{\xi})$ be a vector in $\mathbb{R}^d$ with entries $U_i(\boldsymbol{\xi})\, \delta_{qj}$, where i, j are some fixed indices and $q = 1, \ldots, d$, that is $\boldsymbol{U}^{(i)}(\boldsymbol{\xi})$ has zero coordinates except for the coordinate j which is $U_i(\boldsymbol{\xi})$. The boundary conditions in Eq. 8.92 imply that the entries of $\boldsymbol{U}^{(i)}(\boldsymbol{\xi})$ are $(\bar{\gamma}_{ik}\, \xi_k)\, \delta_{qj}$ for $\boldsymbol{\xi} \in \partial D$. These observations and the Gauss divergence theorem give

$$\begin{aligned} \int_D U_{i,j}(\boldsymbol{\xi})\, d\boldsymbol{\xi} &= \int_D \nabla \cdot \boldsymbol{U}^{(i)}(\boldsymbol{\xi})\, d\boldsymbol{\xi} = \int_{\partial D} \boldsymbol{U}^{(i)}(\boldsymbol{\xi}) \cdot \boldsymbol{n}(\boldsymbol{\xi})\, d\sigma(\boldsymbol{\xi}) \\ &= \int_{\partial D} \bar{\gamma}_{ik}\, \xi_k\, \delta_{qj}\, n_q(\boldsymbol{\xi})\, d\sigma(\boldsymbol{\xi}) = \int_D \frac{\partial(\bar{\gamma}_{ik}\, \xi_k)}{\partial \xi_j}\, d\boldsymbol{\xi} = \int_D \bar{\gamma}_{ij}\, d\boldsymbol{\xi} = v\, \bar{\gamma}_{ij} \end{aligned}$$

so that $\bar{\Gamma}_{ij} = (1/v) \int_D U_{i,j}(\boldsymbol{\xi})\, d\boldsymbol{\xi} = \bar{\gamma}_{ij}$.

We now calculate the average of the strain energy. Let $\tilde{\boldsymbol{U}}(\boldsymbol{\xi})$ be a vector in $\mathbb{R}^d$ with entries $S_{iq}(\boldsymbol{\xi})\, U_i(\boldsymbol{\xi})\, \delta_{qj}$, $q = 1, \ldots, d$, and note that

$$\nabla \cdot \tilde{\boldsymbol{U}}(\boldsymbol{\xi}) = \frac{\partial(S_{ij}(\boldsymbol{\xi})\, U_i(\boldsymbol{\xi}))}{\partial \xi_j} = S_{ij}(\boldsymbol{\xi})\, U_{i,j}(\boldsymbol{\xi})$$

since $S_{ij,j}(\boldsymbol{\xi}) = 0$. The indices i and j are arbitrary but fixed. The average strain energy accumulated in the elastic body is

$$\begin{aligned} \bar{\mathcal{U}} &= \frac{1}{2\,v} \int_D S_{ij}(\boldsymbol{x})\, U_{i,j}(\boldsymbol{x})\, d\boldsymbol{x} = \frac{1}{2\,v} \int_D \nabla \cdot \tilde{\boldsymbol{U}}(\boldsymbol{\xi})\, d\boldsymbol{\xi} \\ &= \frac{1}{2\,v} \int_{\partial D} \tilde{U}_q(\boldsymbol{\xi})\, n_q(\boldsymbol{\xi})\, d\sigma(\boldsymbol{\xi}) = \frac{1}{2\,v} \int_{\partial D} S_{ij}(\boldsymbol{\xi})\, \bar{\gamma}_{ik}\, \xi_k\, n_j(\boldsymbol{\xi})\, d\sigma(\boldsymbol{\xi}) \\ &= \frac{1}{2\,v} \int_D \frac{\partial(S_{ij}(\boldsymbol{\xi})\, \bar{\gamma}_{ik}\, \xi_k)}{\partial \xi_j}\, d\boldsymbol{\xi} = \frac{1}{2}\, \bar{\gamma}_{ij} \int_D S_{ij}(\boldsymbol{\xi})\, d\boldsymbol{\xi} = \frac{1}{2}\, \bar{S}_{ij}\, \bar{\gamma}_{ij}, \end{aligned}$$

where the above equalities follow from the Gauss divergence theorem, boundary conditions, and the definition of the average stress provided that there are no body forces. ■

The **Voigt average**

$$A^{(V)} = \frac{1}{v} \int_D A(x)\, dx \tag{8.94}$$

and the **overall stiffness** $\bar{A}$ in Eq. 8.90 with $D(x) = D$ satisfy the inequality

$$\bar{\gamma}^T \bar{A}\, \bar{\gamma} \le \bar{\gamma}^T A^{(V)}\, \bar{\gamma} \quad \text{for any } \bar{\gamma} = \{\bar{\gamma}_{ij}\}. \tag{8.95}$$

Proof: The Voigt average corresponds to the assumption that the strain field in a heterogeneous medium is uniform, and is analogous to the effective stiffness of a parallel system (Example 8.35). We say that $A^{(V)}$ is an upper bound on $\bar{A}$ since the matrix $A^{(V)} - \bar{A}$ is positive definite (Eq. 8.95).

Let $\bar{A}$ be the overall material stiffness given by Eq. 8.90 in which $D(x)$ coincides with D. The corresponding average stress and strain fields over D are related by $\bar{S}_{ij} = \bar{A}_{ijkl}\, \bar{\Gamma}_{kl}$ so that the average strain energy is $\bar{\mathcal{U}} = \frac{1}{2}\, \bar{\gamma}^T\, \bar{A}\, \bar{\gamma}$ (Eq. 8.93), where $\bar{\gamma}$ is a column vector including all distinct components of the tensor $\bar{\gamma}_{ij}$, that is, the average strains $(\bar{\gamma}_{11}, \bar{\gamma}_{22}, \bar{\gamma}_{33}, \bar{\gamma}_{12}, \bar{\gamma}_{23}, \bar{\gamma}_{31})$. As previously stated, these relationships are of little use for applications because $\bar{A}_{ijkl}$ depends on detailed information on the stress and strain fields (Eqs. 8.90-8.91).

Let $\Gamma_{ij}(x)$ be the strain field corresponding to the equilibrium configuration of a linear elastic random heterogeneous material in D with the boundary conditions in Eq. 8.92. The strain energy corresponding to this field is $\bar{\mathcal{U}} = \frac{1}{2}\, \bar{\gamma}^T\, \bar{A}\, \bar{\gamma}$. Consider also the uniform strain field $\bar{\gamma}_{ij}$ corresponding to a linear elastic homogeneous material in D under the same boundary conditions (Eq. 8.92). We have

$$\begin{aligned}\bar{\mathcal{U}} &= \frac{1}{2}\, \bar{\gamma}^T\, \bar{A}\, \bar{\gamma} = \frac{1}{2\,v} \int_D \Gamma_{ij}(x)\, A_{ijkl}(x)\, \Gamma_{kl}(x)\, dx \le \frac{1}{2\,v} \int_D \bar{\gamma}_{ij}\, A_{ijkl}(x)\, \bar{\gamma}_{kl}\, dx \\ &= \frac{1}{2}\, \bar{\gamma}_{ij}\, \bar{\gamma}_{kl} \left[\frac{1}{2v} \int_D A_{ijkl}(x)\, dx \right] = \frac{1}{2}\, \bar{\gamma}^T\, A^{(V)}\, \bar{\gamma},\end{aligned}$$

since the actual strain field minimizes the strain energy ([68], Section 10.7). ■

Example 8.42: Consider a rod of length $v > 0$ and unit cross section characterized by the random stiffness $A(x)$, $0 \le x \le v$. The displacements of the left and right ends of the rod are zero and $\bar{\gamma} > 0$, respectively. The overall constitutive law is $\bar{S} = \bar{A}\, \bar{\Gamma}$ and

$$\bar{A} = \frac{1}{v} \int_0^v A(x)\, \frac{\Gamma(x)}{\bar{\Gamma}}\, dx \le A^{(V)} = \frac{1}{v} \int_0^v A(x)\, dx,$$

where $\bar{A}$ is defined by Eq. 8.90. ◇

Proof: The average strain energy accumulated in the rod is $\bar{\mathcal{U}} = (1/2)\, \bar{A}\, \bar{\gamma}^2$. According to the principle of minimum potential energy, we have

$$\bar{\mathcal{U}} = \frac{1}{2}\, \bar{A}\, \bar{\gamma}^2 \le \frac{1}{2\,v} \int_0^v A(x)\, \bar{\gamma}^2\, dx = \frac{1}{2}\, \bar{\gamma}^2 \left(\frac{1}{v} \int_0^v A(x)\, dx \right) = \frac{1}{2}\, A^{(V)}\, \bar{\gamma}^2$$

so that $\bar{A} \le A^{(V)}$. ■

8.5.2.2 Stress controlled experiment. Reuss's average

Consider a linearly elastic material in an open bounded subset D of $\mathbb{R}^d$ defined at the beginning of Section 8.5.2. The material is subjected to the boundary traction

$$S_{ij}(\boldsymbol{x})\, n_j(\boldsymbol{x}) = \bar{s}_{ij}\, n_j(\boldsymbol{x}), \quad \boldsymbol{x} \in \partial D, \tag{8.96}$$

where $\bar{s}_{ij} = \bar{s}_{ji}$ are specified constants, $n_j(\boldsymbol{x})$ are the coordinates of the exterior normal $\boldsymbol{n}(\boldsymbol{x})$ to ∂D at $\boldsymbol{x} \in D$, and v denotes the volume of D. Let $\bar{S}_{ij}$, $\bar{\Gamma}_{ij}$, and $\bar{\mathcal{U}}$ be the average of S_{ij}, Γ_{ij}, and $\mathcal{U}$ over D, respectively. It is assumed that the body forces F_i in Eq. 8.88 are zero.

The average stress and the average strain energy are ([194], Section II.B)

$$\bar{S}_{ij} = \frac{1}{v}\int_D S_{ij}(\boldsymbol{\xi})\, d\boldsymbol{\xi} = \bar{s}_{ij},$$
$$\bar{\mathcal{U}} = \frac{1}{2\,v}\int_D S_{ij}(\boldsymbol{\xi})\,\Gamma_{ij}(\boldsymbol{\xi})\, d\boldsymbol{\xi} = \frac{1}{2}\,\bar{s}_{ij}\,\bar{\Gamma}_{ij}. \tag{8.97}$$

Proof: For any stress field we have

$$\int_{\partial D} S_{ij}(\boldsymbol{\xi})\,\xi_k\, n_j(\boldsymbol{\xi})\, d\sigma(\boldsymbol{\xi}) = \int_D \left(S_{ij}(\boldsymbol{x})\,\xi_k\right)_{,j} d\boldsymbol{\xi}$$
$$= \int_D \left(S_{ij,j}(\boldsymbol{\xi})\,\xi_k + S_{ij}(\boldsymbol{\xi})\,\delta_{jk}\right) d\boldsymbol{\xi} = \int_D S_{ik}(\boldsymbol{\xi})\, d\boldsymbol{\xi}$$

by the Gauss divergence theorem and the equilibrium condition $S_{ij,j}(\boldsymbol{x}) = 0$. Because the stress field is constant on the boundary of D, we have

$$\int_D S_{ik}(\boldsymbol{\xi})\, d\boldsymbol{\xi} = \int_{\partial D} S_{ij}(\boldsymbol{\xi})\,\xi_k\, n_j(\boldsymbol{\xi})\, d\sigma(\boldsymbol{\xi})$$
$$= \int_{\partial D} \bar{s}_{ij}\,\xi_k\, n_j(\boldsymbol{\xi})\, d\sigma(\boldsymbol{\xi}) = \int_D \bar{s}_{ik}\, d\boldsymbol{\xi} = v\,\bar{s}_{ik}$$

so that $\bar{S}_{ik} = (1/v)\int_D S_{ik}(\boldsymbol{x})\, d\boldsymbol{x} = \bar{s}_{ik}$.

The average of the strain energy over D is

$$\bar{\mathcal{U}} = \frac{1}{2\,v}\int_D S_{ij}(\boldsymbol{\xi})\, U_{i,j}(\boldsymbol{\xi})\, d\boldsymbol{\xi} = \frac{1}{2\,v}\int_{\partial D} S_{ij}(\boldsymbol{\xi})\, U_i(\boldsymbol{\xi})\, n_j(\boldsymbol{\xi})\, d\sigma(\boldsymbol{\xi})$$
$$= \bar{s}_{ij}\,\frac{1}{2\,v}\int_{\partial D} U_i(\boldsymbol{\xi})\, n_j(\boldsymbol{\xi})\, d\sigma(\boldsymbol{\xi}) = \bar{s}_{ij}\,\frac{1}{2\,v}\int_D U_{i,j}(\boldsymbol{\xi})\, d\boldsymbol{\xi}$$
$$= \bar{s}_{ij}\,\frac{1}{2\,v}\int_D \Gamma_{ij}(\boldsymbol{\xi})\, d\boldsymbol{\xi} = \frac{1}{2}\,\bar{s}_{ij}\,\bar{\Gamma}_{ij}$$

with the notation used to derive $\bar{\mathcal{U}}$ in Eq. 8.93 and boundary conditions. ■

The **Reuss average**

$$C^{(R)} = \frac{1}{v} \int_D C(x)\, dx \tag{8.98}$$

and the **overall compliance** $\bar{C}$ in Eq. 8.91 with $D(x) = D$ satisfy the inequality

$$\bar{s}^T\, C^{(R)}\, \bar{s} \le \bar{s}^T\, \bar{C}\, \bar{s} \quad \text{for any } \bar{s} = (\bar{s}_{ij}). \tag{8.99}$$

Proof: The Reuss average corresponds to the assumption that the stress field is uniform throughout the heterogeneous material, and is analogous to the effective stiffness of a series system (Example 8.35). We say that $C^{(R)}$ is an upper bound on $\bar{C}$ since $C^{(R)} - \bar{C}$ is positive definite (Eq. 8.99).

Let $\bar{C}$ be the overall material compliance given by Eq. 8.91 in which $D(x)$ coincides with D. The corresponding average strains and stresses are related by $\bar{\Gamma}_{ij} = \bar{C}_{ijkl}\, \bar{S}_{kl}$ so that the average strain energy is $\bar{\mathcal{U}} = \frac{1}{2}\, \bar{s}^T\, \bar{C}\, \bar{s}$ (Eq. 8.97), where $\bar{s}$ is a column vector whose entries are the distinct elements of $\bar{s}_{ij}$. As previously mentioned, the calculation of the average compliance coefficients $\bar{C}_{ijkl}$ is impractical.

Let $S_{ij}(x)$ be the stress field corresponding to the equilibrium configuration of a linear elastic random heterogeneous material subjected to the boundary conditions in Eq. 8.96. The average strain energy of this solid is given by $\bar{\mathcal{U}} = \frac{1}{2}\, \bar{s}^T\, \bar{C}\, \bar{s}$. Consider also the constant stress field $\bar{s}_{ij}$ corresponding to a linear elastic homogeneous material in D satisfying Eq. 8.96. We have

$$\begin{aligned} \bar{\mathcal{U}}^c &= \frac{1}{2}\, \bar{s}^T\, \bar{C}\, \bar{s} = \frac{1}{2\,v} \int_D S_{ij}(x)\, C_{ijkl}(x)\, S_{kl}(x)\, dx \le \frac{1}{2\,v} \int_D \bar{s}_{ij}\, C_{ijkl}(x)\, \bar{s}_{kl}\, dx \\ &= \frac{1}{2}\, \bar{s}_{ij}\, \bar{s}_{kl}\, \frac{1}{v} \int_D C_{ijkl}(x)\, dx = \frac{1}{2}\, \bar{s}^T\, C^{(R)}\, \bar{s}, \end{aligned}$$

since the actual stress field minimizes the complementary strain energy and $\bar{s}_{ij}$ satisfies the traction boundary conditions ([68], Section 10.9). ■

The Voigt and Reuss averages can be used to bound the overall stiffness in the sense of the inequalities in Eqs. 8.95 and 8.99. For example, we have seen that $A^{(V)}$ provides an upper bound on $\bar{A}$. We also have that

$$A^{(R)} = (C^{(R)})^{-1} \tag{8.100}$$

is a lower bound on $\bar{A} = (\bar{C})^{-1}$ since $\bar{A} - A^{(R)}$ is positive definite ([166], Appendix A.2).

Example 8.43: Consider the random heterogeneous rod in Example 8.42 and assume that it is subjected to a traction $\bar{s}$ at its ends. The overall constitutive law is $\bar{S} = \bar{A}\, \bar{\Gamma}$ or $\bar{\Gamma} = \bar{C}\, \bar{S}$ and

$$\bar{C} = \frac{1}{v} \int_0^v C(x)\, \frac{S(x)}{\bar{S}}\, dx \le C^{(R)} = \frac{1}{v} \int_0^v C(x)\, dx$$

so that $\bar{A} = 1/\bar{C} \ge A^{(R)} = 1/C^{(R)}$. ◇

Proof: The average complementary strain energy accumulated in the rod is $\bar{\mathcal{U}}^c = (1/2)\,\bar{C}\,\bar{s}^2$. According to the principle of minimum complementary potential energy, we have

$$\bar{\mathcal{U}}^c = \frac{1}{2}\,\bar{C}\,\bar{s}^2 \le \frac{1}{2\,v}\int_0^v C(x)\,\bar{s}^2\,dx = \frac{1}{2}\,\bar{s}^2\left(\frac{1}{v}\int_0^v C(x)\,dx\right) = \frac{1}{2}\,C^{(R)}\,\bar{s}^2$$

so that $\bar{C} \le C^{(R)}$. ■

Example 8.44: Let D be a cube with unit sides containing a multi-phase material consisting of n isotropic homogeneous phases with shear and bulk elastic moduli $G^{(r)}$ and $K^{(r)}$, and volume v_r, $r = 1, \ldots, n$, so that $\sum_{r=1}^{n} v_r = 1$. It is assumed that the dimension of individual constituents is much smaller than unity. Let $\bar{G}$ and $\bar{K}$ denote the effective shear and bulk moduli of the material in D corresponding to uniform pure shear and hydrostatic pressure, respectively. For example, $\bar{G}$ for a uniform pure shear s_0 in the (x_1, x_2)-plane is $\bar{G} = s_0/\bar{\Gamma}_{12}$, where $\bar{\Gamma}_{12} = \int_D \Gamma_{12}(\boldsymbol{x})\,d\boldsymbol{x}$ since D has unit volume (Eq. 8.90). The Voigt and Reuss averages give the bounds

$$\left[\sum_{r=1}^{n}\frac{v_r}{G^{(r)}}\right]^{-1} \le \bar{G} \le \sum_{r=1}^{n} v_r\,G^{(r)} \quad\text{and}\quad \left[\sum_{r=1}^{n}\frac{v_r}{K^{(r)}}\right]^{-1} \le \bar{K} \le \sum_{r=1}^{n} v_r\,K^{(r)}$$

on the effective shear and bulk moduli, $\bar{G}$ and $\bar{K}$. ◇

Proof: Suppose that the cube is sheared in the (x_1, x_2)-plane. The effective shear modulus $\bar{G}$ in this plane is bounded by the Voigt and Reuss averages given by Eq. 8.94 and Eq. 8.100. These equations can also be used to bound the effective bulk modulus $\bar{K}$ corresponding to hydrostatic pressure.

The Reuss averages provide lower bounds on $\bar{G}$ and $\bar{K}$ and correspond to uniform boundary stresses (Eq. 8.100), while the Voigt averages relate to uniform boundary strains (Eq. 8.94). That the Reuss average is smaller than the Voigt average also follows from the Cauchy inequality ([149], Theorem 2, p. 46)

$$\left|\sum_{r=1}^{n} a_r\,b_r\right|^2 \le \left(\sum_{r=1}^{n}|a_r|^2\right)\left(\sum_{r=1}^{n}|b_r|^2\right)$$

applied for $a_r = (v_r\,G_r)^{1/2}$ and $b_r = (v_r/G_r)^{1/2}$. This inequality holds for any a_r, b_r in $\mathbb{R}$ or $\mathbb{C}$. ■

8.5.2.3 Physically based approximations

We have mentioned that the determination of macroscopic material properties from Eqs. 8.90 and 8.91 is computationally intensive because the strain and stress fields, respectively, must be known everywhere in D. The previous two sections provide bounds on the overall stiffness and compliance coefficients $\bar{A}_{ijkl}$ and $\bar{C}_{ijkl}$. An alternative approach is to calculate $\bar{A}_{ijkl}$ and $\bar{C}_{ijkl}$ from Eqs. 8.90 and

8.91 in which the actual strain and stress fields are approximated. The accuracy of the resulting effective moduli $\bar{A}_{ijkl}$ and $\bar{C}_{ijkl}$ depends on the postulated strain and stress fields. We illustrate this approach by two examples. The perturbation and the Neumann series method are applied in the following section to develop alternative approximations for $\bar{A}_{ijkl}$ and $\bar{C}_{ijkl}$.

Example 8.45: Consider the multi-phase material in Example 8.44 and let $\nu^{(r)}$ be the Poisson ratio of phase r, $r = 1, \ldots, n$. Suppose that the material is sheared in the (x_1, x_2)-plane by a deterministic stress s_0. The overall or effective shear modulus in this plane is $\bar{G} = \sum_{r=1}^{n} v_r\, G^{(r)}\, \bar{\Gamma}_{12}^{(r)} / \bar{\Gamma}_{12}$, where $\bar{\Gamma}_{12}^{(r)}$ and $\bar{\Gamma}_{12}$ are averages of the strain field Γ_{12} over D_r and D, respectively, and D_r denotes the subset of D occupied by phase r (Eq. 8.90 and Example 8.44). We also denote by $\bar{K}$ the effective bulk modulus (Example 8.44).

The use of the above expression for the effective shear modulus $\bar{G}$ requires knowledge of the strain field. Two options are available. We can calculate the actual strain field Γ_{12} or we can postulate an approximate strain field Γ_{12} and then calculate $\bar{\Gamma}_{12}$. We use the latter option and take $\bar{\Gamma}_{12}^{(r)}$ to be the strain in a spherical inclusion with the properties of phase r that is embedded in an infinite elastic matrix with the macroscopic properties $\bar{G}$ and $\bar{K}$ of the composite. The resulting stress field is $\bar{\Gamma}_{12}^{(r)} = s_0 / \left(\bar{G} + \bar{\beta}(G^{(r)} - \bar{G})\right)$ where $\bar{\beta} = 2\,(4 - 5\,\bar{\nu})\,(15\,(1 - \bar{\nu}))$, $\bar{\nu} = (3\,\bar{K} - 2\,\bar{G}) / \left(6\,\bar{K} + 2\,\bar{G}\right)$, and $\bar{K} = \sum_{r=1}^{n} v_r\, K^{(r)}\, \bar{\Theta}^{(r)} / \bar{\Theta}$ [34]. The approximate contraction $\bar{\Theta}^{(r)}$ of phase r is $\bar{\Theta}^{(r)} = p / \left(\bar{K} + \bar{\alpha}\,(K^{(r)} - \bar{K})\right)$, where $\bar{\alpha} = (1 + \bar{\nu}) / (3\,(1 - \bar{\nu}))$ and p denotes a deterministic hydrostatic pressure applied to D. The average strains and contraction over D are

$$\bar{\Gamma}_{12} = \sum_{r=1}^{n} v_r\, \bar{\Gamma}_{12}^{(r)} \quad \text{and} \quad \bar{\Theta} = \sum_{r=1}^{n} v_r\, \bar{\Theta}^{(r)}.$$

Figure 8.17 shows histograms of $\bar{K}$ and $\bar{G}$ for a three-phase material with concentrations $(0.2, 0.3, 0.5)$. The corresponding phase shear and bulk moduli are $(1, 2, 3)$ and $(4, 1, 2)$, respectively. Let n_v denote the average number of constituents. A larger n_v means a larger reference volume D. The histograms in the figure are for $n_v = 20$ and $n_v = 200$. Larger values of n_v can be interpreted as larger material specimens. The histograms in Fig. 8.17 are centered on the asymptotic values, 1.9122 and 2.1420, of $\bar{K}$ and $\bar{G}$ corresponding to $n_v \to \infty$. The uncertainty in $\bar{K}$ and $\bar{G}$ decreases with the specimen size. For example, the range of the histograms of $\bar{G}$ is $(1.57, 2.71)$ for $n_v = 20$ and $(1.97, 2.33)$ for $n_v = 200$. The corresponding standard deviations of $\bar{G}$ are 0.1941 and 0.0657 for the small and large specimen, respectively. $\diamond$

Proof: The stress-strain relationship for an isotropic homogeneous material is $S_{ij} = \lambda\, \Gamma_{\alpha\alpha}\, \delta_{ij} + 2\, G\, \Gamma_{ij}$, where λ and G are material constants. The average of the stress

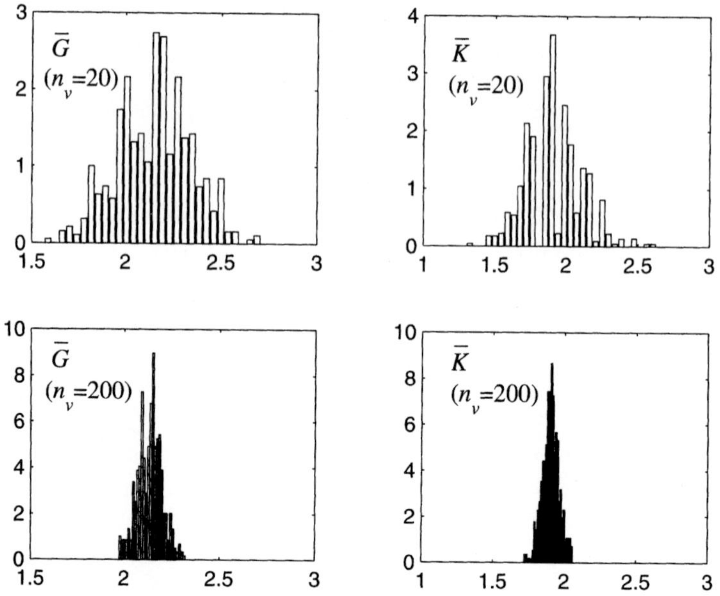

Figure 8.17. Histograms of $\bar{K}$ and $\bar{G}$ for a three-phase material

field in D is (Eq. 8.90 with $D(\boldsymbol{x}) = D$)

$$\bar{S}_{ij} = \frac{1}{v} \int_D S_{ij}(\boldsymbol{x})\, d\boldsymbol{x} = \sum_{r=1}^{n} v_r \left[\delta_{ij}\, \lambda^{(r)} \frac{1}{v_r} \int_{D_r} \Gamma_{\alpha\alpha}(\boldsymbol{x})\, d\boldsymbol{x} + 2\, G^{(r)} \frac{1}{v_r} \int_{D_r} \Gamma_{ij}(\boldsymbol{x})\, d\boldsymbol{x} \right]$$

so that

$$\bar{S}_{12} = \sum_{r=1}^{n} v_r \left[2\, G^{(r)} \frac{1}{v_r} \int_{D_r} \Gamma_{12}(\boldsymbol{x})\, d\boldsymbol{x} \right] = \left[\sum_{r=1}^{n} v_r\, G^{(r)} \frac{\bar{\Gamma}_{12}^{(r)}}{\bar{\Gamma}_{12}} \right] 2\, \bar{\Gamma}_{12},$$

which yields the stated constitutive law since the total shearing strain is equal to $2\bar{\Gamma}_{12}$.

The relationship between pressure and volume change $\Gamma_{\alpha\alpha} = \Gamma_{11} + \Gamma_{22} + \Gamma_{33}$ for an isotropic homogeneous material is $S_{\alpha\alpha} = (3\,\lambda + 2\,G)\,\Gamma_{\alpha\alpha} = K\,\Gamma_{\alpha\alpha}$, where K denotes the bulk modulus. The average $\bar{S}_{\alpha\alpha}$ of $S_{\alpha\alpha}$ over D is

$$\bar{S}_{\alpha\alpha} = \sum_{r=1}^{n} v_r\, K^{(r)} \left[\frac{1}{v_r} \int_{D_r} \Gamma_{\alpha\alpha}(\boldsymbol{x})\, d\boldsymbol{x} \right] = \left[\sum_{r=1}^{n} v_r\, K^{(r)} \frac{\bar{\Gamma}_{\alpha\alpha}^{(r)}}{\bar{\Gamma}_{\alpha\alpha}} \right] \bar{\Gamma}_{\alpha\alpha},$$

so that $\bar{K}$ is as stated.

The expressions of $\bar{G}$, $\bar{K}$, $\bar{\alpha}$, $\bar{\beta}$, and $\bar{\nu}$ form a system of nonlinear algebraic equations whose solution gives $\bar{G}$ and $\bar{K}$. The plots in Fig. 8.17 represent the solutions of these algebraic equations for material samples generated by Monte Carlo simulation.

A multinomial distribution with n distinct outcomes was used to generate samples of the multi-phase material. Let $p_r > 0$, $\sum_{r=1}^{n} p_r = 1$, be the probability of occurrence of phase r in a trial and let X_r denote the number of such occurrences in n_v independent trials, so that

$$P\left(\cap_{r=1}^{n}\{X_r = x_r\}\right) = \frac{n_v!}{x_1!\cdots x_n!}\, p_1^{x_1}\cdots p_n^{x_n}$$

if $\sum_{r=1}^{n} x_r = n_v$ and zero otherwise, where $x_r = 0, 1, \ldots, n_v$ for each phase. The average of the ratio X_r/n_v is p_r, that is, the fractional volume of phase r in D. ■

Example 8.46: A collection of material particles that may have random shape and/or mechanical properties whose interaction is localized at their contacts is referred to as a **granular material**. In contrast to multi-phase media, the particles of a granular material are not included in a matrix. Most studies on granular materials focus on the development of continuum constitutive equations and the analysis of potential changes in material properties, referred to as structure evolution [72, 163]. The analysis of granular materials can be based on (1) concepts of molecular dynamics, (2) continuum smoothing and balance approach, (3) kinetic theory based on the conservation theorem of statistical mechanics and the Maxwell-Bolzman distribution, and (4) mean field hypothesis [72, 108, 148].

We discuss solutions based on the mean field hypothesis. The effective bulk and effective shear moduli, $\bar{\lambda}$ and $2\,\bar{\mu}$, of a random packing of homogeneous and isotropic spheres of the same radius $r > 0$ given by the mean field hypothesis are

$$\bar{\lambda} = \frac{\mu\, k\,(1-\alpha)}{5\,\pi\, r}\left[\frac{a}{1-\nu} - \frac{2\,b}{2-\nu}\right],$$
$$2\,\bar{\mu} = \frac{\mu\, k\,(1-\alpha)}{5\,\pi\, r}\left[\frac{2\,a}{1-\nu} - \frac{6\,b}{2-\nu}\right], \qquad (8.101)$$

where μ and ν are the shear modulus and the Poisson ratio of the spheres, k denotes the average number of contact points for a sphere of the aggregate, a and b are some constants related to the geometry of the contact between two spheres, and α is the aggregate porosity, that is, the ratio of void to total volume. The mean field hypothesis states that (1) each particle has the same number of contacts that is equal to the average number of particle contacts over the entire aggregate and (2) the contacts are equally spaced on the boundary of each particles. These assumptions hold in a perfectly ordered configuration of the aggregate. The use of the mean field hypothesis eliminates the only source of uncertainty, the geometry of packing, and gives an overall relationship between stresses and strains that is deterministic.

The derivation of the effective moduli in Eq. 8.101 involves two steps. First, the constitutive law of the interaction between two particles needs to be established. Second, the mean field hypothesis is applied to eliminate the uncertainty in the coefficients of the equilibrium conditions and establish overall constitutive laws. Details on these derivations can be found in [52]. ◇

Note: We summarize the essential steps of the mean field solution for the special case of an assembly of infinitely long cylindrical particles with circular cross section of radius $r > 0$ and parallel axes. Let S be a plane section through the assembly, which is parallel to the axes of the particles. Consider a rectangular subset S' of S with unit side in the direction of the particle axes. The other side of S' has length $a \gg r$. The area of the intersection of S' with the cylindrical particles of the aggregate is $a^* \simeq \sum_i N_i\,[2\,r\,\sin(\alpha_i)]$, where N_i denotes the number of cylindrical particles intersecting S' with an angle $[\alpha_i, \alpha_i + d\alpha_i)$ (Fig. 8.18). The numbers N_i are random since the packing is not ordered. The interaction

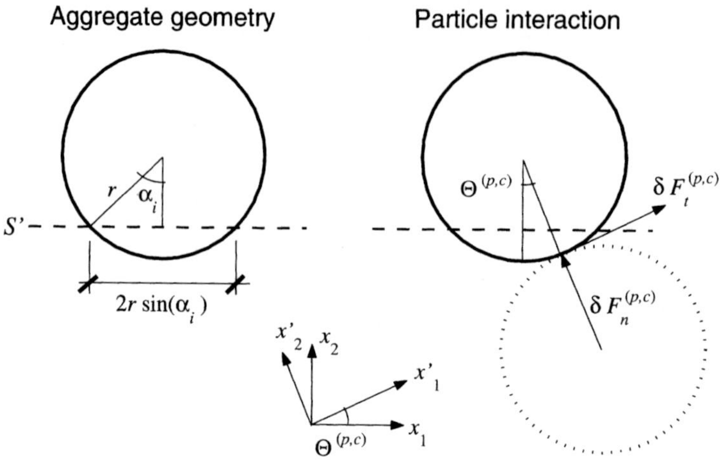

Figure 8.18. Particles of a granular material

between two particles in contact consists of the forces F_n and F_t normal to and contained in the plane tangent to the boundaries of these particles at their contact point. It is assumed that there is no relative movements between particles along their axes so that we can consider that F_t is contained in the (x_1, x_2) plane (Fig. 8.18).

Suppose that the assembly is in equilibrium under equal pressure in the x_1 and x_2 directions. These boundary conditions would generate uniform stress and strain fields in a homogeneous isotropic material replacing the aggregate of particles. Let δs_{ij} be a perturbation of the uniform stress field in this material and $\delta\gamma_{ij}$ denote the corresponding increment in strains. Our objective is to determine the relationship between δs_{ij} and $\delta\gamma_{ij}$, that is, the overall constitutive law or, equivalently, the effective constants of the aggregate. Consider again the set S' and suppose that the stress increment is limited to the component parallel to axis x_2. The total force increment on this section is $a\,\delta s_{22}$. The equilibrium of the assembly is preserved if the original contact forces between particles are slightly altered. Let $\delta F_n^{(p,c)}$ and $\delta F_t^{(p,c)}$ denote changes in the contact forces caused by the stress increment δs_{22}. The assembly is in equilibrium if the condition

$$a\,\delta s_{22} = -\sum_{\text{particles}}\ \sum_{\text{contacts}} \left[\delta F_n^{(p,c)} \cos(\Theta^{(p,c)}) + \delta F_t^{(p,c)} \sin(\Theta^{(p,c)})\right],$$

is satisfied, where $\Theta^{(p,c)}$ is the angle between F_n and axis x_2 (Fig. 8.18). The summation includes contact forces acting on the particles which intersect S' and have their center above this plane. These forces are identified by the superscripts (p, c) in the above equation. The angles $\Theta^{(p,c)}$ are uncertain because of random packing, and therefore $\delta F_n^{(p,c)}$ and $\delta F_t^{(p,c)}$ are random variables.

The analysis of the interaction between two particles can be used to obtain the relationships $\delta F_n = c_n \, \delta\gamma_n$ and $\delta F_t = c_t \, \delta\gamma_t$ between the force increments δF_n and δF_t and corresponding deformation increments, $\delta\gamma_n$ and $\delta\gamma_t$, in the normal and tangential directions, where c_n and c_t are constants. The strain increments in the (x_1', x_2') reference corresponding to these displacements can be calculated simply. Classical formulas of tensor calculus can be used to find the image $\delta\Gamma_{ij}$ of the strain tensor $\delta\Gamma_{ij}'$ in the (x_1, x_2) reference, for example,

$$\delta\Gamma_{22} = \frac{\delta\Gamma_{11}' + \delta\Gamma_{22}'}{2} + \frac{\delta\Gamma_{11}' - \delta\Gamma_{22}'}{2} \cos(2\,\Theta) + \delta\Gamma_{12}' \sin(2\,\Theta).$$

The last two equations and the equilibrium conditions give relations between the stress increment δs_{22} and the strain increments $\delta\Gamma_{ij}$. The above results and the mean field hypothesis yield Eq. 8.101 following the arguments in [52]. ▲

8.5.2.4 Analytically based approximations

The methods in Sec. 8.4.1 can be applied to find moments and other probabilistic properties of the displacement, strain, and stress fields for random heterogeneous media. We show here that some of the methods in this section, for example, the perturbation and Neumann series methods, can be applied to calculate approximately effective material constants and develop bounds on these constants that are tighter than the Voigt and Reuss averages.

It is assumed, as stated at the beginning of Section 8.5.2, that the stiffness coefficients A_{ijkl} are homogeneous, square integrable random fields so that their expectations $E[A_{ijkl}(\boldsymbol{x})] = a_{ijkl}$ exist and are constant. The random part of the stiffness coefficients, $A_{ijkl}(\boldsymbol{x}) - a_{ijkl}$, has mean zero.

Perturbation method. Suppose that the uncertainty in the stiffness coefficients is sufficiently small so that A_{ijkl} has the representation $A_{ijkl}(\boldsymbol{x}) = a_{ijkl} + \varepsilon\, \tilde{R}_{ijkl}(\boldsymbol{x})$, where ε is a small parameter. We will seek solutions for the displacement, strain, and stress fields in power series, for example, the displacement functions are represented by $U_i(\boldsymbol{x}) = U_i^{(0)}(\boldsymbol{x}) + \varepsilon\, U_i^{(1)}(\boldsymbol{x}) + O(\varepsilon^2)$.

If (1) the random fields A_{ijkl} and F_i are ergodic and independent of each other and (2) D is sufficiently large so that sample and ensemble averages of A_{ijkl} and F_i nearly coincide, the constitutive law for the average stress and strain fields can be approximated by $\hat{S}_{ij}(\boldsymbol{x}) = a_{ijkl}\, \hat{\Gamma}_{ij}(\boldsymbol{x})$, where $\hat{S}_{ij}$ and $\hat{\Gamma}_{ij}$ denote approximations to the order ε of $\bar{S}_{ij}$ and $\bar{\Gamma}_{ij}$ in Eq. 8.89, respectively.

Proof: The representation of the displacement field and Eqs. 8.85, 8.87, and 8.88 give

$$\begin{aligned}\frac{1}{2}\,a_{ijkl}\left(U^{(0)}_{k,lj}(\boldsymbol{x})+U^{(0)}_{l,kj}(\boldsymbol{x})\right) &= -F_i,\\ \frac{1}{2}\,a_{ijkl}\left(U^{(1)}_{k,lj}(\boldsymbol{x})+U^{(1)}_{l,kj}(\boldsymbol{x})\right) &= -\frac{1}{2}\,\tilde{R}_{ijkl,j}(\boldsymbol{x})\left(U^{(0)}_{k,l}(\boldsymbol{x})+U^{(0)}_{l,k}(\boldsymbol{x})\right)\\ &\quad -\frac{1}{2}\,\tilde{R}_{ijkl}(\boldsymbol{x})\left(U^{(0)}_{k,lj}(\boldsymbol{x})+U^{(0)}_{l,kj}(\boldsymbol{x})\right)\end{aligned}$$

to order 1 and ε, respectively, provided that $F_i \sim O(1)$. The boundary conditions for the above equations correspond to the terms of order 1 and ε, respectively, of the specified boundary conditions. If the boundary conditions for the equations of order 1 and the body forces F_i are deterministic, the solutions $U_i^{(0)}$ are deterministic. The solutions $U_i^{(1)}$ are random fields irrespective of the properties of F_i and the type of boundary conditions since the differential equations defining these functions are driven by the random part of the stiffness coefficients. The approximate strain and stress fields to the order ε are

$$\begin{aligned}\Gamma_{ij}(\boldsymbol{x}) &= \Gamma^{(0)}_{ij}(\boldsymbol{x})+\varepsilon\,\Gamma^{(1)}_{ij}(\boldsymbol{x})+O(\varepsilon^2),\\ S_{ij}(\boldsymbol{x}) &= a_{ijkl}\,\Gamma^{(0)}_{kl}(\boldsymbol{x})+\varepsilon\left[\tilde{R}_{ijkl}(\boldsymbol{x})\,\Gamma^{(0)}_{kl}(\boldsymbol{x})+a_{ijkl}\,\Gamma^{(1)}_{kl}(\boldsymbol{x})\right]+O(\varepsilon^2)\\ &= S^{(0)}_{ij}(\boldsymbol{x})+\varepsilon\,S^{(1)}_{ij}(\boldsymbol{x})+O(\varepsilon^2),\end{aligned}$$

where $\Gamma^{(q)}_{kl} = \left(U^{(q)}_{k,l}+U^{(q)}_{l,k}\right)/2$ and the expression for S_{ij} is given by the terms of order 1 and ε of $A_{ijkl}(\boldsymbol{x})\,\Gamma_{kl}(\boldsymbol{x})$.

The expectation of the above expression of S_{ij} is

$$E\left[S^{(0)}_{ij}(\boldsymbol{x})+\varepsilon\,S^{(1)}_{ij}(\boldsymbol{x})\right] = a_{ijkl}\,E\left[\Gamma^{(0)}_{ij}(\boldsymbol{x})+\varepsilon\,\Gamma^{(1)}_{ij}(\boldsymbol{x})\right]$$

since $\tilde{R}_{ijkl}$ and $\Gamma^{(0)}_{kl}$ are not related so that the expectation $E\left[\tilde{R}_{ijkl}(\boldsymbol{x})\,\Gamma^{(0)}_{kl}(\boldsymbol{x})\right]$ is equal to $E\left[\tilde{R}_{ijkl}(\boldsymbol{x})\right]E\left[\Gamma^{(0)}_{kl}(\boldsymbol{x})\right]$ and $\tilde{R}_{ijkl}$ has mean zero. If the random fields A_{ijkl} and F_i are ergodic so are the stress and strain fields. Hence, for a sufficiently large D, the expectations $E\left[S^{(0)}_{ij}(\boldsymbol{x})+\varepsilon\,S^{(1)}_{ij}(\boldsymbol{x})\right]$ and $E\left[\Gamma^{(0)}_{ij}(\boldsymbol{x})+\varepsilon\,\Gamma^{(1)}_{ij}(\boldsymbol{x})\right]$ can be approximated by the sample averages of $S^{(0)}_{ij}+\varepsilon\,S^{(1)}_{ij}$ and $\Gamma^{(0)}_{ij}+\varepsilon\,\Gamma^{(1)}_{ij}$, denoted by $\hat{S}_{ij}$ and $\hat{\Gamma}_{ij}$, respectively. These observations give the approximate constitutive law $\hat{S}_{ij} = a_{ijkl}\,\hat{\Gamma}_{kl}$. ■

Neumann series method. Let $R_{ijkl}(\boldsymbol{x}) = A_{ijkl}(\boldsymbol{x}) - a_{ijkl}$ be the random part of A_{ijkl} and assume that the displacement field satisfies the boundary condition in Eq. 8.92. Consider a trial strain field

$$\tilde{\boldsymbol{\Gamma}}^{(m)}(\boldsymbol{x}) = \sum_{k=0}^{m}(-\boldsymbol{G}\,\boldsymbol{R})^k\,\boldsymbol{\gamma}_k(\boldsymbol{x}),$$

where $\boldsymbol{G}$ denotes an integral operator, $\boldsymbol{\gamma}_0 = \bar{\boldsymbol{\gamma}}$ is the uniform strain field used to establish the Voigt bound (Eqs. 8.92 and 8.93), and $\boldsymbol{\gamma}_k$, $k = 1, \ldots, m$, denote unspecified deterministic strain fields.

The average strain energy satisfies the inequality

$$\bar{\mathcal{U}} = \frac{1}{2}\,\bar{\boldsymbol{\gamma}}^T\,\bar{\boldsymbol{A}}\,\bar{\boldsymbol{\gamma}} \le \frac{1}{v}\int_D \tilde{\Gamma}_{ij}^{(m)}(\boldsymbol{x})\,(\boldsymbol{a}+\boldsymbol{R})_{ijkl}(\boldsymbol{x})\,\tilde{\Gamma}_{kl}^{(m)}(\boldsymbol{x})\,d\boldsymbol{x}$$

for any $\boldsymbol{\gamma}_k$. The above upper bound on $\bar{\mathcal{U}}$ can be minimized by an adequate selection of the strain fields $\boldsymbol{\gamma}_k$, $k = 1, \ldots, m$, and provides a tighter bound on the effective stiffness coefficients than the Voigt average.

Proof: The displacement field satisfies the integral equation

$$U_k(\boldsymbol{x}) = U_k^*(\boldsymbol{x}) + \int_D g_{kp}(\boldsymbol{x}, \boldsymbol{\xi})\,[R_{pjkl}(\boldsymbol{\xi})\,U_{k,l}(\boldsymbol{\xi})]_{,j}\,d\boldsymbol{\xi},$$

where g_{kp} are Green's functions defined by

$$a_{ijkl}\,g_{kp,lj}(\boldsymbol{x}, \boldsymbol{\xi}) + \delta_{ip}\,\delta(\boldsymbol{x} - \boldsymbol{\xi}) = 0, \quad \boldsymbol{x} \in D$$

with $g_{kp}(\boldsymbol{x}, \boldsymbol{\xi}) = 0$ for $\boldsymbol{x} \in \partial D$ and U_k^* depends on the boundary conditions ([194], p. 14). The equation for $\boldsymbol{U}$ gives by differentiation $\boldsymbol{\Gamma}(\boldsymbol{x}) = \boldsymbol{\gamma}^*(\boldsymbol{x}) - (\boldsymbol{G}\,\boldsymbol{R}\,\boldsymbol{\Gamma})(\boldsymbol{x})$, where $\boldsymbol{G}$ denotes an integral operator that results from the integral equation for $\boldsymbol{U}$ by differentiation and $\boldsymbol{\gamma}^*$ is the strain field corresponding to U_k^*. An alternative form of the equation for $\boldsymbol{\Gamma}$ is

$$\boldsymbol{\Gamma}(\boldsymbol{x}) = (\boldsymbol{i} + \boldsymbol{G}\,\boldsymbol{R})^{-1}\,\boldsymbol{\gamma}^*(\boldsymbol{x}) = \left[\sum_{k=0}^{\infty}(-\boldsymbol{G}\,\boldsymbol{R})^k\right]\boldsymbol{\gamma}^*(\boldsymbol{x}),$$

where the last equality is valid if the above Neumann series is convergent. Under this condition we can approximate the strain field by the first m terms of the Neumann series expansion of $\boldsymbol{\Gamma}$, that is, by the field

$$\boldsymbol{\Gamma}^{(m)}(\boldsymbol{x}) = \left[\sum_{k=0}^{m}(-\boldsymbol{G}\,\boldsymbol{R})^k\right]\boldsymbol{\gamma}^*(\boldsymbol{x}).$$

The strain energy satisfies the inequality

$$\bar{\mathcal{U}} = \frac{1}{2}\,\bar{\boldsymbol{\gamma}}^T\,\bar{\boldsymbol{A}}\,\bar{\boldsymbol{\gamma}} \le \frac{1}{v}\int_D \Gamma_{ij}^{(m)}(\boldsymbol{x})\,(\boldsymbol{a}+\boldsymbol{R})_{ijkl}(\boldsymbol{x})\,\Gamma_{kl}^{(m)}(\boldsymbol{x})\,d\boldsymbol{x}$$

since $\boldsymbol{\Gamma}^{(m)}$ is an approximate strain field ([194], Section IV).

Note that we can use the strain field $\boldsymbol{\Gamma}^{(m)}$ even if the Neumann series is not convergent since all we need is an approximate strain field, which is consistent with the boundary conditions. Other trial strain fields can be used, for example, the strain field $\tilde{\boldsymbol{\Gamma}}^{(m)}$, since the above bound on the average strain energy holds for any approximate strain field. ■

Example 8.47: Consider a rod with unit cross section and random stiffness $A(x)$, $0 \le x \le v$, subjected to boundary conditions $U(0) = 0$ and $U(v) = \bar{\gamma}\,v$ at its left and right ends, respectively. The average and random parts of A, assumed to be a homogeneous random field in L_2, are $a = E[A(x)]$ and $R = A - a$, respectively. The integral equation for the strain Γ is

$$\Gamma(x) = \bar{\gamma} + \int_0^v G(x, y)\, \Gamma(y)\, dy,$$

where the kernel $G(x, y) = R(y)\,[1 - v\,\delta(x - y)]/(a\,v)$ depends on the average and random parts of A.

Let $\tilde{\Gamma}(x) = \bar{\gamma} + H(x)\,\gamma_1$ be a trial strain field, where γ_1 is a constant and $H(x) = \int_0^v G(x, y)\, dy$. The optimal strain field with this functional form is $\tilde{\Gamma}(x) = \bar{\gamma} + H(x)\,\gamma_1^{\text{opt}}$, where

$$\gamma_1^{\text{opt}} = -\bar{\gamma}\, \frac{\int_0^v A(x)\, H(x)\, dx}{\int_0^v A(x)\, H(x)^2\, dx}.$$

This optimal field provides a tighter bound on $\bar{A}$ than the Voigt average. For example, suppose that A is given by a sum of terms $Y_k\, e^{-\beta\,|x - Z_k|}$, where Y_k are identical copies of an exponential random variable with mean $1/\rho$ and Z_k denote the jumps of a homogeneous Poisson process $N(x)$, $x \geq 0$, with intensity $\lambda > 0$. The average of A is $a = E[A(x)] = 2\,\lambda/(\rho\,\beta)$ so that the expectation of the Voigt average is $E[A^{(V)}] = a = 10$ for $a = 10$, $\lambda = 30$, $\beta = 10$, and $1/\rho = a\,\beta/(2\,\lambda)$. The estimated expectation of the upper bound on $\bar{A}$ corresponding to the optimal strain field is 8.72. This estimate has been calculated from 100 independent samples of A. ◇

Note: The equilibrium condition in terms of displacements is

$$U''(x) = -\frac{1}{a}\,[R'(x)\, U'(x) + R(x)\, U''(x)]$$

and results from the condition $(A(x)\,\Gamma(x))' = 0$, where a prime denotes a differentiation with respect to x. The integration of the above equation and the boundary conditions give

$$U(x) = \left[\bar{\gamma} + \frac{1}{a\,v}\int_0^v M(y)\, dy\right] x - \frac{1}{a}\int_0^x M(y)\, dy,$$

where $M(x) = \int_0^x [R'(y)\, U'(y) + R(y)\, U''(y)]\, dy$. Hence, the strain field is

$$\Gamma(x) = \bar{\gamma} + \frac{1}{a\,v}\int_0^v M(y)\, dy - \frac{1}{a}\, M(x)$$

or, following integration by parts,

$$\Gamma(x) = \bar{\gamma} + \frac{1}{a\,v}\int_0^v R(y)\,(1 - v\,\delta(y - x))\,\Gamma(y)\, dy.$$

Let $G_k(x, y) = \int_0^v G(x, z)\, G_{k-1}(z, y)\, dz$, $k = 2, 3, \ldots$, where $G_1(x, y) = G(x, y)$. If $|v\, G(\cdot, \cdot)| < 1$ a.s., the Neumann series,

$$\Gamma(x) = \bar{\gamma} + \sum_{k=1}^{\infty} \left[\int_0^v G_k(x, y)\, dy\right] \bar{\gamma},$$

is convergent a.s., and represents the strain field in the rod (Eqs. 8.58 and 8.59). We use the above representation of Γ to select trial strain fields $\tilde{\Gamma}$. The Voigt average (Eq. 8.94) corresponds to the trial strain field $\tilde{\Gamma}(x) = \bar{\gamma}$. Since $\bar{A}\,\bar{\gamma}^2 \leq (1/v)\int_0^v A(x)\,\tilde{\gamma}(x)^2\,dx$ holds for any strain field $\tilde{\gamma}$, it can be applied to the strain field $\bar{\gamma} + H(x)\,\gamma_1$. We have selected γ_1^{opt} to minimize the upper bound $(1/v)\int_0^v A(x)\,(\bar{\gamma} + H(x)\,\gamma_1)^2\,dx$ on $\bar{A}\,\bar{\gamma}^2$. ▲

8.6 Evolution and pattern formation

Many physical, economical, biological, and other systems experience notable changes in time because of internal and/or external factors. A change in the attributes of a system beyond a scale factor is called **evolution**. If the evolution ceases as time increases, we say that a **pattern** has emerged. Heterogeneity is a necessary condition for evolution. A homogeneous system cannot evolve. For example, a homogeneous population cannot evolve by natural selection but a heterogeneous population does evolve by natural selection since its individuals have different fitness levels. Randomness of a system property adds another dimension to evolution studies in the sense that probabilistic tools are needed to assess whether evolution and/or pattern formation occurs.

We illustrate the evolution and pattern formation phenomena by examples from elasticity and crystal plasticity (Section 8.6.1 and 8.6.2). Let $\boldsymbol{A}(\boldsymbol{x}, t)$ be a random field specifying a property of an elastic body at location $\boldsymbol{x}$ and time t (Section 8.6.1). We say that this property evolves if $\boldsymbol{A}$ changes in time beyond a scale factor. For example, suppose that $\boldsymbol{A}$ is a homogeneous random field consisting of a superposition of a finite number of waves with fixed length but random amplitude and phase (Section 5.3.1.2). A relative change in time of the ordinates of the spectral density of $\boldsymbol{A}$ indicates a change in the frequency content of this random field so that the field evolves. On the other hand, a proportional increase or decrease in time of all spectral ordinates of $\boldsymbol{A}$ does not mean evolution since such a change affects only the scale but not the frequency content of $\boldsymbol{A}$. If $\boldsymbol{A}$ becomes time-invariant up to a scale factor as $t \to \infty$, we say that a pattern has emerged. Generally, the resulting pattern is simpler than the initial property. For example, the amplitudes of some of the constituent waves of $\boldsymbol{A}$ may decrease to zero in time so that $\boldsymbol{A}$ consists of the superposition of fewer waves at a relatively large time $t > 0$ than at the initial time $t = 0$. Similar considerations apply for crystal plasticity problems (Section 8.6.2).

8.6.1 Elasticity

Let D be an open bounded subset of $\mathbb{R}^d$, $d = 1, 2, 3$, containing a random heterogeneous elastic material. The boundary ∂D of D is subjected to specified displacements varying slowly in time. The stiffness tensor of the material in D is a random field $A_{ijkl}(\boldsymbol{x}, t)$, where $\boldsymbol{x} \in D$, $i, j, k, l = 1, 2, 3$, and $t \geq 0$. Our objective is to assess whether the stiffness A_{ijkl} can evolve and, if it evolves, whether its evolution emerges in a pattern.

Let $\Delta \boldsymbol{U}(\boldsymbol{x}, t)$, $\Delta \Gamma_{ij}(\boldsymbol{x}, t)$, and $\Delta S_{ij}(\boldsymbol{x}, t)$ denote the displacement, strain, and stress increments at $\boldsymbol{x} \in D$ during the time interval $[t, t + \Delta t]$, $\Delta t > 0$, for example, $\Delta \boldsymbol{U}(\boldsymbol{x}, t) = \boldsymbol{U}(\boldsymbol{x}, t + \Delta t) - \boldsymbol{U}(\boldsymbol{x}, t)$. We assume in our discussion the following models for material, kinematics, and external actions.

1. Material. It is assumed that the material:

(a) is **random, heterogeneous, and isotropic** with stiffness [68]

$$A_{ijkl}(\boldsymbol{x}, t) = \Lambda(\boldsymbol{x}, t)\, \delta_{ij}\, \delta_{kl} + \Xi(\boldsymbol{x}, t)\, (\delta_{ik}\, \delta_{jl} + \delta_{ij}\, \delta_{kl}), \tag{8.102}$$

where

$$\Lambda(\boldsymbol{x}, t) = \lambda(t) + \varepsilon\, \tilde{\Lambda}(\boldsymbol{x}, t), \quad \Xi(\boldsymbol{x}, t) = \xi(t) + \varepsilon\, \tilde{\Xi}(\boldsymbol{x}, t), \tag{8.103}$$

λ and ξ are deterministic functions of time, and ε is a small parameter. At the initial time, $t = 0$, $\tilde{\Lambda}$ and $\tilde{\Xi}$ are homogeneous random fields with mean zero. It is assumed that the stiffness random fields Λ and Ξ in Eq. 8.103 are strictly positive a.s., and that these fields are sufficiently smooth so that all the following operations on these fields are valid.

(b) obeys the **evolution rule**

$$\Delta \Lambda(\boldsymbol{x}, t) = c^{(\lambda)} \left[\Delta \Gamma^*_{mn}(\boldsymbol{x}, t)\, \Delta \Gamma^*_{nm}(\boldsymbol{x}, t)\right]^{1/2},$$
$$\Delta \Xi(\boldsymbol{x}, t) = c^{(\xi)} \left[\Delta \Gamma^*_{mn}(\boldsymbol{x}, t)\, \Delta \Gamma^*_{nm}(\boldsymbol{x}, t)\right]^{1/2}, \tag{8.104}$$

where $\Gamma^*_{mn} = \Gamma_{mn} - (1/d)\, \Gamma_{pp}\, \delta_{mn}$ denotes the entry (m, n) of the deviatoric strain tensor and $c^{(\lambda)}$ and $c^{(\xi)}$ are specified constants. This rule can be viewed as an approximation of the crystal plasticity evolution defined later in Section 8.6.2 (Eq. 8.122).

(c) is **locally linear**, that is, the relationship between stress and strain increments is

$$\Delta S_{ij}(\boldsymbol{x}, t) = A_{ijkl}(\boldsymbol{x}, t)\, \Delta \Gamma_{kl}(\boldsymbol{x}, t) \quad , \boldsymbol{x} \in D, \quad t \geq 0. \tag{8.105}$$

2. Kinematics. The deformation is small so that strain increments in $[t, t + \Delta t]$ are

$$\Delta \Gamma_{ij}(\boldsymbol{x}, t) = \frac{1}{2}\left(\Delta U_{i,j}(\boldsymbol{x}, t) + \Delta U_{j,i}(\boldsymbol{x}, t)\right), \quad \boldsymbol{x} \in D, \quad t \geq 0. \tag{8.106}$$

3. Loading. The boundary ∂D of D is subjected to the distortion

$$\frac{1}{2}\left(\Delta U_{i,j}(\boldsymbol{x}, t) + \Delta U_{j,i}(\boldsymbol{x}, t)\right) = \Delta \alpha_{ij}(t) = c_{ij}\, \Delta t, \quad \boldsymbol{x} \in \partial D, \quad t \geq 0, \tag{8.107}$$

in $[t, t+\Delta t]$, where $c_{ij} = c_{ji}$ are some constants. The loading is quasi-static and body forces are zero.

Note: If $\varepsilon = 0$ in Eq. 8.103, the material is deterministic, homogeneous, and isotropic so that $\Delta u_i(\boldsymbol{x}, t) = c_{ik}\, x_k\, \Delta t$ gives the increment of the displacement field in $[t, t+\Delta t]$. The resulting strain increments

$$\Delta\gamma_{ij} = \frac{1}{2}\left(\Delta u_{i,j}(\boldsymbol{x}, t) + \Delta u_{j,i}(\boldsymbol{x}, t)\right) = \Delta\alpha_{ij}(t) = c_{ij}\, \Delta t$$

are space invariant. Because $\Delta\gamma_{ij}/\Delta t = c_{ij}$, we refer to c_{ij} as **strain rates**.

The material model defined by Eqs. 8.102, 8.103, 8.104, and 8.105 is quite simple. We use this model because it allows analytical developments. Such developments are not possible when dealing with more realistic material models, for example, the model for crystal plasticity in Section 8.6.2. We also note that the symbol Δ used in this section to indicate increments of strain, stress, and other functions should not be confused with the Laplace operator used previously in the book. ▲

Our objective is to determine whether the solid defined by Eqs. 8.102 to 8.106 and subjected to the boundary conditions in Eq. 8.107 evolves, that is, whether the frequency content of the random fields $\tilde{\Lambda}$ and/or $\tilde{\Xi}$ (Eqs. 8.102 and 8.103) changes in time. If the material experiences evolution, we will examine whether its evolution emerges into a pattern. To achieve this objective, we need to calculate the properties of the random fields $\tilde{\Lambda}$ and $\tilde{\Xi}$ at all $\boldsymbol{x} \in D$ and $t \geq 0$. The determination of these properties involves two steps. First, we need to find properties of the displacement field $\boldsymbol{U}$. Second, the displacement field and the evolution rule in Eq. 8.104 can be used to find properties of the random fields $\tilde{\Lambda}$ and $\tilde{\Xi}$. The starting point of developments in this section is [116] dealing with deterministic heterogeneous materials, that is, materials that can be described by Eqs. 8.102 and 8.103, in which $\tilde{\Lambda}$ and $\tilde{\Xi}$ are deterministic functions.

The first order perturbation method is used to find properties of the increments $\Delta\boldsymbol{U}$ of the displacement field in the time interval $[t, t+\Delta t]$. Accordingly, these increments are represented by

$$\Delta\boldsymbol{U}(\boldsymbol{x}, t) = \Delta\boldsymbol{u}(\boldsymbol{x}, t) + \varepsilon\, \Delta\tilde{\boldsymbol{U}}(\boldsymbol{x}, t), \tag{8.108}$$

where $\Delta\boldsymbol{u}(\boldsymbol{x}, t)$ is linear in $\boldsymbol{x}$, because it constitutes the increment of the displacement field in a homogeneous and isotropic material with parameters (λ, ξ) (Eq. 8.103) and the boundary conditions in Eq. 8.107. It is convenient to solve the differential equations for $\Delta\tilde{\boldsymbol{U}}(\boldsymbol{x}, t)$ in the frequency domain. Let $\Delta\hat{U}_i(\boldsymbol{q}, t)$, $\hat{\Lambda}(\boldsymbol{q}, t)$, and $\hat{\Xi}(\boldsymbol{q}, t)$ be Fourier transforms of $\Delta\tilde{U}_i(\boldsymbol{x}, t)$, $\Delta\tilde{\Lambda}(\boldsymbol{x}, t)$, and $\Delta\tilde{\Xi}(\boldsymbol{x}, t)$ with respect to the spatial coordinate $\boldsymbol{x}$ at each $t \geq 0$. These definitions hold for almost all samples of $\Delta\tilde{\Lambda}$ and $\Delta\tilde{\Xi}$ by the above properties of these fields.

The Fourier transforms of the differential equations of equilibrium in terms of displacements for order ε are

$$\begin{aligned}&\beta_{ijkl}(t)\,[q_l\,q_j\,\Delta\hat{U}_k(\boldsymbol{q},t) + q_k\,q_j\,\Delta\hat{U}_l(\boldsymbol{q},t)]\\&= \sqrt{-1}\,q_j\,[\hat{\Lambda}(\boldsymbol{q},t)\,\zeta_{ijkl} + \hat{\Xi}(\boldsymbol{q},t)\,\rho_{ijkl}],\end{aligned} \tag{8.109}$$

where

$$\begin{aligned}&\zeta_{ijkl} = \delta_{ij}\,\delta_{kl}, \quad \rho_{ijkl} = \delta_{ik}\,\delta_{jl} + \delta_{il}\,\delta_{kj} \quad \text{and}\\&2\,\beta_{ijkl}(t) = \lambda(t)\,\zeta_{ijkl} + \xi(t)\,\rho_{ijkl}.\end{aligned} \tag{8.110}$$

Proof: We note that the Fourier transforms $\Delta\hat{U}_i(\boldsymbol{q},t)$ of $\Delta\tilde{U}_i(\boldsymbol{x},t)$ in Eq. 8.109 depend on the particular samples of $\Delta\hat{\Lambda}(\boldsymbol{q},t)$ and $\Delta\hat{\Xi}(\boldsymbol{q},t)$ used in this equation so that they are random. The following calculations apply to samples of the random fields $\tilde{\Lambda}$ and $\tilde{\Xi}$. We also note that the functions $\Delta\tilde{U}_i(\boldsymbol{x},t)$, $\Delta\tilde{\Lambda}(\boldsymbol{x},t)$, and $\Delta\tilde{\Xi}(\boldsymbol{x},t)$ are defined only on a subset D of $\mathbb{R}^d$. The Fourier transforms in Eq. 8.109 correspond to extensions of these functions to the entire space $\mathbb{R}^d$. It is common to use periodic extensions.

The equilibrium condition $\Delta S_{ij,j}(\boldsymbol{x},t) = 0$ can be given in the form

$$\left(A_{ijkl}(\boldsymbol{x},t)\,\Delta\Gamma_{kl}(\boldsymbol{x},t)\right)_{,j} = 0$$

or

$$\begin{aligned}&[\,(\,(\lambda(t) + \varepsilon\,\tilde{\Lambda}(\boldsymbol{x},t))\,\zeta_{ijkl} + (\xi(t) + \varepsilon\,\tilde{\Xi}(\boldsymbol{x},t))\,\rho_{ijkl}\,)(\Delta\alpha_{kl}\\&+ \frac{\varepsilon}{2}\,(\Delta\tilde{U}_{k,l}(\boldsymbol{x},t) + \Delta\tilde{U}_{l,k}(\boldsymbol{x},t))\,)\,]_{,j} = 0\end{aligned}$$

since $\Delta\Gamma_{ij}(\boldsymbol{x},t) = \Delta\alpha_{ij}(t) + (\varepsilon/2)\left(\Delta\tilde{U}_{i,j}(\boldsymbol{x},t) + \Delta\tilde{U}_{j,i}(\boldsymbol{x},t)\right)$ (Eqs. 8.106 and 8.108). The resulting condition involving terms of order 1 is satisfied identically and the conditions for terms of order ε gives

$$\beta_{ijkl}(t)\left(\Delta\tilde{U}_{k,lj}(\boldsymbol{x},t) + \Delta\tilde{U}_{l,kj}(\boldsymbol{x},t)\right) = -\left(\tilde{\Lambda}_{,j}(\boldsymbol{x},t)\,\zeta_{ijkl} + \tilde{\Xi}_{,j}(\boldsymbol{x},t)\,\rho_{ijkl}\right)\Delta\alpha_{kl}(t).$$

The result in Eq. 8.109 is the Fourier transform of the above equality. ■

The evolution of material parameters in $[t, t+\Delta t]$ is given by

$$\Delta\lambda = b^{(\lambda)}\,\Delta t, \quad \Delta\xi = b^{(\xi)}\,\Delta t \quad \text{and} \tag{8.111}$$

$$\Delta\tilde{\Lambda}(\boldsymbol{x},t) = b^{(\lambda)}_{mn}\,\Delta\tilde{U}_{m,n}(\boldsymbol{x},t), \quad \Delta\tilde{\Xi}(\boldsymbol{x},t) = b^{(\xi)}_{mn}\,\Delta\tilde{U}_{m,n}(\boldsymbol{x},t), \tag{8.112}$$

where $b^{(\lambda)}$, $b^{(\xi)}$, $b^{(\lambda)}_{mn}$, and $b^{(\xi)}_{mn}$ are constants depending on the average strain rates c_{ij}.

Proof: The change $\Delta\Lambda(\boldsymbol{x},t) = \Lambda(\boldsymbol{x},t+\Delta t) - \Lambda(\boldsymbol{x},t) = \Delta\lambda(t) + \varepsilon\,\Delta\tilde{\Lambda}(\boldsymbol{x},t)$ of the material parameter Λ during the time interval $[t, t+\Delta t]$ depends on the strain increment in this interval (Eq. 8.104). The function $\Delta\Gamma^*_{mn}\,\Delta\Gamma^*_{nm}$ can be approximated to the order ε by

$$\Delta\Gamma^*_{mn}\,\Delta\Gamma^*_{nm} = \Delta\alpha_{mn}\,\Delta\alpha_{nm} - \frac{1}{3}\,(\Delta\alpha_{pp})^2 + 2\,\varepsilon\left(\Delta\alpha_{mn}\,\Delta B_{nm} - \frac{1}{3}\,\Delta\alpha_{pp}\,\Delta B_{qq}\right),$$

where $\Delta B_{pq}(\boldsymbol{x},t) = (\Delta\tilde{U}_{p,q}(\boldsymbol{x},t) + \Delta\tilde{U}_{q,p}(\boldsymbol{x},t))/2$. The arguments of the functions in the above and some of the following equations are not shown for simplicity. The corresponding approximation of the square root of this function is

$$\left(\Delta\Gamma^*_{mn}\,\Delta\Gamma^*_{nm}\right)^{1/2} \simeq \sqrt{a} + \frac{\varepsilon}{\sqrt{a}}\left(\Delta\alpha_{pq} - \frac{1}{3}\,\Delta\alpha_{rr}\,\delta_{qp}\right)\Delta B_{qp}(\boldsymbol{x},t),$$

where $a = \Delta\alpha_{pq}\,\Delta\alpha_{qp} - \frac{1}{3}\,(\Delta\alpha_{rr})^2$ and the function $(y+\varepsilon\,z)^{1/2}$ has been approximated by $y^{1/2} + (\varepsilon\,z)/(2\,y^{1/2})$. We have

$$\Delta\lambda(t) + \varepsilon\,\Delta\tilde{\Lambda}(\boldsymbol{x},t) = c^{(\lambda)}\left[\sqrt{a} + \frac{\varepsilon}{\sqrt{a}}\left(\Delta\alpha_{pq}(t) - \frac{1}{3}\,\Delta\alpha_{rr}(t)\,\delta_{qp}\right)\Delta B_{qp}(\boldsymbol{x},t)\right]$$

so that

$$\Delta\lambda(t) = c^{(\lambda)}\,\sqrt{a},$$
$$\Delta\tilde{\Lambda}(\boldsymbol{x},t) = \frac{c^{(\lambda)}}{\sqrt{a}}\left(\Delta\alpha_{pq}(t) - \frac{1}{3}\,\Delta\alpha_{rr}(t)\,\delta_{qp}\right)\Delta B_{qp}(\boldsymbol{x},t).$$

The results in Eqs. 8.111-8.112 follow since $\Delta\alpha_{ij}(t) = c_{ij}\,\Delta t$ (Eq. 8.107) so that $a^{1/2} = \left(c_{pq}\,c_{qp} - (c_{rr})^2/3\right)^{1/2}\,\Delta t$ and

$$\frac{\Delta\alpha_{mn} - (1/3)\,\Delta\alpha_{mn}\,\delta_{nm}}{\sqrt{a}} = \frac{c_{mn} + c_{nm} - (2/3)\,c_{pp}\,\delta_{nm}}{(c_{pq}\,c_{qp} - (1/3)\,(c_{rr})^2)^{1/2}}.$$

The constants of the first equalities in Eqs. 8.111 and 8.112 are, respectively,

$$b^{(\lambda)} = c^{(\lambda)}\left(c_{pq}\,c_{qp} - (c_{rr})^2/3\right)^{1/2} \quad \text{and} \quad b^{(\lambda)}_{mn} = c^{(\lambda)}\,\frac{c_{mn} + c_{nm} - (2/3)\,c_{pp}\,\delta_{nm}}{(c_{pq}\,c_{qp} - (1/3)\,(c_{rr})^2)^{1/2}}. \tag{8.113}$$

The constants $b^{(\xi)}$ and $b^{(\xi)}_{mn}$ in the second equalities of Eqs. 8.111 and 8.112 can be obtained from $b^{(\lambda)}$ and $b^{(\lambda)}_{mn}$ by multiplication with $c^{(\xi)}/c^{(\lambda)}$. ■

The Fourier transform of Eq. 8.112 is

$$\Delta\hat{\Lambda}(\boldsymbol{q},t) = \sqrt{-1}\,b^{(\lambda)}_{mn}\,q_n\,\Delta\hat{U}_m(\boldsymbol{q},t),$$
$$\Delta\hat{\Xi}(\boldsymbol{q},t) = \sqrt{-1}\,b^{(\xi)}_{mn}\,q_n\,\Delta\hat{U}_m(\boldsymbol{q},t). \tag{8.114}$$

Note: The above equations show that the Fourier transforms of $\Delta\tilde{\Lambda}(\boldsymbol{x}, t)$ and $\Delta\tilde{\Xi}(\boldsymbol{x}, t)$ are linear functions of the Fourier transform of the displacement increments, that is, of the functions $\Delta\hat{U}_m(\boldsymbol{q}, t)$. ▲

The material evolution rules in Eq. 8.111 show that the deterministic part of the stiffness coefficients varies linearly in time and remains space invariant. The evolution of the random part of the stiffness coefficients cannot be obtained directly from Eq. 8.112 since the increments of these coefficients depend on the increments $\Delta\tilde{U}_{m,n}$ of the random part of the displacement field. However, the Fourier transform of this equation (Eq. 8.114) and Eq. 8.109 yield finite difference equations for $\hat{\Lambda}$ and $\hat{\Xi}$. These equations are coupled so that they have to be solved simultaneously. Generally, the coefficients of the equations defining the evolution of $\hat{\Lambda}$ and $\hat{\Xi}$ depend on the wave number $\boldsymbol{q}$ (Eq. 8.109). Hence, the relative magnitude of the amplitudes of the constitutive waves of $\hat{\Lambda}(\boldsymbol{q}, t)$ and $\hat{\Xi}(\boldsymbol{q}, t)$ may change in time and this change implies evolution. Accordingly, the probability law of the random fields $\tilde{\Lambda}(\boldsymbol{x}, t)$ and $\tilde{\Xi}(\boldsymbol{x}, t)$ would also change in time in which case we say that the material properties evolve.

Example 8.48: Consider a rod of length $l > 0$ with stiffness $A(x, t) = a(t) + \varepsilon\,\tilde{A}(x, t)$, where a and $\tilde{A}$ denote the deterministic and random parts of A and ε is a small parameter. It is assumed that the stiffness obeys the evolution rule $\Delta A(x, t) = c^{(a)}\,\Delta\Gamma(x, t)$, where $\Delta\Gamma(x, t) = \Delta U'(x, t)$ and $\Delta A(x, t)$ denote strain and stiffness increments at x during the time interval $[t, t + \Delta t]$, $U(x, t)$ is the displacement function at location x and time t, and $U'(x, t) = \partial U(x, t)/\partial x$. The rod is subjected to an average elongation $\Delta\alpha(t)\,x = c\,x\,\Delta t$, $c > 0$, per unit length during the time interval $[t, t + \Delta t]$.

The stiffness A does not evolve because the spectral density of $\tilde{A}$ at a time $t > 0$ can be obtained from the spectral density of $\tilde{A}$ at time $t = 0$ by scaling. Hence, a very simple system of the type considered in this example cannot evolve although its properties vary randomly in space. We will see in the following two examples that this behavior does not persist for more general systems. ◇

Proof: The increment of the displacement field in the rod during the time interval $[t, t+\Delta t]$ can be approximated by $\Delta U(x, t) = \Delta u(x, t) + \varepsilon\,\Delta\tilde{U}(x, t)$, $x \in D = (0, l)$. The material model and the material evolution rule give (Eqs. 8.102, 8.103, and 8.104)

$$\Delta a(t) + \varepsilon\,\Delta\tilde{A}(x, t) = c^{(a)}\,(\Delta\alpha(t) + \varepsilon\,\Delta\tilde{U}'(x, t)) \quad \text{so that}$$

$$\Delta a(t) = c^{(a)}\,\Delta\alpha(t) = c^{(a)}\,c\,\Delta t \quad \text{and} \quad \Delta\tilde{A}(x, t) = c^{(a)}\,\Delta\tilde{U}'(x, t)$$

are consistent with Eqs. 8.111-8.112.

The evolution of the deterministic part of the stiffness is $a(t) = a(0) + c^{(a)}\,c\,t$. The evolution of the random part of A can be determined from (1) the Fourier transform $\Delta\hat{A}(q, t) = c^{(a)}\,\sqrt{-1}\,q\,\Delta\hat{U}(q, t)$ of the above relationship between $\Delta\tilde{A}$ and $\Delta\tilde{U}'$ (Eq. 8.114) and (2) the term $\tilde{A}'(x, t)\,\Delta\alpha(t) + a(t)\,\Delta U''(x, t) = 0$ of order ε of the equilibrium condition $\Delta S'(x, t) = 0$, whose Fourier transform is

$$\sqrt{-1}\,q\,\hat{A}(q, t)\,\Delta\alpha(t) - a(t)\,q^2\,\Delta\hat{U}(q, t) = 0.$$

The above relationship between $\Delta\hat{U}(q,t)$ and $\hat{A}(q,t)$ and the Fourier transform of the evolution of $\tilde{A}$ give

$$\Delta\hat{A}(q,t) = -\frac{c^{(a)}\,c}{a(t)}\,\hat{A}(q,t)\,\Delta t$$

or, by taking the limit as $\Delta t \to 0$ of the finite difference equation for $\hat{A}$ scaled by Δt,

$$\frac{\partial}{\partial t}\hat{A}(q,t) = -\frac{c^{(a)}\,c}{a(t)}\,\hat{A}(q,t).$$

Hence, $\hat{A}$ at time t is

$$\hat{A}(q,t) = \hat{A}(q,0)\,\exp\left(-\int_0^t \frac{c^{(a)}\,c}{a(s)}\,ds\right) \quad \text{so that}$$

$$\tilde{A}(x,t) = \tilde{A}(x,0)\,\exp\left(-\int_0^t \frac{c^{(a)}\,c}{a(s)}\,ds\right)$$

gives the fluctuating part of the stiffness at an arbitrary time $t \geq 0$. The above equation holds for almost all material samples. If $\tilde{A}(x,0)$ is a homogeneous random field with spectral density $s(q,0)$, the spectral density of $\tilde{A}(x,t)$ is

$$s(q,t) = s(q,0)\,\exp\left(-2\int_0^t \frac{c^{(a)}\,c}{a(s)}\,ds\right)$$

because the spectral density and covariance functions

$$E\left[\tilde{A}(x,t)\,\tilde{A}(y,t)\right] = E\left[\tilde{A}(x,0)\,\tilde{A}(y,0)\right]\,\exp\left(-2\int_0^t \frac{c^{(a)}\,c}{a(s)}\,ds\right)$$

are Fourier pairs. The relationship between the spectral densities of $\tilde{A}$ at the initial and an arbitrary time $t \geq 0$ shows that $\tilde{A}(x;t)$ is a scaled version of $\tilde{A}(x;0)$. Therefore, there is no evolution. ■

Example 8.49: Consider a heterogeneous isotropic material in an open bounded subset D of $\mathbb{R}^2$ specified by Eqs. 8.102 to 8.106. The material is subjected to an average uniform extension and distortion per unit length with increments

$$\Delta\alpha_{11}(t) = c\,\Delta t \quad \text{and} \quad \Delta\alpha_{12}(t) = c\,\mu\,\Delta t$$

during the time interval $[t, t+\Delta t]$, respectively, where $c, \mu > 0$ are some constants (Fig. 8.19).

The change in time of the Fourier transform $\hat{\Lambda}$ of the material parameter $\tilde{\Lambda}$ is given by the finite difference equation

$$\Delta\hat{\Lambda}(\boldsymbol{q},t) = a(\boldsymbol{q};\mu,\lambda,\xi)\,\hat{\Lambda}(\boldsymbol{q},t)\,\Delta t,$$

where $a(\boldsymbol{q};\mu,\lambda,\xi)$ depends on the wave number $\boldsymbol{q}$, the loading parameter μ, and the average material properties λ and ξ. The solution $\hat{\Lambda}(\boldsymbol{q},t)$ of the above finite difference equation shows that the amplitudes of the constituent waves of $\tilde{\Lambda}(\boldsymbol{x},0)$ change in time at different rates depending on the current value of $a(\boldsymbol{q};\mu,\lambda,\xi)$. Hence, the material experiences evolution, which may emerge in a pattern. ◇

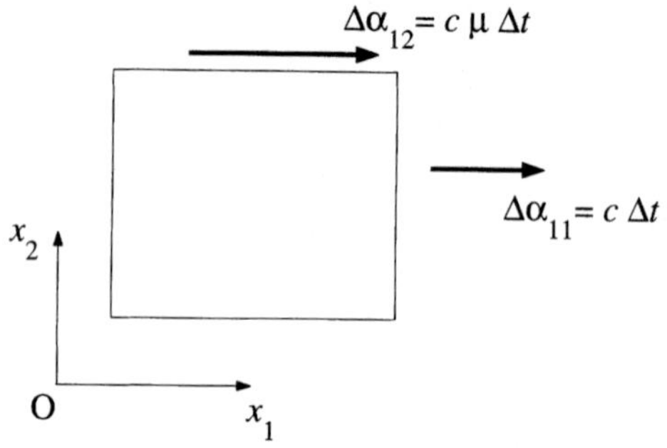

Figure 8.19. Specimen loading

Proof: The increments of the displacements in D during a time interval $[t, t + \Delta t]$ are

$$\Delta U_1(\boldsymbol{x}, t) = \Delta\alpha_{11}\, x_1 + \Delta\alpha_{12}\, x_2 + \varepsilon\, \Delta\tilde{U}_1(\boldsymbol{x}, t) \quad \text{and}$$
$$\Delta U_2(\boldsymbol{x}, t) = \varepsilon\, \Delta\tilde{U}_2(\boldsymbol{x}, t)$$

for $D = (0, 1) \times (0, 1)$ so that

$$\Delta\Gamma_{11}(\boldsymbol{x}, t) = \Delta\alpha_{11} + \varepsilon\, \Delta\tilde{U}_{1,1}(\boldsymbol{x}, t),$$
$$\Delta\Gamma_{22}(\boldsymbol{x}, t) = \varepsilon\, \Delta\tilde{U}_{2,2}(\boldsymbol{x}, t), \quad \text{and}$$
$$\Delta\Gamma_{12}(\boldsymbol{x}, t) = \frac{\Delta\alpha_{12}}{2} + \frac{\varepsilon}{2}(\Delta\tilde{U}_{1,2}(\boldsymbol{x}, t) + \Delta\tilde{U}_{2,1}(\boldsymbol{x}, t))$$

represent the corresponding strain increments at $\boldsymbol{x} \in D$ in $[t, t + \Delta t]$ (Eq. 8.106).

The relationship between stress and strain increments is (Eq. 8.105)

$$\begin{bmatrix} \Delta S_{11}(\boldsymbol{x}, t) \\ \Delta S_{22}(\boldsymbol{x}, t) \\ \Delta S_{12}(\boldsymbol{x}, t) \end{bmatrix} = \begin{bmatrix} \Lambda(\boldsymbol{x}, t) + 2\,\Xi(\boldsymbol{x}, t) & \Lambda(\boldsymbol{x}, t) & 0 \\ \Lambda(\boldsymbol{x}, t) & \Lambda(\boldsymbol{x}, t) + 2\,\Xi(\boldsymbol{x}, t) & 0 \\ 0 & 0 & \Xi(\boldsymbol{x}, t) \end{bmatrix} \begin{bmatrix} \Delta\Gamma_{11}(\boldsymbol{x}, t) \\ \Delta\Gamma_{22}(\boldsymbol{x}, t) \\ \Delta\Gamma_{12}(\boldsymbol{x}, t) \end{bmatrix}$$

so that the quasi-static equilibrium conditions of order ε are

$$(\tilde{\Lambda}_{,1}(\boldsymbol{x}, t) + 2\,\tilde{\Xi}_{,1}(\boldsymbol{x}, t))\, \Delta\alpha_{11} + (\lambda(t) + 2\,\xi(t))\, \Delta\tilde{U}_{1,11}(\boldsymbol{x}, t) + \lambda(t)\, \Delta\tilde{U}_{2,21}(\boldsymbol{x}, t)$$
$$+ \tilde{\Xi}_{,2}(\boldsymbol{x}, t)\, \Delta\alpha_{12}/2 + \xi(t)(\Delta\tilde{U}_{1,22}(\boldsymbol{x}, t) + \Delta\tilde{U}_{2,12}(\boldsymbol{x}, t))/2 = 0 \quad \text{and}$$
$$\tilde{\Xi}_{,1}(\boldsymbol{x}, t)\, \Delta\alpha_{11}/2 + \xi(t)\, (\Delta\tilde{U}_{1,21}(\boldsymbol{x}, t)/2 + \Delta\tilde{U}_{2,11}(\boldsymbol{x}, t))$$
$$+ \tilde{\Xi}_{,2}(\boldsymbol{x}, t)\, \Delta\alpha_{11} + \lambda(t)\, \Delta\tilde{U}_{1,12}(\boldsymbol{x}, t) + (\lambda(t) + 2\,\xi(t))\, \Delta\tilde{U}_{2,22}(\boldsymbol{x}, t) = 0.$$

The Fourier transform of these equations (Eq. 8.109),

$$\begin{bmatrix} q_1^2\,(\lambda(t)+2\,\xi(t))+q_2^2\,\xi(t)/2 & q_1\,q_2\,(\lambda(t)+\xi(t)/2) \\ q_1\,q_2\,(\lambda(t)+\xi(t)/2) & q_2^2\,(\lambda(t)+2\,\xi(t))+q_1^2\,\xi(t)/2 \end{bmatrix} \begin{bmatrix} \Delta\hat{U}_1(\boldsymbol{q},t) \\ \Delta\hat{U}_2(\boldsymbol{q},t) \end{bmatrix}$$
$$= \sqrt{-1}\begin{bmatrix} q_1\,\Delta\alpha_{11}\,(\hat{\Lambda}(\boldsymbol{q},t)+2\,\hat{\Xi}(\boldsymbol{q},t))+q_2\,\Delta\alpha_{12}\,\hat{\Xi}(\boldsymbol{q},t)/2 \\ q_1\,\Delta\alpha_{12}\,\hat{\Xi}(\boldsymbol{q},t)/2+q_2\,\Delta\alpha_{11}\,\hat{\Xi}(\boldsymbol{q},t) \end{bmatrix},$$

constitutes an algebraic system of linear equations for $(\Delta\hat{U}_1, \Delta\hat{U}_2)$ with solution

$$\begin{bmatrix} \Delta\hat{U}_1(\boldsymbol{q},t) \\ \Delta\hat{U}_2(\boldsymbol{q},t) \end{bmatrix} = \frac{\sqrt{-1}\,c\,\Delta t}{a_{11}\,a_{22}-a_{12}^2}\begin{bmatrix} b_{11} & b_{12} \\ b_{21} & b_{22} \end{bmatrix}\begin{bmatrix} \hat{\Lambda}(\boldsymbol{q},t) \\ \hat{\Xi}(\boldsymbol{q},t) \end{bmatrix},$$

where

$$a_{11} = q_1^2\,(\lambda+2\,\xi)+q_2^2\,\xi/2, \quad a_{12}=a_{21}=q_1\,q_2\,(\lambda+\xi/2),$$
$$a_{22} = q_2^2\,(\lambda+2\,\xi)+q_1^2\,\xi/2, \quad b_{11}=q_1\,a_{22}-q_2\,a_{12},$$
$$b_{21} = a_{11}\,q_2-a_{12}\,q_1, \quad b_{12}=(2\,q_1+q_2\,\mu/2)\,a_{22}-q_1\,\mu\,a_{12} \quad \text{and}$$
$$b_{22} = a_{11}\,q_1\,\mu-a_{12}\,(2\,q_1+q_2\,\mu/2).$$

The coefficients a_{ij} and b_{ij} are functions of the wave numbers $\boldsymbol{q}$ and the current average material parameters λ and ξ. The arguments in the above equations are suppressed for simplicity.

The strain function,

$$[\Gamma^*_{mn}(\boldsymbol{x},t)\,\Gamma^*_{nm}(\boldsymbol{x},t)]^{1/2} \simeq \frac{1}{\sqrt{2}}\,[(\Delta\alpha_{11})^2+(\Delta\alpha_{12})^2]^{1/2}$$
$$+\frac{\varepsilon}{\sqrt{2}\,[(\Delta\alpha_{11})^2+(\Delta\alpha_{12})^2]^{1/2}}\,[\Delta\alpha_{11}\,(\Delta\tilde{U}_{1,1}(\boldsymbol{x},t)-\Delta\tilde{U}_{2,2}(\boldsymbol{x},t))$$
$$+\Delta\alpha_{12}\,(\Delta\tilde{U}_{1,2}(\boldsymbol{x},t)+\Delta\tilde{U}_{2,1}(\boldsymbol{x},t))],$$

is used to define the evolution of the material parameters (Eq. 8.104). The resulting equations describing the evolution of the material parameters for the average uniform extension and distortion, $\Delta\alpha_{11} = c\,\Delta t$ and $\Delta\alpha_{12} = c\,\mu\,\Delta t$, are

$$\Delta\lambda = c^{(\lambda)}\,c\,(1+\mu^2)^{1/2}\,\Delta t/\sqrt{2} \quad \text{and} \quad \Delta\xi = c^{(\xi)}\,c\,\mu\,(1+\mu^2)^{1/2}\,\Delta t/\sqrt{2}$$

for terms of order 1, and

$$\Delta\tilde{\Lambda}(\boldsymbol{x},t) = \tilde{c}^{(\lambda)}\,[\Delta\tilde{U}_{1,1}(\boldsymbol{x},t)-\Delta\tilde{U}_{2,2}(\boldsymbol{x},t)+\mu\,(\Delta\tilde{U}_{1,2}(\boldsymbol{x},t)+\Delta\tilde{U}_{2,1}(\boldsymbol{x},t))],$$
$$\Delta\tilde{\Xi}(\boldsymbol{x},t) = \tilde{c}^{(\xi)}\,[\Delta\tilde{U}_{1,1}(\boldsymbol{x},t)-\Delta\tilde{U}_{2,2}(\boldsymbol{x},t)+\mu\,(\Delta\tilde{U}_{1,2}(\boldsymbol{x},t)+\Delta\tilde{U}_{2,1}(\boldsymbol{x},t))],$$

for terms of order ε, where $\tilde{c}^{(\lambda)} = c^{(\lambda)}/\sqrt{2\,(1+\mu^2)}$ and $\tilde{c}^{(\xi)} = c^{(\xi)}/\sqrt{2\,(1+\mu^2)}$. The material parameters λ and ξ vary linearly in time at different rates if $\mu \neq 1$. The evolution of the fluctuating part of material parameters is less simple. The Fourier transforms of the last two equations,

$$\Delta\hat{\Lambda}(\boldsymbol{q},t) = \sqrt{-1}\,\tilde{c}^{(\lambda)}\,c\,[(q_1+\mu\,q_2)\,\Delta\hat{U}_1(\boldsymbol{q},t)+(-q_2+\mu\,q_1)\,\Delta\hat{U}_2(\boldsymbol{q},t)],$$
$$\Delta\hat{\Xi}(\boldsymbol{q},t) = \sqrt{-1}\,\tilde{c}^{(\xi)}\,c\,[(q_1+\mu\,q_2)\,\Delta\hat{U}_1(\boldsymbol{q},t)+(-q_2+\mu\,q_1)\,\Delta\hat{U}_2(\boldsymbol{q},t)],$$

and the linear system of equations giving the Fourier transform of $\boldsymbol{U}$ as a function of the Fourier transforms of $\tilde{\Lambda}$ and $\tilde{\Xi}$ can be used to determine the evolution in time of $\hat{\Lambda}$ and $\hat{\Xi}$. Because the increments of $\hat{\Lambda}$ and $\hat{\Xi}$ in $[t, t+\Delta t]$ are related by

$$\hat{\Lambda}(\boldsymbol{q}, t) = (c^{(\lambda)}/c^{(\xi)})\, \hat{\Xi}(\boldsymbol{q}, t), \quad t \geq 0,$$

it is sufficient to calculate the evolution of, for example, $\hat{\Lambda}$, and use the above equation to find $\hat{\Xi}$. We note that the initial conditions need to be consistent with scale factors relating $\hat{\Lambda}$ and $\hat{\Xi}$ so that $\hat{\Lambda}(\boldsymbol{q}, 0) = (c^{(\lambda)}/c^{(\xi)})\, \hat{\Xi}(\boldsymbol{q}, 0)$.

The evolution equation for $\hat{\Lambda}$ can be given in the form

$$\Delta\hat{\Lambda}(\boldsymbol{q}, t) = -\tilde{c}^{(\lambda)}\, c\, \Delta t\, f(\boldsymbol{q}; \mu, \lambda, \xi)\, \hat{\Lambda}(\boldsymbol{q}, t)$$

or, by taking the limit as $\Delta t \to 0$ of the above finite difference equation for $\hat{\Lambda}$ after division with Δt,

$$\frac{\partial}{\partial t}\hat{\Lambda}(\boldsymbol{q}; t) = -\tilde{c}^{(\lambda)}\, c\, \Delta t\, f(\boldsymbol{q}; \mu, \lambda, \xi)\, \hat{\Lambda}(\boldsymbol{q}; t) = a(\boldsymbol{q}; \mu, \lambda, \xi)\, \hat{\Lambda}(\boldsymbol{q}; t),$$

where

$$f(\boldsymbol{q}; \mu, \lambda, \xi) = \frac{(q_1 + q_2\, \mu)\,(b_{11} + b_{12}\, c^{(\lambda)}/c^{(\xi)}) + (-q_2 + q_1\, \mu)\,(b_{21} + b_{22}\, c^{(\lambda)}/c^{(\xi)})}{a_{11}\, a_{22} - a_{12}^2}.$$

The differential equation $\partial\hat{\Lambda}/\partial t = a(\boldsymbol{q}; \mu, \lambda, \xi)\, \hat{\Lambda}$ shows that the constitutive waves of $\hat{\Lambda}(\boldsymbol{q}; 0)$ change in time at different rates since $a(\boldsymbol{q}; \mu, \lambda, \xi)$ is a function of $\boldsymbol{q}$. That $\hat{\Lambda}$ evolves in time can also be seen from the solution,

$$\hat{\Lambda}(\boldsymbol{q}, t) = \hat{\Lambda}(\boldsymbol{q}, 0)\, \exp\left(\int_0^t a(\boldsymbol{q}; \mu, \lambda(s), \xi(s))\, ds\right),$$

of this equation. ■

Example 8.50: The previous example demonstrates that deterministic heterogeneous materials can evolve. We extend these considerations to random heterogeneous materials, that is, we assume that the material parameters $\tilde{\Lambda}$ and $\tilde{\Xi}$ are homogeneous random fields. We use the spectral densities of the random fields $\tilde{\Lambda}$ and $\tilde{\Xi}$ to assess whether the material parameters exhibit evolution. A similar approach was used in Example 8.48.

Suppose that $\tilde{\Lambda}$ is a homogeneous random field with mean zero and spectral density

$$s(\boldsymbol{q}, 0) = \frac{1}{2}\sum_{k=1}^{n} \sigma_k^2 \left[\delta(\boldsymbol{q} + \boldsymbol{q}^{(k)}) + \delta(\boldsymbol{q} - \boldsymbol{q}^{(k)})\right]$$

at the initial time $t = 0$, where $\sigma_k > 0$ are some positive constants and $\boldsymbol{q}^{(k)} \in \mathbb{R}^2$ denote wave numbers. The field has the spectral representation

$$\tilde{\Lambda}(\boldsymbol{x}, 0) = \sum_{k=1}^{n} \sigma_k \left[A_k \cos(\boldsymbol{q}^{(k)} \cdot \boldsymbol{x}) + B_k \sin(\boldsymbol{q}^{(k)} \cdot \boldsymbol{x})\right],$$

where A_k and B_k are uncorrelated random variables with mean zero and variance 1, and $\boldsymbol{q}^{(k)} \cdot \boldsymbol{x} = q_1^{(k)} x_1 + q_2^{(k)} x_2$ (Section 3.9.4.3).

The spectral density, covariance function, and spectral representation of the random field $\tilde{\Lambda}(\boldsymbol{x}, t)$ at an arbitrary time $t \geq 0$ are

$$s(\boldsymbol{q}, t) = \frac{1}{2} \sum_{k=1}^{n} \gamma_k(t) \left[\delta(\boldsymbol{q} + \boldsymbol{q}^{(k)}) + \delta(\boldsymbol{q} - \boldsymbol{q}^{(k)})\right],$$

$$c(\boldsymbol{\xi}, t) = E\left[\tilde{\Lambda}(\boldsymbol{x}, t)\, \tilde{\Lambda}(\boldsymbol{x} + \boldsymbol{\xi}, t)\right] = \sum_{k=1}^{n} \gamma_k(t) \cos(\boldsymbol{q}^{(k)} \cdot \boldsymbol{\xi}), \quad \text{and}$$

$$\tilde{\Lambda}(\boldsymbol{x}, t) = \sum_{k=1}^{n} \gamma_k(t)^{1/2} \left[A_k \cos(\boldsymbol{q}^{(k)} \cdot \boldsymbol{x}) + B_k \sin(\boldsymbol{q}^{(k)} \cdot \boldsymbol{x})\right],$$

respectively, where $\gamma_k(t) = \sigma_k^2 \exp\left[2 \int_0^t a(\boldsymbol{q}^{(k)}; \mu, \lambda(s), \xi(s))\, ds\right]$. Hence, (1) $\tilde{\Lambda}(\boldsymbol{x}, t)$ is a weakly homogeneous random field with mean zero at each time $t \geq 0$ and (2) the spectral representations of the random fields $\tilde{\Lambda}(\boldsymbol{x}, 0)$ and $\tilde{\Lambda}(\boldsymbol{x}, t)$, $t > 0$, involve waves of the same frequencies but of different amplitudes.

Suppose that the spectral density of $\tilde{\Lambda}(\boldsymbol{x}, t)$ has only four waves with $\boldsymbol{q}^{(1)} = (2, 3)$, $\boldsymbol{q}^{(2)} = (3, 6)$, $\boldsymbol{q}^{(3)} = (5, 7)$, and $\boldsymbol{q}^{(4)} = (10, 0.01)$ and $\sigma_k = 1$, $k = 1, \ldots, 4$. Figure 8.20 shows the evolution in time of a sample path of $\tilde{\Lambda}$ for $\mu = 1$,

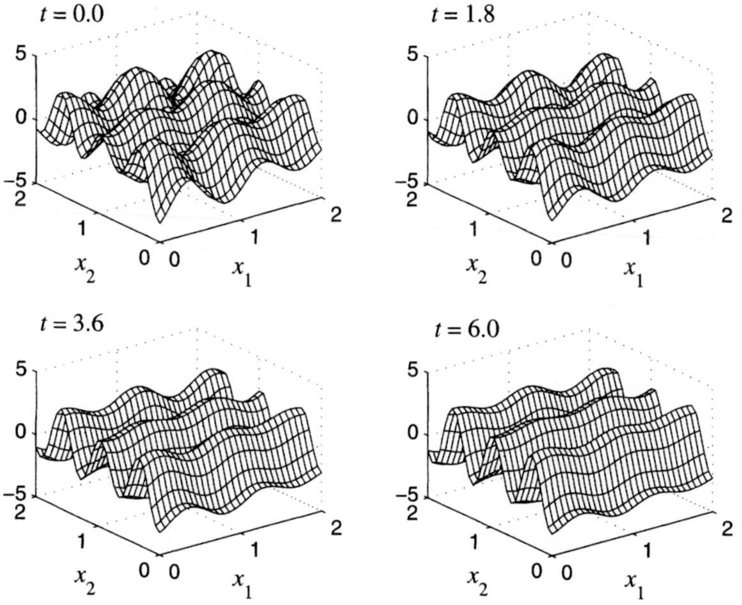

Figure 8.20. A sample path of $\tilde{\Lambda}$

$c = 1$, $c^{(\lambda)} = 1$, $c^{(\xi)} = 1$, $\lambda(0) = 10$, and $\xi(0) = 5$. As time progresses, a single wave becomes dominant. Figure 8.21 shows snapshots of the covariance function

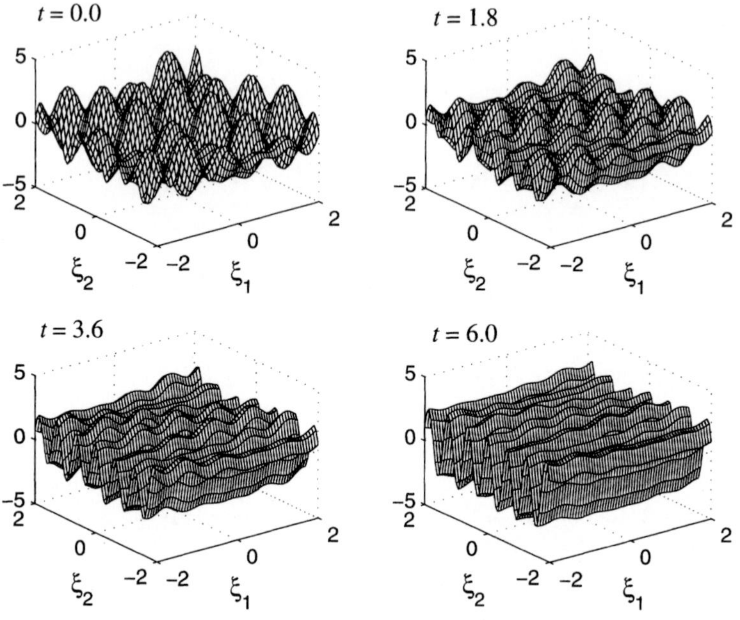

Figure 8.21. Evolution of the covariance function of $\tilde{\Lambda}$

of $\tilde{\Lambda}$ at the initial and later times. These graphs show that material evolves and a pattern emerges as time increases. The resulting pattern is simpler than the initial material structure. ◇

Note: The Fourier transform of a sample $\tilde{\Lambda}(\boldsymbol{x}, 0, \omega)$, $\omega \in \Omega$, of $\tilde{\Lambda}(\boldsymbol{x}, 0)$ is

$$\hat{\Lambda}(\boldsymbol{q}, 0, \omega) = \frac{(2\pi)^2}{2} \sum_{k=1}^{n} \sigma_k \left[\left(A_k(\omega) + \sqrt{-1}\, B_k(\omega) \right) \delta(\boldsymbol{q} + \boldsymbol{q}^{(k)}) \right. \\ \left. + \left(A_k(\omega) - \sqrt{-1}\, B_k(\omega) \right) \delta(\boldsymbol{q} - \boldsymbol{q}^{(k)}) \right].$$

The spectral density of the random field $\tilde{\Lambda}(\boldsymbol{x}, 0)$ and the Fourier transforms of the samples of this field have energy only at frequencies $(\boldsymbol{q}^{(k)}, -\boldsymbol{q}^{(k)})$.

The solution in the previous example gives

$$\hat{\Lambda}(\boldsymbol{q}, t, \omega) = \hat{\Lambda}(\boldsymbol{q}, 0, \omega) \exp\left(\int_0^t a(\boldsymbol{q}; \mu, \lambda(s), \xi(s))\, ds \right)$$

for the Fourier transform of each sample $\omega \in \Omega$ of the material parameters. Because $\hat{\Lambda}(\boldsymbol{q}; 0, \omega)$ has energy only at $(\boldsymbol{q}^{(k)}, -\boldsymbol{q}^{(k)})$, so does $\hat{\Lambda}(\boldsymbol{q}, t, \omega)$. Hence, $\hat{\Lambda}(\boldsymbol{q}, t, \omega)$ and $\hat{\Lambda}(\boldsymbol{q}, 0, \omega)$ have the same frequencies, but the energy associated with these frequencies differ for the two random fields because the ordinates of the spectral density of $\tilde{\Lambda}$ are modulated differently in time depending on the values of $a(\boldsymbol{q}; \mu, \lambda(s), \xi(s))$. The difference in the evolution of the energy associated with each frequency is caused by the dependence of the function $a(\boldsymbol{q}; \mu, \lambda, \xi)$ on $\boldsymbol{q}$.

The sample in Fig. 8.20 is generated from a Gaussian random field. The model is inadequate for representing material stiffness because stiffness is positive and bounded and the Gaussian fields do not have these properties. We use the Gaussian model in Fig. 8.20 only for graphical illustration. ▲

8.6.2 Crystal plasticity

Generally, two scales are used to characterize the evolution of the atomic lattice of polycrystalline metals subjected to time dependent actions: (1) a **microscopic scale**, defined by the grain (crystal) size and (2) a **macroscopic scale**, defined by a sufficiently large assembly of crystals. At macroscopic scale, polycrystalline metals can be described mathematically by continuum models. **Linking hypotheses** are needed to relate material behavior at the microscopic and macroscopic scales. Averaging techniques can be used to go upscale from crystals to continuum properties. Several hypotheses have been proposed to go downscale from continuum to grains.

This section provides basic information on crystal plasticity, presents briefly a numerical method for analyzing the evolution of stresses, strains, and other state variables in a polycrystal, and demonstrates by examples that the orientation of the crystal atomic lattice can evolve in time and that this evolution may emerge in a pattern. Our discussion is based on [50].

8.6.2.1 Planar single crystal

It is assumed that the plastic deformation in a crystal can occur only on a collection of parallel planes, called **slip planes**. A **slip system** is defined by a collection of slip planes and the slip directions in these planes. The orientation of the slip planes is determined by the configuration of the atomic lattice. We limit our discussion to crystals in $\mathbb{R}^2$. More general results can be found in [145].

Two reference frames are considered in crystal plasticity, the **laboratory** or **sample frame** $(\boldsymbol{e}_1, \boldsymbol{e}_2)$, a global system of coordinates attached to the polycrystal, and the **lattice frame** $(\hat{\boldsymbol{e}}_1, \hat{\boldsymbol{e}}_2)$, a local frame attached to the lattice of each crystal (Fig. 8.22). The angle Φ between $\boldsymbol{e}_1$ and $\hat{\boldsymbol{e}}_1$ varies in time during deformation and is a stochastic process for a grain sample if the applied action is a random function of time. If only one crystal is considered, the sample and lattice frames may be taken identical. The geometry of a slip system α is completely defined by the unit vectors $(\boldsymbol{s}^\alpha, \boldsymbol{n}^\alpha)$.

Note: The unit vectors $(\boldsymbol{s}^\alpha, \boldsymbol{n}^\alpha)$ are related to the lattice frame $(\hat{\boldsymbol{e}}_1, \hat{\boldsymbol{e}}_2)$ by

$$\boldsymbol{s}^\alpha = \cos(\theta^\alpha)\,\hat{\boldsymbol{e}}_1 + \sin(\theta^\alpha)\,\hat{\boldsymbol{e}}_2 \quad \text{and} \quad \boldsymbol{n}^\alpha = -\sin(\theta^\alpha)\,\hat{\boldsymbol{e}}_1 + \cos(\theta^\alpha)\,\hat{\boldsymbol{e}}_2,$$

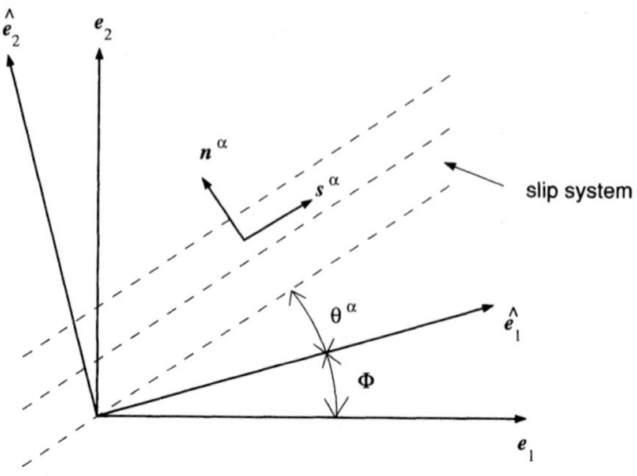

Figure 8.22. Reference frames for crystal plasticity

where θ^α denotes the angle between the direction s^α of the slip system α and $\hat{e}_1$. The **Schmid tensor** in the lattice frame is $t^\alpha = s^\alpha \otimes n^\alpha$ and has the entries $t^\alpha_{ij} = s^\alpha_i\, n^\alpha_j$, $i, j = 1, 2$, where $s^\alpha_1 = \cos(\theta^\alpha)$, $s^\alpha_2 = \sin(\theta^\alpha)$, $n^\alpha_1 = -\sin(\theta^\alpha)$, and $n^\alpha_2 = \cos(\theta^\alpha)$. Note that the trace and the determinant of matrix t^α are zero. We decompose t^α into its symmetric and skew parts denoted by

$$p^\alpha = \mathrm{sym}(t^\alpha) = \begin{bmatrix} -\cos(\theta^\alpha)\sin(\theta^\alpha) & (\cos^2(\theta^\alpha) - \sin^2(\theta^\alpha))/2 \\ (\cos^2(\theta^\alpha) - \sin^2(\theta^\alpha))/2 & \cos(\theta^\alpha)\sin(\theta^\alpha) \end{bmatrix}$$

and

$$q^\alpha = \mathrm{skew}(t^\alpha) = \begin{bmatrix} 0 & 1/2 \\ -1/2 & 0 \end{bmatrix},$$

respectively. The tensor q^α is constant only for planar crystals. ▲

Let S_{ij}, $i, j = 1, 2$, be stresses acting on a crystal defined in the lattice frame. The pressure, the deviatoric stress tensor, and the resolved shear stress on a slip system α are, respectively,

$$S_m = \frac{1}{2}(S_{11} + S_{22}) = \frac{1}{2} S_{pp}, \quad S^*_{ij} = S_{ij} - S_m\,\delta_{ij}, \quad \text{and } S^\alpha = S_{ij}\, s^\alpha_i\, n^\alpha_j. \tag{8.115}$$

An alternative form of the **resolved shear stress** is

$$S^\alpha = t^\alpha_{ij}\, S^*_{ij} = p^\alpha_{ij}\, S^*_{ij} \quad \text{or} \tag{8.116}$$

$$S^\alpha = -2\cos(\theta^\alpha)\sin(\theta^\alpha)\, Z_1 + (\cos^2(\theta^\alpha) - \sin^2(\theta^\alpha))\, Z_2, \tag{8.117}$$

where $Z_1 = (S^*_{11} - S^*_{22})/2 = (S_{11} - S_{22})/2$ and $Z_2 = S^*_{12} = S_{12}$.

Note: The deviatoric stress tensor depends on two parameters, $Z_1 = (S^*_{11} - S^*_{22})/2 = (S_{11} - S_{22})/2$ and $Z_2 = S^*_{12} = S_{12}$. The resolved shear stress corresponding to the pressure tensor $S_m\,\delta_{ij}$ is $S^\alpha = S_m\,\delta_{ij}\,s^\alpha_i\,n^\alpha_j = S_m\,(s^\alpha_i\,n^\alpha_j) = 0$ since $\boldsymbol{s}^\alpha$ and $\boldsymbol{n}^\alpha$ are orthogonal, so that

$$S^\alpha = S^*_{ij}\,s^\alpha_i\,n^\alpha_j = (s^\alpha_i\,n^\alpha_j)\,S^*_{ij} = t^\alpha_{ij}\,S^*_{ij} = p^\alpha_{ij}\,S^*_{ij}$$

by the definition of the Schmid tensor and the symmetry of the deviatoric tensor, where $\delta_{ij} = 1$ for $i = j$ and $\delta_{ij} = 0$ for $i \neq j$.

The relationships between the components $(\boldsymbol{p}^\alpha, \boldsymbol{q}^\alpha)$ and $(\boldsymbol{P}^\alpha, \boldsymbol{Q}^\alpha)$ of the Schmid tensors in the lattice and laboratory frames are

$$\boldsymbol{P}^\alpha = \boldsymbol{A}\,\boldsymbol{p}^\alpha\,\boldsymbol{A}^T \quad \text{and} \quad \boldsymbol{Q}^\alpha = \boldsymbol{A}\,\boldsymbol{q}^\alpha\,\boldsymbol{A}^T, \tag{8.118}$$

where $A_{11} = A_{22} = \cos(\Phi)$ and $A_{12} = -A_{21} = -\sin(\Phi)$ are the entries of the $(2,2)$ matrix $\boldsymbol{A}$. If the deformation of the crystal varies randomly in time, the angle of rotation Φ and the components $(\boldsymbol{P}^\alpha, \boldsymbol{Q}^\alpha)$ of the Schmid tensor are stochastic processes. ▲

The distortion of a crystal subjected to a stress field S_{ij} consists of elastic and plastic strains. Elasticity is neglected in our discussion to simplify the mathematical model [145]. We need to specify a **plastic law** for the crystal characterizing the evolution in time of its slip system. A plastic law is defined by a yield condition, a flow rule, and an evolution rule.

1. The **yield condition**, defining a criterion for the occurrence of plastic flow, is

$$\begin{aligned} &\text{If } |S^\alpha| < S^\alpha_{\text{cr}}, && \text{no slip,} \\ &\text{If } |S^\alpha| = S^\alpha_{\text{cr}}, && \text{slip can occur,} \end{aligned} \tag{8.119}$$

where S^α_{cr} can be interpreted as a strength parameter for the slip system α and represents a material property.

Note: S^α_{cr} may or may not depend on the **shearing rates** $\dot{\Gamma}^\beta$ of the slip systems β in the crystal. Because the resolved shear stress S^α depends linearly on the stress parameters Z_1 and Z_2 (Eq. 8.117), the yield condition is $a^\alpha\,Z_1 + b^\alpha\,Z_2 = \pm S^\alpha_{\text{cr}}$, where a^α and b^α depend on the orientation of the slip system α.

If S^α_{cr} does not depend on $\dot{\Gamma}^\beta$, the yield condition is said to be **rate independent**. Otherwise, we deal with a **visco-plastic** model. A typical visco-plastic model is $S^\alpha_{\text{cr}} = \bar{S}^\alpha\,f(\ldots, \dot{\Gamma}^\beta, \ldots)$, where f is a specified function of the shearing rates $\dot{\Gamma}^\beta$ and $\bar{S}^\alpha$ is a state variable whose time evolution is given by the differential equation $\dot{\bar{S}}^\alpha = g(\bar{S}^\alpha, \ldots, |\dot{\Gamma}^\beta|, \ldots)$, for example,

$$\dot{\bar{S}}^\alpha = \sum_\beta h_{\alpha\beta}(\bar{S}^\alpha)\,|\dot{\Gamma}^\beta|, \tag{8.120}$$

where $h_{\alpha\beta}$ are specified functions. This evolution equation is analogous to a modified version of the Voce macroscopic model [50]. ▲

2. The **flow rule**, giving the velocity gradient $\boldsymbol{L}^p$, is

$$\boldsymbol{L}^p = \sum_\alpha \dot{\Gamma}^\alpha \boldsymbol{t}^\alpha = \sum_\alpha \dot{\Gamma}^\alpha \boldsymbol{p}^\alpha + \sum_\alpha \dot{\Gamma}^\alpha \boldsymbol{q}^\alpha = \boldsymbol{G} + \sum_\alpha \dot{\Gamma}^\alpha \boldsymbol{q}^\alpha, \qquad (8.121)$$

where $\dot{\Gamma}^\alpha$ denotes the shearing rate of the slip system α.

3. The **evolution rule**, giving the rotation rate $\boldsymbol{W}$ of the atomic lattice, is

$$\boldsymbol{W} = \boldsymbol{\Lambda} + \sum_\alpha \dot{\Gamma}^\alpha \boldsymbol{q}^\alpha = \dot{\boldsymbol{R}}\, \boldsymbol{R}^T + \sum_\alpha \dot{\Gamma}^\alpha \boldsymbol{q}^\alpha, \qquad (8.122)$$

where $\boldsymbol{R}$ is a matrix defining the rigid body rotation and $\boldsymbol{\Lambda} = \dot{\boldsymbol{R}}\, \boldsymbol{R}^T$ denotes the lattice spin.

Note: The sum of $\boldsymbol{G} = \sum_\alpha \dot{\Gamma}^\alpha \boldsymbol{p}^\alpha$ and $\boldsymbol{W} - \boldsymbol{\Lambda}$ coincides with the plastic velocity gradient $\boldsymbol{L}^p$ given by Eq. 8.121.

The evolution of the geometry and structure of a crystal experiencing plastic deformation is illustrated in Fig. 8.23. Let D_0 be a subset of $\mathbb{R}^d$ ($d = 2$ in our discussion)

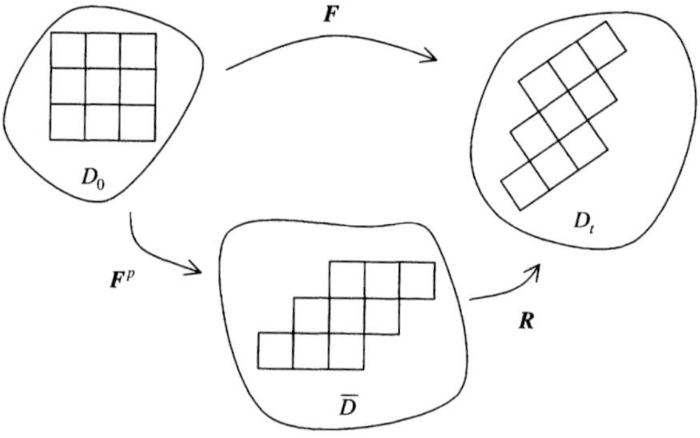

Figure 8.23. Crystal evolution under plastic deformation

giving the geometry of the crystal at the initial time $t = 0$, and let $D_t \subset \mathbb{R}^d$ be its deformed shape at a later time $t > 0$. Denote by $\boldsymbol{x} \mapsto \boldsymbol{y}(\boldsymbol{x}, t)$ the mapping from D_0 to D_t. If the imposed deformation and/or crystal properties are random, D_t is a random subset in $\mathbb{R}^d$. It is assumed that the Jacobian of the mapping from D_0 to D_t is not zero so that the mapping can be inverted. The physical meaning of this assumption is that a non-zero volume in D_0 cannot be mapped into a zero volume in D_t or that there is no subset of D_0 of non-zero Lebesgue measure whose image in D_t has measure zero. The gradients $\boldsymbol{F}(\boldsymbol{x}, t) = \{\partial y_i(\boldsymbol{x}, t)/\partial x_j\}$ of the mapping $\boldsymbol{x} \mapsto \boldsymbol{y}(\boldsymbol{x}, t)$ in the laboratory frame are given by $\boldsymbol{F} = \boldsymbol{R}\, \boldsymbol{F}^p$, where $\boldsymbol{R}$ and $\boldsymbol{F}^p$ correspond to rigid body rotation and lattice distortion, respectively (Fig. 8.23). The velocity gradient $\boldsymbol{L} = \{\partial \dot{y}_i(\boldsymbol{x}, t)/\partial y_j\}$ in the current coordinates can be calculated from $\boldsymbol{L} = \dot{\boldsymbol{F}}\, \boldsymbol{F}^{-1}$ by the definition of $\boldsymbol{F}$ and

$\partial \dot{y}_i/\partial y_j = (\partial \dot{y}_i/\partial x_k)\,(\partial x_k/\partial y_j)$, so that

$$L = \dot{R}\,R^T + R\,\bar{L}^p\,R^T, \quad \text{where } \bar{L}^p = \dot{F}^p\,(F^p)^{-1},$$

since $\dot{F} = \dot{R}\,F^p + R\,\dot{F}^p$, $F^{-1} = (F^p)^{-1}\,R^{-1}$, and $R^{-1} = R^T$. The velocity gradient depends on only three parameters as we neglect the elastic deformation of the crystal, so that there is no volume change ($L_{pp} = 0$). In analysis it is assumed that the deformation is homogeneous at the grain scale, that is, L is constant in each grain [50].

It is common to separate the velocity gradient in two components

$$G = \text{sym}\left(R\,\bar{L}^p\,R^T\right) \quad \text{and} \quad W = \dot{R}\,R^T + \text{skew}\left(R\,\bar{L}^p\,R^T\right)$$

giving the rates of deformation and spin, respectively. The equalities in Eqs. 8.121 and 8.122 follow from the above equation with the notation $\Lambda = \dot{R}\,R^T$ since $L^p = R\,\bar{L}^p\,R^T = \sum_\alpha \dot{\Gamma}^\alpha\,t^\alpha$ (Eq. 8.121). ▲

The evolution of the lattice distortion and orientation in a crystal is controlled by the input velocity gradients, G and W, which need to be specified. The analysis involves two steps. First, the strain rates $\dot{\Gamma}^\alpha$ have to be calculated. There are three distinct cases, crystal with one (rank deficient), two (kinematically determined), and three or more (redundant) slip systems [145]. Second, Eqs. 8.121 and 8.122 can be used to determine the evolution of strains and lattice orientation, respectively. For example, Eq. 8.122 gives

$$\dot{\Phi} = \Lambda_{12} = W_{12} - \frac{1}{2}\sum_\alpha \dot{\Gamma}^\alpha \tag{8.123}$$

for the evolution of the lattice orientation since $q_{11}^\alpha = q_{22}^\alpha = 0$ and $q_{12}^\alpha = -q_{21}^\alpha = 1/2$ (Fig. 8.22).

Note: If the crystal has a single slip system, there is no solution.

If the crystal has two slip systems, the unknown shearing rates $\dot{\Gamma}^\alpha$, $\alpha = 1, 2$, can be calculated from the definition of G in Eq. 8.121. Then Eq. 8.122 yields a differential equation for the angle Φ between the current lattice frame and the laboratory frame.

If the crystal has three or more slip systems, the constitutive laws of the slip systems are needed to find $\dot{\Gamma}^\alpha$. The symmetric part of the velocity rate can be given in the form

$$G = \sum_\alpha \dot{\Gamma}^\alpha\,p^\alpha = \sum_\alpha p^\alpha\,(\dot{\Gamma}^\alpha/S^\alpha)\,S^\alpha = \left[\sum_\alpha (\dot{\Gamma}^\alpha/S^\alpha)\,p^\alpha\,(p^\alpha)^T\right] S^* = M(S^*)\,S^*, \tag{8.124}$$

because the stress field is assumed to be uniform in the crystal so that $S^\alpha = (p^\alpha)^T\,S^*$, where S^* is a vector whose coordinates are the distinct entries of the deviatoric tensor S_{ij}^*. Generally, numerical algorithms are needed to calculate S^* from Eq. 8.124 for an imposed deformation rate G. The resolved shear stress corresponding to S^* (Eq. 8.116),

$$S^\alpha = p_{ij}^\alpha\,S_{ij}^* = (p^\alpha)^T\,S^* = \pm S_{\text{cr}} = \pm \bar{S}_{\text{cr}}\,|\dot{\Gamma}^\alpha/\dot{\Gamma}|^m, \tag{8.125}$$

can be used to determine $\dot{\Gamma}^\alpha$ with $\bar{S}_{\text{cr}}$ obtained, for example, from Eq. 8.120. The evolution of the lattice orientation follows from Eq. 8.122. ▲

Example 8.51: Consider an infinite planar crystal with two slip systems. The crystal is subjected to plastic deformation with velocity gradient $\boldsymbol{L}$. Let $\Phi(t)$ denote the angle between $\hat{\boldsymbol{e}}_1$ and $\boldsymbol{e}_1$ at time t (Fig. 8.22), where $\hat{\boldsymbol{e}}_1$ is aligned with the first slip system, that is, $\hat{\boldsymbol{e}}_1 = \boldsymbol{s}^1$, and θ is the angle between the two slip systems of the crystal. The solution of the differential equation

$$\dot{\Phi}(t) = \frac{L_{12}(t) - L_{21}(t)}{2} - \frac{\sec(2\theta)}{2}[(L_{12}(t) + L_{21}(t))\cos(2\,\Phi(t)) + 2\,L_{11}(t)\,\sin(2\,\Phi(t))],$$

gives the lattice orientation at time $t > 0$ [12, 145]. ◇

Proof: The velocity gradient $\boldsymbol{L}$ depends on only three parameters since $L_{11} + L_{22} = 0$, so that its symmetric part has the entries

$$G_{11} = -G_{22} = L_{11} = -L_{22} = \sum_{\alpha=1}^{2} \dot{\Gamma}^{\alpha}\, P_{11}^{\alpha} \quad \text{and}$$
$$G_{12} = G_{21} = (L_{12} + L_{21})/2 = \sum_{\alpha=1}^{2} \dot{\Gamma}^{\alpha}\, P_{12}^{\alpha}.$$

Because the grain has two slip systems, the slip shearing rates $\dot{\Gamma}^{\alpha}$, $\alpha = 1, 2$, can be determined from the above equations as functions of G_{11} and G_{22}. These expressions of $\dot{\Gamma}^{\alpha}$, $\alpha = 1, 2$, and Eq. 8.123 yield a differential equation for Φ with known coefficients and input.

We also note that the velocity gradient can be decomposed in three parts corresponding to pure compression, pure shear, and spin modes,

$$\boldsymbol{L} = L_{11}(t)\begin{bmatrix} 1 & 0 \\ 0 & -1 \end{bmatrix} + \frac{L_{12}(t) + L_{21}(t)}{2}\begin{bmatrix} 0 & 1 \\ 1 & 0 \end{bmatrix} + \frac{L_{12}(t) - L_{21}(t)}{2}\begin{bmatrix} 0 & 1 \\ -1 & 0 \end{bmatrix}, \tag{8.126}$$

with intensities L_{11}, $(L_{12} + L_{21})/2$, and $(L_{12} - L_{21})/2$, respectively. ■

Example 8.52: Suppose that the crystal in Example 8.51 is subjected to a pure time-invariant compression, that is, $L_{12}(t) = L_{21}(t) = 0$ and L_{11} is a constant. The lattice orientation angle Φ (Fig. 8.22) satisfies the differential equation $\dot{\Phi}(t) = -L_{11}\,\sec(2\theta)\,\sin(2\,\Phi(t))$ and is

$$\Phi(t) = \cot^{-1}\left[\cot(\Phi(0))\, e^{-2\,L_{11}\,t\,\sec(2\theta)}\right],$$

where $\Phi(0)$ denotes the initial state. The fixed points of the differential equation for Φ are $\phi_0 = n\,(\pi/2)$, where n is an integer.

Figure 8.24 shows trajectories of the lattice orientation angle Φ for $L_{11} = 1$, several initial conditions $\Phi(0) \in (-\pi/2, \pi/2)$, and an angle $\theta = \pi/6$ between

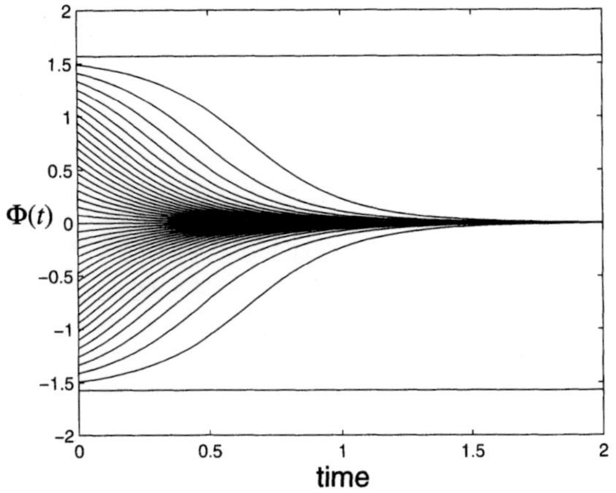

Figure 8.24. Evolution of lattice orientation for pure time-invariant compression

the two slip systems of the crystal. The trajectories of the orientation angle approach the stable equilibrium point $\phi_0 = 0$ as $t \to \infty$ so that the density of $\Phi(t)$ approaches a delta Dirac function centered at zero as time increases, irrespective of the initial state $\Phi(0)$. Hence, atomic lattice evolves in time and its evolution emerges in a pattern [11, 119].

The trajectories of the lattice orientation are less simple if L_{11} is time dependent. For example, suppose that $L_{12} = L_{21} = 0$ and that L_{11} is a stationary Gaussian process with mean zero, variance one, and one-sided spectral density $g(\nu) = 2\beta/[\pi(\nu^2 + \beta^2)]$, $\nu \geq 0$. Figure 8.25 shows a sample of L_{11} for $\beta = 0.1$

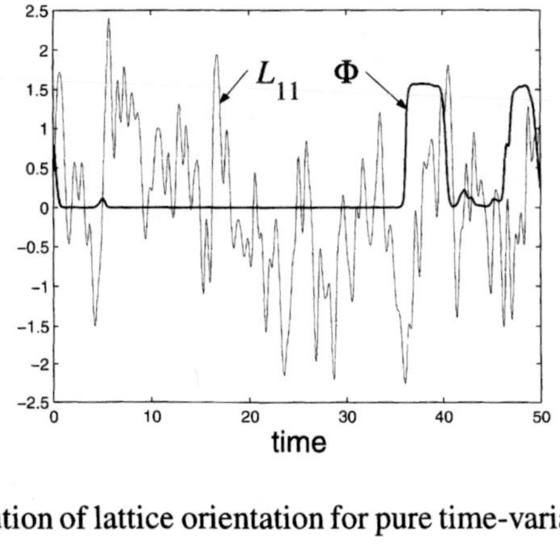

Figure 8.25. Evolution of lattice orientation for pure time-variant random traction

and the corresponding sample of Φ for $\Phi(0) = 0.5$ and $\theta = \pi/6$. The sample path of the lattice orientation Φ oscillates between the fixed points 0 and $\pi/2$ because the equilibrium position depends on the sign of L_{11}, which changes in time [11, 119]. In this case the lattice orientation Φ evolves but no pattern emerges from its evolution. ◇

Note: Let z be a real-valued function defining the state of a physical system specified by the differential equation $\dot{z}(t) = h(z(t), \boldsymbol{\theta})$, where h is a known function and $\boldsymbol{\theta}$ denotes a vector of parameters. The solutions $z_0(\boldsymbol{\theta})$ with the property $h(z_0(\boldsymbol{\theta}), \boldsymbol{\theta}) = 0$ are the **singular** or **fixed points** of the above differential equation. Hence, a singular point is also an equilibrium point since $\dot{z}(t)$ is zero at $z_0(\boldsymbol{\theta})$. A fixed point is said to be **stable** if the solution z starting in a small vicinity of $z_0(\boldsymbol{\theta})$ remains near $z_0(\boldsymbol{\theta})$ as $t \to \infty$. If a fixed point does not have this property, it is said to be **unstable**.

Several methods can be used to determine whether a fixed point is or is not stable. For example, the linearization technique ([104], Chapter 8, [119]) considers the differential equation

$$\begin{aligned}\dot{\tilde{z}}(t) &= h(z_0(\boldsymbol{\theta}), \boldsymbol{\theta}) + h'(z_0(\boldsymbol{\theta}), \boldsymbol{\theta})\,(\tilde{z}(t) - z_0(\boldsymbol{\theta})) \\ &= h'(z_0(\boldsymbol{\theta}), \boldsymbol{\theta})\,(\tilde{z}(t) - z_0(\boldsymbol{\theta}))\end{aligned}$$

derived from the defining equation of z by linearization about $z_0(\boldsymbol{\theta})$. The solution $\tilde{z}$ of this differential equation approximates z in a small vicinity of $z_0(\boldsymbol{\theta})$ and can be used to evaluate the stability of $z_0(\boldsymbol{\theta})$. We note that $y(t) = z(t) - z_0(\boldsymbol{\theta}) = y(0)\, e^{h'(z_0(\boldsymbol{\theta}), \boldsymbol{\theta})\, t}$ since $\dot{y}(t) = h'(z_0(\boldsymbol{\theta}), \boldsymbol{\theta})\, y(t)$. If the real part of $h'(z_0(\boldsymbol{\theta}), \boldsymbol{\theta})$ is strictly negative, y approaches zero as time increases or $\lim_{t\to\infty} z(t) = z_0(\boldsymbol{\theta})$ so that $z_0(\boldsymbol{\theta})$ is stable. ▲

8.6.2.2 Polycrystals

A polycrystal is an ensemble of a large number of crystals interacting with each other. Consider a polycrystal occupying an open bounded subset D of $\mathbb{R}^d$ with boundary ∂D subjected to a specified traction and velocity on ∂D_s and ∂D_v, respectively, where $\partial D_s \cup \partial D_v = \partial D$ and $\partial D_s \cap \partial D_v = \emptyset$. The objective is to find the velocity and stress fields in D. The stress $\boldsymbol{S}$ and velocity $\boldsymbol{V}$ in the polycrystal viewed as a continuum must satisfy (1) equilibrium conditions, $S_{ij,j} + b_i = 0$, where b_i denote body forces, (2) kinematics constraints, $\boldsymbol{L} = \operatorname{grad}(\boldsymbol{V}) = \boldsymbol{G} + \boldsymbol{W}$, (3) a constitutive law, for example, $\boldsymbol{S}^* = \boldsymbol{K}\,\boldsymbol{G}$, where $\boldsymbol{K}$ represents the material stiffness that can be calculated from the crystal properties, and (4) an incompressibility requirement, $\operatorname{div}(\boldsymbol{V}) = V_{i,i} = 0$.

The material stiffness $\boldsymbol{K}$ depends in a complex way on the properties of the constitutive grains and their interaction. Generally, $\boldsymbol{K}$ is determined by the finite element method based on:

(1) The relationships $\boldsymbol{G}^g = \boldsymbol{C}^g\,\boldsymbol{S}^g$ and $\boldsymbol{S}^g = \boldsymbol{K}^g\,\boldsymbol{G}^g$ derived from Eqs. 8.121 and 8.122, where $\boldsymbol{C}^g$ and $\boldsymbol{K}^g$ denote the grain compliance and stiffness, respectively, $\boldsymbol{G}^g$ is the symmetric part of the plastic velocity gradient in a grain, and $\boldsymbol{S}^g$ denotes the deviatoric part of the strain tensor in a grain. The parameters $\boldsymbol{C}^g$ and $\boldsymbol{K}^g$ depend on the resolved shear stress and the Schmid tensor in the grain.

(2) Linking hypotheses relating the microscopic and macroscopic solutions. The Taylor and Sachs hypotheses are frequently used.

- The **Taylor** hypothesis states that (a) the macroscopic and microscopic velocity fields coincide, that is, $\boldsymbol{G} = \boldsymbol{G}^g$ and (b) the macroscopic deviatoric stress tensor $\boldsymbol{S}$ is equal to the average of the deviatoric tensor $\boldsymbol{S}^g$ over the crystal volume, that is, $\boldsymbol{S} = < \boldsymbol{S}^g >$, where $< \cdot >$ denotes volume average. Accordingly, we have the macroscopic constitutive law $\boldsymbol{S} = \boldsymbol{K}\,\boldsymbol{G}$, where $\boldsymbol{K} = < \boldsymbol{K}^g >$. The Taylor hypothesis provides an upper bound analogous to the Voigt average in Eq. 8.94.

- The **Sachs** hypothesis states that (a) the macroscopic and microscopic deviators coincide, that is, $\boldsymbol{S} = \boldsymbol{S}^g$ and (b) the macroscopic velocity is equal to the average velocity over the crystal volume, that is, $\boldsymbol{G} = < \boldsymbol{G}^g >$. Accordingly, we have the macroscopic constitutive law $\boldsymbol{G} = \boldsymbol{C}\,\boldsymbol{S}$, where $\boldsymbol{C} = < \boldsymbol{C}^g >$.

Example 8.53: The calculation of the evolution of the stress, strain, and lattice orientation in a polycrystal involves three steps. First, the constituent grains of the polycrystal need to be characterized. Second, the finite element method can be used to calculate the evolution of the polycrystal state variables. Third, statistical methods need to be used to estimate global features of the polycrystal behavior.

Euler angles measurements of the atomic lattice orientation in a material specimen can be used to develop probabilistic models for the geometrical properties of the grains in a polycrystal. For example, Figs. 8.26 and 8.27 show three-dimensional and contour plots of Φ_1 for aluminum AL 7075, where Φ_1 is one of the three Euler angles defining the atomic lattice orientation. The plots corre-

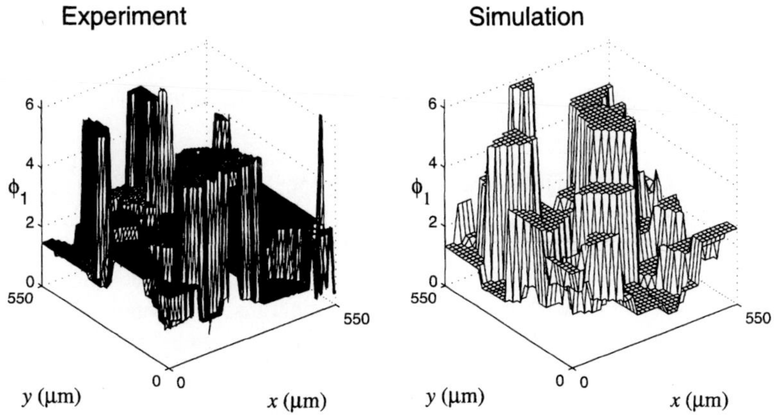

Figure 8.26. Experimental data of Φ_1 and a sample of its translation random field model. Three-dimensional representation

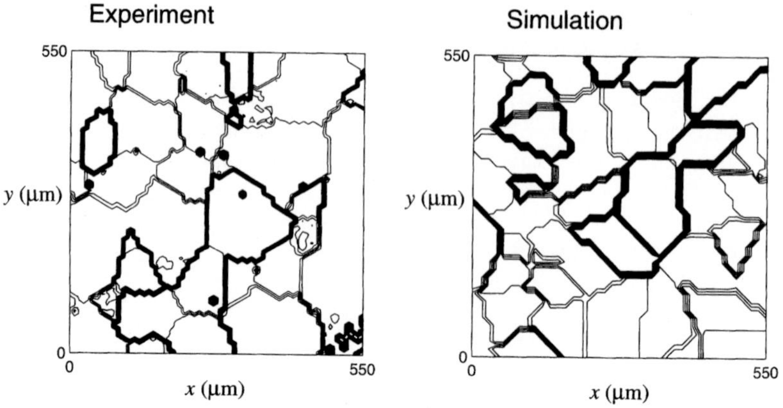

Figure 8.27. Experimental data of Φ_1 and a sample of its translation random field model. Contour lines

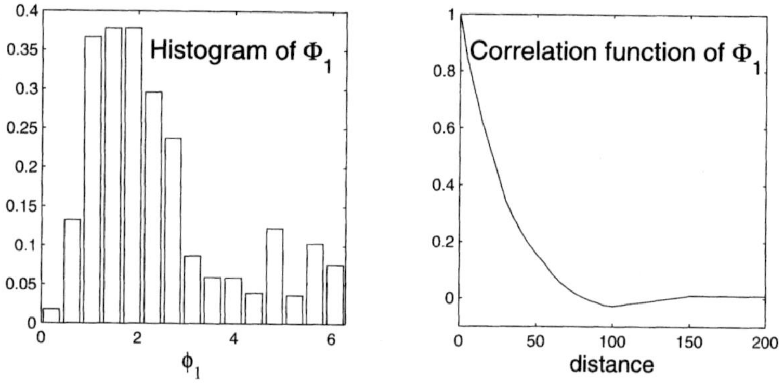

Figure 8.28. Histogram and correlation function of Φ_1

spond to experimental data and a sample of a translation random field calibrated to these data (Section 5.3.3.1). Figure 8.28 shows a histogram and an estimate of the correlation function of Φ_1 based on the experimental data in the previous two figures, where the correlation function of Φ_1 is assumed to be isotropic. The estimates in this figure have been used to calibrate the translation random field model for Φ_1 [11]. These models and statistics on the grain geometry provide sufficient information to capture the essential attributes of a polycrystal.

The finite element method has been used in [188] to find the evolution of a polycrystal consisting of 4096 grains under the assumptions that (1) the grains have a simple geometry, that is, brick grains with the same dimensions, (2) the Euler angles are uniformly distributed random variables in each grain, and (3) the Euler angles in different grains are independent of each other. Numerical results have been obtained for a polycrystal subjected to a 5% initial prestrain and subsequently to strain cycles of amplitude 3%. Figure 8.29 shows the number of grains

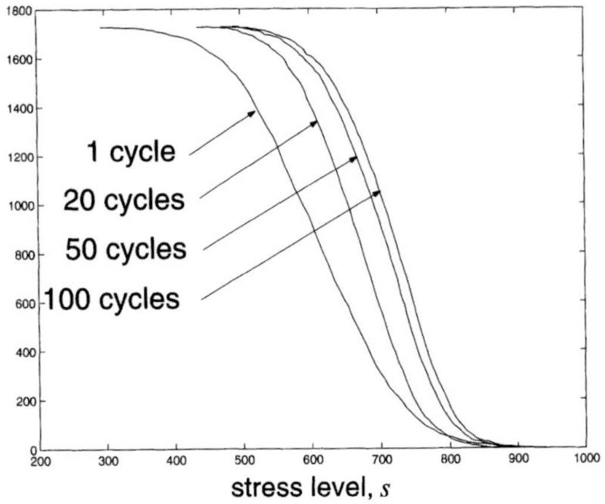

Figure 8.29. Stress distribution in a polycrystal sample

in the polycrystal with stresses exceeding a stress level s after 1, 20, 50, and 100 strain cycles. The plot provides information on the global state of stress in the polycrystal and its evolution in time. $\diamond$

8.7 Stochastic stability

Let $\boldsymbol{X}$ be an $\mathbb{R}^d$-valued stochastic process representing the state of a nonlinear system defined by

$$\dot{\boldsymbol{X}}(t) = \boldsymbol{g}(\boldsymbol{X}(t), \boldsymbol{\alpha} + \boldsymbol{W}(t)), \quad t \geq 0, \tag{8.127}$$

for an initial state $\boldsymbol{X}(0) = \boldsymbol{x}_0$, where $\boldsymbol{\alpha} \in \mathbb{R}^{d'}$ is a parameter, $\boldsymbol{W} \in \mathbb{R}^{d'}$ denotes a white noise process, and $d, d' \geq 1$ are integers. It is assumed that Eq. 8.127 has a stationary solution denoted by $\boldsymbol{X}_s$. If $\boldsymbol{X}_s = \boldsymbol{0}$ satisfies Eq. 8.127, we say that this equation admits a trivial stationary solution. For example, $\boldsymbol{X}_s = \boldsymbol{0}$ is a solution of Eq. 8.127 if this equation is homogeneous.

A main objective of stochastic stability analysis is the determination of subsets of $\mathbb{R}^{d'}$ including all values of $\boldsymbol{\alpha}$ for which solutions $\boldsymbol{X}$ of Eq. 8.127 starting in a small vicinity of a stationary solution $\boldsymbol{X}_s$ converge to $\boldsymbol{X}_s$ as time increases indefinitely. If this behavior is observed for almost all samples of $\boldsymbol{X}$, then $\boldsymbol{X}_s$ is said to be **stable** a.s. Otherwise, $\boldsymbol{X}_s$ is unstable and a **bifurcation**, that is, a qualitative change of this solution, may occur. Stability analysis is relevant for many aerospace, mechanical, and structural systems that exhibit nonlinear behavior [9].

Generally, the functional form of the density of $\boldsymbol{X}_s$ and the Lyapunov exponent are used to assess whether $\boldsymbol{X}_s$ is or is not stable [6, 24, 102, 130, 191]. The functional form of the density of $\boldsymbol{X}_s$ can change significantly as $\boldsymbol{\alpha}$ is perturbed and such a change may be associated with a phase transition in a physical system (Section 9.4.3 in this book, [102]). We define the Lyapunov exponent and present three examples illustrating the stability analysis of the stationary solution for stochastic differential equations driven by Gaussian and Poisson white noise.

Let $\tilde{\boldsymbol{X}} = \boldsymbol{X} - \boldsymbol{X}_s$ be an $\mathbb{R}^d$-valued process giving the difference between a solution $\boldsymbol{X}$ of Eq. 8.127 and a stationary solution $\boldsymbol{X}_s$ of this equation. If $\tilde{\boldsymbol{X}}$ is an ergodic process and $d \geq 2$, then $\ln \| \tilde{\boldsymbol{X}}(t) \| / t$ converges a.s. as $t \to \infty$ to two or more values, referred to as Lyapunov exponents, depending on the initial state $\tilde{\boldsymbol{X}}(0)$. For stability analysis we are interested in the largest Lyapunov exponent, called the **upper** or the **top Lyapunov exponent**, since this exponent controls the rate of growth of $\tilde{\boldsymbol{X}}$. Let λ_{LE} denote the top Lyapunov exponent.

Lyapunov exponent. If

$$\lambda_{\mathrm{LE}} = \lim_{t\to\infty} \frac{1}{t} \ln \| \tilde{\boldsymbol{X}}(t) \| < 0 \text{ and } > 0, \qquad (8.128)$$

then the stationary solution $\boldsymbol{X}_s$ is stable and unstable a.s., respectively. If $\lambda_{\mathrm{LE}} = 0$, then $\boldsymbol{X}_s$ may or may not be stable [9, 10].

Note: Generally, the calculation of the Lyapunov exponent involves two steps. First, a differential equation needs to be developed for the process $\tilde{\boldsymbol{X}} = \boldsymbol{X} - \boldsymbol{X}_s$. An approximate version of this equation obtained by linearization about $\boldsymbol{X}_s$ is used in the analysis. Accordingly, $\tilde{\boldsymbol{X}}$ is the solution of the differential equation

$$d\tilde{X}_i(t) = \sum_{j=1}^{d} a_{ij}\, \tilde{X}_j(t)\, dt + \sum_{k=1}^{d'} \sum_{j=1}^{d} b_{ij}^k\, \tilde{X}_j(t)\, dB_k(t), \quad i = 1, \ldots, d,$$

where B_k, $k = 1, \ldots, d'$, are Brownian motions independent of each other and the coefficients a_{ij}, b_{ij}^k may depend on $\boldsymbol{X}_s$. Hence, $\tilde{\boldsymbol{X}}$ is a diffusion process conditional on $\boldsymbol{X}_s$. Second, the Itô formula can be used to show that $\boldsymbol{S}(t) = \tilde{\boldsymbol{X}}(t)/ \| \tilde{\boldsymbol{X}}(t) \|$ is a diffusion process on the unit sphere in $\mathbb{R}^d$. It can be proved under some mild conditions that the top Lyapunov exponent is given by $\lambda_{\mathrm{LE}} = E[q(\boldsymbol{S}(t))] = \int q(\boldsymbol{s})\, \mu(d\boldsymbol{s})$, where $q(\boldsymbol{s}) = \langle(\tilde{\boldsymbol{a}} - \tilde{\boldsymbol{b}}\, \tilde{\boldsymbol{b}}^T)\, \boldsymbol{s}, \boldsymbol{s}\rangle + (1/2)\, \mathrm{tr}(\tilde{\boldsymbol{b}}\, \tilde{\boldsymbol{b}}^T)$, $\tilde{\boldsymbol{a}} = \{a_{ij}\}$, $\sum_{j=1}^{d} b_{ij}^k\, \tilde{X}_j$ are the entries of $\tilde{\boldsymbol{b}}$, and $\mu(\cdot)$ denotes the marginal probability density function of $\boldsymbol{S}$ assumed to be ergodic [112]. Theoretical considerations on Lyapunov exponents and their use in applications can be found in [6, 7, 13]. ▲

Example 8.54: Let $\boldsymbol{X}(t)$, $t \geq 0$, be an $\mathbb{R}^2$-valued diffusion process defined by

$$\begin{cases} dX_1(t) &= a\, X_1(t)\, dt + \sigma\, \left(X_1(t)\, dB_1(t) + X_2(t)\, dB_2(t)\right) \\ dX_2(t) &= b\, X_1(t)\, dt + \sigma\, \left(X_2(t)\, dB_1(t) - X_1(t)\, dB_2(t)\right), \end{cases}$$

where a, b, σ are some constants and B_1, B_2 denote independent Brownian motion processes. The trivial stationary solution of the above stochastic differential equation is stable a.s. if $(a + b)\, I_0(\beta) + (a - b)\, I_1(\beta) < 0$, where $\beta = (a - b)/(2\,\sigma^2)$ and I_o, I_1, denote modified Bessel functions of the first kind. ◇

Proof: The process $\tilde{X} = X - X_s$ satisfies the same stochastic differential equation as X since the coefficients of the defining equation for X depend linearly on this process. Moreover, $\tilde{X} = X$ because we examine the stability of the trivial stationary solution. The Itô formula in Section 4.6.2 applied to the function $g(\tilde{X}(t)) = \ln \| \tilde{X}(t) \|^2$ gives

$$dR(t) = \left(a\, S_1(t)^2 + b\, S_2(t)^2\right) dt + \sigma\, dB_1(t),$$

where $R(t) = \ln \| \tilde{X}(t) \|$ and $S(t) = \tilde{X}(t)/ \| \tilde{X}(t) \|$. Hence, the function $q(s)$ in the above definition of λ_{LE} is $q(s) = a\, s_1^2 + b\, s_2^2$. We also note that the process S is $S = (\cos(\Phi), \sin(\Phi))$ in polar coordinates, where $\Phi(t) = \tan^{-1}(\tilde{X}_2(t)/\tilde{X}_1(t))$.

The Itô formula applied to the mapping $\tilde{X}(t) \mapsto \Phi(t) = \tan^{-1}(\tilde{X}_2(t)/\tilde{X}_1(t))$ yields

$$d\Phi(t) = (b - a)\, \cos(\Phi(t))\, \sin(\Phi(t))\, dt - \sigma\, dB_2(t),$$

so that Φ is a diffusion process. The stationary density $f(\varphi)$ of Φ satisfies the Fokker-Planck equation

$$\frac{d}{d\varphi}\left((b - a)\, \cos(\varphi)\, \sin(\varphi)\, f(\varphi)\right) = \frac{\sigma^2}{2}\, \frac{d^2 f(\varphi)}{d\varphi^2},$$

and has the expression $f(\varphi) = c\, \exp\left(2\,\beta\, \cos(\varphi)^2\right)$, where $c > 0$ is a constant such that $\int_0^{2\pi} f(\varphi)\, d\varphi = 1$ and f satisfies the boundary condition $f(0) = f(2\pi)$. Elementary calculations give $c = 1/(2\, e^{\beta}\, \pi\, I_0(\beta))$.

The Lyapunov exponent $\lambda_{\text{LE}} = E[q(S(\Phi(t)))] = \int_0^{2\pi} q(S(\varphi)\, f(\varphi)\, d\varphi$ has the expression

$$\lambda_{\text{LE}} = \frac{1}{2\, I_0(\beta)}\, \left[(a + b)\, I_0(\beta) + (a - b)\, I_1(\beta)\right].$$

Hence, the trivial stationary solution is stable a.s. if the stated condition is satisfied. For example, the stationary trivial solution is stable a.s. for $a = -1$, $b = -2$, and $\sigma = 1$ but unstable for $a = 1$, $b = 2$, and $\sigma = 1$. ■

Two additional simple examples are presented. In both examples X is a real-valued process defined by the differential equation

$$dX(t) = (\beta\, X(t) - X(t)^m)\, dt + X(t)\, dS(t), \quad t \geq 0, \tag{8.129}$$

where $m \geq 2$ is an integer, $\alpha, \sigma \in \mathbb{R}$, $\beta = \alpha + \sigma^2/2$, and S denotes an elementary semimartingale noise. The two examples correspond to $S = \sigma\, B$ and $S = C$, where B is a Brownian motion, $C = \sum_{k=1}^{N(t)} Y_k$ denotes a compound Poisson process, N is a Poisson process with intensity $\lambda > 0$, and Y_k are iid random variables with distribution F. An alternative representation of C is $C(t) = \int_{0+}^{t} \int_{\mathbb{R}} y\, \mathcal{M}(ds, dy)$, where $\mathcal{M}$ is a Poisson random measure so that $\mathcal{M}(ds, dy)$ gives the number of jumps of C in $(s, s + ds] \times (y, y + dy]$ and has the expectation $E[\mathcal{M}(ds, dy)] = \lambda\, ds\, dF(y)$. Since the state X has dimension $d = 1$, there is a single Lyapunov exponent so that this exponent is λ_{LE}.

The stationary solution X_s satisfies Eq. 8.129 so that $\tilde{X} = X - X_s$ is the solution of

$$d\tilde{X}(t) = \left[\beta\,\tilde{X}(t) - \left(X(t)^m - X_s(t)^m\right)\right] dt + \tilde{X}(t)\,dS(t) \tag{8.130}$$

that is approximated by

$$d\tilde{X}(t) = \zeta(t)\,\tilde{X}(t)\,dt + \tilde{X}(t)\,dS(t), \tag{8.131}$$

where $\zeta(t) = \beta - m\,X_s(t)^{m-1}$ and $X(t)^m - X_s(t)^m \simeq m\,X_s(t)^{m-1}\,\tilde{X}(t)$. The approximation is acceptable since we consider only small initial deviations from the stationary solution. If λ_{LE} is strictly negative, then almost all samples of the solution of Eq. 8.131 converge to zero as $t \to \infty$, and we say that the stationary solution of Eq. 8.129 is stable a.s.

Example 8.55: If $S = \sigma\,B$ in Eq. 8.129, then

$$\lambda_{\mathrm{LE}} = \begin{cases} \alpha & \text{for } X_s = 0, \\ (1-m)\,\alpha & \text{for } X_s \neq 0, \end{cases}$$

so that the trivial and non-trivial solutions of Eq. 8.129 are stable a.s. for $\alpha < 0$ and $\alpha > 0$, respectively. Figure 8.30 shows five samples of the solution of Eq. 8.129 with $S = \sigma\,B$, $\alpha = \pm 1$, $\sigma = 1$, $m = 3$, and $X(0) = 0.1$. For $\alpha = -1$, the samples of X approach the trivial solution $X_s = 0$ as time increases. The samples

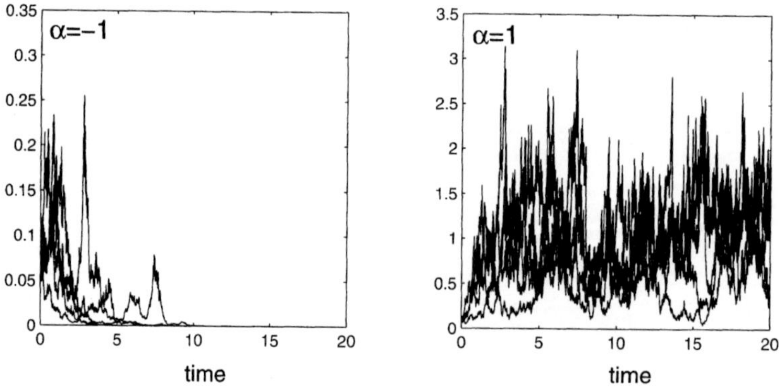

Figure 8.30. Samples of X in Eq. 8.129 with $S = \sigma\,B$ for $\alpha = \pm 1$, $\sigma = 1$, $m = 3$, and $X(0) = 0.1$

of X for $\alpha = 1$ seem to converge as time increases to a stationary process with a mean that is approximately equal to $\sqrt{\beta} = \sqrt{3/2}$, that is, the location of the right well of the potential of X. The samples of X remain in this well for a relatively long time because the noise intensity σ is relatively small. The properties of the samples of X in Fig. 8.30 are consistent with the predicted behavior of X based on λ_{LE}.

The density of X_s is $f_s(x) = c\, x^{(2\alpha/\sigma^2-1)} \exp\left[-2\, x^{m-1}/((m-1)\,\sigma^2)\right]$ for $\alpha > 0$, where $c > 0$ is a constant. This density is the solution of the Fokker-Planck equation for X_s (Section 7.3.1.4). For $\alpha < 0$ the density of X_s is concentrated at zero, that is, $f_s(x) = \delta(x)$. We note that f_s for $\alpha > 0$ has different functional forms for $\alpha < \alpha_{\rm cr}$ and $\alpha > \alpha_{\rm cr}$, where $\alpha_{\rm cr} = \sigma^2/2$. Figure 8.31 shows the density f_s for $m = 3, \sigma = 1$, and several values of α. The thin and heavy lines correspond to values of α smaller and larger than $\alpha_{\rm cr}$.

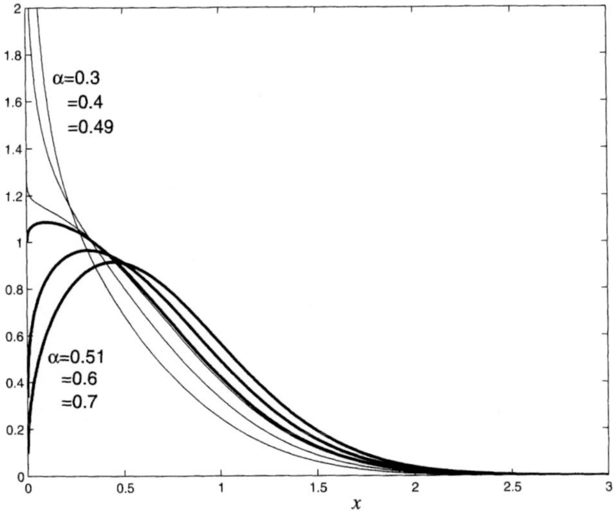

Figure 8.31. Densities of the stationary solution for $m = 3$ and $\sigma = 1$

There are two values of α marking qualitative changes of the stationary solution X_s, $\alpha = 0$ and $\alpha = \sigma^2/2$. These values of α are called **bifurcation** points. The stable stationary solution X_s is trivial and non-trivial for $\alpha < 0$ and $\alpha > 0$, respectively. At $\alpha = \sigma^2/2$ the density f_s of the stable stationary solution undergoes a notable change. The mode of f_s is at zero for $\alpha \in (0, \sigma^2/2)$ but moves at a strictly positive value for $\alpha \in (\sigma^2/2, \infty)$. Hence, the values $\alpha = 0$ and $\alpha = \sigma^2/2$ define a transition from the trivial solution to non-trivial stationary solutions and a transition between two different functional forms of the stationary density f_s, respectively. ◇

Proof: The solution of Eq. 8.131 with $S = \sigma\, B$ for a deterministic function $\zeta(\cdot)$ is a geometric Brownian (Section 4.7.1.1). Accordingly, we have

$$\tilde{X}(t) = \tilde{X}(0) \exp\left(\int_0^t \zeta(s)\, ds - \frac{\sigma^2}{2}\, t + \sigma\, B(t)\right) = \tilde{X}(0)\, e^{t\, Y(t)},$$

where $Y(t) = (1/t) \int_0^t \zeta(s)\, ds - \sigma^2/2 + (\sigma/t)\, B(t)$ or

$$Y(t) = \begin{cases} \alpha + (\sigma/t)\, B(t), & \text{for } X_s = 0 \text{ and} \\ \alpha - m\,(1/t) \int_0^t X_s(u)^{m-1}\, du + (\sigma/t)\, B(t), & \text{for } X_s \neq 0. \end{cases}$$

Note that (1) $B(t)/t \to 0$ a.s. as $t \to \infty$ ([150], p. 496) and (2) the moment of order $m-1$ of the non-trivial stationary solution is $E[X_s(t)^{m-1}] = \alpha$. The Itô formula applied to the function $\ln(X_s(t)^2)$ gives

$$d \ln\left(X_s(t)^2\right) = 2\left((\beta - X_s(t)^{m-1} - \sigma^2/2)\, dt + \sigma\, dB(t)\right)$$

since X_s satisfies the original nonlinear equation defining X, or

$$dE\left[\ln\left(X_s(t)^2\right)\right]/dt = 2\left(\beta - E[X_s(t)^{m-1}] - \sigma^2/2\right)$$

so that $E[X_s(t)^{m-1}] = \beta - \sigma^2/2 = \alpha$ because the left side of the above equation is zero.

If X_s is ergodic, then $(1/t) \int_0^t X_s(u)^{m-1}\, du$ converges a.s. to $E[X_s(t)^{m-1}] = \alpha$ as $t \to \infty$ so that $\lim_{t\to\infty} Y(t) = \alpha$ a.s. for $X_s = 0$ and $\lim_{t\to\infty} Y(t) = \alpha - m\, E[X_s(t)^{m-1}] = (1-m)\,\alpha$ for $X_s \neq 0$. These asymptotic values of Y coincide with the mean of this process at any $t \geq 0$. Since $\tilde{X}(t) = \tilde{X}(0)\, e^{t\, Y(t)}$, the trivial and non-trivial stationary solutions are stable a.s. if $\alpha < 0$ and $(1-m)\,\alpha < 0$, respectively.

We now examine the stability of the stationary solution X_s by calculating the Lyapunov exponent λ_{LE} in Eq. 8.128. Consider the approximate definition of $\tilde{X}$ in Eq. 8.131 with $S = \sigma\, B$. The differential equation for $R(t) = \ln(\tilde{X}(t)^2)$ is (Itô's formula)

$$R(t) - R(0) = 2 \int_0^t \left(\beta - m\, X_s(u)^{m-1} - \sigma^2/2\right) du + 2\sigma\, dB(t) = 2t\, Y(t),$$

or $\ln(|\tilde{X}(t)|/|\tilde{X}(0)|) = t\, Y(t)$, so that $\ln|X(t)|/t$ converges a.s. to $E[Y(t)]$ as $t \to \infty$.

In summary, $X_s = 0$ is a stationary solution for Eq. 8.129 with $S = \sigma\, B$ that is stable a.s. for $\alpha < 0$. If $\alpha > 0$, the trivial solution is not stable but there is a stable non-trivial solution. These conclusions follow from both the analytical solution of Eq. 8.129 and the Lyapunov exponent. In this example, it was not necessary to calculate the Lyapunov exponent λ_{LE} for stability analysis since Eq. 8.131 has a known analytical solution. The Lyapunov exponent has been calculated for illustration. However, in most applications, we do not know $\tilde{X}$ so that the Lyapunov exponent needs to be calculated. ■

Example 8.56: If $S = C$ in Eq. 8.129 and $1 + Y_1 > 0$ a.s., then [81]

$$\lambda_{\text{LE}} = \alpha + \sigma^2/2 + \lambda\, E[\ln(1 + Y_1)].$$

The value of α marking the transition from $X_s = 0$ to a non-trivial stationary solution is $\alpha^*(\lambda) = -\sigma^2/2 - \lambda\, E[\ln(1 + Y_1)]$.

Figure 8.32 shows the boundary $\alpha^*(\lambda)$ between the stability regions for the trivial and non-trivial stationary solutions for λ in the range [5, 40], $\sigma = 1$, $Y_1 \sim U(-a, a)$, and $a = \sigma\sqrt{3/\lambda}$. For these parameter values, we have $\lambda\, E[Y_1^2] = \sigma^2$ so that C is equal in the second moment sense to $\sigma\, B$ in the previous example. The plot in Fig. 8.32 suggests that $\alpha^*(\lambda)$ approaches zero as λ increases, that is,

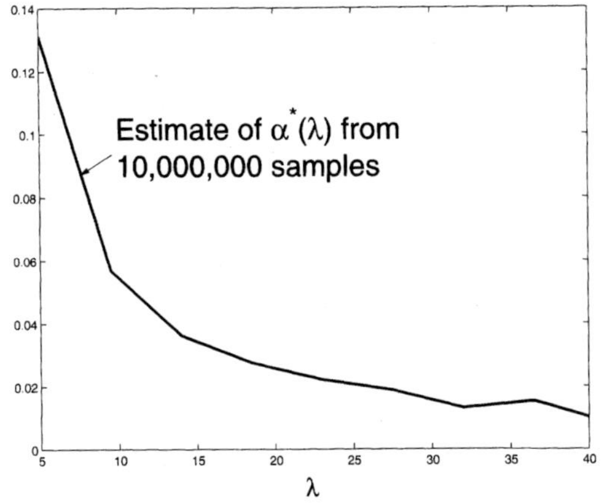

Figure 8.32. Boundary $\alpha^*(\lambda)$ between stability regions for the trivial and non-trivial stationary solutions

it converges to the bifurcation point between the trivial and non-trivial stationary solutions in Example 8.55. The result is not surprising since C becomes a version of the driving noise $\sigma\, B$ as $\lambda \to \infty$.

Figure 8.33 shows samples of X with $X(0) = 0.5$, $\sigma = 1$, $Y_1 \sim U(-a, a)$, $a = \sigma\,\sqrt{3/\lambda}$, $\lambda = 30$, and two values of the parameter α, $\alpha = 0.015 < \alpha^*(30)$ and $\alpha = 1 > \alpha^*(30)$. Consistently with the stability regions in Fig. 8.32, the

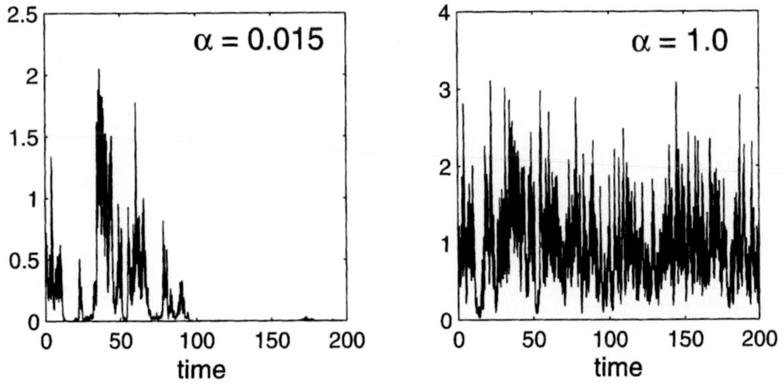

Figure 8.33. Samples of X for two values of α

sample of X corresponding to $\alpha = 0.015$ converges to the trivial solution in time. The sample of X for $\alpha = 1$ seems to approach a stationary solution $X_s \neq 0$ as time increases. ◇

Proof: We first show that the solution of Eq. 8.131 with $S = C$, referred to as the **geometric**

compound Poisson process by analogy with the geometric Brownian motion, is

$$\tilde{X}(t) = \tilde{X}(0) \exp\left(\int_{0+}^{t} \zeta(s-)\,ds + C^*(t)\right) = \tilde{X}(0)\, e^{t\,Z(t)},$$

where $C^*(t) = \sum_{k=1}^{N(t)} \ln(1+Y_k)$ and $Z(t) = (1/t)\int_0^t \zeta(s-)\,ds + C^*(t)/t$. The Itô formula, applied to the function

$$(t,x) \mapsto g(t,x) = \exp\left(\int_{0+}^{t} \zeta(u-)\,du + x\right)$$

whose arguments (t,x) correspond to the $\mathbb{R}^2$-valued process (X_1, X_2) defined by $dX_1(t) = dt$ and $dX_2(t) = dC^*(t)$, gives

$$\begin{aligned} g(t, C^*(t)) - g(0, C^*(0)) &= \int_{0+}^{t} \zeta(s-)\, g(s, C^*(s-))\,ds \\ &+ \sum_{0<s\le t} \left[g(s, C^*(s-) + \Delta C^*(s)) - g(s, C^*(s-))\right] \end{aligned}$$

which, following multiplication with $\tilde{X}(0)$, becomes

$$\tilde{X}(t) - \tilde{X}(0) = \int_{0+}^{t} \zeta(s-)\,\tilde{X}(s-)\,ds + \sum_{0<s\le t} \tilde{X}(s-)\left[e^{\Delta C^*(s)} - 1\right].$$

The differential form of the above equation is

$$d\tilde{X}(t) = \zeta(t-)\,\tilde{X}(t-)\,dt + \tilde{X}(t-)\,dC(t)$$

since $e^{\Delta C^*(s)} - 1$ is Y_k for some $k \ge 1$ and zero if C has a jump Y_k or no jump at time s. Hence, the above expression of $\tilde{X}$ satisfies Eq. 8.131 with $S = C$.

If $\lim_{t\to\infty} Z(t) < 0$ a.s., the stationary solution X_s is stable a.s. as $t \to \infty$. For the trivial stationary solution, $Z(t) = \beta + C^*(t)/t$ so that $\lim_{t\to\infty} Z(t) = \alpha + \sigma^2/2 + \lambda\, E[\ln(1+Y_1)]$ since $C^*(t)/t \to \lambda\, E[\ln(1+Y_1)]$ a.s. as $t \to \infty$ ([150], Theorem 3.3.2, p. 189). Hence, the trivial solution is stable a.s. for $\alpha < \alpha^*(\lambda) = -(\sigma^2/2 + \lambda\, E[\ln(1+Y_1)])$. If a non-trivial stationary solution X_s is an ergodic process, then

$$Z(t) = \beta - m\,\frac{1}{t}\int_{0+}^{t} X_s(u-)^{m-1}\,du + C^*(t)/t$$

converges a.s. to $\beta - m\,E\left[X_s(t-)^{m-1}\right] + \lambda\, E[\ln(1+Y_1)]$ as $t \to \infty$. This solution is stable a.s. for $\alpha > \alpha^*(\lambda)$ because $E[X_s(t-)^{m-1}] = \beta + \lambda\, E[\ln(1+Y_1)]$. To obtain $E[X_s(t)^{m-1}]$, note that the Itô formula applied to the function $\ln(X_s(t)^2)$ gives

$$d\,\ln(X_s(t)^2) = 2\left(\beta - X_s(t-)^{m-1}\right)dt + 2\int_{-\infty}^{\infty} \ln(1+y)\,\mathcal{M}(dt,dy)$$

since X_s satisfies the original nonlinear equation for X so that we have

$$\frac{d}{dt} E\left[\ln(X_s(t)^2)\right] = 2\,\beta - 2\,E[X_s(t-)^{m-1}] + 2\,\lambda\, E[\ln(1+Y_1)]$$

by averaging. Because X_s is a stationary process, the left side of the above equality is zero so that $E[X_s(t-)^{m-1}] = \beta + \lambda\, E[\ln(1+Y_1)]$.

As in the previous example, the Lyapunov exponent is not needed since we know the exact solution, but it is calculated for illustration. The Lyapunov exponent can be found by Itô's formula applied to $R(t) = \ln(\tilde{X}(t)^2)$. This formula gives

$$dR(t) = 2\,\zeta(t-)\,dt + 2\int_{-\infty}^{\infty} \log(1+y)\,\mathcal{M}(dt, dy).$$

so that $|\tilde{X}(t)| = |\tilde{X}(0)|\,\exp\left(\int_0^t \zeta(u-)\,du + C^*(t)\right)$ and

$$\lambda_{\mathrm{LE}} = \lim_{t\to\infty} \frac{1}{t}\left(\int_0^t \zeta(u-)\,du + C^*(t)\right) = \beta - m\, E[X_s(t)^{m-1}] + \lambda\, E[\ln(1+Y_1)]$$

is the Lyapunov exponent. ■

8.8 Localization phenomenon

Consider a system occupying a subset D of $\mathbb{R}^d$. A state variable of the system that has much larger values in one or more small subsets D_k of D than in $D \setminus \cup_k D_k$ is said to experience **localization**.

Localization is frequently observed in applications, and is used in engineering designs. For example, failure of a chain is localized at its weakest link because the links have slightly different capacities and are subjected to the same action. To increase automobile safety, the passenger cabin is designed to be much stiffer than the car ends so that the deformation caused by front and rear collisions will be localized at these ends. Material heterogeneity and geometry are common causes for localization in continuum mechanics. Material and/or geometry randomness is not an essential condition for localization. Uncertainty adds a new dimension to localization, and requires the use of probabilistic tools.

Two examples are used to illustrate the localization phenomenon. The first example shows that earthquake induced strains localize because of soil heterogeneity. Strain localization can cause a phase change in a soil deposit. The second example shows that, while the modes of vibration of a perfectly periodic continuous beam extend over all beam spans, the corresponding modes of a beam with nearly equal spans are concentrated or localized on only a few spans.

8.8.1 Soil liquefaction

Soils are heterogeneous materials whose properties exhibit significant spatial fluctuations. Under external actions, soil deposits can experience a phase transition. For example, sands and other cohesionless soils can change abruptly their consistency from solid to near fluid if subjected to cyclic actions. This phase transition phenomenon, called **liquefaction**, occurs if the pore water pressure exceeds a critical value. Soil liquefaction can have catastrophic effects on building and bridges during seismic events. Generally, relatively small volumes of a soil deposit liquefy during an earthquake. The pockets of liquefied soil appear to have random size and location [144].

Our objective is to (1) assess the potential for liquefaction of cohesionless soil deposits subjected to earthquakes and (2) examine the spatial distribution and the size of liquefied soil pockets. The potential for liquefaction is measured by the pore water pressure. It is assumed that liquefaction occurs at a point $\boldsymbol{x}$ in a soil deposit if the pore water pressure at $\boldsymbol{x}$ exceeds a critical value.

The cone tip resistance Z_1 and the soil classification index Z_2 largely control the occurrence and the extent of liquefaction in a soil deposit. Because Z_1 and Z_2 exhibit a significant spatial variation, it has been proposed to model them by random functions [144]. Let $\boldsymbol{Z} = (Z_1, Z_2)$ be an $\mathbb{R}^2$-valued random field defined in an open bounded subset D of $\mathbb{R}^3$. Because Z_1 and Z_2 are bounded, $\boldsymbol{Z}$ cannot be approximated by a Gaussian random field. We model $\boldsymbol{Z}$ by a translation random field defined by

$$Z_k(\boldsymbol{x}) = F_k^{-1} \circ \Phi(G_k(t)(\boldsymbol{x})) = g_k(G_k(\boldsymbol{x})), \quad k = 1, 2,$$

where F_k denotes the marginal distribution of Z_k, Φ is the distribution of $N(0, 1)$, and $\boldsymbol{G} = (G_1, G_2) : D \times \Omega \to \mathbb{R}^2$ is a homogeneous Gaussian random field defined on a probability space $(\Omega, \mathcal{F}, P)$. The Gaussian field $\boldsymbol{G}$ has mean zero, variance one, and covariance functions $\rho_{kl}(\boldsymbol{\xi}) = E[G_k(\boldsymbol{x})\, G_l(\boldsymbol{x} + \boldsymbol{\xi})]$, $\boldsymbol{x}, \boldsymbol{x} + \boldsymbol{\xi} \in D$. The coordinates of the random field $\boldsymbol{Z}$ can follow any marginal distribution. The correlation functions $E[Z_k(\boldsymbol{x})\, Z_l(\boldsymbol{x} + \boldsymbol{\xi})] = E[g_k(G_k(\boldsymbol{x}))\, g_l(G_l(\boldsymbol{x} + \boldsymbol{\xi}))]$ of $\boldsymbol{Z}$ are determined completely by F_k and ρ_{kl} (Section 5.3.3.1).

Suppose that a soil deposit in D with properties given by the random field $\boldsymbol{Z}$ is subjected to a seismic ground acceleration record $x(t)$, $t \in [0, \tau]$, where τ denotes the earthquake duration. A finite element code, DYNAFLOW, has been used in [144] to calculate the evolution in time of the pore water pressure and other soil states in D under a seismic action $x(t)$. Let $W(\boldsymbol{x}, \omega)$, $\boldsymbol{x} \in D$, be a sample of the pore water pressure at the end of a seismic event corresponding to a sample $\boldsymbol{Z}(\boldsymbol{x}, \omega)$, $\boldsymbol{x} \in D$, of $\boldsymbol{Z}$. We can estimate moments and other probabilistic properties of the pore water pressure random field from a collection of independent samples of W. Information on the finite element code DYNAFLOW can be found in [146].

Example 8.57: Consider a vertical section through a soil deposit D and let $D_v = (0, a) \times (0, b) \subset \mathbb{R}^2$ be a rectangle included in this section with the sides of length a and b oriented in the horizontal and vertical directions, respectively (Fig. 8.34). The soil deposit in D is subjected to the deterministic ground acceleration $x(t)$ in Fig. 8.34. Data analysis shows that the random fields Z_1 and Z_2 follow beta distributions. The parameters of these distributions and the correlation functions of $\boldsymbol{Z}$ are given in [144].

Figure 8.35 shows contour lines for six samples of the pore water pressure random field W corresponding to six independent samples of $\boldsymbol{Z}$. The darker

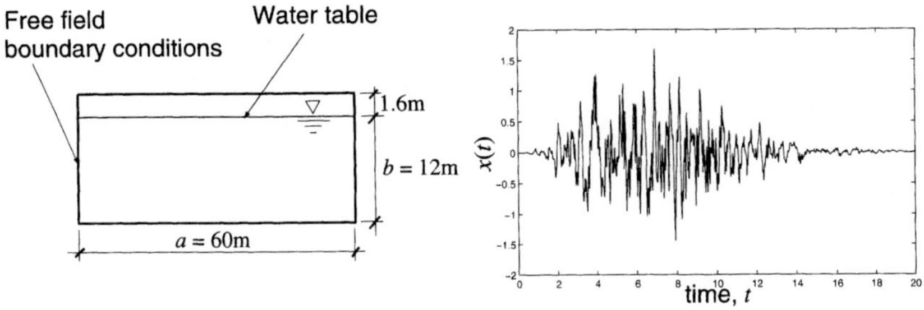

Figure 8.34. Soil deposit and seismic ground acceleration

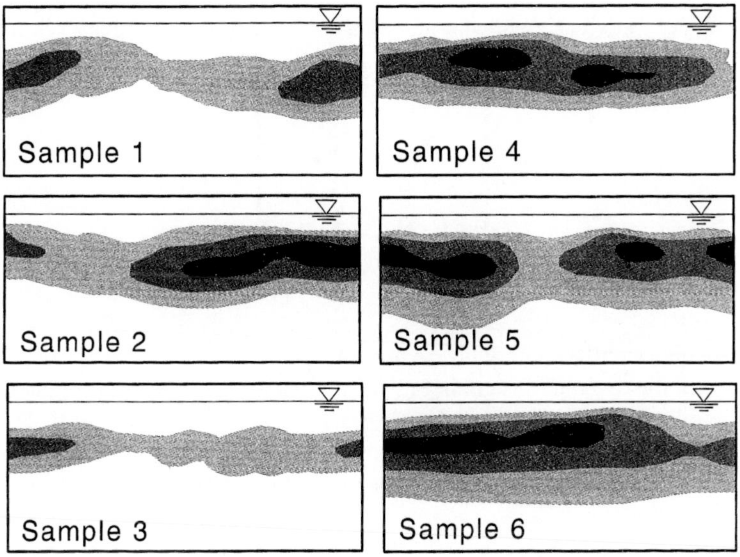

Figure 8.35. Six samples of the pore water pressure random field W

shades in the figure are for larger values of W, and indicate an increased potential for liquefaction. The samples of W are consistent with observed liquefaction patterns [144]. The size and the location of the pockets of liquefaction show a remarkable sample to sample variation. ◇

Note: An extensive data set has been used in [144] to estimate the probability law of the random field $\mathbf{Z}$. If it is assumed that the soil deposit in D is homogeneous, then liquefaction either does not occur or occurs everywhere in D, in contrast with field observations. ▲

8.8.2 Mode localization

Consider the dynamic system in Fig. 8.36 consisting of large number of oscillators connected by springs with the same stiffness k. Suppose first that the oscillators have identical natural frequencies. Then the modes of vibration extend throughout the entire system. Suppose now that the natural frequencies of the oscillators differ slightly and that $k = 0$. The modes and the modal frequencies of the system coincide with the vibrations of individual oscillators and their natural frequencies, respectively. If $k > 0$ is small, the system modes of vibration are localized around individual oscillators and the corresponding modal frequencies are approximately equal to the frequencies of these oscillators. For example, if the system is subjected to a harmonic action of frequency equal to the natural frequency of an oscillator, this oscillator will vibrate with a large amplitude. Its nearest neighbors will be driven off resonance and, since the coupling between the oscillators is weak, will vibrate with much smaller amplitudes. Similarly,

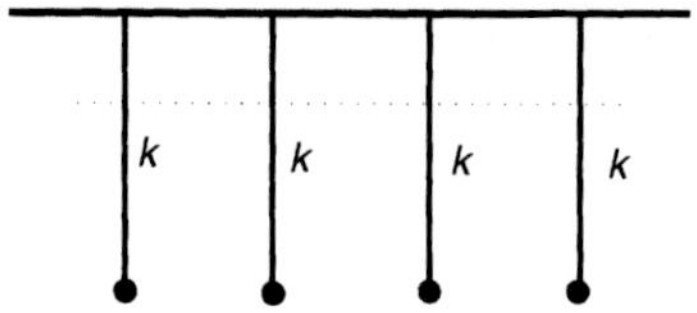

Figure 8.36. A system of simple oscillators

their neighbors will vibrate with even smaller amplitudes, so that the modes of vibration of the dynamic system in Fig. 8.36 with slightly different oscillators and weak couplings will be localized.

It is shown that properties of products of random matrices and the Lyapunov exponents introduced in Section 8.7 can be used to find the rate of decay of the localized modes.

Let $\boldsymbol{X}_n = \boldsymbol{A}_n \boldsymbol{X}_{n-1}$, $n = 1, 2, \ldots$, where $\boldsymbol{A}_1, \boldsymbol{A}_2, \ldots$ are independent copies of a (d, d) matrix $\boldsymbol{A}$ with real-valued entries and $\boldsymbol{X}_0 = \boldsymbol{x}_0 \in \mathbb{R}^d$. Note that the sequence $\boldsymbol{X}_n = \boldsymbol{C}_n \boldsymbol{X}_0$, $n = 1, 2, \ldots$, is an $\mathbb{R}^d$-valued Markov chain, where $\boldsymbol{C}_n = \boldsymbol{A}_n \boldsymbol{A}_{n-1} \cdots \boldsymbol{A}_1$.

We present some properties of matrix $\boldsymbol{C}_n$ and define the upper Lyapunov exponent for the sequence $\boldsymbol{X}_n$.

If $E\left[(\log \| \boldsymbol{A}_1 \|_m)^+\right] < \infty$, the sequence $E[\ln \| \boldsymbol{C}_n \|_m]$ is subadditive, that is,

$$E\left[\ln \| \boldsymbol{C}_{n+p} \|_m\right] \leq E\left[\ln \| \boldsymbol{C}_n \|_m\right] + E\left[\ln \| \boldsymbol{C}_p \|_m\right] \tag{8.132}$$

for $n, p \geq 1$ integers, where $\| \boldsymbol{a} \|_m = \sup\{\| \boldsymbol{a}\,\boldsymbol{x} \| : \boldsymbol{x} \in \mathbb{R}^d, \| \boldsymbol{x} \| = 1\}$ denotes a matrix norm.

Proof: Because $E\left[(\ln \| \boldsymbol{A}_1 \|_m)^+\right] < \infty$, the expectation $E\left[\ln \| \boldsymbol{A}_1 \|_m\right]$ exists but may be $-\infty$. Let $n, p \geq 1$ be integers and note that

$$\begin{aligned} E\left[\ln \| \boldsymbol{C}_{n+p} \|_m\right] &= E\left[\ln \| \boldsymbol{A}_{n+p} \cdots \boldsymbol{A}_1 \|_m\right] \\ &\leq E\left[\ln \| \boldsymbol{A}_{n+p} \cdots \boldsymbol{A}_{n+1} \|_m\right] + E\left[\ln \| \boldsymbol{A}_n \cdots \boldsymbol{A}_1 \|_m\right] \\ &= E\left[\ln \| \boldsymbol{C}_p \|_m\right] + E\left[\ln \| \boldsymbol{C}_n \|_m\right] \end{aligned}$$

by properties of the norm, the definition of $\boldsymbol{C}_n$, and the assumption that the matrices $\boldsymbol{A}_k$ are independent and have the same distribution. The subadditivity of $\ln \| \boldsymbol{C}_n \|$ implies the convergence

$$\frac{1}{n} E\left[\ln \| \boldsymbol{C}_n \|_m\right] \longrightarrow \inf_{q \geq 1} \frac{1}{q} E\left[\ln \| \boldsymbol{C}_q \|_m\right], \quad \text{as } n \to \infty,$$

in $\mathbb{R} \cup \{-\infty\}$ ([24], Example 5.2, p. 13). ■

If $E\left[(\ln \| \boldsymbol{A}_1 \|_m)^+\right] < \infty$ and the independent identically distributed matrices $\boldsymbol{A}_1, \boldsymbol{A}_2, \ldots$ have inverses, the limit

$$\lambda_{\text{LE}} = \lim_{n \to \infty} \frac{1}{n} \ln \| \boldsymbol{C}_n \|_m = \lim_{n \to \infty} \frac{1}{n} \ln \| \boldsymbol{A}_n \cdots \boldsymbol{A}_1 \|_m \tag{8.133}$$

holds a.s. and is called the **upper** or the **top Lyapunov exponent** ([24], Theorem 4.1, p. 11).

Note: The result in Eq. 8.133 is a statement of a theorem by Furstenberg and Keston [69].

Recall that we have used the upper Lyapunov exponent in Section 8.7 to assess the sample stability of the stationary solutions of differential equations driven by Gaussian and Poisson white noise. We have seen that these solutions are stable a.s. if $\lambda_{\text{LE}} < 0$. The Lyapunov exponent in Eq. 8.133 can also be used for the stability analysis of discrete systems with random properties, for example, a system with state $\boldsymbol{X}_n$, $n = 0, 1, \ldots$, defined by the recurrence relationship $\boldsymbol{X}_n = \boldsymbol{A}_n \boldsymbol{X}_{n-1}$, $n = 1, 2, \ldots$, and a specified initial state $\boldsymbol{X}_0$. Such relationships also result from discrete time approximations of stochastic differential equations of the type in Eq. 8.131.

That the Lyapunov exponent gives the rate of growth of the solution is consistent with our intuition. For example, let $\boldsymbol{a}_i$, $i = 1, 2, \ldots$, be deterministic (d, d) matrices and define the sequence $\boldsymbol{x}_n(\boldsymbol{x}_0) = \boldsymbol{a}_n \boldsymbol{a}_{n-1} \cdots \boldsymbol{a}_1 \boldsymbol{x}_0 = \boldsymbol{c}_n \boldsymbol{x}_0$, $n = 1, 2, \ldots$, starting from a deterministic initial state $\boldsymbol{x}_0$. Let $\sigma_{\min}^2$ and $\sigma_{\max}^2$ denote the smallest and largest eigenvalues of $\boldsymbol{c}_n^T \boldsymbol{c}_n$. We have (Section 8.3.2.1)

$$\sigma_{\min}^2 \| \boldsymbol{x}_0 \|^2 \leq \| \boldsymbol{x}_n(\boldsymbol{x}_0) \|^2 = \boldsymbol{x}_0^T \boldsymbol{c}_n^T \boldsymbol{c}_n \boldsymbol{x}_0 \leq \sigma_{\max}^2 \| \boldsymbol{x}_0 \|^2,$$

so that

$$\ln(\sigma_{\min}) + \ln \| \boldsymbol{x}_0 \| \leq \ln \| \boldsymbol{x}_n(\boldsymbol{x}_0) \| \leq \ln(\sigma_{\max}) + \ln \| \boldsymbol{x}_0 \|$$

and $\lim_{n \to \infty} \ln(\sigma_{\min})/n \leq \lambda(\boldsymbol{x}_0) \leq \lim_{n \to \infty} \ln(\sigma_{\max})/n$, where $\lambda(\boldsymbol{x}_0)$ is the limit of $\ln \| \boldsymbol{x}_n(\boldsymbol{x}_0) \| / n$ as $n \to \infty$. If the limits of $\ln(\sigma_{\min})/n$ and $\ln(\sigma_{\max})/n$ as $n \to \infty$ are bounded, so is $\lambda(\boldsymbol{x}_0)$ for every $\boldsymbol{x}_0$ and $\| \boldsymbol{x}_n(\boldsymbol{x}_0) \| \sim e^{n \lambda(\boldsymbol{x}_0)}$ for large values of n so that $\lambda(\boldsymbol{x}_0)$ gives the rate of growth of $\boldsymbol{x}_n(\boldsymbol{x}_0)$ with n. ▲

We have seen in Section 8.7 that systems with dimension $d \geq 2$ have more than one Lyapunov exponent. For stability analysis the upper Lyapunov exponent is relevant. However, for localization we need the smallest positive Lyapunov exponent since it gives the least rate of decay of the state $\boldsymbol{X}_n$, and this rate provides information on the extent of the localization. We denote this Lyapunov exponent by $\tilde{\lambda}_{\mathrm{LE}}$. Additional information on the use of the Lyapunov exponent for mode localization analysis and methods for calculating $\tilde{\lambda}_{\mathrm{LE}}$ in some special cases can be found in [198] and [122] (Chapter 9).

Example 8.58: Consider a perfectly periodic beam with uniformly distributed mass, n spans of equal length l, and torsional springs with the same stiffness s at its supports (Fig. 8.37). Suppose that the spans of this beam are perturbed so

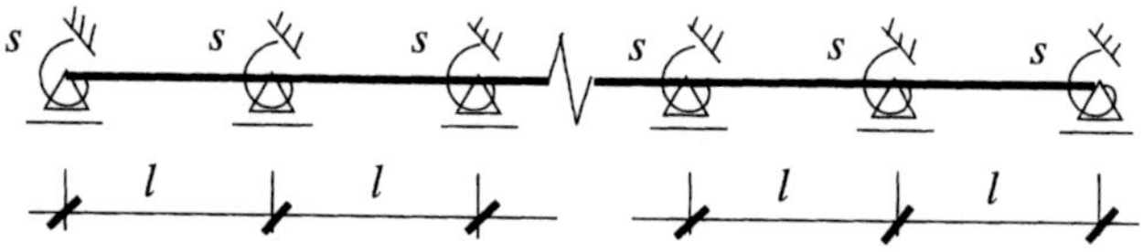

Figure 8.37. A perfectly periodic continuous beam

that the original spans become $L_i = l + R_i$, $i = 1, \ldots, n$, where R_i are independent copies of a random variable R such that $|R| \ll l/2$ a.s. The resulting beam is referred to as a randomly disordered or nearly periodic beam.

Figures 8.38 and 8.39 show the first five modes of vibration of the perfectly periodic and randomly disordered beams, respectively, for $l = 1$, $s = 20$, $n = 100$, $R \sim U(-0.03, 0.03)$, and unit mass per unit of length. The modes of the periodic beam extend over all spans. On the other hand, the modes of the randomly disordered beam are localized on only a few spans. ◇

Note: The modal shapes in Figs. 8.38 and 8.39 were calculated by using the finite element method. Two beam elements were used in each span of the perfectly periodic and randomly disordered beams, so that both beams were modeled by discrete systems with $2n + 1$ degrees of freedom. An algorithm written in MATLAB was used for the modal analysis. The example is taken from [198].

Let Θ_i be the rotation at support $i = 1, \ldots, n+1$ and let $\boldsymbol{\Theta}^{(i)} = (\Theta_i, \Theta_{i+1})$. Then $\boldsymbol{\Theta}^{(i)} = \boldsymbol{T}^{(i)}\, \boldsymbol{\Theta}^{(i-1)}$, where $\boldsymbol{T}^{(i)}$ denotes a transition matrix that can be obtained by classical methods of structural analysis. For the perfectly periodic beam the transition matrices are deterministic and identical. On the other hand, for the randomly disordered beam the matrices $\boldsymbol{T}^{(i)}$ are independent and identically distributed. The recurrence formula for $\boldsymbol{\Theta}$ gives

$$\boldsymbol{\Theta}^{(k)} = \boldsymbol{T}^{(k)} \cdots \boldsymbol{T}^{(2)}\, \boldsymbol{\Theta}^{(1)}$$

so that $\| \boldsymbol{\Theta}^{(k)} \| = \| \boldsymbol{T}^{(k)} \cdots \boldsymbol{T}^{(2)}\, \boldsymbol{\Theta}^{(1)} \|$ and

$$\| \boldsymbol{\Theta}^{(k)} \| = \left(\| \boldsymbol{T}^{(k)} \cdots \boldsymbol{T}^{(2)}\, \boldsymbol{\Theta}^{(1)} / \| \boldsymbol{\Theta}^{(1)} \| \, \| \right) \| \boldsymbol{\Theta}^{(1)} \| \leq \| \boldsymbol{T}^{(k)} \cdots \boldsymbol{T}^{(2)} \|_m \, \| \boldsymbol{\Theta}^{(1)} \|,$$

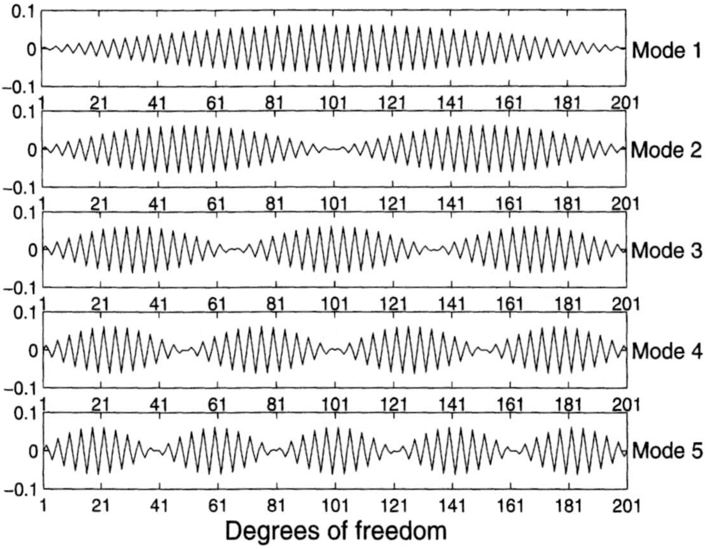

Figure 8.38. First five modes of a perfectly periodic beam

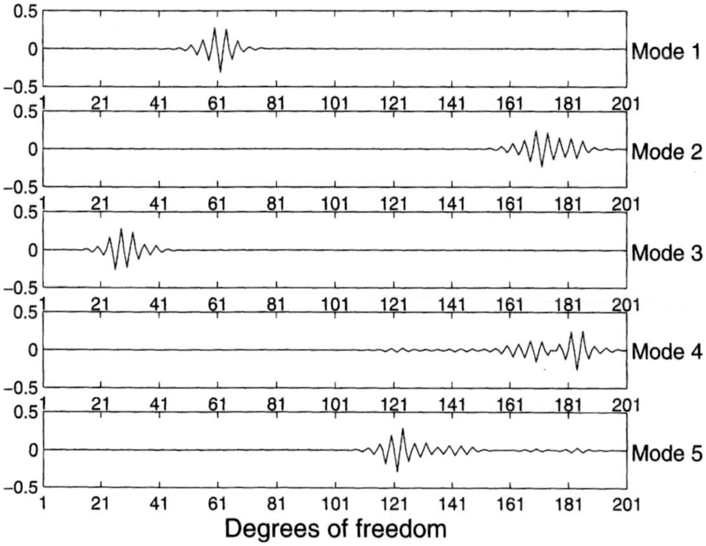

Figure 8.39. First five modes of a randomly disordered beam

so that

$$\| \mathbf{\Theta}^{(k)} \| \sim e^{\tilde{\lambda}_{\mathrm{LE}}(k-1)} \| \mathbf{\Theta}^{(1)} \|, \quad \text{as } k \to \infty,$$

where $\tilde{\lambda}_{\mathrm{LE}}$ is the smallest positive Lyapunov exponent corresponding to the random matrices $\boldsymbol{T}^{(k)}$. The Lyapunov exponent $\tilde{\lambda}_{\mathrm{LE}}$ in the above equation gives the rate of decay of the rotation at supports, and is referred to as a **localization factor**. ▲

8.9 Problems

8.1: Consider a modified version of the partial differential equation in Example 8.1 obtained by replacing its right side with an integrable function $p(\boldsymbol{x})$, $\boldsymbol{x} \in D$. Find the local solution of the resulting partial differential equation. Extend your results to the case in which $p(\boldsymbol{x})$ is a random field.

8.2: Extend the local solution of the partial differential equation in Problem 8.1 to the case in which D is a random set. Assume that the point $\boldsymbol{x}$ at which the local solution is determined is in D a.s.

8.3: Derive approximate expressions for the second moment properties of $\boldsymbol{X}$ in Eq. 8.2 based on a second order Taylor expansion (Eq. 8.9). Use your findings to improve the second moment properties of $\boldsymbol{X}$ calculated in Example 8.4.

8.4: Find the perturbation solution of $\boldsymbol{X}$ in Eq. 8.2 with $\boldsymbol{Y} = \boldsymbol{a} + \varepsilon\, \tilde{\boldsymbol{R}}_1 + \varepsilon^2\, \tilde{\boldsymbol{R}}_2 + O(\varepsilon^3)$, where ε is a small parameter and the random matrices $\tilde{\boldsymbol{R}}_1$ and $\tilde{\boldsymbol{R}}_2$ have mean zero and are independent with each other. Develop expressions for the second moment properties of $\boldsymbol{X}$.

8.5: Solve the problem in Example 8.8 for Y being a shifted exponential random variable with the density $f(y) = \lambda\, e^{-\lambda\,(y-a)}$, $y \geq a$, where $\lambda > 0$ and a are some constants.

8.6: Find the first two moments of X defined by $Y\,X = q$ by both versions of the iteration method, where $Y = a + R$, $R \sim U(-\alpha, \alpha)$, and $0 < \alpha < a$.

8.7: Let $\boldsymbol{S} = (S_{11}, S_{12} = S_{21}, S_{22})$ be a Gaussian vector, where S_{ij} are random stresses. Suppose that the coordinates of $\boldsymbol{S}$ have the mean values $(100, -60, 300)$, are equally correlated with correlation coefficient $\rho = 0.7$, and have the same coefficient of variation $v = 0.3$. Find the first two moments of the eigenvalues and eigenvectors of the stress tensor $\{S_{ij}\}$.

8.8: Consider an infinite thin plate with a through crack of length $2\,a$. The plate is subjected to uniform far field stresses, a tension Q_1 perpendicular to the crack and a shear Q_2. Suppose that (Q_1, Q_2) is a Gaussian vector with mean $(10, 4)$, coefficients of variation $(0.3, 0.3)$, and correlation coefficients $\rho = 0$, 0.7, and 0.9. Let Θ be the direction of crack extension. Find the distribution of Θ based on the relationship $Q_1/Q_2 = [1 - 3\,\cos(\Theta)]/\sin(\Theta)$.

8.9: Find second order Taylor approximations for the eigenvalues and eigenvectors of a square random matrix $\boldsymbol{A}$.

8.10: Consider an (n, n) symmetric matrix $\boldsymbol{A} = \boldsymbol{a} + \boldsymbol{R}$. Denote by λ_i and Λ_i the eigenvalues of the deterministic and random parts, $\boldsymbol{a}$ and $\boldsymbol{R}$, of $\boldsymbol{A}$, respectively. Develop bounds on the probability $P(\Lambda_i \leq x)$.

8.11: Find the second moment properties of the eigenvalues and eigenvectors of the random matrix in Example 8.12 by the iteration method.

8.12: Let $\boldsymbol{G}$ be an $\mathbb{R}^n$-valued Gaussian variable with mean $\boldsymbol{\mu}$ and covariance function $\boldsymbol{\gamma}$. Find the mean zero-crossing rate $\nu(\lambda)$ of the stochastic process $V(\lambda) = \lambda^n + G_1 \lambda^{n-1} + \cdots + G_{n-1} \lambda + G_n$. Use your results to calculate $\nu(\lambda)$ for $n = 3$, $\boldsymbol{\mu} = \boldsymbol{0}$, and $\boldsymbol{\gamma}$ equal to the identity matrix.

8.13: Let U be defined by $dU(t) = -Z\, U(t)\, dt + dB(t)$, $t \geq 0$, where $Z > 0$ a.s., $U(0) = 0$, and B denotes a Brownian motion. Develop a differential equation for the sensitivity factor $V(t) = \partial U(t)/\partial Z$ and calculate the second moment properties of (U, V) for $Z \sim U(1, 3)$.

8.14: Let $\boldsymbol{U}$ be the solution of Eq. 8.45 with $\mathcal{D}$ given by Eq. 8.53, where ε is a small parameter. Find expressions for the mean of $\boldsymbol{U}$ to the order ε^2 and for the correlation function of $\boldsymbol{U}$ to the order ε^4. Calculate the confidence interval $(E[\boldsymbol{U}] - \text{Std}[\boldsymbol{U}], E[\boldsymbol{U}] + \text{Std}[\boldsymbol{U}])$ on $\boldsymbol{U}$ to the order ε^2.

8.15: Let U be the solution of $d^4U(x)/dx^4 + N(x, U(x)) = q$, $x \in (0, l)$, with the boundary conditions $U(0) = 0$, $dU(0)/dx = 0$, $U(l) = 0$, and $d^2U(l)/dx^2 = 0$, where q is constant and $N(x, U(x))$ is a nonlinear random function. Derive equations for the second moments of $U(x)$ for $N(x, U(x)) = A(x)\, U(x) + B(x)\, U(x)^3$, where A and B are square integrable random fields that are uncorrelated with each other.

8.16: Repeat the analysis in Problem 8.15 by using a finite difference approximation for the partial differential equation defining U.

8.17: Let U be the solution of $dU(t)/dt = -Z\, U(t)$, $t \geq 0$, with the initial condition $U(0) = \xi > 0$, where $Z \sim U(1, 2)$. Find U by the Neumann series method, establish conditions for the convergence of this series representation of U, and calculate the first two moments and the marginal distribution of the solution based on the first three terms of the Neumann series.

8.18: Consider the random eigenvalue problem $Z\, \Phi''(x) + \Lambda\, \Phi(x) = 0$, $x \in (0, 1)$, with the boundary conditions $\Phi(0) = \Phi(1) = 0$, where Z is a random variable with positive mean. Find an upper bound on the lowest eigenvalue in this problem.

8.19: Apply the Monte Carlo simulation method to estimate the second moment properties and the distributions of the first three eigenvalues and eigenfunctions of the random eigenvalue problem in Problem 8.18. Assume that Z is uniformly distributed in $(1, 2)$.

8.20: Assume that conductivity σ in Eq. 8.77 is a translation random field defined by $\Sigma(\boldsymbol{x}) = \alpha + \exp(\beta\, G(\boldsymbol{x}))$, $\boldsymbol{x} \in \mathbb{R}^3$, where G is a zero-mean Gaussian field with covariance function $E[G(\boldsymbol{x})\, G(\boldsymbol{x}')] = \exp(-\lambda \;\| \boldsymbol{x} - \boldsymbol{x}' \|)$, and $\alpha, \beta, \lambda > 0$ are some constants. Find the effective conductivity for this random heterogeneous material.

8.21: Generate additional samples of the process in Example 8.56 for different values of λ and α, and determine whether the observed sample properties are consistent with the results in Fig. 8.32.

Chapter 9

Stochastic Systems and Input

9.1 Introduction

In most stochastic problems both the system properties and the input characteristics are random. However, the degree of uncertainty in the input and the system can be very different. For example, it is common in earthquake engineering to assume that the structural system is deterministic because of the very large uncertainty in the seismic ground acceleration. Under this simplifying assumption we can evaluate the seismic performance of structural systems by the methods in Chapter 7. Similarly, developments in Chapters 6 or 8 can be applied to stochastic problems characterized by a negligible uncertainty in both the system and input or the input alone, respectively.

This chapter considers stochastic problems in which both the system and input uncertainty has to be modeled. Methods for analyzing this type of stochastic problems are discussed primarily in Section 9.2. The subsequent sections in this chapter present applications from mechanics, physics, environment and ecology, wave propagation in random media, and seismology. Some brief considerations on model selection conclude the chapter. We start with a summary of the main sources of uncertainty for general stochastic problems.

- **Randomness in system and input properties**. For example, the atomic lattice orientation in metals exhibits random spatial variation (Section 8.6.2.2), identically designed light bulbs and other physical systems have different service life, properties of geological formations are characterized by notable spatial fluctuations, the details of the next earthquake at a site in California are not known.

- **Finite information**. For example, the functional form and the parameters of most mathematical models used in finance, physics, mechanics, and other fields of applied science and engineering need to be estimated from finite records so that they are uncertain and the degree of uncertainty depends on the available sample size.

- **Limited understanding and model simplification**. For example, the underlying principles of a phenomenon may not be well understood and/or may be too complex for modeling and analysis so that simplified representations need to be considered. The prediction of the outcome of a coin tossing experiment is possible in principle because the coin can be viewed as a rigid body subjected to an initial condition. However, the high sensitivity of the outcome of this experiment to initial conditions, properties of the landing surface, and many other parameters render the mathematical modeling of this experiment impractical if not impossible. We need to settle for a global characterization of the coin tossing experiment giving the probability of seeing the head or the tail in a toss.

This chapter examines stochastic problems defined by

$$\mathcal{D}[\mathcal{X}(\boldsymbol{x}, t)] = \mathcal{Y}(\boldsymbol{x}, t), \quad t \geq 0, \quad \boldsymbol{x} \in D \subset \mathbb{R}^q, \tag{9.1}$$

where D is a subset of $\mathbb{R}^q$, $\mathcal{D}$ can be an algebraic, integral, or differential operator with random coefficients, $\mathcal{Y}(\boldsymbol{x}, t)$ is an $\mathbb{R}^m$-valued random function depending on space $\boldsymbol{x} \in D$ and time $t \geq 0$ arguments, the $\mathbb{R}^n$-valued random function $\mathcal{X}$ denotes the output, and $m, n, q \geq 1$ are some integers. The output $\mathcal{X}$ depends on $\mathcal{Y}$, $\mathcal{D}$, and the initial/boundary conditions for Eq. 9.1, which can be deterministic or random. It is difficult to give conditions for the existence and uniqueness of the solution of Eq. 9.1 in this general setting. We will give such conditions for special forms of Eq. 9.1. In many applications it is sufficient to find the solution $\mathcal{X}$ at a finite number of points $\boldsymbol{x}_k \in D$. Then Eq. 9.1 becomes an equation giving the evolution in time of an $\mathbb{R}^d$-valued stochastic process $\boldsymbol{X}(t)$, $t \geq 0$, collecting the functions $\mathcal{X}(t, \boldsymbol{x}_k)$. Similar considerations have been used in Section 7.1. Most of the results in this chapter relate to differential equations, that is, $\mathcal{D}$ in Eq. 9.1 is a differential operator.

Our objective is to calculate second moment and other properties of $\mathcal{X}$ in Eq. 9.1 supplied with deterministic or random initial and boundary conditions, or of its discrete counterpart, the stochastic process $\boldsymbol{X}$. There is no general method that can deliver the probability law of $\mathcal{X}$ or $\boldsymbol{X}$. The next section presents methods for calculating probabilistic properties of $\mathcal{X}$ and $\boldsymbol{X}$. Most of the methods in this section can be viewed as extensions of developments in Chapters 6, 7, and 8. The subsequent sections present applications from various fields of applied science and engineering.

9.2 Methods of analysis

Let $\mathcal{X}$ be the solution of Eq. 9.1. If the coefficients of the operator $\mathcal{D}$ in this equation are not functions of $\mathcal{X}$, we say that $\mathcal{D}$ has **state independent coefficients**. Otherwise, $\mathcal{D}$ has **state dependent coefficients**. We will see that a stochastic problem defined by an operator $\mathcal{D}$ with state dependent coefficients can be recast into another problem defined by an operator $\mathcal{D}^*$ with state independent coefficients. The state independent coefficients of an operator and the input $\mathcal{Y}$ may or may not be defined on the same probability space.

Example 9.1: Let X be the solution of $dX(t) = -R\,X(t)\,dt + \sigma\,dB(t)$, $t \geq 0$, where R is a random variable that is independent of the Brownian motion B and σ is a constant. The operator $\mathcal{D}[X(t)] = (d + R\,dt)[X(t)]$ has state independent coefficients. If the coefficients of the differential equation for X satisfy a.s. the conditions in Section 4.7.1.1, the above stochastic differential equation has a unique strong solution and X is well defined. ◇

Example 9.2: Let X be the solution of

$$m\,\ddot{X}(t) + c\,\dot{X}(t) + \chi(A(t))\,X(t) = Y(t), \quad t \geq 0,$$

where $m, c > 0$ are some constants, Y is a Gaussian process, A is a process defined by the evolution equation $\dot{A}(t) = \alpha\,[k(A(t))\,H(t)]^{\beta}$, $\alpha, \beta > 0$ are some constants, $H(t) = [X(t)^2 + (\dot{X}(t)/\nu(t))^2]^{1/2}$ denotes the envelope of X, $\nu(t) = [\chi(A(t))/m]^{1/2}$, and χ, k are strictly positive functions. The coefficients m, c, α, β in the operator $\mathcal{D} = m\,(d^2/dt^2) + c\,(d/dt) + \chi(A(t))$ and the differential equation for A can be random variables.

The operator $\mathcal{D}$ involves the process A whose value at a time $t \geq 0$ depends on the entire history of X up to this time so that $\mathcal{D}$ has state dependent coefficients. However, the coefficients of the operator $\mathcal{D}^*$ defining the evolution of the augmented vector (X, A) are state independent. ◇

Note: Consider a thin massless rectangular plate that is kept vertical by a continuous support at its top edge. The plate has a mass m attached at its bottom edge and an initial central crack. The process $X(t)$ gives the displacement of the mass m at time $t \geq 0$ under a loading Y, applied at the mass location. The parameter $c > 0$ denotes the plate damping coefficient. The process $A(t)$ denotes the crack length at time t. The constants α, β are material parameters ([175], Section 7.5.2.2). ▲

Two classes of methods are available for finding properties of the solution $\mathcal{X}$ of Eq. 9.1. The methods in the first class constitute direct extensions of some of the results in Chapters 6 and 7, for example, the local solution for some partial differential equation, the Monte Carlo simulation, the conditional analysis, the state augmentation, and the Liouville methods in Sections 9.2.1, 9.2.2, 9.2.3, 9.2.4, and 9.2.5, respectively. These methods involve rather advanced probabilistic tools and can deliver detailed properties of $\mathcal{X}$. The methods in the second class extend some of the results in Chapter 8, for example, the Taylor, perturbation, and Neumann series, the Galerkin, the finite difference, and finite element methods in Sections 9.2.6, 9.2.7, 9.2.8, and 9.3, respectively. Most of these methods involve elementary probabilistic concepts and are particularly useful for finding the second moment properties of $\mathcal{X}$.

9.2.1 Local solutions

We have seen in Sections 8.2 and 8.5.1 that the local solutions in Chapter 6 can be extended to solve a class of stochastic problems with stochastic operator and deterministic input. The methods in Chapter 6 can also be generalized to find local solutions for stochastic problems defined by a class of operators with the functional form of the deterministic problems considered in this chapter but with random coefficients, input, and initial/boundary conditions. The solution of this class of stochastic problems involves two steps. First, the methods in Chapter 6 need to be applied to estimate the required local solution for a collection of samples of the operator $\mathcal{D}$, the input $\mathcal{Y}$, and the initial/boundary conditions. Second, moments and other probabilistic properties can be estimated from the collection of local solutions delivered by the first step. The following example illustrates this generalization of the methods in Chapter 6 by using the random walk method to solve locally a partial differential equation with random coefficients.

Example 9.3: Let U be the solution of the partial differential equation

$$\frac{\partial^2 U(\boldsymbol{x})}{\partial x_1^2} + Z_1 \frac{\partial^2 U(\boldsymbol{x})}{\partial x_2^2} = -Z_2, \quad \boldsymbol{x} \in D = (0,1) \times (0,1)$$

with the boundary conditions $\partial U/\partial x_1 = 0$ on $\{0\} \times (0,1)$, $U = 0$ on $\{1\} \times (0,1)$, $\partial U/\partial x_2 = 0$ on $(0,1) \times \{0\}$, and $U = 3.5349\, x_1^2 + 0.5161\, x_1 + 3.0441$ on $(0,1) \times \{1\}$. The coefficient Z_1 and the input Z_2 are independent random variables that are uniformly distributed in $(3-a, 3+a)$, $0 < a < 3$, and $(16-b, 16+b)$, $0 < b < 16$, respectively.

Let $\boldsymbol{X}(t;\omega)$ be an $\mathbb{R}^2$-valued diffusion process with coordinates

$$\begin{cases} X_1(t,\omega) &= \hat{B}_1(t) + L_1(t), \\ X_2(t,\omega) &= \sqrt{Z_1(\omega)}\,(\hat{B}_2(t) + L_2(t)), \end{cases}$$

where $\hat{B}_i$ are some independent Brownian motions and L_i are local time processes (Section 6.2.3.1). Let $(Z_1, Z_2)(\omega)$ be a sample of (Z_1, Z_2). Denote by $T(\omega) = \inf\{t > 0 : \boldsymbol{X}(t,\omega) \notin D\}$ the first time $\boldsymbol{X}(\cdot,\omega)$ corresponding to this sample (Z_1, Z_2) and starting at $\boldsymbol{x} \in D$ leaves D. The local solution of the above partial differential equation at $\boldsymbol{x} \in D$ for a sample $(Z_1, Z_2)(\omega)$ of (Z_1, Z_2) is

$$U(\boldsymbol{x},\omega) = E^{\boldsymbol{x}}\left[h(\boldsymbol{X}(t,\omega))\right] + \frac{Z_2(\omega)}{2}\, E^{\boldsymbol{x}}\left[T(\omega)\right],$$

where h gives the values of U on the boundaries $\{1\} \times (0,1)$ and $(0,1) \times \{1\}$ of D. Figure 9.1 shows a histogram of $U(\boldsymbol{x})$ for $a = 0.5$, $b = 4$, and $\boldsymbol{x} = (0.25, 0.25)$ that is obtained from 100 samples of $\boldsymbol{Z}$. The estimates of $U(\boldsymbol{x},\omega)$ for each ω have been obtained from 500 samples of $\boldsymbol{X}(\cdot,\omega)$ generated with a time step $\Delta t = 0.001$. The estimated local solution at $\boldsymbol{x} = (0.25, 0.25)$ of the associated deterministic problem ($a = b = 0$) is 4.1808, and nearly coincides with the mean of the histogram in Fig. 9.1. ◇

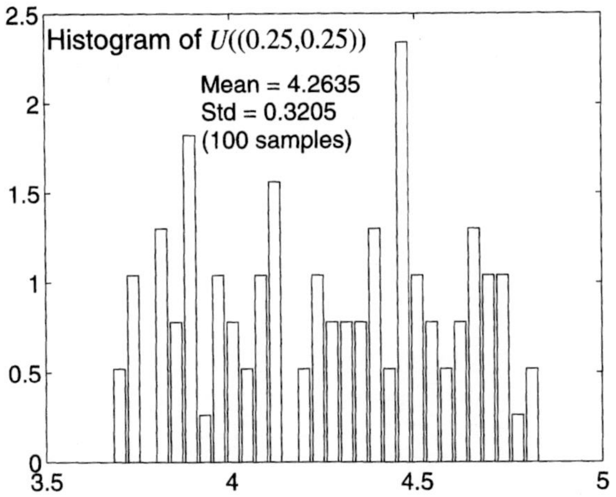

Figure 9.1. Histogram of $U((0.25, 0.25))$

Proof: The Itô formula gives (Section 6.2.3.3)

$$\begin{aligned}
&E^{\boldsymbol{x}}\left[h(\boldsymbol{X}(T(\omega)), \omega))\right] - U(\boldsymbol{x}, \omega) \\
&= \sum_{i=1}^{2} c_i\, E^{\boldsymbol{x}}\left[\int_0^{T(\omega)} \frac{\partial U(\boldsymbol{X}(s, \omega), \omega)}{\partial x_i} \left(d\hat{B}_i(s) + dL_i(s)\right)\right] \\
&+ \frac{1}{2} E^{\boldsymbol{x}}\left[\int_0^{T(\omega)} \left(\frac{\partial^2 U(\boldsymbol{X}(s, \omega), \omega)}{\partial x_1^2} + Z_1(\omega)\, \frac{\partial^2 U(\boldsymbol{X}(s, \omega), \omega)}{\partial x_2^2}\right) ds\right],
\end{aligned}$$

where $c_1 = 1$ and $c_2 = \sqrt{Z_1(\omega)}$. The boundary conditions and the partial differential equation defining U give the above local solution. ■

9.2.2 Monte Carlo simulation

Monte Carlo simulation is the most general method for solving stochastic problems. The method can be used if (1) algorithms are available for solving the deterministic problems corresponding to samples of $\mathcal{D}$, $\mathcal{Y}$, and the initial/boundary conditions and (2) the probability laws of all random variables and functions defining Eq. 9.1 are known. The essential limitation of the Monte Carlo simulation is the computation time, that can be excessive because the time needed to calculate a sample of $\mathcal{X}$ and/or the number of samples needed to find some properties of $\mathcal{X}$ can be very large. The Monte Carlo simulation solution of Eq. 9.1 with state independent coefficients involves the following steps.

1. Generate n_s samples of $\mathcal{Y}$, $\mathcal{D}$, and the initial/boundary conditions for Eq. 9.1.

2. Calculate the solutions $\mathcal{X}(\cdot, \cdot, \omega)$, $\omega = 1, \ldots, n_s$, of Eq. 9.1 for the samples generated in the previous step.

3. Estimate moments and other properties of $\mathcal{X}$ from its samples.

Note: The samples of $\mathcal{X}$ can be used to estimate properties of this random function. For example, $\hat{\boldsymbol{r}}(\boldsymbol{x}, t, \boldsymbol{y}, s) = (1/n_s) \sum_{\omega=1}^{n_s} \mathcal{X}(\boldsymbol{x}, t; \omega)\, \mathcal{X}(\boldsymbol{y}, s; \omega)^T$ is an estimate for the correlation function $E[\mathcal{X}(\boldsymbol{x}, t)\, \mathcal{X}(\boldsymbol{y}, s)^T]$ of $\mathcal{X}$. ▲

If the operator $\mathcal{D}$ in Eq. 9.1 has state dependent coefficients, the following Monte Carlo algorithm can be applied to find properties of $\mathcal{X}$.

1. Generate n_s samples of the input $\mathcal{Y}$, the subset of coefficients of $\mathcal{D}$ that are state independent, and the initial/boundary conditions for Eq. 9.1.

2. Calculate the joint evolution of $\mathcal{X}$ and the state dependent coefficients of $\mathcal{D}$ for the samples generated in the previous step.

3. Calculate moments and other properties of $\mathcal{X}$ from its samples.

The above Monte Carlo simulation algorithms can be applied to generate samples of solutions for the stochastic problems in Examples 9.1 and 9.2, and estimate properties of these solutions. For example, samples of X in Example 9.1 can be obtained by the method in Section 4.7.3 for each sample of R, which can be produced by the algorithms in Section 5.2. This approach can be extended without difficulty to the case in which R is a stochastic process. Samples of (X, A) in Example 9.2 can be obtained from samples of Y and numerical algorithms for integrating nonlinear differential equations, for example, the Runge-Kutta algorithm (Section 4.7.3.2). If the parameters (m, c) and/or (α, β) are random, samples of these parameters need also to be generated. A sample path of (X, A) would correspond in this case to a sample of (m, c), (α, β), and Y.

9.2.3 Conditional analysis

Suppose that $\mathcal{D}$ in Eq. 9.1 has state independent coefficients and let $\boldsymbol{\Theta}$ be a vector collecting all random coefficients of $\mathcal{D}$. The solution of Eq. 9.1 by conditional analysis involves the following steps.

1. Calculate the required properties of $\mathcal{X} \mid \boldsymbol{\Theta}$ by the methods in Chapter 7.

2. Calculate the required properties of $\mathcal{X}$ by eliminating the condition on $\boldsymbol{\Theta}$.

Note: The method is particularly efficient if the dependence of the required properties of $\mathcal{X} \mid \boldsymbol{\Theta}$ is a known function of $\boldsymbol{\Theta}$ and the dimension of $\boldsymbol{\Theta}$ is relatively small. If these conditions are not satisfied, the method can still be applied in conjunction with the Monte Carlo simulation and involves the steps: (1) generate samples $\boldsymbol{\Theta}(\omega)$, $\omega = 1, \ldots, n_s$, of

Θ, (2) calculate the required properties of $\mathcal{X} \mid (\Theta = \Theta(\omega))$, $\omega = 1, \ldots, n_s$, by the methods in Chapter 7, and (3) estimate the required properties of $\mathcal{X}$ from the properties of $\mathcal{X} \mid (\Theta = \Theta(\omega))$, $\omega = 1, \ldots, n_s$, obtained in the previous step.

If $\mathcal{D}$ has state dependent coefficients, the method can be applied to a modified version of Eq. 9.1 defined by an operator $\mathcal{D}^*$ that has state independent coefficients, as we will see later in Section 9.2.4. ▲

Example 9.4: Let X be the process in Example 9.1 with the deterministic initial condition $X(0) = x$. If R and B are independent, then $X|\Theta = X|R$ is a Gaussian process with mean and correlation functions

$$\mu(t \mid R) = E[X(t) \mid R] = x\, e^{-R\,t},$$

$$r(t, s \mid R) = E[X(t)\, X(s) \mid R] = x^2\, e^{-R\,(t+s)} + \frac{\sigma^2}{2\,R}\, e^{-R\,(t+s)} \left(e^{2\,R\,(t \wedge s)} - 1\right).$$

These two functions define the probability law of $X|R$. Properties of X can be obtained from the probability law of $X|R$ by eliminating the condition on R. Numerical integration or Monte Carlo simulation can be used for solution. Figure 9.2 shows the evolution in time of the variance of X for $\sigma = \sqrt{2}$, $R = 1$, and $R \sim U(a - b, a + b)$ with $a = 1$ and $b = 0.7$. ◇

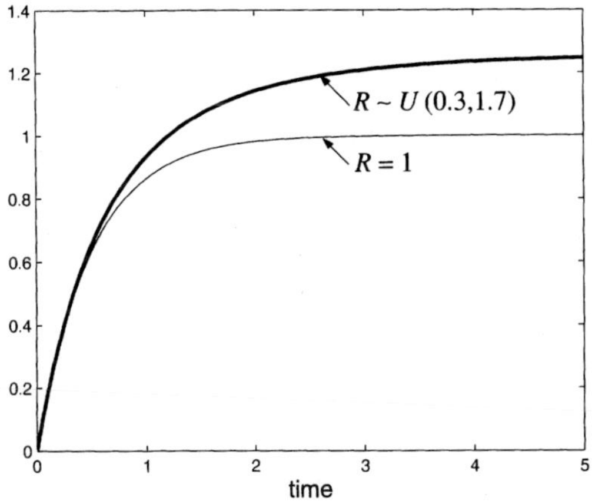

Figure 9.2. Evolution of the variance of X for $\sigma = \sqrt{2}$, $R \sim U(0.3, 1.7)$, and $R = 1$

Proof: The mean and correlation equations in Section 7.2.1.1 can be used to find the second moment properties of $X|R$. Alternatively, we can use direct calculations to obtain the above equations. For example, the mean equation, $\dot{\mu}(t \mid R) = -R\,\mu(t \mid R)$ with initial condition $\mu(0 \mid R) = x$, results by calculating the expectation of the equation for $X|R$. We also have

$$X(t) = x\, e^{-R\,t} + \sigma \int_0^t e^{-R\,(t-u)}\, dB(u),$$

so that the second moment properties of $X \mid R$ are as stated. ■

Example 9.5: Let X be a real-valued stochastic process defined by $dX(t) = -R(t)\,X(t)\,dt + \sigma\,dB(t)$, $t \geq 0$, where $X(0) = x$, σ is a real constant, and R is a stochastic process that is independent of the Brownian motion B. It is assumed that the solution of the above differential equation exists and is unique for almost all samples of R (Section 4.7.1.1). Conditional on R and $X(0) = x$, X is a Gaussian process with mean and correlation functions

$$\mu(t \mid R) = x\,e^{-\int_0^t R(u)\,du},$$
$$r(t, s \mid R) = \mu(t \mid R)\,\mu(s \mid R) + \sigma^2 \int_0^{t \wedge s} e^{-\int_u^t R(\xi)\,d\xi - \int_u^s R(\eta)\,d\eta}\,du.$$

Generally, it is not possible to find simple expressions for $\mu(t \mid R)$ and $r(t, s \mid R)$. However, conditional analysis and Monte Carlo simulation can be used to estimate properties of X. First, formulas need to be developed for the required properties of $X|R$, for example, the mean and correlation functions of $X|R$ given above. Second, these formulas and samples of R can be used to estimate the required properties of X.

Let R be a lognormal translation process defined by $R(t) = a + \exp(G(t))$, where $a > 0$ and G denotes a stationary Gaussian process with mean zero, variance 1, and covariance function $\rho(\tau) = E[G(t)\,G(t+\tau)]$. Figure 9.3 shows ten samples of G with $\rho(\tau) = \exp(-|\tau|)$ generated by the spectral representation method with a cutoff frequency $\nu^* = 5$ and $n = 50$ equally spaced frequencies (Section 5.3.1.1), the corresponding samples of R with $a = 0.5$, the corresponding conditional correlations $r(t, t \mid R)$, and an estimate of $r(t, t)$ based on a hundred samples of R. ◇

Note: The mean and correlation formulas in Section 7.2.1.1 or direct calculations can be used for solution. For example, we can use the solution

$$X(t) = x\,e^{-\int_0^t R(u)\,du} + \sigma \int_0^t e^{-\int_s^t R(u)\,du}\,dB(s)$$

to calculate the second moment properties of $X|R$.

Let $G^{(n)}$ be an approximation of G depending on $2n$ independent Gaussian variables with zero-mean and unit-variance collected in a vector $\mathbf{Z}$ (Section 5.3.1.1). The approximation $R^{(n)}(t) = a + \exp(G^{(n)}(t))$ of the lognormal translation process R depends on the Gaussian vector $\mathbf{Z}$. Denote the conditional correlation function of X corresponding to this representation of R by $r(\cdot, \cdot \mid \mathbf{Z})$. The samples of G in Fig. 9.3 have been obtained from $G^{(n)}$ with $n = 50$ so that $r(t, s \mid \mathbf{Z})$ depends on 100 Gaussian variables. The calculation of $r(t, s)$ by direct integration from

$$r(t, s) = \int_{\mathbb{R}^{100}} r(t, s \mid z) \prod_{i=1}^{100} \left(\frac{1}{\sqrt{2\pi}}\,e^{-z_i^2/2}\,dz_i \right)$$

is not feasible for this value of n. However, $r(t, s)$ can be estimated without difficulty from $r(\cdot, \cdot \mid \mathbf{Z})$ and samples of $\mathbf{Z}$, as demonstrated in Fig. 9.3. ▲

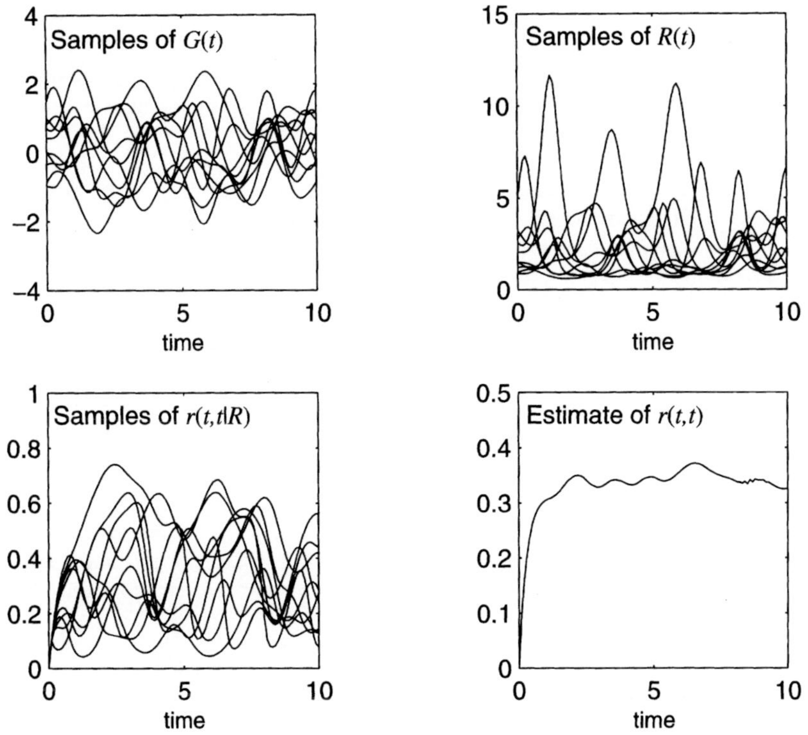

Figure 9.3. Ten samples of G, the corresponding samples of R and of the conditional correlations $r(t, t \mid R)$, and an estimate of $r(t, t)$ based on 100 samples

Example 9.6: Let X be a real-valued process defined by the differential equation $\dot{X}(t) = R(t)\, X(t) + Y(t)$, $t \geq 0$, where R and Y are real-valued stochastic processes. It is assumed that R is independent of Y and has continuous samples. The process X is given by

$$X(t) = H(t, 0)\, X(0) + \int_0^t H(t, s)\, Y(s)\, ds,$$

where the transition matrix $H(t, s)$ satisfies $\partial H(t, s)/\partial t = R(t)\, H(t, s), 0 \leq s \leq t$, $H(s, s) = 1$, so that $H(t, s) = \exp\left(\int_s^t R(u)\, du\right)$. If $X(0) = 0$, the second moment properties of X are

$$E[X(t)] = \int_0^t E[H(t, s)]\, E[Y(s)]\, ds,$$

$$E[X(t)\, X(s)] = \int_0^t \int_0^s E[H(t, u)\, H(s, v)]\, E[Y(u)\, Y(v)]\, du\, dv.$$

The kernel $H(t,s)$ for X in Example 9.4 is $\exp(-R(t-s))\, 1_{\{t \geq s\}}$, where R is a random variable. The expectations $E[H(t,s)]$ and $E[H(t,u)\, H(s,v)]$ in the expressions of the mean and correlation functions of X are given by the moment generating function $E[e^{\xi R}]$ of the random variable R evaluated at $\xi = -(t-s)$ and $\xi = -(t-u+s-v)$, respectively. ◇

Note: Consider a deterministic differential equation $\dot{\boldsymbol{x}}(t) = \boldsymbol{a}(t)\,\boldsymbol{x}(t) + \boldsymbol{y}(t)$, where $\boldsymbol{x} \in \mathbb{R}^d$, $\boldsymbol{a}$ is a (d,d) matrix whose entries are real-valued functions of time, $\boldsymbol{y} \in \mathbb{R}^d$ denotes the input, and the time argument t takes values in a bounded interval $[0,\tau]$. If the entries of $\boldsymbol{a}$ are continuous functions, then (1) the transition matrix $\boldsymbol{h}(\cdot,\cdot)$ is the unique solution of $\partial \boldsymbol{h}(t,s)/\partial t = \boldsymbol{a}(t)\,\boldsymbol{h}(t,s)$, $0 \leq s \leq t$, with $\boldsymbol{h}(s,s)$ equal to the identity matrix and (2) the unique solution of the above differential equation is

$$\boldsymbol{x}(t) = \boldsymbol{h}(t,0)\,\boldsymbol{x}(0) + \int_0^t \boldsymbol{h}(t,s)\,\boldsymbol{y}(s)\,ds$$

in the time interval $[0,\tau]$ ([30], Theorem 1, p. 20, and Theorem 1, p. 40). It is assumed that the samples of R satisfy the condition of these theorems a.s.

The kernels $H(t,s)$, $E[H(t,s)]$, and $E[H(t,u)\, H(s,v)]$ in the above equations can be interpreted as a Green's function for X, the mean of X, and the correlation function of this process. The expectation $E[H(t,u)\, H(s,v)]$ is also referred to as a stochastic Green's function ([3], p. 86). ▲

9.2.4 State augmentation

We have seen that the operator $\mathcal{D}$ in Eq. 9.1 may have state independent or dependent coefficients. If $\mathcal{D}$ has state dependent coefficients, Eq. 9.1 can be replaced by the equation $\mathcal{D}^*[\mathcal{X}^*(\boldsymbol{x},t)] = \mathcal{Y}^*(\boldsymbol{x},t)$, where $\mathcal{X}^*$ and $\mathcal{Y}^*$ are augmented versions of $\mathcal{X}$ and $\mathcal{Y}$, respectively, and $\mathcal{D}^*$ is an operator with state independent coefficients, as illustrated later in this section (Example 9.10). Hence, we can assume that $\mathcal{D}$ in Eq. 9.1 has state independent coefficients without loss of generality. Let $\boldsymbol{\Theta}$ be a vector collecting all random parameters in this equation. The coordinates of $\boldsymbol{\Theta}$ can be random variables or stochastic processes. The state augmentation method involves the following steps.

1. Augment the solution $\mathcal{X}$ of Eq. 9.1 or of its discrete counterpart $\boldsymbol{X}$ with $\boldsymbol{\Theta}$. Let $\mathcal{Z} = (\mathcal{X}, \boldsymbol{\Theta})$ be a random function including the output $\mathcal{X}$ and the random parameters in $\mathcal{D}$.

2. Write a differential equation specifying $\mathcal{Z}$. If $\mathcal{Z}$ is a random process defined by a stochastic differential equation, the methods in Chapter 7 can be used to find moments and other properties of this process.

Example 9.7: Let X be the process in Example 9.4. The operator defining this process has state independent coefficients and depends on a single random parameter $\Theta = R$. The augmented process $\mathcal{Z}$ has the coordinates $(\mathcal{Z}_1 = X, \mathcal{Z}_2 = R)$

and satisfies the stochastic differential equation

$$d\begin{bmatrix} \mathcal{Z}_1(t) \\ \mathcal{Z}_2(t) \end{bmatrix} = \begin{bmatrix} -\mathcal{Z}_1(t)\,\mathcal{Z}_2(t) \\ 0 \end{bmatrix} dt + \begin{bmatrix} \sigma \\ 0 \end{bmatrix} dB(t)$$

with the initial condition $\mathcal{Z}_1(0) = X(0)$ and $\mathcal{Z}_2(0) = R$. Let $\mu(p,q;t) = E[X(t)^p\, R^q]$ be the moment of order $p+q$ of $\mathcal{Z}(t)$, where $p, q \geq 0$ are integers. The moments of $\mathcal{Z}(t)$ satisfy the differential equation

$$\dot{\mu}(p,q;t) = -p\,\mu(p,q+1;t) + \frac{p\,(p-1)\,\sigma^2}{2}\,\mu(p-2,q;t)$$

with the convention that $\mu(p,q;t)$ is zero if at least one of the arguments p, q is strictly negative. Since the moment equations for $\mathcal{Z}(t)$ form an infinite hierarchy, it is not possible to solve exactly these equations.

The Fokker-Planck equation can be used to find the evolution in time of the density f of the diffusion process $(\mathcal{Z}_1, \mathcal{Z}_2)$. For example, the stationary density f_s of $\mathcal{Z}$ is $f_s(z_1, z_2) = c(z_2)\exp\left(-z_1^2\, z_2/\sigma^2\right)$, where $c(z_2) > 0$ is a function of z_2 such that $\int_{\mathbb{R}} dz_1 \int_0^\infty dz_2\, f_s(z_1, z_2) = 1$. ◇

Proof: The differential equation for the second coordinate of $\mathcal{Z}$ states that Θ is time-invariant. The moment equations for $\mu(p,q;t)$ are given by the Itô formula applied to the function $(X(t), R) \mapsto X(t)^p\, R^q$. Although the moments $\mu(0,q;t) = E[R^q]$ are known for any value of q, the moment equations for $\mathcal{Z}(t)$ are not closed. For example, there are two differential equations for $p+q=1$. The equation for $(p=0, q=1)$ provides no information since the moments of R are known, while the equation for $(p=1, q=0)$ involves the moments $\mu(1,0;t)$ and $\mu(1,1;t)$. There are three differential equations for $p+q=2$. The equation for $(p=0, q=2)$ provides no information, while the equations for $(p=2, q=0)$ and $(p=1, q=1)$ involve the moments $\mu(2,0;t)$, $\mu(2,1;t)$, $\mu(1,1;t)$, and $\mu(1,2;t)$. Hence, up to moments of order 2 we have three informative equations and five unknown moments.

The density $f(\cdot,\cdot;t)$ of $(\mathcal{Z}_1(t), \mathcal{Z}_2(t))$ is the solution of the partial differential equation (Section 7.3.1.3)

$$\frac{\partial f}{\partial t} = \frac{\partial}{\partial z_1}\left[z_1\, z_2\, f + \frac{\sigma^2}{2}\frac{\partial f}{\partial z_1}\right].$$

If $R > 0$ a.s., then f converges to a stationary density f_s as $t \to \infty$ so that

$$\frac{\partial}{\partial z_1}\left[z_1\, z_2\, f_s + \frac{\sigma^2}{2}\frac{\partial f_s}{\partial z_1}\right] = 0.$$

Because the expression in the square bracket is a constant and this constant is zero by the boundary conditions at infinity, we have $\ln(f_s(z_1, z_2)) = -z_1^2\, z_2/\sigma + d(z_2)$, where $d(z_2)$ is a constant with respect to the variable of integration z_1, which yields the stated expression for f_s. ■

Example 9.8: Let X be the solution of $dX(t) = -R\,X(t)\,dt + \Gamma\, dB(t)$, where R and Γ are random variables independent of each other and independent of the driving Brownian motion B. The process in Example 9.7 is a special case corresponding to a deterministic rather than random noise intensity. The augmented state $\mathcal{Z}(t) = (\mathcal{Z}_1(t) = X(t), \mathcal{Z}_2(t) = R, \mathcal{Z}_3(t) = \Gamma)$ satisfies the stochastic differential equation

$$d\begin{bmatrix} \mathcal{Z}_1(t) \\ \mathcal{Z}_2(t) \\ \mathcal{Z}_3(t) \end{bmatrix} = \begin{bmatrix} -\mathcal{Z}_1(t)\,\mathcal{Z}_2(t) \\ 0 \\ 0 \end{bmatrix} dt + \begin{bmatrix} \mathcal{Z}_3(t) \\ 0 \\ 0 \end{bmatrix} dB(t)$$

with the initial conditions $\mathcal{Z}_1(0) = X(0)$, $\mathcal{Z}_2(0) = R$, and $\mathcal{Z}_3(0) = \Gamma$. The moments $\mu(p, q, r; t) = E[X(t)^p\, R^q\, \Gamma^r]$ of the state vector $\mathcal{Z}$ are given by the infinite hierarchy of equations

$$\dot{\mu}(p, q, r; t) = -p\,\mu(p, q+1, r; t) + \frac{p\,(p-1)}{2}\,\mu(p-2, q, r+2; t)$$

so that they cannot be obtained exactly from these equations. We use the convention $\mu(p, q, r; t) = 0$ if at least one of the arguments p, q, r is strictly negative.

The Fokker-Planck equation can be used to find the evolution in time of the density f of the augmented state vector $\mathcal{Z}(t)$. If $R > 0$ a.s., f converges to the time-invariant density $f_s(z_1, z_2, z_3) = c(z_2, z_3)\exp\left(-z_1\, z_2/z_3^2\right)$ as $t \to \infty$, where $c(z_2, z_3) > 0$ is a function of z_2 and z_3. $\diamond$

Note: The Itô formula applied to $\mathcal{Z} \mapsto \mathcal{Z}_1^p\, \mathcal{Z}_2^q\, \mathcal{Z}_3^r$ gives the moment equations. The Fokker-Planck equation,

$$\frac{\partial f}{\partial t} = -\frac{\partial}{\partial z_1}\left(-z_1\, z_2\, f\right) + \frac{1}{2}\frac{\partial^2}{\partial z_1^2}\left(z_3^2\, f\right),$$

giving the evolution of the density f of the state vector $\mathcal{Z}$ can be used to establish and solve the differential equation for f_s. ▲

Example 9.9: Let X be the solution of

$$\ddot{X}(t) + 2\,\beta\,\dot{X}(t) + [1 + \sigma_1\, W_1(t)]\, X(t) = \sigma_2\, W_2(t), \quad t \geq 0,$$

where $\beta > 0$, and σ_1, σ_2 are real constants, and W_1 and W_2 denote Gaussian white noise processes defined as the formal derivative of the independent Brownian motion processes B_1 and B_2. The moments $\mu(p, q; t) = E[X_1(t)^p\, X_2(t)^q]$ of $\boldsymbol{X} = (X_1 = X, X_2 = \dot{X})$ satisfy the equations

$$\begin{aligned} \dot{\mu}(p, q; t) &= p\,\mu(p-1, q+1; t) - q\,\mu(p+1, q-1; t) - 2\,q\,\beta\,\mu(p, q; t) \\ &\quad + \frac{q\,(q-1)\,\sigma_1^2}{2}\,\mu(p+2, q-2; t) + \frac{q\,(q-1)\,\sigma_2^2}{2}\,\mu(p, q-2; t) \end{aligned}$$

with the convention in the previous two examples. The moments of order $p+q = 1$ and $p + q = 2$ are given, respectively, by

$$\begin{aligned} &\dot{\mu}(1, 0; t) = \mu(0, 1; t), \\ &\dot{\mu}(0, 1; t) = -\mu(1, 0; t) - 2\,\beta\,\mu(0, 1; t) \quad \text{and} \end{aligned}$$

$$\dot{\mu}(2, 0; t) = 2\,\mu(1, 1; t),$$
$$\dot{\mu}(1, 1; t) = \mu(0, 2; t) - \mu(2, 0; t) - 2\,\mu(1, 1; t),$$
$$\dot{\mu}(0, 2; t) = -2\,\mu(1, 1; t) - 4\,\beta\,\mu(0, 2; t) + \sigma_1^2\,\mu(2, 0; t) + \sigma_2^2.$$

The above equations are closed, so that we can calculate moments of any order of $\boldsymbol{X}$ exactly. ◇

Proof: The process $\boldsymbol{X} = (X_1 = X, X_2 = \dot{X})$ is a diffusion process defined by the stochastic differential equation

$$\begin{cases} dX_1(t) & = X_2(t)\,dt, \\ dX_2(t) & = -[X_1(t) + 2\,\beta\,X_2(t)]\,dt - \sigma_1\,X_1(t)\,dB_1(t) + \sigma_2\,dB_2(t). \end{cases}$$

The average of the Itô formula applied to the function $(X_1\,X_2) \mapsto X_1^p\,X_2^q$ gives the moment equations.

The moment equations of a specified order $p+q$ have the form $\boldsymbol{m}(t) = \boldsymbol{\alpha}\,\boldsymbol{m}(t)$. If the real parts of the eigenvalues of $\boldsymbol{\alpha}$ are negative, the moments $\boldsymbol{m}(t)$ converge to a time-invariant finite value as $t \to \infty$ and we say that the process is asymptotically stable in the moments of order $p+q$. For example, the eigenvalue of $\boldsymbol{\alpha}$ for $p+q = 1$ are $\lambda_{1,2} = -\beta \pm \sqrt{\beta^2 - 1}$ so that $\boldsymbol{X}$ is asymptotically stable in the mean if $\beta > 0$. If the dimension of the vector $\boldsymbol{m}$ is large, it is convenient to use the Hurwitz criterion to assess the sign of the real part of the eigenvalues of $\boldsymbol{a}$ ([30], Theorem 2, p. 55).

The differential equation defining X can be used to model a simply supported beam with constant stiffness χ and length $l > 0$. The beam is subjected to a fluctuating axial force W_1 applied at its ends. The beam deflection $V(x, t)$ at time $t \geq 0$ and coordinate $x \in [0, l]$ satisfies the differential equation

$$\chi\,\frac{\partial^4 V(x,t)}{\partial x^4} = -W_1(t)\,\frac{\partial^2 V(x,t)}{\partial x^2} - m\,\frac{\partial^2 V(x,t)}{\partial t^2},$$

where m is the beam mass per unit length. The representation $V(x, t) = Y(t)\,\sin(\pi\,x/l)$ of V corresponding to the first buckling mode gives

$$\ddot{Y}(t) + \nu^2\,(1 - W(t)/p_{\text{cr}})\,Y(t) = 0,$$

where $\nu^2 = \pi^4\,\chi/(m\,l^4)$ and $p_{\text{cr}} = \pi^2\,\chi/l^2$ is the first buckling load for the beam. This equation has the same form as the equation defining X without the damping term $2\,\beta\,\dot{X}(t)$ and the forcing function $\sigma_2\,W_2$, which have not been considered in the beam model. We also note that the above beam equation can be generalized by considering the possibility that the stiffness χ is random. ■

Example 9.10: Consider the stochastic problem in Example 9.2 characterized by an operator $\mathcal{D}$ with state dependent coefficients and assume that the input Y is a Gaussian white noise with a constant mean μ_y, that is, we have formally $Y(t) = \mu_y + dB(t)/dt$, where B denotes a Brownian motion. The augmented state vector $\mathcal{Z}^* = (\mathcal{Z}_1 = X, \mathcal{Z}_2 = \dot{X}, \mathcal{Z}_3 = A)$ is a diffusion process defined by

$$\begin{cases} d\mathcal{Z}_1(t) & = \mathcal{Z}_2(t)\,dt, \\ d\mathcal{Z}_2(t) & = -\frac{1}{m}\,[\chi(\mathcal{Z}_3(t))\,\mathcal{Z}_1(t) + c\,\mathcal{Z}_2(t)]\,dt + \mu_y\,dt + \frac{1}{m}\,dB(t), \\ d\mathcal{Z}_3(t) & = \alpha\,[k(\mathcal{Z}_3(t))\,H(t)]^\beta\,dt, \end{cases}$$

where $H(t) = [\mathcal{Z}_1(t)^2 + (\mathcal{Z}_2(t)/\nu(t))^2]^{1/2}$, $\nu(t) = [\chi(\mathcal{Z}_3)/m]^{1/2}$ denotes the instantaneous frequency of the plate, and the parameters α, β in the definition of the functions χ and k may be random. The operator $\mathcal{D}^*$ defining $\mathcal{Z}^*$ has state independent coefficients.

The methods in Section 7.3.1 can be applied to find properties of $\mathcal{Z}^*$ conditional on the random parameters in $\mathcal{D}^*$. For example, the stochastic averaging method can be used to derive an approximate evolution equation for the crack length $\mathcal{Z}_3 = A$ ([175], Section 7.5.2.3). ◇

Example 9.11: Let X be the state of a mechanical system

$$dX(t) = a(X(t), \boldsymbol{\theta}_0)\,dt + b(X(t), \boldsymbol{\theta}_0)\,dB(t), \quad t \geq 0,$$

where $\boldsymbol{\theta}_0 \in \mathbb{R}^m$ is a vector depending on some system properties that can be altered by design changes. To assess the usefulness of changing a current value of $\boldsymbol{\theta}_0$, it is necessary to calculate the effects of such changes on X. The functions $V_i(t) = \partial X(t)/\partial \theta_i$, $i = 1, \ldots, m$, evaluated at $\boldsymbol{\theta} = \boldsymbol{\theta}_0$ provide a useful measure of the sensitivity of X to perturbations of $\boldsymbol{\theta}$ about $\boldsymbol{\theta}_0$ and are called **sensitivity factors**.

The sensitivity factors satisfy the differential equations

$$\begin{aligned} dV_i(t) &= \left(\frac{\partial a(X(t), \boldsymbol{\theta}_0)}{\partial X(t)} V_i(t) + \frac{\partial a(X(t), \boldsymbol{\theta}_0)}{\partial \theta_i}\right) dt \\ &+ \left(\frac{\partial b(X(t), \boldsymbol{\theta}_0)}{\partial X(t)} V_i(t) + \frac{\partial b(X(t), \boldsymbol{\theta}_0)}{\partial \theta_i}\right) dB(t), \quad i = 1, \ldots, m, \end{aligned}$$

with coefficient depending on the processes X and V_i. It is assumed that the functions a and b are such that the stochastic differential equations for X and V_i have unique solutions. The methods in Section 7.3.1 can be used to find properties of X and V_i.

Figure 9.4 shows three samples of X and the corresponding samples of the sensitivity factor $V = \partial X/\partial \theta$ for $X(0) = 0$, $a(X(t), \boldsymbol{\theta}_0) = -\theta_0\, X(t)$, $b(X(t), \boldsymbol{\theta}_0) = 1$, and $\theta_0 = 1$. The resulting sensitivity factor V is the solution of the differential equation $dV(t) = -(\theta_0\, V(t) + X(t))\,dt$, and has samples much smoother than the samples of X. ◇

Note: The differential equations for the sensitivity factors result by the differentiation of the defining equation for X with respect to the coordinates θ_i of $\boldsymbol{\theta}$ and then setting $\boldsymbol{\theta}$ equal to $\boldsymbol{\theta}_0$.

Note that the differential equations for sensitivity factors are linear in these factors. Also, the generation of samples of X and V can be performed simultaneously by considering the evolution of the vector (X, V), or it can be performed sequentially by calculating first samples of X and using them to calculate the corresponding samples of V. ▲

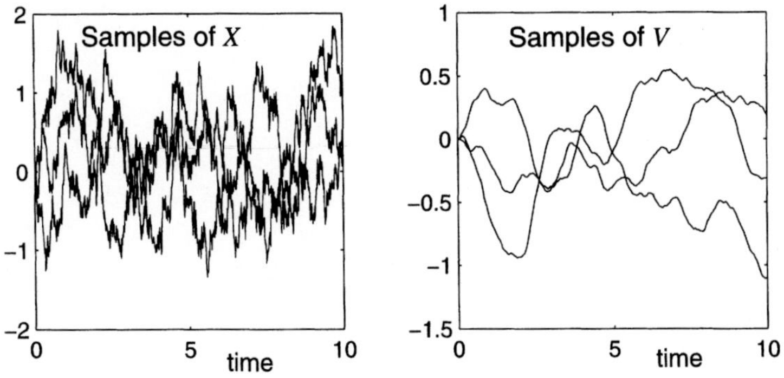

Figure 9.4. Three samples of X and the corresponding samples of V for $\theta_0 = 1$

9.2.5 Liouville equation

Let $\boldsymbol{X}$ be an $\mathbb{R}^d$-valued process defined by the differential equation

$$\dot{\boldsymbol{X}}(t) = \boldsymbol{h}(\boldsymbol{X}(t), t), \quad t \geq 0, \tag{9.2}$$

where $\boldsymbol{h}$ is a deterministic function and $\boldsymbol{X}(0) = \boldsymbol{X}_0$ is an $\mathbb{R}^d$-valued random variable with density f_0. It is assumed that the solution of Eq. 9.2 exists and is unique. Let $\boldsymbol{X}(t) = \boldsymbol{g}(\boldsymbol{X}(0), t)$ be the solution of this equation. The density $f(\cdot, t)$ of $\boldsymbol{X}(t)$, $t \geq 0$, can be obtained from the mapping $\boldsymbol{X}(0) \mapsto \boldsymbol{X}(t)$ or the solution of a special case of the Fokker-Planck equation, called the Liouville equation. We present both methods for finding the density of $\boldsymbol{X}(t)$.

If $\boldsymbol{\xi} \mapsto \boldsymbol{g}(\boldsymbol{\xi}, t)$ has continuous first order partial derivatives with respect to the coordinates of $\boldsymbol{\xi}$ and defines a one-to-one mapping, then

$$f(\boldsymbol{x}; t) = |J| \, f_0(\boldsymbol{g}^{-1}(\boldsymbol{x}, t)), \quad \boldsymbol{x} \in \mathbb{R}^d, \tag{9.3}$$

where $|J|$ and $\boldsymbol{g}^{-1}(\boldsymbol{\xi}, t)$ denote the Jacobian and the inverse of $\boldsymbol{\xi} \mapsto \boldsymbol{g}(\boldsymbol{\xi}, t)$.

Note: The relationship between the random vectors $\boldsymbol{X}(t)$ and $\boldsymbol{X}_0$ gives Eq. 9.3. This equation is useful if the expression for the mapping $\boldsymbol{\xi} \mapsto \boldsymbol{g}(\boldsymbol{\xi}, t)$ can be obtained explicitly (Section 2.11.1 in this book, [176], Theorem 6.2.1, p. 143). The relationship $\boldsymbol{X}(t) = \boldsymbol{g}(\boldsymbol{X}_0, t)$ can also be used to calculate moments and other properties of $\boldsymbol{X}(t)$. ▲

If the solution of Eq. 9.2 exists and is unique, the density $f(\cdot; t)$ of $\boldsymbol{X}(t)$ satisfies the **Liouville** equation

$$\frac{\partial f}{\partial t} = -\sum_{i=1}^{d} \frac{\partial (f\, h_i)}{\partial x_i} \tag{9.4}$$

with the initial and boundary conditions $f(\boldsymbol{x}, 0) = f_0(\boldsymbol{x})$, $\lim_{|x_i|\to\infty} f(\boldsymbol{x}; t) = 0$, $i = 1, \ldots, d$, respectively, so that

$$f(\boldsymbol{x}, t) = f_0(\boldsymbol{g}^{-1}(\boldsymbol{x}, t)) \exp\left[-\int_0^t \sum_{i=1}^{d} \frac{\partial h_i(\boldsymbol{y}, s)}{\partial y_i}\, ds\right], \tag{9.5}$$

where $\boldsymbol{y} = \boldsymbol{g}(\boldsymbol{\xi}, s)$ and $\boldsymbol{\xi} = \boldsymbol{g}^{-1}(\boldsymbol{y}, s)$.

Proof: The proof of the Fokker-Planck equation in Section 7.3.1.3 can be used to establish Eq. 9.4. Let $\varphi(\boldsymbol{u}; t) = E\left[e^{\sqrt{-1}\,\boldsymbol{u}^T \boldsymbol{X}(t)}\right]$ be the characteristic function of $\boldsymbol{X}(t)$ so that $f(\boldsymbol{x}; t) = (1/(2\pi)^d) \int_{\mathbb{R}^d} e^{\sqrt{-1}\,\boldsymbol{u}^T \boldsymbol{x}}\, \varphi(\boldsymbol{u}; t)\, d\boldsymbol{u}$. The derivative with respect to time of φ gives

$$\frac{\partial \varphi}{\partial t} = E\left[\sum_{k=1}^{d} \sqrt{-1}\, u_k\, \dot{X}_k(t)\, e^{\sqrt{-1}\,\boldsymbol{u}^T \boldsymbol{X}(t)}\right] = \sqrt{-1} \sum_{k=1}^{d} u_k\, E\left[\dot{X}_k(t)\, e^{\sqrt{-1}\,\boldsymbol{u}^T \boldsymbol{X}(t)}\right]$$
$$= \sqrt{-1} \sum_{k=1}^{d} u_k\, E\left[h_k(\boldsymbol{X}(t), t)\, e^{\sqrt{-1}\,\boldsymbol{u}^T \boldsymbol{X}(t)}\right].$$

The Fourier transform of the left side of Eq. 9.4 is $\partial f/\partial t$. The Fourier transform of the right side of Eq. 9.4 is $-\sum_{k=1}^{d} \int_{\mathbb{R}^d} [\partial (f\, h_k)/\partial x_k]\, e^{\sqrt{-1}\,\boldsymbol{u}^T \boldsymbol{x}}\, d\boldsymbol{x}$ which gives the right side of the above equation following integration by parts.

The explicit solution of Eq. 9.4 in Eq. 9.5 can be obtained from the associated Lagrange system ([176], Theorem 6.2.2, p. 146) ■

If the function $\boldsymbol{h}$ in Eq. 9.2 depends on a finite number of random parameters $\boldsymbol{\Theta}$, the vector $\boldsymbol{X}$ can be augmented to include these parameters. The evolution of the augmented vector $\tilde{\boldsymbol{X}} = (\tilde{\boldsymbol{X}}_x = \boldsymbol{X}, \tilde{\boldsymbol{X}}_\theta = \boldsymbol{\Theta})$ is defined by a differential equation consisting of Eq. 9.2 and the additional equations $\dot{\tilde{\boldsymbol{X}}}_\theta(t) = \boldsymbol{0}$ with the initial condition $\tilde{\boldsymbol{X}}_\theta(0) = \boldsymbol{\Theta}$.

Example 9.12: Let X be the solution of $\ddot{X}(t) + \nu^2 X(t) = 0$, $t \geq 0$, with the random initial conditions $(X(0) = X_0, \dot{X}(0) = \dot{X}_0)$, where $\nu > 0$ is a constant and f_0 denotes the density of $(X_0, \dot{X}_0)$. The relationship between the values of the state vector $\boldsymbol{X} = (X_1 = X, X_2 = \dot{X})$ at a time $t \geq 0$ and the initial time $t = 0$ is

$$\boldsymbol{X}(t) = \boldsymbol{g}(\boldsymbol{X}(0), t) = \begin{bmatrix} \cos(\nu t) & (1/\nu)\sin(\nu t) \\ -\nu \sin(\nu t) & \cos(\nu t) \end{bmatrix} \boldsymbol{X}(0)$$

so that

$$X(0) = g^{-1}(X(t), t) = \begin{bmatrix} \cos(\nu t) & -(1/\nu) \sin(\nu t) \\ \nu \sin(\nu t) & \cos(\nu t) \end{bmatrix} X(t).$$

The mapping $X(0) \mapsto X(t) = g(X(0), t)$ can be used to calculate moments of $X(t)$ from moments of $X(0)$ and to find the density,

$$f(x; t) = f_0(x_1 \cos(\nu t) - (x_2/\nu) \sin(\nu t), x_1 \nu \sin(\nu t) + x_2 \cos(\nu t))$$

of $X(t)$ from the density of $X(0)$ (Eqs. 9.3 and 9.5). ◇

Note: The Jacobian of $\xi \mapsto g(\xi, t)$ is unity so that $f(x, t) = f_0(g^{-1}(x, t))$ by Eq. 9.3. The evolution equation for X is given by $\dot{X}_1(t) = h_1(X(t), t) = X_2(t)$ and $\dot{X}_2(t) = h_2(X(t), t) = -\nu^2 X_1(t)$ so that $\partial h_1/\partial x_1 = 0$, $\partial h_2/\partial x_2 = 0$, and the exponential function in Eq. 9.5 is unity. The density $f(x; t)$ in the above equation coincides with the result in Eq. 9.3 ([176], Example 6.1, p. 143). ■

Example 9.13: Let X be the solution of $\ddot{X}(t) + C \dot{X}(t) + K X(t) = \alpha \sin(\nu t)$, where $C > 0$ and $K > 0$ are independent random variables and $\alpha, \nu > 0$ are some constants. The augmented vector $\tilde{X}$ with coordinates ($\tilde{X}_1 = X, \tilde{X}_2 = \dot{X}, \tilde{X}_3 = C, \tilde{X}_4 = K$) satisfies the differential equation

$$d\tilde{X}(t) = \begin{bmatrix} \tilde{X}_2(t) \\ -\tilde{X}_4(t)\tilde{X}_1(t) - \tilde{X}_3(t)\tilde{X}_2(t) + \alpha \sin(\nu t) \\ 0 \\ 0 \end{bmatrix} dt$$

so that the density $\tilde{f}$ of $\tilde{X}(t)$ satisfies the Liouville equation (Eq. 9.4)

$$\frac{\partial \tilde{f}}{\partial t} = -\frac{\partial}{\partial \tilde{x}_1}\left(\tilde{x}_2 \tilde{f}\right) - \frac{\partial}{\partial \tilde{x}_2}\left[(-\tilde{x}_4 \tilde{x}_1 - \tilde{x}_3 \tilde{x}_2 + \alpha \sin(\nu t))\, \tilde{f}\right]$$

with solution given by Eq. 9.5. ◇

Note: If in addition the input parameters (α, ν) were random, the corresponding augmented vector $\tilde{X}$ would be an $\mathbb{R}^6$-valued process. The evolution of this process would be defined by the above differential equations and two additional equations stating that the parameters (α, ν) are time-invariant. ▲

Example 9.14: Let X be an $\mathbb{R}^2$-valued process defined by $\dot{X}(t) = A\, X(t)$ with the initial condition $X(0) = X_0$, where the entries (1, 1), (1, 2), (2, 1), and (2, 2) of A are $-A_1$, A_2, A_1, and $-A_2$, respectively, and A_1, A_2 are random variables. Then ([176], Example 8.2, p. 223)

$$X(t) = \frac{1}{A_1 + A_2} \begin{bmatrix} A_2 + A_1 e^{-(A_1+A_2)t} & A_2\left(1 - e^{-(A_1+A_2)t}\right) \\ A_1\left(1 - e^{-(A_1+A_2)t}\right) & A_1 + A_2 e^{-(A_1+A_2)t} \end{bmatrix} X_0.$$

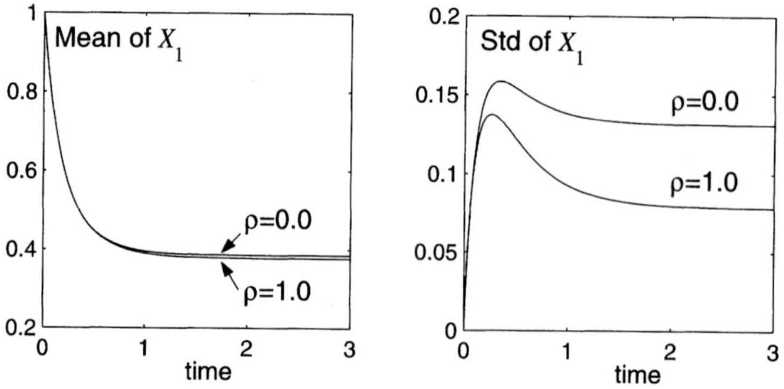

Figure 9.5. Evolution of the mean and standard deviation of X_1 for lognormal random variables (A_1, A_2) and initial condition $\boldsymbol{X}_0 = (1, 0)$

Figure 9.5 shows the evolution in time of estimates of the mean and standard deviation of X_1 for $\boldsymbol{X}_0 = (1, 0)$ and lognormal random variables (A_1, A_2) with means $(1.0, 0.5)$, coefficients of variation $(0.3, 0.4)$, and two values of the correlation coefficient ρ of the Gaussian images of A_1 and A_2. The estimates are based on 500 independent samples of X_1. The first two moments of X_1 converge to the corresponding moments of $A_2/(A_1 + A_2)$ as $t \to \infty$. ◇

Note: The lognormal random variables A_i, $i = 1, 2$, are defined by $A_i = \exp(\mu_{g,i} + \sigma_{g,i}\, G_i)$, where G_i are correlated $N(0, 1)$ variables with correlation coefficient ρ. The parameters $(\mu_{g,i}, \sigma_{g,i})$ relate to the first two moments $(\mu_i, (v_i\, \mu_i)^2)$ of A_i by $\sigma_{g,i}^2 = \ln(1+v_i^2)$ and $\mu_{g,i} = \ln\left(\mu_i/\sqrt{1+v_i^2}\right)$. The correlation coefficients ρ and ξ of (G_1, G_2) and (A_1, A_2) satisfy the condition ([79], pp. 48-52)

$$\xi = \frac{e^{\rho\, \sigma_{g,1}\, \sigma_{g,2}} - 1}{\left[\left(e^{\sigma_{g,1}^2} - 1\right)\left(e^{\sigma_{g,2}^2} - 1\right)\right]^{1/2}}.$$

Because $\dot{X}_1(t) + \dot{X}_2(t) = 0$ by the defining equation for $\boldsymbol{X}$, we have $X_1(t) + X_2(t) = 1$ for the initial condition $\boldsymbol{X}_0 = (1, 0)$, so that $E[X_1(t)] + E[X_2(t)] = 1$ and $\mathrm{Var}[X_1(t)] = \mathrm{Var}[X_2(t)]$ at any time $t \geq 0$ ▲

9.2.6 Taylor, perturbation, and Neumann series methods

We have solved inhomogeneous differential and integral equations with random coefficients in Sections 8.4.1.2, 8.4.1.3, 8.4.1.4, and 8.4.1.5 by the Taylor, perturbation, Neumann series, and the decomposition and iteration methods, respectively. The extension of these methods to the case in which both the operator and the input are stochastic is elementary for the case in which $\mathcal{D}$ has state independent coefficients.

We only present three examples illustrating how some of the methods in Sections 8.4.1.2-8.4.1.5 can be extended to solve Eq. 9.1.

Example 9.15: Let U be the solution of $(d/dt - \alpha + Z\,U(t)^2)\,U(t) = Q\,\sin(\nu\,t)$, where $t \geq 0$, $U(0)$ is specified, (Z, Q) are random variables in L_2 with means (μ_z, μ_q) and variances (σ_z^2, σ_q^2), Z and Q are uncorrelated, and α and $\nu > 0$ are some constants. The solution of this equation, denoted by $U(t; Z, Q)$, is a function of time and depends on the random parameters Z and Q. Let $U(t; \mu_z, \mu_q)$, $V_z(t; \mu_z, \mu_q)$, and $V_q(t; \mu_z, \mu_q)$ denote the functions U, $\partial U/\partial Z$, and $\partial U/\partial Q$, respectively, evaluated at the mean of (Z, Q).

Figure 9.6 shows the deterministic functions $U(t; \mu_z, \mu_q)$, $V_z(t; \mu_z, \mu_q)$, and $V_q(t; \mu_z, \mu_q)$ for $\alpha = 1$, $\mu_z = -0.5$, $\mu_q = 1$, and $\nu = 5$. These functions

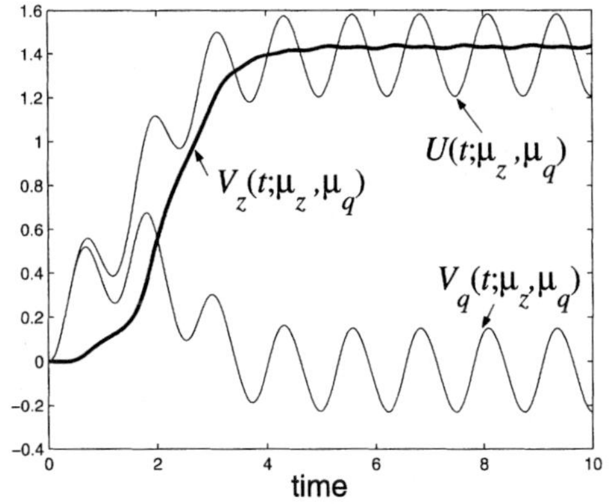

Figure 9.6. Functions $U(t; \mu_z, \mu_q)$, $V_z(t; \mu_z, \mu_q)$, and $V_q(t; \mu_z, \mu_q)$ for $\alpha = 1$, $\mu_z = -0.5$, $\mu_q = 1$, and $\nu = 5$

can be used to find properties of U approximately from

$$U(t; Z, Q) \simeq U(t; \mu_z, \mu_q) + V_z(t; \mu_z, \mu_q)\,(Z - \mu_z) + V_q(t; \mu_z, \mu_q)\,(Q - \mu_q).$$

For example, the approximate mean and variance of $U(t)$ are $\mu(t) \simeq U(t; \mu_z, \mu_q)$ and $\sigma(t)^2 \simeq V_z(t; \mu_z, \mu_q)^2\,\sigma_z^2 + V_q(t; \mu_z, \mu_q)^2\,\sigma_q^2$. ◇

Proof: The function $U(t; \mu_z, \mu_q)$ is the solution of the differential equation for U with (μ_z, μ_q) in place of (Z, Q). The sensitivity factors V_z and V_q can be obtained as in Section 8.4.1.2. For example, the differential equation,

$$(d/dt - \alpha)V_q(t; \mu_z, \mu_q) - 3\,\mu_z\,U(t; \mu_z, \mu_q)^2\,V_q(t; \mu_z, \mu_q) = \sin(\nu\,t),$$

for V_q results by differentiating the defining equation for U with respect to Q and setting (Z, Q) equal to its expectation (μ_z, μ_q). ■

Example 9.16: Let U be the solution of $(d/dt + \beta + \varepsilon\, Y(t))\, U(t) = Q$, where $t \geq 0$, $U(0) = 0$, Y is a stochastic process with mean zero, Q denotes a random variable, and ε is a small parameter.

The first order perturbation solution is $U(t) = U_0(t) + \varepsilon\, U_1(t)$, where $U_0(t) = (Q/\beta)\left(1 - e^{-\beta t}\right)$, $U_1(t) = -(Q/\beta)\, Z(t)\, e^{-\beta t}$, and Z is a process given by $Z(t) = \int_0^t \left(e^{\beta s} - 1\right) Y(s)\, ds$. Moments and other probabilistic properties of U can be approximated from this or any higher order perturbation solution.

Figure 9.7 shows histograms of the exact and first order perturbation solutions for $U(0) = 0$, $\beta = 1$, $t = 5$, $Q \sim U(0.5, 1.5)$, and $Y(t) = Y \sim U(0.5, 1.5)$. The histograms have been calculated from 1,000 samples of these solutions under

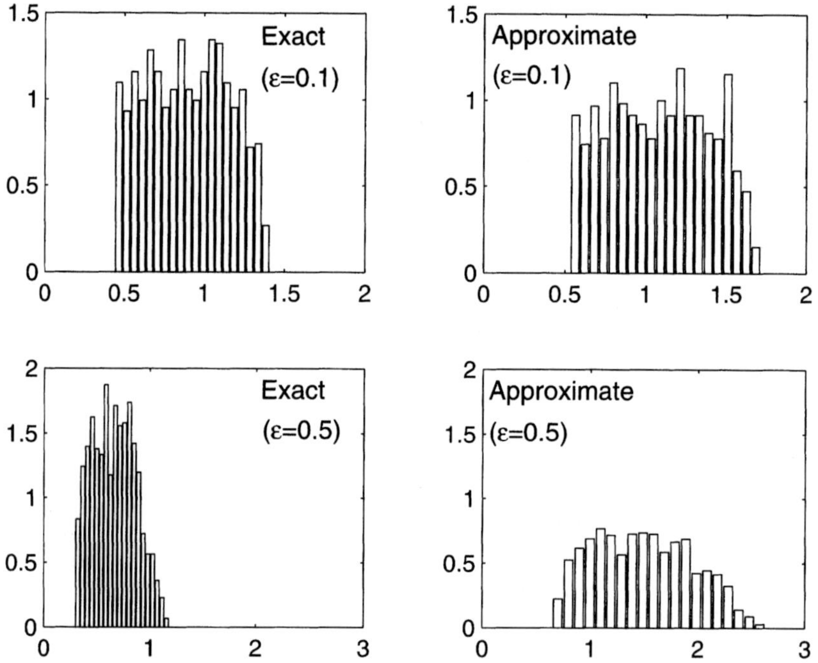

Figure 9.7. Histograms of the exact and first order perturbation solutions for $U(0) = 0$, $\beta = 1$, $t = 5$, $Q \sim U(0.5, 1.5)$, and $Y(t) = Y \sim U(0.5, 1.5)$

the assumption that Q and Y are independent random variables. The histograms of the exact and first order perturbation solutions are similar for $\varepsilon = 0.1$ but differ significantly for $\varepsilon = 0.5$. ◇

Note: The functions U_0 and U_1 satisfy the differential equations $\dot{U}_0(t) + \beta\, U_0(t) = Q$ with $U_0(0) = 0$ and $\dot{U}_1(t) + \beta\, U_1(t) = -Y(t)\, U_0(t)$ with $U_1(0) = 0$, respectively, which yield the stated solutions. The approximate mean is $(E[Q]/\beta)\left(1 - e^{-\beta t}\right) + O(\varepsilon^2)$. The expectation of the product $(U_0(t) + \varepsilon\, U_1(t))\, (U_0(s) + \varepsilon\, U_1(s))$ gives the terms of orders one and order ε but only one of the three terms of order ε^2 (Section 8.4.1.3). ▲

Example 9.17: Consider the differential equation $(d/dt + \beta + Y)\,U(t) = Q$, where $t \in [0, \tau]$, $\beta = 1$, $U(0) = 0$, $Y \sim U(-0.5, 0.5)$, $Q \sim U(0.5, 1.5)$ is independent of Y, and $\tau = 2$. Because $|Y| < 1/\tau$ a.s., the Neumann series representation of U is convergent and

$$U(t) = f(t) - \lambda \int_0^\tau H(t, s; \lambda)\, f(s)\, ds$$

where $\lambda = -1$, $f(t) = (Q/\beta)\,(1 - e^{-\beta t})$, $H(t, s; \lambda) = -\sum_{r=1}^{\infty} \lambda^{r-1} K_r(t, s)$, $K_r(t, s) = \int_0^\tau K(t, \sigma)\, K_{r-1}(\sigma, s)\, d\sigma$ for $r \geq 2$ and $K_1(t, s) = K(t, s) = 1_{[0,t]}(s)\, Y\, e^{-\beta\,(t-s)}$.

Figure 9.8 shows with solid and dotted lines the exact and approximate

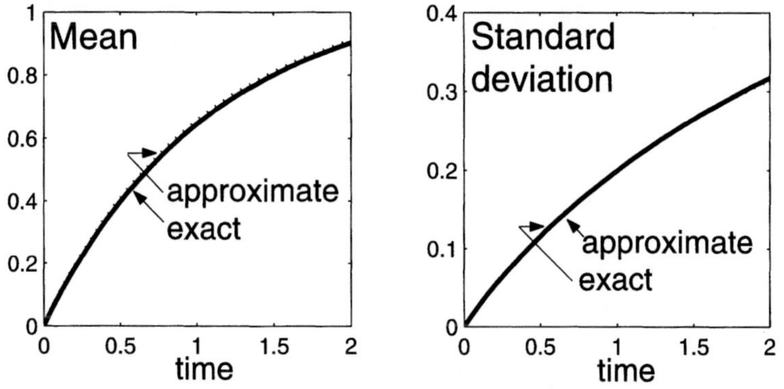

Figure 9.8. Exact and approximate mean and variance of U for $\beta = 1$, $U(0) = 0$, $Y \sim U(-0.5, 0.5)$, $Q \sim U(0.5, 1.5)$, and $\tau = 2$

mean and variance of the solution U in the time interval $[0, 2]$. The approximate solution is given by the first two terms of the above Neumann series. In this case the approximate mean and variance functions are accurate. Generally, a larger number of terms needs to be retained from the Neumann series to obtain satisfactory approximations. ◇

Proof: The integral form of the differential equation for U is

$$\begin{aligned} U(t) &= U(0)\, e^{-\beta t} + \int_0^t e^{-\beta\,(t-s)}\, (-Y\, U(s) + Q)\, ds \\ &= \frac{Q}{\beta}\left(1 - e^{-\beta t}\right) - Y\, e^{-\beta t} \int_0^t e^{\beta s}\, U(s)\, ds \end{aligned}$$

by viewing $-Y\, U(\cdot) + Q$ in the equation for U as the input. Hence, U is the solution of the Fredholm equation of the second kind,

$$U(t) = f(t) + \lambda \int_0^\tau K(t, s)\, U(s)\, ds,$$

with $\lambda = -1$, where $f(t) = (Q/\beta)\,(1 - e^{-\beta t})$ and $K(t, s) = 1_{[0,t]}(s)\, Y\, e^{-\beta\,(t-s)}$ (Section 8.4.1.4). ■

9.2.7 Galerkin method

Let $\mathcal{X}$ be defined by Eq. 9.1 with some initial and boundary conditions. Let $\mathcal{G}$ be a collection of $\mathbb{R}^n$-valued deterministic functions with support D that are square integrable, linearly independent in D, and satisfy the initial and boundary conditions for Eq. 9.1. The members of $\mathcal{G}$ are called **trial functions**. We introduce the inner product $\langle f, g\rangle = \int_D f(\boldsymbol{x})^T\, g(\boldsymbol{x})\, d\boldsymbol{x}$ on $\mathcal{G}$. For computation it is convenient to construct $\mathcal{G}$ such that its members are orthogonal, that is, $\langle f, g\rangle = 0$ for $f, g \in \mathcal{G}$ distinct.

Let $\varphi_p \in \mathcal{G}$, $p = 1, \ldots, m$, and define

$$\tilde{\mathcal{X}}(\boldsymbol{x}, t) = \sum_{p=1}^{m} C_p(t)\, \varphi_p(\boldsymbol{x}), \quad \boldsymbol{x} \in D, \quad t \geq 0, \tag{9.6}$$

where the real-valued coefficients C_p are random and may depend on time. The function $\tilde{\mathcal{X}}$ is a member of the space $\mathcal{S}$ spanned by the trial functions $\varphi_1, \ldots, \varphi_m$. Suppose that we approximate the solution of Eq. 9.1 by a member of $\mathcal{S}$. The **error** or **residual** of this approximation is

$$\mathcal{R}(\boldsymbol{x}, t) = \mathcal{D}[\tilde{\mathcal{X}}(\boldsymbol{x}, t)] - \mathcal{Y}(\boldsymbol{x}, t) = \sum_{p=1}^{m} \mathcal{D}[C_p(t)\, \varphi_p(\boldsymbol{x}, t)] - \mathcal{Y}(\boldsymbol{x}, t), \tag{9.7}$$

where the last equality holds if $\mathcal{D}$ is linear in $\tilde{\mathcal{X}}$.

The **Galerkin solution** for Eq. 9.1 is a member $\tilde{\mathcal{X}}$ of $\mathcal{S}$ whose coordinates C_p are determined from the condition that the residual $\mathcal{R}$ in Eq. 9.7 is orthogonal a.s. to the trial functions $\varphi_1, \ldots, \varphi_m$. The Galerkin solution is (1) useful if the number m of trial functions in Eq. 9.6 is small since, in this case, the solution of Eq. 9.1 is replaced by the solution of a much smaller problem for the unknown coefficients C_p in Eq. 9.6 and (2) accurate if the selected trial functions capture the essential features of the solution of Eq. 9.1. There is no theory for selecting adequate trial functions. Generally, physical and analytical considerations are used to select the trial functions for a particular problem. If there is little information on the class of adequate trial functions, the approximate representation in Eq. 9.6 can be based on, for example, Bernstein polynomials, as illustrated later in this section (Example 9.21).

The coefficients C_p of the Galerkin solution in Eq. 9.6 are given by

$$\sum_{p=1}^{m} \langle \mathcal{D}[C_p(t)\,\varphi_p], \varphi_q \rangle = \langle \mathcal{Y}, \varphi_q \rangle \text{ a.s,} \quad q = 1, \ldots, m, \tag{9.8}$$

provided that $\mathcal{D}$ is linear in $\tilde{\mathcal{X}}$.

Note: We require that almost all samples of $\tilde{\mathcal{X}}$ be orthogonal to the trial functions. This condition gives Eq. 9.8, that is, m algebraic or differential equations for the unknown coefficients C_p.

The coefficients C_p are random because of the uncertainty in the coefficients of $\mathcal{D}$ and the input $\mathcal{Y}$. Generally, these coefficients depend on time so that they are stochastic processes giving the evolution in time of the coordinates of the Galerkin solution $\tilde{\mathcal{X}}$ in the basis defined by the trial functions φ_p. ▲

Under some conditions Eq. 9.8 becomes an algebraic equation with random coefficients for $\boldsymbol{C}$. For example, if $\mathcal{D}$ is a linear differential operator involving only partial derivatives with respect to the coordinates of the space argument $\boldsymbol{x}$, the solution of Eq. 9.1 is time-invariant so that $\tilde{\mathcal{X}}$ in Eq. 9.6 becomes $\tilde{\mathcal{X}}(\boldsymbol{x}) = \sum_{p=1}^{m} C_p\,\varphi_p(\boldsymbol{x})$ and Eq. 9.8 yields the algebraic equation $\boldsymbol{A}\,\boldsymbol{C} = \boldsymbol{Y}$, where $\boldsymbol{A} = \{\langle \mathcal{D}[\varphi_p], \varphi_q \rangle\}$, $p, q = 1, \ldots, m$, $\boldsymbol{C} = (C_1, \ldots, C_m)$, and $\boldsymbol{Y} = \{\langle \mathcal{Y}, \varphi_q \rangle\}$, $q = 1, \ldots, m$. Properties of $\boldsymbol{C}$ can be obtained from the solution of the above algebraic equation.

Example 9.18: Let X be the solution of $\dot{X}(t) = A\,X(t)$, $t \in [0, 1]$, with the initial condition $X(0) = 1$. The exact solution of this equation is $X(t) = \exp(A\,t)$. If A is uniformly distributed in the range (a_1, a_2), the exact moments of X are

$$E[X(t)^m] = E\left[e^{m\,A\,t}\right] = \frac{1}{m\,(a_2 - a_1)\,t}\left(e^{m\,a_2\,t} - e^{m\,a_1\,t}\right).$$

The Galerkin solution with the functional form $\tilde{X}(t) = 1 + C_1\,t + C_2\,t^2$ is

$$\tilde{X}(t) = 1 + \frac{-36\,A\,(2\,A - 5) + 20\,A\,(3\,A - 8)}{3\,A^2 - 12\,A + 20}\,t + \frac{30\,A\,(3\,A - 4) + 40\,A\,(2\,A - 3)}{3\,A^2 - 12\,A + 20}\,t^2.$$

Figure 9.9 shows the mean and standard deviation of X and its approximation $\tilde{X}$ for $A \sim U(-1.5, -0.5)$. The first two moments of $\mathcal{X}$ have been estimated from 500 independent samples of this process. ◇

Note: The functional form considered for $\tilde{X}$ satisfies the initial condition for any values of C_1 and C_2 since $\phi_1(t) = t$ and $\phi_2 = t^2$ are zero at $t = 0$. The conditions in Eq. 9.8 imply $\int_0^1 t\,\mathcal{R}(t)\,dt = 0$ and $\int_0^1 t^2\,\mathcal{R}(t)\,dt = 0$ so that

$$\begin{cases} -A/2 + (1/2 - A/3)\,C_1 + (2/3 - A/4)\,C_2 = 0, \\ -A/3 + (1/2 - A/4)\,C_1 + (1/2 - A/5)\,C_2 = 0, \end{cases}$$

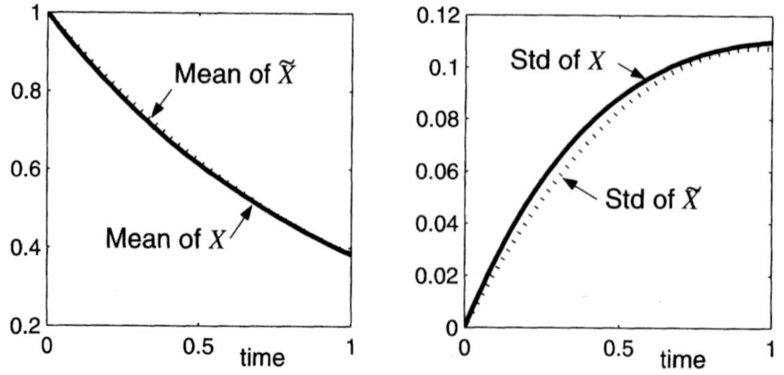

Figure 9.9. Mean and standard deviation of X and $\tilde{X}$ for $A \sim U(-1.5, -0.5)$

where $\mathcal{R}(t) = -A + (1 - A\,t)\,C_1 + t\,(2 - A\,t)\,C_2$ denotes the residual. These equations show that the vector $\boldsymbol{C} = (C_1, C_2)$ is the solution of an algebraic equation with random coefficients and input, and give the above expressions for $\boldsymbol{C}$. If $\boldsymbol{C}$ has a large dimension, it is not possible to obtain analytical expressions for the coordinates of $\boldsymbol{C}$. In this case, the Taylor series, perturbation, and other methods presented in the first part of this chapter can be used for solution ([178], pp. 511-522).

The accuracy of the Galerkin solution depends strongly on the trial functions. For example, the Galerkin solution $\tilde{X}(t) = 1 + C\,t$ with $C = A/(1 - 2\,A/3)$ derived from the orthogonality condition $\int_0^t \mathcal{R}(t)\,t\,dt = 0$ provides unsatisfactory approximations for both the mean and variance of X in $[0, t]$.

We also note that the process $Y(t) = X(t) - 1$ satisfies the differential equation $\dot{Y}(t) = A\,(Y(t) + 1)$ with the initial condition $Y(0) = 0$. It is convenient in some applications to modify the original differential equation so that its new version has homogeneous initial and boundary conditions. ▲

Example 9.19: The Galerkin solution can also be applied to solve nonlinear differential equations with random coefficients and input. Let X be the solution of $\dot{X}(t) = A\,X(t)^2$ for $t \in [0, 1]$ and $X(0) = 1$. Let $\tilde{X}(t) = 1 + \sum_{p=1}^{m} C_p\,\varphi_p(t)$ with $\varphi_p(0) = 0$, $p = 1, \ldots, m$, be a Galerkin solution for this equation. The unknown coefficients of the Galerkin solution satisfy the nonlinear algebraic equation

$$\sum_{p=1}^{m} \alpha_{pk}\,C_p - A \sum_{p,q=1}^{m} \beta_{pqk}\,C_p\,C_q = 0, \quad k = 1, \ldots, m,$$

where $\alpha_{pk} = \int_0^1 \dot{\varphi}_p(t)\,\varphi_k(t)\,dt$ and $\beta_{pqk} = \int_0^1 \varphi_p(t)\,\varphi_q(t)\,\varphi_k(t)\,dt$. ◇

Note: The Galerkin solution $\tilde{X}$ satisfies the initial conditions for any values of the coefficient $C_1, \ldots, C_m$. The residual

$$\mathcal{R}(t) = \sum_{p=1}^{m} C_p\, \dot\varphi_p(t) - A\left(\sum_{p=1}^{m} C_p\, \varphi_p(t)\right)^2$$
$$= \sum_{p=1}^{n} C_p\, \dot\varphi_p(t) - A \sum_{p,q=1}^{m} C_p\, C_q\, \varphi_p(t)\, \varphi_q(t)$$

and the orthogonality conditions $\int_0^1 \mathcal{R}(t)\, \varphi_k(t)\, dt = 0$, $k = 1, \ldots, m$, give the equations for C_p.

Generally, the solutions $C_1, \ldots, C_m$ of the above nonlinear algebraic equations cannot be obtained analytically. Approximate properties of $C_1, \ldots, C_m$ can be obtained by Monte Carlo and other methods for solving algebraic equations with random coefficients (Section 8.3.1). ▲

Example 9.20: Let $\mathcal{X}$ be the solution of Eq. 9.1 with $D = (0, l)$, $l > 0$, and $\mathcal{D} = \partial^4/\partial x^4 + M(x)\, \partial^2/\partial t^2$ depending on a random field M with strictly positive samples in $(0, l)$. The initial and boundary conditions are $\mathcal{X}(x, 0) = 0$, $\mathcal{X}(0, t) = \mathcal{X}(l, t) = 0$, and $\partial \mathcal{X}(0, t)/\partial x^2 = \partial \mathcal{X}(l, t)/\partial x^2 = 0$.

The coefficients C_p of the Galerkin solution

$$\tilde{\mathcal{X}}(x, t) = \sum_{p=1}^{2} C_p(t)\, \varphi_p(x) = C_1(t)\, \sin\left(\frac{\pi\, x}{l}\right) + C_2(t)\, \sin\left(\frac{2\pi\, x}{l}\right)$$

satisfy the differential equations

$$\begin{cases} M_{11}\, \ddot C_1(t) + M_{12}\, \ddot C_2(t) + (\pi/l)^4\, (l/2)\, C_1(t) = \mathcal{Y}_1(t), \\ M_{21}\, \ddot C_1(t) + M_{22}\, \ddot C_2(t) + (2\pi/l)^4\, (l/2)\, C_1(t) = \mathcal{Y}_2(t), \end{cases}$$

where $M_{pq} = \int_0^l M(x)\, \sin(p\,\pi\, x/l)\, \sin(q\,\pi\, x/l)\, dx$, $C_p(0) = 0$, and $\dot C_p(0) = 0$, $p, q = 1, 2$. The terms on the right side of the above equation are $\mathcal{Y}_q(t) = \int_0^l \mathcal{Y}(x, t)\, \sin(q\,\pi\, x/l)\, dx$, $q = 1, 2$.

Suppose that the input is $\mathcal{Y}(x, t) = \mathcal{Y}(t) = \sigma\, Y(t)$, where σ is a constant and Y denotes an Ornstein-Uhlenbeck process defined by $dY(t) = -\rho\, Y(t)\, dt + \sqrt{2\rho}\, dB(t)$, where $\rho > 0$ is a constant and B denotes a Brownian motion. Then the vector $\boldsymbol{C}(t) = (C_1(t), \dot C_1(t), C_2(t), \dot C_2(t), Y(t))$ is an $\mathbb{R}^5$-valued diffusion process conditional on the random coefficients M_{pq}. The methods discussed previously in this chapter and results in Section 7.2.1 can be used to find properties of $\boldsymbol{C}$. ◇

Note: The error of the approximate solution,

$$\mathcal{R}(x, t) = \sum_{p=1}^{2} \left[\left(\frac{p\,\pi}{l}\right)^4 C_p(t) + M(x)\, \ddot C_p(t)\right] \sin\left(\frac{p\,\pi\, x}{l}\right) - \mathcal{Y}(x, t),$$

and the orthogonality conditions give the differential equations for the unknown coefficients C_p. These coefficients and their first derivative are zero at the initial time since $\mathcal{X}(x, 0) = 0$. The coefficients C_p are stochastic processes defined by differential equations with random coefficients and input.

The random function $\mathcal{X}$ gives the deflection of a simply supported beam with unit stiffness, span $l > 0$, and random mass $M(x)$, $x \in [0, l]$, that is subjected to a random load $\mathcal{Y}(x, t)$, $x \in [0, l]$, $t \geq 0$. ▲

Example 9.21: Let $g : [0, 1] \to \mathbb{R}$ be a bounded function that is continuous at $x \in [0, 1]$ and let $b_n(x) = \sum_{k=0}^{n} g(k/n)\, p_{nk}(x)$ be the Bernstein polynomial of order n of g, where $p_{n,k}(x) = [n!/(k!\,(n-k)!)]\, x^k\, (1-x)^{n-k}$. Then $\lim_{n\to\infty} b_n(x) = g(x)$. This limit holds uniformly in $[0, 1]$ if g is continuous in this interval. The Bernstein polynomials can be extended simply to represent real-valued continuous functions defined on $\mathbb{R}^d$, $d > 1$ ([123], p. 51).

Let $G : [0, 1] \times \Omega \to \mathbb{R}$ be a bounded random function on a probability space $(\Omega, \mathcal{F}, P)$, which is uniformly continuous in probability on $[0, 1]$. Then the sequence of **random Bernstein polynomials**,

$$B_n(x) = \sum_{k=0}^{n} G(k/n)\, p_{n,k}(x), \tag{9.9}$$

converges uniformly in probability to G in $[0, 1]$. This property suggests the use of Bernstein polynomials as trial functions in Eq. 9.6. The corresponding Galerkin solution is $\tilde{\mathcal{X}}(x, t) = \sum_{k=0}^{n} C_k(t)\, p_{n,k}(x)$, where C_k are random unknown coefficients, which can be determined from Eq. 9.8.

Random Bernstein polynomials can also be used to develop parametric representations for stochastic processes. These representations are useful in Monte Carlo simulation and analytical studies ([79], Section 4.3.2.3). ◇

Proof: For a given $\varepsilon > 0$ we can find $\delta > 0$ so that $|x - x'| < \delta$ implies $|g(x) - g(x')| < \varepsilon$ by the properties of g. We have

$$|g(x) - b_n(x)| = \left|\sum_{k=0}^{n} (g(x) - g(k/n))\, p_{n,k}(x)\right| \leq \sum_{|k/n - x| < \delta} |g(x) - g(k/n)|\, p_{n,k}(x)$$
$$+ \sum_{|k/n - x| \geq \delta} |g(x) - g(k/n)|\, p_{n,k}(x).$$

The sum indexed by $|k/n - x| < \delta$ is smaller than $\varepsilon \sum_{|k/n-x|<\delta} p_{n,k}(x) \leq \varepsilon$ since $|g(x) - g(k/n)| \leq \varepsilon$ and $\sum_{k=0}^{n} p_{n,k}(x) = 1$. The sum indexed by $|k/n - x| \geq \delta$ is smaller than $2\,\beta \sum_{|k/n-x|\geq\delta} p_{n,k}(x)$, where β bounds g, that is, $\sup_{x\in[0,1]} |g(x)| \leq \beta$. We also have

$$\sum_{k=0}^{n} (k - n\,x)^2\, p_{n,k}(x) = \sum_{k=0}^{n} [k\,(k-1) - (2\,n\,x - 1)\,k + n^2\,x^2]\, p_{n,k}(x) = n\,x\,(1-x),$$

since $\sum_{k=0}^{n} k\, p_{n,k}(x) = n\, x$ and $\sum_{k=0}^{n} k\,(k-1)\, p_{n,k}(x) = n\,(n-1)\, x^2$, and

$$\sum_{|k/n-x|\geq\delta} p_{n,k}(x) \leq \frac{1}{\delta^2} \sum_{|k/n-x|\geq\delta} (k/n-x)^2\, p_{n,k}(x) \leq \frac{n\,x\,(1-x)}{n^2\,\delta^2} \leq \frac{1}{4\,n\,\delta^2}.$$

These results show that $|g(x) - b_n(x)| \leq \varepsilon + \beta/(2\,n\,\delta^2)$ so that $|g(x) - b_n(x)| \leq 2\,\varepsilon$ for $n \geq \beta/(2\,\varepsilon\,\delta^2)$. If g is continuous in $[0, 1]$, then $|g(x) - b_n(x)| \leq 2\,\varepsilon$ holds with δ independent of x so that b_n converges uniformly to g in this interval.

It remains to show that B_n in Eq. 9.9 converges uniformly in probability to G in $[0, 1]$, that is, that for any $\varepsilon, \eta > 0$ there exists $n(\varepsilon, \eta)$ such that $P(|B_n(x) - G(x)| \geq \varepsilon) \leq \eta$ for $n \geq n(\varepsilon, \eta)$, or equivalently that

$$\tilde{d}(B_n, G) = \sup_{x\in[0,1]} d(B_n(x), G(x)) = \sup_{x\in[0,1]} \int_\Omega \frac{|B_n(x) - G(x)|}{1 + |B_n(x) - G(x)|} P(d\omega)$$

can be made as small as desired as $n \to \infty$. Take a fixed $x \in [0, 1]$ and an integer $n \geq 1$. Let $I_i = [(i-1)/n, i/n]$ be the subinterval of $[0, 1]$ containing x. Since G is uniformly continuous in probability in $[0, 1]$, for any $\varepsilon, \eta > 0$ there exists $\delta = \delta(\varepsilon, \eta)$ such that $|x - x'| \leq \delta$ implies $P(|G(x) - G(x')| \geq \varepsilon) \leq \eta$. Hence, we have $P(|G(x) - G(\nu/n)| \geq \varepsilon) \leq \eta$ for $\nu = i-1, i$ if $1/n \leq \delta$. Suppose that n satisfies the condition $1/n \leq \delta$. Then

$$\begin{aligned}
\tilde{d}(B_n, X) &= \sup_{x\in[0,1]} d\,(B_n(t), X(t)) = \sup_{x\in[0,1]} \int_\Omega \frac{|B_n(x) - G(x)|}{1 + |B_n(x) - G(x)|} P(d\omega) \\
&= \sup_{x\in[0,1]} \int_\Omega \frac{|\sum_{\nu=0}^{n} G(\nu/n)\, p_{n,\nu} - G(x) \sum_{\nu=0}^{n} p_{n,\nu}|}{1 + |B_n(x) - G(x)|} P(d\omega) \\
&\leq \sup_{x\in[0,1]} \int_\Omega \frac{\sum_{\nu=0}^{n} |G(\nu/n) - G(x)|\, p_{n,\nu}}{1 + |B_n(x) - G(x)|} P(d\omega) \leq 4\,\beta\,\eta + 3\,\varepsilon + \frac{\beta}{2\,n\,\delta},
\end{aligned}$$

where $\sup_{x\in[0,1]} |G(x)| \leq \beta$ a.s. The above upper bound holds since (1) the integral in the above summation corresponding to $\nu = i-1$ can be separated in two integrals over the events $\{|X((i-1)/n) - X(t)| \geq \varepsilon\}$ and $\{|X((i-1)/n) - X(t)| < \varepsilon\}$ and these integrals are smaller than $2\,\beta\,\eta$ and ε, respectively, (2) similarly, the integral corresponding to $\nu = i$ is smaller than $2\,\beta\,\eta + \varepsilon$, and (3) the remaining sum of integrals satisfies the inequality

$$\int_\Omega \frac{\sum_\nu' |G(\nu/n) - G(x)|\, p_{n,\nu}}{1 + |B_n(x) - B(x)|} P(d\omega) \leq \int_\Omega \sum_{\nu=0}^{n} |G(\nu/n) - G(x)|\, p_{n,\nu}\, P(d\omega)$$

so that it is smaller than $\varepsilon + \beta/(2\,n\,\delta^2)$ by the upper bound on $|g(x) - b_n(x)|$, where the sum $\sum_\nu'$ includes the terms with indices $\nu \neq i-1, i$. We have found that

$$\tilde{d}(B_n, G) \leq 4\,\beta\,\eta + 3\,\varepsilon + \frac{\beta}{2\,n\,\delta^2}$$

for any $\varepsilon, \eta > 0$ and $n \geq n(\varepsilon, \eta)$ so that B_n converges uniformly in probability to G on $[0, 1]$ as $n \to \infty$. ■

9.2.8 Finite difference method

Let $\mathcal{X}$ be the solution of Eq. 9.1 with some initial and boundary conditions. Consider a time sequence $t_i = i\,\Delta t$, $i = 0, 1, \ldots$, $\Delta t > 0$, and a finite collection of points $\boldsymbol{x}_k \in D$. Denote by $\boldsymbol{X}_{k,i} = \mathcal{X}(\boldsymbol{x}_k, t_i)$ the value of $\mathcal{X}$ at the node (k, i) of the finite difference representation for Eq. 9.1. The finite difference solution of this equation involves the following steps.

1. Develop a finite difference representation for Eq. 9.1 and for the initial and boundary conditions of this equation by approximating all derivatives at the nodes (k, i) by finite differences.
2. Find moments and other probabilistic properties of the unknowns $\boldsymbol{X}_{k,i}$ by the methods presented in this and previous chapters.

Note: The derivatives in Eq. 9.1 and in the initial and boundary conditions for this equation can be approximated by finite differences of various accuracy. For example, the partial derivative $\partial\mathcal{X}/\partial t$ can be approximated at $(\boldsymbol{x}, t)$ by $[\mathcal{X}(\boldsymbol{x}, t + \Delta t) - \mathcal{X}(\boldsymbol{x}, t)]/\Delta t$ or $[\mathcal{X}(\boldsymbol{x}, t+\Delta t) - \mathcal{X}(\boldsymbol{x}, t-\Delta t)]/(2\,\Delta t)$. The errors of these finite difference approximations are of orders $O(\Delta t)$ and $O((\Delta t)^2)$, respectively.

If only the spatial derivatives are approximated by finite differences, we obtain differential equations giving the time evolution for a vector with entries $\mathcal{X}(t, \boldsymbol{x}_k)$. The coefficients of these equations depend on values of the random coefficients of $\mathcal{D}$ at the nodes of the finite difference representation for Eq. 9.1. ▲

Example 9.22: Let $\mathcal{X}$ be the solution of

$$\left[U(x)\,\frac{\partial^2}{\partial t^2} - V'(x)\,\frac{\partial}{\partial x} - V(x)\,\frac{\partial^2}{\partial x^2}\right]\mathcal{X}(x, t) = \mathcal{Y}(x, t), \quad x \in (0, 1), \quad t \geq 0,$$

where $U, V > 0$ are random fields, $V'(x) = dV(x)/dx$, and the input $\mathcal{Y}(x, t)$ depends on both time and space arguments. The initial and boundary conditions are $\mathcal{X}(x, 0) = 0$, $\mathcal{X}(0, t) = 0$, and $\partial\mathcal{X}(1, t)/\partial x = 0$.

Take $t_i = i\,\Delta t$ and $x_k = k/n$, $k = 0, 1, \ldots, n$, where $n > 1$ is an integer giving the number of equal intervals considered in $D = (0, 1)$. The finite difference representation of the equation of $\mathcal{X}$ at (x_k, t_i) is

$$X_{k,i+1} = -X_{k,i-1} + A_k\,X_{k-1,i} + B_k\,X_{k,i} + C_k\,X_{k+1,i} + W_k\,Y_{k,i},$$

where $X_{k,i} = \mathcal{X}(x_k, t_i)$, $V_k = V(x_k)$, $V_k' = V'(x_k)$, $U_k = U(x_k)$, and $Y_{k,i} = \mathcal{Y}(x_k, t_i)$. The coefficients in the above finite difference equation are

$$A_k = W_k\left(V_k/(2/n)^2 - V_k'/(2/n)\right),$$

$$B_k = 2\,W_k\left(U_k/((\Delta t)^2) - V_k/(2/n)^2\right), \quad \text{and}$$

$$C_k = W_k\left(V_k'/(2/n) + V_k/(2/n)^2\right),$$

where $W_k = (\Delta t)^2/U_k$. These coefficients are random because they depend on the random variables U_k, V_k, and V'_k. The above finite difference equations for $X_{k,i}$ satisfy the initial and boundary conditions $X_{k,-1} = X_{k,0} = 0$ and $X_{0,i} = 0$, $X_{n+1,i} = X_{n-1,i}$, respectively. ◇

Note: The solution $\mathcal{X}$ gives the elongation of a rod with unit length and cross section area, material density $U(x)$, and modulus of elasticity $V(x)$, subjected to an axial force $Y(x,t)$ at location x and time t. The differential equation for $\mathcal{X}$ follows from (1) the Hooke law stating that the stress in the rod is $S(x,t) = V(x)\,\partial\mathcal{X}(x,t)/\partial x$ and (2) the Newton law giving the condition

$$U(x)\,\frac{\partial^2\mathcal{X}(x,t)}{\partial t^2} = \frac{\partial S(x,t)}{\partial x} + Y(x,t).$$

The rod is at rest at the initial time so that $X_{k,-1} = X_{k,0} = 0$ for all k's. The boundary conditions $X_{0,i} = 0$ and $X_{n+1,i} = X_{n-1,i}$ show that the rod is fixed at $x = 0$ and that the stress is zero at $x = 1$. ▲

We conclude here our discussion on some methods for solving Eq. 9.1. The remaining part of this chapter presents stochastic problems selected from various fields of applied science and engineering. The solutions of the problems in the following sections are based on the methods in Chapters 6, 7, 8, and the first part of Chapter 9. An exception is the discussion of the stochastic finite element method in the following section. This method can be applied to find the solution of Eq. 9.1, but is discussed in the following section because of its wide use in mechanics.

9.3 Mechanics

The **finite element method** is the preferred method for solving deterministic and stochastic problems in mechanics. The method has been applied to analyze problems in physics, chemistry, aerospace, and other fields [28].

The finite element method can be viewed as a mixture of the finite difference and the Rayleigh-Ritz methods. The unknowns in the finite element method are values of $\mathcal{X}$ in Eq. 9.1 at a finite number of points, called **nodes**. The equations for the values of $\mathcal{X}$ at nodes are obtained from variational principles.

We formulate a variational principle used frequently in mechanics, illustrate the Rayleigh-Ritz method by a simple example, give essentials of the finite element method for deterministic problems, extend this formulation to stochastic problems, and present finite element solutions for some stochastic problems.

9.3.1 Variational principles

Consider an elastic solid in an open bounded subset D in $\mathbb{R}^q$, $q = 1, 2, 3$, with boundary ∂D that is in equilibrium under the action of body forces $\boldsymbol{b}(\boldsymbol{x}, t)$,

$\boldsymbol{x} \in D$, boundary traction $\boldsymbol{p}(\boldsymbol{x}, t)$ for $\boldsymbol{x} \in \partial D_\sigma$ and $t \geq 0$, and imposed displacement $\bar{\boldsymbol{u}}(\boldsymbol{x}, t)$ for $\boldsymbol{x} \in \partial D_u$ and $t \geq 0$, where $\partial D = \partial D_\sigma \cup \partial D_u$ and $\partial D_\sigma \cap \partial D_u = \emptyset$. Let $u_i(\boldsymbol{x}, t)$, $s_{ij}(\boldsymbol{x}, t)$, and $\gamma_{ij}(\boldsymbol{x}, t)$ denote the displacement, stress, and strain fields at $\boldsymbol{x} \in D$ at time $t \geq 0$. The vectors $\boldsymbol{s} = (s_{11}, s_{12}, s_{13}, s_{22}, s_{23}, s_{33})$ and $\boldsymbol{\gamma} = (\gamma_{11}, \gamma_{12}, \gamma_{13}, \gamma_{22}, \gamma_{23}, \gamma_{33})$ contain the entire information on stresses and strains since $s_{ij} = s_{ji}$ and $\gamma_{ij} = \gamma_{ji}$. The equilibrium conditions require

$$
\begin{aligned}
&s_{ij,j}(\boldsymbol{x}, t) + b_i(\boldsymbol{x}, t) = 0, \quad \boldsymbol{x} \in D, \\
&s_{ij}(\boldsymbol{x}, t)\, n_j(\boldsymbol{x}, t) - p_i(\boldsymbol{x}, t) = 0, \quad \boldsymbol{x} \in \partial D_\sigma,
\end{aligned} \tag{9.10}
$$

for $t \geq 0$, where $s_{ij,j} = \partial s_{ij}/\partial x_j$, $n_j(\boldsymbol{x}, t)$ denotes the coordinate j of the exterior normal $\boldsymbol{n}(\boldsymbol{x}, t)$ at $\boldsymbol{x} \in \partial D_\sigma$ and time t, and summation is performed on repeated subscripts (Section 8.5.2 in this book, [68], Section 3.4).

Let $\boldsymbol{u}(\boldsymbol{x}, t) = (u_1, u_2, u_3)(\boldsymbol{x}, t)$ be the displacement at $\boldsymbol{x} \in D \subset \mathbb{R}^3$ and $t \geq 0$ and let $\boldsymbol{u}_k$, $k = 1, 2$, be two arbitrary displacement fields in $\bar{D} = D \cup \partial D$ such that $\boldsymbol{u}_k(\boldsymbol{x}, t) = \bar{\boldsymbol{u}}(\boldsymbol{x}, t)$ for $\boldsymbol{x} \in \partial D_u$ and $t \geq 0$. The difference $\delta \boldsymbol{u} = \boldsymbol{u}_1 - \boldsymbol{u}_2$ is said to be a **virtual displacement**. Virtual displacements must vanish on ∂D_u but are arbitrary on ∂D_σ.

The **principle of virtual displacements** states that an elastic solid in $D \subset \mathbb{R}^q$, $q = 1, 2, 3$, is in equilibrium if

$$
\int_D \boldsymbol{s}^T\, \delta \boldsymbol{\gamma}\, d\boldsymbol{x} = \int_D \boldsymbol{b}^T\, \delta \boldsymbol{u}\, d\boldsymbol{x} + \int_{\partial D_\sigma} \boldsymbol{p}^T\, \delta \boldsymbol{u}\, d\sigma(\boldsymbol{x}), \quad \forall \delta \boldsymbol{u}, \tag{9.11}
$$

where $\delta \boldsymbol{\gamma}$ is the first order variation of the strain vector corresponding to $\delta \boldsymbol{u}$ and $d\sigma(\boldsymbol{x})$ is an infinitesimal surface element in ∂D_σ ([28], Section 1.3).

Note: The left and right sides of Eq. 9.11 are the first order variations of the strain energy and external work, respectively. It can be shown that the principle of virtual displacements constitutes an alternative statement of equilibrium. This principle is valid for any material behavior and magnitude of the displacement field ([28], Section 1.3). A dual of the principle of virtual displacements, called principle of virtual forces, can also be established. This principle provides an alternative statement of compatibility ([28], Section 1.5, [68], Section 10.9). ▲

Example 9.23: Let $u(x, t)$ be the solution of

$$
\left(\frac{\partial^4}{\partial x^4} + \frac{m}{\chi} \frac{\partial^2}{\partial t^2} \right) u(x, t) = y(x, t),
$$

for $t \geq 0$, $x \in D = (0, l)$, $m, \chi, l > 0$, the boundary conditions $u(0, t) = u(l, t) = 0$ and $\partial^2 u(0, t)/\partial x^2 = \partial^2 u(l, t)/\partial x^2 = 0$, and the initial conditions $u(x, 0) = g(x)$ and $\partial u(x, 0)/\partial t = h(x)$, where g, h, and y are specified continuous functions. Consider an approximate representation $\tilde{u}(x, t) = \sum_{k=1}^n c_k(t)\, \varphi_k(x)$ of the solution u, where φ_k are specified functions satisfying at least the kinematical boundary conditions and the weights c_k need to be determined. If we take $\varphi_k(x) = \sin(k\,\pi\,x/l)$, the Rayleigh-Ritz method gives

$$\ddot{c}_k(t) + \frac{k^4\,\pi^4\,\chi}{m\,l^4}\,c_k(t) = \frac{2}{m\,l}\int_0^l y(x,t)\,\sin(k,\pi\,x/l)\,dx,$$

where $c_k(0)$ and $\dot{c}_k(0)$ result from $g(x) = \sum_{k=1}^n c_k(0)\,\sin(k\,\pi\,x/l)$ and $h(x) = \sum_{k=1}^n \dot{c}_k(0)\,\sin(k\,\pi\,x/l)$, respectively, and the dots denote differentiation with respect to time. ◇

Proof: The function u represents the displacement at cross section x and time t of a simply supported beam with span l, stiffness χ, and mass m per unit length. The beam strain energy, SE, is

$$\mathrm{SE} = \frac{\chi}{2}\int_0^l \left(\frac{\partial^2 \tilde{u}}{\partial x^2}\right)^2 dx = \frac{\pi^4\,\chi}{4\,l^3}\sum_{k=1}^n k^4\,c_k(t)^2,$$

so that $\delta(\mathrm{SE}) = (\pi^4\,\chi/(2\,l^3))\sum_{k=1}^n k^4\,c_k(t)\,\delta c_k(t)$, where $\delta c_k(t)$ denotes the first order variation of the coefficients $c_k(t)$. The first order variation of the external work, EW, is

$$\begin{aligned}\delta(\mathrm{EW}) &= \int_0^l \left(y(x,t) - m\,\frac{\partial^2 \tilde{u}(x,t)}{\partial t^2})\right)\delta\tilde{u}(x,t)\,dx \\ &= \sum_{k=1}^n \left[\int_0^t y(x,t)\,\sin(k\,\pi\,x/l)\,dx - (m\,l/2)\,\ddot{c}_k(t)\right]\delta c_k(t).\end{aligned}$$

The differential equations for c_k follow from the condition $\delta(\mathrm{EW}) = \delta(\mathrm{SE})$ that must hold for any $\delta c_k(t)$ by the principle of virtual displacements (Eq. 9.11).

The Rayleigh-Ritz and the Galerkin methods consider similar representations for the solution. However, there are two notable differences between these two methods. First, the trial functions in the Galerkin method must satisfy all boundary conditions while the trial functions in the Rayleigh-Ritz can satisfy only the kinematical boundary conditions. Second, the unknown coefficients of the Rayleigh-Ritz and the Galerkin representations are determined from the principle of virtual displacement and an orthogonality condition, respectively. ■

9.3.2 Deterministic problems

Consider an elastic solid in an open bounded subset $D \subset \mathbb{R}^q$, $q = 1, 2, 3$, that is in equilibrium under some external actions. The finite element and the Rayleigh-Ritz methods differ only by the trial functions used in analysis. The support of the trial functions is D for the Rayleigh-Ritz method. On the other hand, the trial functions used in the finite element method are defined locally on non-overlapping subsets of D, called **finite elements**, that partition D ([28], Section 3.2).

Example 9.24: Consider a linear elastic rod of length $l > 0$, constant stiffness $\chi > 0$, unit cross section area, and specific weight $\rho > 0$, that is suspended vertically from one of its ends. Let $x \in D = (0, l)$ denote the distance from the suspension point. We assume small deformations so that the strain field is equal to the first derivative of the displacement function with respect to x. The finite element method involves the following steps.

1. Partition D in finite elements, select nodes, and define node displacements. For example, let $0 = x_1 < x_2 < \cdots < x_{n+1} = l$, partition D in the finite elements $D_k = (x_k, x_{k+1})$, select x_k to be the node coordinates, and denote by u_k the displacement of node x_k, $k = 1, \ldots, n, n+1$, in the direction of coordinate x.

2. Postulate the displacement field in each finite element. For example, take the displacement in the finite element k to be

$$u^{(k)}(x) = \begin{bmatrix} \frac{x_{k+1}-x}{x_{k+1}-x_k} & \frac{x-x_k}{x_{k+1}-x_k} \end{bmatrix} \begin{bmatrix} u_k \\ u_{k+1} \end{bmatrix} = \boldsymbol{\alpha}_k(x)\, \boldsymbol{u}_k, \quad x \in \bar{D}_k = [x_k, x_{k+1}].$$

3. Calculate the corresponding strain field in each finite element from

$$\gamma^{(k)}(x) = \frac{du^{(k)}(x)}{dx} = \begin{bmatrix} -\frac{1}{x_{k+1}-x_k} & \frac{1}{x_{k+1}-x_k} \end{bmatrix} \begin{bmatrix} u_k \\ u_{k+1} \end{bmatrix} = \boldsymbol{\beta}_k\, \boldsymbol{u}_k, \quad x \in \bar{D}_k.$$

4. Apply the principle of virtual displacements for the postulated displacement field, which gives $\sum_{k=1}^{n} \left(\boldsymbol{u}_k^T \boldsymbol{a}_k^T - \boldsymbol{y}_k^T\right) \delta \boldsymbol{u}_k = \boldsymbol{0}$, where $\boldsymbol{a}_k = \int_{x_k}^{x_{k+1}} \chi\, \boldsymbol{\beta}_k^T \boldsymbol{\beta}_k\, dx$ denotes the **element stiffness matrix**, $\boldsymbol{y}_k = \int_{x_k}^{x_{k+1}} \rho\, \boldsymbol{\alpha}_k^T\, dx$ is the **element force matrix**, and $\delta \boldsymbol{u}_k$ is the first variation of the node displacement vector $\boldsymbol{u}_k$. We cannot conclude that $\boldsymbol{u}_k^T \boldsymbol{a}_k^T - \boldsymbol{y}^T = \boldsymbol{0}$ for each $k = 1, \ldots, n$ since the variations $\delta \boldsymbol{u}_k$ are linearly related.

5. Define the **global displacement vector** $\boldsymbol{u} = (\boldsymbol{u}_1, \ldots, \boldsymbol{u}_{n+1})$ and set $\boldsymbol{u}_k = \boldsymbol{j}_k\, \boldsymbol{u}$. The $(2, n+1)$ matrix $\boldsymbol{j}_k$ selects the coordinates k and $k+1$ of $\boldsymbol{u}$. With this notation, the principle of virtual displacement becomes $\left(\boldsymbol{u}^T \boldsymbol{a}^T - \boldsymbol{y}^T\right) \delta \boldsymbol{u} = \boldsymbol{0}$, where $\boldsymbol{a} = \sum_{k=1}^{n} \boldsymbol{j}_k^T \boldsymbol{a}_k^T \boldsymbol{j}_k$ and $\boldsymbol{y} = \sum_{k=1}^{n} \boldsymbol{j}_k^T \boldsymbol{y}_k$ denote the **global stiffness** and **force** matrices. Because the coordinates of $\boldsymbol{u}$ are not related, we have $\boldsymbol{a}\, \boldsymbol{u} = \boldsymbol{y}$. This condition can be used to calculate $\boldsymbol{u}$ and the corresponding stress and strain fields. ◇

Note: Let $u(x)$ denote the rod displacement at $x \in [0, l]$ and $u'(x) = du(x)/dx$. We have $\int_0^l \left[\chi\, u'(x)\, \delta u'(x) - \rho\, \delta u(x)\right] dx = 0$ by Eq. 9.11 under the assumptions that the rod deformation is small and the material is linear elastic. Hence, the rod strain energy and the first order variation of the external work are $\int_0^l (\chi/2) \left(u'(x)\right)^2 dx$ and $\int_0^l \rho\, \delta u(x)\, dx$, respectively. ▲

9.3.3 Stochastic problems

The element and global stiffness and force matrices in Example 9.24 are random if the material properties and the applied actions are uncertain. Therefore, the resulting equilibrium condition is an algebraic equation with random coefficients and input. An extensive review of the stochastic finite element method can be found in [114, 169, 190].

Example 9.25: Consider the problem in Example 9.24 but assume that the rod stiffness is a random field $A(x)$, $x \in [0, l]$, and that the rod is subjected to a random action $Y(x)$, $x \in [0, l]$. If we retain the finite element partition and the functional form of the displacement field in Example 9.24, the element stiffness and force matrices are

$$\boldsymbol{A}_k^* = \int_{x_k}^{x_{k+1}} A(x)\, \boldsymbol{\beta}_k^T \, \boldsymbol{\beta}_k \, dx \quad \text{and } \boldsymbol{Y}_k^* = \int_{x_k}^{x_{k+1}} (\rho + Y(x))\, \boldsymbol{\alpha}_k^T(x)\, dx,$$

respectively. The global stiffness and force matrices $\boldsymbol{A}^*$ and $\boldsymbol{Y}^*$ are linear forms of $\boldsymbol{A}_k^*$ and $\boldsymbol{Y}_k^*$, respectively. Moments and other properties of the random matrices $\boldsymbol{A}_k^*$, $\boldsymbol{Y}_k^*$, $\boldsymbol{A}^*$, and $\boldsymbol{Y}^*$ can be calculated from the definition of these matrices and the probability laws of A and Y. For example, the first two moments of the entries $A_{k,pq}^*$ of the matrices $\boldsymbol{A}_k^*$ are

$$E[A_{k,pq}^*] = \int_{x_k}^{x_{k+1}} E[A(x)] \left(\boldsymbol{\beta}_k^T \, \boldsymbol{\beta}_k\right)_{pq} dx,$$

$$E[A_{k,pq}^* \, A_{l,st}^*] = \int_{x_k}^{x_{k+1}} \int_{x_l}^{x_{l+1}} E[A(x)\, A(y)] \left(\boldsymbol{\beta}_k^T \, \boldsymbol{\beta}_k\right)_{pq} \left(\boldsymbol{\beta}_l^T \, \boldsymbol{\beta}_k\right)_{st} dx\, dy$$

provided that A is in L_2. ◇

The approach in Examples 9.24 and 9.25 can be applied to general solid mechanics problems with random material properties, input, and/or initial and boundary conditions. Consider an elastic solid in an open bounded subset D of $\mathbb{R}^q$, $q = 1, 2, 3$, that is in equilibrium. Let ∂D_σ and ∂D_u denote the parts of the boundary ∂D of D where traction and displacement are specified, respectively. It is assumed that (1) the density of the material in D exhibits a random spatial variation that can be modeled by a random field $R(\boldsymbol{x})$, $\boldsymbol{x} \in D$, (2) the material in D is linear elastic with random properties so that the relationship between the stress and strain fields is given by Hooke's law $\boldsymbol{S}(\boldsymbol{x}, t) = \boldsymbol{A}(\boldsymbol{x})\, \boldsymbol{\Gamma}(\boldsymbol{x}, t)$, where the random field $\boldsymbol{A} = \boldsymbol{A}^T$ denotes the stiffness tensor, and (3) the solid in D is in dynamic equilibrium under the random action $\boldsymbol{Y}(\boldsymbol{x}, t)$, $\boldsymbol{x} \in \partial D_\sigma$, and displacement constraint on ∂D_u for all $t \geq 0$.

Let D_k, $k = 1, \ldots, n$, be a collection of open subsets in D such that $\bar{D} = \cup_{k=1}^n \bar{D}_k$ and $D_k \cap D_l = \emptyset$ for $k \neq l$. The members of this partition of D are called finite elements. We denote by $\partial D_{\sigma,k} = \partial D_\sigma \cap \bar{D}_k$ the common part of the boundary of the finite element k and the boundary of D with specified traction. Let $\boldsymbol{U}_k$ be a vector collecting the displacements of the nodes of the finite element k. The displacement and strain fields in this element are $\boldsymbol{U}^{(k)}(\boldsymbol{x}, t) = \boldsymbol{\alpha}_k(\boldsymbol{x})\, \boldsymbol{U}_k(t)$ and $\boldsymbol{\Gamma}^{(k)}(\boldsymbol{x}, t) = \boldsymbol{\beta}_k(\boldsymbol{x})\, \boldsymbol{U}_k(t)$, where $\boldsymbol{\alpha}_k$ is a specified function and $\boldsymbol{\beta}_k$ results from $\boldsymbol{\alpha}_k$ by differentiation. The vector of global displacements $\boldsymbol{U}$ collects the distinct node displacements in all $\boldsymbol{U}_k$'s and can be related to these vectors by $\boldsymbol{U}_k(t) = \boldsymbol{j}_k\, \boldsymbol{U}(t)$ (Example 9.24).

The vector of global displacements $\boldsymbol{U}$ is the solution of

$$\boldsymbol{M}\,\ddot{\boldsymbol{U}}(t) + \boldsymbol{A}^*\,\boldsymbol{U}(t) = \boldsymbol{Y}^*(t), \quad t \geq 0, \quad \text{where} \tag{9.12}$$

$$\begin{aligned}
\boldsymbol{M} &= \sum_{k=1}^{n} \boldsymbol{j}_k^T \left(\int_{D_k} R(\boldsymbol{x})\, \boldsymbol{\alpha}_k(\boldsymbol{x})^T\, \boldsymbol{\alpha}_k(\boldsymbol{x})\, d\boldsymbol{x} \right) \boldsymbol{j}_k = \sum_{k=1}^{n} \boldsymbol{j}_k^T\, \boldsymbol{M}_k\, \boldsymbol{j}_k, \\
\boldsymbol{A}^* &= \sum_{k=1}^{n} \boldsymbol{j}_k^T \left(\int_{D_k} \boldsymbol{\beta}_k(\boldsymbol{x})^T\, \boldsymbol{A}(\boldsymbol{x})\, \boldsymbol{\beta}_k(\boldsymbol{x})\, d\boldsymbol{x} \right) \boldsymbol{j}_k = \sum_{k=1}^{n} \boldsymbol{j}_k^T\, \boldsymbol{A}_k^*\, \boldsymbol{j}_k, \\
\boldsymbol{Y}^*(t) &= \sum_{k=1}^{n} \boldsymbol{j}_k^T \left(\int_{\partial D_{\sigma,k}} \boldsymbol{Y}(\boldsymbol{x},t)^T\, \boldsymbol{\alpha}_k(\boldsymbol{x})\, d\sigma(\boldsymbol{x}) \right) = \sum_{k=1}^{n} \boldsymbol{j}_k^T\, \boldsymbol{Y}_k^*(t).
\end{aligned} \tag{9.13}$$

Proof: The principle of virtual displacement gives (Eq. 9.11)

$$\begin{aligned}
&\sum_{k=1}^{n} \int_{D_k} \boldsymbol{U}_k(t)^T\, \boldsymbol{\beta}_k(\boldsymbol{x})^T\, \boldsymbol{A}(\boldsymbol{x})\, \boldsymbol{\beta}_k(\boldsymbol{x})\, \delta\boldsymbol{U}_k(t)\, d\boldsymbol{x} \\
&\quad = \sum_{k=1}^{n} \int_{D_k} \left(-R(\boldsymbol{x})\, \boldsymbol{\alpha}_k(\boldsymbol{x})\, \ddot{\boldsymbol{U}}_k(t)\right)^T \boldsymbol{\alpha}_k(\boldsymbol{x})\, \delta\boldsymbol{U}_k(t)\, d\boldsymbol{x} \\
&\quad + \sum_{k=1}^{n} \int_{\partial D_{\sigma,k}} \boldsymbol{Y}(\boldsymbol{x},t)^T\, \boldsymbol{\alpha}_k(\boldsymbol{x})\, \delta\boldsymbol{U}_k(t)\, d\sigma(\boldsymbol{x})
\end{aligned}$$

for $\boldsymbol{b}(t,\boldsymbol{x})$ set equal to the inertia force $-R(\boldsymbol{x})\,\ddot{\boldsymbol{U}}(t,\boldsymbol{x})$. The relationship between the element displacement vectors $\boldsymbol{U}^{(k)}$ and the global displacement vector $\boldsymbol{U}$, the requirement that the above equation be satisfied for all virtual displacements $\delta\boldsymbol{U}(t)$, and elementary matrix manipulations yield Eqs. 9.12 and 9.13. The matrices $\boldsymbol{M}_k$, $\boldsymbol{A}_k^*$, and $\boldsymbol{Y}_k^*$ can be random and are referred to as element mass, stiffness, and force matrices, respectively. The global mass, stiffness, and force matrices are denoted by $\boldsymbol{M}$, $\boldsymbol{A}^*$, and $\boldsymbol{Y}^*$. ■

9.3.4 Methods of analysis

Our objective is to find second moment and other properties of the process $\boldsymbol{U}$ defined by Eq. 9.12. We discuss solutions of Eq. 9.12 based on (1) some of the methods in Section 9.2, referred to as classical solutions, and (2) a method based on polynomial chaos.

9.3.4.1 Classical methods

The methods in Section 9.2 can be used to find second moment and other properties of the solution $\boldsymbol{U}$ of Eq. 9.12. The applications of these methods requires us to find properties of the random matrices $\boldsymbol{M}$, $\boldsymbol{A}^*$, and $\boldsymbol{Y}^*$. These properties can be calculated from their definitions in Eq. 9.13 and the probability laws of the random fields $R(\boldsymbol{x})$, $\boldsymbol{A}(\boldsymbol{x})$, and $\boldsymbol{Y}(\boldsymbol{x}, t)$. For example, the first two moments of the element mass matrices are

$$E[\boldsymbol{M}_k] = \int_{D_k} E[R(\boldsymbol{x})]\, \boldsymbol{\alpha}_k(t)^T\, \boldsymbol{\alpha}_k(\boldsymbol{x})\, d\boldsymbol{x},$$

$$E[M_{k,ij}\, M_{l,pq}] = \int_{D_k \times D_l} E[R(\boldsymbol{x})\, R(\boldsymbol{y})] \left(\boldsymbol{\alpha}_k(\boldsymbol{x})^T\, \boldsymbol{\alpha}_k(\boldsymbol{x})\right)_{ij} \left(\boldsymbol{\alpha}_l(\boldsymbol{y})^T\, \boldsymbol{\alpha}_l(\boldsymbol{y})\right)_{pq} d\boldsymbol{x}\, d\boldsymbol{y}.$$

It is common in applications to calculate properties of the random matrices $\boldsymbol{M}$, $\boldsymbol{A}^*$, $\boldsymbol{Y}^*$, $\boldsymbol{M}_k$, $\boldsymbol{A}_k^*$, and $\boldsymbol{Y}_k^*$ approximately to reduce the computation time. Most approximations are based on the representation of the random fields R, $\boldsymbol{A}$, and $\boldsymbol{Y}$ by random functions that are constant in each finite element, for example, we can take these functions equal in each finite element D_k with (1) one of their values in this element or (2) their spatial average over D_k. There is little justification for these approximations. If we take $R(\boldsymbol{x}) \simeq R(\boldsymbol{\xi}_k)$ for all $\boldsymbol{x} \in D_k$, where $\boldsymbol{\xi}_k$ is a point in D_k, the calculation of the second moment properties of $\boldsymbol{M}_k$,

$$E[\boldsymbol{M}_k] = E[R(\boldsymbol{\xi}_k)] \int_{D_k} \left(\boldsymbol{\alpha}_k(t)^T\, \boldsymbol{\alpha}_k(\boldsymbol{x})\right) d\boldsymbol{x},$$

$$E[M_{k,ij}\, M_{l,pq}] \quad \text{and}$$

$$= E[R(\boldsymbol{\xi}_k)\, R(\boldsymbol{\xi}_l)] \int_{D_k \times D_l} \left(\boldsymbol{\alpha}_k(\boldsymbol{x})^T\, \boldsymbol{\alpha}_k(\boldsymbol{x})\right)_{ij} \left(\boldsymbol{\alpha}_l(\boldsymbol{y})^T\, \boldsymbol{\alpha}_l(\boldsymbol{y})\right)_{pq} d\boldsymbol{x}\, d\boldsymbol{y},$$

is reduced significantly. On the other hand, the calculation efforts required to find the second moment properties of $\boldsymbol{M}_k$ exactly and by the second approximation are similar.

Frequently, the parameters defining constitutive laws in continuum material models are assumed to be random variables and/or fields for use in stochastic finite element studies, for example, the Lamé constants of linear elastic materials are represented by random fields [71, 114]. These representations are not recommended since they can be inconsistent with the material microstructure [136]. The incorporation of the microstructure features in the analysis does not pose any conceptual difficulty. It requires the use of adequate mechanical models and the calibration of the random coefficients of these models to the available information on the material microstructure. We have used such a representation in Section 8.6.2.2 to describe properties of metallic polycrystals.

Example 9.26: Let U be the solution of $A\,U = Y$, where A and Y are independent real-valued random variables with finite mean. This stochastic problem is a special case of Eq. 9.12 defining a static scalar problem. If $(A - E[A])/E[A] < 1$ a.s., then (Section 8.3.1.4)

$$U = \left[1 + \sum_{r=1}^{\infty}\left(-\frac{A - E[A]}{E[A]}\right)^r\right]\frac{Y}{E[A]}.$$

Approximations of the above solution obtained by retaining a finite number of terms from the above series can be used to calculate properties of U. For example, the approximate mean of U based on the first three terms of the above series solution is $(1+\sigma_A^2/\mu_A^2)\,(\mu_Y/\mu_A)$, where μ_A and σ_A^2 denote the mean and variance of A and μ_Y is the mean of Y. ◇

Example 9.27: Let X be the solution of $dX(t) = -(a + \varepsilon\,R)\,X(t)\,dt + \sigma\,dB(t)$, $t \geq 0$, with zero initial conditions, where $a > 0$ and σ are some constants, R is a random variable with mean zero and standard deviation $\sigma_r > 0$, ε is a small parameter, and B denotes a Brownian motion process that is independent of R. The above equation is a special case of Eq. 9.12.

Let $\gamma_{ij}(t \mid R) = E[(X_i(t) - E[X_i(t)])\left(X_j(t) - E[X_j(t)]\right) \mid R]$, $i, j = 0, 1, \ldots$, denote the conditional covariance functions of the processes X_i in the perturbation solution $X(t) = X_0(t) + \varepsilon\,X_1(t) + \cdots$ of the above equation. These functions are

$$\gamma_{00}(t \mid R) = \frac{\sigma^2}{2\,a}\left(1 - e^{-2\,a\,t}\right),$$

$$\gamma_{01}(t \mid R) = -\frac{\sigma^2\,R}{2\,a}\left(\frac{1 - e^{-2\,a\,t}}{2\,a} - t\,e^{-2\,a\,t}\right),$$

$$\gamma_{11}(t \mid R) = \frac{\sigma^2\,R^2}{2\,a}\left(\frac{1 - e^{-2\,a\,t}}{2\,a^2} - \frac{t}{a}\,e^{-2\,a\,t} - t^2\,e^{-2\,a\,t}\right),$$

and approach $\sigma^2/(2\,a)$, $-\sigma^2\,R/(4\,a^2)$, and $\sigma^2\,R^2/(4\,a^3)$ as $t \to \infty$. The stationary variance of X based on the first order perturbation $X_0(t) + \varepsilon\,X_1(t)$ is $\mathrm{Var}[X(t)] = (\sigma/(2\,a))\left(1 + \varepsilon^2\,\sigma_r^2/(2\,a^3)\right)$. ◇

Proof: The $\mathbb{R}^2$-valued process (X_0, X_1) is the solution of

$$d\begin{bmatrix} X_0(t) \\ X_1(t) \end{bmatrix} = \begin{bmatrix} -a & 0 \\ -R & -a \end{bmatrix}\begin{bmatrix} X_0(t) \\ X_1(t) \end{bmatrix} dt + \begin{bmatrix} \sigma \\ 0 \end{bmatrix} dB(t)$$

so that $\dot{\gamma}_{00}(t \mid R) = -2\,a\,\gamma_{00}(t \mid R) + \sigma^2$, $\dot{\gamma}_{01}(t \mid R) = -2\,a\,\gamma_{01}(t \mid R) - R\,\gamma_{00}(t \mid R)$, and $\dot{\gamma}_{11}(t \mid R) = -2\,a\,\gamma_{11}(t \mid R) - 2\,R\,\gamma_{01}(t \mid R)$ (Section 7.2.1.1). The solutions of these equations with zero initial conditions give the above expressions for γ_{ij}. Conditional on R and using the representation $X(t) \simeq X_0(t) + \varepsilon\,X_1(t)$, the approximate second moment of X is $\gamma_{00}(t \mid R) + 2\,\varepsilon\,\gamma_{01}(t \mid R) + \varepsilon^2\,\gamma_{11}(t \mid R)$. The condition on R can be eliminated by averaging with respect to this random variable. The approximate variance of X based on the first order perturbation does not include all terms of order ε^2.

A similar result can be obtained by the first order Taylor method. The processes $\bar{X}$ and V in the first order Taylor approximation $X(t) \simeq \bar{X}(t) + V(t)\, R^*$ of X satisfy the differential equations $d\bar{X}(t) = -a\,\bar{X}(t)\,dt + \sigma\, dB(t)$ and $dV(t) = -a\,V(t)\,dt - \bar{X}(t)\,dt$, where $R^* = \varepsilon\, R$. ■

9.3.4.2 Polynomial chaos

Let $(\Omega, \mathcal{F}, P)$ be a probability space and let G_i, $i = 1, 2, \ldots$, be independent $N(0, 1)$ random variables defined on this space. Denote by $\hat{\mathcal{S}}_p$ the collection of all polynomials of G_i with degree smaller than an integer $p \geq 0$. The family of polynomials in $\hat{\mathcal{S}}_p$ that are orthogonal to every polynomial in $\hat{\mathcal{S}}_l$, $l < p$, is denoted by $\mathcal{S}_p$ and is referred to as the p th **polynomial chaos**. Two polynomials are said to be orthogonal if the expectation of their product is zero. We set $\mathcal{S}_0 = \hat{\mathcal{S}}_0 = \{1\}$.

The polynomials in $\mathcal{S}_p$ can be obtained simply. For example, let $\zeta_0 = 1$, $\zeta_1(G_i) = G_i$, and $\zeta_2(G_k, G_l) = a_0 + a_1\, G_k + a_2\, G_l + a_{12}\, G_k\, G_l$ be members of $\hat{\mathcal{S}}_0$, $\hat{\mathcal{S}}_1$, and $\hat{\mathcal{S}}_2$, respectively. Then $\zeta_1(G_i)$ is also in $\mathcal{S}_1$ because $E[\zeta_1(G_i)\,\zeta_0] = E[G_i] = 0$. The orthogonality conditions,

$$E[\zeta_2(G_k, G_l)\,\zeta_0] = 0 \quad \text{and} \quad E[\zeta_2(G_k, G_l)\,\zeta_1(G_i)] = 0,$$

imply $a_0 + a_{12}\,\delta_{kl} = 0$ and $a_1\,\delta_{ki} + a_2\,\delta_{li} = 0$ so that $\zeta_2(G_k, G_l) = G_k\, G_l - \delta_{kl}$ is in $\mathcal{S}_2$. Similar orthogonality conditions show that the polynomials in $\mathcal{S}_3$ have the form $\zeta_3(G_k, G_l, G_m) = G_k\, G_l\, G_m - G_k\,\delta_{lm} - G_l\,\delta_{km} - G_m\,\delta_{kl}$, where G_k, G_l, and G_m are independent copies of $N(0, 1)$. A table giving polynomial chaos of various orders can be found in ([71], Section 2.4).

If X is a random variable with finite variance defined on $(\Omega, \mathcal{F}, P)$, the series representation,

$$X = \sum_{p \geq 0} \sum_{n_1 + \cdots + n_r = p} \sum_{\lambda_1, \ldots, \lambda_r} a_{\lambda_1, \ldots, \lambda_r}^{n_1, \ldots, n_r}\, \zeta_p(G_{\lambda_1}, \ldots, G_{\lambda_r}), \tag{9.14}$$

of X is convergent in mean square, where $a_{\lambda_1, \ldots, \lambda_r}^{n_1, \ldots, n_r}$ are some real constants ([109], Chapter 6, [71], Section 2.4).

Note: The convergence of the series in Eq. 9.14 indicates that we can approximate a square integrable random variable to any degree of accuracy by polynomials of Gaussian variables. The coefficients $a_{\lambda_1, \ldots, \lambda_r}^{n_1, \ldots, n_r}$ in the above expression of X can be determined simply since the Hermite polynomials ζ_p are orthogonal ([71], Section 2.4).

Approximations $\tilde{X}$ of X defined by finite sums including the first terms in Eq. 9.14, rather than the series representation in this equation, have been employed to solve stochastic problems [71]. The statement in Eq. 9.14 has only been used in these applications to select the functional form for $\tilde{X}$. ▲

Example 9.28: Let G be an $N(0, 1)$ random variable and consider its memoryless transformations $X_1 = \exp(G)$ and $X_2 = |G|$. These random variables can be approximated by

$$\tilde{X}_1 = \sum_{p=0}^{m} a_{p,1}\,\zeta_p(G) \quad \text{and } \tilde{X}_2 = \sum_{p=0}^{m} a_{p,2}\,\zeta_p(G),$$

respectively, where $\zeta_p(G)$ denote Hermite or chaos polynomials of $G \sim N(0, 1)$ and $a_{p,k}$, $k = 1, 2$, are some coefficients. For example, $a_{p,0} = e^{1/2}$ and $a_{p,1} = e^{1/2}/p!$, $p = 1, \ldots, m$, for $\tilde{X}_1$.

Figure 9.10 shows the exact and approximate mappings, $G \mapsto X_k$ and $G \mapsto \tilde{X}_k$, for several values of m. These mappings suggest that the tails of the distributions of the random variables $\tilde{X}_2$ and X_2 may differ significantly. Figure 9.11 showing ratios of approximate to exact variances and kurtosis coefficients confirms this observation [63]. $\diamond$

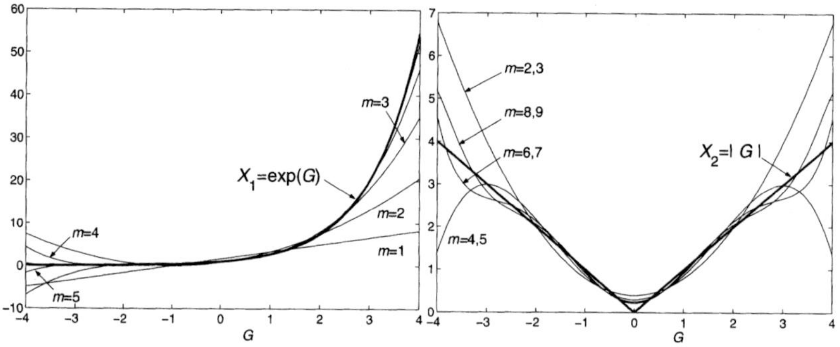

Figure 9.10. Exact and approximate mappings $G \mapsto X_k$ and $G \mapsto \tilde{X}_k$ (adapted from [63])

Proof: Consider an approximation $\tilde{X} = \sum_{p=0}^{m} a_p\,\zeta_p(G)$ of an arbitrary square integrable random variable X suggested by Eq. 9.14. The coefficients a_p of this approximation can be determined from the conditions

$$E[X\,\zeta_q(G)] \simeq E[\tilde{X}\,\zeta_q(G)] = a_q\,E[\zeta_q(G)^2], \quad q = 0, 1, \ldots, m,$$

where the last equality holds since $E[\zeta_p(G)\,\zeta_q(G)] = 0$ for $p \neq q$. We have $a_q \simeq E[X\,\zeta_q(G)]/E[\zeta_q(G)^2]$, $q = 0, 1, \ldots, m$. These equations can be used to calculate the coefficients $a_{p,k}$ in the above approximate representations for X_k, $k = 1, 2$. ■

Example 9.29: Let $X(t) = F^{-1} \circ \Phi(G(t)) = g(G(t))$, $t \in [0, \tau]$, $\tau > 0$, be a translation process, where F is a distribution and G denotes a stationary Gaussian process with mean zero and covariance function $\rho(\tau) = E[G(t + \tau)\,G(t)]$

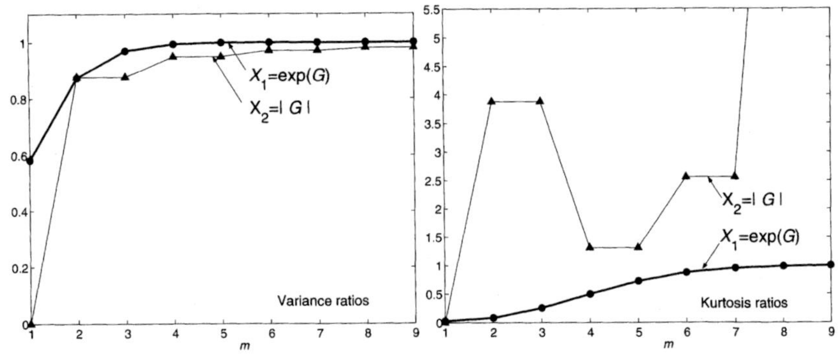

Figure 9.11. Ratios of approximate to exact variances and kurtosis coefficients (adapted from [63])

(Section 3.6.6). The process X can be approximated by

$$\tilde{Y}(t) = \sum_{\nu=0}^{m^*} b_\nu(t)\, \zeta_\nu(G_{i_1}, \ldots, G_{i_\nu}),$$

where $b_\nu(\cdot)$ are some deterministic functions of time depending on the marginal distribution and the correlation function of X, and ζ_ν are Hermite polynomials. $\diamond$

Proof: We can approximate $X(t)$ at a fixed time $t \geq 0$ by $\tilde{X}(t) = \sum_{p=0}^{m} a_p\, \zeta_p(G(t))$ since $X(t)$ is a random variable (Example 9.28). Since X is a stationary process, the coefficients a_p in the representation of $\tilde{X}$ do not depend on time and can be calculated from $a_p \simeq E[X(t)\, \zeta_q(G(t))]/E[\zeta_p(G(t))^2]$. The approximation $\tilde{X}$ represents a nonlinear memoryless transformation of a Gaussian process and is not very useful for calculations. For example, it is difficult to find even the marginal distribution of $\tilde{X}$.

Consider the approximation $\tilde{G}(t) = \sum_{k=1}^{n} G_k\, \varphi_k(t)$ of $G(t)$, where $G_1, \ldots, G_m$ denote independent $N(0, 1)$ variables and φ_k are specified deterministic functions. For example, the spectral representation (Sections 3.9.4.1 and 5.3.1.1) and the Karhunen-Loéve expansion (Section 3.9.4.4) can be used to define $\tilde{G}$. The approximate representation $\tilde{X}$ with $\tilde{G}$ in place of G becomes

$$\tilde{Y}(t) = \sum_{p=0}^{m} a_p\, \zeta_p(\tilde{G}(t)) = \sum_{p=0}^{m} a_p\, \zeta_p\left(\sum_{k=1}^{n} G_k\, \varphi_k(t)\right) = \sum_{\nu=0}^{m^*} b_\nu(t)\, \zeta_\nu(G_{i_1}, \ldots, G_{i_\nu}),$$

where $\zeta_\nu(G_{i_1}, \ldots, G_{i_\nu})$ are Hermite polynomials and $b_\nu(\cdot)$ are functions of time that can be obtained from the above equality [71].

In many applications X is specified partially by its marginal distribution F and correlation function $r(\tau) = E[X(t+\tau)\, X(t)]$. The approximations $\tilde{X}$ and $\tilde{Y}$ of X need to be calibrated such that they match as closely as possible these properties of X. The accuracy of the marginal distribution $\tilde{F}$ of $\tilde{X}$ depends on the number of terms considered in its representation. The marginal distribution of $\tilde{Y}$ is likely to be less accurate than $\tilde{F}$ because its definition is based on the approximation $\tilde{G}$ of G. The correlation function of $\tilde{X}$ is defined by the memoryless mapping $G(t) \mapsto \tilde{X}(t)$ and the correlation function $\rho(\cdot)$ of G. Since $\tilde{X}$ is a translation process, there is no guarantee that there is a correlation function $\rho(\cdot)$ such that $E[\tilde{X}(t+\tau)\,\tilde{X}(t)]$ is a satisfactory approximation of $r(\tau)$ (Section 3.6.6 in this book, [79], Section 3.1.1). Similar comments apply to the approximation $\tilde{Y}$. ■

Example 9.30: Consider a beam with length $l > 0$ and random stiffness. The beam is fixed and free at its left and right ends, respectively, and is subjected to a unit transverse load at its free end. Suppose that the beam stiffness can be modeled by a homogeneous Gaussian random field $Z(x)$, $x \in [0, l]$, with mean $\mu_z > 0$ and covariance function c_z. Let $Z_m(x) = \mu_z + \sum_{q=1}^{m} G_q\, h_q(x)$, $x \in [0, l]$, be an approximate parametric representation of Z, where G_q are independent $N(0, 1)$ random variables and h_q are specified deterministic functions. The above approximation of Z can be obtained, for example, by the spectral representation method (Sections 3.9.4.1 and 5.3.1.1 in this book, [79], Section 4.3.2) or the Karhunen-Loéve series (Section 3.9.4.4 in this book, [71], Section 2.3).

Let $0 = x_1 < x_2 < \cdots < x_{n+1} = l$ be nodes in $[0, l]$ defining the finite elements $D_k = (x_k, x_{k+1})$, $k = 1, \ldots, n$. The displacement vector $\boldsymbol{U}_k$ and the interpolation matrix $\boldsymbol{\alpha}_k(x)$ for the beam element k are

$$\boldsymbol{U}_k = \left[U(x_k), U'(x_k), U(x_{k+1}), U'(x_{k+1})\right]^T,$$
$$\boldsymbol{\alpha}_k(x) = \left[1 - 3\,\xi^2 + 2\,\xi^3,\ l_k\,(\xi - 2\,\xi^2 + \xi^3),\ 3\,\xi^2 - 2\,\xi^3,\ l_k\,(-\xi^2 + \xi^3)\right],$$

where $U'(x) = dU(x)/dx$, $\xi = (x - x_k)/l_k$, $l_k = (x_{k+1} - x_k)$. The corresponding displacement field in this element is $U^{(k)}(x) = \boldsymbol{\alpha}_k(x)\,\boldsymbol{U}_k$.

Figure 9.12 shows estimates of the density f corresponding to the exact tip displacement $U(l)$ and approximations of $U(l)$ given by polynomial chaos of order 1, 2, and 3 based on two terms in the Karhunen-Loeve representation Z_m of Z. The estimates are for $\mu_z = 1$ and $c_z(x_1, x_2) = (0.3)^2 \exp(-|x_1 - x_2|)$ and are based on 10,000 independent samples of the tip displacement. The differences between the Monte Carlo estimates of the density of $U(l)$ based on the exact solution and polynomial chaos solutions are relatively small and seem to decrease with the order of the polynomial chaos expansions. ◇

Note: Let $\boldsymbol{\lambda}_k(x) = d^2\boldsymbol{\alpha}_k(x)/dx^2$. The strain energy in the finite element k is

$$\frac{1}{2}\int_{x_k}^{x_{k+1}} Z_m(x) \left[\frac{d^2 U^{(k)}(x)}{dx^2}\right]^2 dx = \frac{1}{2}\,\boldsymbol{U}_k^T \left(\int_{x_k}^{x_{k+1}} \boldsymbol{\lambda}_k(x)^T\, \boldsymbol{\lambda}_k(x)\, Z_m(x)\, dx\right) \boldsymbol{U}_k$$

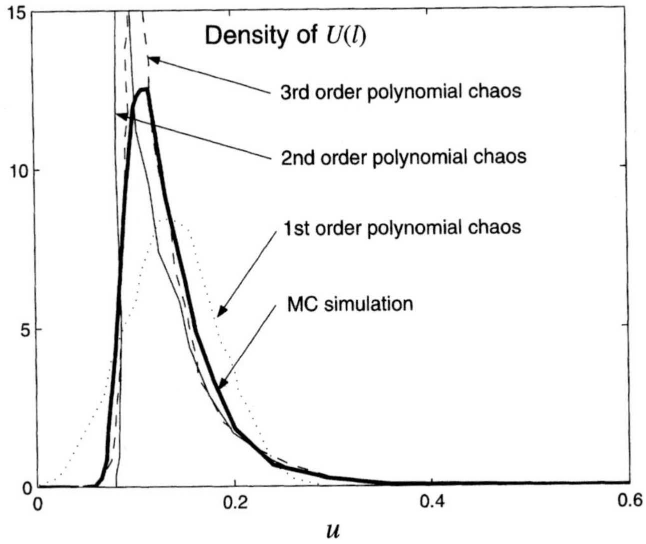

Figure 9.12. Estimates of the density of $U(l)$ by Monte Carlo simulation using the exact solution and polynomial chaos solutions for $\mu_z = 1$ and $c_z(x_1, x_2) = (0.3)^2 \exp(-|x_1 - x_2|)$ (adapted from [71], Fig. 5.36, p. 147)

so that the global stiffness becomes (Eq. 9.13)

$$\begin{aligned}
\boldsymbol{A}^* &= \sum_{k=1}^{n} \boldsymbol{j}_k^T \left(\int_{x_k}^{x_{k+1}} \boldsymbol{\lambda}_k^T(x)\, Z_m(x)\, \boldsymbol{\lambda}_k(x)\, dx \right) \boldsymbol{j}_k \\
&= \mu \sum_{k=1}^{n} \boldsymbol{j}_k^T \left(\int_{x_k}^{x_{k+1}} \boldsymbol{\lambda}_k^T(x)\, \boldsymbol{\lambda}_k(x)\, dx \right) \boldsymbol{j}_k \\
&+ \sum_{q=1}^{m} G_q \sum_{k=1}^{n} \boldsymbol{j}_k^T \left(\int_{x_k}^{x_{k+1}} \boldsymbol{\lambda}_k^T(x)\, h_q(x)\, \boldsymbol{\lambda}_k(x)\, dx \right) \boldsymbol{j}_k = \boldsymbol{a}_0 + \sum_{q=1}^{m} G_q\, \boldsymbol{a}_q.
\end{aligned}$$

The equilibrium condition in Eq. 9.12 gives $\left[\boldsymbol{a}_0 + \sum_{q=1}^{m} G_q\, \boldsymbol{a}_q\right] \boldsymbol{U} = \boldsymbol{y}^*$, where $\boldsymbol{y}^*$ relates to the loading. The solution of this equation by polynomial chaos involves three steps. First, $\boldsymbol{U}$ needs to be represented by a polynomial chaos of the random variables $(G_1, \ldots, G_m)$ used in the representation of Z. Second, the coefficients $a_{\lambda_1,\ldots,\lambda_r}^{n_1,\ldots,n_r}$ of the polynomial chaos representation for $\boldsymbol{U}$ (Eq. 9.14) have to be calculated from the above equilibrium condition and the orthogonality property of the polynomial chaos. Third, the resulting expression of $\boldsymbol{U}$ can be used to approximate probabilistic properties of $\boldsymbol{U}$.

The Gaussian model used for Z is not physically acceptable because the beam stiffness cannot take negative values. It has been used in [71] to simplify calculations. ▲

9.4 Physics

We consider large systems whose macroscopic properties can be described by global variables. The Boltzmann equation, the Ising model, and noise induced transitions are discussed in Sections 9.4.1, 9.4.2, and 9.4.3, respectively. The Boltzmann equation examines systems consisting of a large number of identical particles, and gives the evolution in time of the fraction of particles with some attributes, the Ising model provides a global characterization of magnetism for ferromagnetic materials, and diffusion processes can model the predator-prey relationship in a random environment. In some cases, the qualitative behavior of these systems changes abruptly and this change is referred to as phase transition.

9.4.1 Boltzmann transport equation

Let $\boldsymbol{x} = (x_1, x_2, x_3)$ and $\boldsymbol{p} = (p_1, p_2, p_3)$ denote the position and momentum vectors of a particle with mass m in $\mathbb{R}^3$, where $\boldsymbol{p} = m\,\boldsymbol{v}$ and $\boldsymbol{v} = (v_1, v_2, v_3)$ denotes the particle velocity. The motion of the particle is completely characterized by the evolution of the vector $(\boldsymbol{x}, \boldsymbol{p}) \in \mathbb{R}^6$ in time.

Consider a system consisting of n identical particles. We can study the evolution of this system by tracking the motion of every particle or by finding the evolution in time of a macroscopic system property, for example, the density or temperature. If n is very large, the first approach is not feasible. On the other hand, the second approach applies and delivers macroscopic attributes directly.

Denote by $g(\boldsymbol{x}, \boldsymbol{p}; t)\, d\boldsymbol{x}\, d\boldsymbol{p}$ the number of particles in the infinitesimal subset $(\boldsymbol{x}, \boldsymbol{x} + d\boldsymbol{x}) \times (\boldsymbol{p}, \boldsymbol{p} + d\boldsymbol{p})$ of the space $\mathbb{R}^6$ with coordinates $(\boldsymbol{x}, \boldsymbol{p})$ at a time $t \geq 0$.

If g has continuous partial derivatives, it satisfies the **Boltzmann transport equation**

$$\frac{\partial g}{\partial t} + \frac{1}{m} \sum_{i=1}^{3} p_i \frac{\partial g}{\partial x_i} + \sum_{i=1}^{3} h_i \frac{\partial g}{\partial p_i} = \left(\frac{\partial g}{\partial t} \right)_{\text{coll}} \tag{9.15}$$

where $\boldsymbol{a} = (a_1, a_2, a_3) = d\boldsymbol{v}/dt$, $\boldsymbol{h} = m\,\boldsymbol{a}$, and $(\partial g/\partial t)_{\text{coll}}$ quantifies the effect of particle collisions.

Proof: If there are no collisions, the number of particles in an elementary volume along a flow line is preserved so that $g(\boldsymbol{x}, \boldsymbol{p}; t) = g(\boldsymbol{x} + d\boldsymbol{x}, \boldsymbol{p} + d\boldsymbol{p}; t + dt)$. This conservation condition gives Eq. 9.15 by expanding its right side in Taylor series, taking the limit of the resulting expression as $dt \to 0$, and using the relationships $d\boldsymbol{x} = \boldsymbol{v}\, dt$ and $d\boldsymbol{p} = \boldsymbol{h}\, dt$ that give the change of the position and momentum vectors in $(t, t + dt)$. The Boltzmann equation with no collisions and the Liouville equation in Eq. 9.4 coincide.

If there are collisions, the term $(\partial g/\partial t)_{\text{coll}}$ needs to be specified. Generally, this term is approximated by $(\partial g/\partial t)_{\text{coll}} = -(g - g_0)/\tau_c$, where g_0 is the distribution g at thermal equilibrium and τ_c denotes a relaxation time ([4], Section 9.1). This approximation gives $(g - g_0)_t = (g - g_0)_{t=0} \exp(-t/\tau_c)$ since $\partial g_0/\partial t = 0$ by the definition of the equilibrium distribution. ■

Example 9.31: The steady-state Boltzmann transport equation is given by Eq. 9.15 with $\partial g/\partial t = 0$. If τ_c is a small parameter and $(\partial g/\partial t)_{\text{coll}} = -(g - g_0)/\tau_c$, the first order approximation of this equation is

$$g \simeq g_0 - \tau_c \left(\frac{1}{m} \sum_{i=1}^{3} p_i \frac{\partial g_0}{\partial x_i} + \sum_{i=1}^{3} h_i \frac{\partial g_0}{\partial p_i} \right)$$

with the notation in Eq. 9.15. ◇

Proof: The perturbation solution $g_0 + \tau_c\, g_1 + \cdots$ introduced in the steady-state Boltzmann transport equation yields

$$\tau_c \left(\frac{1}{m} \sum_{i=1}^{3} p_i \frac{\partial}{\partial x_i} + \sum_{i=1}^{3} h_i \frac{\partial}{\partial p_i} \right) (g_0 + \tau_c\, g_1 + \cdots) = -\left[(g_0 + \tau_c\, g_1 + \cdots) - g_0 \right]$$

so that $g_1 = -(1/m) \sum_{i=1}^{3} p_i\, \partial g_0/\partial x_i - \sum_{i=1}^{3} h_i\, \partial g_0/\partial p_i$. ■

The Boltzmann equation is deterministic so that its presentation in this chapter may appear out of place. We give the Boltzmann equation here because it is closely related to the Liouville equation (Eq. 9.4) and its output is a probability density function if normalized. The Boltzmann equation has numerous application in physics and engineering depending on the particular functional form selected for $(\partial g/\partial t)_{\text{coll}}$ ([4], Chapter 9).

9.4.2 Ising model

The ferromagnetism phenomenon can be observed in some metals at temperatures lower than a characteristic value $\mathcal{T}_c$, called the Curie temperature, and is caused by the polarization of a fraction of atomic spins in one direction. At temperatures exceeding $\mathcal{T}_c$ the atomic spins have random orientation so that no macroscopic magnetic field results.

The Ising model provides a simple and approximate representation of the structure of ferromagnetic materials and of the ferromagnetism phenomenon. The model consists of a lattice in $\mathbb{R}^q$, $q = 1, 2, 3$, that has n sites organized in various geometrical shapes. For example, a two-dimensional square lattice is a square with equally spaced sites in the two coordinates. Let $S_i = \pm 1$ denote the atomic spin at site $i = 1, \ldots, n$. The vector $\boldsymbol{S} = (S_1, \ldots, S_n)$ defines a particular lattice configuration.

The energy of the random field Ising model with n sites and configuration $\boldsymbol{S}$ is

$$\mathcal{E} = -\sum_{\langle ij\rangle=1}^{n} \varepsilon_{ij}\, S_i\, S_j - \sum_{i=1}^{n}(h + Z_i)\, S_i, \tag{9.16}$$

where ε_{ij} denotes the interaction energy and $h + Z_i$ is the energy corresponding to an external magnetic field with a deterministic component h and a random component Z_i.

Note: The classical Ising model is given by Eq. 9.16 with $Z_i = 0$ at all sites.

The symbol $\sum_{\langle ij\rangle=1}^{n}$ indicates that summation is performed over the nearest neighbor sites. For example, the summation has at each site of a two-dimensional square lattice four terms corresponding the right/left and above/below neighboring sites ([4], Chapter 11, [103], Chapter 14). The random field Z can be interpreted as a spatially fluctuating external field superposed on h. ▲

If the interaction energy is isotropic, that is, $\varepsilon_{ij} = \varepsilon$, $\varepsilon > 0$, and the fields h and Z_i are zero, the minimum of the resulting lattice energy is $-\varepsilon\,\gamma\, n/2$, where $\gamma\, n/2$ gives the number of terms in the first summation of Eq. 9.16 and $\gamma = 4$ for a square lattice. This minimum value of $\mathcal{E}$ corresponds to a completely polarized lattice, that is, a lattice with all spins of the same orientation. It is not possible to find analytically the minimum of the lattice energy for the general setting in Eq. 9.16.

We give an alternative form of the lattice energy that can be used to develop approximations for the minimum lattice energy, and offers some interesting interpretations of the ferromagnetism phenomenon.

If $\varepsilon_{ij} = \varepsilon$, the energy $\mathcal{E}$ in Eq. 9.16 is

$$\mathcal{E} = -4\,\varepsilon\, N_{++} + 2\,(\varepsilon\,\gamma - h)\, N_{+} + (h - \varepsilon\,\gamma/2)\, n - \sum_{i=1}^{n} S_i\, Z_i, \tag{9.17}$$

where N_{+} and $N_{-} = n - N_{+}$ denote the number of up ($S_i = +1$) and down ($S_i = -1$) spins in the lattice with a configuration $\boldsymbol{S}$, respectively, and N_{++} is the number of nearest neighbor pairs with up spins.

Note: Denote by N_{++}, N_{+-}, and N_{-} the number of nearest neighbor pairs with spins $(+,+)$, $(+,-)$ and $(-,+)$, and $(-,-)$, respectively. It can be shown that $\gamma\, N_{+} = 2\, N_{++} + N_{+-}$ and $\gamma\, N_{-} = 2\, N_{-} + N_{+-}$ so that $N_{+-} = \gamma\, N_{+} - 2\, N_{++}$ and $N_{-} = (\gamma\, N_{-} - N_{+-})/2 = \gamma\, n/2 + N_{++} - \gamma\, N_{+}$ ([103], Section 14.1). Hence, we have

$$\sum_{\langle i,j\rangle=1}^{n} S_i\, S_j = N_{++} + N_{-} - N_{+-} = 4\, N_{++} - 2\,\gamma\, N_{+} + \gamma\, n/2$$

and $\sum_{i=1}^{n} S_i = N_{+} - N_{-} = 2\, N_{+} - n$ ([103], Section 14.1).

The expression of $\mathcal{E}$ in Eq. 9.17 shows that, if $Z_i = 0$ for all $i = 1, \ldots, n$, the lattice energy does not depend on the details of the configuration $\boldsymbol{S}$ but only on the lattice size n and its macroscopic attributes N_+ and N_{++}. Hence, many lattice configurations can have the same energy. ▲

Example 9.32: Suppose that $\varepsilon_{ij} = \varepsilon > 0$ and $Z_i = 0$ for all $i, j = 1, \ldots, n$, so that $\mathcal{E}$ can be expressed in terms of the macroscopic attributes N_+ and N_{++} of the Ising lattice (Eq. 9.17). Let $L = (N_+ - N_-)/n \in [-1, 1]$ denote the fraction of sites with up spin in excess of sites with down spin, that is, the magnetization per particle.

Numerical algorithms are needed to find the minimum of the lattice energy even for the special case in which $Z_i = 0$ at all sites, as considered here. To simplify the analysis, the Bragg-Williams model assumes $N_{++} = (N_+/n)^2\,(\gamma\, n)/2$, which yields the expression $\mathcal{E} = -n\,\left(\varepsilon\,\gamma\, L^2/2 + h\, L\right)$ for the lattice energy. If $h = 0$, then $\mathcal{E} = -n\,\left(\varepsilon\,\gamma\, L^2/2\right)$ with minimum $-n\,\varepsilon\,\gamma/2$ corresponding to $L = 1$. If $h \neq 0$, the determination of the minimum energy is less simple. We use the fact that the value of L minimizing $\mathcal{E}$ maximizes the partition function

$$\mathcal{Q} = \sum_{L=-1}^{1} \frac{n!}{[n\,(1+L)/2]!\,[n\,(1-L)/2]!} \exp[-\beta\,\mathcal{E}]$$

where $\beta = 1/(k\,\mathcal{T})$, k denotes the Boltzmann constant, and $\mathcal{T}$ is the temperature ([103], Section 14.4). If $n \to \infty$, the partition function is maximized at $L = l^*$, where l^* is the solution of $l^* = \tanh\left((h + \gamma\,\varepsilon\, l^*)/(k\,\mathcal{T})\right)$. If $h = 0$, this equation has the solutions $l^* = 0$ and $l^* = \pm l_0$ for $\mathcal{T} > \mathcal{T}_c$ and $\mathcal{T} < \mathcal{T}_c$, respectively, where $\mathcal{T}_c = \varepsilon\,\gamma/k$. Hence, the lattice is ferromagnetic for $\mathcal{T} < \mathcal{T}_c$ but has no magnetization otherwise. ◇

Note: The Bragg-Williams model and arguments similar to those used to obtain Eq. 9.17 yield the expression of the lattice energy.

The partition function is $\mathcal{Q} = \sum_{S_i} \exp(-\beta\,\mathcal{E})$, where the summation is over all possible values of S_i ([103], Section 14.1). Because L is determined by N_+, the number of configurations $(S_1, \ldots, S_n)$ having the same value of L is equal to the number of ways in which we can extract N_+ out of n objects, that is, $n!/(N_+!\,N_-!)$. The equation $l^* = \tanh\left(h/(k\,\mathcal{T}) + \varepsilon\,\gamma\, l^*/(k\,\mathcal{T})\right)$ giving the value of L that maximizes $\mathcal{Q}$ results by using Sterling's formula to approximate the combinatorial coefficient in the expression of $\mathcal{Q}$ and retaining the dominant term of $\mathcal{Q}$ ([103], Section 14.4).

If $h = 0$, we have $l^* = \tanh(\varepsilon\,\gamma\, l^*/(k\,\mathcal{T}))$ so that $l^* = 0$ for $\varepsilon\,\gamma/(k\,\mathcal{T}) < 1$ or $l^* = 0$ and $l^* = \pm l_0$ for $\varepsilon\,\gamma/(k\,\mathcal{T}) > 1$, where $l_0 \in (0, 1)$. The solution $l^* = 0$ under the condition $\varepsilon\,\gamma/(k\,\mathcal{T}) > 1$ is not valid since it corresponds to a minimum rather than a maximum of $\mathcal{Q}$. The solution $l^* = \pm l_0$ is not unique because if $h = 0$ there is no intrinsic difference between up and down spins. ▲

Example 9.33: Consider a lattice with $n = 20 \times 20 = 400$ sites, $\varepsilon_{ij} = \varepsilon = 1$, $h = 0$, and let Z_i, $i = 1, \ldots, n$, be independent Gaussian variables with mean zero and variance σ^2. Let $\mathbf{Z}(\omega)$ be a sample of $\mathbf{Z} = (Z_1, \ldots, Z_n)$, and denote by $\mathcal{E}_{\min}(\omega)$ the minimum of the energy in Eq. 9.16 corresponding to this noise sample.

Figure 9.13 shows estimates of the mean and the coefficient of variation of $\tilde{\mathcal{E}}_{\min} = \mathcal{E}_{\min}/(n\,\sigma)$ for noise intensities σ in the range [0, 10]. The estimates have been calculated from 1,000 samples of $\mathcal{E}_{\min}$ at each noise level. The figure also shows samples of the lattice configuration $\boldsymbol{S}$ at the minimum energy for several

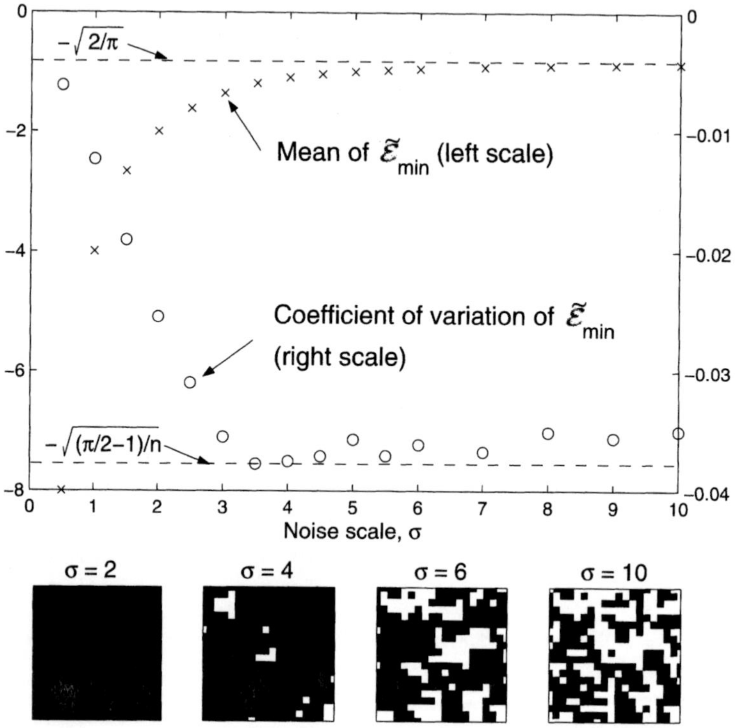

Figure 9.13. Estimates of the mean and standard deviation of $\tilde{\mathcal{E}}_{\min}$

noise levels. For $\sigma = 0$ the minimum energy is $-\gamma\, n/2$ so that its coefficient of variation is zero. The magnitude of the coefficient of variation of $\tilde{\mathcal{E}}_{\min}$ increases with the noise, and approaches an asymptotic value of $-\sqrt{(\pi/2-1)/n}$, which is 0.0378 for $n = 400$. The estimated mean of $\tilde{\mathcal{E}}_{\min}$ converges to $-\sqrt{2/\pi} = -0.0378$ as the noise intensity increases. ◇

Note: An algorithm in Section 5.2.1 has been used to generate independent samples of $\mathbf{Z}$ needed to construct the plots in Fig. 9.13.

If the noise is very large, the atomic spins tend to have the same orientation as the noise so that (1) the contribution of the first term in Eq. 9.16 becomes negligible and (2) the second term is approximately equal in distribution with $-\sum_{i=1}^{n} |Z_i|$. Hence, $\tilde{\mathcal{E}}_{\min}$ can be approximated by the random variable $X = -(1/n) \sum_{i=1}^{n} |G_i|$, where G_i are independent copies of $N(0, 1)$. Because $E[|G_1|] = \sqrt{2/\pi}$ and $E[|G_1|^2] = 1$, the mean and variance of X are $-\sqrt{2/\pi}$ and $(1 - 2/\pi)/n$, respectively, which yields the stated coefficient of variation of $\tilde{\mathcal{E}}_{\min}$ ▲

Example 9.34: The previous example examines the random field Ising model in Eq. 9.16 with $h = 0$ at an arbitrary temperature $\mathcal{T}$. Here we take $\mathcal{T} = 0$ and $h \neq 0$. If the noise Z_i is zero at all sites and $h < 0$ has a large magnitude, then $S_i = -1$ at all sites, that is, all spins point down. As the external field h is increased, it reaches a critical value at which all spins change suddenly their orientation and magnetization jumps from -1 to $+1$ (Fig. 9.14). Hence, the lattice experiences a phase transition since the atomic spins at all its sites change orien-

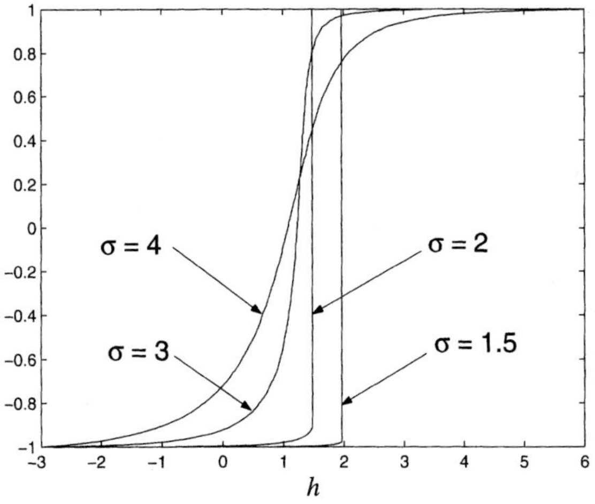

Figure 9.14. Magnetization in the presence of a random external field (adapted from [167])

tation abruptly. Suppose now that the previous experiment is repeated and that the noise in Eq. 9.16 is not zero but consists of independent Gaussian variables Z_i with mean zero and variance σ^2. We can still observe a phase transition in the presence of noise. However, the transition is less abrupt and becomes smoother as the noise intensity increases. The Ising model in this example has been used to study Barkhausen noise in magnetic materials [168]. ◇

Note: For small noise intensities, that is, small values of σ, the couplings between spins dominate and the lattice changes polarization through a large avalanche that sweeps through the entire system as the external field is increased to a particular level. For large values of σ, the coupling between spins is insignificant compared to the local disorder, and the lattice tend to magnetize in many single-spin flips, that is, avalanches contain a single spin. There exists a critical value σ_c of the disorder parameter σ defining a phase transition between these two limit cases. Near the critical point σ_c, avalanches can be described by a power law up to a size depending on the magnitude of $|\sigma - \sigma_c|$ [168]. ▲

9.4.3 Noise induced transitions

Let X be the solution of the stochastic differential equation

$$dX(t) = a(X(t))\,dt + b(X(t))\,dS(t), \quad t \geq 0, \tag{9.18}$$

where S is either a scaled Brownian motion or a compound Poisson process. It is assumed that the functions a, b and the noise S are such that Eq. 9.18 has a unique solution (Sections 4.7.1.1 and 4.7.2). Stochastic problems of the type in Eq. 9.18 have been examined in Section 8.7. We revisit this type of stochastic problems in this section because (1) elementary extensions of Eq. 9.18 obtained, for example by randomizing some of the coefficients of and/or supplying an additive noise to this equation, yield stochastic problems of the type in Eq. 9.1 and (2) the state X in Eq. 9.18 can experience notable phase transitions, encountered in physics, biology, chemistry, and other fields [102].

Our objectives in this section and Section 8.7 differ slightly. In Section 8.7 we have studied the stability of the stationary solution X_s of stochastic differential equations of the type in Eq. 9.18 based on the Lyapunov exponent. A main objective here is to determine whether the density f_s of X_s can exhibit qualitative changes under some conditions. If such changes occur, we say that f_s undergoes **transitions**. If the transitions of f_s are caused by the noise intensity, they are refer to as **noise induced transitions**. Qualitative changes in f_s indicate possible qualitative changes in some macroscopic properties of a large system ([102], Chapter 1).

We illustrate the transition phenomenon by some extensions of the deterministic Verhulst model describing the growth of a biological population. The state $x \in [0, \infty)$ of this model is a measure of the population size, for example, the average population per unit area, and satisfies the differential equation

$$\dot{x}(t) = \rho\, x(t) - x(t)^2, \quad t > 0, \tag{9.19}$$

where $\rho\, x(t)$ is the rate of population increase or decrease depending on the sign of $\rho \in \mathbb{R}$ and the term $-x(t)^2$ accounts for the fact that resources are limited. The parameter ρ is positive and negative for a favorable and hostile environment. The solution of the Verhulst model is

$$x(t) = x(0)\, e^{\rho t} \left[1 + x(0)\,(e^{\rho t} - 1)/\rho\right]^{-1}. \tag{9.20}$$

The stable steady-state solution of the model is $x_s(t) = 0$ for $\rho < 0$ and $x_s(t) = \rho$ for $\rho > 0$. The steady-state solution $x_s(t) = 0$ for $\rho < 0$ becomes unstable at $\rho = 0$ and a new steady-state solution $x_s(t) = \rho$ emerges. The system undergoes a phase transition at $\rho = 0$. Hence, x_s can be viewed as a degenerate random variable with the density $f_s(x) = \delta(x)$ for $\rho < 0$ and $f_s(x) = \delta(x - \rho)$ for $\rho > 0$, where δ denotes the Dirac delta function.

The Verhulst model is a special case of Eq. 9.1 in which $\mathcal{D}$ has deterministic coefficients and $\mathcal{Y}$ is zero. The following two examples consider extensions of the Verhulst model that account for the random variations of the environment. It is shown that the state of these models experience one or more noise induced transitions.

Example 9.35: Suppose that the rate ρ in the Verhulst model is replaced by $\rho + \sigma\, dB(t)/dt$, where B denotes a Brownian motion and σ is a constant giving the scale of the environmental noise. The state X of this version of the Verhulst model is a diffusion process defined by the stochastic differential equation

$$dX(t) = (\rho\, X(t) - X(t)^2)\, dt + \sigma\, X(t)\, dB(t), \quad t > 0.$$

Let X_s be the stationary solution of the above equation. If $\rho > \sigma^2/2$, the density of X_s exists and is $f_s(x) = c\, x^{2(\rho/\sigma^2 - 1)} \exp\left(-2x/\sigma^2\right)$, $x > 0$, where $c > 0$ is a constant. If $\rho < \sigma^2/2$, then $f_s(x) = \delta(x)$. The mean and variance of the non-trivial stationary solution X_s, that is, X_s for $\rho > \sigma^2/2$, are $E[X_s(t)] = \rho$ and $\mathrm{Var}[X_s(t)] = \rho\,\sigma^2/2$, so that the expectation of $X_s(t)$ for $\rho > \sigma^2/2$ coincides with the steady-state solution x_s of the deterministic Verhulst model. Also, $X_s = x_s = 0$ for $\rho < \sigma^2/2$.

Figure 9.15 shows the stationary density f_s for $\sigma = 1$ and ρ in the ranges $(-\infty, \sigma^2/2)$, $(\sigma^2/2, \sigma^2)$, and (σ^2, ∞). The stationary density f_s has transitions at $\rho = \sigma^2/2$ and $\rho = \sigma^2$, in contrast to the Verhulst model that has a single

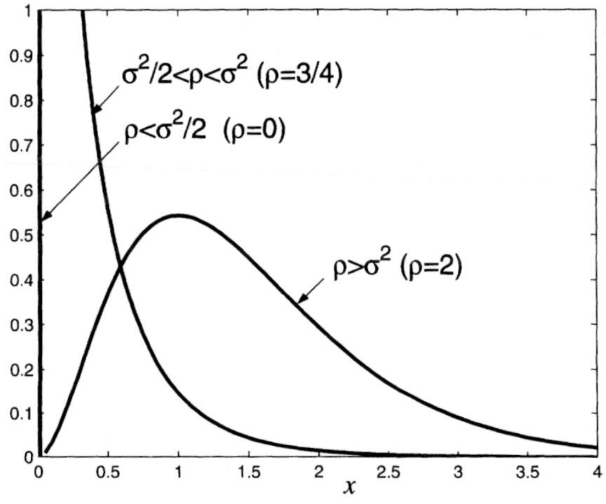

Figure 9.15. The density f_s of the stationary state X_s for $\sigma = 1$, $\rho = 0$, $\rho = 3/4$, and $\rho = 2$

transition at $\rho = 0$. The transitions in Fig. 9.15 represent qualitative changes in the functional form of f_s and are caused by random fluctuations in the environment since $E[\rho\, dt + \sigma\, dB(t)]/dt = \rho$ has the same value as in the deterministic Verhulst model ([102], Section 6.4).

It can be shown that the above differential equation has the solution

$$X(t) = \frac{X(0)\, \exp\left[(\rho - \sigma^2/2)\, t + \sigma\, B(t)\right]}{1 + X(0) \int_0^t \exp\left[(\rho - \sigma^2/2)\, s + \sigma\, B(s)\right]\, ds},$$

where $X(0)$ denotes the initial state ([115], p. 125). Figure 9.16 shows samples of

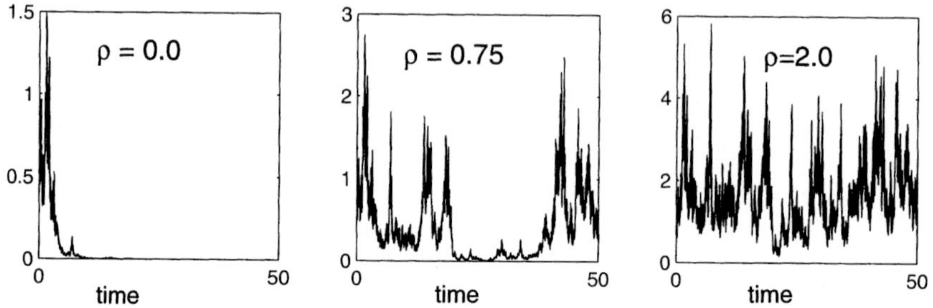

Figure 9.16. Samples of X for $\sigma = 1$ and $\rho = 0$, 3/4, and 2

X for $\sigma = 1$ and $\rho = 0$, 3/4, and 2, that is, the values of ρ in Fig. 9.15. The sample behavior is consistent with the properties of the stationary density in Fig. 9.15. ◇

Proof: The density $f(x; t \mid x_0; 0)$ of $X(t) \mid (X(0) = x_0)$ satisfies the Fokker-Planck equation (Section 7.3.1.3)

$$\frac{\partial f}{\partial t} = -\frac{\partial}{\partial x}\left((\rho\, x - x^2)\, f\right) + \frac{\sigma^2}{2}\frac{\partial^2}{\partial x^2}\left(x^2\, f\right).$$

The stationary density $f_s(x) = \lim_{t\to\infty} f(x; t \mid x_0; 0)$ is the solution of the above equation with $\partial f_s/\partial t = 0$, that is,

$$0 = -\frac{d}{dx}\left((\rho\, x - x^2)\, f_s\right) + \frac{\sigma^2}{2}\frac{d^2}{dx^2}\left(x^2\, f_s\right),$$

so that $(\sigma^2/2)\, d(x^2\, f_s)/dx = (\rho\, x - x^2)\, f_s$ or

$$\frac{df_s(x)}{f_s(x)} = \frac{2}{\sigma^2}\left(\frac{\rho - \sigma^2}{x} - 1\right)$$

since f_s and its first derivative approach zero as $|x| \to \infty$. The above equation gives $\ln(f_s(x)) = 2\,(\rho - \sigma^2)\, \ln(x)/\sigma^2 - 2\,x/\sigma^2 + c_1$ by integration, where c_1 is a constant. The stated expression of f_s results by elementary manipulations.

If $\rho > \sigma^2/2$, the integral $\int_0^\infty x^{2\,(\rho/\sigma^2 - 1)}\, e^{-2\,x/\sigma^2}\, dx$ is bounded so that f_s exists and has the above expression. If $\rho < \sigma^2/2$, f_s is a delta function at $x = 0$, that is,

$f_s(x) = \delta(x)$ ([102], Section 6.4), a result consistent with the expression of $X(t)$ showing that, if $\rho < \sigma^2/2$, then $X(t) \to 0$ a.s. as $t \to \infty$ since $\rho - \sigma^2/2 + \sigma\, B(t)/t$ converges a.s. to $\rho - \sigma^2/2$ as $t \to \infty$ (Section 8.7). Also, note that the first two moments of X_s have the same expression for $\rho > \sigma^2/2$. Hence, the qualitative change in f_s at $\rho = \sigma^2$ cannot be captured by the second moment characterization of X_s.

The solution $x(t) = x(0)\, e^{\rho t} \left[1 + x(0)\,(e^{\rho t} - 1)/\rho\right]^{-1}$ of Eq. 9.19 has the asymptotic value $x_s = \lim_{t\to\infty} x(t) = 0$ and $x_s = \rho$ for $\rho < 0$ and $\rho > 0$, respectively. Hence, $\rho = 0$ constitutes a bifurcation point between two stable solutions. The steady-state solution x_s coincides with the mean of the stationary solution X_s for all values of ρ.

The Lyapunov exponent λ_{LE} obtained by the method in Section 8.7 is $\lambda_{\mathrm{LE}} = \rho - \sigma^2/2$ and $\lambda_{\mathrm{LE}} = (1 - m)\,(\rho - \sigma^2/2)$, $m = 2$, for the trivial and non-trivial stationary solutions, respectively. Since the stationary solution is stable a.s. if $\lambda_{\mathrm{LE}} < 0$, the trivial and non-trivial stationary solutions are stable a.s. for $\rho < \sigma^2/2$ and $\rho > \sigma^2/2$, respectively, in agreement with the results in this section showing that $f_s(x) = \delta(x)$ for $\rho < \sigma^2/2$ and $X_s > 0$ a.s. for $\rho > \sigma^2/2$. The approach in Section 8.7 delivers only the value $\rho = \sigma^2/2$ of ρ at which the stationary solution X_s bifurcates but not the transition of f_s at $\rho = \sigma^2$. However, the Lyapunov exponent method can be extended to detect the transition of f_s at $\rho = \sigma^2$ [10].

We have modeled the uncertainty in the environment by a Gaussian white noise for simplicity. A broadband noise with finite variance would have been more realistic. If such a noise were used to model the environment, the differential equation for X would have been a Stratonovich rather than an Itô equation. The relationship between the Itô and Stratonovich differential equations in Section 4.7.1.2 can be used to recast a Stratonovich equation into an Itô equation. ■

Example 9.36: Let X be defined by the stochastic differential equation in the previous example in which the Gaussian white noise $\sigma\, dB(t)/dt$ is replaced by a Poisson white noise $dC(t)/dt$, where $C(t) = \sum_{k=1}^{N(t)} Y_k$ is a compound Poisson process, N denotes a Poisson process with intensity $\lambda > 0$, and Y_k, $k = 1, 2, \ldots$, are independent copies of a random variable $Y \in L_2$ with mean zero and distribution F. Denote by X_s the stationary solution of the differential equation

$$dX(t) = (\rho\, X(t-) - X(t-)^2)\, dt + X(t-)\, dC(t), \quad t > 0.$$

It is not possible to find analytically the stationary density f_s of the stationary solution X_s. To determine whether f_s exhibits transitions, we consider the characteristic function $\varphi(u; t) = E[\exp(\sqrt{-1}\, u\, X(t))]$. This function is the solution of

$$\begin{aligned} \frac{\partial \varphi(u;t)}{\partial t} &= \rho\, u\, \frac{\partial \varphi(u;t)}{\partial u} + \sqrt{-1}\, \frac{\partial^2 \varphi(u;t)}{\partial u^2} \\ &\quad + \lambda \left[\int_{\mathbb{R}} \varphi(u\,(1+y); t)\, dF(y) - \varphi(u;t) \right]. \end{aligned}$$

The stationary characteristic function $\varphi_s(u) = \lim_{t\to\infty} \varphi(u; t)$ satisfies the above equation in which we set $\partial\varphi/\partial t = 0$. The density f_s of X_s is the Fourier transform of φ_s.

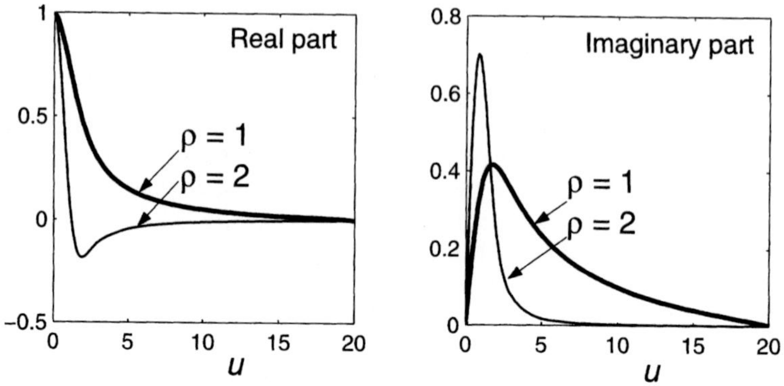

Figure 9.17. Real and imaginary parts of the characteristic function φ_s for $\lambda = 12$, $Y_1 \sim U(-a, a)$, $a = 0.5$, and $\rho = 1, 2$

Figure 9.17 shows the real and imaginary parts of the characteristic function φ_s for $\lambda = 12$, $Y_1 \sim U(-a, a)$, $a = 0.5$, and two values of ρ. The significant difference between the two characteristic functions suggest that f_s may exhibit a transition. The corresponding plots of f_s in Fig. 9.18 show that there is a qualita-

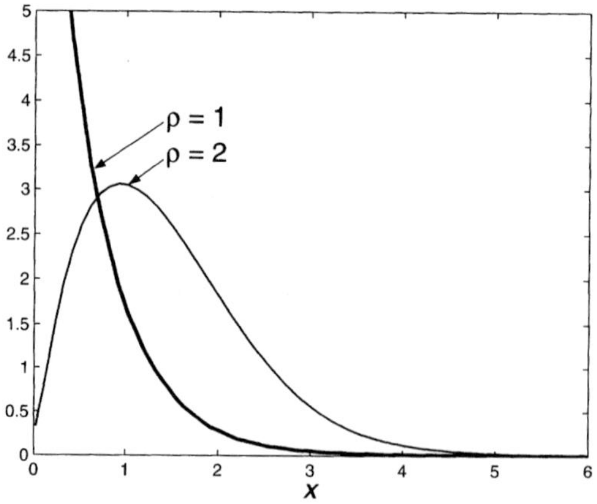

Figure 9.18. The density f_s for $\lambda = 12$, $Y_1 \sim U(-a, a)$, $a = 0.5$, and $\rho = 1, 2$

tive difference between the stationary solutions for $\rho = 1$ and $\rho = 2$. The most likely values of X_s are in a small vicinity of $x = 0$ and away from $x = 0$ for $\rho = 1$ and $\rho = 2$, respectively. We also note that $\text{Var}[C(t)] = \lambda\, t\, E[Y_1^2] = t$

for $\lambda = 12$ and $Y_1 \sim U(-a, a)$ with $a = 0.5$ so that the processes $C(t)$ in this example and $\sigma\, B(t)$ with $\sigma = 1$ in the previous example have the same second moment properties. $\diamond$

Proof: The Itô formula applied to the function $e^{\sqrt{-1}\,u\,X(t)}$ gives

$$e^{\sqrt{-1}\,u\,X(t)} - e^{\sqrt{-1}\,u\,X(0)} = \int_{0+}^{t} \sqrt{-1}\,u\,e^{\sqrt{-1}\,u\,X(s-)}\,dX(s) + \sum_{0<s\le t}\left[e^{\sqrt{-1}\,u\,X(s)} - e^{\sqrt{-1}\,u\,X(s-)} - \sqrt{-1}\,u\,e^{\sqrt{-1}\,u\,X(s-)}\,\Delta X(s)\right],$$

where $\Delta X(s) = X(s) - X(s-)$, or (Section 4.6.1)

$$e^{\sqrt{-1}\,u\,X(t)} - e^{\sqrt{-1}\,u\,X(0)} = \int_{0+}^{t} \sqrt{-1}\,u\,e^{\sqrt{-1}\,u\,X(s-)}\left(\rho\,X(s-) - X(s-)^2\right)ds + \sum_{0<s\le t}\left[e^{\sqrt{-1}\,u\,X(s)} - e^{\sqrt{-1}\,u\,X(s-)}\right].$$

The expectation of the above equation can be performed term by term. The left side gives $\varphi(u;t) - \varphi(u;0)$. The calculation of the expectation of the integral on the right side does not pose difficulties and is not discussed. To calculate the expectation of the summation in the above equation, we note that the term

$$e^{\sqrt{-1}\,u\,X(s)} - e^{\sqrt{-1}\,u\,X(s-)} = e^{\sqrt{-1}\,u\,X(s-)(1+\Delta C(s))} - e^{\sqrt{-1}\,u\,X(s-)}$$

is zero or $e^{\sqrt{-1}\,u\,X(T_k-)(1+Y_k)} - e^{\sqrt{-1}\,u\,X(T_k-)}$ if there is no jump at s or $s = T_k$, respectively, where $T_k = \inf\{t \ge 0 : N(t) = k\}$, $k = 1, \ldots, N(t)$, and $\Delta C(T_k) = C(T_k) - C(T_k-) = Y_k$. Since the expectation of $e^{\sqrt{-1}\,u\,X(s)} - e^{\sqrt{-1}\,u\,X(s-)}$ is

$$E\left\{\sum_{k=1}^{N(t)} E\left[e^{\sqrt{-1}\,u\,X(T_k-)(1+Y_k)} - e^{\sqrt{-1}\,u\,X(T_k-)} \mid N(t)\right]\right\}$$

and the jump times of C are uniformly distributed in $(0, t]$ conditional on $N(t)$, the above conditional expectation is $\int_{\mathbb{R}} dF(y) \int_0^t \varphi(u\,(1+y);\xi)\,d\xi/t - \int_0^t \varphi(u;\xi)\,d\xi/t$ for each $k = 1, \ldots, N(t)$. The stated equation for φ results by differentiating the expectation of the above Itô formula for $e^{\sqrt{-1}\,u\,X(t)}$.

The numerical results in Fig. 9.17 have been obtained by using a finite difference approximation of the equation for φ with the boundary conditions $\varphi(\pm\bar{u};t) = 0$ and $\partial\varphi(\pm\bar{u};t)/\partial u = 0$, where $\bar{u} = 200$. The densities f_s in Fig. 9.18 are the Fourier transforms of the corresponding characteristic functions φ_s. ■

9.5 Environment and ecology

We present environmental measures that are defined by equations with random coefficients and input of the type of Eq. 9.1. For example, the moisture content and pollutant concentration in soil deposits are random functions satisfying partial differential equations with random coefficients and input. This section presents models for the moisture content in soil deposits (Section 9.5.1), water quality indicators in rivers (Section 9.5.2), and pollutant concentration in soil deposits (Section 9.5.3). We also give some numerical results for these environment and ecology measures.

9.5.1 Rainfall runoff model

The spatial variation of soil properties and rainfall intensity can have notable effects on runoff production, river flows, and other hydrological measures [90]. Because both soil properties and rainfall vary randomly in space, the analysis of most hydrology problems involves solutions of partial differential equations with random coefficients and input involving space and time partial derivatives.

Let $\Theta(z, t)$ denote the moisture content in a soil deposit at depth $z \geq 0$ and time $t \geq 0$ and let Θ_s be the saturation water content, where $z = 0$ is the soil surface so that the coordinates x and y are in the plane defined by this surface. Our objective is to find properties of the **ponding time** T_p defined by the condition $\Theta(0, T_p) = \Theta_s$, that is, the time at which the soil surface reaches its saturation water content. It is assumed that (1) soil properties are uniform in the vertical direction z but vary randomly in the horizontal plane, (2) water flow is solely in the z-direction, (3) effects of water accumulation at soil surface and water runoff can be neglected, and (4) the soil saturated hydraulic conductivity K_s is the only soil parameter exhibiting spatial variation.

If the above four assumptions are satisfied, then Θ is the solution of

$$\frac{\partial \Theta}{\partial t} + \frac{\partial}{\partial z}\left(K \frac{d\psi}{d\Theta} \frac{\partial \Theta}{\partial z}\right) + \frac{\partial K}{\partial z} = 0 \tag{9.21}$$

with the initial and boundary conditions $\Theta(z, 0) = \Theta_i(z)$, $z > 0$, and $Q = \partial \Theta(0, t)/\partial z = R$, $t \geq 0$, where R denotes the rainfall rate at (x, y), and the coefficients ψ and K are the suction head and the hydraulic conductivity, respectively ([90], Chapter 5).

Note: The coefficients ψ and K can be related to Θ and K_s by empirical constitutive relationships, so that Eq. 9.21 becomes a partial differential equation for Θ. The functions Θ_i and $Q = \partial \Theta(0, t)/\partial z$ are the moisture content at the initial time $t = 0$ and the water flux in the vertical direction, respectively.

Because of the uncertainty in K_s, Eq. 9.21 has random coefficients. The boundary condition at $z = 0$ can also be random since the rainfall rate R exhibits random spatial and temporal fluctuations. The dependence of the functions Θ, Θ_s, K_s, and R on the coordinates (x, y) is not shown in Eq. 9.21 because we assume that the water can flow only in the z-direction. Therefore, the solution of Eq. 9.21 at a site (x, y) depends only on the soil properties and rainfall rate at this site. ▲

It is not possible to find the probability laws of Θ and T_p analytically. Properties of these random functions can be (1) estimated from samples of Θ given by solutions of Eq. 9.21 corresponding to samples of K_s and R or (2) calculated from approximate analytical solutions of Eq. 9.21 based on heuristic approximations, for example, the approximate expression for T_p in the following equation.

The ponding time can be approximated by

$$T_p \simeq \frac{s^2}{4\,c\,R\,K_s}\left[\frac{R^2}{(R - c\,K_s)^2} - 1\right], \quad \text{for } R > K_s, \tag{9.22}$$

where c and s are some constants ([90], p. 85). If $K_s \geq R$, ponding cannot occur so that $T_p = +\infty$.

Note: The constants c and s depend on Θ_i, Θ_s, soil water diffusion coefficient, soil saturated hydraulic conductivity, and other soil properties ([90], p. 84).

The ponding time in Eq. 9.22 is a nonlinear mapping of K_s and R so that T_p is a random field if K_s and R are random fields. We can calculate properties of T_p from the probability laws of K_s and R. ▲

Example 9.37: Suppose that K_s and R are independent lognormal translation random fields with expectations $E[K_s(x, y)] = 0.018325$ and $E[R] = 0.01665$ measured in centimeters/minute and coefficients of variation $v_k = 0.5$ and $v_r = 0.5$, respectively. For example,

$$K_s(x, y) = \exp\left(\mu_g + \sigma_g\, G(x, y)\right), \quad \boldsymbol{x} = (x, y) \in \mathbb{R}^2, \tag{9.23}$$

where μ_g, σ_g are constants and G denotes a homogeneous Gaussian field with mean zero, variance one, and covariance function $\rho(\boldsymbol{x}_1, \boldsymbol{x}_2) = E[G(\boldsymbol{x}_1)\,G(\boldsymbol{x}_2)]$, $\boldsymbol{x}_1, \boldsymbol{x}_2 \in \mathbb{R}^2$ ([79], Section 3.1). We take $G(x, y) = [G_1(x) + G_2(y)]/\sqrt{2}$, where G_i, $i = 1, 2$, are independent stationary Gaussian processes with mean zero, covariance function $E[G_i(u)\,G_i(u + \zeta)] = \exp(-\alpha\,|\zeta|)$, and $\alpha = 0.5$. The constants μ_g, σ_g can be calculated from the prescribed mean and coefficients of variation of K_s. The rainfall rate R is also modeled by a similar lognormal translation field but the parameter of the covariance function of its Gaussian image is $\alpha = 1$ rather than $\alpha = 0.5$.

Figure 9.19 shows samples of K_s and R in a square $D = (0, a) \times (0, a)$ for $a = 5$. The figure also shows the inverse of the ponding time $1/T_p$ (Eq. 9.22) corresponding to the samples of (K_s, R) and $(K_s, E[R])$ in this figure. The significant spatial variation of the ponding time and the differences between ponding times for random and deterministic rainfall rates indicate that the rainfall runoff model must account for the random spatial variation of both the soil properties and the rainfall rates. ◇

Note: Since the soil conductivity K_s and rainfall rate R are positive, they cannot be modeled by, for example, Gaussian random fields. Lognormal translation fields have been used to represent these parameters.

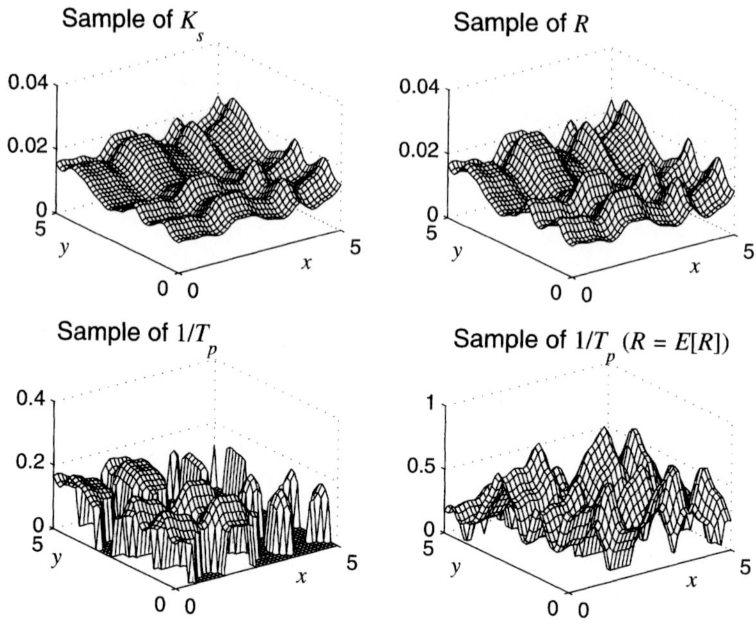

Figure 9.19. Samples of K_s, R, and samples of $1/T_p$ corresponding to (K_s, R) and $(K_s, E[R])$

The lognormal random field K_s is homogeneous with mean and covariance function

$$\mu_k = \exp\left(\mu_g + \sigma_g^2/2\right) \quad \text{and}$$

$$c_k(\boldsymbol{\zeta}) = E[K_s(\boldsymbol{\xi})\, K_s(\boldsymbol{\xi} + \boldsymbol{\zeta})] = \exp\left(2\,\mu_g + \sigma_g^2\,(1 + \rho(\boldsymbol{\zeta}))\right) - \mu_k^2,$$

respectively, so that $\sigma_k^2 = c_k(\mathbf{0})$ and $v_k = \sigma_k/\mu_k$ ([79], p. 49). The finite dimensional distributions of K_s can be obtained from the definition of K_s and the probability law of G (Section 3.6.6 in this book, [79], p. 45). For example, the first order or the marginal distribution of this random field is $P(K_s \leq z) = \Phi((\ln(z) - \mu_g)/\sigma_g)$, where Φ denotes the distribution of $N(0, 1)$. It can also be shown that the scaled covariance function of K_s satisfies the condition $|\xi(\boldsymbol{\zeta})| \leq |\rho(\boldsymbol{\zeta})|$ for all $\boldsymbol{\zeta} \in \mathbb{R}^2$, where $\xi(\boldsymbol{\zeta}) = c_k(\boldsymbol{\zeta})/c_k(\mathbf{0}) = [(1 + v_k^2)^{\rho(\boldsymbol{\zeta})} - 1]/v_k^2$ ([79], p. 49). ▲

Probabilistic models for K_s and R of the type in Example. 9.37 can also be used in Eq. 9.21. Moreover, the model for R can be generalized by allowing the rainfall to vary randomly not only in space but also in time. We have already stated that analytical solutions cannot be obtained for Eq. 9.21, but this equation can be solved by Monte Carlo simulation. First, samples of (K_s, R) need to be generated and the corresponding samples of (Ψ, K) calculated. Second, Eq. 9.21 has to be solved for the samples of (K_s, R, Ψ, K) obtained in the previous step.

Third, the corresponding samples of the ponding time T_p result from its defining equation $\Theta(0, T_p) = \Theta_s$. Fourth, statistics need to be calculated for Θ and T_p from the calculated samples of these random fields.

9.5.2 Water quality in streams

The biochemical oxygen demand (BOD) and the dissolved oxygen (DO) are essential indicators for water quality in streams. The BOD is the total amount of oxygen required by bacteria stabilizing decomposable substances in a stream and the DO is oxygen available in the water. Let v be the average velocity of a stream and denote by $l(t)$ and $c(t)$ the value of BOD and DO, respectively, at a distance $t \geq 0$ downstream from a polluting source. Because the stream velocity is assumed to be constant, time t/v is a scaled distance.

Two classes of stochastic models have been developed for BOD and DO, (1) models based on random walks, birth-death process, or time series and (2) models based on stochastic differential equations. We consider models of the second type ([2], pp. 135-150). One of the simplest deterministic models giving the evolution of the BOD and DO is

$$\begin{cases} \dot{l}(t) &= -(k_1/v)\, l(t), \\ \dot{c}(t) &= -(k_2/v)\, (c(t) - c_s) - (k_1/v)\, l(t), \end{cases} \tag{9.24}$$

where k_1 and k_2 denote the deoxygenation and reaeration rates, respectively, and c_s is the DO saturation concentration ([2], p. 137). The evolution of the state vector (l, c) can be calculated simply from Eq. 9.24 for specified initial conditions. Alternative models have been proposed for the evolution of (l, c) that account for additional processes that may take place in streams ([2], pp. 135-150). Our objectives are to establish a stochastic version for the evolution of BOD and DO and calculate statistics of these water quality measures at any distance $t \geq 0$ downstream from a pollution source.

Consider the differential equation

$$\dot{\boldsymbol{X}}(t) = \boldsymbol{A}\, \boldsymbol{X}(t) + \boldsymbol{Y}(t), \quad t \geq 0, \tag{9.25}$$

where $L(t)$ and $C(t)$ are the coordinates of the state vector $\boldsymbol{X}(t)$, $A_{11} = -(K_1 + K_3)/v$, $A_{12} = 0$, $A_{21} = -K_1/v$, $A_{22} = -K_2/v$, $Y_1(t) = Z(t)/v$, and $Y_2(t) = (K_2\, c_s - K_{2,p})/v$ are the entries of $\boldsymbol{A}$ and $\boldsymbol{Y}(t)$. The upper case letters in Eq. 9.25 are used to denote random variables and functions. The parameters K_1 and K_2 have the same meaning as in Eq. 9.24 but can be random. The parameters K_3, Z, and $K_{2,p}$ denote the rate at which pollution decreases because of sedimentation and absorption, the rate of pollution increase along the river banks caused by small pollution sources, and the rate of DO decrease owing to photosynthesis, respectively ([2], pp. 138, [138]). The parameters K_3, Z, and $K_{2,p}$ may depend on the distance t from the main pollution source. The model in Eq. 9.25 is a linear differential equation with random coefficients, random input, and deterministic or random initial conditions. The model in Eq. 9.25 is an extension of Eq. 9.24 because it (1) considers additional processes, such as sedimentation, absorption, and photosynthesis, and (2) incorporates the uncertainty in the model parameters and input.

The method for calculating probabilistic properties of $\boldsymbol{X}$ in Eq. 9.25 depends on the type of the driving noise. For example, the conditional analysis in Section 9.2.3 and the methods in Chapter 7 can be used if $\boldsymbol{Y}$ is a Brownian motion, a martingale, or a semimartingale. If $\boldsymbol{Y}$ is a random variable rather than a random function, then

$$\boldsymbol{X}(t) = \begin{bmatrix} \Psi_1(t) & 0 \\ \Psi_2(t) & \Psi_3(t) \end{bmatrix} \boldsymbol{X}(0) + \begin{bmatrix} \Xi_1(t) \\ \Xi_2(t) \end{bmatrix} = \boldsymbol{\Psi}(t)\, \boldsymbol{X}(0) + \boldsymbol{\Xi}(t), \tag{9.26}$$

where ([2], pp. 144-145, [138])

$$\begin{aligned}
\Psi_1(t) &= \exp[-(K_1 + K_3)\, t/v], \\
\Psi_2(t) &= K_1\, [\Psi_1(t) - \Psi_3(t)]/(K_1 - K_2 + K_3), \\
\Psi_3(t) &= \exp[-K_2\, t/v], \\
\Xi_1(t) &= Z\, [1 - \Psi_1(t)]/(K_1 + K_3), \\
\Xi_2(t) &= \frac{K_1\, Z}{K_1 - K_2 + K_3} \left[\frac{1 - \Psi_1(t)}{K_1 + K_3} - \frac{1 - \Psi_3(t)}{K_2} \right] \\
&\quad + \frac{c_s\, K_2 - K_{2,p}}{K_2} [1 - \Psi_3(t)].
\end{aligned} \tag{9.27}$$

The probability law of $\boldsymbol{X}$ can be obtained from (1) Eqs. 9.26-9.27 and the properties of the random parameters in these equations by Monte Carlo simulation, (2) the relationship between the densities of functionally dependent random variables (Eqs. 9.3), and (3) the Liouville equation (Eqs. 9.4-9.5). We illustrate the determination of the density of $\boldsymbol{X}(t)$ by Liouville's equation.

If $K_r = k_r$, $r = 1, 2, 3$, are deterministic constants and $(Z, K_{2,p})$ are random variables, the density of the augmented state vector $(\boldsymbol{X}, Z, K_{2,p})$ at $t \geq 0$ is

$$\begin{aligned}
&f(\boldsymbol{x}, z, \chi; t) \\
&= f_0 \left(\frac{x_1 - \xi_1(t)}{\psi_1(t)}, \frac{-\psi_2(t)\,(x_1 - \xi_1(t)) + \psi_1(t)\,(x_2 - \xi_2(t))}{\psi_1(t)\, \psi_3(t)}, z, \chi \right) \\
&\times \exp[(k_1 + k_2 + k_3)\, t/v]
\end{aligned} \tag{9.28}$$

where ξ_q and ψ_p have the expressions of Ξ, $q = 1, 2$, and Ψ_p, $p = 1, 2, 3$, in Eq. 9.27, respectively, with $K_r = k_r$, $r = 1, 2, 3$, and $(Z, K_{2,p}) = (z, \chi)$. The function f_0 is the density of $(\boldsymbol{X}(0), Z, K_{2,p})$.

Note: The augmented vector $\tilde{\boldsymbol{X}} = (\tilde{X}_1 = X_1, \tilde{X}_2 = X_2, \tilde{X}_3 = Z, \tilde{X}_4 = K_{2,p})$ is defined by Eqs. 9.26-9.27 and the differential equations $\dot{\tilde{X}}_3(t) = 0$, $\dot{\tilde{X}}_4(t) = 0$ with the initial conditions $\tilde{X}_3(0) = Z$, $\tilde{X}_4(0) = K_{2,p}$. The density $f(\boldsymbol{x}, z, \chi; t)$ is given by Eq. 9.5 applied to $\tilde{X}$ in which we use $\boldsymbol{X}(0) = \boldsymbol{\Psi}(t)^{-1}[\boldsymbol{X}(t) - \boldsymbol{\Xi}(t)]$ (Eq. 9.26). Generally, the determination of the density $f(\boldsymbol{x}; t) = \int_{\mathbb{R}^2} f(\boldsymbol{x}, z, \chi; t)\, dz\, d\chi$ of $\boldsymbol{X}(t)$ requires numerical integration. ▲

Example 9.38: Suppose that $\boldsymbol{X}(0)$, Z, and $K_{2,p}$ in Eq. 9.26 are random variables and all the other parameters in these equation are deterministic. Let the densities of $X_1(0)$ and $X_2(0)$ be proportional to the densities of $N(6.8, 1)$ and $N(8.7, 0.03)$ in the intervals $[0, \infty)$ and $[0, c_s]$, respectively, and zero outside these intervals. For example, $f_{X_1(0)}(\xi) = c_1\, 1_{[0,\infty)}(\xi)\phi(\xi - 6.8)$, where $c_1 > 0$ is a constant and ϕ denotes the density of $N(0, 1)$. Let Z and $K_{2,p}$ be random variables that are uniformly distributed in the intervals $(0, 0.4)$ and $(0, 0.2)$, respectively. It is also assumed that the random variables $X_1(0)$, $X_2(0)$, Z, and $K_{2,p}$ are mutually independent. Then the joint density of $(X_1(0), X_2(0), Z, K_{2,p})$, that is, the density of the augmented state vector $\tilde{\boldsymbol{X}}(t) = (X_1(t), X_2(t), Z, K_{2,p})$ at $t = 0$, is

$$f_0(\tilde{\boldsymbol{x}}) = c\ \exp[-(\tilde{x}_1 - 6.8)^2/2 - (\tilde{x}_2 - 8.7)^2/0.06]$$

for $\tilde{x}_1 \geq 0$, $\tilde{x}_2 \in [0, c_s]$, $\tilde{x}_3 \in [0, 0.4]$, and $\tilde{x}_4 \in [0, 0.2]$, where $c > 0$ is a normalization constant. If any of these conditions is violated, f_0 is zero. The density of $\tilde{\boldsymbol{X}}(t) = (\boldsymbol{X}(t), Z, K_{2,p})$ at a later time $t \geq 0$ is (Eq. 9.28)

$$f(\tilde{\boldsymbol{x}}; t) = c\ \exp[(k_1 + k_2 + k_3)\, t/v] \exp\left\{ -\frac{1}{2}\left[\frac{\tilde{x}_1 - \xi_1(t)}{\psi_1(t)} - 6.8\right]^2 \right.$$
$$\left. - \frac{1}{0.06}\left[\frac{-\psi_2(t)\,(\tilde{x}_1 - \xi_1(t)) + \psi_1(t)\,(\tilde{x}_2 - \xi_2(t))}{\psi_1(t)\,\psi_3(t)} - 8.7\right]^2 \right\}$$

if $\tilde{x}_1 \in [\xi_1(t), \infty)$, $\tilde{x}_2 \in [c_*, c^*]$, $\tilde{x}_3 \in [0, 0.4]$, and $\tilde{x}_4 \in [0, 0.2]$ and zero if any of these conditions is not satisfied, where $c_* = \max(0, (\psi_2/\psi_1)(\tilde{x}_1 - \xi_1) + \xi_2)$ and $c^* = \min(c_s, c_s\, \psi_3 + (\psi_2/\psi_1)(\tilde{x}_1 - \xi_1) + \xi_2)$ [138].

Figure 9.20 shows the joint density of $\boldsymbol{X}(t)$ at distances $t = 0$ and $t = 5$ miles downstream from a pollution source for $k_1 = 0.35$, $k_2 = 0.75$, $k_3 = 0.20$, $c_s = 10$, and an average stream velocity $v = 7.5$ miles/day. The plots have been calculated from the above expression of the density of $\tilde{\boldsymbol{X}}(t)$ by numerical integration. They show the trend of (BOD, DO) with the distance $t \geq 0$ and the spread about this trend. Such plots can be used to calculate the probability that the DO level does not fall below a critical level DO_{cr} at any site downstream from a pollution source. Values of DO smaller than DO_{cr} can have serious consequences on the aquatic life in a stream. ◇

Note: The densities in Fig. 9.20 have been calculated from the density of $\tilde{\boldsymbol{X}}(t)$ in which the random variables Z and $K_{2,p}$ were assumed to take discrete values. Numerical results are for ten equally spaced values in the range of possible values of these variables.

The relationship between $\boldsymbol{X}(t)$ and $(\boldsymbol{X}(0), Z, K_{2,p})$ in Eq. 9.26 can also be used to calculate moments of BOD and DO. For example the expectations of these pollution

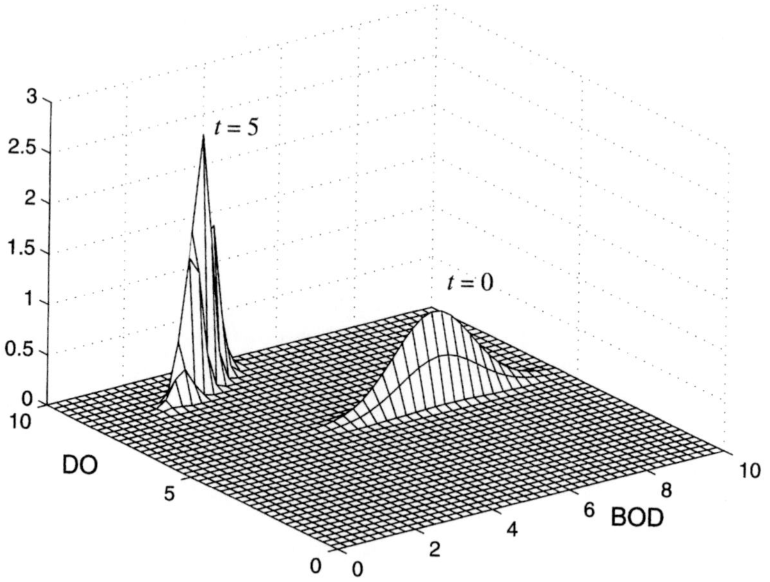

Figure 9.20. The joint density of (BOD, DO) at $t = 0$ and $t = 5$ miles downstream from a pollution source

measures are

$$E[L(t)] = E[X_1(t)] = \psi_1(t)\, E[L(0)] + \frac{1-\psi_1(t)}{k_1+k_3}\, E[Z],$$

$$E[C(t)] = E[X_2(t)] = \psi_2(t)\, E[L(0)] + \psi_3(t)\, E[C(0)] + \frac{k_1}{k_1-k_2+k_3}\left[\frac{1-\psi_1(t)}{k_1+k_3} - \frac{1-\psi_3(t)}{k_2}\right] E[Z] + \frac{c_s\, k_2 - E[K_{2,p}]}{k_2}\,[1-\psi_3(t)]$$

since k_1, k_2, and k_3 are deterministic. ▲

9.5.3 Subsurface flow and transport

Consider a soil deposit in a subset D of $\mathbb{R}^q$, $q = 1, 2, 3$. Let $k(\boldsymbol{x})$ and $d_{ij}(\boldsymbol{x})$ denote the hydraulic conductivity and the dispersion tensor at $\boldsymbol{x} \in D$ and time $t \geq 0$, respectively. The flux and head, $\boldsymbol{q}(\boldsymbol{x})$ and $h(\boldsymbol{x})$, at $\boldsymbol{x} \in D$ are assumed for simplicity to be time-invariant. Let $\boldsymbol{v}(\boldsymbol{x}, t)$ and $c(\boldsymbol{x}, t)$ be the flow velocity and the concentration of a pollutant at $\boldsymbol{x} \in D$ and time $t \geq 0$ corresponding to a specified head and flux on the boundary of D and to a pollutant source.

The spatial and temporal variation of the flow velocity and pollutant concentration depend on the properties of the soil deposit in D, pollutant source, and boundary and initial conditions. Our objective is to find properties of pollutant concentration $c(\boldsymbol{x}, t)$. This function varies randomly in space and time if the soil properties, the pollutant source, and/or initial and boundary conditions are uncertain.

The steady-state head in D is the solution of the partial differential equation

$$\nabla \cdot (k(\boldsymbol{x}) \, \nabla h(\boldsymbol{x})) + s(\boldsymbol{x}) = 0, \quad \boldsymbol{x} \in D, \tag{9.29}$$

with the boundary conditions

$$\begin{aligned} h(\boldsymbol{x}) &= h_0(\boldsymbol{x}), \quad \boldsymbol{x} \in \partial D_u, \quad \text{and} \\ -\boldsymbol{q}(\boldsymbol{x}) \cdot \boldsymbol{n}(\boldsymbol{x}) &= q_0(\boldsymbol{x}), \quad \boldsymbol{x} \in \partial D_\sigma, \end{aligned} \tag{9.30}$$

where the source function s, the head h_0, and the flux q_0 are prescribed functions, ∂D_u and ∂D_σ partition the boundary of D, and $\boldsymbol{n}(\boldsymbol{x})$ is the unit outward normal at $\boldsymbol{x} \in \partial D_\sigma$ ([39], Chapter 20).

Note: The flux in D can be calculated from the Darcy law $\boldsymbol{q}(\boldsymbol{x}) = k(\boldsymbol{x}) \, \nabla h(\boldsymbol{x})$ and the solution of Eqs. 9.29 and 9.30. The velocity $\boldsymbol{v}$ of ground water is $\boldsymbol{v} = \boldsymbol{q}/\phi$, where ϕ denotes the effective soil porosity ([39], Chapter 20). ▲

The concentration $c(\boldsymbol{x}, t)$ of a pollutant at $\boldsymbol{x} \in D$ and time $t \geq 0$ is the solution of the **transport equation**

$$\frac{\partial c}{\partial t} + c \, \nabla \boldsymbol{v} + \boldsymbol{v} \cdot \nabla c - \sum_{i,j=1}^{q} \left(\frac{\partial d_{ij}}{\partial x_i} \frac{\partial c}{\partial x_j} + d_{ij} \frac{\partial^2 c}{\partial x_i \, \partial x_j} \right) = g(\boldsymbol{x}, t), \tag{9.31}$$

where g is a specified pollutant source function and $q = 1, 2, 3$.

Note: Initial and boundary conditions as well as the pollutant source function need to be specified for solution. The conditions on the boundary ∂D of D can be of the Dirichlet and Neumann type specifying the concentration c at and the pollutant flux through ∂D, respectively.

If the flow velocity is constant, the above equation simplifies to

$$\frac{\partial c}{\partial t} + \boldsymbol{v} \cdot \nabla \boldsymbol{c} - \sum_{i,j=1}^{q} \left(\frac{\partial d_{ij}}{\partial x_i} \frac{\partial c}{\partial x_i} + d_{ij} \frac{\partial^2 c}{\partial x_i \, \partial x_j} \right) = g(\boldsymbol{x}, t).$$

If in addition the coefficients d_{ij} are space invariant, the above equation becomes

$$\frac{\partial c}{\partial t} + c \, \nabla \boldsymbol{v} - \sum_{i,j=1}^{q} d_{ij} \frac{\partial^2 c}{\partial x_i \, \partial x_j} = g(\boldsymbol{x}, t).$$

The defining equation for c is a special case of the type of partial differential equation considered in Section 6.1. Hence, the local solutions in Sections 6.2 and 6.3 can be used to find properties of the concentration at a specified location $\boldsymbol{x} \in D$ and time $t \geq 0$. ▲

Example 9.39: Suppose that (1) the ground water flow in an infinite medium has a known constant velocity $v > 0$ along the coordinate x_1, (2) an amount $c_0 > 0$ of pollutant is placed instantaneously at $\boldsymbol{x} = (0, 0, 0)$ and time $t = 0$, and (3) the dispersion tensor $d_{ij} = d\,\delta_{ij}$ is constant. Then c satisfies the partial differential equation $\partial c/\partial t + v\,\partial c/\partial x_1 = d\,\Delta c$, where $\Delta = \sum_{i=1}^{3} \partial^2/\partial x_i^2$, so that

$$c(\boldsymbol{x}, t) = \frac{c_0}{(4\pi\, d\, t)^{3/2}} \exp\left[-\frac{(x_1 - v\,t)^2 + x_2^2 + x_3^2}{4\,d\,t}\right].$$

Figure 9.21 shows contour lines of the pollutant concentration at time $t = 0.1, 1.5$, and 3.0 in a two dimensional medium for $d = 0.2$ and $v = 4$. Two phenomena can be observed. The pollutant eye travels in the x_1 direction with velocity v and the pollutant diffuses in the x_1 and x_2 directions as time increases. ◇

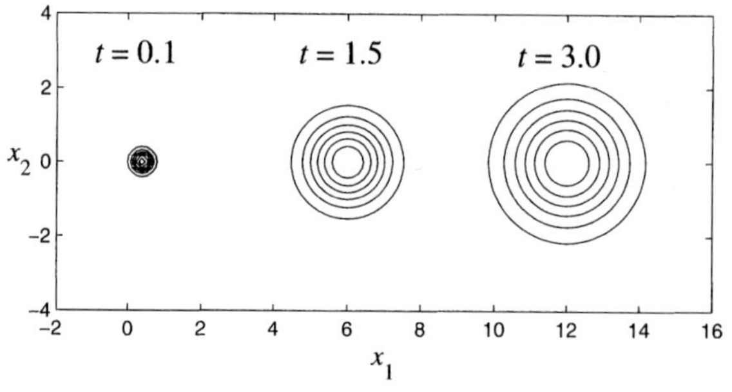

Figure 9.21. Propagation of pollutant in a two-dimensional medium

Note: That the above expression of c is the solution of the transport equation $\partial c/\partial t + v\,\partial c/\partial x_1 = d\,\Delta c$ can be found by direct calculations ([46], Chapter 1). As $t \downarrow 0$, the concentration converges to $c_0\,\delta(\boldsymbol{x})$. The equality $c_0 = \int_{\mathbb{R}^3} c(\boldsymbol{x}, t)\,d\boldsymbol{x}$ holds at each time $t \geq 0$ and expresses the conservation of mass. ▲

Example 9.40: Consider the transport problem in the previous example but assume that the dispersion tensor $D_{ij}(\boldsymbol{x}) = D(\boldsymbol{x})\,\delta_{ij}$ varies randomly in space. The pollutant concentration $C(\boldsymbol{x}, t)$ becomes a random function of space and time defined by (Eq. 9.31)

$$\frac{\partial C}{\partial t} + v\,\frac{\partial C}{\partial x_1} - \sum_{i=1}^{q}\left[\frac{\partial D}{\partial x_i}\frac{\partial C}{\partial x_i} + D\,\frac{\partial^2 C}{\partial x_i^2}\right] = 0,$$

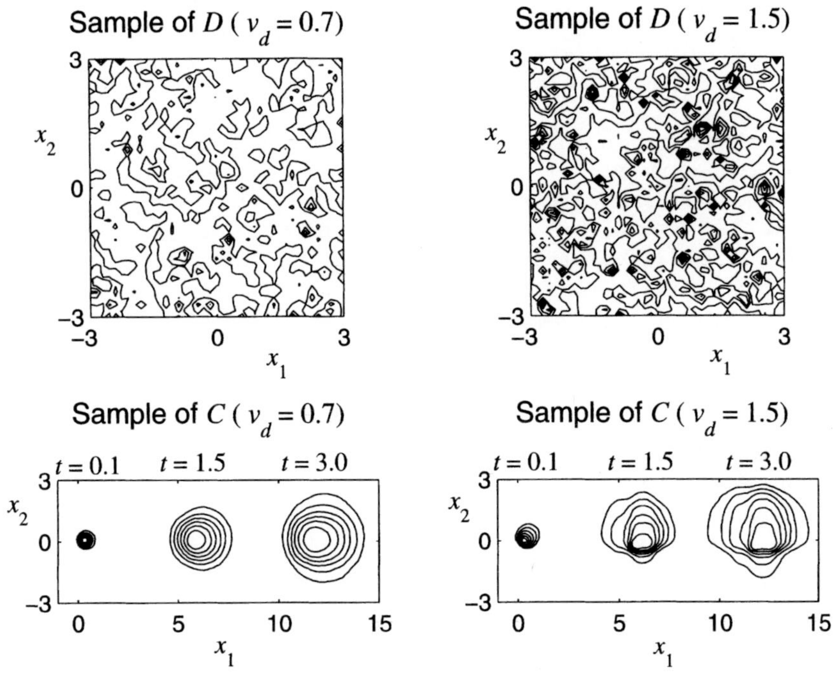

Figure 9.22. Samples of dispersion tensor and contour lines for the corresponding samples of pollutant concentration

with the initial condition $C(\boldsymbol{x}, t) = c_0\,\delta(\boldsymbol{x})$. It is not possible to find the probability law of C analytically so that numerical methods need to be employed.

The top plots in Fig. 9.22 are two samples of the dispersion D assumed to be a lognormal translation field defined by $D(\boldsymbol{x}) = \exp(\mu_g + \sigma_g\, G(\boldsymbol{x}))$, where $G(\boldsymbol{x}) = (G_1(x_1) + G_2(x_2))/\sqrt{2}$, $\boldsymbol{x} \in \mathbb{R}^2$, and G_i, $i = 1, 2$, denote independent Gaussian processes with mean zero and covariance functions $E[G_i(x_i)\, G_i(x_i + \zeta_i)] = \exp(-\alpha\,|\zeta_i|)$, $\alpha = 2$. The expected value of $D(\boldsymbol{x})$ is 0.2, that is, the value of the dispersion in Example 9.39. The samples of D are for coefficients of variation $v_d = 0.7$ and $v_d = 1.5$.

The bottom two plots in Fig. 9.22 are contour lines for the corresponding samples of C at distances $t = 0.1$, 1.5, and 3 downstream from the pollution source for $v = 4$. Figures 9.21 and 9.22 show that the uncertainty in the dispersion tensor can have a significant effect on pollution spread. Also, the sample to sample variation of pollutant concentration can be significant. ◇

Note: The parameters μ_g and σ_g in the definition of $D(\boldsymbol{x})$ can be related to the mean μ_d

and coefficient of variation v_d of this field by

$$\mu_g = \ln\left(\mu_d/\sqrt{1+v_d^2}\right) \quad \text{and} \quad \sigma_g = \sqrt{\ln(1+v_d^2)}.$$

These equations can be used to calculate the values of the parameters μ_g and σ_g for specified values of μ_d and σ_d.

The Monte Carlo simulation algorithm in Section 5.3.3.1 has been used to generate samples of the dispersion tensor D. A MATLAB tool box has been applied to solve the partial differential equation of the pollutant concentration for samples of D.

Note that an extended version of the local solution in Section 6.2 of the type considered in, for example, Section 8.2 can be used to find properties of $C(\boldsymbol{x}, t)$ at an arbitrary location and time. Also, the methods discussed in the first part of this chapter can be used to find moments and other properties of the pollutant concentration approximately. ▲

The calculations in the previous two examples are based on the assumption that the flow velocity is known and deterministic. Generally, the flow velocity is not known and may be a random function of $\boldsymbol{x} \in D$ and $t \geq 0$. In this case, we have to solve also Eq. 9.29 to find the head that gives the flux and flow velocity. There are no simple methods for solving Eqs. 9.29 and 9.31 even in a deterministic setting.

The stochastic version of Eqs. 9.29 and 9.31 results by assuming that the hydraulic conductivity, the dispersion tensor, the input, and the initial and boundary conditions are random functions with known probability law. Generally, the probability law of these functions is not known so that it has to be inferred from observations. Measurements of conductivity and dispersion can be performed at a finite number of points $\boldsymbol{x}_k$ and constitute spatial averages over relatively small volumes centered at $\boldsymbol{x}_k$, where the size of these volumes depends on the measuring device. Higher resolution devices tend to identify smaller scale features of the measured fields so that the resulting records have a larger variability. Methods for incorporating the quality of measurements in the probabilistic models for conductivity and dispersion are discussed in [39] (Chapter 20) and [47].

9.6 Waves in random media

Waves represent disturbances that propagate in a medium and carry energy. These disturbances are functions of the space and time arguments, and are defined by partial differential equations, referred to as the **wave equations**. The coefficients of the wave equations are random because of the uncertainty in the properties of the medium. Propagation of sound and electromagnetic radiation, vibration of mechanical systems, and variation of ocean surface elevation in time are examples of wave propagation.

The stochastic Helmholtz or Schrödinger wave equation,

$$\Delta U(\boldsymbol{x}) + k_0^2\,(1 + \varepsilon\, Z(\boldsymbol{x}))\, U(\boldsymbol{x}) = 0, \quad \boldsymbol{x} \in D \subset \mathbb{R}^3, \tag{9.32}$$

is another example of a wave propagation equation in random media ([173], Section III.13). In Eq. 9.32, ε is a scale parameter, $k_0 > 0$ denotes the wave number corresponding to the deterministic Schrödinger equation ($\varepsilon = 0$), and the random field Z characterizing the uncertainty in the medium properties is such that $1 + \varepsilon\, Z(\boldsymbol{x}) > 0$ a.s. in D. We have obtained in Section 6.2.2.2 a local solution for the deterministic Schrödinger equation based on diffusion processes and the Feynman-Kac formula. This local solution can be extended by considerations in Section 9.2.1 to solve locally the stochastic Schrödinger equation in Eq. 9.32. We note that Eq. 9.32 belongs to the class of stochastic problems defined by Eq. 9.1.

Our objective is to determine properties of waves propagating in random heterogeneous media. There is no general method for calculating moments and other properties of the solution of the stochastic wave equation. In our brief discussion on waves we assume that (1) the differential operator of the wave equation is linear with time-invariant random coefficients (2) the Sommerfeld radiation condition holds, that is, the waves from a point source depart to infinity ([173], p. 27). Even under these assumptions there is no general solution for the stochastic wave equation. However, the analysis simplifies significantly because the superposition principle applies and Green functions exist. Methods discussed in the first part of this chapter can be used to find moments and other properties of the solution of the stochastic wave equation, for example, the perturbation, Neumann series, iteration, and Monte Carlo methods. A comprehensive review of the application of these methods to the solution of various types of wave equations can be found in [173] (Chapter 3). The following two examples consider some simple stochastic wave equations.

Example 9.41: Consider a rod with length $l > 0$, cross section of unit area, and random stiffness $K(x) > 0$ and density $R(x) > 0$, $x \in (0, l)$. The rod is free and fixed at its left and right ends, respectively. It is assumed that a force $Y(t)$ is acting on the left end of the rod and that the rod is at rest at the initial time. Let $U(x, t)$ be the rod displacement at location $x \in (0, l)$ and time $t \geq 0$. Then

$$R(x)\,\frac{\partial^2 U(x,t)}{\partial t^2} = \frac{\partial}{\partial x}\left[K(x)\,\frac{\partial U(x,t)}{\partial x}\right], \qquad x \in (0, l),$$

which becomes

$$\begin{cases} \frac{\partial V(x,t)}{\partial t} &= \frac{1}{R(x)}\,\frac{\partial T(x,t)}{\partial x}, \\ \frac{\partial T(x,t)}{\partial t} &= K(x)\,\frac{\partial V(x,t)}{\partial x}, \end{cases}$$

with the notation $V(x, t) = \partial U(x, t)/\partial t$ and $T(x, t) = K(x)\,\partial U(x, t)/\partial x$. The functions V and T are the rate of change of U in time and the stress in the rod, respectively. The initial and boundary conditions for the above equations are $V(x, 0) = 0$, $T(x, 0) = 0$, $T(0, t) = Y(t)$, and $V(l, t) = 0$.

Figure 9.23 shows contour lines for two samples of U corresponding to $Y(t) = \sin(\nu\, t)$, $\nu = 0.1$, $l = 10$, $R(x) = 1$, and two independent samples of $K(x) = k_0 + \sigma\, \exp(G(x))$, where $k_0 = 2$, $\sigma = 5$, and G is a stationary Gaussian

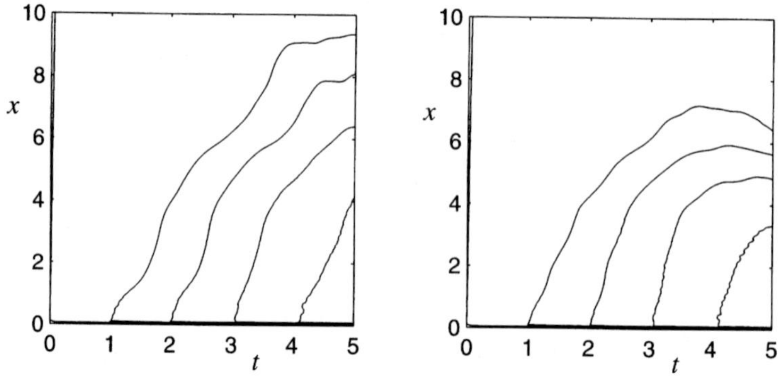

Figure 9.23. Contour lines of two samples of the solution of the wave equation for a rod with random stiffness

field with mean zero and covariance function $E[G(x)\, G(x')] = \exp(-|x - x'|)$. The distinct pattern of the two wave samples is caused by the random fluctuations of the rod stiffness. ◇

Note: Because we assume that the rod is made of a linearly elastic material and its deformation is small, the strain and stress at (x, t) are $\Gamma(x, t) = \partial U(x, t)/\partial x$ and $S(x, t) = K(x)\, \partial U(x, t)/\partial x$, respectively. The equilibrium condition,

$$\partial S(x, t)/\partial x - R(x)\, \partial^2 U(x, t)/\partial t^2 = 0,$$

gives the wave equation.

Let $V_{k,l}$ and $T_{k,l}$ denote the values of V and T at $(x = x_k, t = t_l)$, respectively, where $x_k = k\, h$ are equally spaced points in $(0, l)$, $h > 0$, $k = 0, 1, \ldots, n + 1$, and $t_l = l\, \Delta t$, where $\Delta t > 0$ is the integration time step and $l \geq 0$, $n > 0$ are integers. The $(2\, n, 1)$ vector $\boldsymbol{X}_l = (V_{1,l}, \ldots, V_{n,l}, T_{1,l}, \ldots, T_{n,l})$ satisfies the finite difference equation $\boldsymbol{X}_{l+1} = \boldsymbol{A}\, \boldsymbol{X}_l$, where the entries of the random matrix $\boldsymbol{A}$ are zero except for $A_{i,i} = 1$ for $i = 1, \ldots, 2\, n$, $A_{i,n+i+1} = B_i$ for $i = 1, \ldots, n - 1$, $A_{i,n+i-1} = -B_i$ for $i = 2, \ldots, n$, $A_{n+i,i-1} = -C_i$ for $i = 2, \ldots, n$, and $A_{n+i,i+1} = C_i$ for $i = 1, \ldots, n - 1$, where $B_i = \Delta t/(2\, R(x_i)\, h)$ and $C_i = K(x_i)\, \Delta t/(2\, h)$. This finite difference equation needs to be supplemented by initial and boundary conditions. The evolution in time of moments and other properties of $\boldsymbol{X}$ can be calculated by the methods in Sections 9.2 and 9.3.

If the rod is subjected to an externally applied action $Q(x, t)$, the wave equation becomes

$$R(x)\, \frac{\partial^2 U(x, t)}{\partial t^2} = \frac{\partial}{\partial x} \left[K(x)\, \frac{\partial U(x, t)}{\partial x} \right] + Q(x, t)$$

or

$$\begin{cases} \frac{\partial V(x,t)}{\partial t} & = \frac{1}{R(x)} \left(\frac{\partial T(x,t)}{\partial x} + Q(x, t) \right), \\ \frac{\partial T(x,t)}{\partial t} & = K(x)\, \frac{\partial V(x,t)}{\partial x}, \end{cases}$$

so that the corresponding finite difference equation is $\boldsymbol{X}_{l+1} = \boldsymbol{A}\, \boldsymbol{X}_l + \boldsymbol{Q}_l$, where the entries in the first n rows of $\boldsymbol{Q}_l$ are $Q(x_k, t_l)\, \Delta t$, $k = 1, \ldots, n$, and the last n rows of this matrix are zero.

Let $\mathcal{X}(x, t)$ be an $\mathbb{R}^2$-valued function with coordinates (V, T). Then $\mathcal{X}$ is the solution of Eq. 9.1 with $\mathcal{D} = \boldsymbol{i}\,(\partial/\partial t) - \boldsymbol{\alpha}\,(\partial/\partial x)$ and $\mathcal{Y}(x, t) = \boldsymbol{\beta}\, Q(x, t)/R(x)$, where $\boldsymbol{i}$ is the identity matrix, and the entries of the matrices $\boldsymbol{\alpha}$ and $\boldsymbol{\beta}$ are $\alpha_{11} = \alpha_{22} = 0$, $\alpha_{12} = 1/R(x)$, $\alpha_{21} = K(x)$, $\beta_1 = 1$, and $\beta_2 = 0$. The methods discussed in this chapter can be applied to find properties of $\mathcal{X}$. ▲

Example 9.42: Consider an infinite rod that is homogeneous and deterministic except for a finite interval $(0, l)$, $l > 0$, in which the rod has random heterogeneous properties. The amplitude U of the harmonic waves in the rod is the solution of the differential equation

$$U''(x) + k_0^2\,(1 + \varepsilon\, 1_{(0,l)}(x)\, Z(x))\, U(x) = 0, \quad x \in \mathbb{R},$$

where Z is a random function defined in $(0, l)$.

Suppose an incident wave $e^{-\sqrt{-1}\,k_0\,x}$ propagates toward the random heterogeneous interval of the rod from its right side. Our objective is to find properties of the reflected wave $\Lambda(l)\, e^{\sqrt{-1}\,k_0\,x}$, $x \in (l, \infty)$. The reflection coefficient Λ is viewed as a stochastic process of the argument l that satisfies the differential equation

$$\frac{d\Lambda(l)}{dl} = \varepsilon\, \frac{\sqrt{-1}\,k_0\, Z(l)}{2} \left[e^{\sqrt{-1}\,k_0\,l}\,\Lambda(l) + e^{-\sqrt{-1}\,k_0\,l}\right]^2$$

with the initial condition $\Lambda(0) = 0$. This equation has random coefficients because of the rod properties in $(0, l)$. Figure 9.24 shows the real and imaginary parts of ten samples of Λ in the range $(0, 3)$. The samples have been calculated

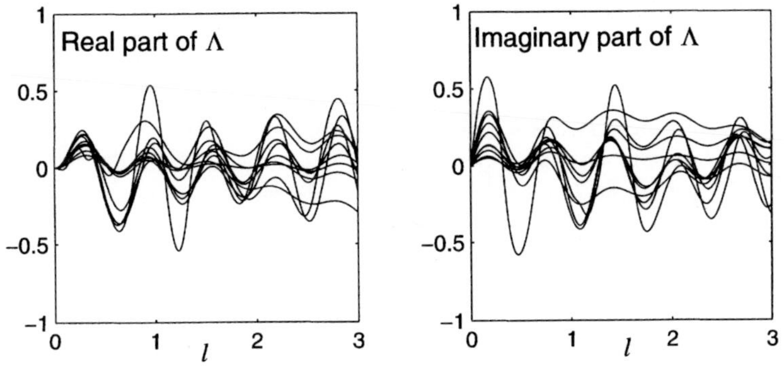

Figure 9.24. Ten samples of Λ in the range $(0, 3)$

from the above differential equation for Λ and the samples of Z in Fig. 9.25 for $k_0 = 5$, $\varepsilon = 0.5$, and $Z(x) = \exp(G(x))$, where G is a stationary Gaussian process with mean zero and covariance function $E[G(x)\, G(x')] = \exp(-|x - x'|)$, $x, x' \in (0, l)$.

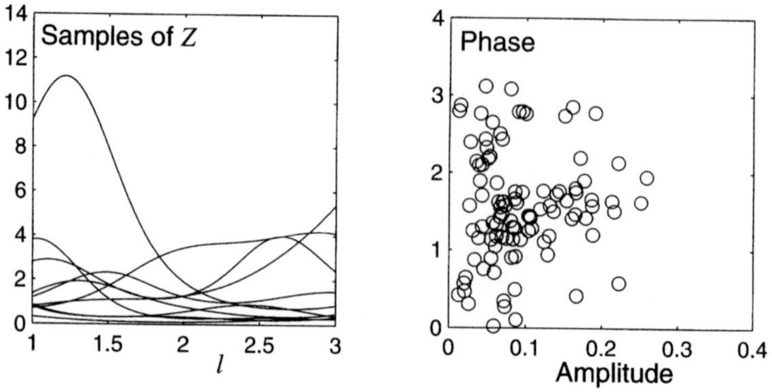

Figure 9.25. Ten samples of the field Z and 100 samples of the amplitude and phase corresponding to 100 samples of Z and $l = 3$

The solution of the wave equation in (l, ∞) is

$$e^{-\sqrt{-1}\,k_0\,x} + A\,e^{\sqrt{-1}\,(k_0\,x+\Phi)},$$

where A and Φ denote the random amplitude and phase of the reflected wave. The right graph in Figure 9.25 shows the amplitude of the reflected waves for 100 samples of Z and the corresponding phase of these waves relative to the incident wave for $l = 3$. The sample to sample variation of the amplitude and phase of the reflected waves is significant. Statistics of A and Φ can be calculated from Monte Carlo simulation results as shown in Fig. 9.25. For example, an estimate of the correlation coefficient $\rho_{A,\Phi}$ of (A, Φ) based on 1000 samples is -0.0195 for $l = 3$. This low value of $\rho_{A,\Phi}$ is consistent with the plot of the amplitude and phase of the reflected wave in Fig. 9.25 showing virtually no linear trend between these two random variables. $\diamond$

Proof: The solution of the wave equation in (l, ∞) is $U(x) = e^{-\sqrt{-1}\,k_0\,x} + \Lambda(l)\,e^{\sqrt{-1}\,k_0\,x}$. Denote by $\tilde{U}$ the solution of the wave equation in $(0, l)$. The continuity conditions at $x = l$ require $\tilde{U}(l) = U(l)$ and $\tilde{U}'(l) = U'(l)$. The ratio of these two conditions is

$$\frac{\tilde{U}'(l)}{\tilde{U}(l)} = \sqrt{-1}\,k_0\,\frac{\Lambda(l)\,e^{\sqrt{-1}\,k_0\,l} - e^{-\sqrt{-1}\,k_0\,l}}{\Lambda(l)\,e^{\sqrt{-1}\,k_0\,l} + e^{-\sqrt{-1}\,k_0\,l}}$$

so that

$$\Lambda(l) = e^{-2\sqrt{-1}\,k_0\,l}\,\frac{\sqrt{-1}\,k_0\,\tilde{U}(l) + \tilde{U}'(l)}{\sqrt{-1}\,k_0\,\tilde{U}(l) - \tilde{U}'(l)}$$

which gives the equation for Λ by differentiation with respect to l and by using continuity conditions at $x = 0$ [140].

We have not imposed any condition on ε in our discussion. If ε is a small parameter, it is possible to derive an evolution equation for the transition probability density of a stochastic process related to the reflection coefficient Λ. The derivation of this asymptotic result can be found in [140]. ■

9.7 Seismology

The energy supplied by the movement of the tectonic plates strains the earth's crust and can cause slips along faults. The fault slips are violent events called earthquakes. Large slips can produce earthquakes that may have devastating social and economical consequences. There are few large earthquakes but many small seismic events.

Seismologists have had a limited success in predicting the occurrence of individual future earthquakes. However, simple global laws for earth dynamics have been found from empirical observations, for example, the Gutenberg-Richter law $\log(n(m)) = a - b\,m$ and the energy-magnitude relationship $m = \alpha + \beta \log(e)$, where $n(m)$ is the number of earthquakes with magnitude larger then m, e denotes the energy released during an earthquake, and a, b, α, β are some constants. These two equations show that $\log(n(m))$ and $\log(e)$ are linearly related.

Large systems with many components tend to evolve into a critical, unstable state that can change abruptly because of minor disturbances. The critical state results solely from the interaction between the system components so that it is **self-organized**. A system may evolve in a critical state if it is large, receives a steady supply of energy, and is dissipative [16]. These conditions are satisfied by the collection of interacting crust faults. The system of tectonic plates extends over large regions, receives a steady supply of energy from the motion of tectonic plates, and dissipates energy through fault slips. This observation suggests that the Gutenberg-Richter law can be a manifestation of the self-organized critical behavior of earth dynamics [16].

We consider two earthquake fault models related to the plate tectonic theory. The first model, referred to as a physical model, consists of a collection of connected blocks that are pulled at a constant speed on a rough surface. The second model, called cellular automata, is a mathematical abstraction. The states of both models satisfy equations with random coefficients and random or deterministic inputs, which can be described by Eq. 9.1.

9.7.1 Physical model

Figure 9.26 shows a collection of equally spaced blocks sitting on a horizontal fixed plate with rough surface. The blocks are connected to each other by coil springs and to a moving plate by flat springs that pressure the blocks to the fixed plate. The velocity of the moving plate is constant and deterministic. If the force acting on a block in the horizontal plane does not exceed the friction between the block and the fixed plate, the block adheres to this plate. Otherwise, the block slips and the released force is transferred to the neighboring blocks.

Suppose that all blocks and springs are identical and the friction coefficient is the same at all points of the fixed plate. Then the blocks are in identical conditions so that either they are at rest or slip at once. If this ideal system is perturbed, then block slips are usually localized, and the energy release associated with slips

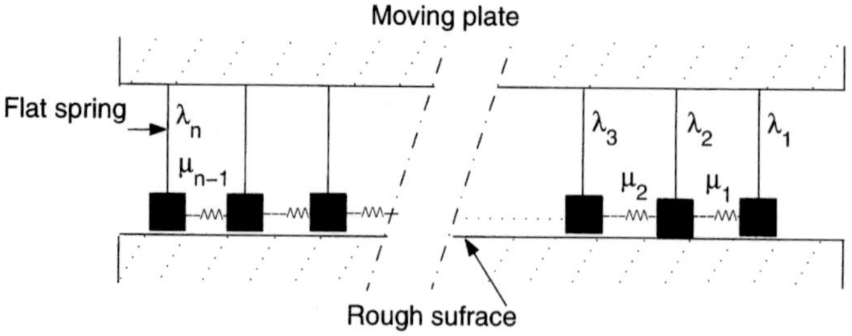

Figure 9.26. Physical model of interacting crust faults (adapted from [35], Fig. 3)

is small. If the friction at a block site is exceeded, the block slips and the released force may cause a single block or a collection of blocks to slip. It is possible that a single block slip will trigger the slip of many neighboring blocks resulting in a large avalanche of block slips and a significant reduction of the system potential energy. Generally, this is not the case, and block slips are confined to small neighborhoods. The block slips resemble the fault slips. The energies released by the slip of one or a few blocks and the slip of many blocks corresponds to small and large seismic events, respectively.

Figure 9.27 shows the time evolution of the potential energy $e(t)$ for the system in Fig. 9.26. Following an initial time interval when potential energy is built in the system, the energy reaches a steady-state regime characterized by small frequent and large infrequent negative jumps of magnitude $X(t) = |e(t) - e(t-)|$. Consider a large time interval in the steady-state regime of the potential energy

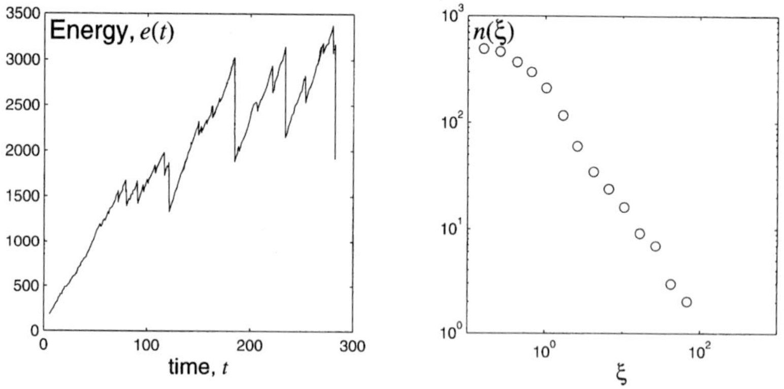

Figure 9.27. Potential energy and frequency-energy relation (adapted from [35],Figs. 4 and 5)

and let $n(\xi)$ denote the number of jumps of X in this time interval with magnitude larger than an arbitrary energy value $\xi > 0$. The resulting relationship between the logarithms of $n(\xi)$ and ξ shown in Fig. 9.27 is nearly linear with a slope of approximately -1.12, in agreement with the Gutenberg-Richter law [35].

The relationship between $n(\xi)$ and ξ in Fig. 9.27 implies that the jump process X associated with the potential energy must have a power-like tail, that is, $P(X(t) > \xi) \sim c\,\xi^{-\rho}$ as $\xi \to \infty$ for some $\rho, c > 0$ ([161], Section 2) because (1) the above power law implies that the logarithms of $P(X(t) > \xi)$ and ξ are linearly related and (2) $n(\xi)$ scaled by $n(0)$ is an estimate of $P(X(t) > \xi)$.

Note: The relationship between $n(\xi)$ and ξ in Fig. 9.27 gives $P(X(t) > \xi) \sim c\,\xi^{-1.12}$ for large values of ξ, and shows that the upper tail of the distribution of $X(t)$ is approximately Cauchy. Let Y be a Cauchy random variable with density $f(y) = 1/[\pi\,(1+y^2)]$ and distribution $F(y) = 1/2 + \arctan(y)/\pi$, $y \in \mathbb{R}$. Hence, $\lim_{y\to\infty} P(Y > y)/y^{-1} = 1/\pi$ so that $P(Y > y) \sim (1/\pi)\,y^{-1}$ as $y \to \infty$ and $\log(P(Y > y)) \sim \log(1/\pi) - \log(y)$ so that $\log(P(Y > y))$ is linearly related to $\log(y)$ and the slope of this linear relationship is -1 for Cauchy distributions.

We also note that the results in Fig. 9.27 are for a stochastic system subjected to a deterministic input since the properties of the rough surface and the velocity of the moving plate in Fig. 9.26 are random and deterministic, respectively. Problems characterized by stochastic systems and deterministic input have been examined in Chapter 8. We discuss the model in Fig. 9.26 here because it can be extended simply to the case in which the velocity of the moving plate is random. ▲

9.7.2 Cellular automata model

Consider a square lattice with n^2 sites and denote by $Z(i, j; t)$ the value of a random field defined on this lattice at time $t = 0, 1, \ldots$ and site (i, j), $i, j = 1, \ldots, n$. We can think of the random field Z as a model for forces acting on hypothetical blocks located at the lattice sites (Fig 9.26).

Suppose that the lattice is in equilibrium at a time $t \geq 0$, that is, $Z(i, j; t) \leq z_c$ at all nodes, where $z_c > 0$ is an integer that can be related to the magnitude of the friction force between the blocks and the fixed plate of the physical model in Fig. 9.26. The effect of the moving plate on the system in Fig. 9.26 can be simulated in cellular automata model by increasing the value of Z with unity at a site selected at random. This site selection attempts to model the spatial variability in the mechanical properties of a fault. If the updated value of Z exceeds z_c, the values of Z have to be modified according to a redistribution rule until the field Z does not exceed z_c at all sites.

The algorithm for evolving the random field Z involves the following three steps [16, 107]. Consider a time t and assume that at this time the lattice is in equilibrium, that is, we have $Z(i, j; t) < z_c$ for all $i, j = 1, \ldots, n$.

1. Select a site (i, j) at random and increase the value of Z at this site by unity, that is, $Z(i, j; t) \mapsto Z(i, j; t) + 1 = \tilde{Z}(i, j; t)$. The values of Z at all the other sites is not changed, that is, $Z(k, l; t) \mapsto Z(k, l; t) = \tilde{Z}(k, l; t)$ for $k \neq i$ and/or $l \neq j$.

2. If $\tilde{Z}(i, j; t) < z_c$, then set $Z(p, q; t + 1) = \tilde{Z}(p, q; t)$, $p, q = 1, \ldots, n$, and return to the previous step. If $\tilde{Z}(i, j; t) \geq z_c$, then the following values of the field are changed according to the rule: $\tilde{Z}(i, j; t) \mapsto \tilde{Z}(i, j; t) - 4$, $Z(i \pm 1, j; t) \mapsto Z(i \pm 1, j; t) + 1$, and $Z(i, j \pm 1; t) \mapsto Z(i, j \pm 1; t) + 1$.

3. Modify Z according to the rules in the previous step at all sites with values larger than z_c till equilibrium has been achieved. The resulting values of Z correspond to time $t + 1$ and are denoted by $Z(i, j; t + 1)$, $i, j = 1, \ldots, n$.

Note: The initial values of Z are taken zero, that is, $Z(i, j; 0) = 0$ at all nodes. The redistribution in the second step differs at boundary sites. For example, suppose that $Z(i, n; t) \geq z_c$ and $i \in \{2, \ldots, n-1\}$. Then $Z(i, n; t)$ is reduced by four, and $Z(i \pm 1, n; t)$, $Z(i, n - 1; t)$ are increased by unity. A unit of energy is allowed to escape through the boundary. The uncertainty in the input can be incorporated by replacing the first step with $Z(i, j; t) \mapsto Z(i, j; t) + Q$, where Q is a $\{1, \ldots, a\}$-valued random variable and $a > 0$ denotes an integer smaller than z_c.

An earthquake is generated during a time interval $(t, t + 1]$ if the application of the above algorithm causes one or more sites to exceed the critical value z_c. If at time t the values of $Z(i, j; t)$ are relatively small with respect to z_c, a local slip associated with an increase of Z at a site (step **1**) is unlikely to propagate in the lattice. On the other hand, if most of the values of Z are nearly equal to z_c, a local slip may trigger an avalanche of slips covering a large region of the lattice. These two extreme events correspond to small and large earthquakes, respectively. ▲

Let $\Theta(t) = \sum_{i,j=1}^{n} S(i, j; t)$, $t = 0, 1, \ldots, \tau$, and $X(t) = |\Theta(t) - \Theta(t - 1)|$, where $\tau > 0$ is an integer. The random variables $\Theta(t)$ and $X(t)$ can be interpreted as potential energy and earthquake magnitude, respectively, at time t. Denote by $N(x) = \#\{X(t) > x : t = 1, \ldots, \tau\}$ the number of earthquakes with magnitude larger than $x > 0$ in the time interval $[0, \tau]$. Then $\tilde{N}(x) = N(x)/N(0)$ is an estimate of the probability $P(X(t) > x)$.

Figure 9.28 shows the evolution in time of two samples of Θ for $n = 50$, $z_c = 3$, and $\tau = 30,000$. The significant difference in the details of the samples of Θ suggests that there is little hope to ever predict details of future seismic events, for example, the time and the magnitude of the next major earthquake. However, the corresponding samples of $\tilde{N}$ versus $\tilde{x} = x/n^2$ are quite similar, indicating that global features of the earthquake dynamics can be predicted. The relation between $\log(\tilde{N})$ and $\tilde{x}$ is nearly linear for both samples with slopes of approximately -1.17 and -1.29. This relationship is similar to the Gutenberg-Richter law introduced at the beginning of Section 9.7.

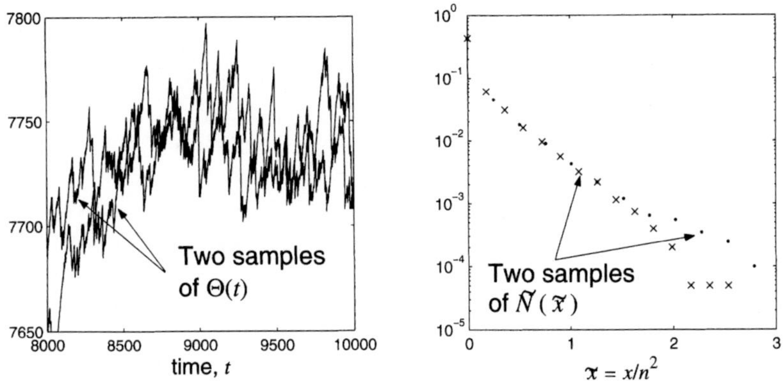

Figure 9.28. Two samples of Θ and the corresponding samples of $\tilde{N}$

9.8 Model selection

Our objective in this chapter and through the entire book has been to solve the **output prediction** problem ([18], Chapter 2), that is, to find properties of the output $\mathcal{X}$ in Eq. 9.1 for known probability law of $\mathcal{Y}$, functional form of $\mathcal{D}$, and probability laws of the random coefficients of $\mathcal{D}$. Generally, the probability law of $\mathcal{Y}$, the functional form of $\mathcal{D}$, and/or the probability laws of the random coefficients of $\mathcal{D}$ are not known. Records and information other than records, referred to in Bayesian inference as **prior information** ([201], Chapter II), can be used to characterize the input $\mathcal{Y}$ and the operator $\mathcal{D}$ of Eq. 9.1. For example, the prior information that the elasticity modulus is positive implies that Gaussian random fields cannot be used to model this material property. In stochastic hydrology it is common to augment the information provided by the record available at a particular river basin with prior information derived from records collected at similar river basins.

The following two sections outline difficulties related to the mathematical formulation of a particular stochastic problem caused by limited information and/or understanding of the input and the system features. The first section examines the sensitivity of the output $\mathcal{X}$ of Eq. 9.1 with a known deterministic operator $\mathcal{D}$ to a class of models for $\mathcal{Y}$ that are consistent with the available information. The last section presents a Bayesian framework for selecting an optimal model for $\mathcal{D}$ and $\mathcal{Y}$ from a collection of competing models. The coefficients of the resulting optimal model are random since the available information is finite.

9.8.1 Partially known input

Suppose that the operator $\mathcal{D}$ in Eq. 9.1 and the initial and boundary conditions for this equation are specified but the information on $\mathcal{Y}$ is limited to its marginal distribution F and correlation function $r(\tau) = E[\mathcal{Y}(t)\,\mathcal{Y}(t+\tau)]$. Our objective is to assess the sensitivity of the output $\mathcal{X}$ of Eq. 9.1 to **equivalent models** for $\mathcal{Y}$, that is, to a class of stochastic processes with marginal distribution F and correlation function r. Sensitivity is measured by the difference between estimates of the marginal distribution, the skewness coefficient, and the kurtosis coefficient of $\mathcal{X}$ corresponding to equivalent models for $\mathcal{Y}$.

In the following two examples we assume that $\mathcal{Y}$ is a stationary process with mean zero, lognormal marginal distribution F, and covariance function $r(\tau) = c(\tau) = \exp(-\lambda\,|\tau|)$, $\lambda > 0$. Two equivalent models are considered for $\mathcal{Y}$, a translation process Y_T and a diffusion process Y_D, defined by

$$\begin{aligned} Y_T(t) &= F^{-1} \circ \Phi(G(t)) \quad \text{and} \\ dY_D(t) &= -\lambda\, Y_D(t)\, dt + b(Y_D(t))\, dB(t), \end{aligned}$$

where G is a stationary Gaussian process with mean zero, variance 1, and covariance function $\rho(\tau) = E[G(t)\, G(t+\tau)]$, Φ denotes the distribution of $N(0,1)$, and B is a Brownian motion. The function ρ is such that Y_T has the specified exponential covariance function (Sections 3.6.6 and 5.3.3.1 in this book, [79], Section 3.1.1) and

$$b(y)^2 = \frac{-2\,\lambda}{f(y)} \int_{y_l}^{y} u\, f(u)\, du,$$

where y_l is the smallest value in the range of the lognormal density $f(y) = dF(y)/dy$ of $\mathcal{Y}$ [36].

Note: The stationary density f_D of Y_D is the solution of the Fokker-Planck equation

$$-\frac{d}{dy}\left(-\lambda\, y\, f_D(y)\right) + \frac{1}{2}\frac{d^2}{dy^2}\left(b(y)^2\, f_D(y)\right) = 0$$

so that $\lambda\, y\, f_D(y) + (1/2)\, d\left(b(y)^2\, f_D(y)\right)/dy = 0$ by the boundary conditions in Section 7.3.1.3. This equation yields the above expression for $b(\cdot)$ by requiring that f_D coincides with the target density f, and solving the resulting equation for $b(\cdot)$. ▲

We denote by X_T and X_D the stationary outputs of Eq. 9.1 to the processes Y_T and Y_D, respectively. In the following two examples the processes Y_T and Y_D modeling $\mathcal{Y}$ follow the same shifted lognormal marginal distribution with zero-mean and unit-variance and the same exponential correlation function with decay parameter $\lambda = 1$, so that they are equivalent.

Example 9.43: Suppose that $\mathcal{D} = d^2/dt^2 + 2\,\zeta\,\nu\, d/dt + \nu^2$. This operator defines a linear oscillator with natural frequency $\nu > 0$ and damping ratio $0 < \zeta < 1$.

The mean and the covariance functions of X_T and X_D coincide since $\mathcal{D}$ is linear and Y_T and Y_D have the same second moment properties (Section 7.2.1.1). However, higher order statistics of X_T and X_D differ. Figure 9.29 shows samples of X_T and X_D corresponding to samples of Y_T and Y_D for $\nu = 10$ and $\zeta = 0.1$.

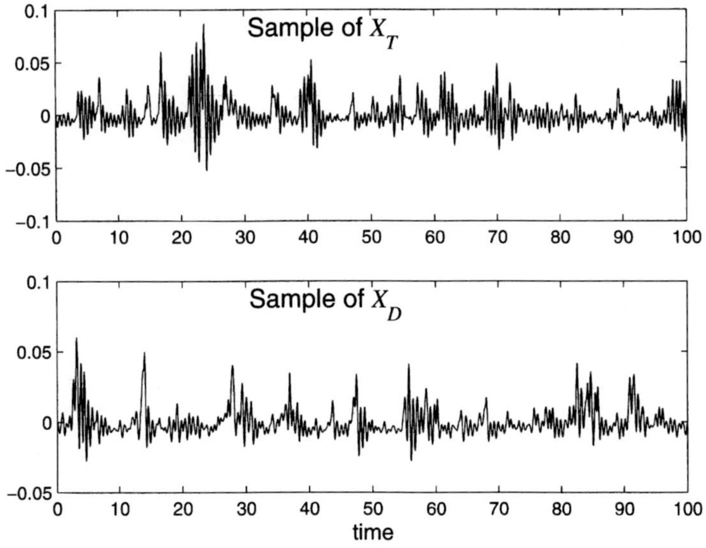

Figure 9.29. Samples of X_T and X_D corresponding to samples of Y_T and Y_D for a linear oscillator with $\nu = 10$ and $\zeta = 0.1$

Estimates of the skewness and kurtosis coefficients, γ_3 and γ_4, are $\hat{\gamma}_3 = 1.34$ and $\hat{\gamma}_4 = 7.6829$ for X_T and $\hat{\gamma}_3 = 1.81$ and $\hat{\gamma}_4 = 8.06$ for X_D. The differences between these estimates are relatively small. However, the differences between the marginal histograms of X_T and X_D and estimates of the probabilities $P(X_T > x)$ and $P(X_D > x)$ in Fig. 9.30 indicate that reliability measures limiting the allowable values of X are sensitive to the particular model used for $\mathcal{Y}$. All estimates are based on 1,000 independent output samples ◇

Example 9.44: Suppose that $\mathcal{D} = d^2/dt^2 + 2\,\zeta\,\nu\,d/dt + k(\cdot)$, where $0 < \zeta < 1$, $k(y) = \nu^2\,y$ if $|y| \leq a$ and $k(y) = \nu^2\,[a + b(1 - \exp(-\beta(|y| - a)))]\,\mathrm{sign}(y)$ if $|y| \geq a$, and $a, b, \nu > 0$ are constants. The operator $\mathcal{D}$ defines a nonlinear oscillator with restoring force $k(\cdot)$ and damping ratio ζ.

The processes X_T and X_D are not equal in the second moment sense since $\mathcal{D}$ is nonlinear. For example, estimates of the variance of these two processes are 0.01 and 0.017, respectively. Figure 9.31 shows samples of X_T and X_D corresponding samples of Y_T and Y_D for $\nu = 10$, $\zeta = 0.1$, $a = 0.04$, and $b = 0.03$. Estimates of the skewness and kurtosis coefficients differ significantly. These estimates are $\hat{\gamma}_3 = 3.03$ and $\hat{\gamma}_4 = 23.24$ for X_T and $\hat{\gamma}_3 = 2.29$ and $\hat{\gamma}_4 = 12$ for X_D. Differences between properties of X_T and X_D can also be seen in Fig. 9.32 showing histograms of X_T and X_D and estimates of the probabilities $P(X_T > x)$ and $P(X_D > x)$. All estimates are based on 1,000 independent output samples. ◇

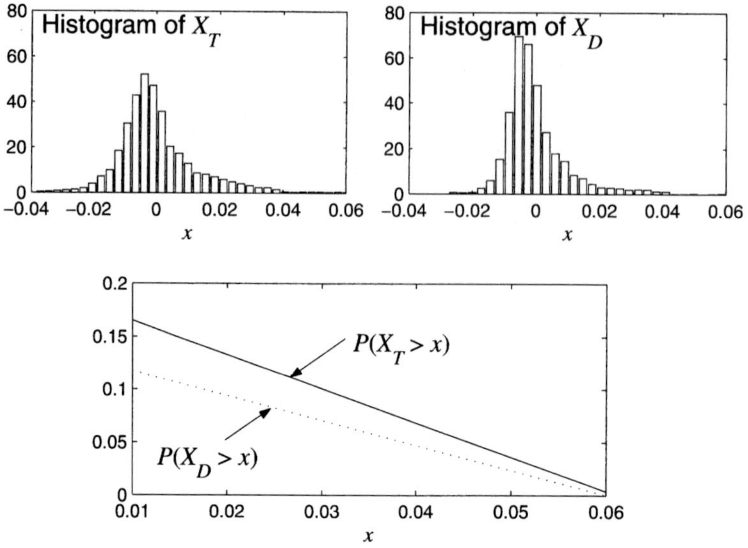

Figure 9.30. Histograms of X_T and X_D and estimates of the probabilities $P(X_T > x)$ and $P(X_D > x)$ for a linear oscillator with $\nu = 10$ and $\zeta = 0.1$

In Examples 9.43 and 9.44 we have assumed that (1) the operator $\mathcal{D}$ in Eq. 9.1 is perfectly known and has deterministic coefficients and (2) the input $\mathcal{Y}$ to this equation is a real-valued stationary process with known marginal distribution and correlation function. We have seen that the properties of the solution $\mathcal{X}$ of Eq. 9.1 depend on the particular model used for $\mathcal{Y}$, although the models considered for $\mathcal{Y}$ are equivalent in the sense that they have the same marginal distribution and correlation function. These results are of concern since the information available in most applications rarely exceeds the information postulated in Examples 9.43 and 9.44, and suggest that, at least for important systems, we need to find properties of $\mathcal{X}$ in Eq. 9.1 for a collection of equivalent models of $\mathcal{Y}$.

9.8.2 Partially known system and input

Consider a stochastic problem of the type in Eq. 9.1 and suppose that the properties of the operator $\mathcal{D}$ and the input $\mathcal{Y}$ are not known. The selection of the operator $\mathcal{D}$, referred to as **system identification** ([18], Chapter 2), requires the specification of the functional form of $\mathcal{D}$ and the coefficients of this functional form. However, it is common to limit the solution of the system identification problem to the estimation of the coefficients of $\mathcal{D}$ assumed to have a known functional form ([18], Chapter 2). This partial solution of the system identification problem can be unsatisfactory.

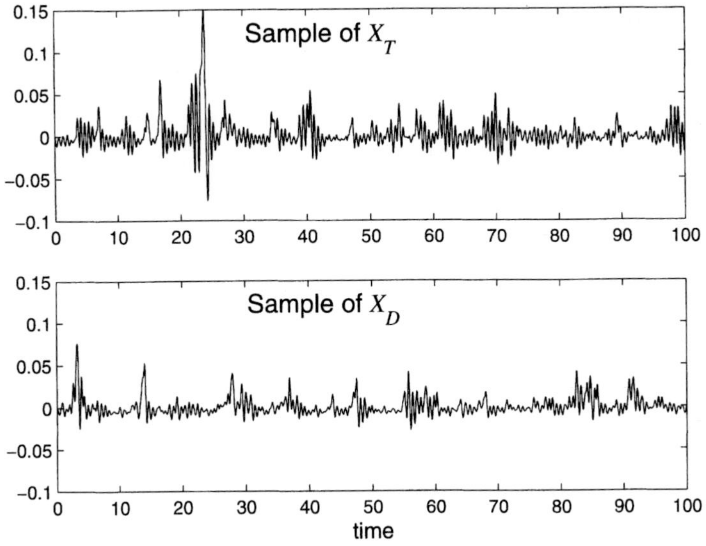

Figure 9.31. Samples of X_T and X_D corresponding to samples of Y_T and Y_D for a nonlinear oscillator with $\nu = 10$, $\zeta = 0.1$, $a = 0.04$, and $b = 0.03$

Example 9.45: Let Eq. 9.1 be a memoryless relation between two random variables, the input $\mathcal{Y}$ and the output $\mathcal{X}$. Suppose that Eq. 9.1 is $\mathcal{X} = \mathcal{D}^{-1}[\mathcal{Y}] = \mathcal{Y}^2$ and we chose arbitrarily to model this equation by a linear regression $\hat{\mathcal{X}} = a\,\mathcal{Y}+b$, where (a, b) are unknown coefficients. If n independent samples (y_i, x_i), $i = 1, \ldots, n$, of these random variables are available, a and b can be estimated by

$$\hat{a} = \frac{\sum_{i=1}^n x_i\, y_i/n - (\sum_{i=1}^n x_i/n)(\sum_{i=1}^n y_i/n)}{\sum_{i=1}^n y_i^2/n - (\sum_{i=1}^n y_i/n)^2} \quad \text{and } \hat{b} = \sum_{i=1}^n \frac{x_i}{n} - \hat{a} \sum_{i=1}^n \frac{y_i}{n}.$$

The selected linear regression model may provide no information on the relation between $\mathcal{X}$ and $\mathcal{Y}$. For example, if $\mathcal{Y} \sim N(0, 1)$, the estimates $(\hat{a}, \hat{b})$ converge to $(0, 1)$ as $n \to \infty$ so that the regression line becomes $\hat{\mathcal{X}} = 1$ under perfect information. ◇

Note: The estimates $\hat{a}$ and $\hat{b}$ minimize the error $\sum_{i-1}^n (a\, y_i + b - x_i)^2$ and have the above expressions. Note that $\hat{a}$ and $\hat{b}$ converge to γ_{xy}/γ_{yy} and $\mu_x - \left(\gamma_{xy}/\gamma_{yy}\right) \mu_y$ as $n \to \infty$, where (μ_x, μ_y) denote the means of $(\mathcal{X}, \mathcal{Y})$, γ_{xy} is the covariance of $(\mathcal{X}, \mathcal{Y})$, and γ_{yy} denotes the variance of $\mathcal{Y}$. If $\mathcal{Y} \sim N(0, 1)$, then $\mu_y = 0$, $\mu_x = E[\mathcal{X}] = E[\mathcal{Y}^2] = 1$, and $\gamma_{xy} = E[\mathcal{X}\,\mathcal{Y}] = E[\mathcal{Y}^3] = 0$, so that $\hat{a} \to 0$ and $\hat{b} \to 1$ as $n \to \infty$. ▲

We now present a method for solving the system identification problem, which selects the optimal models for $\mathcal{D}$ and $\mathcal{Y}$ in Eq. 9.1 from a specified collection of competing models with known functional form but unknown coefficients.

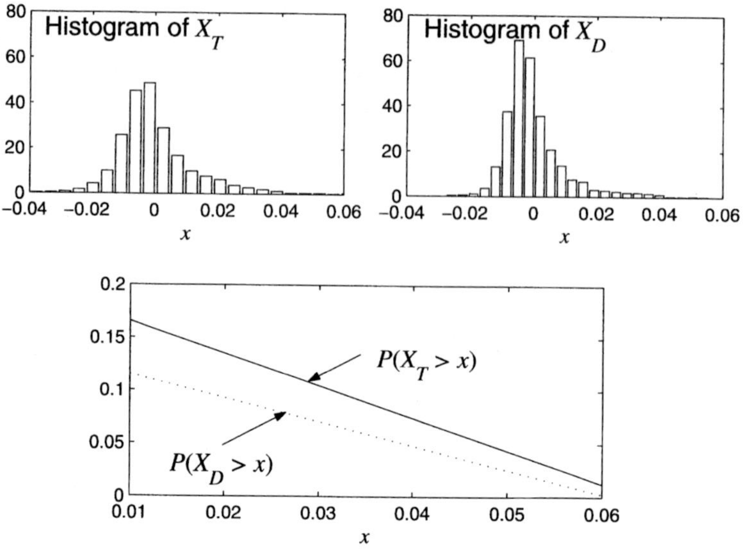

Figure 9.32. Histograms of X_T and X_D and estimates of the probabilities $P(X_T > x)$ and $P(X_D > x)$ for a linear oscillator with $\nu = 10$, $\zeta = 0.1$, $a = 0.04$, and $b = 0.03$

The method is based on a Bayesian inference approach, and is illustrated for the case in which Eq. 9.1 is a Gaussian autoregressive process with unknown order, coefficients, and input noise scale. We denote the output $\mathcal{X}$ and the input $\mathcal{Y}$ in Eq. 9.1 by X and W, respectively.

Let $\mathcal{M}_k$, $k = 1, 2, \ldots, m$, be a collection of autoregressive Gaussian processes with state X defined by the recurrence formula

$$X_{t+1} = \beta_0 + \beta_1 X_t + \cdots + \beta_k X_{t-k+1} + W_t, \tag{9.33}$$

where W_t are independent $N(0, 1/h_k)$ variables, $h_k > 0$, $(\beta_0, \ldots, \beta_k, h_k)$ are unknown coefficients, and $m \geq 1$ is a specified integer defining the size of the collection of competing models.

The ingredients of the Bayesian method for model selection are:

- **Prior information** on the collection of competing models, that is, the value of $m \geq 1$ defining the largest autoregressive model considered in the analysis, the prior densities $f'_k(\boldsymbol{\beta}_k, h_k)$ of the coefficients $\boldsymbol{\beta}_k = (\beta_0, \beta_1, \ldots, \beta_k)$ and h_k in the definition of $\mathcal{M}_k$, $k = 1, \ldots, m$, and the prior probabilities $p'_k > 0$, $\sum_{k=1}^{m} p'_k = 1$, where p'_k is the initial likelihood that $\mathcal{M}_k$ is the correct model. An extensive discussion on the selection of $f'_k(\boldsymbol{\beta}_k, h_k)$ can be found in [201] (Chapters II and III). It is possible to define non-informative prior densities, that is, densities that carry no information regarding the parameters $\boldsymbol{\beta}_k$ and h_k. Physics must also be included in the selection of the competing models. For example, Gaussian models are inadequate for physical parameters known to have bounded values.

- **Measurements** of the state X consisting of n consecutive readings, that is, a record $z = (z_1, \ldots, z_n)$ of $(X_1, \ldots, X_n)$ with size $n > m$, referred to as an observation vector. In our discussion we consider perfect measurements. Noisy measurements can be incorporated simply in the analysis.

- **Utility** function, which is problem specific and may depend on the model use. Utility functions are needed in many applications since the observation vector z is frequently insufficient to provide a reliable solution to the model selection problem. Let $l(\mathcal{M}_k, \mathcal{M}_q)$ be a utility function giving the penalty if the model $\mathcal{M}_k$ is selected and the correct model is $\mathcal{M}_q$ and let $E[\mathcal{M}_k] = \sum_{q=1}^{m} l(\mathcal{M}_k, \mathcal{M}_q)\, p_q''$ denote the expected utility associated with the selection of the model $\mathcal{M}_k$, where p_q'' denotes the posterior probability that $\mathcal{M}_q$ is the correct model and will be calculated later in this section (Eq. 9.37). The **optimal model** minimizes $E[\mathcal{M}_k]$ [89]. We take $l(\mathcal{M}_k, \mathcal{M}_q) = 1 - \delta_{kq}$ in the following example, so that the optimal model will have the highest posterior probability p_k''.

Our analysis delivers (1) an optimal model for $\mathcal{D}$ and $\mathcal{Y}$, that is, the model $\mathcal{M}_{k^*}$ with $E[\mathcal{M}_{k^*}] \geq E[\mathcal{M}_k]$ for $k \neq k^*$ and (2) the posterior densities $f_k''(\boldsymbol{\beta}_k, h_k)$ of the coefficients of the models $\mathcal{M}_k$ and the probabilities p_k'', $k = 1, \ldots, m$. The resulting optimal model $\mathcal{M}_{k^*}$ can be used to estimate extremes, integrals, and other properties of X relevant for reliability studies [89]. The coefficients of $\mathcal{M}_{k^*}$ are random because of the limited information. We now give formulas for calculating f_k'' and p_k''.

Consider the model $\mathcal{M}_k$ in Eq. 9.33 and its regression form

$$\boldsymbol{V}_k = \boldsymbol{a}_k\, \boldsymbol{\beta}_k + \frac{1}{\sqrt{h_k}}\, \boldsymbol{G}_k \tag{9.34}$$

corresponding to an observation vector z, where $\boldsymbol{G}_k$ is an $(n-k, 1)$ matrix with independent $N(0, 1)$ entries, $\boldsymbol{V}_k = (X_{k+1}, \ldots, X_n)$ denotes a column vector,

$$\boldsymbol{a}_k = \begin{bmatrix} 1 & z_k & \cdots & z_1 \\ 1 & z_{k+1} & \cdots & z_2 \\ \vdots & \vdots & & \vdots \\ 1 & z_{n-1} & \cdots & z_{n-k} \end{bmatrix}, \tag{9.35}$$

and $\boldsymbol{\beta}_k = (\beta_0, \beta_1, \ldots, \beta_k)^T$ is a column vector. Given $z = (z_1, \ldots, z_n)$, the vector $\boldsymbol{V}_k$ and the matrix $\boldsymbol{a}_k$ are perfectly known so that Eq. 9.34 becomes a multiple regression model. Methods for regression analysis can be used to find the **posterior density** $f_k''(\boldsymbol{\beta}_k, h_k)$ of the coefficients $\boldsymbol{\beta}_k$ and h_k of $\mathcal{M}_k$ as well as the **posterior probability** p_k'' that $\mathcal{M}_k$ is the correct model, $k = 1, \ldots, m$. The posterior densities f_k'' and probabilities p_k'' incorporate the prior information consisting of f_k' and p_k' and the additional information provided by the observation vector z.

If $f_k'(\boldsymbol{\beta}_k, h_k)$ is a normal-gamma density with parameters $(\boldsymbol{\beta}_k', \boldsymbol{\gamma}_k', \rho_k', \nu_k')$ and $n > k$, then $f_k''(\boldsymbol{\beta}_k, h_k)$ is also a normal-gamma density with parameters

$$\begin{aligned}
(\boldsymbol{\gamma}_k'')^{-1} &= (\boldsymbol{\gamma}_k')^{-1} + \boldsymbol{a}_k^T\, \boldsymbol{a}_k \\
\boldsymbol{\beta}_k'' &= \boldsymbol{\gamma}_k'' \left[(\boldsymbol{\gamma}_k')^{-1}\, \boldsymbol{\beta}_k' + \boldsymbol{a}_k^T\, z_k \right] \\
\rho_k''\, \nu_k'' &= \rho_k'\, \nu_k' + (\boldsymbol{\beta}_k')^T\, (\boldsymbol{\gamma}_k')^{-1}\, \boldsymbol{\beta}_k' - (\boldsymbol{\beta}_k'')^T\, (\boldsymbol{\gamma}_k'')^{-1}\, \boldsymbol{\beta}_k'' + z_k^T\, z_k \\
\nu_k'' &= \nu_k' + n - k,
\end{aligned} \tag{9.36}$$

where $z_k = (z_{k+1}, \ldots, z_n)$ is a column vector.

Note: Consider a model $\mathcal{M}_k$ defined by Eq. 9.33. Under the above assumption on the prior density of $(\boldsymbol{\beta}_k, h_k)$, we have $\boldsymbol{\beta}_k \mid h_k \sim N(\boldsymbol{\beta}_k', \boldsymbol{\gamma}_k'/h_k)$ is a Gaussian vector with mean $\boldsymbol{\beta}_k'$ and covariance matrix $\boldsymbol{\gamma}_k'/h_k$ and $h_k \sim G2(\rho_k', \nu_k')$ follows a gamma-2 distribution with parameters (ρ_k', ν_k'), that is, the prior density of h_k is ([91], p. 226)

$$f_{h_k}'(\xi) = \frac{(\rho_k'\, \nu_k')^{\nu_k'/2}}{\Gamma(\nu_k'/2)}\, \xi^{\nu_k'/2-1}\, e^{-\rho_k'\, \nu_k'\, \xi/2}.$$

A density $f_k'(\boldsymbol{\beta}_k, h_k)$ is said to be a **conjugate prior density** if f_k' and f_k'' have the same functional form, for example, the prior and posterior densities in Eq. 9.36. The posterior density $f_k''(\boldsymbol{\beta}_k, h_k)$ can be obtained by applying the Bayes formula (Section 2.3.4 in this book, [201], Section 3.2.3).

The results in Eq. 9.36 can be used to calculate the **simple predictive distribution**, that is, the distribution of $Y^{(k)} = X_{n+1}$ conditional on $\mathcal{M}_k$ and based on the prior information $f_k'(\boldsymbol{\beta}_k, h_k)$ and the observation vector $z = (z_1, \ldots, z_n)$. We have $Y^{(k)} = \tilde{z}_k^T\, \boldsymbol{\beta}_k + W_n$ from Eq. 9.33 with $t = n$, where $\tilde{z}_k = (1, z_n, \ldots, z_{n-k+1})$ is a column vector. Hence, $Y^{(k)} \mid h_k \sim N\left(\tilde{z}_k^T\, \boldsymbol{\beta}_k'', (1 + \tilde{z}_k^T\, \boldsymbol{\gamma}_k''\, \tilde{z}_k)/h_k\right)$ so that the density of $Y^{(k)}$ is given by

$$\tilde{f}_k(y) = \int_0^\infty \tilde{f}_k(y \mid \xi)\, f_{h_k}''(\xi)\, d\xi,$$

where $\tilde{f}_k(y \mid \xi)$ is the density of $Y^{(k)} \mid (h_k = \xi)$ and f_{h_k}'' denotes the posterior density of h_k. The above integral shows that the random variable $(Y^{(k)} - \tilde{z}_k^T\, \boldsymbol{\beta}_k'')/\sqrt{\nu_k''\, (1 + \tilde{z}_k^T\, \boldsymbol{\gamma}_k''\, \tilde{z}_k)}$ is a standard Student t with ν_k'' degrees of freedom ([201], Section 3.2.4). ▲

The posterior probability that $\mathcal{M}_k$ is the correct model is

$$p_k'' = c\, p_k'\, f_k(z_k), \quad \text{where } c^{-1} = \sum_{k=1}^{m} p_k'\, f_k(z_k) \tag{9.37}$$

$$f_k(z_k) = \int f_k(z_k \mid \boldsymbol{\beta}_k, h_k)\, f_k'(\boldsymbol{\beta}_k, h_k)\, d\boldsymbol{\beta}_k\, dh_k, \tag{9.38}$$

$f_k(z_k \mid \boldsymbol{\beta}_k, h_k)$ is the density of $\boldsymbol{a}_k\, \boldsymbol{\beta}_k + \boldsymbol{G}_k/\sqrt{h_k}$ conditional on $(\boldsymbol{\beta}_k, h_k)$, $z_k = (z_{k+1}, \ldots, z_n)$ (Eq. 9.36), and $f_k'(\boldsymbol{\beta}_k, h_k)$ denotes the prior density of the coefficients $(\boldsymbol{\beta}_k, h_k)$ of $\mathcal{M}_k$.

Note: The vector $\boldsymbol{a}_k\, \boldsymbol{\beta}_k + \boldsymbol{G}_k/\sqrt{h_k}$ is Gaussian with mean $\boldsymbol{a}_k\, \boldsymbol{\beta}_k$ and covariance matrix $\boldsymbol{i}_k/h_k$ conditional on $(\boldsymbol{\beta}_k, h_k)$, where $\boldsymbol{i}_k$ is an identity matrix. Generally, numerical integration is needed to find $f_k(z_k)$. Straightforward calculations show that, if $f_k'(\boldsymbol{\beta}_k, h_k)$ is the normal-gamma density in Eq. 9.36, then

$$f_k(z_k) = (2\pi)^{-(\nu_k''-\nu_k')/2}\, \frac{|\boldsymbol{\gamma}_k''|^{1/2}}{|\boldsymbol{\gamma}_k'|^{1/2}}\, \frac{(\rho_k'\, \nu_k'/2)^{\nu_k'/2}}{(\rho_k''\, \nu_k''/2)^{\nu_k''/2}}\, \frac{\Gamma(\nu_k''/2)}{\Gamma(\nu_k'/2)},$$

where $\Gamma(\cdot)$ denotes the Gamma function. Extensive considerations on the selection of the prior density can be found, for example, in [91] (Chapter 3) and [201] (Section 8.4). ▲

Example 9.46: Two sets of samples of length $n = 10, 20, 50, 100, 150, 200$, and 250 have been generated from an autoregressive Gaussian process X of order $k = 3$ with parameters $\beta_0 = 0$, $\beta_1 = 0.7$, $\beta_2 = -0.5$, $\beta_3 = -0.3$, and $1/h_k = 1$ (Eq. 9.33). We assume that the samples are of unknown origin, consider a collection $\mathcal{M}_k$, $k = 1, 2, 3$, of autoregressive process, and apply the method in this section to find the optimal model $\mathcal{M}_{k^*}$ for X.

Figure 9.33 shows the dependence on the sample size of the posterior probabilities p_k'' in Eq. 9.37 for the two sets of samples generated from $\mathcal{M}_3$. The posterior probabilities p_1'' are nearly zero for all values of n and are not shown. The posterior probabilities p_2'' and p_3'' vary significantly from sample to sample and with the sample size. However, the method delivers the correct model as the available information, that is, the sample size n of the record, increases. The parameters $(\beta_0, \beta_1, \beta_2, \beta_3, h)$ of the optimal model $\mathcal{M}_{k^*} = \mathcal{M}_3$ are random because n is finite. For example, the means of the parameters $(\beta_0, \beta_1, \beta_2, \beta_3)$ of this model are

$$(1.0056, 0.6736, -0.4555, -0.3499) \quad \text{and} \quad (1.1558, 0.6636, -0.4734, -0.3330)$$

for the two samples of X with length $n = 250$ considered in Fig. 9.33. The standard deviations of these parameters are

$$(0.0885, 0.0591, 0.0660, 0.0591) \quad \text{and} \quad (0.0826, 0.0596, 0.0676, 0.0598),$$

respectively. The posterior density of $(\beta_0, \beta_1, \beta_2, \beta_3, h)$ given by Eq. 9.36 can be used to find properties of the output of Eq. 9.1 at this information level.

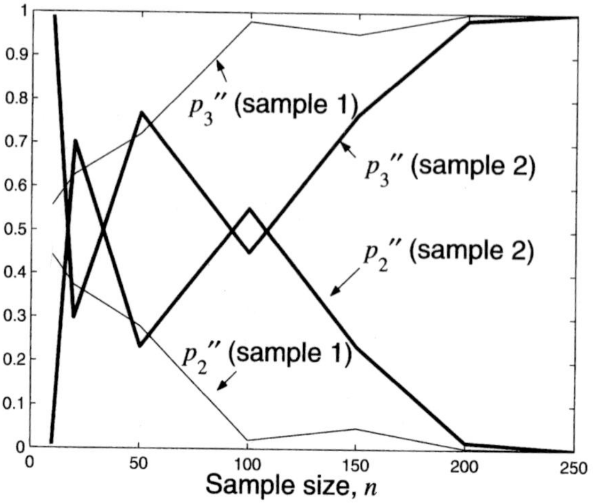

Figure 9.33. Posterior probabilities of second and third order autoregressive Gaussian processes

The evolution of the posterior probabilities in Fig. 9.33 shows that $\mathcal{M}_2$ can be the optimal model if the sample size n is not sufficiently large although (1) the available samples have been generated from $\mathcal{M}_3$ and (2) $\mathcal{M}_2$ is a special case of $\mathcal{M}_3$. That simpler models may be superior to more general models for small samples is not surprising. For small values of n the uncertainty in the coefficients of $\mathcal{M}_3$ is so large that its probability law becomes inconsistent with the observation vector. ◇

Note: The prior information used in the analysis has been derived from five consecutive values of X generated from $\mathcal{M}_3$ with $\beta_0 = 0$, $\beta_1 = 0.7$, $\beta_2 = -0.5$, $\beta_3 = -0.3$, and $1/h_k = 1$. This prior information has virtually no influence on the outcome in Fig. 9.33 since the sample size n considered in analysis is much larger.

Because the absolute values of the roots of the polynomial $1 - \beta_1 z - \beta_2 z^2 - \beta_3 z^3$ with $\beta_1 = 0.7$, $\beta_2 = -0.5$, and $\beta_3 = -0.3$ differ from one, the state of the model $\mathcal{M}_3$ becomes stationary as t increases ([31], Theorem 3.1.3, p. 88). The samples used in Fig. 9.33 correspond to the stationary state of $\mathcal{M}_3$. ▲

9.9 Problems

9.1: Find the local solution of the partial differential equation in Example 9.3 with Z_1 and Z_2 replaced by the random fields $a_1 + \exp(G_1(\boldsymbol{x}))$ and $-a_2 + G_2(\boldsymbol{x})$, respectively, where $a_1, a_2 > 0$ are constants, $\boldsymbol{x} \in D$, G_1 and G_2 are homogeneous Gaussian fields with mean zero and covariance functions $c_k(\boldsymbol{\xi}) = e^{-\lambda_k \|\boldsymbol{\xi}\|}$, $k = 1, 2$, and G_1 is independent of G_2. For numerical calculations take $a_1 = 1$, $a_2 = 5$, and $\lambda_1 = \lambda_2 = 5$.

9.2: Find the local solution of

$$\frac{\partial U(\boldsymbol{x},t)}{\partial t} = A(\boldsymbol{x})^2 \sum_{i=1}^{d} \frac{\partial^2 U(\boldsymbol{x},t)}{\partial x_i^2}, \quad \boldsymbol{x} \in D, \quad t \geq 0,$$

where D is an open bounded subset of $\mathbb{R}^d$ and $A > 0$ denotes a random field. Assume the initial and boundary conditions $U(\boldsymbol{x},0) = \alpha > 0$, $\boldsymbol{x} \in D$, and $U(\boldsymbol{x},t) = 0$ for $\boldsymbol{x} \in \partial D$ and $t \geq 0$, respectively.

9.3: Find U in Example 9.3 by conditional analysis. Use any numerical method to calculate the solution $U \mid (Z_1, Z_2)$.

9.4: Find U in Example 9.3 by the first order Taylor approximation.

9.5: Find the first four moments of X in Example 9.7 by a closure method. Compare the resulting moments with Monte Carlo solutions.

9.6: Find the function $c(z_2)$ in the expression for the stationary density of the state vector $\mathcal{Z}$ in Example 9.7.

9.7: Let X be the solution of $(a + R)\, X(t) = Q(t)$, $t \geq 0$, where R is a real-valued random variable such that $|R/a| < 1$ a.s. and Q is a stochastic process in L_2. Find the Neumann series representation of X and examine its convergence. Calculate the second moment properties of X.

9.8: Find the second moment properties of $\tilde{X}$ in Example 9.8 by one of the closure methods in Section 7.3.1.5.

9.9: Find the function $c(z_2, z_3)$ in the expression for the stationary density of the state vector $\mathcal{Z}$ in Example 9.8.

9.10: Let X be the solution of $\dot{X}(t) = -a\, X(t)^2$ for $a > 0$, $t \geq 0$, and $X(0) = Z$, where Z is a real-valued random variable. Find the density of X at some time $t \geq 0$ by using Eq. 9.3 and Eq. 9.5.

9.11: Complete the solution of the problem in Example 9.20.

9.12: Solve the problem in Example 9.20 by the Galerkin method with Berstein polynomials as trial functions.

9.13: Find the first two moments of U in Example 9.3 by the stochastic finite element method.

9.14: Complete the details of the stochastic finite element analysis in Example 9.30 based on polynomial chaos.

9.15: Find the second order approximation of the solution of the steady-state Boltzmann equation in Example 9.31.

9.16: Calculate the Lyapunov exponent for the stochastic differential equation in Example 9.35. Discuss the stochastic stability of the stationary solution of this equation as a function of ρ.

9.17: Suppose that K_r, $r = 1, 2, 3$, $\boldsymbol{X}(0)$, Z, and K_{2p} in Example 9.38 are real-valued random variables. Find the marginal density of the state $\boldsymbol{X}(t)$ at an arbitrary time $t \geq 0$.

9.18: Find the local solution for Eq. 9.31 and a generalized version of this equation for the case in which the dispersion tensor d_{ij} is random.

9.19: Repeat the calculations in Examples 9.43 and 9.44 for other input processes, which are equivalent in some sense. Also, assume that some of the coefficients of the differential equations in these examples are random.

9.20: Complete the details of the proof presented for Eqs. 9.36 and 9.38.

9.21: Extend results in Example 9.46 to the case in which the observations are imperfect, that is, the sample values z_i are replaced with $z_i + U_i$, where U_i are independent copies of a real-valued random variable with mean zero and known variance.

9.22: Let X be a stationary Gaussian process with mean zero and covariance function $c(\tau) = E[X(t)\, X(t + \tau)]$. The covariance function of X is not given. It is known that c is one of the functions $c_1(\tau) = \sigma^2 \sin(\nu_c \tau)/(\nu_c \tau)$, $c_2(\tau) = \sigma^2 \exp(-\alpha\, |\tau|)$, or $c_3(\tau) = \sigma^2 (1 + \beta\, |\tau|) \exp(-\beta\, |\tau|)$, where $\sigma, \nu_c, \alpha, \beta > 0$ are some unknown constants. Generate a sample of length n of X with $c = c_1$, $\sigma = 1$, and $\nu_c = 30$. Suppose that the resulting sample is of unknown origin. Apply the approach in Example 9.46 to find the optimal model for X from the collection of stationary Gaussian processes with mean zero and covariance functions c_k, $k = 1, 2, 3$.

Bibliography

[1] R. J. Adler. *The Geometry of Random Fields*. John Wiley & Sons, New York, 1981.

[2] G. Adomian, editor. *Applied Stochastic Processes*. Academic Press, New York, 1980.

[3] G. Adomian. *Stochastic Systems*. Academic Press, New York, 1983.

[4] B. K. Agarwal and M. Eisner. *Statistical Mechanics*. John Wiley & Sons, New York, 1988.

[5] T. M. Apostol. *Mathematical Analysis*. Addison-Wesley Publishing Company, Reading, Massachusetts, 1974.

[6] S. T. Ariaratnam. Some illustrative examples of stochastic bifurcation. In J. M. T. Thompson and S. R. Bishop, editors, *Nonlinearity and Chaos in Engineering Dynamics*, pages 267–274. John Wiley & Sons, Ldt., New York, 1994.

[7] S. T. Ariaratnam and N. M. Abdelrahman. Almost sure stochastic stability of viscoelastic plates in supersonic flow. *AIAA Journal*, 39(3):465–472, 2001.

[8] S. T. Ariaratnam and H. N. Pi. On the first-passage time for envelope crossing for a linear oscillator. *International Journal of Control*, 18(1):89–96, 1973.

[9] S. T. Ariaratnam and W. C. Xie. Almost-sure stochastic stability of coupled non-linear oscillators. *International Journal of Non-Linear Mechanics*, 29(2):197–204, 1994.

[10] L. Arnold. A formula connecting sample and moment stability of linear stochastic systems. *SIAM Journal of Applied Mathematics*, 1994.

[11] S. R. Arwade. Probabilistic models for aluminum microstructure and intergranular fracture analysis. Master's thesis, Cornell University, Civil and Environmental Engineering, Ithaca, NY, 1999.

[12] S. R. Arwade. *Stochastic characterization and simulation of material microstructures with applications to alluminum.* PhD thesis, Cornell University, Ithaca, NY, January 2002.

[13] S. F. Asokanthan and S. T. Ariaratnam. Almost-sure stability of a gyropendulum subjected to white noise random support motion. *Journal of Sound and Vibration*, 235(5):801–812, 2000.

[14] M. Avlonis, M. Zaiser, and E. C. Aifantis. Some exactly solvable models for the statistical evolution of internal variables during plastic deformation. *Probabilistic Engineering Mechanics*, 15(2):131–138, 2000.

[15] P. Bak. *How Nature Works.* Springer-Verlag, New York, 1996.

[16] P. Bak and C. Tang. Earthquakes as a self-organized critical phenomenon. *Journal of Geophysical Research*, 94(B11):15,635–15,637, 1989.

[17] N. Bellomo and R. Rigantti. *Nonlinear Stochastic Systems in Physics and Mechanics.* World Scientific, Singapore, 1987.

[18] J. S. Bendat. *Nonlinear System Analysis and Identification from Random Data.* John Wiley & Sons, New York, 1990.

[19] P. Bernard and L. Wu. Stochastic linearization: The theory. *Journal of Applied Probability*, 35:718–730, 1998.

[20] J. Bertoin. *Lévy Processes.* Cambridge University Press, New York, 1998.

[21] A. T. Bharucha-Reid and M. Sambandham. *Random Polynomials.* Academic Press, Inc., New York, 1986.

[22] V. V. Bolotin. *Random Vibration of Elastic Systems.* Martinus Nijhoff Publishers, Boston, 1984.

[23] A. P. Boresi and K. P. Chong. *Elasticity in Engineering Mechanics.* John Wiley & Sons, Inc., New York, 2000.

[24] P. Bougerol and J. Lacroix. *Products of Random Matrices with Applications to Schrödinger Operators.* Birkhäuser, Boston, 1985.

[25] W. E. Boyce. *Probabilistic Methods in Applied Mathematics*, volume 1. Academic Press, New York, 1968.

[26] R. L. Brabenec. *Introduction to Real Analysis.* PWS-KENT Publishing Company, Boston, 1990.

[27] P. Bratley, B.L. Fox, and L. E. Schrage. *A Guide to Simulation.* Springer–Verlag, New York, second edition, 1987.

[28] C. A. Brebbia and J. J. Connor. *Fundamentals of Finite Element Techniques*. Butterworth, London, 1973.

[29] E. O. Brigham. *The Fast Fourier Transform*. Prentice-Hall, Inc., Englewood Cliffs, NJ, 1974.

[30] R. W. Brockett. *Finite Dimensional Linear Systems*. John Wiley & Sons, Inc., New York, 1970.

[31] P. J. Brockwell and R. A. Davis. *Time Series: Theory and Methods*. Springer-Verlag, New York, 1987.

[32] V. Bryant. *Metric Spaces. Iteration and Applications*. Cambridge University Press, 1987.

[33] R. S. Bucy and P. D. Joseph. *Filtering for Stochastic Processes with Applications to Guidance*. Chelsea Publishing Company, New York, 1987.

[34] B. Budiansky. On the elastic moduli of some heterogeneous materials. *Journal of the Mechanics and Physics of Solids*, 13:223–227, 1965.

[35] R. Burridge and L. Knopoff. Model and theoretical seismicity. *Bulletin of the Seismological Society of America*, 57(3):341–371, 1967.

[36] G. Q. Cai and Y. K. Lin. Generation of non–Gaussian stationary stochastic processes. *Physical Review, E*, 54(1):299–303, 1995.

[37] A. S. Cakmak, J. F Botha, and W. G Gray. *Computational and Applied Mathematics for Engineering Analysis*. Springer-Verlag, New York, 1987.

[38] Grigoriu M. Kulkarni S. S. Chati, M. K. and S. Mukherjee. Random walk method for the two- and three-dimensional Laplace, Poisson and Helmholtz equations. *International Journal for Numerical Methods in Engineering*, 51:1133–1156, 2001.

[39] A. H. D. Cheng and C. Y. Yang, editors. *Computational Stochastic Mechanics*. Elsevier Applied Science, New York, 1993.

[40] K. L. Chung. *A Course in Probability Theory*. Academic Press, Inc., New York, 1974.

[41] K. L. Chung. *Green, Brown, and Probability*. World Scientific, New Jersey, 1995.

[42] K. L. Chung and R. J. Williams. *Introduction to Stochastic Integration*. Birkhäuser, Boston, 1990.

[43] R. Courant and D. Hilbert. *Methods of Mathematical Physics*, volume 1. Interscience Publishers, Inc., New York, 1953.

[44] R. Courant and D. Hilbert. *Methods of Mathematical Physics*, volume 2. Interscience Publishers, Inc., New York, 1953.

[45] H. Cramer and M. R. Leadbetter. *Stationary and Related Stochastic Processes*. John Wiley & Sons, Inc., New York, 1967.

[46] G. T. Csanady. *Turbulent Diffusion in the Environment*. D. Reidel Publishing Company, Boston, 1973.

[47] J. H. Cushman. Development of stochastic partial differential equations for subsurface hydrology. *Stochastic Hydrology and Hydraulics*, 1(4):241–262, 1987.

[48] R. E. Cutkosky. A Monte Carlo method for solving a class of integral equations. *Journal of Research of the National Bureau of Standards*, 47(2):113–115, 1951.

[49] W. B. Davenport and W. L. Root. *An Introduction to the Theory of Random Signals and Noise*. McGraw-Hill Book Company, Inc., New York, 1958.

[50] P. R. Dawson and E. B. Martin. Computational mechanics for metal deformation processes using polycrystal plasticity. *Advances in Applied Mechanics*, pages 77–169, 1998.

[51] J. B. Diaz and L. E. Payne. Mean value theorems in the theory of elasticity. In *Proceedings of the Third U. S. National Congress of Applied Mechanics*, pages 293–303. The American Society of Mechanical Engineers, June 11-14 1958.

[52] P. J. Digby. The effective elastic moduli of porous granular rocks. *Journal of Applied Mechanics*, 48:803–808, 1981.

[53] O. Ditlevsen. *Uncertainty Modeling with Applications to Multidimensional Civil Engineering Systems*. McGraw–Hill Inc., New York, 1981.

[54] O. Ditlevsen and G. Lindgren. Empty envelope excursions in stationary Gaussian processes. *Journal of Sound and Vibration*, 122(3):571–587, 1988.

[55] P. Ditlevsen. Observations of α–stable noise induced millennial climate changes from an ice–core record. *Geophysical Research Letters*, 26(10):1441–1444, 1999.

[56] J. Durbin. The first-passage density of a continuous gaussian process to a general boundary. *Journal of Applied Probability*, 22:99–122, 1985.

[57] R. Durrett. *Brownian Motion and Martingales in Analysis*. Wadsworth Advanced Books & Software, Belmont, California, 1984.

[58] R. Durrett. *Stochastic Calculus. A Practical Introduction.* CRC Press, New York, 1996.

[59] R. J. Elliott. *Stochastic Calculus and Applications.* Springer-Verlag, New York, 1982.

[60] R. J. Elliott and P. E. Kopp. *Mathematics of Financial Markets.* Springer-Verlag, New York, 1999.

[61] S. N. Ethier and T. G. Kurtz. *Markov Processes. Characterization and Convergence.* John Wiley & Sons, New York, 1986.

[62] W Feller. *An Introduction to Probability Theory and Its Applications*, volume II. John Wiley & Sons, Inc., New York, second edition, 1971.

[63] R. V. Field, Jr. and M. Grigoriu. A new perspetive on polynomial chaos. In P. D. Spanos and G. Deodatis, editors, *Proceedings of the Fourth International Conference on Computational Stochastic Mechanics, to appear.* Presented at the Fourth International Conference on Computational Stochastic Mechanics, Corfu, Greece, June 9-12, 2002.

[64] A. Frankel. Mapping seismic hazard in the central united and eastern United States. *Seismological Research Letters*, 66(4):8–21, 1995.

[65] B. Friedman. *Lectures on Applications-Oriented Mathematics.* John Wiley & Sons, Inc., New York, 1969.

[66] B. Fristedt and L. Gray. *A Modern Approach to Probability Theory.* Birkhäuser, Boston, 1997.

[67] A. T. Fuller. Analysis of nonlinear stochastic systems by means of the Fokker-Planck equation. *International Journal of Control*, 9(6):603–655, 1969.

[68] Y. C. Fung. *Foundations of Solid Mechanics.* Prentice-Hall, Inc., Englewood Cliffs, NJ, 1965.

[69] H. Furstenberg and H. Keston. Products of random matrices. *Annals od Mathematical Statistics*, 31:457–469, 1960.

[70] C. W. Gardiner. *Handbook of Stochastic Methods for Physics, Chemistry and the Natural Sciences.* Springer-Verlag, New York, second edition, 1985.

[71] R. G. Ghanem and P. D. Spanos. *Stochastic Finite Elements: A Spectral Approach.* Springer–Verlag, New York, 1991.

[72] J. D. Goddard. Continuum modelling of granular assemblies. In H. J. Herrmann, J.-P. Hovi, and S. Luding, editors, *Physics of Dry Granular Media*, pages 1–24. Kluwer Academic Publishers, Boston, 1998.

[73] M. D. Greenberg. *Foundations of Applied Mathematics*. Prentice Hall, Inc., Englewood Cliffs, New Jersey, 1978.

[74] M. Grigoriu. A consistent closure method for nonlinear random vibration. *International Journal of Nonlinear Mechanics*, 26(6):857–866, 1991.

[75] M. Grigoriu. A solution of the random eigenvalue problem by crossing theory. *Journal of Sound and Vibration*, 158(1):58–80, 1992.

[76] M. Grigoriu. Transient response of linear systems to stationary gaussian inputs. *Probabilistic Engineering Mechanics*, 7(3):159–164, 1992.

[77] M. Grigoriu. Simulation of nonstationary Gaussian processes by random trigonometric polynomials. *Journal of Engineering Mechanics, ASCE*, 119(2):328–343, 1993.

[78] M. Grigoriu. Simulation of stationary processes via a sampling theorem. *Journal of Sound and Vibration*, 166(2):301–313, 1993.

[79] M. Grigoriu. *Applied Non-Gaussian Processes: Examples, Theory, Simulation, Linear Random Vibration, and MATLAB Solutions*. Prentice Hall, Englewoods Cliffs, NJ, 1995.

[80] M. Grigoriu. Equivalent linearization for Poisson white noise input. *Probabilistic Engineering Mechanics*, 11(1):45–51, 1995.

[81] M. Grigoriu. Lyapunov exponents for nonlinear systems with Poisson white noise. *Physics Letters A*, 217:258–262, 1996.

[82] M. Grigoriu. Local solutions of Laplace, heat, and other equations by Itô processes. *Journal of Engineering Mechanics, ASCE*, 123(8):823–829, 1997.

[83] M. Grigoriu. Mean and covariance equations for boundary value problems. *Journal of Engineering Mechanics, ASCE*, 123(5):485–488, 1997.

[84] M. Grigoriu. Solution of some elasticity problems by the random walk method. *ACTA Mechanica*, 125:197–209, 1997.

[85] M. Grigoriu. A local solution of the Schrödinger equation. *Journal of Physics A: Mathematical and General*, 31:8669–8676, 1998.

[86] M. Grigoriu. A Monte Carlo solution of transport equations. *Probabilistic Engineering Mechanics*, 13(3):169–174, 1998.

[87] M. Grigoriu. A spectral representation based model for Monte Carlo simulation. *Probabilistic Engineering Mechanics*, 15(4):365–370, October 2000.

[88] M. Grigoriu and S. Balopoulou. A simulation method for stationary Gaussian random functions based on the sampling theorem. *Probabilistic Engineering Mechanics*, 8(3-4):239–254, 1993.

[89] M. Grigoriu, D. Veneziano, and C. A. Cornell. Probabilistic modelling as decision making. *Journal of the Engineering Mechanics Division, ASCE*, 105(EM4):585–596, 1979.

[90] V. K. Gupta, I. Rodriguez-Iturbe, and E. F. Wood, editors. *Scale Problems in Hydrology*. D. Reidel Publishing Company, Boston, 1986.

[91] Raiffa H. and R. Schlaifer. *Applied Statistical Decision Theory*. The M.I.T. Press, Cambridge, Massachusetts, 1961.

[92] P. Hähner. On the foundations of stochastic dislocation dynamics. *Applied Physics A*, 62:473–481, 1996.

[93] J. H. Halton. Sequential monte carlo. In *Proceedings of the Cambridge Philosophical Society (Mathematical and Physical Sciences)*, volume 58, pages 57–78. Cambridge at the University Press, 1962.

[94] J. M. Hammersley and D. C. Handscomb. *Monte Carlo Methods*. Methuen & Co. Ltd., London, 1964.

[95] J. M. Harmmersley and D. C. Handscomb. *Monte Carlo Methods*. Methuen & Co. Ltd., London, 1967.

[96] S. Harmsen and A. Frankel. Deaggregation of probabilistic ground motions in the central and eastern United States. *Bulletin of the Seismological Society of America*, 89(1):1–13, 1999.

[97] A. M. Hasofer and M. Grigoriu. A new perspective on the moment closure method. *Journal of Applied Mechanics*, 62(2):527–532, 1995.

[98] D. B. Hernández. *Lectures on Probability and Second Order Random Fields*. World Scientific, London, 1995.

[99] F. B. Hildebrand. *Methods of Applied Mathematics*. Prentice Hall, Inc., Englewood Cliffs, N. J., 1965.

[100] B. M. Hill. A simple general approach to inference about the tail of a distribution. *The Annals of Statistics*, 3(5):1163–1174, 1975.

[101] E. J. Hinch. *Perturbation Methods*. Cambridge University Press, Cambridge, 1994.

[102] W. Horsthemke and R. Lefever. Phase transitions induced by external noise. *Physics Letters*, 64A(1):19–21, 1977.

[103] K. Huang. *Statistical Mechanics*. John Wiley & Sons, New York, 1987.

[104] I.D. Huntley and R. M. Johnson. *Linear and Nonlinear Differential Equations*. John Wiley & Sons, New York, 1983.

[105] R. A. Ibrahim. *Parametric Random Vibration*. John Wiley & Sons Inc., New York, 1985.

[106] R. Iranpour and P. Chacon. *The Mark Kac Lectures*. Macmillan Publishing Company, New York, 1988.

[107] K. Ito and M. Matsuzaki. Earthquakes as self-organized critical phenomena. *Journal of Geophysical Research*, 95(B5):6853–6860, 1990.

[108] J. T. Jenkins and O. D. L Strack. Mean field inelastic behavior of random arrays of identical spheres. *Mechanics of Materials*, 16:25–33, 1993.

[109] G. Kallianpur. *Stochastic Filtering Theory*. Springer-Verlag, New York, 1980.

[110] R. P. Kanwal. *Linear Integral Equations. Theory and Technique*. Academic Press, New York, 1971.

[111] S. Karlin and H. M. Taylor. *A Second Course in Stochastic Processes*. Academic Press, New York, 1981.

[112] R. Z. Khas'minskii. A limit theorem for the solution of differential equations with random right-hand sides. *Theory of Probability and Its Applications*, XI(3):390–406, 1966.

[113] I. C. Kim and S. Torquato. Determination of the effective conductivity of heterogeneous media by brownian motion simulation. *Journal of Applied Physics*, 68(8):3892–3903, 1990.

[114] M. Kleiber and T. D. Hien. *The Stochastic Finite Element*. John Wiley & Sons, New York, 1992.

[115] P. E. Kloeden and Platen E. *Numerical Solutions of Stochastic Differential Equations*. Springer–Verlag, New York, 1992.

[116] M. A. Koenders. Evolution of spatially structures elastic materials using a harmonic density function. *Physical Review E*, 56(5), 1997.

[117] V. Krishnan. *Nonlinear Filtering and Smoothing: An Introduction to Martingales, Stochastic Integrals and Estimation*. John Wiley & Sons, New York, 1984.

[118] S. Kulkarni, S. Mukherjee, and M. Grigoriu. Applications of the boundary walk method to potential theory and linear elasticity. *Journal of Applied Mechanics, ASME*, 2002. in press.

[119] A. Kumar and P. R. Dawson. The simulation of texture evolution with finite element over orientation space. Application to planar polycrystals. *Applied Mechanics and Engineering*, 130:247–261, 1996.

[120] P. Lancaster and M. Tismenetsky. *The Theory of Matrices*. Academic Press, Inc., New York, second edition, 1985.

[121] M. R. Leadbetter, G. Lindgren, and H. Rootzén. *Extremes and Related Properties of Random Sequences and Processes*. Springer-Verlag, New York, 1983.

[122] Y. K. Lin and G. Q. Cai. *Probabilistic Structural Dynamics. Advanced Theory and Applications*. McGraw–Hill, New York, 1995.

[123] G. G. Lorentz. *Bernstein Polynomials*. Chelsea Publishing Company, New York, 1986.

[124] E. Lukacs. *Characteristic Functions*. Number 5 in Griffin's Statistical Monographs & Courses. Charles Griffin & Company Limited, London, 1960.

[125] K. A. Mal and S. J. Singh. *Deformation of Elastic Solids*. Prentice Hall, 1991.

[126] R. N. Mantegna and H. E. Stanley. *An Introduction to Econophysics. Correlation and Complexity in Finance*. Cambridge University Press, Cambridge, UK, 2000.

[127] L. Meirovitch. *Computational Methods in Structural Dynamics*. Sijthoff & Noordhoff, Rockville, Maryland, 1980.

[128] Yu. A. Melnikov. *Green's Functions in Applied Mathematics*. Computational Mechanics Publications, Boston, 1995.

[129] J. L. Melsa and A. P. Sage. *An Introduction to Probability and Stochastic Processes*. Prentice-Hall, Inc., Englewood Cliffs, NJ, 1973.

[130] C. Meunier and A. D. Verga. Noise and bifurcations. *Journal of Statistical Physics*, 50(1/2):345–375, 1988.

[131] T. Mikosch. *Elementary Stochastic Calculus*. World Scientific, New Jersey, 1998.

[132] A. Naess and J. M. Johnsen. Response statistics of nonlinear, compliant offshore structures by the path integral solution method. *Probabilistic Engineering Mechanics*, 8(2):91–106, 1993.

[133] M. Nicolescu. *Funcţii Reale şi Elemente de Topologie (in Romanian)*. Editura Didactica sı Pedagogica, Bucharest, Romania, 1968.

[134] B. Øksendal. *Stochastic Differential Equations.* Springer-Verlag, New York, 1992.

[135] B. Øksendal. *Stochastic Differential Equations. An Introduction with Applications.* Springer-Verlag, New York, 1998.

[136] M. Ostoja-Starzewski. Micromechanics as a basis of random elastic continuum approximations. *Probabilistic Engineering Mechanics*, 8:107–114, 1993.

[137] M. Ostoja-Starzewski. Micromechanics as a basis of continuum random fields. *Applied Mechanics Reviews*, 47:221–230, 1994.

[138] W. J. Padgett, G. Schultz, and C. P Tsokos. A random differential equation approach to the probability distribution of bod and do in streams. *SIAM Journal of Applied Mathematics*, 32(2):467–483, 1977.

[139] A. S. Papageorgiou. On the characteristic frequencies of acceleration spectra: Patch corner frequency and f-max. *Bulletin of the Seismological Society of America*, 78(2):509–529, 1988.

[140] G. C. Papanicolaou. Wave propagation in a one-dimensional random medium. *SIAM Journal of Applied Mathematics*, 21(1):13–18, 1971.

[141] A. Papoulis. *Probability, Random Variables, and Stochastic Processes.* McGraw-Hill Book Company, New York, 1965.

[142] V. V. Petrov. *Sums of Independent Random Variables.* Springer–Verlag, New York, 1975.

[143] A. J. Pettofrezzo. *Matrices and Transformations.* Dover Publications, Inc., New York, 1966.

[144] R. Popoescu, J. H. Prevost, and G. Deodatis. Effects of spatial variability on soil liquefaction: Some design recommendations. *Géotechnique*, XLVII(5):1019–1036, December 1997.

[145] V. C. Prantil, J. T. Jenkins, and P. R. Dawson. An analysis of texture and plastic spin for planar polycrystals. *Journal of the Mechanics and Physics of Solids*, 42(8):1357–1382, 1993.

[146] J. H. Prevost. A simple plasticity theory for frictional cohesionless soils. *Soil Dynamics and Earthquake Engineering*, 4:9–17, 1985.

[147] P. Protter. *Stochastic Integration and Differential Equations.* Springer–Verlag, New York, 1990.

[148] F. Radjai. Multicontact dynamics. In H. J. Herrmann, J.-P. Hovi, and S. Luding, editors, *Physics of Dry Granular Media.* Kluwer Academic Publishers, Boston, 1998.

[149] K. Rektorys. *Survey of Applicable Mathematics*. The M.I.T. Press, Cambridge, Massachusetts, 1969.

[150] S. I. Resnick. *Adventures in Stochastic Processes*. Birkhäuser, Boston, 1992.

[151] S. I. Resnick. *A Probability Path*. Birkhäuser, Boston, 1998.

[152] J. B. Roberts. *Averaging Methods in Random Vibration*. Number R-245. Technical University of Denmark, Lyngby, Denmark, 1989.

[153] J. B. Roberts and P. D. Spanos. *Random Vibration and Statistical Linearization*. John Wiley & Sons, New York, 1990.

[154] S. M. Ross. *Stochastic Processes*. John Wiley & Sons, New York, 1983.

[155] R. Rubinstein. *Simulation and the Monte Carlo Method*. John Wiley & Sons, New York, NY, 1981.

[156] W. Rudin. *Real and Complex Analysis*. McGraw-Hill, Inc., New York, 1974.

[157] P. A. Ruymgaart and T. T. Soong. *Mathematics of Kalman-Bucy Filtering*. Springer-Verlag, New York, 1988.

[158] I. Rychlik. A note on Durbin's formula for the first-passage density. *Statistics & Probability Letters*, 5:425–428, 1987.

[159] I. Rychlik and G. Lindgren. Crossreg, a computer code for first passage and wave density analysis. Technical report, University of Lund and Lund Institute of Technology, Department of Mathematical Statistics, Lund, Sweden, 1990.

[160] K. K. Sabelfeld. *Monte Carlo Methods in Boundary Value Problems*. Springer–Verlag, New York, 1991.

[161] G. Samorodnitsky. *Long Range Dependence, Heavy Tails and Rare Events*. Lecture notes. MaPhySto, Center for Mathematical Physics and Stochastics. Aarhus, Denmark, 2002.

[162] G. Samorodnitsky and M. S. Taqqu. *Stable Non–Gaussian Random Processes. Stochastic Models with Infinite Variance*. Birkhäuser, New York, 1994.

[163] S. B. Savage. Modeling and granular material boundary value problems. In H. J. Herrmann, J.-P. Hovi, and S. Luding, editors, *Physics of Dry Granular Media*. Kluwer Academic Publishers, Boston, 1998.

[164] J. Scheidt and W. Purkert. *Random Eigenvalue Problem*. North Holland, New York, 1983.

[165] Z. Schuss. *Theory and Applications of Stochastic Differential Equations.* John Wiley & Sons, New York, 1980.

[166] F. C. Schweppe. *Uncertain Dynamic Systems.* Prentice Hall, Inc, Englewood Cliffs, NJ, 1973.

[167] J. P. Sethna, K. Dahmen, S. Kartha, J. A. Krumhansi, B. W. Roberts, and J. D. Shore. Hysteresis and hierarchies: Dynamics of disorder-driven first-order phase transformations. *Physical Review Letters*, 70(21):3347–3350, 1993.

[168] J. P. Sethna, K. Dahmen, and C. R. Myers. Crackling noise. *Nature*, 410:242–250, 2001.

[169] M. Shinozuka. Structural response variability. *Journal of Engineering Mechanics, ASCE*, 113(EM6):825–842, 1987.

[170] M. Shinozuka and Yang J.-N. On the bounds of first excursion probability. *Journal of the Engineering Mechanics Division, ASCE*, 95(EM2):363–377, 1969.

[171] J. G. Simmonds and J. E. Mann. *A First Look at Perturbation Theory.* Robert E. Krieger Publishing Company, Malabar, Florida, 1986.

[172] D. L. Snyder. *Random Point Processes.* John Wiley & Sons, New York, 1975.

[173] K. Sobczyk. *Stochastic Wave Propagation.* Elsevier, New York, 1985.

[174] C. Soize. *The Fokker–Planck Equation for Stochastic Dynamical Systems and Its Explicit Steady State Solution.* World Scientific, New Jersey, 1994.

[175] T. T. Soong and M. Grigoriu. *Random Vibration of Mechanical and Structural Systems.* Prentice Hall, Englewood Cliffs, N.J., 1993.

[176] T.T. Soong. *Probabilistic Modeling and Analysis in Science and Engineering.* John Wiley & Sons, New York, NY, 1981.

[177] P. D. Spanos. Stocahstic linearization in structural dynamics. *Applied Mechanics Reviews*, 34(1):1–8, 1981.

[178] P. D. Spanos and C. A. Brebbia, editors. *Computational Stochastic Mechanics.* Elsevier Applied Science, New York, 1991.

[179] B. F. Spencer and L. A. Bergman. On the numerical solutions of the Fokker-Planck equations for nonlinear stochastic systems. *Nonlinear Dynamics*, 4:357–372, 1993.

[180] D. Stoyan, W. S. Kendall, and J. Mecke. *Stochastic Geometry and Its Applications.* John Wiley & Sons, New York, 1987.

[181] H. Tanaka. Application of importance sampling method to time-dependent system reliability analysis using Girsanov transformation. In N. Shiraishi, M. Shinozuka, and Y. K. Wen, editors, *Structural Safety and Reliability, ICOSSAR'97*, volume 1, Rotterdam, Netherlands, 1998. A. A. Balkema.

[182] S. J. Taylor. Exact asymptotic estimates of brownian path variation. *Duke Mathematical Journal*, 39, 1972.

[183] G. P. Tolstov. *Fourier Series*. Dover Publications, Inc., New York, 1962.

[184] S. Torquato. Thermal conductivity of disordered heterogeneous media form the microstructure. *Reviews in Chemical Engineering*, 4(3 & 4):151–204, 1987.

[185] S. Torquato. *Random Heterogeneous Materials. Microstructure and Macroscopic Properties*. Springer, New York, 2002.

[186] S. Torquato and I. C. Kim. Efficient simulation technique to compute effective properties of heterogeneous media. *Applied Physics Letters*, 55(18):1847–1849, 1989.

[187] F. G. Tricomi. *Integral Equations*. Dover Publications, Inc., New York, 1957.

[188] H. S. Turkmen, R. Loge, P. R Dawson, and M. P. Miller. The micromechanical behavior of a polycrystalline metal during cycling loading, in preparation. 2002.

[189] E. Vanmarcke. On the distribution of the first-passage time for normal stationary random processes. *Journal of Applied Mechanics*, 42:215–220, 1975.

[190] E. Vanmarcke, M. Shinozuka, S. Nakagiri, G. I. Schuëller, and M. Grigoriu. Random fields and stochastic finite elements. *Structural Safety*, 3:143–166, 1986.

[191] W. V. Wedig. *Lyapunov Exponents of Stochastic Systems and Related Bifurcation Problems*. Elsevier Applied Science, 315-327, New York, 1988.

[192] S. Whitaker. Multiphase transport phenomena: matching theory and experiments. In G. Papanicolaou, editor, *Advances in multiphase flow and related problems*, pages 273–295. SIAM, 1986.

[193] J. H. Wilkinson. *The Algebraic Eigenvalue Problem*. Clarendon Press, Oxford, 1988.

[194] J. R. Willis. Variational and related methods for the overall properties of composites. *Advances in Applied Mechanics*, 21:1–78, 1981.

[195] R. Willis and W. W.-G. Yeh. *Groundwater Systems Planning and Management.* Prentice Hall, Englewood Cliffs, NJ, 1987.

[196] S. F. Wojtkiewicz and L. A. Bergman. Numerical solution of high-dimensional Fokker-Planck equations. In A. Kareem, A. Haldar, B. F. Spencer, and Johnson E. A., editors, *Proceedings of the 8'th ASCE Joint Specialty Conference on Probabilistic Mechanics and Structural Reliability, CD-ROM*, July 24-26 2000.

[197] E. Wong and B. Hajek. *Stochastic Processes in Engineering Systems.* Springer–Verlag, New York, 1985.

[198] W-C Xie and S. T. Ariaratnam. Vibration mode localization in large randomly disordered continuous beams. *Fields Institute Communication*, 9:219–238, 1996.

[199] P. Yiou, K. Fuhrer, L. D. Meeker, J. Jouzel, S. Johnsen, and P. A. Mayewski. Paleoclimatic variability inferred from the spectral analysis of Greenland and Antarctic ice-core data. *Journal of Geophysical Research*, 102(C12):26441–26454, 1997.

[200] J. M. Zagajac. *Engineering Analysis over Subdomains.* PhD thesis, Cornell University, Sibley School of Mechanical and Aerospace Engineering, Ithaca, New York, May 1997.

[201] A. Zellner. *An Introduction to Bayesian Inference in Econometrics.* John Wiley & Sons, Inc., New York, 1971.

Index